T0137142

Geometric Algebra Applications Vol. I

Eduardo Bayro-Corrochano

Geometric Algebra Applications Vol. I

Computer Vision, Graphics and Neurocomputing

Springer

Eduardo Bayro-Corrochano
Electrical Engineering and
Computer Science Department
CINVESTAV, Campus Guadalajara
Jalisco, Mexico

ISBN 978-3-030-09085-2 ISBN 978-3-319-74830-6 (eBook)
https://doi.org/10.1007/978-3-319-74830-6

Softcover re-print of the Hardcover 1st edition 2019

Printed on acid-free paper

This Springer imprint is published by the registered company Springer International Publishing AG part of Springer Nature
The registered company address is: Gewerbestrasse 11, 6330 Cham, Switzerland

My Three Dedication Strophes

I. To the social fighters Nelson Mandela who achieved the elimination of the African apartheid and Evo Morales Aima who worked to eliminate the Andean Indian apartheid according the quechua saying Ama Sua, Ama Qhella, Ama Llulla, Ama Llunk'a.

II. To all scientists who do not work for the development of weapons and technology destined to occupy and dominate countries; for all who work for education, health, preservation of the environment and social justice.

III. To my beloved wife Joanna Jablonska and my first wife Mechthild Kaiser for their constant and patient support during all these years of my scientific work; and to my adored children: Esteban, Fabio, Vinzenz, Silvana, Nikolai, Claudio and Gladys.

Preface

This book presents the theory and applications of an advanced mathematical language called *geometric algebra* that greatly helps to express the ideas and concepts and to develop algorithms in the broad domain of robot physics.

In the history of science, without essential mathematical concepts, theories would have not been developed at all. We can observe that in various periods of the history of mathematics and physics, certain stagnation occurred; from time to time, thanks to new mathematical developments, astonishing progress took place. In addition, we see that the knowledge became unavoidably fragmented as researchers attempted to combine different mathematical systems. Each mathematical system brings about some parts of geometry; however, together, these systems constitute a highly redundant system due to an unnecessary multiplicity of representations for geometric concepts. The author expects that due to his persistent efforts to bring to the community geometric algebra for applications as a metalanguage for geometric reasoning, in the near future tremendous progress in computer vision, machine learning, and robotics should take place.

What is geometric algebra? Why is its application so promising? Why should researchers, practitioners, and students make the effort to understand geometric algebra and use it? We want to answer all these questions and convince the reader that becoming acquainted with geometric algebra for applications is a worthy undertaking.

The history of geometric algebra is unusual and quite surprising. In the 1870s, William Kingdon Clifford introduced his geometric algebra, building on the earlier works of Sir William Rowan Hamilton and Hermann Gunther Grassmann. In Clifford's work, we perceive that he intended to describe the geometric properties of vectors, planes, and higher-dimensional objects. Most physicists encounter the algebra in the guise of Pauli and Dirac matrix algebras of quantum theory. Many roboticists or computer graphic engineers use quaternions for 3D rotation estimation and interpolation, as a pointwise approach is too difficult for them to formulate homogeneous transformations of high-order geometric entities. They resort often to tensor calculus for multivariable calculus. Since robotics and engineering make use of the developments of mathematical physics, many beliefs are automatically inherited; for instance, some physicists come away from a study of Dirac theory with the view that Clifford's algebra is inherently quantum mechanical. The goal of this book is to eliminate these kinds of beliefs by giving a clear introduction of geometric algebra and showing this new and promising mathematical framework to multivectors and geometric multiplication in higher dimensions. In this new geometric language, most of the standard matter taught to roboticists and computer science engineers can be advantageously reformulated without redundancies and in a highly condensed fashion. The geometric algebra allows us to generalize and transfer concepts and techniques to a wide range of domains with little extra conceptual work. Leibniz dreamed of a geometric calculus system that deals directly with geometric objects rather than with sequences of numbers. It is clear that by increasing the dimension of the geometric space and the generalization of the transformation group, the invariance of the operations with respect to a reference frame will be more and more difficult. Leibniz's invariance dream is fulfilled for the nD classical geometries using the coordinate-free framework of geometric algebra.

The aim of this book is precise and well planned. It is not merely an expose of mathematical theory; rather, the author introduces the theory and new techniques of geometric algebra, albeit by showing their applications in diverse domains ranging from neurocomputing and robotics to medical image processing.

Guadalajara, Mexico
August 2017

Prof. Eduardo Bayro-Corrochano

Acknowledgements

Eduardo José Bayro-Corrochano would like to thank the Center for Research and Advanced Studies (CINVESTAV, Guadalajara, Mexico) and the Consejo Nacional de Ciencia y Tecnología (SEP-CONACYT, Mexico) for their support of this project. I am also very grateful to my former Ph.D. students Julio Zamora-Esquivel, Nancy Arana-Daniel, Jorge Rivera Rovelo, Leo Reyes Hendrick, Luis Eduardo Falcón, Carlos López-Franco, Rubén Machucho Cadena, Eduardo Ulises Moya-Sáanchez, Eduardo Vázquez, Gehová López-González, Gerardo Altamirano-Gómez, and Oscar Carbajal-Espinoza for fruitful discussions and technical cooperation. Their creative suggestions, criticism, and patient research work were decisive for the completion of this book. In the geometric algebra community, first of all I am indebted to David Hestenes for all his amazing work in developing the modern subject of geometric algebra and his constant encouragement to me for tackling problems in robot physics. Also, I am very thankful to Garret Sobczyk, Rafal Ablamowicz, Anthony Lasenby, Eckard Hitzer, Dietmar Hildebrand, and Joan Lasenby for their support and constructive suggestions. Finally, I am very thankful to the Mexican people, who pay my salary, which made it possible for me to accomplish this contribution to scientific knowledge.

Contents

Part III Image Processing and Computer Vision

How to Use This Book

This section begins by briefly describing the organization and content of the chapters and their interdependency. Then, it explains how readers can use the book for self-study or for delivering a regular graduate course.

Chapter Organization

The book starts with a historical review of cybernetics and gives a modern view of geometric cybernetics. This new era of cybernetics is strong related to the impressive progress and fruitful influence of geometric algebra in science and engineering. Briefly we explain geometric algebra as unified language for mathematics and physics what does offer for geometric computing.

- Part I Fundamentals of Geometric Algebra
  Chapter 2 outlines the history of geometric algebra and gives an introduction to geometric algebra. After preliminary definitions, we explain multivector products and operations and we discuss how to handle linear algebra derivations and simplexes.
  Chapter 3 is devoted to differentiation, linear multivector functions, the determinant, adjoint, and inverses. We study eigenvectors and eigenvalues, symmetric and skew-symmetric transformations and how we formulate expression of tensor calculus involving multivectors.
  Chapter 4 is devoted to geometric calculus. We study the vector manifolds and tangent space, the vector derivative, and multivector fields. We describe the generalized Gaussian map and formulate the second derivative and the Laplace–Beltrami operator. We formulate: projections, shape, curl, and spur as well as grad, div, curl involving monogenic functions and discuss the geometric concept of the blade–Laplace operator. This chapter ends with geometric calculus in 2D and 3D illustrated with the formulation in geometric algebra of the Maxwell equations, spinors, and the Schrödinger, Pauli, and Dirac equations.

Chapter 5 examines Lie group theory, Lie algebra, and the algebra of incidence using the universal geometric algebra generated by reciprocal null cones. We start discussing the Lie groups and Lie algebras of rotors. We analyze the reciprocal null cones and study the general linear group as spin group. We study the $Pin^{sp}(3,3)$, the projective transformations using the null geometric algebras over $\mathbb{R}^{3,3}$ and the Lie groups in $\mathbb{R}^{3,3}$. We study the horosphere and n-dimensional affine space as well as the general linear Lie algebra $gl(\mathcal{N})$ of the general linear Lie group $GL(\mathcal{N})$. This chapter ends with the study of algebra of incidence and describes geometric constraints as flags.

- Part II Euclidean, Pseudo-Euclidean Geometric Algebra, Incidence Algebra, and Conformal Geometric Algebras

  Chapter 6 begins by explaining the geometric algebras for 2D, 3D, and 4D spaces.

  Chapter 7 presents the kinematics of the 2D and 3D spaces. It describes the kinematics of points, lines, and planes using the Euclidean 3D geometric algebra and motor algebra.

  Chapter 8 is devoted to the conformal geometric algebra. We explain the representation of geometric objects and versors. Different geometric configurations are studied using geometric objects as points, lines, planes, circles, spheres, and simplexes. We briefly discuss the use of ruled surfaces.

  Chapter 9 discusses the geometric algebras $G^+_{6,0,2}$, $G_{6,3}$, $G_{9,3}$, and $G^+_{6,0,6}$.

  Chapter 10 discusses the main issues of the implementation of computer programs for geometric algebra and how algorithms can be speeded up using specialized hardware.

- Part III Geometric Computing for Image Processing, Computer Vision, and Neurocomputing

  Chapter 11 presents a complete study of the standard and quaternion and Clifford wavelets and Fourier transforms. We use as kernels quaternion Gabor filters and quaternion atomic functions. Furthermore, we study the quaternion analytic signal, monogenic signal, Hilbert and Riesz transforms.

  Chapter 12 uses geometric algebra techniques to formulate the n-view geometry of computer vision and the formation of 3D projective invariants for both points and lines in multiple images. We show stereo vision and detection of 3D line symmetries of objects using the conformal geometric algebra. We extend these concepts for omnidirectional vision using stereographic mapping onto the unit sphere. We compute invariants in the conformal space and apply for robot navigation. We design a geometric voting scheme for the extraction of lines, circles, and symmetries. The chapter ends with an attempt to model the human visual system in the conformal geometric algebra framework. We use the horopter in our conformal model to build a stereoscopic perception system.

- Part IV Machine Learning

  Chapter 13 presents a complete study of geometric neurocomputing discussing different models for a variety of applications. Geometric feedforward neural networks are described: the quaternion multilayer perceptron, the geometric radial basis function network, spherical radial basis function network, the quaternion wavelet function network. We study Clifford Support Vector Machines for classification and regression. The real- and quaternion-valued spike neural networks are discussed. We explain multiparticle quantum theory, quantum bits and operators action and observables in geometric algebra and the formulation of quaternion quantum gates to build quaternion quantum neural networks. The chapter ends with deep learning and convolutional neural networks formulated in quaternion algebra.

– Part V Applications of Geometric Algebra in Image Processing, Graphics, and Computer Vision
  Chapter 14 shows applications of Lie operators for key point detection, the quaternion Fourier transform for speech recognition, and the quaternion wavelet transform for optical flow estimation. The chapter ends with the application of the Riesz transform for multiresolution image processing.
  In Chap. 15, we use projective invariants for 3D shape and motion reconstruction, and robot navigation using n-view cameras and omnidirectional vision.
  Chapter 16 uses Tensor Voting and geometric algebra for detection of 3D surfaces and the estimation of 3D correspondences and non-rigid motion. We developed the conformal Hough transform for the detection of lines, circles, planes, and sphere. We apply a conformal geometric algebra voting scheme for the extraction in images of circles, lines, and symmetries. We use our conformal model of stereoscopic vision for egomotion, 3D reconstruction, relocalization, and navigation. The chapter ends with 3D pose estimation using conformal geometric algebra.
  Chapter 17 presents the application of marching spheres for 3D medical shape representation and registration.
– Part VI Applications of GA in Machine Learning
  Chapter 18 shows experiments using geometric feedforward neural networks to learn high nonlinear mappings. We illustrate the use of the geometric RBF network for the detection of 2D visual structures and for the detection of the motor the transformation between lines. We present the use of the sphere radial function network for detecting in soft surfaces the underlying 3D structure and to interpolate the surface elasticity. We solve the hand–eye calibration using the geometric radial basis function network. The chapter includes experiments using the Clifford Support Vector Machines for nonlinear classification, object recognition, and multicase interpolation.
  Chapter 19 shows the use of a geometric self-organizing neural net to segment 2D contours and 3D shapes.
  Chapter 20 includes an outline of Clifford algebra. The reader can find concepts and definitions related to classic Clifford algebra and related algebras: Gibbs vector algebra, exterior algebras, and Grassmann–Cayley algebras.

Interdependence of the Chapters

The interdependence of the chapters is shown in Fig. 1. Essentially, there are four groups of chapters:

- Fundamentals of geometric algebra
  - Chapter 1 Geometric Algebra for the Twenty-first Century Cybernetics
  - Chapter 2 Introduction to Geometric Algebra
  - Chapter 3 Differentiation, Linear, and Multilinear Functions in Geometric Algebra
  - Chapter 4 Geometric Calculus
  - Chapter 5 Lie Algebras, Lie Groups, and Algebra of Incidence.
- Useful geometric algebras
  - Chapter 6 2D, 3D, and 4D Geometric Algebras
  - Chapter 7 Kinematics of the 2D and 3D Spaces
  - Chapter 8 Conformal Geometric Algebra
  - Chapter 9 The Geometric Algebras $G^{+}_{6,0,2}$, $G_{6,3}$, $G_{9,3}$, $G^{+}_{6,0,6}$

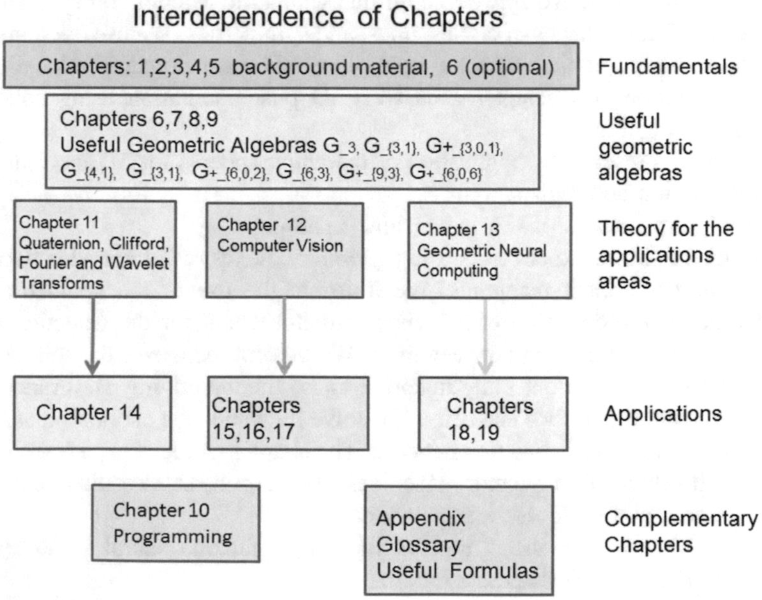

Interdependence of Chapters

Chapters: 1,2,3,4,5 background material, 6 (optional) — Fundamentals

Chapters 6,7,8,9
Useful Geometric Algebras G_3, G_{3,1}, G+_{3,0,1}, G_{4,1}, G_{3,1}, G+_{6,0,2}, G_{6,3}, G+_{9,3}, G+_{6,0,6} — Useful geometric algebras

Chapter 11 Quaternion, Clifford, Fourier and Wavelet Transforms
Chapter 12 Computer Vision
Chapter 13 Geometric Neural Computing — Theory for the applications areas

Chapter 14
Chapters 15,16,17
Chapters 18,19 — Applications

Chapter 10 Programming
Appendix Glossary Useful Formulas — Complementary Chapters

Fig. 1 Chapter interdependence: fundamentals → Theory of applications → Applications → Appendix

- Theory of the applications areas using the geometric algebra framework.
  - Chapter 11 Quaternion Gabor filters and quaternion atomic functions, quaternion analytic signal, monogenic signal, Hilbert and Riesz transforms, Quaternion and Clifford Fourier and wavelet transforms.
  - Chapter 12 Geometric Algebra of Computer Vision
  - Chapter 13 Geometric Neurocomputing
- Applications using real data from cameras, stereo, laser, omnidirectional vision, robot vision
  - Chapter 14 Application of Lie Filters, Quaternion Fourier, and Wavelet Transforms
  - Chapters 15–17 Applications in computer vision
  - Chapters 18 and 19 Applications in neurocomputing
- Complementary material
  - Chapter 1 Geometric cybernetics and geometric algebra
  - Chapter 10 Programming Issue
  - Glossary
  - Useful Formulas for Geometric Algebra
- Chapter 20 Clifford algebras and related algebras: Gibb's vector algebra, exterior algebras, and Grassmann Cayley algebras

The reader should start with the fundamentals of geometric algebra and then select a domain of application, read its theory, and get more insight by studying the related applications. For the more mathematically inclined reader, Chaps. 3–5 are recommended. However, a reader who wants to know more about some definitions and concepts of the classical Clifford algebra can consult Chap. 19 as well. Chapters 1 and 10 provide very useful material to get an overall overview of the advantages of the geometric algebra framework and to know more about the key programming issues and consider our suggestions.

Audience

Let us start with a famous saying of Clifford:

> …for geometry, you know, is the gate of science, and the gate is so low and small that one can only enter it as a little child.
>
> William Kingdon Clifford (1845–1879)

The book is aimed at a graduate level; thus, a basic knowledge of linear algebra and vector calculus is needed. A mind free of biases or prejudices caused by old-fashioned mathematics like matrix algebra and vector calculus is an even better prerequisite to reading this book.

Ideas and concepts in robot physics can be advantageously expressed using an appropriate language of mathematics. Of course, this language is not unique. There are many different algebraic systems, with their own advantages and disadvantages. Furthermore, treating a problem combining excessively different mathematical systems results in a fragmentation of the knowledge. In this book, we present geometric algebra, which we believe is the most powerful available mathematical system to date. It is a unifying language for treating robot physics; as a result, the knowledge is not fragmented and we can produce compact and less redundant mathematical expressions that are prone to be optimized for real-time applications.

One should simply start to formulate the problems using multivectors and versors and forget the formalism of matrices. However, bear in mind that when developing algorithms for particular problems, it is often a matter of techniques integration. For example, one will model the problem with geometric algebra to find geometric constraints and optimize the resulting code for real-time execution; but it is also possible that one may resort for certain computations to other algorithms that were not necessarily developed using the geometric algebra framework, for example Cholesky decomposition or image preprocessing using a morphological filter.

No prior knowledge in computer vision, robotics, wavelets theory, neurocomputing or graphics engineering is indispensable. Acquaintance with some of these topics may be of great help. The book is well suited for a taught course or for self-study at the postgraduate level. For the mathematically more inclined reader, Chaps. 3–5 are complementary. Readers wanting to know more about some definitions and concepts of the classical Clifford algebra should consult Chap. 20.

Exercises

The book offers 212 exercises, so that the reader can gradually learn to formulate equations and compute by hand or using a software package. The exercises were formulated to teach the fundamentals of geometric algebra and also to stimulate the development of the readers' skills in creative and efficient geometric computing.

Chapter 2 Introduction to GA theory and applications: includes 35 exercises for basic geometric algebra computations to learn to get along with the generalized inner product, wedge, join, and meet of multivectors, and some exercises on the computation using simplex.

Chapter 3 includes 15 exercises related to differentiation involving linear and multilinear functions, *k*-vector derivatives of multivectors, computing eigenblades and tensors formulated in geometric algebra.

Chapter 4 includes 16 exercises of geometric calculus involving divergence, curl, shape, spur, coderivative, and the computation of the Gauss map and the blade–Laplace operator.

Chapter 5 presents 32 exercises related to Lie group theory, Lie algebra, and the algebra of incidence using the reciprocal null cones of the universal geometric algebra, directed distance, and the application of flags in the affine plane and includes the computation of versors of the 3D projective group in $G_{3,3}$.

Chapter 6 offers 19 exercises for 2D, 3D, and 4D geometric algebras, particularly to get acquainted with Lie and bivector algebras, power series involving the exterior exponent, and rotor and motor operations for practical use.

Chapter 7 has 20 exercises related to the 2D and 3D kinematics of points, lines, and planes using 3D and 4D geometric algebras. It demands exercises about bivector relations, motion of geometric entities, velocity using motors, and flags for describing configurations of geometric entities attached to an object frame.

Chapter 8 Conformal geometric algebra provides 34 exercises, consisting of basics computations with versor techniques and geometric entities as the point, line, plane, circle, planes, spheres, and computations in the dual space. There are also exercises to prove classical theorems in conformal geometric algebra.

Chapter 9 The Geometric Algebras $G^{+}_{6,0,2}$, $G_{6,3}$, $G^{+}_{9,3}$, $G^{+}_{6,0,6}$ provides 22 exercises to be solved in these mathematical frameworks like modeling sphere, ellipsoids, conoids, cylinders, hyperboloids, and their kinematics and topics in geometry like line geometry, screw, and coscrews, twists and wrenches, kinematics, and dynamics.

Chapter 12 has 15 exercises related to topics in projective geometry, algebra of incidence, projective invariants, and the proof of classical theorems, as well as computations involving tensors of the n-uncalibrated cameras. The computations have to be done using the geometric algebra $G_{3,1}$ with Minkowsky and the 3D Euclidean geometric algebra G_3 for the projective space and projective plane, respectively.

Use of Computer Geometric Algebra Programs

In recent years, we have been using different software packages to check equations, prove theorems, and develop algorithms for applications.

- We recommend using CLICAL for checking equations and proving theorems. CLICAL was developed by Pertti Lounesto's group and is available at http://users.tkk.fi/ppuska/mirror/Lounesto/CLICAL.htm
- For proving equations using symbolic programming, we recommend using the Maple-based package CLIFFORD of Rafal Ablamowicz, which supports $N \leq 9$ and is available at http://math.tntech.edu/rafal
- For practicing and learning geometric algebra computing, and to try a variety of problems in computer science and graphics, we suggest using the C++-based CLUCal of Christian Perwass, available at http://www.perwass.de/cbup/clu.html
- GAIGEN2 of Leo Dorst's group, which generates fast C++ or JAVA sources for low-dimensional geometric algebra, available at http://www.science.uva.nl/ga/gaigen/

- A very powerful multivector software for applications in computer science and physics is the *C++* MV 1.3.0 to 1.6 sources supporting $N \leq 63$ of Ian Bell available at http://www.iancgbell.clara.net/maths/index.htm
- The *C++* GEOMA v1.2 developed by Patrick Stein contains *C++* libraries for Clifford algebra with orthonormal basis. It is available at http://nklein.com/software/geoma
- The reader can also download our *C++* and MATLAB programs, which are being routinely used for applications in robotics, image processing, wavelets transforms, computer vision, neurocomputing, and medical robotics. These are available at http://www.gdl.cinvestav.mx/edb/GAprogramming

Use as a Textbook

The author has taught in the last two decades in various universities a postgraduate course on the topic of geometric algebra and its applications: at the Institut für Informatik und Angewandte Mathematik der Christian Albrechts Universität (WS 1998), Kiel Germany, CINVESTAV Department of Electrical Engineering and Computer Science (annually since 2001) in Guadalajara, Mexico, and as sabbatical faculty at the Informatyk Institut der Karlsruhe Universität (SS 2008, 2 SWS, LV.-No. 24631), in Karlsruhe, Germany and as sabbatical faculty at the Media Lab. MIT (WS2015). The author has been involved with his collaborators in developing concepts and computer algorithms for a variety of application areas. At the same time, he has established a course on geometric algebra and its applications that has proven to be accessible and very useful for beginners. In general, the author follows a sequence of topics based on the material in this book. From the very beginning, students use software packages to try what is taught in the classroom.

- Chapter 2 gives an introduction to geometric algebra.
- Chapter 3 explains differentiation and multivector functions in geometric algebra
- Chapter 4 geometric calculus: discuss various topics, such as vector manifolds, vector derivative, the generalized Gaussian map, second derivative and the Laplace-Beltrami operator, grad, div, curl and monogenic functions; the blade-Laplace operator.
- Chapter 5 Lie algebras, Lie groups of rotors, geometric algebra of the Reciprocal Null Cones, the General Linear Group as Spin Group, $Pin^{sp}(3,3)$, the Horosphere and the affine plane, the general linear Lie algebra $gl(N)$ and the general linear Lie group and the Algebra of Incidence.
- Chapter 6 begins explaining the geometric algebra models for 2D, 3D, and 4D. This allows the student to learn to represent linear transformations as rotors, represent points, lines, and planes, and apply rotors and translators using these geometric entities. The duality principle is explained.

- Chapter 7 presents the kinematics of points, lines, and planes using 3D geometric algebra and motor algebra. The big step here is the use of homogeneous transformations to linearize the rigid motion transformation. Once the student has understood motor algebra, he or she has become proficient in modeling and applying a language of points, lines, and planes to tackle various problems in robotics, computer vision, and mechanical engineering.
- After these experiences, the student should move to treating projective geometry in the 4D geometric algebra $G_{3,1}$ (Minkowski metric). The role of the projective split will be explained. If the student has attended a lecture on computer vision where n-view geometry is taught [128], it is a big advantage to better understand the use of geometric algebra for projective geometry.
- Chapter 8 is devoted to conformal geometric algebra, where the experiences gained with motor algebra and $G_{3,1}$ for projective geometry are key to understanding homogeneous representations, null vectors, and the use of the null cone in the conformal geometric algebra framework. The role of the conformal additive and multiplicative splits is explained. The representation of geometric objects and versors is also shown. Different geometric configurations are studied, including simplexes, flats, plunges, and carriers. We briefly discuss the promising use of ruled surfaces and the role of conformal mapping in computer vision.
- Chapter 9 describes the geometric algebras $G^+_{6,0,2}$, $G_{6,3}$, $G^+_{9,3}$, and $G^+_{6,0,6}$, and it gives some hints for their applications.
- Chapter 10 discusses the programming issues and describes suitable hardware to speed up the computation.

In each theme, the lecturer illustrates the theory with real applications in the domains of computer vision, robotics, neurocomputing, medical robotics, and graphic engineering. For that, we use the chapters in Part III (complementary theory for applications) and Parts IV and V for illustrations of real applications. The analysis of the application examples is a good practice for the student, as they will go through well-designed training illustrations. In this way, they will gain more insight into the potential of geometric algebra, learn how it has to be applied, and finally get algorithms that are directly applicable in their research and engineering projects.

If lecturers around the world want to hold a course on geometric algebra and its applications, they can use this book and a Power Point presentation of our lecturers, which are available for free use at http://www.gdl.cinvestav.mx/edb/GAlecture.

Chapter 1
Geometric Algebra for the Twenty-First Century Cybernetics

The Volume I is devoted to geometric algebra for computer vision, graphics, and machine learning within the broad scope of cybernetics. The Vol II handles the theme geometric algebra for robotics and control. The Vol III presents geometric algebra for integral transforms for science and engineering. As a matter of fact, these topics are fundamental for the ultimate goal of building intelligent machines.

In this chapter, we will discuss the advantages for geometric computing that geometric algebra offers for solving problems and developing algorithms for related themes to geometric cybernetics as computer vision, graphics, and machine learning. We begin with a short tour of the history to find the roots of the fundamental concepts of cybernetics and geometry algebra.

1.1 Cybernetics

This section outlines briefly cybernetics using partially the source of Wikipedia. The word cybernetics stems from the Greek word κυβερνητκή (kyvernitikí, government) and denotes all that are pertinent to κυβερνώ (kyvernó), which means to steer, navigate, or govern; hence, κυβέρνησις (kyvérnisis, government) is the government while κυβερνήτης (kyvernítis) is the governor or the captain. As with the ancient Greek word for pilot, independence of thought is important in the field of cybernetics.

1.1.1 Roots of Cybernetics

The roots of the word cybernetics can be traced back in a work of Plato in The Alcibiades people [154], and it was used in the context of the study of self-governance

E. Bayro-Corrochano, *Geometric Algebra Applications Vol. I*,
https://doi.org/10.1007/978-3-319-74830-6_1

to explain the governance of people. In 1834, the physicist André-Marie Ampére (1775–1836) in his classification system of human knowledge used the French word *cybernétique* to denote the sciences of government. In addition, other studies came into play, the study of teleological mechanisms (from the Greek $\tau\acute{\varepsilon}\lambda o\varsigma$ or telos for end, goal, or purpose) in machines with corrective feedback. This term dates from as far back as the late eighteenth century, when James Watt's steam engine was equipped with a governor, a centrifugal feedback valve for controlling the speed of the engine.

1.1.2 Contemporary Cybernetics

Contemporary cybernetics began as an interdisciplinary study interrelating various fields as control systems, electrical network theory, mechanical engineering, logic modeling, evolutionary biology, and neuroscience in the 1940s. Seminal articles to be mentioned are "Behavior, Purpose and Teleology" by Arturo Rosenblueth, Norbert Wiener, and Julian Bigelow [253] and "A Logical Calculus of the Ideas Immanent in Nervous Activity" by Warren McCulloch and Walter Pitts [207]. Finally, cybernetics as a discipline was firmly established by the visionaries Norbert Wiener, Arturo Rosenbluth, Warren McCulloch, W. Ross Ashby, Alan Turing, John Von Neumann (cellular automata, the universal constructor), and W. Grey Walter who surprisingly was one of the first to build autonomous robots as an aid to the study of animal behavior. Norbert Wiener introduced the neologism *cybernetics* into his scientific theory to denote the study of "teleological mechanisms," and in 1948, this term became popular through his book in Cybernetics, or Control and Communication in the Animal and the Machine [313].

1.1.3 Cybernetics and Related Fields

Cybernetics is a trans-disciplinary approach for exploring regulatory systems, their structures, constraints, interrelations, and possibilities. Cybernetics is relevant to the study of systems, such as mechanical, physical, biological, cognitive, and social systems. Cybernetics is applicable when a system being analyzed is involved in a closed signaling loop. System dynamics, a related field, originated with applications of electrical engineering control theory to other kinds of simulation models. Concepts studied by cyberneticists include, but are not limited to, learning, cognition, adaptation, social control, emergence, communication, efficiency, efficacy, and connectivity. An important characteristic of the new cybernetics is its contribution to bridge the micro–macro gap, for example, linking the individual with the society or quantum computing in cognitive architectures of gravitational machines. Cybernetics is sometimes used as a generic term, which serves as a general framework to include many system-related scientific fields which are listed next. Basic cybernetics studies systems of control as a concept, attempting to discover the basic principles underlying such systems like robot humanoids which use sensors and sophisticated algorithms to perceive, navigate avoiding obstacles, and climb stairs. Basic cybernetics comprised a large list of the fields: artificial intelligence, robotics, computer vision, second-order cybernetics, interactions of actors theory, conversation theory, self-organization in cybernetics. Notably, cybernetics is present in the context of other areas as well as biology, computer science, engineering, management. In mathematics, mathematical cybernetics focuses on the factors of information, interaction of parts in systems, and the structure of systems, i.e., dynamical systems, information theory, systems theory. Other quite different fields where cybernetics is considered are in psychology, sociology, education, art, earth system science and complexity science, and biomechatronics.

1.1.4 Cybernetics and Geometric Algebra

We have seen that cybernetics encompasses a variety of diverse fields, where the common characteristic is the existence of a perception–action cycle system with feedback and a controller, so that the system adapts itself and ultimately augments its degree of awareness and autonomy to interact and maneuver successfully in its environment. Since the modeling of the systems and the processing of entities is key for an efficient system behavior, according to Fourier and Chasles it is more advantageous to model and process data in representations in higher-dimensionality pseudo-Euclidean spaces with a nonlinear computational framework like the horosphere. Geometric entities like lines, planes, circles, and hyperplanes and spheres can be used for screw theory in kinematics and dynamics, geometric neurocomputing, and the design of observers and controllers for perception–action systems. David Hestenes [128] has contributed to developing geometric algebra as a unifying language for mathematics and physics [130, 138] and E. Bayro for perception and action systems [13, 16]. Due to the versatility and the mathematical power of the geometric algebra framework for handling diverse and complex problems in artificial intelligence, robotics, computer vision, second-order cybernetics, self-organization in cybernetics, geometric neural networks, Clifford Fourier and wavelet transforms, fuzzy geometric reasoning, cognitive architectures, quantum computing and humanoids, geometric algebra constitutes undoubtedly an appropriate mathematical system for geometric cybernetics.

1.2 The Roots of Geometry and Algebra

The lengthy and intricate road along the history of mathematics shows that the evolution through time of the two domains *algebra* from the Arabic "alg-jbar" and geometry from the ancient Greek $\gamma\varepsilon\omega\mu\varepsilon\tau\rho\acute{\iota}\alpha$ ($geo = earth, metria =$ measure) started to intermingle early, depending upon certain trends imposed by the different groups and schools of mathematical thought. It was only at the end of the nineteenth century that they become a sort of a clear, integrated mathematical system.

Broadly speaking, on the one hand, algebra is a branch of mathematics concerning the study of structure, relation, and quantity. In addition, it is not restricted to work with numbers, but it also covers the work involving symbols, variables, and set elements. Addition and multiplication are considered general operations, which, in a more general view, lead to mathematic structures such as groups, rings, and fields.

On the other hand, geometry is concerned with essential questions of size, shape, and relative position of figures and with properties of space. Geometry is one of the oldest sciences initially devoted to practical knowledge concerned with lengths, areas, and volumes. In the third century, Euclid put geometry in an axiomatic form, and Euclidean geometry was born. During the first half of the seventeenth century, René Descartes introduced coordinates, and the concurrent development of algebra evolved into a new stage of geometry, because figures such as plane curves could now be represented analytically with functions and equations. In the 1660s, G. Leibniz and Newton, both inventors of infinitesimal calculus, pursued a geometric calculus for dealing with geometric objects rather than with sequences of numbers. The analysis of the intrinsic structure of geometric objects with the works of Euler and Gauss further enriched the topic of geometry and led to the creation of topology and differential geometry. Since the nineteenth century discovery of non-Euclidean geometry, the traditional concept of space has undergone a profound and radical transformation. Contemporary geometry postulates the concept of manifolds, spaces that are greatly more abstract than the classical Euclidean space and that approximately look alike at small scales. These spaces endowed an additional structure: a differentiable structure that allows one to do calculus; that is, a Riemannian metric allows us to measure angles and distances; symplectic manifolds serve as the phase spaces in the Hamiltonian formalism of classical mechanics, and the 4D Lorentzian manifolds serve as a space–time model in general relativity.

Algebraic geometry is considered a branch of the mathematics that combines techniques of abstract algebra with the language and problems of geometry. It plays a central role in modern mathematics and has multiple connections with a variety of fields: complex analysis, topology, and number theory. Algebraic geometry is fundamentally concerned with the study of algebraic varieties that are geometric manifestations of solutions of systems of polynomial equations. One looks for the Gröbner basis of an ideal in a polynomial ring over a field. The most studied classes of algebraic varieties are the *plane, algebraic curves* such as lines, parabolas, lemniscates, and Cassini ovals. Most of the developments of algebraic geometry in the twentieth century were within an abstract algebraic framework studying the intrinsic properties of algebraic properties independent of a particular way of embedding the variety in a setting of a coordinated space as in topology and complex geometry. A key distinction between projective geometry and algebraic geometry is that the former is more concerned with the more geometric notion of the point, whereas the latter puts major emphasis on the more analytical concepts of a regular function and a regular map and extensively draws on sheaf theory.

In the field of mathematical physics, geometric algebra is a multilinear algebra described more technically as a Clifford algebra that includes the geometric product. As a result, the theory, axioms, and properties can be built up in a more intuitive and geometric way. Geometric algebra is a coordinate-free approach to geometry based on the algebras of Grassmann [112] and Clifford [58]. Since the 1960s, David Hestenes [128] has contributed to developing geometric algebra as a unifying language for mathematics and physics [130, 138]. Hestenes also presented a study of projective geometry using Clifford algebra [139] and recently the essential concepts of conformal geometric algebra [185]. Hestenes summarized and precisely defined the role of algebra and geometry in a profound comment emphasizing the role of the capacities of language and spatial perception of the human mind, which, in fact, is the goal of this section. We reproduce it to finalize the section as a prelude to the next section to motivate and justify why geometric algebra can be of great use to build the intelligent machine of which Turing dreamed.

> In his famous survey of mathematical ideas, F. Klein championed the fusing of arithmetic with geometry as a major unifying principle of mathematics. Klein's seminal analysis of the structure and history of mathematics brings to light two major processes by which mathematics grows and becomes organized. They may be aptly referred to as the algebraic and geometric. The one emphasizes algebraic structure, while the other emphasizes geometric interpretation. Klein's analysis shows one process alternatively dominating the other in the historical development of mathematics. But there is no necessary reason that the two processes should operate in mutual exclusion. Indeed, each process is undoubtedly grounded in one of two great capacities of the human mind: the capacity for language and the capacity for spatial perception. From the psychological point of view, then, the fusion of algebra with geometry is so fundamental that one could say, Geometry without algebra is dumb! Algebra without geometry is blind!

- D. Hestenes, 1984

1.3 Geometric Algebra a Unified Mathematical Language

First of all, let us analyze the problems of the community when they use classical mathematical systems to tackle problems in physics or robotics. In this regard, let us resort to an enlightening paper of Hestenes [136] where the author discusses the main issues for modeling the physical reality. The invention of analytical geometry and calculus was essential for Newton to create classical mechanics. On the other hand, the invention of tensor analysis was essential for Einstein to create the theory of relativity. The point here is that without essential mathematical concepts, both theories would not have been developed at all. We can observe in some periods of the history of mathematics certain stagnation and from time to time, thanks to new mathematical developments, astonishing progress in diverse fields. Furthermore, we can notice that as researchers attempt to combine different mathematical systems, unavoidably this attitude in research leads to a fragmentation of knowledge. Each mathematical system brings about some parts of geometry; however, together they constitute a highly redundant system, that is, an unnecessary multiplicity of representations for geometric concepts; see Fig. 1.2. This approach in mathematical physics and in robotics has the following pressing defects:

- Restricted access: The ideas, concepts, methods, and results are unfortunately disseminated across the mathematical systems. Being proficient only in a few of the systems, one does not have access to knowledge formulated in terms of the other mathematical systems.
- Redundancy: Less efficiency due to the repetitive representation of information in different mathematical systems.
- Deficient integration/articulation: Incoherences, incongruences, lack of generalizations, and ineffective integration and articulation of the different mathematical systems.
- Hidden common knowledge: Intrinsic properties and relations represented in different symbolic systems are very difficult to handle.
- Low information density: The information of the problem in question is reduced due to distribution over different symbolic systems.

According to Hestenes [136], the development of a unified mathematical language is, in fact, a problem of the design of mathematical systems based on the following major considerations:

- Optimal encoding of the basic geometric intrinsic characteristics: dimension, magnitude, direction, sense, or orientation.
- Coordinate-free methods to formulate and solve problems in physics and robotics.
- Optimal uniformity of concepts and methods across different domains, so that the intrinsic common structures are made as explicit as possible.
- Smooth articulation between different alternative systems in order to access and transfer information frictionlessly.
- Optimal computational efficiency: The computing programs using the new system should be as or more efficient than any alternative system in challenging applications.

Note that geometric algebra was constructed following these considerations, and in view of the progress of scientific theory, geometric algebra helps greatly to optimize expressions of the key ideas and consequences of the theory.

1.4 What does Geometric Algebra Offer for Geometric Computing?

Next, we will describe the most remarkable features of geometric algebra from the perspective of geometric computing for perception–action systems.

1.4.1 Coordinate-Free Mathematical System

In geometric algebra, one writes coordinate-free expressions to capture concepts and constraints in a sort of high-level geometric reasoning approach. It is expected that the geometric information encoded in the expressions involving geometric products and the actions of operators should not suffer from interference from the reference coordinate frame. As a matter of fact, the results obtained by algebraic computing based on coordinates are geometrically meaningless or difficult to interpret geometrically. This is essentially because they are neither invariant nor covariant under the action of coordinate transformations. In fact, geometric algebra enables us to express fundamental robotics physics in a language that is free from coordinates or indices. The geometric algebra framework gives many equations a degree of clarity that is definitively lost in matrix algebra or tensor algebra.

The introduction of coordinates by R. Descartes (1596–1650) started the algebraization of the geometry. This step in geometry caused a big change from a qualitative description to a qualitative analysis. In fact, coordinates are sequences of numbers and do not have geometric meaning themselves. G. Leibniz (1646–1716) dreamed of a geometric calculus system that deals directly with geometric objects rather than with sequences of numbers. More precisely, in a mathematical system, an element of an expression should have a clear meaning of being a geometric object or a transformation operator for algebraic manipulations such as addition, subtraction, multiplication, and division.

We can illustrate the concept of independence under the action of coordinate transformations by analyzing in the Euclidean plane geometry the following operation with complex numbers $\mathbf{a}, \mathbf{b} \in \mathbb{C}$: $\bar{\mathbf{a}}\mathbf{b} := (a_1, a_2)(b_1, b_2) = (a_1b_1 - a_2b_2, a_1b_2 + a_2b_1)$. This product is not invariant under the Euclidean group; that is, the geometric information encoded in the results of the product cannot be separated from the interference of the reference coordinate frame. However, if we change the complex product to the following product: $\bar{\mathbf{a}}\mathbf{b} := (a_1, -a_2)(b_1, b_2) = (a_1b_1 + a_2b_2, a_1b_2 - a_2b_1)$, then under any rotation $r_{-} : \mathbf{a} \rightarrow \mathbf{a}e^{i\theta}$ centered at the origin, this product remains invariant: $\bar{r}(\mathbf{a})r(\mathbf{b}) := (\mathbf{a}e^{i\theta})(\mathbf{b}e^{i\theta}) = \bar{\mathbf{a}}\mathbf{b}$. Consequently, if the complex numbers are equipped with a scalar multiplication, addition, subtraction, and a geometric product, they turn from a field to a 2D geometric algebra of the 2D orthogonal geometry. It is clear that increasing the dimension of the geometric space and the generalization of the transformation group, the desired invariance will be increasingly difficult. Leibniz's dream is fulfilled for the nD classical geometries using the framework of geometric algebras. In this book, we present the following coordinate-free geometric algebra frameworks: for 2D and 3D spaces with a Euclidean metric, for 4D spaces with a non-Euclidean metric, and for $\mathbb{R}^{n+1,1}$ spaces the conformal geometric algebras.

Assuming that we are handling expression as independently as possible upon a specific coordinate systems, in view of the implementation, one converts these expressions into low-level, coordinate-based that can be directly executed by a fast processor. In general, geometric algebra can be seen as a geometric inference engine of an automated code generator, which is able to take a high-level specification of a physical problem and automatically generate an efficient and executable implementation.

1.4.2 Models for Euclidean and PseudoEuclidean Geometry

When we are dealing with problems in robotics or neurocomputing, an important question is in which metric space we should work. In this book, we are basically concerned with three well-understood space models:

i. Models for 2D and 3D spaces with a Euclidean metric: 2D and 3D are well suited to handle the algebra of directions in the plane and 3D physical space. 3D rotations are represented using rotors (isomorph to quaternions). You can model the kinematics of points, lines, and planes using $G_{3,0,0}$. Rotors can be used for interpolation in graphics and estimation of rotations of rigid bodies.

ii. Models for 4D spaces with non-Euclidean metric: If you are interested in linearizing a rigid motion transformation, you will need a homogeneous representation. For that, we should use a geometric algebra for the 4D space. Here, it is more convenient to choose the motor algebra $G^+_{3,0,1}$ described in Chap. 6. It is the algebra of Plücker lines, which can be used to model the kinematics of points, lines, and planes better than with G_3. Lines belong to the non-singular study 6D quadric and the motors to the 8D Klein quadric. In $G^+_{3,0,1}$, you can formulate a motor-based equation of motion for constant velocity where in the exponent you use a bivector for twists. You can also use motors for interpolation of 3D rigid motion and estimate trajectories using EKF techniques. When you are dealing with problems of projective geometry like in computer vision, again you need a homogeneous coordinate representation, so that the image plane becomes $\mathbb{P}^2$ and the visual space $\mathbb{P}^3$. To handle the so-called n-view geometry [126] based on tensor calculus and invariant theory, you require $G_{3,1}$ (Minkowski metric) for the visual space and $G_{3,0,0}$ for the image plane. This is studied in Chap. 12. Note that the intrinsic camera parameters are modeled with an affine transformation within geometric algebra as part of the projective mapping via a projective split between the projective space and the image plane. Incidence algebra, an algebra of oriented subspaces, can be used in $G_{3,1}$ and $G_{3,0,0}$ to treat problems involving geometric constraints and invariant theory.

iii. Conformal models: If you consider conformal transformations (angle preserving), conformal geometric algebra in Chap. 8 offers a non-Euclidean geometric algebra $G_{(n+1,1)}$ that includes in its multivector basis the null vectors, the origin, and point at infinity. As computational framework, it utilizes the powerful horosphere (the meet between a hyperplane and the null cone). Even though the computational framework uses a nonlinear representation for the geometric entities, one can recover the Euclidean metric. The basic geometric entity is the sphere and you can represent points, planes, lines, planes, circles, and spheres as vectors or in dual forms, the latter being useful to reduce the complexity of algebraic expressions. As you may have noticed, the above-presented geometric algebras can be used either for kinematics in robotics or for projective geometry in computer vision. Provided that you calibrate the digital camera, you can make use of the homogeneous models from conformal geometric algebra to handle problems of robotics and those of computer vision simultaneously, however, without the need to abandon the mathematical framework. Furthermore, incidence algebra of points, lines, planes, circles, and spheres can be used in this framework as well. The topic of omnidirectional vision exploits the model of an image projected on the sphere, whereas all the problems such as rigid motion, depth, and invariant theory can also be handled using conformal geometric algebra (see Chap. 8).

1.4.3 Subspaces as Computing Elements

The wedge product of k basis vectors spans a new entity, the k-vector. A set of all k-vectors spans an oriented subspace $\bigwedge^{k} V^n$. Thus, the entire geometric algebra G_n is given by

$$G_n = \bigwedge^{0} V^n \oplus \bigwedge^{1} V^n \oplus \bigwedge^{2} V^n \oplus \cdots \oplus \bigwedge^{k} V^n \oplus \cdots \oplus \bigwedge^{n} V^n.$$

Geometric algebra uses the subspace structure of the k-vector spaces to construct extended objects ranging from lines, planes, to hyperspheres. If we then represent physical objects in terms of these extended objects, we can model physical phenomena like relativistic particle motion or conformal mappings of the visual manifold into the neocortex using appropriate operators and blade transformations.

1.4.4 Representation of Orthogonal Transformations

Geometric algebra represents orthogonal transformations more efficiently than the orthogonal matrices by reducing the number of coefficients; think of the nine entries of a 3D rotation matrix and the four coefficients of a rotor. In geometric algebra, a versor product is defined as

$$\boldsymbol{O} \rightarrow \boldsymbol{V}\boldsymbol{O}\widetilde{\boldsymbol{V}},$$

where the versor acts on geometric objects of different grade, subspaces, and also on operators, quite a big difference from the matrices. The versor is applied sandwiching the object, because this is the result of successive reflections of the object with respect to hyperplanes. In geometric algebra, a physical object can be described by a $flag$, i.e., a reference using points, lines, planes, circles, and spheres crossing the gravitational center of the object. By applying versors equivalent to an orthogonal transformation on the flag, the relation of the reference geometric entities of the flag will remain invariant, i.e., topology-preserving group action.

1.4.5 Objects and Operators

A geometric object can be represented using multivector basis and wedge products, e.g., a point, line, or a plane. These geometric entities can be transformed for rigid motion, dilation, or reflection with respect to a plane or a sphere. These transformations depend on the metric of the involved space. Thus, we can model the 3D kinematics of such entities in different computational models as in the 3D Euclidean geometric algebra $G_{3,0,0}$, 4D motor algebra $G^*_{3,0,1}$, or the conformal algebra $G_{4,1}$. You can see that the used transformations as versors can be applied to an object regardless the grade of its k-blades. For example, in Fig. 1.1, we can describe the geometry of the arm in terms of points, circles, and spheres and with screw lines the revoluted or prismatic joints and thereafter using direct/inverse kinematics relate these geometric entities from the basis through the joints until the end-effector. Here a pertinent question will be whether or not operators can be located in space like geometric objects. The answer is yes; we can attach to any position the rotors and translators. This is a big difference from matrices; in geometric algebra, the operators or versors can be treated as geometric objects; however, they have a functional characteristic as well. In geometric algebra, objects are specified in terms of basic elements intrinsic to the problem, whereas the operators or versors are constructed depending upon which Lie group we want to use. A versor is built by successive reflections with respect to certain hyperplanes (lines, planes, spheres). In 3D space, an example of a versor is the rotor, which is built by two successive reflections with respect to two planes that intersect the origin.

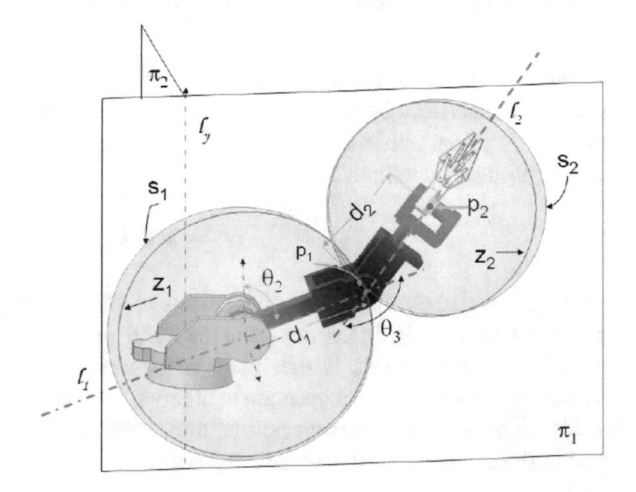

Fig. 1.1 Description of the kinematics of a 5 DOF robot arm using geometric entities of the conformal geometric algebra

A versor is applied sandwiching a geometric object. Since a versor represents a Lie group of the general linear groups, it can also be represented as an exponential form where in the exponent the Lie algebra space is spanned with a bivector basis (Lie generators). The versor and its exponential form are quite effectively represented using bivectors, which is indeed a less redundant representation than that by the matrices. Versor-based techniques can be applied in spaces of arbitrary signature and are particularly well suited for the formulation of Lorentz and conformal transformations.

In tasks of kinematics, dynamics, and modern control theory, we can exploit the Lie algebra representation acting on the bivectorial exponents rather than at the level of Lie group versor representation. In Chaps. 5 and 6, we describe bivector representations of Lie groups.

1.4.6 Extension of Linear Transformations

Linear transformations act on n-D vectors in $\mathbb{R}^n$. Since in geometric algebra a subspace is spanned by vectors, the action of the linear transformation on each individual vector will directly affect the spanned subspace. Subspaces are spanned by wedge products; an outermorphism of a subspace equals the wedge products of the transformed vectors. The outermorphism preserves the grade of any k-blade it acts on; e.g., the unit pseudoscalar must be mapped onto some multiple of itself; this multiple is the determinant of $\underline{f}(I) = \det(\underline{f})I$. We can proceed similarly when we compute the intersection of subspaces via the meet operation. We can transform linearly the result of the meet using duality and then apply an outermorphism. This would be exactly the same if we first transformed the subspaces and then computed the meet. In Chap. 12, we exploit outermorphisms for computing projective invariants in the projective space and image plane.

1.4.7 Signals and Wavelets in the Geometric Algebra Framework

One may wonder why we should interest ourselves in handling n-D signals and wavelets in the geometric algebra framework. The major motivation is that since in image processing, robotics, and control engineering, n-D signals or vectors are corrupted by noise, our filters and observers should smooth signals and extract features but do so as projections on subspaces of the geometric algebra. Thinking in quadrature filters, we immediately traduce this in quaternionic filters or further we can generalize over non-Euclidean metrics the Dirac operator for multidimensional image processing. Thus, by weighting the bivectors with appropriate kernels like Gauss, Gabor, or wavelets, we can derive powerful Clifford transforms to analyze signals using the extended phase concept and carry out convolutions in a certain geometric algebra or on the Riemann sphere. So we can postulate that complex filters can be extended over bivector algebras for computing with Clifford Fourier transforms and Clifford wavelet transforms or even the space and time Fourier transform. In the geometric algebra framework, we gain a major insight and intuition for the geometric processing of noisy n-D signals. In a geometric algebra with a specific non-Euclidean metric, one can compute geometrically coupling, intrinsically different information, e.g., simultaneously process color and thermal images with a multivector derivative or regularize color optical flow with the generalized Laplacian. Chapter 11 is devoted to studying a variety of Clifford Fourier and Clifford wavelet transforms.

1.4.8 Kinematics and Dynamics

In the past, some researchers computed the direct and inverse kinematics of robot arms using matrix algebra and geometric entities like points and lines. In contrast, working in geometric algebra, the repertoire of geometric entities and the use of efficient representation of 3D rigid transformations make the computations easy and intuitive, particularly for finding geometric constraints. In conformal geometric algebra, we can perform kinematic computations using meets of spheres, planes, and screw axes, so that the resulting pair of points yields a realistic solution to the problem. Robot object manipulation together with potential fields can be reformulated in conformal geometry using a language of spheres for planning, grasping, and manipulation. Topics related to kinematics, dynamics, and control are treated in detail in the Volume II: Geometric Algebra for Robotics and Control.

The dynamics of a robot mechanism is normally computed using a Euler–Lagrange equation, where the inertial and Coriolis tensors depend on the degrees of freedom of the robot mechanism. Even though conformal geometry does not have versors for coding affine transformations, we can reformulate these equations, so that the entries of the tensors are projections of the centers of mass points of the limbs with respect to the screw axes of the joints; as a result, we can avoid quadratic entries and facilitate the estimation of the tensor parameters. This is the benefit to handling this kind of problem in either motor algebra or conformal algebra, making use of the algebra of subspaces and versors.

1.5 Solving Problems in Perception and Action Systems

In this section, we will outline how we approach the modeling and design of algorithms to handle tasks of robotic systems acting within the perception–action cycle. In the previous section, we explained what geometric algebra offers as a mathematical system for geometric computing. Here, we will be slightly more concrete

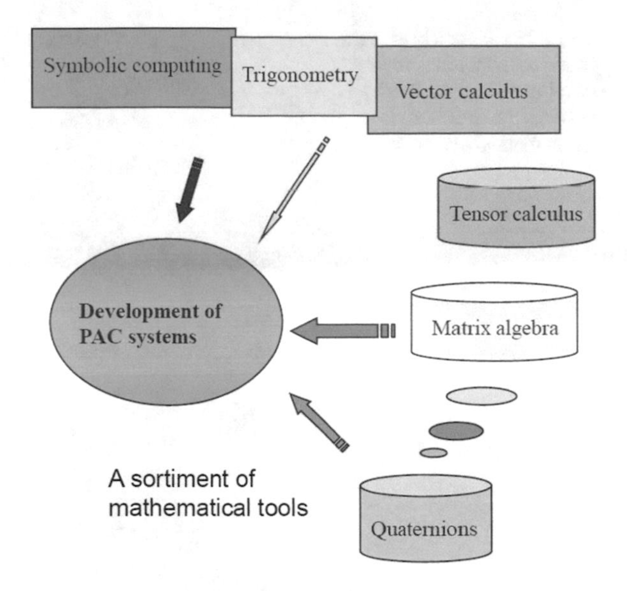

Fig. 1.2 Application of diverse mathematical systems to solve PAC problems

and precise, illustrating the design and implementation of algorithms for real-time geometric computing.

Figure 1.2 shows an abstraction of the attitudes of many researchers and practitioners: How do they approach developing algorithms to solve problems in the domain of PAC systems? Briefly, they split the knowledge across various mathematical systems. As a consequence, as we discussed in Sect. 1.3 above, the ideas, concepts, methods, and results are unfortunately disseminated across the various mathematical systems. Being proficient in only a few of the systems, one does not have access to knowledge formulated in terms of the other mathematical systems. There is a high redundancy due to the repetitive representation of information in different mathematical systems. A deficient articulation of the different mathematical systems degrades their efficiency. The intrinsic properties and relations represented in different symbolic systems are very difficult to handle. The information density is low, because the information from the problem is reduced due to distribution over different symbolic systems.

Bear in mind that geometric algebra was constructed for the optimal encoding of geometric intrinsic characteristics using coordinate-free methods. It ensures an optimal uniformity of concepts and methods across different domains, and it supports a smooth articulation between different alternative systems. The efficient representation of objects and operations guarantees computational efficiency. Since geometric algebra was constructed following these considerations and in view of the progress of scientific theory, geometric algebra is the adequate framework to optimize expressions of the key ideas and consequences of the theory related to perception–action systems; see Fig. 1.3.

Of course, it will not be possibly everything of the PAC problems to be formulated and computed in geometric algebra.We have to ensure that the integration of tech-

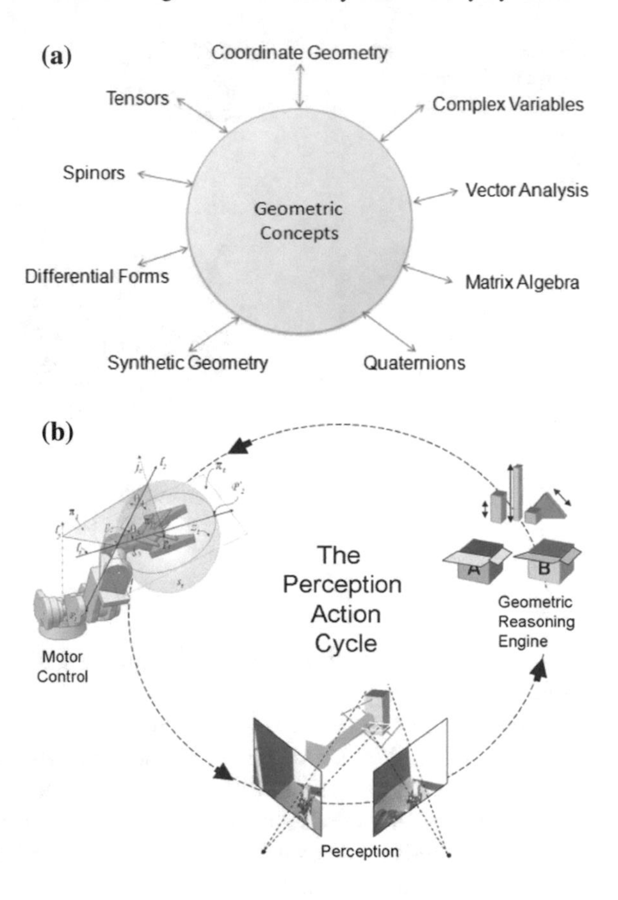

Fig. 1.3 Geometric algebra framework includes for the development of the PAC systems the essential mathematical systems

niques proceeds in a kind of top-down approach. First, we should get acquainted with the physics of the problem aided by the contributions of researchers and paying attention particularly to how they solve the problems. Recall the old Western interpretation of the metaphor *Nanos gigantum humeris insidentes*: one who develops future intellectual pursuits by understanding the research and works created by notable thinkers of the past.

For a start, we should identify where we can make use of geometric algebra. Since geometric algebra is a powerful language for efficient representations and finding geometric constraints, we should first, in a high-symbolic-level approach, postulate formulas (top-down reasoning), symbolically simplify them optimally, and execute them using cost-effective and fast hardware. To close the code generation loop, the conflicts and contradictions caused by our algorithms in their application are fed back in a bottom-up fashion to a negotiation stage in order to ultimately improve our geometric algorithms.

Let us now briefly illustrate this procedure. As a first example, we compute efficiently the inverse kinematic of a robot arm. Figure 1.4 shows the rotation planes of a robot arm. While computing its inverse kinematic, one considers a circle that is the intersection of two reference spheres $z = s_1 \wedge s_2$. We then compute the pair of points

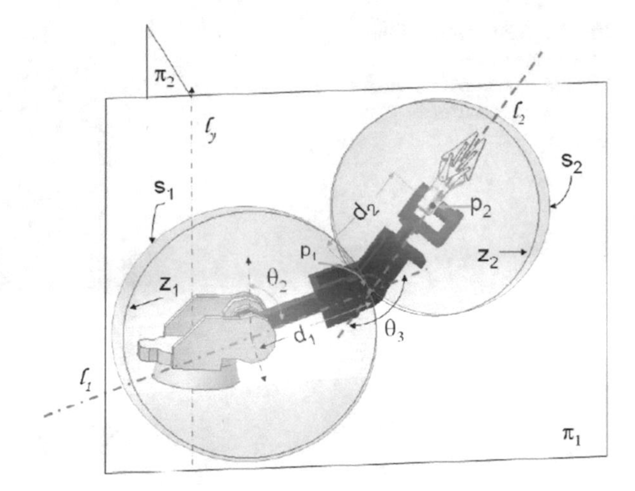

Fig. 1.4 Computing the inverse kinematics of a 5 DOF robot arm

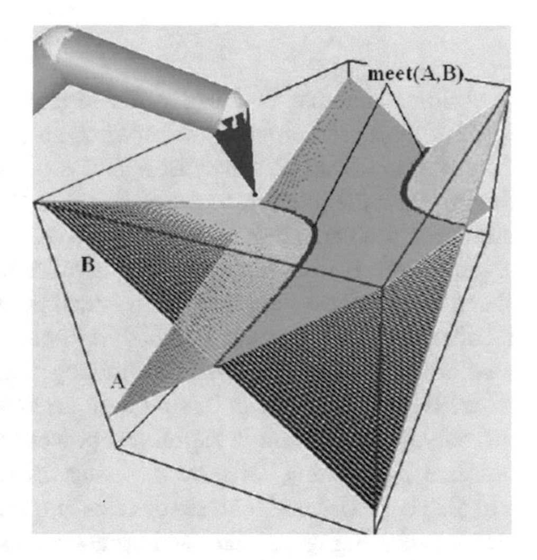

Fig. 1.5 Robot arm welding with laser along the meet of two ruled surfaces

($\boldsymbol{PP}$) as the meet of the swivel plane and the circle: $\boldsymbol{PP} = z \wedge \pi_{swivel}$, so that we can choose one as a realistic position of the elbow. This is a point that lies on the meet of two spheres. After the optimization of the whole equation of the inverse kinematics, we get an efficient representation for computing this elbow position point and other parts of the robot arm, which can all be executed using fast hardware. Note that using a matrix formulation, it will not be possible to generate optimal code.

A second example involves the use of the meet of ruled surfaces. Imagine a laser welding hast to follow the intersection of two highly nonlinear surfaces. You can estimate these using a vision system and model them using the concept of ruled surface. In order to control the motion of the welding laser attached to the end-effector of the robot manipulator, we can follow the contour gained by computing the meet of the ruled surfaces, as depicted in Fig. 1.5.

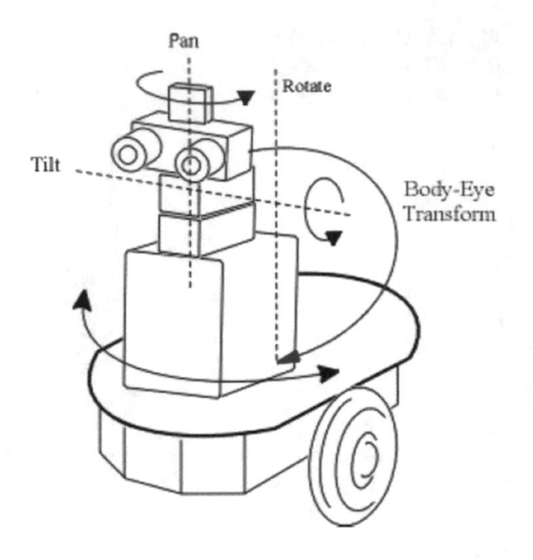

Fig. 1.6 Calibration of the coordinate system of a binocular head with respect to the robot coordinate system

A third example is to solve the sensor-body calibration problem depicted in Fig. 1.6. This problem can be simplified and linearized, exploding the intrinsic relation of screw lines of the sensors and the robot body. The problem is reduced to finding only the unknown motors between these lines. Using matrices will yield a nonlinear problem of the kind $\mathbf{AX} = \mathbf{XB}$. As is evident, a line language of motor algebra suffices to tackle such a problem. A fourth problem entails representing 3D shapes in graphics engineering or medical image processing that traditionally should involve the standard method called *marching cubes*. However, we generalized this method as *marching spheres* [250] using the conformal geometric algebra framework. Instead of using points on simplexes, we use spheres of $G_{4,1}$. In this way, not only do we give more expressive power to the algorithm but we also manage to reutilize the existing software by using, instead of the 2D or 3D vectors, the 4D and 5D vectors, which represent circles and spheres in conformal geometric algebra, respectively. In Fig. 1.7, see the impressive results of carving a 3D shape, where the spheres fill the gaps better than cubes. As a fifth and last example, let us move to geometric computing. Traditionally, neural networks have been vector-based approaches with the burden of applying a coordinate-dependent algorithm for adjusting the weights between neuron layers. In this application, we have to render a contour of a certain noisy shape like a brain tumor. We use a self-organizing neural network called neural gas [97]. Instead of adjusting the weights of the neurons to locate them along the contour or shape, we adjust the exponents of motors. In this way, we are operating in the linear space of the bivector or Lie algebra. The gain is twofold: On the one hand, due to the properties of the tangential Lie algebra space, the approach is linear. On the other hand, we exploit the coordinate-free advantage by working on the Lie algebra manifold using bivectors. Figure 1.8 shows the excellent results from segmenting a section of a brain tumor.

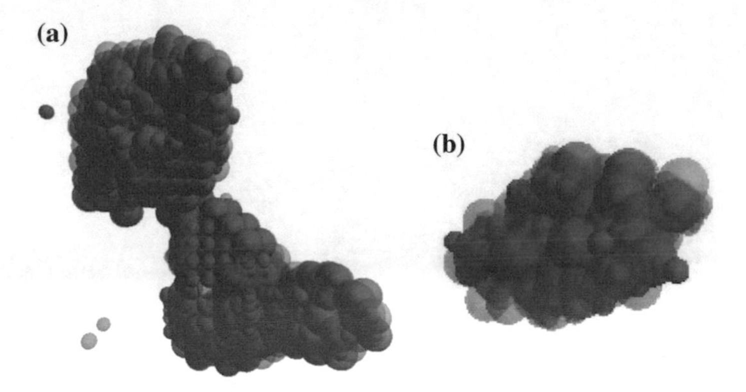

Fig. 1.7 Approximation of shape of three-dimensional objects with marching spheres; **a** approximation of the brain structure extracted from CT images (synthetic data); **b** approximation of the tumor extracted from real patient data

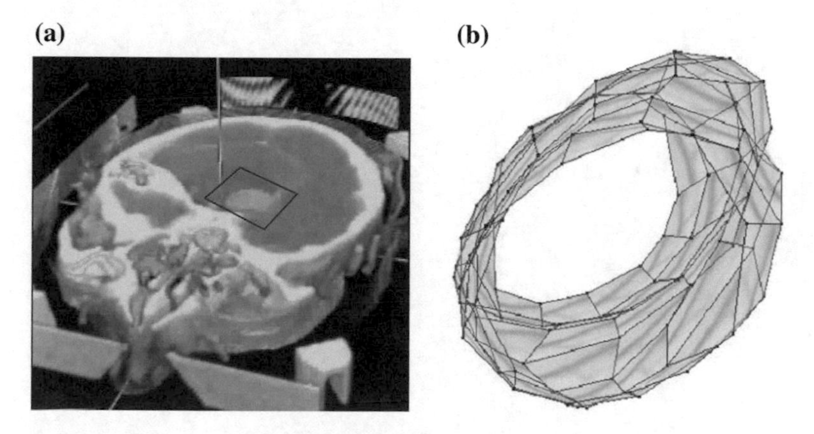

Fig. 1.8 Algorithm for 3D object's shape determination. **a** 3D model of the patient's head containing a section of the tumor in the marked region; **b** final shape after training has finished with a total of 170 versors M (associated with 170 neural units)

Finally, in the chapters of Parts V and Part VI, the reader will find plenty of illustrations using real images and robots where we have used in a smart and creative manner the geometric algebra language. We do hope to encourage readers to use this powerful and promising framework to design new real-time algorithms for perception and action systems.

Part I
Fundamentals of Geometric Algebra

Chapter 2
Introduction to Geometric Algebra

This chapter gives a detailed outline of geometric algebra and explains the related traditional algebras in common use by mathematicians, physicists, computer scientists, and engineers.

2.1 History of Geometric Algebra

Historically, Clifford algebra in its geometric interpretation has constituted the general framework for the embedding and development of ideas of multilinear algebra, multivariable analysis, and the representation theory of Lie groups and Lie algebras. This trend toward geometric algebra started in 300 B.C. with the synthetic geometry of Euclid and has continued to evolve into the present. The analytic geometry of Descartes (1637), the complex algebra of Wessel and Gauss (1798), Hamilton algebra (1843), matrix algebra (Cayley, 1854), exterior algebra (Grassmann, 1844), Clifford algebra (1878), the tensor algebra of Ricci (1890), the differential forms of Cartan (1923), and the spin algebra of Pauli and Dirac (1928) has all contributed to a maturing geometric algebra framework. Geometric algebra offers a multivector concept for representation and a geometric product for multivector computation, which allow for a versatile higher-order representation and computation in domains of different dimensions and metrics. Complex numbers, quaternions, and dual quaternions can all be represented in both rotor and motor bivector algebras. Moreover, double, or hyperbolic, numbers can also be found in geometric algebras of positive signature. Local analysis at tangent spaces, which requires differential operations to enhance the geometric symmetries of invariants, has been done successfully using Lie algebra and Lie theory. Since the Lie algebras are isomorphic with bivector algebras, such differential operations can be advantageously implemented for complex computations of differential geometry, as in the recognition of higher-order symmetries. Projective geometry and multilinear algebra, too, are elegantly reconciled in Clifford

E. Bayro-Corrochano, *Geometric Algebra Applications Vol. I*,
https://doi.org/10.1007/978-3-319-74830-6_2

algebra, providing the resulting algebra of incidence with the duality principle, inner product, and outer morphisms.

In the middle of the nineteenth century, J. Liouville proved, for the three-dimensional case, that any conformal mapping on the whole of $\mathbb{R}^n$ can be expressed as a composition of *inversions in spheres* and *reflections in hyperplanes*, [305]. In particular, *rotation, translation, dilation*, and *inversion* mappings will be obtained with these two mappings. In conformal geometric algebra, these concepts are simplified, because due to the isomorphism between the conformal group on $\mathbb{R}^n$ and the Lorentz group on $\mathbb{R}^{n+1}$, we can easily express, with a linear Lorentz transformation, a nonlinear conformal transformation, and then use *versor* representation to simplify *composition of transformations* with *multiplication of vectors* [185]. Thus, with conformal geometric algebra it is computationally more efficient and simpler to interpret the geometry of the conformal mappings than with matrix algebra. Conformal geometric algebra is the fusion of the Clifford geometric algebra and the non-Euclidean hyperbolic geometry. One of the results of the non-Euclidean geometry demonstrated by Nikolai Lobachevsky in the nineteenth century is that in spaces with hyperbolic structure, we can find subsets that are isomorphic to a Euclidean space. In order to do this, Lobachevsky introduced two constraints, to what is now called the *conformal point* $\boldsymbol{x}_c \in \mathbb{R}^{n+1,1}$. The first constraint is the *homogeneous* representation, *normalizing* the vector $\boldsymbol{x}_c$ such that $\boldsymbol{x}_c \cdot e_\infty = -1$, and the second constraint is such that the vector must be a *null vector*, that is, $\boldsymbol{x}_c^2 = 0$.

Conformal geometric algebra offers enormous advantages by computing in the null cone space. The space $\mathbb{R}^n$ is extended with two null vectors (the origin and the point at infinity), which are used to represent the basic computational unit sphere. In conformal geometric algebra, one represents points, lines, planes, and volumes using unit spheres. Conformal transformations are represented using combinations of versors that are generated via reflections with respect to spheres. Its homogeneous representation allows projective computations; via the inner product one recovers the Euclidean metric. As a result, conformal geometric algebra constitutes a suitable computational framework for dealing with a variety of conformal and projective problems involving time dependence as in causality and space and time analysis. One of the criticisms of conformal geometric algebra is the wrong idea that this system can manipulate only basic entities (points, lines, planes, and spheres), and therefore it won't be useful to model general two- and three-dimensional objects, curves, surfaces, or any other nonlinear entity required to solve a problem of a perception–action system in robotics and computer vision. Surprisingly, ruled surfaces can be treated in conformal geometric algebra for promising applications like the motion guidance of very nonlinear curves, reaching and 3D object manipulation on very nonlinear surfaces.

Our initial attempts to use geometric algebra in have been successful, reinforcing our opinion that there is no need to abandon this framework in order to carry out different kinds of computations. For all of these reasons, we believe that the single unifying language of geometric algebra offers the strongest potential for building perception and action systems; it allows us better to understand links between different fields, incorporate techniques from one field into another, reformulate old procedures, and find extensions by widening their sphere of applicability. Finally, geometric algebra helps to reduce the complexity of algebraic expressions, and as a result improves algorithms both in speed and accuracy.

2.2 What is Geometric Algebra?

The algebras of Clifford and Grassmann are well known to pure mathematicians, but at the beginning they were abandoned by physicists in favor of the vector algebra of Gibbs, the commonly used algebra today in most areas of physics. Geometric algebra is a coordinate-free approach to geometry based on the algebras of Grassmann [112] and Clifford [58]. The geometric approach to Clifford algebra adopted in this book was pioneered in the 1960s by David Hestenes [128], who has since worked on developing his version of Clifford algebra—which will be referred to as *geometric algebra* in this volume—into a unifying language for mathematics and physics [130, 138]. Hestenes also presented a study of projective geometry using Clifford algebra [139] and recently the essential concepts of the conformal geometric algebra [185]. The introductory sections of this chapter will present the basic definitions of geometric algebra.

The reader can find in the Appendix A (Chap. 20), an outline of classical Clifford algebras with explanations of its related subjects and many complementary mathematical definitions which will help to the more mathematical inclined reader to have a broad view of the role of geometric algebra and Clifford algebra. Also a reader with no mathematical background can consult this Appendix to learn more about some concepts and related definitions.

For this mathematical system, we denote scalars with lowercase letters, matrices with uppercase letters, and we use bold lowercase for both vectors in three dimensions and the bivector parts of spinors. Spinors and dual quaternions in four dimensions are denoted by bold uppercase letters.

2.2.1 *Basic Definitions*

Let V^n be a vector space of dimension n. We are going to define and generate an algebra G_n, called a *geometric algebra*. Let $\{e_1, e_2, \ldots, e_n\}$ be a set of orthonormal basis vectors of V^n. The geometric algebra G_n is equipped with a quadratic form Q. The geometric algebra $G(V^n, Q)$ is generated by V^n subject to the condition

$$v^2 = Q(v), \quad \forall v \in V^n. \tag{2.1}$$

This is the fundamental Clifford identity. If the characteristic of the ground field F is not 2, this condition can be rewritten in the following form

$$uv + vu = 2 < u, v >, \quad \forall u, v \in V^n, \tag{2.2}$$

where $< u, v >= \frac{1}{2}(Q(u+v) - Q(u) - Q(v)))$ is the symmetric bilinear form associated to Q. Regarding the quadratic form Q, one can notice that Clifford or geometric algebras are closely related to exterior algebras. As a matter of fact, if $Q = 0$ then the Clifford or geometric algebra $G(V^n, Q)$ is just the exterior algebra $\bigwedge(V^n)$. Since the Clifford or geometric product includes the extra information of the quadratic form Q, the Clifford or geometric product is fundamentally richer than the exterior product. According Eq. (2.2), the product of the orthonormal basis vectors is anticommutative, namely

$$e_j e_k + e_k e_j = 2 < e_j, e_k >= 0 \rightarrow e_j e_k = -e_k e_j, \quad \forall j \neq k. \tag{2.3}$$

The *scalar multiplication* and *sum* in G_n are defined in the usual way of a vector space. The *product* or *geometric product* of elements of the basis of G_n will simply be denoted by juxtaposition. In this way, from any two basis vectors e_j and e_k, a new element of the algebra is obtained and denoted as $e_j e_k \equiv e_{jk}$.

The basis vectors must *square* in $+1$, -1, or 0, this means that there are non-negative integers p, q, and r such that $n = p + q + r$ and

$$e_i e_i = e_i^2 = \begin{cases} +1 & \text{for} \quad i = 1, \ldots, p, \\ -1 & \text{for} \quad i = p+1, \ldots, p+q, \\ 0 & \text{for} \quad i = p+q+1, \ldots, n. \end{cases} \tag{2.4}$$

This *product* will be called the *geometric product* of G_n. With these operations G_n is an associative linear algebra with identity and is called the *geometric algebra* or *Clifford algebra* of dimension $n = p + q + r$, generated by the vector space V^n. It is usual to write G_n for the geometric algebra spanned by the real vector space V^n. The elements of this geometric algebra are called *multivectors*, because they are entities generated by the sum of elements of *mixed grade* of the basis set of G_n, such as

$$\mathbf{A} = \langle \mathbf{A} \rangle_0 + \langle \mathbf{A} \rangle_1 + \cdots + \langle \mathbf{A} \rangle_n, \tag{2.5}$$

where the multivector $\mathbf{A} \in G_n$ is expressed by the addition of its 0-*vector* part (scalar) $\langle \mathbf{A} \rangle_0$, its 1-*vector* part (vector) $\langle \mathbf{A} \rangle_1$, its 2-*vector* part (bivector) $\langle \mathbf{A} \rangle_2$, its 3-*vector* part (trivector) $\langle \mathbf{A} \rangle_3$,..., and its n-*vector* part $\langle \mathbf{A} \rangle_n$. We call an r-*blade* or a *blade of grade* r, to the geometric product of r linearly independent vectors. Because the addition of k-vectors (homogeneous vectors of grade k) is closed and the multiplication of a k-vector by a scalar is another k-vector, the set of all k-vectors is a vector space, denoted $\overset{k}{\bigwedge} V^n$. Each of these spaces is spanned by, $\binom{n}{k}$ k-vectors, where $\binom{n}{k} = \frac{n!}{(n-k)!k!}$.

Thus, our geometric algebra G_n, which is spanned by $\sum_{k=0}^{n} \binom{n}{k} = 2^n$ elements, is a direct sum of its homogeneous subspaces of grades 0, 1, 2,..., n, that is,

$$G_n = \bigwedge^0 V^n \oplus \bigwedge^1 V^n \oplus \bigwedge^2 V^n \oplus \cdots \oplus \bigwedge^k V^n \oplus \cdots \oplus \bigwedge^n V^n, \tag{2.6}$$

where $\bigwedge^0 V^n = R$ is the set of real numbers and $\bigwedge^1 V^n = V^n$. Thus, any multivector of G_n can be expressed in terms of the basis of these subspaces. For instance, in the geometric algebra G_3 with multivector basis

$$\{1, e_1, e_2, e_3, e_{12}, e_{31}, e_{23}, I_3 = e_{123}\}, \tag{2.7}$$

a *typical* multivector u will be of the form

$$\mathbf{u} = \alpha_0 + \alpha_1 e_1 + \alpha_2 e_2 + \alpha_3 e_3 + \alpha_4 e_{12} + \alpha_5 e_{31} + \alpha_6 e_{23} + \alpha_7 I_3, \tag{2.8}$$

where the α_i's are real numbers. When a subalgebra is generated only by multivectors of even grade, it is called the *even* subalgebra of G_n and will be denoted as G_n^+. Note, however, that the set of odd multivectors is not a subalgebra of G_n.

In $G_{p,q,r}$ we have the *pseudoscalar* $I_n \equiv e_{12\cdots n}$, but we can even define a *pseudoscalar* for each set of vectors that square in a positive number, negative number, and zero, that is, $I_p \equiv e_{12\cdots p}$, $I_q \equiv e_{(p+1)(p+2)\cdots(p+q)}$, and $I_r \equiv e_{(p+q+1)(p+q+2)\cdots(p+q+r)}$, where $n = p + q + r$. We say that $G_{p,q,r}$ is degenerated if $r \neq 0$.

2.2.2 Non-orthonormal Frames and Reciprocal Frames

A frame $\{e_1, e_2, \ldots, e_n\}$ span a n-dimensional vector space $\mathbb{R}^n$ and it can be orthonormal or not orthonormal. Consider a set of n linearly independent vectors $\{e_k\}$ which are not orthonormal. Any vector $\boldsymbol{v}$ can be represented uniquely in tensor notation using this set

$$\boldsymbol{v} = v^k e_k. \tag{2.9}$$

To find the components (v^k), we resort to the use of the so-called $reciprocal\ frame$ which fulfills

$$e^i \cdot e_j = \delta^i_j. \tag{2.10}$$

By computing the inner product of $\{e_k\}$ with the vector $\boldsymbol{v}$, one gets the (v^k)

$$e^k \cdot \boldsymbol{v} = e^k \cdot (v^k e_k) = e^i \cdot e_j v^j = v^j \delta^i_j = v^k. \tag{2.11}$$

In order to construct a reciprocal frame, we can use an entirely geometric approach. We begin with the first e^1 which it has to lye outside of the hyperplane $\Pi = e_2 \wedge e_3 \wedge \cdots \wedge e_n\}$. The vector perpendicular to Π is computed via dualisation. For that we use the dualisation operator or Pseudoscalar I

$$e^1 = \alpha e_2 \wedge e_3 \wedge \cdots \wedge e_n I, \tag{2.12}$$

where alpha is some constant, which can be fixed by applying the inner product with e^1

$$\begin{aligned} 1 = e_1 \cdot e^1 &= \alpha e_1 \cdot (e_2 \wedge e_3 \wedge \cdots \wedge e_n I) \\ &= \alpha e_1 \wedge e_2 \wedge e_3 \wedge \cdots \wedge e_n I. \end{aligned} \tag{2.13}$$

Substituting the volume element

$$\boldsymbol{V}_n = e_1 \wedge e_2 \cdots \wedge e_n \tag{2.14}$$

in previous equation, one gets

$$1 = \alpha e_1 \wedge e_2 \cdots \wedge e_n I = \alpha \boldsymbol{V}_n I, \tag{2.15}$$

thus $\alpha = I^{-1} \boldsymbol{V}_n - 1$ which substituted in Eq. (2.12) yields

$$e^1 = e_2 \wedge e_3 \cdots \wedge e_n \boldsymbol{V}_{-1}. \tag{2.16}$$

This result can be extended straightforwardly to compute the other e^k of the reciprocal frame

$$e^k = (-1)^{k+1} e_1 \cdots \wedge \check{e}_2 \wedge \cdots \wedge e_n \boldsymbol{V}_n^{-1}, \tag{2.17}$$

where the check indicates that the term e_k is left out in the wedge product.

2.2.3 Reciprocal Frames with Curvilinear Coordinates

In many situations, one works in non-Cartesian coordinate systems, where a coordinate system is defined by a set of scalar function $\{x^i(\boldsymbol{x})\}$ that is defined over some region. A function $F(\boldsymbol{x})$ can be expressed in terms of such coordinates as $F(x^i)$. If we apply the chain rule,

$$\nabla F = \nabla x^i \partial_i F = e^i \partial_i F, \tag{2.18}$$

this defines the so-called contravariant frame vectors $\{e^i\}$ as follows:

$$e^i = \nabla x^i. \tag{2.19}$$

In Euclidean space, these vectors are fully perpendicular to the surfaces of constant x^i. It is clear that these vectors have vanishing curl:

$$\nabla \wedge e^i = \nabla \wedge (\nabla x^i) = 0. \tag{2.20}$$

The reciprocal frame vectors are the coordinate vectors, which are covariant:

$$e_i = \partial_i x, \tag{2.21}$$

which are formed incrementing the x^i coordinates by keeping all others fixed. These two frames are reciprocal, because they fulfill

$$e_i \cdot e^j = (\partial_i x) \cdot \nabla x^j = \partial_i x^j = \delta^i_j. \tag{2.22}$$

Particularly for the case when the space signature is not Euclidean, we should refrain from using these kind of frame representations restricted to orthogonal frames and compensated with weighting factors as follows:

$$e_i = g_i \hat{e}_i, \qquad e^i = g_i^{-1} \hat{e}_i. \tag{2.23}$$

2.2.4 *Some Useful Formulas*

In [74], introduced the useful result which we will present next and use to recover a rotor in Sect. 6.3.3. Using the basic identity

$$\boldsymbol{x} = \boldsymbol{x}^k e_k = \boldsymbol{x} \cdot e^k e_k = \boldsymbol{x} \cdot e_k e^k, \tag{2.24}$$

many useful results can be derived, consider first

$$e_k e^k \cdot (\boldsymbol{x} \wedge \boldsymbol{y}) = e_k (e^k \cdot \boldsymbol{x} \boldsymbol{y} - e^k \cdot \boldsymbol{y} \boldsymbol{x}) = \boldsymbol{x} \boldsymbol{y} - \boldsymbol{y} \boldsymbol{x} = 2\boldsymbol{x} \wedge \boldsymbol{y}. \tag{2.25}$$

Using induction and utilizing again the Eq. (2.24), this result can be generalized for a p-grade multivector $\boldsymbol{X}_p$

$$e_k e^k \cdot \boldsymbol{X}_p = p \boldsymbol{X}_p. \tag{2.26}$$

Consider now

$$e_k e^k = e_k (e_j \cdot e_k e^j) = e_j \cdot e_k e^k e^j. \tag{2.27}$$

Note that the inner product $e_j \cdot e_k$ is symmetric with respect to j, k, thus one can only pick up the component of $e^k e^j$. This is only a scalar, thus one can only get a scalar contribution to the sum, namely

$$e_k e^k = e_k \cdot e^k = n, \tag{2.28}$$

where n corresponds to the dimension of the space. Now computing the wedge product

$$e_k e^k \wedge \boldsymbol{X}_p = e_k(e^k \boldsymbol{X}_p - e^k \cdot \boldsymbol{X}_p) = (n-p)\boldsymbol{X}_p, \tag{2.29}$$

and combining the above results, one gets finally

$$e_k \boldsymbol{X}_p e^k = (-1)^p e_k(e^k \wedge \boldsymbol{X}_p - e^k \cdot \boldsymbol{X}_p) = (-1)^p(n-2p)\boldsymbol{X}_p. \tag{2.30}$$

2.3 Multivector Products

It will be convenient to define other products between the elements of this algebra, which will allow us to set up several geometric relations (unions, intersections, projections, etc.) between different geometric entities (points, lines, planes, spheres, etc.) in a very simple way.

First, we define the *inner product*, $\boldsymbol{a} \cdot \boldsymbol{b}$, and the *exterior or wedge product*, $\boldsymbol{a} \wedge \boldsymbol{b}$, of any two 1-*vectors* $\boldsymbol{a}, \boldsymbol{b} \in G_n$, as the *symmetric* and *antisymmetric* parts of the geometric product $\boldsymbol{ab}$, respectively. That is, from the expression

$$\boldsymbol{ab} = \overbrace{\frac{1}{2}(\boldsymbol{ab} + \boldsymbol{ba})}^{symmetric\ part} + \overbrace{\frac{1}{2}(\boldsymbol{ab} - \boldsymbol{ba})}^{antisymmetric\ part}, \tag{2.31}$$

we define the *inner product*

$$\boldsymbol{a} \cdot \boldsymbol{b} \equiv \frac{1}{2}(\boldsymbol{ab} + \boldsymbol{ba}) \tag{2.32}$$

and the *outer* or *wedge product*

$$\boldsymbol{a} \wedge \boldsymbol{b} \equiv \frac{1}{2}(\boldsymbol{ab} - \boldsymbol{ba}). \tag{2.33}$$

Thus, we can now express the geometric product of two vectors in terms of these two new operations as

$$\boldsymbol{ab} = \boldsymbol{a} \cdot \boldsymbol{b} + \boldsymbol{a} \wedge \boldsymbol{b}. \tag{2.34}$$

From (2.32) and (2.33), $\boldsymbol{a} \cdot \boldsymbol{b} = \boldsymbol{b} \cdot \boldsymbol{a}$ and $\boldsymbol{a} \wedge \boldsymbol{b} = -\boldsymbol{b} \wedge \boldsymbol{a}$.

The inner product of the vectors $\boldsymbol{a}$ and $\boldsymbol{b}$ corresponds to the standard *scalar* or *dot* product and produces a scalar. In a subsection ahead, we will discuss in detail other complementary definition: of inner product called *contraction* introduced by Lounesto [197]. The outer or wedge product of these vectors is a new quantity,

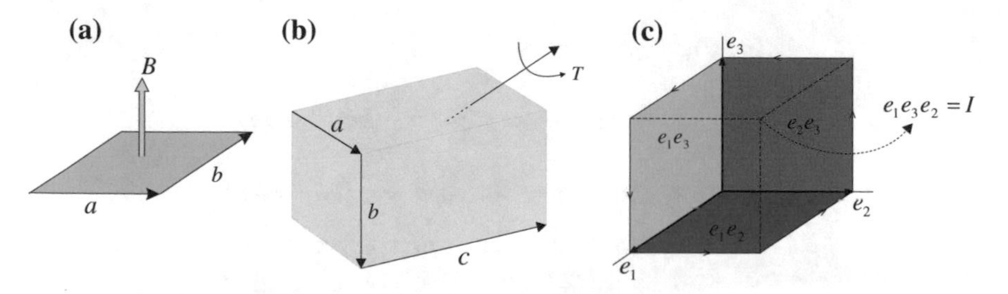

Fig. 2.1 **a** (left) The directed area, or bivector, $\boldsymbol{B} = \boldsymbol{a} \wedge \boldsymbol{b}$; **b** (middle) the oriented volume, or trivector, $\boldsymbol{T} = \boldsymbol{a} \wedge \boldsymbol{b} \wedge \boldsymbol{c}$; **c** (right) multivector basis of the Euclidean 3D space

which we call a *bivector*. We can think of a bivector as an oriented area in the plane containing $\boldsymbol{a}$ and $\boldsymbol{b}$ formed by sweeping $\boldsymbol{a}$ along $\boldsymbol{b}$, see Fig. 2.1a. Thus, $\boldsymbol{b} \wedge \boldsymbol{a}$ will have an opposite orientation, because the wedge product is anticommutative. The outer product is immediately generalizable to higher dimensions: for example, $(\boldsymbol{a} \wedge \boldsymbol{b}) \wedge \boldsymbol{c}$, a *trivector*, is interpreted as the oriented volume formed by sweeping the area $\boldsymbol{a} \wedge \boldsymbol{b}$ along vector $\boldsymbol{c}$, see Fig. 2.1b. Note that the outer product does not determine the spatial forms of the bivector and trivector: it computes only their orientations and values of their involved determinants corresponding to the area and volume respectively. Figure 2.1a, b depict such forms simply to favor the geometric understanding. The Fig. 2.1c depicts the multivector basis of the Euclidean 3D space.

The symmetric part of the geometric product of two vectors $\boldsymbol{a}$ and $\boldsymbol{b}$ is called the *anticommutator product* and the antisymmetric part the *commutatorproduct*. These expressions are denoted as follows:

$$\boldsymbol{a} \overline{\times} \boldsymbol{b} = \frac{1}{2}(\boldsymbol{ab} + \boldsymbol{ba}) \qquad \texttt{anticommutator product}, \tag{2.35}$$

$$\boldsymbol{a} \underline{\times} \boldsymbol{b} = \frac{1}{2}(\boldsymbol{ab} - \boldsymbol{ba}) \qquad \texttt{commutator product}. \tag{2.36}$$

In general, the geometric product of two multivectors $\boldsymbol{A}, \boldsymbol{B} \in G_n$ can be written as the sum of their commutator and anticommutator:

$$\boldsymbol{AB} = \frac{1}{2}(\boldsymbol{AB} + \boldsymbol{BA}) + \frac{1}{2}(\boldsymbol{AB} - \boldsymbol{BA}) = \boldsymbol{A} \overline{\times} \boldsymbol{B} + \boldsymbol{A} \underline{\times} \boldsymbol{B}. \tag{2.37}$$

In the literature, the commutator of two multivectors $\boldsymbol{A}, \boldsymbol{B} \in G_n$ is usually written as $[\boldsymbol{A}, \boldsymbol{B}]$ and the anticommutator as $\{\boldsymbol{A}, \boldsymbol{B}\}$, although that we prefer to emphasize their role within the geometric product by using the operator symbols $\overline{\times}$ and $\underline{\times}$.

The geometric product of a vector and a bivector reads

$$\begin{aligned} a(b\wedge c) &= \frac{1}{2}a(bc - cb) \\ &= (a\cdot b)c - (a\cdot c)b - \frac{1}{2}(bac - cab) \\ &= 2(a\cdot b)c - 2(a\cdot c)b + \frac{1}{2}(bc - ca)a \\ &= 2(a\cdot b)c - 2(a\cdot c)b + (a\wedge c)a. \end{aligned} \tag{2.38}$$

From this result, one gets an extension of the definition of the inner product:

$$a\cdot(b\wedge c) = \frac{1}{2}[(a(b\wedge c) - (b\wedge c)a] = (a\cdot b)c - (a\cdot c)b. \tag{2.39}$$

The remaining part of the product corresponds to a trivector:

$$a\wedge(b\wedge c) = \frac{1}{2}[(a(b\wedge c) + (b\wedge c)a] = a\wedge b\wedge c. \tag{2.40}$$

So the geometric product of a vector and a bivector is

$$a(b\wedge c) = a\cdot(b\wedge c) + a\wedge(b\wedge c) = (a\cdot b)c - (a\cdot c)b + a\wedge b\wedge c. \tag{2.41}$$

The reader can see that in geometric algebra one has to handle expressions with large number of brackets; thus, we should follow an operator ordering convention: in the absence of brackets, inner and outer products take precedence over the geometric products, so we can write

$$(a\cdot b)c = a\cdot bc, \tag{2.42}$$

where the right-hand side is not to be confused with $a\cdot(bc)$.

2.3.1 Further Properties of the Geometric Product

The geometric product of a vector and a bivector extends simply to that of a vector a and a grade-r multivector A_r (composed of r orthogonal blades a_r). To carry out the geometric product, first decompose A_r into blades

$$\begin{aligned} aA_r &= (aa_1)a_2\ldots a_r = (a\cdot a_1)a_2\ldots a_r + (a\wedge a_1)a_2\ldots a_r \\ &= 2(a\cdot a_1)a_2\ldots a_r - (a_1 a)a_2\ldots a_r \\ &= 2a\cdot a_1a_2\ldots a_r - 2a\cdot a_2a_1a_3\ldots a_r + a_1a_2aa_3\ldots a_r \\ &= 2\sum_{k=1}^{r}(-1)^{k+1}a\cdot a_k a_1 a_2\ldots \check{a}_k\ldots a_r + (-1)^r a_1a_2\ldots a_r a, \end{aligned} \tag{2.43}$$

where the check indicates that this term is missing from the series. Each term in the sum has grade $r-1$, so one can define

$$\boldsymbol{a}\cdot\boldsymbol{A}_r =< \boldsymbol{a}\boldsymbol{A}_r >_{r-1}= \frac{1}{2}(\boldsymbol{a}\boldsymbol{A}_r - (-1)^r\boldsymbol{A}_r\boldsymbol{a}) \tag{2.44}$$

and the right term is the remaining antisymmetric part. One can therefore write

$$\boldsymbol{a}\wedge\boldsymbol{A}_r =< \boldsymbol{a}\boldsymbol{A}_r >_{r+1}= \frac{1}{2}(\boldsymbol{a}\boldsymbol{A}_r + (-1)^r\boldsymbol{A}_r\boldsymbol{a}). \tag{2.45}$$

Thus, the geometric product of a vector and a grade-r multivector $\boldsymbol{A}_r$ can be written as

$$\boldsymbol{a}\boldsymbol{A}_r = \boldsymbol{a}\cdot\boldsymbol{A}_r + \boldsymbol{a}\wedge\boldsymbol{A}_r =< \boldsymbol{a}\boldsymbol{A}_r >_{r-1} + < \boldsymbol{a}\boldsymbol{A}_r >_{r+1}. \tag{2.46}$$

Geometric product multiplication by a vector lowers and raises the grade of a multivector by 1. In Eq. (2.43), we assumed that the vectors are orthogonal: we can extend this decomposition for the case of non-orthogonal vectors as follows:

$$\begin{aligned}\boldsymbol{a}\cdot(\boldsymbol{a}_1\wedge\boldsymbol{a}_2\wedge\ldots\boldsymbol{a}_r) &= \frac{1}{2}[\boldsymbol{a} < \boldsymbol{a}_1\boldsymbol{a}_2\ldots\boldsymbol{a}_r >_r -(-1)^r < \boldsymbol{a}_1\boldsymbol{a}_2\wedge\ldots\boldsymbol{a}_r >_r \boldsymbol{a}]\\ &= \frac{1}{2} < \boldsymbol{a}\boldsymbol{a}_1\boldsymbol{a}_2\wedge\ldots\boldsymbol{a}_r - (-1)^r\boldsymbol{a}_1\boldsymbol{a}_2\wedge\ldots\boldsymbol{a}_r\boldsymbol{a} >_{r-1}. \end{aligned} \tag{2.47}$$

In this equation, since we are expecting to lower the r-grade by 1, we can insert $\boldsymbol{a}$ within the brackets. Now, by reutilizing equation (2.43), we can write

$$\begin{aligned}\boldsymbol{a}\cdot(\boldsymbol{a}_1\wedge\boldsymbol{a}_2\wedge\ldots\wedge\boldsymbol{a}_r) &= \Big\langle\sum_{k=1}^{r}(-1)^{k+1}\boldsymbol{a}\cdot\boldsymbol{a}_k\boldsymbol{a}_1\boldsymbol{a}_2\ldots\check{\boldsymbol{a}}_k\ldots\boldsymbol{a}_r\Big\rangle_{r-1}\\ &= \sum_{k=1}^{r}(-1)^{k+1}\boldsymbol{a}\cdot\boldsymbol{a}_k\boldsymbol{a}_1\boldsymbol{a}_2\ldots\check{\boldsymbol{a}}_k\ldots\boldsymbol{a}_r. \end{aligned} \tag{2.48}$$

This result is very useful in practice; see the following examples:

$$\begin{aligned}\boldsymbol{a}\cdot(\boldsymbol{a}_1\wedge\boldsymbol{a}_2) &= \boldsymbol{a}\cdot\boldsymbol{a}_1\boldsymbol{a}_2 - \boldsymbol{a}\cdot\boldsymbol{a}_2\boldsymbol{a}_1,\\ \boldsymbol{a}\cdot(\boldsymbol{a}_1\wedge\boldsymbol{a}_2\wedge\boldsymbol{a}_3) &= \boldsymbol{a}\cdot\boldsymbol{a}_1\boldsymbol{a}_2\wedge\boldsymbol{a}_3 - \boldsymbol{a}\cdot\boldsymbol{a}_2\boldsymbol{a}_1\wedge\boldsymbol{a}_3 + \boldsymbol{a}\cdot\boldsymbol{a}_3\boldsymbol{a}_1\wedge\boldsymbol{a}_2 \end{aligned} \tag{2.49}$$

The generalization of Eq. (2.48) will be discussed in detail in later section.

2.3.2 Projections and Rejections

Let be $\boldsymbol{A}_r$ a non-degenerated r-blade or hyperplane in G_n, then any vector $\boldsymbol{v} \in \mathbb{R}^n$ has a *projection* (parallel component) onto the space of $\boldsymbol{A}_r$ defined by

$$P^{\parallel}_{\boldsymbol{A}_r} = (\boldsymbol{v} \cdot \boldsymbol{A}_r)\boldsymbol{A}_r^{-1}, \tag{2.50}$$

and a *rejection* (perpendicular component) from the space of $\boldsymbol{A}_r$ defined by

$$P^{\perp}_{\boldsymbol{A}_r} = (\boldsymbol{v} \wedge \boldsymbol{A}_r)\boldsymbol{A}_r^{-1}. \tag{2.51}$$

Therefore, $\boldsymbol{v}$ can be *split* in terms of a projection and a *rejection* as follows:

$$\boldsymbol{v} = \boldsymbol{v}_{\parallel} + \boldsymbol{v}_{\perp} = P^{\parallel}_{\boldsymbol{A}_r} + P^{\perp}_{\boldsymbol{A}_r}, \tag{2.52}$$

which is an orthogonal decomposition of $\boldsymbol{v}$ with respect to $\boldsymbol{A}_r$. Figure 2.2 shows the projection, rejection, and inner product of the vector $\boldsymbol{v}$ with respect to the space (plane) $\boldsymbol{A}_2 = \boldsymbol{x} \wedge \boldsymbol{y}$. Note that the vector $\boldsymbol{v} \cdot \boldsymbol{x} \wedge \boldsymbol{y}$ is perpendicular to the vector $P^{\parallel}_{\boldsymbol{A}_2}$.

2.3.3 Projective Split

The idea of the *projective split* was introduced by Hestenes [131] in order to connect *projective geometry* and *metric geometry*. This is done by associating the even subalgebra of G_{n+1} with the geometric algebra of the next lower dimension, G_n. The multivector basis of G_{n+1} and G_n is given by $\gamma_1, \gamma_2, \ldots, \gamma_{n+1}$ and $e_1, e_2, \ldots, e_n$, respectively. One can define a mapping between the spaces by choosing a preferred direction in G_{n+1}, γ_{n+1}. Then, by taking the geometric product of a vector $\mathbf{X} \in G_{n+1}$ and γ_{n+1},

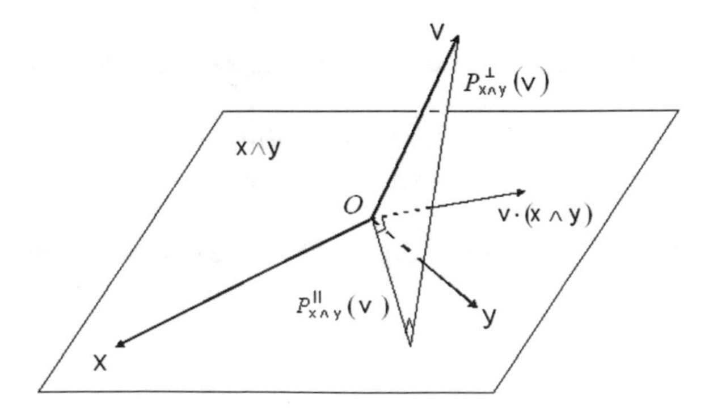

Fig. 2.2 Projection, rejection, and inner product with respect to $\boldsymbol{A}_2 = \boldsymbol{x} \wedge \boldsymbol{y}$

$$\mathbf{X}\gamma_{n+1} = \mathbf{X}\cdot\gamma_{n+1} + \mathbf{X}\wedge\gamma_{n+1} = \mathbf{X}\cdot\gamma_{n+1}\big(1 + \frac{\mathbf{X}\wedge\gamma_{n+1}}{\mathbf{X}\cdot\gamma_{n+1}}\big), \tag{2.53}$$

the vector $\boldsymbol{x} \in G_n$ can be associated with the bivector $\frac{\mathbf{X}\wedge\gamma_{n+1}}{\mathbf{X}\cdot\gamma_{n+1}} \in G_{n+1}$. This result can be projectively interpreted as the pencil of all lines passing though the point γ_{n+1}. In physics, the projective split is called the *space-time split*, and it relates a space-time system $G_{3,1}$ with a Minkowski metric to an observable system G_3 with a Euclidean metric.

2.3.4 Generalized Inner Product

Multivector computing involving inner products is easier if one employs the next equality for the the generalized inner product of two blades $\boldsymbol{A}_r = \boldsymbol{a}_1 \wedge \boldsymbol{a}_2 \wedge \cdots \wedge \boldsymbol{a}_r$, and $\boldsymbol{B}_s = \boldsymbol{b}_1 \wedge \boldsymbol{b}_2 \wedge \cdots \wedge \boldsymbol{b}_s$:

$$\boldsymbol{A}_r \cdot \boldsymbol{B}_s = \begin{cases} ((\boldsymbol{a}_1 \wedge \boldsymbol{a}_2 \wedge \cdots \wedge \boldsymbol{a}_r)\cdot \boldsymbol{b}_1)\cdot(\boldsymbol{b}_2 \wedge \boldsymbol{b}_3 \wedge \cdots \wedge \boldsymbol{b}_s) & \text{if } r \geq s \\ (\boldsymbol{a}_1 \wedge \boldsymbol{a}_2 \wedge \cdots \wedge \boldsymbol{a}_{r-1})\cdot(\mathbf{a}_r \cdot (\boldsymbol{b}_1 \wedge \boldsymbol{b}_2 \wedge \cdots \wedge \boldsymbol{b}_s)) & \text{if } r < s. \end{cases} \tag{2.54}$$

For *right contraction* use the following equation:

$$(\boldsymbol{a}_1 \wedge \boldsymbol{a}_2 \wedge \cdots \wedge \boldsymbol{a}_r)\cdot \boldsymbol{b}_1 = \sum_{i=1}^{r} (-1)^{r-i}\, \mathbf{a}_1 \wedge \cdots \wedge \mathbf{a}_{i-1} \wedge (\boldsymbol{a}_i \cdot \boldsymbol{b}_1) \wedge \boldsymbol{a}_{i+1} \wedge \cdots \wedge \boldsymbol{a}_r \tag{2.55}$$

and for *left contraction*

$$\mathbf{a}_r \cdot (\boldsymbol{b}_1 \wedge \boldsymbol{b}_2 \wedge \cdots \wedge \boldsymbol{b}_s) = \sum_{i=1}^{s} (-1)^{i-1}\, \boldsymbol{b}_1 \wedge \cdots \wedge \boldsymbol{b}_{i-1} \wedge (\boldsymbol{a}_r \cdot \boldsymbol{b}_i) \wedge \boldsymbol{b}_{i+1} \wedge \cdots \wedge \boldsymbol{b}_s. \tag{2.56}$$

From these equations, we can see that the inner product is not commutative for general multivectors. Indeed, if, in particular, $\mathbf{a}$, $\boldsymbol{b}$, and $\mathbf{c}$ are 1-vectors, then from (2.56),

$$\mathbf{a}\cdot(\boldsymbol{b}\wedge\mathbf{c}) = (\mathbf{a}\cdot\boldsymbol{b})\mathbf{c} - (\mathbf{a}\cdot\mathbf{c})\boldsymbol{b}, \tag{2.57}$$

and from Eq. (2.55),

$$(\boldsymbol{b}\wedge\mathbf{c})\cdot\boldsymbol{a} = \boldsymbol{b}(\mathbf{c}\cdot\boldsymbol{a}) - (\boldsymbol{b}\cdot\boldsymbol{a})\mathbf{c} = -(\mathbf{a}\cdot\boldsymbol{b})\mathbf{c} + (\mathbf{a}\cdot\mathbf{c})\boldsymbol{b}. \tag{2.58}$$

Thus, the inner product of a vector $\boldsymbol{a}$ and the bivector $\boldsymbol{B} = \boldsymbol{b}\wedge\boldsymbol{c}$ fulfills

$$\boldsymbol{a}\cdot\boldsymbol{B} = -\boldsymbol{B}\cdot\boldsymbol{a}. \tag{2.59}$$

Let us now consider the inner product for the case $r < s$, given the bivector $\boldsymbol{A}_2 = \boldsymbol{a}_1 \wedge \boldsymbol{a}_2$ and the trivector $\boldsymbol{B}_3 = \boldsymbol{b}_1 \wedge \boldsymbol{b}_2 \wedge b_3$. Then, applying Eq. (2.56) from the left, we should get a vector $\boldsymbol{b}$:

$$
\begin{aligned}
\boldsymbol{A}_2 \cdot \boldsymbol{B}_3 &= < \boldsymbol{A}_2 \cdot \boldsymbol{B}_3 >_{|2-3|} = < \boldsymbol{A}_2 \cdot \boldsymbol{B}_3 >_1 \\
&= (\boldsymbol{a}_1 \wedge \boldsymbol{a}_2) \cdot (\boldsymbol{b}_1 \wedge \boldsymbol{b}_2 \wedge \boldsymbol{b}_3) = \boldsymbol{a}_1 \cdot [\boldsymbol{a}_2 \cdot (\boldsymbol{b}_1 \wedge \boldsymbol{b}_2 \wedge \boldsymbol{b}_3)] \\
&= \boldsymbol{a}_1 \cdot [(\boldsymbol{a}_2 \cdot \boldsymbol{b}_1)\boldsymbol{b}_2 \wedge \boldsymbol{b}_3 - (\boldsymbol{a}_2 \cdot \boldsymbol{b}_2)\boldsymbol{b}_1 \wedge \boldsymbol{b}_3 + (\boldsymbol{a}_2 \cdot \boldsymbol{b}_3)\boldsymbol{b}_1 \wedge \boldsymbol{b}_2] \\
&= (\boldsymbol{a}_2 \cdot \boldsymbol{b}_2)[(\boldsymbol{a}_1 \cdot \boldsymbol{b}_1)\boldsymbol{b}_3 - (\boldsymbol{a}_1 \cdot \boldsymbol{b}_3)\boldsymbol{b}_1] + \\
&\quad +(\boldsymbol{a}_2 \cdot \boldsymbol{b}_3)[(\boldsymbol{a}_1 \cdot \boldsymbol{b}_1)\boldsymbol{b}_2 - (\boldsymbol{a}_1 \cdot \boldsymbol{b}_2)\boldsymbol{b}_1] \\
&= [(\boldsymbol{a}_1 \cdot \boldsymbol{b}_3)[(\boldsymbol{a}_2 \cdot \boldsymbol{b}_2) - (\boldsymbol{a}_1 \cdot \boldsymbol{b}_2)(\boldsymbol{a}_2 \cdot \boldsymbol{b}_3)]\boldsymbol{b}_1 + \\
&\quad +[(\boldsymbol{a}_1 \cdot \boldsymbol{b}_1)[(\boldsymbol{a}_2 \cdot \boldsymbol{b}_3) - (\boldsymbol{a}_1 \cdot \boldsymbol{b}_3)(\boldsymbol{a}_2 \cdot \boldsymbol{b}_1)]\boldsymbol{b}_2 + \\
&\quad +[(\boldsymbol{a}_1 \cdot \boldsymbol{b}_2)[(\boldsymbol{a}_2 \cdot \boldsymbol{b}_1) - (\boldsymbol{a}_1 \cdot \boldsymbol{b}_1)(\boldsymbol{a}_2 \cdot \boldsymbol{b}_2)]\boldsymbol{b}_3 \\
&= \beta_1 \boldsymbol{b}_1 + \beta_2 \boldsymbol{b}_2 + \beta_3 \boldsymbol{b}_3 = \boldsymbol{b}. \qquad (2.60)
\end{aligned}
$$

The projective geometric interpretation of this result tells us that the inner product of a line $\boldsymbol{A}_2$ with the plane $\boldsymbol{B}_3$ should be a vector $\boldsymbol{b}$ or point lying on the plane. Note that the point $\boldsymbol{b}$ is spanned as a linear combination of the vector basis that spans the plane $\boldsymbol{B}_3$. Later we will see that in an algebra of incidence of the projective geometry, the meet operation (inner product with a dual multivector) computes the incidence point of a line through a plane.

Let us consider two other examples of the conformal geometric algebra $G_{4,1}$. Given the Euclidean pseudoscalar $I_E = e_1 \wedge e_2 \wedge e_3$ and the Minkowski plane $E = e_4 \wedge e_5$, let us compute the square of the pseudoscalar $I = I_E \wedge E = e_1 \wedge e_2 \wedge e_3 \wedge e_4 \wedge e_5$:

$$
\begin{aligned}
I^2 &= (I_E \wedge E) \cdot (I_E \wedge E) = -[(I_E \wedge (E \cdot I_E)] \cdot E + [(I_E \cdot I_e) \wedge E] \cdot E \\
&= (I_E^2 \wedge E) \cdot E = ((-1) \cdot E) \cdot E = -E^2 = -1, \qquad (2.61)
\end{aligned}
$$

since E and I_E are orthogonal, $E \cdot I_E = 0$.

Let us now consider Eq. (2.64) $G_{n+1,1}$ given in [185], which claims that the component of the sphere s in $\mathbb{R}^n$ is given by

$$
\boldsymbol{s} = P_E^{\perp}(s) = (s \wedge E)E = s + (s \cdot e_0)e. \qquad (2.62)
$$

Since the Minkowski plane can also be expressed as $E = e_\infty \wedge e_0$, where $e_\infty = e_{n+1} + e_{n+2}$, $e_0 = 0.5(e_{n+1} - e_{n+2})$, and $E^2 = 1$, we proceed by computing the inner product from the right using Eq. (2.55) and the equality $\boldsymbol{B} \cdot (\boldsymbol{a} \wedge \boldsymbol{b}) = \boldsymbol{B} \cdot a \cdot b$, namely,

$$\begin{aligned} s &= s \cdot 1 = (sE)E = (s \cdot E)E + (s \wedge E)E = P_E^{\|}(s) + P_E^{\perp}(s) \\ &= P_E^{\perp}(s) = (s \wedge E)E = (s \wedge e_\infty \wedge e_0) \wedge (e_\infty \wedge e_0) + (s \wedge e_\infty \wedge e_0) \cdot (e_\infty \wedge e_0) \\ &= [(s \wedge e_\infty \wedge e_0) \cdot e_\infty] \cdot e_0 \\ &= [(s \wedge e_\infty) \wedge (e_0 \cdot e_\infty) - s \wedge (e_\infty \cdot e_\infty) \wedge e_0 + (s \cdot e_\infty) \wedge e_\infty \wedge e_0] \cdot e_0 \\ &= -(s \wedge \epsilon_\infty) \cdot e_0 = -\lfloor s \wedge (e_\infty \cdot e_0) - (s \cdot e_0) \wedge e_\infty] \\ &= -[-s - (s \cdot e_0) \wedge e_\infty] = s + (s \cdot e_0) \wedge e_\infty. \end{aligned} \tag{2.63}$$

Let us see how we can compute the meet $\cap$ (incidence algebra intersection) an hand the generalized inner product. Consider the meet of the k-vectors $\boldsymbol{C}_q,\ \boldsymbol{A}_s \in G_n$

$$\boldsymbol{C}_q \cap \boldsymbol{A}_s = \boldsymbol{C}^*_{n-q} \cdot \boldsymbol{A}_s = (\boldsymbol{C}_q I_n^{-1}) \cdot \boldsymbol{A}_s = \boldsymbol{B}_r \cdot \boldsymbol{A}_s, \tag{2.64}$$

where the meet operation is expressed as the inner product of the dual of $\boldsymbol{C}_q$, i.e. $\boldsymbol{C}^*_{n-q} = \boldsymbol{C}_q I_n^{-1} = \boldsymbol{B}_r$, and $\boldsymbol{A}_s$. Thereafter, this inner product can be then easily computed using the generalized inner product of Eq. (2.54).

2.3.5 Geometric Product of Multivectors

In general, any two paired multivectors can be multiplied using the geometric product. Consider two homogeneous multivectors $\boldsymbol{A}_r$ and $\boldsymbol{B}_s$ of grade r and s, respectively. The geometric product of $\boldsymbol{A}_r$ and $\boldsymbol{B}_s$ can be written as

$$\boldsymbol{A}_r \boldsymbol{B}_s = \langle \boldsymbol{AB} \rangle_{r+s} + \langle \boldsymbol{AB} \rangle_{r+s-2} + \ldots + \langle \boldsymbol{AB} \rangle_{|r-s|} \tag{2.65}$$

where $\langle \boldsymbol{AB} \rangle_t$ denotes the t-grade part the multivector $\boldsymbol{A}_r \boldsymbol{B}_s$. Note that

$$\langle \boldsymbol{A}_r \boldsymbol{B}_s \rangle_{|r-s|} \tag{2.66}$$

is a generalized contraction or inner product, and

$$\boldsymbol{A}_r \wedge \boldsymbol{B}_s \equiv \langle \boldsymbol{A}_r \boldsymbol{B}_s \rangle_{r+s}. \tag{2.67}$$

is the *outer* or *wedge product* of the highest grade. The *inner product* of lowest grade $\langle \boldsymbol{AB} \rangle_0$ corresponds to a full contraction, or standard inner product. Since the elements of $\boldsymbol{A}_r \boldsymbol{B}_s$ are of a different grade, $\boldsymbol{A}_r \boldsymbol{B}_s$ is thus an *inhomogeneous multivector.* As illustration, consider $\boldsymbol{ab} = \langle \boldsymbol{ab} \rangle_0 + \langle \boldsymbol{ab} \rangle_2 = \boldsymbol{a} \cdot \boldsymbol{b} + \boldsymbol{a} \wedge \boldsymbol{b}$ and the geometric product of $\boldsymbol{A} = 5e_3 + 3e_1e_2$ and $\boldsymbol{b} = 9e_2 + 7e_3$:

$$\begin{aligned} \boldsymbol{Ab} &= \langle \boldsymbol{Ab} \rangle_0 + \langle \boldsymbol{Ab} \rangle_1 + \langle \boldsymbol{Ab} \rangle_2 + \langle \boldsymbol{Ab} \rangle_3 \\ &= 35(e_3)^2 + 27e_1(e_2)^2 + 45e_3e_2 + 21e_1e_2e_3 \\ &= 35 + 27e_1 - 45e_2e_3 + 21I. \end{aligned} \tag{2.68}$$

In the following sections, expressions of grade 0 will be written ignoring their subindex, that is, $\langle \boldsymbol{ab}\rangle_0 = \langle \boldsymbol{ab}\rangle$.

We can see from Eq. (2.66) that the inner product of a scalar α with a homogeneous multivector $\boldsymbol{A}_r$ has a special treatment: $\boldsymbol{A}\cdot\alpha = \alpha\cdot\boldsymbol{A} = 0$. However, it is not the case for the outer product: $\boldsymbol{A}\wedge\alpha = \alpha\wedge\boldsymbol{A} = \langle\alpha\boldsymbol{A}\rangle_r \equiv \alpha\boldsymbol{A}$. We make the observation that software like GABLE or CLICAL does not have this exceptional case for the inner product. Now, the inner and outer product of any two general multivectors will be obtained applying the left and right distributive laws over their homogeneous parts and then using Eq. (2.66) and (2.67). Note that the definition in (2.66) is not in contradiction to (2.32). Analogously, (2.67) and (2.33) are consistent. From (2.66), it can be said that the inner product $\boldsymbol{A}_r\cdot\boldsymbol{B}_s$ *lowers the grade of* $\boldsymbol{A}_r$ *by s units* when $r \geq s > 0$, and from Eq. (2.67) that the outer product $\boldsymbol{A}_r\wedge\boldsymbol{B}_s$ *raises the grade of* $\boldsymbol{A}_r$ *by s units* for every $r, s \geq 0$.

2.3.6 Contractions and the Derivation

Let us consider a vector and a bivector $\boldsymbol{v}, \boldsymbol{A} \in G_3$. The vector $\boldsymbol{v}$ is tilted by an angle θ out of the plane of a bivector $\boldsymbol{A}$; see Fig. 2.3. Let $\boldsymbol{x}$ be the orthogonal projection of $\boldsymbol{v}$ in the plane, then $|\boldsymbol{x}| = |\boldsymbol{v}|cos\theta$.

In this setting, Lounesto [197] introduced the following notion: The right *contraction* of the bivector $\boldsymbol{A}$ by the vector $\boldsymbol{v}$ is a vector $\boldsymbol{y} = \boldsymbol{A} \vdash \boldsymbol{v}$ that is orthogonal to $\boldsymbol{x}$ and lies on the plane of the bivector $\boldsymbol{A}$ such that

$$\begin{aligned} &|\boldsymbol{y}| = |\boldsymbol{A}||\boldsymbol{x}| \\ &\boldsymbol{y} \perp \boldsymbol{x} \quad \text{and} \quad \boldsymbol{x}\wedge\boldsymbol{y} \parallel \boldsymbol{A}. \end{aligned} \tag{2.69}$$

According to Eq. (2.59),

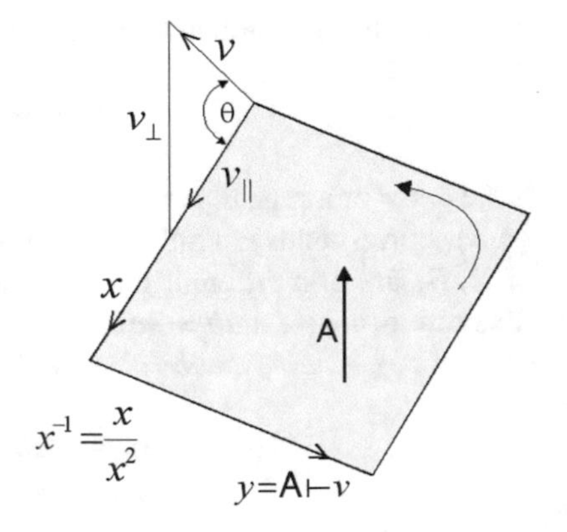

Fig. 2.3 Projection, rejection and inner product with respect to $\boldsymbol{A}_2 = \boldsymbol{a}_1 \wedge \boldsymbol{a}_2$

$$v \rfloor A = -A \lfloor v, \tag{2.70}$$

that is, the left and right contractions have opposite signs. In Fig. refvectspscontract, we see that the area of the rectangle can be computed in terms of the inverse of x, namely, $|A| = |x^{-1}||y|$, where $x^{-1} = \frac{x}{x^2}$. If we write $v_\parallel = x$ and $v_\perp = x - v_\parallel$, the vector v can be expressed in terms of its parallel and perpendicular components:

$$v = P_A^\parallel + P_A^\perp = v_\parallel + v_\perp = (v \rfloor A)A^{-1} + (v \wedge A)A^{-1}, \tag{2.71}$$

where $A^{-1} = A/A^2$, $A^2 = -|A|^2$. According to Eqs. (2.44) and (2.45), the geometric product of v and the bivector A can also be rewritten as

$$vA = v \rfloor A + v \wedge A = \frac{1}{2}(vA - Av) + \frac{1}{2}(vA + Av). \tag{2.72}$$

The left contraction can also be computed using the notion of duality, given the vectors and pseudoscalar $a,\ b,\ I \in G_3$:

$$a \rfloor b = [a \wedge (bI)]I^{-1}, \tag{2.73}$$

this equation means that the left contraction is dual to the wedge product. Later, we will discuss at length the concept of duality and the related operations of meet and join of the incidence algebra.

Now consider in our next formula, the vectors v, b, and the k-vector $A_k \in G_3$:

$$v \rfloor (A_k b) = (v \rfloor A_k)b + (-1)^k A_k (v \rfloor b). \tag{2.74}$$

We see that the left contraction by a vector is a *derivation* in the geometric algebra.

In general, the product of a vector v and an arbitrary r-vector A_r can be decomposed into a sum of the left contraction and the wedge product. Thus using Eqs. (2.44) and (2.45),

$$\begin{aligned} vA_r &= v \rfloor A_r + v \wedge A_r \\ &= \frac{1}{2}(vA_r - (-1)^r A_r v) + \frac{1}{2}(vA_r + (-1)^r A_r v). \end{aligned} \tag{2.75}$$

As we have seen in this section, the right $\rfloor$ and left $\lfloor$ contractions can be computed simply using the Eqs. (2.55) and (2.56) of the generalized inner product (2.54). Therefore in order to avoid using an excessive number of operator symbols, in this book we will not use the operators $\rfloor$ and $\lfloor$ when carrying out contractions.

2.3.7 Hodge Dual

The Hodge star operator establishes a correspondence between the space $\bigwedge^k V^n$ of k-vectors with the space $\bigwedge^{n-k} V^n$ of $(n-k)$-vectors. The dimension of the former space is $\binom{n}{k}$ and that of the latter space is $\binom{n}{n-k}$. Due to the symmetry of the binomial coefficients, the dimensions of both spaces are equal. The mapping between spaces of the same dimension is an isomorphism, which, in the case of the Hodge duality, exploits the inner product and orientation of the vector space. The image of a k-vector under this isomorphism is known as its *Hodge dual* of the k-vector.

The Hodge star operator on an oriented inner product space V^n is a linear operator on the exterior algebra of V^n. This operator interchanges the subspaces of k-vectors and $(n-k)$-vectors, as we show next. Given an oriented orthonormal basis $e_1, e_2, \ldots, e_n$, the Hodge dual of a k-vector is computed as follows

$$\star\,(e_1 \wedge e_2 \wedge \cdots \wedge e_k) = e_{k+1} \wedge e_{k+2} \wedge \cdots \wedge e_n. \tag{2.76}$$

As an illustration, let us apply the Hodge dual to the vector $\boldsymbol{y} \in \mathbb{R}^3$:

$$\begin{aligned} \boldsymbol{Y} &= \star \boldsymbol{y} = \star(y_1 e_1 + y_2 e_2 + y_3 e_3) \\ &= y_1 e_2 \wedge e_3 + y_2 e_3 \wedge e_1 + y_3 e_1 \wedge e_2 \in \bigwedge^2 R^3 \end{aligned} \tag{2.77}$$

and now compute the wedge of a vector $\boldsymbol{x} \in \mathbb{R}^3$ with the bivector $\boldsymbol{Y} = \star \boldsymbol{y} \in \bigwedge^2 R^3$:

$$\boldsymbol{x} \wedge \boldsymbol{Y} = \boldsymbol{x} \wedge \star \boldsymbol{y} = (\boldsymbol{x} \cdot \boldsymbol{y}) e_1 \wedge e_2 \wedge e_3, \tag{2.78}$$

or the wedge product of a bivector $\boldsymbol{X}$ with the Hodge dual of a bivector $\boldsymbol{Y}$:

$$\boldsymbol{X} \wedge \star \boldsymbol{Y} = < \boldsymbol{X}, \boldsymbol{Y} > e_1 \wedge e_2 \wedge e_3. \tag{2.79}$$

Using the Hodge duality, we can establish in $\mathbb{R}^3$ the relationship between the cross product and the wedge product as follows:

$$\begin{aligned} \boldsymbol{x} \wedge \boldsymbol{y} &= \star(\boldsymbol{x} \times \boldsymbol{y}), \\ \boldsymbol{x} \times \boldsymbol{y} &= \star(\boldsymbol{x} \wedge \boldsymbol{y}). \end{aligned} \tag{2.80}$$

In the geometric algebra G_3, the Hodge dual is computed using the reversion of the multivector

$$\star\, \boldsymbol{A} = \widetilde{\boldsymbol{A}} I_3, \tag{2.81}$$

where $I_3 = e_1 \wedge e_2 \wedge e_3$ is the pseudoscalar. The reversion is defined in Eq. (2.92). For example, the Hodge dual of a vector $\boldsymbol{x}$ and a bivector $\boldsymbol{X}$ are computed as

$$\begin{aligned} \star \boldsymbol{x} &= \widetilde{\boldsymbol{x}} I_3 = \boldsymbol{x} I_3 = \boldsymbol{X}, \\ \star \boldsymbol{X} &= \widetilde{\boldsymbol{X}} I_3 = -\boldsymbol{X} I_3 = \boldsymbol{x}. \end{aligned} \tag{2.82}$$

Finally, the relationship between the cross product and the wedge product is written as

$$\begin{aligned} \boldsymbol{x} \wedge \boldsymbol{y} &= \star(\boldsymbol{x} \times \boldsymbol{y}) = (\boldsymbol{x} \times \boldsymbol{y}) I_3, \\ \boldsymbol{x} \times \boldsymbol{y} &= -(\boldsymbol{x} \wedge \boldsymbol{y}) I_3. \end{aligned} \tag{2.83}$$

2.3.8 *Dual Blades and Duality in the Geometric Product*

A fundamental concept algebraically related to the unit pseudoscalar I is that of *duality*. In a geometric algebra G_n, we find dual multivectors and dual operations. The dual of a multivector $\boldsymbol{A} \in G_n$ is defined as follows:

$$\boldsymbol{A}^* = \boldsymbol{A} I_n^{-1}, \tag{2.84}$$

where I_n^{-1} differs from I_n at most by a sign. Note that, in general, I^{-1} might not necessarily commute with $\boldsymbol{A}$.

The multivector bases of a geometric algebra G_n have 2^n basis elements. It can be shown that the second half is the dual of the first half. For example, in G_3 the dual of the scalar is the pseudoscalar, and the dual of a vector is a bivector $e_{23} = I e_1$. In general, the dual of an r-blade is an $(n - r)$-blade.

The operations related directly to the Clifford product are the inner and outer products, which are dual to one another. This can be written as follows:

$$(\boldsymbol{x} \cdot \boldsymbol{A}) I_n = \boldsymbol{x} \wedge (\boldsymbol{A} I_n), \tag{2.85}$$

$$(\boldsymbol{x} \wedge \boldsymbol{A}) I_n = \boldsymbol{x} \cdot (\boldsymbol{A} I_n). \tag{2.86}$$

where $\boldsymbol{x}$ is any vector and $\boldsymbol{A}$ any multivector.

By using the ideas of duality, we are then able to relate the inner product to incidence operators in the following manner. In an n-dimensional space, suppose we have an r-vector $\boldsymbol{A}$ and an s-vector $\boldsymbol{B}$ where the dual of $\boldsymbol{B}$ is given by $\boldsymbol{B}^* = \boldsymbol{B} I^{-1} \equiv \boldsymbol{B} \cdot I^{-1}$. Since $\boldsymbol{B} I^{-1} = \boldsymbol{B} \cdot I^{-1} + \boldsymbol{B} \wedge I^{-1}$, we can replace the geometric product by the inner product alone (in this case, the outer product equals zero, and there can be no $(n+1)$-D vector). Now, using the identity

$$\boldsymbol{A}_r \cdot (\boldsymbol{B}_s \cdot \boldsymbol{C}_t) = (\boldsymbol{A}_r \wedge \boldsymbol{B}_s) \cdot \boldsymbol{C}_t \qquad \text{for} \qquad r + s \leq t, \tag{2.87}$$

we can write

$$\boldsymbol{A} \cdot (\boldsymbol{B} I^{-1}) = \boldsymbol{A} \cdot (\boldsymbol{B} \cdot I^{-1}) = (\boldsymbol{A} \wedge \boldsymbol{B}) \cdot I^{-1} = (\boldsymbol{A} \wedge \boldsymbol{B}) I^{-1}. \tag{2.88}$$

This expression can be rewritten using the definition of the dual as follows:

$$\boldsymbol{A} \cdot \boldsymbol{B}^* = (\boldsymbol{A} \wedge \boldsymbol{B})^*. \tag{2.89}$$

The equation shows the relationship between the inner and outer products in terms of the duality operator. Now, if $r + s = n$, then $\boldsymbol{A} \wedge \boldsymbol{B}$ is of grade n and is therefore a pseudoscalar. Using Eq. (2.88), we can employ the involved pseudoscalar in order to get an expression in terms of a bracket:

$$\begin{aligned} \boldsymbol{A} \cdot \boldsymbol{B}^* &= (\boldsymbol{A} \wedge \boldsymbol{B})^* = (\boldsymbol{A} \wedge \boldsymbol{B}) I^{-1} = ([\boldsymbol{A} \wedge \boldsymbol{B}] I) I^{-1} \\ &= [\boldsymbol{A} \wedge \boldsymbol{B}]. \end{aligned} \tag{2.90}$$

We see, therefore, that the bracket relates the inner and outer products to non-metric quantities.

2.4 Multivector Operations

In this section, we will define a number of right operations that act on multivectors and return multivectors.

2.4.1 *Involution, Reversion and Conjugation Operations*

For an r-grade multivector $\boldsymbol{A}_r = \sum_{i=0}^{r} \langle \boldsymbol{A}_r \rangle_i$, the following operations are defined:

$$\text{Grade involution: } \widehat{\boldsymbol{A}}_r = \sum_{i=0}^{r} (-1)^i \langle \boldsymbol{A}_r \rangle_i, \tag{2.91}$$

$$\text{Reversion: } \widetilde{\boldsymbol{A}}_r = \sum_{i=0}^{r} (-1)^{\frac{i(i-1)}{2}} \langle \boldsymbol{A}_r \rangle_i, \tag{2.92}$$

$$\begin{aligned} \text{Conjugate: } \langle \boldsymbol{A}_r \rangle_k^\dagger &= (\boldsymbol{a}_1 \wedge \boldsymbol{a}_2 \wedge \ldots \wedge \boldsymbol{a}_k)^\dagger \\ &= \boldsymbol{a}_k^\dagger \wedge \boldsymbol{a}_{k-1}^\dagger \wedge \ldots \wedge \boldsymbol{a}_1^\dagger \end{aligned} \tag{2.93}$$

$$\text{Clifford conjugation: } \overline{\boldsymbol{A}}_r = \widetilde{\widehat{\boldsymbol{A}}}_r = \sum_{i=0}^{r} (-1)^{\frac{i(i+1)}{2}} \langle \boldsymbol{A}_r \rangle_i. \tag{2.94}$$

The grade involution simply negates the odd-grade blades of a multivector. The reversion can also be obtained by reversing the order of basis vectors making up the blades in a multivector and then rearranging them in their original order using the anticommutativity of the Clifford product.

Let us explain the case of conjugation, let $\boldsymbol{A} \in \{1, \ldots, p+q\}$ the set of vectors $e_{\boldsymbol{A}} \in G_{p,q}$; the conjugate of $e_{\boldsymbol{A}}$, denoted by $e^{\dagger}_{\boldsymbol{A}}$, is defined as:

$$e^{\dagger}_{\boldsymbol{A}} = (-1)^r \tilde{e}_{\boldsymbol{A}}, \quad r := \texttt{grade of } e_{\boldsymbol{A}}. \tag{2.95}$$

Thus, while $\tilde{\boldsymbol{a}}_i = \boldsymbol{a}_i$, $\boldsymbol{a}^{\dagger}_i$ is not necessarily equal to $\boldsymbol{a}_i$; this is because for $e_i \in G_{p,q}$, $e^{\dagger}_i = -e_i$ if $p < i < p+q$.

The computation of the *magnitude* or modulus of a blade can be done using the reversion operation, as follows:

$$||\boldsymbol{A}|| =< \widetilde{\boldsymbol{A}}\boldsymbol{A} >_0^{\frac{1}{2}} = || < \boldsymbol{A} >_k ||^2. \tag{2.96}$$

Accordingly, the magnitude of a multivector $\boldsymbol{M}$ reads

$$\begin{aligned} ||\boldsymbol{M}|| &= < \widetilde{\boldsymbol{M}}\boldsymbol{M} >_0^{\frac{1}{2}} \\ &= (|| < \boldsymbol{M} >_0 ||^2 + || < \boldsymbol{M} >_1 ||^2 \\ &\qquad + || < \boldsymbol{M} >_2 ||^2 + \cdots + || < \boldsymbol{M} >_n ||^2)^{\frac{1}{2}} \\ &= \sqrt{\sum_{r=0}^{n} || < \boldsymbol{M} >_r ||^2}. \end{aligned} \tag{2.97}$$

In particular, for an *r-vector* $\boldsymbol{A}_r$ of the form $\boldsymbol{A}_r = \mathbf{a}_1 \wedge \boldsymbol{a}_2 \wedge \cdots \wedge \boldsymbol{a}_r$: $\boldsymbol{A}^{\dagger}_r = (\mathbf{a}_1 \cdots \mathbf{a}_{r-1}\mathbf{a}_r)^{\dagger} = \mathbf{a}_r \mathbf{a}_{r-1} \cdots \boldsymbol{a}_1$ and thus $\boldsymbol{A}^{\dagger}_r \boldsymbol{A}_r = \mathbf{a}^2_1\, \mathbf{a}^2_2 \cdots \boldsymbol{a}^2_r$, so, we will say that such an r-vector is null if and only if it has a null vector for a factor. If in such a *factorization* of $\boldsymbol{A}_r$ p, q and s factors square in a positive number, negative, and zero, respectively, we will say that $\boldsymbol{A}_r$ is an r-vector with signature (p, q, s). In particular, if $s = 0$ such a *nonsingular* r-vector has a multiplicative inverse that can also be written using the reversion

$$\boldsymbol{A}^{-1} = (-1)^q \frac{\boldsymbol{A}^{\dagger}}{\boldsymbol{A}\boldsymbol{A}^{\dagger}} = (-1)^q \frac{\boldsymbol{A}^{\dagger}}{|\mathbf{A}|^2} = \frac{\boldsymbol{A}}{\boldsymbol{A}^2}. \tag{2.98}$$

In general, the *inverse* $\boldsymbol{A}^{-1}$ of a multivector $\boldsymbol{A}$, if it exists, is defined by the equation $\boldsymbol{A}^{-1}\boldsymbol{A} = 1$.

2.4.2 Join and Meet Operations

When we work with lines, planes, and spheres, however, it will clearly be necessary to employ operations to compute the *meets* (intersections) or *joins* (expansions) of geometric objects. For this, we will need a geometric means of performing the set-theory operations of intersection, $\cap$, and union, $\cup$. Herewith, $\cup$ and $\cap$ will stand for the algebra of incidence operations of join and meet, respectively.

If in an n-dimensional geometric algebra, the r-vector A and the s-vector B do not have a common subspace (null intersection), one can define the *join* of both vectors as

$$\boldsymbol{C} = \boldsymbol{A} \cup \boldsymbol{B} = \boldsymbol{A} \wedge \boldsymbol{B}, \tag{2.99}$$

so that the join is simply the outer product (an $r + s$ vector) of the two vectors. However, if $\boldsymbol{A}$ and $\boldsymbol{B}$ have common blades, the join would not simply be given by the wedge but by the subspace the two vectors span. The operation join $\cup$ can be interpreted as a *common dividend of lowest grade* and is defined up to a scale factor. The join gives the pseudoscalar if $(r + s) \geq n$. We will use $\cup$ to represent the join only when the blades A and B have a common subspace; otherwise, we will use the ordinary exterior product, $\wedge$, to represent the join.

If there exists a k-vector $\boldsymbol{D}$ such that for $\boldsymbol{A}$ and $\boldsymbol{B}$ we can write $\boldsymbol{A} = \boldsymbol{A}'\boldsymbol{D}$ and $\boldsymbol{B} = \boldsymbol{B}'\boldsymbol{D}$ for some $\boldsymbol{A}'$ and $\boldsymbol{B}'$, then we can define the *intersection* or *meet* using the duality principle as follows:

$$\boldsymbol{D} \cdot I^{-1} = \boldsymbol{D}^* = (\boldsymbol{A} \cap \boldsymbol{B})^* = \boldsymbol{A}^* \cup \boldsymbol{B}^*. \tag{2.100}$$

This is a beautiful result, telling us that the dual of the meet is given by the join of the duals. Since the dual of $\boldsymbol{A} \cap \boldsymbol{B}$ will be taken with respect to the *join* of $\boldsymbol{A}$ and $\boldsymbol{B}$, we must be careful to specify which space we will use for the dual in Eq. (12.37). However, in most cases of practical interest, this join will indeed cover the entire space, and therefore we will be able to obtain a more useful expression for the meet using Eq. (12.33). Thus,

$$\boldsymbol{A} \cap \boldsymbol{B} = ((\boldsymbol{A} \cap \boldsymbol{B})^*)^* = (\boldsymbol{A}^* \cup \boldsymbol{B}^*)I = (\boldsymbol{A}^* \wedge \boldsymbol{B}^*)(I^{-1}I)I = (\boldsymbol{A}^* \cdot \boldsymbol{B}). \tag{2.101}$$

The above concepts are discussed further in [139]. In the theory of Grassman–Cayley algebra, the meet operation is computed using the so-called shuffle formula. This formula will be explained in Sect. 20.2.3 of the Appendix.

2.4.3 Multivector-valued Functions and the Inner Product

A multivector-valued function $f : \mathbb{R}^{p,q} \to G_{p,q}$, where $n = p + q$, has 2^n blade components

$$f(\boldsymbol{x}) = \sum_M f(\boldsymbol{x})_M e_M = \sum_{i=1}^{n} < f(\boldsymbol{x}) >_n, \tag{2.102}$$

where $M \in \{0, 1, 2, \ldots, 12, 23, \ldots, 123, 124, \ldots, 123\cdots(n-1)n\}$ contains all possible 2^n blade subindexes and $f(\boldsymbol{x})_M \in \mathbb{R}$ corresponds to the scalar accompanying each multivector base. Let us consider the complex conjugation in G_3

$$\begin{aligned}\tilde{f}(\boldsymbol{x}) = \sum_M f(\boldsymbol{x})_M \tilde{e}_M &= f(\boldsymbol{x})_0 e_0 + f(\boldsymbol{x})_1 e_1 + f(\boldsymbol{x})_2 e_2 + f(\boldsymbol{x})_3 e_3 - \\ &- f(\boldsymbol{x})_{23} e_{23} - f(\boldsymbol{x})_{32} e_{32} - f(\boldsymbol{x})_{12} e_{12} - f(\boldsymbol{x})_{123} e_{123} \\ &= < f(\boldsymbol{x}) >_0 + < f(\boldsymbol{x}) >_1 - < f(\boldsymbol{x}) >_2 - < f(\boldsymbol{x}) >_3 .\end{aligned} \tag{2.103}$$

Next, we define the inner product of $\mathbb{R}^n \to G_n$ functions f, g by

$$(f, g) = \sum_{\mathbb{R}^n} f(\boldsymbol{x}) \widetilde{g(\boldsymbol{x})} d^n \boldsymbol{x}. \tag{2.104}$$

Let us consider the inner product of two functions in G_3:

$$(f, g)_{L^2(\mathbb{R}^3; G_3)} = \int_{\mathbb{R}^3} f(\boldsymbol{x}) \widetilde{g(\boldsymbol{x})} d^3 \boldsymbol{x} = \sum_{M,N} e_M \tilde{e}_N \int_{\mathbb{R}^3} f_M(\acute{\boldsymbol{x}}) g_N(\boldsymbol{x}) d^3 \boldsymbol{x}, \tag{2.105}$$

where $d^3\boldsymbol{x} = \frac{dx_1 \wedge dx_2 \wedge dx_3}{I_3}$. The scalar part of this equation corresponds to the $L^2 - -norm$

$$||f||^2_{L^2(\mathbb{R}^3; G_3)} =< (f, f)_{L^2(\mathbb{R}^3; G_3)} >_0 = \sum_{\mathbb{R}^3} f(\boldsymbol{x}) * \tilde{f}(\boldsymbol{x}) d^3 \boldsymbol{x} = \int_{\mathbb{R}^3} \sum_M f_M^2(\boldsymbol{x}) d^3 \boldsymbol{x}.$$

The $L^2(\mathbb{R}^n; G_{n,0}) - -norm$ is given by

$$||f||^2 =< (f, f) >_0 = \sum_{\mathbb{R}^n} |f(\boldsymbol{x})|^2 d^n \boldsymbol{x}. \tag{2.106}$$

For the case of square-integrable functions on the sphere $L_2(S^{n-1})$, the inner product and the norm are

$$(f, g)_{L_2} = \int_{S^{n-1}} f(\boldsymbol{x})\widetilde{g(\boldsymbol{x})}dS(\boldsymbol{x}), \tag{2.107}$$

$$||f||^2 = 2^n \int_{S^{n-1}} < f(\boldsymbol{x})\widetilde{f(\boldsymbol{x})} >_0 dS(\boldsymbol{x}), \tag{2.108}$$

where $dS(\boldsymbol{x})$ is the normalized Spin(n)-invariant measure on S^{n-1}.

2.4.4 The Multivector Integral

Let us consider $F(\boldsymbol{x})$ a multivalued function (field) of a vector variable $\boldsymbol{x}$ defined in a certain region of the Euclidean space E^n. If the function is only scalar or vector valued, it will be called a scalar- of a vector-fields respectively. The Riemann integral of a multivector-valued function $F(\boldsymbol{x})$ is defined as follows:

$$\int_{E^n} F(\boldsymbol{x})|d\boldsymbol{x}| = \lim_{\substack{|\Delta x_j| \to 0 \\ n \to \infty}} \sum_{j=1}^{n} F(x_j e_j)\Delta x_j, \tag{2.109}$$

where the quantity in brackets $|d\boldsymbol{x}|$ is used to make the integral grade preserving, because $d\boldsymbol{x}$ is a vector in a geometric algebra G_n. Thus, the integral can be discretized using the sum of quadrature expressions.

2.4.5 Convolution and Correlation of Scalar Fields

The filtering of a continuous signal $f : E^n \to C$ is computed via the convolution of the signal with a filter $h : E^n \to C$ as follows:

$$(h * f)(\boldsymbol{x}) = \int_{E^n} h(\boldsymbol{x}')f(\boldsymbol{x} - \boldsymbol{x}')d\boldsymbol{x}'. \tag{2.110}$$

The spatial correlation is defined by

$$(h \star f)(\boldsymbol{x}) = \int_{E^n} h(\boldsymbol{x}')f(\boldsymbol{x} + \boldsymbol{x}')d\boldsymbol{x}'. \tag{2.111}$$

Note that the convolution can be seen as a correlation with a filter that has been reflected with respect to its center.

2.4.6 Clifford Convolution and Correlation

Given a multivector field $\boldsymbol{F}$ and a multivector-valued filter $\boldsymbol{H}$, their Clifford convolution is defined in terms of the Clifford product of multivectors:

$$(\boldsymbol{H} *_l \boldsymbol{F})(\boldsymbol{x}) = \int_{E^n} \boldsymbol{H}(\boldsymbol{x}')\boldsymbol{F}(\boldsymbol{x}-\boldsymbol{x}')|d\,x'|,$$
$$(\boldsymbol{F} *_r \boldsymbol{H})(\boldsymbol{x}) = \int_{E^n} \boldsymbol{F}(\boldsymbol{x}-\boldsymbol{x}')\boldsymbol{H}(\boldsymbol{x}')|d\,x'|. \tag{2.112}$$

Since the Clifford product is not commutative, we distinguish the application of the filter $\boldsymbol{H}$ from the left and from the right by using the subindex l and r, respectively. For the case of discrete fields, the convolution has to be discretized. If the application of the filter is done from the left, the discretized, left convolution is given by

$$(\boldsymbol{H} *_l \boldsymbol{F})_{i,j,k} = \sum_{i=-d}^{d}\sum_{j=-d}^{d}\sum_{k=-d}^{d} \boldsymbol{H}_{r,s,t}\boldsymbol{F}_{i-r,j-s,k-t}, \tag{2.113}$$

where a 3D uniform grid is used where $i, j, k, r, s, t \in \mathbb{Z}$, (i, j, k) denotes the grid nodes and d^3 is the dimension of the filter grid.

The spatial Clifford correlation can be defined in a similar manner as the Clifford convolution:

$$(\boldsymbol{H} \star_l \boldsymbol{F})(\boldsymbol{x}) = \int_{E^n} \boldsymbol{H}(\boldsymbol{x}')\boldsymbol{F}(\boldsymbol{x}+\boldsymbol{x}')|d\,x'|,$$
$$(\boldsymbol{F} \star_r \boldsymbol{H})(\boldsymbol{x}) = \int_{E^n} \boldsymbol{F}(\boldsymbol{x}+\boldsymbol{x}')\boldsymbol{H}(\boldsymbol{x}')|d\,x'|. \tag{2.114}$$

This correlation formula can be seen simply as a convolution with a filter that has been reflected with respect to its center.

We can also carry out the scalar Clifford convolution in the Fourier domain. Consider the vector fields $\boldsymbol{f}, \boldsymbol{h} : E^3 \rightarrow E^3 \in G_3$. Since

$$(\boldsymbol{h} *_s \boldsymbol{f})(\boldsymbol{x}) = <(\boldsymbol{h} *_l \boldsymbol{f})>_0,$$
$$(\boldsymbol{h} * \boldsymbol{f})_3 = 0, \tag{2.115}$$

the Clifford convolution is

$$\mathcal{F}\{(\boldsymbol{h} *_s \boldsymbol{f})\}(\boldsymbol{u}) = <\mathcal{F}\{\boldsymbol{h}\}, \mathcal{F}\{\boldsymbol{f}\}> + <\mathcal{F}\{\boldsymbol{h}\}, \mathcal{F}\{\boldsymbol{f}\}>_3 . \tag{2.116}$$

Since the Clifford Fourier transform of 3D vector fields contains a vector part and a bivector part, the trivector part $<\mathcal{F}\{\boldsymbol{h}\}, \mathcal{F}\{\boldsymbol{f}\}>_3$ is generally nonzero.

2.5 Linear Algebra

This section presents the geometric algebra approach to the basic concepts of linear algebra and is presented here for completeness. Next chapter will present related topics in more depth such as differentiation, linear and multilinear functions, eigenblades, and tensors formulated in geometric algebra also operators and transformations in common use by mathematicians, physicists, computer scientists, and engineers.

A linear function f maps vectors to vectors in the same space. The extension of f to act linearly on multivectors is possible via the so-called *outermorphism* $\underline{f}$, which defines the action of $\underline{f}$ on r-blades thus:

$$\underline{f}(\boldsymbol{a}_1\wedge\boldsymbol{a}_2\wedge\ldots\wedge\boldsymbol{a}_r) = \underline{f}(\boldsymbol{a}_1)\wedge\underline{f}(\boldsymbol{a}_2)\wedge\ldots\wedge\underline{f}(\boldsymbol{a}_r). \tag{2.117}$$

The function $\underline{f}$ is called an outermorphism because $\underline{f}$ preserves the grade of any r-vector it acts upon. The action of $\underline{f}$ on general multivectors is then defined through linearity. The function $\underline{f}$ must therefore satisfy the following conditions:

$$\begin{aligned} \underline{f}(\boldsymbol{a}_1\wedge\boldsymbol{a}_2) &= \underline{f}(\boldsymbol{a}_1)\wedge\underline{f}(\boldsymbol{a}_2) \\ \underline{f}(A_r) &= \langle\underline{f}(A_r)\rangle_r \\ \underline{f}(\alpha_1\boldsymbol{a}_1 + \alpha_2\boldsymbol{a}_2) &= \alpha_1\underline{f}(\boldsymbol{a}_1) + \alpha_2\underline{f}(\boldsymbol{a}_2). \end{aligned} \tag{2.118}$$

Accordingly, the outermorphism of a product of two linear functions is the product of the outermorphisms—that is, if $f(\boldsymbol{a}) = f_2(f_1(\boldsymbol{a}))$, we write $\underline{f} = \underline{f}_2\underline{f}_1$. The *adjoint* $\overline{f}$ of a linear function $\underline{f}$ acting on the vectors $\boldsymbol{a}$ and $\boldsymbol{b}$ can be defined by the property

$$\underline{f}(\boldsymbol{a})\cdot\boldsymbol{b} = \boldsymbol{a}\cdot\overline{f}(\boldsymbol{b}). \tag{2.119}$$

If $\underline{f} = \overline{f}$, the function is *self-adjoint* and can be represented by a symmetric matrix F ($F = F^T$).

Since the outermorphism preserves grade, the unit pseudoscalar must be mapped onto some multiple of itself; this multiple is called the *determinant* of $\underline{f}$. Thus,

$$\underline{f}(I) = \det(\underline{f})I. \tag{2.120}$$

This is a particularly simple definition of the determinant, from which many properties of the determinants follow straightforwardly.

2.5.1 *Linear Algebra Derivations*

In linear algebra, a number of derivations are carried out using frame contraction, instead, we can use vector derivatives as we show in the next equation:

$$\nabla(\boldsymbol{x} \cdot \boldsymbol{v}) = e^i \partial_i (x^i e_j).\boldsymbol{v} = e^i e_j \cdot \boldsymbol{v} \delta_i^j = e^i e_i \cdot \boldsymbol{v} = \boldsymbol{v}. \qquad (2.121)$$

This shows that when differentiating a function that depends linearly on $\boldsymbol{x}$, it is simply equivalent to carry out contractions over frame indices. To take advantage of this, one introduces a vector variable $\boldsymbol{x}$ and denotes the derivative with respect to $\boldsymbol{x}$ by ∂_x. In this manner, we can write Eqs. (2.26)–(2.30).

$$\begin{aligned} \partial_x \boldsymbol{x} \cdot \boldsymbol{X}_m &= m \boldsymbol{X}_m, \\ \partial_x \boldsymbol{x} \wedge \boldsymbol{X}_m &= (n-m)\boldsymbol{X}_m, \\ \dot{\partial}_x \boldsymbol{X}_m \dot{\boldsymbol{x}} &= (-1)^m (n-2m)\boldsymbol{X}_m. \end{aligned} \qquad (2.122)$$

The trace of a linear function can be expressed in terms of a vector derivative as follows:

$$Tr(f) = \partial_x \cdot f(\boldsymbol{x}). \qquad (2.123)$$

The use of the vector derivatives enables us to write terms of an equation that do not depend on a frame in such a way that reflects this independence. In general, this kind of formulation brings out the intrinsic geometric content of an equation.

2.6 Simplexes

An r-dimensional simplex (r-simplex) in $\mathbb{R}^n$ corresponds to the convex hull of r+1 points, of which at least r are linearly independent. A set of points $\{\boldsymbol{x}_0, \boldsymbol{x}_1, \boldsymbol{x}_2, \ldots, \boldsymbol{x}_r\}$ which define a r-simplex is seen as the *frame* for the r-simplex. One can select $\boldsymbol{x}_0$ as the *base point* or *place* of the simplex. Using the geometric algebra, the related notations for the r-simplex are as follows:

$$\boldsymbol{X}_r \equiv \boldsymbol{x}_0 \wedge \boldsymbol{x}_1 \wedge \boldsymbol{x}_2 \wedge \ldots \wedge \boldsymbol{x}_r = \boldsymbol{x}_0 \wedge \bar{\boldsymbol{X}}_r \qquad (2.124)$$

$$\bar{\boldsymbol{X}}_r \equiv (\boldsymbol{x}_1 - \boldsymbol{x}_0) \wedge (\boldsymbol{x}_2 - \boldsymbol{x}_0) \wedge \cdots \wedge (\boldsymbol{x}_r - \boldsymbol{x}_0) \qquad (2.125)$$

where $\bar{\boldsymbol{x}}_i = (\boldsymbol{x}_i - \boldsymbol{x}_0)$ for $i = 1, \ldots, r$. $\bar{\boldsymbol{X}}_r$ is called the *tangent* of the simplex, because it is tangent to the r-plane in which the simplex lies, see Fig. 2.4a. The tangent determines the directed distance to the simplex $\bar{\boldsymbol{X}}_r / r!$ and it assigns a definite orientation to the simplex. On the other hand, the *volume* of the simplex is given by $(r!)^{-1} |\bar{\boldsymbol{X}}_r|$ which corresponds to area of the triangle formed by $\bar{\boldsymbol{x}}_1$ and $\bar{\boldsymbol{x}}_2$ in Fig. 2.4b.

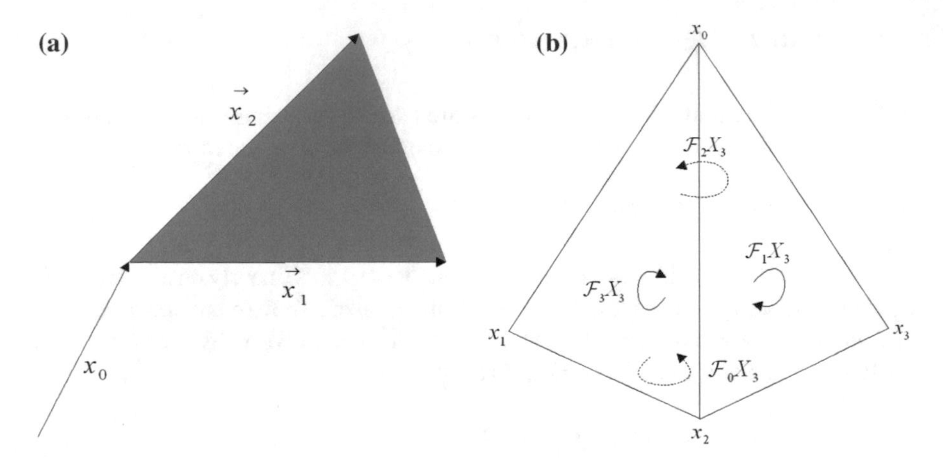

Fig. 2.4 **a** (left) Tangent of the simplex; **b** (right) volume of an 3-hedron

In general, this is the volume of an $(r + 1)$-hedron. Note that $\bar{\boldsymbol{X}}_r$ is independent of the choice of the origin, but $\boldsymbol{X}_r$ is not.

Now, an r-dimensional plane parallel to the subspace of $\boldsymbol{X}_r$ and through the point $\boldsymbol{x}$ is the solution set of

$$(\boldsymbol{y} - \boldsymbol{x})\wedge \boldsymbol{X}_r = 0, \tag{2.126}$$

for $\boldsymbol{y} \in \mathbb{R}^n$. According Eq. (2.126), the equation for the plane of the simplex is given by

$$\boldsymbol{y}\wedge \bar{\boldsymbol{X}}_r = \boldsymbol{x}_0\wedge \bar{\boldsymbol{X}}_r = \boldsymbol{X}_r. \tag{2.127}$$

The term $\boldsymbol{X}_r$ corresponds to the *moment* of the simplex, because is the wedge product of a point touching the simplex with the orientation entity, think in the moment of a line.

Since Eq. (2.125) represents the pseudoscalar for the simplex frame $\{\boldsymbol{x}_i\}$, it determines also a dual frame $\{\boldsymbol{x}^i\}$ given by Eq. (2.10):

$$\boldsymbol{x}^i \cdot \boldsymbol{x}_j = \delta^i_j. \tag{2.128}$$

The *face* opposite $\boldsymbol{x}_i$ in simplex $\boldsymbol{X}_r$ is described by its moment

$$\mathcal{F}^r_i \boldsymbol{X}_r \equiv \boldsymbol{X}^i_r \equiv \boldsymbol{x}^i \cdot \boldsymbol{X}_r = (-1)^{i+1}\boldsymbol{x}_0\wedge \boldsymbol{x}_1\wedge \cdots \wedge \check{\boldsymbol{x}}_i\wedge \cdots \wedge x_r. \tag{2.129}$$

This equation defines a *face* operator $\mathcal{F}_i$, see Fig. 2.4b. Note that the face $\boldsymbol{X}_r^i$ is an $r-1$-dimensional simplex and

$$\boldsymbol{X}_r = \boldsymbol{x}_i \wedge \boldsymbol{X}_r^i, \tag{2.130}$$

for any $0 \leq i \leq r$.

The *boundary* of a simplex $\boldsymbol{X}_r$ is given by the multivector sum as follows:

$$\Omega_b \boldsymbol{X}_r = \sum_{i=0}^{r} \boldsymbol{X}_r^i = \sum_{i=0}^{r} \mathcal{F}_i \boldsymbol{X}_r. \tag{2.131}$$

This boundary operator fulfills

$$\boldsymbol{X}_r = \boldsymbol{x}_i \wedge \Omega_b \boldsymbol{X}_r, \tag{2.132}$$

for any $0 \leq i \leq r$. Taking into account the identity

$$(\boldsymbol{x}_1 - \boldsymbol{x}_0) \wedge (\boldsymbol{x}_2 - \boldsymbol{x}_0) \wedge \cdots \wedge (\boldsymbol{x}_r - \boldsymbol{x}_0) = \sum_{i=0}^{r} (-1)^i \boldsymbol{x}_0 \wedge \cdots \wedge \check{\boldsymbol{x}}_i \wedge \cdots \wedge \boldsymbol{x}_r. \tag{2.133}$$

and Eqs. (2.125) and (2.129), we derive

$$\bar{\boldsymbol{X}}_r = \Omega_b \boldsymbol{X}_r. \tag{2.134}$$

Finally, the following relations hold for the operators $\mathcal{F}_i$ and Ω_b:

$$\mathcal{F}_i \mathcal{F}_i = 0, \tag{2.135}$$
$$\mathcal{F}_i \mathcal{F}_j = -\mathcal{F}_j \mathcal{F}_i \tag{2.136}$$
$$\mathcal{F}_i \Omega_b = -\Omega_b \mathcal{F}_i \tag{2.137}$$
$$\Omega_b \Omega_b = 0. \tag{2.138}$$

Note that these operators relations are strongly analogous to relations one find in algebraic topology.

2.7 Exercises

2.1 Given $\boldsymbol{x}, \boldsymbol{y} \in G_{2,0,0}$, expand the bivector $\boldsymbol{x} \wedge \boldsymbol{y}$ in terms of geometric products. Prove that it anticommutes with both $\boldsymbol{x}$ and $\boldsymbol{y}$, however commutes with any vector outside the plane.

2.2 Prove that the magnitude of bivector $\boldsymbol{x} \wedge \boldsymbol{y}$ is $|\boldsymbol{x}||\boldsymbol{y}|sin(\theta)$.

2.3 Given $\boldsymbol{a} = e_2 + 2e_1e_2$, $\boldsymbol{b} = e_1 + 2e_2$, and $\boldsymbol{c} = 3e_1e_2 \in G_{2,0,0}$, compute $\boldsymbol{ab}$, $\boldsymbol{ba}$, and $\boldsymbol{bac}$.

2.4 Given the coordinates (2,5), (5,7), and (6,0) for the corners of a triangle, compute in $G_{2,0,0}$, the triangle's area, using multivectors.

2.5 Given $\boldsymbol{a} = e_1 - 2e_2$, $\boldsymbol{b} = e_1 + e_2$, and $\boldsymbol{r} = 5e_1 - e_2 \in G_{2,0,0}$, compute α and β for $\boldsymbol{r} = \alpha\boldsymbol{a} + \beta\boldsymbol{b}$.

2.6 Given $\boldsymbol{a} = 8e_1 - e_2$ and $\boldsymbol{b} = 2e_1 + 2e_2 \in G_{2,0,0}$, compute $\boldsymbol{a}_{||}$ and $\boldsymbol{a}_{\perp}$ with respect to $\boldsymbol{b}$.

2.7 For each $\boldsymbol{x} \in G_{2,0,0}$, prove $\boldsymbol{x}\bar{\boldsymbol{x}} = \bar{\boldsymbol{x}}\boldsymbol{x}$ and compute the inverse $\boldsymbol{x}^{-1}$ when $\boldsymbol{x}\bar{\boldsymbol{x}} \neq 0$.

2.8 Prove, using multivectors in $G_{2,0,0}$, the following sinus identity:

$$\frac{sin\alpha}{|a|} = \frac{sin\beta}{|b|} = \frac{sin\gamma}{|c|}.$$

2.9 For a Euclidean unit 2-blade, $I^2 = -1$, interpret this geometrically in terms of versors.

2.10 Prove that in $\mathbb{R}^3$ the cross product is equivalent to the following expressions:

$$\boldsymbol{x} \times \boldsymbol{y} = -I\boldsymbol{x}\wedge\boldsymbol{y} = -\boldsymbol{x} \cdot (I\boldsymbol{y}) = \boldsymbol{y} \cdot (I\boldsymbol{x}).$$

2.11 Interpret geometrically the equations of Exercise 2.10 and establish that the following expressions are true:

$$\boldsymbol{x} \times (\boldsymbol{y} \times \boldsymbol{z}) = -\boldsymbol{x} \cdot (\boldsymbol{y}\wedge\boldsymbol{z}) = -(\boldsymbol{x} \cdot \boldsymbol{y}\boldsymbol{z} - \boldsymbol{x} \cdot \boldsymbol{z}\boldsymbol{y}),$$

and

$$\boldsymbol{x} \cdot (\boldsymbol{y} \times \boldsymbol{z}) = [\boldsymbol{x}, \boldsymbol{y}, \boldsymbol{z}] = \boldsymbol{x}\wedge\boldsymbol{y}\wedge\boldsymbol{z}I^{-1}.$$

2.12 Given the 2-blade (bivector) $\boldsymbol{X} = e_1\wedge(e_2 - e_3)$ that represents a plane, check if the following vectors lie in that plane: (i) e_1; (ii) $e_1 + e_2$; (iii) $e_1 + e_2 + e_3$ and (iv) $e_1 - 2e_2 + e_3$.

2.13 In 3D space, a trivector $\boldsymbol{x}\wedge\boldsymbol{y}\wedge\boldsymbol{z}$ can be written in terms of a determinant:

$$\boldsymbol{x}\wedge\boldsymbol{y}\wedge\boldsymbol{z} = det([\boldsymbol{x}\,\boldsymbol{y}\boldsymbol{z}])e_1\wedge e_2\wedge e_3, \tag{2.139}$$

where $[\boldsymbol{x}\,\boldsymbol{y}\boldsymbol{z}]$ is a matrix with column vectors. Express the trivector not with a matrix but fully in terms of geometric algebra.

2.14 Show in $G_{3,0,0}$ that the pseudoscalar I commutes with e_1, e_2, and e_3. Compute the volume of a parallelepiped spanned by the vectors $\boldsymbol{a} = 3e_1 - 4e_2 + 4e_3$, $\boldsymbol{b} = 2e_1 + 4e_2 - 2e_3$, and $\boldsymbol{c} = 4e_1 - 2e_2 - 3e_3$.

2.15 Compute $\boldsymbol{AB}$ in $G_{3,0,0}$ if $\boldsymbol{A} = 3e_1e_2 + 6e_2e_3$ and $\boldsymbol{B} = \boldsymbol{a} \wedge \boldsymbol{b}$ where $\boldsymbol{a} = 3e_1 - 4e_2 + 5e_3$ and $\boldsymbol{b} = 2e_1 + 4e_2 - 2e_3$.

2.16 Given in $G_{3,0,0}$ the vector $\boldsymbol{a} = 3e_1 + 5e_2 + 6e_3$ and the bivector $\boldsymbol{B} = 3e_2e_3 + 6e_3e_1 + 2e_1e_2$, compute the parallel and orthogonal projections of $\boldsymbol{a}$ with respect to the plane $\boldsymbol{B}$. Then compute the cross product of these projected components and interpret geometrically the dual relationship of the result with respect to $\boldsymbol{B}$.

2.17 Show in $G_{3,0,0}$ that the geometric product of the bivector $\boldsymbol{A}$ and any vector $\boldsymbol{x}$ can be decomposed as follows:

$$\boldsymbol{Ax} = \boldsymbol{A} \cdot \boldsymbol{x} + \frac{1}{2}(\boldsymbol{Ax} - \boldsymbol{xA}) + \boldsymbol{A} \wedge \boldsymbol{x}.$$

2.18 Given $\boldsymbol{x} = 1 + \boldsymbol{a} + \boldsymbol{B}$, where $\boldsymbol{a} \in \mathbb{R}^3$ and $\boldsymbol{B} \in \bigwedge^2 \mathbb{R}^3$.
(a) The outer inverse of $\boldsymbol{x}$ is $\boldsymbol{x}^{\wedge(-1)} = 1 - \boldsymbol{a} - \boldsymbol{B} + \alpha \boldsymbol{a} \wedge \boldsymbol{B}$, where $\alpha \in \mathbb{R}$. Compute α. (*Hint*: Use the power series or $\boldsymbol{x} \wedge \boldsymbol{x}^{\wedge(-1)} = 1$).
(b) The outer square root of $\boldsymbol{x}$ is $\boldsymbol{x}^{\wedge(\frac{1}{2})} = 1 + \frac{1}{2}\boldsymbol{a} + \frac{1}{2}\boldsymbol{B} + \beta \boldsymbol{a} \wedge \boldsymbol{B}$, where $\beta \in \mathbb{R}$. Compute β. (*Hint*: $\boldsymbol{x}^{\wedge(\frac{1}{2})} \wedge \boldsymbol{x}^{\wedge(\frac{1}{2})} = \boldsymbol{x}$).
(c) Give a geometric interpretation of the outer inverse and of the outer square root and suggest their applications.

2.19 In 4D space with an associated orthonormal basis $\{e_i\}_{i=1}^4$, project the 2-blade $\boldsymbol{Y} = (e_1 + e_2) \wedge (e_3 + e_4)$ onto the 2-blade $\boldsymbol{X} = e_1 \wedge e_3$. Compute then the rejection as the difference of $\boldsymbol{Y}$ and its projection. Prove that this is not a blade.

2.20 In $\mathbb{R}^4$, show that the 2-vector $\boldsymbol{X} = e_1 \wedge e_2 + e_3 \wedge e_4$ is not a 2-blade which means that it cannot be expressed as the wedge product of two vectors. Hint: given $\boldsymbol{X} = \boldsymbol{x} \wedge \boldsymbol{y}$, express $\boldsymbol{x}$ and $\boldsymbol{y}$ in terms of the basis vectors, expand the wedge product, and try to solve the resulting scalar equations.

2.21 In Exercise 2.19, check that the 2-vector $\boldsymbol{X} = e_1 \wedge e_2 + e_3 \wedge e_4$ does not contain any vector other than 0. We mean by *contain*, denoted $\boldsymbol{X} \subseteq \boldsymbol{Y}$, the case when all vectors in $\boldsymbol{X}$ are also in $\boldsymbol{Y}$.

2.22 In $G_{2,0,0}$ the multivectors $\boldsymbol{X}$ and $\boldsymbol{Y}$ are given by

$$\boldsymbol{X} = x_0 + x_1e_1 + x_2e_2 + x_3e_1e_2 \qquad \boldsymbol{Y} = y_0 + y_1e_1 + y_2e_2 + y_3e_1e_2,$$

where the basis vectors e_1, e_2 are orthonormals. Compute their geometric product

$$\boldsymbol{XY} = w_0 + w_1e_1 + w_2e_2 + w_3e_1e_2,$$

making explicit w_0, w_1, w_2, w_3 and establish that $\langle \boldsymbol{XY} \rangle = \langle \boldsymbol{YX} \rangle$.

2.23 Expand out the trivector $x \wedge (y \wedge z)$ in terms of geometric products. Is the result antisymmetric on x and y? Justify why one can also write the following equality:

$$x \wedge y \wedge z = \frac{1}{2}(xyz - w \wedge y \wedge x),$$

and prove that the following equation is true:

$$x \wedge y \wedge z = \frac{1}{6}(xyz + zxy + ywx - xzy - yaz - wyx).$$

2.24 Prove using wedge and inner product the formula $x \times (y \times z) = x \cdot (zy) - x \cdot (yz)$, using wedge- and inner-products of the vector calculus formula.

2.25 Compute the area of a parallelogram spanned by the vectors $x = 2e_1 + e_3$ and $y = e1 - e_3$ relative to the area of $e_1 \wedge e_3$.

2.26 In $G_{3,0,0}$, compute the intersection of the non-homogeneous line L_1 with position vector e_3 and direction $e_2 + e_3$ and the line L_2 with position vector $e_2 - e_3$ and direction e_2.

2.27 The projection of a vector v with respect to a plane A_r is given by:

$$v = v_{\|} + v_{\perp} = P^{\|}_{A_r} + P^{\perp}_{A_r},$$

justify why the vector $v \cdot A_r$ is orthogonal to the plane formed by $v_{\|}$ and $v_{\perp}$.

2.28 With a bivector B and the general multivectors X, Y, prove that

$$B \times (XY) = (B \times X)Y + X(B \times Y),$$

hence that

$$B \times (x \wedge X_r) = (B \cdot x) \wedge X_r + x \wedge (B \times X_r).$$

Use these results to establish that the operation of commuting with a bivector is grade preserving.

2.29 Affine transformations: consider the standard orthonormal basis of $\mathbb{R}^{n,0}$ for the shear transformation $g_s : x \rightarrow g_s(x) = x + s(x \cdot e_2)e_1$; compute the transformation matrices $[g_s]$ and $[g_s^*]$ both to act on vectors. Depict your results to see the shear effect of a planar line and its normal vector.

2.30 Simplex: given the three points $x_0 = e_3, x_1 = e_2 + e_3, x_2 = e_1 + e_3$, compute using CLICAL [196], CLUCAL [234] or eClifford [1], the simplex tangent $\bar{X}_2$, the simplex volume $(2!)^{-1}|\bar{X}_2|$ and the simplex moment $X_2 = x_0 \wedge \bar{X}_2$. Show that if $x_0 = 0$ then the simplex passes through the origin and its moment vanishes.

2.31 Simplex: given the four points $\boldsymbol{x}_0 = e_1$, $\boldsymbol{x}_1 = e_2$, $\boldsymbol{x}_2 = e_1 + e_2$ and $\boldsymbol{x}_3 = e_3$, compute using CLICAL [196], CLUCAL [234] or eClifford [1], the four faces $\mathcal{F}_i^3 \boldsymbol{X}_3$ opposite to the points $\boldsymbol{x}_i$ $i = 0, \ldots, 3$.

2.32 Using the points given in Exercise 2.31, compute using CLICAL [196], CLUCAL [234] or eClifford [1], the boundary $\Omega_b X_3$ of the simplex $\boldsymbol{X}_3$.

2.33 Using the points given in Exercise 2.31, check using CLICAL [196], CLUCAL [234] or eClifford [1], if the following relations hold for the operators $\mathcal{F}_i$ and Ω_b:

$$
\begin{aligned}
\mathcal{F}_i \mathcal{F}_i &= 0, \\
\mathcal{F}_i \mathcal{F}_j &= -\mathcal{F}_j \mathcal{F}_i \\
\mathcal{F}_i \Omega_b &= -\Omega_b \mathcal{F}_i \\
\Omega_b \Omega_b &= 0.
\end{aligned}
$$

2.34 Using CLICAL [196], CLUCAL [234] or eClifford [1] compute the join $\boldsymbol{X} \cup \boldsymbol{Y}$ and meet $\boldsymbol{X} \cap \boldsymbol{Y}$ for the following blades: (i) $\boldsymbol{X} = e_2$ and $\boldsymbol{Y} = 3e_1$; (ii) $\boldsymbol{X} = e_1$ and $\boldsymbol{Y} = 4e_1$; (iii) $\boldsymbol{X} = e_1$ and $\boldsymbol{Y} = -e_2$; (iv) $\boldsymbol{X} = e_1 \wedge e_2$ and $\boldsymbol{Y} = 2e_1$; (v) $\boldsymbol{X} = (e_1 + e_2)/\sqrt{2}$ and $\boldsymbol{Y} = e_1$; (vi) $\boldsymbol{X} = e_1 \wedge e_2$ and $\boldsymbol{Y} = 0.00001e_1 + e_2$ and (vii) $\boldsymbol{X} = e_1 \wedge e_2$ and $\boldsymbol{Y} = cos(\theta)e_1 + sin(\theta)e_2$.

Chapter 3
Differentiation, Linear, and Multilinear Functions in Geometric Algebra

This chapter gives a detailed outline of differentiation, linear and multilinear functions, eigenblades, and tensors formulated in geometric algebra and explains the related operators and transformations in common use by mathematicians, physicists, computer scientists, and engineers.

We have learned that readers of the book [138] may have difficulties to understand it and practitioners have difficulties to try the equations in certain applications. For this reason, this chapter reviews concepts and equations most of them introduced by D. Hestenes and G. Sobczyk [138] and Ch. Doran and A. Lasenby [74]. This chapter is written in a clear manner for readers interested in applications in computer science and engineering. The explained equations will be required to understand advanced applications in next chapters.

In [131, 138], David Hestenes showed the advantages of developing the theory of linear and multilinear functions on finite spaces using geometric algebra framework. In this chapter, we will briefly outline the geometric approach to linear and multilinear functions, mappings, and kernels which are important for a number of advanced applications treated throughout of this work.

Operators and transformations are somehow awkward and less efficient in specific computations. For this purpose, matrix algebra is usually used; however, matrix algebra has the disadvantage of requiring a specific selection of frames, which is surprisingly irrelevant for the task in question. Furthermore, the use of matrices is tailored to the theory of linear functions and not envisaged to be applied to the representation of multilinear or nonlinear functions. As we explain in this chapter, as brilliantly shown by D. Hestenes [138], geometric algebra corrects this deficiency and provides other advantages preserving the intrinsic definitions. In particular, geometric algebra representations do not use matrices to avoid redundant matrix entries, and as a result, algorithms can be speed up for real-time applications.

E. Bayro-Corrochano, *Geometric Algebra Applications Vol. I*,
https://doi.org/10.1007/978-3-319-74830-6_3

3.1 Differentiation

This section starts defining the vector derivative and presents its properties; in subsequent subsections, we treat the derivatives on vector manifolds and explain the relation of derivatives to directed integration. The subsections about vector and multivector differentiation is a study based on the formulas proposed by D. Hestenes and G. Sobzcyk [138].

3.1.1 *Differentiation by Vectors*

The derivative of a multivector-valued function $F = F(\tau)$ with respect to a scalar parameter τ is expressed using the Leibniz's notation

$$\partial_\tau F(\tau) = \frac{\partial F(\tau)}{\partial \tau} = \lim_{\Delta\tau \to 0} \frac{F(\tau + \Delta\tau) - F(\tau)}{\Delta\tau}. \tag{3.1}$$

In the study of fields in geometric algebra, we represent a position by a vector spanned in geometric algebra in terms of basis $\{e^k\} \in G_n$, and thus these vectors inherit all the properties of the geometric product. Remarkably, the known concepts of gradient, divergence, and curl, which are developed separately in vector analysis, will be reduced to a single concept of the vector derivative in geometric algebra.

The vector derivative is the derivative ∂ or *nabla* ∇ with respect to the vector position $\boldsymbol{x}$, and it can be written with respect to a fixed frame $\{e^k\}$ with coordinates $x^k = e^k \cdot x$ as follows:

$$\partial = \nabla = \sum_k e^k \frac{\partial}{\partial x^k}. \tag{3.2}$$

Now, let be $F(\boldsymbol{x})$ any multivector-valued function of position, or more generally a position-dependent linear function, if we compute the inner product of ∂ with the vector $\boldsymbol{v}$, we get the *directional derivative* in the $\boldsymbol{v}$ direction or the *first differential* of F. Briefly, the $\boldsymbol{v} - \textit{derivative}$ of F is given by

$$\boldsymbol{v} \cdot \partial F(\boldsymbol{x}) \equiv \left.\frac{\partial F(\boldsymbol{x} + \tau \boldsymbol{v})}{\partial \tau}\right|_{\tau=0} = \lim_{\tau \to 0} \frac{F(\boldsymbol{x} + \tau \boldsymbol{v}) - F(\boldsymbol{x})}{\tau}. \tag{3.3}$$

The derivative (by $\boldsymbol{x}$) of the function $F(\boldsymbol{x})$ is denoted by $\partial_x F(\boldsymbol{x})$. The differential operator ∂_x is defined assuming that posses all the properties of a vector in geometric algebra G_n, and thus we can formulate its basic properties of ∂_x explicitly as follows, for the unit pseudoscalar $I \in G_n$

$$I \wedge \partial_x = 0 \tag{3.4}$$

$$I \partial_x = I \cdot \partial_x \tag{3.5}$$

According to Eq. (2.50), the projection of a r-vector $\boldsymbol{U}_r$ into the hypervolume $\boldsymbol{V}_n$ can be expanded as follows, c.f. ([138], 1-3.16)

$$P(\boldsymbol{U}_r) = (\boldsymbol{U}_r \cdot \boldsymbol{V}_n)\boldsymbol{V}_n^{-1} = \sum_{j_1<...<j_r} \boldsymbol{U}_r \cdot (\boldsymbol{v}_{j_r} \wedge ... \wedge \boldsymbol{v}_{j_1} \boldsymbol{v}^{j_1} \wedge ... \wedge \boldsymbol{v}^{j_r}). \tag{3.6}$$

Like other vectors in G_n, the derivative has the properties

$$\partial_x = P(\partial_x) = \sum_k \boldsymbol{v}^k \boldsymbol{v}_k \cdot \partial_x, \tag{3.7}$$

from this equation, it follows

$$\boldsymbol{v} \cdot \partial_x = P(\boldsymbol{v}) \cdot \partial_{x'}, \tag{3.8}$$

where $x' = f(x)$.

The differential of $F = F(\boldsymbol{x})$ can be expressed using the definition of the inner product in terms of the geometric product, c.f. ([138], 2-1.6)

$$\boldsymbol{v} \cdot \partial F = \frac{1}{2}(\boldsymbol{v}\partial F + \dot{\partial}\boldsymbol{v}\dot{F}), \tag{3.9}$$

where the overdots indicate that F is to be differentiated when $\boldsymbol{v}$ is also regarded to be a function of $\boldsymbol{x}$.

The differential of the identity function $F(\boldsymbol{x}) = \boldsymbol{x}$ reads, c.f. ([138], 2-1.18)

$$\boldsymbol{v} \cdot \partial_x \boldsymbol{x} = P(\boldsymbol{v}) = \partial_x(\boldsymbol{x} \cdot \boldsymbol{v}). \tag{3.10}$$

The left equality of Eq. (3.10) can be computed using the Eqs. (3.1) and (3.8). Let us prove the right equality, note that for the base vectors $\boldsymbol{v}_k$, one has

$$\boldsymbol{v}_k \cdot \partial \boldsymbol{x} = P(\boldsymbol{v}_k) = \boldsymbol{v}_k, \tag{3.11}$$

so using Eq. (3.7), one gets the desired result

$$\partial_x(\boldsymbol{x} \cdot \boldsymbol{v}) = \sum_k \boldsymbol{v}^k \boldsymbol{v}_k \cdot \partial_x(\boldsymbol{x} \cdot \boldsymbol{v}) = \sum_k \boldsymbol{v}^k \boldsymbol{v}_k \cdot \boldsymbol{v} = P(\boldsymbol{v}). \tag{3.12}$$

From Eqs. (3.7) and (3.10), we derive the operator identity

$$\partial_x = P(\partial_x) = \partial_v \boldsymbol{v} \cdot \partial_x. \tag{3.13}$$

Furthermore, It should be mentioned that the operator $\boldsymbol{v} \cdot \partial$ preserves the grade, namely

$$\boldsymbol{v} \cdot \partial < F >_r = < \boldsymbol{v} \cdot \partial F > r. \tag{3.14}$$

Next, we formulate straightforwardly four general linear properties,

$$\begin{aligned} (\boldsymbol{u}+\boldsymbol{v})\cdot\partial F &= \boldsymbol{u}\cdot\partial F+\boldsymbol{v}\cdot\partial F\\ \alpha\boldsymbol{v}\cdot\partial F &= \alpha(\boldsymbol{v}\cdot\partial F)\\ \boldsymbol{v}\cdot(\partial F+\partial G) &= \boldsymbol{v}\cdot\partial F+\boldsymbol{v}\cdot\partial G\\ \boldsymbol{v}\cdot(\partial FG) &= (\boldsymbol{v}\cdot\partial F)G+F(\boldsymbol{v}\cdot\partial G) \end{aligned} \tag{3.15}$$

Another general property of the differentials concerns composite functions. Given $\boldsymbol{x}'=f(\boldsymbol{x})$, to formulate its differential, we expand $F(\boldsymbol{x})=G(f(\boldsymbol{x}))$ utilizing the Taylor expansion

$$f(\boldsymbol{x}+\tau\boldsymbol{v})=f(\boldsymbol{x})+\tau\boldsymbol{v}\cdot\partial f(\boldsymbol{x})+\frac{\tau^2}{2!}(\boldsymbol{v}\cdot\partial)^2 f(\boldsymbol{x})+\cdots \tag{3.16}$$

thus, we can express

$$\begin{aligned} \boldsymbol{v}\cdot\partial_x G(f(\boldsymbol{x}) &= \partial_\tau G(f(\boldsymbol{x}+\tau\boldsymbol{v}))|_{\tau=0}\\ &= \partial_\tau G(f(\boldsymbol{x})+\tau f(\boldsymbol{v}))|_{\tau=0}\\ &= \boldsymbol{v}\cdot\partial f(\boldsymbol{x})\cdot\partial_{x'}G(\boldsymbol{x}')|_{x'=f(x)} \end{aligned} \tag{3.17}$$

There is another version of the chain rule for scalar-valued functions such as $s=s(\boldsymbol{x})$, c.f. ([138], 1-1.15)

$$\boldsymbol{v}\cdot\partial_x F(s(\boldsymbol{x}))=(\boldsymbol{v}\cdot\partial s(\boldsymbol{x}))\partial_s F(s(\boldsymbol{x})). \tag{3.18}$$

The *second differential* of the function $F=F(\boldsymbol{x})$ is defined by, c.f. ([138], 2-1.16.a)

$$F_{uv}=\boldsymbol{v}\cdot\dot{\partial}\boldsymbol{u}\cdot\partial\dot{F}(\boldsymbol{x}). \tag{3.19}$$

From the definition Eq. (3.1) of the differential as a limit, one can derive the *integrability* condition

$$F_{uv}=F_{vu}. \tag{3.20}$$

The geometric product of ∂ and $F(\boldsymbol{x})$ is given by, c.f. ([138], 2-1.21.a)

$$\partial F=\partial\cdot F+\partial\wedge F. \tag{3.21}$$

If $F=\psi(\boldsymbol{x})$ is scalar-valued then $\partial\cdot\psi(\boldsymbol{x})=0$, thus

$$\begin{aligned} \partial\psi(\boldsymbol{x}) &= \partial\cdot\psi(\boldsymbol{x})+\partial\wedge\psi(\boldsymbol{x})\\ &= \partial\wedge\psi(\boldsymbol{x}). \end{aligned} \tag{3.22}$$

$\partial\psi(\boldsymbol{x})$ corresponds to the *gradient* of $\psi(\boldsymbol{x})$. Moreover, we will call $\partial \cdot F$ the *divergence* of F and $\partial \wedge F$ the *curl* of F.

The fundamental problem of calculus on manifolds is to invert a differential equation

$$\partial f = g, \tag{3.23}$$

to obtain the solution

$$f = \partial^{-1} g. \tag{3.24}$$

According to [138], thanks to the use of the fundamental theorem, one gets an explicit expression for the *antiderivative* operator ∂^{-1} which leads to many results of which the generalized Cauchy's integral might be the most important.

If one differentiates the function $F = F(\boldsymbol{x})$ twice, using Eq. (3.13) and the definition (3.19) of the second differential, one obtains, c.f. ([138], 2-1.28.c)

$$\partial_x^2 F(\boldsymbol{x}) = \partial_v \partial_u F_{uv} = (\partial_v \cdot \partial_u + \partial_v \wedge \partial_u) F_{uv} \tag{3.25}$$

Due to the integrability condition, Eq. (3.20), one gets

$$\partial_x \wedge \partial_x F(\boldsymbol{x}) = \partial_v \wedge \partial_u F_{uv} = -\partial_v \wedge \partial_u F_{uv} = 0, \tag{3.26}$$

thus, the operator identity

$$\partial_x \wedge \partial_x = 0, \tag{3.27}$$

expresses the integrability condition for the vector derivative. Conversely, from Eq. (3.27), one can derive the integrability condition (3.20), namely using the inner product between bivectors Eqs. (2.54) and (3.19), one gets

$$(\boldsymbol{u} \wedge \boldsymbol{v}) \cdot (\partial \wedge \partial) F = F_{vu} - F_{uv} = 0. \tag{3.28}$$

Due to the integrability, one can affirm that the second derivative is a scalar differential operator, namely

$$\partial_x^2 = \partial_x \cdot \partial_x, \tag{3.29}$$

which corresponds to the *Laplacian* linear operator.

In the computations besides the basic Eq. (3.13), the derivatives of elementary functions are needed. Next, we list most commonly used derivatives which were proposed by Hestenes, c.f. ([138], 2-(1.32-1.37))

$$\partial|\boldsymbol{v}|^2 = \partial\boldsymbol{v}^2 = 2P(\boldsymbol{v}) = 2\boldsymbol{v}, \tag{3.30}$$

$$\partial\wedge\boldsymbol{v} = 0, \tag{3.31}$$

$$\partial\boldsymbol{v} = \partial\cdot\boldsymbol{v} = n, \tag{3.32}$$

$$\partial|\boldsymbol{v}|^k = k|\boldsymbol{v}|^{k-2}\boldsymbol{v}, \tag{3.33}$$

$$\partial\Big(\frac{\boldsymbol{v}}{|\boldsymbol{v}|^k}\Big) = \frac{n-k}{|\boldsymbol{v}|^k}, \tag{3.34}$$

$$\partial log|\boldsymbol{v}| = \boldsymbol{v}^{-1}. \tag{3.35}$$

If $\boldsymbol{V} = P(\boldsymbol{V}) =< \boldsymbol{V} >_r$, c.f. ([138], 2-(1.38-1.40))

$$\dot{\partial}(\dot{\boldsymbol{v}}\cdot\boldsymbol{V}) = \boldsymbol{V}\cdot\partial\boldsymbol{v} = r\boldsymbol{V}, \tag{3.36}$$

$$\dot{\partial}(\dot{\boldsymbol{v}}\wedge\boldsymbol{V}) = \boldsymbol{V}\wedge\partial\boldsymbol{v} = (n-r)\boldsymbol{V}, \tag{3.37}$$

$$\dot{\partial}\boldsymbol{V}\dot{\boldsymbol{v}} = \sum_k \boldsymbol{a}^k\boldsymbol{V}\boldsymbol{a}_k = (-1)^r(n-2r)\boldsymbol{V}. \tag{3.38}$$

According to [138], these equations can be proved straightforwardly as follows: Eq. (3.30) follows from (3.10) by the product rule. Equation (3.31) follows from (3.30) by the integrability condition (3.27). Equation (3.32) follows from (3.31), (3.7), (3.10) and $v^k \cdot v_j = \delta^k_j$, hence

$$\partial\cdot\boldsymbol{v} = \sum_k \boldsymbol{v}^k\cdot(\boldsymbol{v}_k\cdot\partial\boldsymbol{v}) = \sum_k \boldsymbol{v}^k\cdot\boldsymbol{v}_k = n. \tag{3.39}$$

Equations (3.33), (3.34), and (3.35) are derived by the use of "scalar chain rule" (3.18). The proof of Eqs. (3.36), (3.37) and (3.38) is left as exercises listed at the end of this chapter.

3.1.2 *Differential Identities*

Since the derivative inheritates the same algebraic properties as a vector, one gets several differential identities simply replacing some of their vectors by ∂. Consider the vector-valued functions $\boldsymbol{u} = \boldsymbol{u}(x)$, $\boldsymbol{v} = \boldsymbol{v}(x)$, $\boldsymbol{w} = \boldsymbol{w}(x)$ and using the identity of Eq. (3.37), we get

$$\boldsymbol{v}\cdot(\partial\wedge\boldsymbol{u}) = \boldsymbol{v}\cdot\partial\boldsymbol{u} - \dot{\partial}\dot{\boldsymbol{u}}\cdot\boldsymbol{v}, \tag{3.40}$$

thus

$$\partial(\boldsymbol{u}\cdot\boldsymbol{v}) = \boldsymbol{u}\cdot\partial\boldsymbol{v} + \boldsymbol{v}\cdot\partial\boldsymbol{u} - \boldsymbol{u}\cdot(\partial\wedge\boldsymbol{v}) - \boldsymbol{v}\cdot(\partial\wedge\boldsymbol{u}). \tag{3.41}$$

Using (3.37), the computation of the Lie bracket gives

$$[\boldsymbol{u}, \boldsymbol{v}] = \partial \cdot (\boldsymbol{u} \wedge \boldsymbol{v}) - \boldsymbol{v}\partial \cdot \boldsymbol{u} + \boldsymbol{u}\,\partial \cdot \boldsymbol{v}, \tag{3.42}$$

a result which differs of the conventional one, c.f. ([138], 2-1.45)

$$[\boldsymbol{u}, \boldsymbol{v}] = \boldsymbol{u} \cdot \partial \boldsymbol{v} - \boldsymbol{v} \cdot \partial \boldsymbol{u}. \tag{3.43}$$

The Jacobi identity is given by

$$\boldsymbol{u} \cdot (\boldsymbol{v} \wedge \boldsymbol{w}) + \boldsymbol{v} \cdot (\boldsymbol{w} \wedge \boldsymbol{u}) + \boldsymbol{w} \cdot (\boldsymbol{u} \wedge \boldsymbol{v}) = 0. \tag{3.44}$$

Using the Jacobi identity (3.44), one gets

$$\begin{aligned} \boldsymbol{u} \cdot (\partial \wedge \boldsymbol{v}) &= \dot{\boldsymbol{v}} \cdot (\dot{\partial} \wedge \boldsymbol{u}) + \dot{\partial} \cdot (\boldsymbol{u} \wedge \dot{\boldsymbol{v}}) \\ &= (\boldsymbol{u} \wedge \partial) \cdot \boldsymbol{v} + \boldsymbol{u} \cdot \partial \boldsymbol{v} - \boldsymbol{u}\partial \cdot \boldsymbol{v}. \end{aligned} \tag{3.45}$$

By dotting as Eq. (3.40) from the left instead with the bivector $\boldsymbol{w} \wedge \boldsymbol{v}$, one gets

$$\begin{aligned} (\boldsymbol{w} \wedge \boldsymbol{v}) \cdot (\partial \wedge \boldsymbol{u}) &= \boldsymbol{v} \cdot \dot{\partial} \dot{\boldsymbol{u}} \cdot \boldsymbol{w} - \boldsymbol{w} \cdot \dot{\partial} \dot{\boldsymbol{u}} \cdot \boldsymbol{v} \\ &= \boldsymbol{v} \cdot \partial(\boldsymbol{u} \cdot \boldsymbol{w}) - \boldsymbol{w} \cdot \partial(\boldsymbol{u} \cdot \boldsymbol{v}) + [\boldsymbol{w}, \boldsymbol{v}] \cdot \boldsymbol{u}. \end{aligned} \tag{3.46}$$

From c.f. ([138], 1-1.60.c-d), we get a generalization of Eq. (3.40) and (3.45), namely c.f. ([138], 2-1.49)

$$\begin{aligned} \boldsymbol{U} \times (\partial \wedge \boldsymbol{v}) &= \boldsymbol{U} \cdot \partial \boldsymbol{v} - \dot{\partial} \dot{\boldsymbol{v}} \cdot \boldsymbol{U} \\ &= \boldsymbol{U} \wedge \boldsymbol{\partial} v - \dot{\partial} \dot{\boldsymbol{v}} \wedge \boldsymbol{U}. \end{aligned} \tag{3.47}$$

Hestenes [138] at (2-1.50) derived the following differential identity, given the homogeneous multivector functions $\boldsymbol{F}(\boldsymbol{x}) =< \boldsymbol{F}(\boldsymbol{x}) >_r$ and $\boldsymbol{G}(\boldsymbol{x}) =< \boldsymbol{G}(\boldsymbol{x}) >_s$, reordering the factors

$$\begin{aligned} \dot{\boldsymbol{F}} \wedge \dot{\partial} \wedge \dot{\boldsymbol{G}} &= (-1)^r \partial \wedge (\boldsymbol{F} \wedge \boldsymbol{G}) \\ &= \boldsymbol{F} \wedge \partial \wedge \boldsymbol{G} + (-1)^r (\partial \wedge \boldsymbol{F}) \wedge \boldsymbol{G} \\ &= \boldsymbol{F} \wedge \partial \wedge \boldsymbol{G} + (-1)^{r+s(r+1)} \boldsymbol{G} \wedge \partial \wedge \boldsymbol{F}, \end{aligned} \tag{3.48}$$

its proof is left as exercise.

Given a *simplicial* variable $\boldsymbol{x}_{(r)} = \boldsymbol{x}_1 \wedge ... \wedge \boldsymbol{x}_r$, a simplicial derivative is given by, c.f. ([138], 2-1.36.b)

$$\partial_{(r)} = (r!)^{-1} \partial_{\boldsymbol{x}_r} \wedge ... \wedge \partial_{\boldsymbol{x}_2} \wedge \partial_{\boldsymbol{x}_1}. \tag{3.49}$$

Now, consider a function of a simplicial variable $L(\boldsymbol{X}) = L(\boldsymbol{x}_{(r)}) = L(\boldsymbol{x}_1 \wedge ... \wedge \boldsymbol{x}_r)$, the r-simplicial derivative of $L(\boldsymbol{x}_{(r)})$ is computed as follows, c.f. ([138], 2-3.5)

$$\partial_{(r)} L(\boldsymbol{x}_{(r)}) = \boldsymbol{x}! \partial_{\boldsymbol{x}} L = \partial_r \wedge ... \wedge \partial_2 \wedge \partial_1 L(\boldsymbol{x}_1 \wedge ... \wedge \boldsymbol{x}_r). \tag{3.50}$$

3.1.3 Multivector Derivatives, Differentials, and Adjoints

In the previous section, we study the differentiation of functions of scalars and vector variables; this subsection deals with a more general theory of differentiation of functions in any multivector variable. We review some equations developed in this matter by D. Hestenes and G. Sobczyk in Sect. 2.2-2 [138], which we believe are very useful for this book on applications.

The multivector derivative constitutes a central concept of geometric calculus. The derivative relates with every continuously differentiable function two auxiliary functions: the differential and the adjoint.

Let $F = F(\boldsymbol{X})$ be a function defined in G_n and let $P(\boldsymbol{U}) = (\boldsymbol{U} \cdot I) \cdot I^{\dagger}$ be the projection of a given multivector $\boldsymbol{U}$ into G_n. We define the $\boldsymbol{U} - derivative$ of F (at $\boldsymbol{X}$) as follows

$$\boldsymbol{U} * \partial_X F(\boldsymbol{X}) = \partial_\tau F(\boldsymbol{X} + \tau P(\boldsymbol{U}))|_{\tau=0}. \tag{3.51}$$

If this result is considered as a linear function of $\boldsymbol{U}$, one calls it the differential of F (at $\boldsymbol{X}$) and employs any of these notations

$$\underline{F} = \underline{F}(\boldsymbol{U}) = \underline{F}(\boldsymbol{X}, \boldsymbol{U}) = F_U(\boldsymbol{X}) = \boldsymbol{U} * \partial F. \tag{3.52}$$

From the definition (3.51), the following linear properties can be derived, c.f. ([138], 2-(2.20-2.23))

$$\underline{F}(\boldsymbol{U}) = \underline{F}(P(\boldsymbol{U})), \tag{3.53}$$
$$\boldsymbol{U} * \partial < F >_p = < \boldsymbol{U} * \partial F >_p = < \underline{F}(\boldsymbol{U}) >_p, \tag{3.54}$$
$$\underline{F}(\boldsymbol{U} + \boldsymbol{V}) = \underline{F}(\boldsymbol{U}) + \underline{F}(\boldsymbol{V}), \tag{3.55}$$
$$\underline{F}(\lambda \boldsymbol{U}) = \lambda \underline{F}(\boldsymbol{U}), \quad \texttt{if } \lambda = < \lambda > . \tag{3.56}$$

The sum and products of $G = G(\boldsymbol{X})$ and $F = F(\boldsymbol{X})$ are decomposed as follows, c.f. ([138], 2-(2.7-2.8))

$$\boldsymbol{U} * \partial(F + G) = \boldsymbol{U} * \partial F + \boldsymbol{U} * \partial G, \tag{3.57}$$
$$\boldsymbol{U} * \partial(FG) = (\boldsymbol{U} * \partial F)G + F(\boldsymbol{U} * \partial G). \tag{3.58}$$

The differential of the composite function $F(\boldsymbol{X}) = G(f(\boldsymbol{X}))$ is computed using the chain rule, c.f. ([138], 2-(2.9-2.12))

$$U * \partial F = U * \partial_X G(f(X)) = \underline{f}(U) * \partial G, \tag{3.59}$$

or equivalently

$$\underline{F}(U) = \underline{G}(\underline{f}(U)). \tag{3.60}$$

The second differential of $F = F(U)$ is defined as follows

$$F_{UV} = F_{UV}(X) = V * \dot{\partial} U * \partial \dot{F}, \tag{3.61}$$

which satisfies the integrability condition

$$F_{UV}(X) = F_{VU}(X). \tag{3.62}$$

The differential transformation of a bivector induced by $f(\boldsymbol{v}) = \boldsymbol{v} \cdot F$ is computed from

$$\underline{f}(\boldsymbol{v} \wedge \boldsymbol{u}) = f(\boldsymbol{v}) \wedge f(\boldsymbol{u}) = (\boldsymbol{v} \cdot F) \wedge (\boldsymbol{u} \cdot F) \tag{3.63}$$

and

$$(\boldsymbol{v} \wedge \boldsymbol{u}) \cdot (F \wedge F) = \boldsymbol{v} \cdot [2(\boldsymbol{u} \cdot F) \wedge F] = 2(\boldsymbol{v} \wedge \boldsymbol{u}) \cdot FF + 2(\boldsymbol{v} \cdot F) \wedge (\boldsymbol{u} \cdot F), \tag{3.64}$$

thus c.f. ([138], 3-4.21)

$$\underline{f}(\boldsymbol{v} \wedge \boldsymbol{u}) = \frac{1}{2}(\boldsymbol{v} \wedge \boldsymbol{u}) \cdot (F \wedge F) - (\boldsymbol{v} \wedge \boldsymbol{u}) \cdot (FF), \tag{3.65}$$

see Eq. (3-4.22) in [138] for the differential and adjoint transformation on an arbitrary p-vector V_p.

The derivative $\partial = \partial_X \in G_n$ w.r.t. a variable X posses the same algebraic properties of a multivector in G_n, that means that it can be spanned via a vector basis $\{\boldsymbol{v}_J\}$

$$\partial X = P(\partial_X) = \sum_J \boldsymbol{v}^J \boldsymbol{v}_J * \partial_X, \tag{3.66}$$

where ∂_X can be further represented in terms of p-components

$$\partial_X = \sum_p \partial_{<X>_p}, \tag{3.67}$$

where the derivative w.r.t. a variable $X_p =< X >_p \in G_n^p$ reads

$$\partial_{<X>_p} =< \partial_X >_p= \sum_J < v^J >_p< v_J >_p *\partial_X. \tag{3.68}$$

Also, this equation can be written in an abbreviate form as

$$\partial = \sum_p \partial_{\overline{p}}. \tag{3.69}$$

Given a multivector $U \in G_n$, the U-derivative w.r.t. the variables $U_{\hat{p}} =< U >_p \in G_n^p$ can be computed via the scalar product

$$U * \partial = \sum_p U * \partial_{\hat{p}} = \sum_p U_{\hat{p}} * \partial_{\hat{p}}. \tag{3.70}$$

Note that the derivative can be computed using the identity, namely

$$\partial_X = \partial_U U * \partial_X, \tag{3.71}$$

this result implies

$$\partial F \equiv \partial_X F(X) = \partial_U \underline{F}(X, U) \equiv \underline{\partial}\, \underline{F}. \tag{3.72}$$

These results are now used to define the *adjoint* of $F = F(X)$ as follows

$$\overline{F} = \overline{F}(U') = \underline{\partial}\, \underline{F} * U' = \partial F * U'. \tag{3.73}$$

This leads to a familiar expression used to define the adjoint of a linear function $\underline{F}$.

$$V * \overline{F}(U) = \underline{F}(V) * U. \tag{3.74}$$

According to Sect. 2-2 of [138], the adjoint has the following linear properties, c.f. ([138], 2-(2.20-2.23))

$$P(\overline{F}(U)) = \overline{F}(U), \tag{3.75}$$

$$< \overline{F}(U) >_p = \underline{\partial}_{\overline{p}} \underline{F} * U = \partial_{\overline{p}} F * U, \tag{3.76}$$

$$\overline{F}(U + V) = \overline{F}(U) + \overline{F}(V), \tag{3.77}$$

$$\overline{F}(\lambda U) = \lambda \overline{F}(U), \quad \texttt{if } \lambda =< \lambda >. \tag{3.78}$$

The derivative obeys the sum and product rules, c.f. ([138], 2-(2.24-2.26.b))

$$\underline{\partial}_{\overline{p}}(F + G) = \underline{\partial}_{\overline{p}} F + \underline{\partial}_{\overline{p}} G, \tag{3.79}$$

$$\underline{\partial}_{\overline{p}}(FG) = \dot{\partial}_{\overline{p}} \dot{F} G + \dot{\partial}_{\overline{p}} F \dot{G}, \tag{3.80}$$

$$\dot{F} \dot{\partial}_{\overline{p}} \dot{G} = \dot{F} \dot{\partial}_{\overline{p}} G + F \dot{\partial}_{\overline{p}} \dot{G}. \tag{3.81}$$

Given $F(\boldsymbol{X}) = G(\boldsymbol{X}')$ and $\boldsymbol{X}' = f(\boldsymbol{X})$, the way to compute the transformation of the derivative induced by the change of variables $\boldsymbol{X} \to \boldsymbol{X}'$ is as follows

$$\partial_{\overline{p}} =< \partial_X >_p = \overline{f}(< d_{X'} >_r) = \overline{f}(\partial'_{\overline{p}}). \tag{3.82}$$

The chain rule for the derivative can be written as

$$\partial_{\overline{p}} F = \dot{\partial}_{\overline{p}} G(\dot{f}) = \overline{f}(\dot{\partial}'_{\overline{p}})\dot{G}. \tag{3.83}$$

Consider $\boldsymbol{X} = \boldsymbol{X}(\tau)$ a multivector-valued function of a scalar variable τ which describes a curve; its derivative equals to the usual scalar derivative. Since the argument of $\boldsymbol{X}(\tau)$ is a scalar, hence $< \partial_\tau >_p \boldsymbol{X} = 0$ for $p \neq 0$, then using Eqs. (3.68) and (3.69), one gets

$$\partial_\tau \boldsymbol{X} =< \partial_\tau > \boldsymbol{X} = \frac{d\boldsymbol{X}}{d\tau}. \tag{3.84}$$

The differential of a scalar-valued $\boldsymbol{X} = x(\tau)$ reads, c.f. ([138], 2-2.27.b)

$$\underline{\boldsymbol{X}}(\lambda) = \lambda * \partial \boldsymbol{X} = \lambda \partial_p \boldsymbol{X} = \lambda \frac{d\boldsymbol{X}}{d\tau}. \tag{3.85}$$

We can see that for scalar variables, the differential differs from the derivative just by a multiplicative factor.

The adjoint of $\boldsymbol{X} = \boldsymbol{X}(\tau)$ is given by, c.f. ([138], 2-2.27.c)

$$\overline{\boldsymbol{X}}(\boldsymbol{U}) = \partial_\tau \boldsymbol{X} * \boldsymbol{U} = \left(\frac{d\boldsymbol{X}}{d\tau}\right) * \boldsymbol{U} \tag{3.86}$$

Now, applying the chain rule to the function $F = F(\boldsymbol{X}(\tau))$, one obtains, c.f. ([138], 2-2.27.d)

$$\frac{dF}{d\tau} = \overline{\boldsymbol{X}}(\partial_\tau)F(\tau) = \left(\frac{d\boldsymbol{X}}{d\tau}\right) * \partial_X F(\boldsymbol{X}). \tag{3.87}$$

For a scalar-valued $\boldsymbol{X} = x(\tau)$, only $\alpha =< \boldsymbol{U} >_0$ the scalar value of $\boldsymbol{U}$ plays a role in Eq. (3.86); thus, the adjoint reduces to

$$\overline{\boldsymbol{X}}(\alpha) = \alpha \frac{dx}{d\tau}. \tag{3.88}$$

Thus, the chain rule of Eq. (3.87) can be written as the usual form using a change of scalar variables, namely

$$\frac{dF}{d\tau} = \frac{dx}{d\tau}\frac{dF}{dx}. \tag{3.89}$$

D. Hestenes and G. Sobczyk [138] provide in Chap. 1 p. 57 a list of basic multivector derivatives very useful in applications which are assembled next. We assume that $\boldsymbol{X} = F(\boldsymbol{X})$ is the identity function on some linear subspace of G_n with dimension d. $\boldsymbol{U}^{||}$ denotes the projection of the multivector $\boldsymbol{U}$ into that subspace, c.f. ([138], 2-(2.28.a-2.35))

$$\boldsymbol{U} * \partial_X \boldsymbol{X} = \dot{\partial}_X \dot{\boldsymbol{X}} * \boldsymbol{U} = \boldsymbol{U}^{||} \tag{3.90}$$

$$\boldsymbol{U} * \partial_X \boldsymbol{X}^{\dagger} = \dot{\partial}_X \dot{\boldsymbol{X}}^{\dagger} * \boldsymbol{U} = \boldsymbol{U}^{||\dagger}. \tag{3.91}$$

$$\partial_X \boldsymbol{X} = d, \tag{3.92}$$

$$\partial_X |\boldsymbol{X}|^2 = 2\boldsymbol{X}^{\dagger}, \tag{3.93}$$

$$\boldsymbol{U} * \partial_X \boldsymbol{X}^k = \boldsymbol{U}^{||}\boldsymbol{X}^{k-1} + \boldsymbol{X}\boldsymbol{U}^{||}\boldsymbol{X}^{k-2} + \cdots + \boldsymbol{X}^{k-1}\boldsymbol{U}^{||}, \tag{3.94}$$

$$\partial_X |\boldsymbol{X}|^k = k|\boldsymbol{X}|^{k-2}\boldsymbol{X}^{\dagger}, \tag{3.95}$$

$$\partial_X log|\boldsymbol{X}| = \frac{\boldsymbol{X}^{\dagger}}{|\boldsymbol{X}|^2}, \tag{3.96}$$

$$\boldsymbol{U} * \partial_X \{|\boldsymbol{X}|^k \boldsymbol{X}\} = |\boldsymbol{X}|^k \left\{\boldsymbol{U}^{||} + k\frac{\boldsymbol{U}\boldsymbol{X}^{\dagger}\boldsymbol{X}}{|\boldsymbol{X}|^2}\right\}, \tag{3.97}$$

$$\partial_X \{|\boldsymbol{X}|^k \boldsymbol{X}\} = |\boldsymbol{X}|^k \left\{d + k\frac{\boldsymbol{X}^{\dagger}\boldsymbol{X}}{|\boldsymbol{X}|^2}\right\}. \tag{3.98}$$

The proofs of these formulas are similar as the proofs of corresponding Eqs. (3.30)–(3.38), and we left for the reader as exercises.

3.2 Linear Multivector Functions

A linear function fulfills the superposition property defined as follows

$$f(\alpha\mathbf{x} + \beta\mathbf{y}) = \alpha f(\mathbf{x}) + \beta f(\mathbf{y}), \tag{3.99}$$

for scalars α, β and vectors $\mathbf{x}$, $\mathbf{y}$.

The composition of two linear functions f and g (applied first) results in a third function given by

$$h(\mathbf{x}) = f(g(\mathbf{x})) = fg(\mathbf{x}). \tag{3.100}$$

A linear function f maps vectors to vectors in the same space. The extension of f to act linearly on multivectors is possible via the so-called *outermorphism* $\underline{f}$, which defines the action of $\underline{f}$ on r-blades thus:

$$\underline{f}(\boldsymbol{a}_1 \wedge \boldsymbol{a}_2 \wedge \ldots \wedge \boldsymbol{a}_r) = \underline{f}(\boldsymbol{a}_1) \wedge \underline{f}(\boldsymbol{a}_2) \wedge \ldots \wedge \underline{f}(\boldsymbol{a}_r). \tag{3.101}$$

The function $\underline{f}$ is called an outermorphism because $\underline{f}$ preserves the grade of any r-vector it acts upon. The action of $\underline{f}$ on general multivectors is then defined through linearity. The function $\underline{f}$ must therefore satisfy the following conditions

$$\begin{aligned} \underline{f}(\boldsymbol{a}_1 \wedge \boldsymbol{a}_2) &= \underline{f}(\boldsymbol{a}_1) \wedge \underline{f}(\boldsymbol{a}_2) \\ \underline{f}(\alpha_1 \boldsymbol{a}_1 + \alpha_2 \boldsymbol{a}_2) &= \alpha_1 \underline{f}(\boldsymbol{a}_1) + \alpha_2 \underline{f}(\boldsymbol{a}_2), \end{aligned} \tag{3.102}$$

for the scalars α_1, α_2 and two vectors $\boldsymbol{a}_1, \boldsymbol{a}_2$.

Extended linear functions are grade-preserving

$$\underline{f}(A_r) = \langle \underline{f}(A_r) \rangle_r \tag{3.103}$$

and are also multilinear

$$\underline{f}(\alpha_1 A + \alpha_2 B) = \alpha_1 \underline{f}(A) + \alpha_2 \underline{f}(B), \tag{3.104}$$

which holds for any scalars α_1, α_2 and any multivectors A and B.

The outermorphism of a product of two linear functions is the product of the outermorphisms—that is, if $f(\boldsymbol{a}) = f_2(f_1(\boldsymbol{a}))$, we write $\underline{f} = \underline{f}_2 \underline{f}_1$. Consider the multivector $\boldsymbol{A}_r = \boldsymbol{a}_1 \wedge \boldsymbol{a}_2 \wedge \ldots \wedge \boldsymbol{a}_r$, we see that

$$\begin{aligned} \underline{f}(\boldsymbol{A}_r) = \underline{f}(\boldsymbol{a}_1 \wedge \boldsymbol{a}_2 \wedge \ldots \wedge \boldsymbol{a}_r) &= \underline{f}_2 \underline{f}_1(\boldsymbol{a}_1) \wedge \underline{f}_2 \underline{f}_1(\boldsymbol{a}_2) \cdots \wedge \underline{f}_2 \underline{f}_1(\boldsymbol{a}_r) \\ &= \underline{f}_2(\underline{f}_1(\boldsymbol{a}_1) \wedge \underline{f}_1(\boldsymbol{a}_2) \cdots \wedge \underline{f}_1(\boldsymbol{a}_r)) \\ &= \underline{f}_2 \underline{f}_1(\boldsymbol{a}_1 \wedge \boldsymbol{a}_2 \cdots . \wedge \boldsymbol{a}_r) \\ &= \underline{f}_2 \underline{f}_1(\boldsymbol{A}_r). \end{aligned} \tag{3.105}$$

3.3 The Determinant

Since the outermorphism preserves grade, the unit pseudoscalar or hypervolume $\boldsymbol{I} = e_1 e_2 ... e_n$ must be mapped onto some multiple of itself; this multiple is called the *determinant* of $\underline{f}$. Thus,

$$\underline{f}(I) = \det(\underline{f}) I, \tag{3.106}$$

here the determinant $\det(\overline{f})$ represents the volume of the hypervolume $\boldsymbol{I}$ as a scalar factor. This result agrees with the tensor. This is visualized in Fig. 3.1 for the case of the pseudoscalar of the 3D Euclidean geometric algebra; the orthonormal basis is changed to an affine basis.

Since the pseudoscalar $I \in G_n$ for any n-dimensional space is unique up to scaling and linear functions are grade-preserving, we can define

$$\underline{f}(I) = det(\underline{f}) I. \tag{3.107}$$

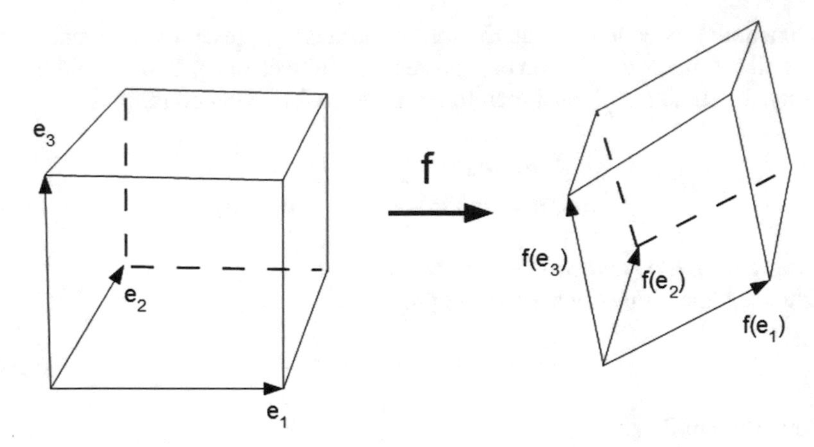

Fig. 3.1 Unit cube is transformed to a parallelepiped. The factor of the parallelepiped volume is the determinant $f(e_1)\wedge f(e_2)\wedge f(e_3) = \underline{f}(I) = \det(f)I$

Now given the product of functions $\underline{f}_3 = \underline{f}_1\underline{f}_2$ and using the equality (3.107) for the functions $\underline{f}_i$ $i = 1, 2, 3$ acting on the pseudoscalar I, we get

$$\begin{aligned}\underline{f}_3(I) &= \underline{f}_1(\underline{f}_2(I)) = \underline{f}_1(\det(\underline{f}_2)I) = \det(\underline{f}_2))\underline{f}_1(I)\\ &= \det(\underline{f}_1)\det(\underline{f}_2)I,\end{aligned} \tag{3.108}$$

since $\underline{f}_3(I) = \det(\underline{f}_3)I$, one conclude that the determinant of the product of two linear functions equals to the product of their determinants

$$\det(\underline{f}_3) = \det(\underline{f}_1)\det(\underline{f}_2). \tag{3.109}$$

It is straightforward to establish that the determinant of the adjoint equals to that of the original function, namely

$$\det(\underline{f}) =< \underline{f}(I)I^{-1} >_0=< \overline{f}(I^{-1})I >_0= \det(\overline{f}). \tag{3.110}$$

3.4 Adjoints and Inverses

The adjoint reverses the action of a linear operator, namely if a function maps from one space to another, the adjoint maps in the opposite direction.

The *adjoint* $\overline{f}$ of a linear function $\underline{f}$ acting on the vectors $\boldsymbol{a}$ and $\boldsymbol{b}$ can be defined by the property

$$\boldsymbol{a}\cdot\underline{f}(\boldsymbol{b}) = \overline{f}(\boldsymbol{a})\cdot\boldsymbol{b}, \tag{3.111}$$

$\forall\, \boldsymbol{a}, \boldsymbol{b}$.

If $\underline{f} = \overline{f}$, the function is *self-adjoint* and can be represented by a symmetric matrix F ($F = F^T$). The $\overline{f}(\boldsymbol{a})$ can be decomposed in an arbitrary frame

$$\overline{f}(\boldsymbol{a}) = \overline{f}(\boldsymbol{a}) \cdot e_k e^k = \boldsymbol{a} \cdot \underline{f}(e_k) e^k. \tag{3.112}$$

Now we can extend the adjoint over the wedge product

$$\begin{aligned}\overline{f}(\boldsymbol{a} \wedge \boldsymbol{b}) &= (\boldsymbol{a} \cdot \underline{f}(e^i) e_i) \wedge (\boldsymbol{b} \cdot \underline{f}(e^j) e_j) = e_i \wedge e_j \boldsymbol{a} \cdot \underline{f}(e^i) \boldsymbol{b} \cdot \underline{f}(e^j) \\ &= \frac{1}{2} e_i \wedge e_j [\boldsymbol{a} \cdot \underline{f}(e^i) \boldsymbol{b} \cdot \underline{f}(e^j) - \boldsymbol{a} \cdot \underline{f}(e^j) \boldsymbol{b} \cdot \underline{f}(e^i)] \\ &= \frac{1}{2} e_i \wedge e_j (\boldsymbol{a} \wedge \boldsymbol{b}) \cdot \underline{f}(e^j \wedge e^i).\end{aligned} \tag{3.113}$$

This shows us that the extension of the adjoint $\overline{f}(\boldsymbol{B}_p)$ equals to the adjoint of the extended function, namely c.f. ([74], 4.138)

$$\boldsymbol{A}_p \cdot \overline{f}(\boldsymbol{B}_p) = \underline{f}(\boldsymbol{A}_p) \cdot \boldsymbol{B}_p. \tag{3.114}$$

An hand this result, we can now formulate the extension of the adjoint for multivectors of different grade; firstly consider the following decomposition equation

$$\begin{aligned}\boldsymbol{a} \cdot \underline{f}(\boldsymbol{b} \wedge \boldsymbol{c}) &= \boldsymbol{a} \cdot \underline{f}(\boldsymbol{b}) \underline{f}(\boldsymbol{c}) - \boldsymbol{a} \cdot \underline{f}(\boldsymbol{c}) \underline{f}(\boldsymbol{b}) = \underline{f}[\overline{f}(\boldsymbol{a}) \cdot \boldsymbol{b}\boldsymbol{c} - \overline{f}(\boldsymbol{a}) \boldsymbol{c}\boldsymbol{b}] \\ &= \underline{f}[\overline{f}(\boldsymbol{a}) \cdot (\boldsymbol{b} \wedge \boldsymbol{c})].\end{aligned} \tag{3.115}$$

Using a direct extension of this postulate, we can finally write the following formulas very useful in the practice, c.f. ([74], 4.142)

$$\underline{f}(\boldsymbol{A}_p) \cdot \boldsymbol{B}_q = \underline{f}[\boldsymbol{A}_p \cdot \overline{f}(\boldsymbol{B}_q)] \quad p \geq q \tag{3.116}$$

$$\boldsymbol{A}_p \cdot \overline{f}(\boldsymbol{B}_q) = \overline{f}[\underline{f}(\boldsymbol{A}_p) \cdot \boldsymbol{B}_q] \quad p \leq q. \tag{3.117}$$

These formulas are now ready to help us to compute the inverse, setting $\boldsymbol{B}_q = \boldsymbol{I}$ in Eq. (3.117), we get

$$\boldsymbol{A}_p det(\underline{f}) \boldsymbol{I} = \overline{f}[\underline{f}(\boldsymbol{A}_p) \boldsymbol{I}] \tag{3.118}$$

rewritten as

$$\begin{aligned}\boldsymbol{A}_p &= \overline{f}[\underline{f}(\boldsymbol{A}_p) I] I^{-1} det(\underline{f})^{-1} \\ &= det(\underline{f})^{-1} I \overline{f}[I^{-1} \underline{f}(\boldsymbol{A}_p)] = \underline{f}^{-1} \underline{f}(\boldsymbol{A}_p),\end{aligned} \tag{3.119}$$

where due the anticommutation $\overline{f}[\underline{f}(\boldsymbol{A}_p) I] I^{-1} = I \overline{f}[I^{-1} \underline{f}(\boldsymbol{A}_p)]$. Since all the terms applied on $\underline{f}(\boldsymbol{A}_p)$ return $\boldsymbol{A}_p$, they represent the inverse of the function in question, namely c.f. ([74], 4.152)

$$\underline{f}^{-1}(\boldsymbol{A}_p) = det(\underline{f})^{-1} I \overline{f}(I^{-1}\boldsymbol{A}_p) \tag{3.120}$$

and similarly

$$\bar{f}^{-1}(\boldsymbol{A}_p) = det(\underline{f})^{-1} I \underline{f}(I^{-1}\boldsymbol{A}_p) \tag{3.121}$$

3.5 Eigenvectors and Eigenblades

In this section, we will make use of the differential and adjoint outermorphisms in the study of invariant spaces and linear transformations. This section is based in the work of D. Hestenes and G. Sobczyk [138]; see Sect. 3-3.

The geometric algebra G_n is spanned by $\sum_{k=0}^{n} \binom{n}{k} = 2^n$ elements. G_n is a direct sum of its homogeneous subspaces of grades 0, 1, 2, ...,k,...,n, that is

$$G_n = \bigwedge^0 \mathbb{R}^n \oplus \bigwedge^1 \mathbb{R}^n \oplus \bigwedge^2 \mathbb{R}^n \oplus \cdots \oplus \bigwedge^k \mathbb{R}^n \oplus \cdots \oplus \bigwedge^n \mathbb{R}^n \tag{3.122}$$

Each of these spaces is spanned by $m = \binom{n}{k} = \frac{n!}{(n-k)!k!}$ k-vectors. Hence, $\bigwedge^0 \mathbb{R}^n =$ R is the set of real numbers, $\bigwedge^1 \mathbb{R}^n = \mathbb{R}^n$ is a n-dimensional linear vector space spanned by a vectors basis, $\bigwedge^2 \mathbb{R}^n$ is a linear m-dimensional vector space spanned by a bivector basis, $\bigwedge^k \mathbb{R}^n$ is a linear $m = \binom{n}{k}$-dimensional vector space spanned by a k-vector basis, ... , $\bigwedge^n \mathbb{R}^n$ is one-dimensional space spanned by the unit pseudoscalar I. The k-vector $\boldsymbol{V} =< \boldsymbol{V} >_k \in \bigwedge^k \mathbb{R}^n$ is called a k-blade if and only if it can be factored into a product of k anticommuting vectors $\boldsymbol{v}_1, \boldsymbol{v}_2, \cdots, \boldsymbol{v}_k$ computed by the outer product of k-vectors, namely $\boldsymbol{V} = \boldsymbol{v}_1 \wedge \boldsymbol{v}_2 \wedge \cdots \wedge \boldsymbol{v}_k$.

An eigenvector $\boldsymbol{v}$ of the linear transformation f with eigenvalues α fulfills

$$f(\boldsymbol{v}) = \alpha \boldsymbol{v}, \tag{3.123}$$

where the eigenvalues α take scalar values. The differential outermorphism $\underline{f}$ will help us for the generalization of the notion of an eigenvector.

In general, if we have a multivector $\boldsymbol{M} \in G_n$

$$\boldsymbol{M} =< \boldsymbol{M} >_1 + < \boldsymbol{M} >_2 + ... + < \boldsymbol{M} >_k + ... + < \boldsymbol{M} >_n, \tag{3.124}$$

we will compute a k-eigenblade just for it's own subspace $\bigwedge^k \mathbb{R}^n$.

Let us consider $V = < V >_k$ be a $k - blade$, we say that the V is a $k - eigenblade$ of f with $k - eigenvalue$ α if

$$\underline{f}(V) = \alpha V \in \bigwedge^k \mathbb{R}^n. \tag{3.125}$$

Note that the computation of a k-eigenblade is carried out strictly under the differential outermorphism $\underline{f}$ in its subspace $\bigwedge^n \mathbb{R}^n$.

For (3.125), V will be called the *left k-eigenblade*. This reduces to (3.123) for the case of a vector $V = < V >_1 = v$.

A k-eigenblade with non-vanishing k-eigenvalue constitutes a subspace of k-vectors which is invariant under the action of $\underline{f}$.

Using the adjoint outermorphism $\overline{f}$, one can formulate a generalization of eigenvector complementary to (3.125), namely

$$\overline{f}(U) = \beta U, \tag{3.126}$$

where U is called the *right k-eigenblade*. If V is both left and right *k-eigenblade*, V is a *proper k-eigenblade* with a proper *k-eigenvalue*.

For the pseudoscalar $I \in G_n$, if I is a n-eigenblade of f, it is also a proper blade of f and the determinant of f is the corresponding n-eigenvalue

$$\underline{f}(I) = \alpha I = det(f)I = \overline{f}(I), \tag{3.127}$$

A k-eigenblade is irreducible under f if it can not be factorized into an outer product of blades which are also eigenblades of f. Consider the factorization of the pseudoscalar I into irreducible eigenblades of $f : V_1, V_2, \cdots V_m$, namely

$$I = V_1 \wedge V_2 \wedge \cdots V_m. \tag{3.128}$$

The factorization (3.128) can be seen as the decomposition of the vector space into a direct sum of m-dimensional invariant subspaces.

$$G_n = \bigwedge^0 \mathbb{R}^n \oplus \bigwedge^1 \mathbb{R}^n \oplus \bigwedge^2 \mathbb{R}^n \oplus \cdots \oplus \bigwedge^m \mathbb{R}^n. \tag{3.129}$$

Given (3.128), it can be proved that the characteristic polynomial $C_{f_k}(\lambda)$ of f restricted to $\bigwedge^k \mathbb{R}^n$ divides the characteristic polynomial of f on $\bigwedge^n \mathbb{R}^n$, where $I \in \bigwedge^n \mathbb{R}^n$. Introducing the auxiliary function $F(x) = f(x) - \lambda x$ and comparing with the Eq. (2.8b) (in [138] Chap. 3 p. 73) for the characteristic polynomial of f with the Eq. (3.127), we get, c.f. ([138], 3-3.6)

$$\underline{F}(I) = C_f(\lambda)I. \tag{3.130}$$

However, it is a bit more difficult to prove that $\boldsymbol{V}_k$ is an eigenblade of F with the eigenvalue C_{fk}, c.f. ([138], 3-3.7)

$$\underline{F}(\boldsymbol{V}_k) = C_{f_k}(\lambda)\boldsymbol{V}_k. \tag{3.131}$$

According to [138], for the proof, it can be shown that

$$\boldsymbol{V}_k \cdot \partial_x F(\boldsymbol{x}) = \boldsymbol{V}_k \cdot \partial_x F[P_k(\boldsymbol{x})] = \boldsymbol{V}_k \cdot \partial_x P_k(F[P_k(\boldsymbol{x})]), \tag{3.132}$$

where $P_k(\boldsymbol{x}) \equiv \boldsymbol{x} \cdot \boldsymbol{V}_k \boldsymbol{V}_k^{-1}$ stands for the projection of $\boldsymbol{x}$ into $\boldsymbol{V}_k$, hence $F[P_k(\boldsymbol{x})]$ is a restriction of F to $\bigwedge^k \mathbb{R}^n$. Thereafter, the proof of (3.130) and (3.131) are basically similar. Now, having established (3.130) and (3.131), the relation $\underline{F}(I) = \underline{F}(\boldsymbol{V}_1) \wedge \underline{F}(\boldsymbol{V}_2) \cdots \wedge \underline{F}(\boldsymbol{V}_m)$ (outermorphism) leads to the factorization of the characteristic polynomial, namely c.f. ([138], 3-3.8)

$$C_f(\lambda) = C_{f_1}(\lambda) C_{f_2}(\lambda) \cdots C_{f_m}(\lambda). \tag{3.133}$$

If $\underline{f}(\boldsymbol{V}) = \alpha \boldsymbol{V}$ and $\overline{f}(\boldsymbol{U}) = \beta \boldsymbol{U}$, then it implies c.f. ([138], 3-3.9.a-b)

$$\beta \underline{f}(\boldsymbol{V} \cdot \boldsymbol{U}) = \alpha \boldsymbol{V} \cdot \boldsymbol{U}, \;\; \texttt{if grade } \boldsymbol{V} \geq \texttt{grade } \boldsymbol{U}, \tag{3.134}$$

$$\alpha \overline{f}(\boldsymbol{V} \cdot \boldsymbol{U}) = \beta \boldsymbol{V} \cdot \boldsymbol{U}, \;\; \texttt{if grade } \boldsymbol{U} \geq \texttt{grade } \boldsymbol{V}. \tag{3.135}$$

If we consider that $\boldsymbol{V} = I$ is the pseudoscalar of G_n, then (3.134) implies that for a non-singular transformation, the dual of a right eigenblade $\boldsymbol{U}$ is a left eigenblade $I \cdot \boldsymbol{U} = \boldsymbol{U}I = \boldsymbol{U}^*$ with eigenvalue α/β. Thus, applying this result to the factorization (3.128), it yields to the *dual factorization*, c.f. ([138], 3-3.10) namely

$$I = \boldsymbol{V}^1 \wedge \boldsymbol{V}^2 \wedge \cdots \boldsymbol{V}^m, \tag{3.136}$$

where we compute $\boldsymbol{V}^k = (-1)^{\epsilon_k} \boldsymbol{V}_1 \wedge \cdots \check{\boldsymbol{V}}_k \wedge \cdots \wedge \boldsymbol{V}_m I^\dagger$, with ϵ_k =(grade $\boldsymbol{V}_k$)(grade $\boldsymbol{V}_1 + \cdots +$ grade $\boldsymbol{V}_{k-1}$) and $\boldsymbol{V}_j \cdot \boldsymbol{V}^k = \delta_k^j$.

Moreover, the eigenvalue α_k of each blade $\boldsymbol{V}_k$ equals to the eigenblade of the dual blade $\boldsymbol{V}^k$, c.f. ([138], 3-3.13), namely

$$\alpha_k = (\boldsymbol{V}_k)^{-1} \underline{f}(\boldsymbol{V}_k) = (\boldsymbol{V}^k)^{-1} \overline{f}(\boldsymbol{V}^k). \tag{3.137}$$

Finally, the projection $\boldsymbol{u}_\parallel = \boldsymbol{u} \cdot \boldsymbol{V}\boldsymbol{V}^{-1}$ of any eigenvector $\boldsymbol{u}$ into the *eigen space* $\bigwedge^k \mathbb{R}^n$ of a proper blade of $\boldsymbol{V}$ with nonzero value corresponds to an eigenvector with the same eigenvalue as $\boldsymbol{u}$, hence if $\underline{f}(\boldsymbol{V}) = \alpha \boldsymbol{V} = \overline{f}(\boldsymbol{V})$ and $\underline{f}(\boldsymbol{u}) = \beta \boldsymbol{u}$, then using (3.134) and (3.135)

$$f(u_{\|}) = f[(u \cdot V) \cdot V^{-1}] = \frac{\beta}{\alpha} f[\overline{f}(u \cdot V) \cdot V^{-1}] \tag{3.138}$$

$$= \frac{\beta}{\alpha}(u \cdot V) \cdot \underline{f}(V^{-1}) = \beta(u \cdot V) \cdot V^{-1} = \beta u_{\|} \tag{3.139}$$

This result can be used for the classification of degenerated eigenvalues.

This section based on c.f. ([138], 3.3) has presented a brief discussion on the eigenblade concept which will be used in the next chapters devoted to the applications in image processing, computer vision, and robotics.

3.6 Symmetric and Skew Symmetric Transformations

Linear transformations $f = f(v)$ can be uniquely decomposed in terms of symmetric and skew-symmetric transformations

$$f(v) = f_{+}(v) + f_{-}(v). \tag{3.140}$$

Utilizing the definitions of the differential and adjoint

$$\begin{aligned} \underline{f}(v) &= v \cdot \partial_u f(u) = \partial_u u \cdot f(v) = f(v), \\ \overline{f}(v) &\equiv \partial_u v \cdot f(u), \end{aligned} \tag{3.141}$$

and the identity $u \cdot (\partial \wedge v) = u \cdot \partial v - \dot{\partial} \dot{v} \cdot u$, in Eq. (3.140), one can easily derive its symmetric and skew-symmetric parts

$$\begin{aligned} f_{+}(v) &= \frac{1}{2}(\underline{f}(v) + \overline{f}(v)) = \partial_v\left(\frac{1}{2} v \cdot f(v)\right), \\ f_{-}(v) &= \frac{1}{2}(\underline{f}(v) - \overline{f}(v)) = \frac{1}{2} v \cdot (\partial \wedge f(v)). \end{aligned} \tag{3.142}$$

A linear transformation is *symmetric* if it fulfills the following equivalent conditions

$$\begin{aligned} &\underline{f}(v) = \overline{f}(v) \ \texttt{or} \ f_{-}(v) = 0 \ \forall \ v \\ &v \cdot f(u) = u \cdot f(v) \ \forall v \ \texttt{and} \ u \\ &\partial \wedge f(v) = 0. \end{aligned} \tag{3.143}$$

3.7 Traction Operators: Contraction and Protraction

Following Sect. 3.9 of [138] for the systematic algebraic manipulation and classification of multiforms, it is required to define an operation called *tractions*. A multiform $F = F(V)$ has a grade q if

$$F(V) =< F(V) >_q . \tag{3.144}$$

A multiform of degree p and grade q is called a $q, p - form$. If $q = p$, the multiform is $grade - preserving$. If a multiform maps p-blades into p-blades, this is a $blade - preserving$ multiform. Clearly, every blade-preserving multiform is grade-preserving.

Given a multiform $F(V) = F(< V >_p)$ of degree p, the operation of contraction determines a multiform of grade $p - 1$ defined by

$$\partial_x \cdot F(\boldsymbol{v} \wedge \boldsymbol{U}), \tag{3.145}$$

where $\boldsymbol{v}$ is a vector and $\boldsymbol{U} =< \boldsymbol{U} >_{p-1}$. Next, we introduce a new term of multiform called *protraction* acting on F

$$\partial_v \wedge F(\boldsymbol{v} \wedge \boldsymbol{U}), \tag{3.146}$$

The contraction of a $q, p - form$ is a $(q - 1), (p - 1)$-form, while its protraction is a $(q + 1), (p - 1)$-form. As you see, the contraction lowers both the degree and grade of the multiform by one unit, while the protraction lowers the degree and raises the grade by one unit.

The *traction* of a multiform F is given by, c.f. ([138], 3-9.8.c)

$$\partial_v F(\boldsymbol{v} \wedge \boldsymbol{U}) = \partial_v \cdot F(\boldsymbol{v} \wedge \boldsymbol{U}) + \partial_v \wedge F(\boldsymbol{v} \wedge \boldsymbol{U}). \tag{3.147}$$

A multiform is called *tractionless* if its traction vanishes. A linear transformation is 1,1-form, and it is symmetric if and only if it is protractionless. In general, symmetric multiforms are not necessarily protractionless, let us see an example of a 2,2-form

$$F(\boldsymbol{v} \wedge \boldsymbol{u}) = \frac{1}{2}(\boldsymbol{v} \wedge \boldsymbol{u}) \cdot (\boldsymbol{V} \wedge \boldsymbol{V}) + (\boldsymbol{v} \wedge \boldsymbol{u}) \cdot (\boldsymbol{V}\boldsymbol{V}), \tag{3.148}$$

where $\boldsymbol{v}, \boldsymbol{u}$ are vectors and $\boldsymbol{V}$ a bivector. This multiform is simply the outermorphism of a skew-symmetric linear transformation $f(\boldsymbol{v}) = \boldsymbol{v} \cdot \boldsymbol{V}$. From Eq. 3.148, one can easily see that $F(\boldsymbol{v} \wedge \boldsymbol{u})$ is symmetric. If $F(\boldsymbol{V}) = F(< \boldsymbol{V} >_p) =< F(\boldsymbol{V}) >_p$ is a linear function of $\boldsymbol{V}$. Using the Eq. (3.36), one gets, c.f. ([138], 3-9.9.b)

$$\partial_v F(\boldsymbol{v} \wedge \boldsymbol{u}) = \frac{3}{2}\boldsymbol{u} \cdot (\boldsymbol{V} \wedge \boldsymbol{V}) + \boldsymbol{u} \cdot (\boldsymbol{V}\boldsymbol{V}), \tag{3.149}$$

thus the protraction is

$$\partial_v \wedge F(\boldsymbol{v}\wedge \boldsymbol{u}) = 2\boldsymbol{u}\cdot(\boldsymbol{V}\wedge \boldsymbol{V}), \tag{3.150}$$

which in general does not vanish. Notwithstanding that a symmetric multiform is not necessarily protectionless, a protectionless multiform is undoubtedly symmetric, if it is grade-preserving, namely

$$\partial_v \wedge F(\boldsymbol{v}\wedge \boldsymbol{U}) = 0 \texttt{ implies } F(\boldsymbol{V}) = \overline{F}(\boldsymbol{V}). \tag{3.151}$$

If $F(\boldsymbol{V}) = F(< \boldsymbol{V} >_p) =< F(\boldsymbol{V}) >_p$, then $F(\boldsymbol{V})$ is a linear function of $\boldsymbol{V}$.

According to Eq. (3.50), we use for further computations the notations for the simplicial variable and simplicial derivative

$$\begin{aligned} \boldsymbol{v}_{(p)} &\equiv \boldsymbol{v}_1 \wedge \boldsymbol{v}_2 \cdots \wedge \boldsymbol{v}_p, \\ \partial_{(p)} &= \frac{1}{r!}\partial_p \wedge \cdots \partial_2 \wedge \partial_1. \end{aligned} \tag{3.152}$$

Let $\boldsymbol{v}$ a vector variable and let $\boldsymbol{Q} =< \boldsymbol{Q} >_k$ be an arbitrary $k-vector$, then their inner and wedge products read, c.f. ([138], 3-9.15.a-b)

$$\begin{aligned} \partial_v \boldsymbol{v}\cdot \boldsymbol{Q} \quad &= k\boldsymbol{Q} = \boldsymbol{Q}\cdot \partial_v \boldsymbol{v} \texttt{ for } k \geq 1, \\ \partial_v \boldsymbol{v}\wedge \boldsymbol{Q} &= (n-k)\boldsymbol{Q} = \boldsymbol{Q}\wedge \partial_v \boldsymbol{v} \texttt{ for } k \geq 0. \end{aligned} \tag{3.153}$$

Let $F = F(\boldsymbol{V})$ be an arbitrary bi-form, the bivector derivative $\partial F = \partial_V F(\boldsymbol{V})$ can be decomposed into 0-vector, 2-vector, and a 4-vector parts as follows

$$\partial F = \partial \cdot F + \partial \times F + \partial \wedge F, \tag{3.154}$$

where $\times$ is the commutator product.

Since the bi-form $F = F(\boldsymbol{V}) = F(\boldsymbol{u}\wedge \boldsymbol{v})$ is a linear function, the bivector derivative is related to the vector derivatives as follows

$$\partial F = \frac{1}{2}\partial_v \wedge \partial_u F(\boldsymbol{u}\wedge \boldsymbol{v}) = \frac{1}{2}\partial_v \partial_u F(\boldsymbol{u}\wedge \boldsymbol{v}). \tag{3.155}$$

According to [138], derivatives of linear bivector functions of homogeneous grade are computed as follows, c.f. ([138], 3-9.18.a-c)

$$\begin{aligned} \partial_V \boldsymbol{V}\cdot \boldsymbol{Q} &= \frac{k(k-1)}{2}\boldsymbol{Q} = \boldsymbol{Q}\cdot \partial_V \boldsymbol{V} \texttt{ for } k \geq 2, \\ \partial_V \boldsymbol{V}\times \boldsymbol{Q} &= k(n-k)\boldsymbol{Q} = \boldsymbol{Q}\times \partial_V \boldsymbol{V} \texttt{ for } k \geq 1, \\ \partial_V \boldsymbol{V}\wedge \boldsymbol{Q} &= \frac{(n-k)(n-k-1)}{2}\boldsymbol{Q} = \boldsymbol{Q}\wedge \partial_V \boldsymbol{V} \texttt{ for } k \geq 0, \end{aligned} \tag{3.156}$$

3.8 Tensors in Geometric Algebra

In this section, we explain how the theory of tensors can be integrated into geometric calculus in a natural and efficient manner. Similarly, with the previous study in detail of functions or mappings acting on multivectors, it is of most importance to clarify the use of tensor calculus in geometric algebra, particularly for the broad domain of applications like dynamics in robotics, affine transformations in control engineering, mappings in computer vision, functions and kernels in signal filtering, Clifford transforms and in neurocomputing. This section is based on the work of D. Hestenes and G. Sobczyk [138]; see Sect. 3-10. We want to explain the more relevant equations of these authors possible in a more comprehensive manner, because the goal of the book is to use such equations for real-time applications.

3.8.1 Tensors

A geometric function of r variables $T = T(\boldsymbol{V}_1, \boldsymbol{V}_2, \cdots, \boldsymbol{V}_r)$ is called a r-linear, if it is a linear function for each argument. Furthermore, T is a tensor of degree r on $\mathbb{R}^n$ if each variable is restricted to some space $\mathbb{R}^n$ defined on a geometric algebra. In a more general case, when each variable is defined on a geometric algebra G_n, the tensor T is called an $extensor\ of\ degree$ r on G_n.

If the values $\boldsymbol{v}_i$ of the tensor $T = T(\boldsymbol{v}_1, \boldsymbol{v}_2, ..., \boldsymbol{v}_r)$ on $\mathbb{R}^n$ are s-vectors in $\bigwedge^s \mathbb{R}^n$, then T has a grade s and degree of r or a s, r-form. A tensor T of grade zero is called a multilinear form. The tensor T determines a multilinear form τ of rank $r + s$, and it is defined as follows

$$\tau(\boldsymbol{v}_1, \boldsymbol{v}_2, ..., \boldsymbol{v}_{r+s}) = (\boldsymbol{v}_{r+1}, \boldsymbol{v}_{r+2}, ..., \boldsymbol{v}_{r+s}) \cdot T(\boldsymbol{v}_1, \boldsymbol{v}_2, ..., \boldsymbol{v}_r) \tag{3.157}$$

Conversely τ determines T as follows

$$\begin{aligned} T(\boldsymbol{v}_1, \boldsymbol{v}_2, ..., \boldsymbol{v}_r) &= (s!)^{-1} \partial_{u_s} \wedge \cdots \partial_{u_1} \tau(\boldsymbol{v}_1, \boldsymbol{v}_2, ..., \boldsymbol{v}_r, \boldsymbol{u}_1, \boldsymbol{u}_2, ..., \boldsymbol{u}_s) \\ &= \partial_U U * T(\boldsymbol{v}_1, \boldsymbol{v}_2, ..., \boldsymbol{v}_r), \end{aligned} \tag{3.158}$$

for $r = s = 1$, T is a linear transformation and τ is its associate bilinear form. Note that the relation (3.158) is expressed using the *simplicial* derivative (3.50).

According to [138], we have maintained the conventional terminology for multilinear functions. This is to help readers to recognize familiar concepts and also to avoid the proliferation of unnecessary mathematical terminology.

In geometric algebra, new tensors can be generated applying the tensors operations of addition, multiplication, expansion via the wedge product and contraction via the inner product.

Given the tensors R and T having the same degree and arguments, the sum is given by

$$S = R(\boldsymbol{v}_1, ..., \boldsymbol{v}_p) + T(\boldsymbol{v}_1, ..., \boldsymbol{v}_q), \tag{3.159}$$

the sum is a tensor S, if only if R and T have the same degree and the same arguments.

The geometric product of tensors R and T is given by

$$Q = RT \equiv R(\boldsymbol{v}_1, ..., \boldsymbol{v}_p)T(\boldsymbol{v}_1, ..., \boldsymbol{v}_q), \tag{3.160}$$

without restrictions on the arguments, Q is a tensor of degree $p+q$. Considering the non-commutativity of the geometric product, RT, $R \cdot T$, $R \wedge T$, $R \times T$ yield in general different tensors of degree $p+q$.

The *contraction* of a tensor T of degree p is a tensor of degree $p-2$ defined by, c.f. ([138], 3-10.4)

$$T(\boldsymbol{v}_1, ..., \boldsymbol{v}_{j-1}, \partial_{\boldsymbol{v}_k}, \boldsymbol{v}_{j+1}, ..., \boldsymbol{v}_p) = \partial_{\boldsymbol{v}_j} \cdot \partial_{\boldsymbol{v}_k} T(\boldsymbol{v}_1, \boldsymbol{v}_2, ..., \boldsymbol{v}_p), \tag{3.161}$$

where (3.161) is the contraction of T in the jth and kth places and

$$\partial_{\boldsymbol{v}_k} T(\boldsymbol{v}_1, \boldsymbol{v}_2, ..., \boldsymbol{v}_p) \tag{3.162}$$

is called the *traction* of the tensor T in the kth place, c.f. ([138], 3-10.5). This tensor can be obtained from the tensor T of degree $p+1$ defined by the product $\boldsymbol{v}_{p+1}T(\boldsymbol{v}_1, \boldsymbol{v}_2, ..., \boldsymbol{v}_p)$ by contracting the variables $\boldsymbol{v}_k$ and $\boldsymbol{v}_{p+1}$. The definition (3.162) generalizes the definition of traction given in Sect. 3.7.

Let us give some illustrations of tensor manipulation. Given the bilinear forms $\alpha(\boldsymbol{v}, \boldsymbol{u})$ and $\beta(\boldsymbol{v}, \boldsymbol{u})$, their composition reads c.f. ([138], 3-10.7)

$$\gamma(\boldsymbol{v}, \boldsymbol{u}) = \partial_x \cdot \partial_y \beta(\boldsymbol{v}, \boldsymbol{x})\alpha(\boldsymbol{y}, \boldsymbol{u}) = \beta(\boldsymbol{v}, \partial_y)\alpha(\boldsymbol{y}, \boldsymbol{u}). \tag{3.163}$$

The bilinear forms determine the tensor of degree 1, namely

$$Q(\boldsymbol{v}) = \partial_u \gamma(\boldsymbol{v}, \boldsymbol{u}), \quad R(\boldsymbol{v}) = \partial_u \alpha(\boldsymbol{v}, \boldsymbol{u}), \quad T(\boldsymbol{v}) = \partial_u \beta(\boldsymbol{v}, \boldsymbol{u}). \tag{3.164}$$

Differentiating (3.163), we obtain the expression for composite linear transformations

$$Q(\boldsymbol{v}) = \beta(\boldsymbol{v}, \partial_u)R(\boldsymbol{u}) = T(\boldsymbol{v}) \cdot \partial_u R(\boldsymbol{u}) = R(T(\boldsymbol{v})). \tag{3.165}$$

An important issue for applications is the conventional covariant formulation of tensor analysis. Next, we restrict us to tensors with scalar values, i.e., multilinear forms. In this regard, (3.160) corresponds essentially to the conventional tensor product.

The covariant components of a rank 3 tensor $\gamma = (\boldsymbol{v}, \boldsymbol{u}, \boldsymbol{w})$ with respect to a frame $\{\boldsymbol{v}_k\}$ for $\mathbb{R}^n$ are given by

$$\gamma_{ijk} = \gamma(\boldsymbol{v}_i, \boldsymbol{v}_j, \boldsymbol{v}_k), \tag{3.166}$$

and the contravariant components by

$$\gamma^{ijk} = \gamma(\boldsymbol{v}^i, \boldsymbol{v}^j, \boldsymbol{v}^k), \tag{3.167}$$

where the $\{\boldsymbol{v}^k\}$ is the reciprocal frame. The next quantities correspond to the mixed components expressed using upper of lower indices to indicate contravariant and covariant components, respectively

$$\gamma_k^{ij} = \gamma(\boldsymbol{v}^i, \boldsymbol{v}^i, \boldsymbol{v}_k), \tag{3.168}$$

Arbitrary expressions of γ can be expressed in terms of components as follows

$$\gamma(\boldsymbol{v}, \boldsymbol{u}, \boldsymbol{w}) = \nu^i \mu^j \omega^k \gamma_{ijk} = \nu_i \mu_j \omega_k \gamma^{ijk} = \nu_i \mu_j \omega^k \gamma_k^{ij}, \tag{3.169}$$

where $\nu_i = \boldsymbol{v}_i \cdot \boldsymbol{v}$, $\mu_j = \boldsymbol{v}_j \cdot \boldsymbol{u}$, $\omega_k = \boldsymbol{v}_k \cdot \boldsymbol{w}$ are, respectively, the covariant components of the vectors $\boldsymbol{u}, \boldsymbol{v}, \boldsymbol{w}$, while $\nu^i = \boldsymbol{v}^i \cdot \boldsymbol{v}$, $\mu^j = \boldsymbol{v}^j \cdot \boldsymbol{u}$, $\omega^k = \boldsymbol{v}^k \cdot \boldsymbol{w}$ are their contravariant components.

The components of the contraction of the tensor $\gamma(\boldsymbol{v}, \boldsymbol{u}, \boldsymbol{w})$ are, c.f. ([138], 3-10.12)

$$\gamma_k^{ij} \equiv \gamma(\boldsymbol{v}^i, \boldsymbol{v}^k, \boldsymbol{v}_k) = \gamma(\boldsymbol{v}^i, \partial_u, \boldsymbol{u}), \tag{3.170}$$

where it used the habitual summation convention and $\partial_u = \boldsymbol{v}^k \boldsymbol{v}_k \cdot \partial_u$ and $\boldsymbol{v}_k \cdot \partial_u \boldsymbol{u} = \boldsymbol{v}_k$. Thus, we can see that the definition of contraction given in Eq. (3.161) certainly corresponds to the conventional one of tensor analysis.

If the Eq. (3.163) is expressed in terms of tensor components, we obtain the familiar rule of matrix multiplication, namely

$$\gamma_k^i \equiv \gamma(\boldsymbol{v}^i, \boldsymbol{v}_k) = \beta(\boldsymbol{v}^i, \boldsymbol{v}_j)\alpha(\boldsymbol{v}^j, \boldsymbol{v}_k) = \beta_j^i \alpha_k^j. \tag{3.171}$$

Let us see how different tensor components are related. Consider any frame $\{\boldsymbol{v}'_k\}$ for $\mathbb{R}^n$ which is related to the frame $\{\boldsymbol{v}_k\}$ via a a linear transformation $\underline{f}$ with components $f_k^j = \boldsymbol{v}^j \cdot \underline{f}(\boldsymbol{v}_k)$, hence c.f. ([138], 3-10.14.a)

$$\boldsymbol{v}'_k = \underline{f}(\boldsymbol{v}_k) = f_k^j \boldsymbol{v}_j. \tag{3.172}$$

On the other hand, the reciprocal frames are related by the adjoint transformation as follows

$$v^j = \overline{f}(v'^k) = f^j_k v'^k. \tag{3.173}$$

Let us now consider the mixed components

$$\gamma^{i'j'}_{k'} \equiv \gamma(v'^i, v'^j, v'^k), \tag{3.174}$$

of the tensor $\gamma(v, u, w)$ with respect to the frame $\{v'_k\}$, the related components γ^{ij}_k w.r.t. $\{v_k\}$ are given by the following equations, c.f. ([138], 3-10.16)

$$\begin{aligned} \gamma^{ij}_k f^k_r &= \gamma(v^i, v^j, \underline{f}(v_r)) = \gamma(\overline{f}(v'^i), \overline{f}(v'^j), \underline{f}(v_r)) \\ &= f^i_p f^j_q \gamma(v'^p, v'^q, v'_r) = f^i_p f^j_q \gamma^{p'q'}_{r'}. \end{aligned} \tag{3.175}$$

Finally, the transformation of a tensor γ induced by the linear transformation $\underline{f}$ of its domain $\mathbb{R}^n$ into another vector space $\mathbb{R}'^n$ is called an active transformation. Given a rank 2 tensor $\gamma(v, u)$ defined in $\mathbb{R}^n$, a tensor $\gamma'(v', u')$ defined in $\mathbb{R}'^n$ is determined simply by the following linear substitution, c.f. ([138], 3-10.17)

$$\gamma'(v', u') = \gamma(\underline{f}^{-1}(v'), \overline{f}(u')). \tag{3.176}$$

Note that the transformation γ into γ' is covariant in the first place (or argument) and contravariant in the second place. The reader can resort to the Sect. 3-10 of [138] to get more insight about the formulation of tensor calculus in the framework of geometric algebra.

3.8.2 Orthonormal Frames and Cartesian Tensors

If the frames consist of orthonormal vectors in Euclidean space, the formulation yields Cartesian tensors which we are more familiar with. This subsection is enriched with the equations given in Sect. 4.5.1 of the book of Ch. Doran and C. A. Lasenby [74].

Since

$$e_i \cdot e^j = e_i \cdot e_j = \delta_{ij}, \tag{3.177}$$

there is no distinction between frames and their reciprocals, so there is no need to distinguish between raised or lowered indices. As an illustration, let us show some simple examples.

The rotation of a frame using a rotor is given by

$$e'_i = \boldsymbol{R} e_i \widetilde{\boldsymbol{R}} = R_{ij} e_j, \tag{3.178}$$

where R_{ij} is a tensor consisting of the components of the rotor $\boldsymbol{R}$

$$R_{ij} = (\boldsymbol{R} e_i \widetilde{\boldsymbol{R}}) \cdot e_j. \tag{3.179}$$

Let us consider now a sequence of two rotations,

$$\begin{aligned} R_{ij} R_{ik} = (\boldsymbol{R} e_i \widetilde{\boldsymbol{R}}) \cdot e_j (\boldsymbol{R} e_i \widetilde{\boldsymbol{R}}) \cdot e_k &= (\widetilde{\boldsymbol{R}} e_j \boldsymbol{R}) \cdot (\widetilde{\boldsymbol{R}} \cdot e_k \boldsymbol{R}) \\ &= \widetilde{\boldsymbol{R}} e_j \cdot e_k \boldsymbol{R} = \widetilde{\boldsymbol{R}} \boldsymbol{R} \delta_{jk} = \delta_{jk}, \end{aligned} \tag{3.180}$$

thus similarly

$$R_{ik} R_{jk} = \delta_{ij}. \tag{3.181}$$

If we change the frame of a vector $v_i = e_i \cdot \boldsymbol{v}$, after a contraction of the indexes j, we get

$$v_i' = e_i' \cdot \boldsymbol{v} = R_{ij} e_j \cdot \boldsymbol{v} = R_{ij} v_j. \tag{3.182}$$

Usually, one contracts raised and lower indexes, in this case since there is no distinction between frames, we have proceeded just contracting the lowered indexes.

The components of a linear function A can be expressed as

$$A_{ij} = e_i \cdot A(e_j), \tag{3.183}$$

this decomposition yields a $n \times n$ matrix. The components of a vector $\boldsymbol{A}(\boldsymbol{x})$ are

$$x_i' = e_i \cdot A(\boldsymbol{x}) = e_i \cdot A(x_j e_j) = A_{ij} x_j, \tag{3.184}$$

this result corresponds to the matrix A acting on column vector $\boldsymbol{x}$.

The components of the product of of two linear functions A and B are given by

$$(AB)_{ij} = AB(e_j) \cdot e_i = B(e_j) \cdot \bar{A}(e_i) = B(e_j) \cdot e_k e_k \bar{A}(e_i) = A_{ik} B_{kj}, \tag{3.185}$$

we recognize the familiar rule of matrix multiplication. If the frame changes to a rotated frame, the components of the transform tensor change to

$$A_{ij}' = R_{ik} R_{jl} A_{kl}. \tag{3.186}$$

Depending upon the k number of indexes, objects are referred to as rank k tensors, e.g., a rank 1 tensor is a vector and a rank 2 a matrix.

3.8.3 *Non-orthonormal Frames and General Tensors*

This subsection is enriched with the equations given in Sect. 4.5.2 of the book of Ch. Doran and C. A. Lasenby [74]. We will discuss the case of arbitrary basis sets in spaces of arbitrary (non-degenerate signature). Non-orthonormal frames arise when one works with curvilinear coordinate systems. Consider the basis vectors $\{e_k\}$ as an arbitrary frame for a n-dimensional space of unspecified signature. Using a reciprocal frame denoted by $\{e^k\}$, we can relate two frames as follows

$$e^i \cdot e_j = \delta^i_j, \tag{3.187}$$

this equation still holds by mixed signature spaces. The vector $\boldsymbol{x}$ has components $(x^1, x^2, ..., x^n)$ in the $\{e_k\}$ frame, and $(x_1, x_2, ..., x_n)$ in the $\{e^k\}$ frame, we can compute the inner product contracting the raised and lower indexes, namely

$$\boldsymbol{x} \cdot \boldsymbol{y} = (x^i e_i) \cdot (y_j e^j) = x^i y_j e_i \cdot e^j = x^i y_j \delta^j_i = x^i y_j. \tag{3.188}$$

Another way to carry out inner products is to resort to the so-called *metric tensor* g_{ij}

$$g_{ij} = e_i \cdot e_j. \tag{3.189}$$

The array of the components of g_{ij} constitutes a $n \times n$ symmetric matrix. The inverse of a metric tensor is written as g^{ij}, and it is given by

$$g^{ij} = e^i \cdot e^j, \tag{3.190}$$

which fulfills

$$g^{ik} g_{kj} = e^i \cdot e^k e_k \cdot e_j = e^i \cdot e_j = \delta^i_j. \tag{3.191}$$

Using the metric tensor, the inner product can be written as

$$\boldsymbol{x}\boldsymbol{y} = x^i y_i = x_i y^i = x^i y^j g_{ij} = x_i y_j g^{ij}. \tag{3.192}$$

Let us now consider rank 2 tensors. Given a linear function A in a general non-orthonormal frame, as previously shown, if A acts on the frame vector e_j, the result is given in terms of the components of the reciprocal frame

$$A_{ij} = e_i \cdot A(e_j), \tag{3.193}$$

the array of the components A_{ij} is a $n \times n$ matrix or a rank 2 tensor. Its entries depend of the choice of the frame. Similar expression can be written for combinations of frame vectors and reciprocal vectors, e.g.,

$$A^{ij} = A(e^i) \cdot e^j. \tag{3.194}$$

One well-known use of the metric tensor is to interchange between these expressions

$$A^{ij} = e^i \cdot A(e^j) = e^i \cdot e^k e_k \cdot A(e_l e^l \cdot e^j) = g^{ik} g^{jl} A_{kl}. \tag{3.195}$$

In the realm of the theory of linear operators, the metric tensor g_{ij} can be seen as the identity operator expressed in a non-orthonormal frame.

If A_{ij} are the components of A in some frame, then the components of $\bar{A}$ are computed as follows

$$\bar{A}_{ij} = \bar{A}(e_j) \cdot e_i = e_j \cdot A(e_i) = A_{ji}. \tag{3.196}$$

In matrix algebra, the components of $\bar{A}$ correspond to the components of transposed of the matrix A.

For mixed index tensors, one needs some care to handle the indices, namely

$$A_i^j = A(e^j) \cdot e_i = e^j \cdot \bar{A}(e_i) = \bar{A}_i^j. \tag{3.197}$$

As in matrix algebra, if A is a symmetric function then $\bar{A} = A$, in this case

$$A_{ij} = A(e_j) \cdot e_i = A(e_i) \cdot (e_j) = A_{ji}, \tag{3.198}$$

where A_{ji} represent a symmetric matrix. Similarly for other indices' cases: $A^{ij} = A^{ji}$ and $A_i^j = A_j^i$.

The components of the product function AB can be made explicit as follows

$$(AB)_{ij} = AB(e_j) \cdot e_i = B(e_j) \cdot \bar{A}(e_i) = B(e_j) \cdot \epsilon_k e^k \bar{A}(e_i) = A_i^k B_{kj}. \tag{3.199}$$

Utilizing correctly the combination of subscript and superscript indices, we can get the usual matrix multiplication. On the other hand, one can work entirely using subscripted indices including the metric tensor

$$(AB)_{ij} = A_{ik} B_{lj} g^{kl}. \tag{3.200}$$

Introducing a second non-orthonormal frame $\{f_\mu\}$, one can relate two frames via a transform matrix $f_{\mu i}$, namely

$$f_{\mu i} = f_\mu \cdot e_i, \quad f^{\mu i} = f^\mu \cdot e^i, \tag{3.201}$$

where the the Latin μ and Greek i indices are used to distinguish the components of one frame from another. Note that these matrices satisfy

$$f_{\mu i} f^{\mu j} = f_{\mu} \cdot e_i f^{\mu} \cdot \epsilon^j = e_i e^j = \delta_i^j, \tag{3.202}$$
$$f_{\mu i} f^{\nu i} = f_{\mu} \cdot e_i f^{\nu} \cdot \epsilon^i = f_{\mu} e^{\nu} = \delta_{\mu}^{\nu}. \tag{3.203}$$

The decomposition of a vector $\boldsymbol{x}$ in terms of these frames is given by

$$\boldsymbol{x} = x^i e_i = x^i f^{\mu} e_i \cdot f_{\mu} = x^i f_{\mu i} f^{\mu}. \tag{3.204}$$

The transformation law for the individual components is thus given by the contraction of the lower and upper index i

$$x_{\mu} = f_{\mu i} x^i. \tag{3.205}$$

These formulas can be extended straightforwardly to include linear functions.

$$A_{\mu\nu} = f_{\mu i} f_{\nu j} A^{ij}, \tag{3.206}$$

and for mixtures of indices

$$A_{\mu}^{\nu} = f_{\mu}^{i} f_{j}^{\nu} A_{i}^{j}. \tag{3.207}$$

If this is expressed in terms of matrix multiplication, we have simply an equivalent transformation.

Let us give some examples of the use of tensor in geometric algebra in the domains of computer vision and kinematics.

Example 3.1 The equation of a point $\boldsymbol{X}^i \in G_{3,1}$ ($\mathbb{P}^3$ or $\mathbb{R}^4$), a tensor of grade 1 and rank 4, projected onto the image plane in G_3 ($\mathbb{P}^2$ or $\mathbb{R}^3$) is a point $\boldsymbol{x}^j \in G_3$, a tensor of grade 2 and rank 3, given given by

$$\boldsymbol{x}^j = \phi_i^j \boldsymbol{X}^i, \tag{3.208}$$

where the $\boldsymbol{x}^j$ is expanded in terms of a bivector basis $\hat{e}_1 = e_2 e_3$, $\hat{e}_2 = e_3 e_1$ and $\hat{e}_3 = e_3 e_1$ and $\phi_i^j = T(\phi_1, \phi_2, \phi_3)$ is a grade 3 rank 4 tensor with the geometric interpretation of a frame transformation in terms of *optical planes* ϕ_j and each represented as a linear combination of the duals of the basis vectors or trivectors

$$\begin{aligned} \phi_j &= t_{j1}(Ie_1) + t_{j2}(Ie_2) + t_{j3}(Ie_3) + t_{j4}(Ie_4) = t_{jk}(Ie_k) \\ &= t_{j1}(e_2 e_3 e_4) + t_{j2}(e_1 e_3 e_4) + t_{j3}(e_1 e_2 e_3) + t_{j4}(e_1 e_2 e_3). \end{aligned} \tag{3.209}$$

Note that the coefficients of the projected point in the image plane are computed by a frame transformation ϕ_i^j (as basis trivectors), i.e., contracting the vectors with the trivectors which in matrix formulation reads

$$\boldsymbol{x} = \begin{bmatrix} x_1 \\ x_2 \\ x_3 \end{bmatrix} = \phi^i X_i = \begin{bmatrix} t_{11} & t_{12} & t_{13} & t_{14} \\ t_{21} & t_{22} & t_{23} & t_{24} \\ t_{31} & t_{32} & t_{33} & t_{34} \end{bmatrix} \begin{bmatrix} X_1 \\ X_2 \\ X_3 \\ X_4 \end{bmatrix} \equiv P\boldsymbol{X},$$

where P is called the projective matrix from the visual space $\mathbb{P}^3$ or $\mathbb{R}^4$ to the image plane $\mathbb{P}^2$ or $\mathbb{R}^3$.

Example 3.2 Next, we consider the projection of world lines onto the image plane. Suppose we have a world line $L = \mathbf{X}_1 \wedge \mathbf{X}_2 \in G_{3,1}$ ($\mathbb{P}^3$) joining the points $\mathbf{X}_1$ and $\mathbf{X}_2$. If $\boldsymbol{x}_1 = (\mathbf{A}_0 \wedge \mathbf{X}_1) \cap \Phi_A$ and $\boldsymbol{x}_2 = (\mathbf{A}_0 \wedge \mathbf{X}_2) \cap \Phi_A$ (i.e., the intersections of the optical rays with the image plane Φ_A), then the projected line in the image plane is clearly given by

$$l = \boldsymbol{x}_1 \wedge \boldsymbol{x}_2.$$

Since we can express l in the bivector basis for the plane, we obtain

$$l = l^j L_j^A,$$

where $L_1^A = \mathbf{A}_2 \wedge \mathbf{A}_3$, $L_2^A = \mathbf{A}_3 \wedge \mathbf{A}_1$ and $L_3^A = \mathbf{A}_1 \wedge \mathbf{A}_2$ and $\boldsymbol{A}_i \in G_{3,1}$ We can also write l as follows:

$$l = \boldsymbol{x}_1 \wedge \boldsymbol{x}_2 = (\mathbf{X}_1 \cdot \phi_A^j)(\mathbf{X}_2 \cdot \phi_A^k) \mathbf{A}_j \wedge \mathbf{A}_k \equiv l^i L_i^A, \tag{3.210}$$

Utilizing the fact that join of the duals is the dual of the meet, we are then able to deduce identities of the following form for each l^j:

$$l_i = (\mathbf{X}_1 \wedge \mathbf{X}_2) \cdot (\phi_A^2 \wedge \phi_A^3) = (\mathbf{X}_1 \wedge \mathbf{X}_2) \cdot (\phi_2^A \cap \phi_3^A)^* = L_i^j \cdot (L_j^A)^*.$$

Thus, the equation of a line $L_j^A \in G_{3,1}$ ($\mathbb{P}^3$ or $\mathbb{R}^4$), a tensor of grade 2 and rank 4 projected onto the image plane in G_3 ($\mathbb{P}^2$ or $\mathbb{R}^3$) is a line $l^j \in G_3$, a tensor of grade 2 and rank 3, given by

$$l_i = L_i^j (L_j^A)^* \equiv L_i^j L_j^A, \tag{3.211}$$

where the $\boldsymbol{l}_i$ is expand in terms of a bivector basis $\hat{e}_1 = e_2 e_3$, $\hat{e}_2 = e_3 e_1$ and $\hat{e}_3 = e_3 e_1$ and $L_i^j = T(L_1, L_2, L_3)$ is a grade 2 rank 4 tensor with the geometric interpretation of a frame transformation in terms of *optical lines* L_j. If we express the world lines and optical lines as bivectors (b_j), $L = \alpha_j b_j + \tilde{\alpha}_j I b_j$ (Plücker coordinates) and $L_A^j = \beta_j b_j + \tilde{\beta}_j I b_j$, respectively , we can write the previous equation as a matrix equation:

$$\boldsymbol{l} = \begin{bmatrix} l^1 \\ l^2 \\ l^3 \end{bmatrix} = \begin{bmatrix} \beta_{11} & \beta_{12} & \beta_{13} & \tilde{\beta}11 & \tilde{\beta}_{12} & \tilde{\beta}_{13} \\ \beta_{21} & \beta_{22} & \beta_{23} & \tilde{\beta}21 & \tilde{\beta}_{22} & \tilde{\beta}_{23} \\ \beta_{31} & \beta_{32} & \beta_{33} & \tilde{\beta}31 & \tilde{\beta}_{32} & \tilde{\beta}_{23} \end{bmatrix} \begin{bmatrix} \alpha_1 \\ \alpha_2 \\ \alpha_3 \\ \tilde{\alpha}_1 \\ \tilde{\alpha}_2 \\ \tilde{\alpha}_3 \end{bmatrix} \equiv P_L \bar{\boldsymbol{l}},$$

where $\bar{l}$ is the vector of *Plücker coordinates* $[\alpha_1, \alpha_2, \alpha_3, \tilde{\alpha}_1, \tilde{\alpha}_2, \tilde{\alpha}_3]$ and the matrix P_L contains the β and $\tilde{beta}$'s, that is, information about the camera configuration.

Example 3.3 Let us represent in the motor algebra $G^+_{3,0,1}$ the kinematics of the geometric object a line expressed in terms of the orientation and momentum or Plücker coordinates: $\boldsymbol{L} = n_x e_{23} + n_y e_{31} + n_z e_{12} + I(m_x e_{23} + m_y e_{31} + m_z e_{12}) = \boldsymbol{n} + \boldsymbol{I}\boldsymbol{m}$. The screw motion is represented with a motor $\boldsymbol{M} = \boldsymbol{T}\boldsymbol{R}$, and thus the kinematics of a line using tensor notation is given by

$$\begin{aligned} \boldsymbol{L} &= \boldsymbol{M}\boldsymbol{L}\widetilde{\boldsymbol{M}} = \boldsymbol{T}\boldsymbol{R}\boldsymbol{L}\widetilde{\boldsymbol{R}}\widetilde{\boldsymbol{T}} \\ L_i &= T_{ij} R_{ij} L_j, \end{aligned} \tag{3.212}$$

where the line L_i is represented as a first-order covariant tensor and R_{ij}, T_{ij} are the second-order tensors consisting of the components of the rotor $\boldsymbol{R}$ and the translator $\boldsymbol{T}$, respectively.

The kinematics of the point $\boldsymbol{X} = 1 + \boldsymbol{I}\boldsymbol{x}$ written in tensor notation reads

$$\begin{aligned} \boldsymbol{X}' &= \boldsymbol{M}\boldsymbol{X}\hat{\widetilde{\boldsymbol{M}}} = \boldsymbol{T}\boldsymbol{R}\boldsymbol{X}\widetilde{\boldsymbol{R}}\boldsymbol{T} \\ X_i &= \hat{T}_{ij} R_{ij} X_j, \end{aligned} \tag{3.213}$$

and that of the plane $\boldsymbol{\Phi} = \boldsymbol{n} + \boldsymbol{I}\boldsymbol{d}$

$$\begin{aligned} \boldsymbol{\Phi}' &= \boldsymbol{M}\boldsymbol{\Phi}\hat{\widetilde{\boldsymbol{M}}} = \boldsymbol{T}\boldsymbol{R}\boldsymbol{\Phi}\widetilde{\boldsymbol{R}}\boldsymbol{T} \\ \Phi'_i &= \hat{T}_{ij} R_{ij} \Phi_j, \end{aligned} \tag{3.214}$$

Note that in the case of the point $\boldsymbol{X}_j$ and the plane $\boldsymbol{\Phi}_j$, we use a covariant representations; this is because the motor algebra is a framework of lines which are covariant tensors.

In the next chapters devoted to applications, we will use the tensor equations formulated in geometric algebra.

3.9 Exercises

3.1 Prove that a vector $v = v(x)$ associated with an interior point x of the manifold $\mathcal{M}$ is a tangent vector at x if and only if $v \wedge I = v(x) \wedge I(x) = 0$.

3.2 Given $F = F(x) = fe_1 + ge_{23} - hI$ a multivector-valued function on $\mathcal{M}$, where I is the pseudoscalar of G_3, compute the derivative of F in the direction $v = ae_1 + be_2 + ce_3 \in G_3$, i.e., $v \cdot \partial F = v(x) \cdot \partial_x F(x)$.

3.3 Compute the vector derivative $\nabla = e_1\partial_x + e_2\partial_y + e_3\partial_z$ acting on the vector $v = ae_1 - be_2 + ce_3$. Then introduce the field $\Psi = ve_{12}$ and prove if Ψ is analytic, i.e., $\nabla\Psi = 0$.

3.4 Given the $F = F(x) = fe_1 + ge_{23} - he_3$, a multivector-valued function on $\mathcal{M}$, where I is the pseudoscalar of G_3, compute the geometric product of $\partial_x = e_1\partial_x + e_2\partial_y + e_3\partial_z$ and $F(x)$.

3.5 Prove c.f. ([138], 2-(1.32-1.37)):

i. $\partial|\boldsymbol{v}|^2 = \partial\boldsymbol{v}^2 = 2P(\boldsymbol{v}) = 2\boldsymbol{v}$.
ii. $\partial\wedge\boldsymbol{v} = 0$.
iii. $\partial\boldsymbol{v} = \partial\cdot\boldsymbol{v} = n$.
iv. $\partial|\boldsymbol{v}|^k = k|\boldsymbol{v}|^{k-2}\boldsymbol{v}$.
v. $\partial\left(\frac{\boldsymbol{v}}{|\boldsymbol{v}|^k}\right) = \frac{n-k}{|\boldsymbol{v}|^k}$.
vi. $\partial log|\boldsymbol{v}| = \boldsymbol{v}^{-1}$.

3.6 If $\boldsymbol{V} = P(\boldsymbol{V}) = <\boldsymbol{V}> r$, prove, c.f. ([138], 2-(1.38-1.40)) :

i. $\dot{\partial}(\dot{\boldsymbol{v}}\cdot\boldsymbol{V}) = \boldsymbol{V}\cdot\partial\boldsymbol{v} = r\boldsymbol{V}$.
ii. $\dot{\partial}(\dot{\boldsymbol{v}}\wedge\boldsymbol{V}) = \boldsymbol{V}\wedge\partial\boldsymbol{v} = (n-r)\boldsymbol{V}$.
iii. $\dot{\partial}\boldsymbol{V}\dot{\boldsymbol{v}} = \sum_k \boldsymbol{a}^k\boldsymbol{V}\boldsymbol{a}_k = (-1)^r(n-2r)\boldsymbol{V}$.

3.7 Assume that $\boldsymbol{X} = F(\boldsymbol{X})$ is the identity function on some linear subspace of G_n with dimension d. $\boldsymbol{U}^{||}$ denotes the projection of the multivector $\boldsymbol{U}$ into that supspace, prove the following equalities, c.f. ([138], 2-(2.28.a-2.35)):

i. $\boldsymbol{U} * \partial_X\boldsymbol{X} = \dot{\partial}_X\dot{\boldsymbol{X}} * \boldsymbol{U} = \boldsymbol{U}^{||}$.
ii. $\boldsymbol{U} * \partial_X\boldsymbol{X}^\dagger = \dot{\partial}_X\dot{\boldsymbol{X}}^\dagger * \boldsymbol{U} = \boldsymbol{U}^{||\dagger}$.
iii. $\partial_X\boldsymbol{X} = d$.
iv. $\partial_X|\boldsymbol{X}|^2 = 2\boldsymbol{X}^\dagger$.
v. $\boldsymbol{U} * \partial_X\boldsymbol{X}^k = \boldsymbol{U}^{||}\boldsymbol{X}^{k-1} + \boldsymbol{X}\boldsymbol{U}^{||}\boldsymbol{X}^{k-2} + \cdots + \boldsymbol{X}^{k-1}\boldsymbol{U}^{||}$.
vi. $\partial_X|\boldsymbol{X}|^k = k|\boldsymbol{X}|^{k-2}\boldsymbol{X}^\dagger$.
vii. $\partial_X log|\boldsymbol{X}| = \frac{\boldsymbol{X}^\dagger}{|\boldsymbol{X}|^2}$.
viii. $\boldsymbol{U} * \partial_X\{|\boldsymbol{X}|^k\boldsymbol{X}\} = |\boldsymbol{X}|^k\left\{\boldsymbol{U}^{||} + k\frac{\boldsymbol{U}\boldsymbol{X}^\dagger\boldsymbol{X}}{|\boldsymbol{X}|^2}\right\}$.
ix. $\partial_X\{|\boldsymbol{X}|^k\boldsymbol{X}\} = |\boldsymbol{X}|^k\left\{d + k\frac{\boldsymbol{X}^\dagger\boldsymbol{X}}{|\boldsymbol{X}|^2}\right\}$.

3.8 The vectors $\boldsymbol{u}, \boldsymbol{v}$ build the plane $\boldsymbol{u}\wedge\boldsymbol{v}$, consider their transformation given by $\boldsymbol{u}' = f(\boldsymbol{u}) = 8\boldsymbol{u} - 6\boldsymbol{v}$ and $\boldsymbol{v}' = f(\boldsymbol{v}) = 6\boldsymbol{u} - 8\boldsymbol{v}$. Firstly, use linear algebra to compute the eigenvectors and their eigenvalues for this transformation. Then, use geometric algebra to compute the determinant, and an eigen bi-blade with its eigenvalue. What can you tell about the geometry of the transformation $\boldsymbol{u}\wedge\boldsymbol{v} \rightarrow \boldsymbol{u}'\wedge\boldsymbol{v}'$. Can you write a geometric algebra expression in terms of bivectors equivalent to this transformation?

3.9 Propose a nontrivial linear map g: $\mathbb{R}^2 \rightarrow \mathbb{R}^2$ which has an eigenvector and an eigen bi-blade, both with eigenvalue 1. Interpret geometrically your results.

3.10 Consider that a linear transformation G has a complex eigenvector $v + iu$ with associated eigenvalue $\alpha + i\beta$. Explain what is the effect of G on the plane $v\wedge u$ and then interpret the action of G in this plane.

3.11 Consider the rotation matrix,

$$R = \begin{pmatrix} \cos(\theta) & \sin(\theta) \\ -\sin(\theta) & \cos(\theta) \end{pmatrix},$$

with following eigenvalues and eigenvectors

$$\{e^{\pm i\theta}\}, \quad \begin{pmatrix} 1 \\ i \end{pmatrix}, \begin{pmatrix} 1 \\ -i \end{pmatrix}.$$

The isomorph operator to R is the 2D rotor $\boldsymbol{R} \in G_{2,0}$ write for $\boldsymbol{R}$ its corresponding eigen bivectors and eigenvalues.

3.12 Consider the rotation matrix $R \in \mathbb{R}^3$. The real eigenvector for the 3D rotation matrix R has a natural interpretation as the axis of rotation. The 3D rotor $\boldsymbol{R} \in G_{3,0}$ is isomorph to this matrix R, give its corresponding eigen bivectors and eigen values. Give a geometric interpretation of your results.

3.13 Normally we compute eigenvectors and their associate eigenvalues using linear algebra. Propose now a linear map $f : \mathbb{R}^3 \rightarrow \mathbb{R}^3$ which you encounter in your work and compute the possibly eigenvectors, the eigen-2-blades, the eigen-3-blades, and their associated eigenvalues.

3.14 Consider the active transformation of a tensor γ by the linear transformation $\underline{f} : \mathbb{R}^n \rightarrow \mathbb{R}^m$; given a rank 2 tensor $\gamma(\boldsymbol{v}, \boldsymbol{u})$ defined in $\mathbb{R}^n$, show that a tensor $\gamma'(\boldsymbol{v}', \boldsymbol{u}')$ defined in $\mathbb{R}^m$ is determined by the linear substitution

$$\gamma'(\boldsymbol{v}', \boldsymbol{u}') = \gamma(\underline{f}^{-1}(\boldsymbol{v}'), \overline{f}(\boldsymbol{u}')).$$

3.15 Cartesian tensors: the rotation of new frame is given by $e'_i = \boldsymbol{R} e_i \tilde{\boldsymbol{R}} = \bigwedge_{ij} e_j$, where $\boldsymbol{R}$ is a rotor and $\bigwedge_{ij}$ are the components of the rotation defined by $\bigwedge_{ij} =$

$(\boldsymbol{R}e_i\tilde{\boldsymbol{R}}) \cdot e_j$. It follows $\bigwedge_{ij}\bigwedge_{ik} = (\boldsymbol{R}e_i\tilde{\boldsymbol{R}}) \cdot e_j(\boldsymbol{R}e_i\tilde{\boldsymbol{R}}) \cdot e_k = (\tilde{\boldsymbol{R}}e_j\boldsymbol{R}) \cdot (\tilde{\boldsymbol{R}}e_k\boldsymbol{R}) = \delta_{jk}$ and similarly $\bigwedge_{ik}\bigwedge_{jk} = \delta_{ij}$. The components of the linear function F are given by $F_{ij} = e_i \cdot F(e_j)$. The vector v has the components $v_i = e_i \cdot v$; under a transform, the components are then given by $v'_i = e'_i \cdot v = \bigwedge_{ij} v_j$. The components of the vector $F(v)$ are given by $e_i \cdot F(v) = e_i \cdot F(v_je_j) = F_{ij}v_j$. Based on these equations, prove the following relations

i. If F and G are a pair of linear functions, check that c.f. ([74], 4.195-4.196)

$$(FG)_{ij} = FG(e_j) \cdot e_i = G(e_j) \cdot \overline{F}(e_i) = G(e_j) \cdot e_k e_k \cdot \overline{F}(e_i) = F_{ik}G_{kj}.$$

ii. If the frame is changed to a new rotate frame, the components of the rank 2 tensor transform is given by
$F'_{ij} = \bigwedge_{ik}\bigwedge_{jl} F_{kl}$

Chapter 4
Geometric Calculus

We have learned that readers of the chapter on *geometric calculus* of the book [138] may have difficulties to understand the subject and practitioners have difficulties to try the equations in certain applications. For this reason, this chapter presents the most relevant equations for applications proposed by D. Hestenes and G. Sobczyk [138]. This study is written in a clear manner for readers interested in applications in computer science and engineering.

This chapter presents an outline of geometric calculus. We are used to the grad, div, curl, and Laplace operators, which are formulated using a single *vector derivative*. The derivative operator is essential in complex analysis and enables the extension of complex analysis to higher dimensions. The synthesis of vector differentiation and geometric algebra is called *geometric calculus* [138]. Next, we will describe these operators in the geometric algebra framework.

4.1 Vector Manifolds and Tangent Space

A differential manifold is a set of points on which operations of differential calculus can be carried out. An m-dimensional manifold $\mathcal{M}$ can be locally parametrized by a set of (scalar) coordinates. These coordinates determine an invertible mapping from $\mathcal{M}$ into an Euclidean space $\mathcal{E}_m$.

A *vector manifold* $\mathcal{M}$ is a set of vectors called points. If the vectors $x, y \in \mathcal{M}$, the vector $x - y$ is called a *chord* from y to x. A limit of a sequence of chords defines a tangent vector. A vector $v(x)$ is seen to be tangent to a point x in $\mathcal{M}$, if exists a curve $\{c(\tau); 0 < \tau < \epsilon\}$ in $\mathcal{M}$ coming from the point $x(0) = x$ such that

$$v(x) = v \cdot \partial x \equiv \frac{\mathrm{d}x(\tau)}{\mathrm{d}\tau}|_{\tau=0} = \lim_{\tau \rightarrow} \frac{x(\tau) - x}{\tau}. \tag{4.1}$$

E. Bayro-Corrochano, *Geometric Algebra Applications Vol. I*,
https://doi.org/10.1007/978-3-319-74830-6_4

The unit pseudoscalar $I = I(\boldsymbol{x})$ is an m-blade-valued function defined on the manifold $\mathcal{M}$. A continuous manifold $\mathcal{M}$ is *orientable*, if its pseudoscalar $I = I(\boldsymbol{x})$ is single-valued and *non − orientable* if $I = I(\boldsymbol{x})$ is double valued. If $I = I(\boldsymbol{x})$ is defined and a continuous function at every point x, then $\mathcal{M}$ is continuous. A manifold $\mathcal{M}$ is differentiable if its pseudoscalar $I = I(\boldsymbol{x})$ is differentiable at each point $\boldsymbol{x}$ of $\mathcal{M}$. One calls $\mathcal{M}$ smooth if $I = I(\boldsymbol{x})$ has derivatives of all orders. A vector $v(\boldsymbol{x})$ associate with an interior point of $\mathcal{M}$ is a tangent vector at $\boldsymbol{x}$ if and only if

$$v \wedge I = v(\boldsymbol{x}) \wedge I(\boldsymbol{x}) = 0. \tag{4.2}$$

The set $T_x\mathcal{M}$ of all vectors tangent to $\mathcal{M}$ is called the *tangent space* at $\boldsymbol{x}$. $\mathcal{M}$ is called a m-dimensional manifold, and $T_x\mathcal{M}$ is m-dimensional vector space. The tangent space $T_x\mathcal{M}$ at an interior point $\boldsymbol{x}$ of $\mathcal{M}$ generates a geometric algebra called the tangent algebra of $\mathcal{M}$, and it is denoted by $G(\boldsymbol{x}) \equiv G(T_x\mathcal{M})$. One assumes that $T_x\mathcal{M}$ is not singular, i.e., it possesses a unit pseudoscalar $I_{\mathcal{M}} = I(\boldsymbol{x})_{\mathcal{M}}$. The pseudoscalar $I(\boldsymbol{x})_{\mathcal{M}}$ determines the projection of $P(\boldsymbol{x}, M)$ of any multivector M into the tangent algebra $G(\boldsymbol{x}) = G(I(\boldsymbol{x})_{\mathcal{M}})$, namely c.f. ([138], 4–1.4)

$$\begin{aligned} P(\boldsymbol{x}, M) &= [M \cdot I(\boldsymbol{x})_{\mathcal{M}}] \cdot I_{\mathcal{M}}^{-1}(\boldsymbol{x}), \\ &\texttt{or} \\ P(M) &= (M \cdot I_{\mathcal{M}}) \cdot I_{\mathcal{M}}^{-1} = I_{\mathcal{M}}^{-1} \cdot (I_{\mathcal{M}} \cdot M). \end{aligned} \tag{4.3}$$

The vector-valued function $x = x(x^1, x^2, ..., x^m)$ represents a patch of $\mathcal{M}^m$ parameterized by scalar coordinates; see Fig. 4.1. The inverse mapping into $\mathbb{R}^m$ is given by the coordinate functions $x^\mu = x^\mu(x)$. Next, we define a coordinate frame $\{e_\mu = e_\mu(x)\}$ as follows

$$e_\mu = \partial_\mu x = \frac{\partial x}{\partial x^\mu} = \lim \frac{\Delta x}{\Delta x^\mu}, \tag{4.4}$$

with a pseudoscalar given by c.f. ([135], Sect. 3)

$$e_{(m)} = e_1 \wedge e_2 \wedge \ldots \wedge e_m = |e_{(m)}| I_m. \tag{4.5}$$

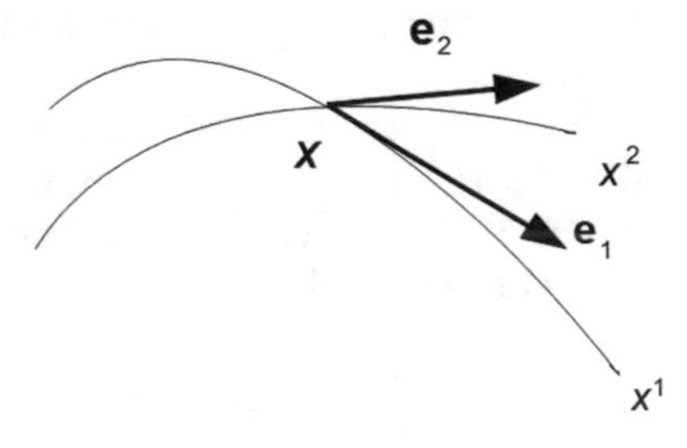

Fig. 4.1 Patch of $\mathcal{M}^m$ parameterized by scalar coordinates

The computations with frames are much easier if one uses reciprocal frames $\{x^\mu\}$, which are defined by the equation $e^\mu \cdot e_\nu = \delta^\mu_\nu$ which have the solution given by

$$e^\mu = (e_1 \wedge \ldots ()_\mu \wedge \ldots \wedge e_m) e_{(m)}^{-1}, \tag{4.6}$$

where the μth vector is omitted from the wedge products. Note that this can be used for a coordinate definition of the vector derivative, i.e., the derivative with respect to the point x:

$$\partial = \partial_x = e^\mu \partial_\mu, \; \texttt{where} \; \partial_\mu = e_\mu \cdot \partial = \frac{\partial}{\partial x^\mu}. \tag{4.7}$$

Thus, the reciprocal vectors can be expressed as gradients, namely c.f. ([135], Sect. 3-1)

$$e^\mu = \partial x^\mu. \tag{4.8}$$

4.1.1 Direct Integrals and the Fundamental Theorem

This section is based on the Hestenes' work ([135], Sect. 4). Let $F = F(x)$ be a multivector function on the manifold $\mathcal{M} = \mathcal{M}^m$ with direct measure $d^m x = |d^m x| I_m(x)$, see Fig. 4.2. The measure can be formulated in terms of coordinates, namely c.f. ([135], Sect. 4)

$$d^m x = d_1 x \wedge d_2 x \wedge \ldots \wedge d_m x = e_1 \wedge e_2 \wedge \ldots \wedge e_m dx^1 dx^2 \ldots dx^m, \tag{4.9}$$

where $d_\mu x = e_\mu(x) d^\mu x$m (no sum). The usual scalar-valued volume element of integration is given by

$$|d^m x| = |e_{(m)}| dx^1 dx^2 \ldots dx^m, \tag{4.10}$$

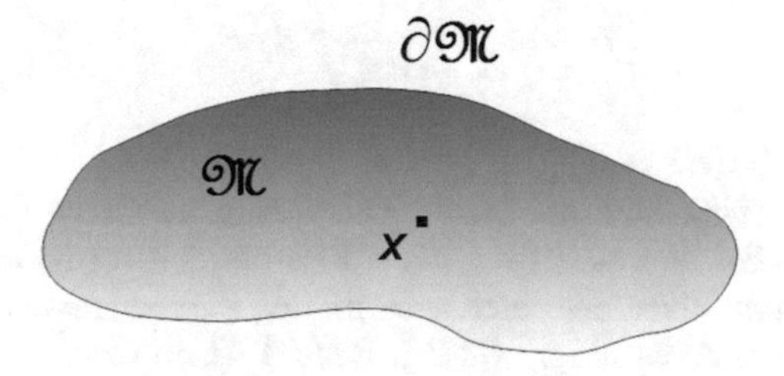

Fig. 4.2 Vector manifold

Thus, the direct integral of F can be now formulated simply as a standard multiple integral, namely c.f. ([135], Sect. 4)

$$\int_{\mathcal{M}} d^m x F = \int_{\mathcal{M}} e_{(m)} dx^1 dx^2 ... dx^m. \tag{4.11}$$

Consider the special case of a manifold embedded in a vector space $\mathcal{M} = \mathcal{M}^m \in V^n$. Let denote $\nabla = \nabla_x$ for the derivative of point in the vectors space V^n. As explained in Sect. 4.3, the derivative of any field $F = F(x)$ can be algebraically decomposed as follows

$$\nabla F = \nabla \cdot F + \nabla \wedge F, \tag{4.12}$$

which unifies the divergence and curl concepts into a single vector derivative. Using geometric calculus, we can formulate the first generalization of the fundamental theorem of calculus as follows, c.f. ([135], Sect. 4-2)

$$\int_{\mathcal{M}} (d^m x) \cdot \nabla F = \int_{\partial \mathcal{M}} d^{m-1} x F. \tag{4.13}$$

The projection of the derivative ∇ to a derivative on the manifold $\mathcal{M}$ is formulated as

$$\partial = \partial_x = I_m^{-1}(I_m \cdot \nabla), \tag{4.14}$$

thus, by writing $d^m x \partial = (d^m x) \cdot \partial m = (d^m x) \cdot \nabla$; hence, the theorem of Eq. (4.13) can be rewritten as follows c.f. ([135], Sect. 4-3)

$$\int_{\mathcal{M}} d^m x \partial F = \int_{\partial \mathcal{M}} d^{m-1} x F, \tag{4.15}$$

note that this result has explicit reference to the embedding space. Hestenes proposed the *tangential derivative*, a coordinate-free definition for the vector derivative w.r.t x in $\mathcal{M}$ without reference to any embedding space, c.f. ([135], Sect. 4-4),

$$\partial F = \lim_{dw \to 0} \frac{1}{dw} \oint d\tau F, \tag{4.16}$$

where $dw = d^m x$ and $d\tau = d^{m-1} x$.

Next, we will show how geometric calculus generalizes Cartan's theory of differential forms. For the case of $k \leq m$, a differential k-form $L = L(d^k x, x)$ on a manifold $\mathcal{M}^m$ is a multivector-valued k-form, i.e., a linear function of the k-vector $d^k x$ at each point x. Accordingly ([135], Sect. 4-4), in Cartan's terminology, the exterior differential of the k-form L is a $(k+1)$-form dL formulated as

$$dL \equiv \dot{L}(d^{k+1}x \cdot \dot{\partial}) = L(d^{k+1}x \cdot \dot{\partial}, \dot{x}), \tag{4.17}$$

note the use of just dL supresses the dependence on the volumen element $d^{k+1}x$ that is explicit in this definition.

Finally, the fundamental theorem of geometric calculus can be expressed in its most general form as, c.f. ([135], Sect. 4-5)

$$\int_{\mathcal{M}} dL = \oint_{\partial\mathcal{M}} L. \tag{4.18}$$

Note that Cartan's theory is rather limited to functions of the form $L =< d^k x F(x) >_0 = (d^k x) \cdot F(x)$, where F is k-vector valued; hence, the exterior differential is given by

$$dL =< d^{k+1}x\ \partial F(x) >_0 =< d^{k+1}x\ \partial \wedge F(x) >= (d^{k+1}x) \cdot (\partial \wedge F). \tag{4.19}$$

Note that Cartan's exterior differential is equivalent to the curl in geometric calculus.

4.2 Vector Derivative

In the study of fields in geometric algebra, we will represent a position by a vector $\boldsymbol{x} \in \mathbb{R}^n$. Shown in Eq. 4.21, the vector derivative ∇ is the derivative with respect to the vector position $\boldsymbol{x}$ and it can be written with respect to a fixed frame $\{e^k\}$ with coordinates $x^k = e^k \cdot x$ as follows:

$$\partial = \nabla = \sum_k e^k \frac{\partial}{\partial x^k}. \tag{4.20}$$

Since the vectors $\{e^k\} \in G_n$, these vectors inherit the full geometric product. For $F(\boldsymbol{x})$ any multivector-valued function of position, or more generally a position-dependent linear function, if we compute the inner product of ∇ with the vector $\boldsymbol{v}$, we get the *directional derivative* in the $\boldsymbol{v}$ direction on F as follows

$$\boldsymbol{v} \cdot \nabla F(\boldsymbol{x}) = \lim_{\tau \to 0} \frac{F(\boldsymbol{x} + \tau \boldsymbol{v}) - F(\boldsymbol{x})}{\tau}. \tag{4.21}$$

In next subsections, we will explain operators that are a result of different actions of the vector derivative ∇ on vectors fields.

4.3 Vector Fields, Grad, Div, Curl, and Monogenic Functions

The vector derivative ∇ acting on a scalar vector field $f(\boldsymbol{x})$ produces the gradient $\nabla f(\boldsymbol{x})$, which is vector whose components in the frame $\{e^k\}$ are the partial derivatives with respect to the x^k-coordinates. In Euclidean space, the vector $\nabla f(\boldsymbol{x})$ points in the direction of the steepest increase of the function $f(\boldsymbol{x})$. However, in spaces of mixed signature, such as in the space with Minkowski signature used for projective geometry, one cannot easily interpret the direction of the vector $\nabla f(\boldsymbol{x})$.

Given a vector field $F(\boldsymbol{x})$, the full vector derivative ∇F comprises of a scalar and a bivector parts as follows

$$\nabla F = \nabla \cdot F + \nabla \wedge F. \tag{4.22}$$

The scalar term corresponds to the *divergence* of $F(\boldsymbol{x})$ which in terms of of the constant basis vectors $\{e_k\}$ can be written as

$$\nabla \cdot F = \frac{\partial}{\partial x^k} e^k \cdot F = \frac{\partial F^k}{\partial x^k} = \partial_k F^k. \tag{4.23}$$

If we compute the divergence of a vector $\boldsymbol{x}$, we get the dimension of the space

$$\nabla \cdot \boldsymbol{x} = \frac{\partial x^k}{\partial x^k} = n. \tag{4.24}$$

The bivector antisymmetric part defines the exterior derivative of the field $F(\boldsymbol{x})$

$$\nabla \wedge F = e^i \wedge (\partial_i F) = e^i \wedge e^j \partial_i F_j. \tag{4.25}$$

The components are the antisymmetric elements of $\partial_i F_j$. In three dimensions, these components are the components of the *curl*, note that $\nabla \wedge F$ is a bivector and not simply an axial vector. The antisymmetric part or bivector is related in 3D with the *curl* via the cross-product as follows:

$$\nabla \wedge F = I_3 \nabla \times F. \tag{4.26}$$

Note that $\nabla \wedge F$ is not an axial vector, but something different, a bivector, which represents an oriented plane. Also, one can see as in the true sense of geometric calculus that one can generalize the curl to arbitrary dimensions.

To get a curl vector one takes the dual of the curl, namely

$$\nabla \times F. = -I_3 \nabla \wedge F. \tag{4.27}$$

Be aware that the exterior derivative generalizes the curl to arbitrary dimensions. As an example, consider the geometric product of ∇ and a vector $\boldsymbol{x}$

$$\nabla \boldsymbol{x} = \nabla \cdot \boldsymbol{x} + \nabla \wedge \boldsymbol{x} = \frac{\partial x^k}{\partial x^k} + e^i \wedge e_i = n + e^i \wedge e^j (e_i \cdot e_j) = n, \tag{4.28}$$

where the symmetric part $e_i \cdot e_j$=0.

Finally, according to the definition of the geometric product between two vectors, the divergence $\nabla \cdot F$ or symmetric part of the geometric product ∇F can be expressed in terms of the geometric product, namely

$$\nabla \cdot F = \frac{1}{2}(\nabla F + \dot{F}\dot{\nabla}), \tag{4.29}$$

and the antisymmetric part or exterior derivative

$$\nabla \wedge F = \frac{1}{2}(\nabla F - \dot{F}\dot{\nabla}). \tag{4.30}$$

F is called divergence-free, iff $\partial \cdot F \equiv 0$ on $\mathcal{M}$ and F is called curl-free $\partial \wedge F \equiv 0$ on $\mathcal{M}$. Example, the identity map $F(\boldsymbol{x}) = \boldsymbol{x}$ is curl-free and its divergence is $\partial \cdot \boldsymbol{x} = m$.

Given F a multivector-valued function, F is called left monogenic iff $\partial F \equiv 0$ on $\mathcal{M}$ and right monogenic iff $F\partial \equiv 0$ on $\mathcal{M}$. Monogenic functions generalize the concept of *holomorphic functions* to arbitrary dimensions and arbitrary manifolds.

4.4 Multivector Fields

The preceding computations can easily be extended to the case of the vector derivative acting on a multivector field $\boldsymbol{F} \in G_n$ as follows:

$$\nabla \boldsymbol{F} = e^k \partial_k \boldsymbol{F}. \tag{4.31}$$

For the m-grade multivector field $\boldsymbol{F}_m$, its inner and wedge products read

$$\nabla \cdot \boldsymbol{F}_m = <\nabla \boldsymbol{F}_m>_{m-1}, \qquad \nabla \wedge \boldsymbol{F}_m = <\nabla \boldsymbol{F}_m>_{m+1}, \tag{4.32}$$

which are known as divergence and curl, respectively. Be aware that divergence of a divergence vanishes,

$$\nabla \cdot (\nabla \cdot \boldsymbol{F}) = 0, \tag{4.33}$$

and also the curl of a curl:

$$\nabla \wedge (\nabla \wedge \boldsymbol{F}) = e^i \wedge \partial_i (e^j \wedge \partial_j \boldsymbol{F}) = e^i \wedge e^j \wedge (\partial_i \partial_j \boldsymbol{F}) = 0, \tag{4.34}$$

by convention here, the inner product of a vector and a scalar is zero. We should define the following conventions: (i) in the absence of brackets, ∇ acts on the object to its immediate right; (ii) when ∇ is followed by an expression in brackets, the derivative acts on all elements of the expression; (iii) when ∇ acts on a multivector to which one is not adjacent, one uses overdots to describe the procedure:

$$\dot{\nabla} \boldsymbol{F} \dot{\boldsymbol{G}} = e^k \boldsymbol{F} \partial_k \boldsymbol{G}. \tag{4.35}$$

According to this notation, one can write

$$\nabla(\boldsymbol{F}\boldsymbol{G}) = \nabla \boldsymbol{F}\boldsymbol{G} + \dot{\nabla} \boldsymbol{F} \dot{\boldsymbol{G}}, \tag{4.36}$$

which is a form of the Leibniz rule,

$$\dot{\nabla} \dot{f}(\boldsymbol{x}) = \nabla f(\boldsymbol{x}) - e^k f(\partial_k \boldsymbol{x}). \tag{4.37}$$

One uses the overdot notation for linear functions, namely given a position-dependent linear function $f(\boldsymbol{x})$, one write

$$\dot{\nabla} \dot{f}(\boldsymbol{x}) = \nabla f(\boldsymbol{x}) - e^k f(\partial_k \boldsymbol{x}), \tag{4.38}$$

note that $\dot{\nabla} \dot{f}(\boldsymbol{a})$ differentiates only the position dependence in the linear function and not in its argument. We can differentiate various multivectors that depend linearly in $\boldsymbol{x}$, e.g.,

$$\nabla \boldsymbol{x} \cdot \boldsymbol{X}_m = e^k e_k \cdot X_m, \tag{4.39}$$

where $\boldsymbol{X} - m$ is a grade-m multivector. Using Eq. (2.122), we can write the divergence and exterior products of $\boldsymbol{X} - m$ as follows

$$\begin{aligned} \nabla \boldsymbol{x} \cdot \boldsymbol{X}_m &= m \boldsymbol{X}_m, \\ \nabla \boldsymbol{x} \wedge \boldsymbol{X}_m &= (n - m) \boldsymbol{X}_m, \\ \dot{\nabla} \boldsymbol{X}_m \dot{\boldsymbol{x}} &= (-1)^m (n - 2m) \boldsymbol{X}_m, \end{aligned} \tag{4.40}$$

n stands for the dimension of the space.

4.5 The Generalized Gaussian Map

The Gauss map can be defined globally if and only if the surface is orientable, in which case its degree is half the *Euler characteristic* [88, 135]. The Gauss map can always be defined locally that means just on a small piece of the surface. It can be easily proved that the Jacobian determinant of the Gauss map is equal to *Gaussian curvature*, and the differential of the Gauss map is called the *shape operator*, which will be discussed below in the geometric algebra framework.

Firstly, we denote by $\bigwedge^{j}\mathbb{R}^{m+k}$ as the exterior algebra of j-blades over $\mathbb{R}^{m+k}$ with the exterior product$\wedge$. $\bigwedge^{j}\mathbb{R}^{m+k}$ is spanned by $r = \begin{pmatrix} m+k \\ j \end{pmatrix} = \frac{m+k!}{(m+k-j)!j!}$ j-vectors. $\bigwedge^{j}\mathbb{R}^{m+k}$ is a linear r-dimensional vector space spanned by a j-vector basis.

Consider the unit pseudoscalar $I_n = e_1 \wedge e_2 \wedge ... e_n$ in $\bigwedge^{j}\mathbb{R}^{m+k}$, where $n = m + k$. The dual of a multivector $V_m \in \bigwedge^{j}\mathbb{R}^{m+k}$ is given by $V_m^* = V_m I_n$. With a nonzero m-vector V_m, it is associated a m-dimensional subspace

$$\mathcal{V}_m = \{v \in \mathbb{R}^{m+k} : V_m \wedge v = 0\}, \tag{4.41}$$

so the dual of V_m, the V_m^* represents the k-dimensional space which is orthogonal to V_m. The orientation of the m-blade V_m has the orientation of the subspace $\mathcal{V}_m$ so that the $\mathcal{V}_m$ is a point in the oriented Grassmannian manifold $\mathbb{G}_m(\mathbb{R}^{m+k})$. Two m-blades U_m and V_m are mapped onto the same subspace if and only if $U_m = s V_m$, where s is some positive scalar. An inverse map can be obtained from the Plücker embedding

$$\begin{aligned} G_m(\mathbb{R}^{m+k}) &\rightarrow \mathbb{P}(\bigwedge^{m}\mathbb{R}^{m+k}) \\ span(v_1, v_2, ..., v_m) &\rightarrow v_1 \wedge v_2 \wedge ... \wedge v_m. \end{aligned} \tag{4.42}$$

This resultant space is projective, because a different choice of basis gives a different exterior product, but the two m-blades differ just in a scalar factor, i.e., the determinat of the basis matrix changes see Sect. 20.2.3.

Consider $[\bigwedge^{m}\mathbb{R}^{m+k}]$ a set of unit-m-blades, the mapping given in (4.42) results in a bijection between normed m-blades and oriented m-subspaces. Since the geometric algebra G_{m+k} of $\mathbb{R}^{m+k}$ is a metric space, the subset of unit-m-blades can be equipped with the subspace topology. In addition, $[\bigwedge^{m}\mathbb{R}^{m+k}]$ is locally homeomorphic to $\mathbb{R}^{m \cdot k}$ and it can be endowed with a differentiable atlas. In general, the $[\bigwedge^{m}\mathbb{R}^{m+k}]$ is a differentiable manifold and from now on, we will call the *unit m-blade manifold*.

We denote by $\mathcal{M}$ an m-dimensional smooth submanifold of the Euclidean vector space $\mathbb{R}^{m+k}$. For any point $\boldsymbol{p} \in \mathcal{M}$, a parallel transport of the tangent space $T_p\mathcal{M}$

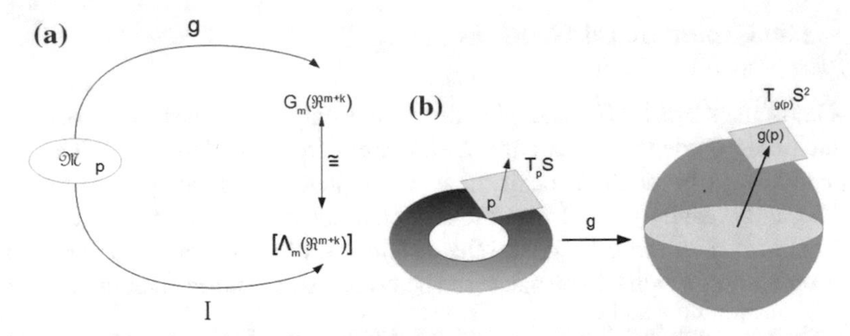

Fig. 4.3 **a** Commuting diagram. **b** Gauss map of $T_p S$ on a 2D manifold to a $T_{n(p)}S^2$ on a sphere

to the origin yields an m-subspace of $\mathbb{R}^{m+k}$, hence a point $\mathrm{g}(\boldsymbol{p})$ in the Grassmanian manifold $\mathbb{G}_m(\mathbb{R}^{m+k})$; see Sect. 20.2.3.

The mapping $\boldsymbol{p} \to g(\boldsymbol{p})$ is called the generalized Gauss map. We can assign a unit-m-blade $I(\boldsymbol{p})$ to the mapping $g(\boldsymbol{p})$ which in the geometric algebra framework will be called the pseudoscalar of the tangent space $T_p\mathcal{M}$. In geometric measure theory [88], this m-vector field was named the orientation of the manifold, i.e., if $\tau_j = \tau_j(\boldsymbol{p})$, $j = 1, ..., m$ is an orthonormal basis of $T_p\mathcal{M}$, then the pseudoscalar of the tangent space $T_p\mathcal{M}$ is given by

$$I(\boldsymbol{p}) = \tau_1(\boldsymbol{p})\wedge\tau_2(\boldsymbol{p})\wedge...\tau_m(\boldsymbol{p}). \tag{4.43}$$

Thus, we will call the following mapping

$$\boldsymbol{p} \to I(\boldsymbol{p}) \tag{4.44}$$

the *Gauss map*. Figure 4.3 depicts the related commuting diagram and a Gauss map of 2D manifold T_pS to a sphere $T_{n(p)}S^2$.

4.6 Second Derivative and the Laplace–Beltrami Operator

Given a multivector-valued function F, the second derivative of F decomposes as follows

$$\begin{aligned}
\partial^2 F &= (\partial\partial)F = \partial(\partial F(\boldsymbol{x})) = \partial(\partial\cdot F + \partial\wedge F)\\
&= \partial\cdot\partial\cdot F + \partial\wedge\partial\cdot F + \partial\cdot\partial\wedge F + \partial\wedge\partial\wedge F\\
&= (\partial\cdot\partial)F + (\partial\wedge\partial)\cdot F + (\partial\wedge\partial)\times F + (\partial\wedge\partial)\wedge F.\\
&= (\partial\cdot\partial)F + (\partial\wedge\partial)\cdot F + (\partial\wedge\partial)\times F + (\partial\wedge\partial)\wedge F.
\end{aligned} \tag{4.45}$$

However, ∂ is considered as a vector and $\partial \wedge \partial$ does not vanish generally, and it has algebraic properties of a bivector; see Sect. 3.1.1.

The double derivative $\partial \cdot \partial$ is a scalar differential operator, namely

$$\partial \cdot \partial = \sum_k \left(e^k \frac{\partial}{\partial x^k} \right) \left(e^k \frac{\partial}{\partial x^k} \right) = \nabla^2, \tag{4.46}$$

which corresponds to the *Laplacian* linear operator.

In Riemannian normal coordinates, the operator $\triangle_M$

$$\triangle_M = \partial \cdot \partial = \sum_k \left(\tau^k \frac{\partial}{\partial x^k} \right) \left(\tau^k \frac{\partial}{\partial x^k} \right) \tag{4.47}$$

is well known in classical differential geometry as the Laplace–Beltrami operator $\triangle_M$ on $\mathcal{M}$, c.f. ([68], p.163) or ([275], Proposition 1.2.1).

A real function ϕ is called *harmonic* iff $\triangle_{\mathcal{M}} \phi \equiv 0$ on the manifold $\mathcal{M}$. An example of harmonic functions is the Euler–Lagrange equation of a certain energy.

4.7 Projection, Shape, Curl, and Spur

The differential geometry of a given vector manifold $\mathcal{M} = \{x\}$ is completely determined by properties of its pseudoscalar $I = I(x)$. For the case of an oriented manifold, the pseudoscalar is a single-valued field which is defined on the manifold $\mathcal{M} = \{x\}$. Figure 4.4a shows that the pseudoscalar $I(x)$ determines the tangent space at each point x.

Let us represent the identity function as $(x) = x$, i.e., it maps each point to itself. The first differential of the identity function is the projection of vector v into the tangent space $\mathcal{M}$ at each point of the manifold, namely

$$v \cdot \partial(x) = P(v) = (v \cdot I) \cdot I^{-1} = v \cdot II^{-1}. \tag{4.48}$$

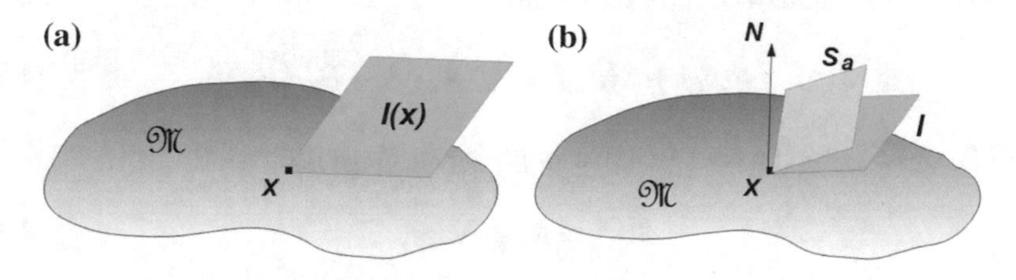

Fig. 4.4 (a) Manifold pseudoscalar. (b) Shape bivector S_a and Spur N

The second differential of the identity is the first differential of the projection, c.f. ([138], 4-2.2a-b)

$$P_u(v) = u \cdot \dot{\partial}\dot{\mathcal{P}}(v) = u \cdot \partial P(v) - P(u \cdot \partial v), \tag{4.49}$$

or more generally

$$P_u(F) = u \cdot \dot{\partial}\dot{\mathcal{P}}(F) = u \cdot \partial P(F) - P(u \cdot \partial F). \tag{4.50}$$

This quantity is well defined at each point of x of $\mathcal{M}$ for any multivector function $F = F(x)$ and a vector $u = u(x)$. The function F can be decomposed as follows

$$F = F_{\|} + F_{\perp}, \tag{4.51}$$

where $F_{\|} = P(F)$ and $F_{\perp} = F - P(F)$ are the projection and reaction of F into the tangent algebra, respectively.

Hestenes and Sobczyk [138] introduced the derivative of the projection operator, ([138], 4-2.14)

$$\mathcal{S}(F) = \dot{\partial}\dot{\mathcal{P}}(F) = \partial_u P_u(F), \tag{4.52}$$

where the projection operator $\mathcal{P}$ is being differentiated. The $S(F)$ is called the *shape* of the function $F = F(x)$ and the operator S is referred as the *shape* operator of the manifold $\mathcal{M}$; see Fig. 4.4b. The shape operator is linear

$$S(\alpha F + \beta G) = \alpha S(F) + \beta S(G); \tag{4.53}$$

where F, G are multivector functions. In case of a vector function $v = v(x)$, we compute

$$S(v_{\|}) = \partial_u \wedge P_u(v), \tag{4.54}$$

$$S(v_{\perp}) = \partial_u \cdot P_u(v). \tag{4.55}$$

Using Eqs. (4.52) and (4.53), we can write

$$S(v) = S(v_{\|}) + S(v_{\perp}) = \partial_u \cdot P_u(v) + \partial_u \wedge P_u(v). \tag{4.56}$$

Eq. (4.55), (4.55) can be rewritten in a single equation as follows, ([138], 4-2.18-20)

$$S(v) = S_v + N \cdot v, \tag{4.57}$$

where, on the one hand, the bivector-valued function

$$S_v \equiv \dot{\partial} \wedge \dot{P}(v) = S(P(v)) \tag{4.58}$$

is the *curl* tensor or the curl of the manifold $\mathcal{M}$, to make clear that it is the curl of the projection of a vector. On the other hand,

$$N \equiv \dot{P}(\dot{\partial}) = \partial_v \cdot S_v = \partial_v S_v \tag{4.59}$$

is called the *spur* of the manifold $\mathcal{M}$ to make clear that it is the spur of the tensor $P_u(v)$; therefore,

$$\partial_u \cdot \partial_v P_u(v) = P_u(\partial_u) = \dot{P}(\dot{\partial}) = N. \tag{4.60}$$

N is everywhere orthogonal to tangent vectors of the manifold $\mathcal{M}$, that is why is called spur as something that sticks out of the manifold; see Fig. 4.4b. $N = N(x)$ is the trace of the second fundamental form at x of $\mathcal{M}$ ([275], Proposition 2.2.2), and by definition, the mean curvature vector at x on $\mathcal{M}$ ([275], p. 68), ([67], p. 301) or ([68], p. 158).

After, we have examined in detail the shape of vector-valued function, we can now establish the general properties of the shape operator defined in Eq. (4.52). The most important properties of the shape are formulated as, c.f. ([130], 4-2.35.a-)

$$S(F_\parallel) = S(P(F)) = \dot{\partial} \wedge \dot{P}(F), \tag{4.61}$$

$$S(F_\perp) = P(S(F)) = \dot{\partial} \cdot \dot{P}(F), \tag{4.62}$$

where

$$S(F) = S(F_\parallel) + S(F_\perp). \tag{4.63}$$

Taking into account that at each point x of $\mathcal{M}$ for any multivector function $F = F(x)$ and any vector function $u = u(x)$

$$P_u(\alpha F + \beta G) = \alpha P_u(F) + \beta P_u(G), \tag{4.64}$$

$$P_u(< F >_k) = < P_u(F) >_k, \tag{4.65}$$

where $P_u(< F >_0) = 0.$, we have then

$$< S(F) >_{k+1} = SP(< F >_k), \tag{4.66}$$

$$< S(F) >_{k-1} = PS(< F >_k). \tag{4.67}$$

Hence, the shape operator is able to distinguish $F_\parallel$ from $F_\perp$ by raising the grade of the former while lowering the grade of the latter.

Accordingly ([138], 4-1.11), the integrability condition

$$P(\partial \wedge \partial) = I^{-1} I \cdot (\partial \wedge \partial) = 0, \tag{4.68}$$

plays a role by considering the second derivative. Let us first consider the difference between the derivative and coderivative, and differentiating the tangency condition $F = P(F)$, we see that for a field $F = F(x)$ the derivative and coderivative are related as follows:

$$\partial F = \nabla F + S(F), \tag{4.69}$$

where $S(F)$ is the shape defined by Eq. (4.52). Due to the Eq. (4.62), we write

$$\partial \wedge F = \nabla \wedge F + S(F), \partial \cdot F = \nabla \cdot F, \quad if \; P(F) = F.$$

Without the condition on F and the notation for coderivative, Eq. (4.62) can be written as follows

$$\begin{aligned} \partial P(F) &= P(\partial P(F)) + S(P(F)), \\ \partial \wedge P(F) &= P(\partial \wedge F) + S(P(F)), \\ \partial \cdot P(F) &= P(\partial \cdot P(F)). \end{aligned} \tag{4.70}$$

Applying P on the previous equation

$$P(\partial P(GF)) = P(\partial \wedge F), \tag{4.71}$$

or equivalently

$$\nabla \wedge F = P(\partial \wedge F). \tag{4.72}$$

The codivergence and cocurl of a field $F = P(F)$ are simply related by duality

$$\nabla \cdot F = [\nabla \wedge (FI)]I^{-1}, \tag{4.73}$$

where is the pseudoscalar of the manifold $\mathcal{M}$. Considering now the integrability condition, taking $F = \partial$, one obtains the following operator equation

$$\partial \wedge \partial = \partial \wedge P(\partial) = P(\partial \wedge \partial) + S(\partial), \tag{4.74}$$

where $S(\partial)$ is the shape of ∂. But, accordingly (4.68), $P(\partial \wedge \partial) = 0$, we obtain then the general relation

$$\partial \wedge \partial F = S(\partial)F. \tag{4.75}$$

4.8 Blade–Laplace Operator

Given F a multivector-valued function, the blade–Laplace operator of F is given by

$$\begin{aligned}\triangle_L F &= \partial \cdot (\partial \wedge F) + \partial \wedge (\partial \cdot F) \\ &= (\partial \cdot \partial) F + (\partial \wedge \partial) \times F = \partial^2 F + (\partial \wedge \partial) \times F.\end{aligned} \tag{4.76}$$

This result can be written in terms of the second left and second right derivatives

$$\triangle_L F = \frac{1}{2}(\partial^2 F + F\partial^2). \tag{4.77}$$

The blade–Laplace operator of Eq. (4.76) can be also rewritten as follows:

$$\triangle_L F = \triangle_{\mathcal{M}} F + (\partial \wedge \partial) \times F, \tag{4.78}$$

where the expression includes the Laplace–Beltrami operator $\triangle_M$ and $\triangle_L F$ is equal to equation ([184], 4.10-iv). Functions F on $\mathcal{M}$ with vanishing blade–Laplace operator $\triangle_L F \equiv 0$ are the monogenic functions.

In terms of the result of Eq. (4.75), we can formulate finally an additional equation of the blade–Laplacian

$$\triangle_L F = \triangle_{\mathcal{M}} F + S(\dot{\partial}) \times \dot{F}. \tag{4.79}$$

Note that the blade–Laplacian differs greatly from the second derivative, namely taking a scalar function $F : \mathcal{M} \to \mathbb{R}$, according to the previous result

$$\partial^2 F = \triangle_L F + \partial \wedge \partial F = \triangle_L F + S(\partial F), \tag{4.80}$$

which is equal to equation ([184], 4.10-iv).

4.9 Geometric Calculus in 2D

The vector derivative relates and combines the algebraic properties of geometric algebra with vector calculus in a natural manner. Let us first see the derivative in 2D. Vectors are written in terms of a right-handed orthonormal frame as

$$\boldsymbol{x} = x^1 e_1 + x^2 e_2 = x e_1 + y e_2, \tag{4.81}$$

where the scalar coordinates are represented with superscripts. The vector derivative reads

$$\nabla = e_1\partial_x + e_2\partial_y = e_1(\partial_x + I_2\partial_y), \tag{4.82}$$

where I_2 is the pseudoscalar of G_2. Consider the vector derivative acting on the vector $\boldsymbol{v} = fe_1 - ge_2$ as follows:

$$\nabla\boldsymbol{v} = (e_1\partial_x + e_2\partial_y)(fe_1 - ge_2) = \left(\frac{\partial f}{\partial x} - \frac{\partial g}{\partial y}\right) - I_2\left(\frac{\partial g}{\partial x} + \frac{\partial f}{\partial y}\right). \tag{4.83}$$

Note that the two terms in the parentheses are the same that vanish in the Cauchy–Riemann equations. One recognizes the close relationship between the complex analysis and the 2D vector derivative. In order to clarify further, let us introduce the complex field Ψ:

$$\Psi = \boldsymbol{v}e_1 = f + I_2 g. \tag{4.84}$$

That Ψ is analytic means that it satisfies the Cauchy–Riemann equations. This statement can be written using the vector derivative as follows:

$$\nabla\psi = 0. \tag{4.85}$$

This fundamental equation can be generalized straightforwardly to higher dimensions. For example, in 3D, Eq. (4.85) with Ψ an arbitrary even-grade multivector defines the spin harmonics, which are fundamental for the Pauli and Dirac theories of electron orbitals (see Sect. 4.11 and the Chap. 20. In 4D, for the space–time algebra, one uses

$$\nabla = e^0\partial_t + e^i\partial_{x^i}, \qquad i = 1, 2, 3. \tag{4.86}$$

Using an even-grade multivector for Ψ, the equation $\nabla\psi = 0$ represents the wave function for a massless fermion (e.g., a neutrino). If $\Psi = F$ is a pure bivector in the space–time algebra, the equation $\nabla F = 0$ encodes all the Maxwell equations for the case of free-field electromagnetism. Finally, one can recognize that all these examples are simply special cases of the same underlying mathematics.

4.10 Electromagnetism: The Maxwell Equations

By using the space–time vector derivative and the geometric product, one can unify all four of the Maxwell equations into a single equation. This result undoubtedly confirms the impressive power of the geometric algebra framework. This section is based on ([74], 7.1). The gradient and curl operators are not invertible, while the vector derivative of the geometric algebra is fully invertible. This helps also to simplify the algebraic equation manipulation enormously. The space–time algebra $G_{1,3}$ has the basis $e_0,\ e_1,\ e_2,\ e_3$. The multivector basis is

$$\underbrace{1}_{scalar}, \underbrace{e_\mu}_{4\ vectors}, \underbrace{e_\mu e_\nu}_{6\ bivectors}, \underbrace{Ie_\mu}_{4\ trivectors}, \underbrace{I}_{pseudoscalar}. \tag{4.87}$$

The pseudoscalar is $I = e_0e_1e_2e_3$, with

$$I^2 = (e_0e_1e_2e_3)(e_0e_1e_2e_3) = -(e_2e_3)(e_2e_3) = -1. \tag{4.88}$$

The space–time vector derivative is

$$\nabla = e^\mu \partial_\mu, \qquad \partial_\mu = \frac{\partial}{\partial x^\mu}, \tag{4.89}$$

where using superscripts, $x^0 = x \cdot e^0 = t$ is the time coordinate, and $x^i = x \cdot e^i$ are the three spatial coordinates: here $e^0 = e_0$ and $e^i = -e_i$. Let us compute the space–time split of the vector derivative,

$$\nabla e_0 = (e^0\partial_t + e^i\partial_i)e_0 = \partial_t - \boldsymbol{\sigma}_i\partial_i = \partial_t - \boldsymbol{\nabla}, \tag{4.90}$$

where the minus sign is due to the Lorentzian metric. We start by considering the four Maxwell equations:

$$\begin{aligned} \boldsymbol{\nabla} \cdot \boldsymbol{B} &= 0, & \boldsymbol{\nabla} \cdot \boldsymbol{E} &= \rho, \\ \boldsymbol{\nabla} \times \boldsymbol{E} &= -\partial_t \boldsymbol{B}, & \boldsymbol{\nabla} \times \boldsymbol{B} &= \boldsymbol{J} + \partial_t \boldsymbol{E}. \end{aligned} \tag{4.91}$$

Let us first find the space–time equation to relate both the electric and magnetic fields. The familiar nonrelativistic form of the Lorentz force law is given by

$$\frac{d\boldsymbol{p}}{dt} = q(\boldsymbol{E} + \boldsymbol{v} \times \boldsymbol{B}), \tag{4.92}$$

where all the vectors are expressed in the frame e_0 and $\boldsymbol{p} = p \wedge e_0$. By multiplying by $v \cdot e_0$, one converts the derivative into one with respect to proper time. Applying it to the first term on the right-hand side, one gets

$$v \cdot e_0 \boldsymbol{E} = v \cdot (\boldsymbol{E} \wedge e_0) - (v \cdot \boldsymbol{E}) \wedge e_0 = (\boldsymbol{E} \cdot v) \wedge e_0, \tag{4.93}$$

where $\boldsymbol{E} \wedge e_0$ vanishes due to the wedge between its terms $\boldsymbol{\sigma}_k = e_k e_0$ and e_0. The magnetic term changes to

$$\begin{aligned} v \cdot e_0 \boldsymbol{v} \times \boldsymbol{B} &= -v \cdot e_0 \boldsymbol{v} \cdot (I\boldsymbol{B}) = -(v \wedge e_0) \times (I\boldsymbol{B}) \\ &= [(I\boldsymbol{B}) \cdot v] \wedge e_0 + [e_0 \cdot (I\boldsymbol{B})] \wedge v = [(I\boldsymbol{B}) \cdot v] \wedge e_0, \end{aligned} \tag{4.94}$$

where we used the result of the Jacobi identity. Using these results, Eq. (4.92) can be written in the following form:

$$\frac{d\boldsymbol{p}}{d\tau} = \dot{p}\wedge e_0 = q[(\boldsymbol{E} + I\boldsymbol{B}) \cdot v]\wedge e_0. \tag{4.95}$$

In this equation, the space–time bivector

$$F = \boldsymbol{E} + I\boldsymbol{B} \tag{4.96}$$

is the *electromagnetic field strength*, also known as the Faraday bivector.

Now, in order to group the two Maxwell source equations, introduce the space–time vector J with

$$\rho = J \cdot e_0, \qquad \boldsymbol{J} = J\wedge e_0. \tag{4.97}$$

Then we form

$$Je_0 = \rho + \boldsymbol{J} = \boldsymbol{\nabla} \cdot \boldsymbol{E} - \partial_t \boldsymbol{E} + \boldsymbol{\nabla} \boldsymbol{\times} \boldsymbol{B}. \tag{4.98}$$

Consider the following algebraic manipulation using the bivector $\boldsymbol{y} = y\wedge e_0$:

$$e_0\wedge(x \cdot \boldsymbol{y}) = e_0\wedge[x \cdot (y\wedge_0)] = e_0\wedge(-x \cdot e_0 y) = x \cdot e_0 y\wedge e_0 = x \cdot e_0 \boldsymbol{y}.$$

In this manner, one can write

$$-\partial_t \boldsymbol{E} = -e_0 \cdot \nabla \boldsymbol{E} = -e_0\wedge(\nabla \cdot \boldsymbol{E}). \tag{4.99}$$

The full $\boldsymbol{E}$-term can then be written as

$$\begin{aligned} \boldsymbol{\nabla} \cdot \boldsymbol{E} - \partial_t \boldsymbol{E} &= (e_0\wedge\nabla) \cdot \boldsymbol{E} - e_0\wedge(\nabla \cdot \boldsymbol{E}) = (\nabla \cdot \boldsymbol{E}) \cdot e_0 + (\nabla \cdot \boldsymbol{E})\wedge e_0 \\ &= (\nabla \cdot \boldsymbol{E})e_0. \end{aligned} \tag{4.100}$$

Now consider the cross-product of two bivectors:

$$\boldsymbol{x} \times \boldsymbol{y} = (xe_0) \times (xe_0) = (x\wedge e_0) \times (x\wedge e_0) = -x\wedge y e_0 e_0. \tag{4.101}$$

Using this result, we rewrite

$$\boldsymbol{\nabla} \boldsymbol{\times} \boldsymbol{B} = -I\nabla \times \boldsymbol{B} = I(\nabla \wedge e_0) \times (\boldsymbol{B}) = -I\nabla\wedge\boldsymbol{B}e_0 = \nabla \cdot (I\boldsymbol{B})e_0. \tag{4.102}$$

Combining Eqs. (4.100) and (4.102), we get

$$Je_0 = \nabla \cdot (\boldsymbol{E} + I\boldsymbol{B})e_0. \tag{4.103}$$

Using the definition of the Faraday bivector

$$F = \boldsymbol{E} + I\boldsymbol{B}, \tag{4.104}$$

we can combine the two equations into one:

$$\nabla \cdot F = J. \tag{4.105}$$

Let us considering the two remaining Maxwell equations,

$$\boldsymbol{\nabla} \cdot \boldsymbol{B} = 0 \qquad \boldsymbol{\nabla} \times \boldsymbol{E} = -\partial_t \boldsymbol{B}. \tag{4.106}$$

The first can be rewritten using the duality principle and $\boldsymbol{E} \wedge e_0 = 0$:

$$\begin{aligned} 0 = \boldsymbol{\nabla} \cdot \boldsymbol{B} &= (e_0 \wedge \nabla) \wedge (I\boldsymbol{B}) = \nabla \wedge (I\boldsymbol{B}) \wedge e_0 = \nabla \wedge (\boldsymbol{E} + I\boldsymbol{B}) \wedge e_0 \\ &= \nabla \wedge F \wedge e_0. \end{aligned} \tag{4.107}$$

Let us now consider the inner product, making use of Eq. (4.90):

$$\begin{aligned} (\nabla \wedge F) \cdot e_0 &= \nabla \wedge (F \cdot e_0) + \partial_t F =< \boldsymbol{\nabla} \boldsymbol{E} e_0 >_2 + \partial_t F \\ &= - < (\partial_t - \boldsymbol{\nabla}) \boldsymbol{E} >_2 + \partial_t (\boldsymbol{E} + I\boldsymbol{B}) \\ &= I(\partial_t \boldsymbol{B} + \boldsymbol{\nabla} \times \boldsymbol{E}) = 0. \end{aligned} \tag{4.108}$$

Since $(\nabla \wedge F) \cdot e_0$ equals zero, we can merge this result with Eq. (4.105) into one single equation using the geometric product

$$\nabla F = J. \tag{4.109}$$

In contrast using tensor calculus, one utilizes two equations:

$$\partial_u F^{\mu\nu} = J^{\nu}, \qquad \partial_u \epsilon^{\mu\nu\rho\sigma} F_{\rho\sigma} = 0. \tag{4.110}$$

4.11 Spinors, Shrödinger–Pauli, and Dirac Equations

As an illustration on the formulation and use of spinor spaces in geometric calculus, let us now consider the spinor representation in the Schrödinger–Pauli equation. See Chap. 5 and Sect. 20.1.6 in the Chap. 20 for further definitions and details on pin and spin groups and spinors. This section is based on section ([197], 4.8).

In the following equation, the wave function sends space–time points to *Pauli spinors*

$$\psi(\mathbf{r}, t) = \begin{pmatrix} \psi_1 \\ \psi_2 \end{pmatrix} \in \boldsymbol{C}^2, \tag{4.111}$$

where $\psi_1, \psi_2 \in \boldsymbol{C}$. If one replaces the Pauli spinors by the square matrix spinors with zero entries in the second column, see

$$\psi(\mathbf{r}, t) = \begin{pmatrix} \psi_1 & 0 \\ \psi_2 & 0 \end{pmatrix}, \tag{4.112}$$

one obtains an isomorphic linear space S. This representation can be expressed as

$$\psi \in \mathtt{Mat}(2, \mathbb{C})f \simeq G_3 f, \qquad f = \begin{pmatrix} 1 & 0 \\ 0 & 0 \end{pmatrix}. \tag{4.113}$$

Such matrices where only the first column has nonzero entries, form a *left ideal* S of G_3, i.e.,

$$x\psi \in S, \;\; \texttt{for all}\; x \in G_3 \;\textit{and}\; \psi \in S \subset G_3. \tag{4.114}$$

The vector bases σ_i of $\mathtt{Mat}(2, \mathbb{C})$ and e_i of G_3 are related as follows: $e_1 \simeq \sigma_1, e_2 \simeq \sigma_2, e_3 \simeq \sigma_3$. The linear space S has a basis $\{f_0, f_1, f_2, f_3\}$, where

$$f_0 = \frac{1}{2}(1 + e_3) \simeq \begin{pmatrix} 1 & 0 \\ 0 & 0 \end{pmatrix}, \quad f_1 = \frac{1}{2}(e_{23} + e_2) \simeq \begin{pmatrix} 0 & 0 \\ i & 0 \end{pmatrix},$$
$$f_2 = \frac{1}{2}(e_{31} - e_1) \simeq \begin{pmatrix} 0 & 0 \\ -1 & 0 \end{pmatrix}, \quad f_3 = \frac{1}{2}(e_{12} + e_{123}) \simeq \begin{pmatrix} i & 0 \\ 0 & 0 \end{pmatrix},$$

where $f = f_0$ is an *idempotent*, i.e., $f^2 = f$. An element $\psi \in S$, expressed in coordinate form in the basis $\{f_0, f_1, f_2, f_3\}$, multiplied by the left with an arbitrary even element $b \in G_3^+$

$$b\psi = (b_0 + b_1 e_{23} + b_2 e_{31} + b_3 e_{12})(\psi_0 f_0 + \psi_1 f_1 + \psi_2 f_2 + \psi_3 f_3), \tag{4.115}$$

corresponds to the following matrix multiplication

$$b\psi \simeq \begin{pmatrix} b_0 & -b_1 & -b_2 & -b_3 \\ b_1 & b_0 & b_3 & -b_2 \\ b_2 & -b_3 & b_0 & b_1 \\ b_3 & b_2 & -b_1 & b_0 \end{pmatrix} \begin{pmatrix} \psi_0 \\ \psi_1 \\ \psi_2 \\ \psi_3 \end{pmatrix}. \tag{4.116}$$

This kind of square matrices corresponding to the left multiplication by even elements constitutes a subring of Mat(4,R), which in turn is an isomorphic image of the quaternion ring $\mathcal{H}$.

4.11.1 Spinor Operators

Until know, spinors have been objects that have been operated upon. This subsection is based on ([197], 4.8) and ([128]). One can replace such passive spinors by *active* spinor operators. Instead of the spinors given by Eq. (4.113) in minimal left ideals, we formulate the following even elements:

$$\Psi = 2\,even(\psi) = \begin{pmatrix} \psi_1 & -\psi_2^* \\ \psi_2 & \psi_1^* \end{pmatrix} \in G_3^+, \tag{4.117}$$

which can also be computed as $\Psi = \psi + \hat{\psi}$ for $\psi \in G_3 f$. In a classical way, one computes the expected values of the components of the spin in terms of the column spinor $\psi \in \boldsymbol{C}^2$ as follows:

$$s_1 = \psi^\dagger \sigma_1 \psi, \quad s_2 = \psi^\dagger \sigma_2 \psi, \quad s_3 = \psi^\dagger \sigma_3 \psi. \tag{4.118}$$

On the other hand, in terms of $\psi \in G_3 f$ this computation can be repeated as follows

$$s_1 = 2 < \psi e_1 \tilde{\psi} >_0, \quad s_2 = 2 < \psi e_2 \tilde{\psi} >_0, \quad s_3 = 2 < \psi e_3 \tilde{\psi} >_0, \tag{4.119}$$

where $< \cdot >_0$ extracts the 0-blade or scalar part of the involved multivector. However, using the active spinor $\Psi \in G_3^+$, we can compute the expected values straightforwardly in a compact form,

$$\boldsymbol{s} = s_1 e_1 + s_2 e_2 + s_3 e_3 = \Psi e_3 \tilde{\Psi}, \tag{4.120}$$

getting the entity $\boldsymbol{s}$ as a whole. Since Ψ acts like an operator here, we will now call it a *spinor operator*.

The matrix algebra Mat(2,ℂ) is an isomorphic image of the geometric algebra G_3 of the Euclidean space $\mathbb{R}^3$. As a result, not only an vectors $x \in R^3$ and rotations in $SO(3)$ be represented in G_3, but also spinor spaces or spinor representations of the rotation group $SO(3)$ can be constructed in G_3. Recall the Schrödinger equation

$$ih\frac{\partial \psi}{\partial t} = -\frac{h^2}{2m}\nabla^2 \psi + W\psi. \tag{4.121}$$

In an electromagnetic field **E**, **B** with potentials V and **A**, this equation becomes

$$\begin{aligned} ih\frac{\partial \psi}{\partial t} &= \frac{1}{2m}[(-ih\nabla - e\mathbf{A})^2]\psi - eV\psi, \\ &= \frac{1}{2m}[(-h^2\nabla^2 + e^2A^2 + ihe(\nabla \cdot \mathbf{A} + \mathbf{A} \cdot \nabla)]\psi - eV\psi. \end{aligned} \tag{4.122}$$

Note that this equation does not yet involve the electron's spin. In 1927, Pauli introduced the spin into quantum mechanics by integrating a new term in the Schrödinger equation. Using the *Pauli spin matrices*, which fulfill the relation $\sigma_j \sigma_k + \sigma_k \sigma_j = 2\delta_{jk} I$ and the generalized momentum $\boldsymbol{\pi} = -ih\nabla - e\mathbf{A} = \mathbf{p} - e\mathbf{A}$ satisfying $\pi_1\pi_2 - \pi_2\pi_1 = iheB_3$ (permutable cyclically for 1,2,3), one can write

$$(\sigma \cdot \pi)^2 = \pi^2 - he(\sigma \cdot \mathbf{B}), \tag{4.123}$$

where $\pi^2 = p^2 + e^2A^2 - e(\mathbf{p} \cdot \mathbf{A} + \mathbf{A} \cdot \mathbf{p})$. In Eq. (4.122), Pauli replaced the term π^2 by $(\boldsymbol{\sigma} \cdot \boldsymbol{\pi})^2$ obtaining this formulation:

$$ih\frac{\partial\psi}{\partial t} = \frac{1}{2m}[\pi^2 - he(\boldsymbol{\sigma} \cdot \mathbf{B})]\psi - eV\psi. \tag{4.124}$$

Note that in this Schrödinger–Pauli, the spin is described by the term $\frac{he}{2m}(\boldsymbol{\sigma} \cdot \mathbf{B})$.

In geometric algebra G_3, the Pauli contribution can be formulated by replacing the dot product $\boldsymbol{\pi} \cdot \boldsymbol{\pi} = \pi^2$ with $\boldsymbol{\pi}^2 = \boldsymbol{\pi}\boldsymbol{\pi} - he\mathbf{B}$, so that the equation changes to

$$ih\frac{\partial\psi}{\partial t} = \frac{1}{2m}[\pi^2 - he\mathbf{B}]\psi - eV\psi, \tag{4.125}$$

where $\mathbf{B} \in \mathbb{R}^3 \subset G_3$ and $\psi(\mathbf{r}, t) \in S = G_3 f,\ f = \frac{1}{2}(1 + e_3)$. Note how remarkable the formulation of this equation is in geometric calculus, because all the arguments and functions now have values in one algebra. As a result, it greatly facilitates the numerical computations. Now let us move a step forward by using the above-explained active spinor operator. The Schrödinger equation using the spinor operator Ψ reads

$$ih\frac{\partial\Psi}{\partial t} = \frac{1}{2m}\pi^2\Psi - \frac{he}{2m}\mathbf{B}\Psi e_3 - eV\Psi, \tag{4.126}$$

and it explicitly shows the quantization direction e_3 of the spin.

The relativistic phenomena can be taken into consideration by starting the analysis from the equation $E^2/c^2 - \mathbf{p} = m^2c^2$. By inserting in this equation the energy and momentum operators, one gets the *Klein–Gordon* equation

$$h^2\Big(-\frac{1}{c^2}\frac{\partial^2}{\partial t^2} + \frac{\partial^2}{\partial x_1^2} + \frac{\partial^2}{\partial x_2^2} + \frac{\partial^2}{\partial x_3^2}\Big)\psi = m^2c^2\psi. \tag{4.127}$$

In 1928, Dirac linearized the Klein–Gordon equation, formulating it as first-order equation

$$ih\Big(e_0\frac{1}{c}\frac{\partial}{\partial t} + e_1\frac{\partial}{\partial x_1} + e_2\frac{\partial}{\partial x_2} + e_3\frac{\partial}{\partial x_3}\Big)\psi = mc\psi, \tag{4.128}$$

where the symbols e_μ satisfy $e_0^2 = I$, $e_1^2 = e_2^2 = e_3^2 = -I$, and $e_\nu e_\mu = -e_\mu e_\nu$ for $\mu \neq \nu$. Dirac found a set of 4×4 matrices that satisfy these relations. Writing $x_0 = ct$ and substituting $\partial_\mu = \frac{\partial}{\partial x^\mu}$ one gets a condensed form of the Dirac equation:

$$ihe^\mu\partial_\mu\psi = mc\psi. \tag{4.129}$$

One can include an interaction with the electromagnetic field $F^{\mu\nu}$ via the space–time potential $(A^0, A^1, A^2, A^3) = (\frac{1}{c}V, A_x, A_y, A_z)$ of $F^{\mu\nu}$ utilizing the replacement $ih\partial^\mu \rightarrow ih\partial^\mu - eA^\mu$. In these terms, one finally gets the conventional formulation of the Dirac equation:

$$e_\mu(ih\partial^\mu - eA^\mu)\psi = mc\psi, \tag{4.130}$$

which takes into account the relativistic phenomena and also the spin. This equation describes spin-$\frac{1}{2}$ particles like the electron. In the equation, the wave function is a column spinor

$$\psi(x) = \begin{pmatrix} \psi_1 \\ \psi_3 \\ \psi_3 \\ \psi_4 \end{pmatrix} \in \mathbb{C}^4 \quad \psi_\alpha \in \mathbb{C}. \tag{4.131}$$

During the years 1966–1974, David Hestenes reformulated the Dirac theory. In this contribution, the role of the column spinors $\psi(x) \in \mathbb{C}^4$ was taken over by operators in the even subalgebra $G^+_{1,3}$ [128].

4.12 Exercises

4.1 Given the multivector field $\boldsymbol{F}(x) = fe_1 + ge_3 + he_{23} + lI \in G_3$, compute the divergence $\nabla \cdot \boldsymbol{F}$ and curl $\nabla \wedge \boldsymbol{F}$.

4.2 Given the multivector field $\boldsymbol{F}$, prove that the divergence of a divergence $\nabla \cdot (\nabla \cdot \boldsymbol{F}) = 0$ and a curl of a curl $\nabla \wedge (\nabla \wedge \boldsymbol{F}) = 0$.

4.3 Given the vector-valued function $v(x)$ on the manifold $\mathcal{M}$ a unitary sphere located at the origin. $v(x)$ is a vector field if its values lie in the tangent space at each point of the manifold. This property is determined by the pseudoscalar I. Compute its projection $v(x)_\parallel$ and rejection $v(x)_\perp$. Compute the differential $P_u(v)$ for a tangent vector $u(x) = e_1$. Then compute the shape tensor $S(v)$ and the shape bivector S_u or called the curl of the manifold $\mathcal{M}$.

4.4 Prove the following theorems

i. $\dot{\partial} \cdot \dot{P}(v) = S(v_\perp) \Rightarrow \dot{\partial} \wedge \dot{P}(v_\parallel) = 0.$
ii. $\dot{\partial} \wedge \dot{P}(v) = S(v_\parallel) \Rightarrow \dot{\partial} \wedge \dot{P}(v_\perp) = 0.$

4.5 Given the vector-valued function $v(x)$ on the manifold $\mathcal{M}$ a unitary sphere located at the origin. Compute the shape tensor and decompose into a bivector and scalar parts: $S(v) = S_v + N \cdot v$, where $S_v \equiv \dot{\partial} \wedge \dot{P}(v) = S[P(v)] = S(v_\parallel)$ is the curl of the manifold $\mathcal{M}$ and the spur $N \equiv \dot{P}(\dot{\partial}) = \partial_v S_v$.

4.6 Compute the action of the covariant derivative or coderivative D_x on the the multivector-valued function $F = F(x)$, $DF \equiv P(\partial F) = D \wedge F + D \cdot F$.

4.7 The derivative and coderivative are related as follows: $\partial F = \nabla F + S(F)$, check this equation using the shape tensor defined by Eq. (4.52).

4.8 Given the vector-valued function $v(x)$ on the manifold $\mathcal{M}$ a unitary sphere located at the origin. $u(x) = e_1 + e_2$ is a vector field. Compute the curvature of the manifold which is given by the shape commutator defined for vectors v and u by $C(v \wedge u) \equiv S_v \times S_u = P(S_v \times S_u) + P_\perp(S_v \times S_u)$. Note that the curvature decomposes in the intrinsic part $P(S_v \times S_u)$ and the extrinsic part $P(S_v \times S_u)$. The latter is the usual Riemann curvature defined by $R(v \wedge u) \equiv P(S_v \times S_u)$.

4.9 Using Eq. (4.15), show that, c.f.([135], Sect. 4)

$$\partial F = 0 \leftrightarrow \int d^{m-1}x F = 0. \tag{4.132}$$

Also show that for $m = 2$, this can be reconiced as Cauchy's theorem for complex variables. This gives a generalization of that theorem for higher dimensions.

4.10 Using Eq. (4.18), derive a generalization of the famous Cauchy integral formula as explained in [140, 141].

4.11 In geometric calculus, given F compute the Gauss map.

4.12 In geometric calculus, given F compute the double derivative $\nabla^2 F = \partial \cdot \partial F$, the second derivative $\partial^2 F$, the Laplace–Beltrami operator $\triangle_{\mathcal{M}} F$.

4.13 In geometric calculus, given F compute projection $F_\parallel$, the reaction $F_\perp$, the shape $S(F)$, the curl $S(P(v))$, and spur $N = \dot{P}\dot{\partial}$.

4.14 In geometric calculus, given F compute the shape $S(F)$, the derivative, and the coderivative.

4.15 In geometric calculus, given a field $F = P(F)$, compute the shape $S(F)$, the codivergence, and cocurl.

4.16 In geometric calculus, given a field $F = P(F)$, compute the blade Laplacian $\triangle_L F$ and the second derivative $\partial^2 F$, compare them. Do they yield the same result?

4.17 Apply separately the blade–Laplace operator and the Beltrami–Laplace operator to a color image using Mathlab and compare and explain the results.

Chapter 5
Lie Algebras, Lie Groups, and Algebra of Incidence

5.1 Introduction

We have learned that readers of the work of D. Hestenes and G. Sobzyk [138] Chap. 8 and a late article of Ch. Doran, D. Hestenes and F. Sommen [72] Sect. IV may have difficulties to understand the subject and practitioners have difficulties to try the equations in certain applications. For this reason, this chapter reviews concepts and equations most of them introduced by D.Hestenes and G. Sobzyk [138] Chap. 8 and the article of Ch. Doran, D. Hestenes and F. Sommen [72] Sect. IV. This chapter is written in a clear manner for readers interested in applications in computer science and engineering. The explained equations will be required to understand advanced applications in next chapters.

In this chapter, we give the fundamentals of Lie algebra and Lie groups using the computational frameworks of the null cone and the n-dimensional affine plane. Using Lie algebra within this computational framework has the advantage that it is easily accessible to the reader because there is a direct translation of the familiar matrix representations to representations using bivectors from geometric algebra. Generally speaking, Lie group theory is the appropriate tool for the study and analysis of the action of a group on a manifold. Since we are interested in purely geometric computations, the use of geometric algebra is appropriate for carrying out computations in a coordinate-free manner by using a bivector representation of the most important Lie algebras. We will represent Lie operators using bivectors for the computation of a variety of invariants. This section begins with a brief introduction to the Lie group theory. We study the $Pin(n)$ and $Spin(n)$ groups, the projective transformations using the null geometric algebra over $\mathbb{R}^{3,3}$.

E. Bayro-Corrochano, *Geometric Algebra Applications Vol. I*,
https://doi.org/10.1007/978-3-319-74830-6_5

5.2 Lie Algebras and Bivector Algebras

In an abstract way, a Lie group is defined as a manifold $\mathcal{M}$ together with a product $\zeta(x, y)$. Points on the manifold can be labeled as vectors which can be seen as lying in a higher-dimensional embedding space. The product $\zeta(x, y)$ encodes the group product, and it takes two points as arguments and returns a third one. The following conditions applied to the product $\zeta(x, y)$ ensure that this product has the correct group properties.

i. Closure: $\zeta(x, y) \in \mathcal{M}, \forall\, x, y \in \mathcal{M}$.
ii. Identity: there exists en element $e \in \mathcal{M}$ such that $\zeta(e, x) = \zeta(x, e) = x$, $\forall\, x \in \mathcal{M}$.
iii. Inverse: for every element $x \in \mathcal{M}$ there exists a unique element $\hat{x}$ such that $\zeta(x, \hat{x}) = \zeta(\hat{x}, x) = e$.
iv. Associativity: $\zeta[\zeta(x, y), z] = \zeta[x, \zeta(y, z)], \forall\, x, y, z \in \mathcal{M}$.

In general, any manifold equipped with a product which fulfills the properties $i - iv$ is called a Lie group manifold. Most of the group properties can be elucidated by analyzing these properties near to the identity element e. The product $\zeta(x, y)$ induces a *Lie bracket* structure on the elements of the tangent space at the identity e. This tangent space is linear, and it is spanned by a set of Lie algebra vectors also called Lie algebra generators. This basis vectors together with their bracket form a Lie algebra.

5.3 Lie Groups and Lie Algebras of Rotors

In the geometric algebra framework, we can reveal conveniently the properties of Lie algebra using bivectors algebras and furthermore bivectors algebra makes possible to represent Lie groups of the general linear group using few coefficients; e.g., a 3D rotation matrix has nine real entries and its isomorph counterpart a rotor only four.

5.3.1 Lie Group of Rotors

This subsection is based on ([74], 11.3.2-11.3.5). Let us analyze the Lie group properties of the 3D geometric algebra G_3 using rotors. Choose a rotor $\boldsymbol{R}_\alpha$ and imagine a family of rotors, $\boldsymbol{R}(\alpha)$, for which

$$\boldsymbol{R}(0) = 1, \qquad \boldsymbol{R}(\alpha) = \boldsymbol{R}_\alpha, \tag{5.1}$$

this shows that the rotor can be obtained from the identity by a continuous set of rotor transformations. In fact, there are many possible paths to connect $\boldsymbol{R}_\alpha$ to the identity, but there is only one path which has the additional property that

$$\boldsymbol{R}(\alpha+\beta)=\boldsymbol{R}(\alpha)\boldsymbol{R}(\beta). \tag{5.2}$$

These properties belong to the one-parameter subgroup of the rotor group, and it represents to all rotations in a fixed oriented plane. Now consider a family of vectors $\boldsymbol{v}(\alpha)=\boldsymbol{R}\boldsymbol{v}_0\tilde{\boldsymbol{R}}$, where $\boldsymbol{v}_0$ is some fixed initial vector, and the differentiation of the relationship $\boldsymbol{R}\tilde{\boldsymbol{R}}=1$

$$\frac{d}{d\alpha}(\boldsymbol{R}\tilde{\boldsymbol{R}})=\boldsymbol{R}'\tilde{\boldsymbol{R}}+\boldsymbol{R}\tilde{\boldsymbol{R}}'=0, \tag{5.3}$$

where the dash symbol stands for the differentiation. By using these relations, we can compute

$$\begin{aligned}\frac{d}{d\alpha}\boldsymbol{v}(\alpha)&=\boldsymbol{R}'\boldsymbol{v}_0\tilde{\boldsymbol{R}}+\boldsymbol{R}\boldsymbol{v}_0\tilde{\boldsymbol{R}}'=(\boldsymbol{R}'\tilde{\boldsymbol{R}})\boldsymbol{v}(\alpha)-\boldsymbol{v}(\alpha)(\boldsymbol{R}'\tilde{\boldsymbol{R}}),\\&=(-2\boldsymbol{R}'\tilde{\boldsymbol{R}})\cdot\boldsymbol{v}(\alpha),\end{aligned} \tag{5.4}$$

here we make use of the inner product between a bivector and a vector. In this equation, the derivative of a vector should yield a vector as well, as given by the inner product of $(-2\boldsymbol{R}'\tilde{\boldsymbol{R}})\cdot\boldsymbol{v}(\alpha)$. Since $\boldsymbol{R}'\tilde{\boldsymbol{R}}$ is a bivector now called $\boldsymbol{B}(\alpha)$

$$-2\boldsymbol{R}'\tilde{\boldsymbol{R}}=\boldsymbol{B}(\alpha)\longrightarrow\frac{d}{d\alpha}(\boldsymbol{R})=\boldsymbol{R}'=-\frac{1}{2}\boldsymbol{B}(\alpha)\boldsymbol{R}. \tag{5.5}$$

This result is valid for any parameterized set of rotors; however restricted to the curve defined by Eq. (5.2), one get the following equation

$$\begin{aligned}\frac{d}{d\alpha}\boldsymbol{R}(\alpha+\beta)&=-\frac{1}{2}\boldsymbol{B}(\alpha+\beta)\boldsymbol{R}(\alpha+\beta)=-\frac{1}{2}\boldsymbol{B}(\alpha+\beta)\boldsymbol{R}(\alpha)\boldsymbol{R}(\beta),\\&=\frac{d}{d\alpha}[\boldsymbol{R}(\alpha)\boldsymbol{R}(\beta)]=-\frac{1}{2}\boldsymbol{B}(\alpha)\boldsymbol{R}(\alpha)\boldsymbol{R}(\beta).\end{aligned} \tag{5.6}$$

This result shows that the bivector $\boldsymbol{B}$ follows constant along this curve. Integrating Eq. (5.5), one get

$$\boldsymbol{R}(\alpha)=e^{-\frac{\alpha\boldsymbol{B}}{2}}. \tag{5.7}$$

Let us find the equivalent bivector expression of a rotated vector

$$\boldsymbol{v}(\alpha)=\boldsymbol{R}(\alpha)\boldsymbol{v}_0\tilde{\boldsymbol{R}}(\alpha)=e^{-\frac{\alpha\boldsymbol{B}}{2}}\boldsymbol{v}_0e^{\frac{\alpha\boldsymbol{B}}{2}}. \tag{5.8}$$

Differentiating it successively,

$$\begin{aligned} \frac{d\boldsymbol{v}}{d\alpha} &= e^{-\frac{\alpha \boldsymbol{B}}{2}} \boldsymbol{v}_0 \cdot \boldsymbol{B} e^{\frac{\alpha \boldsymbol{B}}{2}}, \\ \frac{d^2\boldsymbol{v}}{d\alpha^2} &= e^{-\frac{\alpha \boldsymbol{B}}{2}} (\boldsymbol{v}_0 \cdot \boldsymbol{B}) \cdot \boldsymbol{B} e^{\frac{\alpha \boldsymbol{B}}{2}}, \\ &\cdots \\ &\text{etc.} \end{aligned} \tag{5.9}$$

One see that for every derivative, the rotated vector is doted with one extra $\boldsymbol{B}$. Since this operation is grade-preserving, the resulting rotated vector can be expressed as a useful Taylor expansion, c.f. ([74], 11.65)

$$\boldsymbol{R}(\alpha)\boldsymbol{v}\tilde{\boldsymbol{R}}(\alpha) = e^{-\frac{\alpha \boldsymbol{B}}{2}} \boldsymbol{v} e^{\frac{\alpha \boldsymbol{B}}{2}} = \boldsymbol{v} + \boldsymbol{v} \cdot \boldsymbol{B} + \frac{1}{2|}(\boldsymbol{v} \cdot \boldsymbol{B}) \cdot \boldsymbol{B} + \cdots . \tag{5.10}$$

5.3.2 Bivector Algebra

It is easy to prove that the operation of commuting a multivector with a bivector is always grade-preserving. In particular, the commutator of two bivectors yields a third bivector; thus, it follows that the space of bivectors is closed under the commutator product. This closed algebra is in fact a Lie algebra which preserves most of the properties of the associated Lie group of rotors. Thus, the Lie group of rotors is formed from the bivector algebra by means of the exponentiation operation.

In Lie algebra, elements are acted on by applying the Lie bracket, which is anti-symmetric and satisfies the so-called *Jacobi identity*. In bivector algebra, the Lie bracket is just the commutator of bivectors. Given three bivectors $\boldsymbol{X}$, $\boldsymbol{Y}$, $\boldsymbol{Z}$, the Jacobi identity is given by

$$(\boldsymbol{X} \times \boldsymbol{Y}) \times \boldsymbol{Z} + (\boldsymbol{Z} \times \boldsymbol{X}) \times \boldsymbol{Y} + (\boldsymbol{Y} \times \boldsymbol{Z}) \times \boldsymbol{X} = 0, \tag{5.11}$$

this equation is simple, and it requires the expansion of each bivector cross product.

Given a set of basis bivectors $\{\boldsymbol{B}_i\}$, the commutator of any pair of these bivectors returns a third bivector which can be expanded as a linear combination of this basis bivectors. Therefore, we can express the commutator operation as follows

$$\boldsymbol{B}_j \times \boldsymbol{B}_k = C^i_{jk} \boldsymbol{B}_i, \tag{5.12}$$

the set if C^i_{jk} is called the *structure constants* of the Lie algebra, and it can be used to recover most of the properties of the corresponding Lie group. The classification of all possible Lie algebras can be carried out using the structure constants. The mathematician E. Cartan completed the solution of this problem.

Let us define the *Killing form* of Lie algebra. First of all, the *adjoint* representation of a Lie group is defined entirely in terms of certain functions which maps the Lie algebra onto itself. Each element of Lie group induces an adjoint representation, due its action, on the Lie algebra. Let us illustrate this using the Lie group of the rotor groups G_3^+. Given the bivectors $\boldsymbol{U}, \boldsymbol{V}$, the map of rotor group is

$$\boldsymbol{V} \rightarrow \boldsymbol{R}\boldsymbol{V}\tilde{\boldsymbol{R}} = Ad_R(\boldsymbol{V}), \tag{5.13}$$

this representation fulfills for consecutive rotations

$$Ad_{R_1}(Ad_{R_2}(\boldsymbol{V})) = Ad_{R_1 R_2}(\boldsymbol{V}). \tag{5.14}$$

Now, an adjoint of the rotor group representation induces an adjoint representation $ad_{\frac{U}{2}}$ of the Lie algebra as follows, c.f. ([74], 11.82)

$$ad_{\frac{U}{2}}(\boldsymbol{V}) = \boldsymbol{U} \times \boldsymbol{V}. \tag{5.15}$$

The adjoint representation of an element of the Lie algebra can be seen as linear map in the space of bivectors. Note that the matrix corresponding to the adjoint representation of the bivector $\boldsymbol{V}_j$ is determined by the structure coefficients, namely c.f. ([74], 11.83)

$$(ad_{V_j})^i_k = 2C^i_{jk}. \tag{5.16}$$

Finally, the Killing form for a Lie algebra is determined in terms of the adjoint representation

$$\mathcal{K}(\boldsymbol{U}, \boldsymbol{V}) = tr(ad_U ad_V), \tag{5.17}$$

up to a normalization. Now, the Killing form for a bivector algebra is simply computed using the inner product, c.f. ([74], 11.85), i.e.,

$$\mathcal{K}(\boldsymbol{U}, \boldsymbol{V}) = \boldsymbol{U} \cdot \boldsymbol{V}. \tag{5.18}$$

According to Eq. (5.18), the rotor groups in Euclidean space have a negative-definite Killing form. A Lie algebra with negative-definite Killing form is called to be of compact type; consequently, the corresponding Lie group is compact.

5.4 Complex Structures and Unitary Groups

In geometric algebra, one can represent complex numbers defining one basis vector for the real axis and other which squares to –1 for the imaginary axis. This suggests us that a n-D complex space could have a natural realization in a 2n-D space. In this section, we will explain the Lie algebras and Lie groups of the rotors and discuss the role of the *doubling bivector* $\boldsymbol{J}$ in $2n$-D spaces by the unitary groups. This section is based on ([74], Sect. 11.4.1).

5.4.1 The Doubling Bivector

Suppose that the n-D space has some arbitrary vector basis $\{e_i\}$, which need not to be orthonormal. We introduce a second set of basis vectors $\{f_i\}$ perpendicular to $\{e_i\}$ which fulfill the following properties

$$e_i \cdot e_j = f_i \cdot f_j = \delta_{ij} \qquad e_i \cdot f_j = 0, \qquad \forall i, j = 1, ..., n. \tag{5.19}$$

One introduces a complex structure through the so-called *doubling bivector*, c.f. ([74], 11.87)

$$\boldsymbol{J} = e_1 \wedge f^1 + e_2 \wedge f^2 + e_3 \wedge f^3 + \cdots + e_n \wedge f^n = e_i \wedge f^i, \tag{5.20}$$

where $\{f^i\}$ are the reciprocal vectors to the frame $\{f_i\}$ frame. For this and next equations, we assume that repeated indices are summed from 1,...,n. Let us show that the bivector $\boldsymbol{J}$ is independent of the initial choice of the frame $\{e_i\}$. Introducing a second pair of frames $\{e'_i\}$ and $\{f'_i\}$ to compute a bivector $\boldsymbol{J}'$ as follows

$$\boldsymbol{J}' = e'_i \wedge f'^i = e'_i \cdot e^j e_j \wedge f'^i = f'_i \cdot f^j e_j \wedge f'^i = e_j \wedge f^j = \boldsymbol{J}. \tag{5.21}$$

In the case, if the $\{e_i\}$ frame is chosen to be orthonormal, we obtain

$$\boldsymbol{J} = e_1 f_1 + e_2 f_2 + e_3 f_3 + \cdots + e_n f_n = e_i f_i = \boldsymbol{J}_1 + \boldsymbol{J}_2 + \cdots + \boldsymbol{J}_n. \tag{5.22}$$

This sum consists of n commuting blades of grade two; each bivector $\boldsymbol{J}_i$ plays the role of an imaginary number representing an oriented ith rotation plane. The doubling bivector satisfies the following conditions

$$\begin{aligned} \boldsymbol{J} \cdot f_i &= (e_j \wedge f_j) \cdot f_i = e_j \delta_{ij} = e_i, \\ \boldsymbol{J} \cdot e_i &= (e_j \wedge f_j) \cdot e_i = -f_i. \end{aligned} \tag{5.23}$$

This computation shows the interesting role of the doubling vector $\boldsymbol{J}$ which relates one n-D space with the other. Other useful relations follow

$$\begin{aligned} \boldsymbol{J} \cdot (\boldsymbol{J} \cdot e_i) &= -\boldsymbol{J} \cdot f_i = -e_i, \\ \boldsymbol{J} \cdot (\boldsymbol{J} \cdot f_i) &= \boldsymbol{J} \cdot e_i = -f_i, \end{aligned} \tag{5.24}$$

thus, for any vector $\boldsymbol{v}$ in the 2n-D space

$$\boldsymbol{J} \cdot (\boldsymbol{J} \cdot \boldsymbol{v}) = (\boldsymbol{v} \cdot \boldsymbol{J}) \cdot \boldsymbol{J} = -\boldsymbol{v}, \qquad \forall\, \boldsymbol{v}. \tag{5.25}$$

Similar as Eq. (5.10), the Taylor expansion involving the doubling vector $\boldsymbol{J}$ is the description of a series of coupled rotations with respect to $e_i \wedge f_i$ planes. Using trigonometric expressions, the Taylor expansion of $\boldsymbol{R}\boldsymbol{v}\tilde{\boldsymbol{R}}$ can be written as follows, c.f. ([74], 11.94)

$$\begin{aligned}
\boldsymbol{R}\boldsymbol{v}\tilde{\boldsymbol{R}} &= e^{-\boldsymbol{J}\frac{\theta}{2}}\boldsymbol{v}e^{\boldsymbol{J}\frac{\theta}{2}} = \boldsymbol{v} + \theta\boldsymbol{v} \cdot \boldsymbol{J} + \frac{\theta^2}{2!}(\boldsymbol{v} \cdot \boldsymbol{J}) \cdot \boldsymbol{J} \cdots, \\
&= \left(1 - \frac{\theta^2}{2!} + \frac{\theta^4}{4!} + \cdots\right)\boldsymbol{v} + \left(\theta - \frac{\theta^3}{3!} - \cdots\right)\boldsymbol{v} \cdot \boldsymbol{J}, \\
&= cos\theta\boldsymbol{v} + sin\theta\boldsymbol{v} \cdot \boldsymbol{J}.
\end{aligned} \tag{5.26}$$

5.4.2 Unitary Groups $U(n)$ and Unitary Lie Algebras $u(n)$

This section is based on ([74], Sects. 11.4.1 and 11.4.2). In the study of unitary groups, one focus in the Hermitian inner product as it is left invariant under group actions. Consider a pair complex vectors

$$z_{1i} = x_i + iy_i, \quad z_{2i} = z_i + iw_i \in \mathbb{C}. \tag{5.27}$$

Their Hermitian inner product is

$$< z_1 | z_2 >= z_{1}{}_i^{*} z_2{}^i = x_i z_i + y_i w_i + i(y_i z_i - x_i w_i). \tag{5.28}$$

We are looking for an analog for the 2n-D space. A real vector $\boldsymbol{x}$ in the $2n$-dimensional algebra can be mapped onto a set of complex coefficients $\{x_i\}$. For that let us first introduce the following vectors

$$x_i = \boldsymbol{x} \cdot e_i + i\boldsymbol{x} \cdot f_i, \qquad y_i = \boldsymbol{z} \cdot e_i + i\boldsymbol{w} \cdot f_i. \tag{5.29}$$

The complex inner product reads

$$< \boldsymbol{x} | \boldsymbol{y} > = x^i y_i^* = (\boldsymbol{x} \cdot e^i + i\boldsymbol{x} \cdot f^i)(\boldsymbol{z} \cdot e_i + \boldsymbol{w} \cdot f_i), \tag{5.30}$$

$$\begin{aligned}
&= \boldsymbol{x} \cdot e^i \boldsymbol{y} \cdot e_i + \boldsymbol{x} \cdot f^i \boldsymbol{y} \cdot f_i + i(\boldsymbol{x} \cdot f^i \boldsymbol{y} \cdot e_i - \boldsymbol{x} \cdot e^i \boldsymbol{y} \cdot f_i), \\
&= \boldsymbol{x} \cdot \boldsymbol{y} + i(\boldsymbol{x} \wedge \boldsymbol{y}) \cdot \boldsymbol{J}.
\end{aligned} \tag{5.31}$$

Here $\boldsymbol{x} \cdot \boldsymbol{y}$ is the real part and the imaginary part is an antisymmetric product formed by projecting the bivector $(\boldsymbol{x} \wedge \boldsymbol{y})$ onto $\boldsymbol{J}$.

In order to proof the invariance group of the Hermitian inner product,

$$(\boldsymbol{x}'\wedge\boldsymbol{y}')\cdot\boldsymbol{J}=(\boldsymbol{x}\wedge\boldsymbol{y})\cdot\boldsymbol{J}, \tag{5.32}$$

where $\boldsymbol{x}'=\boldsymbol{R}\boldsymbol{x}\tilde{\boldsymbol{R}}$ and $\boldsymbol{y}'=\boldsymbol{R}\boldsymbol{y}\tilde{\boldsymbol{R}}$. We compute

$$\begin{aligned}(\boldsymbol{x}'\wedge\boldsymbol{y}')\cdot\boldsymbol{J}&=<\boldsymbol{x}'\wedge\boldsymbol{y}'\boldsymbol{J}>=<\boldsymbol{R}\boldsymbol{x}\tilde{\boldsymbol{R}}\boldsymbol{R}\boldsymbol{y}\tilde{\boldsymbol{R}}\boldsymbol{J}>=<\boldsymbol{x}\boldsymbol{y}\tilde{\boldsymbol{R}}\boldsymbol{J}\boldsymbol{R}>,\\&=(\boldsymbol{x}\wedge\boldsymbol{y})\cdot(\tilde{\boldsymbol{R}}\boldsymbol{J}\boldsymbol{R}),\end{aligned} \tag{5.33}$$

this equation holds for all $\boldsymbol{x}$, $\boldsymbol{y}$ of 2n-D space, and thus, the following equality has to be true

$$\boldsymbol{J}=\tilde{\boldsymbol{R}}\boldsymbol{J}\boldsymbol{R}. \tag{5.34}$$

Since we are dealing with the rotor group which leaves the $\boldsymbol{J}$ invariant, this equality is satisfied. Unitary groups are constructed from reflections and rotations which leave $\boldsymbol{J}$ invariant. A reflection which satisfies the invariance constrain would require that a vector generator $\boldsymbol{u}$ fulfills

$$\boldsymbol{u}\boldsymbol{J}\boldsymbol{u}=\boldsymbol{J}. \tag{5.35}$$

This results implies that $\boldsymbol{u}\cdot\boldsymbol{J}=0$; therefore $(\boldsymbol{u}\cdot\boldsymbol{J})\cdot\boldsymbol{J}=-\boldsymbol{u}=0$. Thus, there are no vector generators of reflections and we conclude that all unitary transformations are generated by elements of the spin group. Note that we have not specified so far the underlying signature, so these computing can be applied in the same way to the unitary groups $U(n)$ and $U(p,q)$ which in turn can be represented by using even multivectors of $G_{2n,0}$ and $G_{2p,2q}$, respectively.

A rotor is given by

$$\boldsymbol{R}=e^{-\frac{\boldsymbol{B}}{2}}, \tag{5.36}$$

where the bivector generators of the unitary group satisfies

$$\boldsymbol{B}\times\boldsymbol{J}=0. \tag{5.37}$$

This defines the bivector realization of the Lie algebra of the unitary group, and it is called $u(n)$. In order to construct bivectors satisfying this relation, we resort to the Jacobi identity of Eq. (5.11) to prove that, c.f. ([74], 11.110-111).

$$[(\boldsymbol{x}\cdot\boldsymbol{J})\wedge(\boldsymbol{y}\cdot\boldsymbol{J})]\times\boldsymbol{J}=-(\boldsymbol{x}\cdot\boldsymbol{J})\wedge\boldsymbol{y}+(\boldsymbol{y}\cdot\boldsymbol{J})\wedge\boldsymbol{x}=-(\boldsymbol{x}\wedge\boldsymbol{y})\times\boldsymbol{J}, \tag{5.38}$$

this leads to

$$[\boldsymbol{x}\wedge\boldsymbol{y}+(\boldsymbol{x}\cdot\boldsymbol{J})\wedge(\boldsymbol{y}\cdot\boldsymbol{J})]\times\boldsymbol{J}=0. \tag{5.39}$$

Note that a bivector similar to the form on the left-hand side will commute with J. Now, if we try all the combinations of the pair $\{e_i, f_i\}$, we find the following Lie algebra basis for $u(n)$

$$
\begin{aligned}
\boldsymbol{J}_i &= e_i f_i && (i = 1, ..., n),\\
\boldsymbol{E}_{ij} &= e_i e_j + f_i f_j && (i < j = 1, ..., n),\\
\boldsymbol{F}_{ij} &= e_i f_j - f_i e_j && (i < j = 1, ..., n).
\end{aligned}
\tag{5.40}
$$

All the bivectors belong to the geometric algebra $G_{2n,0}$, and the vectors $\{e_i\}$ and $\{f_i\}$ form an orthonormal basis for this geometric algebra. It is easy to establish the closure of this algebra under the commutator product. The doubling bivector $\boldsymbol{J}$ belongs to this algebra, and it commutes with all other elements of this algebra. The complex structure is generated by the bivector $\boldsymbol{J} = \boldsymbol{J}_1 + \boldsymbol{J}_2 + \cdots + \boldsymbol{J}_n$. Without considering the term $\boldsymbol{J}$, this algebra defines the special unitary group $SU(n)$. Similar analysis can be carried out for the base space of different signature in order ultimately to build a bivector representation of the Lie algebra $u(p, q)$. These groups can be formulated in terms of even multivectors belonging to $G_{2n,0}$ or $G_{2p,2q}$, respectively.

5.5 Geometric Algebra of Reciprocal Null Cones

This section introduces the 2^{2n}-dimensional geometric algebra $G_{n,n}$. This geometric algebra is best understood by considering the properties of two 2^n-dimensional *reciprocal Grassmann subalgebras*. These subalgebras are generated by the vector bases of the reciprocal null cones N and $\overline{N}$. Let us start by explaining the meaning of null cones.

5.5.1 Reciprocal Null Cones

The *reciprocal null cones* N and $\overline{N}$ are real linear n-dimensional vector spaces whose vectors square to null— that is, for $\boldsymbol{x} \in N$, $\boldsymbol{x}^2$=0 and for $\overline{\boldsymbol{x}} \in \overline{N}$, $\overline{\boldsymbol{x}}^2$=0. In that regard, an associative geometric multiplication also equals zero—namely, for $\boldsymbol{x}$, $\boldsymbol{y}$, $\boldsymbol{z} \in N$,

$$
z^2 = (\boldsymbol{x} + \boldsymbol{y})^2 = (\boldsymbol{x} + \boldsymbol{y})(\boldsymbol{x} + \boldsymbol{y}) = \boldsymbol{x}^2 + \boldsymbol{xy} + \boldsymbol{yx} + \boldsymbol{y}^2 = \boldsymbol{xy} + \boldsymbol{yx} = 0. \tag{5.41}
$$

This result indicates that the symmetric part of geometric product of two vectors is simply zero. Thus, the geometric product of two vectors will be equal to the outer product:

$$\boldsymbol{xy} = \boldsymbol{x} \cdot \boldsymbol{y} + \boldsymbol{x} \wedge \boldsymbol{y} = \frac{1}{2}(\boldsymbol{xy} + \boldsymbol{yx}) + \frac{1}{2}(\boldsymbol{xy} - \boldsymbol{yx}) = \frac{1}{2}(\boldsymbol{xy} - \boldsymbol{yx}) = \boldsymbol{x} \wedge \boldsymbol{y}. \tag{5.42}$$

Similarly, for the vectors in the reciprocal null cone,

$$\overline{\boldsymbol{xy}} = \overline{\boldsymbol{x}} \cdot \overline{\boldsymbol{y}} + \overline{\boldsymbol{x}} \wedge \overline{\boldsymbol{y}} = \frac{1}{2}(\overline{\boldsymbol{xy}} + \overline{\boldsymbol{yx}}) + \frac{1}{2}(\overline{\boldsymbol{xy}} - \overline{\boldsymbol{yx}}) = \frac{1}{2}(\overline{\boldsymbol{xy}} - \overline{\boldsymbol{yx}}) = \overline{\boldsymbol{x}} \wedge \overline{\boldsymbol{y}}. \tag{5.43}$$

The *reciprocal vector bases* $\{w\}$ and $\{\overline{w}\}$ span the reciprocal null cones:

$$N = span\{w_1, ..., w_n\} \qquad \overline{N} = span\{\overline{w}_1, ..., \overline{w}_n\}. \tag{5.44}$$

The null spaces N or to the reciprocal and $\overline{N}$ have the following pseudoscalars

$$W_n = w_1 w_2 \ldots w_n \qquad \bar{W}_n = \bar{w}_1 \bar{w}_2 \ldots \bar{w}_n. \tag{5.45}$$

The *neutral pseudo-Euclidean space* $\mathbb{R}^{n,n}$ is the linear space spanned by the null cones:

$$\mathbb{R}^{n,n} = span\{N, \overline{N}\} = \{\boldsymbol{x} + \overline{\boldsymbol{x}} | \boldsymbol{x} \in N, \overline{\boldsymbol{x}} \in \overline{N}\}. \tag{5.46}$$

The basis $\{w_i, \overline{w}_i\}$ is called a *Witt* basis in the theory of the quadratic forms. The reciprocal vector bases satisfy the following defining *inner product* relations:

$$w_i \cdot \overline{w}_j = \overline{w}_j \cdot w_i = \delta_{ij}, \tag{5.47}$$

and according the symmetric relation of the vector inner product, it follows

$$w_i \overline{w}_j + \overline{w}_j w_i = \delta_{ij}, \tag{5.48}$$

also

$$w_i \cdot w_j = 0 \qquad \overline{w}_i \cdot \overline{w}_j = 0, \tag{5.49}$$

for all $i, j = 1, 2, ..., n$. The inner product relations of Eqs. (5.47) and (5.49) tell us that the reciprocal basis vectors w_i and $\overline{w}_i$ are all null vectors and are mutually orthogonal. Because of Eq. (5.47), the reciprocal vector bases are said to be dual, because they satisfy the relationship $\{w\} \cdot \{\overline{w}\}$=id, where id stands for the identity.

The outer product defining relations are

$$w_i \wedge w_j = w_i w_j = -w_j w_i \qquad \overline{w}_i \wedge \overline{w}_j = \overline{w}_i \overline{w}_j = -\overline{w}_j \overline{w}_i. \tag{5.50}$$

Note that the geometric product of two vectors equals the outer product only when both belong either to the null cone N or to the reciprocal cone $\overline{N}$. This is no longer true for arbitrary $\boldsymbol{x}, \boldsymbol{y} \in \mathbb{R}^{n,n}$ owing to the dual relationship of the reciprocal bases expressed by Eqs. (5.49) and (5.50).

5.5.2 The Universal Geometric Algebra $G_{n,n}$

The vector bases of the reciprocal null cones N and $\overline{N}$ generate the 2^n-dimensional subalgebra $G_N = gen\{w_1, ..., w_n\}$ and the 2^n-dimensional reciprocal subalgebra $G_{\overline{N}} = gen\{\overline{w}_1, ..., \overline{w}_n\}$, which have the structure of Grassmann algebras.

The geometric algebra $G_{\bar{n},\bar{n}}$ is built by the direct product of these 2^n-dimensional Grassmann subalgebras:

$$G_{n,n} = G_N \otimes G_{\overline{N}} = gen\{w_1, ..., w_n, \overline{w}_1, ..., \overline{w}_n\}. \tag{5.51}$$

When n is countably infinite, we call $G_{\infty,\infty}$ the *universal geometric algebra*. The universal algebra $G_{\infty,\infty}$ contains all of the algebras $G_{n,n}$ as subalgebras.

The reciprocal bases $\{w\} \in N$ and $\{\bar{w}\} \in \bar{N}$ are also called *dual*, because they fulfill Eq. (5.47)—$\{w\} \cdot \{\bar{w}\}$=id. They generate the k-vector bases of $G_{\bar{n},\bar{n}}$,

$$\Big\{1, \{w\}, \{\bar{w}\}, \{w_i w_j\}, \{\bar{w}_i \bar{w}_j\}, \{w_i \bar{w}_j\}, ..., \{w_{j_1} ... \bar{w}_{j_i} \bar{w}_{j_l} ... w_{j_k}\}, I, \bar{I}\Big\}, \tag{5.52}$$

consisting of scalars, vectors, bivectors, trivectors, ... , and the dual pseudoscalars $I = w_1 \wedge w_2 \wedge w_3 ... \wedge w_n$ and $\bar{I} = \bar{w}_1 \wedge \bar{w}_2 \wedge \bar{w}_3 ... \wedge \bar{w}_n$ which satisfy $I\bar{I} = 1$. Note that the $\begin{bmatrix} 2n \\ k \end{bmatrix}$-dimensional bases of k-vectors $\{w_{j_1} ... w_{j_i} \bar{w}_{j_l} ... w_{j_k}\}$ for the $\begin{bmatrix} 2n \\ k \end{bmatrix}$ sets of indices $1 \leq j_1 < j_2 < \cdots < j_k \leq 2n$ are generated by different combinations of w's and $\bar{w}$'s.

5.5.3 Representations and Operations Using Bivector Matrices

In this subsection, we use a notation which extends the familiar addition and multiplication of matrices of real numbers to matrices consisting of vectors and bivectors. The notation is somewhat similar to the Einstein summation convention of tensor calculus, and it can be used to directly express the close relationships that exist between Clifford algebra and matrix algebra [279].

We begin by writing the *Witt basis* of null vector $\{w\}$ and the corresponding reciprocal basis $\{\bar{w}\}$ of $G_{\bar{n},\bar{n}}$ in *row form* and *column form*, respectively:

$$\{w\} = [w_1 w_2 \ldots w_n], \qquad \{\bar{w}\} = \begin{bmatrix} \bar{w}_1 \\ \bar{w}_2 \\ \cdot \\ \cdot \\ \cdot \\ \bar{w}_n \end{bmatrix}. \tag{5.53}$$

Taking advantage of the usual matrix multiplication between a *row* and a *column* and the properties of the geometric product, we get

$$\{\bar{w}\}\{w\} = \{\bar{w}\} \cdot \{w\} + \{\bar{w}\} \wedge \{w\} = \mathrm{I} + \begin{bmatrix} \bar{w}_1 \wedge w_1 & \bar{w}_1 \wedge w_2 & \ldots & \bar{w}_1 \wedge w_n \\ \bar{w}_2 \wedge w_1 & \bar{w}_2 \wedge w_2 & \ldots & \bar{w}_2 \wedge w_n \\ \ldots & \ldots & \ldots & \ldots \\ \ldots & \ldots & \ldots & \ldots \\ \bar{w}_n \wedge w_1 & \bar{w}_n \wedge w_2 & \ldots & \bar{w}_n \wedge w_n \end{bmatrix}, \tag{5.54}$$

where I is the $n \times n$ identity matrix. Similarly,

$$\{w\}\{\bar{w}\} = \{w\} \cdot \{\bar{w}\} + \{w\} \wedge \{\bar{w}\} = \sum_{i=1}^{n} w_i \cdot \bar{w}_i + \sum_{i=1}^{n} w_i \wedge \bar{w}_i, \tag{5.55}$$

$$= n + \sum_{i=1}^{n} w_i \wedge \bar{w}_i. \tag{5.56}$$

In terms of the null cone bases, a vector $\boldsymbol{x} \in N$ is given by

$$\boldsymbol{x} = \{w\}\boldsymbol{x}_{\{w\}} = (w_1, w_2, \ldots, w_n) \begin{bmatrix} x_1 \\ x_2 \\ \cdot \\ \cdot \\ \cdot \\ x_n \end{bmatrix}, \tag{5.57}$$

$$= (w_1, w_2, \ldots, w_n) \begin{bmatrix} \overline{w}_1 \cdot \boldsymbol{x} \\ \overline{w}_2 \cdot \boldsymbol{x} \\ \cdot \\ \cdot \\ \cdot \\ \overline{w}_n \cdot \boldsymbol{x} \end{bmatrix} = \{w\}(\{\overline{w}\} \cdot \boldsymbol{x}) = \sum_{i=1}^{n} x_i w_i. \tag{5.58}$$

The vectors $\boldsymbol{x} \in N$ behave like column vectors, and the vectors $\overline{\boldsymbol{x}} \in N^*$ like row vectors. This property makes it possible to define the transpose of the vector $\boldsymbol{x}$ as follows:

$$\boldsymbol{x}^T = (\{w\}x_{\{w\}})^T = x_w^T\{\overline{w}\} = (x_1 x_2 \ldots x_n) \begin{bmatrix} \overline{w}_1 \\ \overline{w}_2 \\ \cdot \\ \cdot \\ \overline{w}_n \end{bmatrix}. \tag{5.59}$$

Note that using the transpose operation it is possible to move between the null cones N and N^*.

5.5.4 Bivector Representation of Linear Operators

One important application of the bivector matrix representation is the representation of a linear operator $f \in End(\mathcal{N})$. Recalling the basic Clifford algebra identity

$$(\boldsymbol{a} \wedge \boldsymbol{b}) \cdot \boldsymbol{x} = (\boldsymbol{b} \cdot \boldsymbol{x})\boldsymbol{a} - (\boldsymbol{a} \cdot \boldsymbol{x})\boldsymbol{b}, \tag{5.60}$$

between the bivector $\boldsymbol{a} \wedge \boldsymbol{b}$ and the vector $\boldsymbol{x}$, we can use the reciprocal bases to express the vector $\boldsymbol{x}$ in the form

$$\boldsymbol{x} = \{w\}\boldsymbol{x}_{\{w\}} = (\{w\} \wedge \{\bar{w}\}) \cdot \boldsymbol{x}, \tag{5.61}$$

where $\boldsymbol{x}_{\{w\}}$ represents the *column vector* components of $\boldsymbol{x}$:

$$\boldsymbol{x}_{\{w\}} = \begin{bmatrix} x_1 \\ x_2 \\ \cdot \\ \cdot \\ x_n \end{bmatrix} = \begin{bmatrix} \bar{w}_1 \cdot \boldsymbol{x} \\ \bar{w}_2 \cdot \boldsymbol{x} \\ \cdot \\ \cdot \\ \bar{w}_n \cdot \boldsymbol{x} \end{bmatrix} = \{\bar{w}\} \cdot \boldsymbol{x}. \tag{5.62}$$

This is the *key idea* to the bivector representation of a linear operator. Let $f \in End(\mathcal{N})$, and we then have the following relationships:

$$\begin{aligned} f(\boldsymbol{x}) &= f(\{w\}\boldsymbol{x}_{\{w\}}) = \{w\}\mathcal{F}\boldsymbol{x}_{\{w\}} = \Big((\{w\}\mathcal{F}) \wedge \{\bar{w}\}\Big) \cdot \boldsymbol{x}, \\ &= F \cdot \boldsymbol{x}, \end{aligned} \tag{5.63}$$

where the bivector $F \in G$ is defined by

$$F = (\{w\}\mathcal{F})\wedge\{\bar{w}\} = \sum_{i=1}^{n}\sum_{j=1}^{n} f_{ij} w_i \wedge \bar{w}_j = \{w\}\wedge(\mathcal{F}\{\bar{w}\}). \tag{5.64}$$

Thus, a linear operator $f \in End(\mathcal{N})$ can now be pictured as a linear mapping $f : \mathcal{N} \to \mathcal{N}$ of the null cone $\mathcal{N}$ onto itself. Furthermore, it can be represented in the bivector form $f(\boldsymbol{x}) = F \cdot \boldsymbol{x}$, where $F = (\{w\}\mathcal{F})\wedge\{\bar{w}\}$ is a bivector in the enveloping geometric algebra G. As an example, we can show for the linear operator T its representation as a bivector matrix:

$$T = \begin{bmatrix} t_{11} & t_{12} & t_{13} & t_{14} \\ t_{21} & t_{22} & t_{23} & t_{24} \\ t_{31} & t_{32} & t_{33} & t_{34} \\ t_{41} & t_{42} & t_{43} & t_{44} \end{bmatrix}, \tag{5.65}$$

$$\to \begin{bmatrix} t_{11} w_1 \wedge \bar{w}_1 & t_{12} w_1 \wedge \bar{w}_2 & t_{13} w_1 \wedge \bar{w}_3 & t_{14} w_1 \wedge \bar{w}_4 \\ t_{21} w_2 \wedge \bar{w}_1 & t_{22} w_2 \wedge \bar{w}_2 & t_{23} w_2 \wedge \bar{w}_3 & t_{24} w_2 \wedge \bar{w}_4 \\ t_{31} w_3 \wedge \bar{w}_1 & t_{32} w_3 \wedge \bar{w}_2 & t_{33} w_3 \wedge \bar{w}_3 & t_{34} w_3 \wedge \bar{w}_4 \\ t_{41} w_4 \wedge \bar{w}_1 & t_{42} w_4 \wedge \bar{w}_2 & t_{43} w_4 \wedge \bar{w}_3 & t_{44} w_4 \wedge \bar{w}_4 \end{bmatrix}. \tag{5.66}$$

Now, by considering $f, g \in gl(\mathcal{N})$ in the bivector form $f(\boldsymbol{x}) = F \cdot \boldsymbol{x}$ and $g(\boldsymbol{x}) = G \cdot \boldsymbol{x}$, and by calculating the commutator $[f, g]$, we find

$$[f, g](\boldsymbol{x}) = F \cdot (G \cdot \boldsymbol{x}) - G \cdot (F \cdot \boldsymbol{x}) = (F \times G) \cdot \boldsymbol{x}, \tag{5.67}$$

where the *commutator product of bivectors* $F \times G$ is defined by $F \times G = \frac{1}{2}[FG - GF]$. Thus, the Lie bracket of the linear operators f and g becomes the commutator product of their respective bivectors F and G.

5.5.5 The Standard Bases of $G_{n,n}$

In the previous subsections, we have discussed in some detail the *orthogonal null vector bases* $\{w_i, \bar{w}_i\}$ of the geometric algebra $G_{\bar{n},\bar{n}}$. Using these orthogonal null vector bases, we can construct new *orthonormal* bases $\{e, \overline{e}\}$ of $G_{n,n}$,

$$\begin{aligned} e_i &= (w_i + \bar{w}_i), \\ \overline{e}_i &= (w_i - \bar{w}_i), \end{aligned} \tag{5.68}$$

for i=1, 2, 3, ... , n. According to the properties of Eqs. (5.47)–(5.50), these bases vectors for $i \neq j$ satisfy

$$
\begin{aligned}
&e_i^2 = 1, \ \bar{e}_i^2 = -1, \\
&e_i \cdot e_j = \delta_{i,j}, \ \bar{e}_i \cdot \bar{e}_j = -\delta_{i,j}, \\
&e_i e_j = -e_j e_i, \ \bar{e}_i \bar{e}_j = -\bar{e}_j \bar{e}_i, \\
&e_i \cdot \bar{e}_j = 0, \\
&e_i \bar{e}_j = -e_j \cdot \bar{e}_i.
\end{aligned}
\tag{5.69}
$$

The orthonormal vector basis $\{e_i\}$ spans a real Euclidean vector space $\mathbb{R}^n$, while the reciprocal orthonormal basis $\{\bar{e}_i\}$ spans the *anti*-Euclidean space $\overline{\mathbb{R}}^n$; hence, $\mathbb{R}^{n,n}$ can be expressed as the direct product of the Euclidean and anti-Euclidean spaces

$$
\mathbb{R}^{n,n} = \mathbb{R}^n \otimes \overline{\mathbb{R}}^n. \tag{5.70}
$$

The bases $\{e_i\}$ generate the geometric subalgebra $G_{n,0}$, whereas $\{\bar{e}_i\}$ generates the geometric subalgebra $G_{0,n}$; thus, the geometric algebra $G_{n,n}$ is generated in terms of the bases $\{e_i\}$ and $\{\bar{e}_i\}$

$$
G_{n,n} = gen\{e_1, ..., e_n, \bar{e}_1, ..., \bar{e}_n\}, \tag{5.71}
$$

where the generated k-vectors are

$$
\Big\{1, \{e\}, \{\bar{e}\}, \{e_i e_j\}, \{\bar{e}_i \bar{e}_j\}, \{e_i \bar{e}_j\}, \{e_i e_j \bar{e}_k\}, ..., \{e_{j_1} ... \bar{e}_{j_i} \bar{e}_{j_l} \dots e_{j_k}\}, I, \bar{I}\Big\}, \tag{5.72}
$$

consisting of scalars, vectors, bivectors, trivectors, ..., and the dual pseudoscalars $I = e_1 \wedge e_2 \wedge e_3 ... \wedge e_n$ and $\bar{I} = \bar{e}_1 \wedge \bar{e}_2 \wedge \bar{e}_3 ... \wedge \bar{e}_n$. Note that the $\begin{bmatrix} 2n \\ k \end{bmatrix}$-dimensional bases of k-vectors $\{e_{j_1} ... \bar{e}_{j_i} \bar{e}_{j_l} \dots e_{j_k}\}$ for the $\begin{bmatrix} 2n \\ k \end{bmatrix}$ sets of indices $1 \le j_1 < j_2 < \dots < j_k \le 2n$ are generated by different combinations of e's and $\bar{e}$'s.

Using the basis $\{e_i, \bar{e}_i\}$, one can construct $(p+q)$-blades

$$
E_{p,q} = E_p \bar{E}^\dagger = E_p \wedge \bar{E}^\dagger, \tag{5.73}
$$

where $E_p = e_1 e_2 ... e_p = E_{p,0}$, $\bar{E}_p = \bar{e}_1 \bar{e}_2 ... \bar{e}_p = E_{0,p}$ and $(\cdot)^\dagger$ stands for the reversion operation. Note the each blade determines a projection $\underline{E} \in \mathbb{R}^{n,n}$ into a $(p+q)$-dimensional subspace $\mathbb{R}^{p,q}$ defined by, c.f. ([72], 3.20-24),

$$
\underline{E}_{p,q}(\boldsymbol{x}) = (\boldsymbol{x} \cdot E_{p,q}) E_{p,q}^{-1} = \frac{1}{2}[\boldsymbol{x} - (-1)^{p+q} E_{p,q} E_{p,q} \boldsymbol{x} E_{p,q}^{-1}]. \tag{5.74}
$$

We use an underbar to distinguish linear operators from elements of the geometric algebra; in order to designate the operator by a multivector as in Eq. 5.74, the operator $\underline{E}_{p,q}$ is determined by the blade $E_{p,q}$.

The vector $\boldsymbol{x}$ is in the space $\mathbb{R}^{n,n}$ if and only if the wedge product is null

$$\boldsymbol{x} \wedge E_{p,q} = \boldsymbol{x}E_{p,q} + (-1)^{p+q}E_{p,q}\boldsymbol{x} = 0. \tag{5.75}$$

For the case of $q = 0$, Eq. (5.74) reads

$$\underline{E}_n(\boldsymbol{x}) = \frac{1}{2}(\boldsymbol{x} + \boldsymbol{x}^*), \tag{5.76}$$

and the involution of $\boldsymbol{x}$

$$\boldsymbol{x}^* = (-1)^{(n+1)} E_n \boldsymbol{x} E_n^{-1}. \tag{5.77}$$

5.5.6 Endomorphisms of $G_{n,n}$

This subsection is based on c.f. ([72], Sect. V). The universal algebra is a proper framework to handle the theory of linear transformations and Lie groups. This arises naturally as the endomorphism algebra $\mathcal{E}nd\,(G_n)$ which is the algebra of linear maps of the Euclidean geometric algebra G_n onto itself. This algebra is isomorphic to the algebra of real $2^n \times 2^n$ matrices, i.e.,

$$\mathcal{E}nd\,(G_n) \simeq M\,(2n, \mathbb{R}). \tag{5.78}$$

For an arbitrary multivector $\boldsymbol{M} \in G_n$, left and right multiplication by the basis vectors e_i determine endomorphisms of G_n defined as follows

$$\begin{aligned} \underline{e}_i &: \boldsymbol{M} \rightarrow e_i(\boldsymbol{M}) = \underline{e}_i\boldsymbol{M}, \\ \underline{\bar{e}}_i &: \boldsymbol{M} \rightarrow \underline{\bar{e}}_i(\boldsymbol{M}) = \overline{\boldsymbol{M}}e_i, \end{aligned} \tag{5.79}$$

where the overbar stands for main involution of G_n,

$$\begin{aligned} \bar{e}_i &= -e_i, \\ \overline{\boldsymbol{M}_1\boldsymbol{M}_2} &= \overline{\boldsymbol{M}}_1\overline{\boldsymbol{M}}_2. \end{aligned} \tag{5.80}$$

The operators $\underline{e}_i$ are linear, and they satisfy the following

$$\begin{aligned} \underline{e}_i\underline{e}_j + \underline{e}_j\underline{e}_i &= 2\delta_{ij}, \\ \underline{\bar{e}}_i\underline{\bar{e}}_j + \underline{\bar{e}}_j\underline{\bar{e}}_i &= -2\delta_{ij}, \\ \underline{e}_i\underline{\bar{e}}_j + \underline{\bar{e}}_j\underline{e}_i &= 0. \end{aligned} \tag{5.81}$$

Considering the inner product definition of G_n, these relations are isomorphic to the relations (5.69) for the vector basis $\{e_i, \bar{e}_i\} \in G_n$, and this establishes the algebra isomorphism

$$G_{n,n} \simeq \mathcal{E}nd(G_n). \tag{5.82}$$

The equations to define this isomorphism were first formulated by F. Sommen and N. Van Acker [282] and included in [72].

Next, let us show that the universal algebra contains a $2n$-dimensional subspace $\mathcal{S}_n$ on which all the endomorphisms G_n are faithfully represented by left multiplication with elements belonging to $G_{n,n}$. The space $\mathcal{S}_n$ is known as *spinor space*, and its elements are called *spinors*. $\mathcal{S}_n$ is a *minimal left ideal* of $G_{n,n}$. Its construction can be described after establishing the necessary algebraic relations.

The spinor space is generated by multiplicating by the pseudoscalar I from the left by the entire mother algebra, as follows c.f. ([72], 5.21),

$$\mathcal{S}_n = G_{n,n} I. \tag{5.83}$$

As defined by Eq. (5.73), $E_{n,n} = E_n \bar{E}_n^{\dagger}$ is the unit pseudoscalar for $G_{n,n}$. Let us study linear transformations on $G_{n,n}$ and on its subspaces, particularly orthogonal transformations. The orthogonal transformation $\underline{R}$ is defined if it leaves the inner product invariant, namely

$$(\underline{R}\boldsymbol{x}) \cdot (\underline{R}\boldsymbol{y}) = \boldsymbol{x} \cdot \boldsymbol{y}. \tag{5.84}$$

This linear transformation is called a *rotation* if *det* $\underline{R}$ =1, i.e.,

$$\underline{R}(E_{n,n}) = E_{n,n}. \tag{5.85}$$

The rotations form a group known as the *special orthogonal group* SO(n, n).

From the standard basis (5.68), the null basis vectors are

$$w_i = \frac{1}{2}(e_i + \bar{e}_i), \qquad \bar{w}_i = \frac{1}{2}(e_i - \bar{e}_i). \tag{5.86}$$

Using these null vectors, one constructs the commuting vectors I_i, which can be written as

$$I_i = \bar{w}_i w_i = (1 + e_i \bar{e}_i) = (1 + \boldsymbol{K}_i) = e_i w_i = \bar{w}_i e_i = w_i \bar{e}_i = -\bar{e}_i w_i, \tag{5.87}$$

for i=1,2,...,n. The idempotent and commutativity properties of I_i and I are

$$\begin{aligned} I_i^2 &= I_i, \qquad I_i \times I_j = 0 \\ I^2 &= I. \end{aligned} \tag{5.88}$$

Multiplying I_i from the left or right by e_i, we get the next relations

$$\begin{aligned} e_i I_i &= \bar{e}_i I_i = w_i I_i = w_i = w_i \bar{w}_i w_i, \\ I_i e_i &= -I_i \bar{e}_i = I_i \bar{w}_i = \bar{w}_i = \bar{w}_i w_i \bar{w}_i. \end{aligned} \tag{5.89}$$

Using the I_i, one constructs the *universal idempotent*, c.f. ([72], 5.14),

$$I = I_1 I_2 \dots I_n = \overline{W}_n^{\dagger} W_n = \overline{W}_n W_n^{\dagger}, \tag{5.90}$$

where W_n is the pseudoscalar of the null space N.

Since $e_i I = \bar{e}_i I = w_i I$ and $E_{n,n} I = I$, where the unit pseudoscalar $E_{n,n} = E_n \bar{E}_n^{\dagger}$ for $G_{n,n}$, it follows, , c.f. ([72], 5.22),

$$E_{n,n} e_i I = -e_i I = E_{n,n} \bar{e}_i I = -\bar{e}_i I. \tag{5.91}$$

Thus, multiplication of $\mathcal{S}$ by $E_{n,n}$ corresponds to the main involution in G_n, written as

$$E_{n,n} \mathcal{S}_n \Longleftrightarrow \overline{G}_n. \tag{5.92}$$

Considering the relations of Eq. (5.81) and the definitions of (5.79), it follows

$$\begin{aligned} e_i \mathcal{S}_n &\longleftrightarrow e_i G_n, \\ \bar{e}_i \mathcal{S}_n &\longleftrightarrow \overline{G}_n e_i, \end{aligned} \tag{5.93}$$

where the overbar stands for the main involution, and using (5.91), one gets

$$\bar{e}_i E_{n,n} \mathcal{S}_n \longleftrightarrow G_n e_i. \tag{5.94}$$

One can easily prove that the reversion in $G_{n,n,}$ corresponds to the reversion in G_n.

All these equations determine the interpretation of the spinor space operators as G_n endomorphisms.

5.5.7 The Bivectors of $G_{n,n}$

Let $\boldsymbol{K}$ be any bivector in $\bigwedge^2(\mathbb{R}^{n,n})$ which can be expressed as sum of n commuting blades with unit square, c.f. ([72], 3.25),

$$\boldsymbol{K} = \sum_{i=0}^{n} \boldsymbol{K}_i, \tag{5.95}$$

where

$$K_i \times K_j = 0, \tag{5.96}$$
$$K_i^2 = 1. \tag{5.97}$$

The bivector $\boldsymbol{K}$ determines an automorphism of $\mathbb{R}^{n,n}$

$$\underline{K} : \boldsymbol{x} \to \bar{\boldsymbol{x}} = \underline{K}\boldsymbol{x} = \boldsymbol{x} \times \boldsymbol{K} = \boldsymbol{x} \cdot \boldsymbol{K}. \tag{5.98}$$

This is a mapping of a vector $\boldsymbol{x}$ into a vector $\bar{\boldsymbol{x}}$, which is called the complement of $\boldsymbol{x}$ with respect to $\boldsymbol{K}$. It is easy to prove that $\underline{K}\bar{\boldsymbol{x}} = \underline{K}^2\boldsymbol{x} = \boldsymbol{x}$ or as an operator equation $\underline{K}^2 = 1$; therefore, $\underline{K}$ is an involution. In addition

$$\boldsymbol{x} \cdot \bar{\boldsymbol{x}} = 0,$$
$$\boldsymbol{x}_\pm = \boldsymbol{x} \pm \bar{\boldsymbol{x}} = \boldsymbol{x} \pm \boldsymbol{x} \cdot \boldsymbol{K}, \tag{5.99}$$

are null vectors. As matter of fact, the sets $\{\boldsymbol{x}_+\}$ and $\{\boldsymbol{x}_-\}$ of all such vectors are dual n-dimensional vector spaces; thus, $\boldsymbol{K}$ determines the desired null space decomposition of the form (5.70) without referring to a vector basis.

A suitable bivector $\boldsymbol{K}$ can be computed using for the $\boldsymbol{K}_i$ the basis $\{e_i, \bar{e}_i\}$

$$K_i = e_i\bar{e}_i = e_i \wedge \bar{e}_i. \tag{5.100}$$

This bivector $\boldsymbol{K}$ has the following properties

$$e_i \times \boldsymbol{K} = e_i \cdot \boldsymbol{K} = e_i(e_j \wedge \bar{e}_j) = e_i \cdot (e_j \wedge \bar{e}_j) = \bar{e}_i,$$
$$\bar{e}_i \times \boldsymbol{K} = \bar{e}_i \cdot \boldsymbol{K} = \bar{e}_i(e_j \wedge \bar{e}_j) = -\bar{e}_i \cdot \bar{e}_j e_i = e_i, \tag{5.101}$$

thus

$$(\boldsymbol{x} \cdot \boldsymbol{K}) \cdot \boldsymbol{K} = \boldsymbol{K} \cdot (\boldsymbol{K} \cdot \boldsymbol{x}) = \boldsymbol{x} \qquad \forall \boldsymbol{x}. \tag{5.102}$$

The difference between the doubling bivector $\boldsymbol{J}$ of Sect. 5.4 and $\boldsymbol{K}$ is crucial, because $\boldsymbol{J}$ generate a complex structure and in contrast $\boldsymbol{K}$ generates instead a *null* structure. In order to see clearly this, let us take any vector $\boldsymbol{x} \in G_{n,n}$ and define

$$\boldsymbol{x}_\pm = \boldsymbol{x} \pm \boldsymbol{x} \cdot \boldsymbol{K}, \tag{5.103}$$

taking its square

$$\boldsymbol{x}_\pm^2 = \boldsymbol{x}^2 \pm 2\boldsymbol{x} \cdot (\boldsymbol{x} \cdot \boldsymbol{K}) + (\boldsymbol{x} \cdot \boldsymbol{K})^2 = \boldsymbol{x}^2 - < \boldsymbol{x} \cdot \boldsymbol{K}\boldsymbol{K}\boldsymbol{x} >,$$
$$= \boldsymbol{x}^2 - [(\boldsymbol{x} \cdot \boldsymbol{K}) \cdot \boldsymbol{K}] \cdot \boldsymbol{x} = \boldsymbol{x}^2 - \boldsymbol{x}^2 = 0, \tag{5.104}$$

as expected by a null vector. Recalling the concept of projective split of Sect. 2.3.3, in a bit similar fashion here the bivector $\boldsymbol{K}$ splits vectors in $\boldsymbol{x} \in G^1_{n,n}$ into two separate null vectors, c.f. ([74], 11.118)

$$\boldsymbol{x} = \boldsymbol{x}_+ + \boldsymbol{x}_-, \tag{5.105}$$

where

$$\boldsymbol{x}_+ = \frac{1}{2}(x + \boldsymbol{x} \cdot \boldsymbol{K}), \qquad \boldsymbol{x}_- = \frac{1}{2}(x - \boldsymbol{x} \cdot \boldsymbol{K}). \tag{5.106}$$

We see that the space of vectors $G^1_{n,n}$ decomposes into a direct sum of two null spaces: N, N^*, where $\boldsymbol{x}_+ \in N$. The vectors in N fulfill, c.f. ([74], 11.119),

$$\boldsymbol{x}_+ \cdot \boldsymbol{K} = \boldsymbol{x}_+ \qquad \forall \boldsymbol{x}_+ \in N. \tag{5.107}$$

According to Eq. (5.104), we can see that all vectors $\boldsymbol{x}_+ \in N$ also square to zero. The space N defines a *Grassmann algebra*.

5.5.8 Mappings Acting on the Null Space

Every linear functions $\boldsymbol{x} \rightarrow \mathtt{f}(\boldsymbol{x})$ acting on an n-dimensional vector space can be represented in the null space N using an operator $\boldsymbol{O}$ as follows

$$x_+ \rightarrow \boldsymbol{O}\boldsymbol{x}_+\boldsymbol{O}^{-1}, \tag{5.108}$$

where $\boldsymbol{O}$ is built by the geometric product on an *even* number of unit vectors. Here vectors $\boldsymbol{x} \in G_n$ are mapped to null vectors $\boldsymbol{x}_+$ in $G_{n,n}$ which in turn are acted on by the multivector $\boldsymbol{O}$ so that the following is true

$$\mathtt{f}(\boldsymbol{x}) + \mathtt{f}(\boldsymbol{x}) \cdot \boldsymbol{K} = \boldsymbol{O}(\boldsymbol{x} + \boldsymbol{x} \cdot \boldsymbol{K})\boldsymbol{O}^{-1}. \tag{5.109}$$

This equation defines a map between linear functions $\mathtt{f}(\boldsymbol{x})$ and multivectors $\boldsymbol{O} \in G_{n,n}$. Note that the map is not quite an isomorphism, due to the fact that both $\boldsymbol{O}$ and -$\boldsymbol{O}$ generate the same function; i.e., $\boldsymbol{O}$ forms a $double - cover$ representation. The map will work, only if the action of $\boldsymbol{O}$ does not leave the space N, for that $\boldsymbol{O}$ must fulfill Eq. (5.107)

$$(\boldsymbol{O}\boldsymbol{x}_+\boldsymbol{O}^{-1}) \cdot \boldsymbol{K} = \boldsymbol{O}\boldsymbol{x}_+\boldsymbol{O}^{-1}. \tag{5.110}$$

After simple algebraic manipulation involving inner product between vectors and bivectors, c.f. ([74], 11.123-124),

$$
\begin{aligned}
\boldsymbol{x}_{+} &= \boldsymbol{O}^{-1}(\boldsymbol{O}\boldsymbol{x}_{+}\boldsymbol{O}^{-1}) \cdot \boldsymbol{K}\boldsymbol{O}, \\
&= \boldsymbol{O}^{-1}\frac{1}{2}(\boldsymbol{O}\boldsymbol{x}_{+}\boldsymbol{O}^{-1}\boldsymbol{K} - \boldsymbol{K}\boldsymbol{O}\boldsymbol{x}_{+}\boldsymbol{O}^{-1})\boldsymbol{O}, \\
&= \boldsymbol{x}_{+} \cdot (\boldsymbol{O}^{-1}\boldsymbol{K}\boldsymbol{O}). \qquad (5.111)
\end{aligned}
$$

In this equation, according to Eq. (5.107), it follows that it is required $\boldsymbol{O}^{-1}\boldsymbol{K}\boldsymbol{O} = \boldsymbol{K}$ or

$$
\boldsymbol{O}\boldsymbol{K} = \boldsymbol{K}\boldsymbol{O}. \qquad (5.112)
$$

Since $\boldsymbol{O}$ is built by a product of even number of unit vectors, it follows $\boldsymbol{O}\boldsymbol{O}^{-1} = \pm 1$. The subgroup which fulfills $\boldsymbol{O}\boldsymbol{O}^{-1} = 1$ are rotors in $G_{n,n}$ and their generators (Lie algebra basis elements) are bivectors. Thus, the condition $\boldsymbol{R}\boldsymbol{K}\tilde{\boldsymbol{R}} = \boldsymbol{K}$ is the direct analog of the condition that defined the unitary group expressed in terms of rotors which leaves in this case $\boldsymbol{J}$ invariant; see Eq. (5.34).

5.5.9 The Bivector or Lie Algebra of $G_{n,n}$

The bivector or Lie algebra generators are the set of bivectors that commute with $\boldsymbol{K}$. The Jacobi identity of Eq. (5.11) guarantees that the commutator of two bivectors which commute with $\boldsymbol{K}$ yields a third which in turn also commutes with $\boldsymbol{K}$. Similar as by the case of the unitary group of Eq. (5.11), ([74], 11.125), formulate the following algebraic constraint

$$
[(\boldsymbol{x} \cdot \boldsymbol{K}) \wedge (\boldsymbol{y} \cdot \boldsymbol{K})] \times \boldsymbol{K} = \boldsymbol{x} \wedge (\boldsymbol{y} \cdot \boldsymbol{K}) + (\boldsymbol{x} \cdot \boldsymbol{K}) \wedge y = (\boldsymbol{x} \wedge \boldsymbol{y}) \times \boldsymbol{K}, \qquad (5.113)
$$

after passing the far most right element to the left of the equation

$$
[\boldsymbol{x} \wedge \boldsymbol{y} - (\boldsymbol{x} \cdot \boldsymbol{K})(\boldsymbol{y} \cdot \boldsymbol{K})] \times \boldsymbol{K} = 0. \qquad (5.114)
$$

By using this constraint, we can again try all combinations of $\{e_i, \bar{e}_i\}$ to produce the bivector basis for the Lie algebra $gl(n)$ of the general linear group $GL(n)$, c.f. ([72], 4.29a-c) and c.f. ([74], Table 11.2)

$$
\begin{aligned}
\boldsymbol{K}_i &= \frac{1}{2}\boldsymbol{F}_{ij} = e_i\bar{e}_i, && (5.115) \\
\boldsymbol{E}_{ij} &= e_ie_j - \bar{e}_i\bar{e}_j \qquad (i < j = 1...n), && (5.116) \\
\boldsymbol{F}_{ij} &= e_i\bar{e}_j - \bar{e}_ie_j \qquad (i < j = 1...n). && (5.117)
\end{aligned}
$$

The $\{e_i\}$ basis vectors are orthonormal with positive signature, and the $\{\bar{e}_i\}$ are orthonormal with negative signature and belong to the geometric algebra $G_{n,n,}$. This algebra contains the bivector $\boldsymbol{K} = \boldsymbol{K}_1 + \boldsymbol{K}_2 + \cdots + \boldsymbol{K}_n$ which generates the Abelian subgroup of global dilation. By factoring out this bivector, one gets the special Lie algebra $sl(n)$. The special linear group $SL(n)$ can be characterized as the group of volume and orientation-preserving linear transformations of $\mathbb{R}^n$.

Note that the difference in structure between the Lie algebra $gl(n)$ of the linear group $GL(n)$ and the unitary group $U(n)$ is only due to the different signatures of their underlying spaces. By the case of the conformal geometric algebras presented in Chap. 8, the Lie algebras are related to a space with pseudo-Euclidean signature.

5.6 The General Linear Group as Spin Group

This section is based on the work of D.Hestenes and G. Sobzyk [138] Chap. 8, a late article of Ch. Doran, D. Hestenes, F. Sommen and Van Hacker [72] Sect. IV, and the C. Doran's PhD thesis [73].

As shown in Sect. 5.3.1, every rotor can be expressed in an exponential form

$$\boldsymbol{R} = e^{-\frac{1}{2}\boldsymbol{B}}, \tag{5.118}$$

where the bivector $\boldsymbol{B}$ is called the generator of $\boldsymbol{R}$. Every bivector determines a unique rotation. The bivectors are generators of a spin or rotation Lie group $Spin(n, n)$. The Lie algebra of $SO(n, n)$ and $Spin(n, n)$ is designed by the lower case notations $so(n, n)$. The whole set of bivectors constitutes a bivector basis which spans the linear space $\bigwedge^2 \mathbb{R}^{n,n}$ and under the commutator forms a Lie algebra. Note that every Lie algebra is a subalgebra of $so(n, n)$.

Lie groups are classified according to their invariants. For the case of the classical groups, the invariants are non-degenerated bilinear or quadratic forms. The geometric algebra framework offers us a much more simpler alternative of invariants formulated in terms of multivectors which shape the bilinear forms. According to [138], every bilinear form can be written as, c.f. ([72], 4.19)

$$\boldsymbol{x} \cdot (\underline{Q}\boldsymbol{y}) = \boldsymbol{x} \cdot (\boldsymbol{y} \cdot \underline{Q}) = (\boldsymbol{x} \wedge \boldsymbol{y}) \cdot \boldsymbol{Q}, \tag{5.119}$$

where $\boldsymbol{Q}$ is a bivector and $\underline{Q}$ is the corresponding linear operator and the form is non-degenerated if $\underline{Q}$ is non-singular, i.e., det $\underline{Q} \neq 0$ and

$$\underline{Q}^2 = \pm\underline{1}. \tag{5.120}$$

Due to the fact that $(\underline{R}\boldsymbol{x} \wedge \underline{R}\boldsymbol{y}) \cdot \boldsymbol{Q} = (\boldsymbol{x} \wedge \boldsymbol{y}) \cdot \underline{R}^{\dagger}\boldsymbol{Q}$ an invariance of Eq. (5.119), it is equivalent to the so-called *stability condition*, c.f. ([72], 4.21)

$$\underline{R}^{\dagger} \wedge \boldsymbol{Q} = \boldsymbol{R}^{\dagger}\boldsymbol{Q}\boldsymbol{R} = \boldsymbol{Q}. \tag{5.121}$$

This can explained as that the invariance group of any skew-symmetric bilinear form is the *stability group* of a bivector. According to Eq. (5.121), the generators of the stability group $G(\boldsymbol{Q})$ for $\boldsymbol{Q}$ necessarily must commute with $\boldsymbol{Q}$. In this regard, let us study the commutator of $\boldsymbol{Q}$ with an arbitrary two blade $\boldsymbol{x}\wedge\boldsymbol{y}$. The Jacobi identity, Eq. (5.11), in terms of bivectors of Eq. (5.11) can be rewritten using $\boldsymbol{x}\wedge\boldsymbol{y}$ as follows, c.f. ([72], 4.22-25)

$$(\boldsymbol{x}\wedge\boldsymbol{y}) \times \boldsymbol{Q} = (\boldsymbol{x} \times \boldsymbol{Q})\wedge\boldsymbol{y} + \boldsymbol{x}\wedge(\boldsymbol{y} \times \boldsymbol{Q}) = (\underline{Q}\boldsymbol{x})\wedge\boldsymbol{y} + \boldsymbol{x}\wedge(\underline{Q}\boldsymbol{y}), \tag{5.122}$$

which leads to

$$\begin{aligned}
[(\boldsymbol{x}\wedge\boldsymbol{y}) \times \boldsymbol{Q}] \times \boldsymbol{Q} &= [(\underline{Q}\boldsymbol{x})\wedge\boldsymbol{y} + \boldsymbol{x}\wedge(\underline{Q}\boldsymbol{y})] \times \boldsymbol{Q}, \\
&= (\underline{Q}^2\boldsymbol{x})\wedge\boldsymbol{y} + 2\underline{Q}(\boldsymbol{x}\wedge\boldsymbol{y}) + \boldsymbol{x}\wedge(\underline{Q}^2\boldsymbol{y}),
\end{aligned} \tag{5.123}$$

using the condition of Eq. (5.120); for any bivector $\boldsymbol{B}$ using linearity, we can extend Eq. (5.123) to

$$(\boldsymbol{B} \times \boldsymbol{Q}) \times \boldsymbol{Q} = 2(\underline{Q}\boldsymbol{B} \pm \boldsymbol{B}). \tag{5.124}$$

Henceforth, if $\boldsymbol{B}$ commutes with $\boldsymbol{Q}$ we obtain

$$\underline{Q}\boldsymbol{B} = \mp\boldsymbol{B}, \tag{5.125}$$

note that the signs are opposite to those of Eq. (5.120). This result tells us that the generators of $G(\boldsymbol{Q})$ are simply eigenvectors of $\underline{Q}$ with eigenvalues ∓ 1. We understand invariance under a Lie group rotation expressed as linear operator $\underline{R}$, if the isometry is preserved, c.f. ([72], 4.10-13), namely

$$(\underline{R}\boldsymbol{x}) \cdot (\underline{Q}\underline{R}\boldsymbol{y}) = \boldsymbol{x} \cdot (\underline{Q}\boldsymbol{y}), \tag{5.126}$$

$$\boldsymbol{x} \cdot (\underline{R}^{\dagger})\underline{Q}\underline{R}\boldsymbol{y}) = \boldsymbol{x} \cdot (\underline{Q}\boldsymbol{y}). \tag{5.127}$$

As an operator equation this condition reads

$$\underline{R}^{\dagger}\underline{Q}\underline{R} = \underline{Q} = \underline{R}\underline{Q}\underline{R}^{\dagger}, \tag{5.128}$$

or equivalent

$$\underline{Q}\underline{R} = \underline{R}\underline{Q}. \tag{5.129}$$

So the invariance group of the quadratic form $\underline{Q}$ consists of those rotations which commute with $\underline{Q}$. In order to illustrate, let us consider the bilinear form $\boldsymbol{x} \cdot \boldsymbol{y}^*$ determined by Eq. (5.77) for involution which distinguishes the subspaces $\mathbb{R}^n$ and $\overline{\mathbb{R}}^n$. By applying the rotation group on $E_n = e_1 e_2 \ldots e_n$, we get

$$\underline{R}E_n = \boldsymbol{R}E_n\tilde{\boldsymbol{R}} = E_n. \tag{5.130}$$

Thus, invariance of the bilinear form $\boldsymbol{x} \cdot \boldsymbol{y}^*$ is equivalent to invariance of the n-blade E_n. Consequently, one can construct a basis for the Lie algebra from the vector basis $\{e_i, \bar{e}_i\}$ for the involution operator $*$, namely

$$\begin{aligned} e_{ij} &= e_i e_j, && \texttt{for } i < j = 1, 2, ..., n, \\ \bar{e}_{kl} &= \bar{e}_k \bar{e}_l, && \texttt{for } k < l = 1, 2, ..., n. \end{aligned} \tag{5.131}$$

Any generator $\boldsymbol{B}$ in the geometric algebra can be expressed as follows, c.f. ([72], 4.10-16)

$$\boldsymbol{B} = \sum_{i<j} \alpha^{ij} e_{ij} + \sum_{i<j} \beta^{ij} \bar{e}_{ij}. \tag{5.132}$$

These Lie algebra generators form the corresponding Lie group or a rotor, c.f. ([72], 4.10-18)

$$\boldsymbol{R} = e^B = e^{\frac{1}{2}\sum_{i<j}\alpha^{ij}e_{ij} + \sum_{i<j}\beta^{ij}\bar{e}_{ij}} = e^{\frac{1}{2}\sum_{i<j}\alpha^{ij}e_{ij}} e^{\frac{1}{2}\sum_{i<j}\beta^{ij}\bar{e}_{ij}}. \tag{5.133}$$

The $\boldsymbol{R}$ is the spin representation for the product group SO(n)$\otimes$SO(n). Since it is determined by the invariance of E_n given in Eq. (5.126), $\boldsymbol{R}$ is called to be the stability group of E_n. Here, for its characterization, it is not needed a direct reference to a quadratic form.

As an illustration, we identify Q with the complementation bivector $\boldsymbol{K}$ of Eq. (5.95) and from Eq. (5.97) $\boldsymbol{K}^2 = 1$. We select an orthonormal basis which factors the components blades $\boldsymbol{K}_i$ into orthogonal factors as in Eq. (5.100). Now, using Eq. (5.101), one obtains straightforwardly the generator basis given by Eqs. (5.115)–(5.117) for the stability group of $\boldsymbol{K}$. The stability group of $\boldsymbol{K}$ can be identified as the general linear group GL(n, $\mathbb{R}$); for this proof see [72].

Eqs. (5.115)–(5.117) can be then used to represent every positive, nonsingular linear transformation in a spinor form, c.f. ([72], 4.34), namely

$$\boldsymbol{R} = e^B = e^{\frac{1}{2}\sum_{i<j}\alpha^{ij}E_{ij} + \sum_{i<j}\beta^{ij}F_{ij} + \sum_{i<j}\gamma^{ij}K_{ij}}, \tag{5.134}$$

$$= e^{\frac{1}{2}\sum_{i<j}\alpha^{ij}E_{ij}} e^{\frac{1}{2}\sum_{i<j}\beta^{ij}F_{ij}} e^{\frac{1}{2}\sum_{i<j}\gamma^{ij}K_{ij}}. \tag{5.135}$$

The Eq. (5.135) is an efficient representation of a composition of linear transformations in terms of the product of specific spinors, each one to handle certain transformation. The computation of composite rotations with a *spin representation* is much more efficient than the use of standard matrix-based methods, particularly due to the redundancy in the matrix coefficients. For example, a 3D rotation matrix has nine real coefficients, in contrast a rotor just four. Furthermore, the spinor representation applied to geometric entities conveys a clear geometric interpretation of the operation. From other perspective, for rotations, spinors in exponential form are not optimal for computational purposes; however, they are very attractive for Lie algebra analysis. Since the Lie algebra is a linear space, one can use the exponent variables for efficient algorithms in control engineering and neurocomputing.

5.6.1 The Special Linear Group SL(n, $\mathbb{R}$)

Continuing with our analysis of the group theory, one notes that $\boldsymbol{K}$ commutes with all other elements of the Lie algebra $gl(n, \mathbb{R})$; thus it generates a one dimensional invariant subgroup of $GL(n, \mathbb{R})$. We can remove $\boldsymbol{K}$ from the group by replacing the $\boldsymbol{K}_i$ in Eq. (5.115) by, c.f. ([72], 4.51)

$$\boldsymbol{H}_i = \boldsymbol{K}_i - \boldsymbol{K}_{i+1} \qquad (i = 1, 2, ..., n-1), \tag{5.136}$$

$\boldsymbol{H}_{ij}$ together with $\boldsymbol{E}_{ij}$ Eq. (5.116) and $\boldsymbol{F}_{ij}$ Eq. (5.117), and these bivectors generate the *special linear group* SL(n, $\mathbb{R}$), the subgroup of GL(n, $\mathbb{R}$) for which the determinant is unity.

Next, we compose the complementation operator $\underline{K}$ with the operator (5.77) to formulate an operator $\underline{K}_*$ defined by

$$\underline{K}_*\boldsymbol{x} = \underline{K}\boldsymbol{x}^* = (\boldsymbol{x}^*) \cdot \boldsymbol{K}. \tag{5.137}$$

Consequently

$$\underline{K}_* e_j = \bar{e}_j, \qquad \underline{K}_* \bar{e}_j = -e_j, \tag{5.138}$$

thus

$$\underline{K}_*^2 = -1. \tag{5.139}$$

In analogy with Eq. (5.125), $\underline{K}_*$ defines a new Lie algebra with generators determined by the outermorphism condition

$$\underline{K}_*(\boldsymbol{B}) = -\boldsymbol{B}. \tag{5.140}$$

Using this result, one can construct a generator basis for an invariance group of $\underline{K}_*$ as follows, c.f. ([72], 4.56-57)

$$\boldsymbol{E}_{ij} = e_i e_j - \bar{e}_i \bar{e}_j, \tag{5.141}$$
$$\boldsymbol{F}_{ij} = e_i \bar{e}_j + \bar{e}_i e_j, \tag{5.142}$$

for $i, j = 1, 2, ..., n, i < j$. These are the generators of the *complex orthogonal group* SO(n, $\mathbb{C}$). According to [72] for odd n, $\boldsymbol{K}$ is the only kind of involutory bivector. For the case when n is even, their invariants determine other groups; see [72] Sect. VI.

The general linear group GL(p, q) can be formulated and studied essentially in the same manner as the Euclidean case previously explained; i.e., E_n is replaced by $E(p, q)$ and the corresponding pseudoscalar I_n is replaced by $I_{p,q} = w_1 \ldots w_p \bar{w}_{p+1} \ldots \bar{w}_n$.

5.6.2 The Pin(n) and Spin(n) Groups

It is worth to study the endomorphism correspondence for GL(n,$\mathbb{R}$); we start with the orthogonal group. Considering any unit vectors $u, v, ..., \in G_n$, Eq. (5.93) yields the following correspondences

$$\begin{aligned} u\bar{u}S_n &\longleftrightarrow uG_{0,n}u, \\ v\bar{v}u\bar{u}S_n &\longleftrightarrow vuG_{n,0}uv. \end{aligned} \tag{5.143}$$

Eq. (5.143) is (the outermorphism of) a reflection in $G_{\bar{n}}$, and Eq. (5.143), a double reflection, is a rotation in the $u \wedge v$ plane. The generalization is straightforward, for k-vectors $u_1, u_2, ..., u_k \in G_{\bar{n}}$, it follows, c.f. ([72], 5.29-32)

$$U_{(k)} = u_k u_{k-1} \ldots u_2 u_1 \quad \text{and} \quad \overline{U} = \bar{u}_k \bar{u}_{k-1} \ldots \bar{u}_2 \bar{u}_1, \tag{5.144}$$

thus

$$\begin{aligned} U_{(k)}\overline{U}_{(k)} &= u_k \ldots u_1 \bar{u}_k \ldots \bar{u}_1 = (-1)^{k-1} u_k \ldots u_2 \bar{u}_k \ldots \bar{u}_2 u_1 \bar{u}_1 \ldots, \\ &= \epsilon u_k \bar{u}_k u_{k-1} \ldots u_2 \bar{u}_2 u_1 \bar{u}_1, \end{aligned} \tag{5.145}$$

where $\epsilon = (-1)^{\frac{1}{2}k(k-1)}$. For an odd k value, one gets

$$\epsilon_k U_{(k)}\overline{U}_{(k)} S_n \longleftrightarrow \overline{U}_{(k)} \overline{G}_n U_{(k)}^{\dagger}, \tag{5.146}$$

and for even k value,

$$\epsilon_k U_{(k)}\overline{U}_{(k)} S_n \longleftrightarrow U_{(k)} G_n U_{(k)}^{\dagger}, \tag{5.147}$$

Eqs. (5.146) and (5.147) describe the complete orthogonal group O(n) as an automorphism group of G_n.

The successive reflections using unit vectors as shown in Eq. (5.144) build a multiplicative groups in $G_{\bar{n}}$, exemplified by U_k of Eq. (5.144); this group is called the Pin group G_n and is denoted by Pin(n). The Pin (n) is a double covering of O(n).

For n even, the subgroup is called Spin(n), which is the double covering group of SO(n).

If we use the notation of Eq. (5.119) and considering a pseudoscalar even I, the $U_{(k)}$ in Eq. (5.144) which are continuously connected with the identity can be rewritten in a exponential form as follows, c.f. ([72], 5.33)

$$U_{(k)} = e^{\frac{1}{2}\sum_{i<j}\alpha^{ij}e_{ij}}, \tag{5.148}$$

henceforth

$$\begin{aligned} \epsilon_k U_{(k)}\overline{U}_{(k)} &= e^{\frac{1}{2}\sum_{i<j}\alpha^{ij}e_{ij}} e^{-\frac{1}{2}\sum_{i<j}\alpha^{ij}\bar{e}_{ij}} = e^{\frac{1}{2}\sum_{i<j}\alpha^{ij}(e_{ij}-\bar{e}_{ij})} \\ &= e^{\frac{1}{2}\sum_{i<j}\alpha^{ij}E_{ij}} = e^{E}. \end{aligned} \tag{5.149}$$

The bivector generators $E_{ij} = e_i e_j - \bar{e}_i\bar{e}_j$ represent the actual left and right multiplications in G_n as presented in Eq. (5.146). The exponent of the bivector $E = \frac{1}{2}\sum_{i<j}\alpha^{ij}E_{ij}$ determines the rotor representation of O(n) given previously by Eq. (5.141).

Corollary Previous subsections and Sect. V of [72] show that projections as well as orthogonal and symmetric transformations on $\mathbb{R}^n$ can be represented in $G_{n,n}$ as simply even monomials, i.e., products of an even number of vectors which in turn cause successive reflections. Hence, we can draw a key conclusion: *every linear transformation* in $\mathbb{R}^n$ can be represented in $G_{n,n}$ as an even multivector which commutes with the complementation bivector $\boldsymbol{K}$. This is quite useful, because the composition of linear transformations can be simply formulated in terms of geometric products among idempotents and rotors in Spin(n, n). Many equation manipulations are simplified thanks to the commutativity with $\boldsymbol{K}$, as it is implicit in the reordering of vectors as shown in Eq. (5.145).

5.7 The $Pin^{sp}(3,3)$

In this subsection, we will formulate the Lie group and Lie algebra of the 3D projective geometry using the geometric algebra framework; specifically, we will model the Lie group $SL(4)$ and its Lie algebra $sl(4)$ in $G_{3,3}$.

According to Sect. 5.5.2, the Witt vector bases of the reciprocal null cones N_3 and $\overline{N}_3$ build $\mathbb{R}^{3,3} = N_3 \oplus \overline{N}_3$ and generate the 2^3-dimensional subalgebra $G_N = gen\{w_1, w_2, w_3\}$ and the 2^3-dimensional reciprocal subalgebra $G_{\overline{N}} = gen\{\overline{w}_1, \overline{w}_2, \overline{w}_3\}$, which have the structure of Grassmann algebras.

The geometric algebra $G_{3,3}$ is built by the direct product of these 2^3-dimensional Grassmann subalgebras:

$$G_{3,3} = G_N \otimes G_{\overline{N}} = gen\{w_1, w_2, w_3, \overline{w}_1, \overline{w}_2, \overline{w}_3\}. \tag{5.150}$$

The reciprocal bases $\{w\} \in N$ and $\{\bar{w}\} \in \bar{N}$ are called *dual*, because they fulfill Eq. (5.47) $w_i \cdot \overline{w}_j = \overline{w}_j \cdot w_i = \delta_{ij}$. They generate the k-vector bases of $G_{3,3}$,

$$\Big\{1, \{w\}, \{\bar{w}\}, \{w_i w_j\}, \{\bar{w}_i \bar{w}_j\}, \{w_i \bar{w}_j\}, ..., \{w_{j_1} ... \bar{w}_{j_i} \bar{w}_{j_l} \dots w_{j_k}\}, I, \bar{I}\Big\}, \tag{5.151}$$

consisting of scalars, vectors, bivectors, trivectors, ... , pseudoscalar and the dual pseudoscalars $I = w_1 \wedge w_2 \wedge w_3$ and $\bar{I} = \bar{w}_1 \wedge \bar{w}_2 \wedge \bar{w}_3$ which satisfy $I\bar{I} = 1$.

The Lie algebra $sl(4)$ of the 3D projective transformations is isomorphic with the bivector algebra of $G_{3,3}$. The canonical homomorphism from $SL(4)$ to $Spin(3,3)$ is not surjective, and the projective transformations of negative determinant do not induce orthogonal transformations in the Plücker coordinate space of lines; to overcome this limitations, Hongbo Li et al [188] proposed the group called $Pin^{sp}(3,3)$ with $Pin(3,3)$ as its normal subgroup, in order to quadruple-cover the group of the projective transformations and polarities.

Let us review some geometric algebras for the representation of points, lines, planes, and their associated Lie groups. The conformal geometric algebra $G_{4,1}$ of the 3D space [13] is equipped with $\begin{bmatrix} 5 \\ 2 \end{bmatrix}$ = 10 bivectors which as Lie algebra generators suffice to build the conformal group consisting of $SE(3)$, dilations and inversions. In addition, the motor algebra $G^+_{3,0,1}$ [13, 21] with $\begin{bmatrix} 4 \\ 2 \end{bmatrix}$=6 bivector basis

$$\{e_{01}, e_{02}, e_{03}, e_{23}, e_{31}, e_{12}\}, \tag{5.152}$$

which correspond to the *Plücker* coordinates of the lines. Now, given two homogeneous points $P, Q \in G_{3,0,1}$, we compute a line wedging both points

$$\boldsymbol{L} = P \wedge Q = P \wedge n = (e_0 + \boldsymbol{p}) \wedge \boldsymbol{n} = e_0 \wedge \boldsymbol{n} + \boldsymbol{p} \wedge \boldsymbol{n}, \tag{5.153}$$

where $\boldsymbol{n} = \boldsymbol{P} - \boldsymbol{Q}$ is the line orientation and $\boldsymbol{m} = \boldsymbol{p} \wedge \boldsymbol{n}$ is its moment.

In the space spanned by the bivectors of Eq. (5.152), not all bivectors will represent lines, only lines $\boldsymbol{L} = l_{01}e_{01} + l_{02}e_{02} + l_{03}e_{03} + l_{23}e_{23} + l_{31}e_{31} + l_{12}e_{12}$ which fulfill the so-called Grassmann-Plücker relation

$$M_2^4 : l_{01}l_{23} + l_{02}l_{31} + l_{03}l_{12} = 0, \tag{5.154}$$

this relation defines a quadric called the manifold of Study [16] which is an algebraic variety M_2^4 of degree 2 and dimension 4. Hence, $G^+_{3,0,1}$ is appropriate for line geometry and handle $SE(3)$ and it can be used to model the kinematics of points, lines, and planes, as shown in [13, 21]. As you see to model the Lie group of the projective geometry $SL(4)$, we require more bivectors, that is why we consider $G_{3,3}$ which offers $\begin{bmatrix} 6 \\ 2 \end{bmatrix}$=15 bivectors enough to handle the 15 DOF of the projective transformation of $\mathbb{R}^4$. Thus, we can infer that the points and planes of the geometric algebra $G_{3,0,1}$ for projective geometry and its group action $GL(4)$ upon $\mathbb{R}^4$ can be modeled in terms of null 3-vector representations and the adjoint action of $Pin^{sp}(3,3)$ as we will explain next in much more detail.

5.7.1 *Line Geometry Using Null Geometric Algebra Over* $\mathbb{R}^{3,3}$

Let e_0, e_1, e_2, e_3 be an orthonormal basis of $\mathbb{R}^4$ which expands the affine plane, where the null vector $e_4 \in G_{3,0,1}$ called e_0 is the origin of the Euclidean affine space 3D space $\mathbb{E}^3$. The unit vectors e_1, e_2, e_3 are points at the plane at infinity $< e_1, e_2, e_3 >$, and they represent the directions of three basis projective lines $\{e_{23}, e_{31}, e_{12}\}$. These lines are in turn the intersections (meets) of three planes $\{e_{023}, e_{012}, e_{012}\}$ which cross the origin e_0.

We project the affine plane onto $\mathbb{R}^{3,3} = N_3 \oplus \overline{N}_3$; hence, the basis e_0, e_1, e_2, e_3 induces a Witt basis.

Li and Zhang [187, 188] proposed a new model of 3D projective geometry by taking the null vectors of $\mathbb{R}^{3,3}$ as algebraic generators. They defined points and planes as two connected components of the set of null 3-spaces of $\mathbb{R}^{3,3}$, respectively, and showed that whenever an element of $Spin_o(3,3)$ acts upon $\mathbb{R}^{3,3}$, it induces a projective transformation by means the outermorphism of the action upon the null 3-vectors representing 3D points and planes. Later on, Klawitter [158] along the same lines of thought, he proposed an explicit expression of the spinor inducing a projective transformation in 4×4 matrix form.

Embedding the line basis of Eq. (5.152) via the function

$$\wp : \bigwedge^2(\mathbb{R}^4) \rightarrow \mathbb{R}^{3,3}, \tag{5.155}$$

we get a space spanned by a null basis given by

$$\{w_1, w_2, w_3, \overline{w}_1, \overline{w}_2, \overline{w}_3\}. \tag{5.156}$$

The embedding of lines $\boldsymbol{L} = e_0 \wedge \boldsymbol{n} + \boldsymbol{p} \wedge \boldsymbol{n} \in G_{3,1}$ of $\mathbb{R}^4$ via the function $\wp(\cdot)$ in $\mathbb{R}^{3,3}$ yields

$$\wp : \boldsymbol{L} \longrightarrow n_x w_1 + n_y w_2 + n_z w_3 + m_x \overline{w}_1 + m_y \overline{w}_2 + m_z \overline{w}_3. \tag{5.157}$$

Note in Eq. (7.2), we model a line in G_3 of $\mathbb{R}^3$ using the available k-blades as

$$L = \mathbf{n} + \boldsymbol{m}, \tag{5.158}$$

where $\mathbf{n}$ and $\boldsymbol{m}$ are a vector and a bivector. In motor algebra $G^+_{3,0,1}$, a line is represented in terms of the bivectors $\boldsymbol{n}$ and $\boldsymbol{m}$ (expanded in the dual bivector basis) and the pseudoscalar I which squares to null,

$$L = \boldsymbol{n} + \boldsymbol{m} = \boldsymbol{n} + I\boldsymbol{m}. \tag{5.159}$$

Finally in $G_{3,3}$ of $\mathbb{R}^{3,3}$, we represent the line in terms of null vectors as given in Eq. (5.157).

5.7.2 Projective Transformations Using Null Geometric Algebra Over $\mathbb{R}^{3,3}$

Recently, Dorst [78] proposed bivector generators for 3D projective transformations. In this subsection, we will explain using our previous equations how we derive the affine and projective transformations in terms of versors of $G_{3,3}$. For that we have to embed the projective transformation of $\mathbb{R}^4$ into $\mathbb{R}^{3,3}$ and use a 6×6 bivector matrix similar as Eq. (5.66) and then select the bivectors generators of $\mathbb{R}^{3,3}$ to construct specialized versors for translation, rotation, perspectivity, and affine transformations. Firstly, let us discuss the inner product and the isometry between $\mathbb{R}^4$ and $\mathbb{R}^{3,3}$.

For the metric space of oriented lines in $\mathbb{R}^{3,3}$, we define next an inner product via the *bracket* of $G_{3,1}$ in $\mathbb{R}^4$ a 4D-volume, given $L_1, L_2 \in G_{3,0,1}$ and the lines $l_1, l_2 \in G_{3,3}$

$$L_1 \cdot L_2 = (P_1 \wedge Q_1) \cdot (P_2 \wedge Q_2) = [\wp^{-1}(l_1) \wedge \wp^{-1}(l_2)] = [P_1, Q_1, P_2, Q_2],$$

the inner product of L_1, L_2 gives the volume spanned by the line orientations and their orthogonal distance or fut, when they intersect the bracket or volume is zero.

Consider a transformation $f : \mathbb{R}^4 \longrightarrow \mathbb{R}^4$ for points and $\underline{f}$ its outermorphism on $\bigwedge^2 \mathbb{R}^4$, the corresponding transformation $F : \mathbb{R}^{3,3} \rightarrow \mathbb{R}^{3,3}$ is $F() \equiv \wp(\underline{f})(\wp^{-1}())$ on lines and screws. Let us compute the dot product of two lines

$$\begin{aligned} F(l_1) \cdot F(l_2) &= F(\wp(L_1)) \cdot F(\wp(L_2)) = \wp(\underline{f}(L_1)) \cdot \wp(\underline{f}(L_2)), \\ &= [\underline{f}(L_1) \wedge \underline{f}(L_2)] = [\underline{f}(L_1 \wedge L_2)] = det(f)[L_1 \wedge L_2], \\ &= det(f)\wp(L_1) \cdot \wp(L_2) = det(f) l_1 \cdot l_2. \end{aligned} \tag{5.160}$$

By the case of an orthogonal transformation det(f)=1 and $det(F) = det(f)^3 = 1$. Projective transformations with det(f)=-1 do not induce orthogonal transformations in the Pluckër space of lines.

For the applications of geometric algebra in computer vision and in graphics engineering, it is needed projective transformations, let us find the projective transformations of lines. In the projective space $\mathbb{P}^3$, given the orientation $\mathbf{u} \in \mathbb{R}^3$ and a point $\mathbf{x} \in \mathbb{R}^3$ touching the line, a line $\mathbf{L}$ is computed as follows

$$\mathbf{L} = \begin{bmatrix} \mathbf{x} \\ 1 \end{bmatrix} \wedge \begin{bmatrix} \mathbf{n} \\ 0 \end{bmatrix} = \begin{bmatrix} \mathbf{n} \\ \mathbf{x} \times \mathbf{n} \end{bmatrix} = \begin{bmatrix} \mathbf{n} \\ \mathbf{m} \end{bmatrix}, \tag{5.161}$$

where $\mathbf{m}$ stands for the moment of the line, which is a vector normal to the surface where the line lies, and note that this line belongs to the projective space $\mathbb{P}^5$. An homography $\mathtt{H}$ in $\mathbb{P}^3$ is given by

$$\mathtt{H} = \begin{bmatrix} \mathtt{A} & \mathbf{t} \\ \mathbf{f} & h \end{bmatrix}, \tag{5.162}$$

where $\mathtt{A}$ is an affine transformation matrix, $\mathbf{t}$ a translation vector, and if $\mathrm{h} = 1$, the vector $\mathbf{f}$ corresponds to the perspectivity in the pinhole model of a digital camera [126]. If we apply the homography to the two projective points and compute the wedge, we obtain the line after the action of the homography

$$\begin{aligned}
\mathbf{L}' &= \mathtt{H}\begin{bmatrix} \mathbf{x} \\ 1 \end{bmatrix} \wedge \mathtt{H}\begin{bmatrix} \mathbf{n} \\ 0 \end{bmatrix} = \begin{bmatrix} \mathtt{A}\mathbf{x} + \mathbf{t} \\ \mathbf{f}^T\mathbf{x} + h \end{bmatrix} \wedge \begin{bmatrix} \mathtt{A}\mathbf{n} \\ \mathbf{f}^T\mathbf{n} \end{bmatrix}, \\
&= \begin{bmatrix} \mathtt{A}\mathbf{n}(\mathbf{f}^T\mathbf{x} + h) - (\mathtt{A}\mathbf{x} + \mathbf{t})\mathbf{f}^{\mathbf{T}}\mathbf{n} \\ (\mathtt{A}\mathbf{x} + \mathtt{t}) \times \mathtt{A}\mathbf{n} \end{bmatrix}, \\
&= \begin{bmatrix} h\mathtt{A}\mathbf{n} + (\mathbf{f}^T\mathbf{x})\mathtt{A}\mathbf{n} - \mathtt{A}\mathbf{x}(\mathbf{f}^T\mathbf{n}) - \mathbf{t}(\mathbf{f}^{\mathbf{T}}\mathbf{n}) \\ \mathtt{A}\mathbf{x} \times \mathtt{A}\mathbf{n} + \mathbf{t} \times \mathtt{A}\mathbf{n} \end{bmatrix}, \\
&= \begin{bmatrix} (h\mathtt{A} - \mathbf{t}^T\mathbf{f}) & -\mathtt{A}[\mathbf{f}]_\times \\ [\mathbf{t}]_\times\mathtt{A} & (\det \mathtt{A})\mathtt{A}^{-T} \end{bmatrix} \begin{bmatrix} \mathbf{n} \\ \mathbf{x} \times \mathbf{n} \end{bmatrix} = \mathtt{T}(\mathbf{L}),
\end{aligned} \tag{5.163}$$

here the vector triple product was used in $(\mathbf{f} \cdot \mathbf{x})\mathtt{A}\mathbf{n} - \mathtt{A}\mathbf{x}(\mathbf{f} \cdot \mathbf{n}) = \mathtt{A}(\mathbf{f} \cdot \mathbf{x})$ $\mathbf{n} - (\mathbf{f} \cdot \mathbf{n})\mathtt{A}\mathbf{x} = \mathtt{A}(-\mathbf{x}(\mathbf{f} \cdot \mathbf{n}) + \mathbf{n}(\mathbf{f} \cdot \mathbf{x})) = \mathtt{A}((\mathbf{f} \cdot \mathbf{x})\mathbf{n} - (\mathbf{f} \cdot \mathbf{n})\mathbf{x})) = \mathtt{A}[\mathbf{f} \times \mathbf{n} \times \mathbf{x}]$ $= -\mathtt{A}[\mathbf{f} \times \mathbf{x} \times \mathbf{n}] = -\mathtt{A}[\mathbf{f}] \times (\mathbf{x} \times \mathbf{n})$, where $[\mathbf{f}]_\times$ stands for the operator as an antisymmetric matrix which represents the cross product with $\mathbf{f}$.

5.7.3 *Lie Groups in* $\mathbb{R}^{3,3}$

Equation (5.163) represents the line transformation into the projective space $\mathbb{P}^5$,

$$\mathtt{T} = \begin{bmatrix} (h\mathtt{A} - \mathbf{t}^T\mathbf{f}) & -\mathtt{A}[\mathbf{f}]_\times \\ [\mathbf{t}]_\times\mathtt{A} & (\det \mathtt{A})\mathtt{A}^{-T} \end{bmatrix}, \tag{5.164}$$

which will be used to find the generators of its Lie algebra. The basis to expand the Lie operators of $\mathtt{T}$ in $\mathbb{R}^{3,3}$ is given by the unit basis $e_i, \bar{e}_i$ with $e_i \cdot e_j = \delta_{i,j}$ and $\bar{e}_i \cdot \bar{e}_j = -\delta_{i,j}$ and the null basis $w_i, \bar{w}_i$ with $w_i \cdot \bar{w}_j = \delta_{i,j}$. The inner product on the unit basis gives the signature of the space in $\mathbb{R}^{3,3}$.

Similarly, one obtains the inner product table using the null basis as shown in the tables below.

$\cdot$	e_1	e_2	e_3	$\bar{e}_1$	$\bar{e}_2$	$\bar{e}_3$
e_1	1	0	0	0	0	0
e_2	0	1	0	0	0	0
e_3	0	0	1	0	0	0
$\bar{e}_1$	0	0	0	−1	0	0
$\bar{e}_2$	0	0	0	0	−1	0
$\bar{e}_3$	0	0	0	0	0	−1

$\cdot$	w_1	w_2	w_3	$\bar{w}_1$	$\bar{w}_2$	$\bar{w}_3$
w_1	0	0	0	1	0	0
w_2	0	0	0	0	1	0
w_3	0	0	0	0	0	1
$\bar{w}_1$	1	0	0	0	0	0
$\bar{w}_2$	0	1	0	0	0	0
$\bar{w}_3$	0	0	1	0	0	0

The projective transformation (5.164) is represented using the unit basis or the null basis as follows

$$\mathtt{T}_{e_i,\bar{e}_j} = \begin{bmatrix} a_{11} & a_{12} & a_{13} & 0 & -f_3 & f_2 \\ a_{21} & a_{22} & a_{23} & f_3 & 0 & -f_1 \\ a_{31} & a_{32} & a_{33} & -f_2 & f_1 & 0 \\ 0 & t_3 & -t_2 & -a_{11} & -a_{21} & -a_{31} \\ -t_3 & 0 & t_1 & -a_{12} & -a_{22} & -a_{32} \\ t_2 & -t_1 & 0 & -a_{13} & -a_{23} & -a_{33} \end{bmatrix}, \tag{5.165}$$

$$\mathtt{T}_{w_i,\bar{w}_j} = \begin{bmatrix} 0 & -f_3 & f_2 & -a_{11} & -a_{21} & -a_{31} \\ f_3 & 0 & -f_1 & -a_{12} & -a_{22} & -a_{32} \\ -f_2 & f_1 & 0 & -a_{13} & -a_{23} & -a_{33} \\ a_{11} & a_{12} & a_{13} & 0 & t_3 & -t_2 \\ a_{21} & a_{22} & a_{23} & -t_3 & 0 & t_1 \\ a_{31} & a_{32} & a_{33} & t_2 & -t_1 & 0 \end{bmatrix}. \tag{5.166}$$

The projective transformation of a vector $\boldsymbol{x} \in \mathbb{R}^{3,3}$ can be computed using the Lie group versors, and the generators of Lie algebra in terms of bivectors, where the generators can be expanded as linear combinations of different bivectors, namely

$$\boldsymbol{x}' = \boldsymbol{V}(\boldsymbol{x})\widetilde{\boldsymbol{V}} = e^{\boldsymbol{B}}\boldsymbol{x}e^{-\boldsymbol{B}} = e^{\sum_{i,j} t_{ij}\boldsymbol{B}_{i,j}}\boldsymbol{x}e^{-\sum_{i,j} t_{ij}\boldsymbol{B}_{i,j}}, \tag{5.167}$$

where $t_{ij} \in \mathbb{R}$ are computed based on the entries of the transformation matrix of Eq. (5.165) or of Eq. (5.166), respectively. The bivectors $\boldsymbol{B}_{i,j}$ are computed in terms of the unit basis $e_i \wedge e_j$ or null basis $w_i \wedge w_j$ as shown below in Eqs. (5.168) and (5.169).

In order to compute any Lie generators in terms of bivectors, firstly one takes the derivatives of $\mathtt{T}_{e_i,\bar{e}_j}$ Eq. (5.165) or $\mathtt{T}_{w_i,\bar{w}_j}$ Eq. (5.166) with respect to a specific parameter ν (degree of freedom), and then one evaluates at $\nu = 0$ to get the corresponding generator of certain Lie algebra associate to the projective Lie group $SL(4,\mathbb{R})$. Then one expresses the resulting Lie generators using bivectors corresponding to the transformation of Eq. (5.165) or of Eq. (5.166). Similar as the procedure given in Eq. (5.63) and (5.66), the resulting Lie generator or matrix in terms of unit vector bivectors $e_i \wedge \bar{e}_j$ reads

$$T_{e_i \wedge \bar{e}_j} = \begin{bmatrix} 0 & a_{12}e_1\wedge e_2 & a_{13}e_1\wedge e_3 & 0 & -f_3e_1\wedge\bar{e}_2 & f_2e_1\wedge\bar{e}_3 \\ a_{21}e_2\wedge e_1 & 0 & a_{23}e_2\wedge e_3 & f_3e_2\wedge\bar{e}_1 & 0 & -f_1e_2\wedge\bar{e}_3 \\ a_{31}e_3\wedge e_1 & a_{32}e_3\wedge e_2 & 0 & -f_2e_3\wedge\bar{e}_1 & f_1e_3\wedge\bar{e}_2 & 0 \\ 0 & t_3\bar{e}_1\wedge e_2 & -t_2\bar{e}_1\wedge e_3 & 0 & -a_{21}\bar{e}_1\wedge\bar{e}_2 & -a_{31}\bar{e}_1\wedge\bar{e}_3 \\ -t_3\bar{e}_2\wedge e_1 & 0 & t_1\bar{e}_2\wedge e_3 & -a_{12}\bar{e}_2\wedge\bar{e}_1 & 0 & -a_{32}\bar{e}_2\wedge\bar{e}_3 \\ t_2\bar{e}_3\wedge e_1 & -t_1\bar{e}_3\wedge e_2 & 0 & -a_{13}\bar{e}_3\wedge\bar{e}_1 & -a_{23}\bar{e}_3\wedge\bar{e}_2 & 0 \end{bmatrix}, \tag{5.168}$$

or in terms of null vector bivectors $w_i \wedge \bar{w}_j$.

$$T_{w_i \wedge \bar{w}_j} = \begin{bmatrix} 0 & -f_3w_1\wedge w_2 & f_2w_1\wedge w_3 & -s_1w_1\wedge\bar{w}_1 & -a_{21}w_1\wedge\bar{w}_2 & -a_{31}w_1\wedge\bar{w}_3 \\ f_3w_2\wedge w_1 & 0 & -f_1w_2\wedge w_3 & -a_{12}w_2\wedge\bar{w}_1 & -s_2w_2\wedge\bar{w}_2 & -a_{32}w_2\wedge\bar{w}_3 \\ -f_2w_3\wedge w_1 & f1w_3\wedge w_2 & 0 & -a_{13}w_3\wedge\bar{w}_1 & -a_{23}w_3\wedge\bar{w}_2 & -s_3w_3\wedge\bar{w}_3 \\ s_1\bar{w}_1\wedge w_1 & a_{12}\bar{w}_1\wedge w_2 & a_{13}\bar{w}_1\wedge w_3 & 0 & t_3\bar{w}_1\wedge\bar{w}_2 & -t_2\bar{w}_1\wedge\bar{w}_3 \\ a_{21}\bar{w}_2\wedge w_1 & s_2\bar{w}_2\wedge w_2 & a_{23}\bar{w}_2\wedge w_3 & -t_3\bar{w}_2\wedge\bar{w}_1 & 0 & t_1\bar{w}_2\wedge\bar{w}_3 \\ a_{31}\bar{w}_3\wedge w_1 & a_{32}\bar{w}_3\wedge w_2 & s_3\bar{w}_3\wedge w_3 & t_2\bar{w}_3\wedge\bar{w}_1 & -t_1\bar{w}_3\wedge\bar{w}_2 & 0 \end{bmatrix}. \tag{5.169}$$

Let us start computing the Lie generators for the translation.

$$\mathcal{L}_\nu = \frac{\mathrm{d}\mathtt{T}_{\mathtt{w}_i,\bar{\mathtt{w}}_{j\nu}}}{\mathrm{d}\nu}|_{\nu\to 0}. \tag{5.170}$$

Translation: Taking the derivatives for each degree of freedom $\{\mathtt{t}_x, \mathtt{t}_y, \mathtt{t}_z\}$ of

$$\mathtt{T}_\mathtt{t} = \begin{bmatrix} \mathtt{I} & \mathtt{O} \\ [\mathbf{t}]_\times & \mathtt{O} \end{bmatrix}, \tag{5.171}$$

$$\mathcal{L}_{\mathtt{t}_x} = \frac{\mathrm{d}\mathtt{T}_{\mathtt{w}_i,\bar{\mathtt{w}}_j\,\mathtt{t}_x}}{\mathrm{d}\mathtt{t}_x}|_{\mathtt{t}_x\to 0} = \begin{bmatrix} \mathtt{I} & \mathtt{O} \\ \begin{bmatrix} 0 & 0 & 0 \\ 0 & 0 & 1 \\ 0 & -1 & 0 \end{bmatrix} & \mathtt{O} \end{bmatrix}, \tag{5.172}$$

similarly for $\{\mathtt{t}_y, \mathtt{t}_z\}$, gathering these results we obtain the Lie operator for translation

$$\mathcal{L}_{\mathtt{t}} = \begin{bmatrix} \mathtt{I} & \mathtt{O} \\ \begin{bmatrix} 0 & 1 & -1 \\ -1 & 0 & 1 \\ 1 & -1 & 0 \end{bmatrix} & \mathtt{I} \end{bmatrix}. \tag{5.173}$$

The versor for translation called *translator* can be computed either using the bivectors $e_i \wedge \bar{e}_j = e_i \bar{e}_j$ (orthonormal basis) of Eq. (5.168) or following the similar approach using the bivectors $w_i \wedge \bar{w}_j$ of Eq. (5.169) to get

$$\begin{aligned} \boldsymbol{T} = e^{\boldsymbol{B}_t} &= e^{\frac{t_3 \bar{e}_1 e_2 - t_2 \bar{e}_1 e_3 + t_1 \bar{e}_2 e_3}{2}}, \qquad (5.174) \\ &= e^{\frac{t_3 \bar{w}_1 \wedge \bar{w}_2 - t_2 \bar{w}_1 \wedge \bar{w}_3 + t_1 \bar{w}_2 \wedge \bar{w}_3}{2}}, \\ &= 1 + \frac{1}{2}(t_3 \bar{e}_1 e_2 - t_2 \bar{e}_1 e_3 + t_1 \bar{e}_2 e_3), \\ &= 1 + \frac{1}{2}(t_3 \bar{w}_1 \wedge w_2 - t_2 \bar{w}_1 \wedge \bar{w}_3 + t_1 \bar{w}_2 \wedge \bar{w}_3). \qquad (5.175) \end{aligned}$$

Perspectivity: Taking the derivatives for each degree of freedom $\{\mathtt{f}_x, \mathtt{f}_y, \mathtt{f}_z\}$ of

$$\mathtt{T}_f = \begin{bmatrix} \mathtt{I} & [\mathbf{f}]_\times \\ \mathtt{O} & \mathtt{I} \end{bmatrix}, \tag{5.176}$$

$$\mathcal{L}_{\mathtt{f}_x} = \frac{\mathtt{d}T_{\mathtt{f}_x}}{\mathtt{d}\mathtt{f}_x}|_{\mathtt{f}_x \to 0} = \begin{bmatrix} \mathtt{I} & \begin{bmatrix} 0 & 0 & 0 \\ 0 & 0 & 1 \\ 0 & -1 & 0 \end{bmatrix} \\ \mathtt{O} & \mathtt{I} \end{bmatrix}, \tag{5.177}$$

similarly for $\{\mathtt{f}_y, \mathtt{f}_z\}$, gathering these results we obtain the Lie operator for perspectivity

$$\mathcal{L}_{\mathtt{f}} = \begin{bmatrix} \mathtt{I} & \begin{bmatrix} 0 & -1 & 1 \\ 1 & 0 & -1 \\ -1 & 1 & 0 \end{bmatrix} \\ \mathtt{O} & \mathtt{I} \end{bmatrix}. \tag{5.178}$$

The versor for perspectivity called *perspector* can be computed either using the bivectors $e_i \bar{e}_j$ of Eq. (5.168) or following the similar approach using the bivectors $w_i \wedge \bar{w}_j$ of Eq. (5.169) to get

$$
\begin{aligned}
\boldsymbol{P} = e^{\boldsymbol{B}_f} &= e^{\frac{-f_3 e_1 \bar{e}_2 + f_2 e_1 \bar{e}_3 - f_1 e_2 \bar{e}_3}{2}}, \\
&= e^{\frac{-f_3 w_1 \wedge w_2 + f_2 w_1 \wedge w_3 - f_1 w_2 \wedge w_3}{2}}, \\
&= 1 + \frac{1}{2}(-f_3 e_1 \bar{e}_2 + f_2 e_1 \bar{e}_3 - f_1 e_2 \bar{e}_3), \\
&= 1 + \frac{1}{2}(-f_3 w_1 \wedge w_2 + f_2 w_1 \wedge w_3 - f_1 w_2 \wedge w_3).
\end{aligned} \tag{5.179}
$$

Rotation: Taking the derivatives for each Euler angles $\{\theta, \phi, \psi\}$ of

$$
\mathtt{T}_{\mathtt{R}} = \begin{bmatrix} \mathtt{R} & \mathtt{O} \\ \mathtt{O} & \mathtt{R}^{-T} \end{bmatrix}, \tag{5.180}
$$

we obtain first the Lie operator for the 3D rotation $\mathtt{R}_z$ with respect to the axis z

$$
\mathcal{L}_\theta = \frac{\mathtt{dT}_{\mathtt{R}_z}}{\mathtt{d}\theta}|_{\theta\to 0} = \frac{\mathtt{d}\begin{bmatrix} \mathtt{R}_z & \mathtt{O} \\ \mathtt{O} & \mathtt{R}_z^T \end{bmatrix}}{\mathtt{d}\theta}|_{\theta\to 0}, \tag{5.181}
$$

$$
= \frac{\mathtt{d}\begin{bmatrix} \begin{bmatrix} \cos(\theta) & -\sin(\theta) & 0 \\ \sin(\theta) & \cos(\theta) & 0 \\ 0 & 0 & 1 \end{bmatrix} & \mathtt{O} \\ \mathtt{O} & \begin{bmatrix} \cos(\theta) & \sin(\theta) & 0 \\ -\sin(\theta) & \cos(\theta) & 0 \\ 0 & 0 & 1 \end{bmatrix} \end{bmatrix}}{\mathtt{d}\theta}|_{\theta\to 0},
$$

$$
= \begin{bmatrix} \begin{bmatrix} 0 & -1 & 0 \\ 1 & 0 & 0 \\ 0 & 0 & 0 \end{bmatrix} & \mathtt{O} \\ \mathtt{O} & \begin{bmatrix} 0 & 1 & 0 \\ -1 & 0 & 0 \\ 0 & 0 & 0 \end{bmatrix} \end{bmatrix}, \tag{5.182}
$$

and similarly for those angles ϕ, ψ with respect to axes y and x, gathering these results we obtain the Lie operator for the rotations $\mathtt{R} = \mathtt{R}_z\mathtt{R}_y\mathtt{R}_z$

$$
\mathcal{L}_{\mathtt{R}} = \begin{bmatrix} \begin{bmatrix} 0 & -1 & -1 \\ 1 & 0 & -1 \\ 1 & 1 & 0 \end{bmatrix} & \mathtt{O} \\ \mathtt{O} & \begin{bmatrix} 0 & 1 & 1 \\ -1 & 0 & 1 \\ -1 & -1 & 0 \end{bmatrix} \end{bmatrix}. \tag{5.183}
$$

The versor for rotation called *rotator* can be computed either using the bivectors $e_i\bar{e}_j$ of Eq. (5.168) or following the similar approach using the bivectors $w_i\wedge\bar{w}_j$ of Eq. (5.169) to get

$$\begin{aligned} \boldsymbol{R} &= e^{\boldsymbol{B}_R}, \\ &= e^{\frac{(-e_2e_3+\bar{e}_2\bar{e}_3)+(-e_1e_3+\bar{e}_1\bar{e}_3)+(-e_1e_2+\bar{e}_1\bar{e}_2)}{2}}, \\ &= e^{\frac{(w_2\wedge\bar{w}_3-\bar{w}_2\wedge w_3)+(w_1\wedge\bar{w}_3-\bar{w}_1\wedge w_3)+(w_1\wedge\bar{w}_2-\bar{w}_1\wedge w_2)}{2}}. \end{aligned} \tag{5.184}$$

Lorentz transformation Taking the derivatives for each variable $\{\vartheta, \varphi, \varpi\}$ of

$$\mathtt{T}_{\mathtt{LO}} = \begin{bmatrix} \mathtt{LO} & \mathtt{O} \\ \mathtt{O} & -\mathtt{LO}^{-T} \end{bmatrix}, \tag{5.185}$$

we obtain first the Lie operator for the 3D Lorentz transformation $\mathtt{LO}$ with respect to the axis z and variable ϑ

$$\mathcal{L}_{\mathtt{LO}_\vartheta} = \frac{\mathrm{d}\mathtt{LO}_\vartheta}{\mathrm{d}\vartheta}|_{\vartheta\to 0} = \frac{\mathrm{d}\begin{bmatrix} \mathtt{LO}_\vartheta & \mathtt{O} \\ \mathtt{O} & -\mathtt{LO}_\vartheta^T \end{bmatrix}}{\mathrm{d}\theta}|_{\theta\to 0}, \tag{5.186}$$

$$= \frac{\mathrm{d}\begin{bmatrix} \begin{bmatrix} \cosh(\vartheta) & \sinh(\vartheta) & 0 \\ \sinh(\vartheta) & \cosh(\vartheta) & 0 \\ 0 & 0 & 1 \end{bmatrix} & \mathtt{O} \\ \mathtt{O} & \begin{bmatrix} \cosh(\vartheta) & -\sinh(\vartheta) & 0 \\ -\sinh(\vartheta) & \cosh(\vartheta) & 0 \\ 0 & 0 & 1 \end{bmatrix} \end{bmatrix}}{\mathrm{d}\vartheta}|_{\vartheta\to 0},$$

$$= \begin{bmatrix} \begin{bmatrix} 0 & 1 & 0 \\ 1 & 0 & 0 \\ 0 & 0 & 0 \end{bmatrix} & \mathtt{O} \\ \mathtt{O} & \begin{bmatrix} 0 & -1 & 0 \\ -1 & 0 & 0 \\ 0 & 0 & 0 \end{bmatrix} \end{bmatrix}, \tag{5.187}$$

and similarly for the variables $\{\varphi, \varpi\}$ with respect to axes y and x, gathering these results we obtain the Lie operator for the Lorentz transformation involving the variables $\{\vartheta, \varphi, \varpi\}$

$$\mathcal{L}_{\mathtt{LO}} = \begin{bmatrix} \begin{bmatrix} 0 & 1 & 1 \\ 1 & 0 & 1 \\ 1 & 1 & 0 \end{bmatrix} & \mathtt{O} \\ \mathtt{O} & \begin{bmatrix} 0 & -1 & -1 \\ -1 & 0 & -1 \\ -1 & -1 & 0 \end{bmatrix} \end{bmatrix}. \tag{5.188}$$

The versor for the Lorentz transformation called *Lorentor* can be computed either using the bivectors $e_i\bar{e}_j$ of Eq. (5.168) or following the similar approach using the bivectors $w_i \wedge \bar{w}_j$ of Eq. (5.169) to get

$$\begin{aligned} \boldsymbol{R}_L &= e^{\boldsymbol{B}_L}, \\ &= e^{\frac{(e_2\wedge e_3 - \bar{e}_2\wedge\bar{e}_3)+(e_1\wedge e_3 - \bar{e}_1\wedge\bar{e}_3)+(e_1\wedge e_2 - \bar{e}_1\wedge e_2)}{2}}, \\ &= e^{\frac{(\bar{w}_2\wedge w_3 - w_2\wedge\bar{w}_3)+(\bar{w}_1\wedge w_3 - w_1\wedge\bar{w}_3)+(\bar{w}_1\wedge w_2 - w_1\wedge\bar{w}_2)}{2}}. \end{aligned} \tag{5.189}$$

The next Lie operators will be found using the parameters of the affine transformation matrix $\mathtt{A}$ in the transformation $\mathtt{T}$ of Eq. (5.164) with $h = 1$ and $\det \mathtt{A} = 1$ or equivalently Eq. (5.165)

$$\mathtt{A} = \begin{bmatrix} a_{11} & a_{12} & a_{13} \\ a_{21} & a_{22} & a_{23} \\ a_{31} & a_{32} & a_{33} \end{bmatrix}. \tag{5.190}$$

Shear: Taking the derivatives for each of the variables off the diagonal of $\mathtt{A}$ $\{a_{12}, a_{13}, a_{21}, a_{23}, a_{31}, a_{32}\}$ in

$$\mathtt{T}_\mathtt{A} = \begin{bmatrix} \mathtt{A} & \mathtt{O} \\ \mathtt{O} & -\mathtt{A}^{-T} \end{bmatrix}, \tag{5.191}$$

we obtain first the Lie operator for a partially shear $\mathtt{A}_{a_{23}}$

$$\mathcal{L}_{a_{23}} = \frac{\mathtt{d}\mathtt{T}_{\mathtt{A}_1}}{\mathtt{d}a_{23}}\Big|_{a_{23}\to 0} = \frac{\mathtt{d}\begin{bmatrix} \mathtt{A}_1 & \mathtt{O} \\ \mathtt{O} & -\mathtt{A}_1^T \end{bmatrix}}{\mathtt{d}a_{23}}\Big|_{a_{23}\to 0}, \tag{5.192}$$

$$= \frac{\mathtt{d}\begin{bmatrix} \begin{bmatrix} 1 & 0 & 0 \\ 0 & 1 & a_{23} \\ 0 & 0 & 1 \end{bmatrix} & \mathtt{O} \\ \mathtt{O} & \begin{bmatrix} 1 & 0 & 0 \\ 0 & 1 & 0 \\ 0 & -a_{23} & 1 \end{bmatrix} \end{bmatrix}}{\mathtt{d}a_{23}}\Big|_{a_{23}\to 0},$$

$$= \begin{bmatrix} \begin{bmatrix} 0 & 0 & 0 \\ 0 & 0 & 1 \\ 0 & 0 & 0 \end{bmatrix} & \mathtt{O} \\ \mathtt{O} & \begin{bmatrix} 0 & 0 & 0 \\ 0 & 0 & 0 \\ 0 & -1 & 0 \end{bmatrix} \end{bmatrix}, \tag{5.193}$$

and similarly for the variables $\{a_{13}, a_{21}, a_{23}, a_{31}, a_{32}\}$, gathering these results we obtain the Lie operator for the shears

$$\mathcal{L}_{\mathtt{A}} = \begin{bmatrix} \begin{bmatrix} 0 & 1 & 1 \\ 1 & 0 & 1 \\ 1 & 1 & 0 \end{bmatrix} & \mathtt{O} \\ \mathtt{O} & \begin{bmatrix} 0 & -1 & -1 \\ -1 & 0 & -1 \\ -1 & -1 & 0 \end{bmatrix} \end{bmatrix}. \tag{5.194}$$

If we combine the Lie operators of opposite shears of $\{a_{12}, a_{13}, a_{21}, a_{23}, a_{31}, a_{32}\}$, we obtain the Lie generators to squeeze with respect to the axis z, y, and x, respectively. The versor for the shear transformation called *Shearotor* can be computed either using the bivectors $e_i\bar{e}_j$ of Eq. (5.168) or following the similar approach using the bivectors $w_i\wedge, \bar{w}_j$ of Eq. (5.169) to get

$$\begin{aligned} \boldsymbol{S} &= e^{\boldsymbol{B}_S}, \\ &= e^{\frac{a_{23}(e_2\wedge e_3-\bar{e}_3\wedge\bar{e}_2)+a_{13}(e_1\wedge e_3-\bar{e}_3\wedge\bar{e}_1)+a_{12}(e_1\wedge e_2-\bar{e}_2\wedge\bar{e}_1)}{2}} \\ &\qquad e^{\frac{a_{32}(e_3\wedge e_2-\bar{e}_2\wedge\bar{e}_3)+a_{31}(e_3\wedge e_1-\bar{e}_1\wedge\bar{e}_3)+a_{21}(e_2\wedge e_1-\bar{e}_1\wedge\bar{e}_2)}{2}}, \\ &= e^{\frac{a_{23}(\bar{w}_2\wedge w_3-w_3\wedge\bar{w}_2)+a_{13}(\bar{w}_1\wedge w_3-w_3\bar{w}_1)+a_{12}(\bar{w}_1\wedge w_2-w_2\wedge\bar{w}_1)}{2}} \\ &\qquad e^{\frac{a_{32}(\bar{w}_3\wedge w_2-w_2\wedge\bar{w}_3)+a_{31}(\bar{w}_3\wedge w_1-w_1\bar{w}_3)+a_{21}(\bar{w}_2\wedge w_1-w_1\wedge\bar{w}_2)}{2}}. \end{aligned} \tag{5.195}$$

Dilation: The Lie operators for scaling or dilation will be found by taking the derivative with respect to the parameters of the dilation transformation matrix $\mathtt{D}$ in the transformation $\mathtt{T}$ of Eq. (5.164)

$$\mathtt{D} = \begin{bmatrix} \{a_{11}, e^{\pm s_1}\} & 0 & 0 \\ 0 & \{a_{22}, e^{\pm s_2}\} & 0 \\ 0 & 0 & \{a_{33}, e^{\pm s_3}\} \end{bmatrix}, \tag{5.196}$$

where you can represent certain dilation transformation in terms of the diagonal entries either $a_{i,i}$ or $e^{\pm s_i}$ $(i = 1 \ldots 3)$. Now, for the Lie operator for a directional scaling by a_{11} or e^{s_1} along axis x, we take the derivative with respect to the parameter a_{11} or s_1

$$\begin{aligned} \mathcal{L}_{\mathcal{D}_{\{a_{11},s_1\}}} &= \frac{\mathtt{d}\mathtt{T}_{\mathcal{D}_{\{a_{11},s_1\}}}}{\mathtt{d}\{a_{11}, s_1\}}|_{\{a_{11},s_1\}} \to 0, \\ &= \frac{\mathtt{d} \begin{bmatrix} \begin{bmatrix} a_{11}, e^{s_1} & 0 & 0 \\ 0 & -a_{22}, e^{-s_1} & 0 \\ 0 & 0 & -a_{33}, e^{-s_1} \end{bmatrix} & \mathtt{0} \\ \mathtt{O} & \mathcal{D}_{\{-a_{11},-s_1\}} \end{bmatrix}}{\mathtt{d}\{a_{11}, s_1\}}|_{\{a_{11},s_1\}} \to 0, \end{aligned}$$

$$= \frac{\mathrm{d}\begin{bmatrix} \mathcal{D}_{\{a_{11},s_1\}} & 0 \\ \circ & \begin{bmatrix} -a_{11}, e^{-s_1} & 0 & 0 \\ 0 & a_{22}, e^{s_1} & 0 \\ 0 & 0 & a_{33}, e^{s_1} \end{bmatrix} \end{bmatrix}}{\mathrm{d}\{a_{11}, s_1\}} |_{\{a_{11},s_1\}} \to 0,$$

$$= \begin{bmatrix} \begin{bmatrix} 1,1 & 0 & 0 \\ 0 & 0,-1 & 0 \\ 0 & 0 & 0,-1 \end{bmatrix} & \circ \\ \circ & \begin{bmatrix} -1,-1 & 0 & 0 \\ 0 & 0,1 & 0 \\ 0 & 0 & 0,1 \end{bmatrix} \end{bmatrix}.$$

If we select the sign at the exponents in different combinations, we can generate Lie operators for scaling with respect to the axis z, y and x or also pull along an axis. Gathering the dilation Lie operators, we obtain the dilation Lie operator where the diagonal entries can be zero or ± 1

$$\mathcal{L}_{D_{a_{ii},s_i}} = \begin{bmatrix} \begin{bmatrix} 1/0, \pm 1 & 0 & 0 \\ 0 & 1/0, \pm 1 & 0 \\ 0 & 0 & 1/0, \pm 1 \end{bmatrix} & \circ \\ \circ & \begin{bmatrix} -1/0, \mp 1 & 0 & 0 \\ 0 & -1/0, \mp 1 & 0 \\ 0 & 0 & -1/0, \mp 1 \end{bmatrix} \end{bmatrix}. \tag{5.197}$$

In case of the Dilator since $e_i \wedge e_i = 0$, we will use the bivectors $e_i \wedge \bar{e}_i$ insteadl; thus, the table is accommodated as follows

$$\mathcal{L}_{D_{a_{ii},s_i}} = \begin{bmatrix} \circ & \begin{bmatrix} -1/0, \mp 1 & 0 & 0 \\ 0 & -1/0, \mp 1 & 0 \\ 0 & 0 & -1/0, \mp 1 \end{bmatrix} \\ \begin{bmatrix} 1/0, \pm 1 & 0 & 0 \\ 0 & 1/0, \pm 1 & 0 \\ 0 & 0 & 1/0, \pm 1 \end{bmatrix} & \circ \end{bmatrix}. \tag{5.198}$$

The versor for the dilation transformation called *Dilator* can be computes either using the bivectors $e_i \bar{e}_j$ of Eq. (5.168) or following the similar approach using the bivectors $\bar{w}_i \wedge w_j$ of Eq. (5.169) to get

$$
\begin{aligned}
\boldsymbol{D} = &\ e^{\boldsymbol{B}_D}, \\
= &\ e^{\frac{1}{2}\{(1/0,\pm)\bar{e}_1\wedge e_1+(0/1,\mp)\bar{e}_2\wedge e_2+(0/1,\mp)\bar{e}_3\wedge e_3\}} \\
&\ e^{\frac{1}{2}\{(-1/0,\pm)e_1\wedge\bar{e}_1+(0/-1,\mp)e_2\wedge\bar{e}_2+(0/-1,\mp)e_3\wedge\bar{e}_3\}}, \\
= &\ e^{\frac{1}{2}\{(1/0,\pm)\bar{w}_1\wedge w_1+(0/1,\mp)\bar{w}_2\wedge w_2+(0/1,\mp)\bar{w}_3\wedge w_3\}} \\
&\ e^{\frac{1}{2}\{(-1/0,\pm)w_1\wedge\bar{w}_1+(0/-1,\mp)w_2\wedge\bar{w}_2+(0/-1,\mp)w_3\wedge\bar{w}_3\}}.
\end{aligned}
\tag{5.199}
$$

5.8 2D Projective Geometry in $\mathbb{R}^{3,3}$

The homography $\mathtt{H}$ in $\mathbb{P}^2$ is given by

$$
\mathtt{H}_{3\times 3} = \begin{bmatrix} \mathtt{A}_{2\times 2} & \mathbf{t}_{2\times 1} \\ \mathbf{f}_{1\times 2} & h_{1\times 1}, \end{bmatrix}, \tag{5.200}
$$

which has 8 DOF, as an homography is a projective transformation up to a scalar factor. The projective transformation for the projective plane $\mathbb{P}^2$ can be represented using the unit basis or the null basis as follows

$$
\mathtt{T}_{e_i,\bar{e}_j} = \begin{bmatrix}
a_{11} & a_{12} & 0 & 0 & 0 & f_2 \\
a_{21} & a_{22} & 0 & 0 & 0 & -f_1 \\
0 & 0 & 0 & -f_2 & f_1 & 0 \\
0 & 0 & -t_2 & -a_{11} & -a_{21} & 0 \\
0 & 0 & t_1 & -a_{12} & -a_{22} & 0 \\
t_2 & -t_1 & 0 & 0 & 0 & 0
\end{bmatrix}, \tag{5.201}
$$

$$
\mathtt{T}_{w_i,\bar{w}_j} = \begin{bmatrix}
0 & 0 & f_2 & -a_{11} & -a_{21} & 0 \\
0 & 0 & -f_1 & -a_{12} & -a_{22} & 0 \\
-f_2 & f_1 & 0 & 0 & 0 & 0 \\
a_{11} & a_{12} & 0 & 0 & 0 & -t_2 \\
a_{21} & a_{22} & 0 & 0 & 0 & t_1 \\
0 & 0 & 0 & t_2 & -t_1 & 0
\end{bmatrix}. \tag{5.202}
$$

Following a similar methodology for the versors obtained for the 3D projective geometry in $\mathbb{P}^3$ and using Eqs. (5.201) and (5.202), one can derive the versors for the Lie groups of $\mathbb{P}^2$, namely for the translation (2 DOF), rotation (1 DOF) about the axis of the plane $e_1\wedge e_2$, scaling (2 DOF), perspectivity (2 DOF), and shear (1 DOF). The derivation of these versors are left as exercises. You will see that in all these versors simply the bivectors involving the basis elements e_3, $\bar{e}_3$, w_3, $\bar{w}_3$ are not present. Furthermore, compare these versors with those of Sect. 5.10.5, and you will see that the Lie algebra of the affine plane is exactly the same as the Lie algebra of $\mathbb{P}^2$ in $\mathbb{R}^{3,3}$.

5.9 Horosphere and n-Dimensional Affine Plane

This section explains briefly the meaning of the computational frameworks the horosphere and the n-dimensional affine plane, which are useful in the study of conformal transformations [240]. This kind of transformation preserves angles between tangent vectors at each point. A common conformal transformation is the one defined by any analytic function in the complex plane. Conformal transformations also exist in the *pseudo-Euclidean* space $\mathbb{R}^{p,q}$. Since conformal transformations are nonlinear transformations, it will be desirable to linearize them. One way to do so is by moving up from the affine plane $\mathcal{A}_w(\mathbb{R}^{p,q})$ to the (p, q)-horosphere $\mathcal{H}_w^{p,q}(R^{p+1,q+1})$. Be aware that in the equations of Sects. 5.9 and 5.10, we use the *Witt basis* of null vector $\{w\}$ and the corresponding reciprocal basis $\{\bar{w}\}$ of $G_{n,n}$

5.9.1 The Horosphere

The *n-dimensional affine plane* $\mathcal{A}_w(\mathbb{R}^{p,q})$ is a homogeneous representation of the points $\boldsymbol{x} \in \mathbb{R}^{p,q}$. It extends the $\mathbb{R}^{p,q}$ to a projective space with signature $\mathbb{R}^{p,q,1}$ using a null vector w as follows:

$$\boldsymbol{x}_h = \boldsymbol{x} + w \in \mathcal{A}_w(\mathbb{R}^{p,q}). \tag{5.203}$$

The $(p,\ q)$- horosphere is normally defined by

$$\mathcal{H}_w^{p,q}(R^{p+1,q+1}) = \{\frac{1}{2}\boldsymbol{x}_h\overline{w}\boldsymbol{x}_h | \boldsymbol{x}_h \in \mathcal{A}_w(\mathbb{R}^{p,q})\} \in R^{p+1,q+1}, \tag{5.204}$$

where the space $\mathbb{R}^{p,q}$ has been extended to $\mathbb{R}^{p+1,q+1}$ in order to have available two null vectors, $w = w_{n+1}$ and $\overline{w} = \overline{w}_{n+1}$. The conformal representation $\boldsymbol{x}_c$ of both a point $\boldsymbol{x} \in R^{p,q}$ and a point $\boldsymbol{x}_h \in \mathcal{A}_e(\mathbb{R}^{p,q})$ is given by

$$\begin{aligned}\boldsymbol{x}_c &= \frac{1}{2}\boldsymbol{x}_h\overline{w}\boldsymbol{x}_h = \frac{1}{2}[(\boldsymbol{x}_h \cdot \overline{w})\boldsymbol{x}_h + (\boldsymbol{x}_h \wedge \overline{w})\boldsymbol{x}_h] = \boldsymbol{x}_h - \frac{1}{2}\boldsymbol{x}_h^2\overline{w},\\ &= \boldsymbol{x} - \frac{1}{2}\boldsymbol{x}_h^2\overline{w} + w = exp(\frac{1}{2}\boldsymbol{x}\overline{w})\ w\ \ exp(-\frac{1}{2}\boldsymbol{x}\overline{w}).\end{aligned} \tag{5.205}$$

This equation tells us that all points on $\mathcal{H}_w^{p,q}$ can be obtained by a simple rotation of w with respect to the plane indicated by the bivector $\boldsymbol{x}\overline{w}$.

5.9.2 The n-Dimensional Affine Plane

The points of the horosphere can be projected down into the affine plane by applying the simple formula,

$$\boldsymbol{x}_h = (\boldsymbol{x}_c \wedge \overline{w}) \cdot w, \tag{5.206}$$

and into the space $\mathbb{R}^{p,q}$ by using

$$\boldsymbol{x} = (\boldsymbol{x}_c \wedge \overline{w} \wedge w) \cdot (\overline{w} \wedge w). \tag{5.207}$$

Figure 5.1 depicts

$$\begin{aligned} \mathcal{A}_w^2 &= \{\boldsymbol{x}_h | \boldsymbol{x}_h = \boldsymbol{x} + w, \boldsymbol{x} \in \mathbb{R}^2\}, \\ \mathcal{H}_w^2 &= \{\boldsymbol{x}_c = \frac{1}{2}\boldsymbol{x}_h \overline{w} \boldsymbol{x}_h | \boldsymbol{x}_h \in \mathcal{A}_w^2\}. \end{aligned} \tag{5.208}$$

Since $\boldsymbol{x} \in \mathbb{R}^2 = span\{e_1, e_2\}$, and the null vectors computed as $w = \frac{1}{2}(e_3 + e_4)$, and $\overline{w} = e_3 - e_4$, any point on the horosphere in these terms is given by

$$\begin{aligned} \boldsymbol{x}_c &= \boldsymbol{x} - \frac{1}{2}\boldsymbol{x}^2 \overline{w} + w = \boldsymbol{x} - \frac{1}{2}(\boldsymbol{x}^2 - 1)e_3 + \frac{1}{2}(\boldsymbol{x}^2 + 1)e_4, \\ &= \boldsymbol{x} + x_3 e_3 + x_4 e_4. \end{aligned} \tag{5.209}$$

In order to be able to depict the horosphere in 3D, in Fig. 5.1, we have ignored the coordinate x_3, considering instead the condition that e_4 is orthogonal to e_1, e_2. In Sect. 5.11, we will use the frameworks of the n-dimensional affine plane and horosphere for computations of incidence algebra.

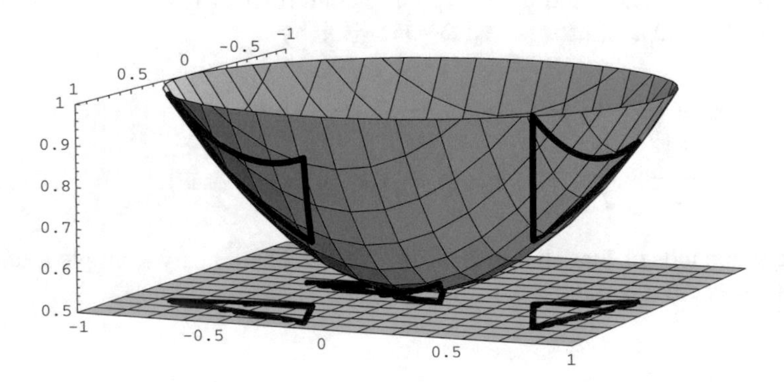

Fig. 5.1 Horosphere of R^2 with triangles of the 2D affine plane projected into the horosphere

5.10 The General Linear Algebra $gl(\mathcal{N})$ of the General Linear Lie Group $GL(\mathcal{N})$

Since each linear operator $f \in End(\mathcal{N})$ can be represented according to Eq. (5.63) in the bivector form, in this section, we study the Lie group $GL(\mathcal{N})$ and the linear Lie algebra $gl(\mathcal{N})$ using the bivector form.

5.10.1 *The General Linear Lie Group $GL(\mathcal{N})$*

The general linear group $GL(\mathcal{N})$ is defined to be the subset of all *endomorphisms* $f \in End(\mathcal{N})$, with the property that $f \in GL(\mathcal{N})$ if and only if $\det(f) \neq 0$ [31]. The determinant of f is defined in the algebra G_N by

$$f(w_1)\wedge f(w_2)\wedge \ldots \wedge f(w_n) = \det(\mathcal{F})w_1\wedge w_2\wedge \ldots \wedge w_n, \tag{5.210}$$

where $\det(\mathcal{F})$ is just the ordinary determinant of the matrix of f with respect to the basis $\{w\}$. Choosing the basis $\{w\}$ makes explicit the isomorphism between the general linear groups $GL(\mathcal{N})$ and $GL(n, \mathbb{C})$. The latter corresponds to the general linear group of all complex $n \times n$ matrices $\mathcal{F}$ with $\det \mathcal{F} \neq 0$. The theory of Lie groups and their corresponding Lie algebras can be considered to be largely the study of the group manifold $GL(n, \mathbb{C})$, since any Lie group is isomorphic to a subgroup of $GL(n, \mathbb{C})$ [99, p. 501].

Since we have referred to $GL(\mathcal{N})$ as a *manifold*, we must be careful to give it the structure of an n^2-dimensional topological metric space. We define the inner product $<f, g>$ of $f, g \in GL(\mathcal{N})$ to be the usual Hermitian positive-definite inner product

$$<f, g>= \sum_{j=1}^{n}\sum_{i=1}^{n} \overline{f_{ij}} g_{ij},$$

where $f_{ij}, g_{ij} \in \mathbb{C}$ are the components of the matrices $\mathcal{F}$ and $\mathcal{G}$ of f and g, respectively, with respect to the basis $\{w\}$. The positive definite norm $|f|$ of $f \in GL(\mathcal{N})$ is defined by

$$|f|^2 =<f, f>= \sum_{j=1}^{n}\sum_{i=1}^{n} \overline{f_{ij}} f_{ij}$$

and is clearly zero if and only if $f = 0$.

The crucial relationship between a Lie group and its corresponding Lie algebra is almost an immediate consequence of the properties of the exponential of a linear operator $f \in End(\mathcal{N})$. The *exponential mapping* may be directly defined by the usual Taylor series

$$e^f = \sum_{i=0}^{\infty} \frac{f^i}{i!},$$

where convergence is with respect to the norm $|f|$. Note that $f^0 = 1$ is the identity operator on $\mathcal{N}$ and that f^k is the composition of f with itself k times.

The logarithm of a linear operator, $\theta_f = \log(f)$, exists and is well defined for any $f \in GL(\mathcal{N})$. The logarithm can also be defined in terms of an infinite series, or more directly, in terms of the *spectral form* of f [277]. Since the logarithm is the inverse function of the exponential function, we can write $f = e^{\theta_f}$ for any $f \in GL(\mathcal{N})$. The logarithmic form $f = e^{\theta_f}$ of $f \in GL(\mathcal{N})$ is useful for defining the *one-parameter group* $\{f_t\}$ of the operator $f \in GL(\mathcal{N})$,

$$f_t(\boldsymbol{x}) = e^{t\theta_f}\boldsymbol{x}. \tag{5.211}$$

The one-parameter group $\{f_t\}$ is *continuously connected to the identity* in the sense that $f_0(\boldsymbol{x}) = \boldsymbol{x}$, and $f_1(\boldsymbol{x}) = f(\boldsymbol{x})$. Note that

$$f_0'(\boldsymbol{x}) = \theta_f e^{t\theta_f}|_{t=0}(\boldsymbol{x}) = \theta_f(\boldsymbol{x}), \tag{5.212}$$

so θ_f is *tangent* to f_t at the identity. The reason that $\{f_t\}$ is called a one-parameter group is because it satisfies the basic additive property

$$f_s f_t = e^{t\theta_f} e^{s\theta_f} = e^{(s+t)\theta_f} = f_{s+t}. \tag{5.213}$$

Since each linear operator $f \in End(\mathcal{N})$ can be represented according to Eq. (5.63) in the bivector form $f(x) = F \cdot \boldsymbol{x}$, we can express the one-parameter group $g_t x = e^{tf} x$ of the skew-symmetric transformation $f(x) = F \cdot x$ in the form

$$g_t x = e^{tf} x \equiv e^{\frac{t}{2}F} x e^{-\frac{t}{2}F}. \tag{5.214}$$

This equation can be proved by showing that the terms of the Taylor series expansion of both sides of Eq. (5.214) are identical at $t = 0$.

We begin with

$$e^{tf} x \doteq e^{\frac{t}{2}F} x e^{-\frac{t}{2}F}. \tag{5.215}$$

Clearly, for $t = 0$, we have

$$e^{0f} x = e^{0F} x e^{0F} = x.$$

Next, taking the first derivative of both sides of Eq. (5.215), we get

$$e^{tf} fx \doteq \frac{1}{2} F e^{\frac{t}{2}F} x e^{-\frac{t}{2}F} - \frac{1}{2} F e^{\frac{t}{2}F} x e^{-\frac{t}{2}F} = e^{\frac{t}{2}F} (F \cdot x) e^{-\frac{t}{2}F}. \tag{5.216}$$

Setting t equal to zero gives the identity $f(x) = F \cdot x$.

Taking the derivative of both sides of (5.216) gives

$$e^{tf} f^2 x \dot{=} \frac{1}{2} F e^{\frac{t}{2}F}(F\cdot x)e^{-\frac{t}{2}F} - \frac{1}{2} F e^{\frac{t}{2}F}(F\cdot x)e^{-\frac{t}{2}F} = e^{\frac{t}{2}F}[F\cdot(F\cdot x)]e^{-\frac{t}{2}F},$$

and setting t equal to zero gives the identity $f^2(x) = F\cdot(F\cdot x)$. Continuing to take successive derivatives of (5.215) gives

$$e^{tf} f^k(x) \dot{=} e^{\frac{t}{2}F}(F^k : x)e^{-\frac{t}{2}F}, \tag{5.217}$$

where $F^k : x$ is defined recursively by $F^1 : x = F\cdot x$ and

$$F^k : x = F\cdot(F^{k-1} : x). \tag{5.218}$$

Finally, setting t equal to zero in Eq. (5.217) gives the identity

$$f^k(x) = F^k : x.$$

Equation (5.218) is interesting because it expresses the powers of a linear operator in terms of "powers" of its defining bivector. It is clear that each bivector defines a unique skew-symmetric linear operator, and conversely, that each skew-symmetric linear operator defines a unique bivector (see Eq. (5.63)). Thus, the study of the structure of a bivector is determined by and uniquely determines the corresponding structure of the corresponding linear operator. The proof of the above theorem is attributable to Marcel Riesz [248].

5.10.2 The General Linear Algebra gl($\mathcal{N}$)

We can now define the *general linear Lie algebra* $gl(\mathcal{N})$ of the general linear Lie group $GL(\mathcal{N})$. As a set, $gl(\mathcal{N}) \equiv End(\mathcal{N})$, which is just the set of all *tangent operators* $\theta_f = \log(f) \in End(\mathcal{N})$ to the one-parameter groups $f_t = e^{t\theta_f}$ defined for each $f \in GL(\mathcal{N})$. But to complete the definition of $gl(\mathcal{N})$, we must specify the algebraic operations of addition and multiplication which allow $End(\mathcal{N})$ to be seen as the Lie algebra $gl(\mathcal{N})$. Addition requires only the ordinary addition of linear operators, but multiplication is defined by the *Lie bracket* $[\theta_f, \theta_g]$ for $\theta_f, \theta_g \in gl(\mathcal{N})$. We will give an analytic definition of the Lie bracket [214, p. 3] which directly ties it to the group structure of $GL(\mathcal{N})$

$$[\theta_f, \theta_g] = \frac{d}{d(t^2)} f_t g_t f_{-t} g_{-t}|_{t=0} = \frac{1}{2t}\frac{d}{dt} f_t g_t f_{-t} g_{-t}|_{t=0}.$$

Evaluating the Lie bracket by using the Taylor series expansions,

$$
\begin{aligned}
[\theta_f, \theta_g] &= \frac{1}{2t}\frac{d}{dt}(f_t g_t f_{-t} g_{-t})\big|_{t=0}, \\
&= \frac{1}{2t}\big(\theta_f f_t g_t f_{-t} g_{-t} + f_t \theta_g g_t f_{-t} g_{-t} - f_t g_t \theta_f f_{-t} g_{-t} - f_t g_t f_{-t} \theta_g g_{-t}\big), \\
&= \left(\frac{1}{2t} f_t(\theta_f g_t - g_t \theta_f) f_{-t} g_{-t}\right)|_{t=0} + \left(\frac{1}{2t} f_t g_t (\theta_g f_{-t} - f_{-t}\theta_g) g_{-t}\right)|_{t=0}, \\
&= \frac{1}{2}(\theta_f\theta_g - \theta_g\theta_f) + \frac{1}{2}(-\theta_g\theta_f + \theta_f\theta_g), \\
&= \theta_f\theta_g - \theta_g\theta_f,
\end{aligned}
\tag{5.219}
$$

we find that

$$g_t = 1 + t\theta_g + \cdots, \quad \text{and} \quad f_{-t} = 1 - t\theta_f + \cdots.$$

We have thus demonstrated that the Lie bracket, defined analytically above, reduces simply to the commutator product of the linear operators θ_f and θ_g in $gl(\mathcal{N})$. As such, it is not difficult to show that it satisfies the Jacobi identity of Eq. (5.11), which is equivalent to the distributive law

$$[\theta_f, [\theta_g, \theta_h]] = [[\theta_f, \theta_g], \theta_h] + [\theta_g, [\theta_f, \theta_h]]. \tag{5.220}$$

When we choose a particular basis $\{w\}$ of $\mathcal{N}$, the isomorphism between the general linear Lie algebra $gl(n, \mathbb{C})$ and $gl(\mathcal{N})$ becomes explicit, and the Lie bracket of linear operators just becomes the Lie bracket of $n \times n$ matrices,

$$[f, g]\{w\} = fg\{w\} - gf\{w\} = \{w\}(\mathcal{F}\mathcal{G} - \mathcal{G}\mathcal{F}) = \{w\}[\mathcal{F}, \mathcal{G}], \tag{5.221}$$

where $[\mathcal{F}, \mathcal{G}]$ are the commutator products of the matrices $\mathcal{F}$ and $\mathcal{G}$. Alternatively, using the bivector representation of Eq. (5.63), the Lie bracket of linear operators is expressed in terms of the Lie bracket of the bivectors of the operators (5.67).

5.10.3 The Orthogonal Groups

The simplest well-known example of an *orthogonal group* is $SO(2)$, which is a subgroup of the general linear group $GL(\mathcal{N}^2)$. As a matrix group, it is generated by all 2×2 matrices of the form

$$X_\theta = \begin{bmatrix} \cos\theta & -\sin\theta \\ \sin\theta & \cos\theta \end{bmatrix}. \tag{5.222}$$

The matrix X_θ generates a counterclockwise rotation in the xy-plane through the angle θ. Using (5.63), we get the corresponding bivector representation

$$\boldsymbol{X}_\theta = \cos(\theta) w_1 \wedge \bar{w}_1 - \sin(\theta) w_1 \wedge \bar{w}_2 + \sin(\theta) w_2 \wedge \bar{w}_1 + \cos(\theta) w_2 \wedge \bar{w}_2. \quad (5.223)$$

For matrices $X_{\theta_1}, X_{\theta_2} \in SO(2)$, the group operation is ordinary matrix multiplication, $X_{\theta_2} X_{\theta_1} = X_{\theta_1+\theta_2}$. For the bivector representation $\boldsymbol{X}_{\theta_1}, \boldsymbol{X}_{\theta_2} \in SO(2)$, the group operation is defined by its *generalized dot product*, that is, for $\boldsymbol{x} \in \mathcal{N}^2$,

$$(\boldsymbol{X}_{\theta_1} : \boldsymbol{X}_{\theta_2}) \equiv \boldsymbol{X}_{\theta_2} \cdot (\boldsymbol{X}_{\theta_1} \cdot \boldsymbol{x}) = \boldsymbol{X}_{\theta_1+\theta_2} \cdot \boldsymbol{x}. \quad (5.224)$$

Note that the bivectors $\boldsymbol{X}_\theta$ are in $G^2_{n,n}$.

Taking the derivatives of X_θ and $\boldsymbol{X}_\theta$ with respect to θ and evaluating at $\theta = 0$ gives the corresponding generators of the associated Lie algebra $so(2)$. As a matrix Lie algebra under the bracket operation of matrices, we find the generator

$$\frac{dX_\theta}{d\theta}|_{\theta\to 0} = \begin{bmatrix} 0 & -1 \\ 1 & 0 \end{bmatrix}. \quad (5.225)$$

As a bivector Lie algebra under the bracket operation of bivectors, we use Eq. (5.223) to find the bivector generator

$$\boldsymbol{B} = \frac{d\boldsymbol{X}_\theta}{d\theta}|_{\theta\to 0} = -w_1 \wedge \bar{w}_2 + w_2 \wedge \bar{w}_1 = -e_1 e_2 + \bar{e}_1 \bar{e}_2. \quad (5.226)$$

The *spinor group Spin*(2) is defined by taking the exponential of the bivector (see Eq. (5.226)),

$$Spin(2) = \{\exp(\frac{1}{2}\theta \boldsymbol{B})|\ \theta \in \mathbb{R}\}.$$

According to Sobczyk [278], the exponential $\exp(\frac{1}{2}\theta\boldsymbol{B})$ can be calculated by noting that the bivector $\boldsymbol{B}$ satisfies the *minimal polynomial*

$$\boldsymbol{B}^3 + 4\boldsymbol{B} = \boldsymbol{B}(\boldsymbol{B} - 2i)(\boldsymbol{B} + 2i) = 0,$$

which implies the decomposition

$$\boldsymbol{B} = 0p_1 + 2ip_2 - 2ip_3,$$

where the mutually annihilating idempotents are defined by

$$p_1 = \frac{\boldsymbol{B}^2 + 4}{4}, \quad p_2 = -\frac{1}{8}\boldsymbol{B}(\boldsymbol{B} + 2i), \quad p_3 = -\frac{1}{8}\boldsymbol{B}(\boldsymbol{B} - 2i).$$

Using this decomposition, we find that

$$\begin{aligned}\exp(\frac{1}{2}\theta\boldsymbol{B}) &= \exp(\frac{0\cdot\theta}{2})p_1 + \exp(i\theta)p_2 + \exp(-i\theta)p_3,\\ &= p_1 + \cos(\theta)(p_2+p_3) + \sin(\theta)i(p_2-p_3),\\ &= p_1 + \cos(\theta)(p_2+p_3) + \frac{\boldsymbol{B}}{2}\sin(\theta).\end{aligned} \tag{5.227}$$

The group action of the *spinor group* $Spin(2)$ is given by

$$\boldsymbol{x}' = \exp(\frac{1}{2}\theta\boldsymbol{B})\boldsymbol{x}\exp(-\frac{1}{2}\theta\boldsymbol{B}),$$

where $\boldsymbol{x} = \{w\}x_{\{w\}} = x_1w_1 + x_2w_2$. We say that $Spin(2)$ is a "double covering" of the orthogonal group $SO(2)$ because the spinors $\pm\exp(\frac{1}{2}\theta\boldsymbol{B})$ represent the same group element. Note that now we can formulate the easy rule for the composition of two group elements, $\exp(\frac{1}{2}\theta_1\boldsymbol{B})$ and $\exp(\frac{1}{2}\theta_2\boldsymbol{B})$,

$$\exp(\frac{1}{2}\theta_1\boldsymbol{B})\exp(\frac{1}{2}\theta_2\boldsymbol{B}) = \exp(\frac{1}{2}(\theta_1+\theta_2)\boldsymbol{B}).$$

If we are only interested in the group $SO(2)$, a more natural place to carry out the calculations is in the Euclidean space $\mathbb{R}^2$. We project the null cone $\mathcal{N}^2$ down to $\mathbb{R}^2$ by using the reciprocal pseudoscalars I_2 and $\overline{I_2}$ defined by

$$I_2 = e_1e_2 \quad \text{and} \quad \overline{I_2} = (2-\sqrt{2})^2(\bar{w}_2+e_2)(\bar{w}_1+e_1).$$

Thus, for $\boldsymbol{x} = \{w\}x_{\{w\}} = x_1w_1 + x_2w_2 \in \mathcal{N}^2$, the projection $\boldsymbol{x}' = P_I(\boldsymbol{x})$ gives

$$\boldsymbol{x}' = P_I(\boldsymbol{x}) = (\boldsymbol{x}\cdot\overline{I})\cdot I = x_1e_1 + x_2e_2 \in \mathbb{R}^2.$$

Note that this projection is invertible, in the sense that we can find $P_{I'}$ such that $\boldsymbol{x} = P_{I'}(\boldsymbol{x}')$. The projection $P_{I'}$ is specified by

$$\boldsymbol{x} = P_{I'}(\boldsymbol{x}') = (\boldsymbol{x}'\cdot\overline{I})\cdot I' = x_1w_1 + x_2w_2, \tag{5.228}$$

where $\overline{I}$ is defined as before and where $I' = w_1w_2$.

In $\mathbb{R}^2$, the generator of rotations is the simple bivector e_2e_1. This bivector can be obtained from the bivector (5.226) in $spin(2,2)$ by the simple projection

$$P_I(\boldsymbol{B}) = I_2^{-1}I_2\cdot\boldsymbol{B} = e_2e_1 = -I_2, \tag{5.229}$$

onto the Lie algebra $so(2)$. For $\boldsymbol{x}' = x_1e_1 + x_2e_2 \in \mathbb{R}^2$, the equivalent rotation is given by

$$\boldsymbol{y}' = \exp(\frac{1}{2}\theta e_2e_1)\boldsymbol{x}'\exp(-\frac{1}{2}\theta e_2e_1). \tag{5.230}$$

The above ideas can be immediately generalized to the general Lie group $GL(\mathcal{N}^n)$ of null cone $\mathcal{N}^n$ and the orthogonal subgroups $SO(p,q)$ where $p+q=n$. The orthogonal group $SO(p,q)$ acts on the space $\mathbb{R}^{p,q}$. Thus, if we wish to work in this Lie group or in the corresponding Lie algebra, we first project the null cone $\mathcal{N}^n$ onto $\mathbb{R}^{p,q}$ by using the reciprocal vector basis elements, then carry out the rotation, and finally return to the null cone by using the inverse projection.

5.10.4 Computing Rigid Motion in the Affine Plane

A rotation in the affine n-plane $\mathcal{A}_w^n = \mathcal{A}_w(\mathbb{R}^n)$, just as in the Euclidean space $\mathbb{R}^n$, is the product of two reflections through two intersecting hyperplanes. If the normal unit vectors to these hyperplanes are m and n, respectively, then the versor of the rotation is given by

$$R = mn = e^{\frac{\theta}{2}B} = \cos(\frac{\theta}{2}) + B\sin(\frac{\theta}{2}), \tag{5.231}$$

where B is the unit bivector defining the plane of the rotation.

A translation of the vector $x_h \in \mathcal{A}_w^n$, along the vector $t \in \mathbb{R}^n$, to the vector $x_h' = x_h + t \in \mathcal{A}_w^n$, is effected by the versor

$$T = \exp(\frac{1}{2}t\bar{w}) = 1 + \frac{1}{2}t\bar{w}, \tag{5.232}$$

when it is followed by the projection $P_A(x') \equiv (x\wedge\bar{w})\cdot w$, which brings the horosphere *back* into the affine plane. Thus, for $x_h \in \mathcal{A}_w^n$, we get

$$\begin{aligned} x' = TxT^{-1} &= \exp(\frac{1}{2}t\bar{w})x_h\exp(-\frac{1}{2}t\bar{w}), \\ &= (1+\frac{1}{2}t\bar{w})x_h(1-\frac{1}{2}t\bar{w}) = x_h + \frac{1}{2}t\bar{w}x_h - \frac{1}{2}x_ht\bar{w} - \frac{1}{4}t\bar{w}x_ht\bar{w}, \\ &= x_h + t + t\cdot(\bar{w}\wedge x^h) - \frac{1}{2}t^2\bar{w}, \\ &= x_h + t - (t\cdot x_h + \frac{1}{2}t^2)\bar{w}. \end{aligned} \tag{5.233}$$

Applying P_A to this result, we get the expected translated vector

$$x_h' = P_A(x') = P_A[x_h + t - (t\cdot x_h + \frac{1}{2}t^2)\bar{w}] = x_h + t. \tag{5.234}$$

The advantage of carrying out translations in the affine plane rather than in the horosphere is that the affine plane is basically still a linear model of Euclidean space, whereas the horosphere is a more complicated *nonlinear* model.

Combining the versors for a rotation and a translation, we get the expression for the versor $M = TR$ of a rigid motion. For $x_h \in \mathcal{A}^n_w$, we then find that

$$x'_h = P_A[Mx_hM^{-1}] = P_A[TRx_hR^{-1}T^{-1}]. \tag{5.235}$$

Equivalently, we will often write $M^{-1} \equiv \widetilde{M}$, expressing M^{-1} in terms of the operation of *conjugation*. Whenever a calculation involves a translation, we must always apply the projection P_A to guarantee that our end result will be in the affine plane. The above calculations can be checked with the Clifford algebra calculator CLICAL 4.0 [196]. Comparisons can also be made to the corresponding calculations made by Hestenes and Li [186] on the horosphere.

5.10.5 The Lie Algebra of the Affine Plane

The Lie algebra of the neutral affine plane $\mathcal{A}_{w_3}(\mathcal{N}^2)$ is useful in the analysis of visual invariants [145], so we will begin with its treatment here. An homography $\mathtt{H}$ in $\mathbb{P}^2$ is given by

$$\mathtt{H} = \begin{bmatrix} \mathtt{A} & \mathbf{t} \\ \mathbf{f} & h \end{bmatrix}, \tag{5.236}$$

where $\mathtt{A}$ is an affine transformation matrix, $\mathbf{t}$ a 2D translation vector, and if h=1, the vector $\mathbf{f} = \mathtt{0}$ corresponds to the well-known matrix representation of the Lie group of affine transformations in the plane, has six independent parameters, or degrees of freedom, and consists of all matrices of the form

$$g(A, \mathbf{v}) = \begin{bmatrix} a_{11} & a_{12} & a \\ a_{21} & a_{22} & b \\ 0 & 0 & 1 \end{bmatrix}, \tag{5.237}$$

where $\det g(A, \mathbf{v}) = \det A \neq 0$.

The one-parameter subgroups are generated by the matrices

$$T_x = \begin{bmatrix} 1 & 0 & x \\ 0 & 1 & 0 \\ 0 & 0 & 1 \end{bmatrix}, \qquad T_y = \begin{bmatrix} 1 & 0 & 0 \\ 0 & 1 & y \\ 0 & 0 & 1 \end{bmatrix},$$
$$D_u = \begin{bmatrix} e^u & 0 & 0 \\ 0 & e^u & 0 \\ 0 & 0 & 1 \end{bmatrix}, \quad R_\theta = \begin{bmatrix} cos(\theta) & -sin(\theta) & 0 \\ sin(\theta) & cos(\theta) & 0 \\ 0 & 0 & 1 \end{bmatrix}, \tag{5.238}$$
$$S_v = \begin{bmatrix} e^v & 0 & 0 \\ 0 & e^{-v} & 0 \\ 0 & 0 & 1 \end{bmatrix}, H_\phi = \begin{bmatrix} cosh(\phi) & sinh(\phi) & 0 \\ sinh(\phi) & cosh(\phi) & 0 \\ 0 & 0 & 1 \end{bmatrix}.$$

Using Eq. (5.212), we obtain the matrix representation of the Lie algebra basis generators by taking the derivative of Eq. (5.238) and evaluating the parameter at zero:

$$\mathcal{L}_x = \begin{bmatrix} 0 & 0 & 1 \\ 0 & 0 & 0 \\ 0 & 0 & 0 \end{bmatrix}, \quad \mathcal{L}_y = \begin{bmatrix} 0 & 0 & 0 \\ 0 & 0 & 1 \\ 0 & 0 & 0 \end{bmatrix},$$
$$\mathcal{L}_s = \begin{bmatrix} 1 & 0 & 0 \\ 0 & 1 & 0 \\ 0 & 0 & 0 \end{bmatrix}, \quad \mathcal{L}_r = \begin{bmatrix} 0 & -1 & 0 \\ 1 & 0 & 0 \\ 0 & 0 & 0 \end{bmatrix}, \tag{5.239}$$
$$\mathcal{L}_b = \begin{bmatrix} 1 & 0 & 0 \\ 0 & -1 & 0 \\ 0 & 0 & 0 \end{bmatrix}, \quad \mathcal{L}_B = \begin{bmatrix} 0 & 1 & 0 \\ 1 & 0 & 0 \\ 0 & 0 & 0 \end{bmatrix}.$$

The above matrix Lie group and matrix Lie algebra can be directly translated into the corresponding Lie group and Lie algebra of the affine plane $\mathcal{A}_{w_3}(\mathcal{N}^2)$. Each of the matrix generators in (5.238) and (5.239) can be replaced by its corresponding bivector representation (5.63). For example, the bivector representations of the generators of the Lie algebra are

$$\begin{aligned}
\boldsymbol{\mathcal{L}}_x &= bivector(\mathcal{L}_x) = w_1 \wedge \bar{w}_3, \\
\boldsymbol{\mathcal{L}}_y &= bivector(\mathcal{L}_y) = w_2 \wedge \bar{w}_3, \\
\boldsymbol{\mathcal{L}}_s &= bivector(\mathcal{L}_s) = w_1 \wedge \bar{w}_1 + w_2 \wedge \bar{w}_2, \\
\boldsymbol{\mathcal{L}}_r &= bivector(\mathcal{L}_r) = w_2 \wedge \bar{w}_1 - w_1 \wedge \bar{w}_2, \\
\boldsymbol{\mathcal{L}}_b &= bivector(\mathcal{L}_b) = w_1 \wedge \bar{w}_1 - w_2 \wedge \bar{w}_2, \\
\boldsymbol{\mathcal{L}}_B &= bivector(\mathcal{L}_\phi) = w_1 \wedge \bar{w}_2 + w_2 \wedge \bar{w}_1.
\end{aligned} \tag{5.240}$$

Expanding these bivector generators in the standard basis (5.68), we get

$$\begin{aligned}
\boldsymbol{\mathcal{L}}_x &= \tfrac{1}{2} e_1 e_3 - \tfrac{1}{2} e_1 \bar{e}_3 - \tfrac{1}{2} e_3 \bar{e}_1 - \tfrac{1}{2} \bar{e}_1 \bar{e}_3, \\
\boldsymbol{\mathcal{L}}_y &= \tfrac{1}{2} e_2 e_3 - \tfrac{1}{2} e_2 \bar{e}_3 - \tfrac{1}{2} e_3 \bar{e}_2 - \tfrac{1}{2} \bar{e}_2 \bar{e}_3, \\
\boldsymbol{\mathcal{L}}_s &= -e_1 \bar{e}_1 - e_2 \bar{e}_2, \\
\boldsymbol{\mathcal{L}}_r &= -e_1 e_2 + \bar{e}_1 \bar{e}_2, \\
\boldsymbol{\mathcal{L}}_b &= -e_1 \bar{e}_1 + e_2 \bar{e}_2, \\
\boldsymbol{\mathcal{L}}_B &= -e_1 \bar{e}_2 - e_2 \bar{e}_1.
\end{aligned} \tag{5.241}$$

Let us see how the Lie algebra of the affine plane can be represented as a Lie algebra of vector fields over the null cone $\mathcal{N}^3$. The *vector derivative* or *gradient* $\partial_x = \frac{\partial}{\partial x}$ at the point $\boldsymbol{x} = xw_1 + yw_2 + zw_3 \in \mathcal{N}^3$ is defined by requiring $\boldsymbol{a} \cdot \partial_x$ to be the *directional derivative* in the direction of $\boldsymbol{a}$. It follows that $\boldsymbol{a} \cdot \partial_x \boldsymbol{x} = \boldsymbol{a}$. We also have

$$\partial_x \boldsymbol{x} = \partial_x \cdot \boldsymbol{x} + \partial_x \wedge \boldsymbol{x} = 3 + \sum_{i=1}^{3} \bar{w}_i \wedge w_i,$$

where $\{w\}$ and $\{\bar{w}\}$ are reciprocal bases for the reciprocal null cones $\mathcal{N}^3$ and $\overline{\mathcal{N}^3}$.

Now, let $\boldsymbol{a} = \boldsymbol{a}(\boldsymbol{x})$ and $\boldsymbol{b} = \boldsymbol{b}(\boldsymbol{x})$ be vector fields in $\mathcal{N}^3$. The Lie bracket $[\boldsymbol{a}, \boldsymbol{b}]$ is defined by

$$[\boldsymbol{a}, \boldsymbol{b}] = \boldsymbol{a} \cdot \partial_x \boldsymbol{b} - \boldsymbol{b} \cdot \partial_x \boldsymbol{a}.$$

Since in $\mathcal{N}^3$, $\partial_x \wedge \partial_x = 0$, we have the important integrability condition that

$$(\boldsymbol{a} \wedge \boldsymbol{b}) \cdot (\partial_x \wedge \partial_x) = [\boldsymbol{a}, \boldsymbol{b}] \cdot \partial_x - [\boldsymbol{a} \cdot \partial_x, \boldsymbol{b} \cdot \partial_x] = 0,$$

where

$$[\boldsymbol{a} \cdot \partial_x, \boldsymbol{b} \cdot \partial_x] = \boldsymbol{a} \cdot \partial_x \boldsymbol{b} \cdot \partial_x - \boldsymbol{b} \cdot \partial_x \boldsymbol{a} \cdot \partial_x,$$

is the Lie bracket or commutator product of the partial derivatives $\boldsymbol{a} \cdot \partial_x$ and $\boldsymbol{b} \cdot \partial_x$. It follows from this identity that

$$[\boldsymbol{a}, \boldsymbol{b}] \cdot \partial_x = [\boldsymbol{a} \cdot \partial_x, \boldsymbol{b} \cdot \partial_x],$$

which relates the Lie bracket of the vector fields $[\boldsymbol{a}, \boldsymbol{b}]$ to the standard Lie bracket of the partial derivatives $[\boldsymbol{a} \cdot \partial_x, \boldsymbol{b} \cdot \partial_x]$.

Let us consider in detail the translation of the Lie algebra of the affine plane to the null vector formulation in the null cone $\mathcal{N}^2$. Recall that the two-dimensional affine plane $\mathcal{A}_e(\mathcal{N}^2)$ in $\mathcal{N}^3$ is defined by

$$\mathcal{A}_e(\mathcal{N}) = \{\boldsymbol{x} \in \mathcal{N}^3 | \ \boldsymbol{x} = x w_1 + y w_2 + w_3\}. \tag{5.242}$$

We have already seen that the Lie algebra of the affine plane can be defined by a Lie algebra of matrices, or by an equivalent Lie algebra of bivectors. We now define this same Lie algebra as a Lie algebra of partial derivatives or as a Lie algebra of vector fields. We have the following correspondences:

$$\mathcal{L}_x = \frac{\partial}{\partial x} = w_1 \cdot \partial_x = \boldsymbol{\mathcal{L}}_x \cdot (\boldsymbol{x} \wedge \partial_x) \leftrightarrow \mathcal{L}_x \boldsymbol{x} = \boldsymbol{\mathcal{L}}_x \cdot \boldsymbol{x} = w_1 = L_x, \tag{5.243}$$

where $\boldsymbol{\mathcal{L}}_x = w_1 \wedge \bar{w}_3$;

$$\mathcal{L}_y = \frac{\partial}{\partial y} = w_2 \cdot \partial_x = \boldsymbol{\mathcal{L}}_y \cdot (\boldsymbol{x} \wedge \partial_x) \leftrightarrow \mathcal{L}_y \boldsymbol{x} = \boldsymbol{\mathcal{L}}_y \cdot \boldsymbol{x} = w_2 = L_y, \tag{5.244}$$

where $\boldsymbol{\mathcal{L}}_y = w_2 \wedge \bar{w}_3$;

$$\begin{aligned}\mathcal{L}_s &= x\frac{\partial}{\partial x} + y\frac{\partial}{\partial y} = (\boldsymbol{x} - w_3)\cdot\partial_{\boldsymbol{x}} = \boldsymbol{\mathcal{L}}_s\cdot(\boldsymbol{x}\wedge\partial_{\boldsymbol{x}})\\ &\leftrightarrow \mathcal{L}_s\boldsymbol{x} = \boldsymbol{\mathcal{L}}_s\cdot\boldsymbol{x} = xw_1 + yw_2 = \boldsymbol{x} - w_3 = L_s,\end{aligned} \tag{5.245}$$

where $\boldsymbol{\mathcal{L}}_s = w_1\wedge\bar{w}_1 + w_2\wedge\bar{w}_2$;

$$\mathcal{L}_r = -y\frac{\partial}{\partial x} + x\frac{\partial}{\partial y} = \boldsymbol{\mathcal{L}}_r\cdot(\boldsymbol{x}\wedge\partial_{\boldsymbol{x}}) \leftrightarrow \mathcal{L}_r\boldsymbol{x} = \boldsymbol{\mathcal{L}}_r\cdot\boldsymbol{x} = L_r, \tag{5.246}$$

where $\boldsymbol{\mathcal{L}}_r = w_2\wedge\bar{w}_1 - w_1\wedge\bar{w}_2$;

$$\mathcal{L}_b = x\frac{\partial}{\partial x} - y\frac{\partial}{\partial y} = \boldsymbol{\mathcal{L}}_b\cdot(\boldsymbol{x}\wedge\partial_{\boldsymbol{x}}) \leftrightarrow \mathcal{L}_b\boldsymbol{x} = \boldsymbol{\mathcal{L}}_b\cdot\boldsymbol{x} = L_b, \tag{5.247}$$

where $\boldsymbol{\mathcal{L}}_b = w_1\wedge\bar{w}_1 - w_2\wedge\bar{w}_2$; and

$$\mathcal{L}_B = y\frac{\partial}{\partial x} + x\frac{\partial}{\partial y} = \boldsymbol{\mathcal{L}}_B\cdot(\boldsymbol{x}\wedge\partial_{\boldsymbol{x}}) \leftrightarrow \mathcal{L}_B\boldsymbol{x} = \boldsymbol{\mathcal{L}}_B\cdot\boldsymbol{x} = L_B, \tag{5.248}$$

where $\boldsymbol{\mathcal{L}}_B = w_1\wedge\bar{w}_2 + w_2\wedge\bar{w}_1$.

Thus, the Lie algebra of the affine plane is generated by the bivectors

$$\mathcal{M}_{bivectors} = \{\boldsymbol{\mathcal{L}}_x, \boldsymbol{\mathcal{L}}_y, \boldsymbol{\mathcal{L}}_s, \boldsymbol{\mathcal{L}}_r, \boldsymbol{\mathcal{L}}_b, \boldsymbol{\mathcal{L}}_B\}, \tag{5.249}$$

or, equivalently, by the vector fields of the form

$$\begin{aligned}\mathcal{M}_{vectorfields} &= \{\boldsymbol{\mathcal{L}}_x\cdot\boldsymbol{x}, \boldsymbol{\mathcal{L}}_y\cdot\boldsymbol{x}, \boldsymbol{\mathcal{L}}_s\cdot\boldsymbol{x}, \boldsymbol{\mathcal{L}}_r\cdot\boldsymbol{x}, \boldsymbol{\mathcal{L}}_b\cdot\boldsymbol{x}, \boldsymbol{\mathcal{L}}_B\cdot\boldsymbol{x}\},\\ &= \{L_x, L_y, L_s, L_r, L_b, L_B\},\end{aligned} \tag{5.250}$$

where $\boldsymbol{\mathcal{L}}\cdot\boldsymbol{x}$ for $\boldsymbol{\mathcal{L}}\in\mathcal{M}_{bivectors}$.

The Lie bracket $[\boldsymbol{\mathcal{L}}_1\cdot\boldsymbol{x}, \boldsymbol{\mathcal{L}}_2\cdot\boldsymbol{x}]$ is given by

$$[\boldsymbol{\mathcal{L}}_1\cdot\boldsymbol{x}, \boldsymbol{\mathcal{L}}_2\cdot\boldsymbol{x}] = \boldsymbol{\mathcal{L}}_2\cdot(\boldsymbol{\mathcal{L}}_1\cdot\boldsymbol{x}) - \boldsymbol{\mathcal{L}}_1\cdot(\boldsymbol{\mathcal{L}}_2\cdot\boldsymbol{x}) = (\boldsymbol{\mathcal{L}}_2\times\boldsymbol{\mathcal{L}}_1)\cdot\boldsymbol{x},$$

where $\boldsymbol{\mathcal{L}}_1\times\boldsymbol{\mathcal{L}}_2 = \frac{1}{2}(\boldsymbol{\mathcal{L}}_1\boldsymbol{\mathcal{L}}_2 - \boldsymbol{\mathcal{L}}_2\boldsymbol{\mathcal{L}}_1)$ is the commutator product of the bivectors $\boldsymbol{\mathcal{L}}_1, \boldsymbol{\mathcal{L}}_2\in\mathcal{M}$.

The Lie algebra of the affine plane is useful for the analysis of motion in the image plane [145]. The vector fields of this Lie algebra are tangent to the flows or integral curves of their group action on the manifold and are presented in Fig. 5.2 as real images.

We have found the generators

$\mathcal{L}_x = \frac{\partial}{\partial x}$	$\mathcal{L}_r = -y\frac{\partial}{\partial x} + x\frac{\partial}{\partial y}$	$\mathcal{L}_B = x\frac{\partial}{\partial x} - y\frac{\partial}{\partial y}$
$\mathcal{L}_y = \frac{\partial}{\partial y}$	$\mathcal{L}_s = x\frac{\partial}{\partial x} + y\frac{\partial}{\partial y}$	$\mathcal{L}_b = y\frac{\partial}{\partial x} + x\frac{\partial}{\partial y}$

(5.251)

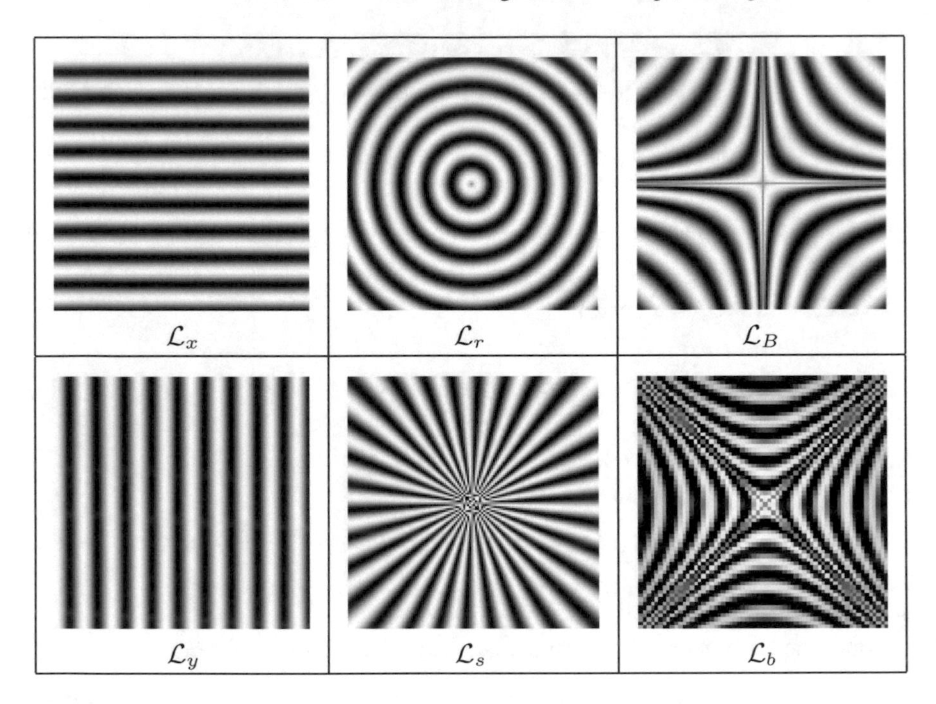

Fig. 5.2 Lie algebra basis in the form of real images

Table 5.1 Lie algebra of the affine plane

$[\cdot,\cdot]$	$\mathcal{L}_x$	$\mathcal{L}_y$	$\mathcal{L}_s$	$\mathcal{L}_r$	$\mathcal{L}_b$	$\mathcal{L}_B$
$\mathcal{L}_x$	0	0	$\mathcal{L}_x$	$\mathcal{L}_y$	$\mathcal{L}_x$	$\mathcal{L}_y$
$\mathcal{L}_y$	0	0	$\mathcal{L}_y$	$-\mathcal{L}_x$	$-\mathcal{L}_y$	$\mathcal{L}_x$
$\mathcal{L}_s$	$-\mathcal{L}_x$	$-\mathcal{L}_y$	0	0	0	0
$\mathcal{L}_r$	$-\mathcal{L}_y$	$\mathcal{L}_x$	0	0	$-2\mathcal{L}_B$	$2\mathcal{L}_b$
$\mathcal{L}_b$	$-\mathcal{L}_x$	$\mathcal{L}_y$	0	$2\mathcal{L}_B$	0	$2\mathcal{L}_r$
$\mathcal{L}_B$	$-\mathcal{L}_y$	$-\mathcal{L}_x$	0	$-2\mathcal{L}_b$	$-2\mathcal{L}_r$	0

of the Lie algebra of the affine plane $\mathcal{A}_{w_3}(\mathcal{N}^2)$ as vector fields along integral curves. Taking the commutator products of these infinitesimal differential generators gives the following multiplication Table 5.1 for this Lie algebra.

Using Table 5.1, we can verify the Jacobi identity for $\mathcal{L}_x$, $\mathcal{L}_s$, and $\mathcal{L}_b$, getting

$$\begin{array}{ccccccl} [\mathcal{L}_x[\mathcal{L}_s\mathcal{L}_b]] & + & [\mathcal{L}_s[\mathcal{L}_b\mathcal{L}_x]] & + & [\mathcal{L}_b[\mathcal{L}_x\mathcal{L}_s]] & = & \\ [\mathcal{L}_x 0] & - & [\mathcal{L}_s\mathcal{L}_y] & + & [\mathcal{L}_b\mathcal{L}_x] & = & \\ 0 & + & \mathcal{L}_y & - & \mathcal{L}_y & = & 0. \end{array} \tag{5.252}$$

Or, equivalently, using CLICAL and the bivector representation for $\mathcal{L}_x$, $\mathcal{L}_r$, and $\mathcal{L}_b$, we calculate

$$
\begin{array}{ccccccc}
[\mathcal{L}_x[\mathcal{L}_r\mathcal{L}_b]] & + & [\mathcal{L}_r[\mathcal{L}_b\mathcal{L}_x]] & + & [\mathcal{L}_b[\mathcal{L}_x\mathcal{L}_r]] & = & \\
2\,[\mathcal{L}_x\mathcal{L}_B] & - & [\mathcal{L}_r\mathcal{L}_y] & + & [\mathcal{L}_b\mathcal{L}_y] & = & \\
2\mathcal{L}_x & - & \mathcal{L}_x & - & \mathcal{L}_x & = 0. &
\end{array} \quad (5.253)
$$

5.11 The Algebra of Incidence

In various applications in robotics, image analysis, and computer vision, the use of projective geometry and the algebra of incidence is extreme useful. Fortunately, these mathematical systems can be efficiently handled within the geometric algebra framework.

In projective geometry, points are represented using homogeneous coordinates of nonzero vectors in the $(n+1)$-dimensional Euclidean space $\mathbb{R}^{n+1}$. These can be seen as *projective rays* identified as points in the n-dimensional projective plane Π^n of $\mathbb{R}^{n+1}$. Furthermore, points, lines, planes, and higher-dimensional k-planes in Π^n are related with 1, 2, 3, and $(k+1)$-dimensional subspaces S^r of $\mathbb{R}^{n+1}$, where $k \leq n$. Since each k-subspace can be associated with a nonzero k-blade A_k of the geometric algebra $G(\mathbb{R}^{n+1})$, it follows that the corresponding $(k-1)$-plane in Π^n can be named by the k-direction of the k-blade A_k.

The *meet* and *join* in Π^n are the principal operations of the algebra of incidence to compute the *intersection* and *union* of the k-planes. Suppose that the set of r points $a_1, a_2, \ldots, a_r \in \Pi^n$ and the set of s points $b_1, b_2, \ldots, b_s \in \Pi^n$ are both in general position (linearly independent vectors in $\mathbb{R}^{n+1}$), then the $(r-1)$-plane in Π^n is specified by the r-blade

$$A_r = a_1 \wedge a_2 \wedge \ldots \wedge a_r \neq 0, \quad (5.254)$$

and the $(s-1)$-plane by the s-blade

$$B_s = b_1 \wedge b_2 \wedge \ldots \wedge b_s \neq 0. \quad (5.255)$$

Considering the a's and b's to be the basis elements of respective subspaces $\mathcal{A}_r$ and $\mathcal{B}_s$, they can be sorted in such a way that

$$\mathcal{A}^r \cup \mathcal{B}^s = span\{a_1, a_2, \ldots a_s, b_{\lambda_1}, \ldots, b_{\lambda_k}\}. \quad (5.256)$$

Supposing that

$$B_s = b_{\lambda_1} \wedge \ldots \wedge b_{\lambda_k} \wedge b_{\alpha_1} \wedge \ldots \wedge b_{\alpha_{s-k}}, \quad (5.257)$$

it follows that the "meet" and "join" of the r-blade A_r and s-blade B_s are, respectively, given by

$$A_r \cup B_s = A_r \wedge b_{\lambda_1} \wedge \ldots \wedge b_{\lambda_k}, \tag{5.258}$$

$$\mathcal{A}^r \cap \mathcal{B}^s = span\{b_{\alpha_1}, \ldots, b_{\alpha_{s-k}}\}. \tag{5.259}$$

Note that if the meet of A_r and $B_s = 0$, their join equals the wedge of the blades $A_r \cup B_s = A_r \wedge B_s$.

After the join of $\mathcal{A}^r$ and $\mathcal{B}^s$ has been computed, the $r + k$-blade

$$\overline{I}_{A_r \cup B_s} = \mathcal{A}^r \cup \mathcal{B}^s, \tag{5.260}$$

can be used for computing the meet of the r- and s-blades A_r and B_s:

$$A_r \cap B_s = A_r \cdot (B_s \cdot I_{A_r \cup B_s}) = (I_{A_r \cup B_s} \cdot A_r) \cdot B_s. \tag{5.261}$$

This expression holds for the positive-definite metric of $\mathbb{R}^{n+1}$. If we use any non-degenerated pseudo-Euclidean space $\mathbb{R}^{p,q}$, where $p + q = n + 1$, we must use instead the reciprocal $r + k$-blade $\overline{I}_{A_r \cup B_s}$, for which the property $I_{A_r \cup B_s} \cdot \overline{I}_{A_r \cup B_s} \neq 0$ is satisfied. For this case, the meet equation reads

$$A_r \cap B_s = A_r \cdot (B_s \cdot \overline{I}_{A_r \cup B_s}) = (\overline{I}_{A_r \cup B_s} \cdot A_r) \cdot B_s. \tag{5.262}$$

Note that if the grade of the blade $A_r \cup B_s$ equals $n = p + q$, we can simply use the inverse of the pseudoscalar, so that $I \cdot \overline{I} = 1$.

In the case of the geometric algebra of the null cone $G(N^{n+1})$, we define the following reciprocal $r + k$-blade for meet Eq. (5.262):

$$\overline{I}_{A_r \cup B_s} = \overline{a}_1 \wedge \overline{a}_2 \wedge \ldots \wedge \overline{a}_s \wedge \overline{b}_{\lambda_1} \wedge \ldots \wedge \overline{b}_{\lambda_k}. \tag{5.263}$$

A more complete discussion of these ideas can be found in [240, 279].

5.11.1 Incidence Relations in the Affine n-Plane

This subsection presents incidence relations between points, lines, planes, and higher-dimensional k-planes using the useful computational framework of the affine n-plane. Let us rewrite Eq. (5.203) in the larger pseudo-Euclidean space $\mathbb{R}^{n+1,1} = \mathbb{R}^n \oplus \mathbb{R}^{1,1}$, where $\mathbb{R}^{1,1} = span\{e_{n+1}, \bar{e}_1\}$:

$$\mathcal{A}_e(\mathbb{R}^n) = \{x_h = x + e| \ x \in \mathbb{R}^n\} \subset \mathbb{R}^{n+1,1}. \tag{5.264}$$

The null vector $e \in \mathbb{R}^{1,1}$ is given by $e = \frac{1}{2}(e_{n+1} + \bar{e}_1)$, and the reciprocal null vector $\bar{e} = (e_{n+1} - \bar{e}_1)$ fulfills the condition $e \cdot \bar{e} = 1$. Now, if we merge the n-affine plane $\mathcal{A}_e(\mathbb{R}^n)$ together with the plane at infinity, we obtain the projective plane Π^n. Each point $\boldsymbol{x} \in \mathcal{A}_e(\mathbb{R}^n)$ is called a *homogeneous representative* of the corresponding point in Π^n. Now points in the affine plane can be represented as rays in the projective space:

$$\mathcal{A}_e^{rays}(\mathbb{R}^n) = \{\boldsymbol{y} |\ \boldsymbol{y} \in \mathbb{R}^{n+1} \text{ and } \boldsymbol{y} \cdot \bar{e} \neq 0\ \} \subset \mathbb{R}^{n+1}. \tag{5.265}$$

Note that in this definition we consider $\boldsymbol{y} \cdot \bar{e} \neq 0$, because rays are directions and they remain the same if we multiply for a scalar. Accordingly, a homogeneous point of the n-affine plane can be uniquely computed from a ray as follows:

$$\frac{\boldsymbol{y}}{\boldsymbol{y} \cdot \bar{e}} \in \mathcal{A}_e(\mathbb{R}^n). \tag{5.266}$$

Now, let us formulate useful incidence relations. If we consider k-points a_1^h, $a_2^h, \ldots, a_k^h \in \mathcal{A}_e^n$, where each $a_i^h = a_i + e$ for $a_i \in \mathbb{R}^n$, and then compute their the outer product, we get the $(k-1)$-plane A^h in Π^n:

$$\begin{aligned} A^h &= a_1^h \wedge a_2^h \wedge \ldots \wedge a_k^h = a_1^h \wedge (a_2^h - a_1^h) \wedge a_3^h \wedge \ldots \wedge a_k^h = \ldots \\ &= a_1^h \wedge (a_2^h - a_1^h) \wedge (a_3^h - a_2^h) \wedge \ldots \wedge (a_k^h - a_{k-1}^h), \\ &= a_1^h \wedge (a_2 - a_1) \wedge (a_3 - a_2) \wedge \ldots \wedge (a_k - a_{k-1}), \\ &= (a_1 + e) \wedge (a_2 - a_1) \wedge (a_3 - a_2) \wedge \ldots \wedge (a_k - a_{k-1}), \\ &= a_1 \wedge a_2 \wedge \ldots \wedge a_k + \\ &\qquad + e \wedge (a_2 - a_1) \wedge (a_3 - a_2) \wedge \ldots \wedge (a_k - a_{k-1}). \end{aligned} \tag{5.267}$$

This equation represents a $(k-1)$-plane in Π^n, but it also belongs to the affine n-plane $\mathcal{A}_e^n$ and thus contains important metrical information which can be extracted by taking the dot product from the left with $\bar{e}$:

$$\begin{aligned} \bar{e} \cdot A^h &= \bar{e} \cdot (a_1^h \wedge a_2^h \wedge \ldots \wedge a_k^h), \\ &= (a_2 - a_1) \wedge (a_3 - a_2) \wedge \ldots \wedge (a_k - a_{k-1}). \end{aligned} \tag{5.268}$$

Interestingly enough, this result, with a little modification, turns out to be the directed content of the $(k-1)$-simplex $A^h = a_1^h \wedge a_2^h \wedge \ldots \wedge a_k^h$ in the affine n-plane:

$$\begin{aligned} \frac{\bar{e} \cdot A^h}{(k-1)!} &= \frac{\bar{e} \cdot (a_1^h \wedge a_2^h \wedge \ldots \wedge a_k^h)}{(k-1)!}, \\ &= \frac{(a_2 - a_1) \wedge (a_3 - a_2) \wedge \ldots \wedge (a_k - a_{k-1})}{(k-1)!}. \end{aligned} \tag{5.269}$$

5.11.2 Directed Distances

Using our previous results, we can propose useful equations in the affine plane to relate points, lines, and planes metrically . The *directed distance* or *foot* from the $(k-1)$-plane $a_1^h\wedge\ldots\wedge a_k^h$ to the point b^h is given by

$$
\begin{aligned}
&d[a_1^h\wedge\ldots a_k^h, b^h] \equiv \\
&[\{\overline{e}\cdot(a_1^h\wedge\ldots\wedge a_k^h)\}\ (\overline{e}\cdot b^h)]^{-1}[\overline{e}\cdot(a_1^h\wedge\ldots\wedge a_k^h\wedge b^h)], \\
&= [a_2-a_1)\wedge\ldots\wedge(a_k-a_{k-1})]^{-1}[(a_2-a_1)\wedge\ldots\wedge(a_k-a_{k-1})\wedge(b-a_k)].
\end{aligned}
\tag{5.270}
$$

In the same sense, the equation of the directed distance between the two lines $a_1^h\wedge a_2^h$ and $b_1^h\wedge b_2^h$ in the affine n-plane reads

$$
\begin{aligned}
&d[a_1^h\wedge a_2^h, b_1^h\wedge b_2^h] \equiv \\
&\quad [\{\overline{e}\cdot(a_1^h\wedge a_2^h)\}\wedge\{\overline{e}\cdot(b_1^h\wedge b_2^h)\}]^{-1}[\overline{e}\cdot(a_1^h\wedge a_2^h\wedge b_1^h\wedge b_2^h)], \\
&\quad = [(a_2-a_1)\wedge(b_2-b_1)]^{-1}[(a_2-a_1)\wedge(b_1-a_2)\wedge(b_2-b_1)].
\end{aligned}
\tag{5.271}
$$

A general equation of the directed distance between the $(r-1)$-plane $A^h = a_1^h\wedge\ldots\wedge a_r^h$ and the $(s-1)$-plane $B^h = b_1^h\wedge\ldots\wedge b_s^h$ in the affine n-plane is similarly given by

$$
\begin{aligned}
&d[a_1^h\wedge\ldots\wedge a_r^h, b_1^h\wedge\ldots\wedge b_s^h] \equiv \\
&\{\overline{e}\cdot(a_1^h\wedge\ldots\wedge a_r^h)\}\wedge\{\overline{e}\cdot(b_1^h\wedge\ldots\wedge b_s^h)\}]^{-1}[\overline{e}\cdot(a_1^h\wedge\ldots\wedge a_r^h\wedge b_1^h\wedge\ldots\wedge b_s^h)], \\
&= [(a_2-a_1)\wedge\ldots\wedge(a_r-a_{r-1})\wedge(b_2-b_1)\wedge\ldots\wedge(b_s-b_{s-1})]^{-1} \\
&[(a_2-a_1)\wedge\ldots\wedge(a_r-a_{r-1})\wedge(b_1-a_r)\wedge(b_2-b_1)\wedge\ldots\wedge(b_s-b_{s-1})].
\end{aligned}
\tag{5.272}
$$

We have to be careful, because if $A^h\wedge B^h = 0$, the directed distance may or may not be equal to zero. If $(a_1^h\wedge\ldots\wedge a_r^h)\wedge(b_1^h\wedge\ldots\wedge b_{s-1}^h) \neq 0$, we can calculate the meet between the $(r-1)$-plane A^h and $(s-1)$-plane B^h,

$$
\begin{aligned}
p &= (a_1^h\wedge\ldots\wedge a_r^h)\cap(b_1^h\wedge\ldots\wedge b_s^h), \\
&= (a_1^h\wedge\ldots\wedge a_r^h)\cdot[(b_1^h\wedge\ldots\wedge b_s^h)\cdot\overline{I}_{A\cup B}],
\end{aligned}
\tag{5.273}
$$

where

$$
\overline{I}_{A\cup B} = \{\overline{e}\cdot[(a_1^h\wedge\ldots\wedge a_r^h)\wedge(b_1^h\wedge\ldots\wedge b_{s-1}^h)]\}\wedge\overline{e}.
$$

It can happen that the point $p = A^h\cap B^h$ may not be in the affine n-plane, but the *normalized* point $p^h = \frac{p}{\overline{e}\cdot p}$ will either be in the affine plane or will be undefined. Finding the "normalized point" is not necessary in many calculations, but is required when the metric plays an important role or in the case of parallel hyperplanes, when it is used as an indicator.

5.11.3 Incidence Relations in the Affine 3-Plane

This subsection presents some algebra of incidence relations for 3D Euclidean space represented in the affine 3-plane $\mathcal{A}_e^3$, with the pseudoscalar $I = e_{123}e$ and the reciprocal pseudoscalar $\bar{I} = \bar{e}e_{321}$ satisfying the condition $I \cdot \bar{I} = 1$. Similar incidence relations were given by Blaschke [38] using dual quaternions, and later by Selig using the 4D degenerate geometric algebra $G_{3,0,1}$ [264]. Unlike the formulas given by these authors, our formulas are generally valid in any dimension and are expressed completely in terms of the meet and join operations in the affine plane. Blaschke and Selig could not exploit the meet and join operations because they were using a geometric algebra with a degenerate metric.

The distance of a point b^h to the line $L^h = a_1^h \wedge a_2^h$ is the *magnitude* or *norm* of the directed distance,

$$|d| = \left|\left[\{\bar{e} \cdot (a_1^h \wedge a_2^h)\} \wedge \{(\bar{e} \cdot (b^h)\}\right]^{-1}\left[\bar{e} \cdot (a_1^h \wedge a_2^h \wedge b^h)\right]\right|. \tag{5.274}$$

The distance of a point b^h to the plane $A^h = a_1^h \wedge a_2^h \wedge a_3^h$ is

$$|d| = \left|\left[\{\bar{e} \cdot (a_1^h \wedge a_2^h \wedge a_3^h)\} \wedge \{(\bar{e} \cdot (b^h)\}\right]^{-1}\left[\bar{e} \cdot (a_1^h \wedge a_2^h \wedge a_3^h \wedge b^h)\right]\right|. \tag{5.275}$$

Let us analyze carefully the incidence relation between the lines $L_1^h = a_1^h \wedge a_2^h$ and $L_2^h = b_1^h \wedge b_2^h$, which are completely determined by their join $I_{L_1^h \cup L_2^h} = L_1^h \cup L_2^h$. The following formulas help to test the incidence relations of the lines.

- If $I_{L_1^h \cup L_2^h}$ is a bivector, the lines coincide and $L_1^h = tL_2^h$ for some $t \in \mathbb{R}$.
- If $I_{L_1^h \cup L_2^h}$ is a 3-vector, the lines are either parallel or intersect in a common point. In this case,

$$p = L_1^h \cap L_2^h = L_1^h \cdot (L_2^h \cdot \bar{I}_{L_1^h \cup L_2^h}), \tag{5.276}$$

  where p is the result of the meet. If $\bar{e} \cdot p = 0$, the lines are parallel; otherwise, they intersect at the point $p_h = \frac{p}{\bar{e} \cdot p}$ in the affine 3-space $\mathcal{A}_e^3$.
- If $I_{L_1^h \cup L_2^h}$ is a 4-vector, the lines are skew. In this case, the distance is given by equation (5.272).

The incidence relation between a line $L^h = a_1^h \wedge a_2^h$ and a plane $B^h = b_1^h \wedge b_2^h \wedge b_3^h$ is also determined by their join, $L^h \cup B^h$. Clearly, if the join is a trivector, the line L^h lies in the plane B^h. The only other possibility is that their join is the pseudoscalar $I = \sigma_{123}e$. In this case,

$$p = L^h \cap B^h = L^h \cdot (B^h \cdot \bar{I}). \tag{5.277}$$

If $\overline{e} \cdot p = 0$, the line is parallel to the plane, with the directed distance determined by Eq. (5.273). Otherwise, their point of intersection in the affine plane is $p_h = \frac{p}{\overline{e} \cdot p}$.

Two planes, $A^h = a_1^h \wedge a_2^h \wedge a_3^h$ and $B^h = b_1^h \wedge b_2^h \wedge b_3^h$, in the affine plane $\mathcal{A}_e^3$ are either parallel, intersect in a line, or coincide. If their join is a trivector, that is, if $A^h = tB^h$ for some $t \in \mathbb{R}^*$, they obviously coincide. If they do not coincide, then their join is the pseudoscalar $I = \sigma_{123}e$. In this case, we calculate the meet as

$$L = A^h \cap B^h = (\overline{I} \cdot A^h) \cdot B^h. \tag{5.278}$$

If $\overline{e} \cdot L = 0$, the planes are parallel, with the directed distance determined by Eq. (5.273). Otherwise, L represents the line of intersection in the affine plane having the direction $\overline{e} \cdot L$.

The equivalent of the above incidence relations was given by Blaschke [38] using dual quaternions, and by Selig [263] utilizing a special or degenerate four-dimensional Clifford algebra. Whereas Blaschke uses only pure quaternions (bivectors) for his representation, Selig uses trivectors for points and vectors for planes. In contrast, in the affine 3-plane, points are always represented by vectors, lines by bivectors, and planes by trivectors. This offers a comprehensive and consistent interpretation which greatly simplifies the underlying conceptual framework. The following equation compares our equations (left side) with those of Blaschke and Selig (right side).

$$\textit{equation } (5.274) \equiv \frac{1}{2}(\tilde{p}l + \tilde{l}p), \tag{5.279}$$

$$\textit{equation } (5.275) \equiv \frac{1}{2}(\tilde{p}\pi + \tilde{\pi}p), \tag{5.280}$$

$$\textit{equation } (5.277) \equiv \frac{1}{2}(\tilde{l}\pi + \tilde{\pi}l). \tag{5.281}$$

5.11.4 Geometric Constraints as Flags

It is often necessary to check a geometric configuration during a rigid motion in Euclidean space, and simple geometric incidence relations can be used for this purpose. For example, a point p is on a line L if and only if

$$p \wedge L = 0. \tag{5.282}$$

Similarly, a point p is on a plane A if

$$p \wedge A = 0. \tag{5.283}$$

A line L will lie in plane A if

$$L \cap A = A. \tag{5.284}$$

Alternatively, the line L can meet the plane A in a single point p, in which case,

$$L \cap A = p,$$

or, if the line L is parallel to the plane A,

$$L \cap A = 0. \tag{5.285}$$

5.12 Conclusion

We have shown how geometric algebra can effectively be used to carry out analysis on a manifold, which is useful in robotics and image analysis. Geometric algebra offers a clear and concise geometric framework of multivectors in which calculations can be carried out. Since the elements and operations in geometric algebra are basis-free, computations are simpler and geometrically more transparent than in more traditional approaches.

Stereographic projection and its generalization to the conformal group and projective geometry have direct application to image analysis from one or more viewpoints. The key idea is that an image is first represented on the null cone and then projected onto affine geometries or onto an n-dimensional affine plane, where the image analysis takes place. Since every Lie algebra can be represented by an appropriate bivector algebra in an affine geometry, it follows that a complete motion analysis should be possible using its bivector representation in geometric algebra. In Chap. 14, we will employ Lie operators expressed in terms of bivectors to detect visual invariants.

5.13 Exercises

5.1 A complex structure is introduced through the doubling bivector,

$$\boldsymbol{J} = e_1 \wedge f^1 + e_2 \wedge f^2 + e_3 \wedge f^3 + \cdots + e_n \wedge f^n = e_i \wedge f^i,$$

if the $\{e_i\}$ frame is chosen to be orthonormal, we obtain

$$\boldsymbol{J} = e_1 f_1 + e_2 f_2 + e_3 f_3 + \cdots + e_n f_n = e_i f_i = \boldsymbol{J}_1 + \boldsymbol{J}_2 + \cdots + \boldsymbol{J}_n.$$

This sum consists of n commuting blades of grade two. Give the geometric interpretation of each bivector $\boldsymbol{J}_i$. What represents each $\boldsymbol{J}_i$ w.r.t. to the rotation plane expressed by $\boldsymbol{J}$.

5.2 The vectors $\{e_i\}$, i = 1,...,3 are the basis for G_3. The bivector algebra of the rotor algebra G_3^+ is spanned by the bivectors e_{23}, e_{31}, e_{12}. Prove that this algebra fulfills the axiom of closure under the commutator product. Find the structure constants of the unitary group as well.

5.3 The bivector algebra of the rotor algebra G_3^+ is spanned by the bivectors e_{23}, e_{31}, e_{12}. Compute the Killing of this bivector algebra.

5.4 The vectors $\{e_i, f_i\}$, i = 1,...,n are the basis for $G_{2n,0}$. The Lie algebra $u(n)$ is defined by the following bivectors:

$$\begin{aligned} J_i &= e_i f_i && (i = 1, ..., n), \\ E_{ij} &= e_i e_j + f_i f_j && (i < j = 1, ..., n), \\ F_{ij} &= e_i f_j - f_i e_j && (i < j = 1, ..., n). \end{aligned}$$

for $i < j = 1, ..., n$. Prove that this algebra fulfills the axiom of closure under the commutator product. Find the structure constants of the unitary group as well.

5.5 Compute the Killing of the Lie algebra $u(n)$ defined by the bivectors given in Exercise 5.3.

5.6 Prove that the quaternion algebra $\mathbb{H}$ isomorphic to the rotor algebra G_3^+ is a ring.

5.7 The $\{e_i\}$ basis vectors are orthonormal with positive signature, and the $\{\bar{e}_i\}$ are orthonormal with negative signature and belong to the geometric algebra $G_{n,n,}$. By using the constraint of Eq. (5.114), given by

$$[\boldsymbol{x} \wedge \boldsymbol{y} - (\boldsymbol{x} \cdot \boldsymbol{K})(\boldsymbol{y} \cdot \boldsymbol{K})] \times \boldsymbol{K} = 0, \tag{5.286}$$

one can again try all combinations of $\{e_i, \bar{e}_i\}$ to produce the bivector basis for the Lie algebra $gl(n)$ of the general linear group $GL(n)$,

$$\begin{aligned} \boldsymbol{K}_i &= \frac{1}{2}\boldsymbol{F}_{ij} = e_i \bar{e}_i, \\ \boldsymbol{E}_{ij} &= e_i e_j - \bar{e}_i \bar{e}_j && (i < j = 1...n), \\ \boldsymbol{F}_{ij} &= e_i \bar{e}_j - \bar{e}_i e_j && (i < j = 1...n). \end{aligned}$$

for $i < j = 1, ..., n$. Prove that this algebra fulfills the axiom of closure under the commutator product. Find the structure constants of the unitary group as well.

5.8 The $\{e_i\}$ basis vectors are orthonormal with positive signature, and the $\{\bar{e}_i\}$ are orthonormal with negative signature and belong to the geometric algebra $G_{n,n,}$. Since $\underline{K}_*$ defines a new Lie algebra with generators determined by the outer morphism condition

$$\underline{K}_*(\boldsymbol{B}) = -\boldsymbol{B},$$

using this result, one can construct a generator basis for an invariance group of $\underline{K}_*$ as follows

$$\begin{aligned} \boldsymbol{E}_{ij} &= e_i e_j - \bar{e}_i \bar{e}_j, \\ \boldsymbol{F}_{ij} &= e_i \bar{e}_j + \bar{e}_i e_j, \end{aligned}$$

for $i, j = 1, 2, ..., n, i < j$. These are the generators of the *complex orthogonal group* SO(n, $\mathbb{C}$). Prove that this Lie algebra fulfills the axiom of closure under the commutator product. Find the structure constants of the unitary group as well. Explain also why for odd n, $\boldsymbol{K}$ is the only kind of involutory bivector. When the case n is even, show which other groups are determined by their invariants.

5.9 Prove how the versor for translation called translator, $\boldsymbol{T} \in G_{3,3}$, is derived in terms of $e_i \bar{e}_j$ and $w_i \wedge \bar{w}_j$ as follows

$$\begin{aligned} \boldsymbol{T} = e^{\boldsymbol{B}_t} &= e^{\frac{t_3 \bar{e}_1 e_2 - t_2 \bar{e}_1 e_3 + t_1 \bar{e}_2 e_3}{2}}, \\ &= e^{\frac{t_3 \bar{w}_1 \wedge \bar{w}_2 - t_2 \bar{w}_1 \wedge \bar{w}_3 + t_1 \bar{w}_2 \wedge \bar{w}_3}{2}}, \\ &= 1 + \frac{1}{2}(t_3 \bar{e}_1 e_2 - t_2 \bar{e}_1 e_3 + t_1 \bar{e}_2 e_3), \\ &= 1 + \frac{1}{2}(t_3 \bar{w}_1 \wedge w_2 - t_2 \bar{w}_1 \wedge \bar{w}_3 + t_1 \bar{w}_2 \wedge \bar{w}_3). \end{aligned}$$

5.10 Prove how the versor for perspectivity called perspector, $\boldsymbol{P} \in G_{3,3}$, is derived in terms of $e_i \bar{e}_j$ and $w_i \wedge \bar{w}_j$ as follows

$$\begin{aligned} \boldsymbol{P} = e^{\boldsymbol{B}_f} &= e^{\frac{-f_3 e_1 \bar{e}_2 + f_2 e_1 \bar{e}_3 - f_1 e_2 \bar{e}_3}{2}}, \\ &= e^{\frac{-f_3 w_1 \wedge w_2 + f_2 w_1 \wedge w_3 - f_1 w_2 \wedge w_3}{2}}, \\ &= 1 + \frac{1}{2}(-f_3 e_1 \bar{e}_2 + f_2 e_1 \bar{e}_3 - f_1 e_2 \bar{e}_3), \\ &= 1 + \frac{1}{2}(-f_3 w_1 \wedge w_2 + f_2 w_1 \wedge w_3 - f_1 w_2 \wedge w_3). \end{aligned}$$

5.11 Prove how the versor for rotation called rotor, $\boldsymbol{R} \in G_{3,3}$, is derived in terms of $e_i \bar{e}_j$ and $w_i \wedge \bar{w}_j$ as follows

$$\begin{aligned}
\boldsymbol{R} &= e^{\boldsymbol{B}_R}, \\
&= e^{\frac{(-e_2e_3+\bar{e}_2\bar{e}_3)+(-e_1e_3+\bar{e}_1\bar{e}_3)+(-e_1e_2+\bar{e}_1\bar{e}_2)}{2}}, \\
&= e^{\frac{(w_2\wedge\bar{w}_3-\bar{w}_2\wedge w_3)+(w_1\wedge\bar{w}_3-\bar{w}_1\wedge w_3)+(w_1\wedge\bar{w}_2-\bar{w}_1\wedge w_2)}{2}}.
\end{aligned}$$

5.12 The bivector $\boldsymbol{B}_{\theta_{23}} = \frac{1}{2}(w_2\wedge\bar{w}_3 - \bar{w}_2\wedge w_3) = \frac{1}{2}(-e_2e_3 + \bar{e}_2\bar{e}_3)$ squares to $\boldsymbol{B}^2_{\theta_{23}} = -\frac{1}{2}(1 + w_2\wedge\bar{w}_3\wedge\bar{w}_2\wedge w_3)$. Since $\boldsymbol{B}^3_{\theta_{23}} = -\boldsymbol{B}_{\theta_{23}}$ prove that

$$e^{\theta\boldsymbol{B}_{\theta_{23}}} = 1 + sin(\theta)\boldsymbol{B}_{\theta_{23}} + (1 - cos(\theta))\boldsymbol{B}^2_{\theta_{23}}. \tag{5.287}$$

See in Appendix B the equations of the exponentials of bivectors.

5.13 In Sect. 5.10.3, we derived for *Spin*(2) the equation

$$\begin{aligned}
\boldsymbol{R}_\theta = \exp(\frac{1}{2}\theta\boldsymbol{B}) &= \exp(\frac{0\cdot\theta}{2})p_1 + \exp(i\theta)p_2 + \exp(-i\theta)p_3, \\
&= p_1 + \cos(\theta)(p_2 + p_3) + \sin(\theta)i(p_2 - p_3), \\
&= p_1 + \cos(\theta)(p_2 + p_3) + \frac{\boldsymbol{B}}{2}\sin(\theta).
\end{aligned}$$

Now, similar as Exercise 5.12, for the rotor

$$\begin{aligned}
\boldsymbol{R} &= e^{\boldsymbol{B}_R}, \\
&= e^{\frac{(-e_2e_3+\bar{e}_2\bar{e}_3)+(-e_1e_3+\bar{e}_1\bar{e}_3)+(-e_1e_2+\bar{e}_1\bar{e}_2)}{2}}, \\
&= e^{\frac{(w_2\wedge\bar{w}_3-\bar{w}_2\wedge w_3)+(w_1\wedge\bar{w}_3-\bar{w}_1\wedge w_3)+(w_1\wedge\bar{w}_2-\bar{w}_1\wedge w_2)}{2}}.
\end{aligned}$$

write $\boldsymbol{R}$ in terms of trigonometric functions. See in Appendix B the equations of the exponentials of bivectors.

5.14 Prove how the versor for Lorentz transformation called Lorentor, $\boldsymbol{R}_L \in G_{3,3}$, is derived in terms of $e_i\bar{e}_j$ and $w_i\wedge\bar{w}_j$ as follows

$$\begin{aligned}
\boldsymbol{R}_L &= e^{\boldsymbol{B}_L}, \\
&= e^{\frac{(e_2\wedge e_3-\bar{e}_2\wedge\bar{e}_3)+(e_1\wedge e_3-\bar{e}_1\wedge\bar{e}_3)+(e_1\wedge e_2-\bar{e}_1\wedge\bar{e}_2)}{2}}, \\
&= e^{\frac{(\bar{w}_2\wedge w_3-w_2\wedge\bar{w}_3)+(\bar{w}_1\wedge w_3-w_1\wedge\bar{w}_3)+(\bar{w}_1\wedge w_2-w_1\wedge\bar{w}_2)}{2}}.
\end{aligned}$$

5.15 Prove how the versor for shear affine transformation called Sheartor, $\boldsymbol{S} \in G_{3,3}$, is derived in terms of $e_i\bar{e}_j$ and $w_i\wedge\bar{w}_j$ as follows

$$\begin{aligned}
\boldsymbol{S} &= e^{\boldsymbol{B}_S}, \\
&= e^{\frac{a_{23}(e_2\wedge e_3-\bar{e}_3\wedge\bar{e}_2)+a_{13}(e_1\wedge e_3-\bar{e}_3\wedge\bar{e}_1)+a_{12}(e_1\wedge e_2-\bar{e}_2\wedge\bar{e}_1)}{2}} \\
&\qquad e^{\frac{a_{32}(e_3\wedge e_2-\bar{e}_2\wedge\bar{e}_3)+a_{31}(e_3\wedge e_1-\bar{e}_1\wedge\bar{e}_3)+a_{21}(e_2\wedge e_1-\bar{e}_1\wedge\bar{e}_2)}{2}}, \\
&= e^{\frac{a_{23}(\bar{w}_2\wedge w_3-w_3\wedge\bar{w}_2)+a_{13}(\bar{w}_1\wedge w_3-w_3\bar{w}_1)+a_{12}(\bar{w}_1\wedge w_2-w_2\wedge\bar{w}_1)}{2}} \\
&\qquad e^{\frac{a_{32}(\bar{w}_3\wedge w_2-w_2\wedge\bar{w}_3)+a_{31}(\bar{w}_3\wedge w_1-w_1\bar{w}_3)+a_{21}(\bar{w}_2\wedge w_1-w_1\wedge\bar{w}_2)}{2}}.
\end{aligned}$$

5.16 Prove how the versor for dilation called Dilator, $\boldsymbol{D} \in G_{3,3}$, is derived in terms of $e_i\bar{e}_j$ and $w_i \wedge \bar{w}_j$ as follows

$$\begin{aligned}
\boldsymbol{D} = &\; e^{\boldsymbol{B}_D}, \\
= &\; e^{\frac{1}{2}\{(1/0,\pm)\bar{e}_1\wedge e_1+(0/1,\mp)\bar{e}_2\wedge e_2+(0/1,\mp)\bar{e}_3\wedge e_3\}} \\
&\; e^{\frac{1}{2}\{(-1/0,\pm)e_1\wedge\bar{e}_1+(0/-1,\mp)e_2\wedge\bar{e}_2+(0/-1,\mp)e_3\wedge\bar{e}_3\}}, \\
= &\; e^{\frac{1}{2}\{(1/0,\pm)\bar{w}_1\wedge w_1+(0/1,\mp)\bar{w}_2\wedge w_2+(0/1,\mp)\bar{w}_3\wedge w_3\}} \\
&\; e^{\frac{1}{2}\{(-1/0,\pm)w_1\wedge\bar{w}_1+(0/-1,\mp)w_2\wedge\bar{w}_2+(0/-1,\mp)w_3\wedge\bar{w}_3\}}.
\end{aligned} \tag{5.288}$$

5.17 Following a similar methodology for the versors obtained for the 3D projective geometry in $\mathbb{P}^3$ and using Eqs. (5.201) and (5.202), derive the versors for the Lie groups of $\mathbb{P}^2$, namely for the translation (2 DOF), rotation (1 DOF) about the axis of the plane $e_1 \wedge e_2$, scaling (2 DOF), perspectivity (2 DOF), and shear (1 DOF).

5.18 Considering the camera pinhole model [126], the projective transformation from the projective space $\mathbb{P}^3$ to the projective plane $\mathbb{P}^2$ is decomposed in a multiplication of an affine transformation K a 3×3 upper triangular matrix representing the intrinsic parameters and a 3×4 matrix which represents extrinsic parameters of the rigid motion $SE(3)$

$$P_{3\times4} = K[R|t], \tag{5.289}$$

using Eq. (5.66) express the entries of matrix in terms of bivectors and then formulate the versor corresponding to the affine transformation of matrix K in the projective plane $\mathbb{P}^2$. Hint: Use the results of Exercise 5.15.

5.19 In Exercise 5.15 you have computed the versor for the matrix K, and the relation of pairs of points in two views is established by the homography H as follows.

$$x_i' = Hx_i, \tag{5.290}$$

for any pair of points $x_i \in \mathbb{P}^2$. Now H can be decomposed as follows

$$H = K[r_1, r_2, t], \tag{5.291}$$

where r_1, r_2 are the column vectors of the rotation matrix R, i.e., $r_3 = r_1 \times r_2$ and t is the translation vector in $P_{3\times4} = K[R|t]$, where $t \cdot r_i = 0$, for $i = 1,2,3$. Using Eq. (5.66), express the entries of the matrix H in terms of bivectors and then formulate the versor for the homography H in the projective plane $\mathbb{P}^2$. Hint: Use the results of Exercise 5.15.

5.20 Prove that $\boldsymbol{x}_c = \frac{1}{2}\boldsymbol{x}_h\overline{w}\boldsymbol{x}_h = exp(\frac{1}{2}\boldsymbol{x}\overline{w})\ e\ exp(-\frac{1}{2}\boldsymbol{x}\overline{w})$, where $\boldsymbol{x}_c \in \mathcal{H}_w^{p,q}$, $x_h \in \mathcal{A}_w^{p,q}$, and $\boldsymbol{x} \in \mathbb{R}^n$.

5.21 The bases of the reciprocal null cones $\{w\} \in N$ and $\{\bar{w}\} \in N$ are called reciprocal or dual bases because they fulfill the relationship $\{w\} \cdot \{\bar{w}\} = id$, where id

is an $n \times n$ identity matrix. The pseudoscalar of $G(N)$ is $I = w_1 \wedge w_2 \wedge w_3 ... w_n$, and of $G(\bar{N})$ is $\bar{I} = \bar{w}_1 \wedge \bar{w}_2 \wedge \bar{w}_3 ... \bar{w}_n$, both of which satisfy the condition $I \cdot \bar{I} = 1$. According to Eq. (5.57), we can express a second basis $\{a\} = \{w\}\mathcal{A} = \{w_1, w_2, w_3, ..., w_n\}A \in N$, where the matrix A is responsible for this change of basis. The hypervolume spanned by the basis $\{a\}$ is $\bigwedge_{i=1}^{n}\{a\} = a_1 \wedge a_2 \wedge \ldots \wedge a_n = \det(\mathcal{A}) w_1 \wedge w_2 \wedge \ldots w_n = \det(\mathcal{A})$I. The *bracket* of $\mathcal{A}$ is simply computed by taking the dot product of this hypervolume and the reciprocal pseudoscalar $\det(\mathcal{A}) = \bigwedge_{i=1}^{n}\{a\} \cdot \bar{I}$. Similar to the standard approach for obtaining a reciprocal basis, it is easy to see that the new reciprocal basis $\{\bar{a}\}$ can be computed by means of the equation

$$\bar{a}_i = (-1)^{i+1} \frac{(a_1 \wedge \ldots \wedge (1)_i \wedge \ldots \wedge a_n) \cdot \bar{I}}{[a_1 \; a_2 \; \ldots \; a_n]},$$

where a_i is left out of the wedge operation in position $(1)_i$. This expression guarantees that $\{\bar{a}\} \cdot \{a\} = id$. Find an expression for computing the inverse of the matrix $(\mathcal{A})$. (*Hint*: Use $\{\bar{a}\} = \mathcal{B}\{\bar{a}\}$.)

5.22 Projections from $G_{3,3}$ to $G_{3,0}$: Consider the Lie algebra $so(3)$ in $G_{3,3}$:

$$\mathcal{L}_x = \begin{bmatrix} 0 & 1 & 0 \\ -1 & 0 & 0 \\ 0 & 0 & 0 \end{bmatrix}, \mathcal{L}_y = \begin{bmatrix} 0 & 0 & 1 \\ 0 & 0 & 0 \\ -1 & 0 & 0 \end{bmatrix}, \mathcal{L}_z = \begin{bmatrix} 0 & 0 & 0 \\ 0 & 0 & -1 \\ 0 & 1 & 0 \end{bmatrix}.$$

Using CLICAL or eClifford, represent this Lie algebra in $G_{3,3}$ using the bivector matrices $\boldsymbol{L}_x$, $\boldsymbol{L}_y$, and $\boldsymbol{L}_z$. Take their projections $P(\boldsymbol{L}_x) = I^{-1}(I \cdot \boldsymbol{L}_x)$, $P(\boldsymbol{L}_y) = I^{-1}(I \cdot \boldsymbol{L}_y)$, and $P(\boldsymbol{L}_z) = I^{-1}(I \cdot \boldsymbol{L}_z)$ using the Euclidean pseudoscalar $I = e_1 e_2 e_3$ and also using the reciprocal pseudoscalar $I = \bar{e}_1 \bar{e}_2 \bar{e}_3$. Explain the dual relation of the results.

5.23 Using CLICAL or eClifford, compute in the 2D affine plane $\mathcal{A}_e^2$ the new position of the point $\boldsymbol{x} = 4e_1 + 2e_2 \in \mathcal{A}_w(\mathbb{R}^2)$ after the translation $\boldsymbol{t} = 6e_1 + 5e_2 \in \mathcal{A}_e(\mathbb{R}^2)$.

5.24 Using CLICAL or eClifford, compute in the 2D affine plane $\mathcal{A}_e^2$ the dilation of the point $\boldsymbol{x} = 3e_1 + 5e_2\mathbb{R}^2$ for

$$\mathcal{D}_u = \begin{bmatrix} e^u & 0 & 0 \\ 0 & e^u & 0 \\ 0 & 0 & 1 \end{bmatrix} = \begin{bmatrix} 1.75 & 0 & 0 \\ 0 & 1.75 & 0 \\ 0 & 0 & 1 \end{bmatrix}.$$

5.25 Using CLICAL or eClifford, compute in the geometric algebra of the null cone $G(N^2)$ the new position of the point $x_0 = 2e_1 + 3e_2 \in \mathcal{A}(\mathbb{R}^2)$ after a rotation of $\theta = \frac{\pi}{6}$. Use Eq. (5.227) with the bivector of the spinor group $Spin(2)$ $\boldsymbol{B} = -w_1 \wedge \bar{w}_2 + -w_2 \wedge \bar{w}_1$. Note that the rotation is not computed with the exponential function but rather with a function depending on the mutually annihilating idempotents.

5.26 Compute in the affine 2-plane $\mathcal{A}_e^2$, with the pseudoscalar $I = e_1 \wedge e_2 \wedge e$ and the reciprocal pseudoscalar $\overline{I} = e_1 \wedge e_2 \wedge \overline{e}$, the meet of the lines $L_1^h = a_1^h \wedge a_2^h$, and $L_2^h = b_1^h \wedge b_2^h$, where $a_1^h = 4e_1 + e$, $a_2^h = 2e_2 + e$, $b_1^h = e$, and $b_2^h = 2e_1 + 3e_2 + e$.

5.27 In the affine 2-plane $\mathcal{A}_e^2$, compute the intersecting point p^h of the lines L_1^h and L_3^h, where L_1^h is the line determined in Exercise 5.7 and L_3^h passes through the point $c_2^h = 4e_1 + 3e_2 + e$ and is orthogonal to the line L_1^h. (*Hint*: Consider the line $L_3^h = p^h \wedge c_2^h$ with the point $p^h = c_2^h + s\, i(a_1 - a_2)$ for $s \in I\!R$, where $i = e_1 e_2$.) Note that $s \neq 0$ can be overlooked, because the line is uniquely defined by the 2-direction of the bivector $p^h \wedge c_2^h$ and not by its magnitude.

5.28 Theorem proving: Let a circle entered at the origin and $\boldsymbol{a}$ and $\boldsymbol{b}$ be the end points of the diameter. Take any point $\boldsymbol{c}$ on the circle and show in the 2-plane $\mathcal{A}_e^2$ that the lines $\boldsymbol{l}_{ac}$ and $\boldsymbol{l}_{cb}$ are perpendicular.

5.29 Theorem proving: Prove the theorem of Desargues's configuration in the 3D-projective plane Π^3. Consider that $x_1,\ x_2,\ x_3$ and $y_1,\ y_2,\ y_3$ are the vertices of two triangles in Π^3 and suppose that $(x_1 \wedge x_2) \cap (y_1 \wedge y_2) = z_3$, $(x_2 \wedge x_3) \cap (y_2 \wedge y_3) = z_1$, and $(x_3 \wedge x_1) \cap (y_3 \wedge y_1) = z_2$. You can claim that $c_1 \wedge c_2 \wedge c_3 = 0$ if and only if there is a point p such that $x_1 \wedge y_1 \wedge p = 0 = x_2 \wedge y_2 \wedge p = x_3 \wedge y_3 \wedge p$. (*Hint*: Express the point as linear combinations of $a_1,\ b_1, a_2,\ b_2$, and $a_3,\ b_3$. The other half of the proof follows by duality of the classical projective geometry.)

5.30 Theorem proving: Consider an arbitrary circumcircled triangle; see Fig. 5.3. From a point $\boldsymbol{d}$ on the circumcircle, draw three perpendiculars to the triangle sides $\boldsymbol{bc}$, $\boldsymbol{ca}$, and $\boldsymbol{ab}$ to meet the circle at points $\boldsymbol{a}_1$, $\boldsymbol{b}_1$, and $\boldsymbol{c}_1$, respectively. Prove that the lines $\boldsymbol{l}_{aa_1}$, $\boldsymbol{l}_{bb_1}$, and $\boldsymbol{l}_{cc_1}$ are parallel. (*Hint*: In the affine 2-plane $\mathcal{A}_e^2$, interpret the geometry of your results according to the grade and the absolute value of the directed distances between the lines.)

Fig. 5.3 Simson's theorem

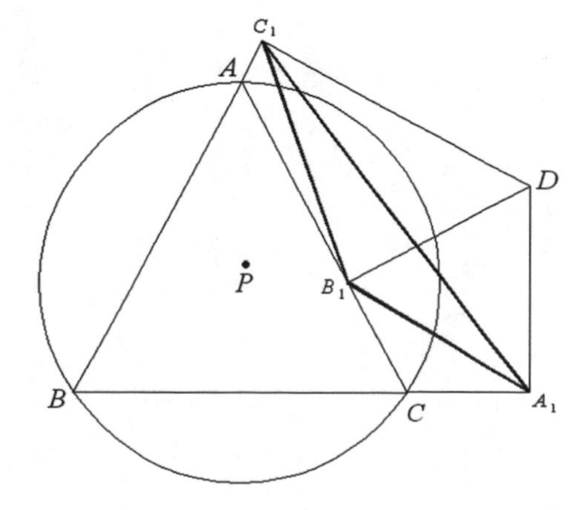

5.31 Consider in the affine 3-plane $\mathcal{A}_e^3$ the points $a_1^h = 3e_1 + 4e_2 + 5e_3 + e$, $a_2^h = 2e_1 - 5e_2 + 2e_3 + e$, and $a_3^h = 1e_1 + 6e_2 + 4e_3 + e$; the line $L_1^h = a_1^h \wedge a_2^h$; and the plane $\phi_1^h = a_1^h \wedge a_2^h \wedge a_3$. Compute, using the Maple packet CLIFFORD 4.0, for (a) one of the points, (b) the line, and (c) the plane, their new positions after the rigid motion. This is defined by a translation of $t_1^h = e_1 + 2e_2 + e_3 + e$ and rotations about the three axes of $\theta_x = \frac{\pi}{5}, \theta_y = \frac{\pi}{3}$, and $\theta_z = \frac{\pi}{6}$. Recall that you compute the translation using the horosphere as an intermediate framework.

5.32 Consider in the affine 3-plane $\mathcal{A}_e^3$ the points $p_0^h = e$, $p_1^h = e_2 + e$, $p_2^h = e_1 + e_2 + e, p_3^h = e_1 + e, p_4^h = e_1 + e_3 + e, p_5^h = e_1 + e_2 + e_3 + e, p_6^h = e_2 + e_3 + e$, and $p_7^h = e_3 + e$; the lines $L_{01} = p_0^h \wedge p_1^h$, $L_{36} = p_3^h \wedge p_6^h$, and $L_{76} = p_7^h \wedge p_6^h$; and the planes $\phi_f = p_0^h \wedge p_1^h \wedge p_3^h$, $\phi_t = p_5^h \wedge p_6^h \wedge p_7^h$, and $\phi_r = p_2^h \wedge p_5^h \wedge p_6^h$. Compute, using CLIFFORD 4.0, the directed distances between p_7^h and ϕ_f, p_5^h and L_{36}, L_{01} and L_{36}, L_{36} and L_{76}, L_{01} and ϕ_t, L_{36} and ϕ_t, ϕ_f and ϕ_t, and ϕ_f and ϕ_r. Interpret the geometry of your results according to the grade and the absolute value of the directed distances.

Part II
Euclidean, Pseudo-Euclidean Geometric Algebras, Incidence Algebra, Conformal and Projective Geometric Algebras

Chapter 6
2D, 3D, and 4D Geometric Algebras

It is the believe that imaginary numbers appeared for the first time around 1540 when the mathematicians Tartaglia and Cardano represented real roots of a cubic equation in terms of conjugated complex numbers. A Norwegian surveyor, Caspar Wessel, was in 1798 the first one to represent complex numbers by points on a plane with its vertical axis imaginary and horizontal axis real. This diagram was later known as the Argand diagram, although the true Argand's achievement was an interpretation of $\mathrm{i} = \sqrt(-1)$ as a rotation by a right angle in the plane. Complex numbers received their name by Gauss and their formal definition as pair of real numbers was introduced by Hamilton in 1835.

6.1 Complex, Double, and Dual Numbers

In a broad sense, the most general complex numbers [156, 316] on the plane can be categorized into three different systems: ordinary complex numbers, double numbers, and dual numbers. In general, a complex number can be represented as a *composed number* $\mathbf{a} = b + \omega c$ using the algebraic operator ω, in which $\omega^2 = -1$ in the case of complex numbers, $\omega^2 = 1$ in the case of double numbers, and $\omega^2 = 0$ in the case of dual numbers. For dual numbers, b represents the real term and c represents the dual term.

For this book, it is useful to recall the notion of a function of a dual variable, in which a differentiable real function $f : \mathcal{R} \rightarrow \mathcal{R}$ with a dual argument $\alpha + \omega\beta$, where $\alpha, \beta \in \mathcal{R}$, can be expanded using a Taylor series. Because $\omega^2 = \omega^3 = \omega^4 = \cdots = 0$, the function reads

E. Bayro-Corrochano, *Geometric Algebra Applications Vol. I*,
https://doi.org/10.1007/978-3-319-74830-6_6

$$
\begin{aligned}
f(\alpha+\omega\beta) &= f(\alpha)+\omega f'(\alpha)\beta+\omega^2 f''(\alpha)\frac{\beta^2}{2!}+\cdots, \\
&= f(\alpha)+\omega f'(\alpha)\beta.
\end{aligned} \tag{6.1}
$$

A useful illustration of this expansion is the exponential function of a dual number,

$$
e^{\alpha+\omega\beta} = e^{\alpha}+\omega e^{\alpha}\beta = e^{\alpha}(1+\omega\beta). \tag{6.2}
$$

In his seminal paper "Preliminary sketch of bi-quaternions" [57], Clifford introduced the use of dual numbers *the motors* or *bi-quaternions* to represent screw motion. Later, Study [290] used dual numbers to represent the relative position of two skew lines in space— that is, $\hat{\theta}=\theta+\omega d$, where $\hat{\theta}$ represents the dual angle, θ for the difference of the line orientation angles, and d for the distance between both lines.

The algebras of complex, double (hyperbolic), and dual numbers are isomorphic to the *center* of certain geometric algebras. For these algebras, we must choose the appropriate multivector basis, so that the unit pseudoscalar squares to 1 for the case of double numbers, to -1 for complex numbers, and to 0 for dual numbers. Note that the pseudoscalar for these numbers maintains its geometric interpretation as a unit hypervolume, and that, as is the case with ω, they are commutative with either vectors or bivectors, depending only upon the type of the geometric algebra used.

In Sect. 6.2, we will consider some examples of composed numbers in geometric algebra: complex numbers in the space $G_{0,1,0}$, double numbers in the space $G_{1,0,0}$, and dual complex numbers in the space $G_{1,0,1}$. We shall describe complex and dual numbers for 2D, 3D, and 4D spaces in some detail. The dual numbers will be used later for the modeling of points, lines, and planes, as well as for the modeling of motion.

6.2 2D Geometric Algebras of the Plane

In this section, we want to illustrate the application of different 2D geometric algebras for the modeling of group transformations on the plane. In doing so, we can also clearly see the geometric interpretation and the use of complex, double, and dual numbers for the cases of rotation, affine, and Lorentz transformations, respectively [251, 316]. We find these transformations in various tasks of image processing. For the modeling of the 2D space, we choose a geometric algebra which has $2^2=4$ elements, given by

$$
\underbrace{1}_{scalar},\quad \underbrace{e_1, e_2}_{vectors},\quad \underbrace{e_1e_2}_{bivector} \equiv I. \tag{6.3}
$$

The highest grade element for the 2D space, called the unit pseudoscalar $I \equiv e_1e_2$, is a bivector. According to the used vector basis, the signature of the geometric algebra will change, yielding complex, double, or dual numbers. Each of these cases is illustrated below.

In the geometric algebra $G_{2,0,0}$, where $I = e_1e_2$ with $I^2 = -1$, we want to represent the rotation of the points (x,y) of the *Euclidean plane*. Here, a rotation of the point $z = xe_1 + ye_2 = r(cos(\alpha)e_1 + sin(\alpha)e_2) \in G_{2,0,0}$ can be computed as the geometric product of the vector and the complex number $e^{I\frac{\theta}{2}} = cos\frac{\theta}{2} + e_1e_2sin\frac{\theta}{2} = (cos\frac{\theta}{2} + Isin\frac{\theta}{2}) \in G^+_{2,0,0}$, or $\in$ Spin(2) (spin group), as follows:

$$
\begin{aligned}
z' &= e^{-I\frac{\theta}{2}} z e^{I\frac{\theta}{2}}, \\
&= e^{-I\frac{\theta}{2}} r(cos(\alpha)e_1 + sin(\alpha)e_2)e^{I\frac{\theta}{2}}, \\
&= \left(cos\frac{\theta}{2} + Isin\frac{\theta}{2}\right)^{-1} r(cos(\alpha)e_1 + sin(\alpha)e_2)\left(cos\frac{\theta}{2} + Isin\frac{\theta}{2}\right), \\
&= r(cos(\alpha+\theta)e_1 + sin(\alpha+\theta)e_2).
\end{aligned} \tag{6.4}
$$

Figure 6.1b illustrates that each point of the 2D image of the die is rotated by θ. Note that this particular form for representing rotation, $e^{I\frac{\theta}{2}} = (cos\frac{\theta}{2} + Isin\frac{\theta}{2})$, can be generalized to higher dimensions (see the algebra of rotors in 3D space in the next section).

Let us now represent the points as dual numbers in the geometric algebra $\mathcal{G}_{1,0,1}$, where $I^2 = 0$. A 2D point can be represented in $G_{1,0,1}$ as $z = xe_1+ye_2 = x(e_1+se_2)$, where $s = \frac{y}{x}$ is the slope. The shear transformation of this point can be computed by applying a unit shear dual number $e^{I\frac{\tau}{2}} = (1 + I\frac{\tau}{2}) \in \mathcal{G}_{1,0,1}$ as follows:

$$
\begin{aligned}
z' &= e^{-I\frac{\tau}{2}} z e^{I\frac{\tau}{2}} = \left(1 - I\frac{\tau}{2}\right)(x(e_1 + se_2))\left(1 + I\frac{\tau}{2}\right), \\
&= x(e_1 + (s+\tau)e_2).
\end{aligned} \tag{6.5}
$$

Note that the overall effect of this transformation is to shear the plane, where the points (x,y) lie parallel to the e_2-axis through the shear τ with a shear angle of $tan^{-1}\tau$. Figure 6.1c depicts the effect of the shear transformation acting on the 2D image of the die.

By using the representation of the double number in $G_{1,1,0}$, where $I^2 = 1$, we can implement the Lorentz transformation of the points. This transformation is commonly used in space–time algebra for special relativity computations, and it has been suggested for use in psychophysics as well [69, 145]. In this context, a 2D point is associated with a double number $z = te_1 + xe_2 = \rho(cosh(\alpha)e_1 + sinh(\alpha)e_2) \in G_{1,1,0}$. The lines $|t| = |x|$ divide the plane into two quadrants with $|t| > |x|$ and two quadrants with $|t| < |x|$. If we apply a 2D unit displacement vector $e^{I\frac{\beta}{2}} = a + Ib = (cosh(\beta) + Isinh(\beta)) \in G_{1,1,0}$ from one of the quadrants, $|t| > |x|$, to an arbitrary point $z = t + Ix$, we get

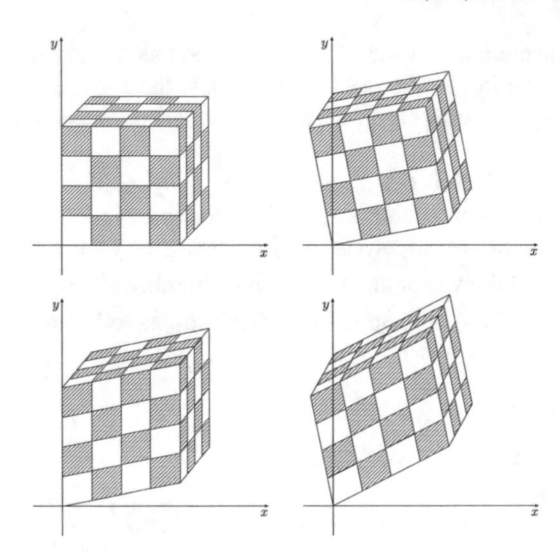

Fig. 6.1 Effects of 2D transformations: **a** original cube, **b** cube after rotation, **c** after shear transformation, and **d** after Lorentz transformation

$$
\begin{aligned}
z' &= e^{-I\frac{\beta}{2}} z e^{I\frac{\theta}{2}}, \\
&= e^{-I\frac{\beta}{2}} \rho(cosh(\alpha)e_1 + sinh(\alpha)e_2) e^{I\frac{\theta}{2}}, \\
&= \left(cosh\frac{\beta}{2} + I sinh\frac{\beta}{2}\right)^{-1} (\rho(cosh\alpha e_1 + sinh\alpha e_2)) \left(cosh\frac{\beta}{2} + I sinh\frac{\beta}{2}\right), \\
&= \rho(cosh(\alpha+\beta)e_1 + sinh(\alpha+\beta)e_2). \qquad (6.6)
\end{aligned}
$$

The point is displaced along a particular hyperbolic path through the interval $\rho\beta$ in $|t| < |x|$. Figure 6.1d illustrates the effect of the Lorentz transformation acting on the 2D image of the die.

6.3 3D Geometric Algebra for the Euclidean 3D Space

For the case of embedding Euclidean 3D space, we choose the geometric algebra $G_{3,0,0}$, which has $2^3 = 8$ elements given by

$$
\underbrace{1}_{scalar}, \quad \underbrace{\{e_1, e_2, e_3\}}_{vectors}, \quad \underbrace{\{e_2e_3, e_3e_1, e_1e_2\}}_{bivectors}, \quad \underbrace{\{e_1e_2e_3\} \equiv I}_{trivector}. \qquad (6.7)
$$

The highest grade algebraic element for the 3D space is a trivector called a unit pseudoscalar $I \equiv e_1e_2e_3$, which squares to -1 and which commutes with the scalars and bivectors in the 3D space. In the algebra of three-dimensional space, we can construct a trivector $\boldsymbol{a} \wedge \boldsymbol{b} \wedge \boldsymbol{c} = \lambda I$, where the vectors $\boldsymbol{a}$, $\boldsymbol{b}$, and $\boldsymbol{c}$ are in general position and $\lambda \in \mathcal{R}$. Note that no 4-vectors exist since there is no possibility of sweeping the volume element $\boldsymbol{a} \wedge \boldsymbol{b} \wedge \boldsymbol{c}$ over a fourth dimension.

Multiplication of the three basis vectors e_1, e_2, and e_3 by I results in the three basis bivectors $e_2 e_3 = I e_1$, $e_3 e_1 = I e_2$, and $e_1 e_2 = I e_3$. These simple bivectors rotate vectors in their own plane by 90°; for example, $(e_1 e_2) e_2 = e_1$, $(e_2 e_3) e_2 = -e_3$. Identifying the unit vectors $\boldsymbol{i}, \boldsymbol{j}, \boldsymbol{k}$ of quaternion algebra with $I e_1, -I\sigma_2, I\sigma_3$ allows us to write the famous Hamilton relations $\boldsymbol{i}^2 = \boldsymbol{j}^2 = \boldsymbol{k}^2 = \boldsymbol{i}\boldsymbol{j}\boldsymbol{k} = -1$. Since the $\boldsymbol{i}, \boldsymbol{j}, \boldsymbol{k}$ are really bivectors, it comes as no surprise that they represent 90° rotations in orthogonal directions and provide a system well suited for the representation of general 3D rotations (see Fig. 2.1c). Rotors are isomorphic with quaternions. The quaternion and rotor follow the left-hand and the right-hand rotation rule, respectively.

6.3.1 The Algebra of Rotors

In geometric algebra, a *rotor* (short name for rotator), $\boldsymbol{R}$, is an even-grade element of the Euclidean algebra of 3D space. If $Q = \{r_0, r_1, r_2, r_3\} \in G_{3,0,0}$ represents a unit quaternion, then the rotor which performs the same rotation is simply given by

$$\boldsymbol{R} = \underbrace{r_0}_{scalar} \underbrace{+ r_1(I e_1) - r_2(I e_2) + r_3(I e_3)}_{bivectors}. \tag{6.8}$$

The rotor algebra $G^+_{3,0,0}$ is therefore a subset of the Euclidean geometric algebra of three-dimensional space.

Consider in $\mathcal{G}_{3,0,0}$ two non-parallel vectors $\boldsymbol{a}$ and $\boldsymbol{b}$ which are referred to the same origin. In general, a rotation operation of a vector $\boldsymbol{a}$ toward the vector $\boldsymbol{b}$ can be performed by two *reflections*, respective to the unit vector axes $\boldsymbol{n}$ and $\boldsymbol{m}$ (see Fig. 6.2). The components of the first reflection are

$$\boldsymbol{a}_{\|} = |\boldsymbol{a}| cos(\alpha) \frac{\boldsymbol{n}}{|\boldsymbol{n}|} = |\boldsymbol{a}||\boldsymbol{n}| cos(\alpha) \frac{\boldsymbol{n}}{|\boldsymbol{n}|^2} = (\boldsymbol{a} \cdot \boldsymbol{n}) \boldsymbol{n}^{-1}, \tag{6.9}$$

$$\begin{aligned} \boldsymbol{a}_{\perp} &= \boldsymbol{a} - \boldsymbol{a}_{\|} = \boldsymbol{a} - (\boldsymbol{a} \cdot \boldsymbol{n}) \boldsymbol{n}^{-1} = (\boldsymbol{a}\boldsymbol{n} - \boldsymbol{a} \cdot \boldsymbol{n}) \boldsymbol{n}^{-1}, \\ &= (\boldsymbol{a} \wedge \boldsymbol{n}) \boldsymbol{n}^{-1}, \end{aligned} \tag{6.10}$$

so the vector $\boldsymbol{a}$ after the first reflection becomes

$$\begin{aligned} \boldsymbol{a}' &= \boldsymbol{a}_{\|} - \boldsymbol{a}_{\perp} = (\boldsymbol{a} \cdot \boldsymbol{n}) \boldsymbol{n}^{-1} - (\boldsymbol{a} \wedge \boldsymbol{n}) \boldsymbol{n}^{-1} = (\boldsymbol{a} \cdot \boldsymbol{n} - \boldsymbol{a} \wedge \boldsymbol{n}) \boldsymbol{n}^{-1}, \\ &= (\boldsymbol{n} \cdot \boldsymbol{a} + \boldsymbol{n} \wedge \boldsymbol{a}) \boldsymbol{n}^{-1} = \boldsymbol{n}\boldsymbol{a}\boldsymbol{n}^{-1}. \end{aligned} \tag{6.11}$$

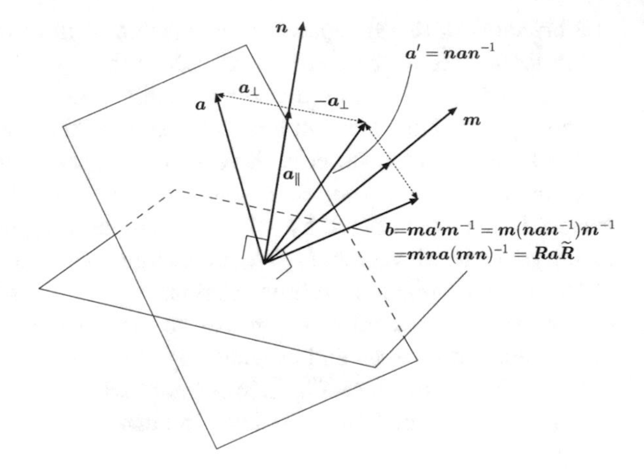

Fig. 6.2 Rotor in the 3D space formed by a pair of reflections

The second reflection respective to the axis unit $\boldsymbol{m}$ completes the vector rotation of $\boldsymbol{a}$ toward $\boldsymbol{b}$, as follows:

$$\begin{aligned} \boldsymbol{b} &= \boldsymbol{m}(\boldsymbol{a}')\boldsymbol{m}^{-1} = \boldsymbol{m}(\boldsymbol{n}\boldsymbol{a}\boldsymbol{n}^{-1})\boldsymbol{m}^{-1} = \boldsymbol{m}\boldsymbol{n}\boldsymbol{a}\boldsymbol{n}^{-1}\boldsymbol{m}^{-1} = (\boldsymbol{m}\boldsymbol{n})\boldsymbol{a}(\boldsymbol{m}\boldsymbol{n})^{-1}, \\ &= \boldsymbol{R}\boldsymbol{a}\boldsymbol{R}^{-1} = \boldsymbol{R}\boldsymbol{a}\tilde{\boldsymbol{R}}. \end{aligned} \tag{6.12}$$

The rotor $\boldsymbol{R}$ composed by these two reflections performs a rotation that is two times greater than the angle between $\boldsymbol{m}$ and $\boldsymbol{n}$. Now, if we consider successive reflections of the vector $\boldsymbol{a}$ with respect to j planes, we get the following resultant transformation

$$\begin{aligned} \boldsymbol{b} &= (\boldsymbol{m}_j \ldots \boldsymbol{m}_2\boldsymbol{m}_1)\boldsymbol{a}(\boldsymbol{m}_j \ldots \boldsymbol{m}_2\boldsymbol{m}_1)^{-1}, \\ &= (\boldsymbol{m}_j\boldsymbol{R}_{(j-1)(j-2)} \ldots \boldsymbol{R}_{32}\boldsymbol{R}_{21})\boldsymbol{a}(\boldsymbol{m}_j\boldsymbol{R}_{(j-1)(j-2)} \ldots \boldsymbol{R}_{32}\boldsymbol{R}_{21})^{-1}, \\ &= \boldsymbol{m}_j\boldsymbol{R}_{1(j-1)}\boldsymbol{a}\boldsymbol{R}_{(j-1)1}\boldsymbol{m}_j, \qquad \texttt{for j odd} \\ &= \boldsymbol{R}_{j1}\boldsymbol{a}\boldsymbol{R}_{1j}, \qquad \texttt{for j even.} \end{aligned} \tag{6.13}$$

Figure 6.3 shows the case for three reflections or j = 3.

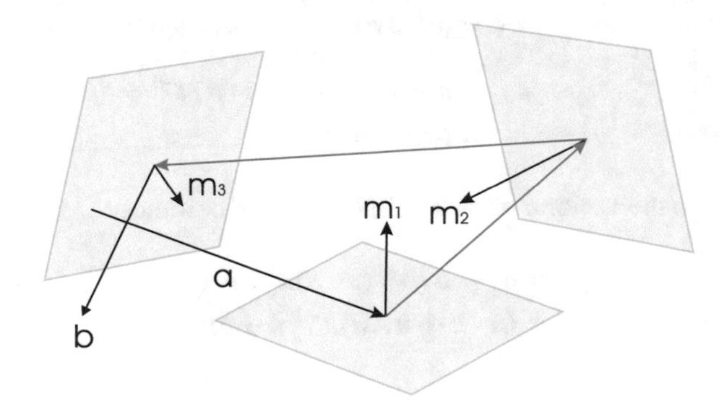

Fig. 6.3 Trajectory of successive reflections

According to Eqs. (2.92)–(2.97), the reversion and magnitude of a rotor $\boldsymbol{R}$ are, respectively, given by

$$\begin{aligned}\tilde{\boldsymbol{R}} &= r_0 - r_1 e_2 e_3 - r_2 e_3 e_1 - r_3 e_1 e_2 = r_0 - \boldsymbol{r},\\ ||\boldsymbol{R}||^2 &= \boldsymbol{R}\tilde{\boldsymbol{R}}.\end{aligned} \tag{6.14}$$

This implies that the unique multiplicative inverse of $\boldsymbol{R}$ is given by

$$\boldsymbol{R}^{-1} = \tilde{\boldsymbol{R}} \parallel \boldsymbol{R} \parallel^{-2}. \tag{6.15}$$

If a rotor $\boldsymbol{R}$ satisfies the equation

$$\boldsymbol{R}\tilde{\boldsymbol{R}} = \parallel \boldsymbol{R} \parallel^2 = r_0^2 - \boldsymbol{r} \cdot \boldsymbol{r} = 1, \tag{6.16}$$

then we say that this rotor is a unit rotor and its multiplicative inverse is simply $\boldsymbol{R}^{-1} = \tilde{\boldsymbol{R}}$, as denoted previously in Eq. (6.12).

Equation (6.12) shows that the unit rotor corresponds to the geometric product of two unit vectors,

$$\boldsymbol{R} = \boldsymbol{m}\boldsymbol{n} = \boldsymbol{m}\cdot\boldsymbol{n} + \boldsymbol{m}\wedge\boldsymbol{n}. \tag{6.17}$$

The components of Eq. (6.17) correspond to the scalar and bivector terms of an equivalent quaternion in $G_{3,0,0}$, and thus $\boldsymbol{R} \in G^+_{3,0,0}$. This even subalgebra corresponds to the algebra of rotors.

Considering the scalar and the bivector terms of the rotor of Eq. (6.17), we can further write the Euler representation of a 3D rotation with angle θ in the left-hand sense, as follows:

$$\begin{aligned}\boldsymbol{R} &= r_0 + \boldsymbol{r} = r_0 + r_1 e_2 e_3 + r_2 e_3 e_1 + r_3 e_1 e_2,\\ &= a_c + a_s \bar{\boldsymbol{r}}_n = cos\left(\frac{\theta}{2}\right) + sin\left(\frac{\theta}{2}\right)\bar{\boldsymbol{r}}_n,\\ &= e^{\frac{\theta}{2}\bar{\boldsymbol{r}}_n},\end{aligned} \tag{6.18}$$

where $\bar{\boldsymbol{r}}_n$ is the unitary rotation-axis vector spanned by the bivector basis e_2e_3, e_3e_1, and e_1e_2, and the scalars a_c and $a_s \in \mathcal{R}$. The polar representation of a rotor given in Eq. (6.18) is possible, because the rotor as a Lie group can be expressed in terms of the Lie algebra of bivectors: The orbits on the Lie group manifold describe the evolution of the actions of rotors. The bivector $\bar{\boldsymbol{r}}_n$ corresponds to the Lie operator tangent to an orbit or geodesic.

The transformation of a rotor $\boldsymbol{p} \mapsto \mathbf{R}\boldsymbol{p}\tilde{\mathbf{R}} = \boldsymbol{p}'$ is a very general way of handling rotations which works for multivectors of any grade and in spaces of any dimension. Rotors combine in a straightforward manner—i.e., a rotor $\boldsymbol{R}_1$ followed by a rotor $\boldsymbol{R}_2$ is equivalent to a total rotor

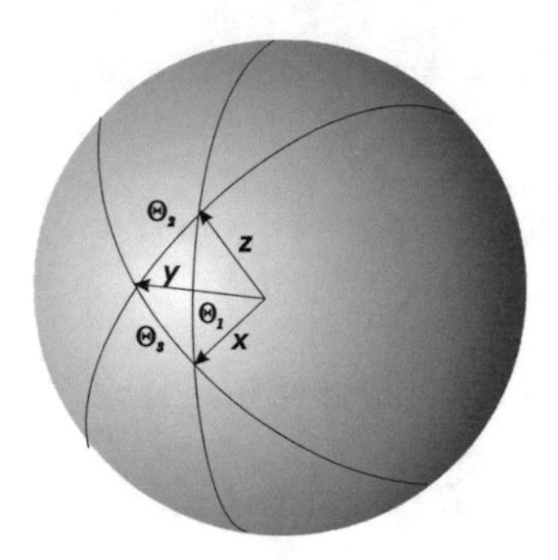

Fig. 6.4 Rotors represented on the sphere

$$R = R_2 R_1. \tag{6.19}$$

The composition of 3D rotations in different planes is described in Eqs. (6.13) and (6.19), the latter can be visualized geometrically in Fig. 6.4, where the half-angle of each rotor is depicted as a directed arc vector called $\boldsymbol{\theta}_i$ confined to a great circle on the unit sphere. The product of rotors $\boldsymbol{R}_1$ and $\boldsymbol{R}_2$ is depicted in Fig. 6.4 by connecting the corresponding arcs at the point z where the two great circles of $\boldsymbol{\theta}_1$ and $\boldsymbol{\theta}_2$ intersect. As a result $\boldsymbol{x} = z\boldsymbol{\theta}_1$ and $\boldsymbol{y} = \boldsymbol{\theta}_2 z$, thus the half-angle of the rotor can be computed depending of a common arc vector: $\boldsymbol{\theta}_1 = z\boldsymbol{x}$ and $\boldsymbol{\theta}_2 = \boldsymbol{y}z$. Combining these results, we get: $\boldsymbol{\theta}_3 = \boldsymbol{\theta}_2\boldsymbol{\theta}_1 = (\boldsymbol{y}z)(z\boldsymbol{x}) = \boldsymbol{y}\boldsymbol{x}$.

A rotor is isomorph with a quaternion. As a result, we can embed quaternions in the more comprehensive mathematical system offered by geometric algebra. Different as in quaternion theory, in geometric algebra the quaternions or rotors have a clear geometric interpretation due to the representation in space of the rotations as described above by using reflections with respect to planes. The Sect. 6.4 is devoted to quaternion algebra including more details useful for applications in image processing.

6.3.2 Orthogonal Rotors

For the rotation of a vector $\boldsymbol{p}$ in the right-hand sense, we simply adopt the rotor with the minus sign to agree with the standard right-hand rule for the direction of the rotation:

$$\begin{aligned} \boldsymbol{R} &= e^{-\frac{\bar{\boldsymbol{r}}_\theta}{2}} = e^{-\frac{\theta}{2}\bar{\boldsymbol{r}}_n}, \\ &= cos(\theta/2) - sin(\theta/2)\bar{\boldsymbol{r}}_n. \end{aligned} \tag{6.20}$$

This rotation operation is depicted in Fig. 6.5. The rotated vector $\boldsymbol{p}'$ is given by

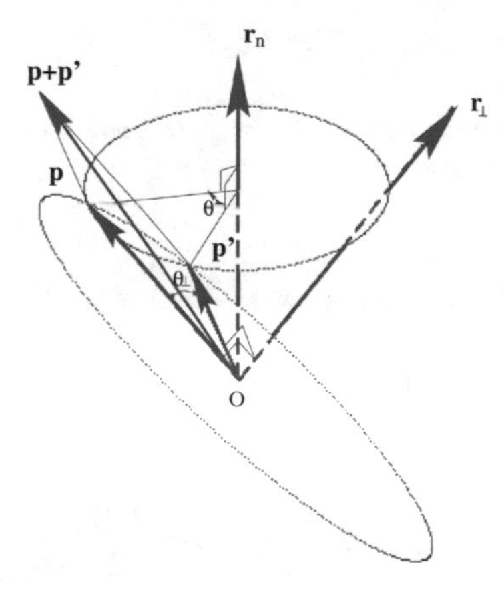

Fig. 6.5 Geometric interpretation of rotation

$$p' = \boldsymbol{R}\boldsymbol{p}\tilde{\boldsymbol{R}} = \Big(cos(\theta/2) - sin(\theta/2)\bar{\boldsymbol{r}}_n\Big)\boldsymbol{p}\Big(cos(\theta/2) + sin(\theta/2)\bar{\boldsymbol{r}}_n\Big). \quad (6.21)$$

Since the rotation path from $\boldsymbol{p}$ and $\boldsymbol{p}'$ is not necessarily unique, neither is the rotor $\boldsymbol{R}$ unique. The shortest path determined by the endpoints $\boldsymbol{p}$ and $\boldsymbol{p}'$ lies on a great circle of a sphere with radius $\| \boldsymbol{p} \|$, and this is called *orthogonal rotation*. The rotor itself is called *orthogonal rotor* $\boldsymbol{R}_\perp$ and it can be calculated using the unit vectors $\frac{\boldsymbol{p}'+\boldsymbol{p}}{\|\boldsymbol{p}'+\boldsymbol{p}\|}$ and $\frac{\boldsymbol{p}}{\|\boldsymbol{p}\|}$, as follows:

$$\begin{aligned}
\boldsymbol{R}_\perp &= -\frac{(\boldsymbol{p}'+\boldsymbol{p})\boldsymbol{p}}{\| \boldsymbol{p}'+\boldsymbol{p} \| \cdot \| \boldsymbol{p} \|} = -\frac{(\boldsymbol{p}'+\boldsymbol{p})\cdot\boldsymbol{p} + (\boldsymbol{p}'+\boldsymbol{p})\wedge\boldsymbol{p}}{\| \boldsymbol{p}'+\boldsymbol{p} \| \cdot \| \boldsymbol{p} \|}, \\
&= -\frac{(\boldsymbol{p}'+\boldsymbol{p})\cdot\boldsymbol{p}}{\| \boldsymbol{p}'+\boldsymbol{p} \| \cdot \| \boldsymbol{p} \|} - \frac{-\boldsymbol{p}\wedge(\boldsymbol{p}'+\boldsymbol{p})}{\| \boldsymbol{p}'+\boldsymbol{p} \| \cdot \| \boldsymbol{p} \|}, \\
&= \boldsymbol{r}_{\perp 0} + \boldsymbol{r}_\perp = cos(\theta_\perp/2) - sin(\theta_\perp/2)\boldsymbol{r}_{n,\perp},
\end{aligned} \quad (6.22)$$

where the rotation-axis bivector $\boldsymbol{r}_\perp$, or the unit rotation-axis bivector $\boldsymbol{r}_{n,\perp}$, are perpendicular to both $\boldsymbol{p}$ and $\boldsymbol{p}'$ and the angle $\theta_\perp/2$ is the angle between the vectors $\boldsymbol{p}$ and $\boldsymbol{p}'+\boldsymbol{p}$. Rotors are isomorphic with quaternions. In signal analysis, quaternions have been used quite often in an operational sense. In contrast, rotors have a clear geometric interpretation and they are used for geometric operations.

6.3.3 Recovering a Rotor

Using previous results, we will derive a procedure to recover a rotor based on two observer measurements. Assume that we have two sets of vectors in 3D $\{e_k\}$ and $\{f_k\}$ which are not necessarily orthonormal and that they are related by a rotation

$$f_k = \boldsymbol{R} e_k \tilde{\boldsymbol{R}}. \tag{6.23}$$

Since the $\boldsymbol{R}$ is unknown, we should derive a simple method to recover $\boldsymbol{R}$. A rotor can be written as

$$\begin{aligned} \boldsymbol{R} &= e^{-\frac{\boldsymbol{B}}{2}}, \\ \tilde{\boldsymbol{R}} &= e^{\frac{\boldsymbol{B}}{2}} = cos\left(\frac{|\boldsymbol{B}|}{2}\right) + sin\left(\frac{|\boldsymbol{B}|}{2}\right)\frac{\boldsymbol{B}}{|\boldsymbol{B}|}. \end{aligned} \tag{6.24}$$

Using the equation

$$\begin{aligned} e_k \tilde{\boldsymbol{R}} e^k &= e_k \left[cos(\frac{|\boldsymbol{B}|}{2}) + sin(\frac{|\boldsymbol{B}|}{2})\frac{\boldsymbol{B}}{|\boldsymbol{B}|} \right] e^k, \\ &= 3cos\left(\frac{|\boldsymbol{B}|}{2}\right) - sin\left(\frac{|\boldsymbol{B}|}{2}\right)\frac{\boldsymbol{B}}{|\boldsymbol{B}|}, \\ &= 4cos\left(\frac{|\boldsymbol{B}|}{2}\right) - \tilde{\boldsymbol{R}}. \end{aligned} \tag{6.25}$$

Reusing Eq. (6.23)

$$f_k e^k = \boldsymbol{R} e_k \tilde{\boldsymbol{R}} e^k = 4cos\left(\frac{|\boldsymbol{B}|}{2}\right)\boldsymbol{R} - 1. \tag{6.26}$$

We can see that the unknown rotor is a scalar multiple of $1 + f_k e^k$, thus one can establish the simple and useful formula

$$\boldsymbol{R} = \frac{1 + f_k e^k}{|1 + f_k e^k|} = \frac{\zeta}{\sqrt{\zeta\tilde{\zeta}}}, \tag{6.27}$$

where $\zeta = 1 + f_k e^k$. Using this formula, one can recover the rotor directly from the frame vectors.

6.4 Quaternion Algebra

The quaternion algebra $I\!H$ was invented by W. R. Hamilton in 1843 [119, 120], when he intended for almost ten years to find an algebraic system which would do for the space $\mathbb{R}^3$ the same as complex numbers do for the space $\mathbb{R}^2$. Interesting enough, the current formalism of vector algebra was simply extracted out from the quaternion product of two vectors by Gibbs in 1901, namely $\boldsymbol{ab} = -\boldsymbol{a} \cdot \boldsymbol{b} + \boldsymbol{a} \times \boldsymbol{b}$.

Hamilton tried to find a multiplications rule for triplets $\boldsymbol{a} = a_1\boldsymbol{i} + a_2\boldsymbol{j} + a_3\boldsymbol{k}$ and $\boldsymbol{b} = b_1\boldsymbol{i} + b_2\boldsymbol{j} + b_3\boldsymbol{k}$, so that $|\boldsymbol{ab}| = |\boldsymbol{a}||\boldsymbol{b}|$ correspond to a multiplicative product of the vectors $\boldsymbol{a}, \boldsymbol{b} \in \mathbb{R}^3$. However, according to a result of Legendre (1830) such bilinear product confined in $\mathbb{R}^3$ does not exist, i.e. no integer of the form $4^a(6b+7)$ with $a \geq 0,\ b \geq 0$ can be obtained as a sum of three squares. Hamilton searched for a generalized complex number system in three dimensions, but no such associative hypercomplex numbers exist in three dimensions. One can see this fact easily by considering two imaginary units $\boldsymbol{i}$ and $\boldsymbol{j}$ such that $\boldsymbol{i}^2 = \boldsymbol{j}^2 = -1$ and furthermore that $1, \boldsymbol{i}, \boldsymbol{j}$ span $\mathbb{R}^3$. So the multiplication has to be of the form $\boldsymbol{i}\boldsymbol{j} = \alpha + \boldsymbol{i}\beta + \boldsymbol{j}\gamma$ for $\alpha, \beta, \gamma \in \mathbb{R}$. Let us prove its consistency, on the one hand

$$\begin{aligned} \boldsymbol{i}(\boldsymbol{i}\boldsymbol{j}) &= \alpha\boldsymbol{i} + -\beta + \gamma(\boldsymbol{i}\boldsymbol{j}) = -\beta + \alpha\boldsymbol{i} + \gamma(\alpha + \boldsymbol{i}\beta + \boldsymbol{j}\gamma), \\ &= -\beta + \alpha\gamma + (\alpha + \beta\gamma)\boldsymbol{i} + \gamma^2\boldsymbol{j}. \end{aligned} \tag{6.28}$$

On the other hand by associativity

$$\boldsymbol{i}(\boldsymbol{i}\boldsymbol{j}) = (\boldsymbol{i}\boldsymbol{i})\boldsymbol{j} = \boldsymbol{i}^2\boldsymbol{j} = -\boldsymbol{j}, \tag{6.29}$$

which is a contradiction with the above equation, since $\gamma^2 \geq 0$ for any $\gamma \in \mathbb{R}$.

The Frobenius theorem proved by Ferdinand Georg Frobenius in 1877 characterizes the finite-dimensional associative algebras over the real numbers. It proves that, if A is a finite-dimensional division algebra over the real numbers R then one of the following cases are true: $A = \mathbb{R}$, $A = \mathbb{C}$ (complex numbers) and A is isomorphic to the quaternion algebra $\mathbb{H}$.

The key idea of Hamilton's discovery was to move to a four-dimensional space and to consider elements of the form $\boldsymbol{q} = s + x\boldsymbol{i} + y\boldsymbol{j} + z\boldsymbol{k}$ for $s, x, y, z \in \mathbb{R}$, where the orthogonal imaginary numbers $\boldsymbol{i}$, $\boldsymbol{j}$ and $\boldsymbol{k}$ obey the following multiplicative rules

$$\boldsymbol{i}^2 = \boldsymbol{j}^2 = -1,\ \ \boldsymbol{k} = \boldsymbol{i}\boldsymbol{j} = -\boldsymbol{j}\boldsymbol{i}\ \ \rightarrow\ \ \boldsymbol{k}^2 = -1. \tag{6.30}$$

Hamilton named his four-component elements *quaternions*. Quaternions form a division ring and are denoted by H in honor of Hamilton. The quaternion algebra is non-associative. Arthur Cayley (1821–1895) was the first person, after Hamilton, to publish a paper on quaternions.

The conjugated of a quaternion is given by

$$\bar{\boldsymbol{q}} = s - x\boldsymbol{i} - y\boldsymbol{j} - z\boldsymbol{k}. \tag{6.31}$$

Arthur Cayley in 1855 discovered the quaternionic representation of four-dimensional rotations, namely

$$\mathbb{R}^4 \rightarrow \mathbb{R}^4, \qquad\qquad \boldsymbol{q}_1 \rightarrow \boldsymbol{q}_2\boldsymbol{q}_1\bar{\boldsymbol{q}}_2, \tag{6.32}$$

where $\mathbb{R}^4$ stands for $\mathbb{H}$ and $\boldsymbol{q}_1, \boldsymbol{q}_1 \in \mathbb{H}$.

For the quaternion $\boldsymbol{q}$, we can compute its partial angles as

$$\arg_i(q) = s\tan 2(x, s), \quad \arg_j(q) = s\tan 2(y, s), \quad \arg_k(q) = s\tan 2(z, s), \tag{6.33}$$

and its partial modules and its projections on its imaginary axes as

$$\begin{aligned} &mod_i(q) = \sqrt{s^2 + x^2}, \; mod_j(s) = \sqrt{s^2 + y^2}, \\ &mod_k(q) = \sqrt{s^2 + z^2}, \\ &mod_i(q)\exp(i \arg_i(q)) = s + xi, \; mod_i(q)\exp(j \arg_j(q)) = s + yj, \\ &mod_k(q)\exp(k \arg_k(q)) = s + zk. \end{aligned} \tag{6.34}$$

In signal analysis, quaternions have been used quite often in an operational sense. In contrast, rotors were introduced for geometric operations. Next, we will provide some definitions for quaternions which will be useful for the analysis and processing of images.

In a similar way as the complex numbers which can be expressed in a polar representation, we can also represent a quaternion in a polar form. The polar representation of a quaternion $\boldsymbol{q} = r + x\boldsymbol{i} + y\boldsymbol{j} + z\boldsymbol{k} \in G^+_{3,0,0}$ is given when the quaternion, seen as a Lie group, is expressed in terms of the Lie algebra of bivectors:

$$\boldsymbol{q} = |\boldsymbol{q}|e^{e_2e_3\phi}e^{e_1e_2\psi}e^{e_1e_3\theta} = |\boldsymbol{q}|e^{\boldsymbol{i}\phi}e^{\boldsymbol{k}\psi}e^{\boldsymbol{j}\theta}, \tag{6.35}$$

in which $(\phi, \theta, \psi) \in [-\pi, \pi[\times[-\frac{\pi}{2}, \frac{\pi}{2}[\times, \pi[- \times [-\frac{\pi}{4}, \frac{\pi}{4}]$.

For a unit quaternion $\boldsymbol{q} = q_0 + q_x i + q_y j + q_z k$, $|\boldsymbol{q}|=1$, its phase can be evaluated first by computing $\psi = -\frac{arcsin(2(q_xq_y - q_0q_z))}{2}$ and then by checking that it adheres to the following rules:

- If $\psi \in]-\frac{\pi}{4}, \frac{\pi}{4}[$, then $\phi = \frac{arg(\boldsymbol{q}\mathcal{I}_j(\bar{\boldsymbol{q}}))}{2}$ and $\theta = \frac{arg(\mathcal{I}_i(\bar{\boldsymbol{q}})\boldsymbol{q})}{2}$.
- If $\psi = \pm\frac{\pi}{4}$, then select either ϕ=0 and $\theta = \frac{arg(\mathcal{I}_k(\bar{\boldsymbol{q}})\boldsymbol{q})}{2}$ or θ=0 and $\phi = \frac{arg(\boldsymbol{q}\mathcal{I}_k(\bar{\boldsymbol{q}}))}{2}$.
- If $e^{\boldsymbol{i}\phi}e^{\boldsymbol{k}\psi}e^{\boldsymbol{j}\theta} = -\boldsymbol{q}$ and $\phi \geq 0$, then $\phi \rightarrow \phi - \pi$.
- If $e^{\boldsymbol{i}\phi}e^{\boldsymbol{k}\psi}e^{\boldsymbol{j}\theta} = -\boldsymbol{q}$ and $\phi < 0$, then $\phi \rightarrow \phi + \pi$.

The reader can find the details of the development of these rules in [44].

The concept of quaternionic Hermitian function is very useful for the computation of the inverse quaternionic Fourier transform using the quaternionic analytic signal, as we will see in an extension of the Hermitian function $f : \mathbb{R} \rightarrow \boldsymbol{C}$ with $f(x) = f^*(-x)$ for every $x \in \mathbb{R}$, we regard $f : \mathbb{R}^2 \rightarrow \mathbb{H}$ as a quaternionic Hermitian function if it fulfills the following non-trivial involution rules [54]:

$$\begin{aligned} f(-x, y) &= -jf(x, y)j = \mathcal{T}_j(f(x, y)), \\ f(x, -y) &= -if(x, y)i = \mathcal{T}_i(f(x, y)), \\ f(-x, -y) &= -if(-x, y)i = -i(-jf(x, y)j)i = (-i - j)f(x, y)(ji), \\ &= -kf(x, y)k = \mathcal{T}_k(f(x, y)). \end{aligned} \tag{6.36}$$

6.5 4D Geometric Algebra for 3D Kinematics

Usually, problems of robotics are treated in algebraic systems of 2D and 3D space. In the case of 3D rigid motion, or Euclidean transformation, we are confronted with a nonlinear mapping; however, if we employ homogeneous coordinates in 4D geometric algebra we can linearize the rigid motion in 3D Euclidean space. That is why we choose three basis vectors which square to one and a fourth vector which squares to zero—to provide dual copies of the multivectors of the 3D space. In other words, we extend the Euclidean geometric algebra $G_{3,0,0}$ to the special or degenerated geometric algebra $G_{3,0,1}$, which is spanned via the following basis:

$$\underbrace{1}_{scalar}, \underbrace{e_k}_{4\ vectors}, \underbrace{e_2e_3, e_3e_1, e_1e_2, e_4e_1, e_4e_2, e_4e_3}_{6\ bivectors}, \underbrace{Ie_k}_{4\ pseudovectors}, \underbrace{I}_{unit\ pseudoscalar}, \quad (6.37)$$

where $e_4^2 = 0$, $e_k^2 = +1$ for $k = 1, 2, 3$. The unit pseudoscalar is $I = e_1e_2e_3e_4$, with

$$I^2 = (e_1e_2e_3e_4)(e_1e_2e_3e_4) = -(e_3e_4)(e_3e_4) = 0. \quad (6.38)$$

The motor algebra $G^+_{3,0,1}$, which is the even subalgebra of $G_{3,0,1}$ can be utilized to obtain linear 4D models of the 3D motion of points, lines, and planes.

6.5.1 *Motor Algebra*

The word *motor* is an abbreviation of "moment and vector." Clifford introduced motors with the name bi-quaternions [57]. Motors are isomorphic to dual quaternions, with the necessary condition $I^2 = 0$. They can be found in the special 4D even subalgebra of $G_{3,0,1}$ introduced in Sect. 6.5. This even subalgebra is denominated by $G^+_{3,0,1}$ and is only spanned via a bivector basis, as follows:

$$\underbrace{1}_{scalar}, \underbrace{e_2e_3, e_3e_1, e_1e_2, e_4e_1, e_4e_2, e_4e_3,}_{6\ bivectors}, \underbrace{I}_{unit\ pseudoscalar}. \quad (6.39)$$

This kind of basis structure also allows us to represent spinors, which are composed of scalar and bivector terms. Motors, then, are also spinors, and as such, they represent a special kind of rotor. Because a Euclidean transformation includes both rotation and translation, we will show in the following subsection a spinor representation for both transformations in the definition of motors. But we must first show the relationship between motors and screw motion theory.

Note that the bivector terms of the basis correspond to the same basis for spanning 3D lines. Note also that the dual of a scalar is the pseudoscalar P and that the duals of the first three basis bivectors are actually the next three bivectors, that is, $(e_2 e_3^*) = I e_2 e_3 = e_4 e_1$.

We said in Sect. 6.3.1 that a rotor relates two vectors in 3D space. According to Clifford [57], a motor operation is necessary to convert the rotation axis of a rotor into the rotation axis of a second rotor. Each rotor can be geometrically represented as a rotation plane with the rotation axis normal to this plane. Figure 6.6a depicts a motor action in detail. Note that the involved rotor axes are represented as line axes. In the figure, we first orient one axis parallel to the other by applying the rotor $\boldsymbol{R}_s$. Then, we slide the rotated axis a distance d along the connecting axis, so that it ends up overlapping the axis of the second rotor. Altogether, this operation can be described as forming a *twist* about a screw with line axis $\boldsymbol{l}$, whose pitch relationship *pitch* equals $\frac{d}{\theta}$ for $\theta \neq 0$. A motor, then, is specified only by its direction and the position of the *screw-axis line*, twist angular magnitude, and pitch. Figure 6.6b shows an action of a motor on a real object. In this case, the motor relates the rotation-axis line of the initial position of the object to the rotation-axis line of its final position. Note that in both figures, the angle and sliding distance indicate how rigid displacement takes place around and along a screw-axis line $\boldsymbol{l}$, respectively. A *degenerated motor* can

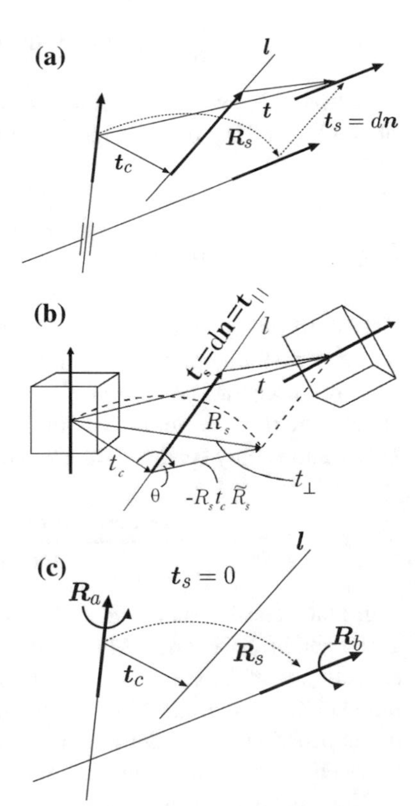

Fig. 6.6 Screw motion about the line axis $\boldsymbol{l}$ ($\boldsymbol{t}_s$: longitudinal displacement by d and $\boldsymbol{R}_s$: rotation angle θ): **a** motor relating two axis lines, **b** motor applied to an object, **c** degenerated motor relating two coplanar rotors. (Note: indicated 3D vectors are represented as bivectors in text)

only rotate and not slide along the line $\boldsymbol{l}$ as Fig. 6.6c shows. In this case, therefore, the two axes are coplanar.

6.5.2 Motors, Rotors, and Translators in $G^+_{3,0,1}$

Since a rigid motion consists of the rotation and translation transformations, it should be possible to split a motor multiplicatively in terms of these two spinor transformations, which we will call a rotor and a *translator*. In the following discussion, we will denote all bivector components of a spinor by bold lowercase letters. Let us now express this procedure algebraically. First of all, let us consider a simple rotor in its *Euler representation* for a rotation with an angle θ,

$$
\begin{aligned}
\boldsymbol{R} &= a_0 + a_1 e_2 e_3 + a_2 e_3 e_1 + a_3 e_1 e_2,\\
&= a_0 + \boldsymbol{a},\\
&= cos\left(\frac{\theta}{2}\right) + sin\left(\frac{\theta}{2}\right)\boldsymbol{n},\\
&= a_c + a_s \boldsymbol{n},
\end{aligned}
\tag{6.40}
$$

where $\boldsymbol{n}$ is the unit 3D bivector of the rotation axis spanned by the bivector basis e_2e_3, e_3e_1, e_1e_2, and a_c, $a_s \in \mathcal{R}$. Now, dealing with the rotor of a screw motion, the rotation-axis vector should be represented as a screw-axis line. For that, we must relate the rotation axis to a reference coordinate system at the distance $\boldsymbol{t}_c$. A 3D translation in motor algebra is represented by a spinor $\boldsymbol{T}_c$ called a translator. If we apply a translator from the left to rotor $\boldsymbol{R}$, and then apply the translator's conjugate from the right, we get a modified rotor,

$$
\begin{aligned}
\boldsymbol{R}_s &= \boldsymbol{T}_c \boldsymbol{R} \widetilde{\boldsymbol{T}}_c,\\
&= \left(1 + I\frac{\boldsymbol{t}_c}{2}\right)(a_0 + \boldsymbol{a})\left(1 - I\frac{\boldsymbol{t}_c}{2}\right),\\
&= a_0 + \boldsymbol{a} + I a_0 \frac{\boldsymbol{t}_c}{2} + I\frac{\boldsymbol{t}_c}{2}\boldsymbol{a} - I a_0 \frac{\boldsymbol{t}_c}{2} - I\boldsymbol{a}\frac{\boldsymbol{t}_c}{2},\\
&= a_0 + \boldsymbol{a} + I\left(\frac{\boldsymbol{t}_c}{2}\boldsymbol{a} - \boldsymbol{a}\frac{\boldsymbol{t}_c}{2}\right),\\
&= a_0 + \boldsymbol{a} + I(\boldsymbol{a} \times \boldsymbol{t}_c).
\end{aligned}
\tag{6.41}
$$

Here, $\boldsymbol{t}_c$ is the 3D vector of translation spanned by the bivector basis e_2e_3, e_3e_1, e_1e_2. Then, expressing the last equation in Euler terms, we get the spinor representation,

$$
\begin{aligned}
\boldsymbol{R}_s &= a_0 + a_s\boldsymbol{n} + Ia_s\boldsymbol{n}\wedge\boldsymbol{t}_c,\\
&= a_c + a_s(\boldsymbol{n} + I\boldsymbol{m}),\\
&= cos\left(\frac{\theta}{2}\right) + sin\left(\frac{\theta}{2}\right)(\boldsymbol{n} + I\boldsymbol{m}),\\
&= cos\left(\frac{\theta}{2}\right) + sin\left(\frac{\theta}{2}\right)\boldsymbol{l}.
\end{aligned} \tag{6.42}
$$

This result is indeed interesting because the new rotor $\boldsymbol{R}_s$ can now be applied with respect to an axis line $\boldsymbol{l}$ expressed in dual terms of direction $\boldsymbol{n}$ and moment $\boldsymbol{m} = \boldsymbol{n} \wedge \boldsymbol{t}_c$. Now, to define the motor finally, let us slide the distance $\boldsymbol{t}_s = d\boldsymbol{n}$ along the rotation-axis line $\boldsymbol{l}$. Since a motor is applied from the left and conjugated from the right, we should use the half of $\boldsymbol{t}_s$ in the spinor expression of $\boldsymbol{T}_s 1$ when we define the motor:

$$
\begin{aligned}
\boldsymbol{M} = \boldsymbol{T}_s\boldsymbol{R}_s &= \left(1 + I\frac{\boldsymbol{t}_s}{2}\right)(a_0 + \boldsymbol{a} + I\boldsymbol{a}\wedge\boldsymbol{t}_c),\\
&= \left(1 + I\frac{d\boldsymbol{n}}{2}\right)(a_c + a_s\boldsymbol{n} + Ia_s\boldsymbol{n}\wedge\boldsymbol{t}_c),\\
&= a_c + a_s\boldsymbol{n} + Ia_s\boldsymbol{n}\wedge\boldsymbol{t}_c + I\frac{d}{2}a_c\boldsymbol{n} - I\frac{d}{2}a_s\boldsymbol{n}\boldsymbol{n},\\
&= \left(a_c - I\frac{d}{2}a_s\right) + \left(a_s + Ia_c\frac{d}{2}\right)(\boldsymbol{n} + I\boldsymbol{n}\wedge\boldsymbol{t}_c),\\
&= \left(a_c - Ia_s\frac{d}{2}\right) + \left(a_s + Ia_c\frac{d}{2}\right)\boldsymbol{l}.
\end{aligned} \tag{6.43}
$$

Note that this expression of the motor makes explicit the unit line bivector of the screw-axis line $\boldsymbol{l}$.

Now let us express a motor using Euler representation. By substituting the constants $a_c = cos\left(\frac{\theta}{2}\right)$ and $a_s = sin\left(\frac{\theta}{2}\right)$ in the motor Eq. (6.43) and using the property of Eq. (6.1), we get

$$
\begin{aligned}
\boldsymbol{M} = \boldsymbol{T}_s\boldsymbol{R}_s &= \left(cos(\frac{\theta}{2}) - Isin(\frac{\theta}{2})\frac{d}{2}\right) + \left(sin(\frac{\theta}{2}) + Icos(\frac{\theta}{2})\frac{d}{2}\right)\boldsymbol{l},\\
&= cos\left(\frac{\theta}{2} + I\frac{d}{2}\right) + sin\left(\frac{\theta}{2} + I\frac{d}{2}\right)\boldsymbol{l},
\end{aligned} \tag{6.44}
$$

which is a dual-number representation of the spinor. Now, let us analyze the resultant expressions,

$$
\begin{aligned}
\boldsymbol{R} &= cos\left(\frac{\theta}{2}\right) + sin\left(\frac{\theta}{2}\right)\boldsymbol{n},\\
\boldsymbol{R}_s &= cos\left(\frac{\theta}{2}\right) + sin\left(\frac{\theta}{2}\right)\boldsymbol{l},\\
\boldsymbol{M} &= cos\left(\frac{\theta}{2} + I\frac{d}{2}\right) + sin\left(\frac{\theta}{2} + I\frac{d}{2}\right)\boldsymbol{l}.
\end{aligned} \tag{6.45}
$$

We can see that the rotation axis $\boldsymbol{n}$ of the simple rotor $\boldsymbol{R}$ is changed to a rotation-axis line, so that $\boldsymbol{R}_s$ now rotates about an axis line. And in the motor expression, the information for sliding distance d is now made explicit in terms of dual arguments of the trigonometric functions. It is also interesting to note that the expression for the motor using dual angles simply extends the expression of $\boldsymbol{R}_s$.

If we expand the exponential function of the dual bivectors using a Taylor series, the result will follow the general expression $e^{\alpha+I\beta} = e^{\alpha} + Ie^{\alpha}\beta = e^{\alpha}(1 + I\beta)$, which is a special case of Eq. (6.1). Once again, we obtain the motor expression as the spinor

$$
e^{l\frac{\theta}{2}+I\frac{t_s}{2}} = \left(1 + I\frac{\boldsymbol{t}_s}{2}\right)e^{l\frac{\theta}{2}} = \boldsymbol{T}_s\boldsymbol{R}_s, \tag{6.46}
$$

where $I\frac{t_s}{2} = I\frac{1}{2}(t_1\sigma_2\sigma_3 + t_2\sigma_3\sigma_1 + t_3\sigma_1\sigma_2) = \frac{1}{2}(t_1\sigma_4\sigma_1 + t_2\sigma_4\sigma_2 + t_3\sigma_4\sigma_3)$.

If we want to express the motor using only rotors in a dual spinor representation, we proceed as follows:

$$
\begin{aligned}
\boldsymbol{M} = \boldsymbol{T}_s\boldsymbol{R}_s &= \left(1 + I\frac{\boldsymbol{t}_s}{2}\right)\mathbf{R}_s,\\
&= \boldsymbol{R}_s + I\frac{\boldsymbol{t}_s}{2}\boldsymbol{R}_s.
\end{aligned} \tag{6.47}
$$

Let us consider carefully the resultant dual part of the motor. This is the geometric product of the bivector $\boldsymbol{t}_s$ and the rotor $\boldsymbol{R}_s$. Since both are expressed in terms of the same bivector basis, their geometric product will be also expressed in this basis, which can be considered as a new rotor $\boldsymbol{R}'_s$. Thus, we can further write

$$
\boldsymbol{M} = \boldsymbol{R}_s + I\frac{\boldsymbol{t}_s}{2}\boldsymbol{R}_s = \boldsymbol{R}_s + I\boldsymbol{R}'_s. \tag{6.48}
$$

In this equation, the line axes of the rotors are skewed (see Fig. 6.6a). That means that they represent the general case of non-coplanar rotors. If the sliding distance $\boldsymbol{t}_s$ is zero, then the motor will degenerate to a rotor

$$
\boldsymbol{M} = \boldsymbol{T}_s\boldsymbol{R}_s = (1 + I\frac{\boldsymbol{t}_s}{2})\boldsymbol{R}_s = (1 + I\frac{0}{2})\boldsymbol{R}_s = \boldsymbol{R}_s. \tag{6.49}
$$

In this case, that is, when the two generating axis lines of the motor are coplanar, we get the so-called *degenerated motor* (see Fig. 6.6c).

Finally, the bivector $\boldsymbol{t}_s$ can be expressed in terms of the rotors using previous results

$$\boldsymbol{R}'_s\widetilde{\boldsymbol{R}}_s = \left(\frac{\boldsymbol{t}_s}{2}\boldsymbol{R}_s\right)\widetilde{\boldsymbol{R}}_s, \tag{6.50}$$

therefore,

$$\boldsymbol{t}_s = 2\boldsymbol{R}'_s\widetilde{\boldsymbol{R}}_s. \tag{6.51}$$

Figure 6.6 shows that the 3D vector $\boldsymbol{t}$, expressed in the bivector basis, is referred to the rotation axis of the rotor, and that $\boldsymbol{t}_s$ is a bivector along the motor-axis line. Thus, $\boldsymbol{t}$, considered here as a bivector, can be computed in terms of the bivectors $\boldsymbol{t}_c$ and $\boldsymbol{t}_s$, as follows:

$$\begin{aligned}
\boldsymbol{t} &= \boldsymbol{t}_\perp + \boldsymbol{t}_\parallel,\\
\boldsymbol{t} &= (\boldsymbol{t}_c - \boldsymbol{R}_s\boldsymbol{t}_c\widetilde{\boldsymbol{R}}_s) + (\boldsymbol{t}\cdot\boldsymbol{n})\boldsymbol{n} = (\boldsymbol{t}_c - \boldsymbol{R}_s\boldsymbol{t}_c\widetilde{\boldsymbol{R}}_s) + d\boldsymbol{n},\\
&= \boldsymbol{t}_c - \boldsymbol{R}_s\boldsymbol{t}_c\widetilde{\boldsymbol{R}}_s + \boldsymbol{t}_s,\\
&= \boldsymbol{t}_c - \boldsymbol{R}_s\boldsymbol{t}_c\widetilde{\boldsymbol{R}}_s + 2\boldsymbol{R}'_s\widetilde{\boldsymbol{R}}_s.
\end{aligned} \tag{6.52}$$

So far, we have analyzed the motor from a geometrical point of view. Next, we will look at the motor's relevant algebraic properties.

6.5.3 Properties of Motors

A general motor can be expressed as

$$\boldsymbol{M}_\alpha = \alpha\boldsymbol{M}, \tag{6.53}$$

where $\alpha \in \mathcal{R}$ and $\boldsymbol{M}$ is a unit motor, as explained in the previous sections. In this section, we will employ unit motors. The norm of a motor $\boldsymbol{M}$ is defined as follows:

$$\begin{aligned}
|\boldsymbol{M}| = \boldsymbol{M}\widetilde{\boldsymbol{M}} = \boldsymbol{T}_s\boldsymbol{R}_s\widetilde{\boldsymbol{R}}_s\widetilde{\boldsymbol{T}}_s &= \left(1 + I\frac{\boldsymbol{t}_s}{2}\right)\boldsymbol{R}_s\tilde{\boldsymbol{R}}_s\left(1 - I\frac{\boldsymbol{t}_s}{2}\right),\\
&= 1 + I\frac{\boldsymbol{t}_s}{2} - I\frac{\boldsymbol{t}_s}{2} = 1,
\end{aligned} \tag{6.54}$$

where $\widetilde{M}$ is the conjugate motor and 1 is the identity of the motor multiplication. Now, using Eq. (6.48) and considering the unit motor magnitude, we find two useful properties, expressed by

$$\begin{aligned} |\boldsymbol{M}| = \boldsymbol{M}\widetilde{\boldsymbol{M}} &= (\boldsymbol{R}_s + I\boldsymbol{R}'_s)(\widetilde{\boldsymbol{R}}_s + I\widetilde{\boldsymbol{R}}'_s), \\ &= \boldsymbol{R}_s\widetilde{\boldsymbol{R}}_s + I(\boldsymbol{R}'_s\widetilde{\boldsymbol{R}}_s + \boldsymbol{R}_s\widetilde{\boldsymbol{R}}'_s) = 1. \end{aligned} \tag{6.55}$$

These equations require the following constraints:

$$\boldsymbol{R}_s\widetilde{\boldsymbol{R}}_s = 1, \tag{6.56}$$

$$\boldsymbol{R}'_s\widetilde{\boldsymbol{R}}_s + \boldsymbol{R}_s\widetilde{\boldsymbol{R}}'_s. \tag{6.57}$$

Now we can show that the combination of two rigid motions can be expressed using two consecutive motors. The resultant motor describes the overall displacement, namely,

$$\begin{aligned} \boldsymbol{M}_c = \boldsymbol{M}_a\boldsymbol{M}_b &= (\boldsymbol{R}_{s_a} + I\boldsymbol{R}'_{s_a})(\boldsymbol{R}_{s_b} + I\boldsymbol{R}'_{s_b}), \\ &= \boldsymbol{R}_{s_a}\boldsymbol{R}_{s_b} + I(\boldsymbol{R}_{s_a}\boldsymbol{R}'_{s_b} + \boldsymbol{R}'_{s_a}\boldsymbol{R}_{s_b}), \\ &= \boldsymbol{R}_{s_c} + I\boldsymbol{R}'_{s_c}. \end{aligned} \tag{6.58}$$

Note that, on the one hand, pure rotations combine multiplicatively, and, on the other hand, the dual parts containing the translation combine additively.

Using Eq. (6.48), let us express a motor in terms of dual spinors:

$$\begin{aligned} \boldsymbol{M} = \boldsymbol{T}_s\boldsymbol{R}_s &= \boldsymbol{R}_s + I\boldsymbol{R}'_s, \\ &= (a_0 + a_1e_2e_3 + a_2e_3e_2 + a_3e_2e_1) + \\ &\qquad I(b_0 + b_1e_2e_3 + b_2e_3e_2 + b_3e_2e_1), \\ &= (a_0 + \boldsymbol{a}) + I(b_0 + \boldsymbol{b}). \end{aligned} \tag{6.59}$$

We can use another notation to enhance the components of the real and dual parts of the motor, as follows:

$$\boldsymbol{M} = (a_0, \boldsymbol{a}) + I(b_0, \boldsymbol{b}). \tag{6.60}$$

Here, each term within the brackets consists of a scalar part and a 3D bivector.

A motor expressed in terms of a translator and a rotor is manipulated similarly as in the case of a rotor, from the left and its conjugate from the right. These left and right operations, called *motor reflections*, are used to build an automorphism equivalent to the screw. Yet, by conjugating only the rotor or only the translator for the second reflection, we can derive different types of automorphisms.

By changing the sign of the scalar and bivector in the real and dual parts of the motor, we get the following variations:

$$\begin{aligned}
\boldsymbol{M} &= (a_0 + \boldsymbol{a}) + I(b_0 + \boldsymbol{b}) = \boldsymbol{T}_s \boldsymbol{R}_s, \\
\widetilde{\boldsymbol{M}} &= (a_0 - \boldsymbol{a}) + I(b_0 - \boldsymbol{b}) = \widetilde{\boldsymbol{R}}_s \widetilde{\boldsymbol{T}}_s, \\
\bar{\boldsymbol{M}} &= (a_0 + \boldsymbol{a}) - I(b_0 + \boldsymbol{b}) = \boldsymbol{R}_s \widetilde{\boldsymbol{T}}_s, \\
\widetilde{\bar{\boldsymbol{M}}} &= (a_0 - \boldsymbol{a}) - I(b_0 - \boldsymbol{b}) = \widetilde{\boldsymbol{R}}_s \boldsymbol{T}_s.
\end{aligned} \tag{6.61}$$

The first, second, and fourth versions will be used for modeling the motion of points, lines, and planes, respectively.

Using Eq. (6.61), it is now a straightforward matter to compute the expressions for the individual components:

$$\begin{aligned}
a_0 &= \frac{1}{4}(\boldsymbol{M} + \widetilde{\boldsymbol{M}} + \bar{\boldsymbol{M}} + \widetilde{\bar{\boldsymbol{M}}}), \\
I b_0 &= \frac{1}{4}(\boldsymbol{M} + \widetilde{\boldsymbol{M}} - \bar{\boldsymbol{M}} - \widetilde{\bar{\boldsymbol{M}}}), \\
\boldsymbol{a} &= \frac{1}{4}(\boldsymbol{M} - \widetilde{\boldsymbol{M}} + \bar{\boldsymbol{M}} - \widetilde{\bar{\boldsymbol{M}}}), \\
I\boldsymbol{b} &= \frac{1}{4}(\boldsymbol{M} - \widetilde{\boldsymbol{M}} - \bar{\boldsymbol{M}} + \widetilde{\bar{\boldsymbol{M}}}).
\end{aligned} \tag{6.62}$$

6.6 4D Geometric Algebra for Projective 3D Space

To this point, we have dealt with transformations in three-dimensional space. When we use homogeneous coordinates, we increase the dimension of the vector space by one. As a result, the transformation of 3D motion becomes linear. Let us now model the projective 3D space P^3. This space corresponds to the homogeneous extended space R^4. In real applications, it is important to regard the signature of the modeled space to facilitate the computations. In the case of the modeling of the projective plane using homogeneous coordinates, we adopt $G_{3,0,0}$ of the ordinary space, E^3, which has the standard Euclidean signature. For the four-dimensional space R^4, we are forced to adopt the same signature as in the case of the Euclidean space. This geometric algebra $G_{1,3,0}$ is spanned with the following basis:

$$\underbrace{1}_{scalar}, \quad \underbrace{e_k}_{4\ vectors}, \quad \underbrace{e_2e_3, e_3e_1, e_1e_2, e_4e_1, e_4e_2, e_4e_3}_{6\ bivectors}, \quad \underbrace{Ie_k}_{4\ pseudovectors}, \quad \underbrace{I}_{unit\ pseudoscalar}, \tag{6.63}$$

where $e^2 = +1$, $e_k^2 = -1$ for k = 1, 2, 3. The unit pseudoscalar is $I = e_1e_2e_3e_4$, with

$$I^2 = (e_1e_2e_3e_4)(e_1e_2e_3e_4) = -(e_3e_4)(e_4) = -1. \tag{6.64}$$

The geometric algebras $G_{3,0,0}$ and $G_{1,3,0}$ will be used for the geometric modeling of the image plane and the visual 3D space. In space–time algebra, the fourth basis vector e_4 of $G_{1,3,0}$ is selected as the time axis for applications of the *projective split* [176]. This helps to associate multivectors of the 4D space with multivectors of the 3D space. The role and use of the projective split for a variety of problems involving the algebra of incidence will also be discussed in Chap. 7.

6.7 Conclusion

This chapter gives an outline of geometric algebra. In particular, it explains the geometric product and the meaning of multivectors. The geometric approach adopted is a version of Clifford algebra, and it is used in the whole book as unifying language for the design of artificial perception action systems. The chapter presents the geometric algebras of the plane and 3D space, where we can find the planar and 3D quaternions. Finally, we introduce 4D geometric algebras useful for computations involving dual quaternions (motors) or projective geometry. The chapter offers various exercises, so that the reader can start to learn to compute in Clifford algebra in an easy manner.

6.8 Exercises

6.1 Using CLICAL [196] or eClifford [1] in $G_{3,0,0}$, rotate the vector $\boldsymbol{r} = 2e_1 + 3e_2 + 2e_3$ about the axis $\boldsymbol{n} = 1.4e_1 + 1.9e_2$ with angle $\theta = |\boldsymbol{n}|$. Since $|\theta| \leq \phi$, the rotation is well defined.

6.2 Two consecutive rotations in $G_{3,0,0}$, one about the axis $\boldsymbol{a}$ with the angle $\alpha = |\boldsymbol{a}|$ and another about the axis $\boldsymbol{b}$ with the angle $\beta = |\boldsymbol{b}|$, are equivalent to the rotation about an axis $\boldsymbol{c}$. Prove this statement using the Rodriguez formula,

$$\boldsymbol{c}' = \frac{\boldsymbol{a}' + \boldsymbol{b}' - \boldsymbol{a}' \times \boldsymbol{b}'}{1 - \boldsymbol{a}' \cdot \boldsymbol{b}'},$$

where $\boldsymbol{a}' = \frac{\boldsymbol{a}}{\alpha} tan(\frac{\alpha}{2})$ and $\alpha = |\boldsymbol{a}|$. (*Hint*: Compare the scalar and bivector terms of $e^{\frac{\gamma}{2}I\boldsymbol{c}} = e^{\frac{\alpha}{2}I\boldsymbol{a}}e^{\frac{\beta}{2}I\boldsymbol{b}}$.)

6.3 Using the vectors $\boldsymbol{a}$ and $\boldsymbol{b}$ of $G_{3,0,0}$, prove that the rotation of $\boldsymbol{a}$ to $\boldsymbol{b}$ can be represented by the rotor

$$\boldsymbol{R} = (\boldsymbol{ab})^{\frac{1}{2}} = \frac{(\boldsymbol{a}+\boldsymbol{b})\boldsymbol{b}}{|\boldsymbol{a}+\boldsymbol{b}|} = \frac{\boldsymbol{a}(\boldsymbol{b}+\boldsymbol{a})}{|\boldsymbol{b}+\boldsymbol{a}|}.$$

Since the norm $|\boldsymbol{a}+\boldsymbol{b}| = [2(1+\boldsymbol{a}\cdot\boldsymbol{b})]^{\frac{1}{2}}$ is not relevant, you can write $\boldsymbol{R}$ as follows:

$$\boldsymbol{R} \doteq (\boldsymbol{a}+\boldsymbol{b})\boldsymbol{b} = \boldsymbol{a}(\boldsymbol{b}+\boldsymbol{a}) = 1 + \boldsymbol{ab}.$$

Also prove this equation. You can interpret the symbol $\doteq$ as a projective identity or an identity up to a scalar factor. (*Hint*: Each rotation can be represented by two reflections.)

6.4 Given the bivector $\boldsymbol{B}$ in the $\boldsymbol{a}\wedge\boldsymbol{b}$ plane and the angle θ between the vectors $\boldsymbol{a}$ and $\boldsymbol{b}$, prove that the rotor

$$\boldsymbol{R} = \frac{1+\boldsymbol{ba}}{|\boldsymbol{a}+\boldsymbol{b}|}$$

can also be written as $e^{-\frac{\theta}{2}\boldsymbol{B}}$.

6.5 Check by hand, if the Eq. 6.45 are correct. Note that the equations have a similar format and that by the equation for the motor the angle is changed to a dual angle and the rotation axis $\boldsymbol{n}$ is changed to a screw line $\boldsymbol{l} = \boldsymbol{n} + I\boldsymbol{m}$.

6.6 Given the rotors $\boldsymbol{R}_1 = e^{-\frac{\alpha}{2}\boldsymbol{B}_1}$ and the rotor $\boldsymbol{R}_2 = e^{-\frac{\alpha}{2}\boldsymbol{B}_2}$, show that their product is

$$\boldsymbol{R}_3 = \boldsymbol{R}_2\boldsymbol{R}_1 = e^{-\frac{\alpha}{2}\boldsymbol{B}_3}.$$

6.7 Shown that for an orthonormal bivector basis, the structure constants C^i_{jk} of the Lie algebra of the 3D rotation group are simply $-\epsilon_{ijk}$.

6.8 Given the bivectors $\boldsymbol{X}$, $\boldsymbol{Y}$, and $\boldsymbol{Z}$ prove the Jacobi identity

$$(\boldsymbol{X}\times\boldsymbol{Y})\times\boldsymbol{Z} + (\boldsymbol{Z}\times\boldsymbol{X})\times\boldsymbol{Y} + (\boldsymbol{Y}\times\boldsymbol{Z})\times\boldsymbol{X} = 0.$$

6.9 Lie algebra: the bivector generators of the unitary group are

$$\begin{aligned}
E_{ij} &= e_i e_j + f_i f_j, \quad (i < j = 1 \ldots n)\\
F_{ij} &= e_i f_j + f_i e_j, \quad (i < j = 1 \ldots n)\\
J_i &= e_i f_i.
\end{aligned}$$

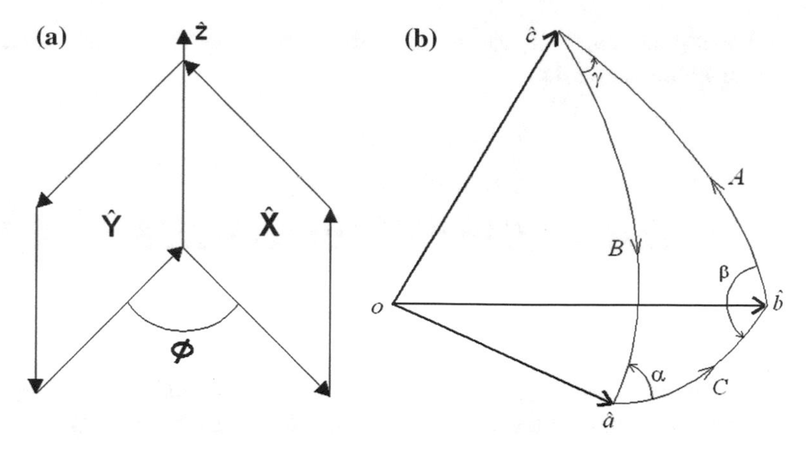

Fig. 6.7 **a** Dihedral angle ϕ between two planes. **b** The spherical triangle

Also show that this algebra is closed under the commutator product and give some examples as well. See in Chap. 5, Sect. 5.4.2.

6.10 Two consecutive motors in $G^+_{3,0,1}$, one about the screw axis $\boldsymbol{L}_a$ with the angle $\alpha = |\boldsymbol{a}|$ and translation $\boldsymbol{t}_a$ and another about the screw axis $\boldsymbol{L}_b$ with the angle $\beta = |\boldsymbol{b}|$ and translation $\boldsymbol{t}_b$, are equivalent to the motor about a screw axis $\boldsymbol{L}_c$. Prove this statement using and extension of the Rodriguez formula given in Exercise 6.2. Draw and interpret your results.

6.11 The dihedral angle ϕ between two planes $\hat{X}$ and $\hat{Y}$ is defined as shown in the Fig. 6.7a shown in the figure bivectors $bl\hat{d}X$ and $\hat{Y}$ and $\hat{z}$ the unit vector along the intersection line, prove that

$$\hat{X}\hat{Y} = e^{Iz},$$

where $z = \theta\hat{z}$.

6.12 Consider the spherical triangle with corners described by three unit vectors coming from the origin to the surface of a unit sphere, see Fig. 6.7b The lengths of the three arcs are given by $|\boldsymbol{A}|$, $|\boldsymbol{B}|$, and $|\boldsymbol{C}|$, prove the following formulas

$$\begin{aligned}\hat{\boldsymbol{a}}\hat{\boldsymbol{b}} &= e^{\boldsymbol{C}},\\ \hat{\boldsymbol{b}}\hat{\boldsymbol{c}} &= e^{\boldsymbol{A}},\\ \hat{\boldsymbol{c}}\hat{\boldsymbol{a}} &= e^{\boldsymbol{B}},\end{aligned}$$

6.13 Following Exercise 6.12, the angles α, β, and γ between the planes are called the dihedral angles, prove that

$$\hat{B}\hat{A} = e^{Ic}, \ \|c\| = \gamma,$$
$$\hat{C}\hat{B} = e^{Ia}, \ \|a\| = \alpha,$$
$$\hat{C}\hat{A} = e^{Ib}, \ \|b\| = \beta.$$

Also prove that

$$e^{C}e^{A}e^{B} = 1, \ e^{Ic}e^{Ib}e^{Ia} = -1.$$

6.14 Following Exercise 6.13, take the scalar part of the equation $e^{-Ic} = -e^{Ib}e^{Ia}$ and prove the cosine law for the angles in spherical trigonometry, namely

$$cos(\gamma) = -cos(\alpha)cos(\beta) + sin(\alpha)sin(\beta)cos(|C|).$$

Since this formulation is more advantageously as traditional formulations, suggest some applications using this equation.

6.15 Using a sphere, draw the rotated rotor $R_2 R_1 \widetilde{R}_2$. What it is geometric meaning of this operation with respect to the action of this sequence of rotors acting on a vector?

6.16 Given the bivector B and X and Y two general multivectors, prove that

$$B \times (XY) = (B \times X)Y + X(B \times Y),$$

hence show that

$$B \times (v \wedge V_r) = (B \cdot v) \wedge V + v \wedge (B \times V_r).$$

By using this result establish the fact that the operation of commuting with a bivector is grade preserving.

6.17 Given a linear function $f(x)$ and an orthonormal frame $\{e_k\}$, one can form the following matrix

$$f_i j = e_i \cdot f(e_j).$$

Prove that the matrix $\bar{f}(x)$ is nothing else as the transport matrix. Furthermore, prove that the product transformations $h = fg$ is determined by the multiplication of f_{ij} and g_{ij}.

6.18 Explain the geometric meaning of the projection $P_R(x) = (x \cdot R)R^{-1}$.

6.19 In $\mathbb{R}^{4,0}$ with an associated orthonormal basis $\{e_i\}_{i=1}^4$, perform a rotation in the plane $e_1 \wedge e_2$, followed with a rotation in the plane $e_3 \wedge e_4$. Compute the rotor of this rotor composition and show that this is the exponent of bivector, not a 2-blade. The resulting rotor is not of a simple form consisting of a scalar plus a 2-blade or even scalar plus a bivector, why?

Chapter 7
Kinematics of the 2D and 3D Spaces

7.1 Introduction

This chapter presents the geometric algebra framework for dealing with 3D kinematics. The reader will see the usefulness of this mathematical approach for applications in computer vision and kinematics. We start with an introduction to 4D geometric algebra for 3D kinematics. Then we reformulate, using 3D and 4D geometric algebras, the classic model for the 3D motion of vectors. Finally, we compare both models, that is, the one using 3D Euclidean geometric algebra and our model, which uses 4D motor algebra.

7.2 Representation of Points, Lines, and Planes Using 3D Geometric Algebra

The modeling of points, lines, and planes in 3D Euclidean space will be done using the Euclidean geometric algebra $G_{3,0,0}$, where the pseudoscalar $I^2 = -1$. A point in 3D space represents a position and thus can be simply spanned using the vector basis of $G_{3,0,0}$:

$$\boldsymbol{x} = xe_1 + ye_2 + ze_3, \tag{7.1}$$

where $x, y, z \in \mathcal{R}$.

In classical vector calculus, a line is described by a position vector $\boldsymbol{x}$ that touches any point of the line and by a vector $\boldsymbol{n}$ for the line direction, that is, $\boldsymbol{l} = \boldsymbol{x} + \alpha\boldsymbol{n}$, where $\alpha \in \mathcal{R}$. In geometric algebra, we employ a multivector concept, and we can thus compactly represent in $G_{3,0,0}$ any line, using a vector $\boldsymbol{n}$ for its direction and a bivector $\boldsymbol{m}$ for the orientation of the plane within which the line lies. Thus,

$$\boldsymbol{l} = \boldsymbol{n} + \boldsymbol{x} \wedge \boldsymbol{n} = \boldsymbol{n} + \boldsymbol{m}. \tag{7.2}$$

E. Bayro-Corrochano, *Geometric Algebra Applications Vol. I*,
https://doi.org/10.1007/978-3-319-74830-6_7

Note that the moment bivector $\boldsymbol{m}$ is computed as the outer product of the position vector $\boldsymbol{x}$ and the line direction vector $\boldsymbol{n}$. We can also compute $\boldsymbol{m}$ as the dual of a vector, that is, $I\boldsymbol{x} \times \boldsymbol{n} = \boldsymbol{m}$.

The representation of the plane is even more striking. The plane is a geometric entity one grade higher than the line, so we would expect that the multivector representation of the plane would be a natural multivector grade extension from that of the line. In classical vector calculus, a plane is described in terms of *Hesse distance*, which represents the distance from the origin to the plane, and a vector which indicates the plane orientation, that is, $\{d, \ \boldsymbol{n}\}$. Note that this description is composed of two separate attributes which come from the equation: $n_x x + n_y y + n_z z - d = \mathbf{n}^T\mathbf{x} - h = 0$. Once again, using geometric algebra we can express the plane more compactly and with clearer geometric sense. In $G_{3,0,0}$, for example, the extension of the line expression to a plane would be expressed in terms of a bivector $\boldsymbol{n}$ and a trivector Id,

$$\boldsymbol{h} = \boldsymbol{n} + \boldsymbol{x} \wedge \boldsymbol{n} = \boldsymbol{n} + Id, \tag{7.3}$$

where the bivector $\boldsymbol{n}$ indicates the plane orientation, and the outer product of the position vector $\boldsymbol{x}$ and the bivector $\boldsymbol{n}$ builds a trivector which can be expressed using Hesse distance, a scalar value, and the unit pseudoscalar I. Note that the trivector represents a volume, whereas the scalar d represents the Hesse distance. Figure 7.1 compares the different representations of points, lines, and planes using classical vector calculus, Euclidean geometric algebra for $G_{3,0,0}$, and motor algebra for $G^+_{3,0,1}$.

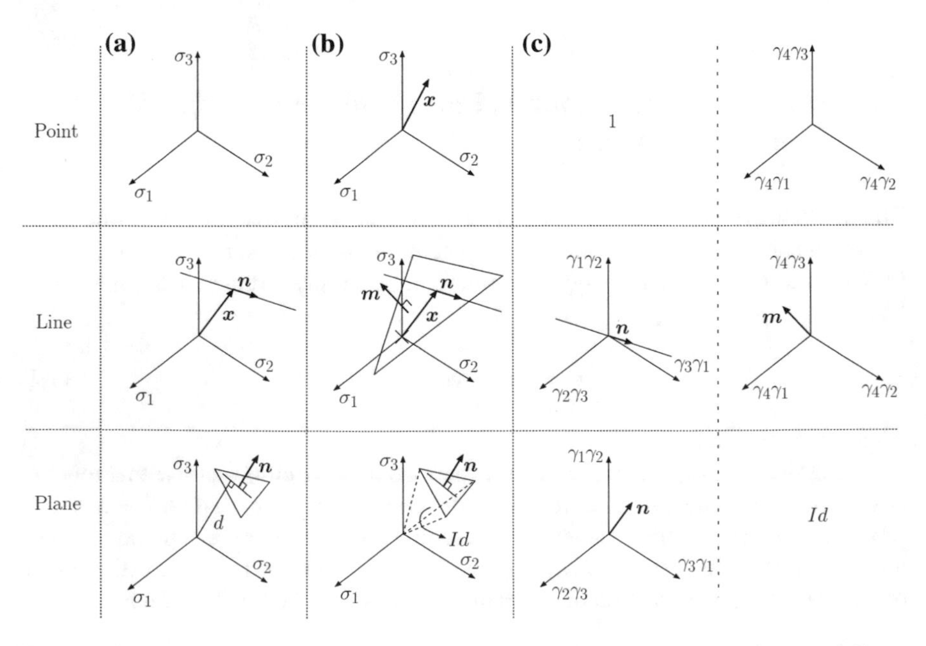

Fig. 7.1 Representations of **a** points, **b** lines, and **c** planes using vector calculus, $\mathcal{G}_{3,0,0}$, and $G_{3,0,1}$

7.3 Representation of Points, Lines, and Planes Using Motor Algebra

In this section, we will model points, lines, and planes in 4D space using the special algebra of the motors $G^+_{3,0,1}$, which spans in 4D the line space using a bivector basis.

For the case of the point representation, we proceed by embedding a 3D point on the hyperplane $X_4 = 1$, so that the equation of the point $\boldsymbol{X} \in G^+_{3,0,1}$ reads

$$\begin{aligned} \boldsymbol{X} &= 1 + x_1 e_4 e_1 + x_2 e_4 e_2 + x_3 e_4 e_3, \\ &= 1 + I(x_1 e_2 e_3 + x_2 e_3 e_1 + x_3 e_1 e_2), \\ &= 1 + I\boldsymbol{x}, \end{aligned} \quad (7.4)$$

or $\boldsymbol{X} = (1, 0) + I(0, \boldsymbol{x})$. We can see that in this expression the real part consists of the scalar 1 and the dual part of a 3D bivector.

Since we are working in the algebra $G^+_{3,0,1}$ spanned only by bivectors and scalars, we see that this special geometric algebra is the most appropriate system for line modeling. Unlike the line representation, the point and the plane are in some sense asymmetric representations with respect to the scalar and bivector parts. Let us now rewrite the line Eq. (7.2) of $G_{3,0,0}$ in the degenerated geometric algebra $G^+_{3,0,1}$. We can express the vector and the dual vector of Eq. (7.2) in $G^+_{3,0,1}$ as a bivector and a dual bivector. Since the product of the unit pseudoscalar $I = e_1 e_2 e_3 e_4$ and any dual bivectors built from the basis $\{e_4 e_1, \ e_4 e_2, \ e_4 e_3\}$ is zero, we must select the bivector basis $\{e_2 e_3, \ e_3 e_1, \ e_1 e_2\}$ for representing the line

$$\boldsymbol{L} = \boldsymbol{n} + I\boldsymbol{m}. \quad (7.5)$$

In this case, the bivectors for the line direction and the moment are computed using two bivector points, $\boldsymbol{x}_1$ and $\boldsymbol{x}_2$, lying on the line, as follows:

$$\begin{aligned} \boldsymbol{n} &= (\boldsymbol{x}_2 - \boldsymbol{x}_1) = (x_{21} - x_{11}) e_2 e_3 + (x_{22} - x_{12}) e_3 e_1 + (x_{23} - x_{13}) e_1 e_2 \\ &= L_{n_1} e_2 e_3 + L_{n_2} e_3 e_1 + L_{n_3} e_1 e_2, \\ \boldsymbol{m} &= \boldsymbol{x}_1 \times \boldsymbol{x}_2, \\ &= (x_{12} x_{23} - x_{13} x_{22}) e_2 e_3 + (x_{13} x_{21} - x_{11} x_{23}) e_3 e_1 + \ldots + \\ &\qquad + (x_{11} x_{22} - x_{12} x_{21}) e_1 e_2, \\ &= L_{m_1} e_2 e_3 + L_{m_2} e_3 e_1 + L_{m_3} e_1 e_2. \end{aligned} \quad (7.6)$$

This line representation using dual numbers is easy to understand and to manipulate algebraically, and it is fully equivalent to the representation in terms of *Plücker coordinates*. Using bracket notation, the line equation becomes $\boldsymbol{L} \equiv (0, \boldsymbol{n}) + I(0, \boldsymbol{m})$, where $\boldsymbol{n}$ and $\boldsymbol{m}$ are spanned with a 3D bivector basis.

For the equation of the plane, we proceed in a similar manner as for Eq. (7.3). We represent the orientation of the plane via the bivector $\boldsymbol{n}$ and the outer product between a bivector touching the plane and its orientation $\boldsymbol{n}$. This outer product results in a quatrivector, which we can express as the Hesse distance $d = (\boldsymbol{x} \cdot \boldsymbol{n})$ multiplied by the unit pseudoscalar,

$$\boldsymbol{H} = \boldsymbol{n} + \boldsymbol{x} \wedge \boldsymbol{n} = \boldsymbol{n} + I(\boldsymbol{x} \cdot \boldsymbol{n}) = \boldsymbol{n} + Id, \tag{7.7}$$

or $\boldsymbol{H} = (0, \boldsymbol{n}) + I(d, 0)$. Note that the plane equation is the dual of the point equation

$$\boldsymbol{H} = (d + I\boldsymbol{n})^* = (I\boldsymbol{n})^* + (d)^* = \boldsymbol{n} + Id, \tag{7.8}$$

where the plane orientation is given by the unit bivector $\boldsymbol{n}$ and the Hesse distance d by the scalar 1.

7.4 Representation of Points, Lines, and Planes Using 4D Geometric Algebra

We can also represent point, lines, and planes using the entire 4D geometric algebra $G_{3,0,1}$. As opposite to the previous representations which use only bivectors, this representation will also use vectors and the trivectors basis. The point expressed in terms of trivectors is given by

$$\boldsymbol{X} = e_1e_2e_3 + x_1e_2e_3e_4 + x_2e_3e_1e_4 + x_3e_1e_2e_4. \tag{7.9}$$

The equation of the line using bivectors basis is exactly the same as that for Eq. (7.5):

$$\begin{aligned} \boldsymbol{L} &= L_{n_1}e_2e_3 + L_{n_1}e_3e_1 + L_{n_3}e_1e_2 + L_{m_1}e_4e_1 + L_{m_2}e_4e_2 + L_{m_3}e_4e_3, \\ &= L_{n_1}e_2e_3 + L_{n_1}e_3e_1 + L_{n_3}e_1e_2 + I(L_{m_1}e_2e_3 + L_{m_2}e_3e_1 + L_{m_3}e_1e_2), \\ &= \boldsymbol{n} + I\boldsymbol{m}. \end{aligned} \tag{7.10}$$

The equation of the plane is spanned using basis vectors in terms of the normal of the plane and the Hessian distance:

$$\boldsymbol{H} = \boldsymbol{n}_x e_1 + \boldsymbol{n}_y e_2 + \boldsymbol{n}_z e_3 + de_4. \tag{7.11}$$

Note that in this equation the multivector basis of the point and plane has been swapped. This equation corresponds to the dual of point Eq. (7.9) and makes use of a vector as the dual of each trivector:

$$\begin{aligned} \boldsymbol{H} &= (de_1e_2e_3 + n_xe_2e_3e_4 + n_ye_3e_1e_4 + n_ze_1e_2e_4)^*, \\ &= d(e_1e_2e_3)^* + n_x(e_2e_3e_4)^* + n_y(e_3e_1e_4)^* + n_z(e_1e_2e_4)^*, \\ &= n_xe_1 + n_ye_2 + n_ze_3 + de_4. \end{aligned} \tag{7.12}$$

Note that the dual operation is actually not carried out via the pseudoscalar $I = e_1e_2e_3e_4$ because this will lead to the square of e_4, which equals zero. In order to explain the relation between Eqs. (7.9) and (7.11), we are simply relating in the dual sense each basis vector with a basis trivector. Since Eqs. (7.4) and (7.7) are the duals of Eqs. (7.9) and (7.11), we can reconsider Fig. 7.1c, now using a trivector coordinate basis for depicting the point Eq. (7.9), and similarly, Fig. 7.1a, now using a vector coordinate basis for the plane Eq. (7.11).

The following sections are concerned with the modeling of the motion of basic geometric entities in 3D and 4D space. By comparing these motion models, we will show the power of geometric algebra in the representation and linearizing of the translation transformation achieved in 4D geometric algebra.

7.5 Motion of Points, Lines, and Planes in 3D Geometric Algebra

The 3D motion of a point $\boldsymbol{x}$ in $G_{3,0,0}$ is given by the following equation:

$$\boldsymbol{x}' = \boldsymbol{R}\boldsymbol{x}\tilde{\boldsymbol{R}} + \boldsymbol{t}. \tag{7.13}$$

Using Eq. (7.2), the motion equation of the line can be expressed as follows:

$$\begin{aligned}
\boldsymbol{l}' &= \boldsymbol{n}' + \boldsymbol{m}' = \boldsymbol{n}' + \boldsymbol{x}' \wedge \boldsymbol{n}',\\
&= \boldsymbol{R}\boldsymbol{n}\tilde{\boldsymbol{R}} + (\boldsymbol{R}\boldsymbol{x}\tilde{\boldsymbol{R}} + \boldsymbol{t}) \wedge (\boldsymbol{R}\boldsymbol{n}\tilde{\boldsymbol{R}}),\\
&= \boldsymbol{R}\boldsymbol{n}\tilde{\boldsymbol{R}} + \boldsymbol{R}\boldsymbol{x}\tilde{\boldsymbol{R}} \wedge \boldsymbol{R}\boldsymbol{n}\tilde{\boldsymbol{R}} + \boldsymbol{t} \wedge \boldsymbol{R}\boldsymbol{n}\tilde{\boldsymbol{R}},\\
&= \boldsymbol{R}\boldsymbol{n}\tilde{\boldsymbol{R}} + \boldsymbol{R}\boldsymbol{x}\tilde{\boldsymbol{R}} \wedge \boldsymbol{R}\boldsymbol{n}\tilde{\boldsymbol{R}} + \frac{\boldsymbol{t}}{2}\boldsymbol{R}\boldsymbol{n}\tilde{\boldsymbol{R}} - \boldsymbol{R}\boldsymbol{n}\tilde{\boldsymbol{R}}\frac{\boldsymbol{t}}{2},\\
&= \boldsymbol{R}\boldsymbol{n}\tilde{\boldsymbol{R}} + \boldsymbol{R}\boldsymbol{n}\frac{\boldsymbol{t}}{2}\tilde{\boldsymbol{R}} + \frac{\boldsymbol{t}}{2}\boldsymbol{R}\boldsymbol{n}\tilde{\boldsymbol{R}} + \boldsymbol{R}\boldsymbol{m}\tilde{\boldsymbol{R}},
\end{aligned} \tag{7.14}$$

where $\boldsymbol{x}'$ stands for the rotated and shifted position vector, $\boldsymbol{n}'$ stands for the rotated orientation vector, and $\boldsymbol{m}'$ for the new line moment.

The model of the motion of the plane in $G_{3,0,0}$ can be expressed in terms of the multivector Hesse Eq. (7.3), as follows:

$$\begin{aligned}
\boldsymbol{h}' &= \boldsymbol{n}' + Id' = \boldsymbol{n}' + \boldsymbol{x}' \wedge \boldsymbol{n}',\\
&= \boldsymbol{R}\boldsymbol{n}\tilde{\boldsymbol{R}} + (\boldsymbol{R}\boldsymbol{x}\tilde{\boldsymbol{R}} + \boldsymbol{t}) \wedge (\boldsymbol{R}\boldsymbol{n}\tilde{\boldsymbol{R}}),\\
&= \boldsymbol{R}\boldsymbol{n}\tilde{\boldsymbol{R}} + \boldsymbol{R}\boldsymbol{x}\tilde{\boldsymbol{R}} \wedge \boldsymbol{R}\boldsymbol{n}\tilde{\boldsymbol{R}} + \boldsymbol{t} \wedge \boldsymbol{R}\boldsymbol{n}\tilde{\boldsymbol{R}},\\
&= \boldsymbol{R}\boldsymbol{n}\tilde{\boldsymbol{R}} + \boldsymbol{t} \wedge \boldsymbol{R}\boldsymbol{n}\tilde{\boldsymbol{R}} + \boldsymbol{R}\boldsymbol{x} \wedge \boldsymbol{n}\tilde{\boldsymbol{R}},\\
&= \boldsymbol{R}\boldsymbol{n}\tilde{\boldsymbol{R}} + \boldsymbol{t} \wedge \boldsymbol{R}\boldsymbol{n}\tilde{\boldsymbol{R}} + \boldsymbol{R}(Id)\tilde{\boldsymbol{R}},\\
&= \boldsymbol{R}\boldsymbol{n}\tilde{\boldsymbol{R}} + \boldsymbol{t}^* \cdot \boldsymbol{R}\boldsymbol{n}\tilde{\boldsymbol{R}} + Id,\\
&= \boldsymbol{R}\boldsymbol{n}\tilde{\boldsymbol{R}} + I(\boldsymbol{t} \cdot \boldsymbol{R}\boldsymbol{n}\tilde{\boldsymbol{R}} + d),
\end{aligned} \tag{7.15}$$

where $\boldsymbol{n}'$ stands for the rotated bivector plane orientation, $\boldsymbol{x}'$ stands for the rotated and shifted position vector, and d' for the new Hesse distance. Here, we use the concept of duality to claim that $\boldsymbol{t} \wedge \boldsymbol{R}\boldsymbol{n}\tilde{\boldsymbol{R}} = \boldsymbol{t}^* \cdot \boldsymbol{R}\boldsymbol{n}\tilde{\boldsymbol{R}} = (I\boldsymbol{t}) \cdot \boldsymbol{R}\boldsymbol{n}\tilde{\boldsymbol{R}}$.

7.6 Motion of Points, Lines, and Planes Using Motor Algebra

The modeling of the 3D motion of geometric primitives using the motor algebra $G^+_{3,0,1}$ takes place in a 4D space where rotation and translation are multiplicative operators that are applied as multiplicative operators, and the result is that the 3D general motion becomes linear. Having created a linear model, we can then compute simultaneously the unknown rotation and translation. This will be useful for the case of the hand–eye problem, or when we apply the motor extended Kalman filter.

For the modeling of the point motion, we use the point representation of Eq. (7.4) and the motor relations given in Eq. (6.62), with $I^2 = 0$:

$$
\begin{aligned}
X' &= 1 + I\boldsymbol{x}' = \boldsymbol{M}X\tilde{\overline{\boldsymbol{M}}} = \boldsymbol{M}(1 + I\boldsymbol{x})\tilde{\overline{\boldsymbol{M}}},\\
&= \boldsymbol{T}_s\boldsymbol{R}_s(1 + I\boldsymbol{x})\tilde{\boldsymbol{R}}_s\boldsymbol{T}_s,\\
&= (1 + I\frac{\boldsymbol{t}_s}{2})\boldsymbol{R}_s(1 + I\boldsymbol{x})\tilde{\boldsymbol{R}}_s(1 + I\frac{\boldsymbol{t}_s}{2}),\\
&= (1 + I\frac{\boldsymbol{t}_s}{2})(1 + I\boldsymbol{R}_s\boldsymbol{x}\tilde{\boldsymbol{R}}_s)(1 + I\frac{\boldsymbol{t}_s}{2}),\\
&= 1 + I\frac{\boldsymbol{t}_s}{2} + I\boldsymbol{R}_s\boldsymbol{x}\tilde{\boldsymbol{R}}_s + I\frac{\boldsymbol{t}_s}{2},\\
&= 1 + I(\boldsymbol{R}_s\boldsymbol{x}\tilde{\boldsymbol{R}}_s + \boldsymbol{t}_s).
\end{aligned} \tag{7.16}
$$

Note that the dual part of this equation in 4D space is fully equivalent to Eq. (7.13), which is in 3D space.

The motion of a 3D line or screw motion can be seen as the rotation of the line about the axis line $\boldsymbol{L}_s$ and its translation along this axis line, as depicted in Fig. 7.2. Note that in the figure the line $\boldsymbol{L}_s$ is shifted a distance $\boldsymbol{t}_c$ from the origin. Now, using line Eq. (7.5) we can express the motion of a 3D line as follows:

$$
\boldsymbol{L}' = \boldsymbol{n}' + I\boldsymbol{m}' = \boldsymbol{M}(\boldsymbol{L})\tilde{\boldsymbol{M}} = \boldsymbol{T}_s\boldsymbol{R}_s(\boldsymbol{n} + I\boldsymbol{m})\tilde{\boldsymbol{R}}_s\tilde{\boldsymbol{T}}_s. \tag{7.17}
$$

This equation can be further expressed purely in terms of rotors, as follows:

$$
\begin{aligned}
\boldsymbol{L}' &= (1 + I\frac{\boldsymbol{t}_s}{\mathbf{2}})\boldsymbol{R}_s(\boldsymbol{n} + I\boldsymbol{m})\tilde{\boldsymbol{R}}_s(1 - I\frac{\boldsymbol{t}_s}{2}),\\
&= (1 + I\frac{\boldsymbol{t}_s}{\mathbf{2}})(\boldsymbol{R}_s\boldsymbol{n}\tilde{\boldsymbol{R}}_s + I\boldsymbol{R}_s\boldsymbol{m}\tilde{\boldsymbol{R}}_s - I\boldsymbol{R}_s\boldsymbol{n}\tilde{\boldsymbol{R}}_s\frac{\boldsymbol{t}_s}{2}),
\end{aligned}
$$

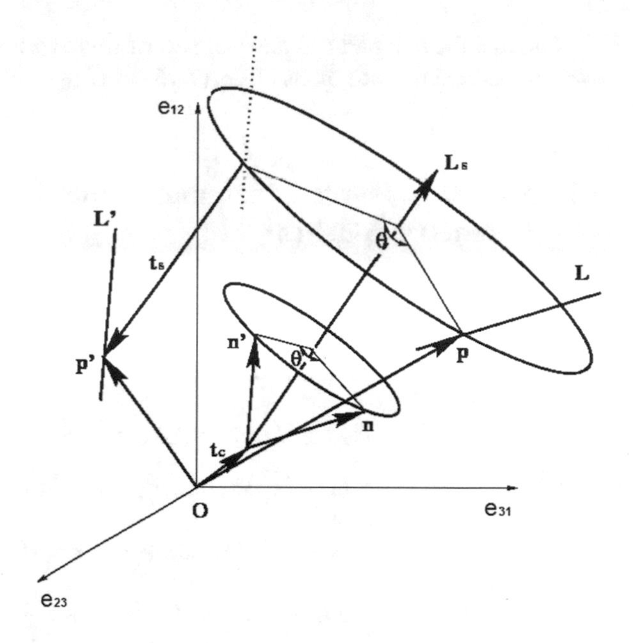

Fig. 7.2 Screw motion of a line

$$= R_s n \tilde{R}_s + I(-R_s n \tilde{R}_s \frac{t_s}{2} + \frac{t_s}{2} R_s n \tilde{R}_s + R_s m \tilde{R}_s),$$
$$= R_s n \tilde{R}_s + I(R_s n \tilde{R}_s{}' + R_s' n \tilde{R}_s + R_s m \tilde{R}_s). \quad (7.18)$$

Note that in this equation, before we merge the bivector $\frac{t_s}{2}$ with the rotor R_s or $\tilde{R}_s$, the real and the dual parts are fully equivalent with the elements of line Eq. (7.14) of $G_{3,0,0}$.

Equation (7.18) is very useful as a linear algorithm to estimate simultaneous rotation and translation, as in the case of hand–eye calibration [20] or for the algorithm for the motor extended Kalman filter.

The transformation of a plane under a rigid motion in $G^+_{3,0,1}$ can be seen as the motion of the dual of the point. Thus, Eq. (7.7) can be utilized to express the motion equation of the plane:

$$H' = n' + Id' = MH\tilde{\bar{M}} = M(n + Id)\tilde{\bar{M}},$$
$$= T_s R_s (n + Id)\tilde{R}_s T_s,$$
$$= (1 + I\frac{t_s}{2})(R_s n \tilde{R}_s + Id)(1 + I\frac{t_s}{2}),$$
$$= R_s n \tilde{R}_s + I(R_s n \tilde{R}_s \frac{t_s}{2} + \frac{t_s}{2} R_s n \tilde{R}_s + d),$$
$$= R_s n \tilde{R}_s + I(t_s \cdot (R_s n \tilde{R}_s) + d). \quad (7.19)$$

Note that the real part and the dual part of this expression are fully equivalent to the bivector and trivector parts of Eq. (7.15) in $G_{3,0,0}$.

7.7 Motion of Points, Lines, and Planes Using 4D Geometric Algebra

For the modeling of the motion of points, lines, and planes in $\mathcal{G}_{3,0,1}$, only the automorphism equivalent of the screw and its conjugate is required.

The motion of the point is given by

$$
\begin{aligned}
\boldsymbol{X}' = \boldsymbol{M}\boldsymbol{X}\tilde{\boldsymbol{M}} &= \boldsymbol{T}_s\boldsymbol{R}_s(\boldsymbol{X})\tilde{\boldsymbol{R}}_s\tilde{\boldsymbol{T}}_s,\\
&= (1+I\frac{\boldsymbol{t}_s}{2})\boldsymbol{R}_s(e_1e_2e_3+I\boldsymbol{x})\tilde{\boldsymbol{R}}_s(1-I\frac{\boldsymbol{t}_s}{2}),\\
&= (1+e_4\frac{\boldsymbol{t}}{2})(e_1e_2e_3+I\boldsymbol{R}_s\boldsymbol{x}\tilde{\boldsymbol{R}}_s)(1-e_4\frac{\boldsymbol{t}}{2}),\\
&= e_1e_2e_3+I\frac{\boldsymbol{t}}{2}+I\boldsymbol{R}_s\boldsymbol{x}\tilde{\boldsymbol{R}}_s+I\frac{\boldsymbol{t}_s}{2},\\
&= e_1e_2e_3+I(\boldsymbol{R}_s\boldsymbol{x}\tilde{\boldsymbol{R}}_s+\boldsymbol{t}_s),
\end{aligned}
\tag{7.20}
$$

where $\boldsymbol{t}_s = t_xe_2e_3+t_ye_3e_1+t_ze_1e_2$, $\boldsymbol{x} = x_xe_1+x_ye_2+x_ze_3$, and $\boldsymbol{t} = t_xe_1+t_ye_2+t_ze_3$. Note that the dual part of this equation in 4D space is fully equivalent to Eq. (7.13) in 3D space.

The equation of the motion of the line is exactly the same as Eq. (7.18).

The motion of the plane is given by

$$
\begin{aligned}
\boldsymbol{H}' &= \boldsymbol{n}'+e_4d' = n'_xe_1+n'_ye_2+n'_ze_3+d'e_4 = \boldsymbol{M}\boldsymbol{H}\tilde{\boldsymbol{M}} = \boldsymbol{M}(\boldsymbol{n}+e_4d)\tilde{\boldsymbol{M}},\\
&= \boldsymbol{T}_s\boldsymbol{R}_s(\boldsymbol{n}+e_4)\tilde{\boldsymbol{R}}_s\tilde{\boldsymbol{T}}_s = (1+I\frac{\boldsymbol{t}_s}{\boldsymbol{2}})(\boldsymbol{R}_s\boldsymbol{n}\tilde{\boldsymbol{R}}_s+e_4d)(1-I\frac{\boldsymbol{t}_s}{\boldsymbol{2}}),\\
&= \boldsymbol{R}_s\boldsymbol{n}\tilde{\boldsymbol{R}}_s+I\frac{\boldsymbol{t}_s}{\boldsymbol{2}}R_s\boldsymbol{n}\tilde{\boldsymbol{R}}_s-\boldsymbol{R}_s\boldsymbol{n}\tilde{\boldsymbol{R}}_sI\frac{\boldsymbol{t}_s}{\boldsymbol{2}}+e_4d,\\
&= \boldsymbol{R}_s\boldsymbol{n}\tilde{\boldsymbol{R}}_s+I(\frac{\boldsymbol{t}_sR_s\boldsymbol{n}\tilde{\boldsymbol{R}}_s-\boldsymbol{R}_s\boldsymbol{n}\tilde{\boldsymbol{R}}_s\boldsymbol{t}_s}{\boldsymbol{2}})+e_4d,\\
&= \boldsymbol{R}_s\boldsymbol{n}\tilde{\boldsymbol{R}}_s+I(\boldsymbol{t}_s\wedge R_s\boldsymbol{n}\tilde{\boldsymbol{R}}_s)+e_4d,\\
&= \boldsymbol{n}'+e_4(I_3\boldsymbol{t}_s)\wedge\boldsymbol{n}'+e_4d = \boldsymbol{n}'+e_4(\boldsymbol{t}_s^*\cdot\boldsymbol{n}')+e_4d,\\
&= \boldsymbol{n}'+e_4(\boldsymbol{t}\cdot\boldsymbol{n}'+d).
\end{aligned}
\tag{7.21}
$$

The real and dual parts of this expression are equivalent in a higher dimension to the bivector and trivector parts of Eq. (7.15) in $G_{3,0,0}$.

7.8 Spatial Velocity of Points, Lines, and Planes

This section begins with the classic formulation of the spatial velocity of a rigid body using matrices, and in a later subsection, we will represent this velocity using instead the more advantageous techniques of motor algebra.

7.8.1 Rigid Body Spatial Velocity Using Matrices

The 3D motion of a rigid body comprises a rotation and a translation and can be computed as a relative motion between a world coordinates fixed frame and a frame attached to the object. In the case of pure rotations, the body frame is set at the origin of the fixed spatial frame. Translations will displace the object frame apart from the spatial frame. The velocity of a single particle of the object is given by

$$v_x(t) = \frac{dx(t)}{dt}. \tag{7.22}$$

Let us first consider the trajectory of a continuous rotational motion given by $R(t)$: $I\!R \rightarrow SO(3)$ that satisfies the constraint

$$R(t)R^T(t) = I. \tag{7.23}$$

By computing the derivative of this equation with respect to time t and passing a term to the right, we get a skew-symmetric matrix $\hat{w}(t)$:

$$\dot{R}(t)R^T(t) + R(t)\dot{R}^T(t) = 0 \Rightarrow \dot{R}(t)R^T(t) = -R(t)\dot{R}^T(t) = \hat{w}(t). \tag{7.24}$$

We get the expression for the derivative of the 3D rotation by multiplying both sides of the last equation by $R(t)$:

$$\dot{R}(t) = \hat{w}(t)R(t). \tag{7.25}$$

Considering that at t_0, there is not yet a rotation, $R(t_0) = I$, the first-order approximation to a rotation matrix is given by

$$\dot{R}(t_0 + dt) \approx I + \hat{w}(t_0)dt. \tag{7.26}$$

The linear space of all skew-symmetric matrices is commonly denoted by

$$so(3) \doteq \{\hat{w} \in I\!R^{3\times 3} | w \in I\!R^3\}. \tag{7.27}$$

In Eq. (7.25), $R(t)$ can be interpreted as the *state transition matrix* of the following linear ordinary differential equation (ODE):

$$\dot{x}(t) = \hat{w}x(t), \quad x(t) \in I\!R^3, \tag{7.28}$$

whose solution is given by

$$x(t) = e^{\hat{w}}x(0), \tag{7.29}$$

and the matrix $e^{\hat{w}t}$ can be expanded as a McClaurin expansion as follows:

$$e^{\hat{w}t} = I + \hat{w}t + \frac{(\hat{w}t)^2}{2!} + \cdots + \frac{(\hat{w}t)^n}{n!} + \cdots. \tag{7.30}$$

According to Rodriguez' formula related to a rotation matrix, the last equation can also be written in a more compact form as

$$e^{\hat{w}t} = I + \frac{\hat{w}}{||w||} sin(||w||) + \frac{\hat{w}^2}{||w||^2}(1 - cos(||w||)). \tag{7.31}$$

Due to the uniqueness of the solution of Eq. (7.28) and assuming $R(0) = I$ is the initial condition of Eq. (7.25), we get

$$R(t) = e^{\hat{w}t}R(0) = e^{\hat{w}t}. \tag{7.32}$$

Interestingly enough, we have derived the exponential relation between the linear spaces of $so(3)$ and $SO(3)$

$$\texttt{exp} : so(3) \rightarrow SO(3) : \quad \hat{w} \mapsto e^{\hat{w}}, \tag{7.33}$$

where the inverse of the exponential map is given by $\hat{w} = log(R)$.

The motion of a rigid body can be represented as a transformation matrix using homogeneous coordinates. The set of all motion matrices forms a group that is not longer Euclidean; thus, the formulation of the rigid body velocity cannot be done as a straightforward extension of Eq. (7.22) of the velocity of a single particle. One can derive the velocity equations of the rigid body based on the parametrization of motion using homogeneous matrices. Let us represent the trajectory of a body as a time-dependent curve using a homogeneous matrix:

$$g(t) = \begin{bmatrix} R(t) & t(t) \\ 0^T & 1 \end{bmatrix} \in I\!R^{4\times4}. \tag{7.34}$$

The inverse of $g(t)$ is given by

$$g^{-1}(t) = \begin{bmatrix} R(t) & t(t) \\ 0^T & 1 \end{bmatrix}^{-1} = \begin{bmatrix} R^T & -R^T t(t) \\ 0^T & 1 \end{bmatrix}. \quad (7.35)$$

In analogy to the case of pure rotation (see Eqs. (7.25–7.28)), we will obtain a matrix consisting of the instantaneous *spatial angular velocity* component expressed as an antisymmetric matrix $\hat{w}$ and the linear velocity component v; for this, we start deriving the following identity with respect to time t

$$\frac{dg(t)g^{-1}(t)}{dt} = \frac{dI}{dt} = 0,$$
$$\dot{g}(t)g^{-1}(t) + g(t)\dot{g}^{-1}(t) = 0. \quad (7.36)$$

Now, passing one term to the right, we get a skew-symmetric matrix $\widehat{V}(t)$

$$\dot{g}(t)g^{-1}(t) = -g(t)\dot{g}^{-1}(t) = \hat{V}(t) \in I\!R^{4\times 4}, \quad (7.37)$$

which equals

$$\widehat{V} = g\dot{(t)}g^{-1}(t) = \begin{bmatrix} \dot{R}(t) & \dot{t}(t) \\ 0^T & 0 \end{bmatrix} \begin{bmatrix} R^T(t) & -R^T(t)t(t) \\ 0^T & 1 \end{bmatrix}, \quad (7.38)$$
$$= \begin{bmatrix} \dot{R}(t)R^T(t) & -\dot{R}(t)R^T(t)t(t) + \dot{t}(t) \\ 0^T & 0 \end{bmatrix} = \begin{bmatrix} \hat{w}(t) & v(t) \\ 0^T & 0 \end{bmatrix}.$$

One gets the expression for the derivative of the 3D motion by multiplying both sides of the last equation by $g(t)$:

$$\dot{g}(t) = (\dot{g}(t)g^{-1}(t))g(t) = \widehat{V}g(t). \quad (7.39)$$

$\widehat{V}$ can be viewed as the tangent vector along the curve of $g(t)$ and can be used to approximate $g(t)$ locally:

$$\dot{g}(t + dt) \approx g(t) + \widehat{V}g(t)dt = (I + \widehat{V}dt)g(t). \quad (7.40)$$

The 4×4 matrix of the form of $\hat{V}$ is called a *twist*. The set of all twists builds a linear space denoted by

$$se(3) \doteq \left\{ \hat{V} \in \begin{bmatrix} \hat{w} & v \\ 0^T & 0 \end{bmatrix} \middle| w \in so(3),\ v \in I\!R^3 \right\} \in I\!R^{4\times 4}. \quad (7.41)$$

$se(3)$ is called the tangent space or Lie algebra of the matrix Lie group $SE(3)$.

If, in Eq. (7.39), $\widehat{V}$ is considered constant, we have a time-invariant linear ordinary equation, which can be integrated to get the following expression:

$$g(t) = e^{\widehat{V}t}g(0). \quad (7.42)$$

If we assume the initial condition $g(0) = I$, we obtain

$$g(t) = e^{\widehat{V}t}. \tag{7.43}$$

As a result, we can claim that the exponential map defines a transformation from $se(3)$ to $SE(3)$, namely

$$exp : se(3) \rightarrow SE(3); \quad \widehat{V} \rightarrow e^{\widehat{V}}. \tag{7.44}$$

This exponential twist can be further expanded using the McClaurin series

$$e^{\widehat{V}t} = I + \widehat{V}t + \frac{(\widehat{V}t)^2}{2!} + \cdots + \frac{\widehat{V}^n}{n!} + \cdots. \tag{7.45}$$

Using the Rodriguez formula (7.31) and certain additional properties of the exponential matrix, one can establish the following relationship:

$$e^{\widehat{V}} = \begin{bmatrix} e^{\hat{w}} & \frac{(I-e^{\hat{w}})\hat{w}v+ww^Tv}{||w||} \\ 0^T & 1 \end{bmatrix}, \texttt{ if } \quad w \neq 0. \tag{7.46}$$

If $w = 0$, the exponential of $\widehat{V}$ becomes

$$e^{\widehat{V}} = \begin{bmatrix} I & v \\ 0^T & 1 \end{bmatrix}. \tag{7.47}$$

Let us know study the rigid motion with respect to a spatial framework X and other Y attached to a rigid body (see Fig. 7.3). Let us again consider the equation of the twist derived above:

$$\widehat{V}^s_{xy} = \begin{bmatrix} \dot{R}_{xy}R^T_{xy} & -\dot{R}_{xy}R^T_{xy}t_{xy} + \dot{t}_{xy} \\ 0^T & 0 \end{bmatrix} = \begin{bmatrix} \hat{w}^s_{xy} & v^s_{xy} \\ 0^T & 0 \end{bmatrix}. \tag{7.48}$$

Note that the physical meaning of the linear velocity component is not very intuitive, due to the antisymmetric matrix $\dot{R}_{xy}R^T_{xy} = \hat{w}^s_{xy}$, there is a component orthogonal to the actual translation and parallel to the velocity with respect to the object frame, namely $(\hat{w}^s_{xy}t_{xy} = w^s_{xy} \times t_{xy}) \,||\, \dot{t}_{xy}$ (see Fig. 7.3). We will see below that only with the bivector notation of the motor algebra, we can achieve a much more intuitive representation of the angular and linear velocities than when we use transformation matrices.

The rigid body spatial velocity V^s_{xy} is a vector comprised of the instantaneous spatial angular velocity w^s_{xy} and the linear component v^s_{xy}:

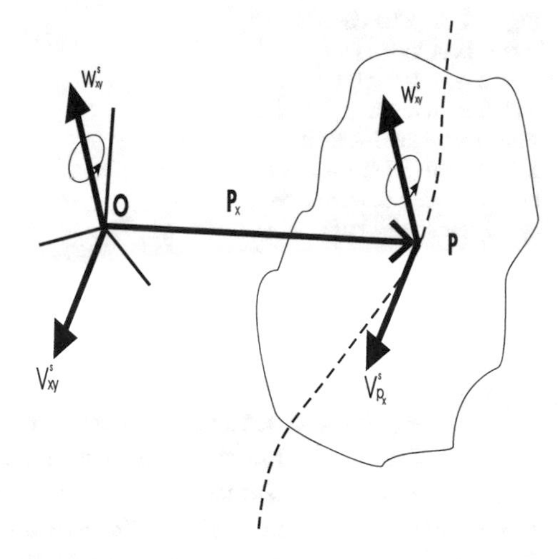

Fig. 7.3 Linear and angular velocities of a body with respect to spatial frame X and body frame Y

$$V^s_{xy} = \begin{bmatrix} w^s_{xy} \\ v^s_{xy} \end{bmatrix}. \tag{7.49}$$

The tangential velocity of a point p_x attached to the rigid body measured with respect to the spatial frame X is given by

$$v^s_{p_x} = \widehat{V}^s_{xy} \begin{bmatrix} p_x \\ 1 \end{bmatrix} = \hat{w}^s_{xy} p_x + v^s_{xy} = w^s_{xy} \times p_x + v^s_{xy}. \tag{7.50}$$

After this review of well-known motion formulas using matrices, we will represent the rigid body spatial velocity now using the more advantageous techniques of motor algebra.

7.8.2 *Angular Velocity Using Rotors*

The angular momentum of a particle with momentum $\boldsymbol{m}$ and position vector $\boldsymbol{x}$ is usually defined in 3D by using the cross-product

$$L = \boldsymbol{x} \times \boldsymbol{m}. \tag{7.51}$$

In geometric algebra, one replaces axial vectors with bivectors: thus, we rewrite the last equation using a bivector:

$$\boldsymbol{L} = \boldsymbol{x} \wedge \boldsymbol{m}, \tag{7.52}$$

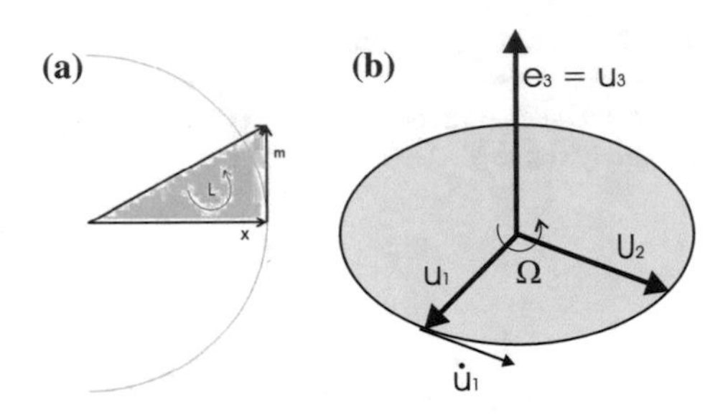

Fig. 7.4 **a** (left) Particle sweeps out the plane $\boldsymbol{L} = \boldsymbol{x} \wedge \boldsymbol{m}$. **b** (right) Rotating frame $\{e_k\}$ with $\boldsymbol{R}$. Bivector angular velocity $\boldsymbol{\Omega} = \boldsymbol{u}_1 \wedge \dot{\boldsymbol{u}}_1$, if the vector ω has the same orientation as $e_1 \wedge e_2$ then $\omega = |\omega| e_3$

see this formulation in the domain of physics in [73]. This formula substitutes the old notion of angular momentum as "an axial vector" with a geometric expression that describes the angular momentum as a particle sweeping out a plane (see Fig. 7.4a).

Since angular momentum is described as a bivector, the angular velocity must be represented as a bivector as well. To do so, we resort to a rotor equation. Suppose the orthonormal frame $\{u_k\}$ is rotating in the 3D space, and it is related to another via a rotor $\boldsymbol{R}$:

$$u_k = \boldsymbol{R}(t) e_k \tilde{\boldsymbol{R}}(t). \tag{7.53}$$

Traditionally, the angular momentum vector $\boldsymbol{w}$ is defined using the cross-product

$$\dot{u}_k = \boldsymbol{w} \times u_k = -I\boldsymbol{w} \wedge u_k = (-I\boldsymbol{w}) \cdot u_k. \tag{7.54}$$

In this equation, the (space) angular velocity bivector is

$$\boldsymbol{\Omega}_S = I_3 \boldsymbol{w}, \tag{7.55}$$

where the pseudoscalar $I_3 \in G_{3,0,0}$. The sign ensures the orientation sense followed by the involved rotor. Now, let us analyze the time dependence with respect to the frame $\{u_k\}$:

$$\dot{u} = \dot{\boldsymbol{R}} e_k \tilde{\boldsymbol{R}} + \boldsymbol{R} e_k \dot{\tilde{\boldsymbol{R}}} = \dot{\boldsymbol{R}} \tilde{\boldsymbol{R}} u_k + u_k \boldsymbol{R} \dot{\tilde{\boldsymbol{R}}}. \tag{7.56}$$

Since $\boldsymbol{R}\tilde{\boldsymbol{R}} = 1$, we derive

$$\partial_t (\boldsymbol{R}\tilde{\boldsymbol{R}}) = \dot{\boldsymbol{R}}\tilde{\boldsymbol{R}} + \boldsymbol{R}\dot{\tilde{\boldsymbol{R}}} = 0, \tag{7.57}$$

which leads to

$$\dot{\boldsymbol{R}}\tilde{\boldsymbol{R}} = -\boldsymbol{R}\dot{\tilde{\boldsymbol{R}}}. \tag{7.58}$$

We substitute Eq. (7.58) into (7.56): the result is simply the inner product of a bivector with a vector:

$$\dot{u} = \dot{\boldsymbol{R}}\tilde{\boldsymbol{R}}u_k - u_k\dot{\boldsymbol{R}}\tilde{\boldsymbol{R}} = (2\dot{\boldsymbol{R}}\tilde{\boldsymbol{R}}) \cdot u_k. \tag{7.59}$$

Comparing Eqs. (7.54) and (7.55), we find an expression for the angular velocity bivector in terms of the rotor:

$$\boldsymbol{\Omega}_S = -2\dot{\boldsymbol{R}}\tilde{\boldsymbol{R}}. \tag{7.60}$$

If we multiply from the right by a rotor and divide by -2, we easily obtain the dynamic equation reduced to a rotor equation

$$\dot{\boldsymbol{R}} = -\frac{1}{2}\boldsymbol{\Omega}_S\boldsymbol{R}, \tag{7.61}$$

or

$$\dot{\tilde{\boldsymbol{R}}} = \frac{1}{2}\tilde{\boldsymbol{R}}\boldsymbol{\Omega}_S. \tag{7.62}$$

The body angular velocity $\boldsymbol{\Omega}_B$ related to a fixed space frame is transformed back as follows

$$\boldsymbol{\Omega}_S = \boldsymbol{R}\boldsymbol{\Omega}_B\tilde{\boldsymbol{R}}. \tag{7.63}$$

Replacing the last equation into Eq. (7.61) gives

$$\dot{\boldsymbol{R}} = -\frac{1}{2}\boldsymbol{R}\boldsymbol{\Omega}_B = -\frac{1}{2}\boldsymbol{\Omega}_S\boldsymbol{R}, \tag{7.64}$$

and into Eq. (7.62) gives

$$\dot{\tilde{\boldsymbol{R}}} = \frac{1}{2}\boldsymbol{\Omega}_B\tilde{\boldsymbol{R}}. \tag{7.65}$$

Assuming that the rotor motion is constant through time, such as a body rotating with a fixed angle, then we fix $\boldsymbol{\Omega}_S$ constant and the rotor Eq. (7.64) can be integrated to give

$$\boldsymbol{R}(t) = e^{-\boldsymbol{\Omega}_S\frac{t}{2}}\boldsymbol{R}(0), \tag{7.66}$$

which represents a rotor that rotates with a constant frequency rotation in the right-hand sense.

7.8.3 Rigid Body Spatial Velocity Using Motor Algebra

In motor algebra, we represent the rigid motion of lines by Eq. (7.17). Lines are represented in terms of two dual bivector bases. Following the same idea followed by the case of the pure rotor motion, we represent the relationship of the moving frames as follows:

$$l'_k(t) = \boldsymbol{M}(t) l_k \widetilde{\boldsymbol{M}}(t), \tag{7.67}$$

where l'_k and l_k are the coefficients of the two frames. Taking its time derivative

$$\dot{l}'_k = \dot{\boldsymbol{M}} l_k \widetilde{\boldsymbol{M}} + \boldsymbol{M} l_k \dot{\widetilde{\boldsymbol{M}}}, \tag{7.68}$$

and substituting $\boldsymbol{M} l_k = l'_k \boldsymbol{M}$, one gets

$$\dot{l}'_k = \dot{\boldsymbol{M}} \widetilde{\boldsymbol{M}} l'_k + l'_k \boldsymbol{M} \dot{\widetilde{\boldsymbol{M}}}. \tag{7.69}$$

Taking the time derivative of the identity $\boldsymbol{M}\widetilde{\boldsymbol{M}} = 1$, we get the useful relations

$$\begin{aligned} \partial_t (\boldsymbol{M}\widetilde{\boldsymbol{M}}) &= 0, \\ \dot{\boldsymbol{M}}\widetilde{\boldsymbol{M}} + \boldsymbol{M}\dot{\widetilde{\boldsymbol{M}}} &= 0, \\ \dot{\boldsymbol{M}}\widetilde{\boldsymbol{M}} &= -\boldsymbol{M}\dot{\widetilde{\boldsymbol{M}}}, \end{aligned} \tag{7.70}$$

which are substituted in Eq. (7.69) to yield

$$\dot{l}_k{}'(t) = \dot{\boldsymbol{M}} \widetilde{\boldsymbol{M}} l'_k - l'_k \dot{\boldsymbol{M}} \widetilde{\boldsymbol{M}}. \tag{7.71}$$

Since the right side of the last equation is the inner product between a bivector and the vector basis coefficients, we rewrite it as follows:

$$\dot{l}_k{}' = (2\dot{\boldsymbol{M}}\widetilde{\boldsymbol{M}}) \cdot l'_k. \tag{7.72}$$

Let us call the bivector

$$\boldsymbol{V}_S = 2\dot{\boldsymbol{M}}\widetilde{\boldsymbol{M}} \tag{7.73}$$

the spatial velocity. We will show next, that it comprises of a bivector for the linear velocity and a bivector for the angular velocity. Recalling that $\boldsymbol{M} = \boldsymbol{T}\boldsymbol{R}$, we proceed as follows

$$
\begin{aligned}
V_S = 2\dot{M}\widetilde{M} &= 2(\dot{T}R + T\dot{R})\widetilde{R}\widetilde{T} = 2\dot{T}R\widetilde{R}\widetilde{T} + 2T\dot{R}\widetilde{R}\widetilde{T}, \\
&= 2\dot{T}\widetilde{T} + 2(1 + I\frac{t_S}{2})(\frac{1}{2}\boldsymbol{\Omega}_S)(1 - I\frac{t_S}{2}), \\
&= I\dot{t}_S(1 - I\frac{t_S}{2}) + (\boldsymbol{\Omega}_S + I\frac{t_S}{2}\boldsymbol{\Omega}_S)(1 - I\frac{t_S}{2}), \\
&= I\dot{t}_S + (\boldsymbol{\Omega}_S + I\frac{t_S}{2}\boldsymbol{\Omega}_S - I\frac{t_S}{2}\boldsymbol{\Omega}_S), \\
&= (\boldsymbol{\Omega}_S + I\dot{t}_S), \\
&= (\boldsymbol{\Omega}_S + I\boldsymbol{v}_S),
\end{aligned}
\tag{7.74}
$$

where the angular velocity bivector $\boldsymbol{\Omega}_S$ is the dual of the linear velocity bivector $\boldsymbol{v}_S$.

Multiplying Eq. (7.73) from the right by M and dividing by -2, we get the dynamic motor equation

$$
\dot{M} = -\frac{1}{2}V_S M. \tag{7.75}
$$

Assuming that the screw motion of the body is constant through time, the dynamic motor Eq. (7.75) can be integrated to give

$$
M(t) = e^{-\frac{V_S}{2}}M(0) = e^{-\frac{(\boldsymbol{\Omega}_S + I\boldsymbol{v}_S)}{2}}M(0). \tag{7.76}
$$

This represents a motor that rotates with a constant frequency rotation in the right-hand sense and has a constant linear velocity as well.

Finally, compare the different algebraic treatments to get Eqs. (7.25), (7.39), (7.61), and (7.75) for thc analysis of the rigid body spatial velocity using matrices, rotor, and motor algebras.

7.8.4 Point, Line, and Plane Spatial Velocities Using Motor Algebra

According to Eq. (7.16), the motion of a point is represented as

$$
X = 1 + Ix = MX_0\widetilde{\widetilde{M}}, \tag{7.77}
$$

where $\widetilde{\widetilde{M}} = \tilde{R}T$. If we take the time derivative of this equation, we get

$$\begin{aligned}
\dot{X} &= \dot{M}X_0\tilde{M} + MX_0\dot{\tilde{M}}, \quad (7.78)\\
&= \dot{T}RX_0\tilde{R}T + T\dot{R}X_0\tilde{R}T + RTX_0\dot{\tilde{R}}T + RTX_0\tilde{R}\dot{T},\\
&= \dot{T}\tilde{T}TRX_0\tilde{R}T + T\dot{R}\tilde{R}RX_0\tilde{R}T + RTX_0\tilde{R}R\dot{\tilde{R}}T + RTX_0\tilde{R}\dot{T}\tilde{T}T,\\
&= \frac{1}{2}[\boldsymbol{\Omega}_S X_0 - X_0\boldsymbol{\Omega}_S] + I\frac{1}{2}[\dot{t}_S X_0 + X_0\dot{t}_S] = \boldsymbol{\Omega}_S \wedge X_0 + (I\boldsymbol{v}_S)\cdot X_0,\\
&= I\boldsymbol{w} + (I\boldsymbol{v}_S)\cdot X_0.
\end{aligned}$$

According to Eqs. (7.5) and (7.7), the motion equations of a line and a plane are represented as

$$\begin{aligned}
L &= n + Im = ML_0\tilde{M},\\
H &= n + Id = MH_0\tilde{\tilde{M}}.
\end{aligned} \quad (7.79)$$

If we take the time derivative of these equations and follow simple algebraic equations similar to the motion equation of the point, we get

$$\dot{L} = \boldsymbol{\Omega}_S \wedge L, \quad (7.80)$$

$$\dot{H} = \boldsymbol{\Omega}_S \wedge H + (I\dot{t}_S)\cdot H = \boldsymbol{\Omega}_S \wedge H + (I\boldsymbol{v}_S)\cdot \boldsymbol{n}. \quad (7.81)$$

Let us consider the velocity V_S when a composition of two motors $M = M_2M_1$. According to Eq. (7.73), we get

$$\begin{aligned}
V_S &= -2\dot{M}\tilde{M} = -2(\dot{M}_2M_1 + M_2\dot{M}_1)(\widetilde{M_2M_1}),\\
&= -2(\dot{M}_2M_1 + M_2\dot{M}_1)\tilde{M}_1\tilde{M}_2 = -2\dot{M}_2\tilde{M}_2 + M_2(-2\dot{M}_1\tilde{M}_1)\tilde{M}_2,\\
&= V_2 + M_2V_1\tilde{M}_2.
\end{aligned} \quad (7.82)$$

In general, for a sequence of n motors, the overall velocity V_S is given as follows:

$$\begin{aligned}
V_S = \; & V_n + M_nM_{n-1}M_{n-2}\cdots M_{n-1}V_{n-1}\tilde{M}_{n-1}\cdots\tilde{M}_{n-1}\tilde{M}_n +\\
& M_{n-1}M_{n-2}M_{n-3}\cdots M_{n-1}V_{n-2}\tilde{M}_{n-1}\cdots\tilde{M}_{n-3}\tilde{M}_{n-2}\tilde{M}_{n-1} +\\
& \cdots + M_3M_2V_2\tilde{M}_3\tilde{M}_2 + M_2V_1\tilde{M}_2,\\
= \; & \sum_{i=1}^{n}\prod_{j=i+1}^{n} M_jV_i\tilde{M}_j.
\end{aligned} \quad (7.83)$$

7.9 Incidence Relations Between Points, Lines, and Planes

The geometric relations between points, lines, and planes expressed in terms of incidence relations are very useful when we are dealing with the geometry of configurations and the relative motion of objects. Blaske introduced the basic relations of incidence using dual quaternions [38]. These relations can be also formulated for the representations of points, lines, and planes in the motor algebra $\mathcal{G}^+_{3,0,1}$ or in the 4D degenerated geometric algebra $G_{3,0,1}$. In general, the incidence relations between a point, a line, and a plane are given by:

$$\widetilde{\boldsymbol{P}}\boldsymbol{L} + \widetilde{\boldsymbol{L}}\boldsymbol{P} = -2I(\boldsymbol{x} \cdot \boldsymbol{n}), \tag{7.84}$$

$$\widetilde{\boldsymbol{P}}\boldsymbol{\Pi} + \widetilde{\boldsymbol{\Pi}}\boldsymbol{P} = -2I(d - \boldsymbol{x} \cdot \boldsymbol{n}_\pi), \tag{7.85}$$

$$\widetilde{\boldsymbol{L}}\boldsymbol{\Pi} + \widetilde{\boldsymbol{\Pi}}\boldsymbol{L} = -2(\boldsymbol{n}_L \cdot \boldsymbol{n}_\pi + \boldsymbol{m}_L \cdot \boldsymbol{n}_\pi), \tag{7.86}$$

where $\widetilde{\boldsymbol{P}}$ and $\widetilde{\boldsymbol{L}}$ are their conjugated (reversion), respectively.

Thus, a point $\boldsymbol{P}$ lying on a line $\boldsymbol{L}$ fulfills the equation

$$\widetilde{\boldsymbol{P}}\boldsymbol{L} + \widetilde{\boldsymbol{L}}\boldsymbol{P} = 0. \tag{7.87}$$

The distance d of a point relative to a plane Π is given by

$$\widetilde{\boldsymbol{P}}\boldsymbol{\Pi} + \widetilde{\boldsymbol{\Pi}}\boldsymbol{P} = de_1e_2e_3e_4. \tag{7.88}$$

If $d = 0$, the point lies on the plane; if d is negative, it is behind the plane; and if d is positive, it is in front of the plane.

The intersection point $\boldsymbol{P}$ of a line $\boldsymbol{L}$ crossing a plane Π can be computed as

$$\widetilde{\boldsymbol{L}}\boldsymbol{\Pi} + \widetilde{\boldsymbol{\Pi}}\boldsymbol{L} = \boldsymbol{P}. \tag{7.89}$$

In this case, if the line is parallel to the plane, the equation equals zero.

Incidence relations are fundamental in projective geometry, and we will study in more detail in next chapters.

7.9.1 Flags of Points, Lines, and Planes

When dealing with geometric configurations as in object modeling or robot navigation planning, it is useful to resort to a type of *geometric indicator*, which allows us to detect whether a geometric condition is fulfilled or not. These kind of indicators are called flags, and they are expressions relating points, lines, and planes that have some common attributes. Flags generate varieties [263].

We can express a point touching a line as the equation of the so-called *point-line flag*:

$$\boldsymbol{F}_{PL} = \boldsymbol{P} + \boldsymbol{L}. \tag{7.90}$$

If $\boldsymbol{P}\widetilde{\boldsymbol{P}} = 1$ and $\boldsymbol{L}\widetilde{\boldsymbol{L}} = 1$, then $\mathcal{F}_{PL}\mathcal{F}^*_{PL} = 1$.

If we have a point which touches a line and also a plane, and in which the orientation of the plane is parallel to the line, then we can represent the *line–plane flag* as

$$\boldsymbol{F}_{L\Pi} = \boldsymbol{L} + \boldsymbol{\Pi}. \tag{7.91}$$

In this equation, if $\boldsymbol{L}\widetilde{\boldsymbol{L}} = 1$ and $\boldsymbol{\Pi}\widetilde{\boldsymbol{\Pi}} = 1$, then $\mathcal{F}_{L\Pi}\widetilde{\mathcal{F}}_{L\Pi} = 1$.

Finally, if a point touches a line and a plane and the line has the same orientation as the normal to the plane, we can assign to this geometry the following *point-line-plane flag*:

$$\boldsymbol{F}_{PL\Pi} = \boldsymbol{P} + \boldsymbol{L} + \boldsymbol{\Pi}. \tag{7.92}$$

Here, if $\boldsymbol{P}\widetilde{\boldsymbol{P}} = 1$, $\boldsymbol{L}\widetilde{\boldsymbol{L}} = 1$, and $\boldsymbol{\Pi}\widetilde{\boldsymbol{\Pi}} = 1$, then $\mathcal{F}_{PL\Pi}\widetilde{\mathcal{F}}_{PL\Pi} = 1$.

7.10 Conclusion

This chapter presents the Clifford or geometric algebra for computations in visually guided robotics. Looking for other suitable ways of representing algebraic relations of geometric primitives, we consider the complex and dual numbers in the geometric algebra framework. It turns out that in this framework the algebra of motors is well suited to express the 3D kinematics. Doing that we can linearize the nonlinear 3D rigid motion transformation. In this chapter, the geometric primitives points, lines, and planes are represented using the 3D Euclidean geometric algebra and the 4D motor algebra. Next the rigid motions of these geometric primitives are elegantly expressed using rotors, motors, and concepts of duality. In the algebra of motors, we extend the 3D Euclidean space representation to a 4D space by means of a dual copy of scalars, vectors, and rotors or quaternions. Finally, we formulate incidence relations between points, lines, and planes using the motor algebra framework.

7.11 Exercises

7.1 Split the vector $\boldsymbol{v} = -2e_1 + 7e_2 + 10e_3$ into the components $\boldsymbol{x} = 3e_1 - 1.5e_2$, $\boldsymbol{y} = 2e_1 + 4e_2 - 2e_3$, and $\boldsymbol{z} = 5e_1 + 3e_2 + 3e_3$. In other words, compute the coefficients α, β, e of the equation $\boldsymbol{v} = \alpha\boldsymbol{x} + \beta\boldsymbol{y} + e\boldsymbol{z}$. Draw in three dimensions the vectors and the volume $\boldsymbol{a}\wedge\boldsymbol{b}\wedge\boldsymbol{c}$.

7.2 The outer square root of a multivector $\boldsymbol{M}$ is $\boldsymbol{x}$, which is the solution of the equation $\boldsymbol{M} = \boldsymbol{x}\wedge\boldsymbol{x}$. If $\boldsymbol{M} = \alpha + \boldsymbol{a}$, where $\alpha \in \boldsymbol{R}$, we can write $\boldsymbol{x} = \sqrt(\alpha)(1 + \frac{\boldsymbol{a}}{\alpha})^{\frac{1}{2}}$. This equation can be expanded using the following series:

$$(1+\frac{\boldsymbol{a}}{\alpha})^{\frac{1}{2}} = 1 + \frac{1}{2}\frac{\boldsymbol{a}}{\alpha} + \frac{\frac{1}{2}(\frac{1}{2}-1)}{2}\cdot\frac{\boldsymbol{a}\wedge\boldsymbol{a}}{\alpha^2} + \frac{\frac{1}{2}(\frac{1}{2}-1)(\frac{1}{2}-2)}{3!}\cdot\frac{\boldsymbol{a}\wedge\boldsymbol{a}\wedge\boldsymbol{a}}{\alpha^3} + ...$$

The quantity of the elements of this series should be at least equal to the dimension of the considered geometric algebra. Generate a program file for CLICAL [196] or eClifford [1] to compute the outer square root of $\boldsymbol{M}$, where Re($\boldsymbol{M}$)>0. Give any $\boldsymbol{M}$ and verify the identity $\boldsymbol{x}\wedge\boldsymbol{x} = \boldsymbol{M}$.

7.3 Rotate in 4D Euclidean space, using CLICAL [196] or eClifford [1], the vector $\boldsymbol{v} = e_1 + e_2 + e_3$, first related to the plane e_1e_2 by the angle $\frac{\phi}{4}$, and then related to the plane e_4e_2 by the angle $\frac{\phi}{5}$.

7.4 Given $\boldsymbol{x} = x_1e_1 + x_2e_2 + x_3e_3$, $\boldsymbol{y} = y_1e_1 + y_2e_2 + y_3e_3$, $\boldsymbol{X} = \boldsymbol{x}e_1e_2e_3$, and $\boldsymbol{Y} = \boldsymbol{y}e_1e_2e_3$, compute the bivectors $\frac{1}{2}(1+\epsilon_1e_2e_3e_4)\boldsymbol{X}$ and $\frac{1}{2}(1-e_1e_2e_3e_4)\boldsymbol{Y}$ and show that they commute.

7.5 Given the bivector $\boldsymbol{B} = \alpha e_1e_2 + \beta e_4e_3$, compute $\boldsymbol{B}\cdot\boldsymbol{B}$, $\boldsymbol{B}\wedge\boldsymbol{B}$, and $\boldsymbol{B}\times\boldsymbol{B}$ and explain what kinds of multivector result.

7.6 For this problem, use the same multivectors $\boldsymbol{X}$ and $\boldsymbol{Y}$ that you used in Exercise 7.4. Given $\boldsymbol{Z} = \frac{1}{2}(1+e_1e_2e_3e_4)\boldsymbol{X} + \frac{1}{2}(1-e_1e_2e_3e_4)\boldsymbol{Y}$, express exp($\boldsymbol{Z}$) using $|\boldsymbol{x}|$ and $|\boldsymbol{y}|$. What are the rotation angles of the rotation $\mathbb{R}^4 \rightarrow \mathbb{R}^4$, $\boldsymbol{u} \rightarrow \boldsymbol{z}\boldsymbol{u}\boldsymbol{z}^{-1}$, where $z = exp(\boldsymbol{Z})$?

7.7 Given in $G_{3,0,0}$ the points $\boldsymbol{a} = 0.5e_1 + 2.0e_2 + 1.3e_3$, $\boldsymbol{b} = 1.5e_1 + 1.2e_2 + 2.3e_3$, and $\boldsymbol{c} = 0.7e_1 + 1.2e_2 - 0.3e_3$, compute the line $\boldsymbol{l}$ crossing $\boldsymbol{a}$ and $\boldsymbol{b}$, and the plane ϕ tangent to the points $\boldsymbol{a}$, $\boldsymbol{b}$, and $\boldsymbol{c}$.

7.8 Using the software packet eClifford [1] and point $\boldsymbol{a}$, line $\boldsymbol{l}$, and plane ϕ from Exercise 7.7, in $G_{3,0,0}$ compute new values for the point, line, and plane after undergoing a rigid motion given by the translation $\boldsymbol{t} = 1.0e_1 - 2.7e_2 + 5.3e_3$. Also, compute the rotor $\boldsymbol{R} = cos(\frac{\theta}{2}) + sin(\frac{\theta}{2})\boldsymbol{n}$, where $\theta = \frac{\pi}{6}$ and $\boldsymbol{n}$ is the unit 3D bivector of the rotation axis given by $\boldsymbol{n} = 0.7e_2e_3 + 1.2e_3e_1 + 0.9e_1e_2$.

7.9 Express point $\boldsymbol{a}$, line $\boldsymbol{l}$, and plane ϕ given in Exercise 7.7 in the algebra of motors $G^+_{3,0,1}$.

7.10 Using the software packet eClifford and the point, line, and plane given in Exercise 7.7, compute their new values after undergoing the rigid motion given by the translator $\boldsymbol{T} = 1 + I\frac{1.0\gamma_2\gamma_3 - 2.7\gamma_3\gamma_1 + 5.3\gamma_1\gamma_2}{2}$ and the rotor $\boldsymbol{R}_s = cos(\frac{\theta}{2}) + sin(\frac{\theta}{2})\boldsymbol{l}$, where $\theta = \frac{\pi}{6}$ and $\boldsymbol{l}$ is the screw-axis line. Compare your results with the results of Exercise 7.8.

7.11 According c.f. ([74], Exercise 3.3), to solve this exercise consider in Sect. 7.8.2 Eq. (7.66).

$$\boldsymbol{R}(t) = e^{-\boldsymbol{\Omega}_S\frac{t}{2}}\boldsymbol{R}(0),$$

Let be a particle in the 3D space which moves along a curve $\boldsymbol{c}(t)$ with constant velocity $|\boldsymbol{v}|$, prove that there exits a bivector Ω such that

$$\dot{\boldsymbol{v}} = \Omega \cdot \boldsymbol{v}.$$

Give an explicit formula for the bivector Ω. Is this bivector unique?

7.12 According c.f. ([74], Exercise 3.4) and Sect. 7.8.2, imagine you measure the components of the position vector $\boldsymbol{x}$ in a rotating frame $\{f_i\}$. Referring this frame to a fixed frame, prove that the components of $\boldsymbol{x}$ are given by

$$x_i = e_i \cdot (\boldsymbol{R}\boldsymbol{x}\widetilde{\boldsymbol{R}}). \tag{7.93}$$

Now, differentiate this expression twice and prove that one can write

$$f_i \ddot{x}_i = \ddot{\boldsymbol{x}} + \Omega \cdot (\Omega \cdot \boldsymbol{x}) + 2\Omega \cdot \dot{\boldsymbol{x}} + \dot{\Omega} \cdot \boldsymbol{x}.$$

Using this, deduce the expressions for the centrifugal, Coriolis and Euler forces in terms of the angular velocity bivector Ω.

7.13 Express the geometric objects of Exercise7.7 in the degenerated algebra $G_{3,0,1}$: the point $\boldsymbol{a}$ using trivectors, the line $\boldsymbol{l}$ using bivectors and the plane ϕ using vectors.

7.14 Using the software packet eClifford and point $\boldsymbol{a}$, line $\boldsymbol{l}$, and plane ϕ from Exercise 7.11, compute new values for the point, line, and plane after undergoing the rigid motion given by the translator $\boldsymbol{T} = 1 + I\frac{1.0\gamma_2\gamma_3 - 2.7\gamma_3\gamma_1 + 5.3\gamma_1\gamma_2}{2}$. Also compute the rotor $\boldsymbol{R}_s = cos(\frac{\theta}{2}) + sin(\frac{\theta}{2})\boldsymbol{l}$, where $\theta = \frac{\pi}{6}$ and $\boldsymbol{l}$ is the screw-axis line. Compare your results with the results of Exercises 7.8 and 7.10.

7.15 Prove the Rodriguez formula for 3D rigid motion using motor algebra. Compare your results with the results from Exercise 2.18. Why is the motor algebra expression superior?

7.16 In the degenerated algebra $G_{3,0,1}$, choose a point $\boldsymbol{P}$ using trivectors and a line $\boldsymbol{L}$ using bivectors, and prove that the following expression,

$$\widetilde{\boldsymbol{P}}\mathbf{L} + \widetilde{\boldsymbol{L}}\boldsymbol{P},$$

equals zero if the point lies on the line. Prove the same equation using representations of the point and line in the motor algebra $G^{+}_{3,0,1}$.

7.17 In the degenerated algebra $G_{3,0,1}$ for the point $\boldsymbol{P}$ using trivectors, and the plane $\boldsymbol{\Pi}$ using vectors, show that the following expression,

$$\widetilde{\boldsymbol{P}}\boldsymbol{\Pi} + \widetilde{\boldsymbol{\Pi}}\boldsymbol{P} = de_1e_2e_3e_4$$

will describe a particular geometric configuration, depending upon the value of d. If d=0, the point lies on the plane; if $d \neq$0, d indicates the distance of the point to the plane; and if d is negative, the point is behind the plane. Using eClifford [1] and some points of the 3D space, check this equation. Check the equation again using the representations of the point and plane in the motor algebra $G^+_{3,0,1}$.

7.18 In the degenerated algebra $G_{3,0,1}$ for the line $\boldsymbol{L}$ using bivectors, the plane $\boldsymbol{\Pi}$ using vectors show that the following expression,

$$\widetilde{\boldsymbol{L}}\boldsymbol{\Pi} + \widetilde{\boldsymbol{L}}\boldsymbol{\Pi},$$

will describe a particular geometric configuration. If this expression equals zero, the line lies on the plane or is parallel to it; if not, the equation yields the intersecting point. This equation can be seen as a kind of meet equation when using a degenerated algebra. Using Clifford and some points of 3D space, check the equation. Check the same equation using the representations of the line and plane in the motor algebra $G^+_{3,0,1}$.

7.19 Three points $\boldsymbol{x}, \boldsymbol{y}, \boldsymbol{z}$ are in general position lying on a 2D simplex (a plane). Any point $\boldsymbol{p}$ on this plane can be represented sum of these points

$$\boldsymbol{p} = s_1\boldsymbol{x} + s_2\boldsymbol{y} + s_3\boldsymbol{z},$$

where the scalars s_i are known as the barycentric coordinates. Compute the scalars s_i in terms of ratios of bivectors which represent areas in the plane for the case when $\boldsymbol{p}$ lies inside the triangle formed by $\boldsymbol{x}, \boldsymbol{y}, \boldsymbol{z}$. Explain this comment. Compute the barycentric coordinates of the gravity center.

7.20 Construct the dual representation of the mid-plane between points $\boldsymbol{x}$ and $\boldsymbol{y}$.

7.21 Flags: flags are very useful when we are interested in relating geometrical points, lines, and planes. Show in the degenerated algebra $G_{3,0,1}$ that the plane perpendicular to the line $\boldsymbol{L}$ passing through the point $\boldsymbol{P}$ is given by

$$\boldsymbol{\Pi}^{\perp} = \frac{1}{2}(\boldsymbol{P}\widetilde{\boldsymbol{L}} - \boldsymbol{L}\widetilde{\boldsymbol{P}}).$$

This equation is a kind of extension which relates the mapping of the flags

$$f_{PL} = \frac{1}{\sqrt{2}}(\boldsymbol{P} + \boldsymbol{L}) \rightarrow f_{P\Pi} = \frac{1}{\sqrt{2}}(\boldsymbol{P} + \boldsymbol{\Pi}^{\perp}).$$

The inverse of this mapping relates the plane to the line perpendicular to the plane passing through the point. That is,

$$\boldsymbol{L}^{\perp} = \frac{1}{2}(\boldsymbol{\Pi}\widetilde{\boldsymbol{P}} - \boldsymbol{P}\widetilde{\boldsymbol{\Pi}}).$$

Check that $(\boldsymbol{L}^{\perp})^{\perp} = \boldsymbol{L}$.

Chapter 8
Conformal Geometric Algebra

8.1 Introduction

The geometric algebra of a 3D Euclidean space $G_{3,0,0}$ has a point basis and the motor algebra $G_{3,0,1}$ a line basis. In the latter geometric algebra, the lines expressed in terms of Plücker coordinates can be used to represent points and planes as well. The reader can find a comparison of representations of points, lines, and planes using $G_{3,0,0}$ and $G_{3,0,1}$ in Chap. 7.

Interestingly enough, in the case of the conformal geometric algebra, we find that the unit element is the sphere, which allows us to represent the other geometric primitives in its terms. To see how this is possible, we begin by giving an introduction in conformal geometric algebra following the same formulation presented in [18, 185] and show how the Euclidean vector space $\mathbb{R}^n$ is represented in $\mathbb{R}^{n+1,1}$. Let $\{e_1, \ldots, e_n, e_+, e_-\}$ be a vector basis with the following properties:

$$e_i^2 = 1, \quad i = 1.., n; \tag{8.1}$$

$$e_\pm^2 = \pm 1, \tag{8.2}$$

$$e_i \cdot e_+ = e_i \cdot e_- = e_+ \cdot e_- = 0, \quad i = 1, \ldots, n. \tag{8.3}$$

Note that this basis is not written in bold. A *null basis* $\{e_0, e_\infty\}$ can be introduced by

$$e_0 = \frac{(e_- - e_+)}{2}, \tag{8.4}$$

$$e_\infty = e_- + e_+, \tag{8.5}$$

with the properties

$$e_0^2 = e_\infty^2 = 0, \quad e_\infty \cdot e_0 = -1. \tag{8.6}$$

A unit pseudoscalar $\boldsymbol{E} \in \mathbb{R}^{1,1}$ that represents the so-called Minkowski plane is defined by

E. Bayro-Corrochano, *Geometric Algebra Applications Vol. I*,
https://doi.org/10.1007/978-3-319-74830-6_8

$$E = e_\infty \wedge e_0 = e_+ \wedge e_- = e_+ e_-, \tag{8.7}$$

having the properties

$$E^2 = 1, \tag{8.8}$$
$$\widetilde{E} = -E, \tag{8.9}$$
$$Ee_\pm = e_\mp = -e_\pm, \tag{8.10}$$
$$Ee_\infty = -e_\infty E = -e_\infty, \quad Ee_0 = -e_0 E = e_0 \quad \text{(absorption)}, \tag{8.11}$$
$$1 - E = -e_\infty e_0, \quad 1 + E = -e_0 e_\infty. \tag{8.12}$$

The dual of E is given by

$$E^* = EI^{-1} = -E\widetilde{I}, \tag{8.13}$$

where I is the pseudoscalar for $\mathbb{R}^{n+1,1}$.

8.1.1 Conformal Split

Euclidean points $\mathbf{x}_e \in \mathbb{R}^n$ can be represented in $\mathbb{R}^{n+1,1}$ in a general way, as

$$x_c = \mathbf{x}_e + \alpha e_0 + \beta e_\infty, \tag{8.14}$$

where α and β are arbitrary scalars. A conformal point $x_c \in \mathbb{R}^{n+1,1}$ can be divided into its Euclidean and conformal parts by an operation called the *additive split*: $\mathbb{R}^{n,1} = \mathbb{R}^n \oplus \mathbb{R}^{1,1}$ [185]. This split is defined by the projection operators P_E (projection) and $P_E^\perp$ (rejection) as follows:

$$P_E(x_c) = (x_c \cdot E)E = \alpha e_0 + \beta e_\infty \in \mathbb{R}^{1,1}, \tag{8.15}$$
$$P_E^\perp(x_c) = (x_c \cdot E^*)\widetilde{E^*} = (x_c \wedge E)E = \mathbf{x}_e \in \mathbb{R}^n, \tag{8.16}$$
$$x_c = P_E(x_c) + P_E^\perp(x_c). \tag{8.17}$$

The names "projection" and "rejection" stem from the geometrical meaning of these operators. The first returns the component of x_c that is parallel to E by a projection (dot product). The latter produces the component of x_c that is orthogonal to E, hence the name (see Fig. 8.1).

Hestenes introduced earlier as the additive split [185] the *multiplicative split*: $\mathbb{R}^{n,1} = \mathbb{R}^n \otimes \mathbb{R}^{1,1}$ which relates conformal and vector space models (covariant Euclidean geometry) [134, 137].

For the case of $G_{4,1}$, the null vector e_0 is assigned to the origin, so that each point lies on the bundle of all lines crossing through the origin. G_3 is the geometry of that

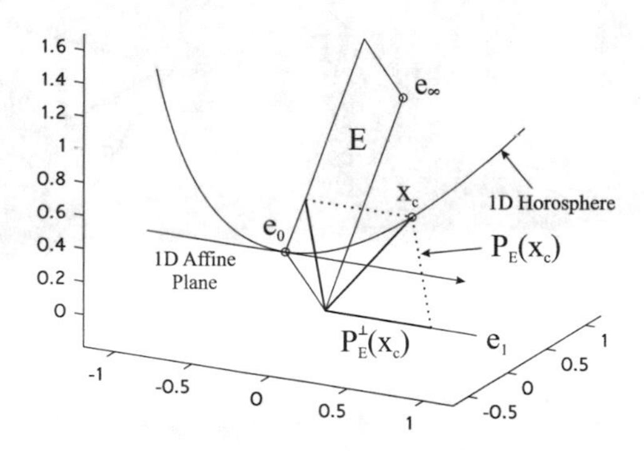

Fig. 8.1 Projection and rejection of vector $x_c \in \mathbb{R}^{2,1}$ from the $\boldsymbol{E}$ plane. The operators are illustrated for the 1D case

bundle, and each point is represented by the vector $\boldsymbol{x} \in \mathbb{R}^3$. The generating basis vectors for G_3 are trivectors in $G_{4,1}$:

$$e_i = e_i \wedge e_\infty \wedge e_0 = e_i(e_\infty \wedge e_0) = e_i \boldsymbol{E} = \boldsymbol{E} e_i, \tag{8.18}$$

$$i = e_1 e_2 e_3 = (e_1 \boldsymbol{E})(e_2 \boldsymbol{E})(e_3 \boldsymbol{E}) = (e_1 e_2 e_3)\boldsymbol{E} = I_c, \tag{8.19}$$

and note that the pseudoscalar is invariant, i.e., $i = I_c$, where $i^2 = -1$. By the case of additive split, the vector basis of $e_1, e_2, e_3 \in G_3$ is not associated with lines or the origin e_0 and its pseudoscalar $I_3 = e_1 e_2 e_3$ is not invariant.

8.1.2 Conformal Splits for Points and Simplexes

We will next give the expressions using conformal splits of the basic geometric entities depicted in Fig. 8.2. The point is given by:

$$\boldsymbol{x} = \left(\mathbf{x} + \frac{1}{2}\mathbf{x}^2 e_\infty + e_0\right)\boldsymbol{E} = \boldsymbol{E}\left(\mathbf{x} - \frac{1}{2}\mathbf{x}^2 e_\infty - e_0\right) = \mathbf{x}\boldsymbol{E} + \frac{1}{2}\mathbf{x}^2 e_\infty - e_0, \tag{8.20}$$

and this fulfills

$$\mathbf{x} = \boldsymbol{x} \wedge e_0 \wedge e_\infty = \boldsymbol{x} \wedge \boldsymbol{E}. \tag{8.21}$$

The line or *spear* is given by:

$$\boldsymbol{L} = \boldsymbol{x} \wedge \boldsymbol{p} \wedge e_\infty = \mathbf{x} \wedge \mathbf{p} e_\infty + (\mathbf{p} - \mathbf{x}) = (\mathbf{d} e_\infty + 1)\mathbf{n}, \tag{8.22}$$

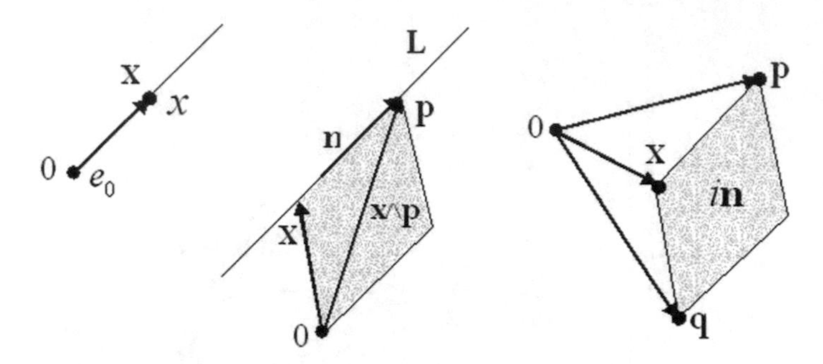

Fig. 8.2 Conformal multiplicative splits: point, line, and a plane

where the Plücker coordinates are given by $\mathbf{n} = (\mathbf{p} - \mathbf{x})$ (tangent), $\mathbf{x} \wedge \mathbf{p} = \mathbf{x} \wedge (\mathbf{p} - \mathbf{x}) = \mathbf{d}\mathbf{n}$ (moment), and $\mathbf{d} = (\mathbf{x} \wedge \mathbf{p})\mathbf{n}^{-1} = \mathbf{x} - (\mathbf{x} \cdot \mathbf{n}^{-1})\mathbf{n}$ (directance). The plane is given by:

$$\boldsymbol{P} = \boldsymbol{x} \wedge \boldsymbol{p} \wedge \boldsymbol{q} \wedge e_{\infty} = \mathbf{x} \wedge \mathbf{p} \wedge \mathbf{q} e_{\infty} + (\mathbf{p} - \mathbf{x}) \wedge (\mathbf{q} - \mathbf{x}) \boldsymbol{E}, \tag{8.23}$$

where the tangent is $(\mathbf{p} - \mathbf{x}) \wedge (\mathbf{q} - \mathbf{x}) = \mathbf{x} \wedge \mathbf{p} + \mathbf{p} \wedge \mathbf{q} + \mathbf{q} \wedge \mathbf{x} = i\mathbf{n}$ and the moment $\mathbf{x} \wedge \mathbf{p} \wedge \mathbf{q} = \mathbf{x} \wedge [(\mathbf{p} - \mathbf{x}) \wedge (\mathbf{q} - \mathbf{x})] = \mathbf{x} \wedge (i\mathbf{n}) = i(\mathbf{x} \cdot \mathbf{n})$.

The dual form of the plane is given by:

$$\boldsymbol{P} = i(\mathbf{x} \cdot \mathbf{n} e_{\infty} + \mathbf{n} \boldsymbol{E}) = i\boldsymbol{n}, \tag{8.24}$$

where

$$\begin{aligned} \boldsymbol{n} = \boldsymbol{x}_2 - \boldsymbol{x}_1 &= (\mathbf{x}_2 - \mathbf{x}_1)\boldsymbol{E} + \frac{1}{2}(\mathbf{x}_2^2 - \mathbf{x}_1^2) e_{\infty}, \\ &= (\mathbf{x}_2 - \mathbf{x}_1)\boldsymbol{E} + \frac{1}{2}(\mathbf{x}_2 + \mathbf{x}_1) \cdot (\mathbf{x}_2 - \mathbf{x}_1) e_{\infty} = \mathbf{n}\boldsymbol{E} + \mathbf{c} \cdot \mathbf{n} e_{\infty}. \end{aligned} \tag{8.25}$$

In this book, we have used more additive split than the multiplicative split; however, the use of the multiplicative split and its representations is matter of our future works.

8.1.3 *Euclidean and Conformal Spaces*

One of the results of the non-Euclidean geometry demonstrated by Nikolai Lobachevsky in the nineteenth century is that in spaces with hyperbolic structure we can find subsets that are isomorphic to a Euclidean space. In order to do this, Lobachevsky introduced two constraints, to what we now call *conformal point* $\boldsymbol{x}_c \in \mathbb{R}^{n+1,1}$; see Fig. 8.3.

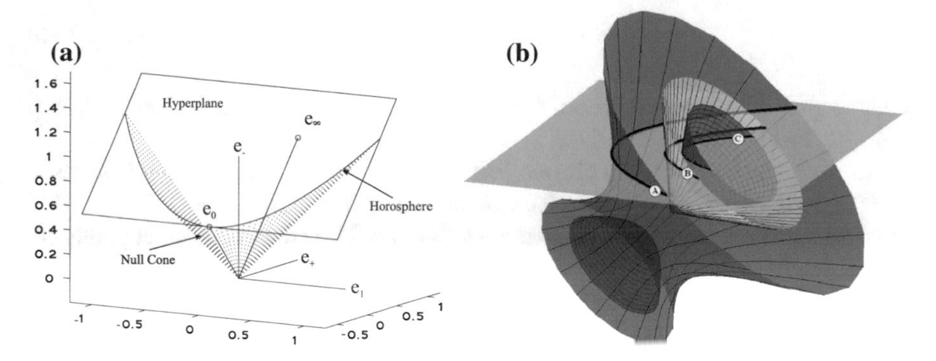

Fig. 8.3 **a** (left) Null cone (dotted lines), the hyperplane $\mathbb{P}+(e_\infty, e_0)$, and the horosphere for 1D. Note that even though the normal of the hyperplane is e_∞, the plane is actually *geometrically parallel* to this vector. **b** (right) Surface levels A, B, and C denoting spheres with a radius positive, zero (null cone), and negative, respectively

The first constraint is the *homogeneous* representation, *normalizing* the vector $\boldsymbol{x}_c$ such that

$$\boldsymbol{x}_c \cdot e_\infty = -1, \tag{8.26}$$

and the second constraint is that the vector must be a *null vector*, that is,

$$\boldsymbol{x}_c^2 = 0. \tag{8.27}$$

Now, recall that a hyperplane $\mathbb{P}(\boldsymbol{n}, \boldsymbol{a}) \in \mathbb{R}^{n+1,1}$ with normal $\boldsymbol{n}$ and passing through the point $\boldsymbol{a}$ is the solution to the equation

$$\boldsymbol{n} \cdot (\boldsymbol{x} - \boldsymbol{a}) = 0, \ \boldsymbol{x} \in \mathbb{R}^{n+1,1}. \tag{8.28}$$

The normalization condition $\boldsymbol{x}_c \cdot e_\infty = e_\infty \cdot e_0 = -1$ is equivalent to the equation

$$e_\infty \cdot (\boldsymbol{x}_c - e_0) = 0, \tag{8.29}$$

which is the equation of a hyperplane $\mathbb{P}(e_\infty, e_0)$. Thus, the normalization condition of Eq. (8.26) constrains the points $\boldsymbol{x}_c$ to lie in a hyperplane passing though e_0 with normal e_∞. Equation (8.26) fixes the scale; however, for the conformal model, another constraint is needed to fix $\boldsymbol{x}_c$ as a unique representation of $\mathbf{x}_e \in \mathbb{R}^n$.

Note that the inner product of two conformal points

$$\begin{aligned}\boldsymbol{x} \cdot \boldsymbol{y} &= \left(\mathbf{x} + \frac{1}{2}\mathbf{x}^2 e_\infty + e_0\right) \cdot \left(\mathbf{y} + \frac{1}{2}\mathbf{y}^2 e_\infty + e_0\right), \\ &= \mathbf{x} \cdot \mathbf{y} - \frac{1}{2}\mathbf{x}^2 - \frac{1}{2}\mathbf{y}^2 = -\frac{1}{2}(\mathbf{x} - \mathbf{y})^2,\end{aligned}$$

$$= -\frac{1}{2}\|\mathbf{x} - \mathbf{y}\|^2, \tag{8.30}$$

yields a quadratic representation of the Euclidean distance between the two Euclidean points: $\mathbf{x}$ and $\mathbf{y}$.

To complete the definition of *generalized homogeneous coordinates* for points in $\mathbb{R}^{n+1,1}$, we resort therefore to the second constraint of Eq. (8.27) $\boldsymbol{x}_c^2 = 0$. The set $\mathbb{N}^{n+1}$ of vectors that square to zero is called the *null cone*. Therefore, conformal points are required to lie in the intersection of the null cone $\mathbb{N}^{n+1}$ with the hyperplane $\mathbb{P}(e_\infty, e_0)$. The resulting surface $\mathbb{N}_e^n$ is called the *horosphere*:

$$\mathbb{N}_e^n = \mathbb{N}^{n+1} \cap \mathbb{P}(e_\infty, e_0) = \{\boldsymbol{x}_c \in \mathbb{R}^{n+1,1} | \boldsymbol{x}_c^2 = 0, \boldsymbol{x}_c \cdot e_\infty = -1\}. \tag{8.31}$$

The homogeneous model horosphere has its origins in the work of F. A: Wachter (1792–1817) a student of Gauss [79].

An illustration of the null cone, the hyperplane, and the horosphere can be seen in Fig. 8.3a. By values of $\boldsymbol{x}_c^2$ that are positive, zero (null cone), and negative, three families of surfaces are obtained as shown in Fig. 8.3b.

The constraints (8.26) and (8.27) now included in (8.31) define an isomorphic mapping between the Euclidean space and the conformal space. Thus, for each conformal point $\boldsymbol{x}_c \in \mathbb{R}^{n+1,1}$ there are a unique Euclidean space point $\mathbf{x}_e \in \mathbb{R}^n$ and unique scalars α, β such that the following mapping is bijective:

$$\mathbf{x}_e \mapsto \boldsymbol{x}_c = \mathbf{x}_e + \alpha e_0 + \beta e_\infty. \tag{8.32}$$

To see how this mapping is obtained, first we see that any point $\boldsymbol{x}_c = \mathbf{x}_e + \alpha e_0 + \beta e_\infty \in \mathbb{N}_e^n$ can be expressed as $\boldsymbol{x}_c = \mathbf{x}_e + k_1 e_+ + k_2 e_-$, for some scalars k_1, k_2, since e_0 and e_∞ are linear combinations of the basis vectors e_+ and e_-. Since $\boldsymbol{E}^2 = 1$, we can apply the conformal split to x_c, and we get

$$\boldsymbol{x}_c = \boldsymbol{x}_c \boldsymbol{E}^2 = (\boldsymbol{x}_c \wedge \boldsymbol{E} + \boldsymbol{x}_c \cdot \boldsymbol{E})\boldsymbol{E} = (\boldsymbol{x}_c \wedge \boldsymbol{E})\boldsymbol{E} + (\boldsymbol{x}_c \cdot \boldsymbol{E})\boldsymbol{E}, \tag{8.33}$$

see Fig. 8.3. Now, recall that $(\boldsymbol{x}_c \wedge \boldsymbol{E})\boldsymbol{E} = \mathbf{x}_e$ is the rejection (see Eq. (8.17)). The expression $(x_c \cdot \boldsymbol{E})\boldsymbol{E}$ can be expanded as

$$(\boldsymbol{x}_c \cdot \boldsymbol{E})\boldsymbol{E} = (\boldsymbol{x}_c \cdot (e_\infty \wedge e_0))\boldsymbol{E} = e_0 + (k_1 + k_2)e_\infty. \tag{8.34}$$

Now, applying the condition that $\boldsymbol{x}_c^2 = 0$, we find from Eq. (8.33) that

$$\begin{aligned} \boldsymbol{x}_c^2 &= ((\boldsymbol{x}_c \wedge \boldsymbol{E})\boldsymbol{E} + (\boldsymbol{x}_c \cdot \boldsymbol{E})\boldsymbol{E})^2, \\ 0 &= (\mathbf{x}_e + e_0 + (k_1 + k_2)e_\infty)^2 = \mathbf{x}_e^2 - (k_1 + k_2), \\ \mathbf{x}_e^2 &= (k_1 + k_2). \end{aligned} \tag{8.35}$$

Finally, using Eq. (8.33), and substituting Eq. (8.35) into (8.34), we get

$$\boldsymbol{x}_c = (\boldsymbol{x}_c \wedge \boldsymbol{E})\boldsymbol{E} + (\boldsymbol{x}_c \cdot \boldsymbol{E})\boldsymbol{E} = \mathbf{x}_e + e_0 + \frac{1}{2}(k_1 + k_2)e_\infty = \mathbf{x}_e + \frac{1}{2}\mathbf{x}_e^2 e_\infty + e_0. \tag{8.36}$$

We can gain further insight into the geometrical meaning of the null vectors by analyzing Eq. (8.36). For instance, by setting $\mathbf{x}_e = 0$, we find that e_0 represents the origin of $\mathbb{R}^n$ (hence the name). Similarly, dividing this equation by $\boldsymbol{x}_c \cdot e_0 = -\frac{1}{2}\mathbf{x}_e^2$ gives

$$\begin{aligned}\frac{\boldsymbol{x}_c}{\boldsymbol{x}_c \cdot e_0} &= -\frac{2}{\mathbf{x}_e^2}\left(\mathbf{x}_e + \frac{1}{2}\mathbf{x}_e^2 e_\infty + e_0\right) = -\frac{2\mathbf{x}_e^2}{\mathbf{x}_e^2}\left(\frac{1}{\mathbf{x}_e} + \frac{1}{2}e_\infty + \frac{e_0}{\mathbf{x}_e^2}\right),\\ &= -2\left(\frac{1}{\mathbf{x}_e} + \frac{1}{2}e_\infty + \frac{e_0}{\mathbf{x}_e^2}\right) \underset{\mathbf{x}_e \to \infty}{\longrightarrow} e_\infty.\end{aligned} \tag{8.37}$$

Thus, we conclude that e_∞ represents the point at infinity.

8.1.4 Stereographic Projection

Conformal geometry is equivalent to stereographic projection in Euclidean space. Generally speaking, a stereographic projection is a mapping, taking points lying on a hypersphere to points lying on a hyperplane and following a simple geometric construction. It is well known that this projection is used in cartography to make maps of the earth; see Fig. 8.4. In that case, the projection plane passes through the equator and the sphere is centered at the origin. To make a projection, a line is drawn from the north pole to each point on the sphere, and the intersection of this line with the projection plane constitutes the stereographic projection.

Next, we will illustrate the equivalence between stereographic projection and conformal geometric algebra in $\mathbb{R}^1$. We will be working in $\mathbb{R}^{2,1}$, with the basis vectors $\{e_1, e_+, e_-\}$ having the usual properties. The projection plane will be the x-axis, and the sphere will be a circle centered at the origin with unitary radius.

Given a scalar $\mathbf{x}_e$ representing a point on the x-axis, we wish to find the point $\boldsymbol{x}_c$ lying on the circle that projects to it (see Fig. 8.4b). The equation of the line passing through the north pole and $\mathbf{x}_e$ is given by $f(\mathbf{x}) = -\frac{1}{\mathbf{x}_e}\mathbf{x} + 1$. The equation of the circle is $\mathbf{x}^2 + f(\mathbf{x})^2 = 1$. Substituting the equation of the line on the circle, we get $\mathbf{x}^2 - 2\mathbf{x}\mathbf{x}_e + \mathbf{x}^2\mathbf{x}_e^2 = 0$, which has the two solutions $\mathbf{x} = 0, \quad \mathbf{x} = 2\frac{\mathbf{x}_e}{\mathbf{x}_e^2+1}$. Only the latter solution is meaningful. Substituting in the equation of the line, we get $f(\mathbf{x}) = \frac{\mathbf{x}_e^2-1}{\mathbf{x}_e^2+1}$. Hence, $\boldsymbol{x}_c$ has coordinates $\boldsymbol{x}_c = \left(2\frac{\mathbf{x}_e}{\mathbf{x}_e^2+1}, \frac{\mathbf{x}_e^2-1}{\mathbf{x}_e^2+1}\right)$, which can be represented in homogeneous coordinates as the vector

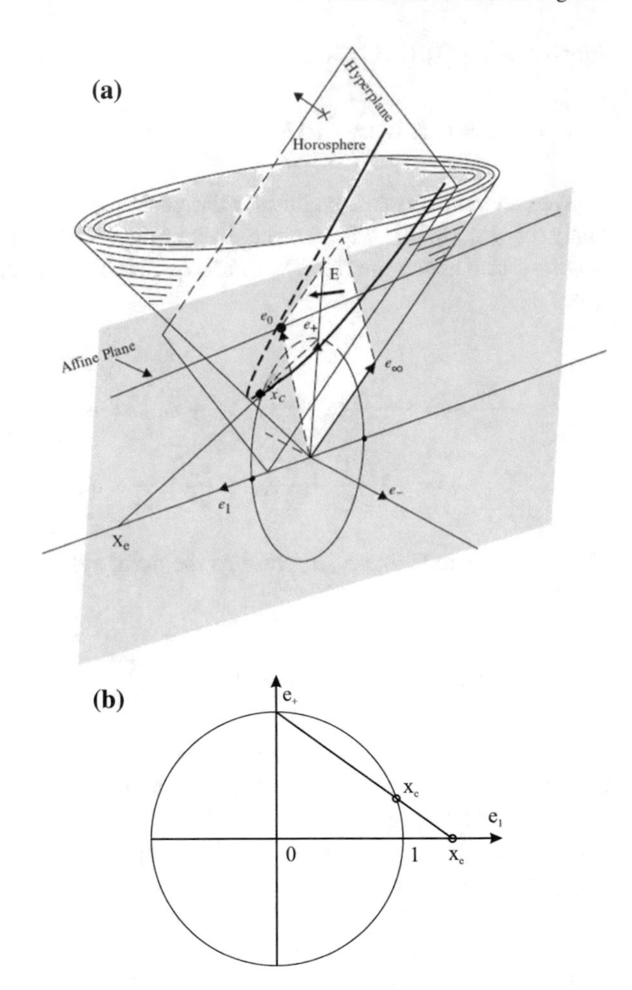

Fig. 8.4 **a** Null cone, hyperplane, horosphere, and affine plane for the 1D case. **b** Stereographic projection for the 1D case

$$\boldsymbol{x}_c = 2\frac{\mathbf{x}_e}{\mathbf{x}_e^2 + 1}e_1 + \frac{\mathbf{x}_e^2 - 1}{\mathbf{x}_e^2 + 1}e_+ + e_-. \tag{8.38}$$

If we take the limits to Eq. (8.38), we get the expected points at infinity and the origin of $\mathbb{R}^n$:

$$\lim_{\mathbf{x}_e \to \infty} \boldsymbol{x}_c = e_+ + e_- = e_\infty, \tag{8.39}$$

$$\lim_{\mathbf{x}_e \to 0} \frac{\boldsymbol{x}_c}{2} = \frac{e_- - e_+}{2} = e_0.$$

This result is a first confirmation that stereographic projection is equivalent to a conformal mapping given by Eq. (8.36). For a second proof, we note that Eq. (8.38) can be rewritten as

$$
\begin{aligned}
x_c &= 2\frac{\mathbf{x}_e}{\mathbf{x}_e^2+1}e_1 + \frac{\mathbf{x}_e^2-1}{\mathbf{x}_e^2+1}e_+ + e_-, \\
&= \frac{2}{\mathbf{x}_e^2+1}\left(\mathbf{x}_e e_1 + \frac{1}{2}(\mathbf{x}_e^2-1)e_+ + \frac{1}{2}(\mathbf{x}_e^2+1)e_-\right).
\end{aligned} \tag{8.40}
$$

Dividing by the scale factor $\frac{2}{\mathbf{x}_e^2+1}$ in order to achieve the constraint imposed by Eq. (8.26)—$x_c \cdot e_\infty = -1$—we arrive at

$$
\begin{aligned}
x_c &= \mathbf{x}_e e_1 + (\mathbf{x}_e^2-1)e_+ + (\mathbf{x}_e^2+1)e_- = \mathbf{x}_e e_1 + \frac{1}{2}\mathbf{x}_e^2 e_\infty + e_0, \\
&= \mathbf{x}_e + \frac{1}{2}\mathbf{x}_e^2 e_\infty + e_0,
\end{aligned} \tag{8.41}
$$

where $\mathbf{x}_e = x_e e_1$, which is precisely Eq. (8.36). Hence, we have demonstrated that conformal geometric algebra is projectively equivalent to a stereographic projection (i.e., *up to a scale factor*).

8.1.5 Inner and Outer Product Null Spaces

For the next definitions, we adopt the classification of geometric entities introduced by Ch. Perwass [233]. We will consider that given a vector $\boldsymbol{x}$ if its inner or wedge product with a blade $\boldsymbol{X}_k$ is zero, we can consider that $\boldsymbol{x}$ with respect to these operations belongs to the null space of the blade $\boldsymbol{X}_k$; see more details of this definition in [232, 233].

Thus, the *Inner Product Null Space* (IPNS) of a blade $\boldsymbol{X}_k \in G_{p,q}^k$, denoted by $\mathbb{NI}(\boldsymbol{X}_k)$, is defined as

$$
\mathbb{NI}(\boldsymbol{X}_k) := \{\boldsymbol{x} \in G_{p,q}^1 : \boldsymbol{x} \cdot \boldsymbol{X}_k = 0\}, \tag{8.42}
$$

regardless if the blade is a null space or not.

Thinking in the dual space, we can define the *Outer Product Null Space* (OPNS) of a blade $\boldsymbol{X}_k \in G_{p,q}^k$, denoted by $\mathbb{NO}(\boldsymbol{X}_k)$, as follows

$$
\mathbb{NO}(\boldsymbol{X}_k) := \{\boldsymbol{x} \in G_{p,q}^1 : \boldsymbol{x} \wedge \boldsymbol{X}_k = 0\}, \tag{8.43}
$$

regardless if the blade is a null space or not.

The dual operation on blades can be best seen considering the dual relation between the IPNS and OPNS of the blade. Let $\boldsymbol{x} \in G_{p,q}^1$, $\boldsymbol{X}_k \in G_{p,q}^k$ with $k \geq 1$ and the pseudoscalar I of $G_{p,q}$, and according to equations given in Sect. 2.4.2, we can formulate the dual relationship as follows

$$
(\boldsymbol{x} \wedge \boldsymbol{X}_k)^* = (\boldsymbol{x} \wedge \boldsymbol{X}_k) \cdot I^{-1} = \boldsymbol{x} \cdot \boldsymbol{X}_k^*, \tag{8.44}
$$

hence,

$$\boldsymbol{x} \wedge \boldsymbol{X}_k = 0 \iff \boldsymbol{x} \cdot \boldsymbol{X}_k^* = 0, \tag{8.45}$$

and thus,

$$\mathbb{NO}(\boldsymbol{X}_k) = \mathbb{NI}(\boldsymbol{X}_k). \tag{8.46}$$

8.1.6 Spheres and Planes

The equation of a sphere of radius ρ centered at point $\mathbf{c}_e \in \mathbb{R}^n$ can be written as

$$(\mathbf{x}_e - \mathbf{c}_e)^2 = \rho^2. \tag{8.47}$$

Since $\boldsymbol{x}_c \cdot \boldsymbol{y}_c = -\frac{1}{2}(\mathbf{x}_e - \mathbf{y}_e)^2$, we can rewrite the above formula in terms of homogeneous coordinates as

$$\boldsymbol{x}_c \cdot \boldsymbol{c}_c = -\frac{1}{2}\rho^2. \tag{8.48}$$

Since $\boldsymbol{x}_c \cdot e_\infty = -1$, we can factor the above expression and then

$$\boldsymbol{x}_c \cdot \left(\boldsymbol{c}_c - \frac{1}{2}\rho^2 e_\infty\right) = 0, \tag{8.49}$$

and this equation corresponds to the IPNS representation and yields finally the simplified equation for the sphere as

$$\boldsymbol{x}_c \cdot \boldsymbol{s} = 0, \tag{8.50}$$

where

$$\boldsymbol{s} = \boldsymbol{c}_c - \frac{1}{2}\rho^2 e_\infty = \mathbf{c}_e + e_0 + \frac{\mathbf{c}_e^2 - \rho^2}{2} e_\infty \tag{8.51}$$

is the equation of the sphere. From this equation and (8.36), we can see that a conformal point is just a sphere with zero radius. The vector s has the properties

$$\boldsymbol{s}^2 = \rho^2 > 0, \tag{8.52}$$

$$e_\infty \cdot \boldsymbol{s} = -1. \tag{8.53}$$

From these properties, we conclude that the sphere $\boldsymbol{s}$ is a point lying on the hyperplane $\boldsymbol{x}_c \cdot e_\infty = -1$, but *outside* the null cone $\boldsymbol{x}_c^2 = 0$. In particular, all points on the hyperplane outside the horosphere determine spheres with positive radius, points lying on the horosphere define spheres of zero radius (i.e., points), and points lying inside the horosphere have imaginary radius. Finally, note that spheres of the same radius form a surface that is parallel to the horosphere.

Alternatively, spheres can be dualized and represented as $(n+1)$-vectors $s^* = sI^{-1}$. Then using the *main conjugation* $\widetilde{I}$ of I defined as

$$\widetilde{I} = (-1)^{\frac{1}{2}(n+2)(n+1)} I = -I^{-1}, \tag{8.54}$$

we can express the constraints of Eqs. (8.52) and (8.53) as

$$\begin{aligned} s^2 &= -\widetilde{s^*} s^* = \rho^2, \\ e_\infty \cdot s &= e_\infty \cdot (s^* I) = (e_\infty \wedge s^*) I = -1. \end{aligned} \tag{8.55}$$

Similar to Eq. (8.50), the equation involving a dual sphere reads

$$\boldsymbol{x}_c \wedge \boldsymbol{s}^* = 0. \tag{8.56}$$

The advantage of the dual form is that the sphere can be directly computed from four points (in 3D) yielding the OPNS representation for the sphere

$$\boldsymbol{s}^* = \boldsymbol{x}_{c_1} \wedge \boldsymbol{x}_{c_2} \wedge \boldsymbol{x}_{c_3} \wedge \boldsymbol{x}_{c_4}. \tag{8.57}$$

If we replace one of these points for the point at infinity, we get

$$\pi^* = \boldsymbol{x}_{c_1} \wedge \boldsymbol{x}_{c_2} \wedge \boldsymbol{x}_{c_3} \wedge e_\infty. \tag{8.58}$$

In standard IPNS form, π is given by

$$\pi = I\pi^* = n + d e_\infty, \tag{8.59}$$

where n is the normal vector and d represents the Hesse distance for the 3D space.

Developing the products, we get the OPNS representation for the plane

$$\begin{aligned} \pi^* &= \boldsymbol{x}_{c_3} \wedge \boldsymbol{x}_{c_1} \wedge \boldsymbol{x}_{c_2} \wedge e_\infty \\ &= \mathbf{x}_{e_3} \wedge \mathbf{x}_{e_1} \wedge \mathbf{x}_{e_2} \wedge e_\infty + ((\mathbf{x}_{e_3} - \mathbf{x}_{e_1}) \wedge (\mathbf{x}_{e_2} - \mathbf{x}_{e_1}))\boldsymbol{E}, \end{aligned} \tag{8.60}$$

which is the equation of the plane passing through the points $\mathbf{x}_{e_1}$, $\mathbf{x}_{e_2}$, and $\mathbf{x}_{e_3}$. We can easily see that $\mathbf{x}_{e_1} \wedge \mathbf{x}_{e_2} \wedge \mathbf{x}_{e_3}$ is a pseudoscalar representing the volume of the parallelepiped with sides $\mathbf{x}_{e_1}$, $\mathbf{x}_{e_2}$, and $\mathbf{x}_{e_3}$. Also, since $(\mathbf{x}_{e_1} - \mathbf{x}_{e_2})$ and $(\mathbf{x}_{e_3} - \mathbf{x}_{e_2})$ are two vectors on the plane, the expression $((\mathbf{x}_{e_1} - \mathbf{x}_{e_2}) \wedge (\mathbf{x}_{e_3} - \mathbf{x}_{e_2}))^*$ is the normal to the plane. Therefore, planes are spheres passing through the point at infinity.

8.1.7 Geometric Identities, Dual, Meet, and Join Operations

A circle z can be regarded as the intersection of two spheres s_1 and s_2. This means that for each point on the circle $x_c \in z$ they lie on both spheres, that is, $x_c \in s_1$ and $x_c \in s_2$. Assuming that s_1 and s_2 are linearly independent, we can write for $x_c \in z$

$$(x_c \cdot s_1)s_2 - (x_c \cdot s_2)s_1 = x_c \cdot (s_1 \wedge s_2) = x_c \cdot z = 0. \tag{8.61}$$

This result tells us that since x_c lies on both spheres, $z = (s_1 \wedge s_2)$ should be the intersection of the spheres or a circle. It is easy to see that the intersection with a third sphere leads to a point pair. We have derived algebraically that the wedge of two linearly independent spheres yields to their intersecting circle (see Fig. 8.5b). This topological relationship between two spheres can be also conveniently described using the dual of the meet operation, namely

$$z = (z^*)^* = (s_1^* \vee s_2^*)^* = s_1 \wedge s_2. \tag{8.62}$$

This new equation says that the dual of a circle can be computed via the meet of two spheres in their dual form. This equation confirms geometrically our previous algebraic computation of Eq. (8.61).

The standard OPNS (dual) form of the circle (in 3D) can be expressed by three points lying on it as

$$z^* = x_{c_1} \wedge x_{c_2} \wedge x_{c_3}, \tag{8.63}$$

see Fig. 8.5a. Similar to the case of planes shown in Eq. (8.58), lines can be defined by circles passing through the point at infinity as

$$l^* = x_{c_1} \wedge x_{c_2} \wedge e_\infty. \tag{8.64}$$

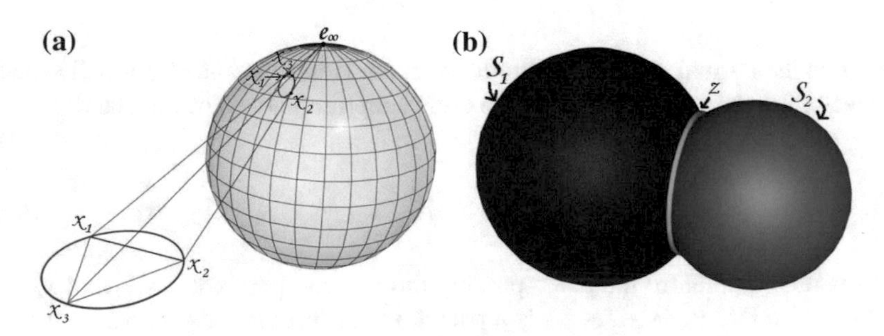

Fig. 8.5 **a** (*left*) Circle computed using three points, note its stereographic projection. **b** (*right*) Circle computed using the meet of two spheres

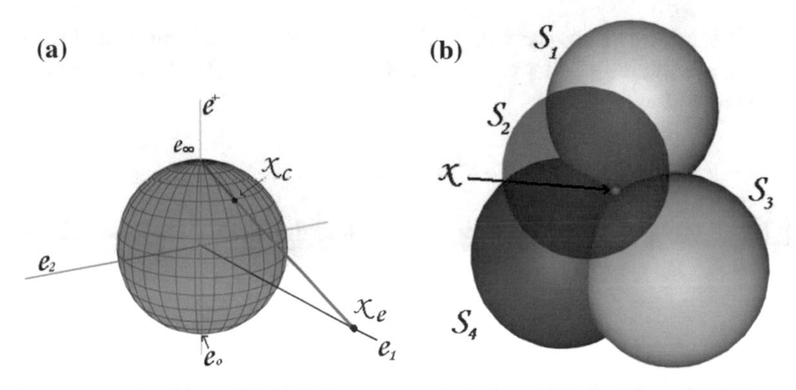

Fig. 8.6 **a** (*left*) Conformal point generated by projecting a point of the affine plane to the unit sphere. Note that we use only the Riemann sphere and the 2D plane. **b** (*right*) Point generated by the meet of four spheres

This can be demonstrated by developing the wedge products as in the case of the planes to yield the OPNS representation of the line:

$$\boldsymbol{x}_{c_1} \wedge \boldsymbol{x}_{c_2} \wedge e_\infty = \mathbf{x}_{e_1} \wedge \mathbf{x}_{e_2} \wedge e_\infty + (\mathbf{x}_{e_2} - \mathbf{x}_{e_1}) \wedge \boldsymbol{E}, \tag{8.65}$$

from where it is evident that the expression $\mathbf{x}_{e_1} \wedge \mathbf{x}_{e_2}$ is a bivector representing the plane where the line is contained and $(\mathbf{x}_{e_2} - \mathbf{x}_{e_1})$ is the direction of the line.

The standard IPNS form of the line can be expressed as

$$L = \boldsymbol{n}\boldsymbol{I}_e - e_\infty \boldsymbol{m}\boldsymbol{I}_e, \tag{8.66}$$

where $\boldsymbol{n}$ and $\boldsymbol{m}$ stand for the line orientation and moment, respectively. The line in the IPNS standard form is a bivector representing the six Plücker coordinates.

The dual of a point $\boldsymbol{x}$ is a sphere $\boldsymbol{s}$. The intersection of four spheres yields the OPNS equation for the point: See Fig. 8.6b. The dual relationships between a point and its dual, the sphere, are

$$\boldsymbol{s}^* = \boldsymbol{x}_{c_1} \wedge \boldsymbol{x}_{c_2} \wedge \boldsymbol{x}_{c_3} \wedge \boldsymbol{x}_{c_4} \leftrightarrow \boldsymbol{x}^* = s_1 \wedge s_2 \wedge s_3 \wedge s_4, \tag{8.67}$$

where the points are denoted as $\boldsymbol{x}_i$ and the spheres $\boldsymbol{s}_i$ for $i = 1, 2, 3, 4$.

There is another very useful relationship between an $(r-2)$-dimensional sphere $\boldsymbol{A}_r$ and the sphere $\boldsymbol{s}^*$ (computed as the dual of a point $\boldsymbol{s}$). If from the sphere $\boldsymbol{A}_r$ we can compute the hyperplane $\boldsymbol{A}_{r+1} \equiv \boldsymbol{e}_\infty \wedge \boldsymbol{A}_r \neq 0$, we can express the meet between the dual of the point $\boldsymbol{s}$ (a sphere) and the hyperplane $\boldsymbol{A}_{r+1}$ as getting the sphere $\boldsymbol{A}_r$ of one dimension lower:

$$(-1)^\epsilon \boldsymbol{s}^* \cap \boldsymbol{A}_{r+1} = (\boldsymbol{s}^* \boldsymbol{I}) \cdot \boldsymbol{A}_{r+1} = \boldsymbol{s}\boldsymbol{A}_{r+1} = \boldsymbol{A}_r. \tag{8.68}$$

This result is telling us an interesting relationship: that the sphere $\boldsymbol{A}_r$ and the hyperplane $\boldsymbol{A}_{r+1}$ are related via the point $\boldsymbol{s}$ (dual of the sphere $\boldsymbol{s}^*$); thus, we then rewrite Eq. (8.68) as follows:

$$\boldsymbol{s} = \boldsymbol{A}_r \boldsymbol{A}_{r+1}^{-1}. \tag{8.69}$$

Using Eq. (8.69) and given the plane π (A_{r+1}) and the circle z $(\boldsymbol{A}_r)$, we can compute the sphere

$$\boldsymbol{s} = z\pi^{-1}. \tag{8.70}$$

Similarly, we can compute another important geometric relationship called the *pair of points* using the Eq. (8.69) directly:

$$\boldsymbol{s} = \boldsymbol{PP}\boldsymbol{L}^{-1}. \tag{8.71}$$

Now, using this result given the line $\boldsymbol{L}$ and the sphere $\boldsymbol{s}$, we can compute the IPNS form of the pair of points $\boldsymbol{PP}$ (see Fig. 8.7b):

$$\boldsymbol{PP} = \boldsymbol{sL} = \boldsymbol{s} \wedge \boldsymbol{L}. \tag{8.72}$$

The OPNS of the pair of points is given by:

$$\boldsymbol{PP}^* = \boldsymbol{x}_{c_1} \wedge \boldsymbol{x}_{c_2}. \tag{8.73}$$

A summary of the standard IPNS and OPNS forms of the basic geometric entities is presented in Table 8.1.

Now, consider the following element in the conformal model

$$\boldsymbol{X} = \alpha(e_0 \wedge \boldsymbol{x}_1 \wedge \ldots \wedge \boldsymbol{x}_k \wedge e_\infty). \tag{8.74}$$

Since $\boldsymbol{x}_i \wedge \boldsymbol{x}_j = \boldsymbol{x}_i \wedge (\boldsymbol{x}_j - \boldsymbol{x}_i)$, the previous equation can be rewritten as follows

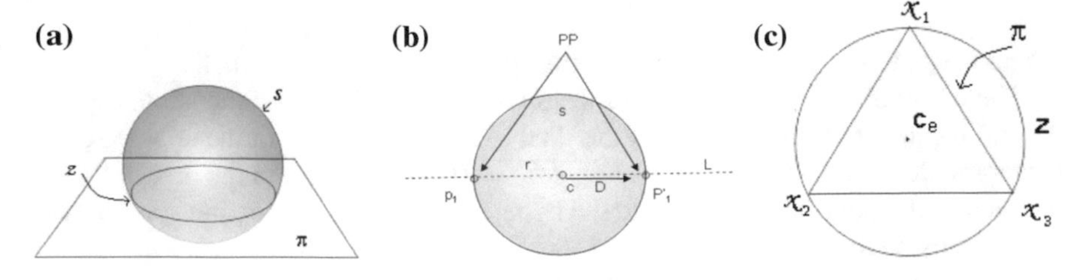

Fig. 8.7 **a** Meet of a sphere and a plane. **b** Pair of points resulting from the meet between a line and a sphere. **c** Center of a circumscribed triangle

Table 8.1 Representation of entities in conformal geometric algebra, G = grade

Entity	IPNS	G	OPNS (dual)	G
Sphere	$s = \mathbf{c} + \frac{1}{2}(\mathbf{c}^2 - \rho^2)e_\infty + e_0$	1	$s^* = x_{c_1} \wedge x_{c_2} \wedge x_{c_3} \wedge x_{c_4}$	4
Point	$x_c = \mathbf{x} + \frac{1}{2}\mathbf{x}^2 e_\infty + e_0$	1	$x^* = s_1 \wedge s_2 \wedge s_3 \wedge s_4$	4
Plane	$\pi = \boldsymbol{n} I_E - d e_\infty$ $\boldsymbol{n} = (\mathbf{x}_{e_1} - \mathbf{x}_{e_2}) \wedge (\mathbf{x}_{e_1} - \mathbf{x}_{e_3})$ $d = (\mathbf{x}_{e_1} \wedge \mathbf{x}_{e_2} \wedge \mathbf{x}_{e_3}) I_E$	1	$\pi^* = e_\infty \wedge x_{c_1} \wedge x_{c_2} \wedge x_{c_3}$	4
Line	$\boldsymbol{L} = \pi_1 \wedge \pi_2$ $\boldsymbol{L} = \boldsymbol{n} I_E - e_\infty \boldsymbol{m} I_E$ $\boldsymbol{n} = (\mathbf{x}_{e_1} - \mathbf{x}_{e_2})$ $\boldsymbol{m} = (\mathbf{x}_{e_1} \wedge \mathbf{x}_{e_2})$	2	$\boldsymbol{L}^* = e_\infty \wedge x_{c_1} \wedge x_{c_2}$	3
Circle	$z = s_1 \wedge s_2 = s_1 \wedge \pi_2$	2	$z^* = x_{c_1} \wedge x_{c_2} \wedge x_{c_3}$	3
Point	$\boldsymbol{PP} = s_1 \wedge s_2 \wedge s_3$	3	$\boldsymbol{PP}^* = x_{c_1} \wedge x_{c_2}$	2
Pair	$\boldsymbol{PP} = s \wedge \boldsymbol{L}$	2		

$$X = \alpha(e_0 \wedge (x_1 - e_0) \wedge \ldots \wedge (x_k - e_0) \wedge e_\infty). \tag{8.75}$$

Now, substituting the conformal representation of the Euclidean points

$$x_i = \mathbf{x}_i + \frac{1}{2}\mathbf{x}_i^2 e_\infty + e_0, \tag{8.76}$$

we obtain

$$X = \alpha\left(e_0 \wedge \left(\mathbf{x}_1 + \frac{1}{2}\mathbf{x}_1^2 e_\infty\right) \wedge \ldots \wedge \left(\mathbf{x}_k + \frac{1}{2}\mathbf{x}_k^2 e_\infty\right) \wedge e_\infty\right). \tag{8.77}$$

Since the wedge product eliminates the extra terms $\frac{1}{2}\mathbf{x}_i^2 e_\infty$, we get

$$X = \alpha(e_0 \wedge \mathbf{x}_1 \wedge \ldots \wedge \mathbf{x}_k \wedge e_\infty). \tag{8.78}$$

Thus, the part involving the vectors $\mathbf{x}_i$ and the weight α can be seen as a purely Euclidean k-blade $\mathbf{X}_k = \alpha \mathbf{x}_1 \wedge \ldots \wedge \mathbf{x}_k$; therefore, this class of blade in the conformal model is equivalent to:

$$X = e_0 \wedge \mathbf{X}_k \wedge e_\infty. \tag{8.79}$$

Finally, we can claim that the general form of a flat k-dimensional Euclidean subspace through the point $\boldsymbol{p}$ is given by:

$$\boldsymbol{p} \wedge \mathbf{X}_k \wedge e_\infty. \tag{8.80}$$

8.1.8 Simplexes and Spheres

In CGA, geometric objects can be computed as the wedge of linearly independent homogeneous points $\boldsymbol{a}_1, \boldsymbol{a}_2, \ldots, \boldsymbol{a}_r$, with $r \leq n$ so that $\boldsymbol{a}_1 \wedge \boldsymbol{a}_2 \wedge \ldots \wedge \boldsymbol{a}_r \neq 0$. This multivector can be expressed in an expanded form as follows:

$$\boldsymbol{a}_1 \wedge \boldsymbol{a}_2 \wedge \ldots \wedge \boldsymbol{a}_r = \mathbf{A}_r + e_0 \mathbf{A}_r^+ + \frac{1}{2} e_\infty \mathbf{A}_r^- - \frac{1}{2} \boldsymbol{E} \mathbf{A}_r^\pm, \tag{8.81}$$

where

$$\begin{aligned}
\mathbf{A}_r &= \mathbf{a}_0 \wedge \mathbf{a}_1 \wedge \ldots \wedge \mathbf{a}_r, \\
\mathbf{A}_r^+ &= \sum_{i=0}^{r} (-1)^i \mathbf{a}_0 \wedge \ldots \wedge \check{\mathbf{a}}_i \wedge j \ldots \wedge \mathbf{a}_r = \big(\mathbf{a}_1 - \mathbf{a}_0\big) \wedge \ldots \wedge \big(\mathbf{a}_r - \mathbf{a}_0\big), \\
\mathbf{A}_r^- &= \sum_{i=0}^{r} (-1)^i \mathbf{a}_i^2 \mathbf{a}_0 \wedge \ldots \wedge \check{\mathbf{a}}_i \wedge \ldots \wedge \mathbf{a}_r, \\
\mathbf{A}_r^\pm &= \sum_{i=0}^{r} \sum_{j=i+1}^{r} (-1)^{i+j} \big(\mathbf{a}_i^2 - \mathbf{a}_j^2\big) \mathbf{a}_0 \wedge \ldots \wedge \check{\mathbf{a}}_i \wedge \ldots \wedge \check{\mathbf{a}}_j \wedge \ldots \wedge \mathbf{a}_r.
\end{aligned} \tag{8.82}$$

The expanded form of Eq. (8.81) gives the following geometric information

- Determine an r-simplex if $\mathbf{A}_r \neq 0$.
- Represent an (r–1)-simplex in a plane which passes through the origin if $\mathbf{A}_r^+ = \mathbf{A}_r^- = 0$.
- Represent an $(r-1)$-sphere if and only if $\mathbf{A}_r^+ \neq 0$.

See [185] for a more detailed study on the expanded form.

In Eq. (8.81), $\mathbf{A}_r$ is the moment of the simplex with a *boundary* (or *tangent*) $\mathbf{A}_r^+$; thus, the corresponding r-simplex can be formulated as follows:

$$e \wedge a_0 \wedge a_1 \wedge \ldots \wedge a_r = e\mathbf{A}_r + \boldsymbol{E}\mathbf{A}_r^+. \tag{8.83}$$

On the other hand, the *volume* (or *content*) of the simplex is $k!|\mathbf{A}_r^+|$, where

$$\begin{aligned}
|\mathbf{A}_r^+|^2 = (\mathbf{A}_r^+)^\dagger \mathbf{A}_r^+ &= -\big(a_r \wedge \ldots \wedge a_0 \wedge e\big) \cdot \big(e \wedge a_0 \wedge\wedge \ldots \wedge a_r\big), \\
&= -(-\frac{1}{2})^r \begin{vmatrix} 0 & 1 & \cdots & 1 \\ 1 & & & \\ \vdots & & d_{ij}^2 & \\ 1 & & & \end{vmatrix},
\end{aligned} \tag{8.84}$$

where $d_{ij} = |\mathbf{a}_i - \mathbf{a}_j|$ is the distance between a pair of points. This determinate is called Cayley–Menger determinant. The directed distance from the origin in $\mathcal{R}^n$ to the plane of the simplex in terms of the points is given by

$$d = \mathbf{A}_r(\mathbf{A}_r^+)^{-1}, \tag{8.85}$$

thus, the square of its absolute value is:

$$|d|^2 = \frac{|\mathbf{A}_r|^2}{|\mathbf{A}_r^+|^2} = \frac{(\mathbf{a}_r \wedge \ldots \wedge \mathbf{a}_0) \cdot (\mathbf{a}_0 \wedge \ldots \wedge \mathbf{a}_r)}{(\bar{\mathbf{a}}_r \wedge \ldots \wedge \bar{\mathbf{a}}_1) \cdot (\bar{\mathbf{a}}_1 \wedge \ldots \wedge \bar{\mathbf{a}}_r)}, \tag{8.86}$$

where $\bar{\mathbf{a}}_i = \mathbf{a}_i - \mathbf{a}_0$ for $i = 1, \ldots, r$. In next section, we present the computing of the directed distance in the 3D affine plane.

8.2 The 3D Affine Plane

In the previous section, we described the general properties of the conformal framework. However, sometimes we would like to use only the projective plane of the conformal framework but not the null cone of this space. This will be the case when we use only rigid transformations, and then we will limit ourselves to the *affine plane*, which is an $n + 1$ dimensional subspace of the hyperplane of reference $\mathbb{P}(e_\infty, e_0)$.

We have chosen to work in the algebra $G_{4,1}$. Since we deal with homogeneous points, the particular choice of null vectors does not affect the properties of the conformal geometry. Points in the affine plane $\boldsymbol{x} \in \mathbb{R}^{4,1}$ are formed as follows:

$$\boldsymbol{x}^a = \mathbf{x}_e + e_0, \tag{8.87}$$

where $\mathbf{x}_e \in \mathbb{R}^3$. From this equation, we note that e_0 represents the origin (by setting $\mathbf{x}_e = 0$) and, similarly, e_∞ represents the point at infinity. Then, the normalization property is expressed as

$$e_\infty \cdot \boldsymbol{x}^a = -1. \tag{8.88}$$

In this framework, the conformal mapping equation is expressed as

$$\boldsymbol{x}_c = \mathbf{x}_e + \frac{1}{2}\mathbf{x}_e^2 e_\infty + e_0 = \boldsymbol{x}^a + \frac{1}{2}\mathbf{x}_e^2 e_\infty. \tag{8.89}$$

For the case when we will be working on the affine plane exclusively, we will be mainly concerned with a simplified version of the *rejection*. Noting that $\boldsymbol{E} = e_\infty \wedge e_0 = e_\infty \wedge e$, we write an equation for rejection as follows:

$$\begin{aligned} P_E^{\perp}(\boldsymbol{x}_c) &= (\boldsymbol{x}_c \wedge \boldsymbol{E})\boldsymbol{E} = (\boldsymbol{x}_c \wedge \boldsymbol{E}) \cdot \boldsymbol{E} = (e_\infty \wedge e_0) \cdot e_0 + (\boldsymbol{x}_c \wedge e_\infty) \cdot e_0, \\ \mathbf{x}_e &= -e_0 + (\boldsymbol{x}_c \wedge e_\infty) \cdot e_0. \end{aligned} \tag{8.90}$$

Now, since the points in the affine plane have the form $\boldsymbol{x}^a = \boldsymbol{x}_e + e_0$, we conclude that

$$\boldsymbol{x}^a = (\boldsymbol{x}_c \wedge e_\infty) \cdot e_0, \tag{8.91}$$

is the mapping from the horosphere to the affine plane.

8.2.1 Lines and Planes

The lines and planes in the affine plane are expressed in a similar fashion to their conformal counterparts as the *join* of two and three points, respectively:

$$\boldsymbol{L}^a = \boldsymbol{x}_1^a \wedge \boldsymbol{x}_2^a, \tag{8.92}$$

$$\Pi^a = \boldsymbol{x}_1^a \wedge \boldsymbol{x}_2^a \wedge \boldsymbol{x}_3^a. \tag{8.93}$$

Note that unlike their conformal counterparts, the line is a *bivector* and the plane is a *trivector*. As seen earlier, these equations produce a moment–direction representation; thus,

$$\boldsymbol{L}^a = e_\infty \mathbf{d} + \boldsymbol{B}, \tag{8.94}$$

where $\mathbf{d}$ is a vector representing the direction of the line and $\boldsymbol{B}$ is a bivector representing the moment of the line. Similarly, we have that

$$\Pi^a = e_\infty \mathbf{n} + \delta e_{123}, \tag{8.95}$$

where $\mathbf{n}$ is the normal vector to the plane and δ is a scalar representing the distance from the plane to the origin. Note that in any case, the direction and normal can be retrieved with $\mathbf{d} = e_\infty \cdot \boldsymbol{L}^a$ and $\mathbf{n} = e_\infty \cdot \Pi^a$, respectively.

In this framework, the intersection or *meet* has a simple expression, too. Let $\boldsymbol{A}^a = a_1^a \wedge \ldots \wedge a_r^a$ and $\boldsymbol{B}^a = b_1^a \wedge \ldots \wedge b_s^a$. Then, the meet is defined as

$$\boldsymbol{A}^a \cap \boldsymbol{B}^a = \boldsymbol{A}^a \cdot (\boldsymbol{B}^a \cdot \bar{I}_{\boldsymbol{A}^a \cup \boldsymbol{B}^a}), \tag{8.96}$$

where $\bar{I}_{\boldsymbol{A}^a \cup \boldsymbol{B}^a}$ is either $e_{12}e_\infty$, $e_{23}e_\infty$, $e_{31}e_\infty$ or $e_{123}e_\infty$, according to which basis vectors span the largest common space of $\boldsymbol{A}^a$ and $\boldsymbol{B}^a$.

8.2.2 Directed Distance

The so-called Hessian normal form is well known from vector analysis to be a convenient representation to specify lines and planes using their distance from the

Fig. 8.8 **a** Line in 2D affine space. **b** Plane in the 3D affine space (note that the 3D space is "lifted" by a null vector e)

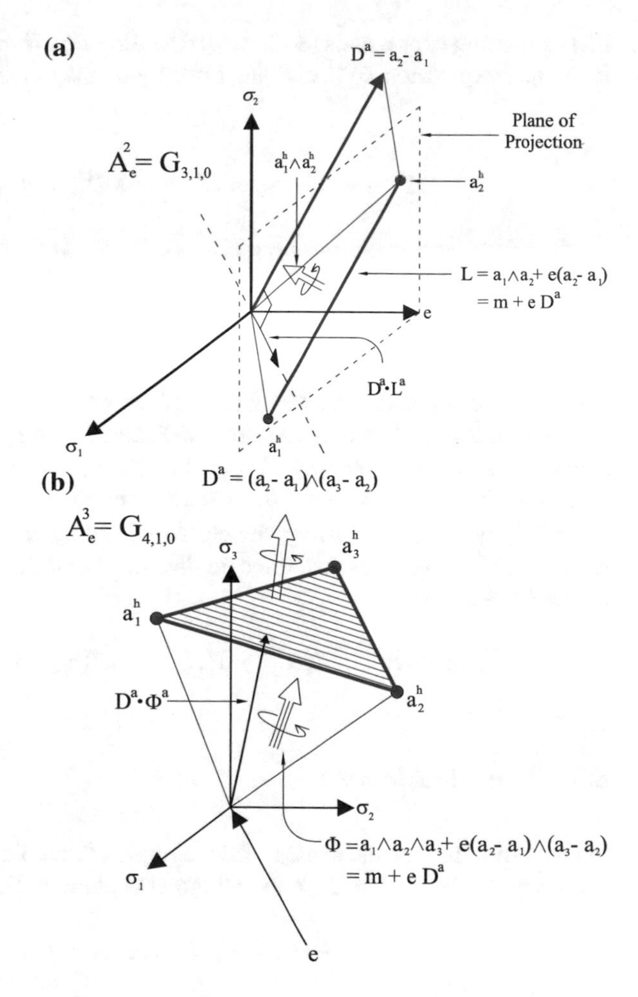

origin (the Hesse distance or directed distance). In this section, we are going to show how CGA can help us to obtain the Hesse distance for more general simplexes and not only for lines and planes. Figure 8.8a, b depicts a line and a plane, respectively, that will help us to develop our equations. Let $\boldsymbol{A}^k$ be a k-line (or plane); then, it consists of a *momentum* M^k of degree k and of a *direction* D^{k-1} of degree $k-1$. For instance, given three Euclidean points a_1, a_2, a_3, their two-simplex defines a dual 3-plane in CGA that can be expressed as

$$\boldsymbol{A}^k \equiv \Phi = M^3 + D^2 e_0 = a_1 \wedge a_2 \wedge a_3 + (a_2 - a_1) \wedge (a_3 - a_1) e_0. \tag{8.97}$$

Then, the directed distance of this plane, denoted as $\boldsymbol{p}^k$, can be obtained taking the inner product between the unit direction D_u^{k-1} and the moment M^k. Indeed, from

(8.1) and using expressions (8.6), we get the direction from $\Phi \cdot e_\infty = D^{k-1}$ and then its unitary expression D_u^{k-1} dividing D^{k-1} by its magnitude. Schematically,

$$\boldsymbol{A}^k \longrightarrow \boldsymbol{A}^k \cdot e_\infty = \boldsymbol{D}^{k-1} \longrightarrow \boldsymbol{D}_u^{k-1} = \frac{\boldsymbol{D}^{k-1}}{|\boldsymbol{D}^{k-1}|}. \tag{8.98}$$

Finally, the directed distance $\boldsymbol{p}^k$ of $\boldsymbol{A}^k$ is

$$\boldsymbol{p}^k = \boldsymbol{D}_u^{k-1} \cdot \boldsymbol{A}^k, \tag{8.99}$$

where the dot operation basically takes place between the direction $\boldsymbol{D}_u^{k-1}$ and the momentum of $\boldsymbol{A}^k$. Obviously, the directed distance vector $\boldsymbol{p}^k$ touches orthogonally the k-plane $\boldsymbol{A}^k$, and as we mentioned at the beginning of this subsection, the magnitude $|\boldsymbol{p}^k|$ equals the Hesse distance. For the sake of simplicity, in Fig. 8.8a, b only $\boldsymbol{D}^{k-1} \cdot \boldsymbol{L}^k$ and $\boldsymbol{D}^{k-1} \cdot \boldsymbol{\Phi}^k$ are, respectively, shown. Now, having this point from the first object, we can use it to compute the directed distance from the k-plane $\boldsymbol{A}^k$ parallel to the object $\boldsymbol{B}^k$ as follows:

$$d[\boldsymbol{A}^k, \boldsymbol{B}^k] = d[\boldsymbol{D}^{k-1} \cdot \boldsymbol{A}^k, \boldsymbol{B}^k] = d[(e_\infty \cdot \boldsymbol{A}^k) \cdot \boldsymbol{A}^k, \boldsymbol{B}^k]. \tag{8.100}$$

8.3 The Lie Algebra

We should introduce the analog of the complex "doubling" bivector for conformal geometry. This corresponds to the Minkowski plane of Eq. (8.7), namely

$$\boldsymbol{E} = e_i \bar{e}_j = e_+ \wedge e_- = e_+ e_-. \tag{8.101}$$

Thus, the bivector generators are the set of bivectors that commute with $\boldsymbol{E}$. We proceed similarly as with the unitary group and formulate an algebraic constraint so for any $\boldsymbol{x}, \boldsymbol{y} \in G_{n+1,1}$

$$[(\boldsymbol{x} \cdot \boldsymbol{E}) \wedge (\boldsymbol{y} \cdot \boldsymbol{E})] \times \boldsymbol{E} = \boldsymbol{x} \wedge (\boldsymbol{y} \cdot \boldsymbol{E}) + (\boldsymbol{x} \cdot \boldsymbol{E}) \wedge y = (\boldsymbol{x} \wedge \boldsymbol{y}) \times \boldsymbol{E}, \tag{8.102}$$

after a simple algebraic manipulation,

$$[\boldsymbol{x} \wedge \boldsymbol{y} - (\boldsymbol{x} \cdot \boldsymbol{E})(\boldsymbol{y} \cdot \boldsymbol{E})] \times \boldsymbol{E} = 0. \tag{8.103}$$

Now, using this constraint, we can again try all combinations of $\{e_i, \bar{e}_i\}$ to produce the bivector basis for the Lie algebra of the conformal group:

$$\boldsymbol{E}_i = e_+ e_-, \tag{8.104}$$

$$\boldsymbol{B}_{ij} = e_i e_j \qquad (i < j = 1...n), \tag{8.105}$$

$$\bar{\boldsymbol{E}}_{ij} = e_i e_\pm \qquad (i = 1, \ldots, n). \tag{8.106}$$

In the next section, we will use this set of Lie algebra operators to define the Lie groups of the conformal geometric algebra.

8.4 Conformal Transformations

In the middle of the nineteenth century, J. Liouville proved, for the three-dimensional case, that any conformal mapping on the whole of $\mathbb{R}^n$ can be expressed as a composition of *inversions in spheres* and *reflections in hyperplanes* [305]. In particular, *rotation, translation, dilation*, and *inversion* mappings will be obtained with these two mappings. In conformal geometric algebra, these concepts are simplified, due to the isomorphism between the conformal group on $\mathbb{R}^n$ and the Lorentz group on $\mathbb{R}^{n+1}$, which helps us to express, with a linear Lorentz transformation, a nonlinear conformal transformation, and then to use *versor* representation to simplify the *composition of transformations* with *multiplication of vectors* [185]. Thus, using conformal geometric algebra, it is computationally more efficient and simpler to interpret the geometry of the conformal mappings, than with matrix algebra. A transformation of geometric figures is said to be *conformal* if it preserves the *shape* of the figures, that is, whether it preserves the angles and hence the shapes of straight lines and circles. In particular, rotation and translation mappings are conformal and are also called *direct motion* transformations. Inversion and reflection mappings preserve the magnitude of the angle but reverse its direction; they are also called *opposite motion* transformations.

Any conformal transformation in $\mathbb{R}^n$,

$$\boldsymbol{x} \longmapsto \boldsymbol{x}\frac{1}{1+\boldsymbol{a}\boldsymbol{x}}, \tag{8.107}$$

can be expressed as a composite of inversions and a translation:

$$\boldsymbol{x} \overset{\longmapsto}{inversion} \quad \frac{\boldsymbol{x}}{\boldsymbol{x}^2}, \tag{8.108}$$

$$\overset{\longmapsto}{translation} \quad \frac{\boldsymbol{x}}{\boldsymbol{x}^2} + \boldsymbol{a}, \tag{8.109}$$

$$\overset{\longmapsto}{inversion} \quad \frac{\frac{\boldsymbol{x}}{\boldsymbol{x}^2} + \boldsymbol{a}}{\left(\frac{\boldsymbol{x}}{\boldsymbol{x}^2} + \boldsymbol{a}\right)\left(\frac{\boldsymbol{x}}{\boldsymbol{x}^2} + \boldsymbol{a}\right)} = \boldsymbol{x}\frac{1}{1+\boldsymbol{a}\boldsymbol{x}}. \tag{8.110}$$

The conformal transformation in conformal geometric algebra uses a versor representation,

$$g(\boldsymbol{x}_c) = \boldsymbol{G}\boldsymbol{x}_c(\boldsymbol{G}^*)^{-1} = \sigma \boldsymbol{x}'_c, \tag{8.111}$$

where $\boldsymbol{x}_c \in \mathbb{R}^{n+1,1}$, $\boldsymbol{G}$ is a versor, and σ is a scalar. $\boldsymbol{G}$ can be expressed in CGA as a composite of versors for transversion, translation, and rotation as follows:

$$\boldsymbol{G} = \boldsymbol{K}_{\mathbf{b}}\boldsymbol{T}_{\mathbf{a}}\boldsymbol{R}_{\alpha}. \tag{8.112}$$

These individual versors will be explained next.

8.4.1 Inversion

By the classical definition, an *inversion* T_S *with respect to a sphere* S (of radius ρ and center $\mathbf{c}$) is such that for any point q at a distance d from $\mathbf{c}$, $T_S(q)$ will be in the same ray from $\mathbf{c}$ to q and at a distance ρ^2/d from $\mathbf{c}$; see Fig. 8.9a. We comment some of the main properties of this transformation: (a) The inverse of a plane through the center of inversion is the plane itself; (b) the inverse of a plane not passing through the center of inversion is a sphere passing through the center of inversion; (c) the inverse of a sphere through the center of inversion is a plane not passing through the center of inversion; (d) the inverse of a sphere not passing through the center of inversion is a sphere not passing through the center of inversion; see Fig. 8.9b, e. Inversion in a sphere maps lines and circles to lines and circles.

In the context of the conformal geometry, the general form of a reflection about a vector is

$$s(\boldsymbol{x}_c) = -\boldsymbol{s}\boldsymbol{x}_c\boldsymbol{s}^{-1} = \boldsymbol{x}_c - 2(\boldsymbol{s}\cdot\boldsymbol{x}_c)\boldsymbol{s}^{-1} = \sigma\boldsymbol{x}'_c, \tag{8.113}$$

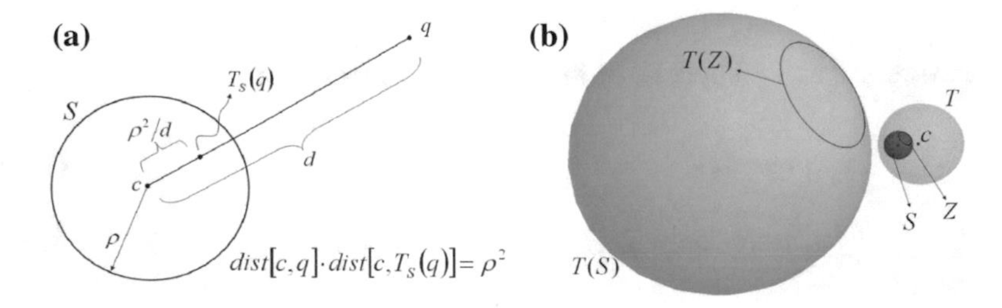

Fig. 8.9 **a** The point $T_S(q)$ is the inverse point of q with respect to the circle S and vice versa. **b** The inversion with respect to a sphere T mapping sphere to sphere and circle to circle, if they are not passing through the center of T

where $\boldsymbol{s}\boldsymbol{x} + \boldsymbol{x}\boldsymbol{s} = 2(\boldsymbol{s} \cdot \boldsymbol{x})$, from the definition of the Clifford product between two vectors. We will now analyze what happens when s represents a sphere. Recall that the equation of a sphere of radius ρ centered at point $\boldsymbol{c}_c$ is the vector

$$\boldsymbol{s} = \boldsymbol{c}_c - \frac{1}{2}\rho^2 e_\infty. \tag{8.114}$$

If $\boldsymbol{s}$ represents the unit sphere centered at the origin, then $\boldsymbol{s}$ and $\boldsymbol{s}^{-1}$ reduce to $e_0 - \frac{1}{2}e_\infty$. Hence, $-2(\boldsymbol{s} \cdot \boldsymbol{x}_c) = \mathbf{x}_e^2 - 1$, and Eq. (8.113) becomes

$$\sigma \boldsymbol{x}_c' = \left(\mathbf{x}_e + \frac{1}{2}\mathbf{x}_e^2 + e_\infty\right) + (\mathbf{x}_e^2 - 1)(e_0 - e_\infty) = \mathbf{x}_e^2(\mathbf{x}_e^{-1} + \mathbf{x}_e^{-2}e_\infty + e_0), \tag{8.115}$$

which is the conformal mapping for $\mathbf{x}_c^{-1}$.

To see how a general sphere inverts a point, we return to Eq. (8.114) to get

$$\boldsymbol{s} \cdot \boldsymbol{x}_c = \boldsymbol{c}_c \cdot \boldsymbol{x}_c - \rho^2 e_\infty \cdot \boldsymbol{x}_c = -[(\mathbf{x}_e - \mathbf{c}_e)^2 - \rho^2]. \tag{8.116}$$

Inserting the equation (8.116) in (8.115) and after a little algebra gives

$$\sigma \boldsymbol{x}_c' = \left(\frac{\mathbf{x}_e - \mathbf{c}_e}{\rho}\right)^2 \left(g(\boldsymbol{x}_e) + \frac{1}{2}g^2(\mathbf{x}_e)e_\infty + e_0\right), \tag{8.117}$$

where

$$g(\mathbf{x}_e) = \frac{\rho^2}{\mathbf{x}_e - \mathbf{c}_e} + \mathbf{c}_e = \frac{\rho^2(\mathbf{x}_e - \mathbf{c}_e)}{(\mathbf{x}_e - \mathbf{c}_e)^2} + \mathbf{c}_e \tag{8.118}$$

is the inversion in $\mathbb{R}^n$.

8.4.2 *Reflection*

The reflection of conformal geometric entities helps us to do any other transformation. The reflection of a point x with respect to the plane π is equal to x minus twice the direct distance between the point and plane (see Fig. (8.10)), that is, $x = x - 2(\pi \cdot x)\pi^{-1}$. To simplify this expression, recall the property of Clifford product of vectors $2(b \cdot a) = ab + ba$ (Fig. 8.10).

The reflection could be written as

$$x' = x - (\pi x - x\pi)\pi^{-1}, \tag{8.119}$$

$$x' = x - \pi x \pi^{-1} - x\pi\pi^{-1}, \tag{8.120}$$

$$x' = -\pi x \pi^{-1}. \tag{8.121}$$

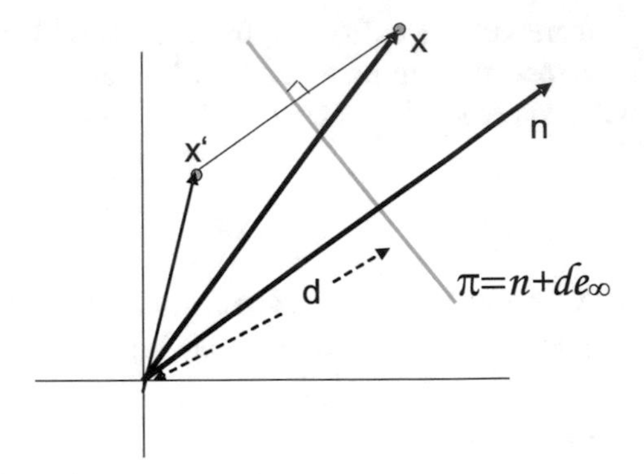

Fig. 8.10 Reflection of a point x with respect to the plane π

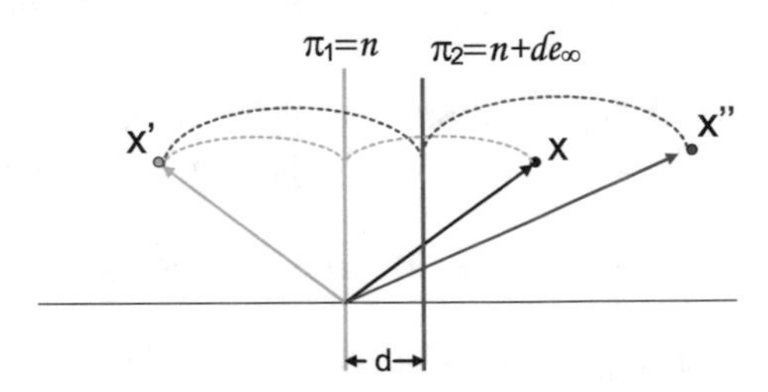

Fig. 8.11 Reflection about parallel planes

For any geometric entity Q, the reflection with respect to the plane π is given by

$$Q' = \pi Q \pi^{-1}. \tag{8.122}$$

8.4.3 Translation

The translation of conformal entities can be obtained by carrying out two reflections in parallel planes π_1 and π_2 (see Fig. (8.11)):

$$Q' = \underbrace{T_a}_{(\pi_2\pi_1)} Q \underbrace{\widetilde{T}_a}_{(\pi_1^{-1}\pi_2^{-1})}, \tag{8.123}$$

$$T_a = (n + de_\infty)n = 1 + \frac{1}{2}ae_\infty = e^{-\frac{a}{2}e_\infty}, \tag{8.124}$$

where $a = 2dn$ (Fig. 8.11).

Since $\mathbf{a}e_\infty = b^i e_i(e_+ + e_-)$, the Lie algebra generators or bivectors for the translator T_a are e_ie_+ and e_ie_- (see Eq. (8.106)).

8.4.4 Transversion

A transversion can be generated from two inversions and a translation. A transversor has the form

$$K_{\mathbf{b}} = e_+ T_{\mathbf{b}} e_+ = (e_\infty - e_0)(1 + \mathbf{b}e_\infty)(e_\infty - e_0) = 1 + \mathbf{b}e_0. \tag{8.125}$$

The transversion generated by $K_{\mathbf{b}}$ can be expressed in various forms:

$$g(\mathbf{x}_e) = \frac{\mathbf{x}_e - \mathbf{x}_e^2\mathbf{b}}{1 - 2\mathbf{b}\cdot\mathbf{x}_e + \mathbf{x}_e^2\mathbf{b}^2} = \mathbf{x}_e(1 - \mathbf{b}\mathbf{x}_e)^{-1} = (\mathbf{x}_e^{-1} - \mathbf{b})^{-1}. \tag{8.126}$$

The last form can be written down directly as an inversion followed by a translation and another inversion. Note that a transversion uses as a null vector the origin e_0, whereas the translator uses the point at infinity e_∞. Since $\mathbf{b}e_0 = b^i e_i \frac{1}{2}(e_- - e_+)$, the Lie algebra generators or bivectors for the transversion T_a are $e_i e_+$ and $e_i e_-$ (see Eq. (8.106)). In fact, the bivector basis is the same for the translator and for the transversion.

8.4.5 Rotation

A rotation is the product of two reflections with respect to two non-parallel planes (see Fig. 8.12):

$$Q' = R_\theta \underbrace{(\pi_2\pi_1)}Q\widetilde{R}_\theta\underbrace{(\pi_1^{-1}\pi_2^{-1})}, \tag{8.127}$$

or, computing the conformal product of the normals of the planes,

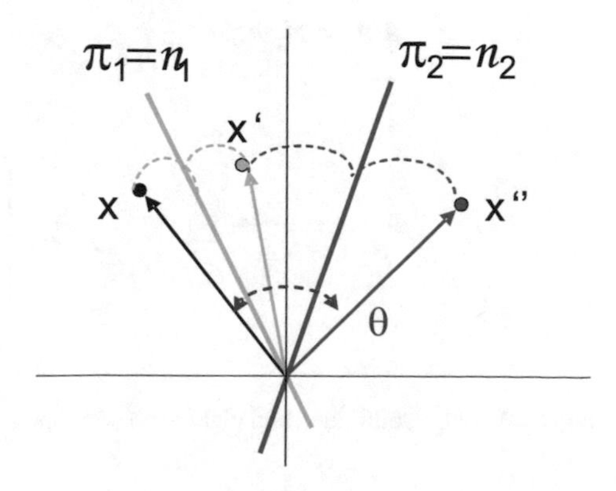

Fig. 8.12 Reflection about non-parallel planes

$$R_\theta = n_2 n_1 = \cos(\frac{\theta}{2}) - \sin(\frac{\theta}{2})l = e^{-\frac{\theta}{2}l}. \quad (8.128)$$

There are three Lie algebra generators or bivectors for the rotors; they are of the type $B_{ij} = e_i e_j$ (see Eq. (8.105)).

With $l = n_2 \wedge n_1$, and θ twice the angle between the planes π_2 and π_1, the screw motion called motor related to an arbitrary axis L is $M = TR\widetilde{T}$:

$$Q' = M_\theta \underbrace{(TR\widetilde{T})}Q\widetilde{M}_\theta\underbrace{((T\widetilde{R}\widetilde{T}))}, \quad (8.129)$$

$$M_\theta = TR\widetilde{T} = \cos(\frac{\theta}{2}) - \sin(\frac{\theta}{2})L = e^{-\frac{\theta}{2}L}. \quad (8.130)$$

The direct kinematics for serial robot arms can be expressed as a succession of motors and surprisingly is valid for points, lines, planes, circles, and spheres.

$$Q' = \prod_{i=1}^{n} M_i Q \prod_{i=1}^{n} \widetilde{M}_{n-i+1}. \quad (8.131)$$

8.4.6 Rigid Motion Using Flags

In Sect. 7.91, the concept of *flag* or *soma* is explained. This helps to embed the frame in a single algebraic as depicted in Fig. 8.13. The flag representation involves a point, line and a plane with a common reference point

$$\boldsymbol{F} = \boldsymbol{x} + \boldsymbol{L}^* + \pi^* = \boldsymbol{x} + I_c\boldsymbol{Q}, \quad (8.132)$$

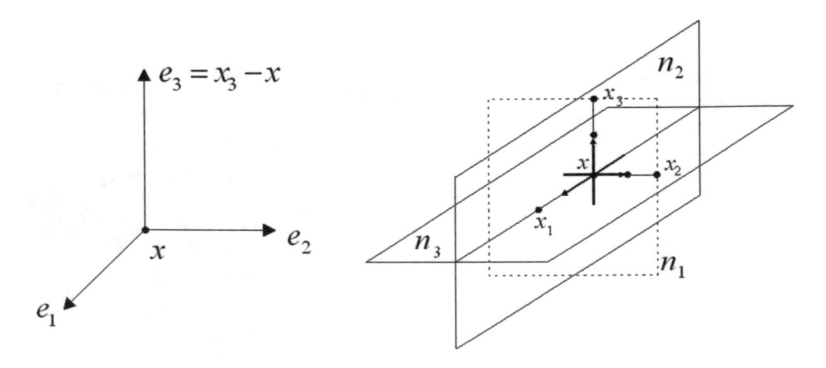

Fig. 8.13 Flag: point, line, and plane with a common point

where

$$L^* = x \wedge x_1 \wedge e_\infty = x \wedge (x_1 - x) \wedge e_\infty = x \wedge e_1 \wedge e_\infty = I_c n_2 n_3,$$
$$\pi^* = x \wedge x_1 \wedge x_2 \wedge e_\infty = x \wedge e_1 \wedge e_2 \wedge e_\infty = e_2 \wedge L^* = I_c n_3,$$

combining these results, one get $Q = n_3 + n_2 n_3 = (1 + n_2) n_3$ which is a conic, namely it fulfills

$$x \wedge F = x \wedge (L^* + \pi^*) = x \wedge (I_c Q) = 0, \tag{8.133}$$

its dual expression is

$$x \cdot Q = 0, \tag{8.134}$$

where

$$Q^2 = (n_3 + n_2 n_3)(n_3 - n_2 n_3) = 0. \tag{8.135}$$

The previous equations describe a point lying on an absolute conic. Recall in projective geometry the equation of homogeneous points lying on the absolute conic.

Finally, the rigid body can be nicely modeled using the flag F as follows, given a motor $M = TR$

$$F' = MF\widetilde{M} = Mx\widetilde{M} + ML^*\widetilde{M} + M\pi^*\widetilde{M}. \tag{8.136}$$

Here, this linearity is due to the congruence theorem.

8.4.7 Dilation

A dilation is the composite of two successive inversions centered at the origin. Using the unit sphere $s_1 = e_0 - \frac{1}{2}e_\infty$, another sphere of arbitrary radius ρ, and $s_2 = e_0 - \frac{1}{2}\rho^2 e_\infty$ as inversors, we get

$$(e_0 - e_\infty)(e_0 - \rho^2 e_\infty) = (1 - E) + (1 + E)\rho^2. \tag{8.137}$$

Normalizing to unity, we have

$$D_\rho = (1 + E)\rho + (1 - E)\rho^{-1} = e^{E\phi}, \tag{8.138}$$

where $\phi = \ln \rho$. To prove that this is indeed a dilation, we note that

$$D_\rho e_\infty D_\rho^{-1} = \rho^{-2} e, \tag{8.139}$$

and similarly,

$$D_\rho e_0 D_\rho^{-1} = \rho^2 e_0. \tag{8.140}$$

Therefore,

$$D_\rho(\mathbf{x}_e + \mathbf{x}_e^2 e_\infty + e_0)D_\rho^{-1} = \rho^2[\rho^{-2}\mathbf{x}_e + (\rho^{-2}\mathbf{x}_e)^2 e_\infty + e_0], \tag{8.141}$$

which is the conformal mapping $g(\boldsymbol{x}_e) = \sigma \boldsymbol{x}'_c$ with $\boldsymbol{x}'_c = \boldsymbol{x}'_e + \boldsymbol{x}'^2 e_\infty + e_0$, where $\boldsymbol{x}'_e = \rho^{-2}\mathbf{x}_e$. The Lie algebra generator or bivector for the dilations is $E = e_+ e_-$ or the Minkowski plane; see Eq. (8.104).

8.4.8 *Involution*

The bivector E (the Minkowski plane of $\mathbb{R}^{1,1}$) represents an operation that corresponds to the main involution, but for an r-blade $\boldsymbol{A}_r$, namely $\bar{\boldsymbol{A}}_r = (-1)^r \boldsymbol{A}_r$. In particular, for vectors $\bar{\mathbf{x}}_e = -\boldsymbol{x}_e$, which can be easily obtained by applying the versor E:

$$\boldsymbol{E}(\mathbf{x}_e + \mathbf{x}_e^2 e_\infty + e_0)\boldsymbol{E} = -\left(-\mathbf{x}_e + \frac{1}{2}\mathbf{x}_e^2 e_\infty + e_0 \right). \tag{8.142}$$

This expression corresponds to the conformal mapping of $-\boldsymbol{x}_e$, thus confirming that the versor E represents the main involution for $\mathbb{R}^n$. This means that the main involution is a reflection via the Minkowski plane $\mathbb{R}^{1,1}$.

8.4.9 *Conformal Transformation*

Finally, using previous results, we can now write a canonical decomposition of a conformal transformation in terms of individual versors:

$$\boldsymbol{G} = \boldsymbol{K}_{\mathbf{b}} \boldsymbol{T}_{\mathbf{a}} \boldsymbol{R}_\alpha, \tag{8.143}$$

where the versors are as follows: for transversion $\boldsymbol{K}_{\mathbf{b}} = e_+ \boldsymbol{T}_{\mathbf{b}} e_+ = 1 + \mathbf{b} e_0$, for translation $\boldsymbol{T}_{\mathbf{a}} = 1 + \frac{1}{2} a e_\infty$, and for rotation $\boldsymbol{R}_\alpha$. This decomposition reveals the structure of the three-parameter group $\{G \in \mathbb{R}_{1,1} | G^* G^\dagger = 1\} \simeq GL_2(\mathbb{R})$. Note that in conformal geometric algebra the conformal transformation is built by successive multiplicative versors: In contrast in $\mathbb{R}^n$, the conformal transformation of Eq. (8.110) is highly nonlinear and its application will require of complicated nonlinear and slow algorithms.

8.5 Ruled Surfaces

Conics, ellipsoids, helicoids, hyperboloids of one sheet are entities that cannot be directly described in CGA, but can be modeled with its multivectors. In particular, a ruled surface is a surface generated by the displacement of a straight line (called *generatrix*) along a directing curve or curves (called a *directrices*). The plane is the simplest ruled surface, but now we are interested in nonlinear surfaces generated as ruled surfaces. For example, a circular cone is a surface generated by a straight line through a fixed point and a point in a circle. It is well known that the intersection of a plane with the cone can generate the conics (see Fig. 8.14). In [252], the cycloidal curves can be generated by two coupled twists. In this section, we are going to see how these and other curves and surfaces can be obtained using only multivectors of CGA.

8.5.1 Cone and Conics

A circular cone is described by a fixed point v_0 (*vertex*), a dual circle $z_0 = a_0 \wedge a_1 \wedge a_2$ (*directrix*), and a rotor $\mathbf{R}(\theta, l)$, $\theta \in [0, 2\pi)$ rotating the straight line $L(v_0, a_0) = v_0 \wedge a_0 \wedge e_\infty$ (*generatrix*) along the axis of the cone $l_0 = z_0 \cdot e_\infty$. Then, the cone w is generated as

$$w = \boldsymbol{R}(\theta, l_0)\, L(v_0, a_0)\widetilde{\boldsymbol{R}}(\theta, l_0)\,, \quad \theta \in [0, 2\pi)\,. \tag{8.144}$$

A conic curve can be obtained with the *meet* of a cone and a plane; see Fig. 8.14.

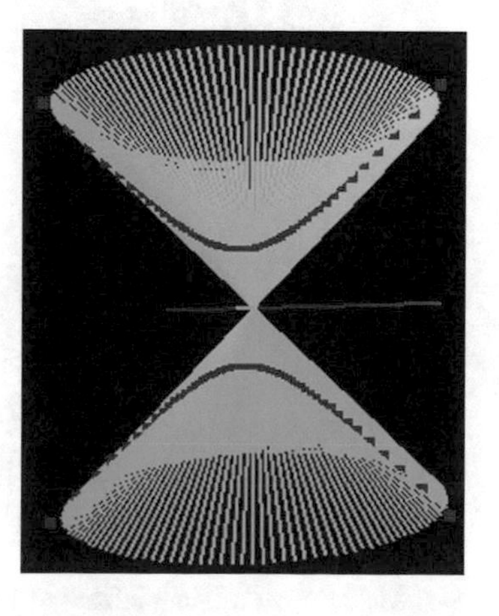

Fig. 8.14 Hyperbola as the *meet* of a cone and a plane

8.5.2 Cycloidal Curves

The family of the cycloidal curves can be generated by the rotation and translation of one or two circles. For example (see Fig. 8.15), the cycloidal family of curves generated by two circles of radius r_0 and r_1 is expressed by the following motor:

$$\boldsymbol{M} = \boldsymbol{T}\boldsymbol{R}_1\widetilde{\boldsymbol{T}}\boldsymbol{R}_2, \tag{8.145}$$

where

$$\boldsymbol{T} = \boldsymbol{T}((r_0 + r_1)(sin(\theta)e_1 + cos(\theta)e_2)), \tag{8.146}$$

$$\boldsymbol{R}_1 = \boldsymbol{R}_1\left(\frac{r_0}{r_1}\theta\right), \tag{8.147}$$

$$\boldsymbol{R}_2 = \boldsymbol{R}_2(\theta). \tag{8.148}$$

Then, each conformal point x is transformed as $\boldsymbol{M}x\widetilde{\boldsymbol{M}}$.

8.5.3 Helicoid

We can obtain the ruled surface called a helicoid rotating a ray segment in a similar way as the spiral of Archimedes. So, if the axis e_3 is the directrix of the rays and is orthogonal to them, then the translator that we need to apply is a multiple of θ, the angle of rotation (see Fig. 8.16a).

Fig. 8.15 A laser welding following a 3D curve: the projection of a cycloidal curve over a sphere

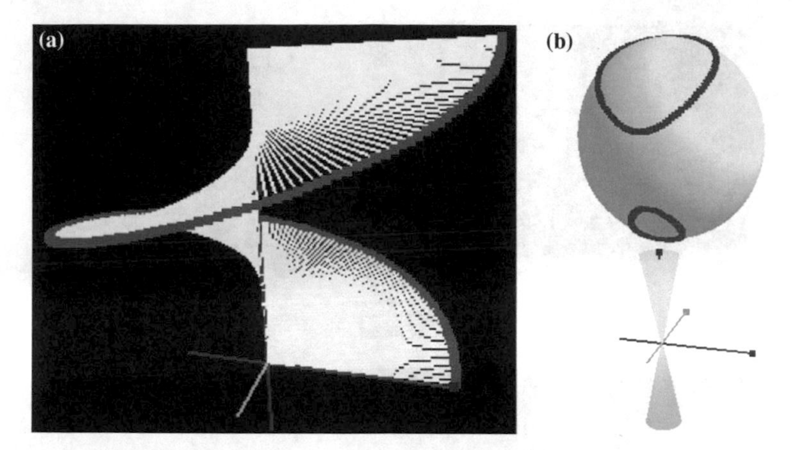

Fig. 8.16 **a** The helicoid is generated by the rotation and translation of a line segment. In CGA, the *motor* is the desired multivector. **b** Intersection as the *meet* of a sphere and a cone

8.5.4 Sphere and Cone

Let us see an example of how the use of algebra of incidence in CGA simplifies the algebraic formulation. The intersection of a cone and a sphere in general position, that is, the axis of the cone does not pass through the center of the sphere, is the three-dimensional curve of all Euclidean points (x, y, z) such that x and y satisfy the quartic equation

$$\left[x^2\left(1+\frac{1}{c^2}\right) - 2x_0x + y^2\left(1+\frac{1}{c^2}\right) - 2y_0y + x_0^2 + y_0^2 + z_0^2 - r^2\right]^2 = 4z_0^2(x^2+y^2)/c^2 \quad (8.149)$$

and x, y, and z the quadratic equation

$$(x-x_0)^2 + (y-y_0)^2 + (z-z_0)^2 = r^2, \quad (8.150)$$

(see Fig. 8.16b). In CGA, the set of points q of the intersection can be expressed as the meet of the dual sphere s and the cone w, Eq. (8.144), defined in terms of its generatrix L, that is

$$q = (s^*) \cdot [\boldsymbol{R}\,(\theta, l_0)\, L(v_0, a_0)\widetilde{\boldsymbol{R}}\,(\theta, l_0)],\ \ \theta \in [0, 2\pi)\,. \quad (8.151)$$

Thus, in CGA, we only need (8.151) to express the intersection of a sphere and a cone; meanwhile in Euclidean geometry, it is necessary to use the complicated Eqs. (8.149) and (8.150).

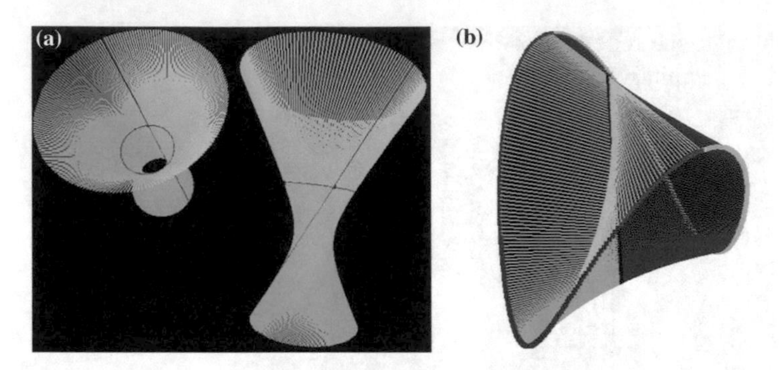

Fig. 8.17 **a** Hyperboloid as the rotor of a line. **b** The Plücker conoid as a ruled surface

8.5.5 *Hyperboloid, Ellipsoids, and Conoid*

The rotation of a line over a circle can generate a hyperboloid of one sheet (see Fig. 8.17a). The ellipse is a curve of the family, of the cycloid; we can obtain an ellipsoid with a translator and a dilator. The cylindroid or Plücker conoid is a ruled surface (see Fig. 8.17b). This ruled surface is like the helicoid, where the magnitude of the translator parallel to the e_3 axis is a multiple of $cos(\theta)sin(\theta)$.

8.6 Exercises

8.1 Formulate the equation of a sphere using conformal geometric algebra.

8.2 Derive the representation of the point pair (0-sphere) $\boldsymbol{PP}$ spanned by the points $\boldsymbol{p}_1$ and $\boldsymbol{p}_2$ at locations e_1 and e_2 with weights 3 and –2.

8.3 Compute the center and radius of $\boldsymbol{PP}$.

8.4 Give the dual representation of $\boldsymbol{PP}$, and compute its center and radius.

8.5 Give the IPNS representation of the circle z through the points $\boldsymbol{p}_1$ and $\boldsymbol{p}_2$ and the unit point e_1. Compute its square radius and center, and the orientation vector of the circle z passing its center.

8.6 Compute the IPNS representation of the sphere $\boldsymbol{s}$ through z and the origin.

8.7 Compute the OPNS representation of the sphere s^*, and read off its center and square radius directly from that dual representation.

8.8 In conformal geometric algebra, write the equations of the direct distances between a point and a line, a point and a plane, and a line and a plane.

8.9 Compute the reflexion of the line $\boldsymbol{L}$ which crosses the points e_1 and e_2 with respect to the unitary sphere which is at the origin.

8.10 The dual circle is the intersection of a plane with a sphere, namely $\boldsymbol{s} \wedge \phi$. Recall that the dual sphere at origin is $\boldsymbol{s}^* = (e_0 - \frac{\rho^2 e_\infty}{2}$. Use CLUCAL to compute the meet of two intersecting dual circles $z_1 = \boldsymbol{T}_{e_1}\big((e_0 - \frac{e_\infty}{2})(-e_3)\big)$ and $z_2 = \boldsymbol{T}_{e_2}\big((e_0 - \frac{e_\infty}{2})(-e_3)\big)$ lying on the $e_1 \wedge e_2$-plane. Here, $\boldsymbol{T}_{e_1}$ and $\boldsymbol{T}_{e_2}$ are the translators of the circles along e_1 and e_2, respectively.

8.11 Explain yourself Eq. 8.68

$$(-1)^{\epsilon} \boldsymbol{s}^* \cap \boldsymbol{A}_{r+1} = (\boldsymbol{s}^* I) \cdot \boldsymbol{A}_{r+1} = \boldsymbol{s} \boldsymbol{A}_{r+1} = \boldsymbol{A}_r.$$

This can be used to formulate a sphere in terms of a circle z and a plane π, i.e.,

$$s = z \pi^{-1}.$$

Using geometric entities like lines, planes, circles, and spheres, draw and compute the projection of a blade A to a blade B, and interpret your resulting blade.

8.12 Prove the flag

$$\boldsymbol{F} = \boldsymbol{x} + \boldsymbol{L}^* + \pi^* = \boldsymbol{x} + I_c \boldsymbol{Q}.$$

8.13 In projective geometry, one applies the homography $H \in \mathbb{P}^3$

$$\mathrm{H} = \begin{bmatrix} \mathrm{R} & \mathbf{t} \\ \mathbf{f} & h \end{bmatrix}$$

to the origin and point at infinity. In conformal geometric algebra, rotate with a rotor $\boldsymbol{R}$ the null vectors the origin e_o and the point at infinity e_∞, explain your results.

8.14 In conformal geometric algebra, translate with a translator $\boldsymbol{T}$ the null vectors the origin e_o and the point at infinity e_∞, explain your results.

8.15 In conformal geometric algebra, write the equations of the direct distances between a sphere and a point, a sphere and a line, and a sphere and a plane.

8.16 In conformal geometric algebra, write the equations of the direct distances between a circle and a point, a circle and a line, and a circle and a plane, and a circle and a sphere.

8.17 In conformal geometric algebra, write the equations of the direct distances between a circle and a point, a circle and a line, a circle and a plane, and a circle and a sphere.

8.18 In conformal geometric algebra, given three points lying on a circle, using the conformal split compute the radio and center of the circle.

8.19 In conformal geometric algebra, given three points a_1, a_2, and s_3 in general position using CLICAL [196], eClifford [1], or CLUCAL [234] compute the expanded form of the plane crossing these tree points.

8.20 Theorem-proving: Prove in conformal geometric algebra the theorem of Desargues configuration. Recall the 3D projective plane Π^3. Consider that x_1, x_2, x_3 and y_1, y_2, y_3 are the vertices of two triangles in Π^3, and suppose that $(x_1 \wedge x_2) \cap (y_1 \wedge y_2) = z_3$, $(x_2 \wedge x_3) \cap (y_2 \wedge y_3) = z_1$, and $(x_3 \wedge x_1) \cap (y_3 \wedge y_1) = z_2$. You can claim that $c_1 \wedge c_2 \wedge c_3 = 0$ if and only if there is a point p such that $x_1 \wedge y_1 \wedge p = 0 = x_2 \wedge y_2 \wedge p = x_3 \wedge y_3 \wedge p$. (*Hint*: Express the the point as linear combinations of a_1, b_1, a_2, b_2, and a_3, b_3. The other half of the proof follows by duality of the classical projective geometry.)

8.21 Theorem-proving: Using conformal geometric algebra, prove the Simpson rule using the join of three projected points; see Fig. 8.18. If the projecting point **D** lies at the circumference, the join of the projected points is zero. Take three arbitrary points lying on a unit circumference and a fourth one nearby. Use CLICAL [196], eClifford [1], or CLUCAL [234] for your computations. (*Hint*: First compute the projected points as the meet of three lines passing by the point **D** orthogonal to the triangle sides. The triangle is formed by three arbitrary points lying on the circumference: **A**, **B**, **C**).

8.22 Prove the Apollonius's theorem using the incidence algebra in conformal geometric algebra; see Fig. 8.19.

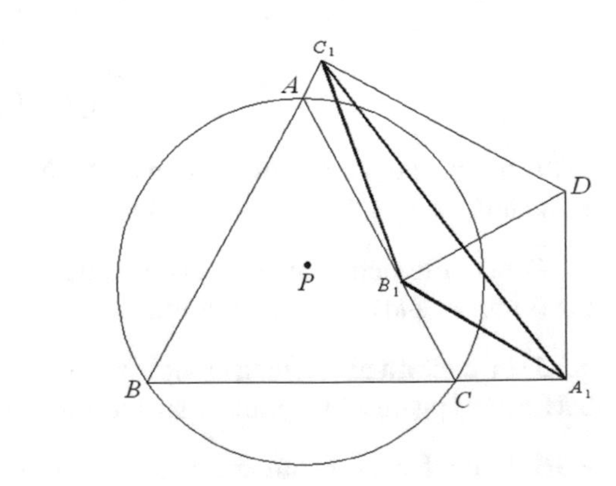

Fig. 8.18 Simson's theorem

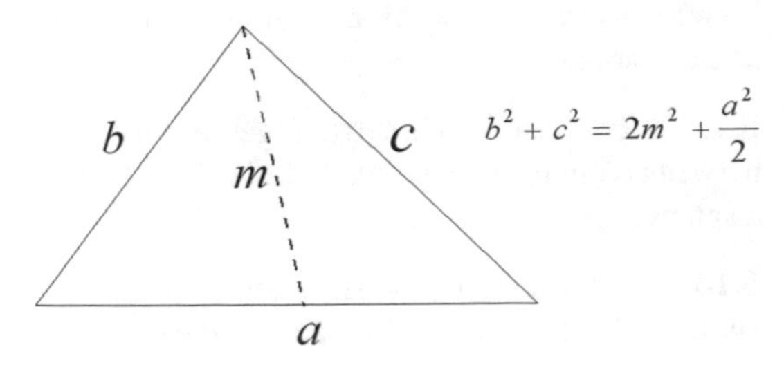

Fig. 8.19 Apollonius's theorem

8.23 Prove the Pascal theorem using the incidence algebra in conformal geometric algebra. Given six points lying in a conic, compute six intersecting lines using the join operation. The meet of these lines should give three intersecting points, and the join of these three intersecting points should be zero. Use CLICAL [196], eClifford [1], or CLUCAL [234] and any six points belonging to a conic of your choice.

8.24 Prove the Desargues theorem using the algebra of incidence in conformal geometric algebra. Let $\boldsymbol{x}_1, \boldsymbol{x}_2, \boldsymbol{x}_3$ and $\boldsymbol{y}_1, \boldsymbol{y}_2, \boldsymbol{y}_3$ be the vertices of two triangles in P^2, and suppose that the meets of the lines fulfill the equations $(\boldsymbol{x}_1 \wedge \boldsymbol{x}_2) \cap (\boldsymbol{y}_1 \wedge \boldsymbol{y}_2) = z_3$, $(\boldsymbol{x}_2 \wedge \boldsymbol{x}_3) \cap (\boldsymbol{y}_2 \wedge \boldsymbol{y}_3) = z_1$, and $(\boldsymbol{x}_3 \wedge \boldsymbol{x}_1) \cap (\boldsymbol{y}_3 \wedge \boldsymbol{y}_1) = z_2$. Then, the join of these points $z_1 \wedge z_2 z_3 = 0$ if and only if a point $\boldsymbol{p}$ exists such that $\boldsymbol{x}_1 \wedge \boldsymbol{y}_1 \wedge \boldsymbol{p} = \boldsymbol{x}_2 \wedge \boldsymbol{y}_2 \wedge \boldsymbol{p} = \boldsymbol{x}_3 \wedge \boldsymbol{y}_3 \wedge \boldsymbol{p} = 0$. In this problem, use CLICAL [196], eClifford [1], or CLUCAL [234] and define two triangles of your choice.

8.25 25 A point $\boldsymbol{p}$ lies on the circumcircle of an arbitrary triangle with vertices $\boldsymbol{x}_1$, $\boldsymbol{y}_1$, and z_1. From the point $\boldsymbol{p}$, draw three perpendiculars to the three sides of the triangle to meet the circle at points $\boldsymbol{x}_2$, $\boldsymbol{y}_2$, and z_2, respectively. Using incidence algebra in conformal geometric algebra, show that the lines $\boldsymbol{x}_1 \wedge \boldsymbol{x}_2, \boldsymbol{y}_1 \wedge \boldsymbol{y}_2$, and $z_1 \wedge z_2$ are parallel.

8.26 In Exercise 7.19, we introduced the barycentric coordinates using a homogeneous model. For this exercise, formulate the expressions for the barycentric coordinates in terms of conformal points.

8.27 Simplex: Show that the weight of $\boldsymbol{x} \wedge \boldsymbol{y} \wedge \boldsymbol{z}/2$ is the area of a triangle $\boldsymbol{xyz}$ and that the weight of $\boldsymbol{x} \wedge \boldsymbol{y} \wedge \boldsymbol{z} \wedge \boldsymbol{w}/3$ is the volume of the tetrahedron $\boldsymbol{xyzw}$.

8.28 The weight of a dual sphere $\boldsymbol{s}^*$ is the weight of its center and equals to $e_\infty \cdot \boldsymbol{s}^*$. Take the dual of this expression to find out that when a sphere passing through the points $\boldsymbol{x}, \boldsymbol{y}, \boldsymbol{z}, \boldsymbol{z}$ becomes zero.

8.29 Prove that the distance measure between a point $\boldsymbol{p}_i$ and the sphere/plane $\boldsymbol{s} = \mathbf{s} + s_4 e_\infty + s_5 e_0$ can be computed in conformal geometry as follows: $\boldsymbol{p}_i \cdot \boldsymbol{s} = \mathbf{p}_i \cdot \mathbf{s} - s_4 - \frac{1}{2} s_5 \mathbf{p}_i^2$ or $\boldsymbol{p}_i \cdot \boldsymbol{s} = \sum_{j=1}^{5} w_{i,j} s_j$, with $w_{i,j} = x_{i,j}$ $j \in \{1, 2, 3\}$, $w_4 = -1$, and $w_5 = -\frac{1}{2}\mathbf{p}_i^2$.

8.30 Use the results of Exercise 8.15 for the sphere to propose the structure of a conformal radial basis function network for data clustering.

8.31 Compute the reflection of the line $\boldsymbol{L}$ through the locations e_1 and e_2 with respect to the unit sphere at the origin.

8.32 Following Exercise 8.30, factorize the result to determine its center and squared radius.

8.33 Compute the reflection of the vector $e_1 + e_2$ in the direction of $3e_3$ with respect to the unit sphere at the origin. Notice the weight in your result.

8.34 Compute the reflection of the line $\boldsymbol{L}$ with respect to the origin.

8.35 Compute the reflection of the line $\boldsymbol{L}$ with respect to the point e_2.

Chapter 9
The Geometric Algebras $G^+_{6,0,2}$, $G_{6,3}$, $G^+_{9,3}$, $G^+_{6,0,6}$

9.1 Introduction

The geometric algebra of a 3D Euclidean space $G_{3,0,0}$ has a point basis and the motor algebra $G^+_{3,0,1}$ a line basis. In the latter, the lines are expressed in terms of Plücker coordinates and the points and planes in terms of bivectors. The reader can find a comparison of representations of points, lines, and planes using vector calculus, $G_{3,0,0}$ and $G^+_{3,0,1}$ in Chap. 7. Extending the degrees of freedom of the mathematical system, in the conformal geometric algebra $G_{4,1}$, using the horosphere framework points, one can model lines, planes, circles, and spheres and also certain Lie groups as versors.

For the applications, we are interested in model not only diverse geometric entities but also their associated Lie groups. The $SE(3)$ group is modeled in $G_{3,0,0}$ in an additive manner using a rotor $\boldsymbol{R}$ and for the translation a vector $\boldsymbol{t}$. In the motor algebra $G^+_{3,0,1}$ [13, 21] with $\begin{bmatrix} 4 \\ 2 \end{bmatrix} = 6$ bivector basis, we represent $SE(3)$ using the versor called motor $\boldsymbol{M} = \boldsymbol{TR}$ which carries out the group action however in a multiplicative manner. As explained in Chap. 8, the conformal geometric algebra $G_{4,1}$ of the 3D space [13], is equipped with $\begin{bmatrix} 5 \\ 2 \end{bmatrix} = 10$ bivectors which as Lie algebra generators suffice to build $SE(3)$ and extend to the conformal group involving $SE(3)$, dilations and inversions. Note that the dilations are simply isotropic transformations an not achieve complete affine transformations. Considering other more involved Lie groups, in the Sect. 5.7, in the framework $G_{3,3}$, we formulated for the 3D projective geometry its Lie group $SL(4)$ and its Lie algebra $sl(4)$ which indeed are useful for complete affine transformations as well.

Along these lines of thought, it would beneficial to extend the degrees of freedom of certain geometric algebras so that we can model the kinematics of a variety of geometric entities together with other Lie groups. For that, we need to extend the degrees of freedom of the geometric algebra in question so that we have enough null vectors for the modeling of the geometric entities and the bivectors for generating

E. Bayro-Corrochano, *Geometric Algebra Applications Vol. I*,
https://doi.org/10.1007/978-3-319-74830-6_9

the required Lie algebras. In the horizon of our work in 3D, we require points, lines, circles, planes and spheres together with the affine and projective groups for example to transform spheres in ellipsoids varying its orientation and position. We have formulated for the 3D projective geometry in $G_{3,3}$ the Lie group $SL(4)$ and its Lie algebra $sl(4)$, so we should now extend the number of the null vectors of the geometric algebras like $G_{3,0,0}$ and $G_{4,1}$ so that we have enough resources for this purpose, so the more natural geometric algebras appear to be $G_{6,3}$, introduced by Zamora [318], which extends to three the number of null vectors and of the 3D Euclidean $G_{3,0,0}$, as the $G_{4,1}$ has just two one null vectors one to represent the origin and another for one point at infinity, in $G_{6,3}$ we can model three orthogonal points at the origin and three at infinity which help to model projective and affine transformations. Another interesting geometric algebra is the $G^+_{6,0,2}$ which in fact is a result of the doubling procedure applied to motor algebra $G^+_{3,0,1}$, and it was proposed by Selig and Bayro [266] to model the kinematics and dynamics of points, lines, and planes. Other interesting geometric algebra is $G_{8,2}$, introduced by Easter [81], which again is the result obtained after the double procedure applied to $G_{4,1}$. In the next subsections, we will briefly describe $G^+_{6,0,2}$, $G_{6,3}$ and $G_{8,2}$ highlighting its attributes which help to overcome the limitations of the motor algebra $G^+_{3,0,2}$ and the conformal geometric algebra $G_{4,1}$.

9.2 The Double Motor Algebra $G^+_{6,0,2}$

In Sect. 6.5.1, we discusses the motor algebra $G^+_{3,0,1}$ a suitable line geometry to model the kinematics of points, lines and planes. Selig and Bayro [266] recast the dynamics of rigid bodies using the double motor algebra $G^+_{6,0,2}$. The authors shown how the velocities, momenta, and inertias can be represented in terms of elements of the algebra

$$G^+_{6,0,2} = \mathrm{gen}\{1, e_i e_j, e_i e, f_i f_j, f_i f, I\}, \tag{9.1}$$

with an orthogonal set of generators e_i and f_i where i = j = 1, 2, 3 and two null vectors e, f; 28 bivectors and the pseudoscalars $I_e = e_1 e_2 e_3 e$, $I_f = f_1 f_2 f_3 f$ and $I = e_1 e_2 e_3 e f_1 f_2 f_3 f = I_e I_f$. Note that, one can transforms a set of 6 bivectors $e_i e_j$, $e_i e$ to its orthogonal set $f_i f_j$, $f_i f$ using the pseudoscalar I, namely

$$f_i f_j = I e_i e_j, \quad f_i f = I e_i e, \tag{9.2}$$

for $i, j = 1, 2, 3$. The Lie algebra of the Lie group of rigid motion $SE(3)$ can be represented as a six-dimensional velocity vectors called twists or sometimes for simplicity screws

$$s = w_x e_2 e_3 + w_y e_3 e_1 + w_z e_1 e_2 + v_x e_1 e + v_y e_2 e + v_z e_3 e = \boldsymbol{w} + I_e \boldsymbol{v}. \tag{9.3}$$

The coscrew expanded with the dual bivector basis reads

$$s_f = w_x f_2 f_3 + w_y f_3 f_1 + w_z f_1 f_2 + v_x f_1 f + v_y f_2 f + v_z f_3 f = \boldsymbol{w} + I_f \boldsymbol{v}. \tag{9.4}$$

The Lie bracket of a pair of screws s_1 and s_2 is computed using their commutator

$$[s_1, s_2] = \frac{1}{2}(s_1 s_2 - s_2 s_1). \tag{9.5}$$

Instantaneously, the body is rotating about an a screw axis $\boldsymbol{L} = \boldsymbol{n} + I_e \boldsymbol{m}$ whit and angular velocity $\boldsymbol{w}$ and simultaneously translating along the same screw axis with linear velocity $\boldsymbol{v}$. Similarly, momenta called coscrews can be written using the dual bivectors as follows

$$\mathcal{P} = p_x f_2 f_3 + p_y f_3 f_1 + p_z f_1 f_2 + q_x f_1 f + q_y f_2 f + q_z f_3 f = \boldsymbol{p} + I_f \boldsymbol{q}. \tag{9.6}$$

The wrenches or coscrews are the dual of the twists or screws. These coscrews can be also used to represent wrenches which consists of torque and force bivectors

$$\boldsymbol{w} = \tau_x e_2 e_3 + \tau_y e_3 e_1 + \tau_z e_1 e_2 + F_x f_3 f_1 + F_y f_3 f_1 + F_z f_1 f_2 = \boldsymbol{\tau} + I_e \boldsymbol{F}. \tag{9.7}$$

For the transformations of screws, we utilize the homogeneous equation (7.34) for the case of the $SE(3)$

$$\mathtt{T} = \begin{bmatrix} \mathtt{R} & \mathtt{O} \\ [\mathtt{t}]_\times \mathtt{R} & \mathtt{R} \end{bmatrix}. \tag{9.8}$$

This transformation of lines can be formulated using motors as explained in Sect. 6.5.2

$$\begin{aligned} \boldsymbol{M} &= \boldsymbol{T}\boldsymbol{R} = e^{\frac{\theta}{2}\boldsymbol{n} + e\frac{\boldsymbol{t}}{2}}, \\ \boldsymbol{M}\widetilde{\boldsymbol{M}} &= 1. \end{aligned} \tag{9.9}$$

Given a Lie algebra element or bivector $\boldsymbol{B}$, the exponential map from the Lie algebra to the Lie group is given by

$$\boldsymbol{M} = e^{\frac{\boldsymbol{B}}{2}} = 1 + (\boldsymbol{B}/2) + \frac{1}{2}(\boldsymbol{B}/2)^2 + \cdots, \tag{9.10}$$

note that the factor $\frac{1}{2}$ is due to that we are working in the double covering group $SE(3)$. We can express a motor $\boldsymbol{M}$ using the basis of $G^+_{3,0,1}$ and $\boldsymbol{M}_f$ using the other set of bivectors of the extended $G^+_{3,0,2}$ as follows

$$
\begin{aligned}
\boldsymbol{M} &= m_0 + m_1 e_2 e_3 + m_2 e_3 e_1 + m_3 e_1 e_2 + m_4 e_1 e + m_5 e_2 e + m_6 e_3 e + m_8 e,\\
\boldsymbol{M}_f &= m_0 + m_1 f_2 f_3 + m_2 f_3 f_1 + m_3 f_1 f_2 + m_4 f_1 f + m_5 f_2 f + m_6 f_3 f + m_8 f.
\end{aligned}
$$

According Eq. (7.75), the dynamic motor equation is

$$
\dot{\boldsymbol{M}} = -\frac{1}{2}\boldsymbol{s}\boldsymbol{M}. \tag{9.11}
$$

Assuming that the screw motion of the body is constant through time, the dynamic motor equation (7.75) can be integrated to give

$$
\boldsymbol{M}(t) = e^{-\frac{s}{2}}\boldsymbol{M}(0) = e^{-\frac{(\boldsymbol{w}+I\boldsymbol{v})}{2}}\boldsymbol{M}(0), \tag{9.12}
$$

where $\boldsymbol{s}$ is the twist. This last equation represents a motor that rotates with a constant frequency rotation in the right-hand sense and has a constant linear velocity as well. The motor expanded with the coscrew is given by

$$
\boldsymbol{M}_f(t) = e^{-\frac{s_f}{2}}\boldsymbol{M}_f(0) = e^{-\frac{(\boldsymbol{w}+I_f\boldsymbol{v})}{2}}\boldsymbol{M}_f(0). \tag{9.13}
$$

The Lie group action $SE(3)$ on lines is given by

$$
\boldsymbol{L}' = \boldsymbol{M}\boldsymbol{L}\widetilde{\boldsymbol{M}} = \boldsymbol{T}\boldsymbol{R}\boldsymbol{L}\widetilde{\boldsymbol{R}}\widetilde{\boldsymbol{T}}, \tag{9.14}
$$

on screws is

$$
\boldsymbol{s}' = \boldsymbol{M}\boldsymbol{s}\widetilde{\boldsymbol{M}}, \tag{9.15}
$$

on coscrews is

$$
\boldsymbol{s}'_f = \boldsymbol{M}_f\boldsymbol{s}_f\widetilde{\boldsymbol{M}}_f. \tag{9.16}
$$

The group action $SE(3)$ on momenta is

$$
\mathcal{P}' = \boldsymbol{M}_f\mathcal{P}\widetilde{\boldsymbol{M}}_f. \tag{9.17}
$$

The kinetic energy of the rigid body is an scalar function computed via the inner product of its momenta coscrew and its velocity screw

$$
\begin{aligned}
E_K &= \frac{1}{2}\mathcal{P}\cdot\boldsymbol{s}, \qquad (9.18)\\
&= \frac{1}{2}(\boldsymbol{w}\cdot\boldsymbol{q} + \boldsymbol{v}\cdot\boldsymbol{p}) = \frac{1}{2}(w_x q_x + w_l dy q_y + w_z q_z + v_x p_x + v_y p_y + v_z p_z).
\end{aligned}
$$

The inertia matrix of a rigid body is given by a 6×6 matrix

$$\mathtt{N} = \begin{bmatrix} \mathtt{I} & m[\mathtt{c}]_\times \\ m[\mathtt{c}]^{\mathtt{T}}_\times & m\mathtt{I}_3 \end{bmatrix}, \quad (9.19)$$

where $\mathtt{I}$ is the 3×3 inertia matrix and $\mathtt{I}_3$ is the identity matrix, m is the mass of the body and $\mathtt{c}$ is the position vector of the center of mass expressed as $[\mathtt{c}]_\times$ an antisymmetric matrix to compute its cross product. Since $G_{6,0,2}$ has not enough null vectors as in $G_{3,3}$ or $G_{6,3}$, we can not express $\mathtt{N}$ as an affine transformation using bivectors as it is shown in Sect. 5.7. The inertia maps velocities to momenta, we will then formulate this mapping in terms of 6D vectors

$$\begin{bmatrix} q_x \\ q_y \\ q_z \\ p_x \\ p_y \\ p_z \end{bmatrix} = \mathtt{N} \begin{bmatrix} w_x \\ w_y \\ w_z \\ v_x \\ v_y \\ v_z \end{bmatrix} = \begin{bmatrix} \mathtt{I}\mathbf{w} + m\mathbf{c} \times \mathbf{v} \\ \mathrm{m}(\mathbf{c} \times \mathbf{w} + \mathbf{v}) \end{bmatrix}, \quad (9.20)$$

which in terms of bivectors can be written as

$$\mathtt{N} : \boldsymbol{s} \rightarrow \mathcal{P} = \boldsymbol{q} + I_f \boldsymbol{p} = (\mathtt{I}\boldsymbol{w} + \mathrm{m}(\boldsymbol{c} \times \boldsymbol{v})) + I_e \mathrm{m}(\boldsymbol{c} \times \boldsymbol{w} + \boldsymbol{v}). \quad (9.21)$$

The transformation properties of the inertia matrix can be inferred based on the fact that the kinetic energy must be invariant, namely

$$\mathtt{N} \rightarrow \mathtt{T}^{-\mathtt{T}}\mathtt{N}\mathtt{T}^{-1}. \quad (9.22)$$

9.2.1 The Shuffle Product

The dual of the join operator for expansion is the meet product for intersection. In non-degenerated geometric algebras, the meet is computed using the pseudoscalar I, namely

$$\boldsymbol{A} \vee \boldsymbol{B} = (\boldsymbol{A}^* \wedge \boldsymbol{B}^*) I = (\boldsymbol{A} I^{-1} \wedge \boldsymbol{B} I^{-1}) I. \quad (9.23)$$

Since $G^+_{6,0,2}$ is degenerated, the I^{-1} doesn't exist, thus we need to resort to the shuffle product of homogeneous k-vectors which yields as result the common vectors existing in both k-vectors. According [308], given $\boldsymbol{A} = a_1 \wedge a_2 \wedge \cdots \wedge a_j$ and $\boldsymbol{B} = b_1 \wedge b_2 \wedge \cdots \wedge b_k$, their shuffle product is computed as follows

$$\boldsymbol{A} \vee \boldsymbol{B} = \sum_\sigma sign(\sigma) \det(a_{\sigma(1)}, \ldots, a_{\sigma(n-k)}, b_1, \ldots, b_k) a_{\sigma(n-k+1)} \wedge \cdots \wedge a_{\sigma(j)}. \quad (9.24)$$

The sum is taken over all permutations σ of 1, 2, ..., j such that $\sigma(1) < \sigma(2) < \cdots < \sigma(n-k)$ and $\sigma(n-k+1) < \sigma(n-k+2) < \cdots < \sigma(j)$. Each a_i can be written as a sum of basis elements

$$a_i = a_{i1}e_1 + a_{i2}e_2 + \cdots + a_{in}e_n. \tag{9.25}$$

The determinant in Eq. (9.24) is the determinant of the matrix whose columns are the coefficients $a_{\sigma(1)i}, a_{\sigma(2)i}, \cdots, b_{\sigma(k)i}$. When $j+k < n$ the shuffle product is zero. As an example consider in $G^+_{6,0,2}$, the shuffle product between $f_1f_2f_3f$ and $f_3f_1e_1e_2e_3e$

$$\begin{aligned} f_1f_2f_3f \vee f_3f_1e_1e_2e_3e &= \det(f_1f_2f_3f_1e_1e_2e_3e)f_3f - \\ &- \det(f_2ff_3f_1e_1e_2e_3e)f_1f_3 + \cdots + \det(f_3ff_3f_1e_1e_2e_3e)f_1f_2, \\ &= f_3f_1. \end{aligned} \tag{9.26}$$

Note that the determinants values are either zero or 1, only the determinant of the middle term is 1.

9.2.2 Equations of Motion

In order to formulate the equation of motion of a rigid body, we need the coadjoint action of a screw on a coscrew, in this way one can compute the Coriolis terms. The operation is dual to the adjoint action of the screws acting on themselves involving a Lie bracket, namely

$$\boldsymbol{s}_1^T\{\boldsymbol{s}_2, \mathcal{P}\} = [\boldsymbol{s}_1, \boldsymbol{s}_2]^T\mathcal{P}, \tag{9.27}$$

which written as vectors read

$$\{\boldsymbol{s}, \mathcal{P}\} = \left\{\begin{pmatrix}\boldsymbol{w}\\ \boldsymbol{v}\end{pmatrix}, \begin{pmatrix}\boldsymbol{q}\\ \boldsymbol{p}\end{pmatrix}\right\} = \begin{pmatrix}\boldsymbol{w}\times\boldsymbol{q} + \boldsymbol{v}\times\boldsymbol{p}\\ \boldsymbol{w}\times\boldsymbol{p}\end{pmatrix}. \tag{9.28}$$

Thus, the dynamic equation can be written as follows

$$\mathtt{N}\frac{d}{dt}\boldsymbol{s} + \{\boldsymbol{s}, N\boldsymbol{s}\} = \mathtt{N}\frac{d}{dt}\boldsymbol{s} + \{\boldsymbol{s}, \mathcal{P}\} = \boldsymbol{\tau}, \tag{9.29}$$

where $\mathtt{N}$ is a 6×6 tensor applied to $\boldsymbol{s}$ and $\boldsymbol{\tau}$ is the external wrench (force/torque coscrew) acting on the body.

Next, we will formulate Eq. (9.29) entirely in the geometric algebra $G^+_{6,0,2}$. The evaluation map of a coscrew on a screw can be formulated using the following invariant element

$$Q = f_2f_3e_1e + f_3f_1e_2e + f_1f_2e_3e + f_1fe_2e_3 + f_2fe_3e_1 + f_3fe_1e_2.$$

The invariance of Q is with respect to the following group action $Q = (\boldsymbol{M}\boldsymbol{M}_f) Q(\widetilde{\boldsymbol{M}}_f\widetilde{\boldsymbol{M}})$. The evaluation map can be formulated using the shuffle operator $\vee$ as follows

$$\mathcal{P}(\boldsymbol{s}) = (\mathcal{P}\wedge Q) \vee \boldsymbol{s} = \mathcal{P} \vee (Q\wedge \boldsymbol{s}). \tag{9.30}$$

Since

$$\begin{aligned} Q\wedge s &= (w_x f_2 f_3 + w_y f_3 f_1 + w_z f_1 f_2 + v_x f_1 f + v_y f_2 f + v_z f_3 f) e_1 e_2 e_3 e, \\ &= (w_x f_2 f_3 + w_y f_3 f_1 + w_z f_1 f_2 + v_x f_1 f + v_y f_2 f + v_z f_3 f) I_e, \end{aligned} \tag{9.31}$$

then evaluating the shuffle operation $\vee$, we get

$$\mathcal{P}(\boldsymbol{s}) = \mathcal{P} \vee (Q\wedge \boldsymbol{s}) = w_x q_x + w_y q_y + w_z q_z + v_x p_x + v_y p_y + v_z p_z. \tag{9.32}$$

The inertia matrix can be transformed in a diagonal one via a suitable rigid body motion. The resulting diagonal matrix rewritten in terms of bivectors reads

$$\boldsymbol{N} = d_x f_1 f e_1 e + d_y f_2 f e_2 e + d_z f_3 f_2 e_3 e + m f_2 f_3 e_2 e_3 + m f_3 f_1 e_3 e_1 + m f_1 f_2 e_1 e_2,$$

where m stands for the body mass and the d_i is the mass times the square of the radius of gyration about the principle axis i.

In order to construct the map from velocities to momenta, we use another invariant, the pseudoscalar $I_f = f_1 f_2 f_3 f$. This element commutes with any $\boldsymbol{M}$, i.e. $I_f = \boldsymbol{M} I_f \widetilde{\boldsymbol{M}}$. The invariance of I_f is with respect to the group action $I_f = (\boldsymbol{M}\boldsymbol{M}_f) I_f (\widetilde{\boldsymbol{M}}_f\widetilde{\boldsymbol{M}})$ as well.

The evaluation map from velocities to momenta is given by

$$\mathcal{P}(\boldsymbol{s}) = (I_f\wedge \boldsymbol{N}) \vee s = I_f \vee (\boldsymbol{N}\wedge \boldsymbol{s}). \tag{9.33}$$

Similar as computing Eq. (9.30) using the bivector inertia one obtains

$$\mathcal{P} = m v_x f_2 f_3 + m v_y f_3 f_1 + m v_z f_1 f_2 + d_x w_x f_1 f + d_y w_y f_2 f + d_z w_z f_3 f. \tag{9.34}$$

Next, let us study the action of $SE(3)$ on the inertia. First, we transform the moment coscrew using $\boldsymbol{M}\boldsymbol{M}_f$

$$\begin{aligned} (\boldsymbol{M}\boldsymbol{M}_f)\mathcal{P}(\widetilde{\boldsymbol{M}}_f\widetilde{\boldsymbol{M}}) &= (\boldsymbol{M}\boldsymbol{M}_f)(I_f \vee (\boldsymbol{N}\wedge s)(\widetilde{\boldsymbol{M}}_f\widetilde{\boldsymbol{M}})) \\ &= (\boldsymbol{M}\boldsymbol{M}_f) I_f (\widetilde{\boldsymbol{M}}_f\widetilde{\boldsymbol{M}}) \vee ((\boldsymbol{M}\boldsymbol{M}_f)\boldsymbol{N}(\widetilde{\boldsymbol{M}}_f\widetilde{\boldsymbol{M}})\wedge(\boldsymbol{M}\boldsymbol{M}_f)\boldsymbol{s}(\widetilde{\boldsymbol{M}}_f\widetilde{\boldsymbol{M}}). \end{aligned} \tag{9.35}$$

Since I_f is invariant, $\mathcal{P}$ commutes with $\boldsymbol{M}$ and $\boldsymbol{s}$ commutes with $\boldsymbol{M}_f$, we obtain

$$\boldsymbol{M}_f \mathcal{P} \widetilde{\boldsymbol{M}}_f \widetilde{\boldsymbol{M}} = I_f \vee (\boldsymbol{M}\boldsymbol{M}_f \boldsymbol{N} \widetilde{\boldsymbol{M}}_f \widetilde{\boldsymbol{M}} \wedge \boldsymbol{M}\boldsymbol{s}\widetilde{\boldsymbol{M}}). \tag{9.36}$$

Thus, we can infer that the inertia, the screw and coscrew must transform according to

$$\begin{aligned} \boldsymbol{N} &\rightarrow (\boldsymbol{M}\boldsymbol{M}_f)\boldsymbol{N}(\widetilde{\boldsymbol{M}}_f\widetilde{\boldsymbol{M}}), \\ \boldsymbol{s} &\rightarrow (\boldsymbol{M}\boldsymbol{M}_f)\boldsymbol{s}(\widetilde{\boldsymbol{M}}_f\widetilde{\boldsymbol{M}}), \\ \mathcal{P} &\rightarrow (\boldsymbol{M}\boldsymbol{M}_f)\mathcal{P}(\widetilde{\boldsymbol{M}}_f\widetilde{\boldsymbol{M}}). \end{aligned} \tag{9.37}$$

Let us consider how the diagonal inertia is transformed under a motor for translation or translator in the y-direction $\boldsymbol{M} = 1 + \frac{1}{2}(t_y e_2 e)$, thus

$$\boldsymbol{M}\boldsymbol{M}_f = \left(1 + \frac{1}{2}t_y e_2 e\right)\left(1 + \frac{1}{2}t_y f_2 f\right) = 1 + \frac{1}{2}t_y f_2 f + \frac{1}{2}t_y e_2 e + \frac{1}{4}t_y^2 f_2 f e_2 e.$$

Computing the action on the diagonal inertia, we get

$$\begin{aligned} (\boldsymbol{M}\boldsymbol{M}_f)\boldsymbol{N}(\widetilde{\boldsymbol{M}}_f\widetilde{\boldsymbol{M}}) &= (d_x + mt_y^2) f_1 f e_1 e + d_y f_2 f e_2 e + (d_z + mt_y^2) f_3 f e_3 e \\ &- mt_y f_3 f e_2 e_3 + mt_y f_1 f e_1 e_2 + mt_y f_1 f_2 e_1 e - mt_y f_2 f_3 e_3 e + \\ &m f_2 f_3 e_2 e_3 + m f_3 f_1 e_3 e_1 + m f_1 f_2 e_1 e_2. \end{aligned} \tag{9.38}$$

We can compare this result expressed in terms of bivectors with the corresponding 6×6 inertia matrix

$$\begin{bmatrix} d_x + mt_y^2 & 0 & 0 & 0 & 0 & mt_y \\ 0 & d_y & 0 & 0 & 0 & 0 \\ 0 & 0 & d_z + mt_y^2 & -mt_y & 0 & 0 \\ 0 & 0 & -mt_y & m & 0 & 0 \\ 0 & 0 & 0 & 0 & m & 0 \\ mt_y & 0 & 0 & 0 & 0 & m \end{bmatrix}. \tag{9.39}$$

The kinetic energy of the rigid body can be written as

$$E_k = \frac{1}{2}\mathcal{P} \vee (\mathcal{Q} \wedge \boldsymbol{s}) = \frac{1}{2}I_f \vee (\mathcal{Q} \wedge \boldsymbol{s}) \vee (\boldsymbol{N} \wedge \boldsymbol{s}), \tag{9.40}$$

note that the bivector

$$I_f \vee (\mathcal{Q} \wedge \boldsymbol{s}) = w_x f_2 f_3 + w_y f_3 f_1 + w_z f_1 f_2 + v_x f_1 f + v_y f_2 f + v_z f_3 f, \tag{9.41}$$

is the same as the screw expanded with the bivectors basis $f_j f$ called $\boldsymbol{s}_f$, thus we can rewrite the previous equation in a more compact manner as follows

$$E_k = \frac{1}{2}\boldsymbol{s}_f \vee (\boldsymbol{N} \wedge \boldsymbol{s}). \tag{9.42}$$

According Eq. (9.13), the group action on the coscrew momenta as function of time can be written as

$$\mathcal{P}'(t) = \boldsymbol{M}_f(t)\mathcal{P}\widetilde{\boldsymbol{M}}_f(t) = e^{t\boldsymbol{s}_f}\mathcal{P}e^{-t\boldsymbol{s}_f}. \tag{9.43}$$

As usual, to get the Lie algebra bracket, if we differentiate the last equation with respect to t and then set $t = 0$, we obtain the coadjoint action of the screw $\boldsymbol{s}$ on the coscrew $\mathcal{P}$,

$$\begin{aligned}\{\boldsymbol{s}_f, \mathcal{P}\} &= \frac{1}{2}(\boldsymbol{s}_f\mathcal{P} - \mathcal{P}\boldsymbol{s}_f),\\ \{\boldsymbol{s}_f, \boldsymbol{I}_f \vee (\boldsymbol{N}\wedge\boldsymbol{s})\} &= \frac{1}{2}(\boldsymbol{s}_f(\boldsymbol{I}_f \vee (\boldsymbol{N}\wedge\boldsymbol{s})) - (\boldsymbol{I}_f \vee (\boldsymbol{N}\wedge\boldsymbol{s}))\boldsymbol{s}_f).\end{aligned} \tag{9.44}$$

According to Newton's second law, the force applied to the body is equal to the rate of change of momentum

$$\frac{d}{dt}\mathcal{P} = \boldsymbol{\mathcal{T}}, \tag{9.45}$$

where $\boldsymbol{\mathcal{T}}$ is the applied wrench. Next, for simplicity we do not include explicitly the variable t in Eq. (9.38) and it is rewritten it in terms of the bivector exponents

$$\boldsymbol{N}' = (\boldsymbol{M}\boldsymbol{M}_f)\boldsymbol{N}(\widetilde{\boldsymbol{M}}_f\widetilde{\boldsymbol{M}}) = e^{\frac{t}{2}(\boldsymbol{s}+\boldsymbol{s}_f)}\boldsymbol{N}e^{-\frac{t}{2}(\boldsymbol{s}_f+\boldsymbol{s}_f)}, \tag{9.46}$$

then differentiating this equation with respect to t and setting $t = 0$, we get

$$\frac{d}{dt}\boldsymbol{N} = \frac{1}{2}(\boldsymbol{s}\boldsymbol{N} - \boldsymbol{N}\boldsymbol{s}) + \frac{1}{2}(\boldsymbol{s}_f\boldsymbol{N} - \boldsymbol{N}\boldsymbol{s}_f). \tag{9.47}$$

According Eq. (9.33), the momentum is given by

$$\mathcal{P}(\boldsymbol{s}) = \boldsymbol{I}_f \vee (\boldsymbol{N}\wedge\boldsymbol{s}). \tag{9.48}$$

Since $\boldsymbol{I}_f$ is an invariant, its derivative is zero and $(\boldsymbol{s}\boldsymbol{N} - \boldsymbol{N}\boldsymbol{s})\wedge\boldsymbol{s} = 0$, thus we can simplify the derivative of $\mathcal{P}$ Eq. (9.45) as follows

$$\frac{d}{dt}\mathcal{P} = \boldsymbol{I}_f \vee (\boldsymbol{N}\wedge\dot{\boldsymbol{s}}) + \frac{1}{2}\boldsymbol{I}_f((\boldsymbol{s}_f\boldsymbol{N} - \boldsymbol{N}\boldsymbol{s}_f)\wedge\boldsymbol{s}). \tag{9.49}$$

Note that the second term on the right-hand side of previous equation is similar to Eq. (9.44).

Furthermore, we can improve the motion equation using the invariant $I_e = e_1e_2e_3e$, since $I_f I_e \vee \boldsymbol{p} = I \vee \boldsymbol{p} = \boldsymbol{p}$ for any element $\boldsymbol{p} \in G_{6,0,2}$, multiplying previous equation by I_e, we obtain

$$\boldsymbol{N} \wedge \dot{\boldsymbol{s}} + \frac{1}{2}((\boldsymbol{s}_f \boldsymbol{N} - \boldsymbol{N} \boldsymbol{s}_f) \wedge \boldsymbol{s}) = \boldsymbol{\tau} I_e. \tag{9.50}$$

9.3 The Geometric Algebra $G_{6,3}$

The $G_{6,3}$, was introduced by Zamora [318], in his work he describes the modeling of the geometries entities and certain Lie groups using versors. In this subsection, we will study in detail the $G_{6,3}$ Note that $G_{6,3}$ includes the 3D geometric algebra, G_3 and the mother algebra of the 3D Euclidean space $G_{3,3}$ useful for projective and affine geometry. Note that $G_{3,0}$ can be extended to the conformal GA $G_{4,1}$ to have a couple of null vectors e_0 and e_∞ and further to $G_{6,3}$ to have three couples of null vectors $\{e_{01}, e_{\infty 1}\}$,$\{e_{02}, e_{\infty 2}\}$ and $\{e_{03}, e_{\infty 3}\}$. The definitions given below are based on concepts of the geometric algebra of the reciprocal null cones explained in Sect. 5.5 and in conformal geometric algebra of Chap. 8.

The unit vector basis of $G_{6,3}$ is $\{e_1, e_2, e_3, e_4, e_5, e_6, e_7, e_8, e_9\}$ with the following properties:

$$\begin{aligned} e_i^2 &= 1, \quad i = 1, 2, 3, \\ e_j^2 &= 1, \quad j = 4, 5, 6, \\ e_k^2 &= -1, \quad k = 7, 8, 9, \\ e_i \cdot e_j &= e_i \cdot e_k = e_j \cdot e_k = 0. \end{aligned} \tag{9.51}$$

Note that this basis is not written in bold. We can form a set of null basis consisting of three couples of null vectors $\{e_{0i}, e_{\infty i}\}$ for $i = 1, 2, 3$ as follows

$$\begin{aligned} e_{01} &= \frac{(e_7 - e_4)}{2}, \quad & e_{\infty 1} &= e_7 + e_4, \\ e_{02} &= \frac{(e_8 - e_5)}{2}, \quad & e_{\infty 2} &= e_8 + e_5, \\ e_{03} &= \frac{(e_9 - e_6)}{2}, \quad & e_{\infty 3} &= e_9 + e_6, \end{aligned} \tag{9.52}$$

with the properties

$$e_{0i}^2 = e_{\infty i}^2 = 0, \quad e_{\infty i} \cdot e_{0i} = -1. \tag{9.53}$$

The 3D Euclidean pseudoscalar I_e and the pseudoscalar I of $G_{6,3}$ are defined by

$$I_e = e_1e_2e_3 \qquad I = e_1e_2e_3e_4e_5e_6e_7e_8e_9. \tag{9.54}$$

The unit pseudoscalars $\boldsymbol{E}_1$, $\boldsymbol{E}_2$, $\boldsymbol{E}_3$ represent three Minkowski planes and their wedge a 6-vector unit pseudoscalar $\boldsymbol{E}_{123} \in \mathbb{R}^{3,3}$ as follows

$$\begin{aligned}
\boldsymbol{E}_1 &= e_{\infty 1} \wedge e_{01} = e_4 \wedge e_7 = e_4 e_7, \\
\boldsymbol{E}_2 &= e_{\infty 2} \wedge e_{02} = e_5 \wedge e_8 = e_5 e_8, \\
\boldsymbol{E}_3 &= e_{\infty 3} \wedge e_{03} = e_6 \wedge e_9 = e_6 e_9, \\
\boldsymbol{E}_{123} &= \boldsymbol{E}_1 \wedge \boldsymbol{E}_2 \wedge \boldsymbol{E}_3 = \boldsymbol{E}_1 \boldsymbol{E}_2 \boldsymbol{E}_3
\end{aligned} \tag{9.55}$$

having the properties

$$\begin{aligned}
\boldsymbol{E}_1^2 &= \boldsymbol{E}_2^2 = \boldsymbol{E}_3^2 = 1, \\
\widetilde{\boldsymbol{E}}_1 &= -\boldsymbol{E}_1, \quad \widetilde{\boldsymbol{E}}_2 = -\boldsymbol{E}_2, \quad \widetilde{\boldsymbol{E}}_3 = -\boldsymbol{E}_3, \\
\boldsymbol{E}_{123}^2 &= 1, \quad \widetilde{\boldsymbol{E}}_{123} = -\boldsymbol{E}_{123}, \\
\boldsymbol{E}_1 e_{\infty 1} &= -e_{\infty 1} \boldsymbol{E}_1 = -e_{\infty_1}, \quad \boldsymbol{E}_1 e_{01} = -e_{01} \boldsymbol{E}_1 = e_{01} \quad \text{(absorption)}, \\
1 - \boldsymbol{E}_1 &= -e_{\infty 1} e_{01}, \quad 1 + \boldsymbol{E}_1 = -e_{01} e_{\infty 1}, \\
\boldsymbol{E}_1 e_{\infty 1} &= -e_{\infty 1} \boldsymbol{E}_1 = -e_{\infty_1}, \quad \boldsymbol{E}_1 e_{01} = -e_{01} \boldsymbol{E}_2 = e_{02}, \\
1 - \boldsymbol{E}_2 &= -e_{\infty 2} e_{02}, \quad 1 + \boldsymbol{E}_2 = -e_{02} e_{\infty 2}, \\
\boldsymbol{E}_3 e_{\infty 3} &= -e_{\infty 3} \boldsymbol{E}_3 = -e_{\infty 3}, \quad \boldsymbol{E}_3 e_{03} = -e_{03} \boldsymbol{E}_3 = e_{03}, \\
1 - \boldsymbol{E}_3 &= -e_{\infty 3} e_{03}, \quad 1 + \boldsymbol{E}_3 = -e_{03} e_{\infty 3}.
\end{aligned} \tag{9.56}$$

The unit pseudoscalar of $G_{6,3}$ can be computed as $I = I_e E_{123}$. The duals of $\boldsymbol{E}_i$ and $\boldsymbol{E}_{123}$ are given by

$$\begin{aligned}
\boldsymbol{E}_i{}^* &= \boldsymbol{E}_i I^{-1} = -\boldsymbol{E}\widetilde{I}, \quad \texttt{for i} = 1, 2, 3 \\
\boldsymbol{E}_{123}^* &= \boldsymbol{E}_{123} I^{-1} = -\boldsymbol{E}\tilde{I}.
\end{aligned} \tag{9.57}$$

Note that $\mathbb{R}^{6,3} = \mathbb{R}^3 \oplus \mathbb{R}^{3,3}$. We define other pseudoscalars which can be used to change between a IPNS representation to an OPNS one

$$I_i = I_e \wedge e_{\infty 1} \wedge e_{\infty 2} \wedge e_{\infty 3} \wedge e_0 \qquad I_o = I_e \wedge e_{01} \wedge e_{02} \wedge e_{03} \wedge e_{\infty}. \tag{9.58}$$

9.3.1 *Additive Split of $G_{6,3}$*

Similar as in conformal geometric algebra of Chap. 8, we can compute the origin and a point at infinity using the stereographic projection with respect to three projection axes formed with three origins and three points at infinity. Given an Euclidean point $\mathbf{x}_e = xe_1 + ye_2 + ze_3 \in \mathbb{R}^3$, we map into $G_{6,3}$ as follows

$$\begin{aligned}\boldsymbol{x}_w = \; & 2\frac{x}{x^2+1}e_1 + \frac{x^2-1}{x^2+1}e_4 + e_7 + \\ & 2\frac{y}{y^2+1}e_2 + \frac{y^2-1}{y^2+1}e_5 + e_8 + 2\frac{z}{z^2+1}e_3 + \frac{z^2-1}{z^2+1}e_6 + e_9.\end{aligned} \tag{9.59}$$

The origin is a linear combination of e_{01}, e_{02}, and e_{03} and it is computed as follows

$$\begin{aligned}e_0 &= \lim_{x,y,x\to 0}(\boldsymbol{x}_w) = \frac{1}{2}(e_7 - e_4) + \frac{1}{2}(e_8 - e_5) + \frac{1}{2}(e_9 - e_6), \\ &= e_{01} + e_{02} + e_{03}.\end{aligned} \tag{9.60}$$

One can interpret this result as the origin e_0 is a linear combination of three null vectors lying in general position on a plane passing thorough the origin e_0.

On the other hand, the point at infinity can be computed as the limit at ∞ of the stereographic projection with respect to three projection axes formed with three origins and three points at infinity, namely

$$\begin{aligned}e_\infty &= \lim_{x,y,x\to\infty}(\boldsymbol{x}_w), \\ &= (e_7 + e_4) + (e_8 + e_5) + (e_9 + e_6) = (e_{\infty 1} + e_{\infty 2} + e_{\infty 3}), \\ &= \frac{1}{3}(e_{\infty 1} + e_{\infty 2} + e_{\infty 3}),\end{aligned} \tag{9.61}$$

since e_∞ is homogeneous, it has been normalized by 3 just to fulfill the Lobachevsky relation $e_\infty \cdot e_0 = 1$. One can interpret this result as the point at infinity e_∞ is the linear combination of three points at infinity lying in general position on the plane at infinity π_∞.

Euclidean points $\mathbf{x}_e \in \mathbb{R}^3$ can be represented in $G_{6,3}$ as

$$\boldsymbol{x}_w = \mathbf{x}_e + \sum_{i=1,3}(\alpha_i e_{0i} + \beta_i e_{\infty i}), \tag{9.62}$$

where α_i and β_i are arbitrary scalars. Now, for an intuitive modeling of other geometric objects like ellipsoids and cylinders, the representation of $\boldsymbol{x}_w$ can rewritten in terms of the origin e_0 and the point at infinity e_∞ as follows

$$\boldsymbol{x}_w = xe_1 + ye_2 + ze_3 + \frac{1}{2}(x^2 e_{\infty 1} + y^2 e_{\infty 2} + z^2 e_{\infty 3}) + e_0. \tag{9.63}$$

A point $\boldsymbol{x}_w \in \mathbb{R}^{6,3}$ can be divided into its Euclidean and additive split parts by an operation called the *additive split*: $\mathbb{R}^{6,3} = \mathbb{R}^3 \oplus \mathbb{R}^{3,3}$ [185]. This additive split is defined by the projection operators $P_E^{||}$ (projection) and $P_E^{\perp}$ (rejection) as follows:

$$P_E^{||}(\boldsymbol{x}_w) = (\boldsymbol{x}_w \cdot \boldsymbol{E}_{123})\boldsymbol{E}_{123} = \sum_{i=1,3} (\alpha_i e_{0i} + \beta_i e_{\infty i}) \in \mathbb{R}^3, \tag{9.64}$$

$$P_E^{\perp}(\boldsymbol{x}_w) = (\boldsymbol{x}_w \cdot \boldsymbol{E}_{123}^*)\widetilde{\boldsymbol{E}_{123}^*} = (\boldsymbol{x}_w \wedge \boldsymbol{E}_{123})\boldsymbol{E}_{123} = \mathbf{x}_e \in \mathbb{R}^{3,3}, \tag{9.65}$$

$$\boldsymbol{x}_w = P_E(\boldsymbol{x}_w) + P_E^{\perp}(\boldsymbol{x}_w). \tag{9.66}$$

9.3.2 Geometric Entities of $G_{6,3}$

Recall that in $G_{4,1}$, the basic entity is the sphere with $SE(3)$ transformation (rotations, translations) and isotropic dilations. Since in $G_{6,3}$ it is allowed to formulate affine transformations, its basic entity of grade one is a quadric surface like ellipsoids, hyperboloid, spheres.

For the modeling of quadric surfaces Q, we will resort to a procedure already illustrated in Sect. 8.1.6, namely a point $\boldsymbol{x}_w$ lying on a surface Q satisfies the following constraint

$$x_w \cdot Q = 0. \tag{9.67}$$

The classic equation of an ellipsoid is

$$\frac{(x-p)^2}{a^2} + \frac{(y-q)^2}{b^2} + \frac{(z-s)^2}{c^2} = 1, \tag{9.68}$$

which can be expanded and factorized as

$$\frac{px}{a^2}e_1 + \frac{qy}{b^2}e_2 + \frac{sz}{c^2}e_3 - \frac{1}{2}(\frac{p^2}{a^2} + \frac{q^2}{b^2} + \frac{s^2}{c^2} - 1) - \frac{1}{2}(\frac{x^2}{2a^2} + \frac{y^2}{2b^2} + \frac{z^2}{2c^2}) = 0. \tag{9.69}$$

Considering the point Eq. (9.63), we can factorize (9.69) of in terms of inner product $\boldsymbol{x}_w \cdot Q = \boldsymbol{x}_w$ where the Q corresponds to the equation of an ellipsoid.

$$Q = \frac{p}{a^2}e_1 + \frac{q}{b^2}e_2 + \frac{s}{c^2}e_3 + \frac{1}{2}\left(\frac{p^2}{a^2} + \frac{q^2}{b^2} + \frac{s^2}{c^2} - 1\right)e_\infty + \\ + \left(\frac{1}{a^2}e_{01} + \frac{1}{b^2}e_{02} + \frac{1}{c^2}e_{03}\right), \tag{9.70}$$

note the Q is a IPNS vector. On the other hand, the 6-vector of the dual Q^* in OPNS reads

$$Q^* = QI = \boldsymbol{x}_{w1} \wedge \boldsymbol{x}_{w2} \wedge \boldsymbol{x}_{w3} \wedge \boldsymbol{x}_{w4} \wedge \boldsymbol{x}_{w5} \wedge \boldsymbol{x}_{w6}, \tag{9.71}$$

is a 6-vector.

Using Eq. (9.70) of the ellipsoid, one can straightforwardly derive the equation of the sphere, namely using the same ratio r for the axes x, y, z and factorizing the projective scalar r^2 we get

$$\begin{aligned} S &= \frac{p}{r^2}e_1 + \frac{q}{r^2}e_2 + \frac{s}{r^2}e_3 + \frac{1}{2}\left(\frac{p^2}{r^2} + \frac{q^2}{r^2} + \frac{s^2}{r^2} - 1\right)e_\infty + \left(\frac{1}{r^2}e_{01} + \frac{1}{r^2}e_{02} + \frac{1}{r^2}e_{03}\right), \\ &= pe_1 + qe_2 + se_3 + \frac{1}{2}(p^2 + q^2 + s^2 - r^2)e_\infty + (e_{01} + e_{02} + e_{03}), \\ &= \boldsymbol{x}_e + \frac{1}{2}(\boldsymbol{x}_e^2 - r^2)e_\infty + e_0, \end{aligned} \tag{9.72}$$

note that the equation for the sphere is equal to the equation for the sphere as in $G_{4,1}$.

Using Eq. (9.70) of the ellipsoid, one can again straightforwardly derive the equation of the cylinder, namely setting the ratio of axis z to infinity, we get

$$C = \frac{p}{a^2}e_1 + \frac{q}{b^2}e_2 + \frac{1}{2}\left(\frac{p^2}{a^2} + \frac{q^2}{b^2} - 1\right)e_\infty + \left(\frac{1}{a^2}e_{01} + \frac{1}{b^2}e_{02}\right). \tag{9.73}$$

For the 6-vector of the dual C^* in OPNS, we set one of the point to infinity,

$$C^* = CI = \boldsymbol{x}_{w1} \wedge \boldsymbol{x}_{w2} \wedge \boldsymbol{x}_{w3} \wedge \boldsymbol{x}_{w4} \wedge \boldsymbol{x}_{w5} \wedge \boldsymbol{e}_{\infty 3}. \tag{9.74}$$

The interpretation is of the cylinder is that an axis z of an ellipsoid touches the plane at infinity. Fig. 9.1 shows the intersection of a cylinder and an hyperboloid.

The equation for a pair of planes PP_{yz} is derived from Eq. (9.68)

$$(x - p)^2 = a^2, \tag{9.75}$$

which can be rewritten in terms of a inner product $\boldsymbol{x}_w \cdot PP_{yz}$. Thus, using Eq. (9.70) of the ellipsoid, one can again straightforwardly derive the equation of the pair of planes, namely setting the ratio of axis y and z to infinity, we get

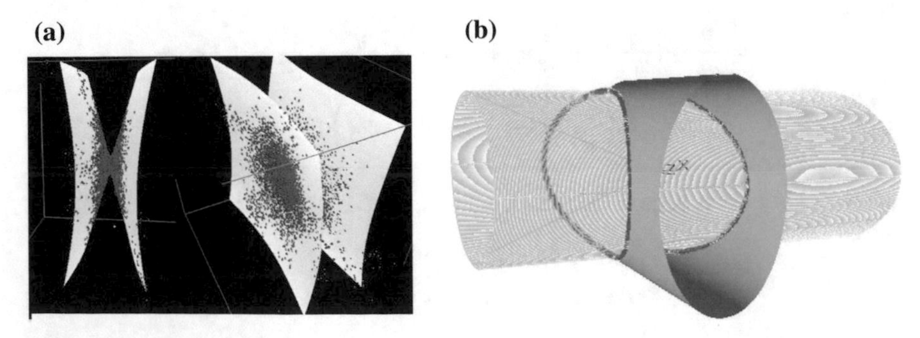

Fig. 9.1 **a** Hyperboloid in $G_{6,3}$. **b** Cylinder and hyperboloid in $G_{6,3}$. Figures provided by Julio Zamora-Esquivel

$$PP_{yz} = \frac{p}{a^2}e_1 + \frac{1}{2}\left(\frac{p^2}{a^2} - 1\right)e_\infty + \frac{1}{a^2}e_{01},$$
$$= pe_1 + \frac{1}{2}(p^2 - a^2)e_\infty + e_{01}. \tag{9.76}$$

The OPNS representation of two planes parallel to the yz planes is formulated setting two points at infinity

$$PP^*_{yz} = \boldsymbol{x}_{w1} \wedge \boldsymbol{x}_{w2} \wedge \boldsymbol{x}_{w3} \wedge \boldsymbol{x}_{w4} \wedge \boldsymbol{e}_{\infty 2} \wedge \boldsymbol{e}_{\infty 3}. \tag{9.77}$$

The equation of the plane equals to the one in $G_{4,1}$ and its a vector

$$\pi = n + de_\infty. \tag{9.78}$$

The OPNS representation of a plane can be computed as the wedge product of three points in general position lying on a plane and three points at infinity or the plane at infinity

$$\pi^* = \boldsymbol{x}_{w1} \wedge \boldsymbol{x}_{w2} \wedge \boldsymbol{x}_{w3} \wedge e_{\infty 1} \wedge \boldsymbol{e}_{\infty 2} \wedge \boldsymbol{e}_{\infty 3} = \boldsymbol{x}_{w1} \wedge \boldsymbol{x}_{w2} \wedge \boldsymbol{x}_{w3} \wedge \pi_\infty. \tag{9.79}$$

The equation of the line equals to the one in $G_{4,1}$ and it is bivector

$$L = nI_e + e_\infty m. \tag{9.80}$$

The OPNS representation of a line can be computed using the wedge of two points and the plane at infinity

$$L^* = \pi^* = \boldsymbol{x}_{w1} \wedge \boldsymbol{x}_{w2} \wedge e_{\infty 1} \wedge \boldsymbol{e}_{\infty 2} \wedge \boldsymbol{e}_{\infty 3} = \boldsymbol{x}_{w1} \wedge \boldsymbol{x}_{w2} \wedge \pi_\infty. \tag{9.81}$$

9.3.3 Intersection of Surfaces

An ellipse $\boldsymbol{q}$ can be regarded as the intersection of two ellipsoids $\boldsymbol{Q}_1$ and $\boldsymbol{Q}_2$. This means that for each point on the ellipse $\boldsymbol{x}_w \in \boldsymbol{q}$ they lie on both ellipsoids, that is, $\boldsymbol{x}_w \in \boldsymbol{Q}_1$ and $\boldsymbol{x}_w \in \boldsymbol{Q}_2$. Assuming that $\boldsymbol{Q}_1$ and $\boldsymbol{Q}_2$ are linearly independent, we can write for $\boldsymbol{x}_w \in \boldsymbol{q}$

$$(\boldsymbol{x}_w \cdot \boldsymbol{Q}_1)\boldsymbol{Q}_2 - (\boldsymbol{x}_w \cdot \boldsymbol{Q}_2)\boldsymbol{Q}_1 = x_w \cdot (\boldsymbol{Q}_1 \wedge \boldsymbol{Q}_2) = \boldsymbol{x}_w \cdot \boldsymbol{q} = 0. \tag{9.82}$$

This result tells us that since $\boldsymbol{x}_w$ lies on both ellipsoids,

$$\boldsymbol{q} = (\boldsymbol{Q}_1 \wedge \boldsymbol{Q}_2), \tag{9.83}$$

thus the intersection of the ellipsoids is the ellipse $\boldsymbol{q}$.

In OPNS, the intersection of surfaces are curves which lie in the intersection of the surfaces, e.g. the intersection of an ellipsoid and a plane is an ellipse. In general by intersecting two surfaces, one can compute very interesting contours which in turn can be used effectively for the applications as geometric constraints. The intersection or meet operation of two surfaces can be computed in OPNS via the wedge of the two surfaces

$$C^* = S_1^* \wedge S_2^*. \tag{9.84}$$

Similarly as Eq. (8.69), the ellipsoid $\boldsymbol{A}_r$ and the hyperplane $\boldsymbol{A}_{r+1}$ are related via the point $\boldsymbol{Q}$ (dual of the ellipsoid $\boldsymbol{Q}^*$), thus, we then rewrite Eq. (8.68) as follows:

$$\boldsymbol{Q} = \boldsymbol{A}_r \boldsymbol{A}_{r+1}^{-1}. \tag{9.85}$$

Using Eq. (9.85) and given the plane π (A_{r+1}) and the ellipse $\boldsymbol{q}$ ($\boldsymbol{A}_r$), we can compute the ellipsoid back

$$\boldsymbol{Q} = \boldsymbol{q}\pi^{-1}. \tag{9.86}$$

Figure 9.2 shows the intersection of two hyperboloids, and a quadric (two planes crossing the origin) and an hyperboloid. Now, using this result given the line $\boldsymbol{L}$ and the sphere $\boldsymbol{s}$ or an ellipsoid $\boldsymbol{Q}$, we can compute the IPNS form of the pair of points $\boldsymbol{PP}$ lying on the sphere or on the ellipsoid

$$\boldsymbol{PP} = \boldsymbol{sL} = \boldsymbol{s} \wedge \boldsymbol{L}. \tag{9.87}$$

$$\boldsymbol{PP} = \boldsymbol{QL} = \boldsymbol{Q} \wedge \boldsymbol{L}. \tag{9.88}$$

The OPNS of the pair the points is then given by

$$\boldsymbol{PP}^* = \boldsymbol{x}_{w_1} \wedge \boldsymbol{x}_{w_2}. \tag{9.89}$$

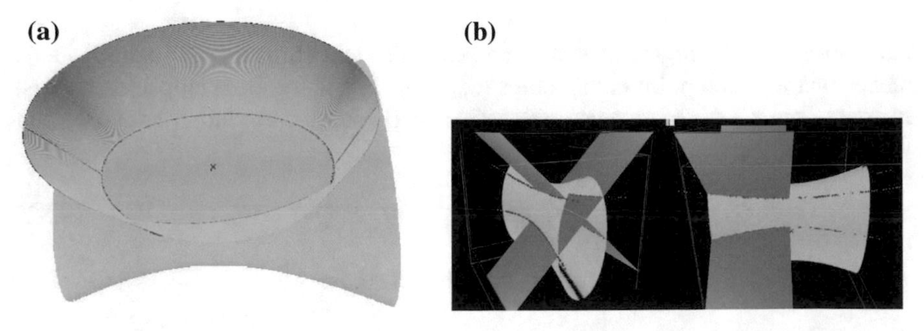

Fig. 9.2 **a** Intersection of two hyperboloids in $G_{6,3}$. **b** Intersection of two quadrics: two planes crossing the origin and a hyperboloid. Figures provided by Julio Zamora-Esquivel

9.3.4 Transformations of $G_{6,3}$

In Sect. 5.7.3, we study the projective groups of $G_{3,3}$ using versors, these transformations can be used for the projective transformation of a line $\boldsymbol{L} \in G_{3,3}$ or $\boldsymbol{L} = \boldsymbol{n} + e_\infty \boldsymbol{m} in G_{6,3}$.

Now, similar as in the case of conformal geometric algebra $G_{4,1}$, for the geometric objects in $G_{6,3}$ we can use the transversion $\boldsymbol{K_b}$ in terms of the origin $e_0 = e_{01} + e_{02} + e_{03}$, the translator $\boldsymbol{T}$ in terms of the point at infinity $e_\infty = \frac{1}{3}(e_{\infty 1} + e_{\infty 2} + e_{\infty 3})$, the rotor using a bivector $\boldsymbol{n}$ and the their combination for the motor $\boldsymbol{M}$ in terms of the screw axis $\boldsymbol{L}$, namely

$$\begin{aligned}
\boldsymbol{K_b} &= 1 + \mathbf{b} e_o, \\
\boldsymbol{T} &= e^{-\frac{t}{2}e_\infty} = 1 + \frac{1}{2}\boldsymbol{t} e_\infty, \\
\boldsymbol{R} &= e^{-\frac{\theta}{2}\boldsymbol{n}} = cos\left(\frac{\theta}{2}\right) - sin\left(\frac{\theta}{2}\right)\boldsymbol{n}, \\
\boldsymbol{M} &= \boldsymbol{T}\boldsymbol{R} = e^{-\frac{\theta}{2}\boldsymbol{L}}.
\end{aligned} \tag{9.90}$$

The versor for perspectivity called *perspector* can be compute either using the bivectors $e_i e_j$ like Eq. (5.168) or following the similar approach using the bivectors $e_{0i} \wedge e_{\infty j}$ like Eq. (5.169) to get

$$\begin{aligned}
\boldsymbol{P} = e^{\boldsymbol{B}_f} &= e^{\frac{f_3 e_4 e_8 - f_2 e_4 e_9 + f_1 e_5 e_9}{2}}, \\
&= e^{\frac{f_3 e_{01} \wedge e_{02} - f_2 e_{01} \wedge e_{03} + f_1 e_{02} \wedge e_{03}}{2}}, \\
&= 1 + \frac{1}{2}(f_3 e_4 e_8 - f_2 e_4 e_9 + f_1 e_5 e_9), \\
&= 1 + \frac{1}{2}(f_3 e_{01} \wedge e_{02} - f_2 e_{01} \wedge e_{03} + f_1 e_{02} \wedge e_{03}).
\end{aligned} \tag{9.91}$$

The versor for the Lorentz transformation called *Lorentor* can be compute either using the bivectors $e_i \wedge e_j$ like Eq. (5.168) or following the similar approach using the bivectors $e_{0i} \wedge e_{\infty j}$ like Eq. (5.169) to get

$$\begin{aligned}
\boldsymbol{R}_L &= e^{\boldsymbol{B}_L}, \\
&= e^{\frac{(e_5 e_6 - e_8 e_9) + (e_4 e_6 - e_7 e_9) + (e_4 e_5 + e_7 e_8)}{2}}, \\
&= e^{\frac{(e_{02} \wedge e_{\infty 3} - e_{02} \wedge e_{\infty 3}) + (e_{\infty 1} \wedge e_{03} - e_{01} e_{\infty 3}) + (e_{\infty 1} \wedge e_{02} - e_{01} \wedge e_{\infty 1})}{2}}.
\end{aligned} \tag{9.92}$$

The versor for the Shear transformation called *sheartor* can be compute either using the bivectors $e_i e_j$ like Eq. (5.168) or following the similar approach using the bivectors $e_{0i} \wedge, e_{\infty j}$ like Eq. (5.169) to get

$$
\begin{aligned}
\boldsymbol{S} &= e^{\boldsymbol{B}_S}, \\
&= e^{\frac{a_{23}(e_5e_6-e_9e_8)+a_{13}(e_4e_6-e_9e_7)+a_{12}(e_4e_5-e_8e_7)}{2}} \\
&\quad e^{-\frac{a_{32}(e_6e_5-e_8e_9)+a_{31}(e_6e_4-e_7e_9)+a_{12}(e_4e_5-e_8e_7)}{2}}, \\
&= e^{\frac{a_{23}(e_{\infty 2}\wedge e_{03}-e_{02}\wedge e_{\infty 3}+a_{13}(e_{\infty 1}\wedge e_{03}-e_{01}e_{\infty 3})+a_{12}(e_{\infty 1}\wedge e_{02}-e_{01}\wedge e_{\infty 2})}{2}} \\
&\quad e^{\frac{a_{32}(e_{\infty 3}\wedge e_{02}-e_{03}\wedge e_{\infty 2})+a_{31}(e_{\infty 3}\wedge e_{01}-e_{03}e_{\infty 1})+a_{21}(e_{\infty 2}\wedge e_{01}-e_{02}\wedge e_{\infty 1}}{2}}.
\end{aligned}
\tag{9.93}
$$

Example, we can apply an affine transformation to a cube C and stretch two faces to rhomboids and the other four faces to rectangles as follows,

$$
\mathcal{R} = \boldsymbol{S}\mathcal{C}\tilde{\boldsymbol{S}}. \tag{9.94}
$$

By considering the fact that the point at infinity is a linear combination of three points at infinity in general position $e_\infty = \frac{1}{3}(e_{\infty 1} + e_{\infty 2} + e_{\infty 3})$, we can formulate the dilator D to scale or dilate a surface in the direction pointed by each of these points at infinity, thus the versor for the dilation transformation $\boldsymbol{D}$ can be formulated as in $G_{3,3}$ either using the bivectors e_ie_j like Eq. (5.168) or following the similar approach using the bivectors $e_0 i\wedge, e_{\infty j}$ like Eq. (5.169) to get

$$
\begin{aligned}
\boldsymbol{D} &= e^{\boldsymbol{B}_D}, \\
&= e^{\frac{(1/0,\pm)e_7e_4+(0/1,\mp)e_8e_5+(0/1,\mp)e_9e_6+(-1/0,\pm)e_4e_7(0/-1,\mp)e_5e_8+(0/-1,\mp)e_6e_9}{2}}, \\
&= e^{\frac{(1/0,\pm)e_{\infty 1}e_{01}+(0/1,\mp)e_{\infty 2}e_{02}+(0/1,\mp)e_{infty3}e_{03}}{2}} \\
&\quad e^{\frac{(-1/0,\pm)e_{01}\wedge e_{\infty 1}+(0/-1,\mp)e_{02}\wedge e_{\infty 2}+(0/-1,\mp)e_{03}\wedge e_{\infty 3}}{2}}.
\end{aligned}
\tag{9.95}
$$

Example to transform an sphere S into an ellipsoid, we apply the dilator as follows,

$$
\mathcal{Q} = \boldsymbol{D}\mathcal{S}\tilde{\boldsymbol{D}}. \tag{9.96}
$$

9.4 The Geometric Subalgebras $G^+_{9,3}$ and $G^+_{6,0,6}$

We are interested to extend the double motor $G^+_{6,0,2}$ to $G^+_{6,0,6}$ in order to augment the null vectors to six and hence to have the necessary bivectors to formulate the affine transformation of lines in terms of versors. This will be a suitable mathematical framework to handle the algebra of lines, screws, and coscrews utilizing versors for $SE(3)$ Lie Group and for the affine transformations of lines.

Let us consider the geometric subalgebra $G^+_{9,3}$ which is generated as follows

$$
G^+_{9,3} = \texttt{gen}\{1, e_ie_j, e_ie, f_if_j, f_if, \ldots, I\}, \tag{9.97}
$$

where the bivectors are computed using an orthogonal set of basis vectors e_i, $f_i = e_{i+3}$ for $i = 1, 2, 3$ and six null vectors $e, f, g, \bar{e}, \bar{f}, \bar{g}$. $G_{9,3}^{+}$ has 66 bivectors like $e_ie_j, e_ie, f_if_j, f_if, e\bar{e}, f\bar{f}, g\bar{g}, \ldots, e_8e_9$ and some pseudoscalars like $I_e = e_1e_2e_3e$, $I_f = f_1f_2f_3f$, $I_g = g_1g_2g_3g$ and $I = e_1e_2e_3ef_1f_2f_3fg_1g_2g_3g = I_eI_fI_g$. Note that, one can transforms a set of three bivectors e_ie_j, e_ie to its orthogonal set f_if_j, f_if using the pseudoscalar $I_{ef} = I_eI_f$ or e_ie_j, e_ie to its orthogonal set g_ig_j, g_ig using the pseudoscalar $I_{eg} = I_eI_g$ or f_if_j, f_if to its orthogonal set g_ig_j, g_ig using the pseudoscalar $I_{fg} = I_fI_g$, namely

$$\begin{aligned} f_if_j &= I_{ef}e_ie_j, \quad f_if = I_{ef}e_ie, \\ g_ig_j &= I_{eg}e_ie_j, \quad g_ig = I_{eg}e_ie, \\ g_ig_j &= I_{fg}f_if_j, \quad g_ig = I_{fg}f_if, \end{aligned} \tag{9.98}$$

for $i, j = 1, 2, 3$. According Eq. (9.53) the set of null basis $G_{9,3}^{+}$ are

$$\begin{aligned} e_{01} &= \frac{(e_7 - e_10)}{2}, \quad e_{\infty 1} = e_7 + e_10, \\ e_{02} &= \frac{(e_8 - e_11)}{2}, \quad e_{\infty 2} = e_8 + e_11, \\ e_{03} &= \frac{(e_9 - e_12)}{2}, \quad e_{\infty 3} = e_9 + e_12, \end{aligned}$$

with the properties

$$e_{0i}^2 = e_{\infty i}^2 = 0, \quad e_{\infty i} \cdot e_{0i} = -1. \tag{9.99}$$

We will rename them as

$$\begin{aligned} \bar{e} &= e_{01}, \quad e = e_{\infty 1}, \\ \bar{f} &= e_{02}, \quad f = e_{\infty 2}, \\ \bar{g} &= e_{03}, \quad g = e_{\infty 3}. \end{aligned} \tag{9.100}$$

The goal to extend of $G_{6,0,2}^{+}$ to $G_{6,0,6}^{+}$ is to get the double motor algebra but equipped with six null vectors $\bar{e}, \bar{f}, \bar{g}, e, f, g$. Note that $\bar{e}, \bar{f}, \bar{g}$ represent the null vectors lying at the plane crossing the origin and e, f, g three null vectors called points at infinity lying at the plane at infinity.

Now, the representations of the screws $\boldsymbol{s}$, coscrews, and momenta $\mathcal{P}$ are equal as in $G_{6,0,2}^{+}$. Since in $G_{6,0,6}^{+}$, we have three null vectors of lying on null cone e, f, g and other three $\bar{e}, \bar{f}, \bar{g}$ on the dual null cone, we can formulate in terms of bivectors (generated by these null vectors) the affine transformation of the inertia matrix $\mathtt{N}$ as it is explained next. Since $G_{6,0,2}^{+}$ is a subalgebra of $G_{9,3}^{+}$, we can identify $G_{6,0,6}^{+}$ as the subalgebra of $G_{9,3}^{+}$ as well, consisting of bivectors in terms of six vectors basis $e_1, e_2, e_3, e_4 = f_1, e_5 = f_2, e_6 = f_3$ and the six null vectors named as $\bar{e}, \bar{f}, \bar{g}, e, f, g$. Thus, we can use these six null vectors of $G_{6,0,6}^{+}$ to represent the inertia matrix $\mathtt{N}$.

According Eq. (9.19), the inertia matrix of a rigid body is given by a 6×6 matrix

$$\mathtt{N} = \begin{bmatrix} \mathtt{I} & m[\mathtt{c}]_\times \\ m[\mathtt{c}]_\times^{\mathtt{T}} & m\mathtt{I}_3 \end{bmatrix}, \tag{9.101}$$

where $\mathtt{I}$ is the 3×3 inertia matrix and $\mathtt{I}_3$ is the identity matrix, m is the mass of the body, and $\mathtt{c}$ is the position vector of the center of mass expressed as $[\mathtt{c}]_\times$ an antisymmetric matrix to compute its cross product. Since $G_{6,0,2}$ has not enough null vectors as in $G_{3,3}$, $G^+_{9,3}$ or $G^+_{6,0,6}$, we can now express $\mathtt{N}$ in $G^+_{6,0,6}$ as an affine transformation using bivectors as it is shown in Sect. 5.5.

Note that Eq. (9.101) corresponds to the line transformation into the projective space $\mathbb{P}^5$. Similarly as shown in Sect. 5.7.3 for Lie groups in $\mathbb{R}^{3,3}$, Eq. (9.101) can be represented using the tables of the inner product of the unit basis or the null basis

$\cdot$	e_1	e_2	e_3	e_7	e_8	e_9
e_1	1	0	0	0	0	0
e_2	0	1	0	0	0	0
e_3	0	0	1	0	0	0
e_7	0	0	0	−1	0	0
e_8	0	0	0	0	−1	0
e_9	0	0	0	0	0	−1

$\cdot$	$\bar{e}$	$\bar{f}$	$\bar{g}$	e	f	g
$\bar{e}$	0	0	0	1	0	0
$\bar{f}$	0	0	0	0	1	0
$\bar{g}$	0	0	0	0	0	1
e	1	0	0	0	0	0
f	0	1	0	0	0	0
g	0	0	1	0	0	0

.

Using these tables, we can multiply the entries of the inertia matrix with bivectors as follows

$$\mathtt{T}_{\mathtt{e}_1,\mathtt{e}_2,\mathtt{e}_3,\mathtt{e}_7,\mathtt{e}_8,\mathtt{e}_9} = \begin{bmatrix} i_{11} & i_{12} & i_{13} & 0 & -mc_3 & mc_2 \\ i_{21} & i_{22} & i_{23} & mc_3 & 0 & -mc_1 \\ i_{31} & i_{32} & i_{33} & -mc_2 & mc_1 & 0 \\ 0 & mc_3 & -mc_2 & m & 0 & 0 \\ -mc_3 & 0 & mc_1 & 0 & m & 0 \\ mc_2 & -mc_1 & 0 & 0 & 0 & m \end{bmatrix}, \tag{9.102}$$

$$\mathtt{T}_{\bar{\mathtt{e}},\bar{\mathtt{f}},\bar{\mathtt{g}},\mathtt{e},\mathtt{f},\mathtt{g}} = \begin{bmatrix} 0 & -mc_3 & mc_2 & m & 0 & 0 \\ mc_3 & 0 & -mc_1 & 0 & m & 0 \\ -mc_2 & mc_1 & 0 & 0 & 0 & m \\ i_{11} & i_{12} & i_{13} & 0 & mc_3 & -mc_2 \\ i_{21} & i_{22} & i_{23} & -mc_3 & 0 & mc_1 \\ i_{31} & i_{32} & i_{33} & mc_2 & -mc_1 & 0 \end{bmatrix}. \tag{9.103}$$

Inertia and m$\mathtt{I}_3$: similar as in Sect. 5.7.3, we can compute the Lie operator for the inertia $\mathtt{I}$ and $m\ \mathtt{I}_3$

$$\mathcal{L}_{\mathtt{I},\mathtt{mI}_3} = \begin{bmatrix} 0 & \begin{bmatrix} 1\,0\,0 \\ 0\,1\,0 \\ 0\,0\,1 \end{bmatrix} \\ \begin{bmatrix} 1\,1\,1 \\ 1\,1\,1 \\ 1\,1\,1 \end{bmatrix} & 0 \end{bmatrix}. \tag{9.104}$$

The versor for inertia I and $\mathtt{mI}_3$ called *inertor* can be computed either using the bivectors $e_i \wedge e_j$ as Eq. (5.168) or following the similar approach using the bivectors combining $e, f, g, \bar{e}, \bar{f}, \bar{g}$ as Eq. (5.169) to get

$$\begin{aligned} \boldsymbol{I}_I &= e^{\boldsymbol{B}_I}, \\ &= e^{-\frac{i_{11}e\wedge\bar{e}+i_{12}e\wedge\bar{f}+i_{13}e\wedge\bar{g}+i_{21}f\wedge\bar{e}+i_{22}f\wedge\bar{f}+i_{23}f\wedge\bar{g}+i_{31}g\wedge\bar{e}+i_{32}g\wedge\bar{f}+i_{33}g\wedge\bar{g}+m\bar{e}\wedge e+m\bar{f}\wedge f+m\bar{g}\wedge g}{2}}. \end{aligned} \tag{9.105}$$

$\mathbf{m}[c]_x$ **and** $\mathbf{m}[c]_x^T$: similar as in Sect. 5.7.3, we can computed the Lie operator simultaneously for both $\mathrm{m}[c]_x$ and $\mathrm{m}[c]_x^T$

$$\mathcal{L}_{m[c]_x, m[c]_x^T} = \begin{bmatrix} \begin{bmatrix} 0 & -1 & 1 \\ 1 & 0 & -1 \\ -1 & 1 & 0 \end{bmatrix} & 0 \\ 0 & \begin{bmatrix} 0 & 1 & -1 \\ -1 & 0 & 1 \\ 1 & -1 & 0 \end{bmatrix} \end{bmatrix}. \tag{9.106}$$

The versor for $m[c]_x$ and $\mathrm{m}[c]_x^T$ can be computed either using the bivectors $e_i \wedge e_j$ as Eq. (5.168) or following the similar approach using the bivectors combining $e, f, g, \bar{e}, \bar{f}, \bar{g}$ as Eq. (5.169) to get

$$\begin{aligned} \boldsymbol{I}_{mc} &= e^{\boldsymbol{B}_{mc}}, \\ &= e^{m\frac{c_3\bar{e}\wedge\bar{f}-c_2\bar{e}\wedge\bar{g}+c_3\bar{f}\wedge\bar{g}-c_1 f\wedge g+c_2 e\wedge g+c_3 e\wedge f}{2}}. \end{aligned} \tag{9.107}$$

Thus, the resulting versor for the inertia matrix called *inertor* is computed in terms of the last versors as follows

$$\boldsymbol{N} = \boldsymbol{I}_I \boldsymbol{I}_{mc} = e^{\boldsymbol{B}_I} e^{\boldsymbol{B}_{mc}} = e^{\boldsymbol{B}_I + \boldsymbol{B}_{mc}}. \tag{9.108}$$

Having formulated the inertia matrix as a versor, we can now write in $G_{6,0,6}^{+}$ or $G_{9,3}^{+}$ the dynamic equation (9.29) entirely using versors, screws, and coscrews. Recalling Eq. (9.28)

$$\{s, \mathcal{P}\} = \left\{ \begin{pmatrix} \boldsymbol{w} \\ \boldsymbol{v} \end{pmatrix}, \begin{pmatrix} \boldsymbol{q} \\ \boldsymbol{p} \end{pmatrix} \right\} = \begin{pmatrix} \boldsymbol{w} \times \boldsymbol{q} + \boldsymbol{v} \times \boldsymbol{p} \\ \boldsymbol{w} \times \boldsymbol{p} \end{pmatrix}, \tag{9.109}$$

and the dynamic equation (9.29) which was written as follows

$$\mathrm{N}\frac{d}{dt}\boldsymbol{s} + \{\boldsymbol{s}, \boldsymbol{N}\boldsymbol{s}\} = \mathrm{N}\frac{d}{dt}\boldsymbol{s} + \{\boldsymbol{s}, \mathcal{P}\} = \boldsymbol{\tau}. \tag{9.110}$$

With the versor formulation of the inertia matrix, an affine transformation in $\mathcal{P}^6$, we can rewrite Eq. (9.110) entirely using versors, screws and coscrews, namely

$$\boldsymbol{N}\dot{\boldsymbol{s}}\tilde{\boldsymbol{N}} + \{\boldsymbol{s}, \boldsymbol{N}\boldsymbol{s}\tilde{\boldsymbol{N}}\} = \boldsymbol{N}\dot{\boldsymbol{s}}\tilde{\boldsymbol{N}} + \{\boldsymbol{s}, \mathcal{P}\} = \boldsymbol{\tau}. \tag{9.111}$$

9.5 Exercises

9.1 Proof the invariance of the inertia matrix expressed in Eq. (9.22).

9.2 Show the relation between the shuffle product of Eq. (9.24) and the generalized inner product given by the Eqs. (2.55) and (2.56). How we can use the generalized inner product to get same result as the shuffle product yields?

9.3 Proof in $G^+_{6,0,2}$, the invariance of the inertia matrix N expressed in Eq. (9.22), but now in terms of bivectors $\boldsymbol{N}$ Eq. (9.33) and the transformation T as motors $\boldsymbol{M}$.

9.4 Proof that we can infer that the inertia, the screw and coscrew must transform according to

$$\begin{aligned}
\boldsymbol{N} &\rightarrow (\boldsymbol{M}\boldsymbol{M}_f)\boldsymbol{N}(\widetilde{\boldsymbol{M}}_f\widetilde{\boldsymbol{M}}),\\
\boldsymbol{s} &\rightarrow (\boldsymbol{M}\boldsymbol{M}_f)\boldsymbol{s}(\widetilde{\boldsymbol{M}}_f\widetilde{\boldsymbol{M}}),\\
\mathcal{P} &\rightarrow (\boldsymbol{M}\boldsymbol{M}_f)\mathcal{P}(\widetilde{\boldsymbol{M}}_f\widetilde{\boldsymbol{M}}).
\end{aligned}$$

9.5 Proof that the equation in terms of quatrivectors

$$\begin{aligned}
(\boldsymbol{M}\boldsymbol{M}_f)\boldsymbol{N}(\widetilde{\boldsymbol{M}}_f\widetilde{\boldsymbol{M}}) = {} & (d_x + mt_y^2) f_1 f e_1 e + d_y f_2 f e_2 e + (d_z + mt_y^2) f_3 f e_3 e\\
& - mt_y f_3 f e_2 e_3 + mt_y f_1 f e_1 e_2 + mt_y f_1 f_2 e_1 e - mt_y f_2 f_3 e_3 e +\\
& m f_2 f_3 e_2 e_3 + m f_3 f_1 e_3 e_1 + m f_1 f_2 e_1 e_2.
\end{aligned}$$

corresponds to the 6×6 inertia matrix

$$\begin{bmatrix}
d_x + mt_y^2 & 0 & 0 & 0 & 0 & mt_y\\
0 & d_y & 0 & 0 & 0 & 0\\
0 & 0 & d_z + mt_y^2 & -mt_y & 0 & 0\\
0 & 0 & -mt_y & m & 0 & 0\\
0 & 0 & 0 & 0 & m & 0\\
mt_y & 0 & 0 & 0 & 0 & m
\end{bmatrix}.$$

9.6 One extends $G_{4,1}$ to $G_{6,3}$ to get three null vectors as origins and three null vectors as three points at infinity. Show with equations, why these null vectors can transform spheres to ellipsoids.

9.7 Using the Eq. (9.70) of the ellipsoid, one can derive the equation of the cylinder, namely setting the ratio of axis z to infinity, we get

$$C = \frac{p}{a^2}e_1 + \frac{q}{b^2}e_2 + \frac{1}{2}\left(\frac{p^2}{a^2} + \frac{q^2}{b^2} - 1\right)e_\infty + \left(\frac{1}{a^2}e_{01} + \frac{1}{b^2}e_{02}\right).$$

Derive the equation of an hyperboloid depicted in Fig. 9.1b.

9.8 In $G_{6,3}$, compute with eClifford and Maple [1] the intersection of a cylinder with a hyperboloid. Plot with Maple your results.

9.9 In $G_{6,3}$, compute with eClifford and Maple [1] the intersection of a sphere with a hyperboloid. Plot with Maple your results.

9.10 In $G_{6,3}$, compute with eClifford and Maple [1] the intersection of a ellipsoid with a plane. Plot with Maple your results.

9.11 Compute with eClifford and Maple [1] the transformation of an ellipsoid using the following versors

$$\begin{aligned}
\boldsymbol{K}_{\mathbf{b}} &= 1 + \mathbf{b}e_o,\\
\boldsymbol{T} &= e^{-\frac{t}{2}e_\infty} = 1 + \frac{1}{2}\boldsymbol{t}e_\infty,\\
\boldsymbol{R} &= e^{-\frac{\theta}{2}\boldsymbol{n}} = cos\left(\frac{\theta}{2}\right) - sin\left(\frac{\theta}{2}\right)\boldsymbol{n},\\
\boldsymbol{M} &= \boldsymbol{TR} = e^{-\frac{\theta}{2}\boldsymbol{L}}.
\end{aligned} \tag{9.112}$$

Plot with Maple your results.

9.12 In $G_{6,3}$, compute with eClifford and Maple [1] the rigid motion of an hyperboloid and the transformation of a sphere to an ellipsoid. Plot with Maple your results.

9.13 In $G_{6,3}$, using eClifford and Maple [1] apply the Perspector to cube and a Lorentor to a cube. Plot with Maple and interpret your results the effect of the Perspectivity and the transformation in space and time by the Lorentz transformation.

9.14 In $G^+_{9,3}$, using eClifford and Maple [1] apply the Sheartor to a cube and a Dilator to a cube. Plot with Maple and interpret your results.

9.15 In $G^+_{9,3}$, using eClifford and Maple [1] apply the Dilator to transform a sphere into an ellipsoid. Plot with Maple and interpret your results.

9.16 In $G^+_{9,3}$, write the equations of a screw line $\boldsymbol{L}$, a twist $\boldsymbol{s}$, and using the inertor a momentum $\boldsymbol{p}$, then a wrench $\boldsymbol{w}$. Explain your self that these geometric entities are needed to write the Newton–Euler dynamic equation of serial manipulators.

9.17 In $G^+_{9,3}$ compute the Lie bracket $[\boldsymbol{s}_i, \boldsymbol{s}_j]$ of twists $\boldsymbol{s}_i$ and $\boldsymbol{s}_j$.

9.18 In $G^+_{9,3}$ compute the Lie cobracket $\{\boldsymbol{s}_i, \boldsymbol{w}_j\}$ of twist $\boldsymbol{s}_i$ and wrench $\boldsymbol{w}_j$.

9.19 In $G^+_{9,3}$, write the direct kinematics of a n-DOF serial manipulator.

9.20 In $G^+_{9,3}$, write the direct kinematics of a four legs Steward platform.

9.21 In $G^+_{9,3}$, using Eq. (9.111), write of a Newton–Euler procedure to compute the dynamics of a double spherical pendulum.

9.22 In $G^+_{9,3}$, Using Eq. (9.111), write of a Newton–Euler procedure to compute the dynamics of a robot manipulator with three rotational joints.

Chapter 10
Programming Issues

In this chapter, we will discuss the programming issues to compute in the geometric algebra framework. We will explain the technicalities for the programming which you have to take into account to generate a sound source code. At the end we will discuss the use of specialized hardware as FPGA and NVidia CUDA to improve the efficiency of the code processing for applications in real time.

10.1 Main Issues for an Efficient Implementation

In the last decade were many attempts to develop software for Clifford algebra and geometric algebra computing some for computer science and numerical computing [75, 197, 234] and others for symbolic computing [1, 6, 77]. The efficient implementation is still an issue which depends greatly on the algorithmic complexity of the required algorithms and also the kind of accelerating cards being used. Fast hardware for running geometric algebra algorithms was recently proposed by [103, 215, 235, 287]. Nevertheless, it is still a long way to go in order to have an efficient and fast software to run geometric algorithms for theorem proving and for real-time applications. We have to admit, however, that the current developments are good contributions to achieve this main goal.

According to the majority of the Clifford or geometric algebra software developers, it seems that there are already three well-established issues for an efficient implementation: the basic computing entities or multivectors, the role of metric, and the computational burden caused by the number of involved basic operations. We will next discuss these issues from our point of view and including useful suggestions of other authors [32, 76].

- 1: Multivectors: the geometric entities of a geometric algebra G_n are represented as multivectors which are expanded in a multivector basis of length 2^n. It is not prudent to represent all multivectors in a frame including many zeros, rather specialize the representation fixing the length to accomplish the needed representation and geometric products ignoring all the unnecessary multivector components. This

E. Bayro-Corrochano, *Geometric Algebra Applications Vol. I*,
https://doi.org/10.1007/978-3-319-74830-6_10

implies a more limited use of grades or basis blades which consequently reduces the storage and processing time.
- 2: When we formulate representations and algorithms for a certain problem, we need to identify exactly the appropriate geometric algebra with a suitable metric. This is not necessarily a trivial task, to answer the question in which metric space $\mathbb{R}^{m,n}$ we should work, it will depend how well we are acquainted with the problem in question and the manner how we should tackle the problem. For instance, when we work in classification using neurocomputing it is more advantageous to treat the feature space in higher dimensions in order to linearize the classification problem. Thus, we should not develop a program for a general geometric algebra and apply it for a particular problem. All indicates that we should have a specialized implementation for every metrically different geometric algebra.
- 3: The number of basic operations on multivectors compared with those in vector calculus (linear transformations, inner or cross products, norms, etc.) is quit large. Usual multivector expressions demand of multiple products and operations which can be reduced in number and complexity by folding them into one calculation instead to execute them one by one by calling a series of generic functions. This tell us that we should rather program things out explicitly for each operation restricted to each multivector type. It is appropriate to avoid a combinatorial explosion due to the number of multivector blade components and variables.

The analysis of these three issues suggests us not to implement a program for a general geometric algebra rather in a metric-defined geometric algebra, we have to break it up and specialize both the representation and the operation computations. Bear in mind that multivectors are unfortunately too big, the role of metrics constrains indeed the world representation but not necessarily enough for an efficient computation.

The operations on multivectors are, in general, simple and universal but unfortunately too slow if they are not specialized, optimized, and formulated in such a way so that one can also take advantage of cost-effective hardware to speed up the computations.

10.1.1 Specific Aspects for the Implementation

The specific considerations for an efficient implementation can be summarized as follows:

- Do not waste computer memory. Try to store each variable as compactly as possible. Keep circumstantial results of functions in between which can reused later on.
- Implement the generic functions over the algebra in a efficient manner, such as factorization, meet. For that, one should consider various practical details:
  - (i) Reduce unpredictable memory access.
  - (ii) Similar to the FFT, find zero coordinates and avoid to process them.

– (iii) Optimize non-trivial geometric algebra expressions, particularly being careful not to implement them as a series of function calls.
– (iv) Conditional branches are prohibitive in order to avoid the loss of processor cycles due to an incorrectly predicted conditional branch.
– (v) Whenever possible, unroll loops, that is, let them run only over the dimension of the algebra or the subspaces determined by grade of the blade of the k-vectors.

10.2 Implementation Practicalities

As you may have noticed, we have not mentioned a particular programming language. You should consider as potential programming languages C++, JAVA, Python, Mathlab and for symbolic calculations Mathematica and Maple with eClifford [1]. In the last section, we give some comments on some existing multivector software packets.

In general, you should consider the specification of the algebra to work in, the implementation of the general multivector class, and the generation of optimized functions. Next, we discuss these aspects in some detail.

10.2.1 *Specification of the Geometric Algebra $G_{p,q}$*

The program requires first to know

- the dimension n of the algebra G_n.
- The metric of the algebra $G_{p,q}$, type of quadratic form expressing the space signature.
- The definition of the specialized types, list of basis blades, and versors.
- The definition of the constants. This is referred to as constant multivector values, which should be considered constant, for example basis vectors, e_0 and the pseudoscalar I_n. By including the constants in the algebra specifications, they can be easily recognized and optimized away by the code generator or compiler.

10.2.2 *The General Multivector Class*

For the program, the implementation of a general multivector is more appropriate. This is because many equations of geometric algebra are computed regardless of the multivector type of the input and expansions and contractions vary the output multivector length as well. On the other hand due to noisy entries, the multivector type of variables may change unexpectedly at run time.

We should care for the administration of the coordinates of the general multivector, because during the computation we can naively try to store all of the 2^n multivector components, some of which could in fact be zero or perhaps not all are needed to accomplish certain computational goal. Therefore, in order to increase the computational performance, we should avoid to store coordinates which are zero. Paying attention to the grade of blades, we can select only the k-blades of a multivector which necessary are involved in a certain computation, rejecting the irrelevant ones. Since versors are even or odd, we can at least leave the half of their grade counterparts. Note that for low-dimensional spaces, selecting the involved k-vectors reduces the dependency of a slow conditional branching by looking at the coordinate level. However reduction by grade still misses some good opportunities at higher level of complexity due to the inherent characteristics of the computation. This applies to entities which have common characteristics for intervening in the computation and depend heavily on the specific use of the algebra, for example select only flat points in conformal geometric algebra. In general, alerting the processor with some proper constraints of the problem in question could help to group zero coordinates of blades of the same grade, therefore alleviating even further the problem caused by unnecessary computations. Symbolic simplification and factorization of multivector functions will help to reduce their processing time, particularly if they are called in loops.

For each multivector type used in the program, it is desirable to generate specialized multivector classes, bearing in mind that the class provides storage for the non-constant coordinates and some functionality to convert back and forth between the general and the specialized multivector classes. The rest of the functionality will be provided by the functions over the entire algebra.

10.2.3 Optimization of Multivector Functions

The optimization of multivector functions over elements of the algebra is extremely important to speed up the processing and to guarantee high performance of the computer program. To start, the functions are formulated based on their high-level definitions. However, the precise syntax of the definition depends greatly on the generative programming method used. Using programming language C^{++} a kind of $meta-programming\,approach$ the functions are formulated as a C^{++} template functions. These functions are instantiated with specific type of multivectors. A reasonable procedure for generating code and possible optimized from a high-level definition should take into consideration the following commonsense advices

- The types of the function arguments are substituted with efficient specializations.
- The expressions in the function are formulated explicitly on a basis.
- Simplify the expressions symbolically, so that most of the operations and products are executed at the basis level; unnecessary computations are removed; identical terms are fused, reducing length looking at coordinate level, grade of k-vectors, group level of complex geometric entities, etc.

- Avoid unnecessary branching, conditionals and loops. For multivector classification and factorization, it is advisable to unroll loops in order to ensure that each variable has a fixed multivector type. Here via blade factorization after loops are unrolled, all variables in the algorithm get a fixed type; consequently, most conditional branches can be removed.
- The algorithms for meet and join of blades cannot be optimized by using specialized multivectors. This is because in the algorithm the multivector type of several variables is normally not fixed. Also there are several other unavoidable conditional branches.
- The return type of the function is specified.

10.2.4 Factorization

Many researchers have explored ways to factorize representations [32, 75, 94]. Common multiplicative representations in geometric algebra include blades as wedge products of vectors and versors as geometric product of vectors (reflections for rotations, reversions, translations, etc). A k-blade corresponds to the wedge product of k vectors, and similarly a versor is the result of the wedge of certain number n of vectors. After factorization of a k-blade or a versor, for both the resulting amount of vectors can be stored as a list of vectors. Note that the storage requirements of blades and versors becomes $O(n^2)$ instead of $O(2^n)$, when we factorize them using the basis-of-blades standard method. Recall that multivectors are often not homogeneous, i.e., k-blades or versors, and are expanded as a sum of multiple blades or versors of different grade; thus, this requires more storage.

As illustration, we include the following procedure for factorizing a blade which was proposed by [32, 75, 94]:

- Input: a nonzero-blade $\boldsymbol{X}$ of grade r.
- Compute its norm $\alpha = |\boldsymbol{X}|$.
- For $\boldsymbol{X}$, find the basis blade $\check{\boldsymbol{X}}$ with the largest coordinate and then determine the r basis vectors e_i that span $\check{\boldsymbol{X}}$.
- Normalize the input blade $\boldsymbol{U} = \boldsymbol{X}/\alpha$.
- For all the r basis vectors $e_i \in$ of $\check{\boldsymbol{X}}$ compute:
  - Project e_i onto $\boldsymbol{U}$, i.e., $g_i = (e_i \cdot \boldsymbol{U})\boldsymbol{U}^{-1}$.
  - Normalize g_i and store it into the list of factors.
  - Update $\boldsymbol{U} \leftarrow g_i^{-1} \cdot \boldsymbol{U}$.
- Obtain the last factor $g_r = \boldsymbol{U}$ and normalize it.
- Output: the list the factors g_i and the scale α.

Further research work has to be done for more efficient factorization methods specially for high-dimensional ($n < 10$) algebras. New insights into this regard can be found in the Fontjine's Ph.D thesis [94].

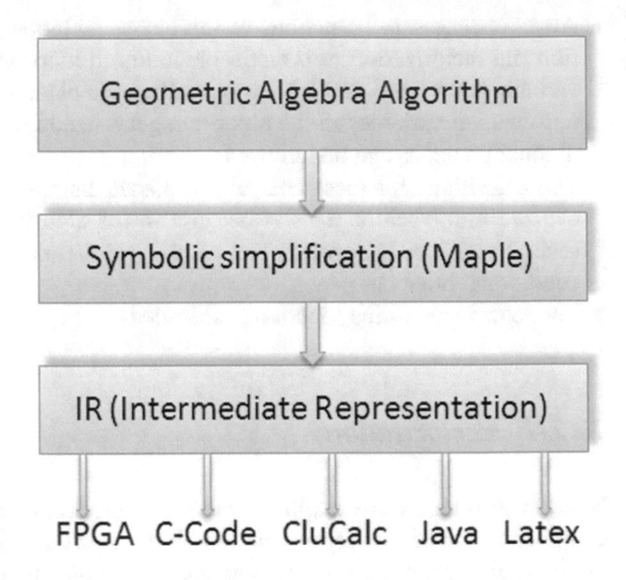

Fig. 10.1 Computing the pair of points as the meet of the swivel plane and the circle

10.2.5 Speeding up Geometric Algebra Expressions

To reduce the complexity of multivector expressions, we can resort to a gradual symbolic simplification which can then let run using cost-effective hardware; see Fig. 10.1. The main idea here is that after we formulated an algorithm in geometric algebra terms for a certain task, we reduce its complexity using symbolic simplification Maple with eClifford [1]. This produces a generic Intermediate Representation (IR) that can be used for the generation of different output formats such as C-code, FPGA descriptions (Verilog language), or CLUCalc code in order to visualize the results. For this purpose, an architecture called *Gaalop* was developed by Hildebrand et al. [142]. As an example, let us consider a problem of inverse kinematic.

Figure 10.2 shows the rotation planes of robot arm elbow. We consider a circle which is the intersection of two reference spheres $z = s_1 \wedge s_2$. We compute then the pair of points ($\boldsymbol{P}\boldsymbol{P}$) as the meet of the swivel plane and the circle: $\boldsymbol{P}P = z \wedge \pi_{swivel}$, so that we can choose a realistic position of the elbow. This is a point which lies on the meet of two spheres. This position is given by the 3D coordinates $\boldsymbol{p}_{ex}$, $\boldsymbol{p}_{ey}$, and $\boldsymbol{p}_{ez}$. After the optimization of the equation of the inverse kinematics, we get an efficient representation for the computing of this elbow position point. You can see for $\boldsymbol{p}_{ex}$ in Fig. 10.3 its data flow and its pipeline structure. Using this implementation, the authors reported for the whole inverse kinematics algorithm an remarkable acceleration of more than 130 times faster than the speed of an unoptimized algorithm [143].

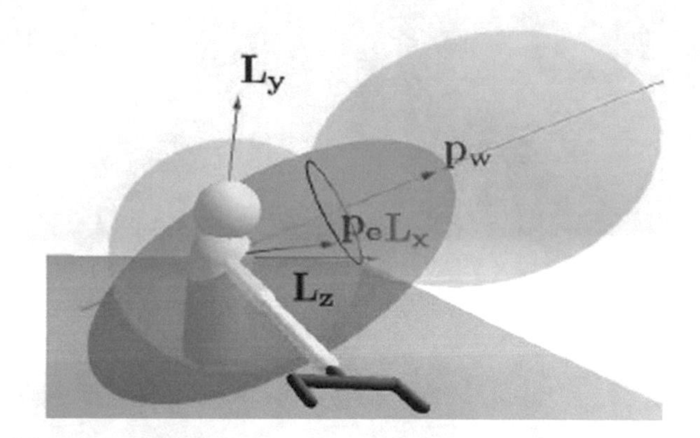

Fig. 10.2 Computing the pair of points as the meet of the swivel plane and the circle

10.2.6 Multivector Software Packets

The existing software packets for multivector programming are:
Fortran-based CLICAL developed by the group of Pertti Lounesto. It is very useful for fast computation and theorem proving. It is available at
http://users.tkk.fi/ppuska/mirror/Lounesto/CLICAL.htm
Mathlab-based geometric algebra tutorial called GABLE supports $N \leq 3$ and it is available at
http://staff.science.uva.nl/~leo/GABLE/index.html
Maple-based eCLIFFORD for computing in any dimension, because the computing of the Clifford product is carried out using Welsh functions [197]. It is very useful for symbolic programming and theorem proving, and it is available at
http://math.tntech.edu/rafal
$C{++}$-based CLUCal. It is very useful for practicing and learning geometric algebra computations in 2D and 3D particularly for visualization, computer vision, and crystallography. It is available at
http://www.perwass.de/cbup/clu.html.
GAIGEN2 generates fast $C{++}$ or JAVA sources for low-dimensional geometric algebra. Very useful for practicing and learning geometric algebra computing and to try a variety of problems of computer science and graphics. It is available at
http://www.science.uva.nl/ga/gaigen/
$C{++}$ MV 1.3.0 sources supporting $N \leq 63$. The author Ian Bell has developed up to 1.6 with significant functionality extensions and bug fixes. Very powerful multivector software for applications in computer science and physics. It is available at
http://www.iancgbell.clara.net/maths/index.htm
The $C{++}$ GEOMA v1.2 developed by Patrick Stein contains $C{++}$ libraries for Clifford algebra with orthonormal basis. It is available at
http://nklein.com/software/geoma

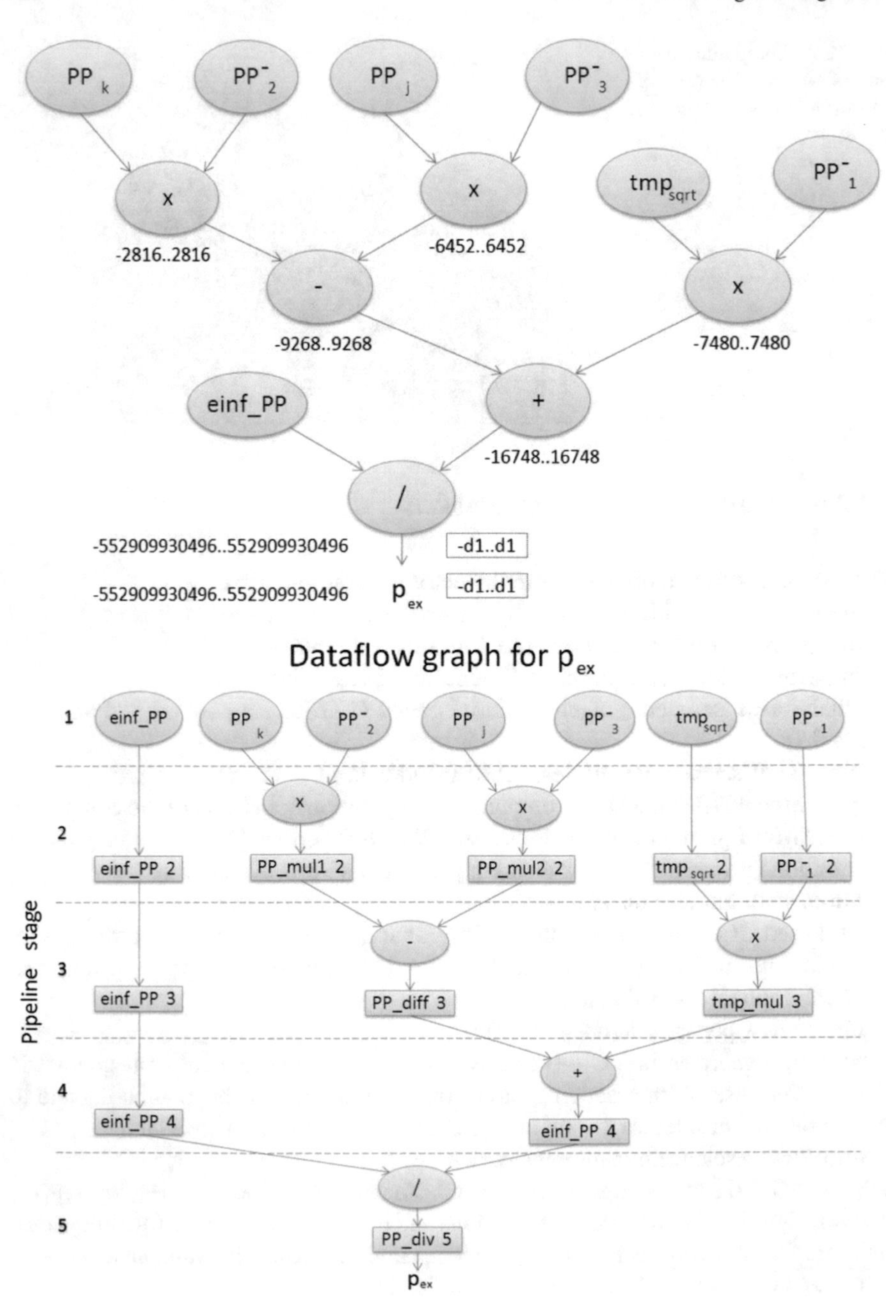

Fig. 10.3 Data flow and pipeline structure for a fast computing of the elbow position point

The reader can also download our $C++$ programs which are being routinely used for applications in robotics, image processing, wavelets transforms, computer vision, neurocomputing, and medical robotics. These are available at

http://www.gdl.cinvestav.mx/edb/GAprogramming.

Readers who want to develop their own programs for Clifford or geometric algebra applications should consider the advice given in this section, learn from the above-cited multivector software packets, and adopt and integrate their new developments. To learn geometric algebra CLUcal, GAIGEN2 and eClifforf [1] are highly recommended. To write a $C++$ geometric algebra program, one should start to look the GEOMA, the MV 1.3.0, and the code generator of GAIGEN or download the programs of our homepage. For symbolical computing and proof of theorems, CLICAL [197] and eCLIFFORD [1] are extremely useful.

10.2.7 Specialized Hardware to Speed up Geometric Algebra Algorithms

There are many works using dedicated hardware or FPGA to accelerate geometric algebra algorithms

- Dedicated hardware [95, 283]
- The use of FPGA for geometric algebra computing [286, 287]
- The use of NVidia CUDA for geometric algebra computing

In Chap. 16 and Sect. 16.5, we illustrate the implementation of computer vision algorithms using conformal geometric algebra and speed up using FPGA. Currently, we are accelerating interpolation in conformal geometric algebra using FPGA and Nvidia CUDA.

Part III
Image Processing and Computer Vision

Chapter 11
Quaternion–Clifford Fourier and Wavelet Transforms

11.1 Introduction

This chapter presents the theory and use of the Clifford Fourier transforms and Clifford wavelet transforms. We will show that using the mathematical system of the geometric algebra, it is possible to develop different kinds of Clifford Fourier and wavelet transforms which are very useful for image filtering, pattern recognition, feature detection, image segmentation, texture analysis, and image analysis in frequency and wavelet domains. These techniques are fundamental for automated visual inspection, robot guidance, medical image processing, analysis of image sequences, as well as for satellite and aerial photogrammetry.

First, we review the traditional one- and two-dimensional Fourier transforms. Then, the complex and quaternionic Fourier transforms are explained, and together with quaternionic Gabor filters, their role for phase analysis is clarified. The quaternionic phase concept helps us to disentangle possible symmetries of 2D signals. In Chap. 14 dedicated to the applications, various illustrations of the use of the quaternionic phase concept will show the power of analysis in the quaternionic frequency domain. As an extension of these transforms, the space and time Fourier transform and the n-dimensional Clifford Fourier transform are developed straightforwardly using the geometric algebra framework.

Additionally we present the extension of the real and complex wavelet transform to the quaternion wavelet transform and its applicability. Finally as a natural consequence, we derive the n-dimensional Clifford wavelet transform as well.

A chapter devoted to image analysis would be not complete if it lacked an analysis in the frequency domain, and so here we present the *quaternionic Fourier transform* (QFT) and the quaternionic Gabor filters. During the 1990s, a number of different attempts were made to use the algebra of quaternions for computations in the frequency domain. Chernov [54] used quaternions to speed up the evaluation of the 2D discrete, complex-valued, Fourier transform. Ell [85] introduced the QFT and applied it to the analysis of 2D linear, time-invariant, partial-differential systems. He used the ψ phase component of the polar representation of the QFT, Eq. (11.31), as

E. Bayro-Corrochano, *Geometric Algebra Applications Vol. I*,
https://doi.org/10.1007/978-3-319-74830-6_11

an indicator of the stability of the system. Later on, Bülow focused on the development of a quaternionic phase concept and clarified many theoretical and practical aspects of the QFT [44]. Sangwine [260] utilized the QFT for color image processing, assigning the individual RGB components to the quaternion imaginary components. The images are thus transformed holistically, as opposed to the more simplified approach in which each color channel is separately transformed utilizing the 2D Fourier transform. Using the Clifford algebra framework, we will also show that the highest level of the hierarchy of harmonic transformations is occupied by the n-dimensional Clifford Fourier transform. First, we review briefly the 1D and 2D Fourier transforms.

11.1.1 The One-Dimensional Fourier Transform

For a continuous and integrable function $f \in L^2(\mathbb{R})$, the Fourier transform of f is defined by the function $\mathcal{F}\{f\} : \mathbb{R} \to \mathbb{C}$ given by

$$F(u) \triangleq \mathcal{F}[f(x)] = \int_{-\infty}^{\infty} f(x)e^{-i2\pi ux}dx, \tag{11.1}$$

where $i^2 = -1$ is the unit imaginary. The function $F(u)$ can be expressed in terms of its complex parts or in polar form as follows:

$$F(u) = F_r(u) + iF_i(u) = S(u)e^{i2\pi\phi(u)}, \tag{11.2}$$

where $S(u)$ is the Fourier spectrum of $f(x)$, $S^2(u)$ is the power spectrum, and $\phi(u)$ is its phase angle.

If $F(u) \in L^2(\mathbb{R})$ and $f \in L^2(\mathbb{R})$, the inverse Fourier transform is given by

$$f(x) \triangleq \mathcal{F}^{-1}[F(u)] = \int_{-\infty}^{\infty} F(u)e^{i2\pi ux}du. \tag{11.3}$$

Table 11.1 presents a summary of some basic properties of the Fourier transform.

11.1.2 The Two-Dimensional Fourier Transform

The one-dimensional Fourier transform, Eq. (11.1), can be easily extended to treat a function $f(x, y)$ of two variables, namely a continuous and integrable function $f \in L^2(\mathbb{R}^2)$, and the Fourier transform of f is defined by the function $\mathcal{F}\{f\} : \mathbb{R}^2 \to \mathbb{C}$ given by

$$F(u, v) \triangleq \mathcal{F}[f(x, y)] = \int_{-\infty}^{\infty}\int_{-\infty}^{\infty} f(x, y)e^{-i2\pi(xu+yv)}dxdy. \tag{11.4}$$

Table 11.1 Properties of the one-dimensional Fourier transform

Property	Function	Fourier transform
Linearity	$\alpha f(x) + \beta g(x)$	$\alpha F(u) + \beta G(u)$
Delay	$f(x - t)$	$e^{-i2\pi ut} F(u)$
Shifting	$e^{i2\pi u_0 x} f(x)$	$F(u - u_0)$
Scaling	$f(\alpha x)$	$\frac{1}{\lvert\alpha\rvert} F(\frac{u}{\alpha})$
Convolution	$(f \star g)(x)$	$F(u)G(u)$
Derivative	$f^n(x)$	$(i2\pi u)^n F(u)$
Parseval theorem	$\int_{-\infty}^{\infty} \lvert f(x)\rvert^2 dx$	$\int_{-\infty}^{\infty} \lvert F(u)\rvert^2 du$

Table 11.2 Properties of the two-dimensional Fourier transform

Property	Function	Fourier transform
Rotation	$f(\pm x, \pm y)$	$F(\pm u, \pm v)$
Linearity	$\alpha f(x) + \beta g(x)$	$\alpha F(u, v) + \beta G(u, v)$
Conjugation	$f^*(x, y)$	$F^*(-u, -v)$
Separability	$f_1(x) f_2(y)$	$F_1(u) F_2(v)$
Scaling	$f(\alpha x, \beta y)$	$\frac{1}{\lvert\alpha\beta\rvert} F(\frac{u}{\alpha}, \frac{v}{\beta})$
Shifting	$f(x \pm x_0, y \pm y_0)$	$e^{\pm i2\pi(x_0 u + y_0 v)} F(u, v)$
Modulation	$e^{\pm i2\pi(u_0 x + v_0 y)} f(x, y)$	$F(u \mp u_0, v \mp v_0)$
Convolution	$h(x, y) = f(x, y) \star g(x, y)$	$H(u, v) = F(u, v)G(u, v)$
Multiplication	$h(x, y) = f(x, y)g(x, y)$	$H(u, v) = F(u, v) \star G(u, v)$
Spatial correlation	$h(x, y) = f(x, y) \odot g(x, y)$	$H(u, v) = F(-u, -v)G(u, v)$
Inner product	$I = \int_{-\infty}^{\infty} \int_{-\infty}^{\infty} f(x, y) g^*(x, y) dxdy$	$I = \int_{-\infty}^{\infty} \int_{-\infty}^{\infty} F(u, v) G^*(u, v) dudv$

The two-dimensional inverse Fourier transform is given by

$$f(x, y) \triangleq \mathcal{F}^{-1}[F(u, v)] = \int_{-\infty}^{\infty} \int_{-\infty}^{\infty} F(u, v) e^{i2\pi(ux+vy)} dudv. \qquad (11.5)$$

Similar to the one-dimensional case, the Fourier spectrum, and phase and power spectrum are respectively given by the following relations:

$$S(u, v) = |F(u, v)| = |F_r^2(u, v) + F_i^2(u, v)|^{\frac{1}{2}}, \qquad (11.6)$$

$$\phi((u, v)) = tan^{-1}\Big[\frac{F_i(u, v)}{F_r(u, v)}\Big], \qquad (11.7)$$

$$P(u, v) = S(u, v)^2 = F_r^2(u, v) + F_i^2(u, v). \qquad (11.8)$$

Table 11.2 presents a summary of some basic properties of the two-dimensional Fourier transform.

Fig. 11.1 Upper row: Original images. Second row: the Fourier or global phase of each image was change by the other image phase. See Eq. (11.10)

11.1.3 Fourier Phase, Instantaneous Phase, and Local Phase

There exist three different types of signal phase: the Fourier also known as global phase, the instantaneous phase, and the local phase. One advantage of the phase information is that it allows us to obtain an invariant or equivariant response [90]. For instance, it has been shown that the phase has an invariant response to image brightness and it can also be invariant to the rotations [90, 110, 164].

The Fourier phase is the most common phase computation and corresponds to the angular phase of the complex Fourier transform of a signal, and this phase indicates the relative position of the frequency components [44]. In order to show the role of global phase versus the amplitude in Fig. 11.1, the global (Fourier) phase of each image was computed and mixed with the amplitude of the other image, i.e.,

$$output1 = ifft\left(|I1|e^{i\phi_2}\right), \tag{11.9}$$

$$output2 = ifft\left(|I2|e^{i\phi_1}\right). \tag{11.10}$$

According to [35, 110] and clearly shown in Fig. 11.1, the global phase of an image carries the main information of the image. In contrast, the instantaneous and the local phase are used when it is needed another kind of information, instead of search for the phase of a certain frequency component, one wants to know what is the phase at a certain position in a signal in space domain. In different fields of signal processing such as telecommunications, geophysics [165, 183, 291], the computation of local amplitude and local phase corresponds to the analytical signal.

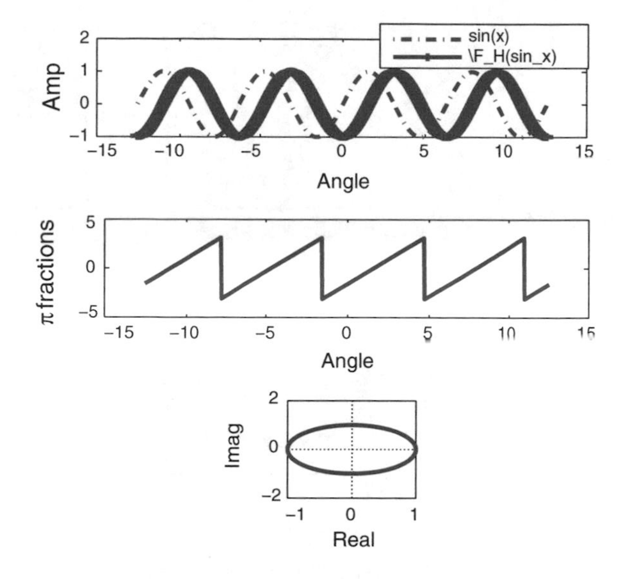

Fig. 11.2 (From top to bottom). The original signal ($sin(x)$), and its Hilbert transform. The instantaneous phase and the relation between the real (original signal) and the imaginary part (the Hilbert transform of the original signal)

For 1D signals, the phase computation is carried out using of the analytical signal involving the Hilbert transform ($f_{\mathbf{H}}(x) = \frac{1}{\pi x}$). The analytical signal is given by [44]

$$f_A(f(x)) = f(x) + i f_{\mathbf{H}}(x), \tag{11.11}$$

$$f_A(f(x)) = |A|e^{i\theta}, \tag{11.12}$$

where $|A| = \sqrt{f(x)^2 + f_{\mathbf{H}}(x)^2}$ and $\theta = arctan(\frac{f(x)}{f_{\mathbf{H}}(x)})$ permit us to extract the magnitude and phase independently. As an example of 1D analytical signal, Fig. 11.2 shows the original signal $\sin(x)$, its Hilbert transform, the phase $theta$, and the plot of the real part $real(f_A(f(x))) = \sin(x)$ versus the imaginary part $imag(f_A(f(x))) = f_{\mathbf{H}}(\sin(x))$. The circle of Fig. 11.2 represents the angle in the complex plane. Figure 11.3 shows at the upper left an odd part of the signal $sin(x)$ and at the upper right an even part, and at the bottom left the real part (x-$axis$) and at the bottom right the imaginary part (y-$axis$) which are related to the even and odd parts of the signal. An edge (odd) with negative slope has a phase range $(0, \pi)$, whereas a positive slope edge has $(\pi, 2\pi)$. The even part of the signal is related to the line; if the line has positive values, the phase range is $(\frac{3\pi}{4}, \frac{\pi}{2})$; meanwhile, if the line has negative values, the phase range is $(\frac{\pi}{2}, \frac{3\pi}{4})$.

In image processing, the instantaneous phase components characterize different image primitives such as lines or edges [110]. The Figure 11.4 compares the global and the instantaneous phase (using the Hilbert transform) of an image. It highlights the main difference between the global phase and the instantaneous phase of a image. The global phase is more sensitive to the change of brightness than the instantaneous phase. The instantaneous phase response is tightly related to the odd and even responses, i.e., to the lines and edges.

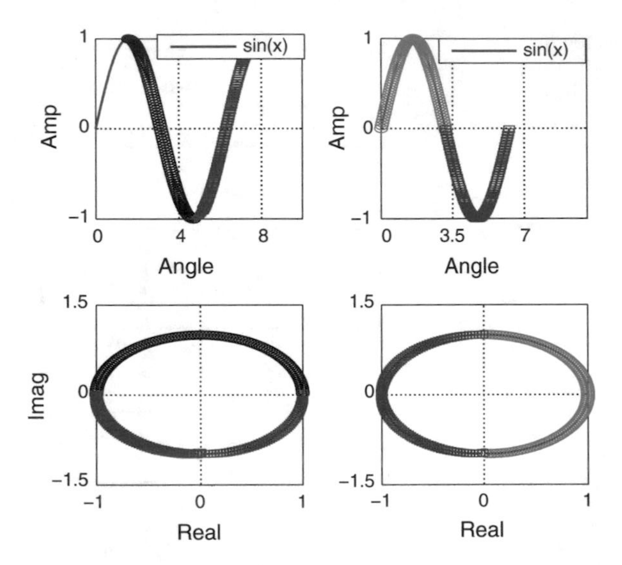

Fig. 11.3 (Upper left) The odd part of the signal. (Upper right) the even part. (Bottom left) real part (x-axis) and imaginary part (y-axis) which are related to the even and odd parts of the signal

Fig. 11.4 (From left to right) Original image, global magnitude, global phase, instantaneous magnitude, and instantaneous phase

11.1.4 Quaternionic Fourier Transform

Like the 2D Fourier transform and the Hartley transform, the quaternionic Fourier transform (QFT) is a linear and invertible transformation. The QFT is limited to 2D signals. Clifford algebra Fourier transforms (CFTs) can deal with transformations of higher-dimensional signals. The complex Fourier transform and the QFT can be seen as special cases of CFTs. The *2D Fourier transform* (FT) is given by

$$\begin{aligned} F_c(u) &= \int_{-\infty}^{\infty}\int_{-\infty}^{\infty} T(i2\pi ux)f(\boldsymbol{x})T(i2\pi vy)d^2\boldsymbol{x}, \\ &= \int_{-\infty}^{\infty}\int_{-\infty}^{\infty} e^{(-i2\pi ux)}f(\boldsymbol{x})e^{(-i2\pi vy)}d^2\boldsymbol{x}, \end{aligned} \tag{11.13}$$

where $\boldsymbol{x} = (x, y)$, $\boldsymbol{u} = (u, v)$, and $f(\boldsymbol{x})$ is a 2D-real-valued function. $T(i2\pi ux)$ and $T(i2\pi vy)$ are Fourier kernels applied to both axes of the 2D signal. If we assign, instead, Fourier kernels which depend upon the quaternion bases $i = \sigma_1\sigma_2$ and

$j = \sigma_2\sigma_3$ $(ij = -k = \sigma_1\sigma_3)$, we then get a straightforward expression for the QFT:

$$F_q(u) = \int_{-\infty}^{\infty}\int_{-\infty}^{\infty} T(i2\pi ux) f(\boldsymbol{x}) T(j2\pi vy) d^2\boldsymbol{x},$$
$$F_q(u) = \int_{-\infty}^{\infty}\int_{-\infty}^{\infty} e^{(-i2\pi ux)} f(\boldsymbol{x}) e^{(-j2\pi vy)} d^2\boldsymbol{x}, \tag{11.14}$$

where $\boldsymbol{x} = (x, y)$, $\boldsymbol{u} = (u, v)$, and $f(\boldsymbol{x})$ is a real-, complex-, or quaternion-valued 2D function.

The *inverse quaternionic Fourier transform* (IQFT) is given by

$$f(\boldsymbol{x}) = \int_{-\infty}^{\infty}\int_{-\infty}^{\infty} \tilde{T}(i2\pi ux) F_q(u) \tilde{T}(j2\pi vy) d^2\boldsymbol{u},$$
$$= \int_{-\infty}^{\infty}\int_{-\infty}^{\infty} e^{(i2\pi ux)} F_q(\boldsymbol{u}) e^{(j2\pi vy)} d^2\boldsymbol{u}, \tag{11.15}$$

where $T(i2\pi ux)\tilde{T}(i2\pi ux) = 1$ and $T(j2\pi vy)\tilde{T}(j2\pi vy) = 1$.

Let us now give the concept of *Hermitian function* in the quaternionic domain. As an extension of the Hermitian function $f : \mathbb{R} \to \mathbb{C}$ with $f(x) = f^*(-x)$ for every $x \in \mathbb{R}$, we regard $f : \mathbb{R}^2 \to \mathbb{H}$ as a quaternionic Hermitian function if it fulfills the following non-trivial involution rules [54]:

$$\begin{aligned} f(-x, y) &= -jf(x, y)j = \mathcal{T}_j(f(x, y)), \\ f(x, -y) &= -if(x, y)i = \mathcal{T}_i(f(x, y)), \\ f(-x, -y) &= -if(-x, y)i = -i(-jf(x, y)j)i = (-i - j)f(x, y)(ji), \\ &= -kf(x, y)k = \mathcal{T}_k(f(x, y)). \end{aligned} \tag{11.16}$$

The concept of quaternionic Hermitian function is very useful for the computation of the inverse QFT using the quaternionic analytic signal.

11.1.5 2D Analytic Signals

This subsection discusses different points of views found in the literature concerning the fundamental concept of the analytical signal of 2D signals. The reader can find a detailed analysis of this issue in [44, 110]. The particular formulation of the 2D analytical signal used influences the final representation of the signal in the four quadrants of the frequency domain. Thus, in order to recover the entire real 2D signal using the inverse Fourier transform, one must take this final representation into account. The same thought process applies when one computes the magnitude and phase of the 2D signal.

The analytical signal of a 2D signal is given by

$$f_A(\boldsymbol{x}) = f(\boldsymbol{x}) \star \left(\delta^2(\boldsymbol{x}) + \frac{i}{\pi^2 xy}\right),$$
$$= f(\boldsymbol{x}) + if(\boldsymbol{x}) \star \left(\frac{i}{\pi^2 xy}\right) = f(\boldsymbol{x}) + if_{H_i}(\boldsymbol{x}), \quad (11.17)$$

where $f_{H_i}(\boldsymbol{x})$ is called the total *Hilbert transform* [288] and the symbol $\star$ stands for a 2D convolution operation. In the frequency domain, the signal is split among the four quadrants according to the following equation:

$$F_A(\boldsymbol{u}) = F(\boldsymbol{u})\Big(1 - isign(u)sign(v)\Big) \quad (11.18)$$

(see Fig. 11.5a).

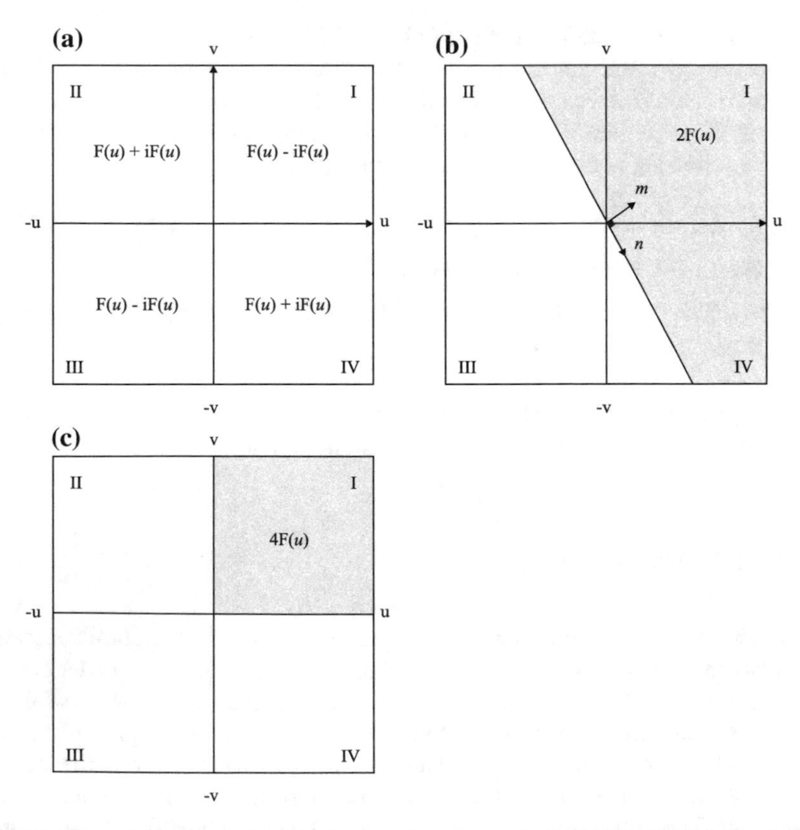

Fig. 11.5 2D analytic signals in the frequency domain: **a** standard 2D analytical signal $F_A(\boldsymbol{u})$, **b** partial analytical signal $F_{A_p}(\boldsymbol{u})$, **c** Hahn's 2D analytical signal $F_{A_{h1}}(\boldsymbol{u})$, and **d** quaternionic 2D analytical signal

The *partial analytical signal* locates the signal energy in the frequency domain on one side of a reference line indicated by the normal unit vector of the line $\boldsymbol{m} = (cos(\theta), sin(\theta))$, which is perpendicular to the line direction $\boldsymbol{n}$. The partial analytical signal in spatial domain is given by the equation

$$f_{A_p}(\boldsymbol{x}) = f(\boldsymbol{x}) \star \Big(\delta(\boldsymbol{x} \cdot \boldsymbol{m}) + \frac{i}{\pi \boldsymbol{x} \cdot \boldsymbol{m}}\Big)\delta(\boldsymbol{x} \cdot \boldsymbol{n}), \tag{11.19}$$

and in frequency domain by the equation

$$F_{A_p}(\boldsymbol{u}) = F(\boldsymbol{u})\Big(1 + sign(\boldsymbol{u} \cdot \boldsymbol{m})\Big) \tag{11.20}$$

(see Fig. 11.5b).

In another approach, Hahn [116, 117] introduced the following notion of the analytic signal:

$$\begin{aligned} f_{A_h1}(\boldsymbol{x}) &= f(\boldsymbol{x}) \star \Big(\delta(x) + \frac{i}{\pi x}\Big)\Big(\delta(y) + \frac{i}{\pi y}\Big), \\ &= f(\boldsymbol{x}) - f(\boldsymbol{x}) \star \Big(\frac{1}{\pi^2 xy}\Big) + if(\boldsymbol{x}) \star \Big(\frac{\delta(y)}{\pi x}\Big) + if(\boldsymbol{x}) \star \Big(\frac{\delta(x)}{\pi y}\Big), \\ &= f(\boldsymbol{x}) - f_{H_i}(\boldsymbol{x}) + i\Big(f_{H_{i_a}}(\boldsymbol{x}) + f_{H_{i_b}}(\boldsymbol{x})\Big), \end{aligned} \tag{11.21}$$

where $f_{H_i}(\boldsymbol{x})$ is the total Hilbert transform and $f_{H_{i_a}}(\boldsymbol{x})$, $f_{H_{i_b}}(\boldsymbol{x})$ are partial Hilbert transforms which only reference the x- and y-axes, respectively. In the frequency domain, the analytical signal is localized only in the first quadrant and is multiplied four times according to

$$F_{A_h1}(\boldsymbol{u}) = \Big(1 + sign(u)\Big)\Big(1 + sign(v)\Big)F(\boldsymbol{u}), \tag{11.22}$$

(see Fig. 11.5c). In this approach, owing to the Hermitian symmetry, only one-half of the plane of the frequency spectrum is redundant, so that a second analytical signal with its spectrum located in the second quadrant is required:

$$\begin{aligned} f_{A_h2}(\boldsymbol{x}) &= f(\boldsymbol{x}) + f_{H_i}(\boldsymbol{x}) - i\Big(f_{H_{i_a}}(\boldsymbol{x}) - f_{H_{i_b}}(\boldsymbol{x})\Big), \\ F_{A_h2}(\boldsymbol{u}) &= \Big(1 + sign(-u)\Big)\Big(1 + sign(v)\Big)F(\boldsymbol{u}). \end{aligned} \tag{11.23}$$

The entire 2D real signal can only be recovered by taking into account both analytic signals, F_{A_h1} and F_{A_h2} (for more details, see [117]).

As Fig. 11.5d shows, the QFT of a 2D real signal is a quaternionic hermitian and thus the total information of the signal is not lost. This suggests that the quaternionic analytical signal may be defined by modifying the Hahn's equation (11.21), utilizing instead the quaternion basis

$$
\begin{aligned}
f_{A_q}(\boldsymbol{x}) &= f(\boldsymbol{x}) \star \Big(\delta(x) + \frac{i}{\pi x}\Big)\Big(\delta(y) + \frac{j}{\pi y}\Big), \\
&= f(\boldsymbol{x}) + i f(\boldsymbol{x}) \star \Big(\frac{1}{\pi^2 xy}\Big) + j f(\boldsymbol{x}) \star \Big(\frac{\delta(y)}{\pi x}\Big) + ij f(\boldsymbol{x}) \star \Big(\frac{\delta(x)}{\pi y}\Big), \\
&= f(\boldsymbol{x}) + i f_{H_{i_a}}(\boldsymbol{x}) + j f_{H_{i_b}}(\boldsymbol{x}) - k f_{H_i}(\boldsymbol{x}),
\end{aligned} \tag{11.24}
$$

where $f_{H_{i_a}}(\boldsymbol{x})$ and $f_{H_{i_b}}(\boldsymbol{x})$ are the partial Hilbert transforms and $f_{H_i}(\boldsymbol{x})$ is the total Hilbert transform. In the frequency domain, the quaternionic analytical signal is given by

$$
F_{A_q}(\boldsymbol{u}) = \Big(1 + sign(u)\Big)\Big(1 + sign(v)\Big) F_q(\boldsymbol{u}) \tag{11.25}
$$

(see Fig. 11.5d).

We will now show that we can indeed obtain $f(x, y)$ by utilizing the first quadrant of the frequency spectrum $F_q(u, v)$ four times. To do so, we employ the simple property of quaternions, $Re(\boldsymbol{q}) = Re(-i\boldsymbol{q}i) = Re(-j\boldsymbol{q}j) = Re(-k\boldsymbol{q}k)$, such that

$$
\begin{aligned}
f(x, y) &= 4Re\Big(\int_0^\infty \int_0^\infty e^{(i2\pi ux)} F_q(\boldsymbol{u}) e^{(j2\pi vy)} d^2\boldsymbol{u}\Big), \\
&= Re\Big(\int_0^\infty \int_0^\infty e^{(i2\pi ux)} F_q(\boldsymbol{u}) e^{(j2\pi vy)} d^2\boldsymbol{u}\Big) \\
&\quad + Re\Big(\int_0^\infty \int_0^\infty e^{(i2\pi ux)} - i F_q(\boldsymbol{u}) i e^{(j2\pi vy)} d^2\boldsymbol{u}\Big) \\
&\quad + Re\Big(\int_0^\infty \int_0^\infty e^{(i2\pi ux)} - j F_q(\boldsymbol{u}) j e^{(j2\pi vy)} d^2\boldsymbol{u}\Big) \\
&\quad + Re\Big(\int_0^\infty \int_0^\infty e^{(i2\pi ux)} - k F_q(\boldsymbol{u}) k e^{(j2\pi vy)} d^2\boldsymbol{u}\Big), \\
&= Re\Big(\int_0^\infty \int_0^\infty e^{(i2\pi ux)} F_q(\boldsymbol{u}) e^{(j2\pi vy)} d^2\boldsymbol{u}\Big) \\
&\quad + Re\Big(\int_0^\infty \int_0^\infty e^{(i2\pi ux)} F_q(u, -v) e^{(j2\pi vy)} d^2\boldsymbol{u}\Big) \\
&\quad + Re\Big(\int_0^\infty \int_0^\infty e^{(i2\pi ux)} F_q(-u, v) e^{(j2\pi vy)} d^2\boldsymbol{u}\Big) \\
&\quad + Re\Big(\int_0^\infty \int_0^\infty e^{(i2\pi ux)} F_q(-u, -v) e^{(j2\pi vy)} d^2\boldsymbol{u}\Big), \\
&= Re\Big(\int_0^\infty \int_0^\infty e^{(i2\pi ux)} F_q(\boldsymbol{u}) e^{(j2\pi vy)} dudv\Big) \\
&\quad + Re\Big(\int_{-\infty}^0 \int_0^\infty e^{(i2\pi ux)} F_q(u, v) e^{(j2\pi vy)} dudv\Big) \\
&\quad + Re\Big(\int_0^\infty \int_{-\infty}^0 e^{(i2\pi ux)} F_q(u, v) e^{(j2\pi vy)} dudv\Big) \\
&\quad + Re\Big(\int_{-\infty}^0 \int_{-\infty}^0 e^{(i2\pi ux)} F_q(u, v) e^{(j2\pi vy)} dudv\Big), \\
&= Re\Big(\int_{-\infty}^\infty \int_{-\infty}^\infty e^{(i2\pi ux)} F_q(\boldsymbol{u}) e^{(j2\pi vy)} dudv\Big).
\end{aligned} \tag{11.26}
$$

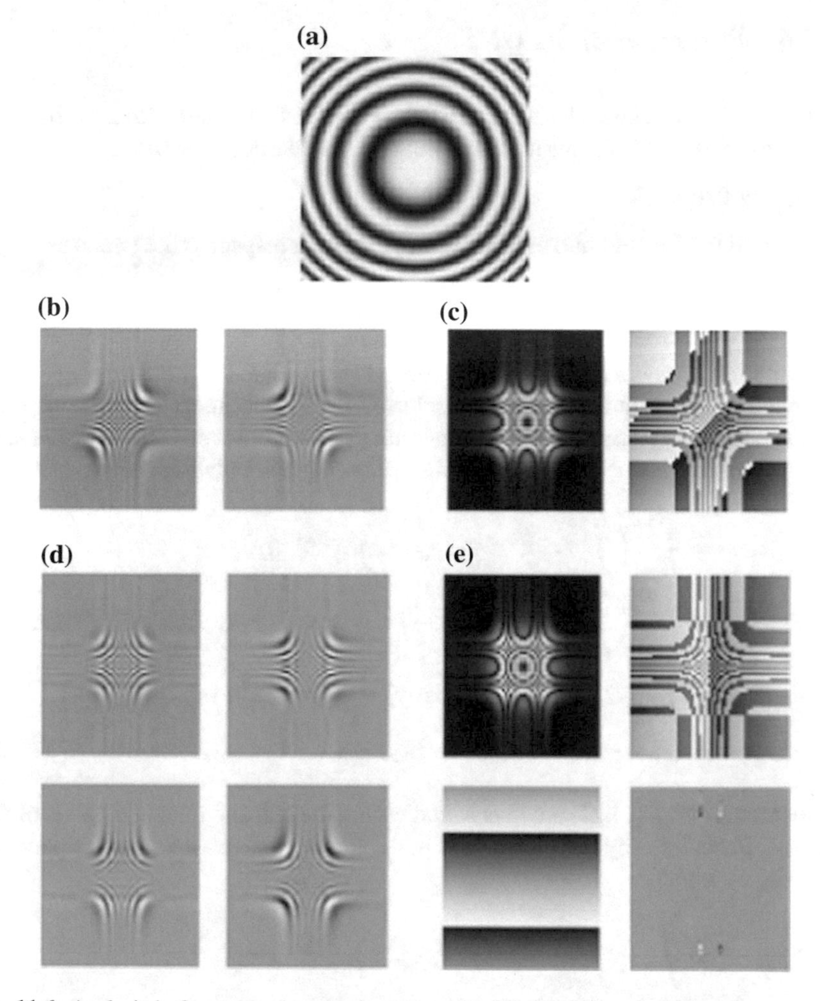

Fig. 11.6 Analysis in frequency domain: (upper row) **a** 2D signal, **b** real and imaginary parts of FT, **c** magnitude and phase of FT, (lower row) **d** real and imaginary parts of QFT, **e** magnitude and phases of QFT

Note that Hahn's approach to the analytical signal fails to recover a 2D complex Hermitian signal using the first quadrant of the frequency spectrum. That is why Hahn introduced two complex signals for each of the real 2D signal equations, (11.22) and (11.23). By contrast, using the quaternionic analytical signal the entire real 2D signal can be recovered from the first quadrant of the frequency spectrum. This is due to the Hermitian symmetry properties of the quaternionic analytical signal. Figure 11.6 shows the complex and quaternionic Fourier transforms of a 2D signal.

11.1.6 Properties of the QFT

Next, we will give some relevant properties of the QFT. We start with a treatment of the symmetries of 2D signals using the Fourier and Hartley transforms.

Symmetry Properties

A 2D signal can be split into even and odd parts with respect to the two axes:

$$f(\boldsymbol{x}) = f_{ee}(\boldsymbol{x}) + f_{oo}(\boldsymbol{x}) + f_{eo}(\boldsymbol{x}) + f_{oe}(\boldsymbol{x}), \tag{11.27}$$

where the subindexes e (for even) and o (for odd) are related to the x-axis and y-axis, respectively. This split depends on the selected origin and image orientation. Since the 2D Fourier transform has real and imaginary parts, it is not possible to disentangle the four components of Eq. (11.27), as the following computation shows:

$$\begin{aligned} F(\boldsymbol{u}, \boldsymbol{v}) &= \int_{-\infty}^{\infty} \Big(\int_{-\infty}^{\infty} e^{-i2\pi ux} f(x, y) dx \Big) e^{-i2\pi vy} dy, \\ &= \int_{-\infty}^{\infty} \Big(\int_{-\infty}^{\infty} (cos(2\pi ux) - isin(2\pi ux)) f(x, y) dx \Big) e^{-i2\pi vy} dy, \\ &= \int_{-\infty}^{\infty} (H_e - iH_o) \Big(cos(2\pi vy) - isin(2\pi vy) \Big) dy, \\ &= (H_{ee} - H_{oo}) - i(H_{oe} + H_{eo}) \in \boldsymbol{C}. \end{aligned} \tag{11.28}$$

The elements of Eq. (11.27) are intermixed within the real and imaginary parts of this equation. In contrast, by computing the QFT following the quaternion multiplication $ij = -k$, we obtain

$$\begin{aligned} F_q(\boldsymbol{u}, \boldsymbol{v}) &= \int_{-\infty}^{\infty} \Big(\int_{-\infty}^{\infty} e^{-i2\pi ux} f(x, y) dx \Big) e^{-j2\pi vy} dy, \\ &= \int_{-\infty}^{\infty} \Big(\int_{-\infty}^{\infty} (cos(2\pi ux) - isin(2\pi ux)) f(x, y) dx \Big) e^{-j2\pi vy} dy, \\ &= \int_{-\infty}^{\infty} (H_e - iH_o) e^{-j2\pi vy} dy, \\ &= \int_{-\infty}^{\infty} (H_e - iH_o) \Big(cos(2\pi vy) - jsin(2\pi vy) \Big) dy, \\ &= \int_{-\infty}^{\infty} \Big(H_e cos(2\pi vy) - iH_o cos(2\pi vy) - jH_e sin(2\pi vy) + \\ &\qquad +(ij) H_o sin(2\pi vy) \Big) dy, \\ &= H_{ee} - iH_{oe} - jH_{eo} - kH_{oo} \in I\!H. \end{aligned} \tag{11.29}$$

Note that here the QFT separates the four components of Eq. (11.27) by projecting them into the orthonormal quaternionic basis. By contrast, the Hartley transform (HT) cannot split these parts, as the following computation shows:

$$
\begin{aligned}
H(\boldsymbol{u}) &= \int_{-\infty}^{\infty}\int_{-\infty}^{\infty} f(\boldsymbol{x})\Big\{cos(2\pi\boldsymbol{u}\cdot\boldsymbol{x}) + sin(2\pi\boldsymbol{u}\cdot\boldsymbol{x})\Big\}d^2\boldsymbol{x},\\
&= \int_{-\infty}^{\infty}\int_{-\infty}^{\infty} f(\boldsymbol{x})\Big\{cos(2\pi ux)cos(2\pi vy) - sin(2\pi ux)sin(2/ivy) +\\
&\quad +cos(2\pi ux)sin(2\pi vy) + sin(2\pi ux)cos(2\pi vy)\Big\}d^2\boldsymbol{x},\\
&= H_{ee}(\boldsymbol{u}) + H_{oo}(\boldsymbol{u}) + H_{eo}(\boldsymbol{u}) + H_{oe}(\boldsymbol{u}) \in R.
\end{aligned}
\tag{11.30}
$$

Polar Representation of the QFT

The polar representation of the QFT is given by

$$
F_q(\boldsymbol{u}) = \Big|F_q(\boldsymbol{u})\Big|e^{i\phi(\boldsymbol{u})}e^{k\psi(\boldsymbol{u})}e^{k\theta(\boldsymbol{u})},
\tag{11.31}
$$

and the evaluation of its phase follows the quaternion rules given in Sect. 6.4. This kind of representation is very helpful for image analysis if we use the phase concept.

Shift Property

With regard to the polar representation of the QFT, it is appropriate to mention the shift property of the QFT:

$$
\begin{aligned}
F_q(\boldsymbol{u})_T &= \int_{-\infty}^{\infty}\int_{-\infty}^{\infty} e^{(-i2\pi ux)} f(\boldsymbol{x}-\boldsymbol{d})e^{(-j2\pi vy)}d^2\boldsymbol{x},\\
&= e^{(-i2\pi ud_1)}F_q(\boldsymbol{u})e^{(-j2\pi ud_2)},\\
&= \Big|F_q(\boldsymbol{u})_T\Big|e^{i(\phi(\boldsymbol{u})-2\pi ud_1)}e^{k\psi(\boldsymbol{u})}e^{j(\chi(\boldsymbol{u})-2\pi vd_2)}.
\end{aligned}
\tag{11.32}
$$

Modulation

The modulation of the 2D signal causes a shifting in the 2D frequency space, which is the result of the modulation in the space domain of the 2D signal from both sides by two orthogonal carriers with frequencies u_0 and v_0:

$$
f_m(x, y) = e^{(i2\pi u_0 x)} f(x, y)e^{(j2\pi v_0 y)}.
\tag{11.33}
$$

By taking the QFT of $f_m(x, y)$, we obtain

$$
F_q\Big\{f_m(x, y)\Big\}(u, v) = F_q(u - u_0, v - v_0).
\tag{11.34}
$$

Convolution

Consider the real 2D signals f_1 and f_2 with their quaternionic Fourier transforms F_{1q} and F_{2q}, respectively. The convolution of these 2D signals can be carried out in the frequency domain as a sort of multiplication of their QFTs. This can be easily proved, by either integrating with respect to the Fourier kernel $e^{-(j2\pi yv)}$:

$$
\begin{aligned}
F_q &= \int_{-\infty}^{\infty}\int_{-\infty}^{\infty} e^{-(i2\pi xu)}\Big(f_1(\boldsymbol{x}) \star f_2(\boldsymbol{x})\Big)e^{-(j2\pi yv)}d^2\boldsymbol{x},\\
&= \int_{-\infty}^{\infty}\int_{-\infty}^{\infty} e^{-(i2\pi xu)}\Big(\int_{-\infty}^{\infty}\int_{-\infty}^{\infty} f_1(\boldsymbol{x}')f_2(\boldsymbol{x}-\boldsymbol{x}')d^2\boldsymbol{x}'\Big)e^{-(j2\pi yv)}d^2\boldsymbol{x},\\
&= \int_{-\infty}^{\infty}\int_{-\infty}^{\infty} e^{-(i2\pi x'u)} f_1(\boldsymbol{x}')F_{2q}(\boldsymbol{u})e^{-(j2\pi y'v)}d^2\boldsymbol{x}',\\
&= \int_{-\infty}^{\infty}\int_{-\infty}^{\infty} e^{-(i2\pi x'u)} f_1(\boldsymbol{x}')cos(-2\pi y'v)F_{2q}(\boldsymbol{u})d^2\boldsymbol{x}'\\
&\quad + \int_{-\infty}^{\infty}\int_{-\infty}^{\infty} e^{-(i2\pi x'u)} f_1(\boldsymbol{x}')jsin(-2\pi y'v)\Big(-jF_{2q}(\boldsymbol{u})j\Big)d^2\boldsymbol{x}',\\
&= F_{1q_e}(\boldsymbol{u})F_{2q}(\boldsymbol{u}) + F_{1q_o}(\boldsymbol{u})\Big(-jF_{2q}(\boldsymbol{u})j\Big),\\
&= F_{1q_e}(\boldsymbol{u})F_{2q}(\boldsymbol{u}) + F_{1q_o}(\boldsymbol{u})\mathcal{T}_j(F_{2q}(\boldsymbol{u})),
\end{aligned} \tag{11.35}
$$

or by integrating with respect to the Fourier kernel $e^{-(i2\pi xv)}$:

$$
\begin{aligned}
F_q &= F_{1q}(\boldsymbol{u})F_{2q_e}(\boldsymbol{u}) + \Big(-iF_{1q}(\boldsymbol{u})i\Big)F_{2q_o}(\boldsymbol{u}),\\
&= F_{1q}(\boldsymbol{u})F_{2q_e}(\boldsymbol{u}) + \mathcal{T}_i(F_{1q}(\boldsymbol{u}))F_{2q_o}(\boldsymbol{u}).
\end{aligned} \tag{11.36}
$$

Note that if one of the functions is even with respect to at least one of the kernel arguments, the convolution is then equal to the product of the individual spectra:

$$
F_q = F_{1q}(\boldsymbol{u})F_{2q}(\boldsymbol{u}). \tag{11.37}
$$

As with the method to discretize the FT and its inverse, we can proceed similarly with the QFT and its inverse. Given a discrete two-dimensional signal of size $M \times N$ with quaternionic components $f_{mn} \in I\!H$, the *discrete quaternionic Fourier transform* (DQFT) and the *inverse discrete quaternionic Fourier transform* (IDQFT) are given, respectively, by

$$
F_{q_{uv}} = \sum_{m=0}^{M-1}\sum_{n=0}^{N-1} e^{(-\frac{i2\pi um}{M})} f_{mn} e^{(-\frac{j2\pi vn}{N})}, \tag{11.38}
$$

$$
f_{mn} = \frac{1}{MN}\sum_{m=0}^{M-1}\sum_{n=0}^{N-1} e^{(\frac{i2\pi um}{M})} F_{q_{uv}} e^{(\frac{j2\pi vn}{N})}. \tag{11.39}
$$

The reader can then easily implement a computer program to determine the QFT and IQFT. Using the fast Fourier transform (FFT), we first compute the one-dimensional DFT of f_{mn} in a row-wise sense, as follows:

$$f_{un} = \sum_{m=0}^{M-1} e^{(-\frac{i2\pi um}{M})} f_{mn} = Real(f_{un}) + Imag(f_{un}). \tag{11.40}$$

This spectrum f_{un} is then divided into real and imaginary parts, which in turn are now transformed in a columnwise sense, once again using a one-dimensional DFT, to give

$$F_{uvr} = \sum_{n=0}^{N-1} \Big(Real(f_{un})\Big) e^{(-\frac{i2\pi vn}{M})}, \tag{11.41}$$

$$F_{uvi} = \sum_{n=0}^{N-1} \Big(Imag(f_{un})\Big) e^{(-\frac{i2\pi vn}{M})}. \tag{11.42}$$

Using these results, we can finally compose the DQFT:

$$F_{uvq} = Real(F_{uvr}) + iReal(F_{uvi}) + jImag(F_{uvr}) + kImag(F_{uvi}). \tag{11.43}$$

Note that the first, row-wise computation outputs complex numbers (Eq. (11.40)) and that the second, columnwise computation rotates their real and imaginary parts spatially by 90°. In this way, the procedure yields the orthogonal bivector components of the DQFT of Eq. (11.43). This method is an implementation of the procedure to build quaternions from complex numbers, namely, by using the *doubling technique* [316]. Current FFT routines and hardware can be reutilized with few changes to implement a fast quaternionic Fourier transform (FQFT). The reader can find more details about the implementation of an FQFT in [89].

11.2 Gabor Filters and Atomic Functions

In the field of image processing, Gabor filters and atomic functions have proved to be a very useful bandpass filters with the beneficial property that they are optimally localized both in space and in the frequency domains [101]. Applications for Gabor filters and atomic functions include pattern recognition, classification, texture analysis, local phase estimation, and frequency estimation. In the following subsections, the 2D Gabor filter, the atomic function, and the phase concept are explained in detail.

11.2.1 2D Gabor Filters

A two-dimensional *complex Gabor filter* is a *linear shift-invariant* filter with the impulse response of a complex carrier modulated by a Gauss function:

$$h_c(\boldsymbol{x}; \boldsymbol{u}_0, \sigma, \alpha, \phi) = g(x', y')exp(2\pi i(u_0 x + v_0 y)). \tag{11.44}$$

Here, the Gauss function is given by

$$g(x, y) = Cexp\Big(-\frac{x^2 + (\alpha y)^2}{\sigma^2}\Big), \tag{11.45}$$

where α is the aspect radio and $C = \frac{\alpha}{2\pi\sigma^2}$ the normalizing factor so that $\int_{\mathbb{R}} g(x, y)$ $dxdy = 1$. The coordinates of $g(x', y')$ have been rotated about the origin

$$\begin{matrix} x' \\ y' \end{matrix} = \begin{matrix} \cos\theta & \sin\theta x, \\ -\sin\theta & \cos\theta y. \end{matrix} \tag{11.46}$$

In frequency domain, the 2D Gabor filter has the following transfer function:

$$\begin{aligned} h_c(\boldsymbol{x}; \boldsymbol{u}_0, \sigma, \alpha, \phi) &\rightarrow H_c(\boldsymbol{u}; \boldsymbol{u}_0, \sigma, \alpha, \phi), \\ H_c(\boldsymbol{u}; \boldsymbol{u}_0, \sigma, \alpha, \phi) &= exp(-2\pi^2\sigma^2[(u' - u'_0)^2 + (v' - v'_0)^2/\alpha]). \end{aligned} \tag{11.47}$$

Figure 11.7 shows a complex Gabor filter whose center in frequency and orientation are given by $f_0 = \sqrt{u_0^2 + v_0^2}$ and $\theta = atan(\frac{v_0}{u_0})$, respectively. Choosing $\theta = \phi$, the principal axis of the Gauss function is aligned with the orientation of ϕ.

Now, by assigning two orthogonal complex carriers to the axes of the 2D signal, we can further extend the complex Gabor filter into the quaternionic Gabor filter:

$$\begin{aligned} h_q(\boldsymbol{x}; \boldsymbol{u}_0, \sigma, \alpha, \phi = 0) &= g(\boldsymbol{x}; \sigma, \alpha)exp(i2\pi u_0 x)exp(j2\pi v_0 y), \\ &= g(\boldsymbol{x}; \sigma, \alpha)exp\left(i\frac{s_1 w_1 x}{\sigma}\right)exp\left(j\frac{s_2 \alpha w_2 y}{\sigma}\right). \end{aligned} \tag{11.48}$$

Note that the complex carriers are dependent on the quaternion bases i and j and that for simplicity we do not use a rotated Gauss function.

The transfer function of a quaternionic Gabor filter is a direct interpretation of the modulation theorem of the quaternionic Fourier transform (QFT) explained in Sect. 11.1. It consists of a shifted Gaussian function in the quaternionic frequency domain:

$$\begin{aligned} h_q(\boldsymbol{x}; \boldsymbol{u}_0, \sigma, \alpha, \phi = 0) &\rightarrow H_q(\boldsymbol{u}; \boldsymbol{u}_0, \sigma, \alpha, \phi = 0), \\ H_q(\boldsymbol{u}; \boldsymbol{u}_0, \sigma, \alpha, \phi = 0) &= exp(-2\pi^2\sigma^2[(u - u_0)^2 + (v - v_0)^2/\alpha^2]), \end{aligned} \tag{11.49}$$

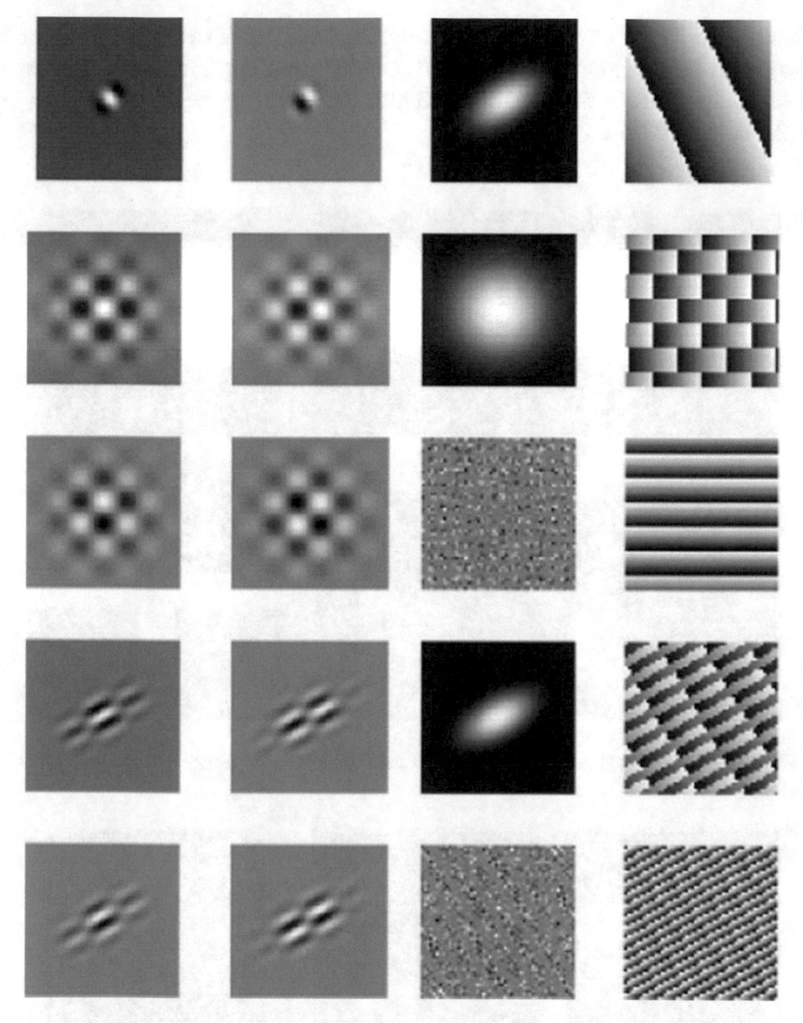

Fig. 11.7 Complex and quaternion Gabor filters: (upper left) real and imaginary parts (r,i) of a complex Gabor 50 × 50-pixel filter with $\sigma_1 = 8$, $\sigma_2 = 4$, $s = 2$, and $\alpha = 30°$; (upper right) magnitude and phase (ϕ) for this filter; (middle left) real and imaginary parts (i,j,k) of a quaternionic Gabor 50 × 50-pixel filter with $\sigma_1 = 10$, $\sigma_2 = 10$, $s_1 = 4$, $s_2 = 4$, and $\alpha = 0$; (middle right) magnitude and phases (ϕ, ψ, and θ) for this filter; (lower left) real and imaginary parts (i,j,k) of a quaternionic 50 × 50-pixel Gabor filter with $\sigma_1 = 8$, $\sigma_2 = 4$, $s_1 = 2$, $s_2 = 4$, and $\alpha = 30°$; (lower right) magnitude and phases (ϕ, ψ, and θ) for this filter

by which the greater amount of the Gabor filter's energy is preserved for positive frequencies u_0 and v_0 in the upper-right quadrant. In this regard, the convolution of a real image with a quaternionic Gabor filter approximates a quaternionic analytic signal. Figure 11.7 shows two examples of quaternion Gabor filters.

Figure 11.8 shows a convolved medical image using a 50 × 50-pixel Gabor filter with $\sigma_1 = 7$, $\sigma_2 = 2$, $s_1 = 2$, $s_2 = 2$, and $\alpha = 90°$, and Fig. 11.9 shows a convolved medical image using a 50 × 50-pixel Gabor filter with $\sigma_1 = 7$, $\sigma_2 = 2$, $s_1 = 2$, $s_2 = 2$, and $\alpha = 67.5°$.

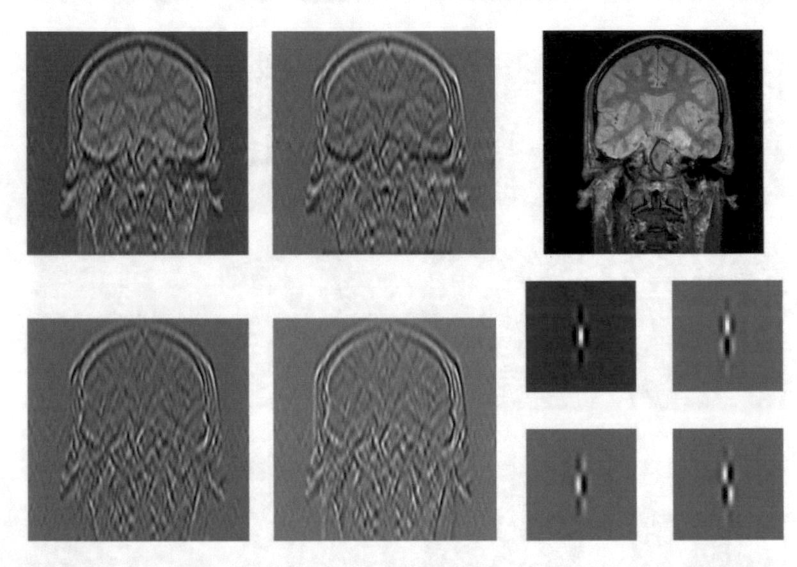

Fig. 11.8 Real and imaginary parts of convolved quaternionic image using qgabor(50,7,2,2, 2, 90)

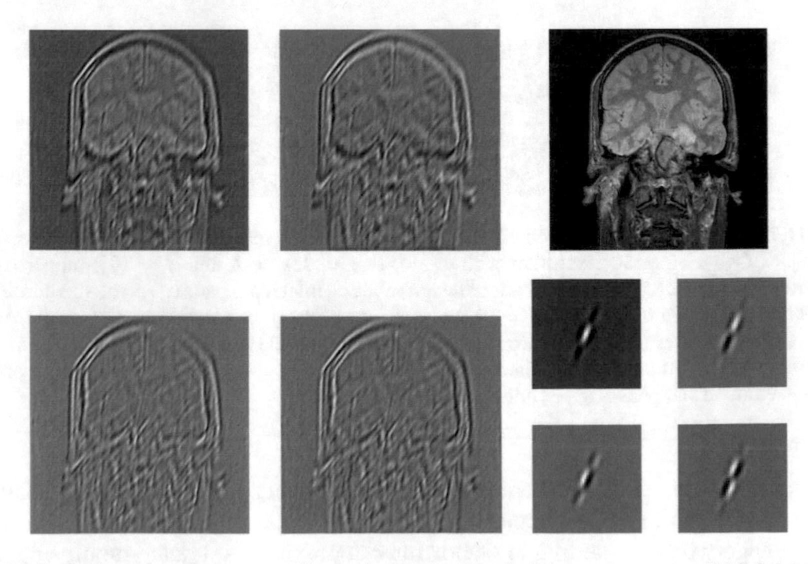

Fig. 11.9 Real and imaginary parts of convolved quaternionic image using qgabor(50,7,2,2,2,67.5)

11.2.2 Atomic Functions

The atomic functions (AF) were first developed in the 1970s, jointly by V. L. and V. A. Rvachev [161].

The atomic functions (AF) are compactly supported, infinitely differentiable solutions of differential functional equations with a shifted argument [231], i.e.,

$$Lf(x) = \lambda \sum_{k=1}^{M} c(k) f(ax - b(k)), |a| > 1, b, c, \lambda \in N, \tag{11.50}$$

where $L = \frac{d^n}{dx^n} + a_1 \frac{d^{n-1}}{dx^{n-1}} + \cdots + a_n$ is a linear differential operator with constant coefficients. Among other AFs, the atomic function $up(x)$ is the simplest and, at the same time, the most useful primitive function to generate other kinds of atomic functions [231]. It satisfies the equation

$$f(x)' = 2\left(f(2x+1) - f(2x-1)\right), \tag{11.51}$$

with compact support. The main characteristics of this function are well described in [115, 161, 231]. The function $up(x)$ is infinitely differentiable, $up(0) = 1$, $up(-x) = up(x)$. Other types of AF satisfying Eq. (11.50) are $fup_n(x)$, $\Xi_n(x)$, $h_a(x)$ [115]. In this work, we only use the functions $up(x)$ and $dup(x)$, see Eq. (11.54).

The atomic function $up(x)$ is generated by infinite convolutions of rectangular impulses. The function $up(x)$ has the following representation in terms of the Fourier transform [161, 168, 231]:

$$up(x) = \frac{1}{2\pi} \int_{-\infty}^{\infty} \prod_{k=1}^{\infty} \frac{\sin(\nu 2^{-k})}{\nu 2^{-k}} e^{i\nu x} d\nu, \tag{11.52}$$

$$= \frac{1}{2\pi} \int_{-\infty}^{\infty} \hat{up}(\nu) e^{i\nu x} d\nu. \tag{11.53}$$

Some properties of the AF that we take advantage of have been reported in [161, 168, 170, 231]; these properties include the following attributes:

- The $up(x)$ function is a compactly supported function in the space domain. Therefore, it can obtain good local characteristics.
- Since derivatives of any order can be represented in terms of shifts, any derivative can be represented as an operator, and the n-order derivatives are defined by

$$d^{(n)} up(x) = 2^{n(n+1)/2} \sum_{k=1}^{2^n} \delta_k up(2^n x + 2^n + 1 - 2k), \tag{11.54}$$

where, $\delta_1 = 1$, $\delta_{2k} = -\delta_k$, $\delta_{2k-1} = \delta_k$.

- The AFs are infinitely differentiable (C^∞). As a result, the AFs and their Fourier transforms are rapidly decreasing functions. Therefore, their Fourier transforms decrease on the real axis faster than any power function.
- The AF windows were compared with classic ones by means of parameters such as the equivalent noise bandwidth, the 50% overlapping region correlation, the parasitic modulation amplitude, the maximum conversion losses (in decibels), the maximum side-lobe level (in decibels), the asymptotic decay rate of the side lobes (in decibels per octave), the window width at the six-decibel level, the coherent gain, etc. All atomic windows exceed classic ones in terms of the asymptotic decay rate [231].
- The Fourier approximations of the $up(x)$ are defined by

$$up(x) = \frac{1}{2}\sum_{k=1}^{n} cos(\pi kx)\prod_{i=1}^{m}\frac{\sin(\pi k2^{-i})}{\pi k2^{-i}}. \quad (11.55)$$

A natural extension of the $up(x)$ function to the case of many variables is based on the usual tensor product of 1D $up(x)$ [231]. As a result, we have

$$up(x, y) = up(x)up(y). \quad (11.56)$$

In Fig. 11.10, we show a 2D atomic function on the space and frequency domains. A radial atomic function was mentioned in [231] as $up(\sqrt{x^2+y^2})$ (see Fig. 11.11). However in [169], the function $Plop(x, y)$ was defined as a radial infinite differentiable function with compact support (see Fig. 11.11), i.e.,

$$Pl\hat{op}(\nu, \upsilon) = \prod_{h=0}^{\infty}\sum_{k=0}^{\infty}\frac{[-(u^2+\nu^2)]^k}{3^{2k(h+1)}[(k+1)!]^2}. \quad (11.57)$$

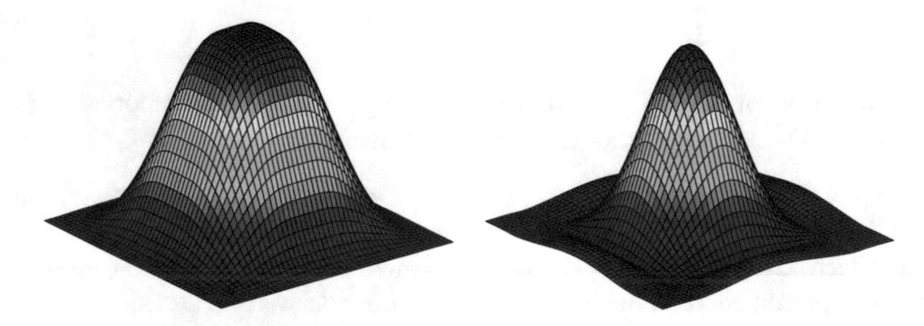

Fig. 11.10 Left: $up(x, y)$; right: Fourier transform $\hat{up}(\nu, \upsilon)$

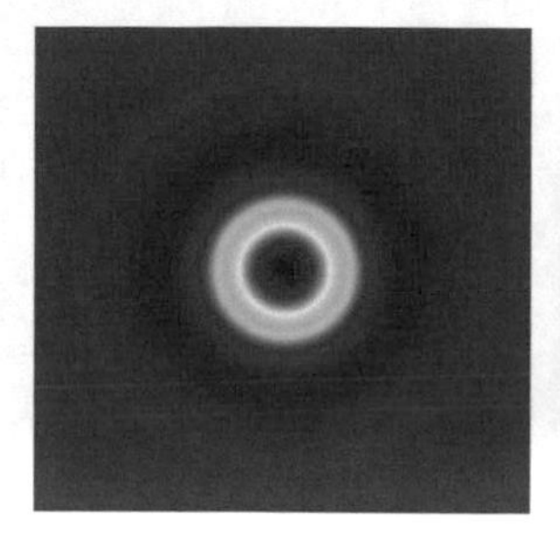
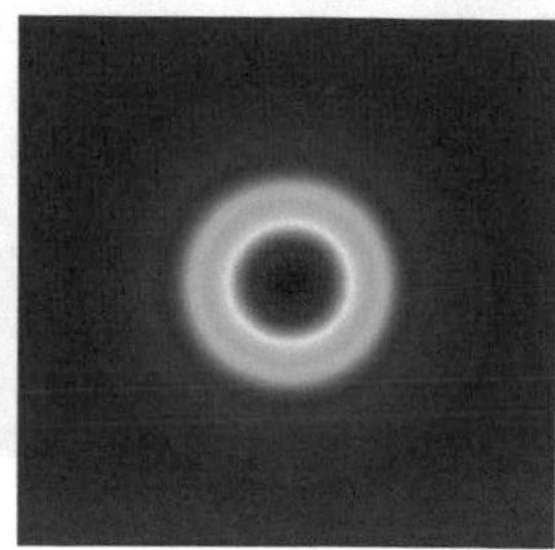

Fig. 11.11 Left: $\hat{up}(\sqrt{\nu^2 + \upsilon^2})$; right: $Pl\hat{o}p(\nu, \upsilon)$

11.2.3 The $dup(x)$

Some mask operators, such as the Sobel, Prewitt, and Kirsch, are used to process the images. A common drawback to these approaches is that it is impossible to ensure the required characteristics over a wide range of the working band of the processed signals and difficult to retune to adapt to the signal parameters [168]. This means that adaptation of the differential operator to the behavior of the input signal by broadening or narrowing its band is desirable, in order to ensure a maximum signal-to-noise ratio [168]. The $dup(x)$ function can be retuned to be adapted to the signal parameters. The $up(x)$ function satisfies Eq. (11.50) as follows:

$$dup(x) = 2up(2x + 1) - 2up(2x - 1). \tag{11.58}$$

The problem of adaptation of the differential operator reduces to the synthesis of infinitely differentiable finite functions with small-diameter carriers that are used for constructing the weighting windows [115, 168, 245]. One of the most effective solutions is obtained with the help of the atomic functions [168]. The AF can be used in two ways: for the construction of a window in a certain frequency region to improve the properties of the impulse response or the direct synthesis based in Eq. (11.50) [168]. Therefore, the function $up(x)$ satisfies the functional equation (11.53), and if we compute the Fourier transform of Eq. (11.53), we obtain

$$i\nu F[up(2x)] = \left(e^{i\nu} - e^{-i\nu}\right) F[up(2x)]), \tag{11.59}$$

$$F(dup(x)) = 2i \sin(\nu) F(up(2x)). \tag{11.60}$$

Figure 11.12 shows the $dup(x)$ and its Fourier transform $\hat{dup}(\nu)$. The function $dup(x)$ provides a good window in the spatial frequency regions because the side lobe has been completely eliminated [168]. Similar to Eq. (11.56), we can get a 2D expression of each derivative: $dup(x, y)_x = dup(x)up(y)$, $dup(x, y)_y = up(x)dup(y)$, $dup(x, y)_{x,y} = dup(x)dup(y)$.

The function $up(x)$ satisfies Eq. (11.51) and also satisfies Eq. (11.50), as mentioned. Upon differentiating Eq. (11.53) term by term, we obtain the derivatives of $up(x)$ Eq. (11.54) $d^{(n)}up(x)$.

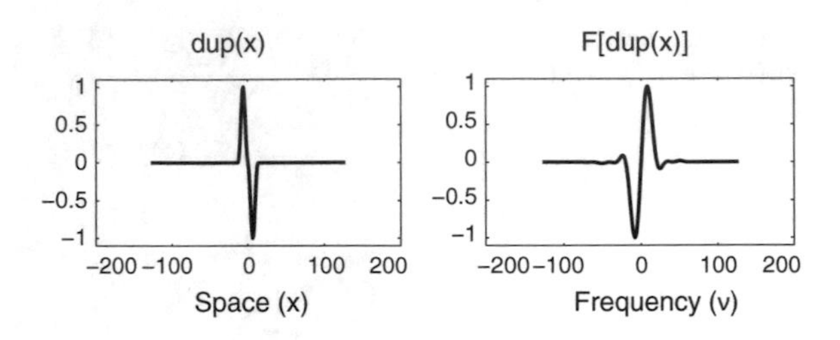

Fig. 11.12 Left: derivative of atomic function $dup(x)$; right: Fourier transform $\check{dup}(\nu)$

The function $dup(x)$ provides a good window in the spatial frequency regions since, in this case (frequency), the side lobes have been completely eliminated [108]. Figure 11.13 illustrates the first derivative, dup (Eq. (11.51)) in 1D and 2D, and also shows the 2D convolution with a test image. This differentiator, dup, can also be used as an oriented line detector via simple rotation as well.

11.2.4 Quaternion Atomic Function $Qup(x)$

The $up(x)$ function is easily extensible to two dimensions. Since a 2D signal can be split into even (e) and odd (o) parts [44],

$$f(x, y) = f_{ee}(x, y) + f_{oe}(x, y) + f_{eo}(x, y) + f_{oo}(x, y), \tag{11.61}$$

one can then separate the four components of Eq. (11.53) and represent them as a quaternion as follows:

$$\begin{aligned} Qup(x, y) = up(x, y)[&\cos(w_x)\cos(w_y) + i\sin(w_x)\cos(w_y) + \\ &+ j\cos(w_x)\sin(w_y) + k\sin(w_x)\sin(w_y)], \\ &= Qup_{ee}(x, y) + iQup_{oe}(x, y) + jQup_{eo}(x, y) + kQup_{oo}(x, y), \\ &= \Phi^q(x, y) + i\Psi_i^q(x, y) + j\Psi_j^q(x, y) + k\Psi_k^q(x, y). \end{aligned} \tag{11.62}$$

One can compute the components as follows:

$$\begin{aligned} \Phi^q(x, y) &= \phi^i(x)\phi^j(y) = s_x s_y up(x, y)\cos(w_x)\cos(w_y), \\ \Psi_i^q(x, y) &= \phi^i(x)\psi^j(y) = s_x s_y up(x, y)\cos(w_x)\ \sin(w_y), \\ \Psi_j^q(x, y) &= \psi^i(x)\phi^j(y) = s_x s_y up(x, y)\sin(w_x)\cos(w_y), \\ \Psi_k^q(x, y) &= \psi^i(x)\psi^j(y) = s_x s_y up(x, y)\sin(w_x)\sin(w_y), \end{aligned} \tag{11.63}$$

where $\phi^i(x) = s_x up(x)\cos(x)$ and $\psi(x)^i = s_x up(x)\sin(x)$ are the 1D complex filters applied along the rows and columns, respectively.

Note that in Eq. (11.62) we rename Qup as Φ, because we want to use similar notation used for multiresolution wavelet pyramids. Figure 11.14 shows a quaternion

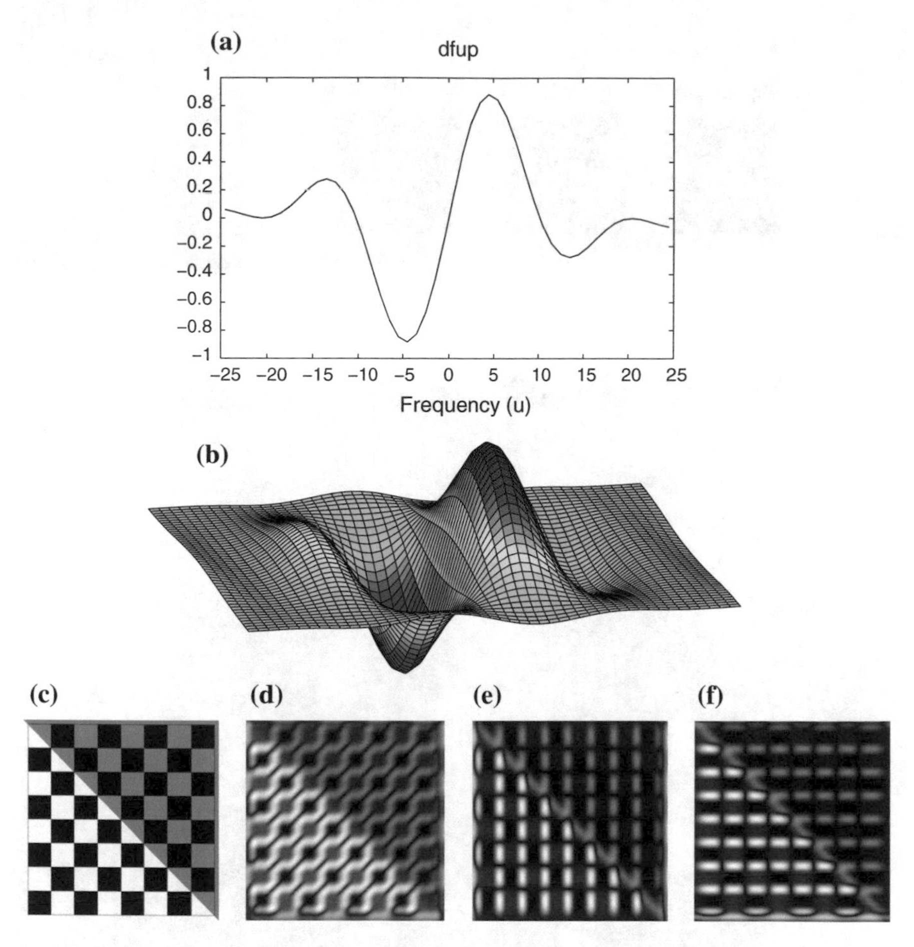

Fig. 11.13 Convolution of $dup(x, y)$ with the test image: (upper row) **a** $dup(x)$; **b** $dup(x, y)$. (lower row) **c** test image; results of convolutions with $dup(x, y)$ oriented in: **d** 0°; **e** 45°; **f** 135°

atomic function Qup in the space and frequency domains with its four components: the real part Qup_{ee} and the imaginary parts Qup_{eo}, Qup_{oe}, and Qup_{oo}. We can clearly see the differences in each part of our filter.

Next, we show the performance of Qup to detect edge changes using the phase concept. In Fig. 11.15, we see the phase changes along a transversal cut of an image with straight edges. Figure 11.16 shows the quaternionic phases by three transverse lines along a circle, a square, and a group of squares. We can see that the phases yield very useful information about shape contour changes.

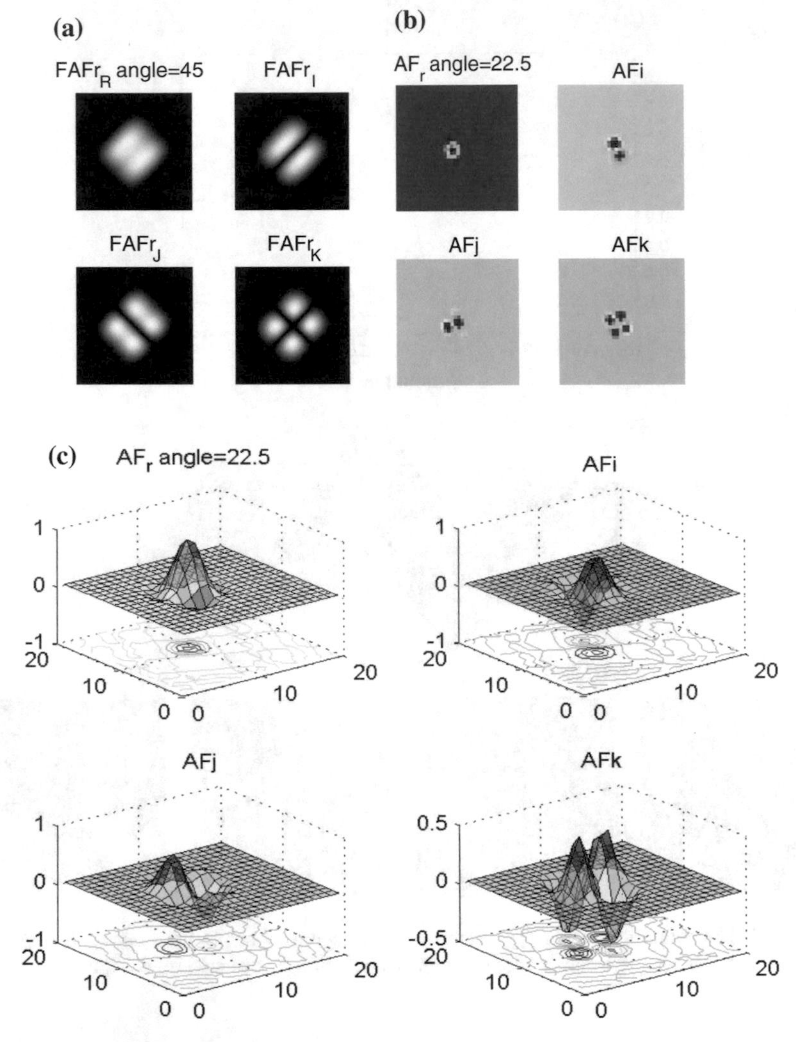

Fig. 11.14 Quaternion atomic function; elongated by $s_x = 0.3$, $s_y = 0.25$ (see Eq. (11.63)), and oriented at 45°. **a** In the frequency domain, the magnitude and the three imaginary quaternionic components; **b** in the spatial domain, the magnitude and the three imaginary quaternionic components; **c** 3D shapes of the quaternionic components in the spatial domain

11.3 Quaternionic Analytic Signal, Monogenic Signal, Hilbert Transform, and Riesz Transform

This section is based in the work published by E.U. Moya and E. Bayro-Corrochano [217, 219]. The *local quaternionic phase* of a two-dimensional signal can be measured using the angular phase of the filter response of a quaternionic Gabor filter. The

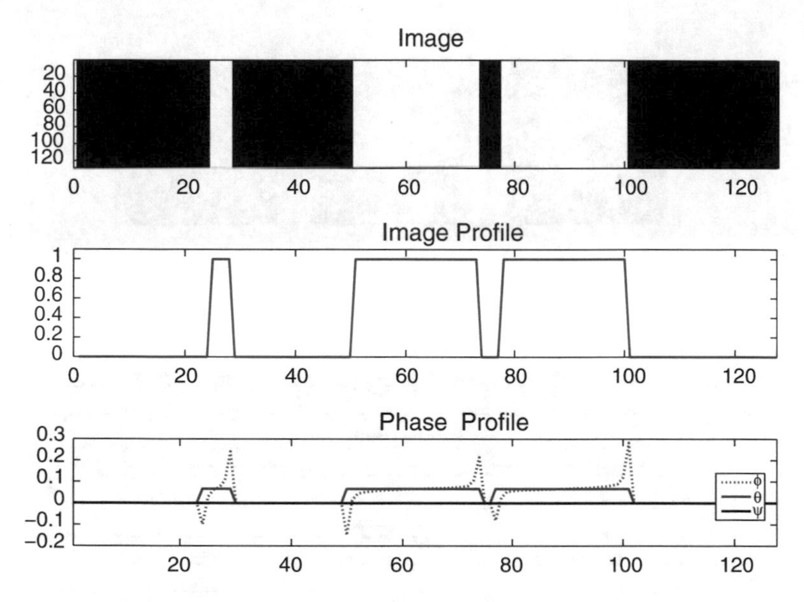

Fig. 11.15 (From up to bottom) Image; image of a transverse cut; quaternionic phases with respect to the transverse cut

evaluation of the angular phase is carried out according to the rules of the quaternion phase presented in Sect. 6.4.

The phase concept can be used for 3D reconstruction using interferometry techniques. Figure 11.17 shows the measurement of the phase change of 3D objects: a cube and a bell shape form which were illuminated with a light grid. In order to complete the whole 3D object representation, we can use a stereoscopic system to get the 3D information of some key points of the illuminated object, and together with the unwrapped phase, we can compute for each phase object point its corresponding 3D value.

11.3.1 Local Phase Information

In this work, we are interested in the local phase as it helps to separate the signal structure into even and odd part, for instance lines or edges [37, 110, 218]. One understands the term local phase as the computation of the instantaneous phase in a restricted bandwidth of the real signal. In general, the local phase is obtained by computing the instantaneous phase using the analytic signal or monogenic signal and a bandpass filter.

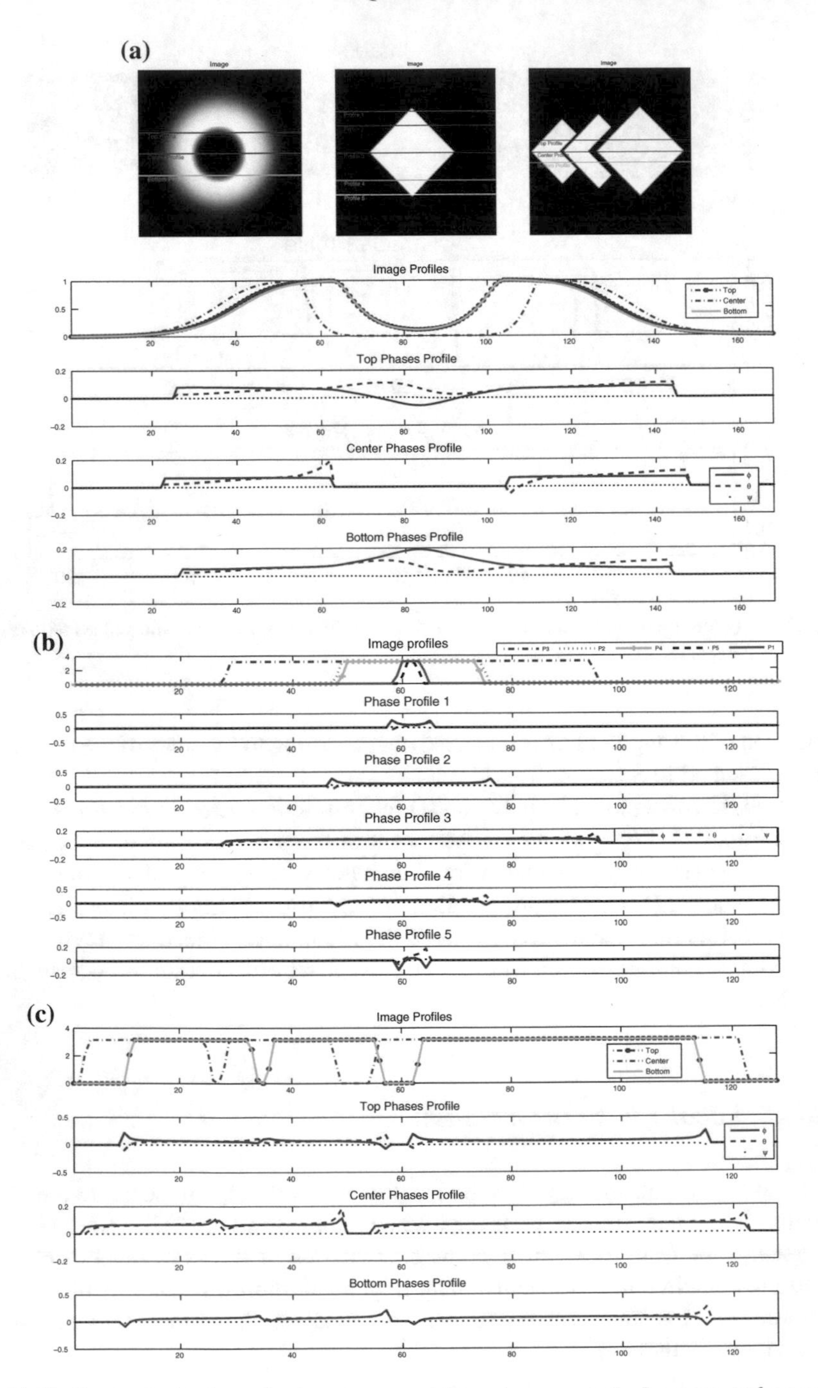

Fig. 11.16 Quaternionic phases by three transversal lines along a **a** circle; **b** square, and a **c** group of squares

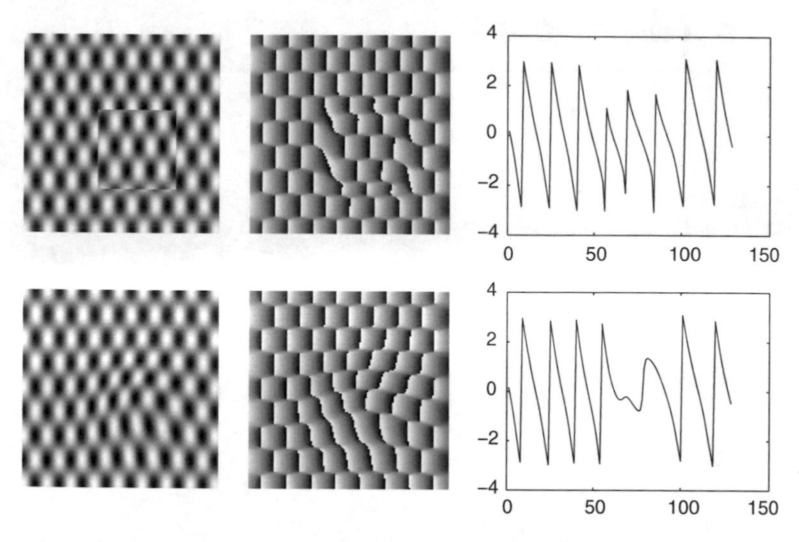

Fig. 11.17 Left images: original 150 × 150 images which are convolved with qgabor(25,7,7,2,2,90). Middle images: one of phase images. Right images: phase changes modulo 2π along the row 64

The main approaches to obtain the local phase in images the quaternion analytic function and the monogenic function use the quaternionic Gabor filter (with Hilbert and partial Hilbert transform) [44] and Log-Gabor filter (with Riesz transform) [90] in a multiscale approach, respectively. However, in our work instead of using a Gabor or Log-Gabor filter, we use the $AF\ up(x, y)$ a compact support window with good local properties, which is easy to derivate with dyadic operations. The $up(x, y)$ can be used as a filter with the Hilbert and Riesz transforms.

In order to explain why we use the $up(x, y)$ instead of the Gabor or the Log-Gabor filters, let us consider first the uncertainty principle for 1D signals, the Gabor filter is located on the curve of optimal uncertainty $\Delta u \Delta x = 1/4\pi$ below the general quadrature filter, and a Poisson filter lies above on its curve which is $\sqrt{2}$ times the curve of the Gabor [90]. An atomic filter $up(x)$ due to its compactness has a lower uncertainty in space than the Gabor, and it lies a bit lower and to the right, see Fig. 11.18. The Fourier transform yields an infinite uncertainty in space $\Delta x = \infty$ and lies at the top of the ordinate, and conversely, the Hilbert transform yields an infinity uncertainty in frequency domain $\Delta u = \infty$ and at the far right of the abscissa.

Thus, one can claim that the use of compact support filters in space results in a localized and much better processing of the signal in space.

Since the atomic filter is compact in the space domain and introduces less errors due to truncation as the Gabor or Log-Gabor filters, it will definitively guarantee a more accurate phase analysis than these filters and also much better than the global phase computed just with the Fourier transform.

It is known that in order to increase the accuracy of the phase estimation, it is needed more filters of reduced bandwidth to cover a certain passband and also more

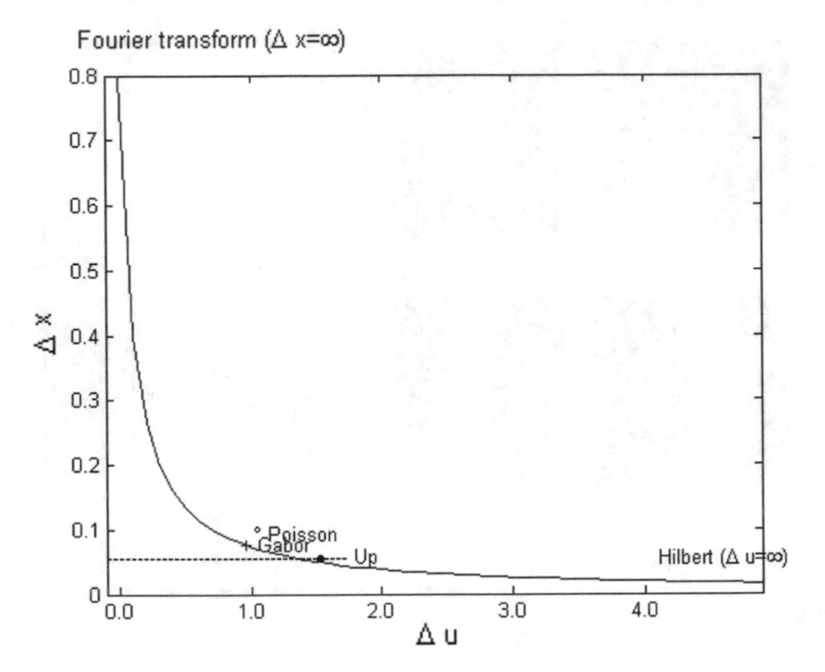

Fig. 11.18 Uncertainties for the Gabor, Poisson and *up* filters

samples are required at the local orientation. Since the atomic function is compact in space, it will cover with less uncertainty the expected passband. In the experimental analysis of Sect. 14.6.1, we will compare global phase analysis versus a multi-scale local phase analyses using the Gabor, Log-Gabor, and the atomic function filter.

11.3.2 Quaternionic Analytic Signal

The quaternionic analytic signal in the space domain is defined as [44]

$$f_A^q(x, y) = f(x, y) + i f_{\mathbf{H}i}(x, y) + j f_{\mathbf{H}j}(x, y) + k f_{\mathbf{H}k}(x, y), \tag{11.64}$$

where $f_{\mathbf{H}i}(x, y) = f(x, y) \star \frac{1}{\pi x}$ and $f_{\mathbf{H}j}(x, y) = f(x, y) \star \frac{1}{\pi y}$ are the partial Hilbert transforms and $f_{\mathbf{H}k}(x, y) = f(x, y) \star \frac{1}{\pi^2 xy}$ is the Hilbert transform, which involves both axes x and y. Bülow has shown that the QFT kernel is expressed in terms of the Hilbert transforms. Given a quaternion $\boldsymbol{q}$, its phases can be computed by the rules given in Sect. 6.4.

11.3.3 Monogenic Signal

The monogenic signal was proposed by M. Felsberg and G. Sommer [90], and it extends the analytic signal in 1D to the nD approach. The monogenic signal for 2D signals is represented by

$$f_{\mathbf{M}}(\mathbf{x}) = f(\mathbf{x}) + (i, j) f_{\mathbf{R}}(\mathbf{x}) = f(\mathbf{x}) + (i, j) f(\mathbf{x}) * \frac{\mathbf{x}}{2\pi |\mathbf{x}|^3}. \tag{11.65}$$

The magnitude of the signal is computed by $|f_{\mathbf{M}}(\mathbf{x})| = \sqrt{(i, j) f_{\mathbf{R}}^2 + f(\mathbf{x})^2}$. Since the monogenic signal is constructed from the original signal and its Riesz transform, we can express the local phase ψ and the local orientation θ as follows:

$$\psi = \arctan\left(\frac{|(i, j) f_{\mathbf{R}} * f(\mathbf{x})|}{f(\mathbf{x})}\right), \tag{11.66}$$

$$\theta = \arctan\left(\frac{f_{\mathbf{R}j} * f(\mathbf{x})}{f_{\mathbf{R}i} * f(\mathbf{x})}\right), \tag{11.67}$$

where the subindexes i, j stand for the i and j partial Hilbert transforms of the $(i, j) f_{\mathbf{R}}$.

11.3.4 Hilbert Transform Using AF

The Hilbert transform and the partials Hilbert transforms can be expressed in terms of the up and its derivative as follows:

$$g_{\mathbf{H_i}} f(x, y) = f(x, y) * \left(dup(x, y)_x * -\frac{1}{\pi} \log(|x|)\right), \tag{11.68}$$

$$g_{\mathbf{H_j}} f(x, y) = f(x, y) * \left(dup(x, y)_y * -\frac{1}{\pi} \log(|y|)\right), \tag{11.69}$$

$$g_{\mathbf{H_k}} f(x, y) = f(x, y) * \left(dup(x, y)_{xy} * -\frac{1}{\pi^2} \log(|x|) \log(|y|)\right). \tag{11.70}$$

Proof The Hilbert transform in terms of the convolution is defined as

$$f_{\mathbf{H}i}(x) = f(x) * \frac{1}{\pi x}, \tag{11.71}$$

by using the association and the distribution properties of the convolution operation

$$f * (g * h) = (f * g) * h, \tag{11.72}$$

$$\nabla(f * g) = \nabla f * g = f * \nabla g, \tag{11.73}$$

where $f, g \in \Re^2$, $\nabla = e_1\frac{\partial}{\partial x} + e_2\frac{\partial}{\partial y}$ allows us to rewrite the Hilbert transform in terms of derivatives [291]. If $g(x, y) = -\frac{1}{\pi}\log(|x|)$, which is the fundamental solution of Laplace's equation, it is easy to show that the Hilbert transform and the partial Hilbert transforms can be written as follows:

$$f_{\mathbf{H_i}} f(x, y) = \frac{\partial f(x, y)}{\partial x} * -\frac{1}{\pi}\log(|x|), \tag{11.74}$$

$$f_{\mathbf{H_j}} f(x, y) = \frac{\partial f(x, y)}{\partial y} * -\frac{1}{\pi}\log(|y|), \tag{11.75}$$

$$f_{\mathbf{H_k}} f(x, y) = \frac{\partial^2 f(x, y)}{\partial x \partial y} * \frac{1}{\pi^2}\log(|x|)\log(|y|). \tag{11.76}$$

We can get the Hilbert transform in terms of $dup(x)$, such as

$$g_{\mathbf{H_i}} f(x, y) = f(x, y) * \left(dup(x, y)_x * -\frac{1}{\pi}\log(|x|)\right), \tag{11.77}$$

$$g_{\mathbf{H_j}} f(x, y) = f(x, y) * \left(dup(x, y)_y * -\frac{1}{\pi}\log(|y|)\right), \tag{11.78}$$

$$g_{\mathbf{H_k}} f(x, y) = f(x, y) * \left(dup(x, y)_{xy} * -\frac{1}{\pi^2}\log(|x|)\log(|y|)\right) \tag{11.79}$$

and the local quaternionic analytic signal can be rewritten as

$$f_A^q(x, y) = f(x, y) * up(x, y) + i g_{\mathbf{H}i}(x, y) + j g_{\mathbf{H}j}(x, y) + k g_{\mathbf{H}k}(x, y). \tag{11.80}$$

On the left of Fig. 11.19.a shows three curves, the function $up(x)$ (dash-line), its Hilbert transform $f_{\mathbf{H_i}} up(x)$ (solid-line), and the computation of $g_{\mathbf{H_i}} f(x)$, Eq. (11.77). Figure 11.19b shows the real and the imaginary parts of the analytic signal. The real part corresponds to signal $up(x)$, and the imaginary part is defined by the Hilbert transform of the signal (solid-line). The dash-line curve shows the same computation using Eq. (11.77).

11.3.5 Riesz Transform Using AF

For a $2D$ signal $f(\mathbf{x})$, the Riesz transform can be rewritten in terms of $Plop(\mathbf{x})$ as follows:

$$g_{\mathbf{R}}(\mathbf{x}) = f(\mathbf{x}) * \left(\nabla^2 Plop(\mathbf{x})_{xy} * -\frac{1}{2\pi} sign(|\mathbf{x}|) log(|\mathbf{x}|)\right). \tag{11.81}$$

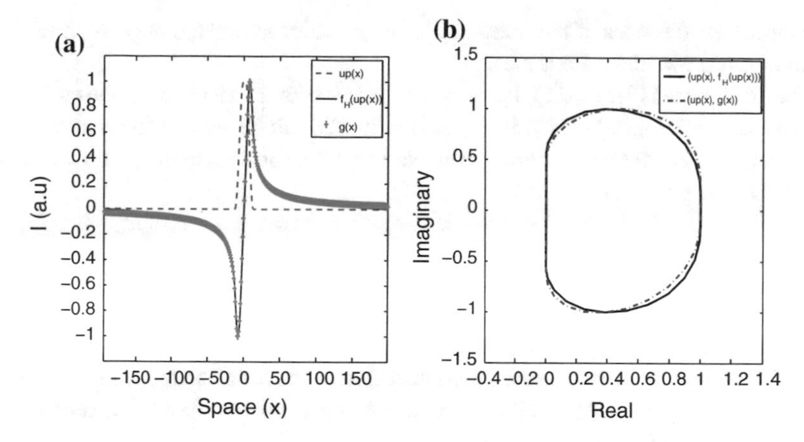

Fig. 11.19 **a** $up(x)$ (dashed-line), its Hilbert transform (solid-line) and Eq. (11.77) $g_{\mathbf{H_i}}up(x)$ (circles). **b** $up(x)$ (real part) versus the Hilbert transform (imaginary part). The dashed-line $up(x)$ versus the imaginary part of $g_{\mathbf{H_i}}f(x)$

Proof As the last theorem, we can rewrite the Riesz transform as

$$f_{\mathbf{R}}(\mathbf{x}) = f(\mathbf{x}) * -\frac{\mathbf{x}}{2\pi|\mathbf{x}|^3}, \tag{11.82}$$

$$\begin{aligned}
f_{\mathbf{R}}(\mathbf{x}) &= \nabla^2 f(\mathbf{x}) * \left(\frac{1}{2\pi} sign(|\mathbf{x}|) log(|\mathbf{x}|)\right), \\
f_{\mathbf{R}}(\mathbf{x}) &= f(\mathbf{x}) * \nabla^2 \left(\frac{1}{2\pi} sign(|\mathbf{x}|) log(|\mathbf{x}|)\right), \\
f_{\mathbf{R}}(\mathbf{x}) &= f(\mathbf{x}) * \nabla \left(\frac{1}{2\pi|\mathbf{x}|}\right), \\
f_{\mathbf{R}}(\mathbf{x}) &= f(\mathbf{x}) * -\frac{\mathbf{x}}{2\pi|\mathbf{x}|^3}.
\end{aligned} \tag{11.83}$$

Finally, we have

$$g_{\mathbf{R}}(\mathbf{x}) = f(\mathbf{x}) * \left(\nabla^2 Plop(\mathbf{x}) * -\frac{1}{2\pi} sign(|\mathbf{x}|) log(|\mathbf{x}|)\right). \tag{11.84}$$

11.4 Radon Transform of Functionals

The aim of this section is to put in context the use of Radon transform (RT) theory together with the QWT for the detection of contours of arbitrary shapes using either grayscale or color images. Some essential concepts are outlined briefly; however, we give enough references of the literature for any reader interested to go in much more detail about these theoretic issues.

The Radon transform (RT), introduced by J. Radon in 1917, describes a function in terms of its (integral) projections [243]. The RT can be seen as the mapping from the function onto the projections of the Radon transform. The original formulation of the RT was given by

$$\mathcal{R}\{I\}(d,\phi) = \int_{\mathbb{R}} I(d\cos\phi - s\sin\phi, d\sin\phi + s\cos\phi)ds, \tag{11.85}$$

which represents the projections of the function I along the lines $c_l(d,\phi)$. The inverse of the RT corresponds to the back-reconstruction of the function from the projections. Furthermore, one can use the RT to detect the shape; for this purpose, one reformulates Eq. (11.85) to detect lines

$$\mathcal{R}\{I\}(d,\phi) = \int_{x \ on \ c_l(d,\phi)} I(x,y)dxdy = \int_{\mathbb{R}} \delta(x\cos\phi + y\sin\phi - d)dxdy. \tag{11.86}$$

The generalization of the RT to detect arbitrary shape's contours $c_l(d,\phi)$ appears now to be straightforward, so let us consider the following formulas:

$$\begin{aligned} \mathcal{R}_{c(p)}\{I\}(\boldsymbol{p}) &= \int_{\boldsymbol{x} \ on \ c(p)} I(\boldsymbol{x})d\boldsymbol{x} = \int_{\mathbb{R}^N} I(c(\boldsymbol{s};\boldsymbol{p}))\|\frac{\partial c(\boldsymbol{s})}{\partial \boldsymbol{s}}\|d\boldsymbol{s}, \\ &= \int_{\mathbb{R}^D} I(\boldsymbol{x})\delta(C(\boldsymbol{x};\boldsymbol{p}))d\boldsymbol{x}. \end{aligned} \tag{11.87}$$

The first formulation is aimed at assigning votes at the point $\boldsymbol{p}$ of the Radon parameter space based on the integral of $I(\boldsymbol{x})$ for points lying on the shape's contour $c(\boldsymbol{p})$. The second formulation is the same following the absolute value of the gradient $\frac{\partial c(s)}{\partial s}$ along the segment of contour $c(\boldsymbol{s})$, and the third votes for points along a contour described by the null constraint formulated in terms of a Dirac delta function; for more details, see [102]. The RT builds in the Radon parameter space a function $P(\boldsymbol{p})$ having peaks for those parameter vectors $\boldsymbol{p}$ for which the corresponding shape $c(\boldsymbol{p})$ is present in the image. As a result, the problem of shape detection has been simplified to a task of peak detection. The third formulation of the RT stresses the importance of the use of generalized functions. In fact, we can recognize the form of a linear integral operator (Fredholm operator) $\mathcal{L}_C$ with kernel C:

$$(\mathcal{L}_C I)(\boldsymbol{p}) = \int_{\mathbb{R}^D} C(\boldsymbol{p},\boldsymbol{x})I(\boldsymbol{x})d\boldsymbol{x}. \tag{11.88}$$

In Eq. (11.87), the kernel C is of the form $C(\boldsymbol{p},\boldsymbol{x}) = \delta(C(\boldsymbol{x};\boldsymbol{p}))$. With respect to shape detection, the role of the operator $\mathcal{L}_C$ is to compute via the inner product the match between the image shape and a template C for a given parameter set $\boldsymbol{p}$. Note the close connection between the Radon transform and template matching. Since the matching criterion is the inner product between the template $T = \delta(C(\boldsymbol{x};\, \boldsymbol{p}))$ and the image I, Eq. (11.87) can be rewritten with respect to a parameter vector $\boldsymbol{p}$ as a parameter response function $P(\boldsymbol{p})$ as follows:

$$P(\boldsymbol{p}) = \int_{\mathbb{R}^D} T(\boldsymbol{x}, \boldsymbol{p}) I(\boldsymbol{x}) d\boldsymbol{x}. \tag{11.89}$$

Although this technique can be used to detect gray-value blobs in I, usually this equation is applied to an edge/line map $E(\boldsymbol{x})$, which contains the contours of the shapes instead of being applied directly to the image $I(x, y)$. In this regard, Eq. (11.89) reads

$$P(\boldsymbol{p}) = \int_{(x,y,\varphi) \ \ on \ \ c^{[\varphi]}(\boldsymbol{x};\boldsymbol{p})} I^{[\varphi]}(x, y, \varphi) dx dy d\varphi, \tag{11.90}$$

where φ indicates the edge orientation. Finally, if we use the quaternionic wavelet atomic function for curves parameterized in terms of the quaternionic parameters $c^{[\varphi(\phi,\theta)]}(s;\, \boldsymbol{p}) = (x(s;\, \boldsymbol{p}), y(s;\, \boldsymbol{p}), \varphi(\phi(s;\, \boldsymbol{p}), \theta(s;\, \boldsymbol{p})))$, the standard RD equation can be formulated using the 2D quaternionic phase concept as follows:

$$P(\boldsymbol{p}) = \int_{(x,y,\phi) \ \ on \ \ c^{[\varphi(\phi,\theta)]}(\boldsymbol{x};\boldsymbol{p})} I^{[\varphi(\phi,\theta)]}(x, y, \varphi) dx dy d\varphi. \tag{11.91}$$

In fact, Eq. (11.89) is equivalent to Eq. (11.85) of the RT of the $\mathbb{H}$-embedded Riesz transform $f_R(\boldsymbol{x}) = (i, j) f_r(\boldsymbol{x})$ of a 2D signal $f(\boldsymbol{x})$ given by the Hilbert transform $(h_1(t))$ of the RT of $f(\boldsymbol{x})$ according to

$$\mathcal{R}\{I_R\}(d, \theta) = (i, j)\mathbf{n}_\theta h_1(t) * \mathcal{R}\{f\}(t, \theta), \tag{11.92}$$

where the I_R is the Riesz transform of the image. This equation is the RT of the $\mathbb{H}$-embedded Riesz transform and is computed with respect to a line with orientation θ and Hesse distance d.

11.5 Clifford Fourier Transforms

The real Fourier transform given in Eq. (11.1) can be straightforwardly extended to Clifford Fourier transforms for different dimensions and metrics, where we have to consider carefully the role of the involved pseudoscalar of $I_n \in G_n = G_{n,0}$. In next subsections, we present the 3D CFT, the space–time CFT, and the n-dimensional CFT.

11.5.1 Tridimensional Clifford Fourier Transform

The Tridimensional Clifford Fourier transform (3D CFT) for vector fields was first introduced first by Bernard Jancewicz [151], then used by Ebling and Sheuermann [83], and recently the work on the $\mathcal{C}_{3,0}$, Clifford Fourier Transform by Hitzer and Mawardi [144, 205] enriched even more the study of this transform. In the tridimensional geometric algebra, the pseudoscalar is a trivector with signature $I_3^2 = (\boldsymbol{e}_1\boldsymbol{e}_2\boldsymbol{e}_3)^2 = -1$. Similar as in the case of the real Fourier transform given in Eq. (11.1), we can use the trivector instead of the imaginary number i with $i^2 = -1$.

We know that the scalar and the pseudoscalar build a complex number which in turn becomes the centrum of the G_3. Thus for any tridimensional multivector field $f : \mathbb{R}^3 \rightarrow G_3$, the Tridimensional Clifford Fourier transform of f is defined by the function given by

$$F(\boldsymbol{w})_{G_3} \triangleq \mathcal{F}[f(\boldsymbol{x})] = \int_{\mathbb{R}^3} f(\boldsymbol{x}) e^{-I_3 \boldsymbol{w}\cdot\boldsymbol{x}} d^3\boldsymbol{x}, \tag{11.93}$$

where $\boldsymbol{w} = w_1\boldsymbol{e}_1 + w_2\boldsymbol{e}_2 + w_3\boldsymbol{e}_3$, $\boldsymbol{x} = x_1\boldsymbol{e}_1 + x_2\boldsymbol{e}_2 + x_3\boldsymbol{e}_3$ and $d^3\boldsymbol{x} = \frac{dx_1 \wedge dx_2 \wedge dx_3}{I_3}$. Since the pseudoscalar I_3 commutes with any element of G_3, the Clifford Fourier kernel $e^{-I_3\boldsymbol{w}\cdot\boldsymbol{x}}$ will commute with every element of G_3 as well. The *inverse* 3D Clifford Fourier transform is computed as follows:

$$f(\boldsymbol{x}) \triangleq \mathcal{F}^{-1}[F(\boldsymbol{w})_{G_3}] = \frac{1}{(2\pi)^3} \int_{\mathbb{R}^3} F(\boldsymbol{w}) e^{I_3 \boldsymbol{w}\cdot\boldsymbol{x}} d^3\boldsymbol{w}. \tag{11.94}$$

The vectors $\boldsymbol{x}$ and $\boldsymbol{w}$ represent position in spatial and frequency domains, respectively.

Table 11.3 presents a summary of some basic properties of the 3D Clifford Fourier transform.

In order to reduce the computational complexity of the CFT, we can first group conveniently the multivector elements of function $f(\boldsymbol{x}) \in G_3$ and then apply the traditional two-dimensional Fast Fourier Transform (FFT). Let us first rewrite the

Table 11.3 Properties of the 3D Clifford Fourier transform

Property	Function	3D Clifford Fourier transform
Linearity	$\alpha f(\boldsymbol{x}) + \beta g(\boldsymbol{x})$	$\alpha F(\boldsymbol{w}) + \beta G(\boldsymbol{w})$
Delay	$f(\boldsymbol{x} - \boldsymbol{t})$	$e^{-I_3\boldsymbol{w}\cdot\boldsymbol{t}} F(\boldsymbol{w})$
Shifting	$e^{I_3\boldsymbol{w}_0\cdot\boldsymbol{x}} f(\boldsymbol{x})$	$F(\boldsymbol{w} - \boldsymbol{w}_0)$
Scaling	$f(\alpha\boldsymbol{x})$	$\frac{1}{\lvert\alpha^3\rvert} F(\frac{\boldsymbol{w}}{\alpha})$
Convolution	$(f \star g)(\boldsymbol{x})$	$F(\boldsymbol{w})G(\boldsymbol{w})$
Derivative	$f^n(\boldsymbol{x})$	$(I_3\boldsymbol{w})^n F(\boldsymbol{w})$
Parseval theorem	$\int_{\mathbb{R}^3} \lvert f(\boldsymbol{x})\rvert^2 d^3\boldsymbol{x}$	$\int_{\mathbb{R}^3} \lvert F(\boldsymbol{w})\rvert^2 d^3\boldsymbol{w}$
Gaussian	$e^{(\alpha\boldsymbol{x})^2/2}$	$\lvert\alpha\rvert^{-3} e^{-(\boldsymbol{w}/\alpha)^2/2}$

tridimensional multivector-valued function $f(\boldsymbol{x}) \in \mathbb{R}^3 \in G_3$ in terms of four complex signals; using instead of the imaginary number, i the trivector I_3 as follows:

$$\begin{aligned} f(\boldsymbol{x}) &= [f(\boldsymbol{x})]_0 + [f(\boldsymbol{x})]_1 + [f(\boldsymbol{x})]_2 + [f(\boldsymbol{x})]_3, \\ &= f(\boldsymbol{x})_0 + f(\boldsymbol{x})_1 \boldsymbol{e}_1 + f(\boldsymbol{x})_2 \boldsymbol{e}_2 + f(\boldsymbol{x})_3 \boldsymbol{e}_3 + \\ &\quad + f(\boldsymbol{x})_{23} \boldsymbol{e}_{23} + f(\boldsymbol{x})_{31} \boldsymbol{e}_{31} + f(\boldsymbol{x})_{12} \boldsymbol{e}_{12} + f(\boldsymbol{x})_{I_3} I_3, \\ &= (f(\boldsymbol{x})_0 + f(\boldsymbol{x})_{I_3} I_3) + (f(\boldsymbol{x})_1 + f(\boldsymbol{x})_{23} I_3) \boldsymbol{e}_1 + \\ &\quad + (f(\boldsymbol{x})_2 + f(\boldsymbol{x})_{31} I_3) \boldsymbol{e}_2 + (f(\boldsymbol{x})_3 + f(\boldsymbol{x})_{12} I_3) \boldsymbol{e}_3, \end{aligned} \tag{11.95}$$

where bivectors are expressed as dual vectors, e.g., $I_3 \boldsymbol{e}_2 = \boldsymbol{e}_3 \boldsymbol{e}_1$. Now taking into account the linearity property of the CFT, 3D CFT can be written as follows:

$$\begin{aligned} F(\boldsymbol{w}) = \mathcal{F}[f(\boldsymbol{x})] &= \mathcal{F}[(f(\boldsymbol{x})_0 + f(\boldsymbol{x})_{I_3} I_3)] + \mathcal{F}[(f(\boldsymbol{x})_1 + f(\boldsymbol{x})_{23} I_3)] \boldsymbol{e}_1 + \\ &\quad + \mathcal{F}[(f(\boldsymbol{x})_2 + f(\boldsymbol{x})_{31} I_3)] \boldsymbol{e}_2 + \mathcal{F}[(f(\boldsymbol{x})_3 + f(\boldsymbol{x})_{12} I_3)] \boldsymbol{e}_3. \end{aligned} \tag{11.96}$$

This kind of separation can be applied to multivector fields of arbitrary dimension D; as a result, Clifford Fourier transformations can be computed by carrying out several standard Fourier transformations. For the 2D and 3D cases, one require two FFTs and 4 FFTs, respectively.

11.5.2 Space and Time Geometric Algebra Fourier Transform

As we show in Chap. 12, in computer vision the 3D projective space is treated in the geometric algebra of Minkowski metric $G_{3,1}$ and it is related to the projective plane (image) G_3 via the projective split. Since in Sect. 11.5.1 we study the 3D Clifford Fourier transform in G_3, now it will be very interesting to formulate a Clifford Fourier transform using the framework of the space–time geometric algebra $G_{3,1}$. In $G_{3,1}$, the space–time vectors are given by $\boldsymbol{x} = x e_1 + y e_2 + z e_3 + t e_4 = \mathbf{x} + t e_4$ and the space–time frequency vectors by $\boldsymbol{w} = u e_1 + v e_2 + w e_3 + s e_4 = \mathbf{w} + s e_4$. Given a 16D space–time algebra functions $f : \mathbb{R}^{3,1} \rightarrow G_3$, the Space–Time Geometric Algebra Fourier Transform (ST-GAFT) is computed as follows:

$$F(\boldsymbol{w})_{ST-CFT} = \int_{\mathbb{R}^{3,1}} e^{-e_4 t s} f(\boldsymbol{x}) e^{-I_3 \mathbf{x} \cdot \mathbf{w}} d^4 \boldsymbol{x}, \tag{11.97}$$

where the space–time volume $d^4 \boldsymbol{x} = dt dx dy dz$. The part $\int_{\mathbb{R}^3} f(\boldsymbol{x}) e^{-I_3 \mathbf{x} \cdot \mathbf{w}} d^3 \mathbf{x}$ corresponds to the 3D CFT. The Inverse Space–Time Geometric Algebra Fourier Transform (IST GAFT) is given by

$$f(\boldsymbol{x}) = \int_{\mathbb{R}^{3,1}} e^{e_4 ts} F(\boldsymbol{w})_{ST-CFT} e^{I_3 \mathbf{x}\cdot\mathbf{w}} d^4\boldsymbol{w}, \tag{11.98}$$

where the space–time frequency volume $d^4\boldsymbol{w} = dsdudvdw$.

11.5.3 n-Dimensional Clifford Fourier Transform

Equation (11.1) for the real Fourier transform can be straightforwardly extended in the n-dimensional geometric algebra G_n framework. Given a multivector-valued function $f : \mathbb{R}^n \rightarrow G_n$, its n-dimensional Clifford Fourier transform (nD CFT) is given by

$$F(w)_{G_n} = \int_{\mathbb{R}^n} f(\boldsymbol{x}) e^{-I_n \boldsymbol{w}\cdot\boldsymbol{x}} d^n\boldsymbol{x}, \tag{11.99}$$

where I_n stands for the pseudoscalar of $G_n = G_{n,0}$ for dimensions $n = 2, 3$ (mod) 4 or also possible for the geometric algebra $G_{0,n}$ for $n = 1, 2, 3$ (mod 4). In next subsections, we present the 3D CFT and the space–time CFT.

11.6 From Real to Clifford Wavelet Transforms for Multiresolution Analysis

The word *wavelet* was used for the first time in the thesis of Alfred Haar in 1909. Surprisingly, the wavelet transform (WT) has become a useful signal processing tool only in the last few decades, mainly due to the contributions in the areas of applied mathematics and signal processing [155, 157, 200, 203, 216]. Generally speaking, the WT is an approach that definitely overcomes the shortcomings of the window Fourier transform. Thanks to the development of the Quaternion Fourier transform (QFT) [44], the generalization of the real and complex wavelet transforms was straightforward to the hypercomplex wavelet transform.

In this next subsections, we review and explain the real, complex, and quaternion wavelet transforms.

11.6.1 Real Wavelet Transform

The real wavelet transform (RWT) implements a meaningful decomposition of a signal $f(x)$ onto a family of functions. Such functions are dilations and translations of a unique function called a *mother wavelet* ψ that fulfills

$$\int_{-\infty}^{\infty} \psi(t)dt = 0. \tag{11.100}$$

The mother wavelet is normalized $||\psi|| = 1$ and is centered with respect to a certain neighborhood ($t = 0$).

In general, the wavelet family is generated by affine transformations. In the one-dimensional case, these wavelets are given by

$$\psi_{s,t}(x) = \sqrt{(s)}\psi(s(x-t)), \tag{11.101}$$

where $(s,t) \in R^2$ represents the scale and translation, respectively.

Considering $\mathbf{L}^2(\mathbb{R})$ the vector space of measurable and square-integrable one-dimensional functions of f, we can define a wavelet transform for functions $f(x) \in \mathbf{L}^2(\mathbb{R})$ in terms of a certain wavelet ψ as follows:

$$Wf(s,t) = \int_{-\infty}^{+\infty} f(x)\psi_{s,t}(x)dx = \langle f(x), \psi_{s,t}(x)\rangle. \tag{11.102}$$

We should interpret a wavelet transform simply as a decomposition of a signal $f(x)$ into a set of frequency channels that have the same bandwidth on a logarithmic scale. If certain requirements are fulfilled, the inverse of the wavelet transform exists [157, 200, 202] and thus $f(x)$ can be reconstructed.

In many image processing tasks like feature extraction or the design of matching algorithms, it is highly desirable for the used filters to exhibit good locality, in both space and frequency. The key advantage of the wavelet transform is the adaptability of its parameters s and t.

11.6.2 Discrete Wavelets

The discretization of the continuous wavelet transform helps to eliminate any redundancies. For that the functions $\psi_{s,t}(x)$ of Eq. (11.102) are discretized via the parameters s and t:

$$s = \frac{1}{2^j}, \qquad t = k, \tag{11.103}$$

for the integers $(j,k) \in \mathbb{Z}^2$. The resulting functions are called dyadic discrete wavelets:

$$\psi_{j,k}(x) = \frac{1}{\sqrt{2^j}}\psi\left(\frac{x-k}{2^j}\right), \tag{11.104}$$

for $(j,k) \in \mathbb{Z}^2$.

In the dyadic scale space of Eq. (11.104), if we denote $A_j f$ as the approximation of a given $f(x)$ at the scale $s = \frac{1}{2^j}$, we express the difference between two successive approximations as

$$D_j f = A_{j-1} f - A_j f, \tag{11.105}$$

where j stands for the number of a certain level. In practice, we consider a limited number of levels $j = 1, 2, \ldots, n$. The coarsest level is n, and A_0 is an identity operator. In that regard, the function $f(x)$ can be decomposed as

$$\begin{aligned} f(x, y) &= A_1 f + D_1 f, \\ &= A_2 f + D_2 f + D_1 f, \\ &\vdots \\ &= A_n f + \sum_{j=1}^{n} D_j f. \end{aligned} \tag{11.106}$$

As it has been shown [202], the multiresolution analysis of the one-dimensional function $f(x)$ can be carried out in terms of a scaling function ϕ and its associated wavelet function ψ as follows:

$$A_j f(x) = \sum_{k=-\infty}^{+\infty} \langle f(u), \phi_{j,k}(u) \rangle \phi_{j,k}(x), \tag{11.107}$$

$$D_j f(x) = \sum_{k=-\infty}^{+\infty} \langle f(u), \psi_{j,k}(u) \rangle \psi_{j,k}(x),$$

where

$$\psi_{j,k}(x) = \frac{1}{\sqrt{2^j}} \psi\left(\frac{x-k}{2^j}\right), \qquad \phi_{j,k}(x) = \frac{\sqrt{1}}{\sqrt{2^j}} \phi\left(\frac{x-k}{2^j}\right), \tag{11.108}$$

for $(j, k) \in \mathbb{Z}^2$. The relation between these functions is clearly described in the frequency domain,

$$\hat{\psi}(2w) = e^{-iw} \overline{H(e^{-iw})} \hat{\phi}(w), \tag{11.109}$$

where $\hat{\phi}$ stands for the Fourier transform of ϕ, H stands for the transfer function of ϕ, and $\bar{H}$ is its complex conjugate.

The multiresolution analysis of two-dimensional functions $f(x, y)$ can be formulated straightforwardly as an extension of the above equations:

$$
\begin{aligned}
f(x, y) &= A_1 f + D_{1,1} f + D_{1,2} f + D_{1,3} f, \\
&= A_2 f + D_{2,1} f + D_{2,2} f + D_{2,3} f + D_{1,1} f + D_{1,2} f + D_{1,3} f, \\
&\vdots \\
&= A_n f + \sum_{j=1}^{n} [D_{j,1} f + D_{j,2} f + D_{j,3} f]. \qquad (11.110)
\end{aligned}
$$

We can characterize each approximation function $A_j f(x, y)$ and the difference components $D_{j,p} f(x, y)$ for $p = 1, 2, 3$ by means of a 2D scaling function $\Phi(x, y)$ and its associated wavelet functions $\Psi_p(x, y)$ as follows:

$$
\begin{aligned}
A_j f(x, y) &= \sum_{k=-\infty}^{+\infty} \sum_{l=-\infty}^{+\infty} a_{j,k,l} \Phi_{j,k,l}(x, y), \qquad (11.111) \\
D_{j,p} f(x, y) &= \sum_{k=-\infty}^{+\infty} \sum_{l=-\infty}^{+\infty} d_{j,p,k,l} \Psi_{j,p,k,l}(x, y),
\end{aligned}
$$

where

$$
\begin{aligned}
\Phi_{j,k,l}(x, y) &= \frac{1}{2^j} \Phi\left(\frac{x-k}{2^j}, \frac{y-l}{2^j}\right), \quad (j, k, l) \in Z^3, \qquad (11.112) \\
\Psi_{j,p,k,l}(x, y) &= \frac{1}{2^j} \Psi_p\left(\frac{x-k}{2^j}, \frac{y-l}{2^j}\right),
\end{aligned}
$$

and

$$
\begin{aligned}
a_{j,k,l}(x, y) &=< f(x, y), \Phi_{j,k,l}(x, y) >, \qquad (11.113) \\
d_{j,p,k,l} &=< f(x, y), \Psi_{j,p,k,l}(x, y) > . \qquad (11.114)
\end{aligned}
$$

In order to carry out a separable multiresolution analysis, we decompose the scaling function $\Phi(x, y)$ and the wavelet functions $\Psi_p(x, y)$ as follows:

$$
\begin{aligned}
\Phi(x, y) &= \phi(x)\phi(y), \\
\Psi_1(x, y) &= \phi(x)\psi(y), \\
\Psi_2(x, y) &= \psi(x)\phi(y), \\
\Psi_3(x, y) &= \psi(x)\psi(y), \qquad (11.115)
\end{aligned}
$$

where ϕ is a 1D scale function and ψ is its associated wavelet function. The functions Ψ_1, Ψ_2, Ψ_3 are expected to extract the details of the y-axis, x-axis, and diagonal directions, respectively.

11.6.3 Wavelet Pyramid

Equation (11.110) represents in a compact manner a pyramid structure for the processing of an image $f(x, y)$. Given a 2D image $f(x, y)$, sampled at $x = 1, 2, \ldots, m_x$ and $y = 1, 2, \ldots m_y$, we establish a pyramidal processing by computing the coefficients $a_{j,k,l}$ and $d_{j,p,k,l}$ for each level of $j = 1, 2, \ldots, n$. For each level j of the pyramid, we group these coefficients into the matrices A_j, $D_{j,p}$ for $p = 1, 2, 3$ as follows:

$$A_j = (a_{j,k,l})_{(k=1,2,\ldots,\frac{m_x}{2^j};l=1,2,\ldots,\frac{m_y}{2^j})}, \tag{11.116}$$

$$D_{j,p} = (d_{j,p,k,l})_{(k=1,2,\ldots,\frac{m_x}{2^j};l=1,2,\ldots,\frac{m_y}{2^j})}. \tag{11.117}$$

The coefficients $a_{j,k,l}$ and $d_{j,p,k,l}$ can be obtained simply via an iterative computation of the impulse responses h and g of the filters ϕ and ψ. This procedure is known as Mallat's algorithm [202].

11.6.4 Complex Wavelet Transform

Despite the advantage of the locality in both the spatial and frequency domains, the wavelet pyramid of real-valued wavelets unfortunately has the drawback of being neither translation-invariant nor rotation-invariant. As a result, no procedure can yield phase information. This is one of the important reasons why researchers are interested in hypercomplex wavelet transforms like complex or the quaternion wavelet transforms.

Since the translation in the spatial domain is represented as a rotation in the complex domain, the complex-valued wavelet can be used for multiscale phase analysis of the image signals. This allows the interpolability of the wavelet transform at subpixel accuracy through the wavelet pyramid levels.

There are many kinds of complex wavelets. Lina [193] extended the Daubechies wavelets to the complex ones. The complex wavelet, briefly outlined here, was developed by Magarey and Kingsbury [200] and used for motion estimation of video frames.

It is worth pointing out that the efficiency of the matching strategies and the similarity distance measures are highly dependent on how well the wavelet is designed. By satisfying an image-matching principle, the wavelet filter pair (h, g) (impulse responses of the filters ϕ and ψ) must be compactly supported in the spatial domain; i.e., they should show regularity (differentiable to a high order) and symmetry (leading to a linear phase). In practice, the orthogonality cannot be strictly fulfilled.

As an illustration, we will describe the complex wavelet designed by Magarey and Kingsbury [157, 200], which was also utilized by Pan [225]. The impulse responses of the scale function h and the wavelet function g are a complex pair of even-lengthed modulated windows

$$h(k) = \tilde{b}_1 \tilde{w}_1(k+0.5) e^{i\tilde{w}_1(k+0.5)}, \qquad g(k) = b_1 w_1(k+0.5) e^{i w_1(k+0.5)}, \tag{11.118}$$

for $k = -n_w, -n_w + 1, \ldots, n_w - 1$; b_1 and $\tilde{b}_1$ are complex constants. w_1 and $\tilde{w}_1$ are a pair of real-valued windows of width $2n_w$, symmetric with respect to $k = 0$, and their magnitudes decay to zero at both ends. The commonly used low-pass filter is the Gauss filter

$$\tilde{w}_1(k) = e^{-\frac{k^2}{2\tilde{\sigma}_1^2}}, \qquad w_1(k) = e^{-\frac{k^2}{2\sigma_1^2}}. \tag{11.119}$$

In order to satisfy the trade-off of good locality of matching and information sufficiency, the minimum width of the window function should be 4, consequently, $n_w = 2$. It is also required to fulfill the complementarity of the modulation frequencies w_1 and w_2 in the frequency range $[0, \phi]$ as follows:

$$w_1 + w_2 = \pi. \tag{11.120}$$

Figure 11.20 shows a complex wavelet designed by Magarey and Kingsbury [200] and also used by Pan [225]. This complementary constraint for the pair of the low-pass filter ϕ and the high-pass filter ψ imposes that $w_1 > \tilde{w}_2$. In Fig. 11.20c, the Fourier transforms of h and g have a conjugated symmetry in regard to their modulation frequencies w_1 and w_2. Due to the fact that one-dimensional signals have conjugated symmetric spectra, one can neglect the negative half-spectrum $[-\pi, o]$ without losing information on the real 1D input signal. In order to cover the frequency range $[0, \pi]$ completely without significant gaps and causing minimal overlap, one can adjust the filters at each level j according to the following relationship:

$$w_j = 3\tilde{w}_j. \tag{11.121}$$

Thus, according to Eq. (11.120), the modulation frequencies at the first (finest) level are

$$w_1 = \frac{5\pi}{6}, \qquad \tilde{w}_1 = \frac{\pi}{6} \tag{11.122}$$

(see Fig. 11.20). In practice, the modulation frequencies through the levels are subdivided as $w_j = \frac{w_{j-1}}{2}$.

For the case of 2D wavelet analysis, we can use the 1D complex wavelets of Eq. (11.118) and achieve the separability described by Eq. (11.115). These 2D wavelet filters are predominantly first quadrant filters in the frequency domain.

Since real, discrete images contain significant information in the first and second frequency quadrants, one has to use the complex conjugated filters $\bar{h}$ and $\hat{g}$ in addition to h and g. Thus, in order to include in the computation the information from the second quadrant, one has to calculate the matrices $\tilde{D}_{j,p}$ p $= 1, 2, 3$ of the difference of coefficients, for each j-level. Kingsbury [157] called this algorithm the dual-tree complex wavelet transform which means that one uses a mirror processing tree to

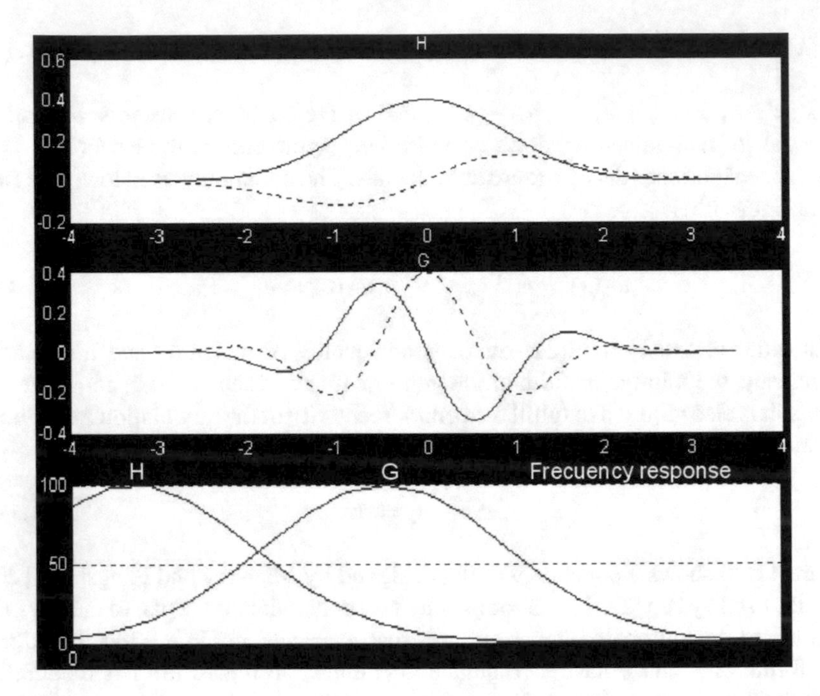

Fig. 11.20 Complex wavelets: (from above) **a** scaling function ϕ, **b** wavelet function ψ, **c** both functions in the frequency domain

include the second quadrant. Thus, the complex wavelet analysis by the transition from level $j-1$ to j transforms two complex approximation submatrices to eight complex approximation and difference submatrices as follows:

$$\{A_{j-1}, \tilde{A}_{j-1}\} \rightarrow \{A_j, \tilde{A}_j, D_{j,p}, \tilde{D}_{j,p}, \ p=1,2,3\}, \tag{11.123}$$

where $\tilde{A}_j$ is the mirror of A_j and $\tilde{D}_{j,p}$ is the mirror of $D_{j,p}$.

11.6.5 Quaternion Wavelet Transform

The quaternion wavelet transform is a natural extension of the real and complex wavelet transform, taking into account the axioms of the quaternion algebra, the quaternionic analytic signal [44], and the separability property described by Eq. (11.115). The QWT is applied for signals of two or more dimensions.

Multiresolution analysis can also be straightforwardly extended to the quaternionic case; we can therefore improve the power of the phase concept, which in the real wavelets is not possible, and in the case of the complex, it is limited to only one phase. Thus, in contrast to the similarity distance used in the complex wavelet pyramid [225], we favor the quaternionic phase concept for top-down parameter estimation.

For the quaternionic versions of the wavelet scale function h and the wavelet function g, we choose two quaternionic modulated Gabor filters in quadrature as follows:

$$h^q = g(x, y, \sigma_1, \varepsilon) \exp\left(\mathbf{i}\frac{c_1\omega_1 x}{\sigma_1}\right) \exp\left(\mathbf{j}\frac{c_2\varepsilon\omega_2 y}{\sigma_1}\right),$$

$$= h^q_{ee} + h^q_{oe}\mathbf{i} + h^q_{eo}\mathbf{j} + h^q_{oo}\mathbf{k}, \tag{11.124}$$

$$g^q = g(x, y, \sigma_2, \varepsilon) \exp\left(\mathbf{i}\frac{\tilde{c}_1\tilde{\omega}_1 x}{\sigma_2}\right) \exp\left(\mathbf{j}\frac{\tilde{c}_2\varepsilon\tilde{\omega}_2 y}{\sigma_2}\right),$$

$$= g^q_{ee} + g^q_{oe}\mathbf{i} + g^q_{eo}\mathbf{j} + g^q_{oo}\mathbf{k}, \tag{11.125}$$

where the parameters σ_1, σ_2, c_1, c_2, $\tilde{c_1}$, $\tilde{c_2}$, w_1, w_2, $\tilde{w}_1$, $\tilde{w}_2$ are selected to fulfill the requirements of Eqs. (11.120) and (11.122). Note that the horizontal axis x is related to $\mathbf{i}$ and the vertical axis y is related to $\mathbf{j}$, and both imaginary numbers of the quaternion algebra fulfill the equation $\mathbf{k} = \mathbf{ji}$.

The right parts of Eqs. (11.124) and (11.125) obey a natural decomposition of a quaternionic analytic function: The subindex *ee* (even-even) stands for a symmetric filter, *eo* (even-odd) or *oe* (odd-even) both stand for asymmetrical filters, and *oo* (odd-odd) stands for an asymmetrical filter as well. Thus, we can clearly see that h^q and g^q of Eqs. (11.124) and (11.125) are powerful filters to disentangle the symmetries of the 2D signals.

At this point, we can show the disadvantageous property of a complex wavelet, i.e., the merging of important information in its two filter components. When $\epsilon = 1$, the even and odd parts of the complex wavelet merge information from two components of the quaternionic wavelet rather than separating this information:

$$h_e(x, y) = g(x, y) \cos(\omega_1 x + \omega_2 y), \tag{11.126}$$

$$= g(x, y)(\cos(\omega_1 x) \cos(\omega_2 y) - \sin(\omega_1 x) \sin(\omega_2 y)),$$

$$= h^q_{ee}(x, y) - h^q_{oo}(x, y),$$

$$h_o(x, y) = g(x, y) \sin(\omega_1 x + \omega_2 y), \tag{11.127}$$

$$= g(x, y)(\cos(\omega_1 x) \sin(\omega_2 y) - \sin(\omega_1 x) \cos(\omega_2 y)),$$

$$= h^q_{oe}(x, y) + h^q_{eo}(x, y). \tag{11.128}$$

This indicates also that for image analysis, using the phase concept, complex wavelets can use only one phase, whereas the quaternionic wavelets offer three phases.

It is also possible to steer quaternionic wavelets. Kingsbury [157] computes six complex filters combining the real and imaginary parts of complex wavelets at each level of its quaternion wavelet pyramid. In the quaternionic wavelet pyramid, one can also generate these six selective filters as shown in Fig. 11.21 simply by applying

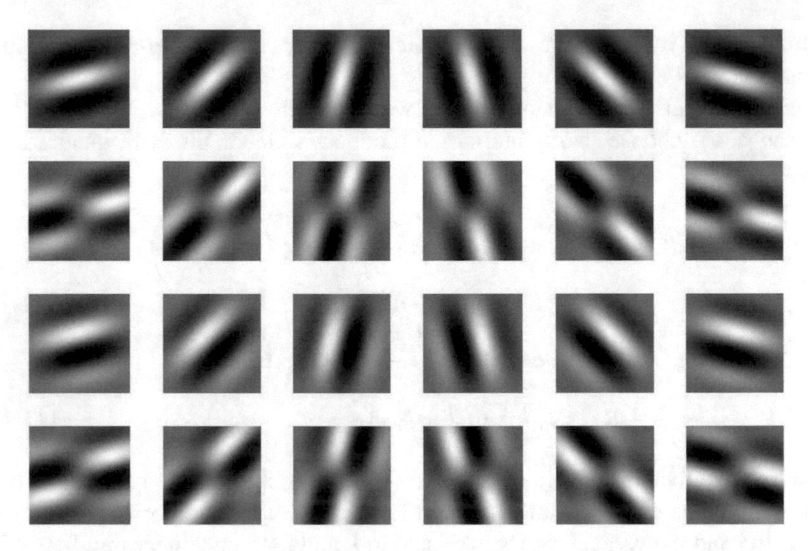

Fig. 11.21 Quaternion wavelet filters with selective orientations (from the left): 15, 45, 75, −75, −45, −15 degrees

the automorphisms of Eq. (11.16) to a basic quaternion filter. The advantage of our selective quaternion wavelets is that they provide three phases.

Next, we present an important theorem that tells us how we can recover the whole image's energy by using only the energy contained in the first quadrant.

Theorem *Assuming that the shift-invariant wavelet filters are used in the procedure, the 2D signal* $f(x, y)$ *can be fully reconstructed from the quaternionic wavelet transform* $Wf(u, v)^q$ *only by considering the first quadrant. Note that the perfect reconstruction is carried out by using well-designed kernels:*

$$\begin{array}{ll} \textit{low-pass} & \overline{\Phi(u,v)} = \overline{\phi(u,v)} e^{(i2\pi\tilde{u}x)} e^{(j2\pi\tilde{v}y)}, \\ \textit{and} & \\ \textit{high-pass} & \overline{\Psi(u,v)} = \overline{\psi(u,v)} e^{(i2\pi ux)} e^{(j2\pi vy)}. \end{array}$$

Proof Using the automorphisms of the quaternion algebra, $Re(\boldsymbol{q}) = Re(-\boldsymbol{i}\boldsymbol{q}\boldsymbol{i}) = Re(-\boldsymbol{j}\boldsymbol{q}\boldsymbol{j}) = Re(-\boldsymbol{k}\boldsymbol{q}\boldsymbol{k})$ (see Eq. (11.16)) and taking four times the signal energy of the first quadrant, we can reconstruct $f(x, y)$ as follows:

$$\begin{aligned} f(x, y) &= f(x, y)_l + f(x, y)_h, \\ &= 4Re\Big(\int_0^\infty \int_0^\infty (\overline{\Phi(u, v)} + \overline{\Psi(u, v)}) Wf(u, v)^q du dv\Big). \end{aligned}$$

Let us consider a signal reconstructed by the low-pass kernel:

$$
\begin{aligned}
f(x,y)_l &= 4Re\Big(\int_0^\infty\int_0^\infty \overline{\Phi(u,v)}Wf(u,v)^q dudv\Big),\\
&= 4Re\Big(\int_0^\infty\int_0^\infty \overline{\phi(u,v)e^{(i2\pi ux)}}Wf(u,v)^q e^{(j2\pi vy)}dudv\Big),\\
&= Re\Big(\int_0^\infty\int_0^\infty \overline{\phi(u,v)e^{(i2\pi ux)}}Wf(u,v)^q e^{(j2\pi vy)}dudv\Big),\\
&\quad +Re\Big(\int_0^\infty\int_0^\infty \overline{\phi(u,v)e^{(i2\pi ux)}} - iWf(u,v)ie^{(j2\pi vy)}dudv\Big)\\
&\quad +Re\Big(\int_0^\infty\int_0^\infty \overline{\phi(u,v)e^{(i2\pi ux)}} - jWf(u,v)^q je^{(j2\pi vy)}dudv\Big)\\
&\quad +Re\Big(\int_0^\infty\int_0^\infty \overline{\phi(u,v)e^{(i2\pi ux)}} - kWf(u,v)^q ke^{(j2\pi vy)}dudv\Big),\\
&= Re\Big(\int_0^\infty\int_0^\infty \overline{\phi(u,v)e^{(i2\pi ux)}}Wf(u,v)^q e^{(j2\pi vy)}dudv\Big)\\
&\quad +Re\Big(\int_{-\infty}^0\int_0^\infty \overline{\phi(u,v)e^{(i2\pi ux)}}Wf(\boldsymbol{u})^q e^{(j2\pi vy)}dudv\Big)\\
&\quad +Re\Big(\int_0^\infty\int_{-\infty}^0 \overline{\phi(u,v)e^{(i2\pi ux)}}Wf(u,v)^q e^{(j2\pi vy)}dudv\Big)\\
&\quad +Re\Big(\int_{-\infty}^0\int_{-\infty}^0 \overline{\phi}(u,v)e^{(i2\pi ux)}Wf(u,v)^q e^{(j2\pi vy)}dudv\Big),\\
&= Re\Big(\int_{-\infty}^\infty\int_{-\infty}^\infty \overline{\phi(u,v)e^{(i2\pi ux)}}Wf^q(u,v)e^{(j2\pi vy)}dudv\Big).
\end{aligned}
\tag{11.129}
$$

Note that one can prove in the same way the case of the high-pass kernel $\overline{\Psi(u,v)} = \psi(u,v)e^{(i2\pi ux)}e^{(j2\pi vy)}$. Thus, the complete 2D signal is equal to the sum of the outputs of these two kernels considering only the energy of the first quadrant:

$$
\begin{aligned}
f(x,y) &= 4Re\Big(\int_0^\infty\int_0^\infty \overline{\Phi(u,v)}Wf(u,v)^q dudv\Big)\\
&\quad +4Re\Big(\int_0^\infty\int_0^\infty \overline{\Psi(u,v)}Wf(u,v)^q dudv\Big),\\
&= Re\Big(\int_{-\infty}^\infty\int_{-\infty}^\infty \overline{\phi(u,v)e^{(i2\pi ux)}}Wf^q(u,v)e^{(j2\pi vy)}dudv\Big)\\
&\quad +Re\Big(\int_{-\infty}^\infty\int_{-\infty}^\infty \overline{\psi(u,v)e^{(i2\pi ux)}}Wf^q(u,v)e^{(j2\pi vy)}dudv\Big),\\
&= f(x,y)_l + f(x,y)_h.
\end{aligned}
\tag{11.130}
$$

The implications of this theorem are very important, particularly when we design a processing schema for multiresolution analysis, which is explained next.

11.6.6 Quaternionic Wavelet Pyramid

The quaternionic wavelet multiresolution analysis can be easily formulated from the basic ideas given in Sects. 11.6.2 and 11.6.3. For the 2D image function $f(x, y)$, a quaternionic wavelet multiresolution will be written as

$$f(x, y) = A_n^q f + \sum_{j=1}^{n} [D_{j,1}^q f + D_{j,2}^q f + D_{j,3}^q f]. \tag{11.131}$$

The upper index q stands for indicating quaternion 2D signal. We can characterize each approximation function $A_j^q f(x, y)$ and the difference components $D_{j,p}^q f(x, y)$ for $p = 1, 2, 3$ by means of a 2D scaling function $\Phi^q(x, y)$ and its associated wavelet functions $\Psi_p^q(x, y)$ as follows:

$$A_j^q f(x, y) = \sum_{k=-\infty}^{+\infty} \sum_{l=-\infty}^{+\infty} a_{j,k,l} \Phi_{j,k,l}^q(x, y), \tag{11.132}$$

$$D_{j,p}^q f(x, y) = \sum_{k=-\infty}^{+\infty} \sum_{l=-\infty}^{+\infty} d_{j,p,k,l} \Psi_{j,p,k,l}^q(x, y),$$

where

$$\Phi_{j,k,l}^q(x, y) = \frac{1}{2^j} \Phi^q \left(\frac{x-k}{2^j}, \frac{y-l}{2^j} \right), \quad (j, k, l) \in Z^3, \tag{11.133}$$

$$\Psi_{j,p,k,l}^q(x, y) = \frac{1}{2^j} \Psi_p^q \left(\frac{x-k}{2^j}, \frac{y-l}{2^j} \right),$$

and

$$a_{j,k,l}(x, y) =< f(x, y), \Phi_{j,k,l}^q(x, y) >, \tag{11.134}$$

$$d_{j,p,k,l} =< f(x, y), \Psi_{j,p,k,l}^q(x, y) > .$$

In order to carry out a separable quaternionic multiresolution analysis, we decompose the scaling function $\Phi^q(x, y)_j$ and the wavelet functions $\Psi_p^q(x, y)_j$ for each level j as follows:

$$\begin{aligned} \Phi^q(x, y)_j &= \phi^i(x)_j \phi^j(y)_j, \\ \Psi_1^q(x, y)_j &= \phi^i(x)_j \psi^j(y)_j, \\ \Psi_2^q(x, y)_j &= \psi^i(x)_j \phi^j(y)_j, \\ \Psi_3^q(x, y)_j &= \psi^i(x)_j \psi^j(y)_j, \end{aligned} \tag{11.135}$$

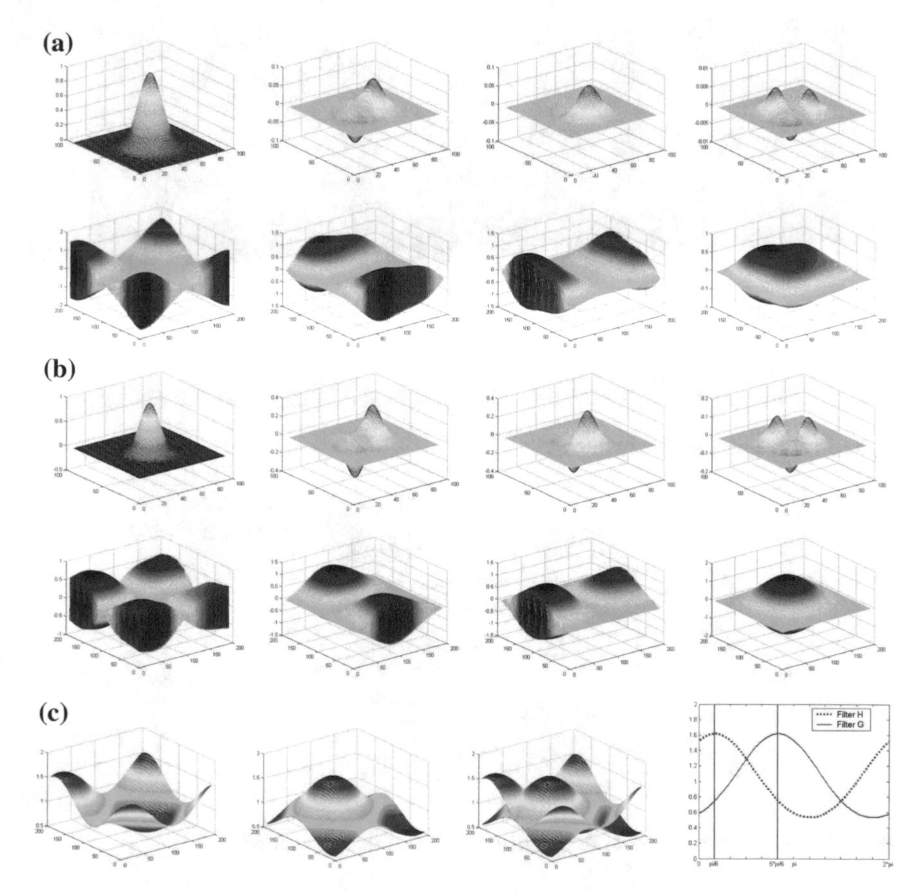

Fig. 11.22 Quaternionic Gabor filters in the space and frequency domains. Note that in this domain for visualizing, the transfer function was sampled at a higher rate and we erased the central part of the spectrum. **a** (first and second upper rows) approximation filter Φ, **b** (third and four rows) detail filter Ψ_3 (diagonal), **c** (fifth row from the left) magnitude of the filter in the frequency domain: (c.1) approximation, (c.2) detail (c.3) depiction of both, (c.4) magnified cross section with orientation of 45°. Note that the approximation and the detail filters are located at the frequencies $\frac{\pi}{6}$ and $\frac{5\pi}{6}$, respectively, fulfilling the requirements of Eqs. (11.120) and (11.122)

where $\phi^{\boldsymbol{i}}(x)_j$ and $\psi(x)^{\boldsymbol{i}}_j$ are 1D complex filters applied along the rows and columns, respectively. Note that in ϕ and ψ, we use the imaginary number $\boldsymbol{i}$, $\boldsymbol{j}$ of quaternions that fulfill $\boldsymbol{ji} = \boldsymbol{k}$.

In Fig. 11.22, we show the quaternion scaling or approximation filter $\Phi^q(x, y)_j$ and the quaternion wavelet function $\Psi_3^q(x, y)_j$, which is designed for detecting diagonal details. Note that, as in Fig. 11.20, the approximation and the detail filters are located at the frequencies $\frac{\pi}{6}$ and $\frac{5\pi}{6}$, respectively, fulfilling the requirements of Eqs. (11.120) and (11.122). This figure also shows the quaternion filters in the quaternionic frequency domain, which were transformed using the Quaternion Fourier Transform (QFT) [44] (Fig. 11.23).

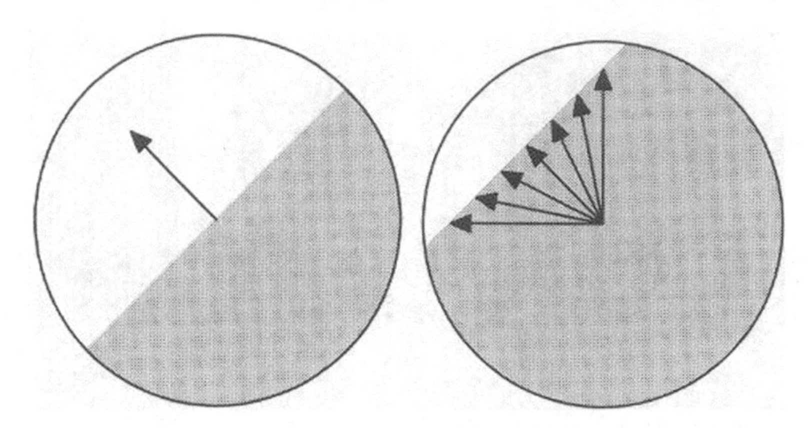

Fig. 11.23 The aperture problem

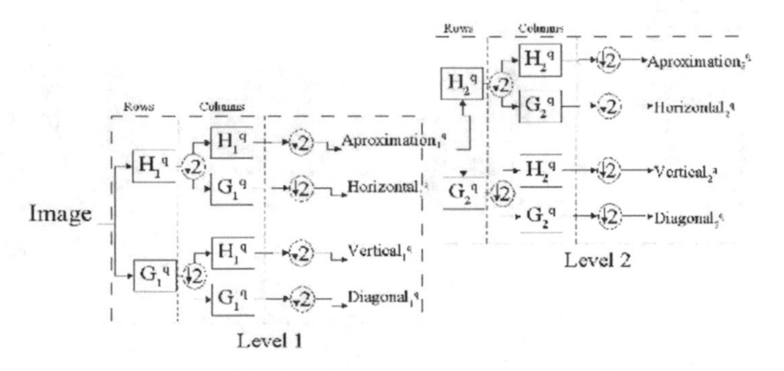

Fig. 11.24 Abstraction of two levels of the quaternionic wavelet pyramid

By using these formulas, we can build quaternionic wavelet pyramids. Figure 11.24 shows the two primary levels of the pyramid (fine to coarse). According to Eq. (11.135), the approximation after the first level $A_1^q f(x, y)$ is the output of $\Phi^q(x, y)_1$, and the differences $D_{1,1}^q f$, $D_{1,2}^q f$, $D_{1,3}^q f$ are the outputs of $\Psi_{1,1}^q(x, y)$, $\Psi_{1,2}^q(x, y)$ and $\Psi_{1,3}^q(x, y)$. The procedure continues through the j levels decimating the image at the outputs of the levels (indicated in Fig. 11.24 within the circle).

The quaternionic wavelet analysis from level $j-1$ to level j corresponds to the transformation of one quaternionic approximation to a new quaternionic approximation and three quaternionic differences, i.e.,

$$\{A_{j-1}^q\} \rightarrow \{A_j^q,\ D_{j,p}^q,\ p = 1, 2, 3\}. \tag{11.136}$$

Note that we do not use the idea of a mirror tree expressed in Eq. (11.123). As a result, the quaternionic wavelet tree is the compact and economic processing structure to be used for the case of n-dimensional multi-resolution analysis.

The procedure of quaternionic wavelet multiresolution analysis depicted partially in Fig. 11.24 is as follows:

(i) convolve the 2D real signal at level j and convolve it with the scale and wavelet filters H_j^q and G_j^q along the rows of the 2D signal.
(ii) H_j^q and G_j^q are convolved with the columns of the previous responses of the filters H_j^q and G_j^q.
(iii) subsample the responses of these filters by a factor of two ($\downarrow 2$).
(iv) the real part of the approximation at level j is taken as input at the next level j. This process continues through the all levels $j = 1, \ldots, n$ repeating the steps $1 \to 4$.

The theorem of Sect. 11.6.5 indicates that we do not need to take into consideration any other quadrant than the first one; thus, we do not need to create a mirror architecture or a *dual tree* as in the case of complex wavelet multiresolution analysis [157]. In Kingsbury's paper, the multi-resolution architecture outputs four-element "complex" vectors $\{a, b, c, d\} = a + bi_1 + ci_2 + di_1$ that, according to the author, are not quaternions, as they have different algebraic properties. This structure is generated based on concepts of the signal and filter theory. Even though the dual-tree wavelet schema works correctly, the wavelet quaternion transform leads to an architecture that merges two branches of the dual tree into one single quaternionic tree. The big advantage of the quaternionic wavelet tree is that, with the same amount of computational resources, it offers three phases for the analysis using the phase concept. The way of applying the quaternionic phase concept will be explained and illustrated with real experiments in Chap. 14.

11.6.7 The Tridimensional Clifford Wavelet Transform

In this section, we will use the similitude group $SIM(3)$ denoted by $G_s = \mathbb{R}^+ \times SO(3) \otimes \mathbb{R}^3 = \{(s, r_\theta, \boldsymbol{t}) | s \in \mathbb{R}^+, r_\theta \in SO(3), \boldsymbol{t} \in \mathbb{R}^3\}$, where s stands for the dilation parameter, $\boldsymbol{t}$ for a translation vector, and $\boldsymbol{\theta}$ for the $SO(3)$ rotation parameters, see Mawardi and Hitzer [206] for and study of the Clifford algebra $Cl_{3,0}$-valued wavelet transform.

The group action G_s on $\mathbb{R}^3$ in G_3 will be represented in terms of rotors as follows:

$$G_s : \quad \mathbb{R}^3 \to \mathbb{R}^3, \tag{11.137}$$

$$\boldsymbol{x} \to s\boldsymbol{R}\boldsymbol{x}\widetilde{\boldsymbol{R}} + \boldsymbol{t}. \tag{11.138}$$

The left Haar measure on G_s is given by

$$d\lambda(s, \boldsymbol{\theta}, \boldsymbol{t}) = d\mu(s, \boldsymbol{\theta})\, d^3\boldsymbol{t}, \tag{11.139}$$

where $d\mu(s,\boldsymbol{\theta}) = \frac{ds\,d\boldsymbol{\theta}}{s^4}$ and $d\boldsymbol{\theta} = \frac{1}{8\pi^2} sin(\theta_1)d\theta_1 d\theta_2 d\theta_3$.

Now, let us define in the 3D geometric algebra G_3 framework a mother wavelet that can be transformed by the action of the similitude group G_s. The group action on the mother wavelet can be formulated in terms of a unitary linear operator

$$\begin{aligned} U_{s,\boldsymbol{\theta},\boldsymbol{t}} : L^2(\mathbb{R}^3; G_3) \to \quad & L^2(G_s; G_3), \\ \psi(\boldsymbol{x}) \to \quad & U_{s,\boldsymbol{\theta},\boldsymbol{t}}\psi(\boldsymbol{x}) = \psi_{s,\boldsymbol{\theta},\boldsymbol{t}}(\boldsymbol{x}), \\ & = \frac{1}{s^{\frac{3}{2}}}\psi\left(r_{\boldsymbol{\theta}}^{-1}\left(\frac{\boldsymbol{x}-\boldsymbol{t}}{s}\right)\right). \end{aligned} \tag{11.140}$$

The family of wavelets $\psi_{s,\boldsymbol{\theta},\boldsymbol{t}}$ is known as daughter Clifford wavelets. Note that the normalization constant $s^{\frac{3}{2}}$ guarantees that the norm of $\psi_{s,\boldsymbol{\theta},\boldsymbol{t}}$ is independent of s, namely

$$||\psi_{s,\boldsymbol{\theta},\boldsymbol{t}}||_{L^2(\mathbb{R}^3;G_3)} = ||\psi||_{L^2(\mathbb{R}^3;G_3)}. \tag{11.141}$$

This can be proved easily:

$$\begin{aligned} ||\psi_{s,\boldsymbol{\theta},\boldsymbol{t}}||^2_{L^2(\mathbb{R}^3;G_3)} &= \int_{\mathbb{R}^3} \sum_M \frac{1}{s^3}\psi_M^2\left(r_{\boldsymbol{\theta}}^{-1}\left(\frac{\boldsymbol{x}-\boldsymbol{t}}{s}\right)\right)d^3\boldsymbol{x}, \\ &= \frac{1}{s^3}\int_{\mathbb{R}^3}\sum_M \psi_M^2(\boldsymbol{u})s^3 det(r_{\boldsymbol{\theta}})d^3\boldsymbol{u} = \int_{\mathbb{R}^3}\sum_M \psi_M^2(\boldsymbol{u})d^3\boldsymbol{u}. \end{aligned}$$

In the G_3 Clifford Fourier domain, Eq. (11.140) can be represented as follows:

$$\mathcal{F}\{\psi_{s,\boldsymbol{\theta},\boldsymbol{t}}\}(\boldsymbol{w}) = e^{-I_3\boldsymbol{t}\cdot\boldsymbol{w}} s^{\frac{3}{2}}\hat{\psi}(s r_{\boldsymbol{\theta}}^{-1}(\boldsymbol{w})). \tag{11.142}$$

Substituting $(\boldsymbol{x} - \boldsymbol{t})/s = \boldsymbol{y}$ for the argument of Eq. (11.140) under the Clifford Fourier integral, we get

$$\begin{aligned} \mathcal{F}\{\psi_{s,\boldsymbol{\theta},\boldsymbol{t}}\}(\boldsymbol{w}) &= \int_{\mathbb{R}^3}\frac{1}{s^{\frac{3}{2}}}\psi(r_{\boldsymbol{\theta}}^{-1}\boldsymbol{y})e^{-I_3\boldsymbol{w}(\boldsymbol{t}+s\boldsymbol{y})}s^3 d^3\boldsymbol{y}, \\ &= e^{-I_3\boldsymbol{t}\cdot\boldsymbol{w}}s^{\frac{3}{2}}\int_{\mathbb{R}^3}\psi(r_{\boldsymbol{\theta}}^{-1}\boldsymbol{y})e^{-I_3 s\boldsymbol{w}\cdot\boldsymbol{y}}d^3\boldsymbol{y}, \\ &= e^{-I_3\boldsymbol{t}\cdot\boldsymbol{w}}s^{\frac{3}{2}}\hat{\psi}(s r_{\boldsymbol{\theta}}^{-1}(\boldsymbol{w})). \end{aligned} \tag{11.143}$$

We will call $\psi \in L^2(\mathbb{R}^3; G_3)$ an admissible wavelet if

$$C_\psi = \int_{\mathbb{R}^+}\int_{SO(3)} s^3\{\hat{\psi}(s r_{\boldsymbol{\theta}}^{-1}(\boldsymbol{w}))\}^{\sim}\hat{\psi}(s r_{\boldsymbol{\theta}}^{-1}(\boldsymbol{w}))d\mu \tag{11.144}$$

is an invertible multivector constant and infinite at any $\boldsymbol{w} \in \mathbb{R}^3$. We will see later that the admissibility condition is important to guaranteeing that the Clifford wavelet transform is invertible. Note that for $\boldsymbol{w} = 0$, $\hat{\psi}(0) = \int_{\mathbb{R}^3} \psi(\boldsymbol{x}) e^{I_3 0 \cdot \boldsymbol{x}} d^3\boldsymbol{x} = 0$ for the scalar part of C_ψ to be finite. Thus, similar to the classical real-valued wavelets, an admissible Clifford-valued mother wavelet $\psi \in L^2(\mathbb{R}^3; G_3)$ ought to satisfy

$$\int_{\mathbb{R}^3} \psi(\boldsymbol{x}) d^3\boldsymbol{x} = \int_{\mathbb{R}^3} \psi_M(\boldsymbol{x}) e_M d^3\boldsymbol{x} = 0, \tag{11.145}$$

where $\psi_M(\boldsymbol{x})$ are real-valued wavelets. It means that the integral of very component ψ_i of the Clifford mother wavelet is zero, i.e., $\int_{\mathbb{R}^3} \psi_M(\boldsymbol{x}) d^3\boldsymbol{x} = 0$.

The 3-Dimensional Clifford Wavelet Transform (3D-CWT) with respect to the mother wavelet $\psi \in L^2(\mathbb{R}^3; G_3)$ is given by

$$\begin{aligned} T_\psi : L^2(\mathbb{R}^3; G_3) &\to L^2(G_s; G_3), \\ f \to T_\psi f(s, \boldsymbol{\theta}, \boldsymbol{t}) &= \int_{\mathbb{R}^3} f(\boldsymbol{x}) \widetilde{\psi_{s,\boldsymbol{\theta},\boldsymbol{t}} \boldsymbol{d}^3}\boldsymbol{x}, \\ &= (f, \psi_{s,\boldsymbol{\theta},\boldsymbol{t}})_{L^2(\mathbb{R}^3; G_3)}. \end{aligned} \tag{11.146}$$

The Clifford wavelet transform of Eq. (11.146) has a Clifford Fourier representations given by the following expression:

$$T_\psi f(s, \boldsymbol{\theta}, \boldsymbol{t}) = \frac{1}{(2\pi)^3} \int_{\mathbb{R}^3} \hat{f}(\boldsymbol{w}) s^{\frac{3}{2}} \{\hat{\psi}(s r_\theta^{-1}(\boldsymbol{w}))\}^{\sim} e^{I_3 \boldsymbol{t} \cdot \boldsymbol{w}} \boldsymbol{d}^3\boldsymbol{w}. \tag{11.147}$$

Finally, the Inverse Tridimensional Clifford Wavelet Transform (3D-ICWT) is given by the following expression:

$$\begin{aligned} f(\boldsymbol{x}) &= \int_{G_s} T_\psi f(s, \boldsymbol{\theta}, \boldsymbol{t}) \psi_{s,\boldsymbol{\theta},\boldsymbol{t}} C_\psi^{-1} d\mu \boldsymbol{d}^3\boldsymbol{t}, \\ &= \int_{G_s} (f, \psi_{s,\boldsymbol{\theta},\boldsymbol{t}})_{L^2(\mathbb{R}^3; G_3)} \psi_{s,\boldsymbol{\theta},\boldsymbol{t}} C_\psi^{-1} d\mu \boldsymbol{d}^3\boldsymbol{t}, \end{aligned} \tag{11.148}$$

where C_ψ is given by Eq. (11.158).

11.6.8 The Continuous Conformal Geometric Algebra Wavelet Transform

In this subsection, we present the continuous conformal geometric algebra wavelet transform (CGAWT) on the sphere S^{n-1} based on the conformal group of the sphere G_c. A possible description of the group is done in terms of a projective identification of the points of the Euclidean space $\mathbb{R}^n$ with rays in the null cone in $\mathbb{R}^{n+1,1}$. As shown in Sect. 8.4, in conformal geometric algebra $G_{n+1,1}$, in general, the conformal transformation can be expressed as a composite of versors for transversion, translation, and rotation as follows:

$$\boldsymbol{G} = D_\rho \boldsymbol{K}_{\mathbf{b}} \boldsymbol{T}_{\mathbf{a}} \boldsymbol{R}_\alpha, \tag{11.149}$$

and as a result, due to the multiplicative nature of the versors, we can avoid complex nonlinear algorithms. In contrast, Cerejeiras et al. [51] used the Möbius transformation in $\mathbb{R}^n$ for the formulation of the continuous wavelet transform and wavelet frames on the sphere. This transformation is expressed in a nonlinear manner as a ratio:

$$m_a(\boldsymbol{x}) = \frac{(\boldsymbol{x} - \boldsymbol{a})}{(1 + \boldsymbol{a}\boldsymbol{x})}, \quad \boldsymbol{a} \in \mathbb{R}^n, \quad |\boldsymbol{a}| < 1. \tag{11.150}$$

In this study, we will consider the space of the square-integrable multivector-valued functions on the sphere, the space $L_2(S^{n-1})$. In this space, the inner product and the norm are defined as follows:

$$\begin{aligned} < f, g >_{L_2} &= \int_{S^{n-1}} \overline{f(\boldsymbol{x})} g(\boldsymbol{x}) dS(\boldsymbol{x}), \\ ||f||^2 &= 2^n \int_{S^{n-1}} < \overline{f(\boldsymbol{x})} f(\boldsymbol{x}) >_0 dS(\boldsymbol{x}), \end{aligned} \tag{11.151}$$

where $< \cdot >_0$ stands for the scalar component of the multivector and $dS(\boldsymbol{x})$ is the normalized Spin(n)-invariant measure on S^{n-1}. We consider the following unitary operators acting on the multivector function $\psi \in L_2(S^{n-1})$: the rotor and the dilator $\boldsymbol{R}, \boldsymbol{D}_\rho \in G_{n+1,1}$,

$$\psi \rightarrow \boldsymbol{D}_\rho \, \psi(\boldsymbol{R}\boldsymbol{x}\tilde{\boldsymbol{R}}). \tag{11.152}$$

In general, the Continuous Conformal Geometric Algebra Wavelet Transform (CCGAWT) with respect to the mother wavelet $\psi(\boldsymbol{x}) \in L_2(S^{n-1})$ is given by

$$\begin{aligned} T_\psi : L^2(S^{n-1}; G_{n+1,1}) &\rightarrow L^2(G_c; G_{n+1,1}), \\ f \rightarrow T_\psi f(\rho, \boldsymbol{\theta}) &= \int_{S^{n-1}} f(\boldsymbol{x}) \widetilde{\psi_{\rho,\boldsymbol{\theta}}} \boldsymbol{d} S(\boldsymbol{x}), \\ &= (f, \psi_{\rho,\boldsymbol{\theta}})_{L^2(G_c; G_{n+1,1})}. \end{aligned} \tag{11.153}$$

Wiaux et al. [310] proved the correspondence principle between spherical wavelets and Euclidean wavelets by applying the inverse stereographic projection of a wavelet on the plane. Thus, typical functions and wavelets can be carried onto the 2-sphere such as the 2D Gauss function and the 2D Gabor function:

$$\psi_{Gauss}(\mathbf{x}) = e^{|\mathbf{x}|^2} \rightarrow \psi(\theta, \varphi)_{Gauss} = e^{(-tan^2(\frac{\varphi}{2}))}, \tag{11.154}$$

$$\psi_G(\mathbf{x}) = e^{i\mathbf{k}_0 \cdot \mathbf{x} - |\mathbf{x}|^2} \rightarrow \psi_G(\theta, \varphi) = \frac{e^{i\mathbf{k}_0} tan\frac{\theta}{2} cos(\varphi_0 - \varphi) e^{-\frac{1}{2} tan^2 \frac{\theta}{2}}}{1 + cos(\theta)}.$$

11.6.9 The n-Dimensional Clifford Wavelet Transform

The similitude group of $\mathbb{R}^n$, $SIM(n)$, $G_s = \mathbb{R}^+ \times SO(n) \otimes \mathbb{R}^n = \{(s, r_\theta, \boldsymbol{t}) | s \in \mathbb{R}^+, r_\theta \in SO(n), \boldsymbol{t} \in \mathbb{R}^n\}$, where s stands for the dilation parameter, $\boldsymbol{t}$ for a translation vector, and $\boldsymbol{\theta}$ the $SO(n)$ rotation parameters. The n-Dimensional Clifford Wavelet Transform (nD-CWT) with respect to the mother wavelet $\psi \in L^2(\mathbb{R}^n; G_n)$ is given by

$$
\begin{aligned}
T_\psi : L^2(\mathbb{R}^n; G_n) &\rightarrow L^2(G_s; G_n), \\
f \rightarrow T_\psi f(s, \boldsymbol{\theta}, \boldsymbol{t}) &= \int_{\mathbb{R}^n} f(\boldsymbol{x}) \widetilde{\psi_{s,\boldsymbol{\theta},\boldsymbol{t}}} \boldsymbol{d}^n \boldsymbol{x}, \\
&= (f, \psi_{s,\boldsymbol{\theta},\boldsymbol{t}})_{L^2(\mathbb{R}^n; G_n)}. \qquad (11.155)
\end{aligned}
$$

The Clifford wavelet transform of Eq. (11.155) has a Clifford Fourier representation given by the following expression:

$$
T_\psi f(s, \boldsymbol{\theta}, \boldsymbol{t}) = \frac{1}{(2\pi)^n} \int_{\mathbb{R}^n} \hat{f}(\boldsymbol{w}) s^{\frac{n}{2}} \{\hat{\psi}(s r_\theta^{-1}(\boldsymbol{w}))\} \, e^{I_n \boldsymbol{t} \cdot \boldsymbol{w}} \boldsymbol{d}^n \boldsymbol{w}. \qquad (11.156)
$$

Finally, the Inverse n-dimensional Clifford Wavelet Transform (nD-ICWT) is given by the following expression:

$$
\begin{aligned}
f(\boldsymbol{x}) &= \int_{G_s} T_\psi f(s, \boldsymbol{\theta}, \boldsymbol{t}) \psi_{s,\boldsymbol{\theta},\boldsymbol{t}} C_\psi^{-1} d\mu \boldsymbol{d}^n \boldsymbol{t}, \\
&= \int_{G_s} (f, \psi_{s,\boldsymbol{\theta},\boldsymbol{t}})_{L^2(\mathbb{R}^3; G_3)} \psi_{s,\boldsymbol{\theta},\boldsymbol{t}} C_\psi^{-1} d\mu \boldsymbol{d}^n \boldsymbol{t}, \qquad (11.157)
\end{aligned}
$$

where C_ψ is similar to Eq. (11.158) above but for a n dimension

$$
C_\psi = \int_{\mathbb{R}^+} \int_{SO(n)} s^n \{\hat{\psi}(s r_\theta^{-1}(\boldsymbol{w}))\} \tilde{\ } \hat{\psi}(s r_\theta^{-1}(\boldsymbol{w})) d\mu. \qquad (11.158)
$$

11.7 Conclusion

This chapter has shown that low-level image processing improves if signal representation and processing are carried out in a system of rich algebraic properties like geometric algebra. In this system, an n-dimensional representation of 2D signals unveils properties which are otherwise obscured with the use of algorithms developed by applying matrix algebra over the real or complex field. Strikingly, the bivector algebra of the geometric algebras allows us to disentangle the symmetries of 2D signals, as in the case of the quaternionic Fourier and wavelet transforms. The chapter presents also different Clifford Fourier and Wavelet transforms in various dimensions and metrics. This opens up an area for the design and implementation of new filters, convolution on the sphere , and estimators for the analysis of nD signals in a much wider and scope.

Chapter 12
Geometric Algebra of Computer Vision

12.1 Introduction

This chapter presents a mathematical approach based on *geometric algebra* for the computation of problems in computer vision. We will show that geometric algebra is a well-founded and elegant language for expressing and implementing those aspects of linear algebra and projective geometry that are useful for computer vision. Since geometric algebra offers both geometric insight and algebraic computational power, it is useful for tasks such as the computation of projective invariants, camera calibration, and the recovery of shape and motion. We will mainly focus on the geometry of multiple uncalibrated cameras and omnidirectional vision.

The following section introduces 3D and 4D geometric algebras and formulates the aspects of projective geometry relevant for computer vision within the geometric algebra framework. Given this background, in Sects. 12.2.4–12.2.5 we will look at the concepts of projective transformations and projective split. Section 12.3 presents the algebra of incidence, and Sect. 12.4 the algebra in projective space of points, lines, and planes. An analysis of monocular, binocular, and trinocular geometries is given in Sect. 12.6. We dedicate the following sections to omnidirectional vision using however the conformal geometric algebra framework. The motivation to resort to this framework is because mirrors can be represented using parameterized spheres; thus, the computing can be greatly simplified.

Conclusions follow in the final section.

In this chapter, vectors will be notated in boldface type (except for basis vectors) and multivectors will appear in bold italics. Lowercase letters are used to denote vectors in 3D Euclidean space, and uppercase letters to denote vectors in 4D projective space. We will also denote a geometric algebra $G_{p,q,r}$, which refers to an n-dimensional geometric algebra in which p-basis vectors square to $+1$, q-basis vectors to -1, and r-basis vectors to 0, so that $p + q + r = n$.

E. Bayro-Corrochano, *Geometric Algebra Applications Vol. I*,
https://doi.org/10.1007/978-3-319-74830-6_12

12.2 The Geometric Algebras of 3D and 4D Spaces

The need for a mathematical framework to understand and process digital camera images of the 3D world prompted researchers in the late 1970 s to use *projective geometry*. By using homogeneous coordinates, we were able to embed both 3D Euclidean visual space in the projective space P^3 or R^4, and the 2D Euclidean space of the image plane in the projective space P^2 or R^3. As a result, inherently nonlinear projective transformations from 3D space to the 2D image space now became linear, and points and directions could be differentiated rather than being represented by the same quantity. The use of projective geometry was indeed a step forward. However, there is still a need [131] for a mathematical system which reconciles projective geometry with multilinear algebra. Indeed, in most of the computer vision literature these mathematical systems are divorced from one another. Depending on the problem at hand, researchers typically resort to different systems, for example, dual algebra [48] for incidence algebra and the Hamiltonian formulation for motion estimation [321]. Here, we suggest the use of a system which offers all of these mathematical facilities. Unlike matrix and tensor algebra, geometric algebra does not obscure the underlying geometry of the problem. We will, therefore, formulate the main aspects of such problems using geometric algebra, starting with the modeling of 3D visual space and the 2D image plane.

12.2.1 3D Space and the 2D Image Plane

To introduce the basic geometric models in computer vision, we consider the imaging of a point $\mathbf{X} \in R^4$ into a point $\boldsymbol{x} \in R^3$. We will assume that the reader is familiar with the basic concepts of using homogeneous coordinates, which will be discussed in greater detail in later sections. The optical center C of the camera may be different from the origin of the world coordinate system O, as depicted in Fig. 12.1.

In the standard matrix representation, the mapping $P : \mathbf{X} \longrightarrow \boldsymbol{x}$ is expressed by the homogeneous transformation matrix

$$P = \begin{bmatrix} t_{11} & t_{12} & t_{13} & t_{14} \\ t_{21} & t_{22} & t_{23} & t_{14} \\ t_{31} & t_{32} & t_{33} & t_{34} \end{bmatrix}, \tag{12.1}$$

which may be decomposed into a product of three matrices,

$$P = K P_0 M_0^c, \tag{12.2}$$

where P_0, K, and M_0^c will now be defined. P_0 is the 3×4 matrix,

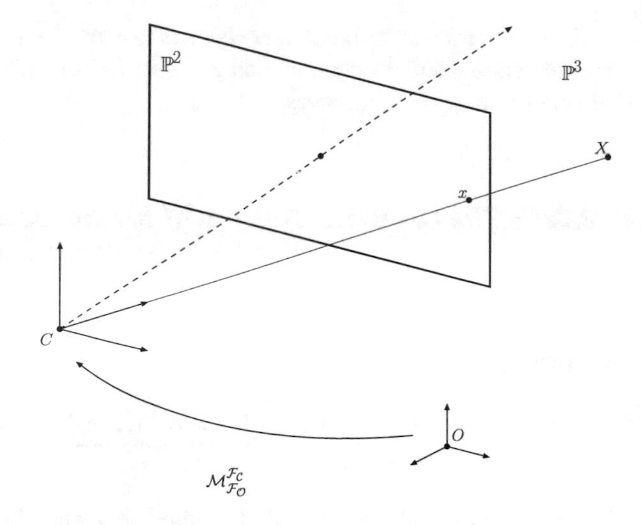

Fig. 12.1 Pinhole camera model

$$\begin{bmatrix} 1\,0\,0\,0 \\ 0\,1\,0\,0 \\ 0\,0\,1\,0 \end{bmatrix}, \tag{12.3}$$

which simply projects down from 4D to 3D and represents a projection from homogeneous coordinates of space to homogeneous coordinates of the image plane (Fig. 12.1).

M_0^c represents the 4×4 matrix containing the rotation and translation which takes the world frame $\mathcal{F}_0$ to the camera frame $\mathcal{F}_c$ and is given explicitly by

$$M_0^c = \begin{bmatrix} R & \boldsymbol{t} \\ \boldsymbol{0}^T & 1 \end{bmatrix}. \tag{12.4}$$

This Euclidean transformation is described by the *extrinsic parameters* of rotation (3×3 matrix R) and translation (3×1 vector $\boldsymbol{t}$). Finally, the 3×3 matrix K expresses the assumed camera model as an affine transformation between the camera plane and the image coordinate system, so that K is an upper triangular matrix. In the case of the perspective (or pinhole) camera, the matrix K which we now call K_p is given by

$$K_p = \begin{bmatrix} \alpha_u & \gamma & u_0 \\ 0 & \alpha_v & v_0 \\ 0 & 0 & 1 \end{bmatrix}. \tag{12.5}$$

The five parameters in K_p represent the camera parameters of scaling, shift, and rotation in the camera plane. In this case, the distance from the optical center to the image plane is finite. In later sections, we will formulate the perspective camera in the geometric algebra framework.

One important task in computer vision is to estimate the matrix of intrinsic camera parameters K_p and the rigid motion given in M_0^c, in order to be able to reconstruct 3D data from image sequences.

12.2.2 The Geometric Algebra of 3D Euclidean Space

The 3D space is spanned by three basis vectors $\{e_1, e_2, e_3\}$, with $e_i^2 = +1$ for all $i = 1, 2, 3$, and the 3D geometric algebra generated by these basis vectors has $2^3 = 8$ elements given by

$$\underbrace{1}_{scalar}, \underbrace{\{e_1, e_2, e_3\}}_{vectors}, \underbrace{\{e_2e_3, e_3e_2, e_1e_2\}}_{bivectors}, \underbrace{\{e_1e_2e_3\} \equiv I}_{trivector}. \tag{12.6}$$

Here, bivectors can be interpreted as oriented areas and trivectors as oriented volumes. Note that we will not use bold for these basis vectors. The highest grade element is a trivector called the unit *pseudoscalar*. It can easily be verified that the pseudoscalar $e_1e_2e_3$ squares to -1 and commutes with all multivectors (a multivector is a general linear combination of any of the elements in the algebra) in the 3D space. The unit pseudoscalar I is crucial when discussing duality. In a three-dimensional space, we can construct a trivector $\boldsymbol{a} \wedge \boldsymbol{b} \wedge \boldsymbol{c}$, but no 4-vectors exist, since there is no possibility of sweeping the volume element $\boldsymbol{a} \wedge \boldsymbol{b} \wedge \boldsymbol{c}$ over a fourth dimension.

The three basis vectors e_i multiplied by I give the following basis bivectors:

$$Ie_1 = e_2e_3 \qquad Ie_2 = e_3e_1 \qquad Ie_3 = e_1e_2. \tag{12.7}$$

If we identify the $\boldsymbol{i}, \boldsymbol{j}, \boldsymbol{k}$ of the quaternion algebra with e_2e_3, $-e_3e_2$, and e_1e_2, we can recover the famous *Hamilton relations*:

$$\boldsymbol{i}^2 = \boldsymbol{j}^2 = \boldsymbol{k}^2 = \boldsymbol{i}\boldsymbol{j}\boldsymbol{k} = -1. \tag{12.8}$$

In geometric algebra, a *rotor* $\boldsymbol{R}$ is an even-grade element of the algebra which satisfies the equation $\boldsymbol{R}\tilde{\boldsymbol{R}} = 1$. The relation between quaternions and rotors is as follows: If $\mathcal{Q} = \{q_0, q_1, q_2, q_3\}$ represents a *quaternion*, then the rotor which performs the same rotation is simply given by

$$\boldsymbol{R} = q_0 + q_1(Ie_1) - q_2(Ie_2) + q_3(Ie_3). \tag{12.9}$$

The quaternion algebra is therefore seen to be a subset of the geometric algebra of three-dimensional space.

12.2.3 A 4D Geometric Algebra for Projective Space

For the modeling of the image plane, we use $G_{3,0,0}$, which has the standard *Euclidean signature*. We will show that if we choose to map between projective space and 3D Euclidean space via the projective split (see Sect. 12.2.5), we are then forced to use the 4D geometric algebra $G_{1,3,0}$ for $\mathbb{P}^3$. The Lorentzian metric we are using here has no adverse effects in the operations we outline in this chapter. However, we will briefly discuss in a later section how a $\{++++\}$ metric for our 4D space and a different split is being favored in recent research.

The Lorentzian 4D algebra has as its vector basis e_1, e_2, e_3, e_4, where $e_4^2 = +1$ and $e_i^2 = -1$ for $i = 1, 2, 3$. This then generates the following multivector basis:

$$\underbrace{1}_{scalar}, \underbrace{e_k}_{4\ vectors}, \underbrace{e_2e_3, e_3e_1, e_1e_2, e_4e_1, e_4e_2, e_4e_3}_{6\ bivectors}, \underbrace{Ie_k}_{4\ trivectors}, \underbrace{I}_{pseudoscalar}. \quad (12.10)$$

The pseudoscalar is $I = e_1e_2e_3e_4$, with

$$I^2 = (e_1e_2e_3e_4)(e_1e_2e_3e_4) = -(e_3e_4)(e_3e_4) = -1. \quad (12.11)$$

The fourth basis vector, e_4, can also be seen as a selected direction for the *projective split* [23] operation in 4D. We will see shortly that by carrying out the geometric product via e_4, we can associate bivectors of our 4D space with vectors of our 3D space. The role and use of the projective split operation will be treated in more detail in a later section.

12.2.4 Projective Transformations

Historically, the success of homogeneous coordinates has partly been due to their ability to represent a general displacement as a single 4×4 matrix and to linearize nonlinear transformations [87].

The following equation indicates how a projective transformation may be linearized by going up one dimension in the GA framework. In general, a point (x, y, z) in the 3D space is projected onto the image via a transformation of the form:

$$x' = \frac{\alpha_1 x + \beta_1 y + \delta_1 z + \epsilon_1}{\tilde{\alpha} x + \tilde{\beta} y + \tilde{\delta} z + \tilde{\epsilon}}, \quad y' = \frac{\alpha_2 x + \beta_2 y + \delta_2 z + \epsilon_2}{\tilde{\alpha} x + \tilde{\beta} y + \tilde{\delta} z + \tilde{\epsilon}}. \quad (12.12)$$

This transformation, which is expressed as the ratio of two linear transformations, is indeed nonlinear. In order to convert this nonlinear transformation in $\mathbb{E}^3$ into a linear transformation in R^4, we define a linear function $\underline{f}_p$ by mapping vectors onto vectors in R^4 such that the action of $\underline{f}_p$ on the basis vectors $\{e_i\}$ is given by

$$\underline{f}_p(e_1) = \alpha_1 e_1 + \alpha_2 e_2 + \alpha_3 e_3 + \tilde{\alpha} e_4,$$
$$\underline{f}_p(e_2) = \beta_1 e_1 + \beta_2 e_2 + \beta_3 e_3 + \tilde{\beta} e_4,$$
$$\underline{f}_p(e_3) = \delta_1 e_1 + \delta_2 e_2 + \delta_3 e_3 + \tilde{\delta} e_4,$$
$$\underline{f}_p(e_4) = \epsilon_1 e_1 + \epsilon_2 e_2 + \epsilon_3 e_3 + \tilde{\epsilon} e_4. \tag{12.13}$$

When we use homogeneous coordinates, a general point P in $\mathbb{E}^3$ given by $\boldsymbol{x} = xe_1 + ye_2 + ze_3$ becomes the point $\mathbf{X} = (Xe_1 + Ye_2 + Ze_3 + We_4)$ in R^4, where $x = X/W$, $y = Y/W$ and $z = Z/W$. Now, using $\underline{f}_p$, the linear map of $\mathbf{X}$ onto $\mathbf{X}'$ is given by

$$\mathbf{X}' = \sum_{i=1}^{3}\{(\alpha_i X + \beta_i Y + \delta_i Z + \epsilon_i W)e_i\} + (\tilde{\alpha} X + \tilde{\beta} Y + \tilde{\delta} Z + \tilde{\epsilon} W)e_4. \tag{12.14}$$

The coordinates of the vector $\boldsymbol{x}' = x'e_1 + y'e_2 + z'e_3$ in $\mathbb{E}^3$ which correspond to $\mathbf{X}'$ are given by

$$x' = \frac{\alpha_1 X + \beta_1 Y + \delta_1 Z + \epsilon_1 W}{\tilde{\alpha} X + \tilde{\beta} Y + \tilde{\delta} Z + \tilde{\epsilon} W} = \frac{\alpha_1 x + \beta_1 y + \delta_1 z + \epsilon_1}{\tilde{\alpha} x + \tilde{\beta} y + \tilde{\delta} z + \tilde{\epsilon}}, \tag{12.15}$$

and similarly,

$$y' = \frac{\alpha_2 x + \beta_2 y + \delta_2 z + \epsilon_2}{\tilde{\alpha} x + \tilde{\beta} y + \tilde{\delta} z + \tilde{\epsilon}}, \qquad z' = \frac{\alpha_3 x + \beta_3 y + \delta_3 z + \epsilon_3}{\tilde{\alpha} x + \tilde{\beta} y + \tilde{\delta} z + \tilde{\epsilon}}. \tag{12.16}$$

If the above represents projection from the world onto a camera image plane, we should take into account the focal length of the camera. This would require that $\alpha_3 = f\tilde{\alpha}$, $\beta_3 = f\tilde{\beta}$, etc. Thus, we can define $z' = f$ (focal length) independent of the point chosen. The nonlinear transformation in $\mathbb{E}^3$ then becomes a linear transformation, $\underline{f}_p$, in R^4. The linear function $\underline{f}_p$ can then be used to prove the invariant nature of various quantities under projective transformations [178].

12.2.5 The Projective Split

The idea of the *projective split* was introduced by Hestenes [131] in order to connect *projective geometry* and *metric geometry*. This is done by associating the even subalgebra of G_{n+1} with the geometric algebra of next lower dimension, G_n. One can define a mapping between the spaces by choosing a preferred direction in G_{n+1}, e_{n+1}. Then, by taking the geometric product of a vector $\mathbf{X} \in G_{n+1}$ and e_{n+1},

$$\mathbf{X}e_{n+1} = \mathbf{X}\cdot e_{n+1} + \mathbf{X}\wedge e_{n+1} = \mathbf{X}\cdot e_{n+1}\left(1 + \frac{\mathbf{X}\wedge e_{n+1}}{\mathbf{X}\cdot e_{n+1}}\right), \tag{12.17}$$

the vector $\boldsymbol{x} \in G_n$ can be associated with the bivector $\frac{\mathbf{X} \wedge e_{n+1}}{\mathbf{X} \cdot e_{n+1}} \in G_{n+1}$. This result can be projectively interpreted as the pencil of all lines passing though the point e_{n+1}. In physics, the projective split is called the *space–time split*, and it relates a space–time system G_4 with a Minkowski metric to an observable system G_3 with a Euclidean metric.

In computer vision, we are interested in relating elements of projective space with their associated elements in the Euclidean space of the image plane. Optical rays (bivectors) are mapped to points (vectors), optical planes (trivectors) are mapped to lines (bivectors), and optical volumes (4-vectors) to planes (trivector or pseudoscalar).

Suppose we choose e_4 as a selected direction in R^4. We can then define a mapping which associates the bivectors $e_i e_4, i = 1, 2, 3$ in R^4 with the vectors $e_i, i = 1, 2, 3$ in $\mathbb{E}^3$:

$$e_1 \equiv e_1 e_4, \quad e_2 \equiv e_2 e_4, \quad e_3 \equiv e_3 e_4. \tag{12.18}$$

Note that in order to preserve the Euclidean structure of the spatial vectors e_i (i.e., $e_i^2 = +1$) we are forced to choose a non-Euclidean metric for the basis vectors in R^4. That is why we select the basis $e_1^2 = +1, \; e_i = -1, \; i = 1, 2, 3$ for $G_{1,3,0}$. This is precisely the metric structure of Lorentzian space–time used in studies of relativistic physics. We note here that although we have chosen to relate our spaces via the projective split, it is possible to use a Euclidean metric $\{+ + ++\}$ for our 4D space and define the split using reciprocal vectors [232]. It is becoming apparent that this is the preferred procedure since it generalizes nicely to splits from higher-dimensional spaces. However, for the problems discussed in this chapter, we encounter no problems by using the projective split.

Let us now see how we associate points via the projective split. For a vector $\mathbf{X} = X_1 e_1 + X_2 e_2 + X_3 e_3 + X_4 e_4$ in R^4 the projective split is obtained by taking the geometric product of $\mathbf{X}$ and e_4:

$$\boldsymbol{X} e_4 = \boldsymbol{X} \cdot e_4 + \boldsymbol{X} \wedge e_4 = X_4 \left(1 + \frac{\boldsymbol{X} \wedge e_4}{X_4} \right) \equiv X_4 (1 + \boldsymbol{x}). \tag{12.19}$$

According to Eq. (12.18), we can associate $\boldsymbol{X} \wedge e_4 / X_4$ in R^4 with the vector $\boldsymbol{x}$ in $\mathbb{E}^3$. Similarly, if we start with a vector $\boldsymbol{x} = x_1 e_1 + x_2 e_2 + x_3 e_3$ in $\mathbb{E}^3$, we represent it in R^4 by the vector $\mathbf{X} = X_1 e_1 + X_2 e_2 + X_3 e_3 + X_4 e_4$ such that

$$\begin{aligned} \boldsymbol{x} &= \frac{\boldsymbol{X} \wedge e_4}{X_4} = \frac{X_1}{X_4} e_1 e_4 + \frac{X_2}{X_4} e_2 e_4 + \frac{X_3}{X_4} e_3 e_4, \\ &= \frac{X_1}{X_4} e_1 + \frac{X_2}{X_4} e_2 + \frac{X_3}{X_4} e_3, \end{aligned} \tag{12.20}$$

which implies $x_i = \frac{X_i}{X_4}$ for $i = 1, 2, 3$. This manner of representing $\boldsymbol{x}$ in a higher-dimensional space can therefore be seen to be equivalent to using *homogeneous coordinates* $\mathbf{X}$ for $\boldsymbol{x}$.

Let us now look at the representation of a line L in R^4. A line is given by the outer product of two vectors:

$$\begin{aligned} L &= \boldsymbol{A} \wedge \boldsymbol{B}, \\ &= (L^{14} e_1 e_4 + L^{24} e_2 e_4 + L^{34} e_3 e_4) + (L^{23} e_2 e_3 + L^{31} e_3 e_1 + L^{12} e_1 e_2), \\ &= (L^{14} e_1 e_4 + L^{24} e_2 e_4 + L^{34} e_3 e_4) - I(L^{23} e_1 e_4 + L^{31} e_2 e_4 + L^{12} e_3 e_4), \\ &= \boldsymbol{n} - I\boldsymbol{m}. \end{aligned} \tag{12.21}$$

The six quantities $\{n_i, m_i\}$ $i = 1, 2, 3$ are precisely the Plücker coordinates of the line. The quantities $\{L^{14},\ L^{24},\ L^{34}\}$ are the coefficients of the *spatial part* of the bivector which represents the line direction $\boldsymbol{n}$. The quantities $\{L^{23},\ L^{31},\ L^{12}\}$ are the coefficients of the *non-spatial part* of the bivector which represents the *moment* of the line $\boldsymbol{m}$.

Let us now see how we can relate this line representation to an $\mathbb{E}^3$ representation via the projective split. We take a line L, joining points $\mathbf{A}$ and $\mathbf{B}$,

$$L = \mathbf{A} \wedge \mathbf{B} = \langle \boldsymbol{AB} \rangle_2 = \langle \boldsymbol{A} e_4 e_4 \boldsymbol{B} \rangle_2. \tag{12.22}$$

Here, the notation $\langle M \rangle_k$ tells us to take the grade-k part of the multivector M. Now, using our previous expansions of $\boldsymbol{X} e_4$ in the projective split for vectors, we can write

$$L = (\mathbf{A} \cdot e_4)(\mathbf{B} \cdot e_4)\langle (1 + \boldsymbol{a})(1 - \boldsymbol{b}) \rangle_2, \tag{12.23}$$

where $\boldsymbol{a} = \frac{\mathbf{A} \wedge e_4}{\mathbf{A} \cdot e_4}$ and $\boldsymbol{b} = \frac{\mathbf{B} \wedge e_4}{\mathbf{B} \cdot e_4}$ are the $\mathbb{P}^3$ representations of $\mathbf{A}$ and $\mathbf{B}$. Writing $A_4 = \mathbf{A} \cdot e_4$ and $B_4 = \mathbf{B} \cdot e_4$ then gives us

$$\begin{aligned} L &= A_4 B_4 \langle 1 + (\boldsymbol{a} - \boldsymbol{b}) - \boldsymbol{ab} \rangle_2, \\ &= A_4 B_4 \{(\boldsymbol{a} - \boldsymbol{b}) + \boldsymbol{a} \wedge \boldsymbol{b}\}. \end{aligned} \tag{12.24}$$

Let us now "normalize" the spatial and non-spatial parts of the above bivector:

$$\begin{aligned} L' &= \frac{L}{A_4 B_4 |\boldsymbol{a} - \boldsymbol{b}|} = \frac{(\boldsymbol{a} - \boldsymbol{b})}{|\boldsymbol{a} - \boldsymbol{b}|} + \frac{(\boldsymbol{a} \wedge \boldsymbol{b})}{|\boldsymbol{a} - \boldsymbol{b}|}, \qquad (12.25) \\ &= (n_x e_1 + n_y e_2 + n_z e_3) + (m_x e_2 e_3 + m_y e_3 e_1 + m_z e_1 e_2), \\ &= (n_x e_1 + n_y e_2 + n_z e_3) + I_3(m_x \sigma_1 + m_y e_2 + m_z e_3) = \boldsymbol{n}' + I_3 \boldsymbol{m}'. \end{aligned}$$

Here, $I_3 = e_1 e_2 e_3 \equiv I_4$. Note that in $\mathbb{E}^3$ the line has two components, a vector representing the direction of the line and the dual of a vector (bivector) representing the moment of the line. This kind of representation completely encodes the position of the line in 3D space by specifying the plane in which the line lies and the perpendicular distance of the line from the origin. Finally, for the plane $\pi = \boldsymbol{A} \wedge \boldsymbol{B} \wedge \boldsymbol{C}$ the expected result should be $\pi' = \boldsymbol{n}' + I_3 d$ (left as exercise 9.6).

12.3 The Algebra of Incidence

In this section, we will discuss the use of geometric algebra for the *algebra of incidence* [139]. First, we will define the concept of the bracket; then, we will discuss *duality*; and, finally, we will show that the basic projective geometry operations of meet and join can be expressed easily in terms of standard operations within the geometric algebra. We also briefly discuss the linear algebra framework in GA, indicating how it can be used within projective geometry. One of the main reasons for moving to a projective space is that lines, planes, etc., may be represented as real geometric objects so that operations of intersection, etc., can be performed using simple manipulations (rather than sets of equations, as in the Euclidean space $\mathbb{E}^3$).

12.3.1 The Bracket

In an n-D space, any pseudoscalar will span a hypervolume of dimension n. Since, up to scale, there can only be one such hypervolume, all pseudoscalars P are multiples of the unit pseudoscalar I, such that $P = \alpha I$, with α being a scalar. We compute this scalar multiple by multiplying the pseudoscalar P and the inverse of I:

$$PI^{-1} = \alpha I I^{-1} = \alpha \equiv [P]. \tag{12.26}$$

Thus, the *bracket* $[P]$ of the pseudoscalar P is its magnitude, arrived at by multiplication from the right by I^{-1}. This bracket is precisely the bracket of the Grassmann–Cayley algebra. The sign of the bracket does not depend on the signature of the space, and as a result, it has been a useful quantity for non-metrical applications of projective geometry.

The bracket of n-vectors $\{\boldsymbol{x}_i\}$ is

$$\begin{aligned}[\boldsymbol{x}_1\boldsymbol{x}_2\boldsymbol{x}_3 ... \boldsymbol{x}_n] &= [\boldsymbol{x}_1 \wedge \boldsymbol{x}_2 \wedge \boldsymbol{x}_3 \wedge ... \wedge \boldsymbol{x}_n], \\ &= (\boldsymbol{x}_1 \wedge \boldsymbol{x}_2 \wedge \boldsymbol{x}_3 \wedge ... \wedge \boldsymbol{x}_n) I^{-1}. \end{aligned} \tag{12.27}$$

It can also be shown that this bracket expression is equivalent to the definition of the determinant of the matrix whose row vectors are the vectors $\boldsymbol{x}_i$.

To understand how we can express a bracket in projective space in terms of vectors in Euclidean space, we can expand the pseudoscalar P using the projective split for vectors:

$$P = \mathbf{X}_1 \wedge \mathbf{X}_2 \wedge \mathbf{X}_3 \wedge \mathbf{X}_4 = \langle \mathbf{X}_1 e_4 e_4 \mathbf{X}_2 \mathbf{X}_3 e_4 e_4 \mathbf{X}_4 \rangle_4,$$
$$= W_1 W_2 W_3 W_4 \langle (1+\boldsymbol{x}_1)(1-\boldsymbol{x}_2)(1+\boldsymbol{x}_3)(1-\boldsymbol{x}_4) \rangle_4,$$

where $W_i = \mathbf{X}_i \cdot e_4$ from Eq. (12.19). A pseudoscalar part is produced by taking the product of three spatial vectors (there are no spatial bivector $\times$ spatial vector terms), such that

$$\begin{aligned} P &= W_1 W_2 W_3 W_4 \langle -\boldsymbol{x}_1\boldsymbol{x}_2\boldsymbol{x}_3 - \boldsymbol{x}_1\boldsymbol{x}_3\boldsymbol{x}_4 + \boldsymbol{x}_1\boldsymbol{x}_2\boldsymbol{x}_4 + \boldsymbol{x}_2\boldsymbol{x}_3\boldsymbol{x}_4 \rangle_4, \\ &= W_1 W_2 W_3 W_4 \langle (\boldsymbol{x}_2 - \boldsymbol{x}_1)(\boldsymbol{x}_3 - \boldsymbol{x}_1)(\boldsymbol{x}_4 - \boldsymbol{x}_1) \rangle_4, \\ &= W_1 W_2 W_3 W_4 \{ (\boldsymbol{x}_2 - \boldsymbol{x}_1) \wedge (\boldsymbol{x}_3 - \boldsymbol{x}_1) \wedge (\boldsymbol{x}_4 - \boldsymbol{x}_1) \}. \end{aligned} \tag{12.28}$$

If $W_i = 1$, we can summarize the above relationships between the brackets of four points in R^4 and $\mathbb{E}^3$ as follows:

$$\begin{aligned} [\mathbf{X}_1\mathbf{X}_2\mathbf{X}_3\mathbf{X}_4] &= (\mathbf{X}_1 \wedge \mathbf{X}_2 \wedge \mathbf{X}_3 \wedge \mathbf{X}_4) I_4^{-1}, \\ &\equiv \{ (\boldsymbol{x}_2 - \boldsymbol{x}_1) \wedge (\boldsymbol{x}_3 - \boldsymbol{x}_1) \wedge (\boldsymbol{x}_4 - \boldsymbol{x}_1) \} I_3^{-1}. \end{aligned} \tag{12.29}$$

12.3.2 The Duality Principle and Meet and Join Operations

In order to introduce the concepts of *duality* which are so important in projective geometry, we must first define the dual A^* of an r-vector A as

$$A^* = AI^{-1}. \tag{12.30}$$

This notation, A^*, relates the ideas of duality to the notion of a *Hodge dual* in differential geometry. Note that, in general, I^{-1} might not necessarily commute with $\boldsymbol{A}$.

We see, therefore, that the dual of an r-vector is an $(n-r)$-vector. For example, in 3D space the dual of a vector ($r = 1$) is a plane or bivector ($n - r = 3 - 1 = 2$).

By using the ideas of duality, we are then able to relate the inner product to incidence operators in the following manner. In an n-D space, suppose we have an r-vector A and an s-vector B, where the dual of B is given by $B^* = BI^{-1} \equiv B \cdot I^{-1}$. Since $BI^{-1} = B \cdot I^{-1} + B \wedge I^{-1}$, we can replace the geometric product by the inner product alone (in this case, the outer product equals zero, and there can be no $(n+1)$-D vector). Now, using the identity

$$A_r \cdot (B_s \cdot C_t) = (A_r \wedge B_s) \cdot C_t \quad \text{for} \quad r + s \leq t, \tag{12.31}$$

we can write

$$A \cdot (BI^{-1}) = A \cdot (B \cdot I^{-1}) = (A \wedge B) \cdot I^{-1} = (A \wedge B) I^{-1}. \tag{12.32}$$

This expression can be rewritten using the definition of the dual as follows:

$$A \cdot B^* = (A \wedge B)^*. \tag{12.33}$$

This equation shows the relationship between the inner and outer products in terms of the duality operator. Now, if $r + s = n$, then $A \wedge B$ is of grade n and is therefore a pseudoscalar. Using Eq. (12.26), it follows that

$$\begin{aligned} A \cdot B^* &= (A \wedge B)^* = (A \wedge B)I^{-1} = ([A \wedge B]I)I^{-1}, \\ &= [A \wedge B]. \end{aligned} \tag{12.34}$$

We see, therefore, that the bracket relates the inner and outer products to non-metric quantities. It is via this route that the inner product, normally associated with a metric, can be used in a non-metric theory such as projective geometry. It is also interesting to note that since duality is expressed as a simple multiplication by an element of the algebra, there is no need to introduce any special operators or any concept of a different space.

When we work with lines and planes, however, it will clearly be necessary to employ operations for computing the intersections, or *joins*, of geometric objects. For this, we will require a means of performing the set-theory operations of intersection, $\cap$, and union, $\cup$.

If in an n-dimensional geometric algebra the r-vector A and the s-vector B do not have a common subspace (null intersection), one can define the *join* of both vectors as follows:

$$J = A \cup B = A \wedge B, \tag{12.35}$$

so that the join is simply the outer product (an $r + s$-vector) of the two vectors. However, if A and B have common blades, the join would not simply be given by the wedge but by the subspace the two vectors span. The operation join J can be interpreted as a *common dividend of lowest grade* and is defined up to a scale factor. The join gives the pseudoscalar if $(r + s) \geq n$. We will use $\cup$ to represent the join only when the blades A and B have a common subspace; otherwise, we will use the ordinary exterior product, $\wedge$, to represent the join.

If there exists a k-vector C such that for A and B we can write $A = A'C$ and $B = B'C$ for some A' and B', then we can define the *intersection* or *meet* using the duality principle as follows:

$$(A \cap B)^* = A^* \cup B^*. \tag{12.36}$$

This is a beautiful result, telling us that the dual of the meet is given by the join of the duals. Since the dual of $A \cap B$ will be taken with respect to the *join* of A and B, we must be careful to specify which space we will use for the dual in Eq. (12.36). However, in most cases of practical interest this join will indeed cover the entire space, and therefore, we will be able to obtain a more useful expression for the meet using Eq. (12.33). Thus,

$$A \cap B = ((A \cap B)^*)^* = (A^* \cup B^*)I = (A^* \wedge B^*)(I^{-1}I)I = (A^* \cdot B). \quad (12.37)$$

The above concepts are discussed further in [139].

12.4 Algebra in Projective Space

Having introduced duality, defined the operations of meet and join, and given the geometric approach to linear algebra, we are now ready to carry out geometric computations using the algebra of incidence.

Consider three non-collinear points, $P_1,\ P_2,\ P_3$, represented by vectors $\boldsymbol{x}_1,\ \boldsymbol{x}_2,\ \boldsymbol{x}_3$ in $\mathbb{E}^3$ and by vectors $\mathbf{X}_1,\ \mathbf{X}_2,\ \mathbf{X}_3$ in R^4. The line L_{12} joining points P_1 and P_2 can be expressed in R^4 by the bivector

$$L_{12} = \mathbf{X}_1 \wedge \mathbf{X}_2. \quad (12.38)$$

Any point P, represented in R^4 by $\mathbf{X}$, on the line through P_1 and P_2, will satisfy the equation

$$\mathbf{X} \wedge L_{12} = \mathbf{X} \wedge \mathbf{X}_1 \wedge \mathbf{X}_2 = 0. \quad (12.39)$$

This is therefore the equation of the line in R^4. In general, such an equation is telling us that $\mathbf{X}$ belongs to the subspace spanned by $\mathbf{X}_1$ and $\mathbf{X}_2$—that is, that

$$\mathbf{X} = \alpha_1 \mathbf{X}_1 + \alpha_2 \mathbf{X}_2 \quad (12.40)$$

for some $\alpha_1,\ \alpha_2$. In computer vision, we can use this equation as a geometric constraint to test whether a point $\mathbf{X}$ lies on L_{12}.

The plane Φ_{123} passing through points $P_1,\ P_2,\ P_3$ is expressed by the following trivector in R^4:

$$\Phi_{123} = \mathbf{X}_1 \wedge \mathbf{X}_2 \wedge \mathbf{X}_3. \quad (12.41)$$

In 3D space, there are generally three types of intersections we wish to consider: the intersection of a line and a plane, a plane and a plane, and a line and a line. To compute these intersections, we will make use of the following general formula [138], which gives the inner product of an r-blade, $A_r = \boldsymbol{a}_1 \wedge \boldsymbol{a}_2 \wedge \ldots \wedge \boldsymbol{a}_r$, and an s-blade, $B_s = \boldsymbol{b}_1 \wedge \boldsymbol{b}_2 \wedge \ldots \wedge \boldsymbol{b}_s$ (for $s \leq r$):

$$B_s \cdot (\boldsymbol{a}_1 \wedge \boldsymbol{a}_2 \wedge \ldots \wedge \boldsymbol{a}_r) = \sum_{j} \epsilon(j_1 j_2 \cdots j_r) B_s \cdot (\boldsymbol{a}_{j_1} \wedge \boldsymbol{a}_{j_2} \wedge \ldots \wedge \boldsymbol{a}_{j_s}) \boldsymbol{a}_{j_s+1} \wedge \ldots \wedge \boldsymbol{a}_{j_r}. \quad (12.42)$$

In the equation, we sum over all the combinations $\boldsymbol{j} = (j_1, j_2, ..., j_r)$ such that no two j_k's are the same. If $\boldsymbol{j}$ is an even permutation of $(1, 2, 3, ..., r)$, then the expression $\epsilon(j_1 j_2 ... j_r) = +1$, and it is an odd permutation if $\epsilon(j_1 j_2 ... j_r) = -1$.

12.4.1 Intersection of a Line and a Plane

In the space R^4, consider the line $A = \mathbf{X}_1 \wedge \mathbf{X}_2$ intersecting the plane $\Phi = \mathbf{Y}_1 \wedge \mathbf{Y}_2 \wedge \mathbf{Y}_3$. We can compute the intersection point using a *meet* operation, as follows:

$$A \cap \Phi = (\mathbf{X}_1 \wedge \mathbf{X}_2) \cap (\mathbf{Y}_1 \wedge \mathbf{Y}_2 \wedge \mathbf{Y}_3) = A \cap \Phi = A^* \cdot \Phi. \quad (12.43)$$

Here, we have used Eq. (12.37), and we note that in this case the join covers the entire space.

Note also that the pseudoscalar I_4 in $G_{1,3,0}$ for R^4 squares to -1, that it commutes with bivectors but anticommutes with vectors and trivectors, and that its inverse is given by $I_4{}^{-1} = -I_4$. Therefore, we can claim that

$$A^* \cdot \Phi = (AI^{-1}) \cdot \Phi = -(AI) \cdot \Phi. \quad (12.44)$$

Now, using Eq. (12.43), we can expand the meet, such that

$$\begin{aligned} A \cap \Phi &= -(AI) \cdot (\mathbf{Y}_1 \wedge \mathbf{Y}_2 \wedge \mathbf{Y}_3), \\ &= -\{(AI) \cdot (\mathbf{Y}_2 \wedge \mathbf{Y}_3)\} \mathbf{Y}_1 + \{(AI) \cdot (\mathbf{Y}_3 \wedge \mathbf{Y}_1)\} \mathbf{Y}_2 + \\ &\quad +\{(AI) \cdot (\mathbf{Y}_1 \wedge \mathbf{Y}_2)\} \mathbf{Y}_3. \end{aligned} \quad (12.45)$$

Noting that $(AI) \cdot (\mathbf{Y}_i \wedge \mathbf{Y}_j)$ is a scalar, we can evaluate Eq. 12.45 by taking scalar parts. For example, $(AI) \cdot (\mathbf{Y}_2 \wedge \mathbf{Y}_3) = \langle I(\mathbf{X}_1 \wedge \mathbf{X}_2)(\mathbf{Y}_2 \wedge \mathbf{Y}_3) \rangle = I(\mathbf{X}_1 \wedge \mathbf{X}_2 \wedge \mathbf{Y}_2 \wedge \mathbf{Y}_3)$. From the definition of the bracket given earlier, we can see that if $P = \mathbf{X}_1 \wedge \mathbf{X}_2 \wedge \mathbf{Y}_2 \wedge \mathbf{Y}_3$, then $[P] = (\mathbf{X}_1 \wedge \mathbf{X}_2 \wedge \mathbf{Y}_2 \wedge \mathbf{Y}_3) I_4{}^{-1}$. If we therefore write $[\mathbf{A}_1 \mathbf{A}_2 \mathbf{A}_3 \mathbf{A}_4]$ as a shorthand for the magnitude of the pseudoscalar formed from the four vectors, then we can readily see that the meet reduces to

$$A \cap \Phi = [\mathbf{X}_1 \mathbf{X}_2 \mathbf{Y}_2 \mathbf{Y}_3] \mathbf{Y}_1 + [\mathbf{X}_1 \mathbf{X}_2 \mathbf{Y}_3 \mathbf{Y}_1] \mathbf{Y}_2 + [\mathbf{X}_1 \mathbf{X}_2 \mathbf{Y}_1 \mathbf{Y}_2] \mathbf{Y}_3, \quad (12.46)$$

thus giving the intersection point (vector in R^4).

12.4.2 Intersection of Two Planes

The *line of intersection of two planes*, $\Phi_1 = \mathbf{X}_1 \wedge \mathbf{X}_2 \wedge \mathbf{X}_3$ and $\Phi_2 = \mathbf{Y}_1 \wedge \mathbf{Y}_2 \wedge \mathbf{Y}_3$, can be computed via the meet of Φ_1 and Φ_2:

$$\Phi_1 \cap \Phi_2 = (\mathbf{X}_1 \wedge \mathbf{X}_2 \wedge \mathbf{X}_3) \cap (\mathbf{Y}_1 \wedge \mathbf{Y}_2 \wedge \mathbf{Y}_3). \tag{12.47}$$

As in the previous section, this expression can be expanded as

$$\begin{aligned} \Phi_1 \cap \Phi_2 &= \Phi_1{}^* \cdot (\mathbf{Y}_1 \wedge \mathbf{Y}_2 \wedge \mathbf{Y}_3), \\ &= -\{(\Phi_1 I) \cdot \mathbf{Y}_1\}(\mathbf{Y}_2 \wedge \mathbf{Y}_3) + \{(\Phi_1 I) \cdot \mathbf{Y}_2\}(\mathbf{Y}_3 \wedge \mathbf{Y}_1) + \\ &\quad + \{(\Phi_1 I) \cdot \mathbf{Y}_3\}(\mathbf{Y}_1 \wedge \mathbf{Y}_2). \end{aligned}$$

Once again, the join covers the entire space and so the dual is easily formed. Following the arguments of the previous section, we can show that $(\Phi_1 I) \cdot \mathbf{Y}_i \equiv -[\mathbf{X}_1 \mathbf{X}_2 \mathbf{X}_3 \mathbf{Y}_i]$, so that the meet is

$$\begin{aligned} \Phi_1 \cap \Phi_2 &= [\mathbf{X}_1 \mathbf{X}_2 \mathbf{X}_3 \mathbf{Y}_1](\mathbf{Y}_2 \wedge \mathbf{Y}_3) + [\mathbf{X}_1 \mathbf{X}_2 \mathbf{X}_3 \mathbf{Y}_2](\mathbf{Y}_3 \wedge \mathbf{Y}_1) + \\ &\quad + [\mathbf{X}_1 \mathbf{X}_2 \mathbf{X}_3 \mathbf{Y}_3](\mathbf{Y}_1 \wedge \mathbf{Y}_2), \end{aligned} \tag{12.48}$$

thus producing a line of intersection or bivector in R^4.

12.4.3 Intersection of Two Lines

Two lines will intersect only if they are coplanar. This means that their representations in R^4, $A = \mathbf{X}_1 \wedge \mathbf{X}_2$, and $B = \mathbf{Y}_1 \wedge \mathbf{Y}_2$ will satisfy the equation

$$A \wedge B = 0. \tag{12.49}$$

This fact suggests that the computation of the intersection should be carried out in the 2D Euclidean space which has an associated 3D projective counterpart R^3. In this plane, the intersection point is given by

$$\begin{aligned} A \cap B &= A^* \cdot B = -(A I_3) \cdot (\mathbf{Y}_1 \wedge \mathbf{Y}_2), \\ &= -\{((A I_3) \cdot \mathbf{Y}_1)\mathbf{Y}_2 - ((A I_3) \cdot \mathbf{Y}_2)\mathbf{Y}_1\}, \end{aligned} \tag{12.50}$$

where I_3 is the pseudoscalar for R^3. Once again, we evaluate $((A I_3) \cdot \mathbf{Y}_i)$ by taking scalar parts:

$$(A I_3) \cdot \mathbf{Y}_i = \langle \mathbf{X}_1 \mathbf{X}_2 I_3 \mathbf{Y}_i \rangle = I_3 \mathbf{X}_1 \mathbf{X}_2 \mathbf{Y}_i = -[\mathbf{X}_1 \mathbf{X}_2 \mathbf{Y}_i]. \tag{12.51}$$

The meet can therefore be written as

$$A \cap B = [\mathbf{X}_1\mathbf{X}_2\mathbf{Y}_1]\mathbf{Y}_2 - [\mathbf{X}_1\mathbf{X}_2\mathbf{Y}_2]\mathbf{Y}_1, \tag{12.52}$$

where the bracket $[\mathbf{A}_1\mathbf{A}_2\mathbf{A}_3]$ in R^3 is understood to mean $(\mathbf{A}_1 \wedge \mathbf{A}_2 \wedge \mathbf{A}_3)I_3{}^{-1}$. This equation is often an impractical means of performing the intersection of two lines. (See [232] for a method which creates a plane and intersects one of the lines with this plane; see also [77] for a discussion of what information can be gained when the lines do not intersect. See Sect. 5.11 for a complete treatment of the incidence relations between points, lines, and planes in the n-affine plane.)

12.4.4 Implementation of the Algebra

In order to implement the expressions and procedures outlined so far in this chapter, we have used a computer algebra package written for Maple. The program can be found in [174] and works with geometric algebras of $G_{1,3,0}$ and $G_{3,0,0}$; a more general version of this program, which works with a user-defined metric on an n-D algebra, is in the public domain [6]. Using these packages, we are easily able to simulate the situation of several cameras (or one moving camera) looking at a world scene and to do so entirely in projective (4D) space. Much of the work described in subsequent sections has been tested in Maple.

12.5 Projective Invariants

In this section, we will use the framework established in this chapter to show how standard invariants can be expressed both elegantly and concisely using geometric algebra. We begin by looking at algebraic quantities that are invariant under projective transformations, arriving at these invariants using a method which can be easily generalized from one dimension to two and three dimensions.

12.5.1 The 1D Cross-Ratio

The *fundamental projective invariant* of points on a line is the so-called cross-ratio, ρ, defined as

$$\rho = \frac{AC}{BC}\frac{BD}{AD} = \frac{(t_3 - t_1)(t_4 - t_2)}{(t_4 - t_1)(t_3 - t_2)},$$

where $t_1 = |PA|$, $t_2 = |PB|$, $t_3 = |PC|$, and $t_4 = |PD|$. It is fairly easy to show that for the projection through O of the collinear points A, B, C, and D onto any line, ρ remains constant. For the 1D case, any point q on the line L can be written as $\boldsymbol{q} = te_1$ relative to P, where e_1 is a unit vector in the direction of L. We can then move up a dimension to a 2D space, with basis vectors (e_1, e_2), which we will call R^2 and in which $\boldsymbol{q}$ is represented by the following vector $\mathbf{Q}$:

$$\mathbf{Q} = Te_1 + Se_2. \tag{12.53}$$

Note that, as before, $\boldsymbol{q}$ is associated with the bivector, as follows:

$$\boldsymbol{q} = \frac{\mathbf{Q}\wedge e_2}{\mathbf{Q}\cdot e_2} = \frac{T}{S}e_1e_2 \equiv \frac{T}{S}e_1 = te_1. \tag{12.54}$$

When a point on line L is projected onto another line L', the distances t and t' are related by a projective transformation of the form

$$t' = \frac{\alpha t + \beta}{\tilde{\alpha} t + \tilde{\beta}}. \tag{12.55}$$

This nonlinear transformation in $\mathbb{E}^1$ can be made into a linear transformation in R^2 by defining the linear function $\underline{f}_1$ which maps vectors onto vectors in R^2:

$$\underline{f}_1(e_1) = \alpha_1 e_1 + \tilde{\alpha} e_2,$$
$$\underline{f}_1(e_2) = \beta_1 e_1 + \tilde{\beta} e_2.$$

Consider two vectors $\mathbf{X}_1$ and $\mathbf{X}_2$ in R^2. Now form the bivector

$$\mathcal{S}_1 = \mathbf{X}_1\wedge\mathbf{X}_2 = \lambda_1 I_2,$$

where $I_2 = e_1e_2$ is the pseudoscalar for R^2. We can now look at how $\mathcal{S}_1$ transforms under $\underline{f}_1$:

$$\mathcal{S}_1' = \mathbf{X}_1'\wedge\mathbf{X}_2' = \underline{f}_1(\mathbf{X}_1\wedge\mathbf{X}_2) = (\det\underline{f}_1)(\mathbf{X}_1\wedge\mathbf{X}_2). \tag{12.56}$$

This last step follows is a result of a linear function, which must map a pseudoscalar onto a multiple of itself, the multiple being the determinant of the function. Suppose that we now select four points of the line L, whose corresponding vectors in R^2 are $\{\mathbf{X}_i\}$, $i = 1, ..., 4$, and consider the ratio $\mathcal{R}_1$ of two wedge products:

$$\mathcal{R}_1 = \frac{\mathbf{X}_1\wedge\mathbf{X}_2}{\mathbf{X}_3\wedge\mathbf{X}_4}. \tag{12.57}$$

Then, under $\underline{f}_1$, $\mathcal{R}_1 \rightarrow \mathcal{R}_1'$, where

$$\mathcal{R}'_1 = \frac{\mathbf{X}'_1 \wedge \mathbf{X}'_2}{\mathbf{X}'_3 \wedge \mathbf{X}'_4} = \frac{(\det \underline{f}_1)\mathbf{X}_1 \wedge \mathbf{X}_2}{(\det \underline{f}_1)\mathbf{X}_3 \wedge \mathbf{X}_4}. \tag{12.58}$$

$\mathcal{R}_1$ is therefore invariant under $\underline{f}_1$. However, we want to express our invariants in terms of distances on the 1D line. To do this, we must consider how the bivector $\mathcal{S}_1$ in R^2 projects down to $\mathbb{E}^1$:

$$\begin{aligned}
\mathbf{X}_1 \wedge \mathbf{X}_2 &= (T_1 e_1 + S_1 e_2) \wedge (T_2 e_1 + S_2 e_2),\\
&= (T_1 S_2 - T_2 S_1) e_1 e_2,\\
&\equiv S_1 S_2 (T_1/S_1 - T_2/S_2) I_2,\\
&= S_1 S_2 (t_1 - t_2) I_2.
\end{aligned} \tag{12.59}$$

In order to form a projective invariant which is independent of the choice of the arbitrary scalars S_i, we must now consider *ratios* of the bivectors $\mathbf{X}_i \wedge \mathbf{X}_j$ (so that $\det \underline{f}_1$ cancels), and then *multiples* of these ratios (so that the S_i's cancel). More precisely, consider the following expression:

$$Inv_1 = \frac{(\mathbf{X}_3 \wedge \mathbf{X}_1) I_2^{-1} (\mathbf{X}_4 \wedge \mathbf{X}_2) I_2^{-1}}{(\mathbf{X}_4 \wedge \mathbf{X}_1) I_2^{-1} (\mathbf{X}_3 \wedge \mathbf{X}_2) I_2^{-1}}. \tag{12.60}$$

Then, in terms of distances along the lines, under the projective transformation $\underline{f}_1$, Inv_1 goes to Inv'_1, where

$$Inv_1 = \frac{S_3 S_1 (t_3 - t_1) S_4 S_2 (t_4 - t_2)}{S_4 S_1 (t_4 - t_1) S_3 S_2 (t_3 - t_2)} = \frac{(t_3 - t_1)(t_4 - t_2)}{(t_4 - t_1)(t_3 - t_2)}, \tag{12.61}$$

which is independent of the S_i's and is indeed the 1D classical projective invariant, the cross-ratio. Deriving the cross-ratio in this way allows us to easily generalize it to form invariants in higher dimensions.

12.5.2 2D Generalization of the Cross-Ratio

When we consider points in a plane, we once again move up to a space with one higher dimension, which we shall call $\mathbb{R}^3$. Let a point P in the plane M be described by the vector $\boldsymbol{x}$ in $\mathbb{E}^2$, where $\boldsymbol{x} = x e_1 + y e_2$. In R^3, this point will be represented by $\mathbf{X} = X e_1 + Y e_2 + Z e_3$, where $x = X/Z$ and $y = Y/Z$. As described in this chapter, we can define a general projective transformation via a linear function $\underline{f}_2$ by mapping vectors to vectors in R^3, such that

$$\underline{f}_2(e_1) = \alpha_1 e_1 + \alpha_2 e_2 + \tilde{\alpha} e_3,$$
$$\underline{f}_2(e_2) = \beta_1 e_1 + \beta_2 e_2 + \tilde{\beta} e_3, \tag{12.62}$$
$$\underline{f}_2(e_3) = \delta_1 e_1 + \delta_2 e_2 + \tilde{\delta} e_3.$$

Now, consider three vectors (representing non-collinear points) $\mathbf{X}_i$, $i = 1, 2, 3$, in R^3, and form the trivector

$$\mathcal{S}_2 = \mathbf{X}_1 \wedge \mathbf{X}_2 \wedge \mathbf{X}_3 = \lambda_2 I_3, \tag{12.63}$$

where $I_3 = \epsilon_1 e_2 e_3$ is the pseudoscalar for R^3. As before, under the projective transformation given by $\underline{f}_2$, $\mathcal{S}_2$ transforms to $\mathcal{S}_2'$, where

$$\mathcal{S}_2' = \det \underline{f}_2 \mathcal{S}_2. \tag{12.64}$$

Therefore, the ratio of any trivector is invariant under $\underline{f}_2$. To project down into $\mathbb{E}^2$, assuming that $\mathbf{X}_i e_3 = Z_i(1 + \boldsymbol{x}_i)$ under the projective split, we then write

$$\begin{aligned} \mathcal{S}_2 I_3{}^{-1} &= \langle \mathbf{X}_1 \mathbf{X}_2 \mathbf{X}_3 I_3{}^{-1} \rangle, \\ &= \langle \mathbf{X}_1 e_3 e_3 \mathbf{X}_2 \mathbf{X}_3 e_3 e_3 I_3{}^{-1} \rangle, \\ &= Z_1 Z_2 Z_3 \langle (1 + \boldsymbol{x}_1)(1 - \boldsymbol{x}_2)(1 + \boldsymbol{x}_3) e_3 I_3{}^{-1} \rangle, \end{aligned} \tag{12.65}$$

where the $\boldsymbol{x}_i$ represent vectors in $\mathbb{E}^2$. We can only get a scalar term from the expression within the brackets by calculating the product of a vector, two spatial vectors, and $I_3{}^{-1}$, i.e.,

$$\begin{aligned} \mathcal{S}_2 I_3{}^{-1} &= Z_1 Z_2 Z_3 \langle (\boldsymbol{x}_1 \boldsymbol{x}_3 - \boldsymbol{x}_1 \boldsymbol{x}_2 - \boldsymbol{x}_2 \boldsymbol{x}_3) e_3 I_3{}^{-1} \rangle, \\ &= Z_1 Z_2 Z_3 \{ (\boldsymbol{x}_2 - \boldsymbol{x}_1) \wedge (\boldsymbol{x}_3 - \boldsymbol{x}_1) \} I_2{}^{-1}. \end{aligned} \tag{12.66}$$

It is therefore clear that we must use multiples of the ratios in our calculations, so that the arbitrary scalars Z_i cancel. In the case of four points in a plane, there are only four possible combinations of $Z_i Z_j Z_k$ and it is not possible to cancel all the Z's by multiplying two ratios of the form $\mathbf{X}_i \wedge \mathbf{X}_j \wedge \mathbf{X}_k$ together. For five coplanar points $\{\mathbf{X}_i\}$, $i = 1, ..., 5$, however, there are several ways of achieving the desired cancelation. For example,

$$Inv_2 = \frac{(\mathbf{X}_5 \wedge \mathbf{X}_4 \wedge \mathbf{X}_3) I_3^{-1} (\mathbf{X}_5 \wedge \mathbf{X}_2 \wedge \mathbf{X}_1) I_3^{-1}}{(\mathbf{X}_5 \wedge \mathbf{X}_1 \wedge \mathbf{X}_3) I_3^{-1} (\mathbf{X}_5 \wedge \mathbf{X}_2 \wedge \mathbf{X}_4) I_3^{-1}}.$$

According to Eq. (12.66), we can interpret this ratio in $\mathbb{E}^2$ as

$$
\begin{aligned}
Inv_2 &= \frac{(\boldsymbol{x}_5-\boldsymbol{x}_4)\wedge(\boldsymbol{x}_5-\boldsymbol{x}_3)I_2^{-1}(\boldsymbol{x}_5-\boldsymbol{x}_2)\wedge(\boldsymbol{x}_5-\boldsymbol{x}_1)I_2^{-1}}{(\boldsymbol{x}_5-\boldsymbol{x}_1)\wedge(\boldsymbol{x}_5-\boldsymbol{x}_3)I_2^{-1}(\boldsymbol{x}_5-\boldsymbol{x}_2)\wedge(\boldsymbol{x}_5-\boldsymbol{x}_4)I_2^{-1}},\\
&= \frac{A_{543}A_{521}}{A_{513}A_{524}},
\end{aligned}
\tag{12.67}
$$

where $\frac{1}{2}A_{ijk}$ is the area of the triangle defined by the three vertices $\boldsymbol{x}_i, \boldsymbol{x}_j, \boldsymbol{x}_k$. This invariant is regarded as the 2D generalization of the 1D cross-ratio.

12.5.3 3D Generalization of the Cross-Ratio

For general points in $\mathbb{E}^3$, we have seen that we move up one dimension to compute in the 4D space R^4. For this dimension, the point $\boldsymbol{x} = xe_1 + ye_2 + ze_3$ in $\mathbb{E}^3$ is written as $\mathbf{X} = Xe_1 + Ye_2 + Ze_3 + We_4$, where $x = X/W,\ y = Y/W,\ z = Z/W$. As before, a nonlinear projective transformation in $\mathbb{E}^3$ becomes a linear transformation, described by the linear function $\underline{f}_3$ in R^4.

Let us consider 4-vectors in R^4, $\{\mathbf{X}_i\}$, $i = 1, ..., 4$ and form the equation of a 4-vector:

$$
\mathcal{S}_3 = \mathbf{X}_1\wedge\mathbf{X}_2\wedge\mathbf{X}_3\wedge\mathbf{X}_4 = \lambda_3 I_4,
\tag{12.68}
$$

where $I_4 = e_1e_2e_3e_4$ is the pseudoscalar for R^4. As before, $\mathcal{S}_3$ transforms to $\mathcal{S}_3'$ under $\underline{f}_3$:

$$
\mathcal{S}_3' = \mathbf{X}_1'\wedge\mathbf{X}_2'\wedge\mathbf{X}_3'\wedge\mathbf{X}_4' = \det\underline{f}_3\mathcal{S}_3.
\tag{12.69}
$$

The ratio of any two 4-vectors is therefore invariant under $\underline{f}_3$, and we must take multiples of these ratios to ensure that the arbitrary scale factors W_i cancel. With five general points, we see that there are five possibilities for forming the combinations $W_iW_jW_kW_l$. It is then a simple matter to show that one cannot consider multiples of ratios such that the W factors cancel. It is, however, possible to do this if we have six points. One example of such an invariant might be

$$
Inv_3 = \frac{(\mathbf{X}_1\wedge\mathbf{X}_2\wedge\mathbf{X}_3\wedge\mathbf{X}_4)I_4^{-1}(\mathbf{X}_4\wedge\mathbf{X}_5\wedge\mathbf{X}_2\wedge\mathbf{X}_6)I_4^{-1}}{(\mathbf{X}_1\wedge\mathbf{X}_2\wedge\mathbf{X}_4\wedge\mathbf{X}_5)I_4^{-1}(\mathbf{X}_3\wedge\mathbf{X}_4\wedge\mathbf{X}_2\wedge\mathbf{X}_6)I_4^{-1}}.
\tag{12.70}
$$

Using the arguments of the previous sections, we can now write

$$
\begin{aligned}
&(\mathbf{X}_1\wedge\mathbf{X}_2\wedge\mathbf{X}_3\wedge\mathbf{X}_4)I_4^{-1} \equiv \\
&\qquad W_1W_2W_3W_4\{(\boldsymbol{x}_2-\boldsymbol{x}_1)\wedge(\boldsymbol{x}_3-\boldsymbol{x}_1)\wedge(\boldsymbol{x}_4-\boldsymbol{x}_1)\}I_3^{-1}.
\end{aligned}
\tag{12.71}
$$

We can therefore see that the invariant Inv_3 is the 3D equivalent of the 1D cross-ratio and consists of ratios of volumes,

$$Inv_3 = \frac{V_{1234}V_{4526}}{V_{1245}V_{3426}}, \tag{12.72}$$

where V_{ijkl} is the volume of the solid formed by the four vertices $\boldsymbol{x}_i, \boldsymbol{x}_j, \boldsymbol{x}_k, \boldsymbol{x}_l$.

Conventionally, all of these invariants are well known, but we have outlined here a general process which is straightforward and simple for generating projective invariants in any dimension.

12.6 Visual Geometry of n-Uncalibrated Cameras

In this section, we will analyze the constraints relating the geometry of n-uncalibrated cameras. First, the pinhole camera model for one view will be defined in terms of lines and planes. Then, for two and three views, the epipolar geometry is defined in terms of bilinear and trilinear constraints. Since the constraints are based on the coplanarity of lines, we will only be able to define relationships expressed by a single tensor for up to four cameras. For more than four cameras, the constraints are linear combinations of bilinearities, trilinearities, and quadrilinearities.

12.6.1 Geometry of One View

We begin with the monocular case depicted in Fig. 12.2. Here, the image plane is defined by a vector basis of three arbitrary non-collinear points $\mathbf{A}_1$, $\mathbf{A}_2$, and $\mathbf{A}_3$, with the optical center given by $\mathbf{A}_0$ (all vectors in R^4). Thus, $\{\mathbf{A}_i\}$ can be used as a coordinate basis for the image plane $\Phi_A = \mathbf{A}_1 \wedge \mathbf{A}_2 \wedge \mathbf{A}_3$, so that any point $\mathbf{A}'$ lying in Φ_A can be written as

$$\mathbf{A}' = \alpha_1 \mathbf{A}_1 + \alpha_2 \mathbf{A}_2 + \alpha_3 \mathbf{A}_3. \tag{12.73}$$

We are also able to define a bivector basis of the image plane $\{L_i^A\}$ spanning the lines in Φ_A:

$$L_1^A = \mathbf{A}_2 \wedge \mathbf{A}_3, \qquad L_2^A = \mathbf{A}_3 \wedge \mathbf{A}_1, \qquad L_3^A = \mathbf{A}_1 \wedge \mathbf{A}_2. \tag{12.74}$$

The bivectors $\{L_i^A\}$ together with the optical center allow us to define three planes, ϕ_i^A, as follows:

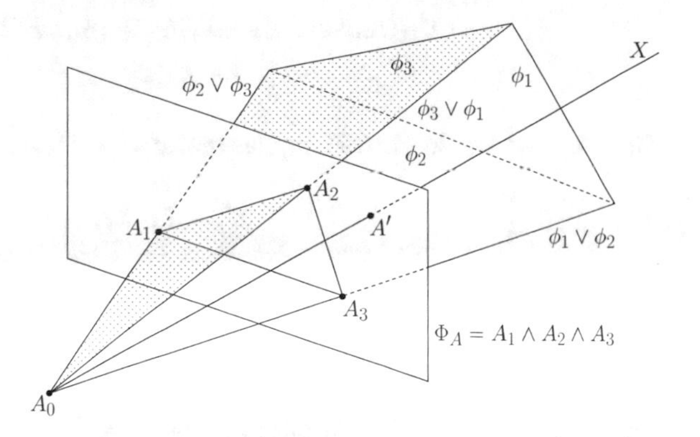

Fig. 12.2 Projection into a single camera: the monocular case

$$\begin{aligned}
\phi_1^A &= \mathbf{A}_0 \wedge \mathbf{A}_2 \wedge \mathbf{A}_3 = \mathbf{A}_0 \wedge L_1^A, \\
\phi_2^A &= \mathbf{A}_0 \wedge \mathbf{A}_3 \wedge \mathbf{A}_1 = \mathbf{A}_0 \wedge L_2^A, \\
\phi_3^A &= \mathbf{A}_0 \wedge \mathbf{A}_1 \wedge \mathbf{A}_2 = \mathbf{A}_0 \wedge L_3^A.
\end{aligned} \tag{12.75}$$

We will call the planes ϕ_j^A *optical planes*. Clearly, each is a trivector and can be written as

$$\phi_j^A = t_{j1}(Ie_1) + t_{j2}(Ie_2) + t_{j3}(Ie_3) + t_{j4}(Ie_4) \equiv t_{jk}(Ie_k), \tag{12.76}$$

since there are four basis trivectors in our 4D space. These optical planes also clearly intersect the image plane in the lines $\{L_j^A\}$. Furthermore, the intersections of the optical planes also define a bivector basis which spans the pencil of *optical rays* (rays passing through the optical center of the camera) in R^4. Thus,

$$\begin{aligned}
L_{A1} &= \phi_2 \cap \phi_3 \equiv \mathbf{A}_0 \wedge \mathbf{A}_1, \\
L_{A2} &= \phi_3 \cap \phi_1 \equiv \mathbf{A}_0 \wedge \mathbf{A}_2, \\
L_{A3} &= \phi_1 \cap \phi_2 \equiv \mathbf{A}_0 \wedge \mathbf{A}_3,
\end{aligned} \tag{12.77}$$

so that any optical ray resulting from projecting a world point $\mathbf{X}$ onto the image plane can be written as

$$\mathbf{A}_0 \wedge \mathbf{X} = x_j L_{Aj}.$$

We can now interpret the camera matrices, used so widely in computer vision applications, in terms of the quantities defined in this section.

The projection of any world point $\mathbf{X}$ onto the image plane is notated $\boldsymbol{x}$ and is given by the intersection of line $\mathbf{A}_0 \wedge \mathbf{X}$ with the plane Φ_A. Thus,

$$\boldsymbol{x} = (\mathbf{A}_0 \wedge \mathbf{X}) \cap (\mathbf{A}_1 \wedge \mathbf{A}_2 \wedge \mathbf{A}_3) = X_\mu \{(\mathbf{A}_0 \wedge e_\mu) \cap (\mathbf{A}_1 \wedge \mathbf{A}_2 \wedge \mathbf{A}_3)\}, \tag{12.78}$$

where μ is summed over 1 to 4. We can now expand the meet given by Eq. (12.78) to get

$$\boldsymbol{x} = X_j\{[\mathbf{A}_0\wedge e_j\wedge\mathbf{A}_2\wedge\mathbf{A}_3]\mathbf{A}_1 + [\mathbf{A}_0\wedge e_j\wedge\mathbf{A}_3\wedge\mathbf{A}_1]\mathbf{A}_2 + [\mathbf{A}_0\wedge e_j\wedge\mathbf{A}_1\wedge\mathbf{A}_2]\mathbf{A}_3\}. \tag{12.79}$$

Since $\boldsymbol{x} = x^k\mathbf{A}_k$, Eq. (12.79) implies that $\boldsymbol{x} = X_j P_{jk}\mathbf{A}_k$ and therefore that

$$x^k = P_{jk}X_j,$$

where

$$P_{jk} = [\mathbf{A}_0\wedge e_j\wedge L_k^A] \equiv [\phi_k\wedge e_j] = -t_{kj}, \tag{12.80}$$

since $Ie_j\wedge e_k = -I\delta_{jk}$. The matrix P takes $\mathbf{X}$ to $\boldsymbol{x}$ and is therefore the standard camera projection matrix. If we define a set of vectors $\{\phi_A^j\}$, $j = 1, 2, 3$, which are the duals of the planes $\{\phi_j^A\}$—that is, $\phi_A^j = \phi_j^A I^{-1}$—it is then easy to see that

$$\phi_A^j = -\phi_j^A I = I\phi_j^A = -[t_{j1}e_1 + t_{j2}e_2 + t_{j3}e_3 + t_{j4}e_4]. \tag{12.81}$$

Thus, we see that the projected point $\boldsymbol{x} = x^j\mathbf{A}_j$ may be given by

$$x^j = \mathbf{X}\cdot\phi_A^j \qquad \text{or} \qquad \boldsymbol{x} = (\mathbf{X}\cdot\phi_A^j)\mathbf{A}_j. \tag{12.82}$$

That is, the coefficients in the image plane are formed by projecting $\mathbf{X}$ onto the vectors formed by taking the duals of the optical planes. This is, of course, equivalent to the matrix formulation

$$\boldsymbol{x} = \begin{bmatrix} x_1 \\ x_2 \\ x_3 \end{bmatrix} = \begin{bmatrix} \phi_A^1 \\ \phi_A^2 \\ \phi_A^3 \end{bmatrix} X = \begin{bmatrix} t_{11}\ t_{12}\ t_{13}\ t_{14} \\ t_{21}\ t_{22}\ t_{23}\ t_{24} \\ t_{31}\ t_{32}\ t_{33}\ t_{34} \end{bmatrix} \begin{bmatrix} X_1 \\ X_2 \\ X_3 \\ X_4 \end{bmatrix} \equiv \boldsymbol{P}\boldsymbol{X}. \tag{12.83}$$

The elements of the camera matrix are therefore simply the coefficients of each optical plane in the coordinate frame of the world point. They encode the intrinsic and extrinsic camera parameters as given in Eq. (12.2).

Next, we consider the projection of world lines in R^4 onto the image plane. Suppose we have a world line $L = \mathbf{X}_1\wedge\mathbf{X}_2$ joining the points $\mathbf{X}_1$ and $\mathbf{X}_2$. If $\boldsymbol{x}_1 = (\mathbf{A}_0\wedge\mathbf{X}_1)\cap\Phi_A$ and $\boldsymbol{x}_2 = (\mathbf{A}_0\wedge\mathbf{X}_2)\cap\Phi_A$ (i.e., the intersections of the optical rays with the image plane), then the projected line in the image plane is clearly given by

$$l = \boldsymbol{x}_1\wedge\boldsymbol{x}_2.$$

Since we can express l in the bivector basis for the plane, we obtain

$$l = l^j L_j^A,$$

where $L_1^A = \mathbf{A}_2 \wedge \mathbf{A}_3$, etc., as defined in Eq. (12.74). From our previous expressions for projections given in Eq. (12.82), we see that we can also write l as follows:

$$l = \boldsymbol{x}_1 \wedge \boldsymbol{x}_2 = (\mathbf{X}_1 \cdot \phi_A^j)(\mathbf{X}_2 \cdot \phi_A^k)\mathbf{A}_j \wedge \mathbf{A}_k \equiv l^p L_p^A, \tag{12.84}$$

which tells us that the *line coefficients* $\{l^j\}$ are

$$\begin{aligned} l^1 &= (\mathbf{X}_1 \cdot \phi_A^2)(\mathbf{X}_2 \cdot \phi_A^3) - (\mathbf{X}_1 \cdot \phi_A^3)(\mathbf{X}_2 \cdot \phi_A^2), \\ l^2 &= (\mathbf{X}_1 \cdot \phi_A^3)(\mathbf{X}_2 \cdot \phi_A^1) - (\mathbf{X}_1 \cdot \phi_A^1)(\mathbf{X}_2 \cdot \phi_A^3), \\ l^3 &= (\mathbf{X}_1 \cdot \phi_A^1)(\mathbf{X}_2 \cdot \phi_A^2) - (\mathbf{X}_1 \cdot \phi_A^2)(\mathbf{X}_2 \cdot \phi_A^1). \end{aligned} \tag{12.85}$$

Using the identity in Eq. (12.36) and utilizing the fact that the join of the duals is the dual of the meet, we are then able to deduce identities of the following form for each l^j:

$$l^1 = (\mathbf{X}_1 \wedge \mathbf{X}_2) \cdot (\phi_A^2 \wedge \phi_A^3) = (\mathbf{X}_1 \wedge \mathbf{X}_2) \cdot (\phi_2^A \cap \phi_3^A)^* = L \cdot (L_1^A)^*.$$

We therefore obtain the general result,

$$l^j = L \cdot (L_j^A)^* \equiv L \cdot L_A^j, \tag{12.86}$$

where we have defined L_A^j to be the dual of L_j^A. Thus, we have once again expressed the projection of a line L onto the image plane by contracting L with the set of lines dual to those formed by intersecting the optical planes.

We can summarize the two results derived here for the projections of points ($\mathbf{X}_1$ and $\mathbf{X}_2$) and lines ($L = \mathbf{X}_1 \wedge \mathbf{X}_2$) onto the image plane:

$$\begin{aligned} \boldsymbol{x}_1 &= (\mathbf{X}_1 \cdot \phi_A^j)\mathbf{A}_j, \qquad \boldsymbol{x}_2 = (\mathbf{X}_2 \cdot \phi_A^j)\mathbf{A}_j, \\ l &= (L \cdot L_A^j)L_j^A \equiv l^k L_k^A. \end{aligned} \tag{12.87}$$

Having formed the sets of dual planes $\{\phi_A^j\}$ and dual lines $\boldsymbol{L}_A^j$ for a given image plane, it is then conceptually very straightforward to project any point or line onto that plane.

If we express the world and image lines as bivectors, $L = \alpha_j e_j + \tilde{\alpha}_j I e_j$ and $L_A^p = \beta_j e_j + \tilde{\beta}_j I e_j$, we can write Eq. (12.87) as a matrix equation:

$$\boldsymbol{l} = \begin{bmatrix} l^1 \\ l^2 \\ l^3 \end{bmatrix} = \begin{bmatrix} u_{11} & u_{12} & u_{13} & u_{14} & u_{15} & u_{16} \\ u_{21} & u_{22} & u_{23} & u_{24} & u_{25} & u_{26} \\ u_{31} & u_{32} & u_{33} & u_{34} & u_{35} & u_{36} \end{bmatrix} \begin{bmatrix} \alpha_1 \\ \alpha_2 \\ \alpha_3 \\ \tilde{\alpha}_1 \\ \tilde{\alpha}_2 \\ \tilde{\alpha}_3 \end{bmatrix} \equiv P_L \bar{\boldsymbol{l}}, \tag{12.88}$$

where $\bar{l}$ is the vector of *Plücker coordinates* $[\alpha_1, \alpha_2, \alpha_3, \tilde{\alpha}_1, \tilde{\alpha}_2, \tilde{\alpha}_3]$ and the matrix P_L contains the β and $\tilde{beta}$'s, that is, information about the camera configuration.

When we back-project a point $\boldsymbol{x}$ or line l in the image plane, we produce their duals, that is, a line l_x or a plane ϕ_l, respectively. These back-projected lines and planes are given by the following expressions:

$$l_x = \mathbf{A}_0 \wedge \boldsymbol{x} = (\mathbf{X}\cdot\phi_A^j)\mathbf{A}_0 \wedge \mathbf{A}_j = (\mathbf{X}\cdot\phi_A^j)\boldsymbol{L}_j^A, \tag{12.89}$$

$$\phi_l = \mathbf{A}_0 \wedge l = (L\cdot L_A^j)\mathbf{A}_0 \wedge L_j^A = (L\cdot L_A^j)\phi_j^A. \tag{12.90}$$

12.6.2 Geometry of Two Views

In this and subsequent sections, we will work in projective space R^4, although a return to 3D Euclidean space will be necessary when we discuss invariants in terms of image coordinates; this will be done via the projective split. Figure 12.3 shows a world point $\mathbf{X}$ projecting onto points $\mathbf{A}'$ and $\mathbf{B}'$ in the two image planes ϕ_A and ϕ_B, respectively.

The so-called epipoles $\boldsymbol{E}_{AB}$ and $\boldsymbol{E}_{BA}$ correspond to the intersections of the line joiningthe optical centers with the image planes. Since the points $\mathbf{A}_0$, $\mathbf{B}_0$, $\mathbf{A}'$, $\mathbf{B}'$ are

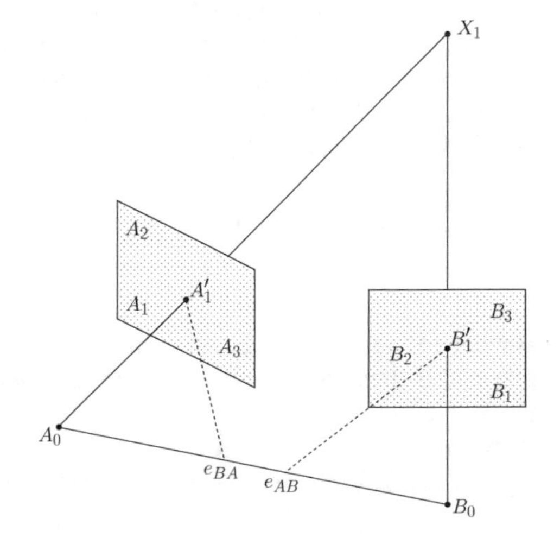

Fig. 12.3 Sketch of binocular projection of a world point

coplanar, we can formulate the bilinear constraint by taking advantage of the fact that the outer product of these four vectors must disappear. Thus,

$$\mathbf{A}_0 \wedge \mathbf{B}_0 \wedge \mathbf{A}' \wedge \mathbf{B}' = 0. \tag{12.91}$$

Now, if we let $\mathbf{A}' = \alpha_i \mathbf{A}_i$ and $\mathbf{B}' = \beta_j \mathbf{B}_j$, then Eq. (12.91) can be written as

$$\alpha_i \beta_j \{\mathbf{A}_0 \wedge \mathbf{B}_0 \wedge \mathbf{A}_i \wedge \mathbf{B}_j\} = 0. \tag{12.92}$$

Defining $\tilde{F}_{ij} = \{\mathbf{A}_0 \wedge \mathbf{B}_0 \wedge \mathbf{A}_i \wedge \mathbf{B}_j\} I^{-1} \equiv [\mathbf{A}_0 \mathbf{B}_0 \mathbf{A}_i \mathbf{B}_j]$ gives us

$$\tilde{F}_{ij} \alpha_i \beta_j = 0, \tag{12.93}$$

which corresponds in R^4 to the well-known relationship between the components of the *fundamental matrix* [199] or the *bilinear constraint* in E^3, F, and the image coordinates [199]. This suggests that $\tilde{F}$ can be seen as a linear function mapping two vectors onto a scalar:

$$\tilde{F}(\mathbf{A}, \mathbf{B}) = \{\mathbf{A}_0 \wedge \mathbf{B}_0 \wedge \mathbf{A} \wedge \mathbf{B}\} I^{-1}, \tag{12.94}$$

so that $\tilde{F}_{ij} = \tilde{F}(\mathbf{A}_i, \mathbf{B}_j)$. Note that viewing the fundamental matrix as a linear function means that we have a coordinate-independent description. Now, if we use the projective split to associate our point $\mathbf{A}' = \alpha_i \mathbf{A}_i$ in the image plane with its $\mathbb{E}^3$ representation $\boldsymbol{a}' = \delta_i \boldsymbol{a}_i$, where $\boldsymbol{a}_i = \frac{\mathbf{A}_i \wedge e_4}{\mathbf{A}_i \cdot e_4}$, it is not difficult to see that the coefficients are expressed as follows:

$$\alpha_i = \frac{\mathbf{A}' \cdot e_4}{\mathbf{A}_i \cdot e_4} \delta_i. \tag{12.95}$$

Thus, we are able to relate our 4D fundamental matrix $\tilde{F}$ to an *observed* fundamental matrix F in the following manner:

$$\tilde{F}_{kl} = (\mathbf{A}_k \cdot e_4)(\mathbf{B}_l \cdot e_4) F_{kl}, \tag{12.96}$$

so that

$$\alpha_k \tilde{F}_{kl} \beta_l = (\mathbf{A}' \cdot e_4)(\mathbf{B}' \cdot e_4) \delta_k F_{kl} \epsilon_l, \tag{12.97}$$

where $\boldsymbol{b}' = \epsilon_i \boldsymbol{b}_i$, with $\boldsymbol{b}_i = \frac{\mathbf{B}_i \wedge e_4}{\mathbf{B}_i \cdot e_4}$. F is the standard fundamental matrix that we would form from observations.

12.6.3 Geometry of Three Views

The so-called trilinear constraint captures the geometric relationships existing between points and lines in three camera views. Figure 12.4 shows three image planes ϕ_A, ϕ_B, and ϕ_C with bases $\{\mathbf{A}_i\}$, $\{\mathbf{B}_i\}$, and $\{\mathbf{C}_i\}$ and optical centers $\mathbf{A}_0$, $\mathbf{B}_0$, $\mathbf{C}_0$. Intersections of two world points $\mathbf{X}_i$ with the planes occur at points $\mathbf{A}'_i$, $\mathbf{B}'_i$, $\mathbf{C}'_i$, $i = 1, 2$. The line joining the world points is $L_{12} = \mathbf{X}_1 \wedge \mathbf{X}_2$, and the projected lines are denoted by L'_A, L'_B, and L'_C.

We first define three planes:

$$\Phi'_A = \mathbf{A}_0 \wedge \mathbf{A}'_1 \wedge \mathbf{A}'_2, \quad \Phi'_B = \mathbf{B}_0 \wedge \mathbf{B}'_1 \wedge \mathbf{B}'_2, \quad \Phi'_C = \mathbf{C}_0 \wedge \mathbf{C}'_1 \wedge \mathbf{C}'_2. \tag{12.98}$$

It is clear that L_{12} can be formed by intersecting Φ'_B and Φ'_C:

$$L_{12} = \Phi'_B \cap \Phi'_C = (\mathbf{B}_0 \wedge L'_B) \cap (\mathbf{C}_0 \wedge L'_C). \tag{12.99}$$

If $L_{A_1} = \mathbf{A}_0 \wedge \mathbf{A}'_1$ and $L_{A_2} = \mathbf{A}_0 \wedge \mathbf{A}'_2$, then we can easily see that L_1 and L_2 intersect with L_{12} at $\mathbf{X}_1$ and $\mathbf{X}_2$, respectively. We therefore have

$$L_{A_1} \wedge L_{12} = 0 \qquad \text{and} \qquad L_{A_2} \wedge L_{12} = 0, \tag{12.100}$$

which can then be written as

$$(\mathbf{A}_0 \wedge \mathbf{A}'_i) \wedge \{(\mathbf{B}_0 \wedge L'_B) \cap (\mathbf{C}_0 \wedge L'_C)\} = 0 \quad \text{for } i = 1, 2. \tag{12.101}$$

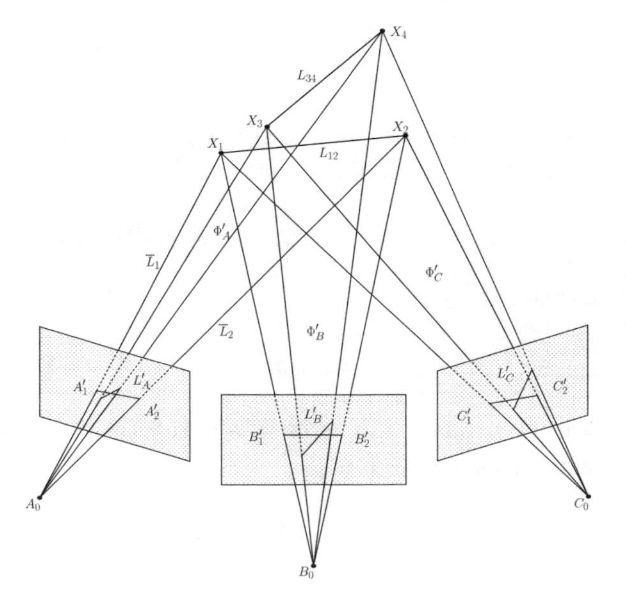

Fig. 12.4 Model of the trinocular projection of the visual 3D space

This suggests that we should define a linear function T which maps a point and two lines onto a scalar as follows:

$$T(\mathbf{A}', L'_B, L'_C) = (\mathbf{A}_0 \wedge \mathbf{A}') \wedge \{(\mathbf{B}_0 \wedge L'_B) \cap (\mathbf{C}_0 \wedge L'_C)\}. \tag{12.102}$$

Now, using the line bases of the planes B and C in a similar manner as was used for plane A in Eq. (12.74), we can write

$$\mathbf{A}' = \alpha_i \mathbf{A}_i, \quad L'_B = l^B_j L^B_j, \quad L'_C = l^C_k L^C_k. \tag{12.103}$$

If we define the components of a tensor as $T_{ijk} = T(\mathbf{A}_i, L^B_j, L^C_k)$, and if $\mathbf{A}'$, L'_B, and L'_C are all derived from projections of the same two world points, then Eq. (12.101) tells us that we can write

$$T_{ijk} \alpha_i l^B_j l^C_k = 0. \tag{12.104}$$

T is the *trifocal tensor* [125, 270] and Eq. (12.104) is the *trilinear constraint*. In [122, 270] this constraint was arrived at by consideration of camera matrices; here, however, Eq. (12.104) is arrived at from purely geometric considerations, namely, that two planes intersect in a line, which in turn intersects with another line. To see how we relate the three projected *lines*, we express the line in image plane ϕ_A joining $\mathbf{A}'_1$ and $\mathbf{A}'_2$ as the intersection of the plane joining $\mathbf{A}_0$ to the world line L_{12} with the image plane $\Phi_A = \mathbf{A}_1 \wedge \mathbf{A}_2 \wedge \mathbf{A}_3$:

$$L'_A = \mathbf{A}'_1 \wedge \mathbf{A}'_2 = (\mathbf{A}_0 \wedge L_{12}) \cap \Phi_A. \tag{12.105}$$

Considering L_{12} as the meet of the planes $\Phi'_B \cap \Phi'_C$ and using the expansions of L'_A, L'_B, and L'_C given in Eq. (12.103), we can rewrite this equation as

$$l^A_i L^A_i = \Big((\mathbf{A}_0 \wedge \mathbf{A}_i) \wedge l^B_j l^C_k \{(\mathbf{B}_0 \wedge L^B_j) \cap (\mathbf{C}_0 \wedge L^C_k)\} \Big) \cap \Phi_A. \tag{12.106}$$

Using the expansion of the meet given in Eq. (12.48), we have

$$l^A_i L^A_i = [(\mathbf{A}_0 \wedge \mathbf{A}_i) \wedge l^B_j l^C_k \{(\mathbf{B}_0 \wedge L^B_j) \cap (\mathbf{C}_0 \wedge L^C_k)\}] L^A_i, \tag{12.107}$$

which, when we equate coefficients, gives

$$l^A_i = T_{ijk} l^B_j l^C_k. \tag{12.108}$$

Thus, we obtain the familiar equation which relates the projected lines in the three views.

12.6.4 Geometry of n-Views

If we have n-views, let us choose four of these views and denote them by A, B, C, and N. As before, we assume that $\{\mathbf{A}_j\}$, $\{\mathbf{B}_j\}$, ... etc., $j = 1, 2, 3$ define the image planes (Fig. 12.5).

Let $\Phi_{Ai} = \mathbf{A}_0 \wedge \mathbf{A}_i \wedge \mathbf{A}'$, $\Phi_{Bi} = \mathbf{B}_0 \wedge \mathbf{B}_i \wedge \mathbf{B}'$, etc., where $\mathbf{A}'$, $\mathbf{B}'$, etc., are the projections of a world point P onto the image planes. The expression $\Phi_{Aj} \vee \Phi_{Bk}$ represents a line passing through the world point P, as does the equation $\Phi_{Cl} \cap \Phi_{Nm}$. Since these two lines intersect, we have the condition

$$\{\Phi_A j \cap \Phi_B k\} \wedge \{\Phi_C l \cap \Phi_N m\} = 0. \tag{12.109}$$

Consider also the world line $L = \mathbf{X}_1 \wedge \mathbf{X}_2$ which projects down to l_a, l_b, l_c, l_n in the four image planes. We know from the previous sections that it is possible to write L in terms of these image lines as the meet of two planes in various ways—for example,

$$L = (\mathbf{A}_0 \wedge l_a) \cap (\mathbf{B}_0 \wedge l_b), \tag{12.110}$$

$$L = (\mathbf{C}_0 \wedge l_c) \cap (N_0 \wedge l_n). \tag{12.111}$$

Now, since $L \wedge L = 0$, we can consider $l_a = \ell_a^i L_i^A$, etc., and then write

$$\ell_a^i \ell_b^j \ell_c^k \ell_n^m [(\mathbf{A}_0 \wedge L_i^A) \cap (\mathbf{B}_0 \wedge L_j^B)] \wedge [(\mathbf{C}_0 \wedge L_k^C) \cap (N_0 \wedge L_m^N)] = 0, \tag{12.112}$$

which can be further expressed as

$$\ell_a^i \ell_b^j \ell_c^k \ell_n^m Q_{ijkm} = 0. \tag{12.113}$$

Here, Q is the so-called *quadrifocal tensor* and Eq. 12.113 is the quadrilinear constraint recently discussed in [125]. The above constraint in terms of lines is straightforward, but it is also possible to find a relationship between point coordinates

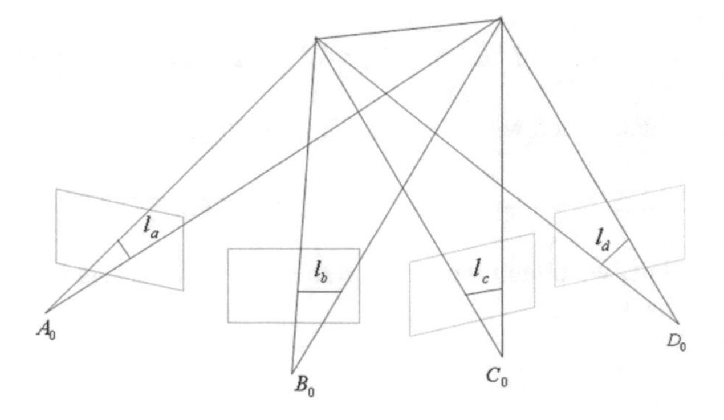

Fig. 12.5 Model of the tetraocular projection of the visual 3D space

and Q. To do this, we expand Eq. (12.109) as follows:

$$\alpha_r \beta_s \delta_t \eta_u \{[(\mathbf{A}_0 \wedge L^A_{jr}) \cap (\mathbf{B}_0 \wedge L^B_{ks})] \wedge [(\mathbf{C}_0 \wedge L^C_{lt}) \cap (N_0 \wedge L^N_{mu})]\} = 0, \quad (12.114)$$

where we have used the notation $L^A_{jr} = \mathbf{A}_j \wedge \mathbf{A}_r \equiv \epsilon_{ijr} L^A_i$. Thus, we can also write the above equation as

$$\alpha_r \beta_s \delta_t \eta_u \epsilon_{i_1 jr} \epsilon_{i_2 ks} \epsilon_{i_3 lt} \epsilon_{i_4 mu} Q_{i_1 i_2 i_3 i_4} = 0, \quad (12.115)$$

for any $\{i, j, k, m\}$.

12.7 Calibrated Camera Model and Stereo Vision Using the Conformal Geometric Algebra Framework

In Sect. 12.2, we have used projective geometry to model the n-view geometry as we considered the case of an uncalibrated camera. Nowadays, there are off-line robust methods to calibrate a camera. The computed intrinsic parameters do not change even if the camera fastened to an end-effector suffer of mechanical vibrations. Thus, after calibration, what remains is to estimate the rigid motion $SE(3)$. This simplifies the modeling of the camera which can be done easily in the conformal geometric algebra.

To introduce the basic geometric models in computer vision, we consider the imaging of a point $\mathbf{X} \in R^4$ into a point $\mathbf{x} \in \mathbf{R}^3$. The optical center C of the camera may be different from the origin of the world coordinate system O, as depicted in Fig. 12.6a.

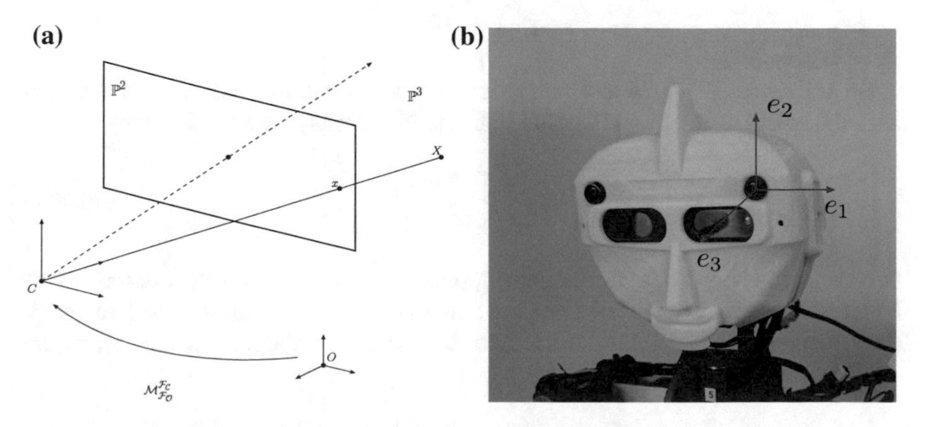

Fig. 12.6 **a** Pan–tilt unit. **b** The frame reference is fixed in left camera

In the standard matrix representation, the mapping $P : \mathbf{X} \longrightarrow \boldsymbol{x}$ is expressed by the homogeneous transformation matrix

$$P = \begin{bmatrix} p_{11} & p_{12} & p_{13} & p_{14} \\ p_{21} & p_{22} & p_{23} & p_{14} \\ p_{31} & p_{32} & p_{33} & p_{34} \end{bmatrix}, \tag{12.116}$$

which may be decomposed into a product of three matrices,

$$P = K P_0 M_0^c, \tag{12.117}$$

where P_0, K, and M_0^c will now be defined. P_0 is the 3×4 matrix,

$$P_0 = \begin{bmatrix} 1 & 0 & 0 & 0 \\ 0 & 1 & 0 & 0 \\ 0 & 0 & 1 & 0 \end{bmatrix}, \tag{12.118}$$

The 3×3 matrix K expresses the assumed camera model as an affine transformation between the camera plane and the image coordinate system, so that K is an upper triangular matrix. In the case of the perspective (or pinhole) camera, the matrix K, which we now call K_p and it is given by

$$K_p = \begin{bmatrix} \alpha_u & \gamma & u_0 \\ 0 & \alpha_v & v_0 \\ 0 & 0 & 1 \end{bmatrix}. \tag{12.119}$$

After calibration, the matrix K_p becomes the identity and the projective mapping, and Eq. (12.117) is simplified to

$$P = P_0 M_0^c, \tag{12.120}$$

where M_0^c represents the 4×4 matrix containing the rotation and translation which takes the world frame $\mathcal{F}_0$ to the camera frame $\mathcal{F}_c$, and is given explicitly by

$$M_0^c = \begin{bmatrix} R & \boldsymbol{t} \\ \mathbf{0}^T & 1 \end{bmatrix}. \tag{12.121}$$

This Euclidean transformation or rigid motion corresponds to $SE(3)$ described by the *extrinsic parameters* of rotation (3×3 matrix R) and translation (3×1 vector $\boldsymbol{t}$). Thus, the projective mapping consists of a stack of three orthogonal planes π_1, π_2, π_3 also called optical and it reads

$$P = [\mathtt{R}|\mathtt{t}] = \begin{bmatrix} r_{11} & r_{12} & r_{13} & t_{14} \\ r_{21} & r_{22} & r_{23} & t_{14} \\ r_{31} & r_{32} & r_{33} & t_{34} \end{bmatrix} = \begin{bmatrix} \mathtt{n}_1 & h_1 \\ \mathtt{n}_2 & h_2 \\ \mathtt{n}_3 & h_3 \end{bmatrix} = \begin{bmatrix} \pi_1 \\ \pi_2 \\ \pi_3 \end{bmatrix}, \tag{12.122}$$

where for $i_1, ..., 3$ $\mathtt{n_i}$ and h_i are the plane normals and the plane Hesse distances, respectively. The mapping given by the matrix P projects down homogeneous points from 4D to 3D and represents a projection from homogeneous coordinates of space to homogeneous coordinates of the image plane as follows

$$\mathbf{x} = \begin{bmatrix} x_1 \\ x_2 \\ x_3 \end{bmatrix} = P\mathbf{X} = [\mathrm{R}|\mathtt{t}]\,\mathbf{X} = \begin{bmatrix} r_{11}\ r_{12}\ r_{13}\ t_{14} \\ r_{21}\ r_{22}\ r_{23}\ t_{14} \\ r_{31}\ r_{32}\ r_{33}\ t_{34} \end{bmatrix} \begin{bmatrix} X_1 \\ X_2 \\ X_3 \\ X_4 \end{bmatrix},$$
$$= \begin{bmatrix} \mathtt{n}_1\ h_1 \\ \mathtt{n}_2\ h_2 \\ \mathtt{n}_3\ h_3 \end{bmatrix} \mathbf{X} = \begin{bmatrix} \pi_1 \\ \pi_2 \\ \pi_3 \end{bmatrix} \mathbf{X} = \begin{bmatrix} \pi_1 \cdot \mathbf{X} \\ \pi_2 \cdot \mathbf{X} \\ \pi_3 \cdot \mathbf{X} \end{bmatrix}, \qquad (12.123)$$

where the point $\mathbf{X}$ is projected on the optical planes π_1, π_2, π_3.

The mapping of projective lines down from 4D to 3D is given by

$$\mathbf{l} = \begin{bmatrix} l_1 \\ l_2 \\ l_3 \end{bmatrix} = P_L\mathbf{L} = \begin{bmatrix} l_{11}\ l_{12}\ l_{13}\ l_{14}\ l_{15}\ l_{16} \\ l_{21}\ l_{22}\ l_{23}\ l_{24}\ l_{25}\ l_{26} \\ l_{31}\ l_{32}\ l_{33}\ l_{34}\ l_{35}\ l_{36} \end{bmatrix} \begin{bmatrix} l_1 \\ l_2 \\ l_3 \\ l_4 \\ l_3 \\ l_4 \end{bmatrix},$$
$$= \begin{bmatrix} \begin{bmatrix} \mathtt{n}_1 \\ \mathtt{m}_1 \end{bmatrix} \cdot \begin{bmatrix} \mathtt{n} \\ \mathtt{m} \end{bmatrix} \\ \begin{bmatrix} \mathtt{n}_2 \\ \mathtt{m}_2 \end{bmatrix} \cdot \begin{bmatrix} \mathtt{n} \\ \mathtt{m} \end{bmatrix} \\ \begin{bmatrix} \mathtt{n}_3 \\ \mathtt{m}_3 \end{bmatrix} \cdot \begin{bmatrix} \mathtt{n} \\ \mathtt{m} \end{bmatrix} \end{bmatrix} = \begin{bmatrix} \mathtt{L}_1 \cdot \mathbf{L} \\ \mathtt{L}_2 \cdot \mathbf{L} \\ \mathtt{L}_3 \cdot \mathbf{L} \end{bmatrix}, \qquad (12.124)$$

where $\mathtt{L1}, \mathtt{L}_2, \mathtt{L}_3$ are the so-called *optical lines* expressed in Plücker coordinates which correspond to the intersection by pairs of the optical planes π_1, π_2, π_3. Note that a line can be also written in terms of the orientation and momentum of the line, i.e., $\mathtt{L_i} = [\mathtt{n_i}, \mathtt{m_i}]^{\mathtt{T}}$.

Since in the conformal geometric algebras of the 3D space and image plane $G_{4,1}$ and $G_{3,1}$, respectively, we cannot formulate the projective Lie groups like in $G_{3,3}$ as shown in Sect. 5.7, we will simply compute these mappings, in the affine plane and not on the horosphere, using the inner products between optical planes, $\pi_i + e_\infty d_i$ for $i = 1, ..., 3$ and points $\boldsymbol{x}_a = \boldsymbol{x}_e + e_0$, and between optical lines, $\boldsymbol{L}_i = \boldsymbol{n}_i + e_\infty \boldsymbol{m}_i$ for $i = 1, ..., 3$ and lines $\boldsymbol{L} = \boldsymbol{n} + e_\infty \boldsymbol{m}$ as follows

$$\boldsymbol{x}_a = \begin{bmatrix} \pi_1 \cdot \boldsymbol{x}_a \\ \pi_2 \cdot \boldsymbol{x}_a \\ \pi_3 \cdot \boldsymbol{x}_a \end{bmatrix}, \tag{12.125}$$
$$\boldsymbol{l}_a = \begin{bmatrix} \boldsymbol{L}_1 \cdot \boldsymbol{L} \\ \boldsymbol{L}_2 \cdot \boldsymbol{L} \\ \boldsymbol{L}_3 \cdot \boldsymbol{L} \end{bmatrix}.$$

Note that the computing in the affine planes is equivalent as the computing of the projective mappings of points and lines from the projective space to the image plane using $G_{3,1}$ and $G_{3,0}$.

12.7.1 Stereo and Detection of the 3D Symmetry Line of an Object

In a stereo camera rig, the cameras are calibrated, i.e., the camera calibration matrices left K_l and right K_r and the rigid transformation between both cameras are known, thus computing the correspondence between the image plane points, we can compute the disparity map straightforwardly.

Once the image segmentation is obtained in each camera, it is converted to an edge image using an appropriate edge filter, and after that, the fast reflectional symmetry detection [189] is applied, obtaining a parameterization of the line given by its angle and the radius or distance perpendicular to the line symmetry (θ, r) (see Fig. 12.7). More practical details about this procedure are given in Sect. 16.7.

Using the intrinsic parameters obtained from camera calibration (from both cameras), the line symmetry is transformed from the image coordinates to camera coordinates, i.e., from $\mathbb{R}^2 \rightarrow \mathbb{R}^3$ and then create the line in conformal space as follows

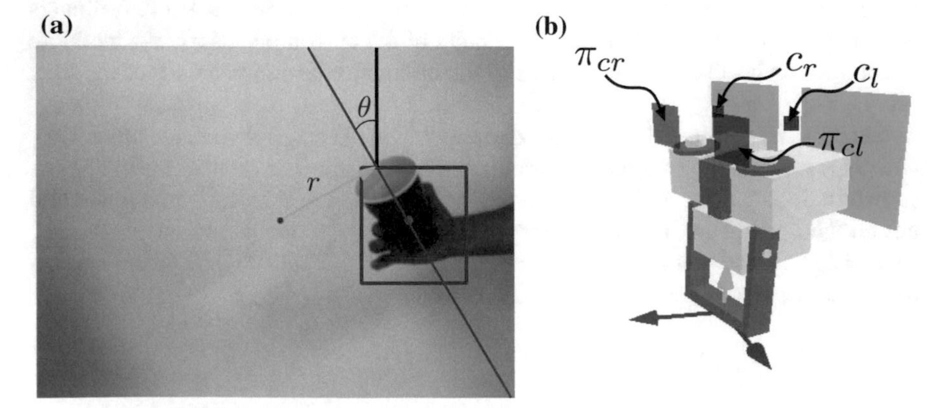

Fig. 12.7 **a** Parameters obtained from fast reflectional symmetry detection. **b** Geometric entities in pan–tilt unit

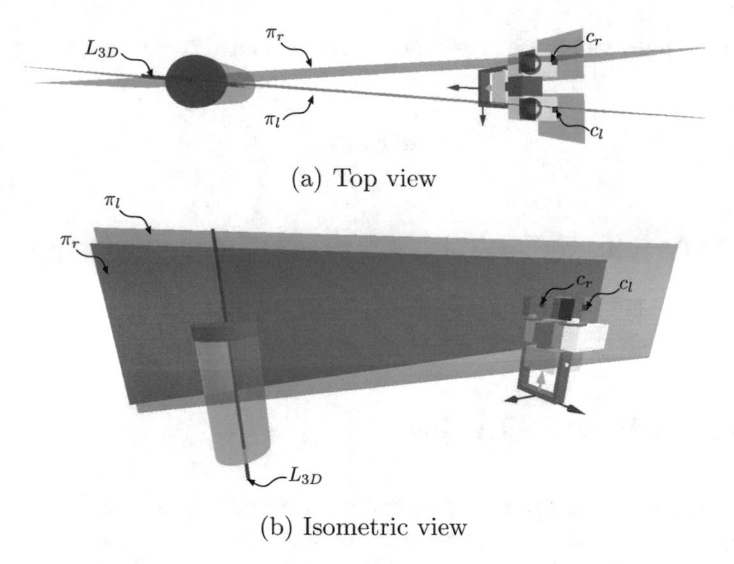

Fig. 12.8 Schematic representation of the estimation of the 3D line symmetry of a cylinder

$$L = \cos(\theta)e_{23} + \sin(\theta)e_{31} + re_3 \wedge e_\infty - fe_\infty(\cos(\theta)e_2 - \sin(\theta)e_1). \quad (12.126)$$

where f is the focal distance from the origin of the coordinate frame of the camera and the camera plane.

Notice that from Eq. 12.126, the line L lies in a parallel plane to xy plane. Since a rigid transformation $[R, t]$ relates both cameras, we need to define the line in both camera planes π_{cl} and π_{cr} (Fig. 12.7b). For that, a motor M is constructed with this transformation that relates the cameras. The lines are then defined as follow:

$$L_l = L, \quad (12.127)$$

$$L_r = ML\widetilde{M}, \quad (12.128)$$

where L_l is the line of the left camera plane π_{cl} and L_r is from the right camera plane π_{cr}. Figure 12.8 shows a general scheme of the presented idea.

To obtain the 3D line symmetry of an object, it is necessary to get the points at camera center which can be calculated from calibration matrix. Without loss of generality, it is possible to define the camera center of the left camera c_l as the origin and the right camera center c_r as a rigid transformation defined before applied to the point at the origin, i.e.,

$$c_l = e_0, \quad (12.129)$$

$$c_r = Me_0\widetilde{M}. \quad (12.130)$$

With the lines and the image center points obtained, we create two planes as follow

$$\pi_l^* = c_l \wedge L_l^*, \tag{12.131}$$
$$\pi_r^* = c_r \wedge L_r^*. \tag{12.132}$$

Finally, the symmetry line in 3D L_{3D} is created by intersecting the planes

$$L_{3D} = \pi_l \wedge \pi_r. \tag{12.133}$$

12.8 Omnidirectional Vision

We know that traditional perspective cameras have a narrow field of view. One effective way to increase the visual field is to use a catadioptric sensor, which consists of a conventional camera and a convex mirror [33]. In order to be able to model the catadioptric sensor geometrically, it must satisfy the restriction that all the measurements of light intensity pass through only one point in space (effective viewpoint). The complete class of mirrors that satisfy such restriction was analyzed by Baker and Nayar [9].

In [105], an unifying theory for central catadioptric systems was introduced. They showed that central catadioptric projection is equivalent to a projective mapping from a sphere to a plane. This is done by projecting the points on the unit sphere (centered at the origin) with respect to the point $\mathbf{N} = (0, 0, \lambda)$ onto a plane orthogonal to the Z-axis and with Hesse distance μ. The parameters λ and μ are in function of the mirror parameters p and d (see Table 12.1), where the *latus rectum* or focal chord [42] is $4p$ and d is the distance between the two focal points.

Table 12.1 Mirror parameters λ and μ of the unified catadioptric projection

Mirror	λ	μ
Parabolic	1	$2p - 1$
Hyperbolic	$\frac{1}{\sqrt{d^2+4p^2}}$	$\frac{d(1-2p)}{\sqrt{d^2+4p^2}}$
Elliptical	$\frac{1}{\sqrt{d^2+4p^2}}$	$\frac{d(1-2p)}{\sqrt{d^2+4p^2}}$

12.8.1 Omnidirectional Vision and Geometric Algebra

The central catadioptric projection in terms of conformal geometric algebra was first introduced in [24]. In this work, the authors showed how the unified theory of catadioptric projection can be handled effectively and easily by the conformal geometric algebra framework.

For the catadioptric image formation, we only need three entities (see Fig. 12.9). The first one is a unit sphere S (not necessarily centered at the origin of the coordinate system). The second one is a point N, which is at a distance λ from the sphere center. Finally, the third entity is a plane Π, which is orthogonal to the line $S \wedge N \wedge e$ and at a distance μ from the sphere center. Observe that this is a more general definition of the unified model. Recall that the plane equation $\Pi = \hat{\mathbf{n}} + \delta e$ has two unknowns: the vector $\mathbf{n} \in \mathbb{R}^3$ and the scalar δ. The vector $\mathbf{n}$ can be extracted from the orthogonal line to the plane; thus,

$$\mathbf{n} = (S \wedge N \wedge e) \cdot I_3^{-1} , \tag{12.134}$$

and $\widehat{\mathbf{n}} = \mathbf{n}/|\mathbf{n}|$. The distance from the sphere (center) to the plane can be calculated with $S \cdot \Pi$, since we know that the distance from the sphere to the plane is μ; then, we have $S \cdot \Pi = \mu$. Thus,

$$S \cdot \Pi = (\mathbf{c} + \frac{1}{2}(\mathbf{c}^2 - \rho^2)e + e_0) \cdot (\widehat{\mathbf{n}} + \delta e) = \mathbf{c} \cdot \widehat{\mathbf{n}} - \delta = \mu , \tag{12.135}$$

and then $\delta = \mathbf{c} \cdot \widehat{\mathbf{n}} - \mu$. Therefore, the equation of the plane is

$$\Pi = \widehat{\mathbf{n}} + \delta e = \widehat{\mathbf{n}} + (\mathbf{c} \cdot \widehat{\mathbf{n}} - \mu)e = \widehat{\mathbf{n}} + (S \cdot \widehat{\mathbf{n}} - \mu)e. \tag{12.136}$$

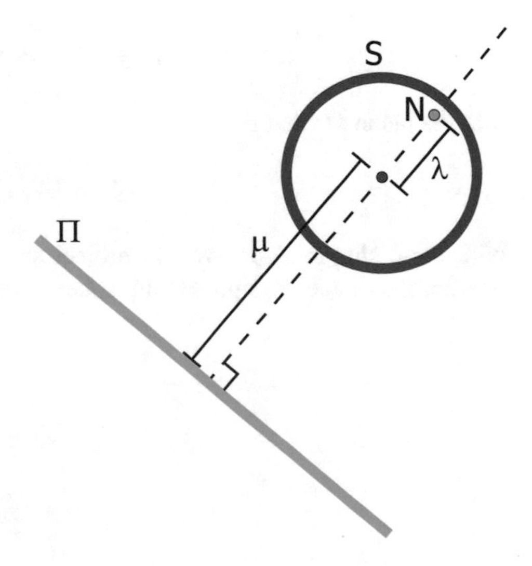

Fig. 12.9 Catadioptric unifying model expressed with conformal geometric algebra entities

An interesting thing that we must note in the definition of this model is that we never talk about any coordinate system. This is because the conformal geometric algebra is a free coordinate system framework, and it allows us to define our model without referring to any other.

In the next section, we will see how points in the space are projected to the catadioptric image through this model.

12.8.2 Point Projection

A point $\mathbf{x} \in \mathbb{R}^3$, represented with X in the conformal space, is projected to a point Y on the catadioptric image in a two-step projection. The first step is the projection of the point X onto the sphere S; this means that we must find the line on which the point X and the center of the sphere lie. This can easily be done using the line

$$L_1^* = X \wedge S \wedge e_\infty \,, \tag{12.137}$$

and then the intersection of the line with the sphere is

$$Z^* = S \cdot L_1^* \,, \tag{12.138}$$

which is a *point pair* ($Z^* = P_1 \wedge P_2$). From it, we take the nearest point with

$$P_1 = \frac{Z^* + |Z^*|}{Z^* \cdot e} \,. \tag{12.139}$$

The second step is the projection of the point P_1 to the catadioptric image plane. This is done by intersecting the line

$$L_2^* = N \wedge P_1 \wedge e_\infty, \tag{12.140}$$

with the plane Π; that is,

$$Q = L_2^* \cdot \Pi \,. \tag{12.141}$$

With these simple steps, we can project any point in the 3D visual space to the catadioptric image through the unit sphere (see Fig. 12.10).

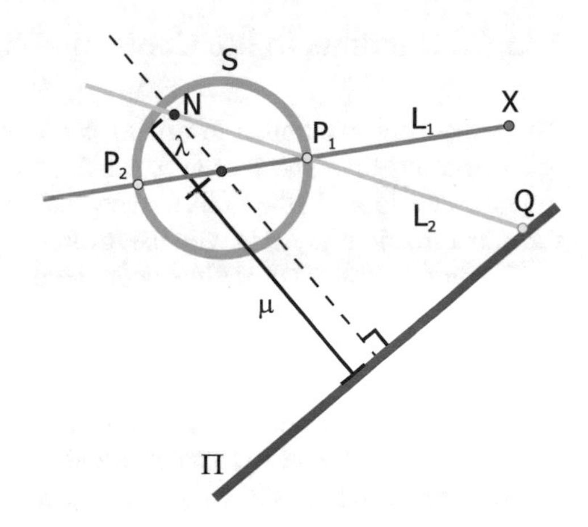

Fig. 12.10 Point projection in the catadioptric unify model, which is expressed in the conformal geometric algebra framework

12.8.3 Inverse Point Projection

In the previous section, we saw how a point in space is projected onto the catadioptric image. Now, in this section, we will see the inverse process: Given a point on the catadioptric image, how can we recover the original 3D point? First, let Q be a point in the catadioptric image; then, the first step is the projection of the point Q onto the sphere S. This is done by intersecting the line

$$L_2^* = Q \wedge N \wedge e_\infty, \tag{12.142}$$

with the sphere S, that is,

$$Z = L_2^* \cdot S\,. \tag{12.143}$$

From the *point pair* Z, we extract the point P_1 using

$$P_1 = \frac{Z^* + |Z^*|}{Z^* \cdot e}\,, \tag{12.144}$$

which is nearest to Q. The second step is to find the original line L_1^* with

$$L_1^* = S \wedge P_1 \wedge e_\infty\,. \tag{12.145}$$

The original 3D point X lies on the line L_1^*, but the exact point cannot be found because a single view does not allow us to know the projective depth.

12.9 Invariants in the Conformal Space

The previously mentioned invariants can also be calculated using the conformal geometric algebra. The reason to do this is because our omnidirectional system model is developed in the CGA framework. Thus, it would be nice if we could relate the omnidirectional system with the invariants theory.

To compute the 1D cross-invariant, we first embed the 2D point $\mathbf{x} \in \mathcal{G}_2$ with the conformal point $X \in \mathcal{G}_{3,1}$. Now, recall that the outer product of r conformal points can be described with (8.81). If you observe the $\mathbf{A}_r$ term (8.82), you will see that it encodes the outer product of the Euclidean points embedded by the conformal points. Therefore, the outer product of r conformal points, besides representing geometric objects, can also be used to calculate projective invariants. Consider the outer product of two conformal points $X_1, X_2 \in \mathcal{G}_{3,1}$; that is,

$$\begin{aligned} X_1 \wedge X_2 &= \mathbf{A}_r - \mathbf{A}_r^+ e_0 - \frac{1}{2}\mathbf{A}_r^- e_\infty - \frac{1}{2}\mathbf{A}_r^\pm E, \qquad (12.146)\\ &= (\mathbf{x}_1 \wedge \mathbf{x}_2) - (\mathbf{x}_2 - \mathbf{x}_1)e_0 - \frac{1}{2}(\mathbf{x}_1^2\mathbf{x}_2 - \mathbf{x}_2^2\mathbf{x}_1)e_\infty - \frac{1}{2}(\mathbf{x}_2^2 - \mathbf{x}_1^2)E. \end{aligned}$$

Note that it represents a *point pair* or 1D sphere. Also note that the term $\mathbf{A}_r$ contains the outer product of the Euclidean points $\mathbf{x}_1$ and $\mathbf{x}_2$. Now, to extract the $\mathbf{A}_r$ term from a conformal geometric entity, we can do the following:

$$\begin{aligned} X_1 \wedge X_2 \wedge E &= \mathbf{A}_r \wedge E - \mathbf{A}_r^+ e_0 \wedge E - \frac{1}{2}\mathbf{A}_r^- e_\infty \wedge E - \frac{1}{2}\mathbf{A}_r^\pm E \wedge E,\\ &= \mathbf{A}_r \wedge E\ , \qquad (12.147) \end{aligned}$$

since $e_0 \wedge E = e_\infty \wedge E = E \wedge E = 0$. Thus, the outer product of a conformal geometric entity with E gives us the term $\mathbf{A}_r = \mathbf{x}_1 \wedge \mathbf{x}_2 = \delta e_1 e_2$ multiplied by E, which is

$$\mathbf{A}_r \wedge E = \mathbf{x}_1 \wedge \mathbf{x}_2 \wedge E = \delta e_1 e_2 e_+ e_- = \delta e_{12+-}\ , \qquad (12.148)$$

where $I_{3,1} = e_{12+-}$ is the pseudoscalar of $\mathcal{G}_{3,1}$. If we multiply the above equation by $I_{3,1}^{-1}$, we obtain $\delta = (\mathbf{A}_r \wedge E)I_{3,1}^{-1}$. Therefore, the 1D cross-ratio using four conformal points X_1, X_2, X_3, X_4 can be calculated with

$$\mathcal{C}_1(X_1, X_2, X_3, X_4) == \frac{(\mathbf{X}_1 \wedge \mathbf{X}_2 \wedge E)I_{3,1}^{-1}(\mathbf{X}_3 \wedge \mathbf{X}_4 \wedge E)I_{3,1}^{-1}}{(\mathbf{X}_1 \wedge \mathbf{X}_3 \wedge E)I_{3,1}^{-1}(\mathbf{X}_2 \wedge \mathbf{X}_4 \wedge E)I_{3,1}^{-1}}. \qquad (12.149)$$

A similar formulation can be done for 2D and 3D cross-ratios. The 2D cross-ratio can be calculated with

$$\mathcal{C}_2(X_1, X_2, X_3, X_4, X_5) = \frac{(\mathbf{X}_5 \wedge \mathbf{X}_4 \wedge \mathbf{X}_3 \wedge E)I_{4,1}^{-1}(\mathbf{X}_5 \wedge \mathbf{X}_2 \wedge \mathbf{X}_1 \wedge E)I_{4,1}^{-1}}{(\mathbf{X}_5 \wedge \mathbf{X}_1 \wedge \mathbf{X}_3 \wedge E)I_{4,1}^{-1}(\mathbf{X}_5 \wedge \mathbf{X}_2 \wedge \mathbf{X}_4 \wedge E)I_{4,1}^{-1}}, \tag{12.150}$$

where $I_{4,1} = e_{123+-}$ is the pseudoscalar for $\mathcal{G}_{4,1}$. Observe that in this case we are working with circles (or 2D spheres) instead of *point pairs*, since the outer product of three points leads to a circle. The 3D cross-ratio can be calculated with

$$\mathcal{C}_3(X_1, X_2, X_3, X_4, X_5, X_6) = \frac{(\mathbf{X}_2 \wedge \mathbf{X}_3 \wedge \mathbf{X}_4 \wedge \mathbf{X}_5 \wedge E)I_{5,1}^{-1}(\mathbf{X}_1 \wedge \mathbf{X}_4 \wedge \mathbf{X}_5 \wedge \mathbf{X}_6 \wedge E)I_{5,1}^{-1}}{(\mathbf{X}_1 \wedge \mathbf{X}_4 \wedge \mathbf{X}_2 \wedge \mathbf{X}_5 \wedge E)I_{5,1}^{-1}(\mathbf{X}_3 \wedge \mathbf{X}_4 \wedge \mathbf{X}_6 \wedge \mathbf{X}_5 \wedge E)I_{5,1}^{-1}} . \tag{12.151}$$

12.9.1 Invariants and Omnidirectional Vision

We have seen how to calculate projective invariants from circles using the conformal geometric algebra. Furthermore, in Sect. 12.8, we also saw how to model the omnidirectional vision system using conformal geometric algebra. Now, we will see how to combine both ideas.

The first thing that we must know is that the projective invariants do not hold in the catadioptric image. However, they hold on the image sphere. Thus, if we project the points on the catadioptric image to the unit sphere, we can recover the projective invariant. To clarify this, we will explain the 1D projective case, which can be seen as a cross section of the 2D case. First, let S be the unit sphere centered at the origin, defined as

$$S = e_0 - \frac{1}{2}e_\infty . \tag{12.152}$$

Also, let the Euclidean points $\mathbf{q}_1, \mathbf{q}_2, \ldots \mathbf{q}_n$ with conformal representation

$$Q_i = \mathbf{q}_i + \frac{1}{2}\mathbf{q}_i^2 e_\infty + e_0, \quad \text{for } i = 1, \cdots, n , \tag{12.153}$$

camera model
be points in the catadioptric image plane. Remember that the number of points (n) needed in the 1D and 2D cases are four and five, respectively. Note that in the 1D cases the points Q_i lie on the intersection of the plane e_3 (i.e., the plane passing through the origin and with normal e_3) with the catadioptric image plane (Fig. 12.11).

Using Eqs. (12.142)–(12.144), we project the points Q_i in the catadioptric image onto the sphere to get the points P_i (in the 1D case, the points lie on a circle that is the intersection of the sphere with the e_3-plane); see Fig. 12.11.

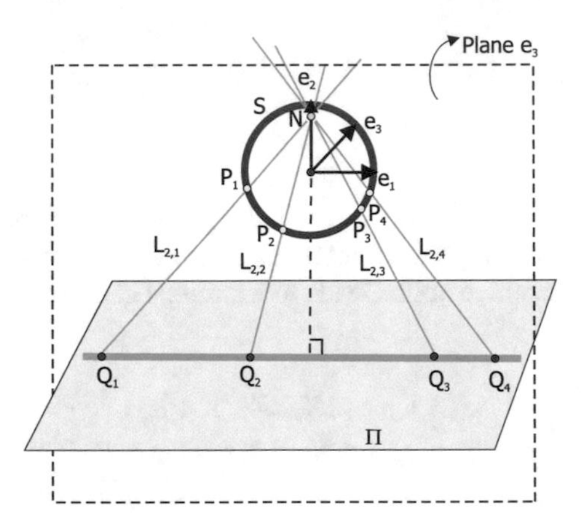

Fig. 12.11 Projection of the points Q_i in the catadioptric image to the points P_i onto the sphere

To compare the invariants on the sphere with the invariants on the projective plane, we define the projective plane Π_p as

$$\Pi_p = e_2 + e_\infty \,, \tag{12.154}$$

which is the plane with normal e_2 and Hesse distance equal to 1. Now, to project the points P_i onto the projective image plane (see Fig. 12.12), we find the line $L_{1,i}$ that passes through the center of the sphere and the point P_i with Eq. 12.145. Then, the lines $L_{1,i}$ are intersected with the projective plane to find the points U_i with

$$U_i = L^*_{1,i} \cdot \Pi_p, \quad \text{for } i = 1, \ldots, n. \tag{12.155}$$

The point U_i is called a *flat point*, which is the outer product of a conformal point with the null vector e_∞ (the point at infinity). To obtain the conformal point from the *flat point*, we can use

$$V_i = \frac{U_i \wedge e_0}{(-U_i \cdot E)E} + \frac{1}{2}\left(\frac{U_i \wedge e_0}{(-U_i \cdot E)E}\right)^2 e_\infty + e_0 \,. \tag{12.156}$$

Once we have the points P_i on the sphere S and the points V_i on the plane Π_p, we calculate their respective 1D invariant with Eqs. (12.60) and (12.149); thus (Fig. 12.13),

$$\delta = \mathcal{C}_1(P_1, P_2, P_3, P_4) = \mathcal{C}_1(V_1, V_2, V_3, V_4) \,. \tag{12.157}$$

In the 2D case, we do the same, but instead we use (12.150) to calculate the invariants (Figs. 12.14 and 12.15). Therefore, we now know that if we project the points on the catadioptric image onto the sphere, we can compute the projective invariants.

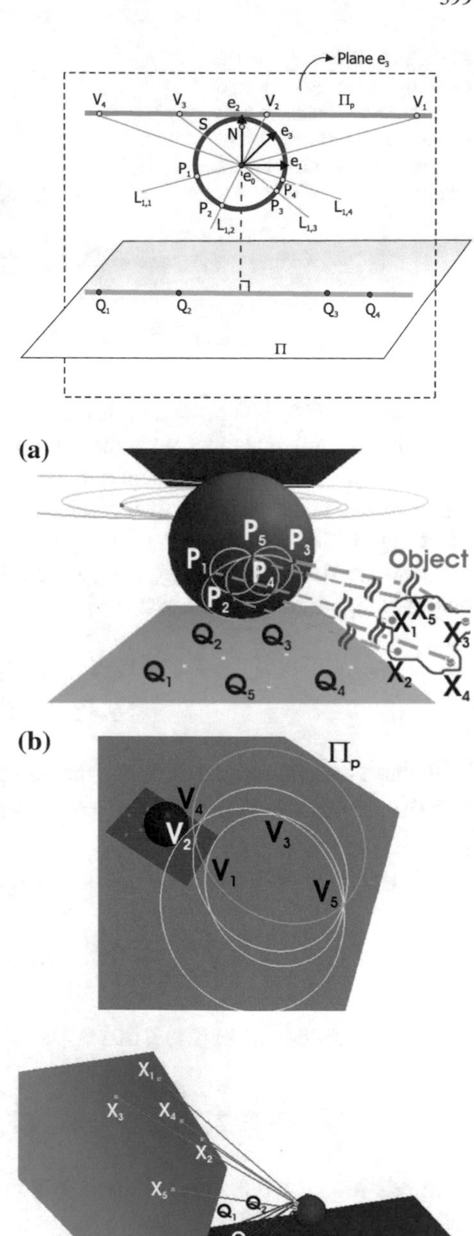

Fig. 12.12 Points on the sphere projected onto the projective image plane

Fig. 12.13 **a** Points Q_i on the catadioptric image projected onto the sphere as the points P_i. These points define the four circles necessary to calculate the 2D invariant on the sphere. **b** Points P_i projected onto the plane Π_p as the points V_i and the four circles formed with them to calculate the 2D invariant on the plane

Fig. 12.14 How five coplanar points on the space are projected in the catadioptric image plane through a sphere

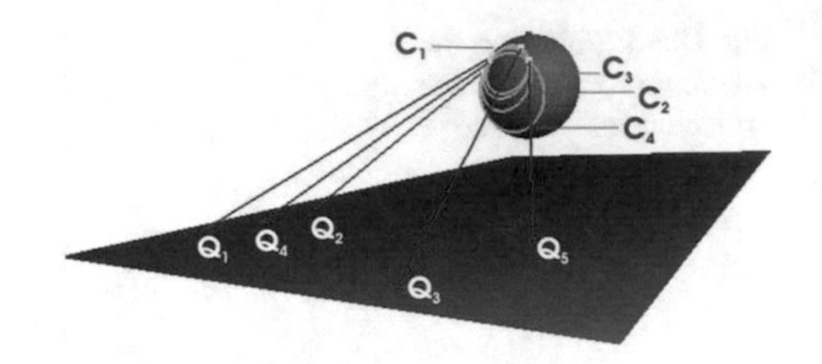

Fig. 12.15 Points Q_i on the catadioptric image projected onto the sphere; these points define the four circles necessary to calculate the 2D invariant on the sphere

12.9.2 Projective and Permutation p^2-Invariants

Equation (12.150) is a projective invariant; however, it is permutation-sensitive. In [211], the authors introduce what they call projective and permutation p^2-*invariants*. Besides removing the dependence of labeling, the p^2-*invariants* use a redundant representation that significantly increases the tolerance to positional errors and allows the design of less sensitive correspondence algorithms.

Given the five points $(P_1, P_2, \ldots, P_5)$ on the sphere, we calculate two independent projective invariants as follows:

$$\begin{aligned}\gamma_1 &= \mathcal{C}_2(P_1, P_2, P_3, P_4, P_5),\\ \gamma_2 &= \mathcal{C}_2(P_2, P_1, P_3, P_4, P_5).\end{aligned} \tag{12.158}$$

With these two values, we calculate the components α_i of a five-dimensional vector $v = \alpha_1 e_1 + \alpha_2 e_2 + \alpha_3 e_3 + \alpha_4 e_4 + \alpha_5 e_5 \in \mathcal{G}_5$ as

$$\begin{aligned}&\alpha_1 = \mathcal{J}(\gamma_1), \quad \alpha_2 = \mathcal{J}(\gamma_2), \quad \alpha_3 = \mathcal{J}\left(\frac{\gamma_1}{\gamma_2}\right),\\ &\alpha_4 = \mathcal{J}\left(\frac{\gamma_2 - 1}{\gamma_1 - 1}\right), \quad \alpha_5 = \mathcal{J}\left(\frac{\gamma_1[\gamma_2 - 1]}{\gamma_2[\gamma_1 - 1]}\right),\end{aligned} \tag{12.159}$$

where $\mathcal{J}$ is defined as in [211], that is,

$$\mathcal{J}(\gamma) = \frac{2\gamma^6 - 6\gamma^5 + 9\gamma^4 - 8\gamma^3 + 9\gamma^2 - 6\gamma + 2}{\gamma^6 - 3\gamma^5 + 3\gamma^4 - \gamma^3 + 3\gamma^2 - 3\gamma + 1}. \tag{12.160}$$

In this way, the obtained v-invariant is independent of the order of the points to calculate it (Fig. 12.13).

Next, we describe a procedure to obtain the 3D pose estimation for tracking and manipulation of the object. The perception system consists of a stereo vision system (SVS) mounted in a pan–tilt unit (PTU) as shown in Fig. 16.32. First, a calibration process for the SVS is realized. This process consists of retrieving the position and orientation of the principal axis of the right camera with respect to the principal axis of the left camera.

12.10 Conformal Geometric Algebra Voting Scheme

In this section, we introduce the generalized version of our voting method, the particularities for extraction of specific kind of features like lines, circles, and symmetries. The essential components of the algorithm involve two phases: the information representation using CGA and then to communicate this information by means of a local and a global voting step. This section is based in the works of G. Altamirano-Gómez and E. Bayro-Corrochano [2, 3].

12.10.1 Representation of Information Using CGA

Let $\mathbb{R}^n$ be a real n-dimensional vector space; then, a token, denoted by t, and a geometric structure, denoted by F, are represented as multivectors of CGA $G_{n+1,1}$. The possible combinations between tokens and geometric structures can constitute *flags*; see Sect. 7.9.1. Then, a token t on a geometric structure F satisfies: $F \cdot t = 0$. Consequently, for a set of tokens $\{t_1, t_2, \cdots, t_N\}$, the flag, F, that satisfies $F \cdot t_i = 0$, for $i = 1, 2, \ldots, N$ defines a minimum for:

$$\frac{1}{N} \sum_{i=1}^{N} W(a_1, \cdots, a_m, t_i) \frac{(F \cdot t_i)^2}{|F|^2}, \tag{12.161}$$

where W is a function that maps a set of parameters $\{a_1, \cdots, a_m, t_i\}$ to a scalar value. We call W a *perceptual saliency function*, since it is used to codify perceptual properties according to Gestalt principles.

Equation (12.161) is the key to generalize voting schemes, since the inner product relates a set of tokens with a geometric structure via an incidence relationship; moreover, each product is limited by a function W, which introduces perceptual restrictions on F. So that, the geometric structure F that minimizes (12.161) satisfies geometric and perceptual properties.

For example, we can express a point, p, touching a line, L, as the equation of the so-called point-line flag:

$$F = p + L, \tag{12.162}$$

and the incidence relationship can be expressed in IPNS as:

$$L \cdot p = 0, \tag{12.163}$$

and its dual, OPNS representation is:

$$L^* \wedge p = 0. \tag{12.164}$$

In the same way, the point-plane flag represents a point touching a plane :

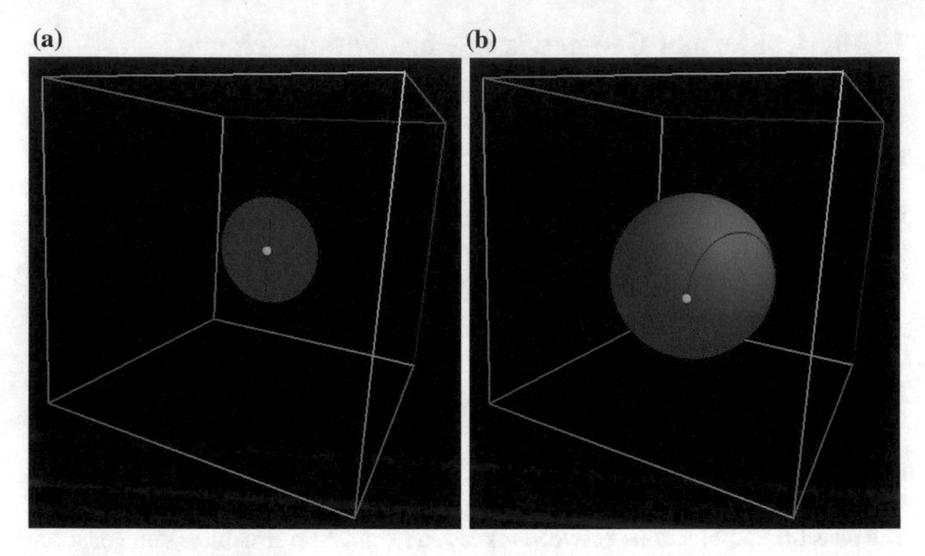

Fig. 12.16 **a** A point-plane and a point-line flags; the plane is shown as a unitary circle. **b** A point-circle and a point-sphere flags

$$F = p + \Pi, \tag{12.165}$$

and its IPNS and OPNS representation can be summarized as:

$$\Pi \cdot p = \Pi^* \wedge p = 0. \tag{12.166}$$

In CGA, the equation that represents a line is equal to a circle with infinite radius, in the same way, a plane is a sphere with infinite radius, and the same principle is valid for hyperplanes and hyperspheres in higher dimensions. Hence, a point-hypersphere flag is defined as follows:

$$F = p + S, \tag{12.167}$$

and using IPNS and OPNS representations we can summarize the incidence relationship as:

$$S \cdot p = S^* \wedge p = 0, \tag{12.168}$$

which is shown in Fig. 12.16.

More over, in (12.161) we have a scalar field, $W(a_1, \cdots, a_m, t_i)$, applied over the geometric entities; the role of this function is to relate a flag with its perceptual properties. The simplest case is when $W = 1$; then, each point on the flag is equally salient. This is shown in Fig. 12.16, where a set of flags have the same saliency value. A different example is shown in Fig. 16.20, where we apply a saliency decay function [209] over point-line and point-circle flags. Thus, we have to design flags and perceptual saliency functions according to the kind of information that we want to extract; in the following sections, we show two applications.

Finally, Eq. (12.161) is a generalization of a geometric voting scheme to handle the use of data, tokens, and flags of any dimension, which can be constructed from any kind of information like gray-value or RGB images, RGB-D data.

12.10.2 Communication of Information

This stage has two parts: a local voting process, which extracts salient geometric entities supported in a local neighborhood, and a global voting process, which clusters the output obtained by the local voting process.

Local Voting Given a perceptual saliency function W, and a set of tokens $T = \{t_1, t_2, \cdots, t_N\}$, the local voting step consists in: selecting a token, $t_0 \in T$, defining a subset, T_0, that contains all tokens in the neighborhood of t_0, and computing the geometric structure, F, that minimizes (12.161) for the tokens in subset T_0. To find F, we apply a voting methodology to take the outliers out, and compute F using the rest of the tokens.

The voting procedure is performed as follows: Let t_0 be the selected token, then each token on its neighborhood casts a vote in the form of a perceptually salient geometric element.

Definition 12.1 *A perceptually salient geometric element is a set of points together with a function, that assigns a scalar value to each element of the set.*

Using CGA, a geometric element with density is represented by:

$$\bar{F} = \{p_c : p_c \cdot F = 0\}, \ W : \mathcal{R}^m \rightarrow \mathcal{R}, \tag{12.169}$$

where p_c and F are a point and a geometric element, respectively, in conformal representation, whereas $\bar{F}$ is a set of conformal points, and W is a function that assigns a scalar value to each element of set $\bar{F}$. Therefore, F codifies the geometric structure of an object, while the function W codifies its perceptual saliency.

Thus, the voting consists in mapping each token on the neighborhood of t_0 to a perceptually salient geometric element. In the voting space, each geometric structure F is a point that has associated a perceptual saliency value or density. Then, we use DBSCAN algorithm [259] to separate votes, and cluster those that has similar geometric structure (e.g., hyperplanes with the same normal and Hesse distance, or hyperspheres with the same center and radius).

Next, we compute the perceptual saliency of each cluster:

$$\bar{W} = \sum_i W_i, \tag{12.170}$$

where W_i is the perceptual saliency of each geometric element in the cluster. Finally, we select clusters that surpass a threshold value, or the cluster with highest perceptual saliency.

In contrast to the Tensor Voting algorithm, which extracts only the feature with maximum support, the use of a clustering technique in our algorithm allows us to extract several features at the same time.

Global Voting The local voting process delivers a set of salient geometric structures for each token. The global voting process consists in grouping similar geometric structures using DBSCAN algorithm. Then, one computes the perceptual saliency of each cluster and selects such clusters which surpass a threshold value. In Sect. 16.5, we present experiments of the conformal geometric algebra voting scheme using synthetic and real images.

12.11 Conformal Model for Stereoscopic Perception Systems

Now we explain how we developed a conformal model to represent 3D information in visual space. We start by discussing the role of conformal image mapping for modeling the human visual system. Then, we explain how we use the horopter in our conformal model to build a stereoscopic perception system. In next section, we describe how we implement a real-time system using our model.

12.11.1 Conformal Image Mapping for Modeling the Human Visual System

In the last few decades, there has been a lot of research to elucidate image mapping in the visual systems of primates and humans. Figure 12.17a shows the human visual system. The visual cues excite nodes of the retina, which generate biological signals. These signals traverse through optical fibers and finally are merged at the neocortex to produce a disparity map. Our brain uses this map to recreate a space and time impression of the visual space. Figure 12.17b depicts the so-called Vieth-Müller circle, or geometric locus, of the 3D points 1 to 5, which causes the same disparity on the eye's retinas; see the lines passing the nodal points arriving at points 1 to 5 on the retinas. If one suffers from myopia, the circle, also called the theoretical horopter, is deformed to an ellipsoid or empirical horopter. In the ideal case, we can regard the eyeball as perfectly symmetric. To build an artificial vision system, we have to achieve almost a circular horopter. Figure 12.17c depicts the human stereo vision using Cartesian coordinates, and Fig. 12.17d depicts it using polar coordinates. In a series of outstanding papers, Erick L. Schwartz shows the developments concerning the image mapping of visual information to the neocortex. He claims that "to simulate

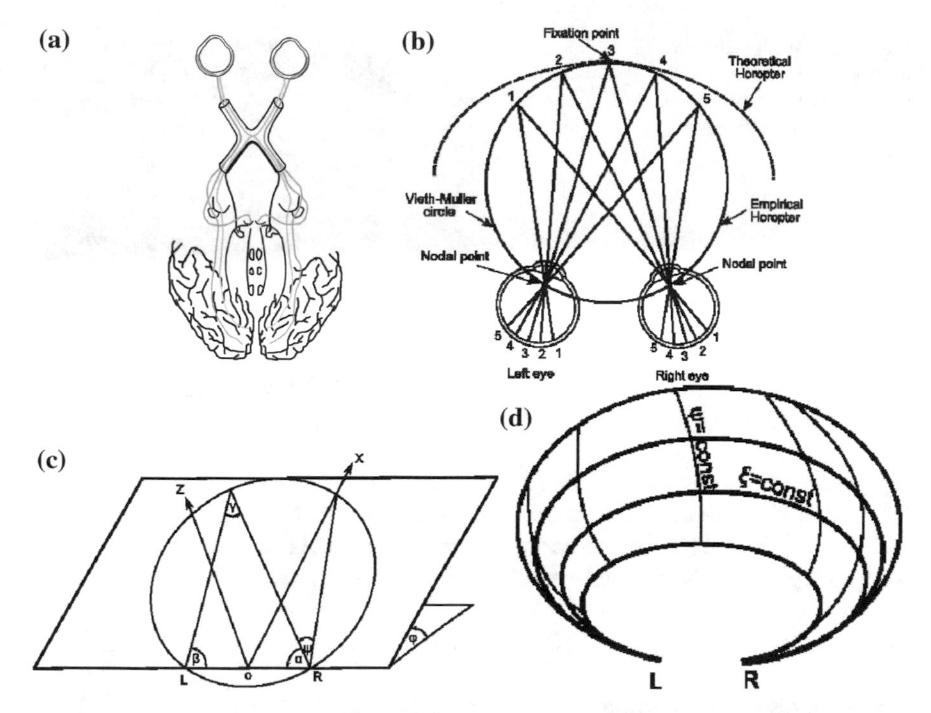

Fig. 12.17 **a** Human vision system. **b** Horopter and the Vieth-Müller circle [KCN2]. **c** Cartesian representation of the stereo vision system. **d** Polar representation of the stereo vision system

the image properties of the human visual system (and perhaps other sensory systems) conformal image mapping is a necessary technique" [96]. The mapping function

$$w = k\,log(z + a), \tag{12.171}$$

is a widely accepted approximation to the topographic structure of the primate V1 foveal and parafoveal regions. An extension of it by simply adding an additional parameter captures the full-field topographic map in terms of the *dipole map* function

$$w = k\,log\frac{(z + a)}{(z + b)}. \tag{12.172}$$

However, these models are still unsatisfactory, as they cannot describe *topografic shear* due to the fact that they are both explicitly complex-analytic or conformal. Balasubramanian et al. [7] suggested a very simple procedure for topographic shear in V1, V2, and V3 assuming that cortical topographic shear is rotational (a compression along iso-eccentricity contours). The authors model the constant rotational with a quasi-conformal mapping called the *wedge mapping*. Using five independent parameters, this mapping yields an approximation to the V1, V2, and V3

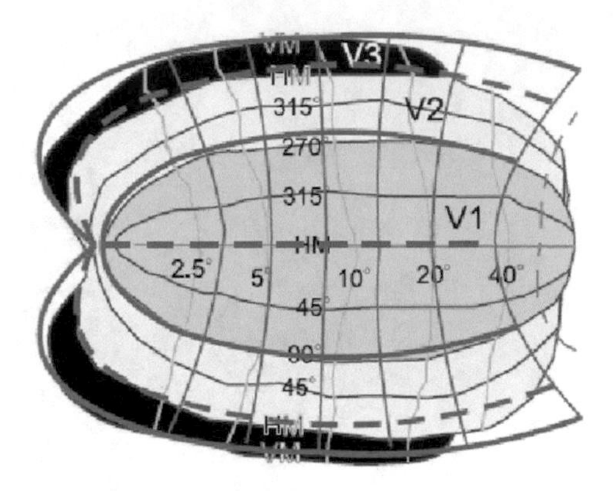

Fig. 12.18 Wedge–dipole model semiqualitatively superimposed on the topography of the human visual areas

topographic structures unifying these three areas into a single V1-V2-V3 complex as follows: First, we represent any point in the visual hemifield with the complex variable $z = r\, e^{i\theta}$, where r and θ denote the eccentricity and polar angle, respectively. The wedge map for the three visual areas Vk, $k = 1, 2, 3$, is the map

$$\eta_k(r\, e^{i\theta}) = r\, e^{i\Theta_k(\theta)}, \tag{12.173}$$

where the respective functions for V1, V2, and V3 are given by

$$\begin{aligned}
\Theta_1(\theta) &= \alpha_1\theta, \\
\Theta_2(\theta) &= \begin{cases} -\alpha_2(\theta - \frac{\pi}{2}) + \Theta_1(+\frac{\pi}{2})\ if & 0^+ \leq \theta \leq \frac{\pi}{2}, \\ -\alpha_2(\theta + \frac{\pi}{2}) + \Theta_1(-\frac{\pi}{2})\ if & -\frac{\pi}{2} \leq \theta \leq 0^-, \end{cases} \\
\Theta_3(\theta) &= \begin{cases} \alpha_3\theta + \Theta_2(0^+) \quad if & 0^+ \leq \theta \leq \frac{\pi}{2}, \\ \alpha_3(\theta) + \Theta_2(0^-)\ if & -\frac{\pi}{2} \leq \theta \leq 0^-. \end{cases}
\end{aligned} \tag{12.174}$$

The wedge warps three copies of V1, V2, and V3 of the visual hemifield and localizes them into a pie form, where each one is compressed by an amount α_k in the azimuthal direction, thus resulting in a rotational shear in each of the wedges. Finally, the wedge map is further modified via a dipole map using Eq. (12.172). The result is the full *wedge–dipole model* depicted semiqualitatively in Fig. 12.18.

Many visual cortex architectures of the primates and the human have an important feature responsible for the procedure of mixing the visual data of the left and right eyes. It has been shown that ocular dominance columns represent thin strips (5–10 min of arc) alternating the left- and right-eye input to the brain. According to Yeshurfun and Schwartz [317], such an architecture, when operated upon with a *cepstral filter*, provides a strong cue for binocular *stereopsis*. The creature can sense depth using this visual cue.

In our work, we have a different motivation; we extend the 2D horopter concept [145] to the 3D horopter sphere for fusing the left and right stereoscopic images in a sphere (see Fig. 12.19c). This is basically a 3D representation using polar coordinates of the 3D visual space. This representation occurs after the stereopsis has

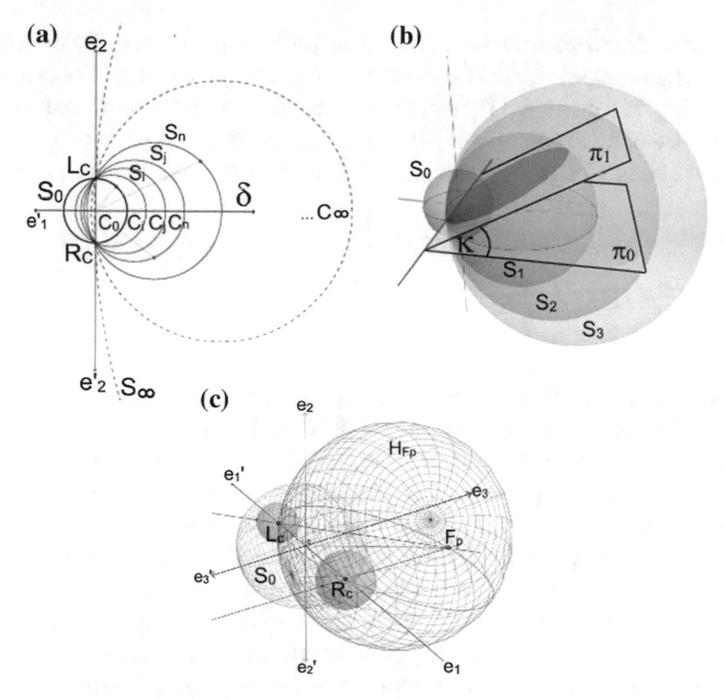

Fig. 12.19 **a** Horopter and spheres center by bisector line when the depth δ grows. **b** Spherical horopter and the unit sphere, where the horopter depends on the azimuth angle κ. **c** Conformal horopter configuration. The left camera center L_c, the right camera center R_c, and the fixation point F_p define a circle, which, by means of varying κ, defines a family of spheres or horopter spheres

been computed. When one traverses the spheres outwardly, one gets the sense of *directed depth*. In conformal geometric algebra, the directed depth is a vector pointing outward from the egocenter of our horopter. Its magnitude is the scalar value of the depth.

Why do we believe that this representation is useful? Let us answer this question by first showing the advantages of our mathematical system. In conformal geometric algebra, the computational unit is the sphere. One can map all the 3D visual information onto a family of spheres using the 3D horopter concept. As the 3D visual information is mapped on spheres, we can start to use this representation, however, in the 5D space of conformal geometric algebra. We can explode then all the computational advantages of this mathematical system, such as applying incidence algebra among circles, planes, and spheres and utilizing linear transformations like translators, rotors, and dilators in terms of spinors. In this way, the 3D visual information on the sphere can be treated more efficiently in the conformal geometric algebra framework for various humanoid tasks like recognition, reasoning, planning, and conducting autonomous actions.

We can also claim that all the efforts on conformal mapping are restricted to the mapping on the primate and human visual areas; however, we are introducing a mathematical framework in order to have an artificial way to fuse in 3D the images of the left and right cameras for the depth sensing necessary for recognition, representation, reasoning, and planning. Furthermore, using the powerful conformal geometric algebra framework, we can relate quite advantageously the algebra of visual primitives of perception with the kinematics and dynamics of robot mechanisms.

12.11.2 Horopter and the Conformal Model

The horopter is the 3D geometric locus in space where an object has to be placed in order to stimulate exactly two points in correspondence in the left and right retinas of a biological binocular vision system [289]. In Fig. 12.19c, we see a horopter depending of an azimuth angle κ. In other words, the horopter represents a set of points that cause minimal (almost zero) disparity on the retinas. We draw the horopter tracing an arc through the fixation point and the nodal points of the two retinas; see Fig. 12.19a. The theoretical horopter is known as the Vieth-Müller circle. Note that each fixation distance has its own Vieth-Müller circle. According to this theoretical view, the following assumptions can be made: Each retina may be a perfect circle; both retinas are of the same size; corresponding points are perfectly matched in their retina locations; and points in correspondence are evenly spaced across the nasal and temporal retinas of the right and left eyes.

If an object is located on either side of the horopter, a small amount of disparity is caused by the eyes. The brain analyzes this disparity and computes the relative distance of the object with respect to the horopter. In a narrow range near the horopter, the stereopsis does not exist. That is due to very small disparities that are not enough to stimulate stereopsis. Empirical horopter measurements (even done using the *Nonius* method) do not agree with the Vieth-Müller circle. There are two obvious reasons for this inconsistency, either due to irregularities in the distribution of visual directions in the two eyes or a result of the optical distortion in the retinal image. There are various physiological reasons why the horopter can be distorted. Another cause of distortion is the asymmetric distribution of oculocentric visual distributions. In addition to a regional asymmetry in local signs in one eye, the distribution between the two eyes may not be congruent (correspondence problem), which may be another cause of horopter distortion. Asymmetric mapping from the retina to the neocortex in both eyes also causes a deviation of the horopter from the Vieth-Müller circle.

The simple configuration of the horopter shown in Fig. 12.19a is nothing more than a very naive geometric representation using polar coordinates of the geometric locus of the visual space. In contrast, using the tools of conformal geometric algebra, we can claim that binocular vision can be reformulated advantageously using a spherical retina. Now we show how we find the horopter in the sphere of conformal geometry. Actually, we are dealing with a bunch of spheres intersecting the centers of the cameras L_C and R_C; see Fig. 12.19b. This is the pencil of spheres in the projective space of spheres. Note that the L_C and R_C camera's centers are Poncelet points. Since a stereo system only sees in front of it, we consider the spheres emerging toward the front. When the space locus of objects expands, the centers of the spheres move along the bisector line of the stereo rig. This is when the depth δ grows; see Fig. 12.19a. From now on, we will use the term "horopter sphere" rather than "horopter circle," because when we change the azimuth of the horopter circle, we are simply selecting a different circle of a particular horopter sphere s_i; see Fig. 12.19b. As a result, we can consider that all the points of the visual space are lying on the pencil of the horopter sphere. Let us translate this description in terms of equations of conformal geometric algebra.

We call the unit horopter sphere s_0 the one whose center is the egocenter of the stereo rig (see Fig. 12.19c) and that has the sphere equation $s_0 = p - \frac{1}{2}\rho^2 e_\infty = \mathbf{c}_0 + \frac{1}{2}(\mathbf{c}_0^2 - \rho^2)e_\infty + e_0$, where its center (egocenter) is attached to the true origin of space of conformal geometric algebra and the radius is the half the stereo rig length $\rho_0 = \frac{1}{2}|L_C - R_C|$. The center $\mathbf{c}_i$ of any horopter sphere s_i moves toward the point at infinity as $\mathbf{c}_i = \mathrm{c}_i + \frac{1}{2}(\mathrm{c}_i^2)e_\infty + e_0$, where c_i is the Euclidean 3D vector. Thus, we can write the equation of the sphere s_i as $s_i = \mathrm{c}_i + \frac{1}{2}(\mathrm{c}_i^2 - \rho_i^2)e_\infty + e_0$, where the radius is computed in terms of the stereopsis depth $\rho_i = \frac{1}{2}(1 + \delta)$.

Consider the figure of the model for the visual human system in Fig. 12.19c. We see that the horopter circles lie on a pencil of planes π_i. We can obtain the same circles z_i simply by intersecting in our conformal model such a pencil of planes with the pencil of tangent spheres as depicted in Fig. 12.19c. The intersection is computed using the meet operation of the duals of the plane and sphere and taking the dual of the result as $z_i = \pi_i \wedge s_i$. Now, taking the meet of any two horopter spheres, we gain a circle that lies on the front parallel plane with respect to the digital camera's common plane $z = s_i \wedge s_j$. Later, taking the meet of this circle with the unit horopter, we regain the Poncelet points L_C and R_C, which in our terms are called the point pair $\mathbf{PP}_{LR} = \mathbf{z} \wedge \mathbf{s}_0 = \mathbf{L} \wedge (\mathbf{s}_0^*)$ (note that the second part of the equation computes the point pair wedging the dual of sphere s_0^* with the line crossing the camera centers L_C and R_C).

If we further consider Fig. 12.19b, the intersecting plane π_i cuts the horopter spheres, generating a geometric locus on the plane as depicted in Fig. 12.19b, c. These horopter circles fulfill an interesting property. If one takes an inversion of all horopter circles with centers on the line l, we get the radial lines of the polar diagram; see the right in Fig. 12.20a. Now, since the plane π_i (by varying angle κ) intersects the family of horopter spheres producing the horopter circles of Fig. 12.20a, whose inverse is a 2D log-polar diagram, we can conclude that the inverse of the arrangement of horopter spheres and Poncelet points is equivalent to a 3D log-polar diagram, as depicted in Fig. 12.20b. To understand this better, let us take any radial line of the 2D log-polar diagram and express it in conformal geometric algebra: $\mathbf{L} = \mathbf{X} \wedge \mathbf{Y} \wedge e_\infty$. Now, applying an inversion to this line, we get a circle; i.e., $z = e_4\mathbf{L}e_4$. Note that this inversion is implemented as a reflection on e_4. The 3D log-polar diagram is an extraordinary result, because contrary to the general belief that conformal image processing takes places in a 2D log-polar diagram, we can consider that the visual processing rather takes place in a 3D log-polar diagram. This claim is novel as well

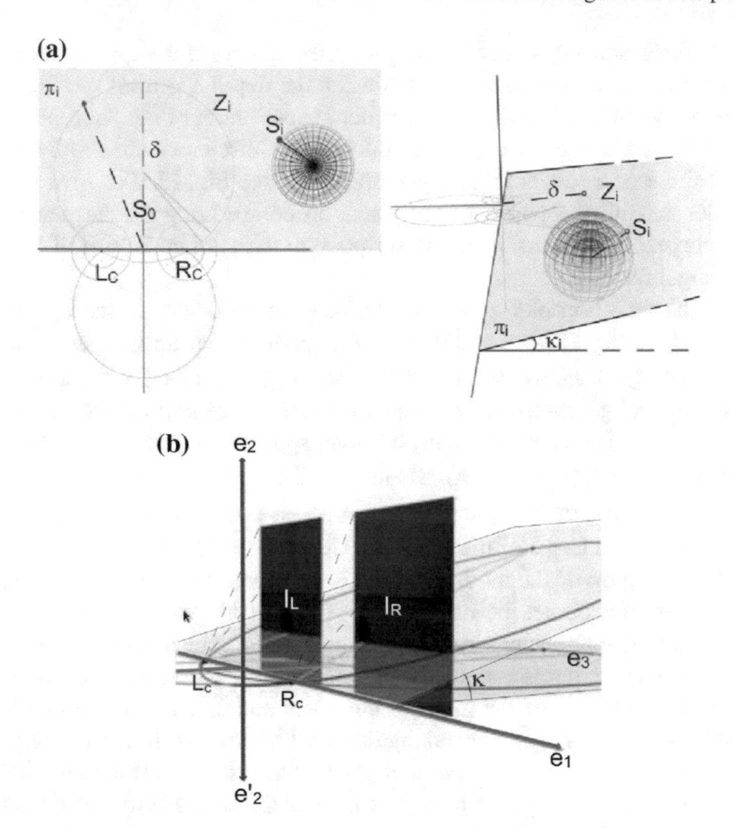

Fig. 12.20 **a** Log-polar representation of visual space. **b** Nodal points L_C-R_C (camera's centers), image planes I_L-I_R, and the azimuth angle κ

as promising, because this framework can be used for 3D flow estimation, as opposed to the use of one view or even an arrangement of two log-polar cameras.

12.12 Conclusion

This chapter has outlined the use of geometric algebra as a framework for analysis and computation in computer vision. In particular, the framework for projective geometry was described and the analysis of tensorial relations between multiple camera views was presented in a wholly geometric fashion. The projective geometry operations of meet and join are easily expressed analytically and easily computed in geometric algebra. Indeed, it is the ease with which we can perform the algebra of incidence (intersections of lines, planes, etc.) that simplifies many of the otherwise complex tensorial relations. The concept of duality has been discussed and used specifically in projecting down from the world to image planes; in geometric algebra, duality is a particularly simple concept and one in which the non-metric properties of the inner product become apparent.

In this chapter, omnidirectional vision has been formulated in the conformal geometric algebra which seems a perfect framework to handle the parametrization of different mirrors as reflecting spheres; in this way, the awkward formulas of the projective geometry of catadrioptic cameras are greatly simplified. We compute projective invariants using projected spheres, and in the image plane, we compute the using circle relations. We reformulated these projective invariants using the concept of the permutation p^2-*invariants*. Besides removing the dependence of labeling, the p^2-*invariants* use a redundant representation that significantly increases the tolerance to positional errors and allows the design of less sensitive correspondence algorithms.

Traditional methods for detection of geometric entities in images resort to voting schemes like Hough transform and Tensor Voting. In this chapter, we presented a generalization of these methods, which allows to represent complex structures in terms of basic geometric primitives. Our algorithm combines the mathematical framework of conformal geometric algebra and perceptual grouping principles of the Gestalt psychology theory, so that, we can detect structures defined by their form, as well as by their perceptual properties. In addition, our method is applied in a hierarchical way, allowing it to detect structures that are supported with local data, as well as with global information.

In addition, in this chapter we introduced a novel conformal model for humanlike vision. The standard concept of horopter circles is extended to a horopter sphere, which leads to a 3D log-polar representation of the visual space. This representation of the visual space can be used for a robot to track features and to detect 3D pose of itself and surrounding objects.

12.13 Exercises

12.1 Compute in P^2 the intersecting point of the lines $\boldsymbol{A} = \boldsymbol{x}_1 \wedge \boldsymbol{x}_2$ and $\boldsymbol{B} = \boldsymbol{y}_1 \wedge \boldsymbol{y}_2$. The homogeneous coordinates of the involved points are $\boldsymbol{x}_1 = (3, 1, -1)$, $\boldsymbol{x}_2 = (-1, 1, 1)$, $\boldsymbol{y}_1 = (2, -1, 1)$, and $\boldsymbol{y}_2 = (-2, 0, 1)$.

12.2 Points and planes are in P^3. Find in P^3 the relative positions of the points $\boldsymbol{P} = (-2, -4, -4, 1)$ and $\boldsymbol{Q} = (2, -4, 5, 1)$, respective to the plane $\boldsymbol{E} = \boldsymbol{A}_1 \wedge \boldsymbol{A}_2 \wedge \boldsymbol{A}_3$, where $\boldsymbol{A}_1 = (2, -1, 1, -6)$, $\boldsymbol{A}_2 = (1, 1, -1, 0)$, and $\boldsymbol{A}_3 = (-1, 0, 0, 4)$. Explain these results. (*Hint*: The join $\boldsymbol{P} \wedge \boldsymbol{E} = dI$ spans a quatrivector, where the scalar d corresponds to the Hesse distance or foot. Thus, you must compute the brackets $[\boldsymbol{PE}]$ and $[\boldsymbol{QE}]$. When the bracket is positive, the point is at the right, and when it is negative, the point is at the left.)

12.3 In P^3 compute the relative orientation of the lines $\boldsymbol{L}_1 = \boldsymbol{A}_1 \wedge \boldsymbol{A}_2$ and $\boldsymbol{L}_2 = \boldsymbol{A}_3 \wedge \boldsymbol{A}_4$, where $\boldsymbol{A}_1 = (2, 1, -3, -1)$, $\boldsymbol{A}_2 = (1, 3, 5, 2)$, $\boldsymbol{A}_3 = (1, 2, 1, 4)$, and $\boldsymbol{A}_4 =$

$(-3,-1,2,4)$. (*Hint*: The join $\boldsymbol{L}_1\wedge\boldsymbol{L}_2 = dI$ spans a quatrivector, where the scalar d corresponds to the Hesse distance or foot. Thus, you must compute the bracket $[\boldsymbol{L}_1\boldsymbol{L}_2] > 0$ and interpret your result using the right-hand rule.)

12.4 In P^3 compute the intersection of the following planes: $\boldsymbol{E}_1 = \boldsymbol{A}_1\wedge\boldsymbol{A}_2\wedge\boldsymbol{A}_3$ and $\boldsymbol{E}_2 = \boldsymbol{B}_1\wedge\boldsymbol{B}_2\wedge\boldsymbol{B}_3$, where $\boldsymbol{A}_1 = (2,1,-3,-1)$, $\boldsymbol{A}_2 = (1,3,5,2)$, $\boldsymbol{A}_3 = (1,2,1,4)$, and $\boldsymbol{B}_1 = (4,2,-6,-2)$, $\boldsymbol{B}_2 = (4,3,5,1)$, $\boldsymbol{B}_3 = (3,6,3,12)$.

12.5 In P^2 compute the resulting intersecting point $\boldsymbol{p}$ of the intersecting lines $\boldsymbol{a}\wedge b$ and the line passing by the point $\boldsymbol{c}$ which lies off the line $\boldsymbol{a}\wedge b$ for the following points: $\boldsymbol{a} = (2,3,1)$, $\boldsymbol{b} = (10,10,1)$, and $\boldsymbol{c} = (5,9,1)$. (*Hint*: First compute the line passing by point $\boldsymbol{c}$ using the direction orthogonal to the line $\boldsymbol{a}\wedge\boldsymbol{b}$; then, compute the meet of these lines.)

12.6 Derive the mapping of a plane $\pi = \boldsymbol{A}\wedge\boldsymbol{B}\wedge\boldsymbol{C}$ of the projective space to the projective plane. The expected result should be $\pi' = \boldsymbol{n}' + I_3 d$.

12.7 Prove the Simpson rule using the join of three projected points. If the projecting point $\boldsymbol{p}$ lies at the circumference, the join of the projected points is zero. Take three arbitrary points lying on a unit circumference and a fourth one nearby. Use CLICAL or eClifford for your computations. (*Hint*: First compute the projected points as the meet of three lines passing by the point $\boldsymbol{p}$ orthogonal to the triangle sides. The triangle is formed by three arbitrary points lying on the circumference.)

12.8 Prove the Pascal theorem using the incidence algebra in P^2. Given six points lying in a conic, compute six intersecting lines using the join operation. The meet of these lines should give three intersecting points, and the join of these three intersecting points should be zero. Use eClifford and any six points belonging to a conic of your choice.

12.9 Prove the Desargues theorem using the algebra of incidence. Let $\boldsymbol{x}_1$, $\boldsymbol{x}_2$, $\boldsymbol{x}_3$ and $\boldsymbol{y}_1$, $\boldsymbol{y}_2$, $\boldsymbol{y}_3$ be the vertices of two triangles in P^2, and suppose that the meets of the lines fulfill the equations $(\boldsymbol{x}_1\wedge\boldsymbol{x}_2)\cap(\boldsymbol{y}_1\wedge\boldsymbol{y}_2) = z_3$, $(\boldsymbol{x}_2\wedge\boldsymbol{x}_3)\cap(\boldsymbol{y}_2\wedge\boldsymbol{y}_3) = z_1$, and $(\boldsymbol{x}_3\wedge\boldsymbol{x}_1)\cap(\boldsymbol{y}_3\wedge\boldsymbol{y}_1) = z_2$. Then, the join of these points $z_1\wedge z_2\wedge z_3 = 0$ if and only if a point $\boldsymbol{p}$ exists such that $\boldsymbol{x}_1\wedge\boldsymbol{y}_1\wedge\boldsymbol{p} = \boldsymbol{x}_2\wedge\boldsymbol{y}_2\wedge\boldsymbol{p} = \boldsymbol{x}_3\wedge\boldsymbol{y}_3\wedge\boldsymbol{p} = 0$. In this problem, use eClifford and define two triangles of your choice.

12.10 A point $\boldsymbol{p}$ lies on the circumcircle of an arbitrary triangle with vertices $\boldsymbol{x}_1$, $\boldsymbol{y}_1$, and z_1. From the point $\boldsymbol{p}$, draw three perpendiculars to the three sides of the triangle to meet the circle at points $\boldsymbol{x}_2$, $\boldsymbol{y}_2$, and z_2, respectively. Using incidence algebra, show that the lines $\boldsymbol{x}_1\wedge\boldsymbol{x}_2$, $\boldsymbol{y}_1\wedge\boldsymbol{y}_2$, and $z_1\wedge z_2$ are parallel.

12.11 Consider a world point $\boldsymbol{X}\in P^3$ projected onto two image planes (see Fig. 12.3). Show that the bilinear constraint can be expressed as $\alpha_i^T\tilde{F}_{ij}\beta_j{=}0$, where α_i^T and β_j are tensor notations. (*Hint*: Use the geometric constraint $\boldsymbol{A}_0\wedge\boldsymbol{B}_0\wedge\boldsymbol{A}'\wedge\boldsymbol{B}'{=}0$, where $\boldsymbol{A}_0$ and $\boldsymbol{B}_0$ are the optical centers of the images and $\boldsymbol{A}' = \alpha_i\boldsymbol{A}_i$ and $\boldsymbol{B}' = \beta_j\boldsymbol{B}_j$ are the image points spanned using three arbitrary image points $\boldsymbol{A}_i$ or $\boldsymbol{B}_i$. The epipolar plane is given by $\boldsymbol{A}_0\wedge\boldsymbol{B}_0\wedge\boldsymbol{A}'$.)

12.12 Consider two non-intersecting world lines $\boldsymbol{L}_{12}$ and $\boldsymbol{L}_{34}$. Their projecting lines intersect in the points α_i and β_j. Show that α_i lies on the epipolar line passing by β_j.

12.13 Consider a world line $\boldsymbol{L}_{12} \in P^3$ projected onto three image planes (see Fig. 12.4). The projected line onto the first image plane Φ_A is given by

$$l_i^A \boldsymbol{L}_{A_i} = \left[\boldsymbol{A}_0 \wedge \left(l_j^B l_k^C \{(\boldsymbol{B}_0 \wedge \boldsymbol{L}_j^B) \cap (\boldsymbol{C}_0 \wedge \boldsymbol{L}_k^C)\}\right)\right] \cap \boldsymbol{\Phi}_A.$$

Geometrically, this equation means that the optical planes $\boldsymbol{B}_0 \wedge l_j^B \boldsymbol{L}_j^B$ and $\boldsymbol{C}_0 \wedge l_j^C \boldsymbol{L}_j^C$ intersect in the line $\boldsymbol{L}_{12}$. Now, the join operation of $\boldsymbol{L}_{12}$ with $\boldsymbol{A}_0$ builds the optical plane that intersects the image plane $\boldsymbol{\Phi}_A$ by the line $l_{A_i} \boldsymbol{L}_{A_i}$ (linear combination of three arbitrary image lines $\boldsymbol{L}_{A_i}$). Expand the equation to get the coefficients l_i^A in terms of the trilinear constraint or trifocal tensor as follows:

$$l_i^A = T_{ijk} l_j^B l_k^C.$$

12.14 Write two sets of equations for the Pascal theorem, one set involving two cameras and the second set involving three. Note that the brackets relate the cameras via the bilinear or trilinear constraints. See the following chapter for a discussion of conics and the Pascal theorem.

12.15 Write a simulation in the motor algebra $G_{3,0,1}^+$ using Maple and eClifford to model the 3D kinematics of the eye; see [15].

Part IV
Machine Learning

Chapter 13
Geometric Neurocomputing

13.1 Introduction

It appears that for biological creatures, the external world may be internalized in terms of intrinsic geometric representations. We can formalize the relationships between the physical signals of external objects and the internal signals of a biological creature by using extrinsic vectors to represent those signals coming from the world and intrinsic vectors to represent those signals originating in the internal world. We can also assume that external and internal worlds employ different reference coordinate systems. If we consider the acquisition and coding of knowledge to be a distributed and differentiated process, we can imagine that there should exist various domains of knowledge representation that obey different metrics and that can be modeled using different vectorial bases. How it is possible that nature should have acquired through evolution such tremendous representational power for dealing with such complicated signal processing [160]. In a stimulating series of articles, Pellionisz and Llinàs [228, 229] claim that the formalization of geometrical representation seems to be a dual process involving the expression of extrinsic physical cues built by intrinsic central nervous system vectors. These vectorial representations, related to reference frames intrinsic to the creature, are covariant for perception analysis and contravariant for action synthesis. The geometric mapping between these two vectorial spaces can thus be implemented by a neural network which performs as a metric tensor [229].

Along this line of thought, we can use Clifford, or geometric, algebra to offer an alternative to the tensor analysis that has been employed since 1980 by Pellionisz and Llinàs for the perception and action cycle (PAC) theory. Tensor calculus is covariant, which means that it requires transformation laws for defining coordinate-independent relationships. Clifford, or geometric, algebra is more attractive than tensor analysis because it is coordinate-free, and because it includes spinors, which tensor theory does not. The computational efficiency of geometric algebra has also been confirmed in various challenging areas of mathematical physics [73]. The other mathematical system used to describe neural networks is matrix analysis. But, once again, geometric algebra better captures the geometric characteristics of the problem

E. Bayro-Corrochano, *Geometric Algebra Applications Vol. I*,
https://doi.org/10.1007/978-3-319-74830-6_13

independent of a coordinate reference system, and it offers other computational advantages that matrix algebra does not, e.g., bivector representation of linear operators in the null cone, incidence relations (meet and join operations), and the conformal group in the horosphere.

Initial attempts at applying geometric algebra to neural geometry have already been described in earlier papers [17, 19, 132, 133]. In this chapter, we demonstrate that standard feedforward networks in geometric algebra are generalizable. We present the geometric multilayer perceptron and the geometric radial basis function. Moreover, instead of using radial basis function at the hidden layer, we use quaternion wavelets for the Quaternion Wavelet Function Network. The chapter introduces also the Clifford Support Vector Machines (CSVM) as a generalization of the real- and complex-valued support vector machines using the Clifford geometric algebra. In this framework, we handle the design of kernels involving the Clifford or geometric product. In this approach, one redefines the optimization variables as multivectors. This allows us to have a multivector as output. Therefore, we can represent multiple classes according to the dimension of the geometric algebra in which we work. We show that one can apply CSVM for classification and regression and interpolation. The CSVM is an attractive approach for the MIMO processing of high-dimensional geometric entities. The use of feedforward neural networks and the SV machines within the geometric algebra framework widens their sphere of applicability and furthermore expands our understanding of their use for multidimensional learning. The experimental analysis confirms the potential of geometric neural networks and Clifford-valued SVMs for a variety of real applications using multidimensional representations, such as in graphics, augmented reality, machine learning, computer vision, medical image processing, and robotics.

In this work, we also show the quaternionic spike neural networks in conjunction with the training algorithm, working on quaternion algebra, outperform the real-valued spike neural networks. They are ideal for applications in the area of neurocontrol of manipulators with angular movements.

The progress in quantum computing and neurocomputing indicates that the formulation of real-time operators in terms of neural networks seems a promising approach particularly as it is inspired in neuroscience as well. We formulate the algorithms for the Rotor or Quaternion Quantum Neural Networks (RQNN) or Quaternion Quantum Neural Networks (QQNN) as models for real-time computing superior by far as traditional quantum computing using a Von Neumann computing based on digital gates.

The revolution of the convolutional neural networks is basically due to the increase of the amount of hidden filter convolutional layers, the inclusion of the pooling and ReLU layers. In this chapter, we propose as special case of the Geometric Algebra CNN the Quaternion CNN, simply extending the representation, convolution, and products in the quaternion algebra framework.

13.2 Real-Valued Neural Networks

The approximation of *nonlinear mappings* using neural networks is useful in various aspects of signal processing, such as in pattern classification, prediction, system modeling, and identification. This section reviews the fundamentals of standard real-valued feedforward architectures.

Cybenko [62] used for the approximation of a continuous function, $g(\mathbf{x})$, the superposition of weighted functions:

$$y(\mathbf{x}) = \sum_{j=1}^{N} w_j s_j(\mathbf{w}_j^T \mathbf{x} + \theta_j), \tag{13.1}$$

where $s(.)$ is a continuous discriminatory function like a sigmoid, $w_j \in \mathcal{R}$ and $\mathbf{x}, \theta_j, \mathbf{w}_j \in \mathcal{R}^n$. Finite sums of the form of Eq. (13.1) are dense in $C^0(I_n)$, if $|g_k(\mathbf{x}) - y_k(\mathbf{x})| < \varepsilon$ for a given $\varepsilon > 0$ and all $\mathbf{x} \in [0, 1]^n$. This is called a *density theorem* and is a fundamental concept in approximation theory and nonlinear system modeling [62, 146].

A structure with k outputs y_k, having several layers using logistic functions, is known as the *multilayer perceptron* (MLP) [255]. The output of any neuron of a hidden layer or of the output layer can be represented in a similar way:

$$o_j = f_j\left(\sum_{i=1}^{N_i} w_{ji} x_{ji} + \theta_j\right) y_k = f_k\left(\sum_{j=1}^{N_j} w_{kj} o_{kj} + \theta_k\right), \tag{13.2}$$

where $f_j(\cdot)$ is logistic and $f_k(\cdot)$ is logistic or linear. Linear functions at the outputs are often used for pattern classification. In some tasks of pattern classification, a hidden layer is necessary, whereas in some tasks of automatic control, two hidden layers may be required. Hornik [146] showed that standard multilayer feedforward networks are able accurately to approximate any measurable function to a desired degree. Thus, they can be seen as *universal approximators*. In the case of a training failure, we should attribute any error to inadequate learning, an incorrect number of hidden neurons, or a poorly defined deterministic relationship between the input and output patterns.

Poggio and Girosi [238] developed the *radial basis function* (RBF) network, which consists of a superposition of weighted Gaussian functions:

$$y_j(\mathbf{x}) = \sum_{i=1}^{N} w_{ji} G_i\big(\mathrm{D}_i(\mathbf{x} - \mathbf{t}_i)\big), \tag{13.3}$$

where y_j is the j-output, $w_{ji} \in \mathcal{R}$, G_i is a Gaussian function, D_i is an $N \times N$ dilatation diagonal matrix, and $\mathbf{x}, \mathbf{t}_i \in \mathcal{R}^n$. The vector $\mathbf{t}_i$ is a translation vector. This architecture is supported by the regularization theory.

13.3 Complex MLP and Quaternionic MLP

An MLP is defined to be in the complex domain when its weights, activation function, and outputs are complex-valued. The selection of the activation function is not a trivial matter. For example, the extension of the sigmoid function from $\mathcal{R}$ to $\mathcal{C}$,

$$f(z) = \frac{1}{(1 + e^{-z})}, \tag{13.4}$$

where $z \in \mathcal{C}$, is not allowed, because this function is analytic and unbounded [104]; this is also true for the functions $\tanh(z)$ and e^{-z^2}. We believe these kinds of activation functions exhibit problems with convergence in training due to their singularities. The necessary conditions that a complex activation $f(z) = a(x, y) + ib(x, y)$ has to fulfill are: $f(z)$ must be nonlinear in x and y, the partial derivatives a_x, a_y, b_x, and b_y must exist ($a_x b_y \not\equiv b_x a_y$), and $f(z)$ must not be entire. Accordingly, Georgiou and Koutsougeras [104] proposed the formulation

$$f(z) = \frac{z}{c + \frac{1}{r}|z|}, \tag{13.5}$$

where $c, r \in \mathcal{R}^+$. These authors thus extended the traditional real-valued backpropagation learning rule to the complex-valued rule of the *complex multilayer perceptron* (CMLP).

Arena et al. [5] introduced the quaternionic multilayer perceptron (QMLP), which is an extension of the CMLP. The weights, activation functions, and outputs of this net are represented in terms of quaternions [119]. Arena et al. chose the following non-analytic bounded function

$$\begin{aligned} f(\boldsymbol{q}) &= f(q_0 + q_1 i + q_2 j + q_3 k), \\ &= (\frac{1}{1 + e^{-q_0}}) + (\frac{1}{1 + e^{-q_1}})i + (\frac{1}{1 + e^{-q_2}})j + (\frac{1}{1 + e^{-q_3}})k, \end{aligned} \tag{13.6}$$

where $f(\cdot)$ is now the function for quaternions. These authors proved that superpositions of such functions accurately approximate any continuous quaternionic function defined in the unit polydisc of $\mathcal{C}^n$. The extension of the training rule to the CMLP was demonstrated in [5].

13.4 Geometric Algebra Neural Networks

Real, complex, and quaternionic neural networks can be further generalized within the geometric algebra framework, in which the weights, the activation functions, and the outputs are now represented using multivectors. For the real-valued neural networks discussed in Sect. 13.2, the vectors are multiplied with the weights, using the scalar product. For geometric neural networks, the scalar product is replaced by the geometric product.

13.4.1 The Activation Function

The activation function of Eq. (13.5), used for the CMLP, was extended by Pearson and Bisset [227] for a type of Clifford MLP by applying different Clifford algebras, including quaternion algebra. We propose here an activation function that will affect each multivector basis element. This function was introduced independently by the authors [19] and is in fact a generalization of the function of Arena et al. [5]. The function for an n-dimensional multivector $\boldsymbol{m}$ is given by

$$\begin{aligned} \boldsymbol{f}(\boldsymbol{m}) &= \boldsymbol{f}(m_0 + m_i e_i + m_j e_j + m_k e_k + \cdots + m_{ij} e_i \wedge e_j + \cdots + \\ &\quad + m_{ijk} e_i \wedge e_j \wedge e_k + \cdots + m_n e_1 \wedge e_2 \wedge \cdots \wedge e_n), \\ &= (m_0) + f(m_i)e_i + f(m_j)e_j + f(m_k)e_k + \cdots + f(m_{ij})e_i \wedge e_j + \\ &\quad + \cdots + f(m_{ijk})e_i \wedge e_j \wedge e_k + \cdots + f(m_n)e_i \wedge e_j \wedge \ldots \wedge e_n, \end{aligned} \qquad (13.7)$$

where $\boldsymbol{f}(\cdot)$ is written in bold to distinguish it from the notation used for a single-argument function $f(\cdot)$. The values of $f(\cdot)$ can be of the sigmoid or Gaussian type.

13.4.2 The Geometric Neuron

The *McCulloch–Pitts neuron* uses the scalar product of the input vector and its weight vector [255]. The extension of this model to the *geometric neuron* requires the substitution of the scalar product with the Clifford or geometric product, i.e.,

$$\mathbf{w}^T\mathbf{x} + \theta \qquad \Rightarrow \qquad \boldsymbol{wx} + \boldsymbol{\theta} = \boldsymbol{w} \cdot \boldsymbol{x} + \boldsymbol{w} \wedge \boldsymbol{x} + \boldsymbol{\theta}. \qquad (13.8)$$

Figure 13.1 shows in detail the McCulloch–Pitts neuron and the geometric neuron. This figure also depicts how the input pattern is formatted in a specific geometric algebra. The geometric neuron outputs a richer kind of pattern. We can illustrate this with an example in $G_{3,0,0}$:

$$\begin{aligned} \boldsymbol{o} &= \boldsymbol{f}(\boldsymbol{wx} + \boldsymbol{\theta}), \qquad (13.9) \\ &= \boldsymbol{f}(s_0 + s_1 e_1 + s_2 e_2 + s_3 e_3 + s_4 e_1 e_2 + s_5 e_1 e_3 + s_6 e_2 e_3 + s_7 e_1 e_2 e_3), \\ &= f(s_0) + f(s_1)e_1 + f(s_2)e_2 + f(s_3)e_3 + f(s_4)e_1 e_2 + \cdots + \\ &\quad + f(s_5)e_1 e_3 + f(s_6)e_2 e_3 + f(s_7)e_1 e_2 e_3, \end{aligned}$$

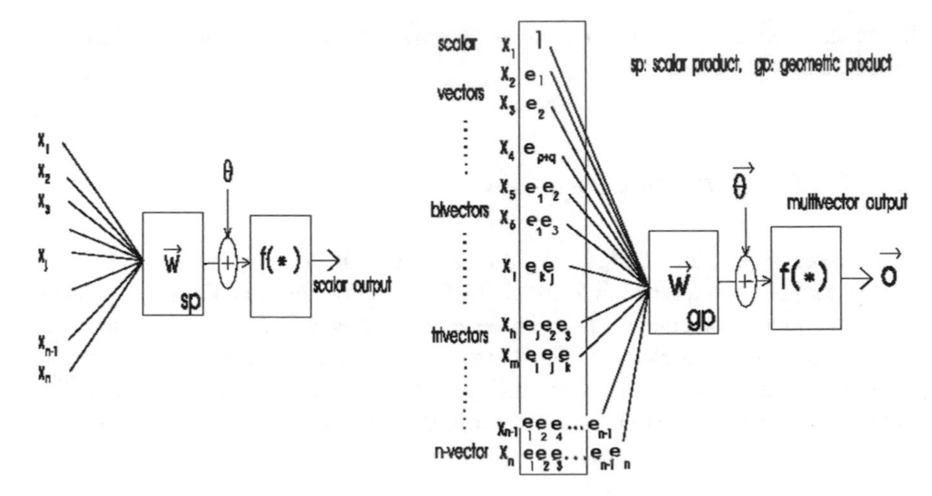

Fig. 13.1 McCulloch–Pitts neuron and geometric neuron

where f is the activation function defined in Eq. (13.7) and $s_i \in \mathcal{R}$. If we use the McCulloch–Pitts neuron in the real-valued neural network, the output is simply the scalar given by

$$o = f(\sum_i^N w_i x_i + \theta). \tag{13.10}$$

The geometric neuron outputs a signal with more geometric information

$$\boldsymbol{o} = \boldsymbol{f}(\boldsymbol{w}\boldsymbol{x} + \boldsymbol{\theta}) = \boldsymbol{f}(\boldsymbol{w} \cdot \boldsymbol{x} + \boldsymbol{w} \wedge \boldsymbol{x} + \boldsymbol{\theta}). \tag{13.11}$$

It has both a scalar product like the McCulloch–Pitts neuron

$$\boldsymbol{f}(\boldsymbol{w} \cdot \boldsymbol{x} + \theta) = f(s_0) \equiv f(\sum_i^N w_i x_i + \theta), \tag{13.12}$$

and also the outer product given by

$$\begin{aligned} \boldsymbol{f}(\boldsymbol{w} \wedge \boldsymbol{x} + \boldsymbol{\theta} - \theta) = f(s_1)e_1 + f(s_2)e_2 + f(s_3)e_3 + f(s_4)e_1e_2 + \cdots + \\ + f(s_5)e_1e_3 + f(s_6)e_2e_3 + f(s_7)e_1e_2e_3. \end{aligned} \tag{13.13}$$

Note that the outer product gives the scalar cross-products between the individual components of the vector, which are nothing more than the multivector components of points or lines (vectors), planes (bivectors), and volumes (trivectors). This characteristic can be used for the implementation of geometric preprocessing in the extended geometric neural network. To a certain extent, this kind of neural network resembles the higher-order neural networks of [230]. However, an extended

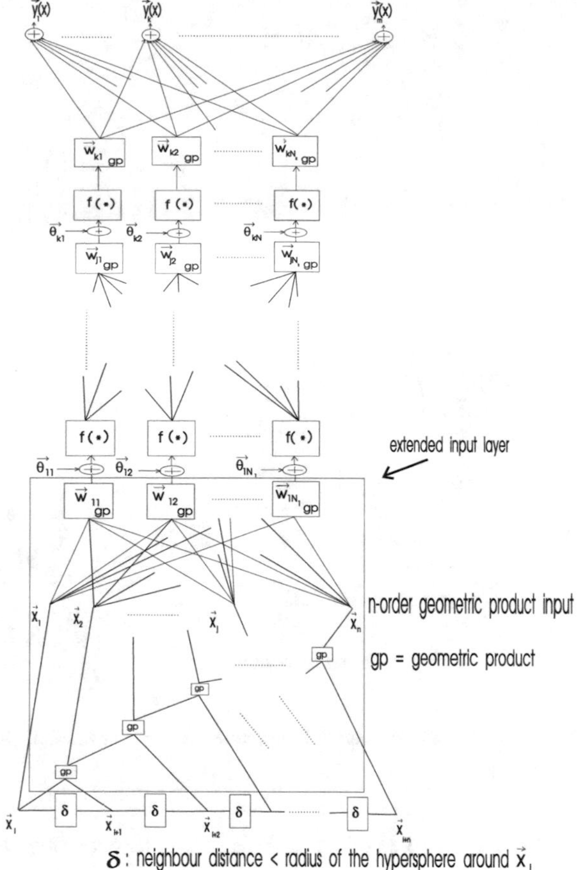

Fig. 13.2 Geometric neural network with extended input layer

geometric neural network uses not only a scalar product of higher order, but also all the necessary scalar cross-products for carrying out a *geometric cross-correlation*. Figure 13.2 shows a geometric network with its extended first layer.

In conclusion, a geometric neuron can be seen as a kind of *geometric correlation operator*, which, in contrast to the McCulloch–Pitts neuron, offers not only points but also higher-grade multivectors such as planes, volumes, and hypervolumes for interpolation.

13.4.3 Feedforward Geometric Neural Networks

Figure 13.3 depicts standard neural network structures for function approximation in the geometric algebra framework. Here, the inner vector product has been extended to the geometric product and the activation functions are according to (13.7).

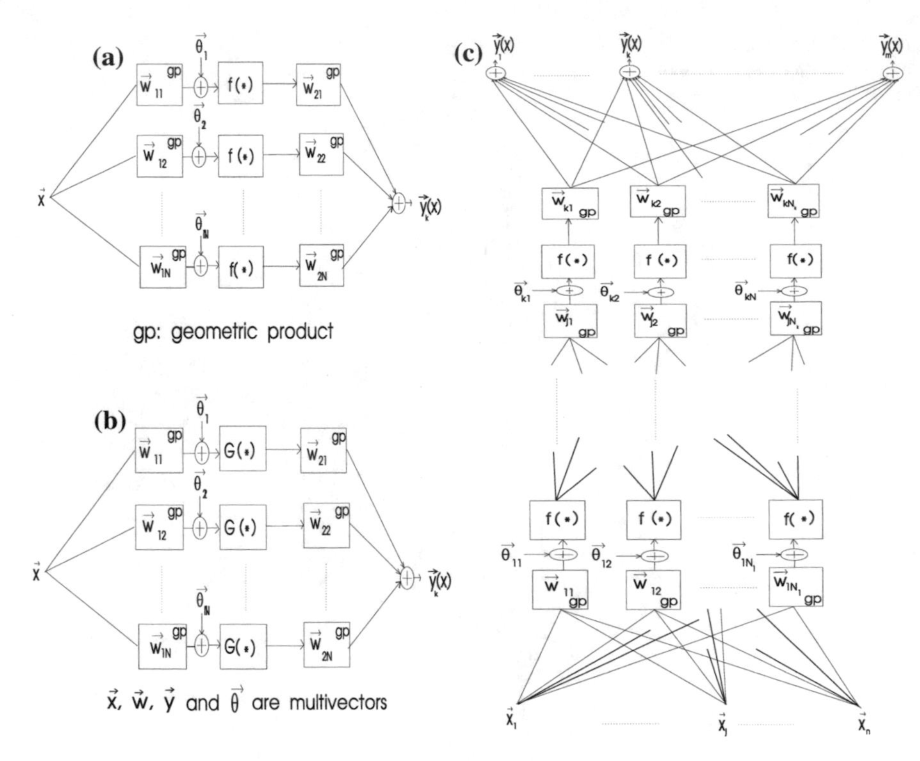

Fig. 13.3 Geometric network structures for approximation: **a** Cybenko's, **b** GRBF network, **c** GMLP$_{p,q,r}$

Equation (13.1) of Cybenko's model in geometric algebra is

$$\boldsymbol{y}(\boldsymbol{x}) = \sum_{j=1}^{N} \boldsymbol{w}_j \boldsymbol{f}(\boldsymbol{w}_j \cdot \boldsymbol{x} + \boldsymbol{w}_j \wedge \boldsymbol{x} + \boldsymbol{\theta}_j). \tag{13.14}$$

The extension of the MLP is straightforward. The equations using the geometric product for the outputs of hidden and output layers are given by

$$\begin{aligned} \boldsymbol{o}_j &= \boldsymbol{f}_j(\sum_{i=1}^{N_i} \boldsymbol{w}_{ji} \cdot \boldsymbol{x}_{ji} + \boldsymbol{w}_{ji} \wedge \boldsymbol{x}_{ji} + \theta_j), \\ \boldsymbol{y}_k &= \boldsymbol{f}_k(\sum_{j=1}^{N_j} \boldsymbol{w}_{kj} \cdot \boldsymbol{o}_{kj} + \boldsymbol{w}_{kj} \wedge \boldsymbol{o}_{kj} + \theta_k). \end{aligned} \tag{13.15}$$

In radial basis function networks, the dilatation operation, given by the diagonal matrix $\mathtt{D_i}$, can be implemented by means of the geometric product with a dilation $\boldsymbol{D}_i = e^{\alpha \frac{\bar{i}i}{2}}$ [138], i.e.,

$$\mathrm{D}_i(\mathbf{x} - \mathbf{t}_i) \Rightarrow \boldsymbol{D}_i(\boldsymbol{x} - \boldsymbol{t}_i)\tilde{\boldsymbol{D}}_i, \tag{13.16}$$

$$\boldsymbol{y}_k(\boldsymbol{x}) = \sum_{j=1}^{N} \boldsymbol{w}_{kj} \boldsymbol{G}_j(\boldsymbol{D}_j(\boldsymbol{x}_{ji} - \boldsymbol{t}_j)\tilde{\boldsymbol{D}}_j). \tag{13.17}$$

Note that in the case of the geometric RBF we are also using an activation function according to (13.7). Equation (13.17) with $\boldsymbol{w}_{kj} \in R$ represents the equation of an RBF architecture for multivectors of 2^n dimension, which is isomorph to a real-valued RBF network with 2^n-dimensional input vectors. In Sect. 13.8, we will show that we can use support vector machines for the automatic generation of an RBF network for multivector processing.

13.4.4 Generalized Geometric Neural Networks

One major advantage to use Clifford geometric algebra in neurocomputing is that the formulated neural network work for all types of multivectors: real, complex, double (or hyperbolic), and dual, as well as for different types of computing models, like horospheres and null cones (see [27, 134]. The chosen multivector basis for a particular geometric algebra $G_{p,q,r}$ defines the signature of the involved subspaces. The signature is computed by squaring the pseudoscalar: If $I^2 = -1$, the net will use complex numbers; if $I^2 = 1$, the net will use double or hyperbolic numbers; and if $I^2 = 0$, the net will use dual numbers (a degenerated geometric algebra). For example, for $G_{0,2,0}$, we can have a quaternion-valued neural network; for $G_{1,1,0}$, a hyperbolic MLP; for $G_{0,3,0}$, a hyperbolic (double) quaternion-valued RBF; or for $G_{3,0,0}$, a net which works in the entire Euclidean three-dimensional geometric algebra; for $G^+_{4,1,0}$, a net which works in the horosphere; or, finally, for $G^+_{3,3,0}$, a net which uses only the bivector null cone.

The conjugation involved in the training learning rule depends on whether we are using complex, hyperbolic, or dual-valued geometric neural networks, and varies according to the signature of the geometric algebra (see Eqs. (13.21–13.23)).

13.4.5 The Learning Rule

This section demonstrates the multidimensional generalization of the gradient descent learning rule in geometric algebra. This rule can be used for training the geometric MLP (GMLP) and for tuning the weights of the geometric RBF (GRBF). Previous learning rules for the real-valued MLP, complex MLP [104], and quaternionic MLP [5] are special cases of this extended rule.

13.4.6 Multidimensional Backpropagation Training Rule

The *norm of a multivector* $\boldsymbol{x}$ for the learning rule is given by

$$|\boldsymbol{x}| = (\boldsymbol{x}|\boldsymbol{x})^{\frac{1}{2}} = \Big(\sum_A [\boldsymbol{x}]_A^2\Big)^{\frac{1}{2}}. \tag{13.18}$$

The geometric neural network with n inputs and m outputs approximates the target mapping function

$$\mathcal{Y}_t : (G_{p,q,r})^n \rightarrow (G_{p,q,r})^m, \tag{13.19}$$

where $(G_{p,q,r})^n$ is the n-dimensional module over the geometric algebra $G_{p,q,r}$ [227]. The error at the output of the net is measured according to the metric

$$E = \frac{1}{2}\int_{x \in X} |\mathcal{Y}_w - \mathcal{Y}_t|^2, \tag{13.20}$$

where $\boldsymbol{X}$ is some compact subset of the Clifford module $(G_{p,q,r})^n$ involving the product topology derived from Eq. (13.188) for the norm and where $\mathcal{Y}_w$ and $\mathcal{Y}_t$ are the learned and target mapping functions, respectively. The *backpropagation algorithm* [255] is a procedure for updating the weights and biases. This algorithm is a function of the negative derivative of the error function (Eq. 13.190) with respect to the weights and biases themselves. The computing of this procedure is straightforward, and here, we will only give the main results. The updating equation for the multivector weights of any hidden j-layer is

$$\boldsymbol{w}_{ij}(t+1) = \eta\Big[\big(\sum_k^{N_k} \delta_{kj} \otimes \overline{\boldsymbol{w}_{kj}}\big) \odot \boldsymbol{F}'(\boldsymbol{net}_{ij})\Big] \otimes \overline{\boldsymbol{o}_i} + \alpha \boldsymbol{w}_{ij}(t), \tag{13.21}$$

for any k-output with a nonlinear activation function

$$\boldsymbol{w}_{jk}(t+1) = \eta\Big[(\boldsymbol{y}_{k_t} - \boldsymbol{y}_{k_a}) \odot \boldsymbol{F}'(\boldsymbol{net}_{jk})\Big] \otimes \overline{\boldsymbol{o}_j} + \alpha \boldsymbol{w}_{jk}(t), \tag{13.22}$$

and for any k-output with a linear activation function

$$\boldsymbol{w}_{jk}(t+1) = \eta(\boldsymbol{y}_{k_t} - \boldsymbol{y}_{k_a}) \otimes \overline{\boldsymbol{o}_j} + \alpha \boldsymbol{w}_{jk}(t). \tag{13.23}$$

In the above equations, $\boldsymbol{F}$ is the activation function defined in Eq. (13.7), t is the update step, η and α are the *learning rate* and the momentum, respectively, $\otimes$ is the Clifford or geometric product, $\odot$ is the scalar product, and $\overline{(\cdot)}$ is the *multivector anti-involution* (reversion or conjugation).

In the case of the non-Euclidean $G_{0,3,0}$, $\overline{(\cdot)}$ corresponds to the simple conjugation. Each neuron now consists of p+q+r units, each for a multivector component. The biases are also multivectors and are absorbed as usual in the sum of the activation signal, here defined as $\boldsymbol{net}_{ij}$. In the learning rules, Eqs. (13.21–13.23), the computation of the geometric product and the anti-involution varies depending on the geometric algebra being used [239]. To illustrate, the conjugation required in the learning rule for quaternion algebra is $\bar{\boldsymbol{x}} = x_0 - x_1e_1 - x_2e_2 - x_3e_1e_2$, where $\boldsymbol{x} \in G_{0,2,0}$.

13.4.7 Simplification of the Learning Rule Using the Density Theorem

Given $\boldsymbol{X}$ and $\boldsymbol{Y}$ as compact subsets belonging to $(\mathcal{G}_{p,q})^n$ and $(G_{p,q})^m$, respectively, and considering $\mathcal{Y}_t : \boldsymbol{X} \rightarrow \boldsymbol{Y}$ a continuous function, we are able to find some coefficients $w_1, w_2, w_3, ..., w_{N_j} \in \mathcal{R}$ and some multivectors $\boldsymbol{y}_1, \boldsymbol{y}_2, \boldsymbol{y}_3, ..., \boldsymbol{y}_{N_j} \in G_{p,q}$ and $\boldsymbol{\theta}_1, \boldsymbol{\theta}_2, \boldsymbol{\theta}_3, ..., \boldsymbol{\theta}_{N_j} \in G_{p,q}$ so that the following inequality $\forall \epsilon > 0$ is valid:

$$E(\mathcal{Y}_t, \mathcal{Y}_w) = sup\big[|\mathcal{Y}_t(\boldsymbol{x}) - \sum_{j=1}^{N_j} w_j \boldsymbol{f}_j(\sum_{i=1}^{N_i} \boldsymbol{w}_i\boldsymbol{x} + \boldsymbol{\theta}_i)|\boldsymbol{x} \in \boldsymbol{X}\big] < \epsilon, \quad (13.24)$$

where $\boldsymbol{f}_j$ is the multivector activation function of Eq. (13.7). Here, the approximation given by

$$S = \sum_{j=1}^{N_j} w_j \boldsymbol{f}_j(\sum_{i=1}^{N_i} \boldsymbol{w}_i\boldsymbol{x} + \boldsymbol{\theta}_i) \quad (13.25)$$

is the subset of the class of functions $C^0(G_{p,q})$ with the norm

$$|\mathcal{Y}_t| = sup_{\boldsymbol{x}\in X}|\mathcal{Y}_t(\boldsymbol{x})|. \quad (13.26)$$

And finally, since Eq. (13.24) is true, we can say that S is dense in $C^0(G_{p,q})$. The density theorem presented here is the generalization of the one used for the quaternionic MLP by Arena et al. [5].

The density theorem shows that the weights of the output layer for the training of geometric feedforward networks can be real values. Therefore, the training of the output layer can be simplified; i.e., the output weight multivectors can be the scalars of the blades of grade k. This k-grade element of the multivector is selected by convenience (see Eq. (2.65)).

13.4.8 Learning Using the Appropriate Geometric Algebras

The primary reason for processing signals within a geometric algebra framework is to have access to representations with a strong geometric character and to take advantage of the geometric product. It is important, however, to consider the type of

Table 13.1 Clifford or geometric algebras up to dimension 16

	q	→				
p	$\mathcal{R}$	$\mathcal{C}$	$\mathcal{H}$	${}^2\mathcal{H}$	$\mathcal{H}(2)$	$\mathcal{C}(4)$
	${}^2\mathcal{R}$	$\mathcal{R}(2)$	$\mathcal{C}(2)$	$\mathcal{H}(2)$	${}^2\mathcal{H}(2)$	$\mathcal{H}(4)$
↓	$\mathcal{R}(2)$	${}^2\mathcal{R}(2)$	$\mathcal{R}(4)$	$\mathcal{C}(4)$	$\mathcal{H}(4)$	${}^2\mathcal{H}(4)$
	$\mathcal{C}(2)$	$\mathcal{R}(4)$	${}^2\mathcal{R}(4)$	$\mathcal{R}(8)$	$\mathcal{C}(8)$	$\mathcal{H}(8)$
	$\mathcal{H}(2)$	$\mathcal{C}(4)$	$\mathcal{R}(8)$	${}^2\mathcal{R}(8)$	$\mathcal{R}(16)$	$\mathcal{C}(16)$

geometric algebra that should be used for any specific problem. For some applications, the decision to use the model of a particularly geometric algebra is straightforward. However, in other cases, without some a priori knowledge of the problem, it may be difficult to assess which model will provide the best results. If our preexisting knowledge of the problem is limited, we must explore the various network topologies in different geometric algebras. This requires some orientation in the different geometric algebras that could be used. Since each geometric algebra is either isomorphic to a matrix algebra of $\mathcal{R}$, $\mathcal{C}$, or $\mathcal{H}$, or simply the tensor product of these algebras, we must take great care in choosing the geometric algebras. Porteous [239] showed the *isomorphisms*

$$G_{p+1,q} = \mathcal{R}_{p+1,q} \cong G_{q+1,p} = \mathcal{R}_{q+1,p}, \tag{13.27}$$

and presented the following expressions for completing the *universal table of geometric algebras*:

$$\begin{aligned} G_{p,q+4} &= \mathcal{R}_{p,q+4} \cong \mathcal{R}_{p,q} \otimes \mathcal{R}_{0,4} \cong \mathcal{R}_{0,4} \cong \mathbb{H}(2), \\ G_{p,q+8} &= \mathcal{R}_{p,q+8} \cong \mathcal{R}_{p,q} \otimes \mathcal{R}_{0,8} \cong \mathcal{R}_{0,8} \cong \mathcal{R}(16), \end{aligned} \tag{13.28}$$

where $\otimes$ stands for the real tensor product of two algebras. Equation (13.28) is known as the periodicity theorem [239]. We can use Table 13.1, which presents the Clifford, or geometric, algebras up to dimension 16, to search for the appropriate geometric algebras. The entries of Table 13.1 correspond to the p and q of the $G_{p,q}$, and each table element is isomorphic to the geometric algebra $G_{p,q}$.

Examples of this table are the geometric algebras $\mathcal{R} \cong G_{0,0}$, $\mathcal{R}_{0,1} \cong \mathcal{C} \cong G_{0,1}$, $\mathcal{H} \cong \mathcal{G}_{0,2}$ and $\mathcal{R}_{1,1} \cong {}^2\mathcal{R} \cong G_{1,1}$, $\mathcal{C}(2) \cong \mathcal{C} \otimes \mathcal{R}(2) \cong G_{3,0} \cong G_{1,2}$ for the 3D space, and $\mathbb{H}(2) \cong \mathcal{G}_{1,3}$ for the 4D space.

13.5 Geometric Radial Basis Function Networks

A real-valued RBFN has a feedforward structure consisting of a single hidden layer of m locally tuned units, which are fully interconnected to an output layer of L linear units. Each connection is weighted by a real value. All hidden units receive the n-dimensional real-valued input vector X. Each hidden unit output z_j represents the closeness of the input X to an n-dimensional parameter vector c_j associated with the jth hidden unit. The response of the jth hidden unit ($j = 1, 2, ..., m$) is computed as

$$z_j = k\left(\frac{||X - c_j||}{\sigma^2{}_j}\right), \tag{13.29}$$

where $k(\cdot)$ is a strictly positive radially symmetric function (kernel) with a unique maximum at its "center" c_j and which drops off rapidly to zero away from the center and has an appreciable value only when the distance $||X - c_j||$ is smaller than σ_j. The parameter σ_j is the width of the respective field in the input space from unit j. Given an input vector X, the output of an RBFN is the L-dimensional activity vector Y whose lth component ($l = 1, 2, \ldots L$) is given by

$$Y_l(X) = \sum_{m=1}^{M} w_{lj} z_j(X), \tag{13.30}$$

where w_{lj} are weights in the network. The accuracy of these networks is controlled by three parameters: the number of basis functions, their location, and their width. These values and weights applied to the RBF function outputs are determined by the training process. K-means clustering is one of many methods to find the cluster centers of the RBF functions. It is convenient to use a gradient descent method to find the weights of the output layer.

13.5.1 Geometric RBF Networks

Let us define the mapping of the geometric RBF network as $g(P, X)$, where P is the set of parameters of our network, X is the input, and $g(\cdot)$ is the network depicted in Fig. 13.4), which is defined by hypercomplex-valued neurons. By adaptive learning, P is tuned so that the training data fit the network model as closely as possible. $g(\cdot)$ uses Gaussian basis functions to define the closeness of X to the geometric centers $\hat{\mathbf{c}}$. Our scheme uses rotors ($\hat{\mathbf{R}}$) as weights of the output layer, which are combined linearly to define the output Y of the network. $X, Y, \hat{\mathbf{R}}, \hat{\mathbf{c}}$ are in $\mathcal{G}_3$ and are normalized. These parameters and the σ_i values of each center define P. The geometric entity X, which can be a line, point, plane, or any other geometric entity in the appropriate $\mathcal{G}_n$, will be transformed by rotors considering each Gaussian basis function result, each of which represents a closeness factor of the geometric entity X to each geometric center. These factors will be useful during the geometric transformation of centers by rotors in a phased manner until the network output is close to the expected output. These factors help in the training step because they define the output error by applying rotors to the geometric centers depending on the closeness of X to each center. Given that each neuron is defined by geometric entities and operators (addition, distance, rotor), each neuron can work with hypercomplex entities, which can be defined by $\mathcal{G}_n$.

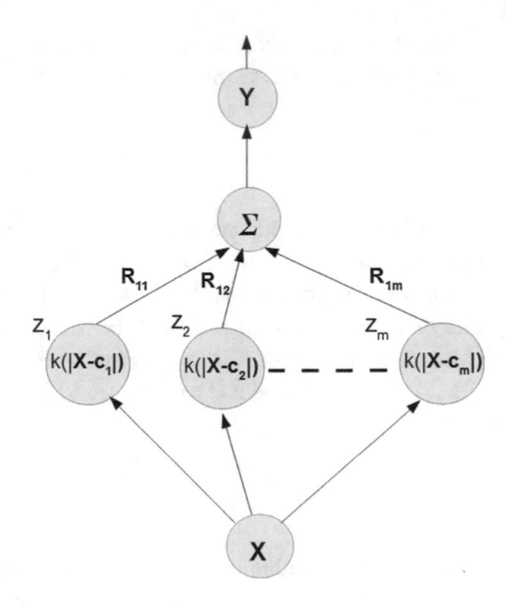

Fig. 13.4 Hypercomplex-valued RBF network

13.5.2 Training of Geometric RBF Networks

The training set is a labeled pair X_i, Y_i that represents associations of a given mapping. Given a number of centers, the adaptive learning process starts when the first training input pair is presented to the network. A geometric radial basis function network uses the K-means clustering algorithm to define the centers of each Gaussian basis function and the Euclidean distance to indicate the nearness of X_i to each center c_i. The parameter σ_i is defined for each Gaussian unit to equal the maximum distance between centers. To determine the $\hat{\mathbf{R}}$ that best approximates the mapping between X_i and Y_i, a GRFBN uses an ordinary least-squares method (LMS) for training. Our scheme updates the $\hat{R}$. To rotate the center c_i, we use the associated R_i (weight) and the respective radial basis function output. As we can see, the K-means clustering algorithm and LMS algorithm proceed with their own individual computations concurrently, which accelerates the training process. Our GRBFN works iteratively as follows:

(1) $Initialization$. Choose random values for initial centers $c_k(0)$.
(2) Iterate until no noticeable changes are observed in the centers c_k.

   (a) $Sampling$. Draw a sample input **x** coded by a multivector from the input space I. **x** is input into the algorithm at iteration n.
   (b) *Similarity definition. k(x)* denotes the index of the best-matching center for input **x**. Find *k(x)* at iteration n by using the minimum-distance Euclidean criterion or directance:
   $k(x) = \underset{k}{\text{argmin}}||x(n) - c_k(n)||, k = 1, 2..., m,$
   $c_k(n)$ is the center of the kth radial function at the nth iteration. Note that

the dimension of the k-vector x is the same as that of the k-vector c_k and in this case is 3.

(c) *Updating*. Adjust the centers of the radial basis functions:

$$c_k(n+1) = \begin{cases} c_k(n) + \rho[x(n>) - c_k(n)], & \text{if } k = k(x), \\ c_k(n), & \textit{otherwise} \end{cases}$$

ρ is a *learning-rate parameter* in the range $0 < \rho < 1$.

(d) *Error*.
$e(n) = Y(n) - \sum_{i=1}^{m} R_i(n)\,(z_i(n)c_i(n))\,\widetilde{R}_i(n)$.

(e) *Rotor vector updating*.
$R_i(n+1) = R_i(n) + \eta z_i(n)e(n)$.
η is a *learning-rate parameter* in the range $0 < \eta < 1$.

(f) *Continuation*. Increment n by 1,

where $z_i(n) \in \mathbb{R}$ and $R_i(n), c_i(n), X_i(n), Y_i(n) \in \mathcal{G}_3$. As we can observe, the GRBFN uses the Clifford product.

Note that the network is defined by

$$g(P, X, n) = \sum_{i=1}^{M} R_i\,(z_i(n)c_i(n).)\,\widetilde{R}_i. \tag{13.31}$$

In this equation, one can see that the factors $z_i(n)$ help in the training step because they determine the closeness of X to the geometric centers $\hat{\mathbf{c}}$ in such a way that they help determine the network output error via applying rotors to the geometric entity of the center $c_i(n)$. At the end, the network output is a geometric entity: the linear combination of the rotation of each geometric center using the factors.

13.6 Spherical Radial Basis Function Network

Returning to the idea of the RBFN, we propose the use of hyperspheres (circles, spheres) as activation functions of the neurons in the hidden layer. There are several of these activation functions in the literature; more references can be found in [192]. The most classical RBF uses the Gaussian function and is known as a universal approximator [238].

Here, the network will strictly use the circles or spheres. Our motivation to use circles or spheres is due to the fact that the sphere is the computational unity of the conformal geometric algebra. In the literature, no work has reported that it uses this kind of activation function. In the geometric algebra framework, the circle and sphere are represented in the IPNS space as vectors—quite a big difference instead of using cumbersome quadratic equations to represent circles or spheres. The problem of locating circles or spheres depends on the knowledge of their radius and centroid when using the representation in the conformal GA. Adjusting the parameters is similar to that when adjusting Gaussian functions.

During the training of the network, given a set of N points, the circles or spheres are added one by one until the mean square error function reaches an expected lower value. The maximum number of neurons is N, and the allowed maximum number of epochs is M. Since the network uses the IPNS vector representation for the spheres, each allocated sphere is a sort of support vector to reconstruct the surface. This network is called the spherical radial basis function network (SRBFN). Following the layer of the spherical neurons or support vectors, which is aimed to reconstruct a surface, comes a coupled layer that is useful for a smooth interpolation of the elasticity of the surface. This layer uses the general regression neural networks (GRNN), a special type of probabilistic networks using density probability functions

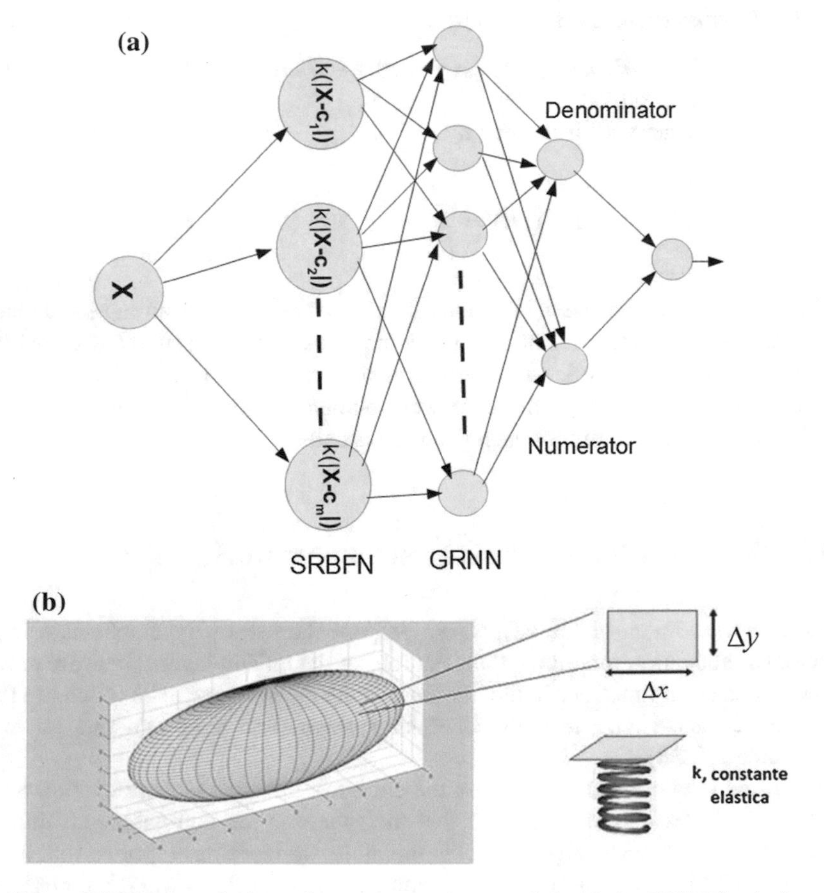

Fig. 13.5 **a** SRBFN network, for surface reconstruction, coupled with the GRNN for the interpolation of the elasticity; **b** surface and associated elasticity at the bump

as activation functions, for the interpolation; see [30, 285]. These networks are widely used in control, in time series, and for the approximation of functions. The use of the SRBFN together with the GRNN can be of great use as an interface between the sensor domain and the robotic mechanism, because the SRBFN reconstructs the surface and, in addition, the GRNN interpolates its elasticity, which varies at each point of the surface. A depiction of the SRBFN together with the GRNN is shown in Fig. 13.5. From a biological perspective, the use of the networks helps to fuse visual data perceived by our eyes and the flexibility data of the surface perceived by our sense of touch.

13.7 Quaternion Wavelet Function Network

Inspired in the radial basis function network, in the past, Bayro [17] proposed for the first time the wavelet neural network. Now, we will formulate the Quaternion Wavelet Function Network (QWFN) as $g(P, X)$, where P is the set of parameters of our network, X is the input, and $g(\cdot)$ is the network depicted in Fig. 13.6, which is defined by quaternion wavelet valued at each hidden neuron. By adaptive learning, P is tuned so that the training data fit the network model as closely as possible. $g(\cdot)$ uses quaternion wavelet basis functions in each neuron to define the closeness of X to the affine transformation of the wavelets, for that a learning rule has to adjust the quaternion wavelet centers $\hat{\mathbf{c}}$ and their wide. Our model uses quaternions (rotors) as weights at the output layer to rotate the output quaternion of the hidden layer. These rotated outputs are combined linearly to define the output Y of the network. The

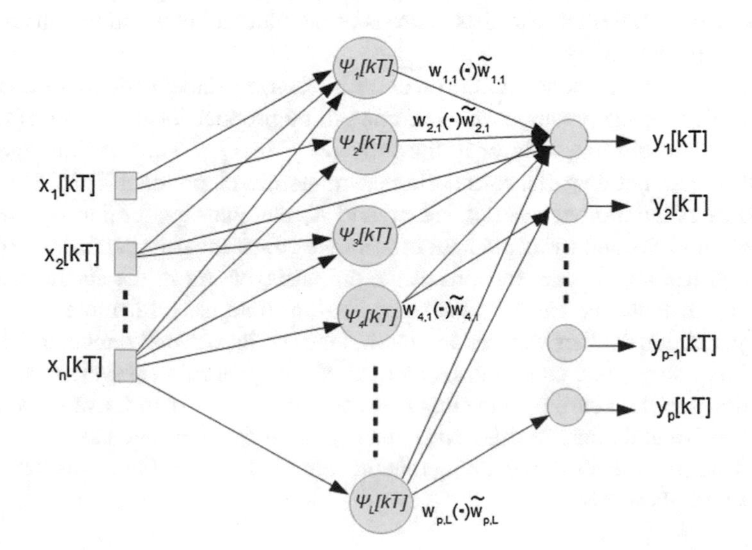

Fig. 13.6 Quaternion wavelet function network

quaternion wavelet parameters for each neuron and the quaternion for rotation of the output layer define P. These factors will be useful during the parameter adaptation like the centers and wide of the quaternion wavelets and the quaternion (rotors) $\boldsymbol{q}_{ij}(\boldsymbol{o}_i)\tilde{\boldsymbol{q}}_{ij}$ acting on the output layer or outstar. The parameter adaptation runs until the network output is close to the expected output. Since each neuron is defined by a quaternion wavelet, the quaternion inputs are convolved with the quaternion wavelet per each neuron and consequently filtered. The resulting quaternion outputs are further rotated and gather to compute the quaternion outputs.

13.7.1 Training of the Quaternion Wavelet Function Network

The training set is a labeled pair X_i, Y_i that represents associations of a given mapping. Similar as in a standard RBF network, given a number of centers, the adaptive learning process starts adjusting to the centers of the mother quaternion wavelets of each quaternion neuron of the hidden layer and then adjusting to the parameters of their quaternion wavelet daughters. A quaternion wavelet function network uses the K-means clustering algorithm to define the center of each mother quaternion wavelet, and the Euclidean distance indicates the nearness of X_i to each center c_i. The indexed parameters of the daughter quaternion wavelets for each neuron are computed so that the coming quaternion signal for each hidden quaternion neuron is well expanded by this quaternion wavelet basis. For that basically, one expands the incoming quaternion signal with the quaternion wavelet transform. To determine the best approximated mapping between X_i and Y_i, a QWFN uses an ordinary least-squares method (LMS) for training. Our scheme updates the center and indexes of the quaternion wavelets neurons at the hidden layer and the quaternions (rotors) at the output layer.

Since the each component of the quaternion neuron is independent, we apply the 1D real valued to each component. The quaternion product for the rotations will be applied only at the output layer in the form of $\boldsymbol{q}_{ij}(\boldsymbol{o}_i)\tilde{\boldsymbol{q}}_{ij}$. The training algorithm consists of mainly two parts: considering the quaternion input data $X_i \in \mathbb{H}$ we fixed the centers of the mother wavelet and expand K_i daughter quaternion wavelets for each i-neuron. Second using gradient descent we adjust the four parameters of each quaternion (rotors) for the rotations at the output layer. As in the standard RBFN on the top of that, one can pass backpropagation to adjust a bit more the centers of the quaternion mother wavelet and the indexes of its wavelet daughters. Also at this training stage, one can vary the amount K_i of quaternion wavelets at each i-quaternion wavelet neuron. At the first pass, one fix $K_h = K$ to a fix value, and then in the iterative updating, one can leave this degree of freedom open and the QWFN will find the suitable K_i. The learning rule procedure of the QWFN works iteratively, and it is described next.

The X_i for N inputs are quaternion signals, and the quaternion signals at the M hidden quaternion neurons are expanded in each quaternion neuron to K_i quaternion wavelets. Since the four components of the quaternion neuron are independent of each other and denoted by the index $r = 1, ..., 4$, we will expand each component simply using a real-valued wavelet expansion; the mother wavelet ψ_l^r of type RASP1 [64], with associate parameters of dilation and contraction $a_l^r[kT]$ and translation $b_{p,l}^r$ $(r = 1, ..., 4)$, is as follows:

$$\tau_l^r[kT] = \frac{\sqrt{\left(\sum_{p=1}^{P}(u_p^r[kT] - b_{p,l}^r[kT])^2\right)}}{a_l^r[kT]},$$
$$\psi^r[kT] = \frac{\tau_l^r[kT]}{(\tau_l^r[kT]+1)^2}, \quad (13.32)$$

where the discrete partial derivative with respect to $b_{p,l}^r[kT]$ is given by

$$\frac{\partial\psi_l^r[kT]}{\partial b_{p,l}^r[kT]} = \frac{3\tau_l^r[kT]^2 - 1}{a_l^r[kT](\tau_l^{r2}[kT]+1)^3}. \quad (13.33)$$

The outputs $y_p^r[kT]$ for $p = 1, ..., P$ are given by

$$y_p[kT] = \sum_{i=1}^{M}(\boldsymbol{w}_{ip}\psi_i\tilde{\boldsymbol{w}}_{ip}), \quad (13.34)$$

where $\boldsymbol{w}_{ip}$ are unit quaternions to rotate the quaternion wavelets of the hidden quaternion neurons.

The parameters $a_l^r[kT]$, $b_{p,l}^r[kT]$ and $w_i^r[kT]$ are represented as a vectors. For the updating the parameters, we resort to the descent gradient using the least mean square (LMS) function

$$E[kT] = [E_1[kT]E_2[kT]...E_p[kT]...E_p[kT]]^T,$$
$$= \frac{1}{2}\sum_{k}^{T}\sum_{r=1}^{4}(e_p^r)^2(k), \quad (13.35)$$

where $e_p^r = y_p^r[kT] - \hat{y}_p^r[kT]$ is the error quaternion componentweise. We apply the descent gradient

$$\Delta W^r[kT] = -\frac{\partial E[kT]}{\partial W^r[kT]} = -e^r_p[kT](\dot{w}^r_l \psi^r_l w^r_l[kT] + w^r_l \psi^r_l \dot{\tilde{w}}^r_l[kT]),$$
$$\Delta B^r[kT] = -\frac{\partial E[kT]}{\partial B^r[kT]} = -e^r_p[kT]\frac{\partial \psi^r_l[kT]}{\partial b^r_{p,l}[kT]} w^r_{p,l},$$
$$\Delta A^r[kT] = -\frac{\partial E[kT]}{\partial A^r[kT]} = \tau_l \frac{\partial E[kT]}{\partial B^r[kT]}. \quad (13.36)$$

The updating of the parameters follows:

$$W^r[(k+1)T] = W^r[kT] + \eta \Delta W^r[kT], \quad (13.37)$$
$$B^r[(k+1)T] = B^r[kT] + \eta \Delta B^r[kT],$$
$$A^r[(k+1)T] = A^r[kT] + \eta \Delta A^r[kT],$$

for $r = 1, ..., 4$.

13.8 Support Vector Machines in Geometric Algebra

The *support vector machine* (SV machine) developed by Vladimir N. Vapnik [303] applies optimization methods for learning. Using SV machines, we can generate a type of two-layer network and RBF networks, as well as networks with other kernels. Our idea is to generate neural networks by using SV machines in conjunction with geometric algebra and thereby in the neural processing of multivectors. We will call our approach the *support multivector machine* (SMVM). We shall review SV machines briefly and then explain the SMVM.

The SV machine maps the input space R^d into a high-dimensional *feature space* H, given by $\Phi : R^d \Rightarrow H$, satisfying a kernel $K(\boldsymbol{x}_i, \boldsymbol{x}_j) = \Phi(x_i) \cdot \Phi(x_j)$, which fulfills *Mercer's condition* [303]. The SV machine constructs an optimal hyperplane in the feature space that divides the data into two clusters.

SV machines build the mapping

$$f(\boldsymbol{x}) = sign\Big(\sum_{supportvectors} y_i \alpha_i K(\boldsymbol{x}_i, \boldsymbol{x}) - b \Big). \quad (13.38)$$

The coefficients α_i in the separable case (and analogously in the non-separable case) are found by maximizing the functional based on Lagrange coefficients:

$$W(\alpha) = \sum_{i=1}^{l} \alpha_i - \frac{1}{2} \sum_{i,j}^{l} \alpha_i \alpha_j y_i y_j K(x_i, x_j), \quad (13.39)$$

subject to the constraints $\sum_{i=1}^{l} \alpha_i y_i = 0$, where $\alpha_i \geq 0$, i=1,2,...,l. This functional coincides with the functional for finding the optimal hyperplane.

Examples of SV machines include

$$K(\boldsymbol{x}, \boldsymbol{x}_i) = [(\boldsymbol{x} \cdot \boldsymbol{x}_i) + 1]^d \text{ (polynomial learning machines),} \tag{13.40}$$

$$K_\gamma(|x - x_i|) = exp\{-\gamma|\boldsymbol{x} - \boldsymbol{x}_i|^2\} \text{ (radial basis functions machines),} \tag{13.41}$$

$$K(\boldsymbol{x}, \boldsymbol{x}_i) = S\Big(v(\boldsymbol{x} \cdot \boldsymbol{x}_i) + c\Big) \text{ (two-layer neural networks).} \tag{13.42}$$

13.9 Linear Clifford Support Vector Machines for Classification

For the case of the Clifford SVM for classification, we represent the data set in a certain Clifford algebra $\mathcal{G}_n$, where $n = p + q + r$, where any multivector base squares to 0, 1 or -1 depending on whether they belong to p, q, or r multivector bases, respectively. We consider the general case of an input comprising D multivectors and one multivector output; i.e., each data ith-vector has D multivector entries $\boldsymbol{x}_i = [\boldsymbol{x}_{i1}, \boldsymbol{x}_{i2}, ..., \boldsymbol{x}_{iD}]^T$, where $\boldsymbol{x}_{ij} \in G_n$ and D is its dimension. Thus, the ith-vector dimension is D$\times 2^n$; then, each data ith-vector $\boldsymbol{x}_i \in G_n^D$. This ith-vector will be associated with one output of the 2^n possibilities given by the following multivector output:

$$\boldsymbol{y}_i = y_{i\,s} + y_{i\,e_1} + y_{i\,e_2} + \cdots + y_{i\,I} \in \{\pm 1 \pm e_1 \pm e_2 \ldots \pm I\},$$

where the first subindex s stands for scalar part, e.g., $2^2 = 4$ outputs for quaternions (or $G_{0,2,0}$) $\boldsymbol{y}_i = y_{i\,s} + y_{i\,e_2e_3} + y_{i\,e_3e_1} + y_{i\,e_1e_2} \in \{\pm 1 \pm e_2e_3 \pm e_3e_1 \pm e_1e_2\}$. For the classification, the CSVM separates these multivector-valued samples into 2^n groups by selecting a good enough function from the set of functions

$$\boldsymbol{f}(\boldsymbol{x}) = \boldsymbol{w}^{\dagger^T} \boldsymbol{x} + \boldsymbol{b}, \tag{13.43}$$

where $\boldsymbol{x}, \boldsymbol{w} \in G_n^D$ and $\boldsymbol{f}(\boldsymbol{x}), \boldsymbol{b} \in G_n$. An entry of the optimal hyperplane

$$\boldsymbol{w} = [\boldsymbol{w}_1, \boldsymbol{w}_2, ..., \boldsymbol{w}_k, ..., \boldsymbol{w}_D]^T, \tag{13.44}$$

is given by

$$\boldsymbol{w}_k = w_{k\,s} + \ldots + w_{k\,e_1e_2} e_1e_2 + \cdots + w_{k\,I} I \in G_n. \tag{13.45}$$

Let us see in detail the last function equation:

$$\begin{aligned} f(x) &= w^{\dagger^T} x + b, \\ &= [w_1^\dagger, w_2^\dagger, ..., w_D^\dagger]^T [x_1, x_2, ..., x_D] + b, \\ &= \sum_{i=1}^{D} w_i^\dagger x_i + b, \end{aligned} \tag{13.46}$$

where $w_i^\dagger x_i$ corresponds to the Clifford product of two multivectors and $w_i^\dagger$ is the reversion of the multivector w_i.

Next, we introduce a structural risk similar to the real-valued SVM for classification. By using a loss function similar to Vapnik's ξ-*insensitive* one, we utilize the following linear constraint quadratic programming for the *primal equation*:

$$\begin{aligned} &min\ L(w, b, \xi)_P = \frac{1}{2} w^{\dagger^T} w + C \sum_{i,j} \xi_{ij}, \\ &subject\ to \\ &y_{ij}(f(x_i))_j = y_{ij}(w^{\dagger^T} x_i + b)_j >= 1 - \xi_{ij}, \\ &\xi_{ij} >= 0 \texttt{ for all } i, j, \end{aligned} \tag{13.47}$$

where ξ_{ij} stands for the slack variables, i indicates the data ith-vector and j indexes the multivector component, i.e., $j = 1$ for the coefficient of the scalar part, $j = 2$ for the coefficient of e_1,..., $j = 2^n$ for the coefficient of I.

By using Lagrange multipliers techniques [45, 92], we obtain the *Wolfe* dual programming of Eq. (13.47):

$$max\ L(w, b, \xi)_D = \sum_{i,j} \alpha_{ij} - \frac{1}{2} w^{\dagger^T} w, \tag{13.48}$$

subject to $a^T \cdot \mathbf{1} = 0$, and all the Lagrange multipliers for each entry w_k should fulfill $0 \leq (\alpha_{k_s})_j \leq C$, $0 \leq (\alpha_{k_{e_1}})_j \leq C$, ..., $0 \leq (\alpha_{k_{e_1 e_2}})_j \leq C$, ..., $0 \leq (\alpha_{k_I})_j \leq C$ for $j = 1, ..., l$. In $a^T \cdot \mathbf{1} = 0$, $\mathbf{1}$ denotes a vector of all ones and the entries of the vector

$$\begin{aligned} a = \ & [[a_{1_s}, a_{1_{e_1}}, a_{1_{e_2}}, \ldots, a_{1_{e_1 e_2}}, \ldots, a_{1_I}], \ldots, \\ & , [a_{k_s}, a_{k_{e_1}}, a_{k_{e_2}}, \ldots, a_{k_{e_1 e_2}}, \ldots, a_{k_I}], \ldots, \\ & , [a_{D_s}, a_{D_{e_1}}, a_{D_{e_2}}, \ldots, a_{D_{e_1 e_2}}, \ldots, a_{D_I}]. \end{aligned} \tag{13.49}$$

are given by

$$
\begin{aligned}
\boldsymbol{a}_{k_s}^T &= [(\alpha_{k_s})_1(y_{k_s})_1, (\alpha_{k_s})_2(y_{k_s})_2, ..., (\alpha_{k_s})_l(y_{k_s})_l],\\
\boldsymbol{a}_{k_{e_1}}^T &= [(\alpha_{k_{e_1}})_1(y_{k_{e_1}})_1, \ldots, (\alpha_{k_{e_1}})_l(y_{k_{e_1}})_l],\\
&\cdot\\
&\cdot\\
&\cdot\\
\boldsymbol{a}_{k_I}^T &= [(\alpha_{k_I})_1(y_{k_I})_1, (\alpha_{k_I})_2(y_{k_I})_2, ..., (\alpha_{k_I})_l(y_{k_I})_l],
\end{aligned}
\tag{13.50}
$$

note that the vector $\boldsymbol{a}^T$ has dimension $(D \times 2^n \times l) \times 1$.

Consider the optimal weight vector $\boldsymbol{w}$ in the equation (13.48), an entry of it is given by $\boldsymbol{w}_k$, see Eq. (13.45). Each of the components of $\boldsymbol{w}_k$ is computed applying the KKT conditions to the Lagrangian of Eq. (13.47) using l multivector samples as follows:

$$
\begin{aligned}
w_{ks} &= \sum_{j=1}^{l} \Big((\alpha_s)_j(y_s)_j\Big)(x_{ks})_j,\\
w_{ke_1} &= \sum_{j=1}^{l} \Big((\alpha_{e_1})_j(y_{e_1})_j\Big)(x_{ke_1})_j,\\
&\cdots\\
w_{kI} &= \sum_{j=1}^{l} \Big((\alpha_I)_j(y_I)_j\Big)(x_{iI})_j,
\end{aligned}
\tag{13.51}
$$

where $(\alpha_s)_j, (\alpha_{e_1})_j, ..., (\alpha_I)_j,\ j = 1, ..., l$, are the Lagrange multipliers.

The threshold $\boldsymbol{b} \in G_n$ can be computed by using the KKT conditions with the Clifford support vectors as follows:

$$
\begin{aligned}
\boldsymbol{b} &= b_s + b_{e_1}e_1 + \cdots + b_{e_1e_2}e_1e_2 + \cdots + b_I I,\\
&= \sum_{j=1}^{l}(\boldsymbol{y}_j - \boldsymbol{w}^{\dagger^T}\boldsymbol{x}_j)/l.
\end{aligned}
\tag{13.52}
$$

However, it is desirable to formulate a compact representation of the ensuing *Gramm matrix* involving multivector components; this will certainly help in the programing of the algorithm. For that, let us first consider the Clifford product of $\boldsymbol{w}^{\dagger^T}\boldsymbol{w}$, which can be expressed as follows:

$$
\boldsymbol{w}^{\dagger^T}\boldsymbol{w} = \langle \boldsymbol{w}^{\dagger^T}\boldsymbol{w}\rangle_s + \langle \boldsymbol{w}^{\dagger^T}\boldsymbol{w}\rangle_{e_1} + \langle \boldsymbol{w}^{\dagger^T}\boldsymbol{w}\rangle_{e_2} + \ldots + \langle \boldsymbol{w}^{\dagger^T}\boldsymbol{w}\rangle_I. \tag{13.53}
$$

Since $\boldsymbol{w}$ has the components presented in (13.51), Eq. (13.53) can be rewritten as follows:

$$
\begin{aligned}
w^{\dagger^T} w &= a_s^T \langle x^{\dagger^T} x\rangle_s a_s + \cdots + a_s^T \langle x^{\dagger^T} x\rangle_{e_1e_2} a_{e_1e_2} + \\
&+ \cdots + a_s^T \langle x^{\dagger^T} x\rangle_I a_I + a_{e_1}^T \langle x^{\dagger^T} x\rangle_s a_s + \ldots + \\
&+ a_{e_1}^T \langle x^{\dagger^T} x\rangle_{e_1e_2} a_{e_1e_2} + \cdots + a_{e_1}^T \langle x^{\dagger^T} x\rangle_I a_I + \ldots + \\
&+ a_I^T \langle x^{\dagger^T} x\rangle_s a_s + a_I^T \langle x^{\dagger^T} x\rangle_{e_1} a_{e_1} + \cdots + \\
&+ a_I^T \langle x^{\dagger^T} x\rangle_{e_1e_2} a_{e_1e_2} + \cdots + a_I^T \langle x^{\dagger^T} x\rangle_I a_I.
\end{aligned}
\tag{13.54}
$$

Renaming the matrices of the t-grade parts of $\langle x^{\dagger^T} x\rangle_t$, we rewrite previous the equation as:

$$
\begin{aligned}
w^{\dagger^T} w &= a_s^T H_s a_s + a_s^T H_{e_1} a_{e_1} + a_s^T H_{e_1e_2} a_{e_1e_2} + \\
&+ \cdots + a_s^T H_I a_I + a_{e_1}^T H_s a_s + a_{e_1}^T H_{e_1} a_{e_1} + \cdots + \\
&+ a_{e_1}^T H_{e_1e_2} a_{e_1e_2} + \cdots + a_{e_1}^T H_I a_I + \cdots + \\
&+ a_I^T H_s a_s + a_I^T H_{e_1} a_{e_1} + \cdots + a_I^T H_{e_1e_2} a_{e_1e_2} + \\
&+ \cdots + a_I^T H_I a_I.
\end{aligned}
\tag{13.55}
$$

We gather the submatrices of the t-grade parts of $\langle x^{\dagger^T} x\rangle_t$ in a positive semidefinite matrix $\boldsymbol{H}$ that is the expected generalized $Gram$ matrix:

$$
\boldsymbol{H} = \begin{bmatrix}
H_s H_{e_1} H_{e_2} \ldots\ldots\ldots\ldots\ldots H_{e_1e_2} \ldots H_I \\
H_{e_1}^T H_s \ldots H_{e_4} \ldots\ldots H_{e_1e_2} \ldots H_I H_s \\
H_{e_2}^T H_{e_1}^T H_s \ldots H_{e_1e_2} \ldots H_I H_s H_{e_1} \\
\cdot \\
\cdot \\
\cdot \\
H_I^T \ldots H_{e_1e_2}^T \ldots\ldots\ldots\ldots H_{e_2}^T H_{e_1}^T H_s
\end{bmatrix}.
\tag{13.56}
$$

Note that the diagonal entries are equal to H_s, and since H is a symmetric matrix, the lower matrices are transposed.

Finally, using the previous definitions and equations, the Wolfe dual programming can be written as follows:

$$
max\ L(\boldsymbol{w}, \boldsymbol{b}, \boldsymbol{\xi})_D = \sum_{i,j} \alpha_{ij} - \frac{1}{2}\boldsymbol{a}^T \boldsymbol{H} \boldsymbol{a}
\tag{13.57}
$$

subject to $\boldsymbol{a}^T \cdot \mathbf{1} = 0, 0 \leq (\alpha_{k_s})_j \leq C, 0 \leq (\alpha_{k_{e_1}})_j \leq C, ..., 0 \leq (\alpha_{k_{e_1e_2}})_j \leq C, ..., 0 \leq (\alpha_{k_I})_j \leq C$ for $j = 1, ..., l$. where $\boldsymbol{a}$ is given by Eq. (13.49).

13.10 Nonlinear Clifford Support Vector Machines for Classification

For the nonlinear Clifford-valued classification problems, we require a Clifford algebra-valued kernel $K(\boldsymbol{x}, \boldsymbol{y})$. In order to fulfill the Mercer theorem, we resort to a componentwise Clifford algebra-valued mapping

$$\boldsymbol{x} \in G_n \stackrel{\phi}{\longrightarrow} \Phi(\boldsymbol{x}) = \Phi_s(x) + \Phi_{e_1} e_1 + \Phi_{e_1} e_2(x) e_2 + \cdots + I\Phi_I(x) \in G_n. \quad (13.58)$$

In general, we build a Clifford kernel $K(x_m, x_j)$ by taking the Clifford product between the conjugated of $\boldsymbol{x_m}$ and $\boldsymbol{x}_j$ as follows:

$$K(\boldsymbol{x}_m, \boldsymbol{x}_j) = \boldsymbol{\Phi}(\boldsymbol{x})^{\dagger}\boldsymbol{\Phi}(\boldsymbol{x}). \quad (13.59)$$

Next, as an illustration we present kernels using different geometric algebras. According to the Mercer theorem, there exists a mapping $\boldsymbol{u} : G \rightarrow \mathcal{F}$, which maps the multivectors $\boldsymbol{x} \in G_n$ into the complex Euclidean space:

$$\boldsymbol{x} \stackrel{v}{\rightarrow} u(\boldsymbol{x}) = u_r(\boldsymbol{x}) + I u_I(\boldsymbol{x}). \quad (13.60)$$

Recall that the center of a geometric algebra, i.e., $\{s, I = e_1 e_2\}$, is isomorphic with $\mathbb{C}$:

$$\begin{aligned}
K(\boldsymbol{x}_m, \boldsymbol{x}_n) &= u(\boldsymbol{x}_m)^{\dagger} u(\boldsymbol{x}_n), \\
&= (u(\boldsymbol{x}_m)_s u(\boldsymbol{x}_n)_s + u(\boldsymbol{x}_m)_I u(\boldsymbol{x}_n)_I) + \\
&\quad + I(u(\boldsymbol{x}_m)_s u(\boldsymbol{x}_n)_I - u(\boldsymbol{x}_m)_I u(\boldsymbol{x}_n)_s), \\
&= (k(\boldsymbol{x}_m, \boldsymbol{x}_n)_{ss} + k(\boldsymbol{x}_m, \boldsymbol{x}_n)_{II}) + \\
&\quad + I(k(\boldsymbol{x}_m, \boldsymbol{x}_n)_{Is} - k(\boldsymbol{x}_m, \boldsymbol{x}_n)_{sI}), \\
&= \boldsymbol{H_r} + I\boldsymbol{H_I}.
\end{aligned} \quad (13.61)$$

For the quaternion-valued Gabor kernel function, we use $\boldsymbol{i} = e_2 e_3$, $\boldsymbol{j} = -e_3 e_1$, $\boldsymbol{k} = e_1 e_2$. The Gaussian window Gabor kernel function reads

$$K(\boldsymbol{x_m}, \boldsymbol{x_n}) = g(\boldsymbol{x_m}, \boldsymbol{x_n}) exp^{-\boldsymbol{i}\mathbf{w}_0^T(\boldsymbol{x_m} - \boldsymbol{x_n})}, \quad (13.62)$$

where the normalized Gaussian window function is given by

$$g(\boldsymbol{x_m}, \boldsymbol{x_n}) = \frac{1}{\sqrt{2\pi}\rho} exp^{-\frac{\|\boldsymbol{x_m} - \boldsymbol{x_n}\|^2}{2\rho^2}} \quad (13.63)$$

and the variables $\boldsymbol{w_0}$ and $\boldsymbol{x_m} - \boldsymbol{x_n}$ stand for the frequency and space domains, respectively.

Unlike the Hartley transform or the 2D complex Fourier transform, this kernel function nicely separates the even and odd components of the given signal, i.e.,

$$\begin{aligned}K(\boldsymbol{x_m}, \boldsymbol{x_n}) &= K(\boldsymbol{x_m}, \boldsymbol{x_n})_s + K(\boldsymbol{x_m}, \boldsymbol{x_n})_{e_2e_3} + \cdots \\ &+ K(\boldsymbol{x_m}, \boldsymbol{x_n})_{e_3e_1} + K(\boldsymbol{x_m}, \boldsymbol{x_n})_{e_1e_2}, \\ &= g(\boldsymbol{x_m}, \boldsymbol{x_n})cos(\mathbf{w}_0^T \boldsymbol{x_m})cos(\mathbf{w}_0^T \boldsymbol{x_m}) + \cdots \\ &+ g(\boldsymbol{x_m}, \boldsymbol{x_n})cos(\mathbf{w}_0^T \boldsymbol{x_m})sin(\mathbf{w}_0^T \boldsymbol{x_m})\boldsymbol{i} + \cdots \\ &+ g(\boldsymbol{x_m}, \boldsymbol{x_n})sin(\mathbf{w}_0^T \boldsymbol{x_m})cos(\mathbf{w}_0^T \boldsymbol{x_m})\boldsymbol{j} + \cdots \\ &+ g(\boldsymbol{x_m}, \boldsymbol{x_n})sin(\mathbf{w}_0^T \boldsymbol{x_m})sin(\mathbf{w}_0^T \boldsymbol{x_m})\boldsymbol{k}.\end{aligned}$$

Since $g(\boldsymbol{x_m}, \boldsymbol{x_n})$ fulfills Mercer's condition, it is straightforward to prove that the $k(\boldsymbol{x_m}, \boldsymbol{x_n})_u$ in the above equations satisfy these conditions as well.

After defining these kernels, we can proceed in the formulation of the SVM conditions. We substitute the mapped data $\Phi_i(\boldsymbol{x}) = \sum_{u=1}^{n} < \Phi_i(\boldsymbol{x}) >_u$ into the linear function $f(\boldsymbol{x}) = \boldsymbol{w}^{\dagger^T} \boldsymbol{\Phi}(\boldsymbol{x}) + \boldsymbol{b}$. The problem can be stated in a similar fashion to (13.57)–(13.52). In fact, we can replace the kernel function in (13.57) to accomplish the Wolfe dual programming and thereby to obtain the kernel function group for nonlinear classification:

$$\begin{aligned}\boldsymbol{H_s} &= \left[K_s(\boldsymbol{x_m}, \boldsymbol{x_j})\right]_{m,j=1,\ldots,l}, \\ \boldsymbol{H_{e_1}} &= \left[K_{e_1}(\boldsymbol{x_m}, \boldsymbol{x_j})\right]_{m,j=1,\ldots,l}, \\ &\vdots \\ \boldsymbol{H_{e_n}} &= \left[K_{e_n}(\boldsymbol{x_m}, \boldsymbol{x_j})\right]_{m,j=1,\ldots,l}, \\ \boldsymbol{H_I} &= \left[K_I(\boldsymbol{x_m}, \boldsymbol{x_j})\right]_{m,j=1,\ldots,l}.\end{aligned} \tag{13.64}$$

In the same way, we can use the kernel functions to replace the scalar product of the input data in (13.56). Now, for the valency state classification, one uses the output function of the nonlinear Clifford SVM given by

$$\boldsymbol{y} = csign_m\Big[f(\boldsymbol{x})\Big] = csign_m\Big[\boldsymbol{w}^{\dagger^T} \Phi(\boldsymbol{x}) + \boldsymbol{b}\Big], \tag{13.65}$$

where m stands for the state valency.

13.11 Clifford SVM for Regression

The representation of the data set for the case of Clifford SVM for regression is the same as for Clifford SVM for classification; we represent the data set in a certain Clifford algebra G_n. Each data ith-vector has multivector entries $\boldsymbol{x}_i = [\boldsymbol{x}_{i1}, \boldsymbol{x}_{i2}, ..., \boldsymbol{x}_{iD}]^T$, where $\boldsymbol{x}_{ij} \in G_n$ and D is its dimension. Let $\boldsymbol{x}_1$, $\boldsymbol{y}_1$),($\boldsymbol{x}_2$,

$\boldsymbol{y}_2$),...,($\boldsymbol{x}_j$, $\boldsymbol{y}_j$),...,($\boldsymbol{x}_l$, $\boldsymbol{y}_l$) be the training set of independently and identically distributed multivector-valued sample pairs, where each label is given by $\boldsymbol{y}_i = y_{si} + y_{e_1 i}e_1 + y_{e_2 i}e_2 + \cdots + y_{I_i}I$. The regression problem using multivectors is to find a multivector valued function $f(\boldsymbol{x})$ that has at most an ε-deviation from the actually obtained targets $\boldsymbol{y}_i \in G_n$ for all the training data and, at the same time, is as smooth as possible. We will use a multivector-valued ε-insensitive loss function and arrive at the formulation of Vapnik [303]:

$$
\begin{aligned}
& min\ L(\boldsymbol{w}, \boldsymbol{b}, \boldsymbol{\xi}) = \frac{1}{2}\boldsymbol{w}^{\dagger^T}\boldsymbol{w} + C\sum_{i,j}(\xi_{ij} + \bar{\xi}_{ij}) \\
& subject\ to \\
& (\boldsymbol{y}_i - \boldsymbol{w}^{\dagger^T}\boldsymbol{x}_i - \boldsymbol{b})_j <= (\epsilon + \xi_{ij}), \\
& (\boldsymbol{w}^{\dagger^T}\boldsymbol{x}_i + \boldsymbol{b} - \boldsymbol{y}_i)_j <= (\epsilon + \bar{\xi}_{ij}), \\
& \xi_{ij} >= 0,\ \bar{\xi}_{ij} >= 0 \texttt{ for all } i, j,
\end{aligned}
\tag{13.66}
$$

where $\boldsymbol{w}, \boldsymbol{x} \in G_n^D$ and $(\cdot)_j$ extracts the scalar accompanying a multivector base. Next, we proceed as in Sect. 13.9, and since the expression of the orientation of the optimal hyperplane is similar to that of Eq. (13.44), the components of an entry $\boldsymbol{w}_k$ of the optimal hyperplane $\boldsymbol{w}$ are computed using l multivector samples as follows:

$$
w_{k_s} = \sum_{j=1}^{l}\Big((\alpha_{k_s})_j - (\bar{\alpha}_{k_s})_j\Big)(x_{k_s})_j, \tag{13.67}
$$

$$
w_{k_{e_1}} = \sum_{j=1}^{l}\Big((\alpha_{k_{e_1}})_j - (\bar{\alpha}_{k_{e_1}})_j\Big)(x_{k_{e_1}})_j, \ldots,
$$

$$
w_{k_{e_2e_3}} = \sum_{j=1}^{l}\Big((\alpha_{k_{e_2e_3}})_j - (\bar{\alpha}_{k_{e_2e_3}})_j\Big)(x_{k_{e_2e_3}})_j, \ldots,
$$

$$
w_{k_I} = \sum_{j=1}^{l}\Big((\alpha_{k_I})_j - (\bar{\alpha}_{k_I})_j\Big)(x_{k_I})_j.
$$

We can now redefine the entries of the vector $\boldsymbol{a}$ of Eq. (13.49) for a vector of D multivectors as follows:

$$
\begin{aligned}
\boldsymbol{a} = \Big[& [\hat{\boldsymbol{a}}_{1_s}, \hat{\boldsymbol{a}}_{1_{e_1}}, \hat{\boldsymbol{a}}_{1_{e_2}}, \ldots, \hat{\boldsymbol{a}}_{1_I}], \ldots, [\hat{\boldsymbol{a}}_{k_s}, \hat{\boldsymbol{a}}_{k_{e_1}}, \hat{\boldsymbol{a}}_{k_{e_2}}, \ldots, \hat{\boldsymbol{a}}_{k_I}], \\
& \ldots, [\hat{\boldsymbol{a}}_{D_s}, \hat{\boldsymbol{a}}_{D_{e_1}}, \hat{\boldsymbol{a}}_{D_{e_2}}, \ldots, \hat{\boldsymbol{a}}_{D_I}]\Big].
\end{aligned}
\tag{13.68}
$$

and the entries for the k element are computed using l samples as follows:

$$
\begin{aligned}
\hat{\boldsymbol{a}}_{k_s}^T &= [(\alpha_{k_s 1} - \overline{\alpha}_{k_s 1}),\ (\alpha_{k_s 2} - \overline{\alpha}_{k_s 2}), \cdots, (\alpha_{k_s l} - \overline{\alpha}_{k_s l})],\\
\hat{\boldsymbol{a}}_{k_{e_1}}^T &= [(\alpha_{k_{e_1} 1} - \overline{\alpha}_{k_{e_1}}, \cdots, (\alpha_{k_{e_1} l} - \overline{\alpha}_{k_{e_1} l})],\\
&\ , \ldots,\\
\hat{\boldsymbol{a}}_{k_I}^T &= [(\alpha_{k_I 1} - \overline{\alpha}_{k_I 1}), (\alpha_{k_I 2} - \overline{\alpha}_{k_I 2}), \cdots, (\alpha_{k_I l} - \overline{\alpha}_{k_I l})].
\end{aligned}
\tag{13.69}
$$

Now, we can rewrite the Clifford product $\boldsymbol{w}^{\dagger T}\boldsymbol{w}$, as we did in (13.53)–(13.56) and rewrite the primal problem as follows:

$$
\begin{aligned}
& min\ \frac{1}{2}\boldsymbol{a}^T \boldsymbol{H} \boldsymbol{a} + C(\boldsymbol{\xi} + \bar{\boldsymbol{\xi}}) \\
& subject\ to \\
& (\boldsymbol{y} - \boldsymbol{w}^{\dagger}\boldsymbol{x} - \boldsymbol{b})_j\ \leq (\epsilon + \xi)_j,\\
& (\boldsymbol{w}^{\dagger}\boldsymbol{x} + \boldsymbol{b} - \boldsymbol{y})_j\ \leq (\epsilon + \bar{\xi})_j,\\
& \xi_{ij} >= 0,\ \bar{\xi}_{ij} >= 0 \texttt{ for all } i, j.
\end{aligned}
\tag{13.70}
$$

Thereafter, we write straightforwardly the dual of (13.70) for solving the regression problem:

$$
\begin{aligned}
& max\ \ -\bar{\boldsymbol{\alpha}}^T(\bar{\boldsymbol{\epsilon}} + \boldsymbol{y}) - \boldsymbol{\alpha}^T(\boldsymbol{\epsilon} - \boldsymbol{y}) - \frac{1}{2}\boldsymbol{a}^T \boldsymbol{H}\boldsymbol{a}, \\
& subject\ to \\
& \sum_{j=1}^{l}(\alpha_{s\,j} - \overline{\alpha}_{s\,j}) = 0,\ \sum_{j=1}^{l}(\alpha_{e_1\,j} - \overline{\alpha}_{e_1\,j}) = 0, \ldots,\\
& \sum_{j=1}^{l}(\alpha_{I\,j} - \overline{\alpha}_{I\,j}) = 0,\\
& 0 \leq (\alpha_s)_j \leq C,\ 0 \leq (\alpha_{e_1})_j \leq C, ...,\\
& 0 \leq (\alpha_{e_1 e_2})_j \leq C, ..., 0 \leq (\alpha_I)_j \leq C,\ j = 1,\ ..., l,\\
& 0 \leq (\bar{\alpha}_s)_j \leq C,\ 0 \leq (\bar{\alpha}_{e_1})_j \leq C, ...,\\
& 0 \leq (\bar{\alpha}_{e_1 e_2})_j \leq C, ..., 0 \leq (\bar{\alpha}_I)_j \leq C,\ j = 1,\ ..., l.
\end{aligned}
\tag{13.71}
$$

As explained in Sect. 13.10, for nonlinear regression we utilize a particular kernel for computing $k(\boldsymbol{x}_m, \boldsymbol{x}_n) = \Phi(\boldsymbol{x}_m)^{\dagger}\Phi(\boldsymbol{x}_m)$. We can use the kernels described in Sect. 13.10. By the use of other loss functions, like the Laplace, complex or polynomial, one can extend Eq. (13.71) to include extra constraints.

13.12 Real- and Quaternion-Valued Spike Neural Networks

The mathematical models representing the functioning of the third-generation artificial neurons can be divided into two major categories, namely behavior and threshold models [314]. The abstraction level is directly related with the computational properties of the corresponding model and depends on the inclusion or not of biological aspects, such as cell's ions channels, which is aimed to describe the neuron's behavior [150]. The behavior models include biological aspects, while the threshold ones do not and they assume a special behavior of the firing signal considering a low threshold voltage, such that the neuron's firing occurs when the membrane's potential exceeds the given threshold value. The Hodgkin–Huxley's model is one example of behavior model, while the Perfect Integrate-and-Fire (PIF) model is the most simple example of threshold model, which is considered in this paper.

13.12.1 *Threshold and Firing Models*

The Perfect Integrate-and-Fire (PIF) and Leaky Integrate-and-Fire (LIF) models are included as examples of integrate-and-fire models, and they have been extensively implemented in several research papers [302], due to their simple implementation and low computational cost. The threshold models represent dynamics with a high abstraction level, and they are based on the summation of all the contributions of the presynaptic neurons to the membrane's potential that, in case of exceeding the previously given threshold value, the neuron fires.

The first-order and ordinary differential equation describing the behavior of the integrate-and-fire neurons does not completely specify the behavior of the third-generation artificial neural networks, since it only embraces the dynamic starting from the resting potential conditions until the instant of time just before the neuron fires (the precise instant of time where the receptor neuron emits the output pulse), so that the conditions of firing must be posed.

$$t^{(f)} : u(t^{(f)}) = \theta \ \ y \ \ \frac{du(t)}{dt}|_{t=t^{(f)}} > 0. \tag{13.72}$$

Equation (13.72) expresses this fact by saying that $t^{(f)}$ is the precise moment where the given threshold θ starts to be exceeded and that at this instant of time the membrane's potential is an increasing function.

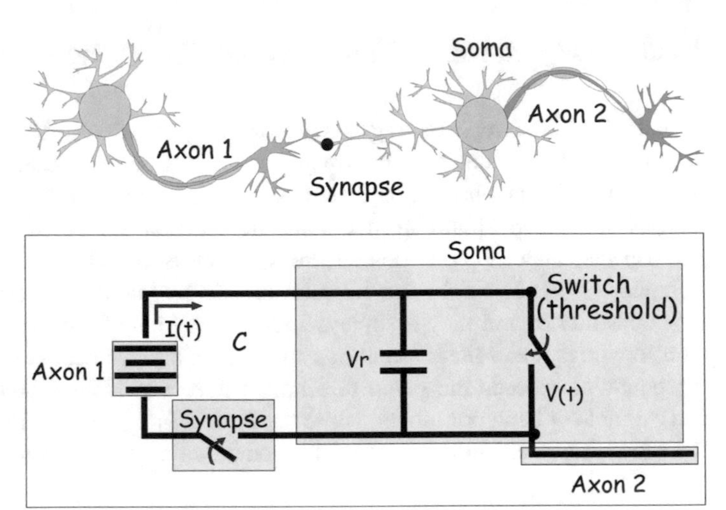

Fig. 13.7 Perfect integrate-and-fire neuron and its representation through an electric circuit with external input

13.12.2 Perfect Integrate-and-Fire Model

This is a model defined by a capacitive circuit, where the input current $I(t)$ accumulates a charge in the capacitor, which produces a potential difference $V(t)$ when the switches, called synapse and threshold in Fig. 13.7, are closed and open, respectively. After the potential $V(t)$ arrives to some previously given threshold value, then the switches synapse and threshold are respectively open and closed in such a way that the capacitor decreases its charge which, at the same time, diminishes the potential $V(t)$. The solution of the ordinary differential equation describing the system that Fig. 13.7 shows is given as follows:

$$V(t) = V_r + \frac{1}{C}\int_{t_o}^{t} I(\tau)d\tau, \tag{13.73}$$

where $V(t)$ is the voltage produced by the accumulated charge in the capacitor, V_r is the value of the capacitor's resting potential (produced by the charge in the capacitor when both switches are open), C is the capacitance of the capacitor, and $I(t)$ is the current produced by the source at time t.

13.12.3 Learning Method

The network architecture for this method necessarily involves a feedforward network of spiking neurons with multiple delayed synaptic terminals, see Fig. 13.8. Formally,

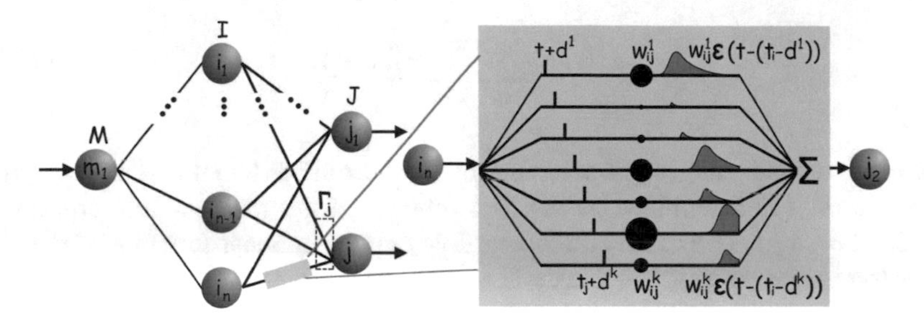

Fig. 13.8 Feedforward spiking neural network and its respective multiple delayed synaptic terminals

a neuron j, having a set Γj of immediate predecessors, receives a set of spikes with firing times t_i, $i \in \Gamma_j$. Any neuron generates at most one spike during the simulation interval, and fires when the internal state variable reaches a threshold θ. The dynamics of the internal state variable $X_j(t)$ are determined by the impinging spikes, whose impact is described by the spike-response function $\epsilon(t)$ weighted by the synaptic efficacy ("weight") w_{ij}:

$$x_j(t) = \sum_{i \in \Gamma_j} w_{ij}\epsilon(t - t_i). \tag{13.74}$$

The spike-response function in Eq. (13.74) effectively models the unweighted postsynaptic potential (PSP) of a single spike impinging on a neuron. The height of the PSP is modulated by the synaptic weight w_{ij} to obtain the effective postsynaptic potential. The spike-response function $\epsilon(t)$ who describes a standard PSP is given by

$$\epsilon(t) = \frac{t}{\tau}e^{1-\frac{t}{\tau}}. \tag{13.75}$$

The delay d^k of a synaptic terminal k is defined by the difference between the firing time of the presynaptic neuron and the time the postsynaptic potential starts rising. The presynaptic spike at a synaptic terminal k is described as a PSP of standard height with delay d^k . The unweighted contribution of a single synaptic terminal to the state variable is then given by

$$y_i^k(t) = \epsilon(t - t_i - d^k), \tag{13.76}$$

with $\epsilon(t)$ a spike-response function shaping a PSP, with $\epsilon(t) = 0$ for $t < 0$. The time t_i is the firing time of presynaptic neuron i, and d^k is the delay associated with the synaptic terminal k.

Extending Eq. (13.74) to include multiple synapses per connection and inserting Eq. (13.76), the state variable x_j of neuron j receiving input from all neurons i can then be described as the weighted sum of the presynaptic contributions

$$x_j(t) = \sum_{i\in\Gamma_j}\sum_{k=1}^{m} w_{ij}^k y_i^k(t), \tag{13.77}$$

where w_{ij}^k denotes the weight associated with synaptic terminal k. The firing time t_j of neuron j is determined as the first time when the state variable crosses the threshold $\theta : x_j(t) \geq \theta$. Thus, the firing time t_j is a nonlinear function of the state variable $x_j : t_j = t_j(x_j)$. The threshold θ is constant and equal for all neurons in the network.

13.12.4 Error Backpropagation

The error backpropagation method called Spike-Propagation [173] is derived in the same way as the derivation of Rumelhart et al. [255]. The target of the algorithm is to learn a set of firing times, denoted t_j^d, at the output neurons $j \in J$ for a given set of patterns $P[t_1...t_h]$, where $P[t_1...t_h]$ defines a single input pattern described by single spike times for each neuron $h \in H$. In this learning method is chosen the least mean square error function as error function. Given desired spike times t_j^d and actual firing times t_j^a, this error function is defined by

$$E = \frac{1}{2}\sum_{j\in J}(t_j^a - t_j^d)^2. \tag{13.78}$$

For the error backpropagation, each synaptic terminal is treated as a separate connection k whit weight w_{ij}^k. Hence, for a backprop rule, we need to calculate the increasing weights

$$\Delta w_{ij}^k = -\eta\frac{\partial E}{\partial w_{ij}^k}, \tag{13.79}$$

where η is the learning rate and w_{ij}^k the weight of connection k from neuron i to neuron j. As t_j is a function of x_j, which depends on the weights w_{ij}^k, using the chain rule in Eq. (13.79), one finds the relationship between the error and the vector of weights (w_{ij}^k).

$$\frac{\partial E}{\partial w_{ij}^k} = \frac{\partial E}{\partial t_j}\frac{\partial t_j}{\partial x_j(t)}\frac{\partial x_j(t)}{\partial w_{ij}^k}. \tag{13.80}$$

The first factor in (13.80), the derivative of E with respect to t_j , is simply

$$\frac{\partial E}{\partial t_j} = (t_j^a - t_j^d). \tag{13.81}$$

For the second factor in (13.80), we assume that for a small enough region around $t = t_j^a$, the function x_j can be approximated by a linear function of t. For such a small region, we approximate the threshold function $\partial t_j(x_j) = -\partial x_j(t_j)/\alpha$. with $\partial t_j/\partial x_j(t)$ the derivate of the inverse function of $x_j(t)$. The value α equals the local derivative of $x_j(t)$ with respect to t, that is, $\alpha = \partial x_j(t)/\partial t(t_j^a)$, which means

$$\frac{\partial t_j}{\partial x_j(t)} = \left.\frac{\partial t_j(x_j)}{\partial x_j(t)}\right|_{x_j=\theta} = \frac{-1}{\alpha},$$
$$= \frac{-1}{\partial x_j(t)/\partial t(t_j^a)} = \frac{-1}{\sum_{i,l} w_{ij}^l(\partial y_i^l(t)/\partial t(t_j^a))}, \tag{13.82}$$

finally we have

$$\frac{\partial x_j(t)}{\partial w_{ij}^a} = \frac{\partial \sum_{n\in\Gamma_j}\sum_l w_{nj}^l y_n^l(t_j^a)}{\partial w_{ij}^k} = y_i^k(t_j^a). \tag{13.83}$$

When we combine these results, Eq. (13.79) becomes to

$$\Delta w_{ij}^k = -\eta\frac{y_i^k(t_j^a)(t_j^d - t_j^a)}{\sum_{i,l} w_{ij}^l(\partial y_i^l(t)/\partial t(t_j^a))}. \tag{13.84}$$

13.12.5 Quaternion Spike Neural Networks

Quaternion algebra (H) has been invented by W.R. Hamilton in 1843 in order to extend in the 3D space the properties of complex numbers. A quaternion can in fact be defined as an extended complex number with three imaginary parts [65, 119, 120]. In this section, a new Spike Neural structure defined in Quaternion Algebra is introduced and a suitable learning algorithm for such a structure is also reported. Let us define an QSNN (Quaternion Spike Neural Network) as a multilayer Spike Neural Network in which inputs and outputs values, weights, and biases are quaternions. Figure 13.9 depicts the extension of the real-valued SNN to a quaternionic Spike Neural Network.

According the generalized multivector training rule explained in Sect. 13.4.5, we use Eq. 13.21 for the rotor algebra G_3^+ which is isomorph with the quaternion algebra $\mathbb{H}$; thus, we use instead the quaternion basis $\boldsymbol{i} = e_2e_3$, $\boldsymbol{j} = -e_3e_1$, $\boldsymbol{k} = e_1e_2$. Accordingly, the learning rule algorithm has been developed for quaternions

$$\Delta W_{ml}^n = -\eta\frac{\partial E}{\partial W_{ml}^n}. \tag{13.85}$$

The error of Eq. (13.85) now belongs to the algebra Hamilton ($E \in H$), and W_{ml}^n is a quaternionic weights vector of the nth delay, between the presynaptic neuron m and

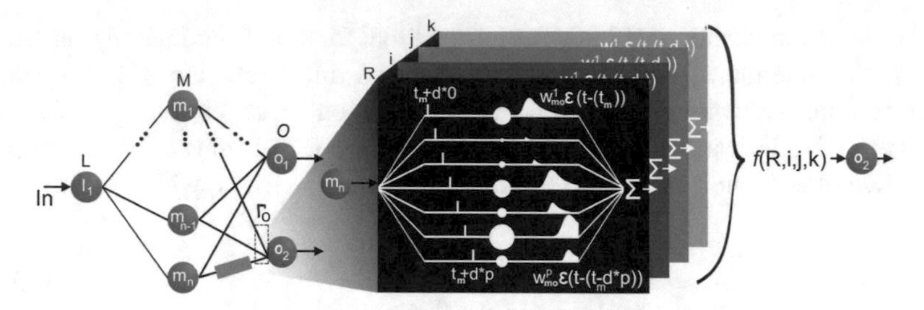

Fig. 13.9 Quaternion feedforward spiking neural network and its respective multiple delayed synaptic terminals

the postsynaptic neuron l, so Eq. (13.85) extends as

$$\frac{\partial E}{\partial W^n_{ml}} = \frac{\partial E}{\partial Wr^n_{ml}} + \frac{\partial E}{\partial Wi^n_{ml}} i + \frac{\partial E}{\partial Wj^n_{ml}} j + \frac{\partial E}{\partial Wk^n_{ml}} k. \tag{13.86}$$

Using the chain rule in (13.86) for each quaternion element, we obtain

$$\begin{aligned}\frac{\partial E}{\partial Wr^n_{ml}} &= \frac{\partial E}{\partial tr^n_{ml}} \frac{\partial tr^n_{ml}}{\partial F^n_{ml}} \frac{\partial F^n_{ml}}{\partial Wr^n_{ml}} + \\ &\frac{\partial E}{\partial ti^n_{ml}} \frac{\partial ti^n_{ml}}{\partial F^n_{ml}} \frac{\partial F^n_{ml}}{\partial Wr^n_{ml}} + \frac{\partial E}{\partial tj^n_{ml}} \frac{\partial tj^n_{ml}}{\partial F^n_{ml}} \frac{\partial F^n_{ml}}{\partial Wr^n_{ml}} + \\ &\frac{\partial E}{\partial tk^n_{ml}} \frac{\partial tk^n_{ml}}{\partial F^n_{ml}} \frac{\partial F^n_{ml}}{\partial Wr^n_{ml}},\end{aligned} \tag{13.87}$$

$$\begin{aligned}\frac{\partial E}{\partial Wi^n_{ml}} &= \frac{\partial E}{\partial tr^n_{ml}} \frac{\partial tr^n_{ml}}{\partial F^n_{ml}} \frac{\partial F^n_{ml}}{\partial Wi^n_{ml}} + \\ &\frac{\partial E}{\partial ti^n_{ml}} \frac{\partial ti^n_{ml}}{\partial F^n_{ml}} \frac{\partial F^n_{ml}}{\partial Wi^n_{ml}} + \frac{\partial E}{\partial tj^n_{ml}} \frac{\partial tj^n_{ml}}{\partial F^n_{ml}} \frac{\partial F^n_{ml}}{\partial Wi^n_{ml}} + \\ &\frac{\partial E}{\partial tk^n_{ml}} \frac{\partial tk^n_{ml}}{\partial F^n_{ml}} \frac{\partial F^n_{ml}}{\partial Wi^n_{ml}},\end{aligned} \tag{13.88}$$

$$\begin{aligned}\frac{\partial E}{\partial Wj^n_{ml}} &= \frac{\partial E}{\partial tr^n_{ml}} \frac{\partial tr^n_{ml}}{\partial F^n_{ml}} \frac{\partial F^n_{ml}}{\partial Wj^n_{ml}} + \\ &\frac{\partial E}{\partial ti^n_{ml}} \frac{\partial ti^n_{ml}}{\partial F^n_{ml}} \frac{\partial F^n_{ml}}{\partial Wj^n_{ml}} + \frac{\partial E}{\partial tj^n_{ml}} \frac{\partial tj^n_{ml}}{\partial F^n_{ml}} \frac{\partial F^n_{ml}}{\partial Wj^n_{ml}} + \\ &\frac{\partial E}{\partial tk^n_{ml}} \frac{\partial tk^n_{ml}}{\partial F^n_{ml}} \frac{\partial F^n_{ml}}{\partial Wj^n_{ml}},\end{aligned} \tag{13.89}$$

$$\begin{aligned}\frac{\partial E}{\partial Wk^n_{ml}} &= \frac{\partial E}{\partial tr^n_{ml}}\frac{\partial tr^n_{ml}}{\partial F^n_{ml}}\frac{\partial F^n_{ml}}{\partial Wk^n_{ml}} + \\ &\frac{\partial E}{\partial ti^n_{ml}}\frac{\partial ti^n_{ml}}{\partial F^n_{ml}}\frac{\partial F^n_{ml}}{\partial Wk^n_{ml}} + \frac{\partial E}{\partial tj^n_{ml}}\frac{\partial tj^n_{ml}}{\partial F^n_{ml}}\frac{\partial F^n_{ml}}{\partial Wk^n_{ml}} + \\ &\frac{\partial E}{\partial tk^n_{ml}}\frac{\partial tk^n_{ml}}{\partial F^n_{ml}}\frac{\partial F^n_{ml}}{\partial Wk^n_{ml}}.\end{aligned} \tag{13.90}$$

Developing each of the partial derivatives, we get

$$\begin{aligned}\frac{\partial E}{\partial W^n_{ml}} &= \frac{Er(\text{tr} + \text{ti}\, i + \text{tj}\, j + \text{tk}\, k)}{\sum_l Wr\frac{\partial F^n_{ml}}{\partial tr^n_{ml}}} + \\ &+\frac{\partial E}{\partial tk^n_{ml}}\frac{\partial tk^n_{ml}}{\partial F^n_{ml}}\frac{Ei(-\text{ti} + \text{tr}\, i + \text{tk}\, j + -\text{tj}\, k)}{\sum_l Wi\frac{\partial F^n_{ml}}{\partial ti^n_{ml}}} + \\ &+\frac{\partial E}{\partial tk^n_{ml}}\frac{\partial tk^n_{ml}}{\partial F^n_{ml}}\frac{Ej(-\text{tj} + -\text{tk}\, i + \text{tr}\, j + \text{ty}\, k)}{\sum_l Wj\frac{\partial F^n_{ml}}{\partial tj^n_{ml}}} + \\ &+\frac{\partial E}{\partial tk^n_{ml}}\frac{\partial tk^n_{ml}}{\partial F^n_{ml}}\frac{Ek(-\text{tk} + \text{tj}\, i + -\text{ti}\, j + \text{tr}\, k)}{\sum_l Wk\frac{\partial F^n_{ml}}{\partial tk^n_{ml}}}.\end{aligned} \tag{13.91}$$

Developing each of the partial derivatives and using the dot product and the cross product between quaternions, we can rewrite Eq. (13.91) in a compact form as follows

$$\frac{\partial E}{\partial W^n_{ml}} = \left(E \odot \frac{1}{\sum_l W\frac{\partial F^n_{ml}}{\partial tr^n_{ml}}}\right) \otimes t^*_l. \tag{13.92}$$

13.12.6 Comparison of SNN Against QSNN

In this subsection, SNN against QSNN neural networks are compared. The comparison is performed using the same parameters of the neuronal model. For both cases, a LIF neuronal model is used, with a value of resistance equal to 730 Oms, a capacitance of 20 μf, a theta $= -30$ mv threshold, a learning rate of 0.1, the weights were initialized arbitrarily at 1.75, and 10 delays of 50 ms and a time window of 1000 ms were used.

The input vector to the neuron SNN is: $IN = [518\ 534\ 477\ 580]$, while the input vector to the neuron QSNN is: $QIN = [518\ 534\boldsymbol{i}\ 477\boldsymbol{j}\ 580\boldsymbol{k}]$ and 6 tests within the range of 0 to 1000 are conducted to test the efficiency of QSNN against SNN. The results obtained are shown in Fig. 13.10. It can be seen in all cases as spike neural networks based on quaternions converge faster than conventional neural networks spike.

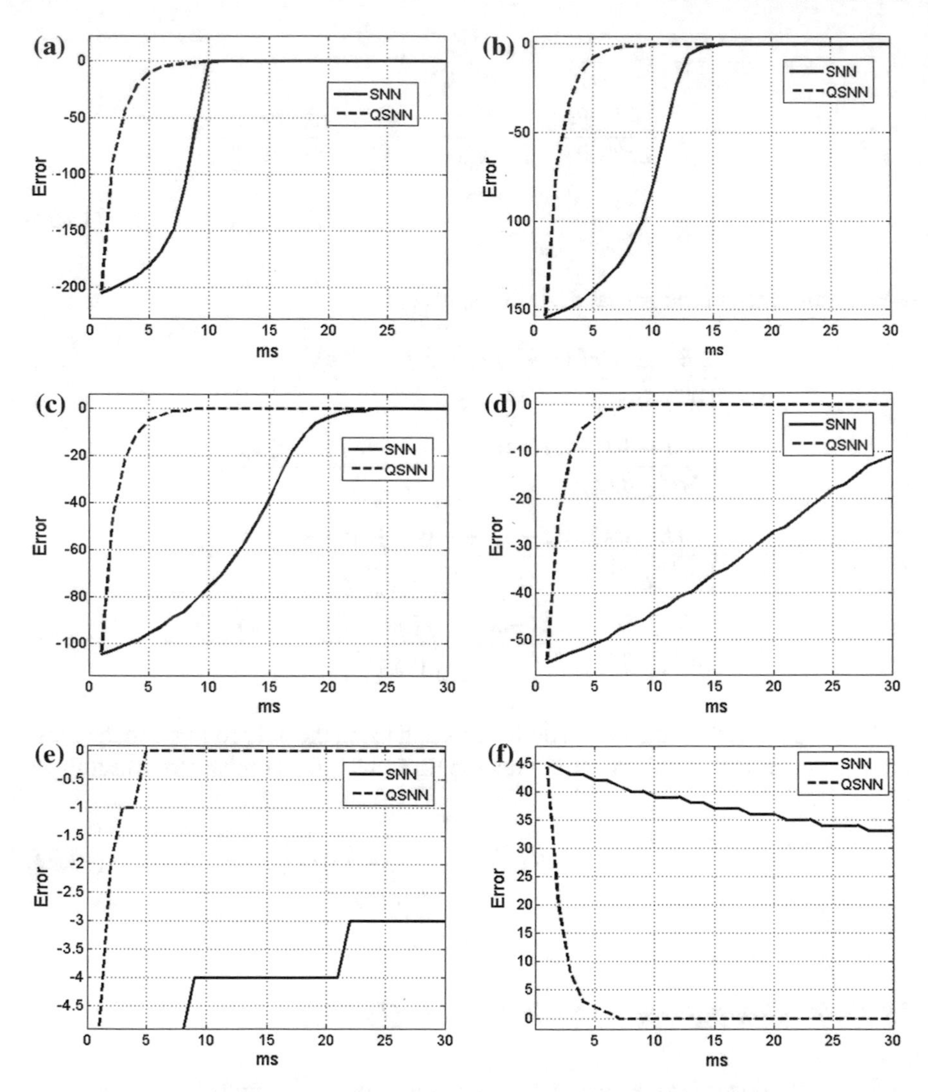

Fig. 13.10 Error curves: The comparison between the SNN (solid line) and QSNN (dashed line), under six different set points, are displayed: In the figure the error peformances **a** the set point is 680 ms, **b** 630 ms, **c** 580 ms, **d** 530 ms, **e** 480 ms, and the set point for the figure **f** was 430 ms

13.12.7 Kinematic Control of a Manipulator of 6 DOF Using QSNN

In this subsection, the kinematic control 6 DOF manipulator is presented, using six neurons in parallel, each neuron controlling a degree of freedom. The manipulator parameters are: lengths of links 1, 2, and 3 equal to 10 cm, 10 cm, and 5 cm respectively. The initial angles are issued by the response of the presynaptic neurons when weights are initialized. In Fig. 13.11 can be observed the evolution of the angles

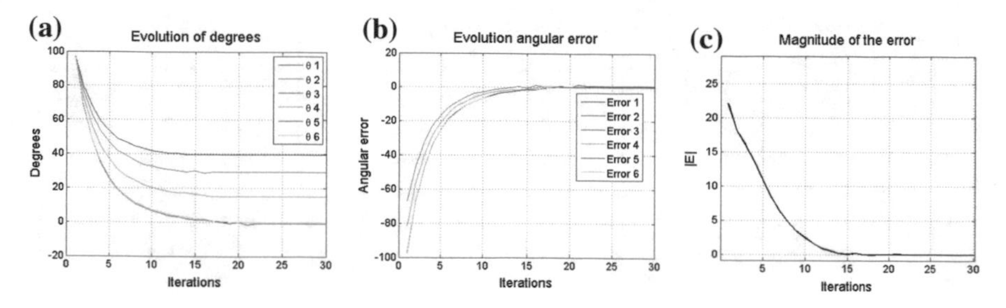

Fig. 13.11 **a** The evolution of the 6 angles. **b** The evolution of the error of each of the 6 degrees of freedom. **c** The evolution of magnitude of the error, and this magnitude is between the set point and the end-effector of the manipulator

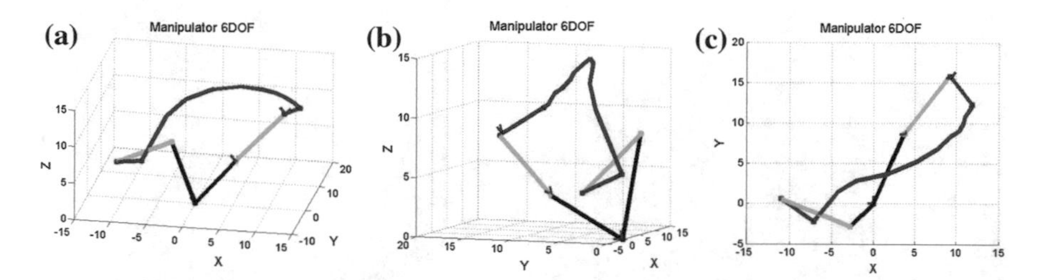

Fig. 13.12 Three different views, the initial configuration of the manipulator, the trajectory of the end-effector (blue line), and the final configuration of 6 DOF manipulator. In black, the first link; green, the second link; and red, the third link

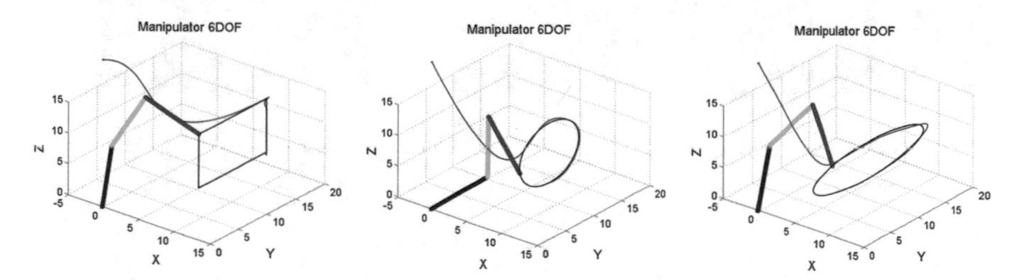

Fig. 13.13 Trajectory paths of the end-effector (blue line) along a square, circle and ellipsoid (in black) of the 6 DOF manipulator

proposed by neurons to evolve from the initial configuration of the manipulator to one of the settings required for the end-effector be in the set point ([12x 12y 10z]). Figure 13.12 shows the motion of the end-effector.

In Figs. 13.13, 13.14, we show the control of a 6 DOF manipulator following the contours of a square, circle, and ellipsoid, respectively.

Next, we present the application of the QSNN for the control of a real humanoid robot arm. Figure 13.15a shows the humanoid robot called Mexone. The length of the shoulder to the elbow is 18.5 cm and from the elbow to the wrist is 12 cm. Figure 13.15b shows the QSNN controller applied to the the robot arm of the humanoid Mexone, where SP is a 6D vector: of six angle references. The difference between the actual motor angles and the desired angles is computed as the error vector. The

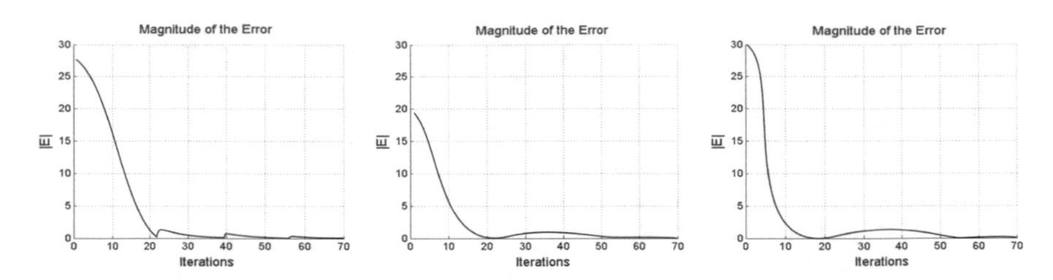

Fig. 13.14 Errors of the trajectories of the end-effector along a square, circle, and ellipsoid

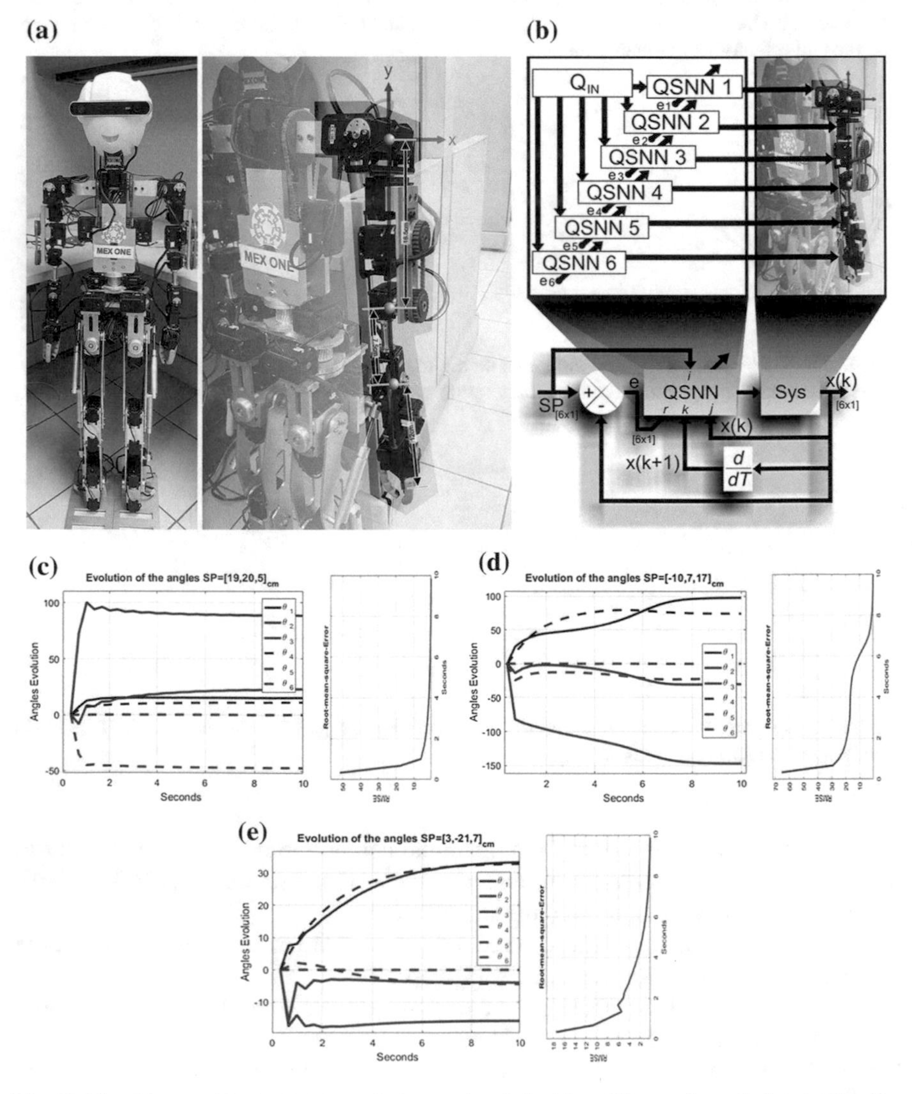

Fig. 13.15 **a** Humanoid robot Mexone. **b** Quaternion Spike Neural Network applied as a controller of the robot humanoid arm. **c–e** trajectories of the six robot arm angles for three position of the hand [19, 25, 5] cm, [-10, 7, 17] cm and [3, -21, 7] cm and error curves during the adaptation

input quaternion (Q_{IN}) consists of the actual angle error and its derivative which adjusts the weights of each neuron. In the superior part of the drawing, we see the outputs of the QS connected to each motor driver.

Figures 13.13c–e show the performance of the QSNN based control to locate the robot hand to three different positions in 3D: [19, 25, 5] cm, [-10, 7, 17] cm and [3, -21, 7] cm. To reach these positions the QSNN controller adjust the six angles needed for computing the inverse kinematics and reach the goal reducing the error in less of 8 sec. Be aware, that the speed is due to the slow motion of the robot arm movement, as the time for the QSNN adaptation takes less of 4 msec. Note that the QSNN are acting as adaptive controllers without the need of an explicit knowledge of the dynamics of the plant.

13.13 Rotor or Quaternion Quantum Neural Networks

In this section, we formulate the quantum gates to build neural networks working in the quaternion algebra framework.

13.13.1 Quantum Computing

The classical Von Neumann computers carry out the digital operations between the logical bits 0 and 1 using logic gates such as NOT, OR, AND, NAND, XOR, NOR, and so on. In quantum computing instead of the bits, the qubit is used. Two possible states for a qubit are the state $|0>$ and $|1>$, which might correspond to the states 0 and 1 for classical bit. The Stern–Gerlach experiment [201] demonstrates that the property of spin has the adequate properties to represent a two-state quantum bit. A measurement returns an up or a down spin represented by $|0>$ (parallel to the field) and $|1>$ (antiparallel), and these are seen as the up and down orientation of these basis states. Before the measurements, a dipole exists in a superposition of these states. The wave function of the qbit is expressed as

$$|\psi = \alpha|0> + \beta|1>, \tag{13.93}$$

where $\alpha, \beta \in \mathbb{C}$ and $|\alpha|^2 + |\beta|^2$=1.

A quantum bit is a two-level entity, represented in the 2D Hilbert space $\mathbb{H}_2$ which is equipped with a fixed basis $B_{\mathbb{H}_2} = \{|0>, |1>\}$, where the $|0>$ and $|1>$ are called basis states.

In quantum computing, the information is extracted from the quantum states, even though there is theoretically an infinite amount of information held in α and β; after the measurement takes place, the information is lost and it can be obtained just a $|0>$ and a $|1>$ quantum state measured with probability $|\alpha|^2$ and $|\beta|^2$, respectively; after the measurement, the α and β are reset to either 0 or 1. This procedure is one of the following five properties which distinguish quantum computing from classical computing.

(i) Superposition: A quantum system unlike a classical system can be in a superposition of $|0>$ and $|1>$ basis states.
(ii) Entanglement: Given two qubits in the state $|\psi>=|0>0>+|1>1>$, there is no way that this can be written as $|\phi>|\chi>$, and this indicates that the two states are intimately entangled.
(iii) Reversible unitary evolution: The Schrödinger's equation is $\hat{H}|\psi>=i\hbar\partial|\psi>/\partial t$ and it integration gives $|\psi(t)>=\hat{U}|\psi(0)>$, where $\hat{U}(t)$ is an unitary operator given by $\hat{U}(t)=\hat{P}\exp[(i/\hbar)\int_o^t dt'\hat{H}(t')]$, where $\hat{P}$ is an operator tangent to the path. The reversion is $\hat{U}\dagger$.
(iv) Irreversibility, measurement, and decoherence: All interactions with the environment are irreversible, whether they are due to measurements or the system settles to thermal equilibrium with its environment. These interactions disturb the quantum system. This process is known decoherence, and it destroys the quantum properties of the system.
(v) No-cloning: The irreversibility of the measurement leads to the inability to copy a state without disturbing it in some way. Thus, it cannot be implemented a general copying routine.

13.13.2 Multiparticle Quantum Theory in Geometric Algebra

In the orthodox formulation of quantum mechanics, the tensor product is used for the construction of both multiparticle states and many of the operators acting on these states. It is a notational formulation for explicitly isolating the Hilbert spaces of different particles. The geometric algebra formalism provides an alternative representation of the tensor product in terms of the geometric product involving multivectors. Tensor product has no neat geometric interpretation and visualization, while the geometric product and the k-vectors (points, lines, planes, and volumes) have a clear interpretation. Entangled quantum states are replaced by multivectors with again clear geometric interpretation that are nothing as bags of shapes like points, lines e_j, squares e_ie_j, cubes $e_ie_je_k$, and so on. For more details of quantum theory in the geometric framework, see [46, 53, 121].

It is commonly believed that the complex space notions and the imaginary unit $i_{\mathbb{C}}$ are fundamental in quantum mechanics. But, using the space–time algebra (STA) [128], the 4D geometry algebra with Minkowski metric, $G_{3,1}$, authors [70, 280, 281] has shown how the $i_{\mathbb{C}}$ which appears in the Dirac, Pauli, and Schrödinger's equations has a geometric interpretation in terms of rotations in real space–time [129]. In this fact, it is even more clear by the geometric algebra of a relativistic configuration space, called multiparticle space and time algebra (MSTA) [70, 71, 175, 280].

Motivated by the undoubtedly usefulness of the STA formalism in describing a single-particle quantum mechanics, the MSTA approach to multiparticle quantum mechanics in both non-relativistic and relativistic settings was originally introduced as an attempt that it would provide both computational and most important interpretation advances in multiparticle quantum theory. The formulation of quantum mechanics using geometric algebra variables appears to be a solution to the Einsten–Podolsky–Rosen paradox [84]. The MSTA approach implies a separate copy of the time dimension for each particle, as well as the three spatial dimensions. In fact, it is an attempt to construct a solid conceptual framework for a multitime approach to quantum theory. Cafaro and Mancini [46] presented an explicit GA characterization of 1- and 2-qubit quantum states together with a GA characterization of a universal set of quantum gates for quantum computation and explained the universality of quantum gates when formulated in the geometric algebra framework. In this section, our contribution is to formulate the quantum gates for the Lie Group SO(3) in terms of neural networks using a subalgebra G_3^+ of the STA. Since G_3^+ is isomorph to the quaternion algebra $\mathbb{H}$, we use the quaternion algebra framework for quantum computing. Our work conciliates the different relevant fields which at present are split due to certain dogmatic concepts which hinders the unification of fields, and just by working in geometric algebra and quaternion neural networks, we manage to integrate the involved fields in an unique mathematical framework which uses as hardware neural networks, see Fig. 13.16. Note that the above-cited authors also do not offer a concept for the whole hardware for the computation of quantum states, they just propose the theory for quantum gates. The progress in quantum computing and neurocomputing indicates that the formulation of real-time operators in terms of neural networks seems a promising approach particularly as it is inspired in neuroscience as well. Considering the progress in MSTA using GA and the quaternion neural networks, it appears logically to formulate the Quaternion Quantum Neural Networks (QQNN) or also called by us Rotor Quantum Neural Networks (RQNN) as models for real-time computing superior by far as traditional quantum computing using a Von Neumann computing based on digital gates.

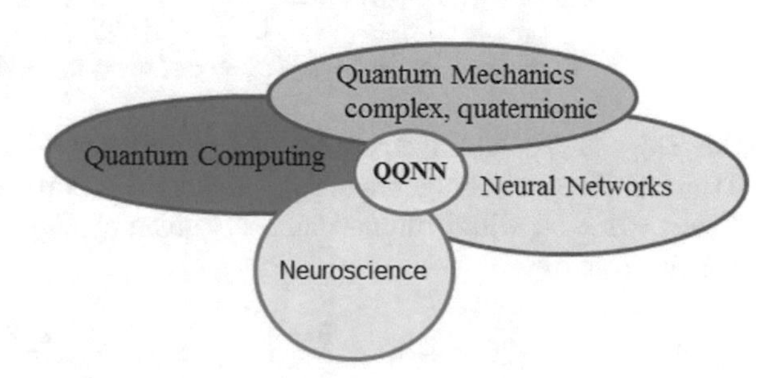

Fig. 13.16 Unification of fields for quantum computinng in geometric algebra

13.13.3 Quantum Bits in Geometric Algebra

Two possible states for a qubit are the state $|0>$ and $|1>$, which might correspond to the states 0 and 1 for classical bit. Here, we use the quantum mechanics notation $|>$ called *Dirac* notation. The difference between bits and qubits is that a qubit can be in a state other than $|0>$ and $|1>$.

In geometric algebra, the pseudoscalar I in G_3 squares to -1 and commutes with all the other elements G_3 and it has identical properties with the unit imaginary i; thus, there is an isomorphism between the bivector basis $Ie_1 = e_{23}$, $Ie_3 = e_{31}$, $Ie_3 = e_{12}$ and the Pauli matrices and the quaternion algebra basis $\boldsymbol{i} = e_{23}$, $\boldsymbol{j} = -e_{31}$, $\boldsymbol{k} = e_{12}$.

The tensor product is the way to put together vector spaces to form large vector spaces, and this is crucial to understand the quantum mechanics of multiparticle systems. If it is allowed N qubits to interact, the state generated has a possible 2^N basis states. This forms a combined Hilbert space written as the tensor product $\mathbb{H}_N = \mathbb{H}_2 \otimes \mathbb{H}_2 \otimes \cdots \otimes \mathbb{H}_2$, where the order of each term matters. For example, a system with two quantum bits is a 4D Hilbert space $\mathbb{H}_4 = \mathbb{H}_2 \otimes \mathbb{H}_2$ with the orthonormal basis $\mathcal{B}_{\mathbb{H}_2^2} = |00>, |01>, |10>, |11>$. The following notations are equivalent for the given state

$$\begin{aligned} |0>_1 \otimes |1>_2 \otimes |0>_3 \otimes \cdots |0>_N &\equiv |0>_1 |1>_2 |0>_3 \cdots |0>_3 \cdots |0>_N, \\ &\equiv | \overset{0}{1} \overset{1}{2}, \overset{0}{3}, ..., \overset{0}{N} > \equiv |010 \cdots 0>, \end{aligned} \quad (13.94)$$

where $|010 \cdots 0>$ means qubit '1' is in state $|0>$, qubit '2' is in state $|1>$, and qubit '3' is in state $|0>$.

For the 3D Euclidean geometric algebra G_3, the $\{e_k^n\}$ generates the direct-product space $[G_3]^n \overset{\text{def}}{=} G_3 \otimes G_3 \otimes \cdots G_3$ of n copies of the 3D Euclidean geometric algebra G_3. In order to express quantum states in G_3, we resort to the $1 \leftrightarrow 1$ mapping [74], defined as follows:

$$\begin{aligned} | > \phi = \alpha|0> + \beta|1> &= \begin{bmatrix} a_0 + ia_3 \\ -a_2 + ia_1 \end{bmatrix} \leftrightarrow \psi, \\ \psi = a_0 + a_1 Ie_1 + a_2 Ie_2 + a_3 Ie_3 &= a_0 + a_1 \boldsymbol{i} - a_2 \boldsymbol{j} + a_3 \boldsymbol{k}, \end{aligned} \quad (13.95)$$

where $a_i \in \mathbb{R}$. The multivectors $\{1, Ie_1, Ie_2, Ie_3\}$ are the computational basis states of the real 4D even subalgebra or rotor algebra G_3^+, corresponding to the dimensional Hilbert space $\mathbb{H}_2^1$ with orthonormal basis given by $\mathcal{B}_{\mathbb{H}_2^1} = \{|0>, |1>\}$. In GA, this basis is given by

$$|0> \leftrightarrow \psi_{|0>} \overset{\text{def}}{=} 1, \quad |1> \leftrightarrow \psi_{|1>} \overset{\text{def}}{=} -Ie_2. \quad (13.96)$$

13.13.4 A Spinor-Quaternion Map

A qubit can be represented by any point on the surface of the Bloch sphere [39], where the north and south poles correspond to the pure states $|0>$ and $|1>$, respectively, see Fig. 13.17. Given point (θ, ϕ) on the sphere and considering the usual spherical coordinates), the corresponding qubit is given by

$$\psi = e^{-i\frac{\phi}{2}} \cos\frac{\theta}{2}|0> + e^{i\frac{\phi}{2}} \sin\frac{\theta}{2}|1>, \tag{13.97}$$

where $i \in \mathbb{C}$. The global phase factor is not encoded in a qubit, as ψ and $\exp(i\alpha)\psi$ correspond to the same state.

A spinor is defined as

$$|\chi > \begin{bmatrix} \alpha \\ \beta \end{bmatrix}, \tag{13.98}$$

with α, $\beta \in \mathbb{C}$, and χ is normalized $< \chi|\chi >= 1$. The multiplication of $|\chi >$ by $\exp(i\alpha)$ gives a different spinor, however one that corresponds to the same qubit. Note that it is not a unique way to decompose $|\chi >$ into three angles (θ, ϕ, α), where θ and ϕ correspond to the location angles of the qubit on the Bloch sphere, e.g., trying to allocate this spinor

$$|\chi >= e^{i\alpha} \begin{pmatrix} \cos(\frac{\theta}{2})e^{-i\frac{\phi}{2}} \\ \sin(\frac{\theta}{2})e^{i\frac{\phi}{2}} \end{pmatrix}, \tag{13.99}$$

results in a singularity for qubits on the z-axis which allows many values for α. This shows us that $|\chi >$ represents indeed a point on a 3D sphere, but the geometry of the 3D sphere can not be represented as a phase 2D sphere. Thus, if α cannot be globally defined, it cannot be removed in some way without certain consequences. This can be circumvent if the spinor $|\chi >$ is rewritten using a unit quaternion ($|\boldsymbol{q}|$=1) via an

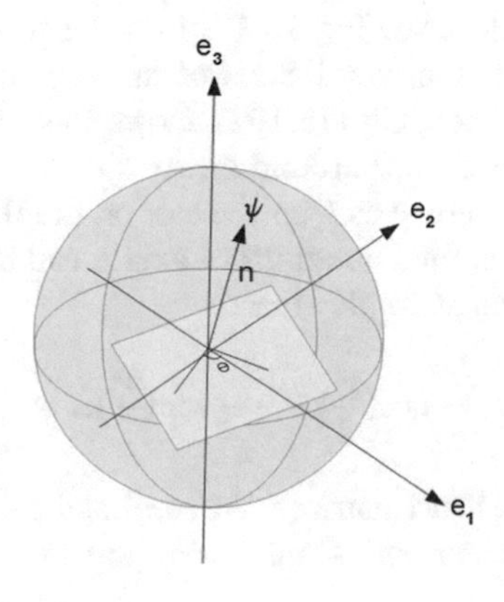

Fig. 13.17 Qubit on the Bloch sphere

invertible mapping $M_i : (|\chi) \rightarrow \boldsymbol{q}$ as follows:

$$M_i : (|\chi) = \boldsymbol{q} = \alpha + \boldsymbol{j}(\beta) = a + \boldsymbol{i}b + \boldsymbol{j}c + \boldsymbol{k}d, \tag{13.100}$$

where $\alpha = a + ib$, $\beta = c + id \in \mathbb{C}$. It is clear now that the space of all unit quaternions lies on a unit 3D sphere; thus, it does the space of all normalized spinors as well.

The ambiguity of α can be clarified using the unit quaternion representation of (13.99)

$$\boldsymbol{q} = e^{\boldsymbol{i}\frac{\alpha}{2}} e^{\boldsymbol{j}\frac{\theta}{2}} e^{-\boldsymbol{i}\frac{\phi}{2}}, \tag{13.101}$$

where for $\theta = 0$ the resulting combination $(\alpha - \frac{\phi}{2})$ can be only assigned a one value. Even though of this ambiguity, Eq. (13.101) can be always used to compute the corresponding Bloch sphere unit vector $\boldsymbol{q}$ in spherical coordinates.

Alternatively, there is another method to find the Bloch unit vector $\mathbf{q}$ which does not use the spinor representation, and it is via the use of pure quaternion $\hat{\boldsymbol{q}}$ with a scalar part zero, namely

$$\hat{\boldsymbol{q}} = \boldsymbol{q}\boldsymbol{i}\tilde{\boldsymbol{q}}, \tag{13.102}$$

$$\mathbf{q} = g(\hat{\boldsymbol{q}}) \equiv q_z e_1 - q_y e_2 + q_x e_3, \tag{13.103}$$

where q_x, q_y, q_z are the quaternion components and the function $g(\cdot)$ is invertible. Equation (13.102) is not invertible; however inserting $\boldsymbol{q}$ of Eq. (13.101) into Eq. (13.102), one sees that the global phase α always disappears exactly, because given two unit quaternions $\boldsymbol{i}$, $\boldsymbol{u}$, then $\boldsymbol{q} = \boldsymbol{u}\boldsymbol{i}\tilde{\boldsymbol{u}}$. Since $Re\{\boldsymbol{q}\} = 0$, then $\boldsymbol{q}$ is always a pure unit quaternion.

The rotations can also be mapped on the Bloch sphere itself. For that, we represent $\boldsymbol{q}$ in the similar form as a rotor of G_3^+

$$\boldsymbol{q} = e^{\frac{\vartheta}{2}\boldsymbol{n}}, \tag{13.104}$$

where $\boldsymbol{n}$ is a unit quaternion, see Fig. 13.17. In order to interpret rotations on the Bloch sphere, we compute the coordinates of the vector $\mathbf{q}$ as follows: $\mathbf{q} = g(\boldsymbol{q})$, $e_3 = g(\boldsymbol{i})$ and $\mathbf{n} = g(\boldsymbol{n})$. Then, Eq. (13.102) shows that the Bloch vector $\mathbf{q}$ can be computed rotating the axes e_3 in ϑ around the axes $\boldsymbol{n}$.

For a spinor $|\psi>$ represented as a unit vector on the Bloch sphere, a rotation of that vector by an angle γ around an arbitrary axis $\mathbf{n}$ can be computed by means of an operation of the following complex matrix

$$\mathtt{R}_{\mathbf{n}} = \cos(\frac{\gamma}{2})\mathtt{I} - i_{\mathbb{C}} \sin(\frac{\gamma}{2})\mathbf{n}\cdot\boldsymbol{\sigma}, \tag{13.105}$$

where $\boldsymbol{\sigma}$ is the vector of the Pauli matrices. We see that there is an isomorphy between $\mathtt{R}_{\mathbf{n}}$ and an exponential quaternion. Considering the map $M_i(|\chi)$ of Eq. (13.100),

Table 13.2 Single-qubit gates presented in terms of a right-multiplied (RM) quaternion using the map $M_i(|\chi)$. For $\pm\pi$-rotations, the two possible directions yield a different sign outcome

Gate	*Matrix operator*	*Equivalent RM quaternion*
Pauli X-gate	$\pm i \begin{bmatrix} 0 & 1 \\ 1 & 0 \end{bmatrix}$	$e^{\pm \boldsymbol{k}\frac{\pi}{2}} = \pm \boldsymbol{k}$
Pauli Y-gate	$\mp i \begin{bmatrix} 0 & -i \\ i & 0 \end{bmatrix}$	$e^{\pm \boldsymbol{j}\frac{\pi}{2}} = \pm \boldsymbol{j}$
Pauli Z-gate	$\pm i \begin{bmatrix} 1 & 0 \\ 0 & -1 \end{bmatrix}$	$e^{\pm \boldsymbol{i}\frac{\pi}{2}} = \pm \boldsymbol{i}$
Phase shift gate	$\begin{bmatrix} e^{-i\theta/2} & 0 \\ 0 & e^{i\theta/2} \end{bmatrix}$	$e^{-\boldsymbol{i}\frac{\theta}{2}}$
Hadamar gate	$\frac{\pm i}{\sqrt{2}} \begin{bmatrix} 1 & 1 \\ -1 & 1 \end{bmatrix}$	$e^{\pm \frac{\boldsymbol{i}+\boldsymbol{k}}{\sqrt{2}}\frac{\pi}{2}} = \pm \frac{\boldsymbol{i}+\boldsymbol{k}}{\sqrt{2}}$

a right multiplication by $exp(\boldsymbol{k}\frac{\gamma}{2})$, $exp(\boldsymbol{j}\frac{\gamma}{2})$, or $exp(\boldsymbol{i}\frac{\gamma}{2})$ on a unit quaternion $\boldsymbol{q}$ rotates the corresponding Bloch sphere vector by an angle γ around the positive axes e_1, e_2, e_3 respectively. Table 13.2 lists some useful special rotations, corresponding to quantum gates; here, we assume the use of the $M_i(|\chi)$ from spinors to quaternions.

13.13.5 Quantum Bit Operators Action and Observables in Geometric Algebra

The action of the conventional Pauli operators $\{\hat{\sum}_k, i_{\mathbb{C}}\hat{I}\}$ expressed in GA is

$$\begin{aligned} &\hat{\sum}_k|\psi> \leftrightarrow e_k\psi e_3 \texttt{ with } k = 1, 2, 3, \\ &i_{\mathbb{C}}|\psi> \leftrightarrow \psi I e_3. \end{aligned} \tag{13.106}$$

Note that the role of the imaginary unit i of the conventional quantum theory is played by the right multiplication by Ie_3. Let us see this in the following computations:

$$\begin{aligned} &\hat{\sum}_1|\psi> = \begin{bmatrix} -a_2 + ia_1 \\ a_0 + ia_3 \end{bmatrix} \leftrightarrow -a_2 + a_3Ie_1 - a_0Ie_2 + a_1Ie_3 = e_1(a_0 + a_kIe_k)e_3, \\ &\hat{\sum}_2|\psi> = \begin{bmatrix} a_1 + ia_2 \\ -a_3 + ia_0 \end{bmatrix} \leftrightarrow a_1 + a_0Ie_1 + a_3Ie_2 + a_2Ie_3 = e_2(a_0 + a_kIe_k)e_3, \\ &\hat{\sum}_3|\psi> = \begin{bmatrix} a_0 + ia_3 \\ a_0 - ia_1 \end{bmatrix} \leftrightarrow a_0 - a_1Ie_1 - a_2Ie_2 + a_3Ie_3 = e_3(a_0 + a_kIe_k)e_3. \end{aligned} \tag{13.107}$$

Doran and Lasenby [74] introduced the following observables, and the quantum n-particle correlator E_n is defined as follows. A complete basis for two-particle states is provided by

$$|0>|0>\leftrightarrow E,\ |0>|1>\leftrightarrow -Ie_2^2E,$$
$$|1>|0>\leftrightarrow -Ie_2^1E,\ |1>|1>\leftrightarrow Ie_2^1Ie_2^2E. \tag{13.108}$$

Further, it is defined

$$J = EIe_3^1 = EIe_3^2 = \frac{1}{2}(Ie_3^1 + Ie_3^2), \tag{13.109}$$

so that

$$J^2 = -E. \tag{13.110}$$

All that is required is to find a *quantum correlator* which satisfies

$$E_n Ie_3^k = E_n Ie_3^l = J_n, \texttt{ for all } k, l. \tag{13.111}$$

E_n is constructed by picking out the $k = 1$ space and correlating all the other spaces to do this, so that

$$E_n = \prod_{k=2}^{n} \frac{1}{2}(1 - Ie_3^1 Ie_3^k). \tag{13.112}$$

Note that the E_n is independent of which the n space is singled out and correlated to.

In the second observable, the complex structure is defined by

$$J_n = E_n Ie_3^k, \tag{13.113}$$

with $J_n^2 = -E_n$, where Ie_3^k can be chosen from any of the n spaces.

In order to reformulate conventional quantum mechanics, the right multiplication by all n-copies of Ie_3 in the n-particle algebra must fulfill the following:

$$\psi Ie_3^1 = \psi Ie_3^2 = \cdots = \psi Ie_3^{n-1} = \psi Ie_3^n. \tag{13.114}$$

These relations are obtained by using the n-particle correlator E_n of Eq. (13.112), which satisfies $E_n Ie_3^k = E_n Ie_3^l = J_n;\ \forall k, l$.

Right multiplication by the quantum correlator E_n is a projection operation that reduces the number of *real* degrees of freedom $4^n = dim_{\mathbb{R}}[G_3^+]^n$ to the expected $2^{n+1} = dim_{\mathbb{R}}\mathbb{H}_2^n$. One can interpret physically the projection as locking the phases of the various particles together. The extension to multiparticle systems involves a separate copy of the STA for each particle, and the standard imaginary unit i induces a correlation between these particles spaces.

13.13.6 Measurement of Probabilities in Geometric Algebra

The overlap probability between two states ψ and ϕ in the N-particle case is formulated by Doran and Lasenby: [74]

$$P(\psi, \phi) = 2^{N-2} < \psi E \psi^{\dagger} \phi E \phi^{\dagger} >_0 - 2^{N-2} < \psi E \psi^{\dagger} \phi J \phi^{\dagger} >_0, \quad (13.115)$$

where the angle bracket extracts the scalar part of the k-vector expression. We have two observables $\psi E \psi^{\dagger}$ and $\phi J \phi^{\dagger}$ and the expression

$$E_n = \prod_{k=2}^{n} \frac{1}{2}(1 - Ie_3^1 Ie_3^k) = \frac{1}{2^{N-1}}\Big(1 + \sum_{n=1}^{[\frac{N-1}{2}]} (-1)^n C_{2n}^N (Ie_3^i)\Big), \quad (13.116)$$

where $C_r^N(Ie_3^i)$ represents all possible combinations of N items taken r at a time, acting on the objects inside the bracket, e.g., $C_2^3(Ie_3^i) = Ie_3^1 Ie_3^2 + Ie_3^1 Ie_3^3 + Ie_2^3 + Ie_3^2 Ie_3^3$, and the number of items is given by

$$C_r^N = \frac{N!}{r!(N-r)!}. \quad (13.117)$$

The second observable can be formulated as follows:

$$J = EIe_3^1 = \frac{1}{2^{N-1}}\Big(1 + \sum_{n=1}^{[\frac{N+1}{2}]} (-1)^{n+1} C_{2n-1}^N (Ie_3^i)\Big). \quad (13.118)$$

For the case of 2 particles $N = 2$, these observables are

$$J = \frac{1}{2}(Ie_3^1 + Ie_3^2), \quad (13.119)$$

$$E = \frac{1}{2}(1 - Ie_3^1 Ie_3^2). \quad (13.120)$$

For a given N-particle state ψ, one encodes the measurement directions, one intends to use, in an auxiliary state ϕ, and then one calculates the overlap probability according to Eq. (13.115).

13.13.7 The 2-Qubit Space–Time Algebra

Cafaro and Mancini [46] described the 2-Qubit STA; for the sake of completeness, we describe it as well. In the 2-particle algebra, there are two bivectors Ie_3^1 and Ie_3^2 playing the role of $i_{\mathbb{C}}$. According to Eq. (13.114),

$$\psi I e_3^1 = \psi I e_3^2, \tag{13.121}$$

From algebraic manipulation of Eq. (13.121), one gets $\psi = \psi E$, where

$$E = \frac{1}{2}(1 - I e_3^1 I e_3^2), \quad E^2 = E. \tag{13.122}$$

The right multiplication by E is seen as a projection operation.

The multivectorial basis $B_{G_3^+ \otimes G_3^+}$ spanning the 16D geometric algebra $G_3^+ \otimes G_3^+$ is given by

$$B_{G_3^+ \otimes G_3^+} = 1, I e_l^1, I e_k^2, I e_l^1 I e_k^2, \tag{13.123}$$

for $k, l = 1, 2, 3$. After a right multiplication of the multivectors in $B_{G_3^+ \otimes G_3^+}$ by the projection operator E, one gets

$$B_{G_3^+ \otimes G_3^+} E = \{E, I e_l^1 E, I e_k^2 E, I e_l^1 I e_k^2 E\}, \tag{13.124}$$

in which after simple algebraic manipulation of each term, one gets

$$\begin{aligned} &E = -I e_3^1 I e_3^2 E,\ I e_1^2 E = -I e_3^1 I e_2^2 E,\ I e_2^2 E = I e_3^1 I e_1^2 E,\ I e_3^2 E = I e_3^1 E, \\ &I e_1^1 E = -I e_2^1 I e_3^2 E,\ I e_1^1 I e_1^2 E = -I e_2^1 I e_2^2 E,\ I e_1^1 I e_2^2 E = I e_2^1 I e_1^2 E, \\ &I e_1^1 I e_3^2 E = I e_2^1 E. \end{aligned} \tag{13.125}$$

By canceling E, we obtain a suitable 8D reduced subalgebra

$$B_{[G_3^+ \otimes G_3^+]/E} = \{1, I e_1^2, I e_2^2, I e_3^2, I e_1^1, I e_1^1 I e_1^2, I e_1^1 I e_2^2, I e_1^1 I e_3^2\}. \tag{13.126}$$

Equation (13.126) shows the basis which spans $B_{[G_3^+ \otimes G_3^+]/E}$. This basis is analog to a suitable standard complex basis which spans the complex Hilbert space $\mathbb{H}_2^2$.

Note that, in the case of the space–time algebra, the representation of the direct-product 2-particle Pauli spinor is formulated as $\psi^1 \phi^2 E$ in terms of the bivector-based spinors ψ^1 and ϕ^2 acting on their own spaces, i.e., $|\psi, \phi >\rightarrow \psi^1 \phi^2 E$.

The 2-particle spin states in $B_{[G_3^+ \otimes G_3^+]/E}$ are given by

$$\begin{aligned} &|0> \otimes |0> = \begin{pmatrix} 1 \\ 0 \end{pmatrix} \otimes \begin{pmatrix} 1 \\ 0 \end{pmatrix} \leftrightarrow E,\ |0> \otimes |1> = \begin{pmatrix} 1 \\ 0 \end{pmatrix} \otimes \begin{pmatrix} 0 \\ 1 \end{pmatrix} \leftrightarrow -I e_2^2 E, \\ &|1> \otimes |0> = \begin{pmatrix} 0 \\ 1 \end{pmatrix} \otimes \begin{pmatrix} 1 \\ 0 \end{pmatrix} \leftrightarrow -I e_2^1 E,\ |1> \otimes |1> = \begin{pmatrix} 0 \\ 1 \end{pmatrix} \otimes \begin{pmatrix} 0 \\ 1 \end{pmatrix} \leftrightarrow -I e_2^1 I e_2^2 E. \end{aligned} \tag{13.127}$$

As illustration, the entangled state between a pair of 2-level systems so-called a *spin singlet state* is formulated as follows:

$$|\psi_{singlet}> = \frac{1}{2}\left\{\begin{pmatrix} 1 \\ 0 \end{pmatrix} \otimes \begin{pmatrix} 0 \\ 1 \end{pmatrix} - \begin{pmatrix} 0 \\ 1 \end{pmatrix} \otimes \begin{pmatrix} 1 \\ 0 \end{pmatrix}\right\} = \frac{1}{\sqrt{2}}(|01> - |10>) \in \mathbb{H}_2^2. \tag{13.128}$$

$$|\psi_{singlet} > \in \mathbb{H}_2^2 \leftrightarrow \psi_{singlet}^{GA} \in [G_3^+ \otimes G_3^+], \tag{13.129}$$

where

$$\psi_{singlet}^{GA} = \frac{1}{2^{\frac{3}{2}}}(Ie_2^1 - Ie_2^2)(1 - Ie_3^1 Ie_3^2). \tag{13.130}$$

Following the work of Cafaro and Mancini [46], the multiplication by the quantum imaginary $i_{\mathbb{C}}$ for 2-particle states is taken by the multiplication with J from the right

$$J = EIe_3^1 = EIe_3^2 = \frac{1}{2}(Ie_3^1 + Ie_3^2), \tag{13.131}$$

henceforth $J^2 = -E$. The action of 2-particle Pauli operators reads

$$\hat{\sum}_k \otimes \hat{I}|\psi > \leftrightarrow -Ie_k^1\psi J, \ \hat{\sum}_k \otimes \hat{\sum}_l |\psi > \leftrightarrow -Ie_k^1 Ie_l^2 \psi E, \ \hat{I} \otimes \hat{\sum}_k |\psi > \leftrightarrow -Ie_k^2 \psi J. \tag{13.132}$$

For example, since

$$\hat{\sum}_l^2 |\psi > \leftrightarrow e_l^2 \psi e_3^2 = e_l^2 \psi E e_3^2 = -e_l^2 \psi EIIe_3^2 = -Ie_l^2 \psi EIe_3^2 = -Ie_l^2 \psi J, \tag{13.133}$$

one can then write

$$\hat{\sum}_k \otimes \hat{\sum}_l |\psi > \leftrightarrow (-Ie_k^1)(-Ie_l^2)\psi J^2 = -Ie_k^1 Ie_l^2 \psi E. \tag{13.134}$$

Since $i_{\mathbb{C}} \hat{\sum}_k |\psi > \leftrightarrow Ie_k \psi$, then we can claim that

$$i_{\mathbb{C}} \hat{\sum}_k \otimes \hat{I}|\psi > \leftrightarrow Ie_k^1 \psi, \ \text{and} \ \hat{I} \otimes i_{\mathbb{C}} \hat{\sum}_k \otimes \hat{I}|\psi > \leftrightarrow Ie_k^2 \psi. \tag{13.135}$$

13.13.8 Gates in Geometric Algebra

Our goal is to use neural networks with neurons which represent gates acting on n-qubits and working in the geometric algebra framework. Thus, we need first to formulate a convenient universal set of quantum gates and then implement them as processing units of neural networks. According to [223], a set of quantum gates $\{\hat{U}_i\}$ is seen as universal if any logical operator $\hat{U}_L$ can be written as

$$\hat{U}_L = \prod_{\hat{U}_L \in \{\hat{U}_i\}} \hat{U}_l. \tag{13.136}$$

Next, according to the proposed formulations by [46, 53, 136], we will represent simple circuit models of quantum computation with 1-qubit quantum gates in the geometric algebra framework. This subsection is based on the work of Cafaro and Mancini [46].

Quantum NOT Gate (or Bit Flip Quantum Model). A nontrivial reversible operation applied to a single qubit is done by means of the NOT operation gate denoted here by $\hat{\sum}_1$. Let us apply on a 1-qubit quantum gate given by $\psi^{GA}_{|q} = a_0 + a_2 I e_2$. Then, the $\hat{\sum}_1$ is defined as

$$\hat{\sum}_1 |q > \overset{\text{def}}{=} |q \oplus 1 > \leftrightarrow \psi^{GA}_{|q\oplus 1>} \overset{\text{def}}{=} e_1(a_0 + a_2 I e_2) e_3.$$

Since $I e_i = e_i I$ and $e_i e_j = e_i \wedge e_j$, we obtain

$$\hat{\sum}_1 |q > \overset{\text{def}}{=} |q \oplus 1 > \leftrightarrow \psi^{GA}_{|q\oplus 1>} = -(a_2 + a_0 I e_2). \tag{13.137}$$

The action of the unitary quantum gate $\hat{\sum}_1^{GA}$ on the basis $\{1, Ie_1, Ie_2, Ie_3\} \in G_3^+$ is as follows:

$$\hat{\sum}_1^{GA} : 1 \to -Ie_2, \hat{\sum}_1^{GA} : Ie_1 \to Ie_3, \hat{\sum}_1^{GA} : Ie_2 \to -1, \hat{\sum}_1^{GA} : Ie_3 \to Ie_1.$$

Quantum Phase Flip Gate. The reversible operation to a single qubit is the phase flip gate denoted by $\hat{\sum}_3$. In the GA framework, the action of the unitary quantum gate $\hat{\sum}_3^{GA}$ on the multivector $\psi^{GA}_{|q|} = a_0 + a_2 I e_2$ is given by

$$\hat{\sum}_3 |q > \overset{\text{def}}{=} (-1)^q |q > \leftrightarrow \psi^{GA}_{(-1)^q|q>} \overset{\text{def}}{=} e_3(a_0 + a_2 I e_2) e_3 = a_0 - a_2 I e_2.$$

The unitary quantum gate $\hat{\sum}_3^{GA}$ acts on the basis $\{1, Ie_1, Ie_2, Ie_3\} \in G_3^+$ as follows:

$$\hat{\sum}_3^{GA} : 1 \to 1, \hat{\sum}_3^{GA} : Ie_1 \to -Ie_1, \hat{\sum}_3^{GA} : Ie_2 \to -Ie_2, \hat{\sum}_3^{GA} : Ie_3 \to Ie_3.$$

Quantum Bit and Phase Flip Gate. A combination of two reversible operations $\hat{\sum}_1^{GA}$ and $\hat{\sum}_3^{GA}$ results in another reversible operation to be applied on a single qubit. This will be denoted by $\hat{\sum}_2 \overset{\text{def}}{=} i_{\mathbb{C}} \hat{\sum}_1 \circ \hat{\sum}_3$, and its action on $\psi^{GA}_{|q>} = a_0 + a_2 I e_2$ is given by

$$\hat{\sum}_2 |q > \overset{\text{def}}{=} i_{\mathbb{C}}(-1)^q |q \oplus 1 > \leftrightarrow \psi^{GA}_{(-1)^q|q\oplus 1>} \overset{\text{def}}{=} e_2(a_0 + a_2 I e_2) e_3 = (a_2 - a_0 I e_2) I e_3.$$

The unitary quantum gate $\hat{\sum}_2^{GA}$ acts on the basis $\{1, Ie_1, Ie_2, Ie_3\} \in G_3^+$ as follows:

$$\hat{\sum}_2^{GA} : 1 \to Ie_1, \hat{\sum}_2^{GA} : Ie_1 \to 1, \hat{\sum}_2^{GA} : Ie_2 \to Ie_3, \hat{\sum}_2^{GA} : Ie_3 \to Ie_2.$$

Hadamar Quantum Gate. The GA formulation of the Walsh–Hadamard quantum gate $\hat{H} \stackrel{\text{def}}{=} \frac{\hat{\sum}_1+\hat{\sum}_3}{\sqrt{2}}$ named $\hat{H}^{GA}$ acts on $\psi^{GA}_{|q>} = a_0 + a_2 I e_2$ as follows:

$$\hat{H}|q> \stackrel{\text{def}}{=} \frac{1}{\sqrt{2}}[|q \oplus 1> +(-1)^q|q>] \leftrightarrow \psi^{GA}_{\hat{H}|q} \stackrel{\text{def}}{=} \Big(\frac{e_1+e_3}{\sqrt{2}}\Big)(a_1 + a_2 I e_2)e_3$$
$$= \frac{a_0}{\sqrt{2}}(1 - Ie_2) - \frac{a_2}{\sqrt{2}}(1 + Ie_2).$$

The Hadamard transformations of the states $|+>$ and $|->$ are given as follows:

$$|+> \stackrel{\text{def}}{=} \frac{|0> +|1>}{\sqrt{2}} \leftrightarrow \psi^{GA}_{|+>} = \frac{1 - Ie_2}{\sqrt{2}},$$
$$|-> \stackrel{\text{def}}{=} \frac{|0> -|1>}{\sqrt{2}} \leftrightarrow \psi^{GA}_{|->} = \frac{1 + Ie_2}{\sqrt{2}}. \tag{13.138}$$

The unitary quantum gate $\hat{H}^{GA}$ acts on the basis $\{1, Ie_1, Ie_2, Ie_3\} \in G_3^+$ as follows:

$$\hat{H}^{GA} : 1 \rightarrow \frac{1 - Ie_2}{\sqrt{2}}, \;\; \hat{H}^{GA} : Ie_1 \rightarrow \frac{-Ie_1 + Ie_3}{\sqrt{2}},$$
$$\hat{H}^{GA} : Ie_2 \rightarrow -\frac{1 + Ie_2}{\sqrt{2}}, \;\; \hat{H}^{GA} : Ie_3 \rightarrow \frac{Ie_1 + Ie_3}{\sqrt{2}}. \tag{13.139}$$

Rotation Quantum Gate. The action of rotation gates $\hat{R}^{GA}_\theta$ acts on $\psi^{GA}_{|q>} = a_0 + a_2 I e_2$ as follows:

$$\hat{R}_\theta|q> \stackrel{\text{def}}{=} \Big(\frac{1 + \exp(i_{\mathbb{C}}\theta)}{2} + (-1)^q \frac{1 - \exp(i_{\mathbb{C}}\theta)}{2}\Big)|q>$$
$$\leftrightarrow \psi^{GA}_{\hat{R}_\theta|q} \stackrel{\text{def}}{=} a_0 + a_2 I e_2(\cos\theta + Ie_3 \sin\theta). \tag{13.140}$$

The unitary quantum gate $\hat{R}^{GA}_\theta$ acts on the basis $\{1, Ie_1, Ie_2, Ie_3\} \in G_3^+$ in following manner:

$$\hat{R}^{GA}_\theta : 1 \rightarrow 1, \;\; \hat{R}^{GA}_\theta : Ie_1 \rightarrow Ie_1(\cos\theta + Ie_3 \sin\theta),$$
$$\hat{R}^{GA}_\theta : Ie_2 \rightarrow Ie_2(\cos\theta + Ie_3 \sin\theta), \;\; \hat{R}^{GA}_\theta : Ie_3 \rightarrow Ie_3. \tag{13.141}$$

Table 13.3 presents a summary of the most relevant quantum gates in the geometric algebra framework to act on the basis states $\{1, Ie_1, Ie_2, Ie_3\} \in G_3^+$.

Table 13.3 Quantum gates to act on the basis states $\{1, Ie_1, Ie_2, Ie_3\} \in G_3^+$, NOT (N), Phase Flip (PF), Bit and Phase Flip (BPF)

1-*Qubit state*	*N*	*PF*	*BPF*	*Hadamard*	*Rotation*
1	$-Ie_2$	1	Ie_1	$\frac{1-Ie_2}{\sqrt{2}}$	1
Ie_1	Ie_3	$-Ie_1$	1	$\frac{-Ie_1+Ie_3}{\sqrt{2}}$	$Ie_1(\cos\theta + Ie_3 \sin\theta)$
Ie_2	-1	$-Ie_2$	Ie_3	$-\frac{1+Ie_2}{\sqrt{2}}$	$Ie_2(\cos\theta + Ie_3 \sin\theta)$
Ie_3	Ie_1	Ie_3	Ie_2	$\frac{Ie_1+Ie_3}{\sqrt{2}}$	Ie_3

13.13.9 Two-Qubit Quantum Computing

In this subsection, we study simple circuit models with 2-qubit quantum gates using the geometric algebra framework. This subsection is based on the work of Cafaro and Mancini [46]. We will show that the set of maximally entangled 2-qubits Bell states can be represented in geometric algebra. The Bell states are an interesting example of maximally entangled quantum states, and they form an orthonormal basis $\mathcal{B}_{Bell}$ in the product Hilbert space $\mathbb{C}^2 \otimes \mathbb{C}^2 \cong \mathbb{C}^4$. Given the 2-qubit computational basis $\mathcal{B}_c = \{|00>, |01>, |10>, |01>\}$, according to [223] the four Bell states can be constructed in the following way:

$$\begin{aligned}
|0>\otimes|0>\to|\psi_{Bell_1} &\stackrel{\text{def}}{=} \left[\hat{U}_{CNOT}\circ(\hat{U}\otimes\hat{I})\right](|0>\otimes|0>) = \frac{1}{\sqrt{2}}(|0>\otimes|0>+|1>\otimes|1>),\\
|0>\otimes|1>\to|\psi_{Bell_2} &\stackrel{\text{def}}{=} \left[\hat{U}_{CNOT}\circ(\hat{U}\otimes\hat{I})\right](|0>\otimes|1>) = \frac{1}{\sqrt{2}}(|0>\otimes|1>+|1>\otimes|0>),\\
|1>\otimes|0>\to|\psi_{Bell_3} &\stackrel{\text{def}}{=} \left[\hat{U}_{CNOT}\circ(\hat{U}\otimes\hat{I})\right](|1>\otimes|0>) = \frac{1}{\sqrt{2}}(|0>\otimes|0>-|1>\otimes|1>),\\
|1>\otimes|1>\to|\psi_{Bell_4} &\stackrel{\text{def}}{=} \left[\hat{U}_{CNOT}\circ(\hat{U}\otimes\hat{I})\right](|1>\otimes|1>) = \frac{1}{\sqrt{2}}(|0>\otimes|1>-|1>\otimes|0>),
\end{aligned} \tag{13.142}$$

where $\hat{H}$ and $\hat{U}_{CNOT}$ stand for the Hadamard and the CNOT gates, respectively. The Bell basis in $\mathbb{C}^2 \otimes \mathbb{C}^2 \cong \mathbb{C}^4$ is given by

$$\mathcal{B}_{Bell} \stackrel{\text{def}}{=} \{|\psi_{Bell_1}>, |\psi_{Bell_2}>, |\psi_{Bell_3}>, |\psi_{Bell_4}>\}. \tag{13.143}$$

According to Eq. (13.142), we obtain

$$\begin{aligned}
|\psi Bell_1> &= \frac{1}{\sqrt{2}}\begin{pmatrix}1\\0\\0\\1\end{pmatrix}, \; |\psi Bell_2> = \frac{1}{\sqrt{2}}\begin{pmatrix}0\\1\\1\\0\end{pmatrix},\\
|\psi Bell_3> &= \frac{1}{\sqrt{2}}\begin{pmatrix}1\\0\\0\\-1\end{pmatrix}, \; |\psi Bell_4> = \frac{1}{\sqrt{2}}\begin{pmatrix}0\\1\\-1\\0\end{pmatrix}.
\end{aligned} \tag{13.144}$$

According to Eqs. (13.127) and (13.142), the formulation of the Bell states in geometric algebra is as follows:

$$|\psi Bell_1 > \leftrightarrow \psi^{GA}_{Bell_1} = \frac{1}{2^{\frac{3}{2}}}(1 + Ie^1_2 Ie^2_2)(1 - Ie^1_3 Ie^2_3)),$$
$$|\psi Bell_2 > \leftrightarrow \psi^{GA}_{Bell_2} = -\frac{1}{2^{\frac{3}{2}}}(Ie^1_2 + Ie^2_2)(1 - Ie^1_3 Ie^2_3)),$$
$$|\psi Bell_3 > \leftrightarrow \psi^{GA}_{Bell_3} = \frac{1}{2^{\frac{3}{2}}}(1 - Ie^1_2 Ie^2_2)(1 - Ie^1_3 Ie^2_3)),$$
$$|\psi Bell_4 > \leftrightarrow \psi^{GA}_{Bell_4} = \frac{1}{2^{\frac{3}{2}}}(Ie^1_2 - Ie^2_2)(1 - Ie^1_3 Ie^2_3)). \quad (13.145)$$

2-Qubit CNOT Quantum gate. According to [223], a CNOT quantum gate can be written as

$$\hat{U}^{12}_{CNOT} = \frac{1}{2}\Big[\Big(\hat{I}^1 + \hat{\sum}^1_3\Big) \otimes \hat{I}^2 + \Big(\hat{I}^1 - \sum^1_3\Big) \otimes \hat{\sum}^2_1\Big], \quad (13.146)$$

where $\hat{U}^{12}_{CNOT}$ is the CNOT gate from qubit 1 to qubit 2, and thus

$$\hat{U}^{12}_{CNOT}|\psi >= \frac{1}{2}\Big[\hat{I}^1 \otimes \hat{I}^2 + \hat{\sum}^1_3 \otimes \hat{I}^2 + \hat{I}^1 \otimes \hat{\sum}^2_1 - \hat{\sum}^1_3\Big] \otimes \hat{\sum}^2_1\Big]|\psi > .$$

Using Eqs. (13.147) and (13.132), it follows

$$\hat{I}^1 \otimes \hat{I}^2|\psi > \leftrightarrow \psi, \; \hat{\sum}^1_3 \otimes \hat{I}^2|\psi > \leftrightarrow -Ie^1_3\psi J, \; \hat{I}^1 \otimes \hat{\sum}^2_1|\psi > \leftrightarrow -Ie^2_1\psi J,$$
$$-\hat{\sum}^1_3 \otimes \hat{\sum}^2_1|\psi > \leftrightarrow Ie^1_3 Ie^2_1 \psi E. \quad (13.147)$$

Now using Eq. (13.132), the CNOT gate of Eq. (13.147) is formulated in GA as follows:

$$\hat{U}^{12}_{CNOT}|\psi > \leftrightarrow \frac{1}{2}(\psi - Ie^1_3\psi J - Ie^2_1\psi J + Ie^1_3 Ie^2_1\psi E). \quad (13.148)$$

2-Qubit Controlled-Phase Gate. According to [223], the action of the controlled-phase gate $\hat{U}^{12}_{CP}$ on $|\psi > \in \mathbb{H}^2_2$ can be formulated as

$$\hat{U}^{12}_{CP}|\psi >= \frac{1}{2}\Big[\hat{I}^1 \otimes \hat{I}^2 + \hat{\sum}^1_3 \otimes \hat{I}^2 + \hat{I}^1 \otimes \hat{\sum}^2_3 - \hat{\sum}^1_3 \otimes \hat{\sum}^2_3\Big]|\psi > . \quad (13.149)$$

Using Eqs. (13.149) and (13.132), it follows

$$\hat{I}^1 \otimes \hat{I}^2|\psi> \leftrightarrow \psi,\ \hat{\sum}^1_3 \otimes \hat{I}^2|\psi> \leftrightarrow -Ie_3^1\psi J,\ \hat{I}^1 \otimes \hat{\sum}^2_3|\psi> \leftrightarrow -Ie_3^2\psi J,$$
$$-\hat{\sum}^1_3 \otimes \hat{\sum}^2_3|\psi> \leftrightarrow Ie_3^1Ie_3^2\psi E. \tag{13.150}$$

Using Eqs. (13.132) and (13.150), one can formulate in geometric algebra the controlled-phase quantum gate as follows:

$$\hat{U}^{12}_{CP}|\psi> \leftarrow \frac{1}{2}(\psi - Ie_3^1\psi J - Ie_3^2\psi J + Ie_3^1Ie_3^2\psi E). \tag{13.151}$$

2-Qubit SWAP Gate. According to [223], the action of the SWAP gate $\hat{U}^{12}_{SWAP}$ on $|\psi> \in \mathbb{H}_2^2$ can be formulated as

$$\hat{U}^{12}_{SWAP}|\psi> = \frac{1}{2}\left[\hat{I}^1 \otimes \hat{I}^2 + \hat{\sum}^1_1 \otimes \hat{\sum}^2_1 + \hat{\sum}^1_2 \otimes \hat{\sum}^2_2 + \hat{\sum}^1_3 \otimes \hat{\sum}^2_3\right]|\psi>. \tag{13.152}$$

Using Eqs. (13.152) and (13.132), it follows

$$\hat{I}^1 \otimes \hat{I}^2|\psi> \leftrightarrow \psi,\ \hat{\sum}^1_1 \otimes \hat{\sum}^2_1|\psi> \leftrightarrow -Ie_2^1Ie_2^2\psi E,$$
$$\hat{\sum}^1_2 \otimes \hat{\sum}^2_2|\psi> \leftrightarrow -Ie_2^1Ie_2^2\psi E,\ \hat{\sum}^1_3 \otimes \hat{\sum}^2_3|\psi> \leftrightarrow -Ie_3^1Ie_3^2\psi E.$$

Using Eqs. (13.132) and (13.153), the SWAP quantum gate in GA reads

$$\hat{U}^{12}_{SWAP}|\psi> \leftarrow \frac{1}{2}(\psi - Ie_1^1Ie_1^2\psi E - Ie_1^2Ie_2^2\psi E - Ie_3^1Ie_3^2\psi E). \tag{13.153}$$

Table 13.4 summarizes the most important 2-qubit quantum gates formulated in the geometric algebra basis $\mathcal{B}_{[G_3^+ \otimes G_3^+]/E}$

Table 13.4 2-Qubit Quantum gates: CNOT, controlled-phase, and SWAP gates to act on the basis $\mathcal{B}_{[G_3^+ \otimes G_3^+]/E}$

2-Q Gates	*2-Qubit states*	*Gate action on states*
CNOT	ψ	$\frac{1}{2}(\psi - Ie_3^1\psi J - Ie_1^2\psi J + Ie_3^1Ie_1^2\psi E)$
CP	ψ	$\frac{1}{2}(\psi - Ie_3^1\psi J - Ie_3^2\psi J + Ie_3^1Ie_3^2\psi E)$
SWAP	ψ	$\frac{1}{2}(\psi - Ie_1^1Ie_1^2\psi E - Ie_1^2Ie_2^2\psi E - Ie_3^1Ie_3^2\psi E)$

13.13.10 *Quaternion Quantum Neurocomputing in Geometric Algebra*

This section is devoted to formulate the Quaternion or Rotor Quantum Gates as Quantum Processing Units for a quaternion quantum neural network and the learning training rule. We will highlight the resemblance of the quaternion or rotor quantum neural network with a quaternion multilayer perceptron.

Quaternion Quantum Perceptron. Altaisky was the first to introduce a quantum perceptron [4]. We extend this model using quaternions. A Quaternion Quantum Perceptron (QQP) process its quaternion (rotor) input qubits. It is modeled by the quaternion quantum updating function

$$|\boldsymbol{y}(t)>= \hat{F}\sum_{i=1}^{m}\hat{\boldsymbol{w}}_{ij}(t)|\boldsymbol{x}_i>, \tag{13.154}$$

implemented by an arbitrary quaternion (rotor) quantum gate operator $\hat{F}$ and $\boldsymbol{w}_{ij}$ quaternion operators representing the synaptic weights, both working on the m quaternion (rotor) input qubits. The quaternion quantum perceptron can be trained by the quaternion quantum equivalent of the Hebbian rule

$$\boldsymbol{w}_{ij}(t+1) = \boldsymbol{w}_{ij}(t) + \eta(|\boldsymbol{d}> - |\boldsymbol{y}(t)>) < \boldsymbol{x}_i|, \tag{13.155}$$

where $|\boldsymbol{d}>$ is the target state and $|\boldsymbol{y}(t)>$ the state of the neuron j at the discrete time step t and $\eta \in [0, 1]$ is the learning rate. Note that the quaternion quantum learning rule is unitary in order to preserve the total probability of the system. Since the our approach is multiparticle of STA, each particle is defined in an quaternion space or Hilbert space $\mathbb{H} \otimes \mathbb{H}$ and its output of a QQP is set to a sum of quaternions from other Hilbert spaces $\mathbb{H} \otimes \mathbb{H}$; thus, it entails the quantum property of superposition.

Quaternion Rotation Gate. A single-qubit quaternion rotation gate can be defined as a unit quaternion or (rotor)

$$\boldsymbol{R} = \cos(\frac{\theta}{2}) + \boldsymbol{n}cos(\frac{\theta}{2}) = e^{\frac{\theta}{2}\boldsymbol{n}}. \tag{13.156}$$

Let the quaternion quantum state be $|\psi>_1 = \cos\theta_0 + \boldsymbol{n}_1 \sin\theta_0$. This state can be transformed as follows:

$$|\psi>_2 = \boldsymbol{R}|\psi>_1 \tilde{\boldsymbol{R}} = \cos(\theta_0+\theta) + \boldsymbol{n}_2\sin(\theta_0+\theta). \tag{13.157}$$

The rotor $\boldsymbol{R}$ shifts the phase of $|\psi>_1$ and changes its unit quaternion $\boldsymbol{n}_1 = n_x\boldsymbol{i} + n_y\boldsymbol{j} + n_z\boldsymbol{k}$ to $\boldsymbol{n}_2$, which is point in another location on the 2D Bloch sphere.

Unitary Operators and Tensor Products. A quaternion quantum neural network involves unitary operators and tensor products to process multiparticles.

An unitary quaternion operator $\boldsymbol{U}$ has to fulfill

$$\boldsymbol{U}\tilde{\boldsymbol{U}} = 1. \tag{13.158}$$

The tensor product is the way to put together vector spaces to form large vector spaces. If it is allowed N qubits to interact, the state generated has a possible 2^N basis states. This forms a combined Hilbert space written as the tensor product $\mathbb{H}_N = \mathbb{H}_2 \otimes \mathbb{H}_2 \otimes \cdots \otimes \mathbb{H}_2$, where the order of each term matters. The following notations are equivalent for the given state

$$\begin{aligned} |0>_1 \otimes |1>_2 \otimes |0>_3 \otimes \cdots |0>_N &\equiv |0>_1 |1>_2 |0>_3 \cdots |0>_3 \cdots |0>_N, \\ &\equiv |\overset{0}{1}\overset{1}{2}, \overset{0}{3}, ..., \overset{0}{N}> \equiv |010 \cdots 0>, \end{aligned} \tag{13.159}$$

where $|010 \cdots 0>$ means qubit '1' is in state $|0>$, qubit '2' is in state $|1>$, and qubit '3' is in state $|0>$. For the 3D Euclidean geometric algebra G_3, the $\{e_k^n\}$ generates the direct-product space $[G_3]^n \overset{\text{def}}{=} G_3 \otimes G_3 \otimes \cdots G_3$ of n copies of the 3D Euclidean geometric algebra G_3.

Quaternion Multi-qubits Controlled-NOT Gate. The quaternion two-qubit controlled-NOT gate has two input qubits, namely the quaternion qubit and the target quaternion qubit, respectively, see a neuron processing unit of this gate in Fig. 13.18. The action of this gate is as follows: According to Eq. (13.137), if the control quaternion qubit is set to 0, the target quaternion qubit is left alone; otherwise, if the control quaternion is set to 1, the target quaternion qubit is flipped. In the quaternion two-qubit controlled-NOT gate, how to condition an action on a single-qubit set is obvious; hence, this condition can be generalized to the quaternion multi-qubits controlled-NOT gate. This condition can be generalized to the quaternion multi-qubits controlled-NOT gate C^n_{CNOT}, and it is defined as

$$C^n_{CNOT} = (x_1 x_2 \cdots x_n > |\psi>)|X^{x_1 x_2 x_3 \cdots x_n |\psi}, \tag{13.160}$$

where the set of quaternions $x_1 x_2 \cdots x_n$ in the exponent of X corresponds to the product of the qubits $x_1 x_2 \cdots x_n$. Figure 13.19 shows a diagram of the quaternion multi-qubits controlled-NOT gate C^n_{CNOT}. That means that the operator X is applied to last one quaternion qubit or target qubit, if the first n quaternion qubits are all equal one, otherwise no action is taken. Given the control quaternion qubits $|x_i> = a_i|0> + \boldsymbol{n}b_i|1>$ for $i = 1, 2, ..., n$, where the quaternion target qubit $psi> = 0>$, the output of the quaternion multi-qubits controlled-NOT gate can be described as

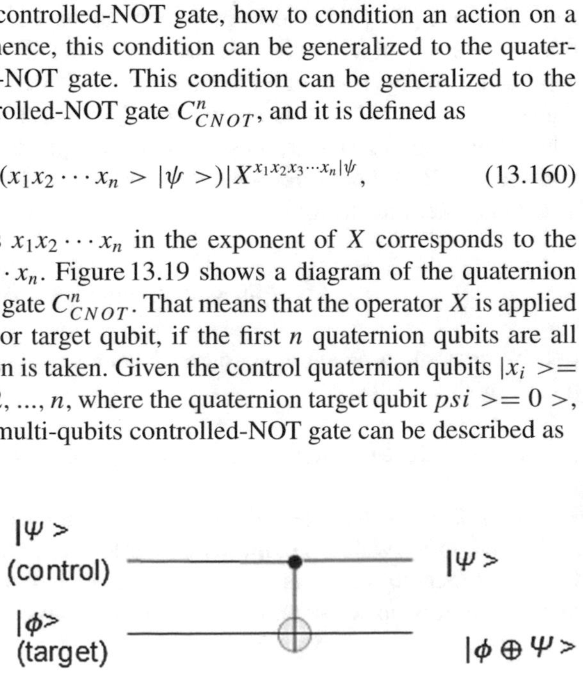

Fig. 13.18 Controlled-NOT gate

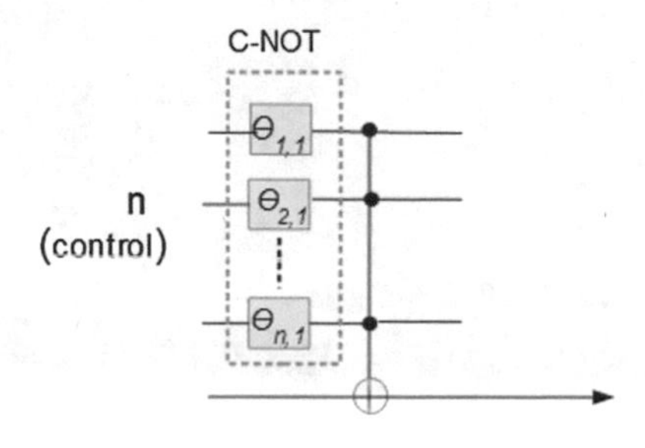

Fig. 13.19 Quaternion multi-qubits controlled-NOT gate

$$C^n_{CNOT}(|x_1> \otimes \cdots \otimes |x_n> \otimes |0> - b_1 b_2 \cdots b_n | \overbrace{11 \cdots 1}^{n} 0> +$$
$$+ b_1 b_2 \cdots b_n | \overbrace{11 \cdots 11}^{n} 1> . \tag{13.161}$$

The meaning of Eq. (13.161) is that the output of C^n_{CNOT} is in the entangled state of $n+1$ quaternion qubits, and the probability of the target quaternion qubit state, in which $|1>$ is observed, equals to $(b_1 b_2 \cdots b_n)^2$.

13.13.11 Quaternion Quantum Neural Network

Cao and Li [50] introduced the quantum-inspired neural network. We extend this work the quantum neural network to the Quaternion Quantum Neural Network (QQNN). At the input of the QQNN, rotor gates rotate the quaternion inputs. Figure 13.20 shows the model of Rotor Gate. Next, we explain the operations and outputs through the layers of the QQNN as shown in Fig. 13.21. The k feature vector is $\hat{X}^k = [\hat{x}^k_1, \hat{x}^k_1, \hat{x}^k_2, ..., \hat{x}^k_n]^T \in \mathbb{R}^n$, where $k = 1, ..., K$ and K is the number of samples. The corresponding quaternion quantum representation reads

$$|X^k >= [|x^k_1 >, |x^k_2 >, ..., |x^k_n >]^T, \tag{13.162}$$

where the quaternion state is

$$|x^k_i >= \cos\left(\frac{2\pi(\hat{x}^k_i - min_i)}{max_i - min_i}\right) + \boldsymbol{n}_i \sin\left(\frac{2\pi(\hat{x}^k_i - min_i)}{max_i - min_i}\right), \tag{13.163}$$

$max_i = max(x^1_i, x^2_i, ..., x^K_i)$, $min_i = min(x^1_i, x^2_i, ..., x^K_i)$.

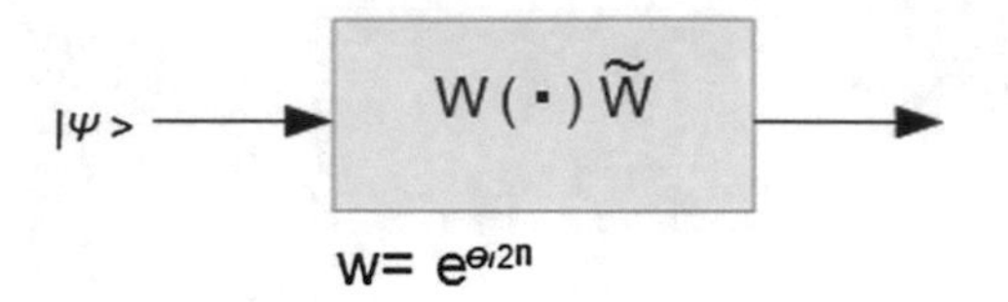

Fig. 13.20 Rotor gate

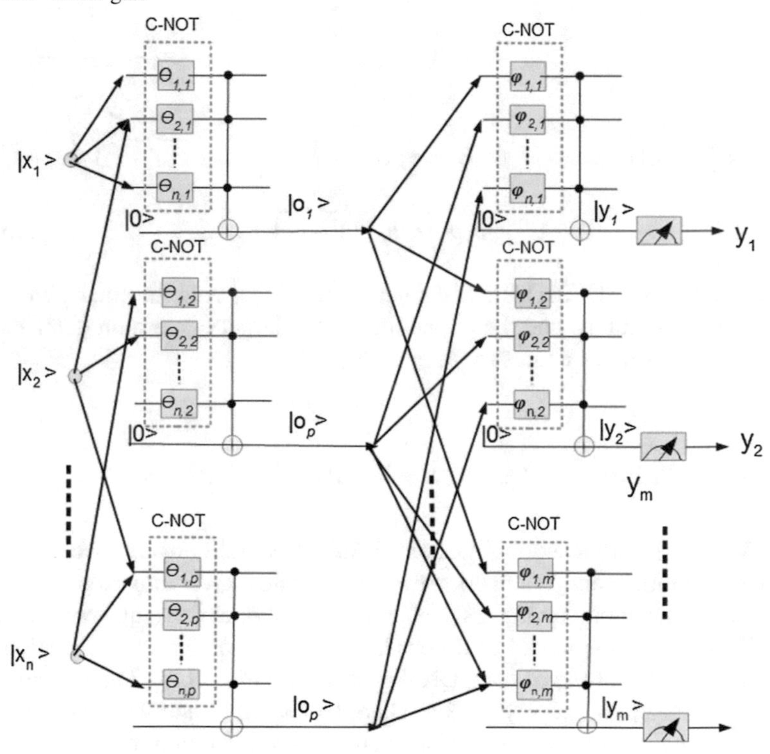

Fig. 13.21 Quaternion quantum neural network

Let the quaternion inputs be

$$|x_i >= \cos\theta_i|0> +\boldsymbol{n}_i \sin\theta_i|1>, \tag{13.164}$$

where $i = 1, 2, ..., n$, and according to Eqs. (13.157–13.161), the outputs of each layer of the QQN can be written as

$$\begin{aligned} |o_j> &= \cos(\varphi_j)|0> +\boldsymbol{n}_j \sin(\varphi_j)|1>, \\ |y_k> &= \cos(\xi_k)|0> +\boldsymbol{n}_k \sin(\xi_k)|1>, \end{aligned} \tag{13.165}$$

where $j = 1, 2, ..., p$ and $k = 1, 2, ..., m$;

$$\varphi_j = \arcsin\Big(\prod_{i=1}^{p}\sin(\theta_i + \theta_{ij})\Big), \quad \boldsymbol{n}_j = \frac{\sum_{i=1}^{p}\boldsymbol{n}_i}{\sqrt{\sum_1^p |\boldsymbol{n}_i|^2}},$$

$$\xi_k = \arcsin\Big(\prod_{j=1}^{m}\sin(\varphi_j + \varphi_{jk})\Big), \quad \boldsymbol{n}_k = \frac{\sum_{i=1}^{m}\boldsymbol{n}_i}{\sqrt{\sum_1^m |\boldsymbol{n}_i|^2}}. \tag{13.166}$$

Henceforth, the outputs of the QQN are written as follows:

$$o_j = \prod_{i=1}^{n}\sin(\theta_i + \theta_{ij}),$$

$$y_k = \prod_{j=1}^{m}\sin\Big(\arcsin(\prod_{i=1}^{n}\sin\big(\theta_i + \theta_{ij})\big) + \varphi_{jk}\Big). \tag{13.167}$$

We extend the complex-valued training rule of Cao and Li [50] to quaternion-valued one. In every layer of the QQNN, the parameters which has to be updated are the rotation angles of the quantum rotation gates. Given the desired normalized outputs $\tilde{y}_1, \tilde{y}_2, ..., \tilde{y}_m$, the evaluation function is defined as

$$E = max_{1\leq l\leq K}\, max_{1\leq k\leq m}|e_k^l| = max_{1\leq l\leq K}\, max_{1\leq k\leq m}|\tilde{y}_k^l - y_k^l|. \tag{13.168}$$

If one uses the gradient descent algorithm, the rotation angles in each layer can be computed as follows:

$$\frac{\partial e_k^l}{\partial\theta_{ij}} = -y_k\, ctg\,(\varphi_j + \varphi_{jk})o_j^l ctg(\theta_i + \theta_{ij}^r)/\sqrt{1-(h_j^l)}^2, \tag{13.169}$$

where $o_j^l = \prod_{i=1}^{n}\sin(\theta_i^l + \theta_{ij})$.

$$-\frac{\partial e_k^l}{\partial\varphi_{jk}} = -y_k\, ctg\,(\varphi_j + \varphi_{jk}), \tag{13.170}$$

and

$$\frac{\partial e_k^l}{\partial \boldsymbol{n}_k^r}, \tag{13.171}$$

where $r = 1, 2, 3$ stands for the three components of the bivector or $\boldsymbol{i}, \boldsymbol{j}, \boldsymbol{k}$ of the unit quaternion $\boldsymbol{n}_k$. However, the gradient calculation is pretty involved, the standard descent algorithm does not converge easily; often, it gets trapped in a local minimum; alternatively, it can be used to adjust the parameters the Levenberg–Marquardt (L-M) algorithm [182] or even evolutionary computing which guarantees global minimum. For the former, let us denote $\mathtt{P}$ for the parameter vector, **e** for the error, and $\mathtt{J}$ for the Jacobian matrix. They are formulated as follows:

$$
\begin{aligned}
\mathtt{P}^T &= [\theta_{1,1}, \theta_{1,2}, ..., \theta_{n,p}, \varphi_{1,1}, \varphi_{1,2}, ..., \varphi_{p,m}, n^1_{1,1}, n^2_{1,1}, n^3_{1,1}, n^1_{1,2}, n^2_{1,2}, n^3_{1,2}, ..., n^1_{p,m}, n^2_{p,m}, n^3_{p,m}],\\
e^T(\mathtt{P}) &= [e^1, e^2, ..., e^l_m, ..., e^K_1, e^K_2, ..., e^K_m],
\end{aligned}
\tag{13.172}
$$

$$
\mathtt{J}(\mathtt{P}) = \begin{bmatrix}
\frac{\partial e^1_1}{\partial \theta_{1,1}} & \cdot & \frac{\partial e^1_1}{\partial \theta_{n,p}} & \cdot & \frac{\partial e^1_1}{\partial \varphi_{1,1}} & \cdot & \frac{\partial e^1_1}{\partial \varphi_{p,m}} & \cdot & \frac{\partial e^1_1}{\partial n^1_{1,1}} & \frac{\partial e^1_1}{\partial n^2_{p,m}} & \frac{\partial e^1_1}{\partial n^3_{p,m}} \\
\vdots & \cdot & \vdots & \cdot & \vdots & \cdot & \vdots & \cdot & \vdots & \cdot & \vdots \\
\frac{\partial e^m_1}{\partial \theta_{1,1}} & \cdot & \frac{\partial e^1_m}{\partial \theta_{n,p}} & \cdot & \frac{\partial e^1_m}{\partial \varphi_{1,1}} & \cdot & \frac{\partial e^1_m}{\partial \varphi_{p,m}} & \cdot & \frac{\partial e^1_m}{\partial n^1_{1,1}} & \frac{\partial e^1_m}{\partial n^2_{p,m}} & \frac{\partial e^1_m}{\partial n^3_{p,m}} \\
\vdots & \cdot & \vdots & \cdot & \vdots & \cdot & \vdots & \cdot & \vdots & \cdot & \vdots \\
\frac{\partial e^L_1}{\partial \theta_{1,1}} & \cdot & \frac{\partial e^L_1}{\partial \theta_{n,p}} & \cdot & \frac{\partial e^L_1}{\partial \varphi_{1,1}} & \cdot & \frac{\partial e^L_1}{\partial \varphi_{p,m}} & \cdot & \frac{\partial e^L_1}{\partial n^1_{1,1}} & \frac{\partial e^L_1}{\partial n^2_{p,m}} & \frac{\partial e^L_1}{\partial n^3_{p,m}} \\
\vdots & \cdot & \vdots & \cdot & \vdots & \cdot & \vdots & \cdot & \vdots & \cdot & \vdots \\
\frac{\partial e^1_1}{\partial \theta_{1,1}} & \cdot & \frac{\partial e^1_1}{\partial \theta_{n,p}} & \cdot & \frac{\partial e^L_m}{\partial \varphi_{1,1}} & \cdot & \frac{\partial e^L_m}{\partial \varphi_{p,m}} & \cdot & \frac{\partial e^L_m}{\partial n^1_{1,1}} & \frac{\partial e^L_m}{\partial n^2_{p,m}} & \frac{\partial e^L_m}{\partial n^3_{p,m}}
\end{bmatrix}.
$$

The iterative L-M algorithm for the QQNN is written as follows:

$$
\mathtt{P}_{t+1} = \mathtt{P}_t - (\mathtt{J}^T(\mathtt{P}_t)\mathtt{J}(\mathtt{P}_t) + \mu_t I)^{-1}\mathtt{J}^T(\mathtt{P}_t)e(\mathtt{P}_t), \tag{13.173}
$$

where t denotes the iterative steps, I stands for the unit matrix, and μ is a small positive real number to ensure that the matrix $(\mathtt{J}^T(\mathtt{P}_t)\mathtt{J}(\mathtt{P}_t) + \mu_t I)$ is invertible.

13.13.12 Quaternion Qubit Neural Network

In the previous subsection, we formulated a multiple quaternion qubit neural network in the space and time algebra. Next, we will propose a Quaternion Quantum Neural Network (QQNN) using one quaternion qubit neurons. Kouda et al. [166] introduced a multilayered feedforward qubit neuron model. Next, we extend this model to the Quaternion Qubit Neural Network.

Qubit Neuron Model. The firing and nonfiring states of the quaternion qubit neuron corresponding to the qubit are $|0>$ and $|1>$, and the quantum superposing states is used as the representation of the quaternion neural states. The quaternion qubit state is given by

$$
f(\theta) = e^{\boldsymbol{n}\theta} = \cos(\theta) + \boldsymbol{n}\sin(\theta), \tag{13.174}
$$

where the bivector $\boldsymbol{n} = n_x\boldsymbol{i} + n_y\boldsymbol{j} + n_z\boldsymbol{k} \equiv n_x e_{23} + n_y e_{31} + n_z e_{12}$. The real part and the bivector correspond to the probability amplitudes of $|0>$ and $|1>$, respectively.

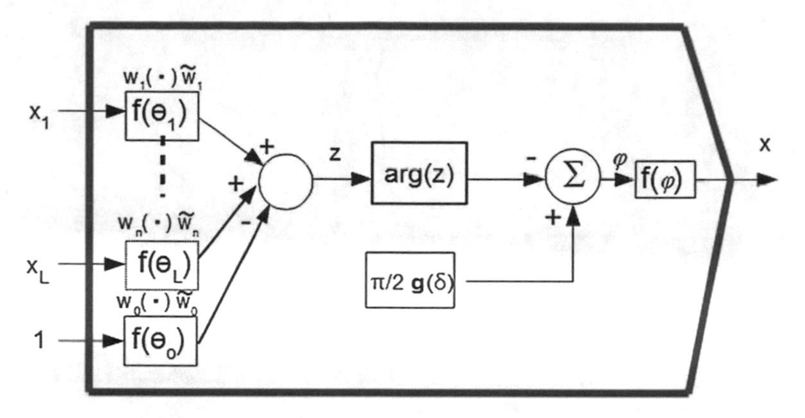

Fig. 13.22 Quaternion qubit neuron

The angle θ represents the quantum phase between the components. The square absolute value of each probability amplitude gives the probability with which the respective state is observed. Figure 13.22 depicts the model of the quaternion qubit neuron. The neuron is fed with the outputs x_l from previous L neuronss and it state is formulated using the Eq. (13.174) as follows:

$$x = f(y). \tag{13.175}$$

The input phase is computed using a 2-qubit controlled-NOT Gate, and it is given by

$$\varphi = \frac{\pi}{2} g(\delta) - arg(z), \tag{13.176}$$

this representation is generalized by introducing the inversion parameter δ to make the inversion smoothly variable using a sigmoid function

$$g(x) = \frac{1}{1 + exp(-x)}. \tag{13.177}$$

The internal state of the quaternion qubit neuron is computed using 1-qubit Rotation Gate, and it is given by

$$\begin{aligned} z &= \sum_{l=1}^{L} f(\theta_l) x_l - f(\theta_0) = \sum_{l=1}^{L} e^{\frac{\theta_l}{2}} f(\varphi_l) e^{\frac{\theta_l}{2}} - f(\theta_0), \\ &= \sum_{l=1}^{L} e^{\frac{\theta_l}{2}} e^{\varphi_l \boldsymbol{n}_l} e^{\frac{\theta_l}{2}} - f(\theta_0) = \sum_{l=1}^{L} e^{\frac{\theta_l}{2}} (\cos(\varphi_l) + \boldsymbol{n}_l \sin(\varphi) e^{-\frac{\theta_l}{2}} - f(\theta_0), \end{aligned} \tag{13.178}$$

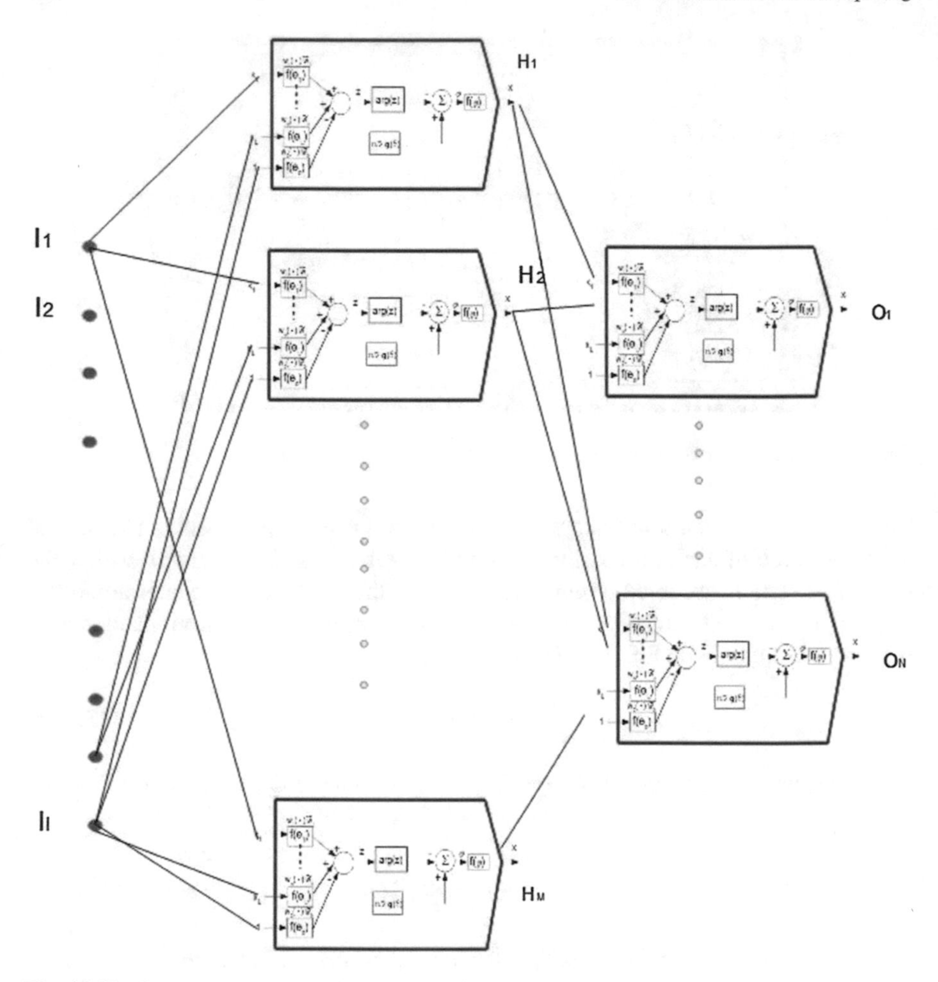

Fig. 13.23 Quaternion qubit feedforward neural network

where θ_l is the phase corresponding to the connecting weight for input x_l and θ_0 is the phase corresponding to a neuron bias or threshold. Note that Eq. (13.178) corresponds to the spatial summation as a feature of the conventional neuron model where the weights are replaced now with quaternion or rotor for the rotations, i.e., $e^{\frac{\theta_l}{2}}(\cdot)e^{\frac{\theta_l}{2}}$.

Quaternion Qubit Neural Network. The feedforward neural network based on the Quaternion Qubit Neuron (QQN) is shown in Fig. 13.23, where the set of inputs at the input layer is $\{I_l\}$ for $l = 1, 2, ..., L$, the set of outputs at the hidden layer is $\{H_m\}$ for $m = 1, 2, ..., M$, and the set of outputs at the output layer is $\{O_n\}$ for $n = 1, 2, ..., N$.

The input–output relation in the input layer neuron is given by

$$y_l^I = \frac{\pi}{2} input_l, \tag{13.179}$$

$$x_l^I = f(y_l^I), \tag{13.180}$$

here y_l^I is the input phase to the lth QQN in the input layer I. x_l^I is the state of the QQN with quantum phase y_l^I, and $input_l$ is the data input data in the range of [0, 1]. The nth output, $output_n$, of the Q-Qubit-NN is defined as the probability that the firing state $|1>$ of the nth QQN x_n^O in the output layer O is observed, that is, the square of the absolute value of the bivector part ($< \cdot >_2$), namely

$$output_n = | < x_n^O >_2 |^2. \tag{13.181}$$

Learning Rule for the Quaternion Qubit Neural Network. Normally to derive the learning rule, one uses the square error function and the steepest descent method to update the parameters. The learning rule comprises of the following equations:

$$E_{total} = \frac{1}{2} \sum_p^K \sum_n^N (t_{n,p} - output_{n,p})^2, \tag{13.182}$$

$$\theta_{l,k}^{new} = \theta_{l,k}^{old} - \eta \frac{\partial E_{total}}{\partial \theta_{l,k}^{old}},$$

$$\theta_k^{new} = \theta_k^{old} - \eta \frac{\partial E_{total}}{\partial \theta_k^{old}},$$

$$\delta_k^{new} = \delta_k^{old} - \eta \frac{\partial E_{total}}{\partial \delta_k^{old}},$$

where K is the number of target patterns to be learned, $output_{n,p}$ is the nth output value for the Quaternion Qubit Neural Network for the input pattern p, $t_{n,p}$ is the target value for that output, and η is the learning rate coefficient. Note as opposite as for the QNNN, in the learning rule for the Quaternion Qubit Neural Network, we do not need to adjust the bivector $\boldsymbol{n}$ of the quaternion 1-qubit neuron. Note that alternatively one can use evolutionary computing as learning rule.

Using the Quaternion Qubit Neural Network, Fig. 13.24a shows the result of solving the XOR problem, Fig. 13.24b shows the 2D clustering of two sets separate by a circle, and Fig. 13.24c shows the 3D clustering of two sets separated by a espiral.

13.14 Deep Learning Using Geometric Algebra

In the previous sections, we have presented different approaches for geometric neurocomputing and the Clifford Support Vector Machines. Most of the cases are feedforward network topologies involving just one hidden layer. So the next step

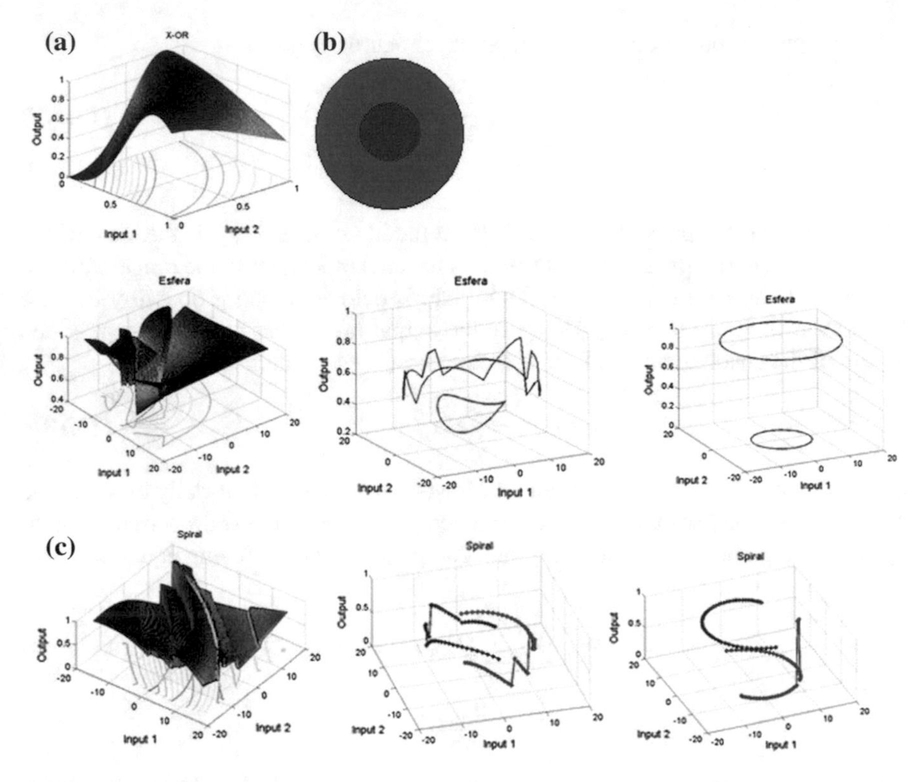

Fig. 13.24 Quaternion qubit neural network: **a** XOR problem: **b** 2D clustering; **c** 3D clustering. The left images show the QQNN output levels after a threshold

is to extend the number of hidden layers to build more complex geometric neural networks for the so-called *deep learning*.

13.14.1 Deep Learning

A pioneer work on the area of deep learning is the model called neocognitron developed by Kuniko Fukushima [100]. This work is strongly influenced by the work of Hubel and Wiesel [149] who discovered that the visual cortex of monkeys is organized in small subregions which are sensitive to specific patterns. These regions are called receptive fields which overlap with each other to cover the whole visual field. Note that the cells act as local filters and they have spatial correlation as well. At the end of the eighties, LeCun [180] proposed different models to increase the learning capacity of neural networks. This effort established a new architecture in neurocomputing the so-called convolutional neural networks (CNNs). This combines three key concepts: local receptive fields, a weight-sharing technique, and a subsampling layer. Interestingly enough, these 2D filters behave as convolvers or kernels to process the signals through the layers; as a result, one can use the FFT to accelerate the image processing through the layers. The learning procedure will be responsible to tune the filter parameters. G. Hinton applied Rectified Linear Units (RELU) to improve restricted Boltzmann machines and as training rule the stochastic gradient descendant [41]. Other remarkable contribution is IMAGET by Fei et al. [66], a CNN for large-scale hierarchical image database.

In a CNN, each layer is composed of several convolutional filters or kernels. By using a big set of kernels in each layer, multiple features can be extracted in which their complexity and sophistication increase as much as if more in-depth signals through the layer flow. Lecun proposed the use of *pooling layers* which compute the maximum value of the average of a set of locally connected inputs. This technique ensures invariance to small translations of features in the input images. With the increase of computational power, the CNN could afford more hidden layers and even process large amounts of image data [66, 82, 256]; as a result, the concept of deep learning was finally consolidated. In fact, a common CNN adjusts several millions of parameters using large data sets. Here, the learning procedure uses some clever techniques and the stochastic/mini-batch gradient descent methods [41, 181, 254]. In these kinds of learning techniques, the weights of the network are updated after each input/mini-batch data is applied to the network, instead of updating the weights after the gradients have been accumulated over the whole training data set or batch. In this way, this method reduces the computational time required for the training, because it avoids redundant computations which may be caused by the repetition of redundant features along diverse patterns.

In the last decade, we eye-witnessed impressive results of CNN computing in natural language processing, computer vision, time series analysis, and robot vision and robot control [171, 276, 292, 319]. On the other hand, researchers developed feedforward neural networks using complex numbers or quaternions and Clifford-valued Support Vector Machines, see Sect. 13.8. Researchers developed CNN using complex numbers as well like LeCun et al. [43], Guberman [113], Hänsch and Helwich [118], and Wilmanski et al. [312]. In this section, we will show how we can extend the previous geometric neural networks for deep learning. Note that the Clifford or Quaternion Support Vector Machine can we connected at the output of a Clifford or Quaternion CNN.

13.14.2 The Quaternion Convolutional Neural Network

The multivector-valued convolutional neural network (CNN) is nothing else as an Geometric Neural Networks with more hidden layers. For that, we can extend the generalized geometric neural networks including more hidden layers and use the multidimensional backpropagation training rule which works in a certain geometric algebra $G_{p,q,r}$ as explained in Sect. 13.4.4.

In this chapter, we have introduced quaternion-valued feedforward neural network using as transfer function either sigmoid functions, radial basis functions or quaternion wavelet functions, hence we will first describe the Quaternion

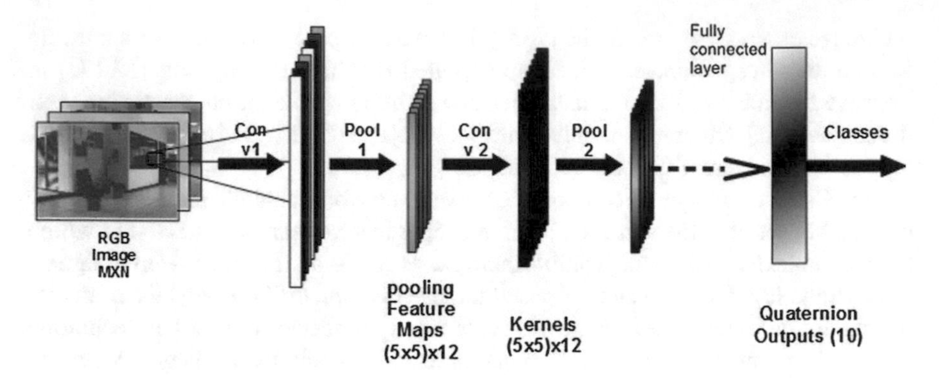

Fig. 13.25 Quaternion convolutional neural network

Convolutional Neural Network (QCNN) using sigmoid functions. Figure 13.25 depicts the QCNN. Note that the algebra G_3^+ is isomorph to the quaternion algebra $\mathbb{H}$; thus, we use instead of the bivectors of G_3^+ the quaternion basis $\boldsymbol{i} = e_2e_3$, $\boldsymbol{j} = -e_3e_1$ and $\boldsymbol{k} = e_1e_2$.

13.14.3 Quaternion Convolution Layers and Max-Pooling Layers

The input, weights, and outputs through the hidden and output layers are represented as quaternions, $\boldsymbol{x}_{i,j}$, $\boldsymbol{w}^k_{i,j}$, $\boldsymbol{o}_{i,j}$, respectively. According to the product in the geometric neuron of Sect. 13.4.2, instead of the inner product as in the real-valued neuron, the incoming signal, a set of quaternions, $\boldsymbol{x}_{i,j}$ and the weight, a set of quaternions, $\boldsymbol{w}(i, j)$ which constitute a quaternion filter or kernel are multiplied via the quaternion left convolution. For QCNN for 1D signals, for example for the application of QCNN to implement quaternion neurocontrollers using deep learning, the signal and weight convolution reads

$$(\boldsymbol{x} *_l \boldsymbol{w})_i = \sum_{p=0}^{M} = (\boldsymbol{x}_p \boldsymbol{w}_{i-p}), \tag{13.183}$$

where the product $(\boldsymbol{x}_p \boldsymbol{w}_{i-p})$ is the quaternion multiplication, $i, p \in \mathbb{N}$ and $M = max[N, L]$, and N, L are the sizes of quaternion input and filter signals, respectively. For 2D signals like images $\boldsymbol{x}_{i,j}$ for $i, j = 1, ..., N$, the quaternion convolution reads

$$(\boldsymbol{x} *_l \boldsymbol{w})_{i,j} = \sum_{p=-K}^{K} \sum_{q=-K}^{K} (\boldsymbol{x}_{p,q} \boldsymbol{w}_{i-p,j-q}), \tag{13.184}$$

where the product is the quaternion multiplication and the $i, j, p, r \in \mathbb{N}$ and K is the size of the quaternion filter template. The kernel is applied along the image. In order to accelerate the quaternion convolution, the quaternion fast Fourier transform can be used, see section for the quaternion Fourier transform.

The quaternion activation signal $\boldsymbol{net}$ of the quaternion neuron is processed by a quaternion activation-function componentwise, namely

$$\begin{aligned} \boldsymbol{F}(\boldsymbol{net}) &= \boldsymbol{F}(\boldsymbol{net}_0 + \boldsymbol{net}_x \boldsymbol{i} + \boldsymbol{net}_y \boldsymbol{j} + \boldsymbol{net}_z \boldsymbol{k}), \\ &= \frac{1}{1+e^{-\boldsymbol{net}_0}} + \frac{1}{1+e^{-\boldsymbol{net}_x}} \boldsymbol{i} + \frac{1}{1+e^{-\boldsymbol{net}_y}} \boldsymbol{j} + \frac{1}{1+e^{-\boldsymbol{net}_z}} \boldsymbol{k}. \end{aligned} \tag{13.185}$$

Hence, according to the geometric neuron, the output of the quaternion k-neuron is given by

$$\boldsymbol{o}_k = F\big((\boldsymbol{x} \star_l \boldsymbol{w})_{i,j} + \boldsymbol{\theta}_k\big), \tag{13.186}$$

where $\boldsymbol{\theta}_k$ is a bias quaternion.

In the QCNN, the output of quaternion max-pooling layers is computed as an extension of the real-valued max-pooling output of the real-valued CNN and it is given by

$$\hat{\boldsymbol{o}} = max_{\boldsymbol{o}_l}(|\boldsymbol{o}_{i-K,j-K}|, ..., |\boldsymbol{o}_{i,j}|, ..., |\boldsymbol{o}_{i+K,j+K}|), \tag{13.187}$$

where $\|\boldsymbol{o}_l\| = \sqrt{o_{l,0}^2 + o_{l,x}^2 + o_{l,y}^2 + o_{l,z}^2}$. The max-pooling output is a real number which indicates the largest magnitude of the winner quaternion l-neuron within a region of the feature map from a previous hidden layer. There are different methods proposed for the CNN [43] to compute more robust estimators of the statistics for a winner neuron; they can be extended straightforwardly for the QCNN. In Sect. 13.8, we introduced the Clifford and Quaternion Support Vector Machines; similarly as with the CNN where SVM can be used at the output, one can connect at the output of a Geometric Algebra CNN or a QCNN a Clifford or Quaternion Support Vector Machines as well.

13.14.4 Quaternion Gradient Descend for QCNN

According to the generalized multivector training rule explained in Sect. 13.4.5, we use Eq. 13.21 for the rotor algebra G_3^+ which is isomorph with the quaternion algebra $\mathbb{H}$; thus, we use instead the quaternion basis $\boldsymbol{i} = e_2e_3$, $\boldsymbol{j} = -e_3e_1$, $\boldsymbol{k} = e_1e_2$.

The *norm of a quaternion* $\boldsymbol{q}$ for the learning rule is given by

$$|\boldsymbol{q}| = (\boldsymbol{q}|\boldsymbol{q})^{\frac{1}{2}} = \Big(\sum_A [\boldsymbol{q}]_A^2\Big)^{\frac{1}{2}}. \tag{13.188}$$

The Quaternion CNN with n inputs and m outputs approximates the target mapping function

$$\mathcal{Y}_t : \underbrace{(\mathbb{H} \otimes \cdots \otimes \mathbb{H})}_{n} \rightarrow \underbrace{(\mathbb{H} \otimes \cdots \otimes \mathbb{H})}_{m}. \tag{13.189}$$

The error at the output of the net is measured according to the metric

$$E = \frac{1}{2} \int_{x \in X} |\mathcal{Y}_w - \mathcal{Y}_t|^2, \tag{13.190}$$

where $\boldsymbol{X}$ is some compact subset of $(\mathbb{H})^n$ involving the product topology derived from Eq. (13.188) for the norm and where $\mathcal{Y}_w$ and $\mathcal{Y}_t$ are the learned and target mapping functions, respectively. The *backpropagation algorithm* [255] is a procedure for updating the weights and biases. This algorithm is a function of the negative derivative of the error function (Eq. 13.190) with respect to the weights and biases themselves. The computing of this procedure is straightforward, and here, we will only give the main results. The updating equation for the quaternion weights of any hidden j-layer is

$$\boldsymbol{w}_{ij}(t+1) = \eta\big[(\sum_{k}^{N_k} \delta_{kj} \otimes \overline{\boldsymbol{w}_{kj}}) \odot \boldsymbol{F}'(\boldsymbol{net}_{ij})\big] \otimes \overline{\boldsymbol{o}_i} + \alpha \boldsymbol{w}_{ij}(t), \tag{13.191}$$

for any k-output with a nonlinear activation function

$$\boldsymbol{w}_{jk}(t+1) = \eta\big[(\boldsymbol{y}_{k_t} - \boldsymbol{y}_{k_a}) \odot \boldsymbol{F}'(\boldsymbol{net}_{jk})\big] \otimes \overline{\boldsymbol{o}_j} + \alpha \boldsymbol{w}_{jk}(t), \tag{13.192}$$

and for any k-output with a linear activation function

$$\boldsymbol{w}_{jk}(t+1) = \eta(\boldsymbol{y}_{k_t} - \boldsymbol{y}_{k_a}) \otimes \overline{\boldsymbol{o}_j} + \alpha \boldsymbol{w}_{jk}(t). \tag{13.193}$$

In the CNN, there are the so-called ReLU activation function [222] which is linear; hence, its derivative equals to 1. The QCNN uses also a linear activation for the quaternion ReLUs.

In the above equations, $\boldsymbol{F}$ is the activation function defined in Eq. (13.185), t is the update step, η and α are the *learning rate* and the momentum, respectively, $\otimes$ is the quaternion product, $\odot$ is the scalar product, and $\overline{(\cdot)}$ is the quaternion conjugation.

13.14.5 Implementation Details

There are many well-developed software packets for CNN computing like Caffe [43], Torch [298], tensor flow [297], and Hinton's work [222] which are equipped with linear algebra libraries. Thus, they can be reutilized to represent quaternions and the quaternion product. However, since these packets were not developed for Clifford- or quaternion-valued CNNs, it is still a bit awkward to reutilize them; thus, it is worthy to write the own code.

Next, we present a simple example to process 32× 32 color images for the 10 handwritten numbers and image pixels are coded using the vector part of the quaternion, i.e., $\boldsymbol{x}(i, j) = r(i, j)\boldsymbol{i} + g(i, j)\boldsymbol{j} + b(i, j)\boldsymbol{k}$. The QCNN architecture is:

- Layer 1: [8 quaternion convolution kernels of size 5×5]
- Layer 2: [quaternion pooling layer]
- Layer 3: [12 quaternion convolutional kernels of size 5×5]
- Layer 4: [quaternion pooling layer]
- Layer 5: [12 quaternion convolution kernels of size 5×5]
- Layer 6: [quaternion pooling layer]
- Layer 7: [16 quaternion convolutional kernels of size 5×5]
- Layer 8: [quaternion pooling layer]
- Layer 9: [16 quaternion convolution kernels of size 5×5]
- Layer 10: [quaternion pooling layer]
- Layer 11: [quaternion fully connected layer with 10 outputs]
- Layer 12: [ReLU layer]

The result of the QCNN is comparable with the CNN of Lecun [40]. Thus, we believe that the use of QCNN should be target when quaternion rotations and affine Lie groups are required, like color image processing where the quaternion wavelet transform [14] plays a role and it could be used for the convolutional kernel layer. Other promising application of the QCNN is for the implementation of 3D pose detectors, for observers, controllers, and deep reinforcement learning for quadrotors and hexapods, robot manipulators and bipeds. We are currently working on these topics.

13.15 Conclusion

According to the literature, there are basically two mathematical systems used in neurocomputing: tensor algebra and matrix algebra. In contrast, the authors have chosen to use the coordinate-free system of Clifford or geometric algebra for the analysis and design of feedforward neural networks. Our work shows that real-, complex-, and quaternion-valued neural networks are simply particular cases of geometric algebra multidimensional neural networks and that some can be generated using support multivector machines.

In this chapter, the real-valued SVM is generalized to Clifford-valued SVM and is used for classification, regression, and interpolation. In particular, the generation of RBF networks in geometric algebra is easier using an SVM, which allows one to find the optimal parameters automatically. The CSVM accepts multiple multivector inputs and multivector outputs, like a MIMO architecture, that allow us to have multiclass applications. We can use CSVM over complex, quaternion, or hypercomplex numbers according to our needs.

The use of feedforward neural networks and the SV machines within the geometric algebra framework widens their sphere of applicability and furthermore expands our understanding of their use for multidimensional learning. The experimental analysis confirms the potential of geometric neural networks and Clifford-valued SVMs for a variety of real applications using multidimensional representations, such as in graphics, augmented reality, machine learning, computer vision, medical image processing, and robotics.

In this work, we also show that the quaternionic spike neural networks in conjunction with the training algorithm, working on quaternion algebra, outperform the real-valued spike neural networks. They are ideal for applications in the area of neurocontrol of manipulators with angular movements, and this lies in the close geometric relationship of the quaternion imaginary axes with rotation axes. Note that for this new spike neural network, we are taking into account two relevant ideas the use of spike neural network which is the best model for oculomotor control and the role of geometric computing.

The progress in quantum computing and neurocomputing indicates that the formulation of real-time operators in terms of neural networks seems a promising approach particularly as it is inspired in neuroscience as well. Considering the progress in multispace and time algebra using GA and the quaternion neural networks, we formulate the algorithms for the Rotor or Quaternion Quantum Neural Networks (RQNN) or Quaternion Quantum Neural Networks (QQNN) as models for real-time computing superior by far as traditional quantum computing using a Von Neumann computing based on digital gates. We are utilizing the QQNN for image processing and to develop neurocontrollers for robotics.

Finally, this chapter is closed with the formulation of Geometric Algebra Deep Learning. The revolution of the convolutional neural networks is basically due to the increase of the amount of hidden filter—convolutional layers—the inclusion of the pooling and ReLU layers. In this work, we have proposed as special case of the Geometric Algebra CNN the Quaternion CNN, simply extending the representation, convolution, and products in the quaternion algebra framework. Previous concepts proposed in previous sections can be reutilized in the QCNN formulation, as the modeling of the quaternion layers, the quaternion gradient descent, the use of the quaternion wavelet transform which can be utilized along the hidden layers, and at the output the possibility to plug the Quaternion Support Vector Machine. By using quaternions to represent 1D or 2D signals (images), we are increasing in roughly four times the amount of parameters to be handled during the training. Despite this increment, the QCNNN appears promising for task involving 3D rotations, affine transformations, 3D pose, the treatment of color (R, G, B) images and in applications for visual guided robotics as quaternion observers, controllers and in quaternion deep reinforcement.

Part V
Applications of GA in Image Processing, Graphics and Computer Vision

Chapter 14
Applications of Lie Filters, Quaternion Fourier, and Wavelet Transforms

This chapter first presents Lie operators for key points detection working in the affine plane. This approach is stimulated by certain evidence of the human visual system; therefore, these Lie filters appear to be very useful for implementing in near future of an humanoid vision system.

The second part of the chapter presents an application of the quaternion Fourier transform for preprocessing for neurocomputing. In a new way, the 1D acoustic signals of French spoken words are represented as 2D signals in the frequency and time domains. These kinds of images are then convolved in the quaternion Fourier domain with a quaternion Gabor filter for the extraction of features. This approach allows to greatly reduce the dimension of the feature vector. Two methods of feature extraction are tested. The feature vectors were used for the training of a simple MLP, a TDNN, and a system of neural experts. The improvement in the classification rate of the neural network classifiers is very encouraging, which amply justifies the preprocessing in the quaternion frequency domain. This work also suggests the application of the quaternion Fourier transform for other image processing tasks.

The third part of the chapter presents the theory and practicalities of the quaternion wavelet transform. This work generalizes the real and complex wavelet transforms and derives a quaternionic wavelet pyramid for multiresolution analysis using the quaternionic phase concept. As an illustration, we present an application of the discrete QWT for optical flow estimation. For the estimation of motion through different resolution levels, we use a similarity distance evaluated by means of the quaternionic phase concept and a confidence mask.

14.1 Lie Filters in the Affine Plane

This section carries out the computations in the affine plane $\mathcal{A}_{e_3}(\mathcal{N}^2)$ for image analysis. We utilize the Lie algebra of the affine plane explained in Sect. 5.10.5 for the design of image filters to detect visual invariants. As an illustration, we apply these filters for the recognition of hand gestures.

E. Bayro-Corrochano, *Geometric Algebra Applications Vol. I*,
https://doi.org/10.1007/978-3-319-74830-6_14

14.1.1 The Design of an Image Filter

In the experiment, we used simulated images of the *optical flow* for two motions: a rotational and a translational motion (see Fig. 14.1a), and a dilation and a translational motion (see Fig. 14.2a). The experiment uses only bivector computations to determine the type of motion, the axis of rotation, and/or the center of the dilation.

To study the motions in the affine plane, we used the Lie algebra of bivectors in the geometric algebra $\mathcal{A}_{e_3}(\mathcal{N}^2)$. The computations were carried out with the help of a computer program which we wrote in C^{++}. Each *flow vector* at any point $\mathbf{x}$ of the image was coded $\mathbf{x} = xe_1 + ye_2 + e_3 \in \mathcal{N}^3$. At each point of the flow image, we applied the commutator product of the six bivectors of the Eq. (5.249). Using the resultant coefficients of the vectors, the computer program calculated which was type of differential invariant or motion was present.

Figure 14.1b shows the result of convolving, via the geometric product, the bivector with a *Gaussian kernel* of size 5×5. Figure 14.1c presents this result using the output of the kernel. The white center of the image indicates the lowest magnitude. Figure 14.2 shows the results for the case of a flow which is expanding. Comparing

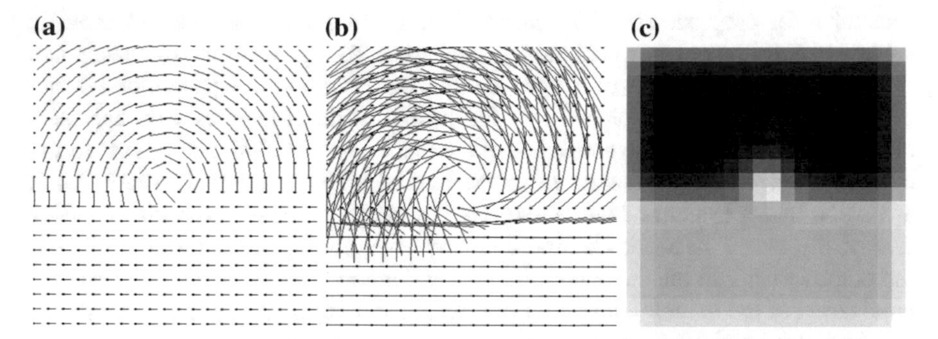

Fig. 14.1 Detection of visual invariants: **a** rotation (L_r) and translational flow (L_x) fields; **b** convolving via geometric product with a Gaussian kernel; **c** magnitudes of the convolution

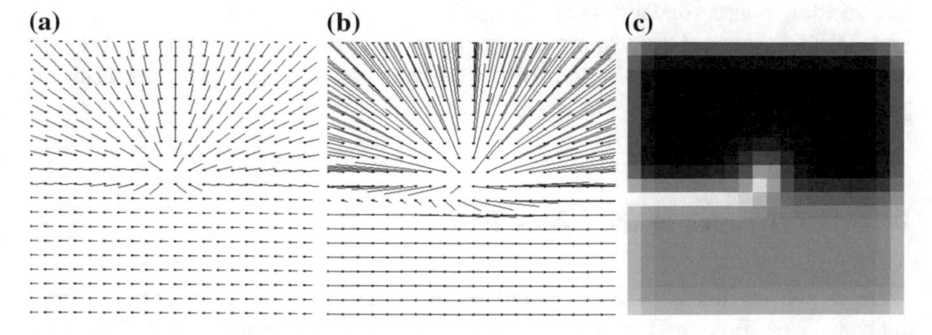

Fig. 14.2 Detection of visual invariants: **a** expansion (L_s) and translational flow (L_x) fields; **b** convolving via the geometric product with a Gaussian kernel; **c** magnitudes of the convolution

Fig. 14.1c with Fig. 14.2c, we note the duality of the differential invariants; the center point of the rotation is invariant, and the invariant of the expansion is a line.

14.1.2 Recognition of Hand Gestures

Another interesting application, suggested by the seminal paper of Hoffman [145], is to recognize a gesture using the key points of an image along with the previous Lie operators arranged in a detection structure, as depicted in Fig. 14.3. These Lie filters may be seen to be *perceptrons*, which play an important role in image preprocessing in the human visual system. It is believed [145] that during the first years of human life, some kinds of Lie operators combined to build the higher-dimensional Lie algebra SO(4,1).

In this sense, we assume that the outputs of the Lie operators are linearly combined with an *outstar output* according to the following equation:

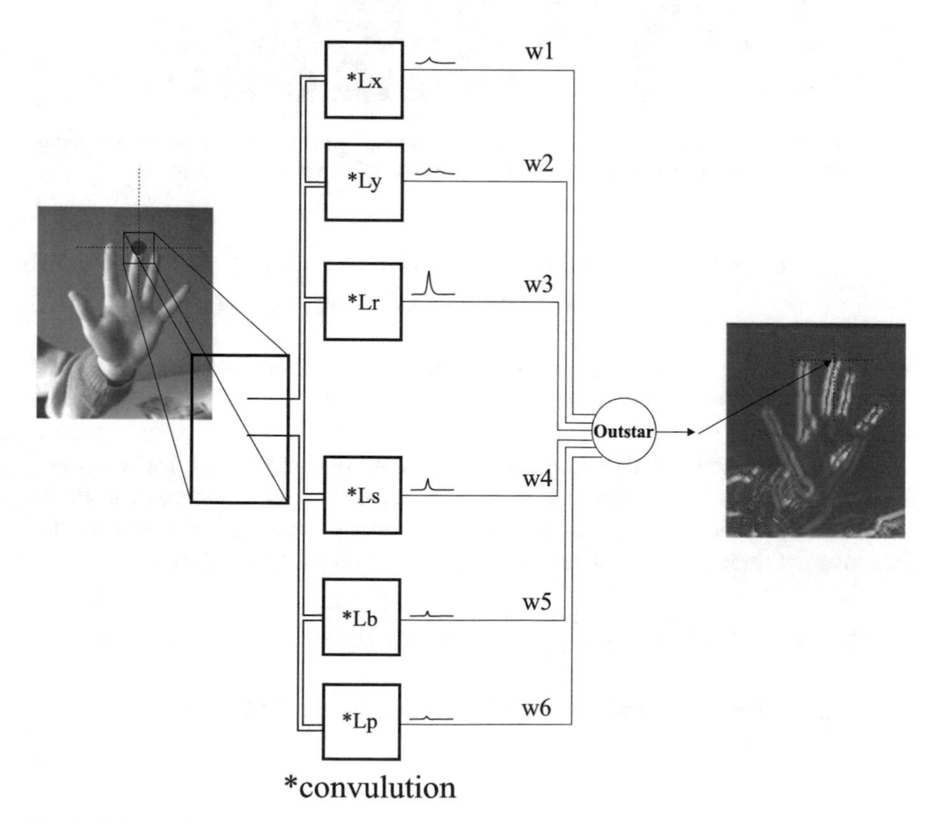

Fig. 14.3 Lie perceptrons arrangement for feature detection

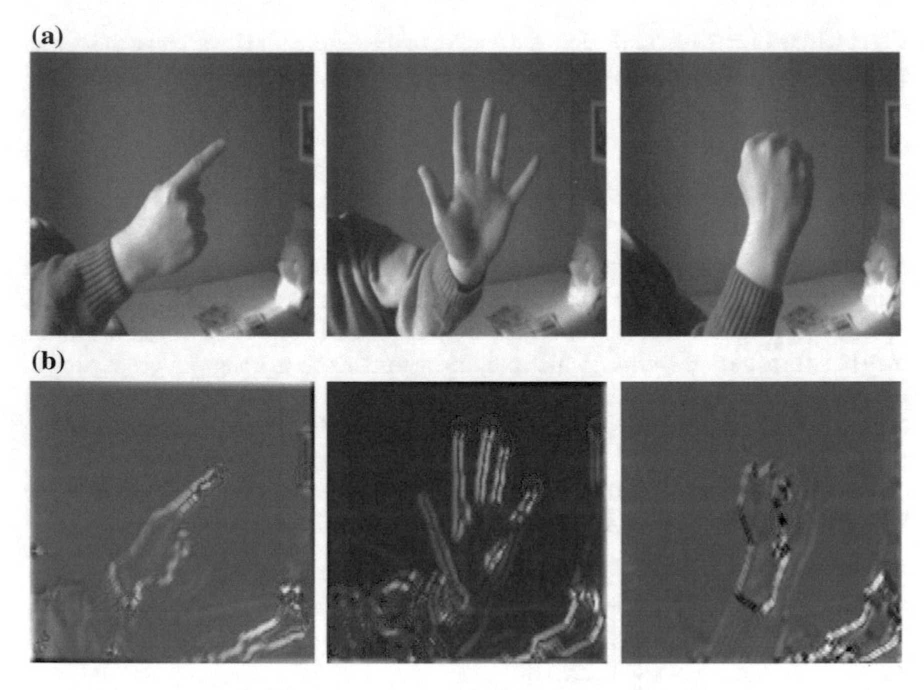

Fig. 14.4 Gestures detection: **a** (top images) gestures for robot guidance (follow, stop, and explore); and **b** (lower images) detected gestures by the robot vision system using Lie operators

$$O_\alpha(x, y) = w_1\mathcal{L}_x(x, y) + w_2\mathcal{L}_y(x, y) + w_3\mathcal{L}_r(x, y) + \\ +w_4\mathcal{L}_s(x, y) + w_5\mathcal{L}_b(x, y) + w_6\mathcal{L}_B(x, y), \qquad (14.1)$$

where the weights w_i can be adjusted if we apply a supervised training procedure. If the desired feature or key point α at the point (i, y) is detected, the output $O_\alpha(x, y)$ goes to zero.

Figure 14.4a shows hand gestures given to a robot. By applying the Lie perceptrons arrangement to the hand region (see Fig. 14.4b), a robot can detect whether it should *follow*, *stop*, or *move in circles*. Table 14.1 presents the weights w_i necessary for detecting the three gestures. Detection tolerance is computed as follows:

$$O_\alpha(x, y) \leq min + \frac{(max - min)Tolerance}{100} \rightarrow detection\ of\ a\ feature\ type,$$

where *min* and *max* correspond to the minimal and maximal Lie operator outputs.

Table 14.1 Weights of the Lie perceptrons arrangement for the detection of hand gestures

Hand gesture	$\mathcal{L}_x$	$\mathcal{L}_y$	$\mathcal{L}_r$	$\mathcal{L}_s$	$\mathcal{L}_b$	$\mathcal{L}_B$	Tolerance (%)
Fingertip	0	0	9	−4	11	−9	10
Stop	0	0	−3	1	1	4	10
Fist	0	0	−2	2	2	−1	10

14.2 Representation of Speech as 2D Signals

In this work, we use the psycho-acoustical model of a loudness meter suggested by E. Zwicker [322]. This meter model is depicted in Fig. 14.5a. The outputs are a 2D representation of sound loudness over time and frequency. The motivation of this work is to use the *loudness image* in order to take advantage of the variation in time of the frequency components of the sound. A brief explanation of this meter model follows.

The sound pressure is picked up by a microphone and converted to an electrical signal, which in turn is amplified. Thereafter, the signal is attenuated to produce the same loudness in a diffuse and free-sound field. In order to take into account the frequency dependence of the sound coming from the exterior and passing through the outer ear, a transmission factor is utilized. The signal is then filtered by a filter bank with filter bands dependent on the critical band rate (Barks). In this work, we have taken only the first 20 Barks, because this study is restricted to the band of speech signals. At the output of the filters, the energy of each filter signal is calculated to obtain the maximal critical band level varying with time. Having 20 of these outputs, a 2D sound representation can be formed as presented in Fig. 14.5b. At each filter output, the loudness is computed taking into account temporal and frequency effects according to the following equation:

$$N' = 0.068 \frac{E_{TQ}}{s \cdot E_0}^{0.25} [(1 - s + s \cdot \frac{E}{E_{TQ}})^{0.25} - 1] \frac{sone}{Bark},$$

where E_{TQ} stands for the excitation level at threshold in quiet, E is the main excitation level, E_0 is the excitation that corresponds to the reference intensity $I_0 = 10^{-12} \frac{W}{m^2}$, s stands for the masking index, and finally one sone is equivalent to 40 phones. To obtain the main loudness value, the specific loudness of each critical band is added. Figure 14.5c depicts the time-loudness evolution in 3D, where the loudness levels are represented along the z-coordinate.

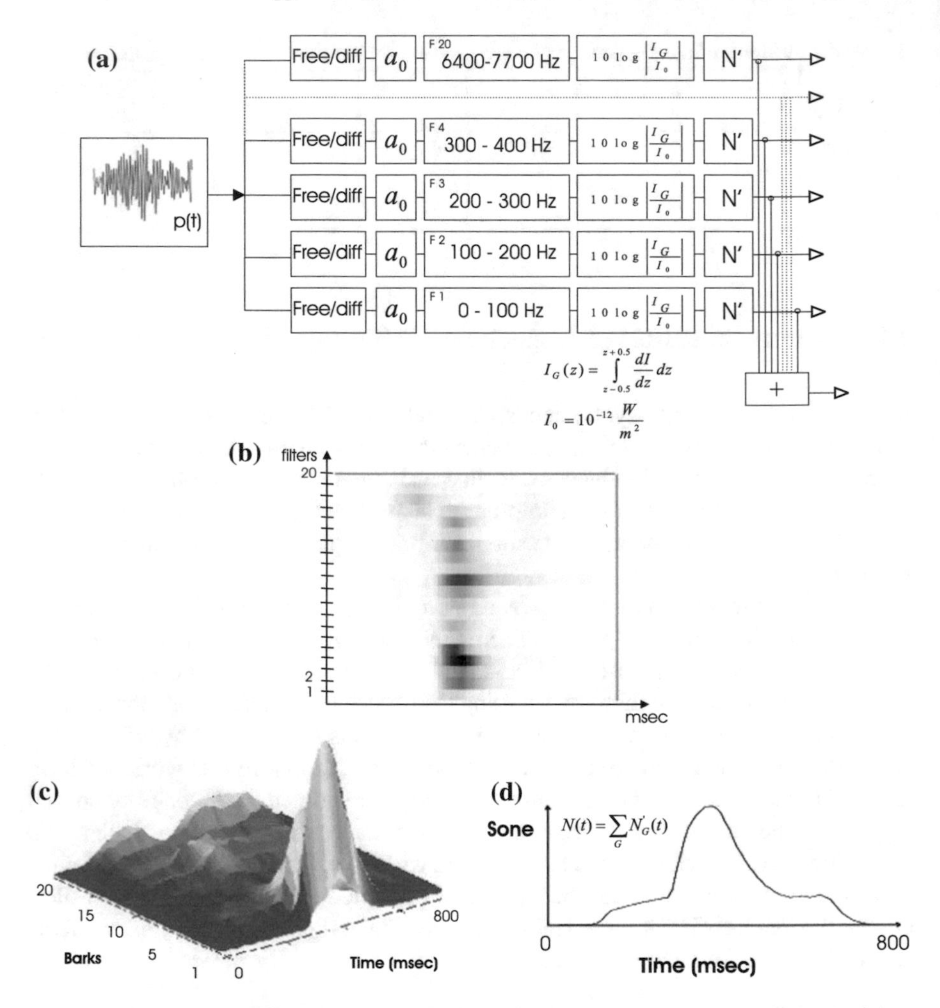

Fig. 14.5 From the top: **a** The psycho-acoustical model of loudness meter suggested by E. Zwicker; **b** 2D representation (vertical outputs of the 20 filters, horizontal axis is the time); **c** a 3D energy representation of **b** where the energy levels are represented along the z-coordinate; **d** main loudness signal or total output of the psycho-acoustical model (the sum of the 20 channels)

14.3 Preprocessing of Speech 2D Representations Using the QFT and Quaternionic Gabor Filter

This section presents two methods of preprocessing. The first is a simple one that could be formulated; however, we will show that the extracted features do not yield a high classification rate, because the sounds of the consonants of the phonemes are not well recognized.

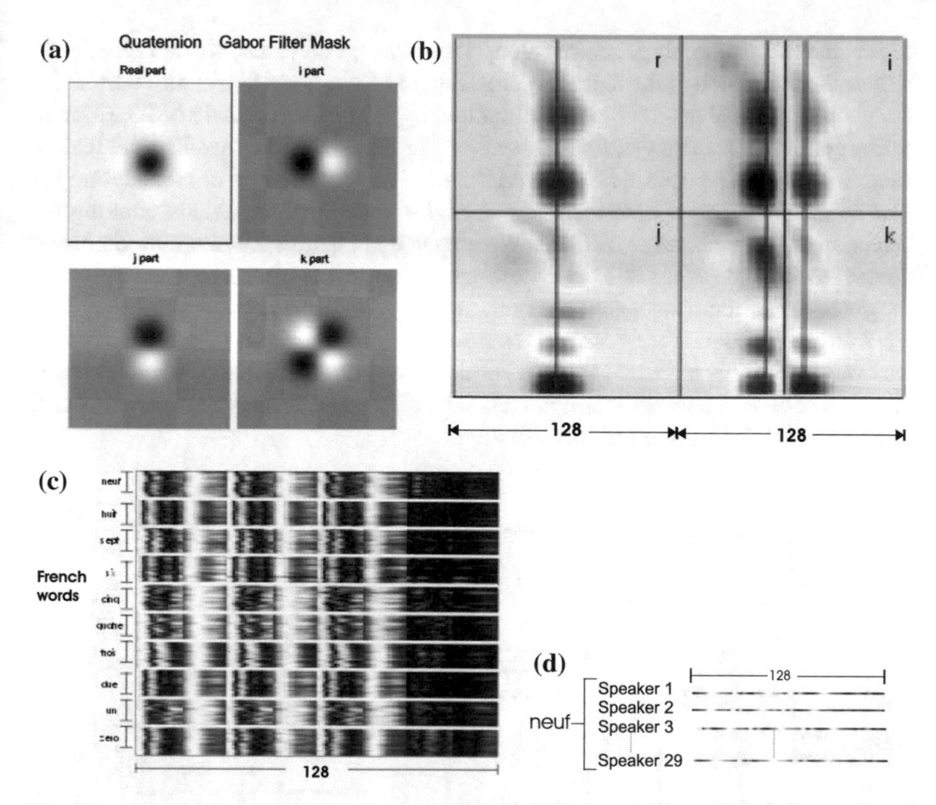

Fig. 14.6 Method 1:(from upper left corner) **a** quaternion Gabor filter; **b** selected 128 features according to energy level (16 channels and 8 analysis lines in the four r, $\boldsymbol{i}, \boldsymbol{j}, \boldsymbol{k}$ images, $16\times 8 = 128$); **c** stack of feature vectors for 10 numbers and 29 speakers (the ordinate axis shows the words of the first 10 french numbers and the abscissa the time in milliseconds); **d** the stack of the French word *neuf* spoken by 29 speakers (also presented in **c**)

14.3.1 Method 1

A quaternion Gabor filter is used for the preprocessing (see Fig. 14.6a). This filter is convolved with an image of 80×80 pixels $\Big((5 \times 16[\textit{channels}]) \times (800/10[\text{msec}])\Big)$ using the quaternion Fourier transform. In order to have for the QFT a 80×80 square matrix for each of the 16 channels, we copied the rows 5 times, for 80 rows. We used only the first approximation of the psycho-acoustical model, which comprises only the first 16 channels. The feature extraction is done in the quaternion frequency domain by searching features along the lines of expected maximum energy (see Fig. 14.6b). After an analysis of several images, we found the best place for these lines in the four images of the quaternion image. Note that this approach first expands the original image $80 \times 80 = 1600$ to $4 \times 1600 = 6400$ and then reduces this result to a feature vector of length 16 [channels]$\times$ 8 [analysis lines] $= 128$. This clearly explains the motivation of our approach; we use a four-dimensional representation of the image for searching features along these lines; as a result, we can effectively reduce the dimension of the feature vector. Figure 14.6c shows the feature vectors for the phonemes of 10 decimal numbers spoken by 29 different female or 20 different male speakers. For example, for the *neuf* of Fig. 14.6c, we have stacked the feature vectors of the 29 speakers, as Fig. 14.6d shows. The consequence of considering the first 16 channels was a notorious loss in high-frequency components, making the detection of the consonants difficult. We also noticed that in the first method, using a wide quaternion Gabor filter, even though higher levels of energy were detected, the detection of the consonants did not succeed. Conversely, using a narrow filter, we were able to detect the consonants, but the detected information of the vowels was very poor. This indicated that we should filter only those zones where changes of sound between consonants and vowels take place. The second method, presented next, is the implementation of this key idea.

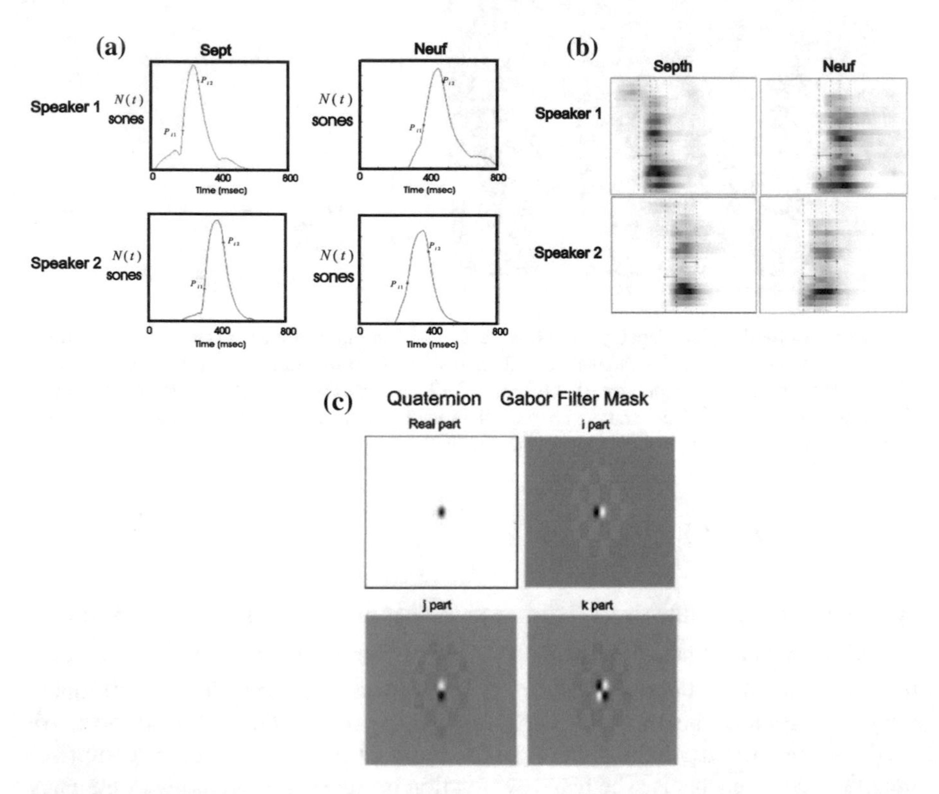

Fig. 14.7 **a** Method 2: the main loudness signals for *sept* and *neuf* spoken by two speakers; **b** determination of the analysis strips using the lines of the inflection points (20 filter responses at the ordinate axis and the time in milliseconds at the abscissa axis); **c** narrow band quaternion Gabor filter

14.3.2 Method 2

The second method does not convolve the whole image with the filter. This method uses all the 20 channels of the main loudness signal. First, we detect the inflection points where the changes of sounds take place, particularly those between consonants and vowels. These inflection points are found by taking the first derivative with respect to the time of the main loudness signal (see Fig. 14.7a for the *sept*). Let us imagine that someone says *ssssseeeeeeepppppth*, with the inflection points, and one detects two transition regions of 60 msec: one for *se* and another for *ept* (see Fig. 14.7b). By filtering these two regions with a narrow quaternion Gabor filter, we split each region to another two, separating *s* from *e* (first region) and *e* from *pth* (for the second region). The four strips represent what happens before and after the vowel *e* (see Fig. 14.8a). The feature vector is built by tracing a line through the maximum levels of each strip. We obtain four feature columns: column 1 for *s*, column 2 for *e*, column 3 for *e*, and column 4 for *pth* (see Fig. 14.8b). Finally, one builds one feature vector of length 80 by arranging the four columns (20 features each). Note that the second method reduces the feature vector length even more, from 128 to 80.

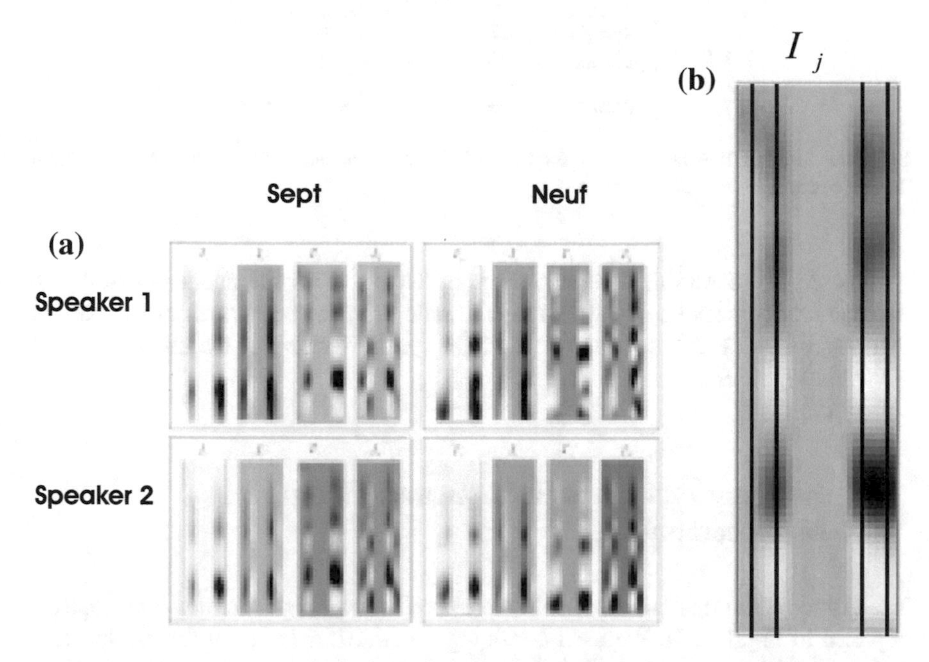

Fig. 14.8 Method 2: from the left **a** selected strips of the quaternion images for the words *sept* and *neuf* spoken by two speakers. **b** zoom of a strip of the component j of the quaternionic image and the selected 4 columns for feature extraction (20 channels × 4 lines = 80 features)

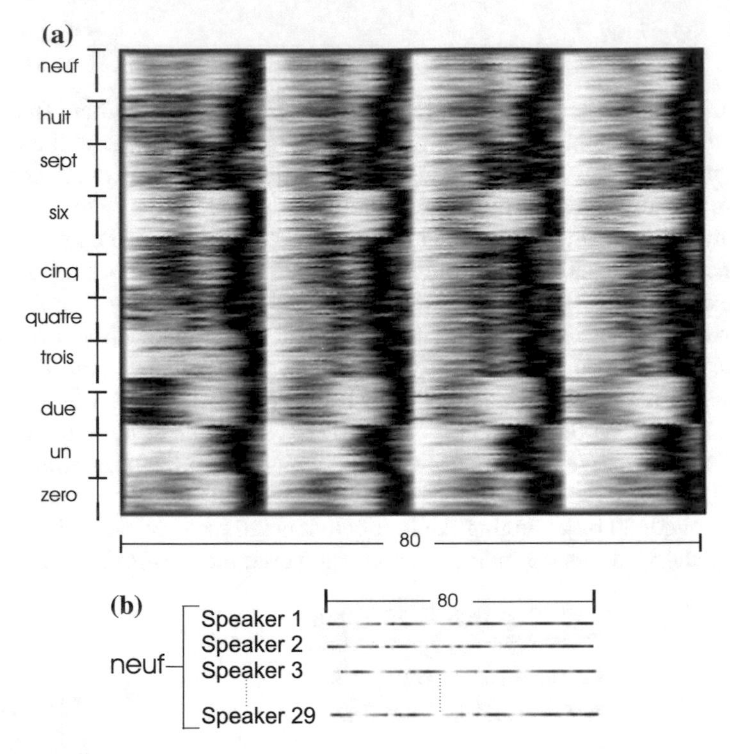

Fig. 14.9 Method 2: **a** stack of feature vectors of 29 words spoken by 29 speakers **b** the stack for the word *neuf* spoken by 29 speakers

Figure 14.9a shows the feature vectors of length 80 for the phonemes of the 10 decimal numbers spoken by 29 different female or male speakers. For example, for the *neuf* of Fig. 14.7b, we have stacked the feature vectors of the 29 speakers, as shown in Fig. 14.9b.

14.4 Recognition of French Phonemes Using Neurocomputing

The extracted features using method 1 were used for training a multilayer perceptron, depicted in Fig. 14.10a. We used a training set of 10 male and 9 female speakers each one spoke the first 10 French numbers *zero*, *un*, *deux*, *trois*, *quatre*, *cinq*, *seize*, *sept*, *huit*, *neuf*, and thus, the training set comprises 190 samples. After the training, a set of 100 spoken words was used for testing method 1 (10 numbers spoken by 5 male and 5 female speakers makes 100 samples). The recognition percentage achieved was 87%. For the second approach, the features were extracted using method 2. The structure used for recognition consists of an assembly of three neural experts

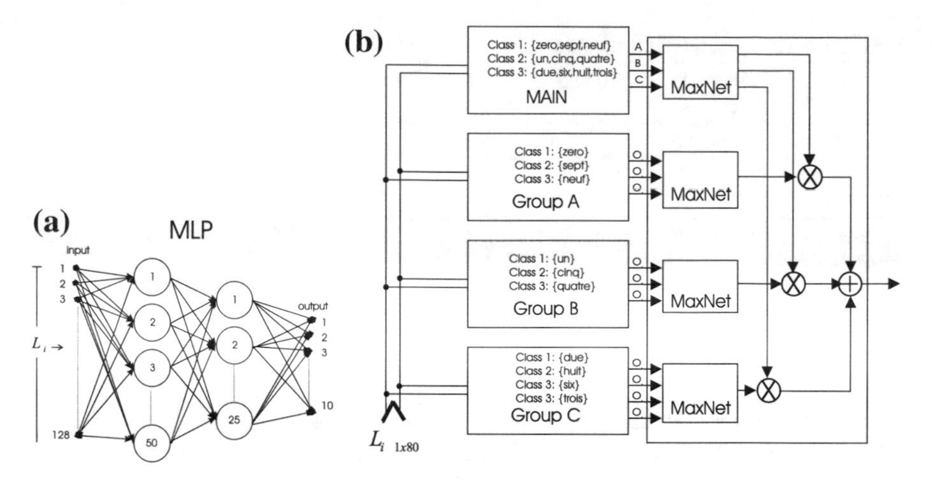

Fig. 14.10 **a** Neural network used for method 1, **b** group of neural networks used for method 2

regulated by a neural network arbitrator. In Fig. 14.10b, the arbitrator is called main and each neural expert is dedicated to recognize a certain group of spoken words. The recognition percentage achieved was 98%. The great improvement in the recognition rate is mainly due to the preprocessing of method 2 and the use of a neural expert system. We carried out a similar test using a set of 150 training samples (10 male and 5 female speakers) and 150 recall samples (5 male and 10 female speakers); the recognition rate achieved was a bit lower than 97%. This may be due to the lower number of female speakers during training and their higher number in the recall. That means that the system specializes itself better for samples spoken by males.

In order to compare with a standard method used in speech processing, we resorted to the Time Delay Neural Network (TDNN) [212]. We used the simulator Stuttgart Neural Network Simulator (SNNS). The input data in the format required for the SNNS was generated using Mathlab. The selected TDNN architecture was as follows: (input layer) 20 inputs for the 20 features that code the preprocessed spoken numbers, 4 delays of length 4; (hidden layer) 10 units, 2 delays of length 2; (output layer) 10 units for the 10 different numbers. We trained the TDNN using 1500 samples, 150 samples for each spoken number. We used the SNNS learning function *timedelaybackprop* and for the updating the function *Timedelay_order*. During the learning, we carried out 1000 cycles, i.e., 1000 iterations for each spoken number. We trained the neural network in two ways: (i) one TDNN with 10 outputs; and (ii) a set of three TDNNs, where each neural network was devoted to learn a small disjoint set of spoken numbers. Of the two methods, the best result was obtained using the first TDNN, which managed to get a 93.8% rate of recognition success. Table 14.2 summarizes the test results. The letters F and M stand for female and male speakers, respectively.

Table 14.2 Comparison of the methods (M = male and F = female)

Method	Training speakers	Test speakers	Samples/speaker	Test samples	Rate (%)
(MLP)	10 M, 9 F	5 M, 5 F	1	100	87
(Neural experts)	10 M, 9 F	5 M, 5 F	1	100	98
(Neural experts)	10 M, 5 F	5 M, 10 F	10	1500	97
(TDNN)	10 M, 5 F	5 M, 10 F	10	1500	93.8

Table 14.2 shows that the best architecture was the one composed of a set of neural experts (method 2). Clearly, the TDNN performed better that the MLP. Taking into account the performance of method 2 and of the TDNN, we find that the preprocessing using the quaternion Fourier transform played a major role.

14.5 Application of QWT

The motion estimation using quaternionic wavelet filters is inferred by means of the measurements of phase changes of the filter outputs. The accuracy of such an estimation depends on how well our algorithm deals with the correspondence problem, which can be seen as a generalization of the aperture problem depicted in Fig. 11.23.

This section deals with the estimation of the optical flow in terms of the estimation of the image disparity. The disparity of a couple of images $f(x,y)_1$ and $f(x,y)_2$ is computed by determining the local displacement, which satisfies $f(x,y)_1 = f(x+d_x, y+d_y)_2 = f(\mathbf{x}+\mathbf{d})$, where $\mathbf{d} = (d_x, d_y)$. The range of $\boldsymbol{d}$ has to be small compared with the image size; thus, the observed features always have to be within a small neighborhood in both images. In order to estimate the optical flow using the quaternionic wavelet pyramid, first we compute the quaternionic phase at each level, then the confidence measure, and finally the determination of the optical flow. The first two will be explained in the next subsections. Thereafter, we will show two examples of optical flow estimation.

14.5.1 *Estimation of the Quaternionic Phase*

The estimation of the disparity using the concept of local phase begins with the assumption that a couple of successive images are related as follows:

$$f_1(x) = f_2(x + \mathbf{d}(x)), \tag{14.2}$$

where $\mathbf{d}(x)$ is the unknown vector. Assuming that the phase varies linearly (here the importance of shifting-invariant filters), the displacement $\mathbf{d}(x)$ can be computed as

$$d_x(x) = \frac{\phi_2(x) - \phi_1(x) + n(2\pi + k)}{2\pi u_{ref}}, \qquad d_y(x) = \frac{\theta_2(x) - \theta_1(x) + m\pi}{2\pi v_{ref}}, \tag{14.3}$$

with reference frequencies (u_{ref}, v_{ref}) that are not known a priori. Here $\phi(x)$ and $\theta(x)$ are the first two components of the quaternionic local phase of the quaternionic filter. We choose $n, m \in \mathbb{Z}$, so that d_x and d_y are within a valid range. Depending on m, k is defined as

$$k = \begin{cases} 0, & \text{if } m \text{ is even}, \\ 1, & \text{if } m \text{ is odd}, \end{cases} \tag{14.4}$$

A good disparity estimation is achieved if (u_{ref}, v_{ref}) are chosen well. There are two methods of dealing with the problem: (i) the constant model, where u_{ref} and v_{ref} are chosen as the central frequencies of the filters; (ii) the model for the complex case called the local model, which supposes that the phase takes the same value $\Phi_1(x) = \Phi_2(x + d)$ in two corresponding points of both images. Thus, one estimates d by approximating Φ_2 via a first-order Taylor's series expansion about $\mathbf{x}$:

$$\Phi_2(\mathbf{x} + \mathbf{d}) \approx \phi_2(\mathbf{x}) + (\mathbf{d} \cdot \nabla)\phi_2(\mathbf{x}), \tag{14.5}$$

where we call $\Phi = (\phi, \theta)$. Solving Eq. (14.5) for $\mathbf{d}$, we obtain the estimated disparity of the local model. In our experiments, we assume that ϕ varies along the $\mathbf{x}$-direction and θ along y. Using this assumption, the disparity Eq. (14.3) can be estimated using the following reference frequencies

$$u_{ref} = \frac{1}{2\pi}\frac{\partial \phi_1}{\partial x}(x), \qquad v_{ref} = \frac{1}{2\pi}\frac{\partial \theta_1}{\partial y}(x). \tag{14.6}$$

In the locations where u_{ref} and v_{ref} are equal to zero, Eq. (14.6) is undefined. One can neglect these localities using a sort of confidence mask. This is explained in the next subsection.

14.5.2 Confidence Interval

In neighborhoods where the energy of the filtered image is low, one cannot estimate the local phase monotonically; similarly, at points where the local phase is zero, the estimation of the disparity is impossible because the disparity is computed by the phase difference divided by the local frequency. In that regard, we need a confidence measurement that indicates the quality of the estimation at a given point. For this purpose, we can design a simple binary confidence mask. Using complex filters, this can be done depending on whether the filter response is reasonable.

In the case of quaternionic filters, we need two confidence measurements for ϕ and θ. According to the multiplication rule of the quaternions, we see that the first two components of the quaternionic phase are defined for almost each point as

$$\phi(\mathbf{x}) = \arg_i(k^q(\mathbf{x})\beta(k^q(\mathbf{x}))), \quad \theta(\mathbf{x}) = \arg_j(\alpha(k^q(\mathbf{x}))k^q(\mathbf{x})), \tag{14.7}$$

where k^q is the quaternionic filter response, and $\arg_i$ and $\arg_j$ were defined in Eq. (6.34). The projections of the quaternionic filter are computed according to Eq. (6.36). Using these angles and projections of the response of the quaternionic filter, we can now extend the well-known confidence measurement of the complex filters

$$Conf(\mathbf{x}) = \begin{cases} 1, & \texttt{if } |\texttt{k(x)}| > \tau, \\ 0, & \texttt{otherwise} \end{cases} \tag{14.8}$$

to the quaternionic case:

$$C_h(k^q(\mathbf{x})) = \begin{cases} 1 & \texttt{if} \quad mod_i(k^q(bfx)\beta(k^q(\mathbf{x}))) > \tau, \\ 0 & \texttt{otherwise} \end{cases} \tag{14.9}$$

$$C_v(k^q(\mathbf{x})) = \begin{cases} 1 & \texttt{if} \quad mod_j(\alpha(k^q(\mathbf{x}))k^q(\mathbf{x}) > \tau, \\ 0 & \texttt{otherwise.} \end{cases} \tag{14.10}$$

Given the outputs of the quaternionic filters k_1^q and k_2^q of two images, we can implement the following confidence masks:

$$Conf_h(\mathbf{x}) = C_h(k_1^q(\mathbf{x}))C_h(k_2^q(\mathbf{x})), \tag{14.11}$$

$$Conf_v(\mathbf{x}) = C_v(k_1^q(\mathbf{x}))C_v(k_2^q(\mathbf{x})), \tag{14.12}$$

where $Conf_h$ and $Conf_v$ are the confidence measurements for the horizontal and vertical disparity, respectively.

However, one cannot use these measurements, because they are fully identical. This can easily be seen due to the simple identity for any quaternion q

$$mod_i(q\beta(\bar{q})) = mod_j(\alpha(\bar{q})q), \tag{14.13}$$

which can be checked straightforwardly by making explicit both sides of the equation. We can conclude, using the previous formulas, that for the responses of two quaternionic filters of either image, the horizontal and vertical confidence measurements will be identical:

$$Conf_h(\mathbf{x}) = Conf_v(\mathbf{x}). \tag{14.14}$$

For a more detailed explanation of the quaternionic confidence interval, the reader can resort to the Ph.D. thesis of Bülow [44].

14.5.3 Discussion on Similarity Distance and the Phase Concept

In the case of the complex conjugated wavelet analysis, the *similarity distance* $S_j((x, y)(x', y'))$ for any pair of image points on the reference image $f(x, y)$ and the matched image (after a small motion) $f'(x', y')$ is defined using the six differential components $D_{j,p}, \tilde{D}_{j,p}$, $p = 1, 2, 3$, of Eq. (11.123). The best match will be achieved by finding u, v, which minimize

$$min_{u,v} S_j((k, l), (k' + u, l' + v)). \tag{14.15}$$

The authors [200, 225] show that, with the continuous interpolation of Eq. (14.15), one can get a quadratic equation for the similarity distance:

$$S_{j,p}((k, l), (k' + u, l' + v)) = s_1(u - u_o)^2 + s_3(u - u_o)(v - v_o) + s_4, \tag{14.16}$$

where u_0, v_0 is the minimum point of the similarity distance surface $S_{j,p}$; s_1, s_2, s_3 are the curvature directions; and s_4 is the minimum value of the similarity distance S_j. The parameters of this approximate quadratic surface provide a subpixel-accurate motion estimate and an accompanying confidence measure. By its utilization in the complex-valued discrete wavelet transform hierarchy, the authors claim to handle the aperture problem successfully.

In the case of the quaternion wavelet transform, we directly use the motion information captured in the three phases of the detail filters. The approach is linear, due to the linearity of the polar representation of the quaternion filter. The confidence throughout the pyramid levels is assured by the bilinear confidence measure given by Eq. (14.12) and the motion is computed by the linear evaluation of the disparity Eq. (14.3). The local model approach helps to estimate the u_{ref} and v_{ref} by evaluating Eq. (14.6). The estimation is not of a quadratic nature, like Eq. (14.16). An extension to this kind of quadratic estimation of motion constitutes an extra venue for further improvement of the application of the QWT for multiresolution analysis.

Fleet [91] claims that the phase-based disparity estimation is limited to the estimation of the components of the disparity vector that is normal to an oriented structure in the image. These authors believe that the image is intrinsically one-dimensional almost everywhere. In contrast to the case of the quaternion confidence measure (see Eq. (14.12)), we see that it singles out those regions where horizontal and vertical displacement can reliably be estimated simultaneously. Thus, by using the quaternionic phase concept, the full displacement vectors are evaluated locally at those points where the aperture problem can be circumvented.

14.5.4 Optical Flow Estimation

In this final part, we will show the estimation of the optical flow of the Rubik cube and Hamburg taxi image sequences. We used the following scaling and wavelet quaternionic filters

$$h^q = g(x, \sigma_1, \varepsilon) \exp\big(i\frac{c_1\omega_1 x}{\sigma_1}\big) \exp\big(j\frac{c_2\varepsilon\omega_2 y}{\sigma_1}\big), \tag{14.17}$$

$$g^q = g(x, \sigma_2, \varepsilon) \exp\big(i\frac{\tilde{c}_1\tilde{\omega}_1 x}{\sigma_2}\big) \exp\big(j\frac{\tilde{c}_2\varepsilon\tilde{\omega}_2 y}{\sigma_2}\big), \tag{14.18}$$

with $\sigma_1 = \frac{\pi}{6}$ and $\sigma_2 = \frac{5\pi}{6}$ so that the filters are in quadrature and $c_1 = \tilde{c}_1 = 3$, $\omega_1 = 1$ and $\omega_2 = 1, \varepsilon = 1$. The resulting quaternionic mask will also be subsampled through the levels of the pyramid. For the estimation of the optical flow, we use two successive images of the image sequence. Thus, two quaternionic wavelet pyramids are generated. For our examples, we computed four levels. According to Eq. (11.135), at each level of each pyramid, we obtain 16 images, accounting for the four quaternionic outputs (approximation Φ and the details Ψ_1 (horizontal), Ψ_2 (vertical),Ψ_3 (diagonal)). The phases are evaluated according to the Eq. (6.4). Figure 14.12 shows the magnitudes and phases obtained at level j using two successive Rubik's images (Fig. 14.11).

After we have computed the phases, we proceed to estimate the disparity images using Eq. (14.3), where the reference frequencies u and v are calculated according to Eq. (14.6). We apply the confidence mask according to the guidelines given in Sect. 14.5.2 and shown in Fig. 14.13a.

After the estimation of the disparity has been filtered by the confidence mask, we proceed to estimate the optical flow at each point computing a velocity vector in terms of the horizontal and vertical details. Now, using the information from the diagonal detail, we adjust the final orientation of the velocity vector. Since the procedure starts from the higher level (top-down), the resulting matrix of the optical flow vectors is expanded in size equal to the next level, as shown for one level in Fig. 14.11. The algorithm estimates the optical flow at the new level. The result is compared with the one of the expanded previous level. The velocity vectors of the previous level fill gaps in the new level.

This procedure is continued until the bottom level. In this way, the estimation is refined smoothly, and the well-defined optical flow vectors are passed from level to level, increasing the confidence of the vectors at the finest level. It is unavoidable for some artifacts to survive at the final stage. A final refinement can be applied, imposing a magnitude thresholding and certainly deleting isolated small vectors. Figure 14.13b presents the computed optical flow for an image couple of the Rubik's image sequence.

Next, we present the computations of optical flow of the Hamburg taxi, using the QWT. Figure 14.14a shows a couple of images; Fig. 14.14b shows the confidence matrices at the four levels; and Fig. 14.14c presents the fused horizontal and vertical

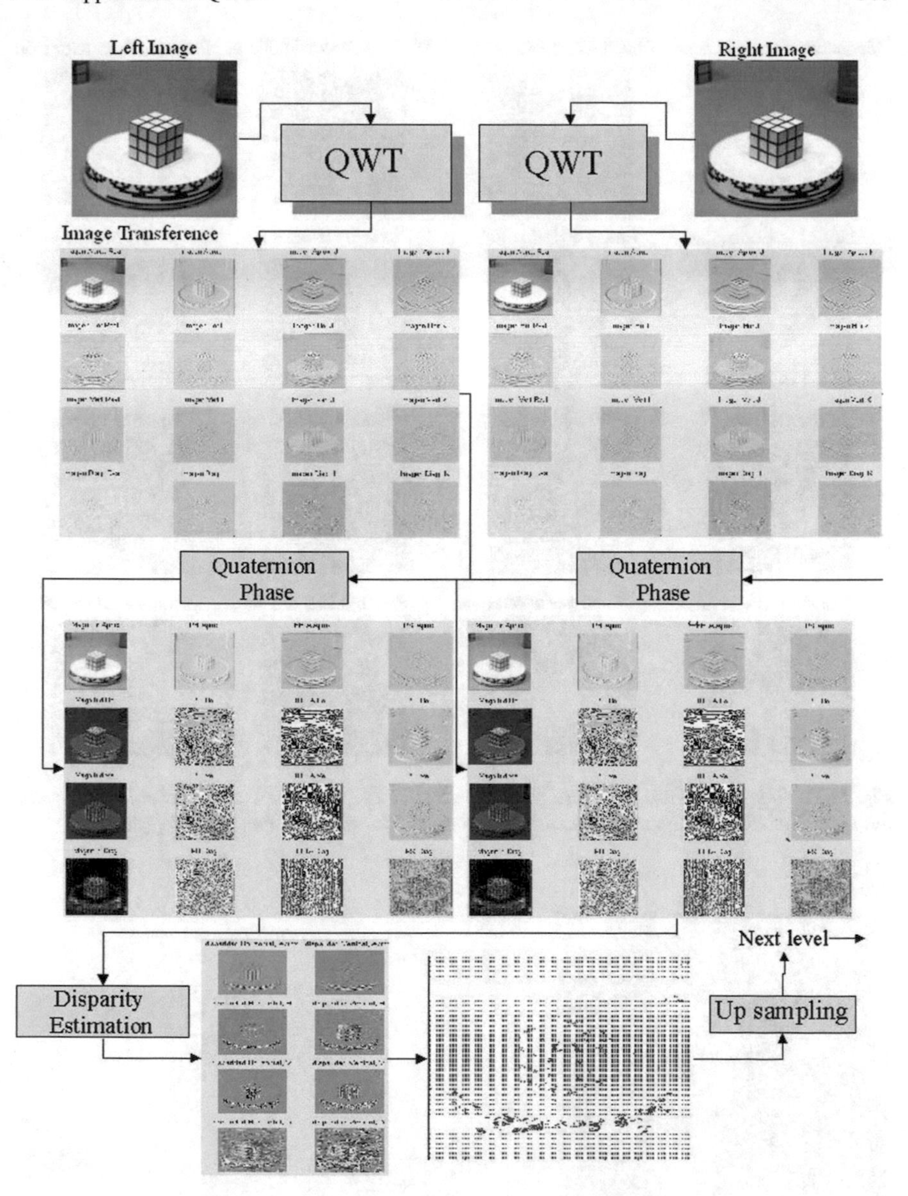

Fig. 14.11 Procedure for the estimation of the disparity at one level

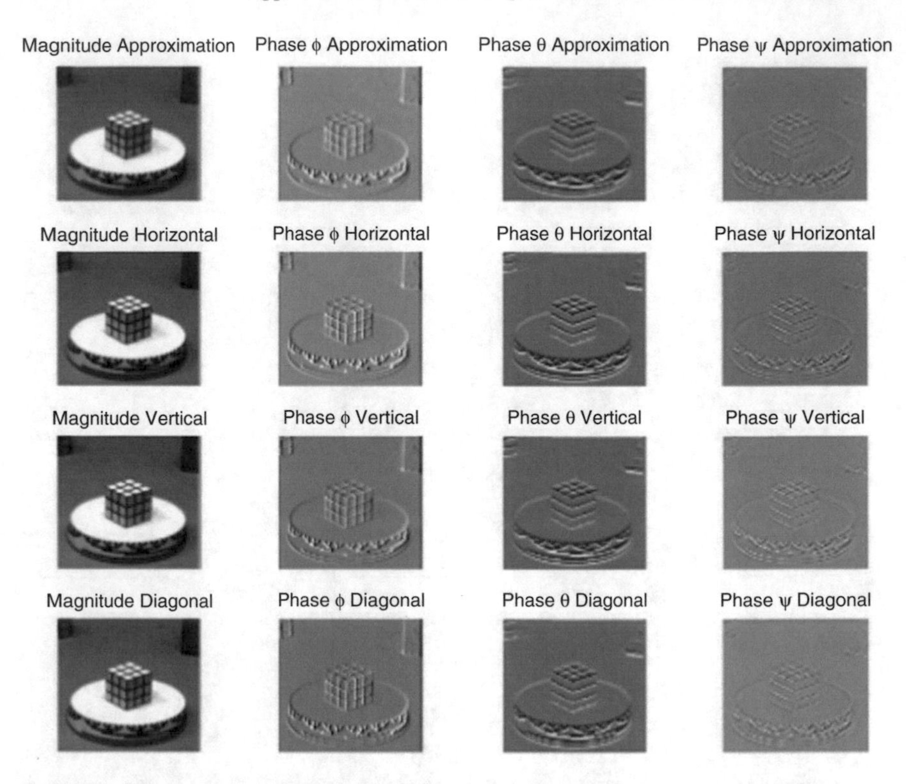

Fig. 14.12 Magnitudes and phase images for the Rubik's sequence at a certain level j: (upper row) the approximation Φ and (next rows) the details Ψ_1 (horizontal), Ψ_2 (vertical), Ψ_3 (diagonal)

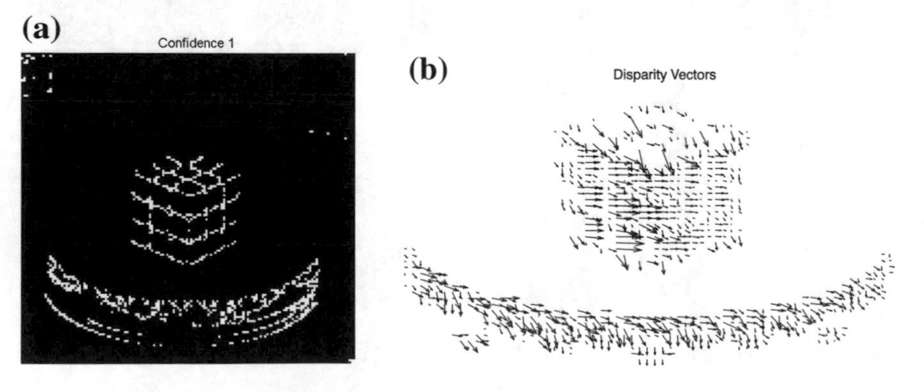

Fig. 14.13 **a** Confidence mask. **b** Estimated optical flow

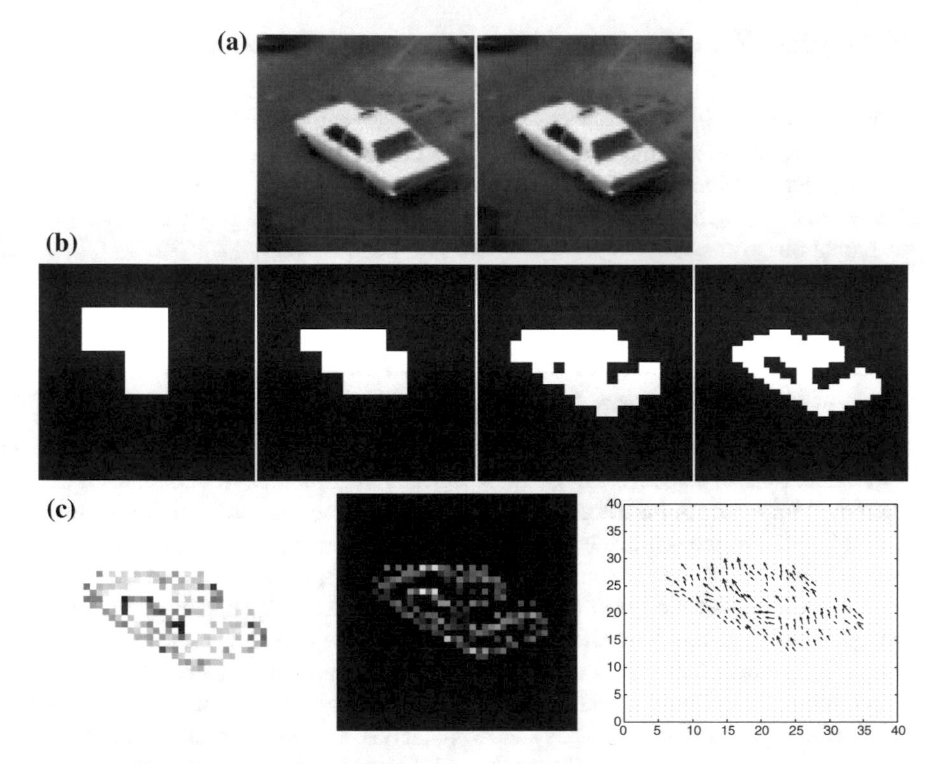

Fig. 14.14 **a** (top row) A couple of successive images; **b** (middle row) Confidence matrices at four levels; **c** (bottom) Horizontal and vertical disparities and optical flow

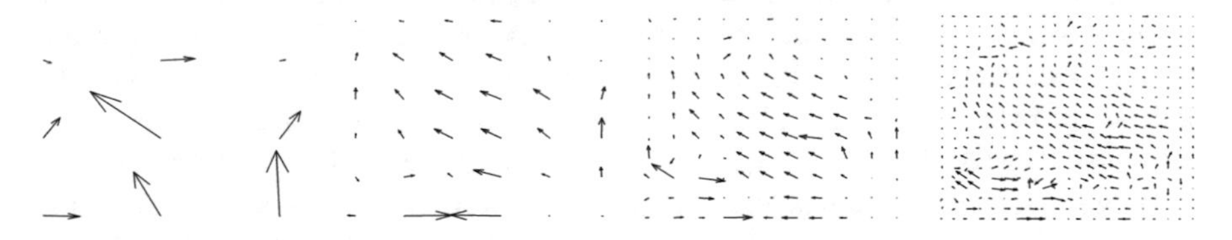

Fig. 14.15 Optical flow computed using discrete real-valued wavelets. (top) Optical flow at fourth level. (bottom) Optical flow at third, second, and first levels

disparities and the resulting optical flow at a high level of resolution. For comparison, we utilized the method of Bernard [34], which uses real-valued discrete wavelets. In Fig. 14.15, we can see that our method yields better results. Based on the results, we can conclude that our procedure using a quaternionic wavelet pyramid and the phase concept for the parameter estimation works very well in both experiments. We believe that the computation of the optical flow using the quaternionic phase concept should be considered by researchers and practitioners as an effective alternative for multiresolution analysis.

14.6 Riesz Transform and Multiresolution Processing

In this section, we illustrate the use of the atomic function for local phase computations using the Riesz transform. Firstly, we compare the Gabor, Log-Gabor, and atomic function filters using Riesz transform in multiresolution pyramid. We discuss also why an appropriate filter yields better results for the local phase analysis than the global phase analysis.

14.6.1 Gabor, Log-Gabor, and Atomic Function Filters Using Riesz Transform in Multiresolution Pyramid

For the local phase analysis, we use three kernels to implement monogenic signals whereby the aperture of the filters is varied equally at three scales of a multiresolution pyramid, whereby the same reference frequency and the same aperture rate variation are used for the three monogenic signals. You can appreciate at these scales the better consistency of the $up(x)$ kernel particularly to detect by means of the phase concept the corners and borders diminishing the expected blurring by a check board image. We suspect that this is an effect of the compactness in space of the $up(x)$ kernel; the Moirè effects are milder specially if you observe the images at scales 2 and 3, whereby in both in the Gauss and LogGauss kernelindexkernel!, Gauss-based monogenic phases, we can see the presence of a sort of bubbles or circular spots. These are often a result of Moire effects. In contrast, the phases of the $up(x)$-based monogenic still preserve through the scales the edges of the check board.

Summarizing, the detection of features is very poor in quality if we would use just the instantaneous phase; in contrast, the use of filters for local phase in a multi-scale analysis permits to detect better features like corners and borders shown in the example of the check board image. Now, as discussed in Sect. 11.3.1 and comparing the pyramid multiresolution layers of Figs. 14.16, 14.17 and 14.18, we can infer that the atomic filter performs better than the Gabor and Log-Gabor filters; this is due to the atomic filter compactness and the lower uncertainty in the space domain.

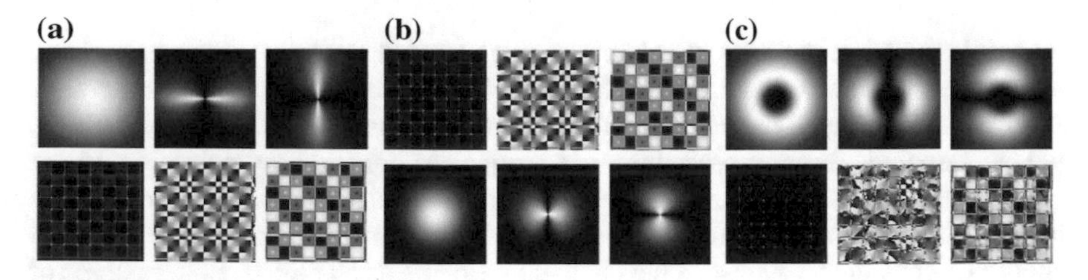

Fig. 14.16 First layer: scale and quadrature filters and monogenic signal (magnitude, local phase ψ, local orientation θ) at the first level of filtering using (rows from top to bottom) **a** Up filter; **b** Log-Gabor filter; **c** Gabor filter

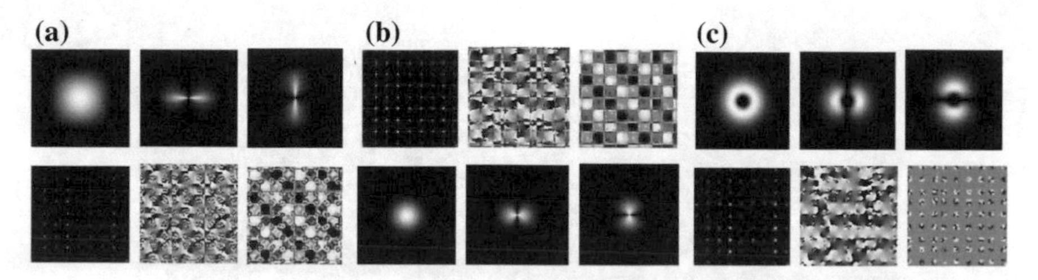

Fig. 14.17 Second layer: scale and quadrature filters and monogenic signal (magnitude, local phase ψ, local orientation θ) at the second level of filtering using (rows from top to bottom) **a** Up filter; **b** Log-Gabor filter; **c** Gabor filter

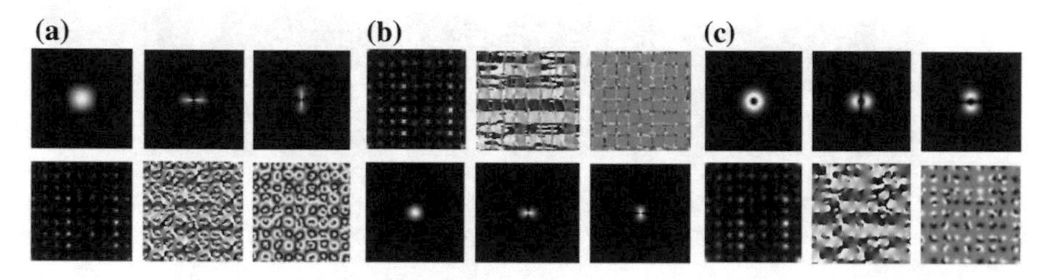

Fig. 14.18 Third layer: scale and quadrature filters and monogenic signal (magnitude, local phase ψ, local orientation θ) at the third level of filtering using (rows from top to bottom) **a** Up filter; **b** Log-Gabor filter; **c** Gabor filter

14.6.2 Multiresolution Analysis Using the Quaternion Wavelet Atomic Function

The *Qup* kernel was used as the mother wavelet in the multiresolution pyramid. Figure 14.19 presents the three quaternionic phases at two scale levels of the pyramid. The bottom row shows the phases after thresholding to enhance the phase structure. You can see how vertical lines and crossing points are highlighted.

The *Qup* mother wavelet kernel was steered: elongation $s_x = 0.3$ and $s_y = 0.25$ and angles $\{0^o, 22.5^o, 45^o, 77.25, 90\}$ through the multiresolution pyramid. Figure 14.20 shows the detected structure.

14.6.3 Radon Transform for Circle Detection by Color Images Using the Quaternion Atomic Functions

First the circles image was filtered to round the sharp contour, then Gaussian noise was added to distort the circle contours, and salt and pepper noise was added to each of the r,g and b color image components. The image was first filtered by steering

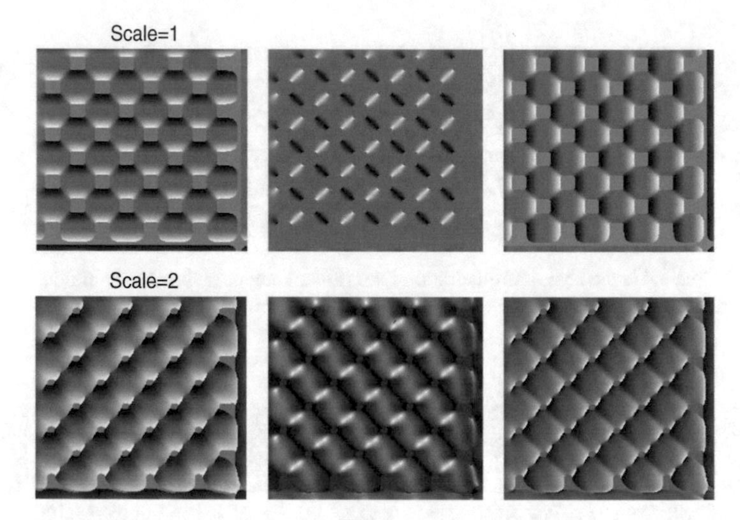

Fig. 14.19 (Upper row) Thresholded quaternionic phases (ϕ, θ, φ) at the first scale of resolution. (second row) Thresholded quaternionic phases (ϕ, θ, φ) at the next scale

a quaternion wavelet atomic function, and at a certain level of the multiresolution pyramid, we applied the Radon transform. Figure 14.21 presents at the upper row the images of two colon circles in color and grayscale level. The next four images at the middle represent the magnitude $|I_q|$ and the quaternionic phases of the filtered image (ϕ, θ, φ) and the next four images the quaternion components after the convolution $I \star AF_r$, $I \star AF_i$, $I \star AF_j$ and $I \star AF_k$. At the lower row, we show the original grayscale level image, its log FFT, and the filtered contour of the cyan circle. You can see that our algorithm has extracted only the contour with respect the color cyan.

Finally, one can improve performance of the multiresolution approach by applying the Riesz transform from coarse to fine levels to identify possible shape contours in the upper level. Then the shapes are oversampled and overlapped with the next lower level to improve the definition of the shapes. At the finest level, one gets the parameters of all well-supported contours.

14.6.4 Feature Extraction Using Symmetry

This section presents the application of the $Qup(x)$ filter for feature extraction using symmetries detected in a multiresolution scheme. Through the levels of the multiresolution pyramid, we detect symmetries of phases using the atomic function-based Riesz transform. The image of symmetries is then treated to extract the feature signature of the object shape by means of the Radon transform.

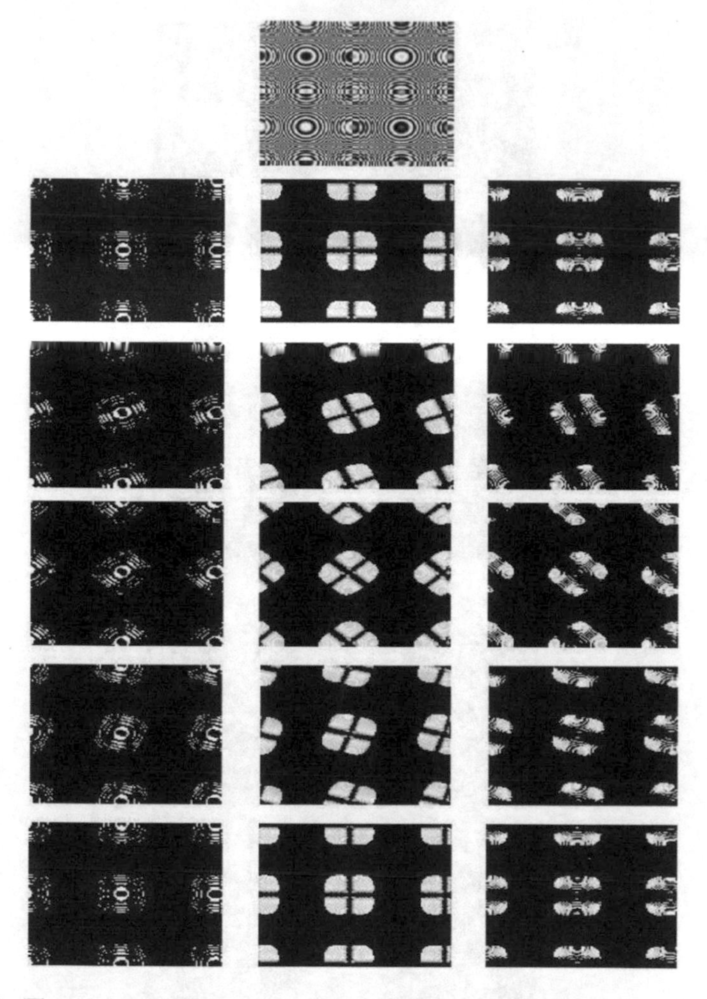

Fig. 14.20 (Three columns) thresholded quaternionic phases (ϕ, θ, φ), filter elongation $s_x = 0.3$, $s_y = 0.25$, steering angles for each row are: 0°, 22.5^o, 45°, 77.25, 90

Procedure. The original image is processed with the atomic function-based Riesz transform. In Figure 14.22 the filters used for the Riesz transform at two scales are shown. Since in this approach the information of the two phases ϕ and θ is immune to changes of the contrast due to high illumination, the phase information can be used

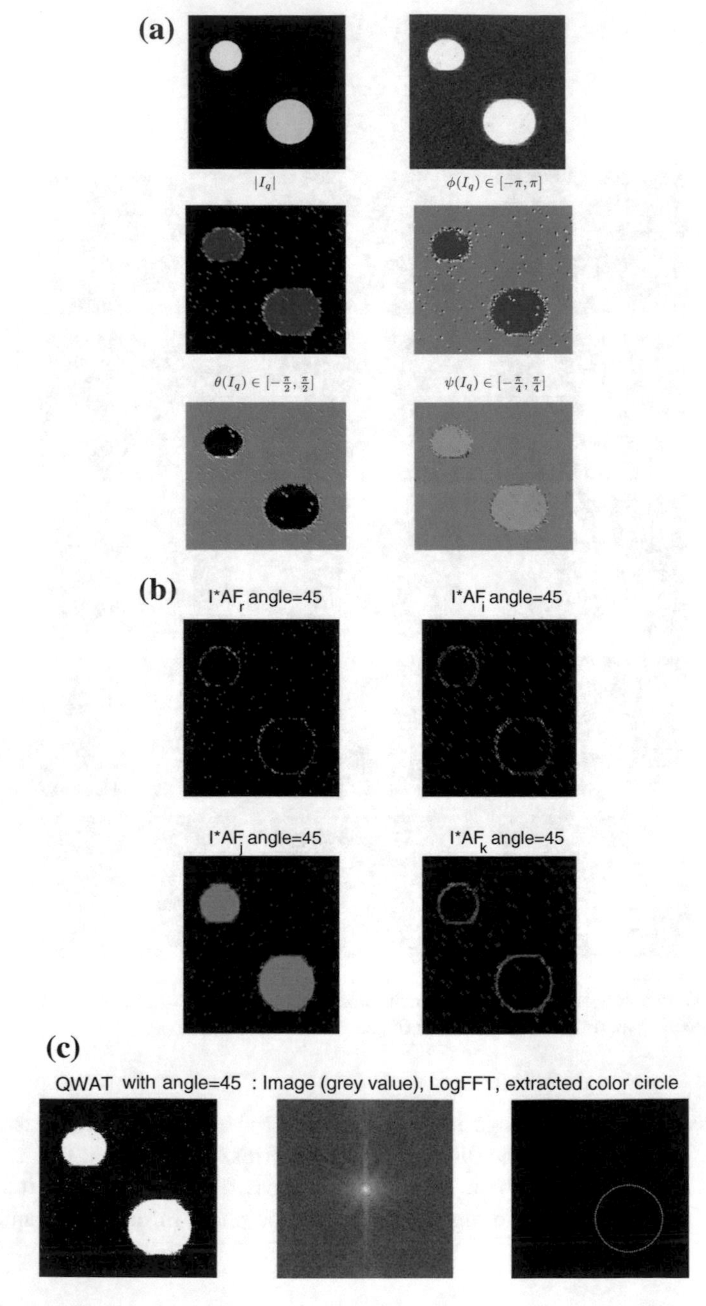

Fig. 14.21 **a** (First three upper rows) Image of two circles in color and grayscale level; **b** (at the middle the first four images: the magnitude $|I_q|$ and the quaternionic phases of the filtered image (ϕ, θ, φ) and the next four images the quaternion components after the convolution $I \star AF_r$, $I \star AF_i$,$I \star AF_j$ and $I \star AF_k$; **c** (lower row) grayscale level image, its log FFT, and the filtered contour of the cyan circle

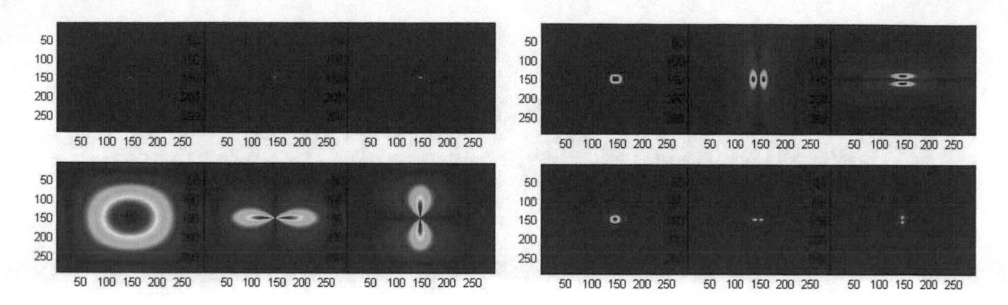

Fig. 14.22 Atomic function filter in (upper row) spatial and (lower row) frequency domains at two pyramid levels

to localize and extract symmetry information independently of contrast changes. The resulting symmetry image is firstly centered at the object centroid C and rotated with respect to the object main axes l, so that the object is aligned with the reference coordinate system which is needed for the application of the Radon transform. Then we use the clockwise-rotated line l and integrate the symmetry values along the line. We get then a value for each angle of rotation. The set of these values builds a feature signature of the object shape symmetry. Why we have to ingrate along the line, this is because the symmetry values varies from the centroid toward the contour of the object. Note that here we are not just computing the symmetry based on the object contour. The changes in the symmetry values within the object shape are the powerful way of computing the symmetries proposed by Kovesi [163, 164]. Using this method at each level of the multiresolution pyramid, the *Qup* filter is steered in 6 orientations to detect along the symmetry. The pyramid runs through 4 scales. The algorithm fuses the symmetry characteristics of the layers yielding a final symmetry image. Different as Kovesi‘s method, in our approach we do not use the energy of the signal; the symmetry is computed using a linear combination of the absolute of two phases, and this result is added to a buffer which after the final pass of the multiresolution pyramid outputs the symmetry image.

The next step is to compute a feature signature vector using this symmetry image by growing slowly the analysis region of symmetry from the centroid toward the contour. The procedure is carried out by integrating along the line to quantify the symmetry changes and most important to change a 2D symmetry description into a 1D feature signature description. You can see that in a natural manner comes into play the role of the use of the Radon transform [243].

Feature signature extraction. Figure 14.22 shows the atomic function filter in spatial and frequency domains at two levels of the multiresolution pyramid. Figure 14.23a shows the image, the symmetry image and the same with its centroid C and its main axis l. Figure 14.23b shows the levels of the symmetry image and phases; note that they are not flat like by a binarised image of a shape, the levels change from the center toward the shape contour. Figure 14.23c shows the feature signatures of the symmetry image and phases. The presence of corners c at the contour correspond to the place of the peaks at the feature signature. Figure 14.23d presents the feature

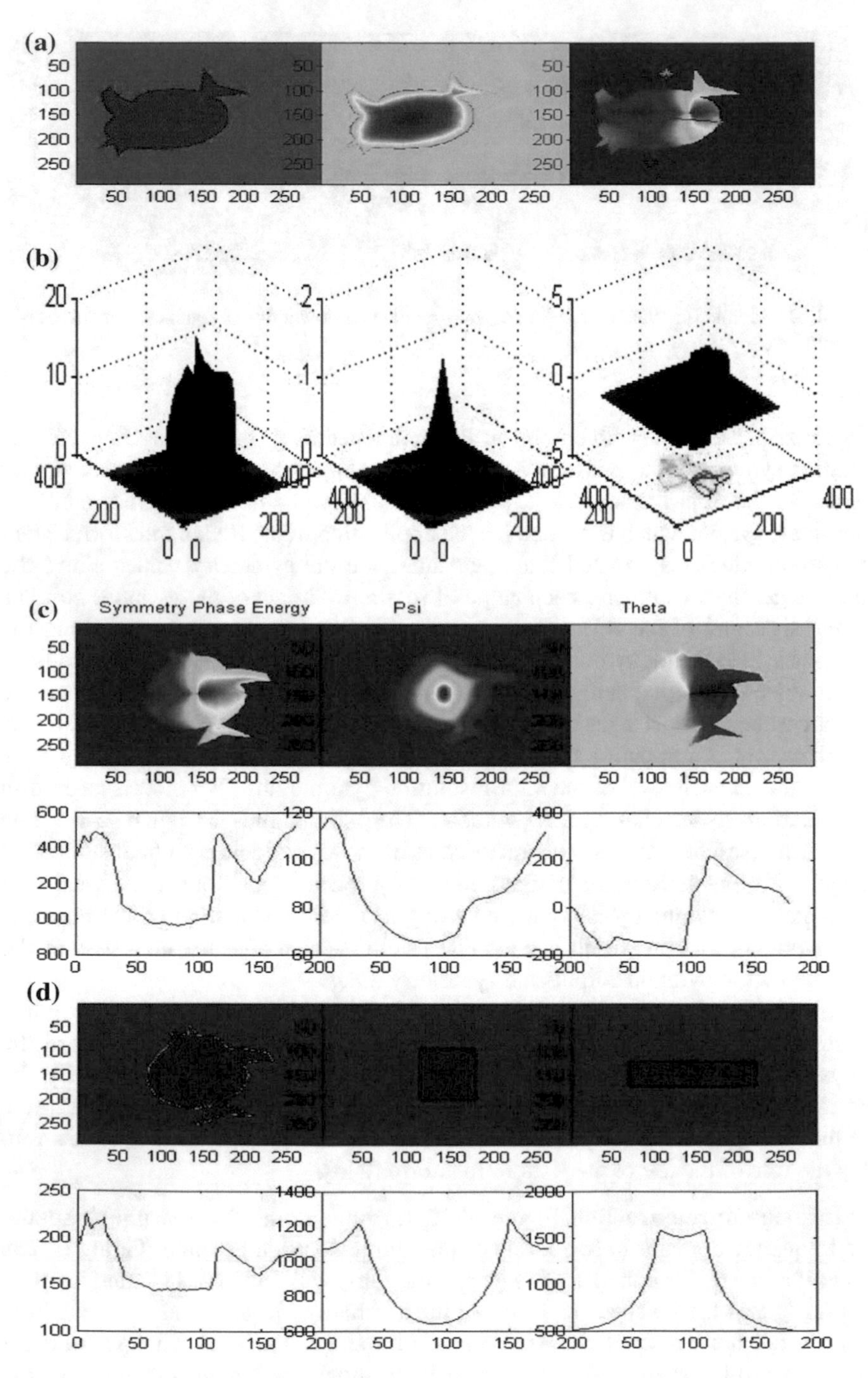

Fig. 14.23 **a** Image, edge image, symmetry image with centroid and main axis. **b** Extracted feature signature: using (left) the symmetry image, (middle) the phase ϕ and (right) the phase θ. **c** Feature signatures of binarised images

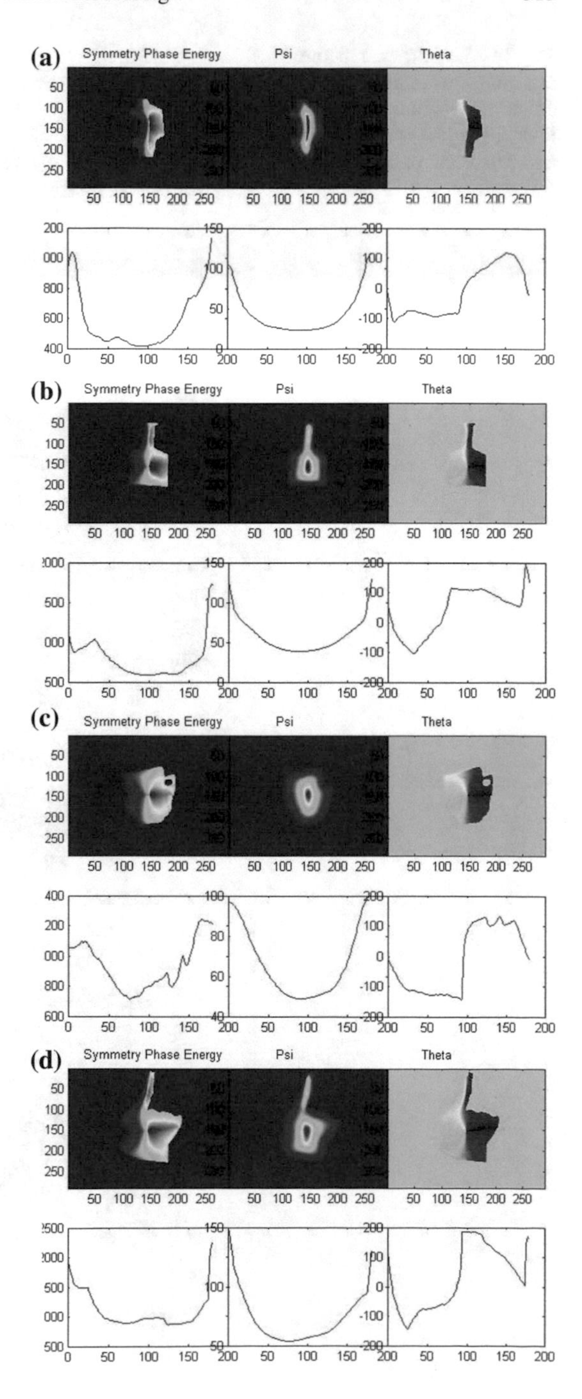

Fig. 14.24 Extracted feature signatures: using (left column) the symmetry image, (middle) the phase ϕ and (right) the phase θ

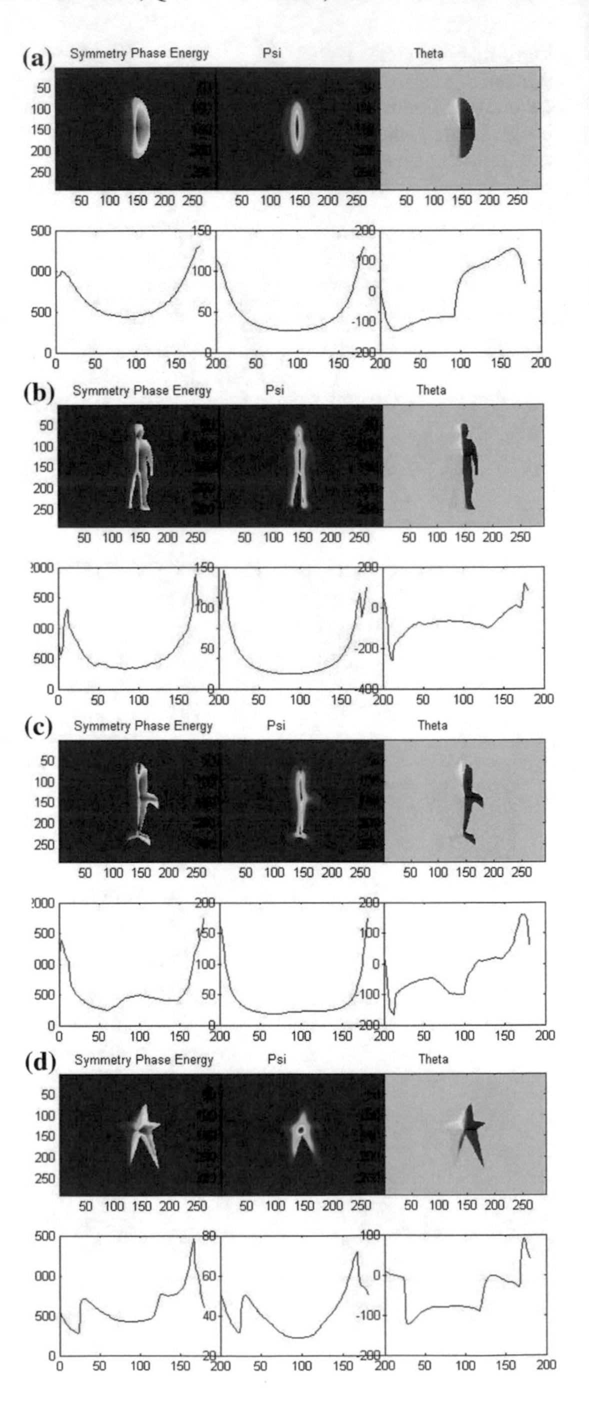

Fig. 14.25 Extracted feature signatures: using (left column) the symmetry image, (middle) the phase $\{\phi$ and (right) the phase $\theta\}$

signatures using the Radon transform of just binarised images. Comparing Fig. 14.23c–d, you can see that the feature signatures of the symmetry and phases images convey more information than that of just the binarised images. Figures 14.24 and 14.25 show the feature signatures of different shapes. By analyzing carefully those feature signatures, one can recognize the discriminative power of the method to obtain feature signatures, namely the feature signatures are quite different from each other. Having the feature signatures this property, they can be used for training classifiers to recognize successfully object shapes.

14.7 Conclusion

In the first part, the chapter shows the application of feature detectors using Lie perceptrons in the 2D affine plane. The filters are not only applicable to optic flow, they may also be used for the detection of key points. Combinations of Lie operators yield to complex structures for detecting more sophisticated visual geometry.

The second part presents the preprocessing for neurocomputing using the quaternion Fourier transform. We applied this technique for the recognition of the first 10 French numbers spoken by different speakers. In our approach, we expanded the signal representation to a higher-dimensional space, where we search for features along selected lines, allowing us to surprisingly reduce the dimensionality of the feature vector. The results also show that the method manages to separate the sounds of vowels and consonants, which is quite rare in the literature of speech processing. For the recognition, we used a neural experts architecture.

The third part of this chapter introduced the theory and practicalities of the QWT, so that the reader can apply it to a variety of problems making use of the quaternionic phase concept. We extended Mallat's multiresolution analysis using quaternion wavelets. These kernels are more efficient than the Haar quaternion wavelets. A big advantage of our approach is that it offers three phases at each level of the pyramid, which can be used for a powerful top-down parameter estimation. As an illustration in the experimental part, we apply the QWT for optical flow estimation.

The fourth part presents the application of the $Qup(x)$ filter for feature extraction using symmetries detected in a multiresolution scheme. Through the levels of the multiresolution pyramid, we detect symmetries of phases using the atomic function-based Riesz transform. The image of symmetries is then treated to extract the feature signature of the object shape by means of the Radon transform.

We believe that this chapter can be very useful for researchers and practitioners interested in understanding and applying the quaternion Fourier and wavelet transforms and the Riesz transform.

Chapter 15
Invariants Theory in Computer Vision and Omnidirectional Vision

15.1 Introduction

This chapter will demonstrate that geometric algebra provides a simple mechanism for unifying current approaches in the computation and application of projective invariants using n-uncalibrated cameras. First, we describe Pascal's theorem as a type of projective invariant, and then the theorem is applied for computing camera-intrinsic parameters. The fundamental projective invariant cross-ratio is studied in one, two, and three dimensions, using a single view and then n views. Next, by using the observations of two and three cameras, we apply projective invariants to the tasks of computing the view-center of a moving camera and to simplified visually guided grasping. The chapter also presents a geometric approach for the computation of shape and motion using projective invariants within a purely geometric algebra framework [138, 139]. Different approaches for projective reconstruction have utilized projective depth [269, 284], projective invariants [61], and factorization methods [237, 296, 300] (factorization methods incorporate projective depth calculations). We compute projective depth using projective invariants, which depend on the use of the fundamental matrix or trifocal tensor. Using these projective depths, we are then able to initiate a projective reconstruction procedure to compute shape and motion. We also apply the algebra of incidence in the development of geometric inference rules to extend 3D reconstruction.

The geometric procedures here presented contribute to the design and implementation of *perception action cycle* (PAC) systems, as depicted in Fig. 15.1.

At the last section, we present a robust technique for landmark identification using omnidirectional vision and the projective and permutation p^2-invariants. Here, the use of permutation p^2-invariant makes more robust the identification of projective invariants in the projective space.

This chapter is organized as follows: Sect. 15.2 briefly explains conics and Pascal's theorem. Section 15.3 demonstrates a method for computing intrinsic camera parameters using Pascal's theorem. Section 15.4 studies in detail the projective invariant cross-ratio in 1D, 2D, and 3D, as well as the generation of projective invariants in 3D.

E. Bayro-Corrochano, *Geometric Algebra Applications Vol. I*,
https://doi.org/10.1007/978-3-319-74830-6_15

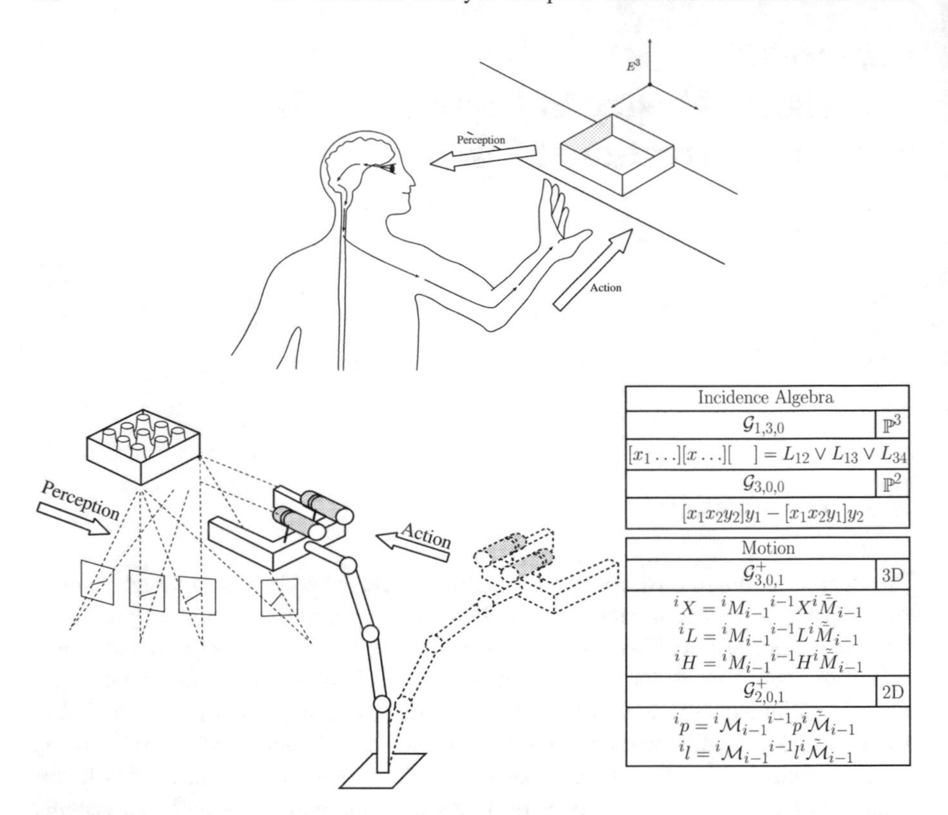

Fig. 15.1 Abstraction of biologic and artificial PACs systems

Section 15.5 presents the theory of 3D projective invariants using n-uncalibrated cameras. Section 15.6 illustrates the use of projective invariants for a simplified task of visually guided robot grasping. The problem of camera self-localization is discussed in Sect. 15.7. Computation of projective depth using projective invariants in terms of the trifocal tensor is given in Sect. 15.8. The treatment of projective reconstruction and the role of the algebra of incidence in completing the 3D shape is given in Sect. 15.9. Section 15.10 presents landmark identification using omnidirectional vision and projective invariants. Section eleven is devoted to the conclusions.

15.2 Conics and Pascal's Theorem

The role of conics and quadrics is well known in projective geometry [267]. This knowledge led to the solution of crucial problems in computer vision [220]. In the last decade, Kruppa's equations, which rely on the conic concept, have been used to compute intrinsic camera parameters [204]. In the present work, we further explore the conics concept and use Pascal's theorem to establish an equation system with clear geometric transparency. We then explain the role of conics and that of Pascal's theorem in relation to fundamental projective invariants. Our work here is based primarily on that of Hestenes and Ziegler [139], that is, on an interpretation of linear algebra together with projective geometry within a Clifford algebra framework.

In order to use projective geometry in computer vision, we utilize homogeneous coordinate representations, which allows us to embed both 3D Euclidean visual space in 3D projective space P^3 or R^4, and the 2D Euclidean space of the image plane in 2D projective space P^2 or R^3. Using the geometric algebra framework, we select for P^2 the 3D Euclidean geometric algebra $G_{3,0,0}$, and for P^3 the 4D geometric algebra $G_{1,3,0}$. The reader should refer to Chap. 12 for more details relating to the geometry of n cameras. Any geometric object of P^3 will be linear projective mapped to P^2 via a projective transformation. For example, the projective mapping of a quadric at infinity in the projective space P^3 results in a conic in the projective plane P^2.

Let us first consider a *pencil of lines* lying on the plane. Any pencil of lines may be well defined by the bivector addition of two of its lines: $\boldsymbol{l} = \boldsymbol{l}_a + s\boldsymbol{l}_b$ with $s \in R \cup \{-\infty, +\infty\}$. If two pencils of lines $\boldsymbol{l}$ and $\boldsymbol{l}' = \boldsymbol{l}'_a + s'\boldsymbol{l}'_b$ can be related one to one so that $\boldsymbol{l} = \boldsymbol{l}'$ for $s = s'$, we say that they are in projective correspondence. Using this idea, we will show that the set of intersecting points of lines in projective correspondence build a conic. Since the intersecting points $\boldsymbol{x}$ of the pencils of lines $\boldsymbol{l}$ and $\boldsymbol{l}'$ fulfill for $s = s'$, the following constraints are met:

$$\begin{aligned} \boldsymbol{x}\wedge\boldsymbol{l} &= \boldsymbol{x}\wedge\boldsymbol{l}_a + s\boldsymbol{x}\wedge\boldsymbol{l}_b = 0, \\ \boldsymbol{x}\wedge\boldsymbol{l}' &= \boldsymbol{x}\wedge\boldsymbol{l}'_a + s\boldsymbol{x}\wedge\boldsymbol{l}'_b = 0. \end{aligned} \tag{15.1}$$

The elimination of the scalar s yields a second-order geometric product equation in $\boldsymbol{x}$,

$$(\boldsymbol{x}\wedge\boldsymbol{l}_a)(\boldsymbol{x}\wedge\boldsymbol{l}'_b) - (\boldsymbol{x}\wedge\boldsymbol{l}_b)(\boldsymbol{x}\wedge\boldsymbol{l}'_a) = 0. \tag{15.2}$$

We can also derive the parameterized conic equation, by simply computing the intersecting point $\boldsymbol{x}$ by means of the meet of the pencils of lines, as follows:

$$\boldsymbol{x} = (\boldsymbol{l}_a + s\boldsymbol{l}_b) \cap (\boldsymbol{l}'_a + s\boldsymbol{l}'_b) = \boldsymbol{l}_a \cap \boldsymbol{l}'_a + s(\boldsymbol{l}_a \cap \boldsymbol{l}'_b + \boldsymbol{l}_b \cap \boldsymbol{l}'_a) + s^2\boldsymbol{l}_b \cap \boldsymbol{l}'_b. \tag{15.3}$$

Let us, for now, define the involved lines as a wedge of points, $\boldsymbol{l}_a = \boldsymbol{a}\wedge\boldsymbol{b}$, $\boldsymbol{l}_b = \boldsymbol{a}\wedge\boldsymbol{b}'$, $\boldsymbol{l}'_a = \boldsymbol{a}'\wedge\boldsymbol{b}$, and $\boldsymbol{l}'_b = \boldsymbol{a}'\wedge\boldsymbol{b}'$, such that $\boldsymbol{l}_a \cap \boldsymbol{l}'_a = \boldsymbol{b}$ and $\boldsymbol{l}_b \cap \boldsymbol{l}'_b = \boldsymbol{b}'$ (see Fig. 15.2a). By substituting $\boldsymbol{b}'' = \boldsymbol{l}_a \cap \boldsymbol{l}'_b + \boldsymbol{l}_b \cap \boldsymbol{l}'_a = \boldsymbol{d} + \boldsymbol{d}'$ into Eq. (15.3), we get

$$\boldsymbol{x} = \boldsymbol{b} + s\boldsymbol{b}'' + s^2\boldsymbol{b}', \tag{15.4}$$

which represents a non-degenerated conic for $\boldsymbol{b}\wedge\boldsymbol{b}''\wedge\boldsymbol{b}' = \boldsymbol{b}\wedge(\boldsymbol{d} + \boldsymbol{d}')\wedge\boldsymbol{b}' \neq 0$. Now, using this equation, let us recompute the original pencils of lines. By

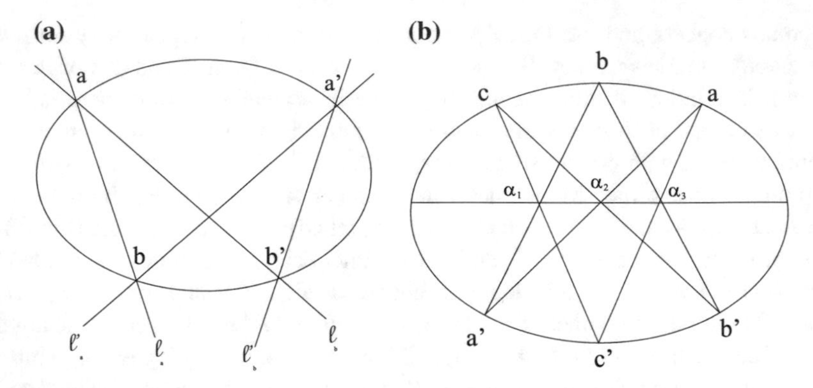

Fig. 15.2 Pencils of lines related to a conic: **a** two projective pencils of lines used to generate the conic; **b** Pascal's theorem

defining $\boldsymbol{l}_1 = \boldsymbol{b}'' \wedge \boldsymbol{b}'$, $\boldsymbol{l}_2 = \boldsymbol{b}' \wedge \boldsymbol{b}$, and $\boldsymbol{l}_3 = \boldsymbol{b} \wedge \boldsymbol{b}''$, we can use Eq. (15.4) to compute the projective pencils of lines:

$$\begin{aligned} \boldsymbol{b} \wedge \boldsymbol{x} &= s\boldsymbol{b} \wedge \boldsymbol{b}'' + s^2 \boldsymbol{b} \wedge \boldsymbol{b}' = s(\boldsymbol{l}_3 - s\boldsymbol{l}_2), \\ \boldsymbol{b}' \wedge \boldsymbol{x} &= \boldsymbol{b}' \wedge \boldsymbol{b} + s\boldsymbol{b}' \wedge \boldsymbol{b}'' = \boldsymbol{l}_2 - s\boldsymbol{l}_1. \end{aligned} \tag{15.5}$$

By considering the points $\boldsymbol{a}, \boldsymbol{a}', \boldsymbol{b}$, and $\boldsymbol{b}'$ and some other point $\boldsymbol{y}$ lying in the conic depicted in Fig. 15.2a, and by using Eq. (15.1) for $s = Is'$ slightly different to s', we get the bracketed expression

$$\begin{aligned} [\boldsymbol{yab}][\boldsymbol{ya'b'}] - I[\boldsymbol{yab'}][\boldsymbol{ya'b}] = 0, \\ I = \frac{[\boldsymbol{yab}][\boldsymbol{ya'b'}]}{[\boldsymbol{yab'}][\boldsymbol{ya'b}]}, \end{aligned} \tag{15.6}$$

for some scalar $I \neq 0$. This equation is well known and represents a projective invariant, a concept which has been used quite a lot in real applications of computer vision. Sections 15.4 and 15.5 show this invariant using brackets of points, bilinearities, and the trifocal tensor (see also Bayro [22] and Lasenby [177]). We can evaluate I of Eq. (15.6) in terms of some other point $\boldsymbol{c}$ to develop a conic equation that can be fully expressed within brackets:

$$[\boldsymbol{yab}][\boldsymbol{ya'b'}] - \frac{[\boldsymbol{cab}][\boldsymbol{ca'b'}]}{[\boldsymbol{cab'}][\boldsymbol{ca'b}]}[\boldsymbol{yab'}][\boldsymbol{ya'b}] = 0,$$

$$[\boldsymbol{yab}][\boldsymbol{ya'b'}][\boldsymbol{ab'c'}][\boldsymbol{a'bc'}] - [\boldsymbol{yab'}][\boldsymbol{ya'b}][\boldsymbol{abc'}][\boldsymbol{a'b'c'}] = 0. \tag{15.7}$$

Again, the resulting equation is well known, which tells us that any conic is uniquely determined by five points in general positions $\boldsymbol{a}$, $\boldsymbol{a}'$, $\boldsymbol{b}$ and $\boldsymbol{b}'$, and $\boldsymbol{c}$. Now, considering Fig. 15.2b, we can identify three collinear intersecting points α_1, α_2, and α_3. By using the collinearity constraint and the pencils of lines in projective correspondence, we can now write a very useful equation:

$$\underbrace{((\boldsymbol{a}'\wedge\boldsymbol{b})\cap(\boldsymbol{c}'\wedge\boldsymbol{c}))}_{\alpha_1}\wedge\underbrace{((\boldsymbol{a}'\wedge\boldsymbol{a})\cap(\boldsymbol{b}'\wedge\boldsymbol{c}))}_{\alpha_2}\wedge\underbrace{((\boldsymbol{c}'\wedge\boldsymbol{a})\cap(\boldsymbol{b}'\wedge\boldsymbol{b}))}_{\alpha_3}=0. \quad (15.8)$$

This expression is a geometric formulation, using brackets, of *Pascal's theorem*, which says that the three intersecting points of the lines which connect opposite vertices of a hexagon circumscribed by a conic are collinear. Equation (15.8) will be used in the following section for computing intrinsic camera parameters.

15.3 Computing Intrinsic Camera Parameters

This section presents a technique within the geometric algebra framework for computing intrinsic camera parameters. In the previous section, it was shown that Eq. (15.7) can be reformulated to express the constraint of Eq. (15.8), known as Pascal's theorem. Since Pascal's equation fulfills a property of any conic, we should also be able to use this equation to compute intrinsic camera parameters. Let us consider three intersecting points which are collinear and fulfill the parameters of Eq. (15.8).

Figure 15.3 shows the first camera image, where the projected rotated points of the conic at infinity $R^T\boldsymbol{A}$, $R^T\boldsymbol{B}$, $R^T\boldsymbol{A}'$, $R^T\boldsymbol{B}'$, and $R^T\boldsymbol{C}'$ are $\boldsymbol{a}=K[R|0]R^T\boldsymbol{A}=K\boldsymbol{A}$, $\boldsymbol{b}=K[R|0]R^T\boldsymbol{B}=K\boldsymbol{B}$, $\boldsymbol{a}'=K[R|0]R^T\boldsymbol{A}'=K\boldsymbol{A}'$, $\boldsymbol{b}'=K[R|0]R^T\boldsymbol{B}'=K\boldsymbol{B}'$, and $\boldsymbol{c}'=K[R|0]\boldsymbol{C}'=K\boldsymbol{C}'$. The point $\boldsymbol{c}=KK^T\boldsymbol{l}_c$ is dependent upon the camera-intrinsic parameters and upon the line $\boldsymbol{l}_c$ tangent to the conic, computed in terms of the epipole $=[p_1,p_2,p_3]^T$ and a point lying at infinity upon the line of the first camera $\boldsymbol{l}_c=[p_1,p_2,p_3]^T\times[1,\tau,0]^T$. Now, using this expression for $\boldsymbol{l}_c$, we can simplify Eq. (15.8) to obtain the equations for the α's in terms of brackets,

$$\big([\boldsymbol{a}'\boldsymbol{b}\boldsymbol{c}']\boldsymbol{c}-[\boldsymbol{a}'\boldsymbol{b}\boldsymbol{c}]\boldsymbol{c}'\big)\wedge\big([\boldsymbol{a}'\boldsymbol{a}\boldsymbol{b}']\boldsymbol{c}-[\boldsymbol{a}'\boldsymbol{a}\boldsymbol{c}]\boldsymbol{b}'\big)\wedge\big([\boldsymbol{c}'\boldsymbol{a}\boldsymbol{b}']\boldsymbol{b}-[\boldsymbol{c}'\boldsymbol{a}\boldsymbol{b}]\boldsymbol{b}'\big)=0\Leftrightarrow \quad (15.9)$$

$$\begin{aligned}
&\big([K\boldsymbol{A}'K\boldsymbol{B}K\boldsymbol{C}']KK^T\boldsymbol{l}_c-[K\boldsymbol{A}'K\boldsymbol{B}KK^T\boldsymbol{l}_c]K\boldsymbol{C}'\big)\wedge\big([K\boldsymbol{A}'K\boldsymbol{A}K\boldsymbol{B}']KK^T\boldsymbol{l}_c-\\
&[K\boldsymbol{A}'K\boldsymbol{A}KK^T\boldsymbol{l}_c]K\boldsymbol{B}'\big)\wedge\big([K\boldsymbol{C}'K\boldsymbol{A}K\boldsymbol{B}']K\boldsymbol{B}-[K\boldsymbol{C}'K\boldsymbol{A}K\boldsymbol{B}]K\boldsymbol{B}'\big)=0\\
&\Big(det(K)K\big([\boldsymbol{A}'\boldsymbol{B}\boldsymbol{C}']K^T\boldsymbol{l}_c-[\boldsymbol{A}'\boldsymbol{B}K^T\boldsymbol{l}_c]\boldsymbol{C}'\big)\Big)\wedge\\
&\wedge\Big(det(K)K\big([\boldsymbol{A}'\boldsymbol{A}\boldsymbol{B}']K^T\boldsymbol{l}_c-[\boldsymbol{A}'\boldsymbol{A}K^T\boldsymbol{l}_c]\boldsymbol{B}'\big)\Big)\wedge\\
&\wedge\Big(det(K)K\big([\boldsymbol{C}'\boldsymbol{A}\boldsymbol{B}']\boldsymbol{B}-[\boldsymbol{C}'\boldsymbol{A}\boldsymbol{B}]\boldsymbol{B}'\big)\Big)=0\Leftrightarrow
\end{aligned} \quad (15.10)$$

$$
\begin{aligned}
&det(K)^3 K\Big(\big([\boldsymbol{A}'\boldsymbol{B}\boldsymbol{C}']K^T\boldsymbol{l}_c - [\boldsymbol{A}'\boldsymbol{B}K^T\boldsymbol{l}_c]\boldsymbol{C}'\big)\wedge\big([\boldsymbol{A}'\boldsymbol{A}\boldsymbol{B}']K^T\boldsymbol{l}_c - \\
&-[\boldsymbol{A}'\boldsymbol{A}K^T\boldsymbol{l}_c]\boldsymbol{B}'\big)\wedge\big([\boldsymbol{C}'\boldsymbol{A}\boldsymbol{B}']\boldsymbol{B} - [\boldsymbol{C}'\boldsymbol{A}\boldsymbol{B}]\boldsymbol{B}'\big)\Big) = 0 \Leftrightarrow \qquad (15.11)\\
&\underbrace{\big([\boldsymbol{A}'\boldsymbol{B}\boldsymbol{C}']K^T\boldsymbol{l}_c - [\boldsymbol{A}'\boldsymbol{B}K^T\boldsymbol{l}_c]\boldsymbol{C}'\big)}_{\alpha_1}\wedge\underbrace{\big([\boldsymbol{A}'\boldsymbol{A}\boldsymbol{B}']K^T\boldsymbol{l}_c - [\boldsymbol{A}'\boldsymbol{A}K^T\boldsymbol{l}_c]\boldsymbol{B}'\big)}_{\alpha_2}\wedge \\
&\wedge\underbrace{\big([\boldsymbol{C}'\boldsymbol{A}\boldsymbol{B}']\boldsymbol{B} - [\boldsymbol{C}'\boldsymbol{A}\boldsymbol{B}]\boldsymbol{B}'\big)}_{\alpha_3} = 0. \qquad (15.12)
\end{aligned}
$$

Note that in Eq. (15.11) the scalars $det(K)^3$ and K are canceled out, thereby simplifying the expression for the αs. The computation of the intrinsic parameters should take into account two possible situations: the intrinsic parameters remain stationary

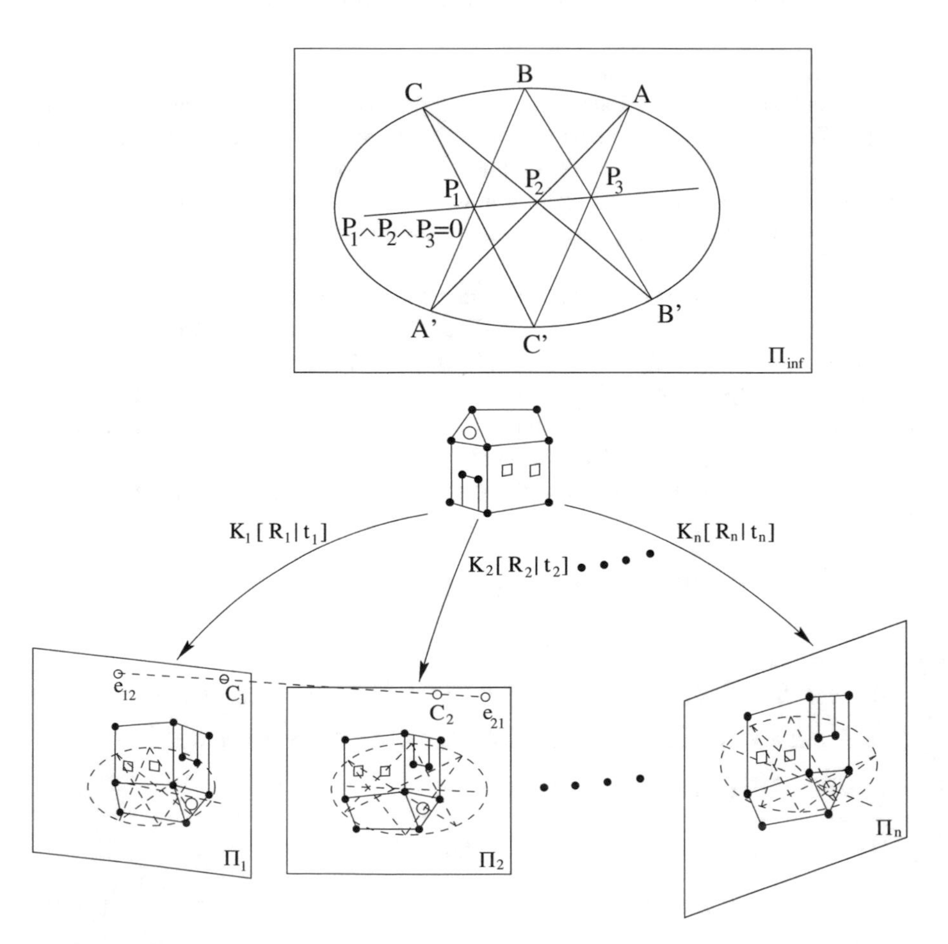

Fig. 15.3 A demonstration of Pascal's theorem at conics at infinity and at images of n-uncalibrated cameras

when the camera is in motion, or they will vary. By using Eq. (15.12), we are able to develop after one camera movement a set of eight quadratic equations, from which we can compute four intrinsic camera parameters (see [26] for the technical details of the computer algorithm) (Fig. 15.3).

15.4 Projective Invariants

In this section, we will use the framework established in Chap. 12 to show how standard invariants can be expressed both elegantly and concisely using geometric algebra. We begin by looking at algebraic quantities that are invariant under projective transformations, arriving at these invariants using a method which can be easily generalized from one dimension to two and three dimensions.

15.4.1 The 1D Cross-Ratio

The *fundamental projective invariant* of points on a line is the so-called cross-ratio, ρ, defined as

$$\rho = \frac{AC}{BC}\frac{BD}{AD} = \frac{(t_3 - t_1)(t_4 - t_2)}{(t_3 - t_2)(t_4 - t_1)},$$

where $t_1 = |PA|$, $t_2 = |PB|$, $t_3 = |PC|$, and $t_4 = |PD|$. It is fairly easy to show that for the projection through O of the collinear points A, B, C, and D onto any line, ρ remains constant. For the 1D case, any point q on the line L can be written as $\boldsymbol{q} = t\sigma_1$ relative to P, where σ_1 is a unit vector in the direction of L. We can then move up a dimension to a 2D space, with basis vectors (γ_1, γ_2), which we will call R^2 and in which $\boldsymbol{q}$ is represented by the following vector $\mathbf{Q}$:

$$\mathbf{Q} = T\gamma_1 + S\gamma_2. \tag{15.13}$$

Note that, as before, $\boldsymbol{q}$ is associated with the bivector, as follows:

$$\boldsymbol{q} = \frac{\mathbf{Q}\wedge\gamma_2}{\mathbf{Q}\cdot\gamma_2} = \frac{T}{S}\gamma_1\gamma_2 \equiv \frac{T}{S}\sigma_1 = t\sigma_1. \tag{15.14}$$

When a point on line L is projected onto another line L', the distances t and t' are related by a projective transformation of the form

$$t' = \frac{\alpha t + \beta}{\tilde{\alpha} t + \tilde{\beta}}. \tag{15.15}$$

This nonlinear transformation in $\mathcal{E}^1$ can be made into a linear transformation in R^2 by defining the linear function $\underline{f}_1$ which maps vectors onto vectors in R^2:

$$\underline{f}_1(\gamma_1) = \alpha_1\gamma_1 + \tilde{\alpha}\gamma_2,$$
$$\underline{f}_1(\gamma_2) = \beta_1\gamma_1 + \tilde{\beta}\gamma_2.$$

Consider two vectors $\mathbf{X}_1$ and $\mathbf{X}_2$ in R^2. Now form the bivector

$$\mathcal{S}_1 = \mathbf{X}_1 \wedge \mathbf{X}_2 = \lambda_1 I_2,$$

where $I_2 = \gamma_1\gamma_2$ is the pseudoscalar for R^2. We can now look at how $\mathcal{S}_1$ transforms under $\underline{f}_1$:

$$\mathcal{S}_1' = \mathbf{X}_1' \wedge \mathbf{X}_2' = \underline{f}_1(\mathbf{X}_1 \wedge \mathbf{X}_2) = (\det \underline{f}_1)(\mathbf{X}_1 \wedge \mathbf{X}_2). \tag{15.16}$$

This last step follows a result of a linear function, which must map a pseudoscalar onto a multiple of itself, the multiple being the determinant of the function. Suppose that we now select four points of the line L, whose corresponding vectors in R^2 are $\{\mathbf{X}_i\}$, $i = 1, ..., 4$, and consider the ratio $\mathcal{R}_1$ of two wedge products:

$$\mathcal{R}_1 = \frac{\mathbf{X}_1 \wedge \mathbf{X}_2}{\mathbf{X}_3 \wedge \mathbf{X}_4}. \tag{15.17}$$

Then, under $\underline{f}_1$, $\mathcal{R}_1 \rightarrow \mathcal{R}_1'$, where

$$\mathcal{R}_1' = \frac{\mathbf{X}_1' \wedge \mathbf{X}_2'}{\mathbf{X}_3' \wedge \mathbf{X}_4'} = \frac{(\det \underline{f}_1)\mathbf{X}_1 \wedge \mathbf{X}_2}{(\det \underline{f}_1)\mathbf{X}_3 \wedge \mathbf{X}_4}. \tag{15.18}$$

$\mathcal{R}_1$ is therefore invariant under $\underline{f}_1$. However, we want to express our invariants in terms of distances on the 1D line. To do this, we must consider how the bivector $\mathcal{S}_1$ in R^2 projects down to $\mathcal{E}^1$:

$$\begin{aligned}
\mathbf{X}_1 \wedge \mathbf{X}_2 &= (T_1\gamma_1 + S_1\gamma_2) \wedge (T_2\gamma_1 + S_2\gamma_2),\\
&= (T_1 S_2 - T_2 S_1)\gamma_1\gamma_2,\\
&\equiv S_1 S_2 (T_1/S_1 - T_2/S_2) I_2,\\
&= S_1 S_2 (t_1 - t_2) I_2.
\end{aligned} \tag{15.19}$$

In order to form a projective invariant which is independent of the choice of the arbitrary scalars S_i, we must now consider *ratios* of the bivectors $\mathbf{X}_i \wedge \mathbf{X}_j$ (so that

$\det \underline{f}_1$ cancels), and then *multiples* of these ratios (so that the S_i's cancel). More precisely, consider the following expression:

$$Inv_1 = \frac{(\mathbf{X}_3 \wedge \mathbf{X}_1) I_2^{-1} (\mathbf{X}_4 \wedge \mathbf{X}_2) I_2^{-1}}{(\mathbf{X}_4 \wedge \mathbf{X}_1) I_2^{-1} (\mathbf{X}_3 \wedge \mathbf{X}_2) I_2^{-1}}. \tag{15.20}$$

Then, in terms of distances along the lines, under the projective transformation $\underline{f}_1$, Inv_1 goes to Inv_1', where

$$Inv_1' = \frac{S_3 S_1 (t_3 - t_1) S_4 S_2 (t_4 - t_2)}{S_4 S_1 (t_4 - t_1) S_3 S_2 (t_3 - t_2)} = \frac{(t_3 - t_1)(t_4 - t_2)}{(t_4 - t_1)(t_3 - t_2)}, \tag{15.21}$$

which is independent of the S_i's and is indeed the 1D classical projective invariant, the cross-ratio. Deriving the cross-ratio in this way allows us to easily generalize it to form invariants in higher dimensions.

15.4.2 2D Generalization of the Cross-Ratio

When we consider points in a plane, we once again move up to a space with one higher dimension, which we shall call R^3. Let a point P in the plane M be described by the vector $\boldsymbol{x}$ in $\mathcal{E}^2$, where $\boldsymbol{x} = x\sigma_1 + y\sigma_2$. In R^3, this point will be represented by $\mathbf{X} = X\gamma_1 + Y\gamma_2 + Z\gamma_3$, where $x = X/Z$ and $y = Y/Z$. As described in Chap. 12, we can define a general projective transformation via a linear function $\underline{f}_2$ by mapping vectors to vectors in R^3, such that

$$\begin{aligned} \underline{f}_2(\gamma_1) &= \alpha_1\gamma_1 + \alpha_2\gamma_2 + \tilde{\alpha}\gamma_3, \\ \underline{f}_2(\gamma_2) &= \beta_1\gamma_1 + \beta_2\gamma_2 + \tilde{\beta}\gamma_3, \\ \underline{f}_2(\gamma_3) &= \delta_1\gamma_1 + \delta_2\gamma_2 + \tilde{\delta}\gamma_3. \end{aligned} \tag{15.22}$$

Now, consider three vectors (representing non-collinear points) $\mathbf{X}_i$, $i = 1, 2, 3$, in R^3, and form the trivector

$$\mathcal{S}_2 = \mathbf{X}_1 \wedge \mathbf{X}_2 \wedge \mathbf{X}_3 = \lambda_2 I_3, \tag{15.23}$$

where $I_3 = \gamma_1\gamma_2\gamma_3$ is the pseudoscalar for R^3. As before, under the projective transformation given by $\underline{f}_2$, $\mathcal{S}_2$ transforms to $\mathcal{S}_2'$, where

$$\mathcal{S}_2' = \det \underline{f}_2 \mathcal{S}_2. \tag{15.24}$$

Therefore, the ratio of any trivector is invariant under $\underline{f}_2$. To project down into $\mathcal{E}^2$, assuming that $\mathbf{X}_i\gamma_3 = Z_i(1 + \boldsymbol{x}_i)$ under the projective split, we then write

$$
\begin{aligned}
\mathcal{S}_2 I_3{}^{-1} &= \langle \mathbf{X}_1 \mathbf{X}_2 \mathbf{X}_3 I_3{}^{-1} \rangle, \\
&= \langle \mathbf{X}_1 \gamma_3 \gamma_3 \mathbf{X}_2 \mathbf{X}_3 \gamma_3 \gamma_3 I_3{}^{-1} \rangle, \\
&= Z_1 Z_2 Z_3 \langle (1 + \boldsymbol{x}_1)(1 - \boldsymbol{x}_2)(1 + \boldsymbol{x}_3) \gamma_3 I_3{}^{-1} \rangle,
\end{aligned} \tag{15.25}
$$

where the $\boldsymbol{x}_i$ represent vectors in $\mathcal{E}^2$. We can only get a scalar term from the expression within the brackets by calculating the product of a vector, two spatial vectors, and $I_3{}^{-1}$, i.e.,

$$
\begin{aligned}
\mathcal{S}_2 I_3{}^{-1} &= Z_1 Z_2 Z_3 \langle (\boldsymbol{x}_1 \boldsymbol{x}_3 - \boldsymbol{x}_1 \boldsymbol{x}_2 - \boldsymbol{x}_2 \boldsymbol{x}_3) \gamma_3 I_3{}^{-1} \rangle, \\
&= Z_1 Z_2 Z_3 \{ (\boldsymbol{x}_2 - \boldsymbol{x}_1) \wedge (\boldsymbol{x}_3 - \boldsymbol{x}_1) \} I_2{}^{-1}.
\end{aligned} \tag{15.26}
$$

It is therefore clear that we must use multiples of the ratios in our calculations, so that the arbitrary scalars Z_i cancel. In the case of four points in a plane, there are only four possible combinations of $Z_i Z_j Z_k$ and it is not possible to cancel all the Z's by multiplying two ratios of the form $\mathbf{X}_i \wedge \mathbf{X}_j \wedge \mathbf{X}_k$ together. For five coplanar points $\{\mathbf{X}_i\}$, $i = 1, ..., 5$, however, there are several ways of achieving the desired cancelation. For example,

$$
Inv_2 = \frac{(\mathbf{X}_5 \wedge \mathbf{X}_4 \wedge \mathbf{X}_3) I_3^{-1} (\mathbf{X}_5 \wedge \mathbf{X}_2 \wedge \mathbf{X}_1) I_3^{-1}}{(\mathbf{X}_5 \wedge \mathbf{X}_1 \wedge \mathbf{X}_3) I_3^{-1} (\mathbf{X}_5 \wedge \mathbf{X}_2 \wedge \mathbf{X}_4) I_3^{-1}}.
$$

According to Eq. (15.26), we can interpret this ratio in $\mathcal{E}^2$ as

$$
\begin{aligned}
Inv_2 &= \frac{(\boldsymbol{x}_5 - \boldsymbol{x}_4) \wedge (\boldsymbol{x}_5 - \boldsymbol{x}_3) I_2^{-1} (\boldsymbol{x}_5 - \boldsymbol{x}_2) \wedge (\boldsymbol{x}_5 - \boldsymbol{x}_1) I_2^{-1}}{(\boldsymbol{x}_5 - \boldsymbol{x}_1) \wedge (\boldsymbol{x}_5 - \boldsymbol{x}_3) I_2^{-1} (\boldsymbol{x}_5 - \boldsymbol{x}_2) \wedge (\boldsymbol{x}_5 - \boldsymbol{x}_4) I_2^{-1}}, \\
&= \frac{A_{543} A_{521}}{A_{513} A_{524}},
\end{aligned} \tag{15.27}
$$

where $\frac{1}{2} A_{ijk}$ is the area of the triangle defined by the three vertices $\boldsymbol{x}_i, \boldsymbol{x}_j, \boldsymbol{x}_k$. This invariant is regarded as the 2D generalization of the 1D cross-ratio.

15.4.3 3D Generalization of the Cross-Ratio

For general points in $\mathcal{E}^3$, we have seen that we move up one dimension to compute in the 4D space R^4. For this dimension, the point $\boldsymbol{x} = x\sigma_1 + y\sigma_2 + z\sigma_3$ in $\mathcal{E}^3$ is written as $\mathbf{X} = X\gamma_1 + Y\gamma_2 + Z\gamma_3 + W\gamma_4$, where $x = X/W$, $y = Y/W$, $z = Z/W$. As before, a nonlinear projective transformation in $\mathcal{E}^3$ becomes a linear transformation, described by the linear function $\underline{f}_3$ in R^4.

Let us consider 4-vectors in R^4, $\{\mathbf{X}_i\}$, $i = 1, ..., 4$ and form the equation of a 4-vector:

$$\mathcal{S}_3 = \mathbf{X}_1 \wedge \mathbf{X}_2 \wedge \mathbf{X}_3 \wedge \mathbf{X}_4 = \lambda_3 I_4, \tag{15.28}$$

where $I_4 = \gamma_1\gamma_2\gamma_3\gamma_4$ is the pseudoscalar for R^4. As before, $\mathcal{S}_3$ transforms to $\mathcal{S}_3'$ under $\underline{f}_3$:

$$\mathcal{S}_3' = \mathbf{X}_1' \wedge \mathbf{X}_2' \wedge \mathbf{X}_3' \wedge \mathbf{X}_4' = \det \underline{f}_3 \mathcal{S}_3. \tag{15.29}$$

The ratio of any two 4-vectors is therefore invariant under $\underline{f}_3$, and we must take multiples of these ratios to ensure that the arbitrary scale factors W_i cancel. With five general points, we see that there are five possibilities for forming the combinations $W_i W_j W_k W_l$. It is then a simple matter to show that one cannot consider multiples of ratios such that the W factors cancel. It is, however, possible to do this if we have six points. One example of such an invariant might be

$$Inv_3 = \frac{(\mathbf{X}_1 \wedge \mathbf{X}_2 \wedge \mathbf{X}_3 \wedge \mathbf{X}_4) I_4^{-1} (\mathbf{X}_4 \wedge \mathbf{X}_5 \wedge \mathbf{X}_2 \wedge \mathbf{X}_6) I_4^{-1}}{(\mathbf{X}_1 \wedge \mathbf{X}_2 \wedge \mathbf{X}_4 \wedge \mathbf{X}_5) I_4^{-1} (\mathbf{X}_3 \wedge \mathbf{X}_4 \wedge \mathbf{X}_2 \wedge \mathbf{X}_6) I_4^{-1}}. \tag{15.30}$$

Using the arguments of the previous sections, we can now write

$$(\mathbf{X}_1 \wedge \mathbf{X}_2 \wedge \mathbf{X}_3 \wedge \mathbf{X}_4) I_4^{-1} \equiv \\ W_1 W_2 W_3 W_4 \{(\boldsymbol{x}_2 - \boldsymbol{x}_1) \wedge (\boldsymbol{x}_3 - \boldsymbol{x}_1) \wedge (\boldsymbol{x}_4 - \boldsymbol{x}_1)\} I_3^{-1}. \tag{15.31}$$

We can therefore see that the invariant Inv_3 is the 3D equivalent of the 1D cross-ratio and consists of ratios of volumes,

$$Inv_3 = \frac{V_{1234} V_{4526}}{V_{1245} V_{3426}}, \tag{15.32}$$

where V_{ijkl} is the volume of the solid formed by the four vertices $\boldsymbol{x}_i, \boldsymbol{x}_j, \boldsymbol{x}_k, \boldsymbol{x}_l$.

Conventionally, all of these invariants are well known, but we have outlined here a general process which is straightforward and simple for generating projective invariants in any dimension.

15.4.4 Generation of 3D Projective Invariants

Any 3D point may be written in $G_{1,3,0}$ as $\mathbf{X}_n = X_n\gamma_1 + Y_n\gamma_2 + Z_n\gamma_3 + W_n\gamma_4$, and its projected image point written in $G_{3,0,0}$ as $\boldsymbol{x}_n = x_n\sigma_1 + y_n\sigma_2 + z_n\sigma_3$, where $x_n = X_n/W_n$, $y_n = Y_n/W_n$, *and* $z_n = Z_n/W_n$. The 3D projective basis consists of four base points and a fifth point for normalization:

$$X_1 = \begin{pmatrix} 1 \\ 0 \\ 0 \\ 0 \end{pmatrix}, X_2 = \begin{pmatrix} 0 \\ 1 \\ 0 \\ 0 \end{pmatrix}, X_3 = \begin{pmatrix} 0 \\ 0 \\ 1 \\ 0 \end{pmatrix}, X_4 = \begin{pmatrix} 0 \\ 0 \\ 0 \\ 1 \end{pmatrix}, X_5 = \begin{pmatrix} 1 \\ 1 \\ 1 \\ 1 \end{pmatrix}.$$

Any other point $\mathbf{X}_i \in \langle G_{1,3,0} \rangle_1$ can then be expressed by

$$\mathbf{X}_i = X_i \mathbf{X}_1 + Y_i \mathbf{X}_2 + Z_i \mathbf{X}_3 + W_i \mathbf{X}_4, \tag{15.33}$$

with (X_i, Y_i, Z_i, W_i) as homogeneous projective coordinates of $\mathbf{X}_i$ in the base $\{\mathbf{X}_1, \mathbf{X}_2, \mathbf{X}_3, \mathbf{X}_4, \mathbf{X}_5\}$.

The first four base points, projected to the projective plane, can be used as a projective basis of $\langle G_{3,0,0} \rangle_1$ if no three of them are collinear:

$$\boldsymbol{x}_1 = \begin{pmatrix} 1 \\ 0 \\ 0 \end{pmatrix}, \quad \boldsymbol{x}_2 = \begin{pmatrix} 0 \\ 1 \\ 0 \end{pmatrix}, \quad \boldsymbol{x}_3 = \begin{pmatrix} 0 \\ 0 \\ 1 \end{pmatrix}, \quad \boldsymbol{x}_4 = \begin{pmatrix} 1 \\ 1 \\ 1 \end{pmatrix}.$$

Using this basis, we can express, in bracketed notation, the 3D projective coordinates X_n, Y_n, Z_n of any 3D point, as well as its 2D projected coordinates x_n, y_n:

$$\frac{X_n}{W_n} = \frac{[234n][1235]}{[2345][123n]}, \quad \frac{Y_n}{W_n} = \frac{[134n][1235]}{[1345][123n]}, \quad \frac{Z_n}{W_n} = \frac{[124n][1235]}{[1245][123n]}. \tag{15.34}$$

$$\frac{x_n}{w_n} = \frac{[23n][124]}{[234][12n]}, \quad \frac{y_n}{w_n} = \frac{[13n][124]}{[134][12n]}. \tag{15.35}$$

These equations show projective invariant relationships, and they can be used, for example, to compute the position of a moving camera (see Sect. 15.7).

The projective structure and its projection on the 2D image can be expressed according to the following geometric constraint, as presented by Carlsson [49]:

$$\begin{pmatrix} 0 & w_5 Y_5 & -y_5 Z_5 & (y_5 - w_5) W_5 \\ w_5 X_5 & 0 & -x_5 Z_5 & (x_5 - w_5) W_5 \\ 0 & w_6 Y_6 & -y_6 Z_6 & (x_5 - w_5) W_5 \\ 0 & w_6 Y_6 & -y_6 Z_6 & (y_6 - w_6) W_6 \\ w_6 X_6 & 0 & -x_6 Z_6 & (x_6 - w_6) W_6 \\ 0 & w_7 Y_7 & -y_7 Z_7 & (y_7 - w_7) W_7 \\ w_7 X_7 & 0 & -x_7 Z_7 & (x_7 - w_7) W_7 \\ \cdot & \cdot & \cdot & \cdot \\ \cdot & \cdot & \cdot & \cdot \\ \cdot & \cdot & \cdot & \cdot \end{pmatrix} \begin{pmatrix} X_0^{-1} \\ Y_0^{-1} \\ Z_0^{-1} \\ W_0^{-1} \end{pmatrix} = 0, \tag{15.36}$$

where X_0, Y_0, Z_0, W_0 are the coordinates of the view-center point. Since the matrix is of rank < 4, any determinant of four rows becomes a zero. Considering $(X_5, Y_5, Z_5, W_5) = (1, 1, 1, 1)$ as a normalizing point and taking the determinant formed by the first four rows of Eq. (15.36), we get a geometric constraint equation involving six points (see Quan [242]):

$$\begin{aligned}(w_5y_6 - x_5y_6)X_6Z_6 + (x_5y_6 - x_5w6)X_6W_6 + (x_5w_6 - y_5w6)X_6Y_6 +\\ +(y_5x_6 - w_5x6)Y_6Z_6 + (y_5w_6 - y_5x6)Y_6W_6 +\\ +(w_5x_6 - w_5y_6)Z_6W_6 = 0.\end{aligned} \quad (15.37)$$

Carlsson [49] showed that Eq. (15.37) can also be derived using *Plücker–Grassmann relations*, which is then computed as the *Laplace expansion* of the 4×8 rectangular matrix involving the same six points as above:

$$\begin{aligned}&[\boldsymbol{X}_1, \boldsymbol{X}_2, \boldsymbol{X}_3, \boldsymbol{X}_4, \boldsymbol{X}_5, \boldsymbol{X}_5, \boldsymbol{X}_6, \boldsymbol{X}_7] = [\boldsymbol{X}_0, \boldsymbol{X}_1, \boldsymbol{X}_2, \boldsymbol{X}_3]\\ &[\boldsymbol{X}_4, \mathbf{X}_5, \boldsymbol{X}_6, \boldsymbol{X}_7] - [\boldsymbol{X}_0, \boldsymbol{X}_1, \boldsymbol{X}_2, \boldsymbol{X}_4][\boldsymbol{X}_3, \boldsymbol{X}_5, \boldsymbol{X}_6, \boldsymbol{X}_7] +\\ &+[\boldsymbol{X}_0, \boldsymbol{X}_1, \boldsymbol{X}_2, \boldsymbol{X}_5][\boldsymbol{X}_3, \boldsymbol{X}_4, \boldsymbol{X}_6, \boldsymbol{X}_7] - [\boldsymbol{X}_0, \boldsymbol{X}_1, \boldsymbol{X}_2, \boldsymbol{X}_6]\\ &[\boldsymbol{X}_3, \mathbf{X}_4, \boldsymbol{X}_5, \boldsymbol{X}_7] + [\boldsymbol{X}_0, \boldsymbol{X}_1, \boldsymbol{X}_2, \boldsymbol{X}_7][\boldsymbol{X}_3, \boldsymbol{X}_4, \boldsymbol{X}_5, \boldsymbol{X}_6] = 0.\end{aligned} \quad (15.38)$$

By using four functions like Eq. (15.38) in terms of the permutations of six points as indicated by their subindices in the table below,

$\boldsymbol{X}_0$	$\boldsymbol{X}_1$	$\boldsymbol{X}_2$	$\boldsymbol{X}_3$	$\boldsymbol{X}_4$	$\boldsymbol{X}_5$	$\boldsymbol{X}_6$	$\boldsymbol{X}_7$
0	1	5	1	2	3	4	5
0	2	6	1	2	3	4	6
0	3	5	1	2	3	4	5
0	4	6	1	2	3	4	6

we get an expression in which bracketed terms having two identical points vanish:

$$\begin{aligned}&[0152][1345] - [0153][1245] + [0154][1235] = 0,\\ &[0216][2346] - [0236][1246] + [0246][1236] = 0,\\ &[0315][2345] + [0325][1345] + [0345][1235] = 0,\\ &[0416][2346] + [0426][1346] - [0436][1246] = 0.\end{aligned} \quad (15.39)$$

It is easy to prove that the bracketed terms of image points can be written in the form $[\boldsymbol{x}_i\boldsymbol{x}_j\boldsymbol{x}_k] = w_iw_jw_k[K][\boldsymbol{X}_0\boldsymbol{X}_i\boldsymbol{X}_j\boldsymbol{X}_k]$, where [K] is the matrix of the intrinsic parameters [220]. Now, if we substitute in Eq. (15.39) all the brackets which have the point $\boldsymbol{X}_0$ with image points, and if we then organize all the products of brackets as a 4×4 matrix, we end up with the singular matrix

$$\begin{pmatrix} 0 & [125][1345] & [135][1245] & [145][1235]\\ [216][2346] & 0 & [236][1246] & [246][1236]\\ [315][2345] & [325][1345] & 0 & [345][1235]\\ [416][2346] & [426][1346] & [436][1246] & 0 \end{pmatrix}. \quad (15.40)$$

Here, the scalars $w_i w_j w_k[K]$ of each matrix entry cancel each other. Now, after taking the determinant of this matrix and rearranging the terms conveniently, we obtain the following useful *bracket! polynomial*:

$$\begin{aligned}
&[125][346]\Big[1236\Big]\Big[1246\Big]\Big[1345\Big]\Big[2345\Big] - \\
&[126][345]\Big[1235\Big]\Big[1245\Big]\Big[1346\Big]\Big[2346\Big] + \\
&[135][246]\Big[1236\Big]\Big[1245\Big]\Big[1346\Big]\Big[2345\Big] - \\
&[136][245]\Big[1235\Big]\Big[1246\Big]\Big[1345\Big]\Big[2346\Big] + \\
&[145][236]\Big[1235\Big]\Big[1246\Big]\Big[1346\Big]\Big[2345\Big] - \\
&[146][235]\Big[1236\Big]\Big[1245\Big]\Big[1345\Big]\Big[2346\Big] = 0.
\end{aligned} \tag{15.41}$$

Surprisingly, this bracketed expression is exactly the *shape constraint* for six points given by Quan [242],

$$i_1 I_1 + i_2 I_2 + i_3 I_3 + i_4 I_4 + i_5 I_5 + i_6 I_6 = 0, \tag{15.42}$$

where

$$\begin{aligned}
&i_1 = [125][346],\ i_2 = [126][345],\ ...,\ i_6 = [146][235], \\
&I_1 = [1236][1246][1345][2345],\ I_2 = [1235][1245][1346][2346],\ ..., \\
&I_6 = [1236][1245][1345][2346],
\end{aligned}$$

are the relative linear invariants in P^2 and P^3, respectively. Using the shape constraint, we are now ready to generate invariants for different purposes.

Let us illustrate this with an example (see Fig. 15.4). According to the figure, a configuration of six points indicate whether or not the end-effector is grasping properly. To test this situation, we can use an invariant generated from the constraint of Eq. (15.41). In this particular situation, we recognize two planes, thus [1235] = 0 and [2346] = 0. By substituting the six points into Eq. (15.41), we can cancel out some brackets, thereby reducing the equation to

$$\begin{aligned}
&[125][346]\Big[1236\Big]\Big[1246\Big]\Big[1345\Big]\Big[2345\Big] - \\
&-[135][246]\Big[1236\Big]\Big[1245\Big]\Big[1346\Big]\Big[2345\Big] = 0,
\end{aligned} \tag{15.43}$$

$$[125][346]\Big[1246\Big]\Big[1345\Big] - [135][246]\Big[1245\Big]\Big[1346\Big] = 0, \tag{15.44}$$

Fig. 15.4 Action of grasping a box

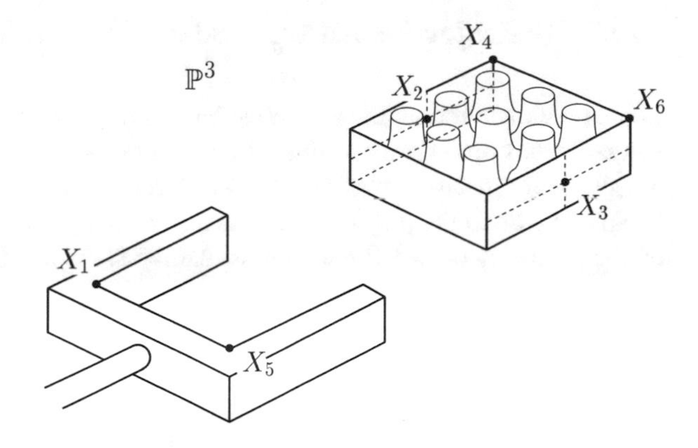

or

$$Inv = \frac{(\mathbf{X}_1 \wedge \mathbf{X}_2 \wedge \boldsymbol{X}_4 \wedge \mathbf{X}_5) I_4^{-1} (\boldsymbol{X}_1 \wedge \mathbf{X}_3 \wedge \mathbf{X}_4 \wedge \mathbf{X}_6) I_4^{-1}}{(\mathbf{X}_1 \wedge \mathbf{X}_2 \wedge \boldsymbol{X}_4 \wedge \mathbf{X}_6) I_4^{-1} (\boldsymbol{X}_1 \wedge \boldsymbol{X}_3 \wedge \mathbf{X}_4 \wedge \mathbf{X}_5) I_4^{-1}},$$
$$= \frac{(\boldsymbol{x}_1 \wedge \boldsymbol{x}_2 \wedge \boldsymbol{x}_5) I_3^{-1} (\boldsymbol{x}_3 \wedge \boldsymbol{x}_4 \wedge \boldsymbol{x}_6) I_3^{-1}}{(\boldsymbol{x}_1 \wedge \boldsymbol{x}_3 \wedge \boldsymbol{x}_5) I_3^{-1} (\boldsymbol{x}_2 \wedge \boldsymbol{x}_4 \wedge \boldsymbol{x}_6) I_3^{-1}}. \tag{15.45}$$

In this equation, any bracket of P^3 after the projective mapping becomes

$$(\mathbf{X}_1 \wedge \mathbf{X}_2 \wedge \mathbf{X}_4 \wedge \mathbf{X}_5) I_4^{-1} \tag{15.46}$$
$$\equiv W_1 W_2 W_4 W_5 \{(\boldsymbol{x}_2 - \boldsymbol{x}_1) \wedge (\boldsymbol{x}_4 - \boldsymbol{x}_1) \wedge (\boldsymbol{x}_5 - \boldsymbol{x}_1)\} I_3^{-1}.$$

The constraint (15.41) ensures that the $W_i W_j W_k W_l$ constants are always canceled. Furthermore, we can interpret the invariant Inv, the equivalent of the 1D cross-ratio, as a ratio of volumes in P^3, and as a ratio of triangle areas in P^2:

$$Inv = \frac{V_{1245} V_{1346}}{V_{1246} V_{1345}} = \frac{A_{125} A_{346}}{A_{135} A_{246}}. \tag{15.47}$$

In other words, we can interpret this invariant in P^3 as the relation of 4-vectors, or volumes, which in turn are built by points lying on a quadric. After they are projected in P^2, they represent an invariant relating areas of triangles encircled by conics.

Then, utilizing this invariant, we can check whether or not the grasper is holding the box correctly. Note that by using the observed 3D points in the image, we can compute this invariant and see if the relation of the triangle areas corresponds well with the parameters for firm grasping. In other words, if the points of the grasper are located at some distance away from the object, the invariant will have a different value that will be the case if the points $\boldsymbol{X}_1$, $\boldsymbol{X}_5$ of the grasper are nearer to the points $\boldsymbol{X}_2$, $\boldsymbol{X}_3$ of the object.

15.5 3D Projective Invariants from Multiple Views

In the previous section, the projective invariant was explained within the context of homogeneous projective coordinates derived from a single image. Since, in general, objects in 3D space are observed from different positions, it would be convenient to be able to extend the projective invariant in terms of the linear constraints imposed by the geometry of two, three, or more cameras.

15.5.1 Projective Invariants Using Two Views

Let us consider a 3D *projective invariant* derived from Eq. (15.41):

$$\boldsymbol{Inv_3} = \frac{[\mathbf{X}_1\mathbf{X}_2\mathbf{X}_3\mathbf{X}_4][\mathbf{X}_4\mathbf{X}_5\mathbf{X}_2\mathbf{X}_6]}{[\mathbf{X}_1\mathbf{X}_2\mathbf{X}_4\mathbf{X}_5][\mathbf{X}_3\mathbf{X}_4\mathbf{X}_2\mathbf{X}_6]}. \tag{15.48}$$

The computation of the bracket

$$[1234] = (\mathbf{X}_1 \wedge \mathbf{X}_2 \wedge \mathbf{X}_3 \wedge \mathbf{X}_4)\boldsymbol{I}_4^{-1} = ((\mathbf{X}_1 \wedge \mathbf{X}_2) \wedge (\mathbf{X}_3 \wedge \mathbf{X}_4))\boldsymbol{I}_4^{-1}$$

of four points from R^4, mapped onto camera images with optical centers $\mathbf{A}_0$ and $\mathbf{B}_0$, suggests the use of a binocular model based on incidence algebra techniques, as discussed in Chap. 12. Defining the lines

$$\begin{aligned}\boldsymbol{L}_{12} &= \mathbf{X}_1 \wedge \mathbf{X}_2 = (\mathbf{A}_0 \wedge \boldsymbol{L}_{12}^A) \cap (\mathbf{B}_0 \wedge \boldsymbol{L}_{12}^B),\\ \boldsymbol{L}_{34} &= \mathbf{X}_3 \wedge \mathbf{X}_4 = (\mathbf{A}_0 \wedge \boldsymbol{L}_{34}^A) \cap (\mathbf{B}_0 \wedge \boldsymbol{L}_{34}^B),\end{aligned}$$

where lines $\boldsymbol{L}_{ij}^A$ and $\boldsymbol{L}_{ij}^B$ are mappings of the line $\boldsymbol{L}_{ij}$ onto the two image planes, results in the expression

$$[1234] = [\mathbf{A}_0\mathbf{B}_0\mathbf{A}'_{1234}\mathbf{B}'_{1234}]. \tag{15.49}$$

Here, $\mathbf{A}'_{1234}$ and $\mathbf{B}'_{1234}$ are the points of intersection of the lines $\boldsymbol{L}_{12}^A$ and $\boldsymbol{L}_{34}^A$ or $\boldsymbol{L}_{12}^B$ and $\boldsymbol{L}_{34}^B$, respectively. These points, lying on the image planes, can be expanded using the mappings of three points $\mathbf{X}_i$, say, $\mathbf{X}_1$, $\mathbf{X}_2$, $\mathbf{X}_3$, to the image planes. In other words, considering $\mathbf{A}_j$ and $\mathbf{B}_j$, $j = 1, 2, 3$, as projective bases, we can expand the vectors

$$\begin{aligned}\mathbf{A}'_{1234} &= \alpha_{1234,1}\mathbf{A}_1 + \alpha_{1234,2}\mathbf{A}_2 + \alpha_{1234,3}\mathbf{A}_3,\\ \mathbf{B}'_{1234} &= \beta_{1234,1}\mathbf{B}_1 + \beta_{1234,2}\mathbf{B}_2 + \beta_{1234,3}\mathbf{B}_3.\end{aligned}$$

Then, using Eq. (12.97) from Chap. 12, we can express

$$[1234] = \sum_{i,j=1}^{3} \tilde{F}_{ij}\alpha_{1234,i}\beta_{1234,j} = \boldsymbol{\alpha}_{1234}^T \tilde{F}\boldsymbol{\beta}_{1234}, \tag{15.50}$$

where $\tilde{F}$ is the fundamental matrix given in terms of the projective basis embedded in R^4, and $\boldsymbol{\alpha}_{1234} = (\alpha_{1234,1}, \alpha_{1234,2}, \alpha_{1234,3})$ and $\boldsymbol{\beta}_{1234} = (\beta_{1234,1}, \beta_{1234,2}, \beta_{1234,3})$ are corresponding points.

The ratio

$$\boldsymbol{Inv_{3F}} = \frac{(\boldsymbol{\alpha}^T{}_{1234}\tilde{F}\boldsymbol{\beta}_{1234})(\boldsymbol{\alpha}^T{}_{4526}\tilde{F}\boldsymbol{\beta}_{4526})}{(\boldsymbol{\alpha}^T{}_{1245}\tilde{F}\boldsymbol{\beta}_{1245})(\boldsymbol{\alpha}^T{}_{3426}\tilde{F}\boldsymbol{\beta}_{3426})} \tag{15.51}$$

is therefore seen to be an invariant using the views of two cameras [48]. Note that Eq. (15.51) is invariant for whichever values of the γ_4 components of the vectors $\mathbf{A}_i, \mathbf{B}_i, \mathbf{X}_i$, etc., are chosen. If we attempt to express the invariant of Eq. (15.51) in terms of what we actually observe, we may be tempted to express the invariant in terms of the homogeneous Cartesian image coordinates $\boldsymbol{a}_i's$, $\boldsymbol{b}_i's$ and the fundamental matrix F calculated from these image coordinates. In order to avoid this, it is necessary to transfer the computations of Eq. (15.51) carried out in R^4 to R^3. Thus, if we define $\tilde{F}$ by

$$\tilde{F}_{kl} = (\mathbf{A}_k \cdot \gamma_4)(\mathbf{B}_l \cdot \gamma_4)F_{kl} \tag{15.52}$$

and consider the relationships $\boldsymbol{\alpha}_{ij} = \frac{\mathbf{A}'_i \cdot \gamma_4}{\mathbf{A}_j \cdot \gamma_4}\boldsymbol{a}_{ij}$ and $\boldsymbol{\beta}_{ij} = \frac{\mathbf{B}'_i \cdot \gamma_4}{\mathbf{B}_j \cdot \gamma_4}\boldsymbol{b}_{ij}$, we can claim

$$\alpha_{ik}\tilde{F}_{kl}\beta_{il} = (\mathbf{A}'_i \cdot \gamma_4)(\mathbf{B}'_i \cdot \gamma_4)\boldsymbol{a}_{ik}F_{kl}\boldsymbol{b}_{il}. \tag{15.53}$$

If F is subsequently estimated by some method, then $\tilde{F}$ as defined in Eq. (15.52) will also act as a *fundamental matrix* or *bilinear constraint* in R^4. Now, let us look again at the invariant $\boldsymbol{Inv_{3F}}$. As we demonstrated earlier, we can write the invariant as

$$\boldsymbol{Inv_{3F}} = \frac{(\boldsymbol{a}^T{}_{1234}F\boldsymbol{b}_{1234})(\boldsymbol{a}^T{}_{4526}F\boldsymbol{b}_{4526})\phi_{1234}\phi_{4526}}{(\boldsymbol{a}^T{}_{1245}F\boldsymbol{b}_{1245})(\boldsymbol{a}^T{}_{3426}F\boldsymbol{b}_{3426})\phi_{1245}\phi_{3426}}, \tag{15.54}$$

where $\phi_{pqrs} = (\mathbf{A}'_{pqrs} \cdot \gamma_4)(\mathbf{B}'_{pqrs} \cdot \gamma_4)$. Therefore, we see that the ratio of the terms $\boldsymbol{a}^T F\boldsymbol{b}$, which resembles the expression for the invariant in R^4 but uses only the observed coordinates and the estimated fundamental matrix, will not be an invariant. Instead, we need to include the factors ϕ_{1234}, etc., which do not cancel. They are formed as follows (see [22]): Since $\boldsymbol{a}'_3$, $\boldsymbol{a}'_4$, and $\boldsymbol{a}'_{1234}$ are collinear, we can write $\boldsymbol{a}'_{1234} = \mu_{1234}\boldsymbol{a}'_4 + (1 - \mu_{1234})\boldsymbol{a}'_3$. Then, by expressing $\mathbf{A}'_{1234}$ as the intersection of the line joining $\mathbf{A}'_1$ and $\mathbf{A}'_2$ with the plane through $\mathbf{A}_0$, $\mathbf{A}'_3$, $\mathbf{A}'_4$, we can use the projective split and equate terms, so that

$$\frac{(\mathbf{A}'_{1234}\cdot\gamma_4)(\mathbf{A}'_{4526}\cdot\gamma_4)}{(\mathbf{A}'_{3426}\cdot\gamma_4)(\mathbf{A}'_{1245}\cdot\gamma_4)} = \frac{\mu_{1245}(\mu_{3426}-1)}{\mu_{4526}(\mu_{1234}-1)}. \tag{15.55}$$

Note that the values of μ are readily obtainable from the images. The factors $\mathbf{B}'_{pqrs}\cdot\gamma_4$ are found in a similar way, so that if $\boldsymbol{b}'_{1234} = \lambda_{1234}\boldsymbol{b}'_4 + (1-\lambda_{1234})\boldsymbol{b}'_3$, etc., the overall expression for the invariant becomes

$$\boldsymbol{Inv}_{3F} = \frac{(\boldsymbol{a}^T_{1234}F\boldsymbol{b}_{1234})(\boldsymbol{a}^T_{4526}F\boldsymbol{b}_{4526})}{(\boldsymbol{a}^T_{1245}F\boldsymbol{b}_{1245})(\boldsymbol{a}^T_{3426}F\boldsymbol{b}_{3426})}\cdot\frac{\mu_{1245}(\mu_{3426}-1)}{\mu_{4526}(\mu_{1234}-1)}\frac{\lambda_{1245}(\lambda_{3426}-1)}{\lambda_{4526}(\lambda_{1234}-1)}. \tag{15.56}$$

In conclusion, given the coordinates of a set of six corresponding points in two image planes, where these six points are projections of arbitrary world points in general position, we can form 3D projective invariants, provided we have some estimate of F.

15.5.2 Projective Invariant of Points Using Three Uncalibrated Cameras

The technique used to form the 3D projective invariants for two views can be straightforwardly extended to give expressions for invariants of three views. Considering four world points $\mathbf{X}_1, \mathbf{X}_2, \mathbf{X}_3, \mathbf{X}_4$ or two lines $\mathbf{X}_1\wedge\mathbf{X}_2$ and $\mathbf{X}_3\wedge\mathbf{X}_4$ projected onto three camera planes, we can write

$$\mathbf{X}_1\wedge\mathbf{X}_2 = (\mathbf{A}_0\wedge\boldsymbol{L}^A_{12}) \cap (\mathbf{B}_0\wedge\boldsymbol{L}^B_{12}),$$
$$\mathbf{X}_3\wedge\mathbf{X}_4 = (\mathbf{A}_0\wedge\boldsymbol{L}^A_{34}) \cap (\mathbf{C}_0\wedge\boldsymbol{L}^C_{34}).$$

Once again, we can combine the above expressions so that they give an equation for the 4-vector $\mathbf{X}_1\wedge\mathbf{X}_2\wedge\mathbf{X}_3\wedge\mathbf{X}_4$,

$$\begin{aligned}\mathbf{X}_1\wedge\mathbf{X}_2\wedge\mathbf{X}_3\wedge\mathbf{X}_4 &= ((\mathbf{A}_0\wedge\boldsymbol{L}^A_{12}) \cap (\mathbf{B}_0\wedge\boldsymbol{L}^B_{12}))\wedge((\mathbf{A}_0\wedge\boldsymbol{L}^A_{34}) \cap (\mathbf{C}_0\wedge\boldsymbol{L}^C_{34})),\\ &= (\mathbf{A}_0\wedge\mathbf{A}_{1234})\wedge((\mathbf{B}_0\wedge\boldsymbol{L}^B_{12}) \cap (\mathbf{C}_0\wedge\boldsymbol{L}^C_{34})).\end{aligned} \tag{15.57}$$

Then, by rewriting the lines $\boldsymbol{L}^B_{12}$ and $\boldsymbol{L}^C_{34}$ in terms of the line coordinates, we get $\boldsymbol{L}^B_{12} = \sum_{j=1}^{3} l^B_{12,j}\boldsymbol{L}^B_j$ and $\boldsymbol{L}^C_{34} = \sum_{j=1}^{3} l^C_{34,j}\boldsymbol{L}^C_j$. As has been shown in Chap. 12, the components of the *trifocal tensor* (which takes the place of the fundamental matrix for three views) can be written in geometric algebra as

$$\tilde{T}_{ijk} = [(\mathbf{A}_0\wedge\mathbf{A}_i)\wedge((\mathbf{B}_0\wedge\boldsymbol{L}^B_j) \cap (\mathbf{C}_0\wedge\boldsymbol{L}^C_k))], \tag{15.58}$$

so that by using Eq. (15.57) we can derive

$$[\mathbf{X}_1 \wedge \mathbf{X}_2 \wedge \mathbf{X}_3 \wedge \mathbf{X}_4] = \sum_{i,j,k=1}^{3} \tilde{T}_{ijk} \alpha_{1234,i} l^B_{12,j} l^C_{34,k} = \tilde{T}(\boldsymbol{\alpha}_{1234}, \boldsymbol{L}^B_{12}, \boldsymbol{L}^C_{34}). \quad (15.59)$$

The invariant $\boldsymbol{Inv}_3$ can then be expressed as

$$\boldsymbol{Inv}_{3T} = \frac{\tilde{T}(\boldsymbol{\alpha}_{1234}, \boldsymbol{L}^B_{12}, \boldsymbol{L}^C_{34}) \tilde{T}(\boldsymbol{\alpha}_{4526}, \boldsymbol{L}^B_{25}, \boldsymbol{L}^C_{26})}{\tilde{T}(\boldsymbol{\alpha}_{1245}, \boldsymbol{L}^B_{12}, \boldsymbol{L}^C_{45}) \tilde{T}(\boldsymbol{\alpha}_{3426}, \boldsymbol{L}^B_{34}, \boldsymbol{L}^C_{26})}. \quad (15.60)$$

Note that the factorization must be done so that the same line factorizations occur in both the numerator and denominator. We have thus developed an expression for invariants in three views, that is a direct extension of the expression for invariants using two views. In calculating the above invariant from observed quantities, we note, as before, that some correction factors will be necessary: Eq. (15.60) is given above in terms of R^4 quantities. Fortunately, this correction is quite straightforward. By extrapolating from the results of the previous section, we simply consider the $\alpha's$ terms in Eq. (15.60) as unobservable quantities, and conversely the line terms, such as $\boldsymbol{L}^B_{12}, \boldsymbol{L}^C_{34}$, are indeed observed quantities. As a result, the expression must be modified, by using to some extent the coefficients computed in the previous section. Thus, for the unique four combinations of three cameras their invariant equations can be expressed as

$$\boldsymbol{Inv}_{3T} = \frac{T(\boldsymbol{a}_{1234}, \boldsymbol{l}^B_{12}, \boldsymbol{l}^C_{34}) T(\boldsymbol{a}_{4526}, \boldsymbol{l}^B_{25}, \boldsymbol{l}^C_{26})\, \mu_{1245}(\mu_{3426} - 1)}{T(\boldsymbol{a}_{1245}, \boldsymbol{l}^B_{12}, \boldsymbol{l}^C_{45}) T(\boldsymbol{a}_{3426}, \boldsymbol{l}^B_{34}, \boldsymbol{l}^C_{26})\, \mu_{4526}(\mu_{1234} - 1)}. \quad (15.61)$$

15.5.3 Comparison of the Projective Invariants

This section presents simulations for the computation of invariants (implemented in Maple) using synthetic data, as well as computations using real images.

The computation of the bilinearity matrix F and the trilinearity focal tensor T was done using a linear method. We believe that for test purposes, this method is reliable. Four different sets of six points $S_i = \{\boldsymbol{X}_{i1}, \boldsymbol{X}_{i2}, \boldsymbol{X}_{i3}, \boldsymbol{X}_{i4}, \boldsymbol{X}_{i5}, \boldsymbol{X}_{i6}\}$, where $i = 1, ..., 4$, were considered in the simulation, and the only three possible invariants were computed for each set $\{I_{1,i}, I_{2,i}, I_{3,i}\}$. Then, the invariants of each set were represented as 3D vectors ($\mathbf{v}_i = [I_{1,i}, I_{2,i}, I_{3,i}]^T$). For the first group of images, we computed four of these vectors that corresponded to four different sets of six points, using two images for the F case and three images for the T case. For the second group of images, we computed the same four vectors, but we used two new images for the F case or three new images for the T case. The comparison of the invariants was

Invariants using F				Invariants using T			
0.000	0.590	0.670	0.460	0.000	0.590	0.310	0.630
	0	0.515	0.68		0	0.63	0.338
		0.59	0			0.134	0.67
			0.69				0.29

0.063	0.650	0.750	0.643	0.044	0.590	0.326	0.640
	0.67	0.78	0.687		0	0.63	0.376
		0.86	0.145			0.192	0.67
			0.531				0.389

0.148	0.600	0.920	0.724	0.031	0.100	0.352	0.660
	0.60	0.96	0.755		0.031	0.337	0.67
		0.71	0.97			0.31	0.67
			0.596				0.518

0.900	0.838	0.690	0.960	0.000	0.640	0.452	0.700
	0.276	0.693	0.527		0.063	0.77	0.545
		0.98	0.59			0.321	0.63
			0.663				0.643

Fig. 15.5 Distance matrices showing performance of invariants after increasing Gaussian noise σ (σ=0.005, 0.015, 0.025, and 0.04)

done using Euclidean distances of the vectors $d(\mathbf{v}_i, \mathbf{v}_j) = (1 - |\frac{\mathbf{v}_i \cdot \mathbf{v}_j}{||\mathbf{v}_i|| ||\mathbf{v}_j||}|)^{\frac{1}{2}}$, which is the same method used in [124].

Since in $d(\mathbf{v}_i, \mathbf{v}_j)$ we normalize the vectors $\mathbf{v}_i$ and $\mathbf{v}_j$, the distance $d(\mathbf{v}_i, \mathbf{v}_j)$ for any of them lies between 0 and 1, and the distance does not vary when $\mathbf{v}_i$ or $\mathbf{v}_j$ is multiplied by a nonzero constant. Figure 15.5 shows a comparison table where each (i, j)th entry represents the distance $d(\mathbf{v}_i, \mathbf{v}_j)$ between the invariants of set S_i, which are the points extracted from the first group of images, and those of set S_j, the points from the second group of images. In the ideal case, the diagonal of the distance matrices should be zero, which means that the values of the computed invariants should remain constant regardless of which group of images they were used for. The entries off the diagonal are comparisons for vectors composed of different coordinates ($\mathbf{v}_i = [I_{1,i}, I_{2,i}, I_{3,i}]^T$) and thus are not parallel. Accordingly, these entries should be larger than zero, and if they are very large, the value of $d(\mathbf{v}_i, \mathbf{v}_j)$ should be approximately 1. The figure clearly shows that the performance of the invariants based on trilinearities is much better than that of the invariants based on bilinearities, since the diagonal values for T are in general closer to zero than is the case for F, and since T entries off the diagonal are in general bigger values than is the case for F entries.

In the case of real images, we used a sequence of images taken by a moving robot equipped with a binocular head. Figure 15.6 shows these images for the left and right eye, respectively. We took image couples, one from the left and one from the right, for the invariants using F, and two from the left and one from the right for the invariants

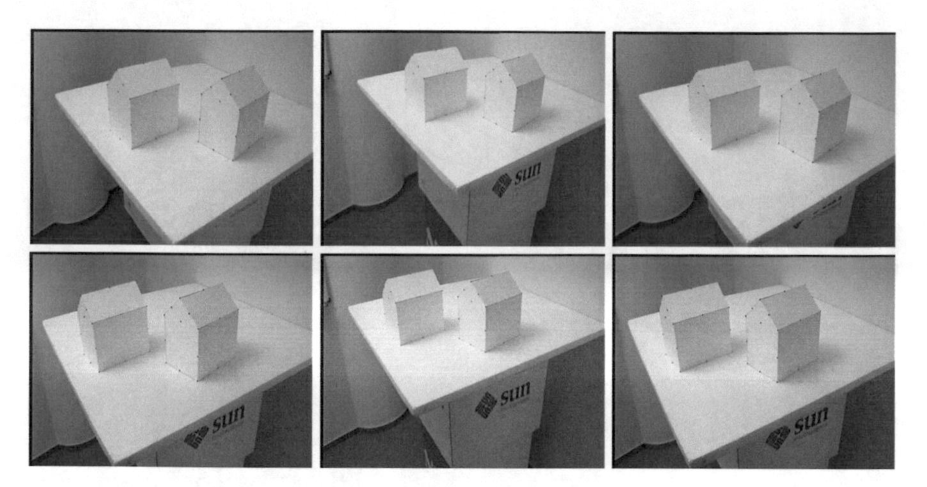

Fig. 15.6 Image sequence taken during navigation by the binocular head of a mobile robot (left camera images are shown in upper row; right camera images in lower row)

Fig. 15.7 Distance matrices show the performance of the computed invariants using bilinearities (top) and trilinearities (bottom) for the image sequence

using F

0.04	0.79	0.646	0.130	0.679	0.89
	0.023	0.2535	0.278	0.268	0.89
		0.0167	0.723	0.606	0.862
			0.039	0.808	0.91
				0.039	0.808
					0.039

using T

0.021	0.779	0.346	0.930	0.759	0.81
	0.016	0.305	0.378	0.780	0.823
		0.003	0.83	0.678	0.97
			0.02	0.908	0.811
				0.008	0.791
					0.01

using T. From the image, we selected thirty-eight points semiautomatically, and from these we chose six sets of points. In each set, the points are in general position. Three invariants of each set were computed, and comparison tables were constructed in the same manner as for the tables of the previous experiment (see Fig. 15.7).

The data show once again that computing the invariants using a trilinear approach is much more robust than using a bilinear approach, a result which is also borne out in theory.

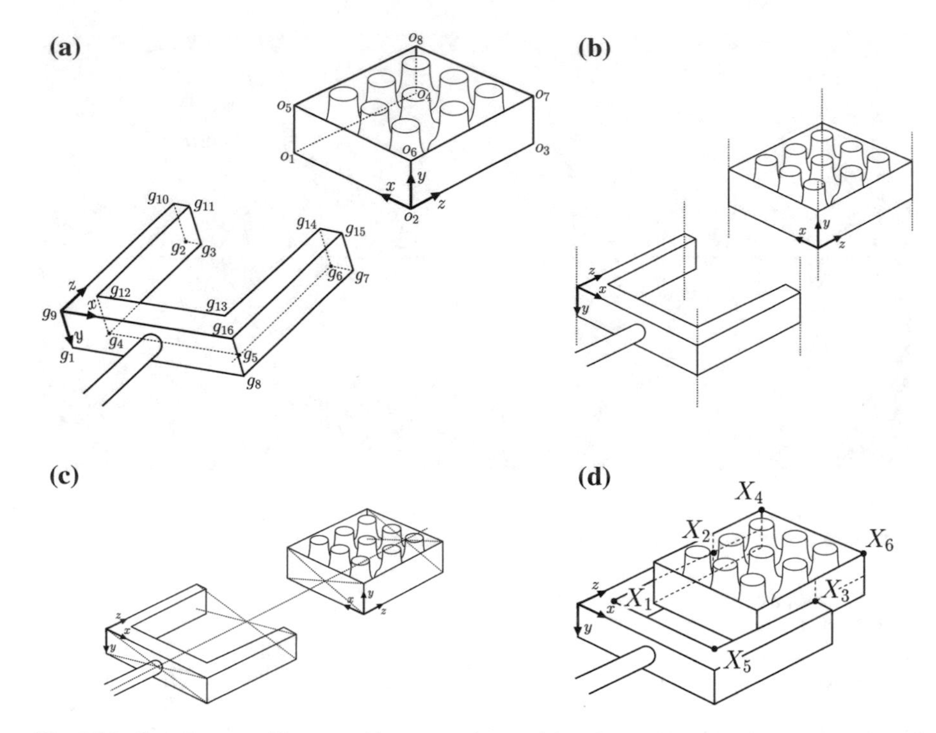

Fig. 15.8 Grasping an object: **a** arbitrary starting position; **b** parallel orienting; **c** centering; **d** grasping and holding

15.6 Visually Guided Grasping

This section presents a practical use of projective invariants using three views. The results will show that despite a certain noise sensitivity in the projective invariants, they can be used for various tasks regardless of camera calibration or coordinate system.

We will apply simple geometric rules using meet or join operations, invariants, and points at infinity to the task of grasping, as depicted in Fig. 15.8a. The grasping procedure uses only image points and consists basically of four steps.

15.6.1 Parallel Orienting

Let us assume that the 3D points of Fig. 15.8 are observed by three cameras A, B, C. The mapped points in the three cameras are $\{\boldsymbol{o}_{A_i}\}$, $\{g_{A_i}\}$, $\{\boldsymbol{o}_{B_i}\}$, $\{g_{B_i}\}$, and $\{\boldsymbol{o}_{C_i}\}$, $\{g_{C_i}\}$. In the projective 3D space P^3, the three points at infinity $\boldsymbol{V}_x$, $\boldsymbol{V}_y$, $\boldsymbol{V}_z$ for the orthogonal corners of the object can be computed as the meet of two parallel lines. Similarly, in the images planes, the *points at infinity* $\boldsymbol{v}_x$, $\boldsymbol{v}_y$, $\boldsymbol{v}_z$ are also computed as the meet of the two projected parallel lines:

$$\begin{aligned} \boldsymbol{V}_x &= (\boldsymbol{O}_1 \wedge \boldsymbol{O}_2) \cap (\boldsymbol{O}_5 \wedge \boldsymbol{O}_6) \rightarrow \boldsymbol{v}_{j_x} = (\boldsymbol{o}_{j_1} \wedge \boldsymbol{o}_{j_2}) \cap (\boldsymbol{o}_{j_5} \wedge \boldsymbol{o}_{j_6}), \\ \boldsymbol{V}_y &= (\boldsymbol{O}_1 \wedge \boldsymbol{O}_5) \cap (\boldsymbol{O}_2 \wedge \boldsymbol{O}_6) \rightarrow \boldsymbol{v}_{j_y} = (\boldsymbol{o}_{j_1} \wedge \boldsymbol{o}_{j_5}) \cap (\boldsymbol{o}_{j_2} \wedge \boldsymbol{o}_{j_6}), \\ \boldsymbol{V}_x &= (\boldsymbol{O}_1 \wedge \boldsymbol{O}_4) \cap (\boldsymbol{O}_2 \wedge \boldsymbol{O}_3) \rightarrow \boldsymbol{v}_{j_z} = (\boldsymbol{o}_{j_1} \wedge \boldsymbol{o}_{j_4}) \cap (\boldsymbol{o}_{j_2} \wedge \boldsymbol{o}_{j_3}), \end{aligned} \quad (15.62)$$

where $j \in \{A, B, C\}$. The parallelism in the projective space P^3 can be checked in two ways. First, if the orthogonal edges of the grasper are parallel with the edges of the object, then

$$(\boldsymbol{G}_1 \wedge \boldsymbol{G}_8) \wedge \boldsymbol{V}_x = 0, \quad (\boldsymbol{G}_1 \wedge \boldsymbol{G}_9) \wedge \boldsymbol{V}_y = 0, \quad (\boldsymbol{G}_1 \wedge \boldsymbol{G}_2) \wedge \boldsymbol{V}_z = 0. \quad (15.63)$$

In this case, the conditions of Eq. (15.63), using the points obtained from a single camera, can be expressed as

$$[g_{i_1} g_{i_8} \boldsymbol{v}_{i_x}] = 0, \quad [g_{i_1} g_{i_9} \boldsymbol{v}_{i_y}] = 0, \quad [g_{i_1} g_{i_2} \boldsymbol{v}_{i_z}] = 0. \quad (15.64)$$

The second way to check the parallelism in the projective space P^3 is to note whether the perpendicular planes of the grasper and those of the object are parallel. If they are, then

$$[\boldsymbol{G}_1 \boldsymbol{G}_8 \boldsymbol{O}_1 \boldsymbol{O}_2] = 0, \quad [\boldsymbol{G}_{15} \boldsymbol{G}_{16} \boldsymbol{O}_5 \boldsymbol{O}_8] = 0, \quad [\boldsymbol{G}_{12} \boldsymbol{G}_{13} \boldsymbol{O}_3 \boldsymbol{O}_4] = 0. \quad (15.65)$$

In this case, the conditions of Eq. (15.65) can be expressed in terms of image coordinates by using either the points obtained from two cameras (the bilinear constraint) or those obtained from three cameras (the trifocal tensor):

$$\begin{aligned} \boldsymbol{x}^T_{j_{g_1 g_8 o_1 o_2}} F_{ij} \boldsymbol{x}_{i_{g_1 g_8 o_1 o_2}} &= 0, \\ \boldsymbol{x}^T_{j_{g_{15} g_{16} o_5 o_8}} F_{ij} \boldsymbol{x}_{i_{g_{15} g_{16} o_5 o_8}} &= 0, \\ \boldsymbol{x}^T_{j_{g_{12} g_{13} o_3 o_4}} F_{ij} \boldsymbol{x}_{i_{g_{12} g_{13} o_3 o_4}} &= 0, \end{aligned} \quad (15.66)$$

$$\begin{aligned} T_{ijk} \boldsymbol{x}_{i_{g_1 g_8 o_1 o_2}} \boldsymbol{l}_{j_{g_1 g_8}} \boldsymbol{l}_{k_{o_1 o_2}} &= 0, \\ T_{ijk} \boldsymbol{x}_{i_{g_{15} g_{16} o_5 o_8}} \boldsymbol{l}_{j_{g_{15} g_{16}}} \boldsymbol{l}_{k_{o_5 o_8}} &= 0, \\ T_{ijk} \boldsymbol{x}_{i_{g_{12} g_{13} o_3 o_4}} \boldsymbol{l}_{j_{g_{12} g_{13}}} \boldsymbol{l}_{k_{o_3 o_4}} &= 0. \end{aligned} \quad (15.67)$$

If the trinocular geometry is known, it is always more accurate to use Eq. (15.67).

15.6.2 Centering

After an initial movement the grasper should be parallel to and centered in front of the object (see Fig. 15.8b). The center points of the grasper and object can be computed as follows:

$$\boldsymbol{C}_o = (\boldsymbol{O}_1 \wedge \boldsymbol{O}_6) \cap (\boldsymbol{O}_2 \wedge \boldsymbol{O}_5), \quad \boldsymbol{C}_g = (\boldsymbol{G}_1 \wedge \boldsymbol{G}_{16}) \cap (\boldsymbol{G}_8 \wedge \boldsymbol{G}_9). \tag{15.68}$$

We can then check whether the line crossing through these center points eventually encounters the point at infinity $\boldsymbol{V}_z$, which is the intersecting point of the parallel lines $\boldsymbol{O}_{j_1} \wedge \boldsymbol{O}_{j_4}$ and $\boldsymbol{O}_{j_2} \wedge \boldsymbol{O}_{j_3}$. For that, we use the constraint which posits that a point is true if it lies on a line such that

$$\boldsymbol{C}_o \wedge \boldsymbol{C}_g \wedge \boldsymbol{V}_z = 0. \tag{15.69}$$

This equation, computed using the image points of a single camera, is given by

$$[\boldsymbol{c}_{i_o} \boldsymbol{c}_{i_g} \boldsymbol{v}_{i_z}] = 0. \tag{15.70}$$

15.6.3 Grasping

We can evaluate the exactitude of grasping when the plane of the grasper touches the plane of the object. This can be done by checking the following coplanar plane condition:

$$[\boldsymbol{C}_o \boldsymbol{C}_g \boldsymbol{o}_1 \boldsymbol{o}_2] = 0. \tag{15.71}$$

Since we want to use image points, we can compute this bracket straightforwardly by using the points obtained from either two or three cameras, employing, respectively, either the bilinear or the trilinear constraint:

$$\begin{aligned} \boldsymbol{x}^T_{j_{c_o c_g o_1 o_2}} F_{ij} \boldsymbol{x}_{i_{c_o c_g o_1 o_2}} &= 0, \\ T_{ijk} \boldsymbol{x}_{i_{c_o c_g o_1 o_2}} \boldsymbol{l}_{j_{c_o c_g}} \boldsymbol{l}_{k_{o_1 o_2}} &= 0. \end{aligned} \tag{15.72}$$

If the epipolar or trinocular geometry is known, it is always more accurate to use Eq. (15.72).

15.6.4 Holding the Object

The final step is to hold the object correctly (see Fig. 15.8d). This can be checked using the invariant in terms of the trifocal tensor given by Eq. (15.61). In this particular problem, an example of a perfect condition would be if the invariant had an approximate value of $\frac{3}{4}$, which would then change to perhaps $\frac{6}{8}$ or $\frac{5}{6}$ when the grasper is distanced a bit from the control point $\boldsymbol{X}_2$, $\boldsymbol{X}_3$. Note, too, that the invariant elegantly relates volumes, indicating a particular relationship between the points of the grasper and those of the object.

15.7 Camera Self-localization

We will now use Eq. (15.34) to compute the 3D coordinates for a moving, uncalibrated camera. For this problem, we first select as a projective basis five fixed points in 3D space, $\boldsymbol{X}_1$, $\boldsymbol{X}_2$, $\boldsymbol{X}_3$, $\boldsymbol{X}_4$, $\boldsymbol{X}_5$, and we consider the unknown point $\boldsymbol{X}_6$ to be the optical center of the moving camera (see Fig. 15.9). Assuming that the camera does not move on a plane, and the projection of the optical center $\boldsymbol{X}_6$ of the first camera position should correspond to the epipole in any of the subsequent views.

We can now compute the moving optical center using points from either two cameras,

$$I_x^F = \frac{X_6}{W_6} = \frac{(\boldsymbol{\delta}^T{}_{2346}\boldsymbol{F}\epsilon_{2346})(\boldsymbol{\delta}^T{}_{1235}\boldsymbol{F}\epsilon_{1235})}{(\boldsymbol{\delta}^T{}_{2345}\boldsymbol{F}\epsilon_{2345})(\boldsymbol{\delta}^T{}_{1236}\boldsymbol{F}\epsilon_{1236})}\frac{\mu_{2345}\lambda_{2345}\mu_{1236}\lambda_{1236}}{\mu_{2346}\lambda_{2346}\mu_{1235}\lambda_{1235}}, \tag{15.73}$$

or three cameras,

$$\begin{aligned} I_x^T &= \frac{X_6}{W_6}, \\ &= \frac{(\boldsymbol{T}_{ijk}^{ABC}\alpha_{2346,i}l^B{}_{23,j}l^C{}_{46,k})(\boldsymbol{T}_{mnp}^{ABC}\alpha_{1235,m}l^B{}_{12,n}l^C{}_{35,p})}{(\boldsymbol{T}_{qrs}^{ABC}\alpha_{2345,q}l^B{}_{23,r}l^C{}_{45,s})(\boldsymbol{T}_{tuv}^{ABC}\alpha_{1236,t}l^B{}_{12,u}l^C{}_{36,v})}\frac{\mu_{2345}\mu_{1236}}{\mu_{2346}\mu_{1235}}. \end{aligned} \tag{15.74}$$

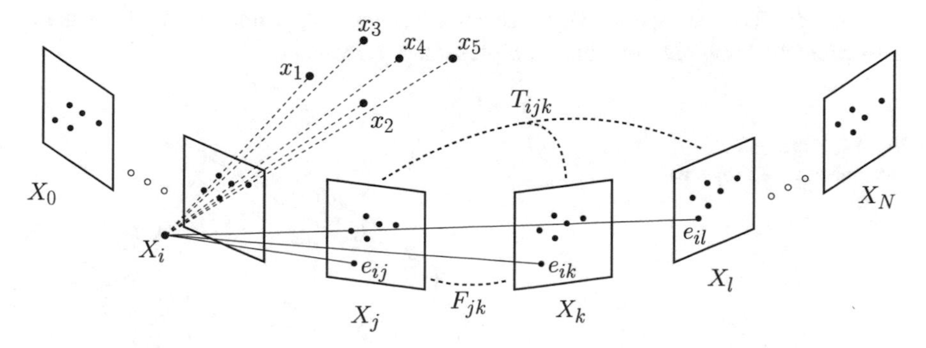

Fig. 15.9 Computing the view-centers of a moving camera

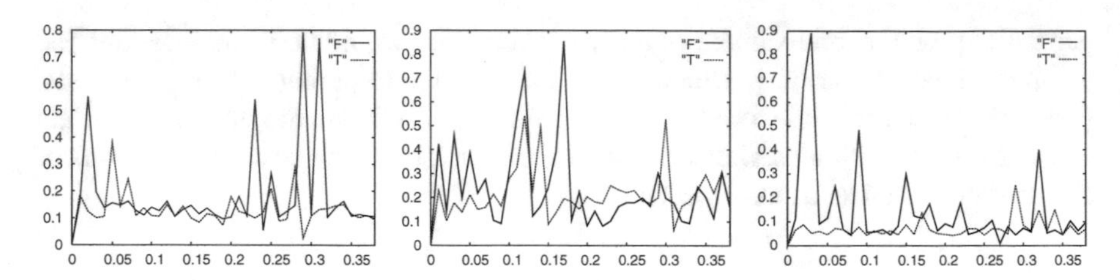

Fig. 15.10 Computing performance of any three view-centers using F (higher spikes) and T (lower spikes); range of additive noise, 0–0.4 pixels

Similarly, by permuting the six points, as in Eq. (15.35), we compute I_y^F, I_y^T and I_z^F, I_z^T. The compensating coefficients for the invariants I_y and I_z vary due to the permuted points. We also simulated the computation of the invariants by increasing noise. Figure 15.10 shows the deviation of the true optical center for three consecutive positions of a moving camera, using two views and three views.

The figure demonstrates that trinocular computation renders more accurate results than binocular computation. The Euclidean coordinates of the optical centers are calculated by applying a transformation, which relates the projective basis to its a priori Euclidean basis.

15.8 Projective Depth

In a geometric sense *projective depth* can be defined as the relation between the distance from the view-center of a 3D point $\boldsymbol{X}_i$ and the focal distance f, as depicted in Fig. 15.11.

We can derive projective depth from a projective mapping of 3D points. According to the pinhole model explained in Chap. 12, the coordinates of any point in the image plane are obtained from the projection of the 3D point to the three optical planes ϕ_A^1, ϕ_A^2, ϕ_A^3. They are spanned by a trivector basis $\gamma_i, \gamma_j, \gamma_k$ and the coefficients t_{ij}. This projective mapping in a matrix representation reads as

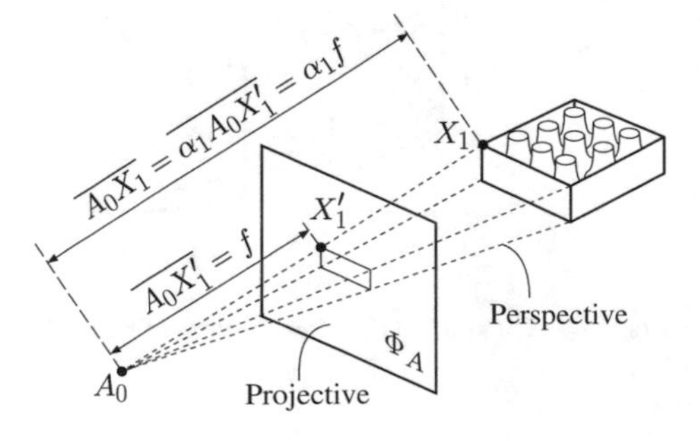

Fig. 15.11 Geometric Interpretation of Projective Depth

$$\lambda \boldsymbol{x} = \begin{bmatrix} x \\ y \\ 1 \end{bmatrix} = \begin{bmatrix} \phi_A^1 \\ \phi_A^2 \\ \phi_A^3 \end{bmatrix} \boldsymbol{X} = \begin{bmatrix} t_{11} & t_{12} & t_{13} & t_{14} \\ t_{21} & t_{22} & t_{23} & t_{24} \\ t_{31} & t_{32} & t_{33} & t_{34} \end{bmatrix} \begin{bmatrix} X \\ Y \\ Z \\ 1 \end{bmatrix},$$

$$= \begin{bmatrix} f & 0 & 0 \\ 0 & f & 0 \\ 0 & 0 & 1 \end{bmatrix} \begin{bmatrix} r_{11} & r_{12} & r_{13} & t_x \\ r_{21} & r_{22} & r_{23} & t_y \\ r_{31} & r_{32} & r_{33} & t_z \\ 0 & 0 & 0 & 1 \end{bmatrix} \begin{bmatrix} X \\ Y \\ Z \\ 1 \end{bmatrix}, \tag{15.75}$$

where the projective scale factor is called λ. Note that the projective mapping is further expressed in terms of a f, rotation, and translation components. Let us attach the world coordinates to the view-center of the camera. The resultant projective mapping becomes

$$\lambda \boldsymbol{x} = \begin{bmatrix} f & 0 & 0 & 0 \\ 0 & f & 0 & 0 \\ 0 & 0 & 1 & 0 \end{bmatrix} \begin{bmatrix} X \\ Y \\ Z \\ 1 \end{bmatrix} \equiv P\boldsymbol{X}. \tag{15.76}$$

We can then straightforwardly compute

$$\lambda = Z. \tag{15.77}$$

The method for computing the projective depth ($\equiv \lambda$) of a 3D point appears simple using invariant theory, i.e., using Eq. (15.34). For this computation, we select a basis system, taking four 3D points in general position $\boldsymbol{X}_1$, $\boldsymbol{X}_2$, $\boldsymbol{X}_3$, $\boldsymbol{X}_5$, and the optical center of camera at the new position as the four-point $\boldsymbol{X}_4$, and $\boldsymbol{X}_6$ as the 3D point which has to be reconstructed. This process is shown in Fig. 15.12.

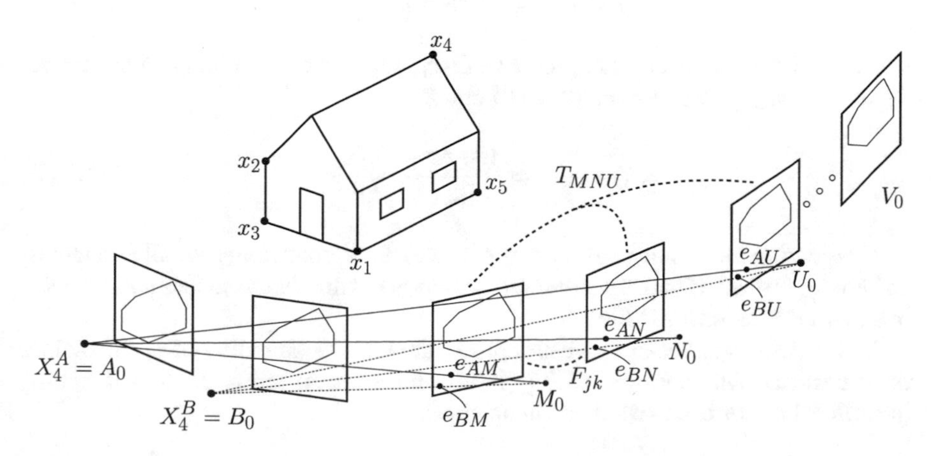

Fig. 15.12 Computing the projective depths of n cameras

Since we are using mapped points, we consider the *epipole* (mapping of the current view-center) to be the fourth point and the mapped sixth point to be the point with unknown depth. The other mapped basis points remain constant during the procedure.

According to Eq. (15.34), the tensor-based expression for the computation of the third coordinate, or projective depth, of a point $\boldsymbol{X}_j$ $(= \boldsymbol{X}_6)$ is given by

$$\lambda_j = \frac{Z_j}{W_j} = \frac{T(\boldsymbol{a}_{124j}, \boldsymbol{l}_{12}^B, \boldsymbol{l}_{4j}^C)T(\boldsymbol{a}_{1235}, \boldsymbol{l}_{12}^B, \boldsymbol{l}_{35}^C)}{T(\boldsymbol{a}_{1245}, \boldsymbol{l}_{12}^B, \boldsymbol{l}_{45}^C)T(\boldsymbol{a}_{123j}, \boldsymbol{l}_{12}^B, \boldsymbol{l}_{3j}^C)} \cdot \frac{\mu_{1245}\mu_{123j}}{\mu_{124j}\mu_{1235}}. \tag{15.78}$$

In this way, we can successively compute the projective depths λ_{ij} of the j-points relating to the i-camera. We will use λ_{ij} in Sect. 15.9, in which we employ the join-image concept and singular value decomposition (SVD) for *singular value decomposition* 3D reconstruction.

Since this type of invariant can also be expressed in terms of the quadrifocal tensor [178], we are also able to compute projective depth based on four cameras.

15.9 Shape and Motion

The orthographic and paraperspective *factorization method for shape and motion* using the affine camera model was developed by Tomasi, Kanade, and Poelman [237, 296]. This method works for cameras viewing small and distant scenes and thus for all scale factors of projective depth $\lambda_{ij} = 1$. In the case of perspective images, the scale factors λ_{ij} are unknown. According to Triggs [300], all λ_{ij} satisfy a set of consistency reconstruction equations of the so-called $join - image$. One way to compute λ_{ij} is by using the epipolar constraint. If we use a matrix representation, this is given by

$$F_{ik}\lambda_{ij}\boldsymbol{x}_{ij} = \boldsymbol{e}_{ik} \wedge \lambda_{kj}\boldsymbol{x}_{kj}, \tag{15.79}$$

which, after computing an inner product with $\boldsymbol{e}_{ik} \wedge x_{kj}$, gives the relation of projective depths for the j-point between camera i and k:

$$\lambda'_{kj} = \frac{\lambda_{kj}}{\lambda_{ij}} = \frac{(\boldsymbol{e}_{ik} \wedge \boldsymbol{x}_{kj})F_{ik}\boldsymbol{x}_{ij}}{||\boldsymbol{e}_{ik} \wedge \boldsymbol{x}_{kj}||^2}. \tag{15.80}$$

Considering the i-camera as a reference, we can normalize λ_{kj} for all k-cameras and use λ'_{kj} instead. If that is not the case, we can normalize between neighbor images in a chained relationship [300].

In Sect. 15.8, we presented a better procedure for the computing of λ_{ij} involving three cameras. An extension of Eq. (15.80), however, in terms of the trifocal or quadrifocal tensor is awkward and unpractical.

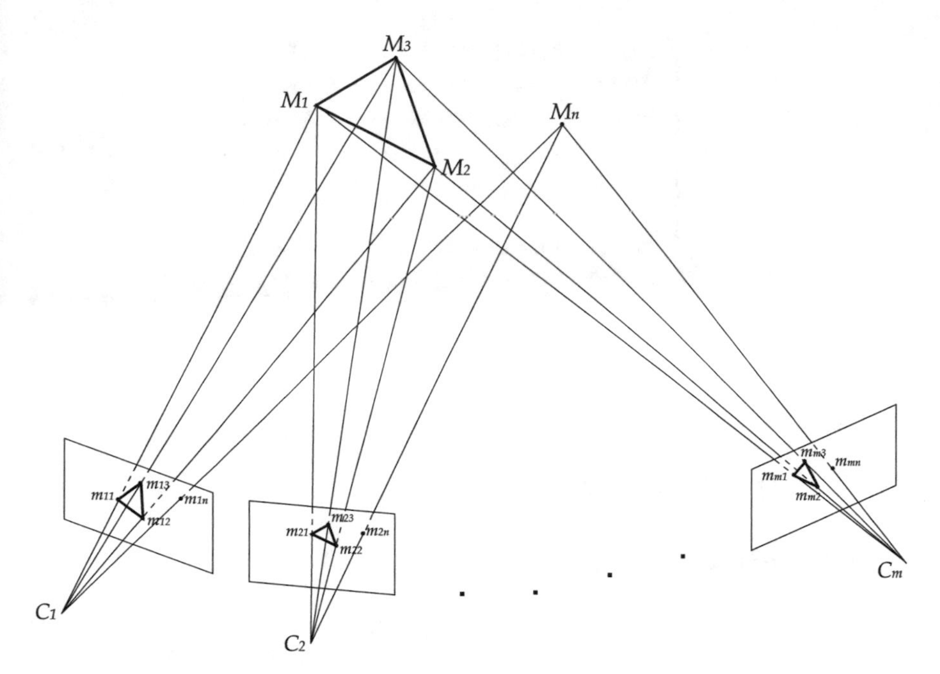

Fig. 15.13 Geometry of the join-image

15.9.1 The Join-Image

The *Join-image* $\mathcal{J}$ is nothing more than the intersections of optical rays and planes with points or lines in 3D projective space, as depicted in Fig. 15.13. The interrelated geometry can be linearly expressed by the fundamental matrix and trifocal and quadrifocal tensors. The reader will find more information about these linear constraints in Chap. 12.

In order to take into account the interrelated geometry, the *projective reconstruction* procedure should bring together all the data of the individual images in a geometrically coherent manner. We do this by considering the points $\boldsymbol{X}_j$ for each i-camera,

$$\lambda_{ij}\boldsymbol{x}_{ij} = P_i\boldsymbol{X}_j, \tag{15.81}$$

as the i-row points of a matrix of rank 4. For m cameras and n points, the $3m \times n$ matrix $\mathcal{J}$ of the join-image is given by

$$\mathcal{J} = \begin{pmatrix} \lambda_{11}\boldsymbol{x}_{11} & \lambda_{12}\boldsymbol{x}_{12} & \lambda_{13}\boldsymbol{x}_{13} & \ldots & \lambda_{1n}\boldsymbol{x}_{1n} \\ \lambda_{21}\boldsymbol{x}_{21} & \lambda_{22}\boldsymbol{x}_{22} & \lambda_{23}\boldsymbol{x}_{23} & \ldots & \lambda_{2n}\boldsymbol{x}_{2n} \\ \lambda_{31}\boldsymbol{x}_{31} & \lambda_{32}\boldsymbol{x}_{32} & \lambda_{33}\boldsymbol{x}_{33} & \ldots & \lambda_{3n}\boldsymbol{x}_{3n} \\ . & . & . & \ldots & . \\ . & . & . & \ldots & . \\ . & . & . & \ldots & . \\ \lambda_{m1}\boldsymbol{x}_{m1} & \lambda_{m2}\boldsymbol{x}_{m2} & \lambda_{m3}\boldsymbol{x}_{m3} & \ldots & \lambda_{mn}\boldsymbol{x}_{mn} \end{pmatrix}. \tag{15.82}$$

For the affine reconstruction procedure, the matrix is of rank 3. The matrix $\mathcal{J}$ of the join-image is therefore amenable to a singular value decomposition for the computation of the shape and motion [237, 296].

15.9.2 The SVD Method

The application of SVD to $\mathcal{J}$ gives

$$\mathcal{J}_{3m\times n} = U_{3m\times r} S_{r\times r} V^T_{n\times r}, \tag{15.83}$$

where the columns of matrix $V^T_{n\times r}$ and $U_{3m\times r}$ constitute the orthonormal base for the input (cokernel) and output (range) spaces of $\mathcal{J}$. In order to get a decomposition in motion and shape of the projected point structure, $S_{r\times r}$ can be absorbed into both matrices, $V^T_{n\times r}$ and $U_{3m\times r}$, as follows:

$$\mathcal{J}_{3m\times n} = (U_{3m\times r} S^{\frac{1}{2}}_{r\times r})(S^{\frac{1}{2}}_{r\times r} V^T_{n\times r}) = \begin{pmatrix} P_1 \\ P_2 \\ P_3 \\ . \\ . \\ . \\ P_m \end{pmatrix}_{3m\times 4} (\boldsymbol{X}_1 \boldsymbol{X}_2 \boldsymbol{X}_3 \ldots \boldsymbol{X}_n)_{4\times n}. \tag{15.84}$$

Using this method to divide $S_{r\times r}$ is not unique. Since the rank of $\mathcal{J}$ is 4, we should use the first four biggest singular values for $S_{r\times r}$. The matrices P_i correspond to the projective mappings or *motion* from the projective space to the individual images, and $\boldsymbol{X}_j$ represents the point structure or *shape*. We can test our approach by using a simulation program written in Maple. Using the method described in Sect. 15.8, we first compute the projective depth of the points of a wire house observed with nine cameras, and we then use SVD to obtain the house's shape and motion. The reconstructed house, after the Euclidean readjustment for the presentation, is shown in Fig. 15.14.

We note that the reconstruction preserves the original form of the model quite well.

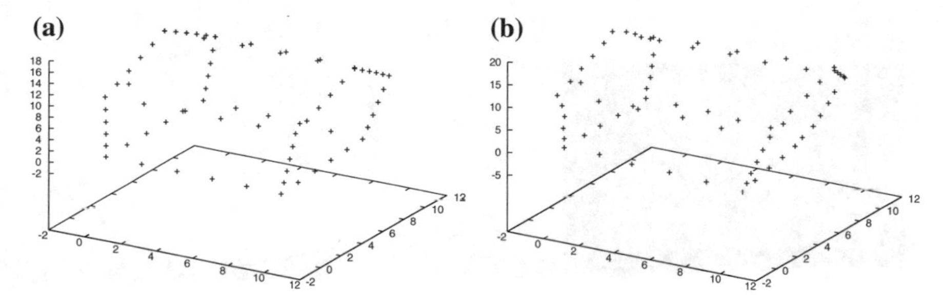

Fig. 15.14 Reconstructed house using **a** noise-free observations and **b** noisy observations

In the following section, we will show how to improve the shape of the reconstructed model by using the geometric expressions $\cap$ (meet) and $\wedge$ (join) from the algebra of incidence along with particular tensor-based invariants.

15.9.3 Completion of the 3D Shape Using Invariants

Projective structure can be improved in one of two ways: (1) by adding points on the images, expanding the join-image, and then applying the SVD procedure; or (2) after the reconstruction is done, by computing new or occluded 3D space points. Both approaches can use, on the one hand, geometric inference rules based on symmetries, or on the other, concrete knowledge about the object. Using three real views of a similar model house with its rightmost lower corner missing (see Fig. 15.15b), we computed in each image the virtual image point of this 3D point. Then we reconstructed the scene, as shown in Fig. 15.15c. We also tried using geometric incidence operations to complete the house, employing space points as depicted in Fig. 15.15d. The figures show that creating points in the images yields a better reconstruction of the occluded point. Note that in the reconstructed image, we transformed the projective shape into a Euclidean one for the presentation of the results. We also used lines to connect the reconstructed points but only so as to make the form of the house visible. Similarly, we used the same procedures to reconstruct the house using nine images; see Fig. 15.16a–d.

The figure shows that the resulting reconstructed point is virtually the same in both cases, which allows us to conclude that for a limited number of views the join-image procedure is preferable, but for the case of several images, an extension of the point structure in the 3D space is preferable.

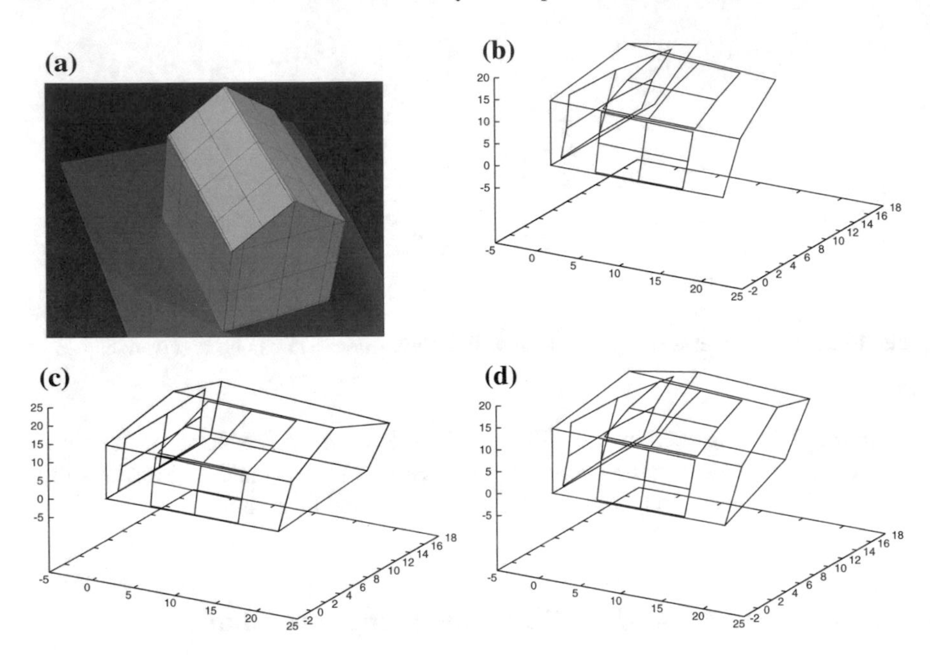

Fig. 15.15 3D reconstruction using three images: **a** one of the three images; **b** reconstructed incomplete house using three images; **c** extending the join-image; **d** completing in the 3D space

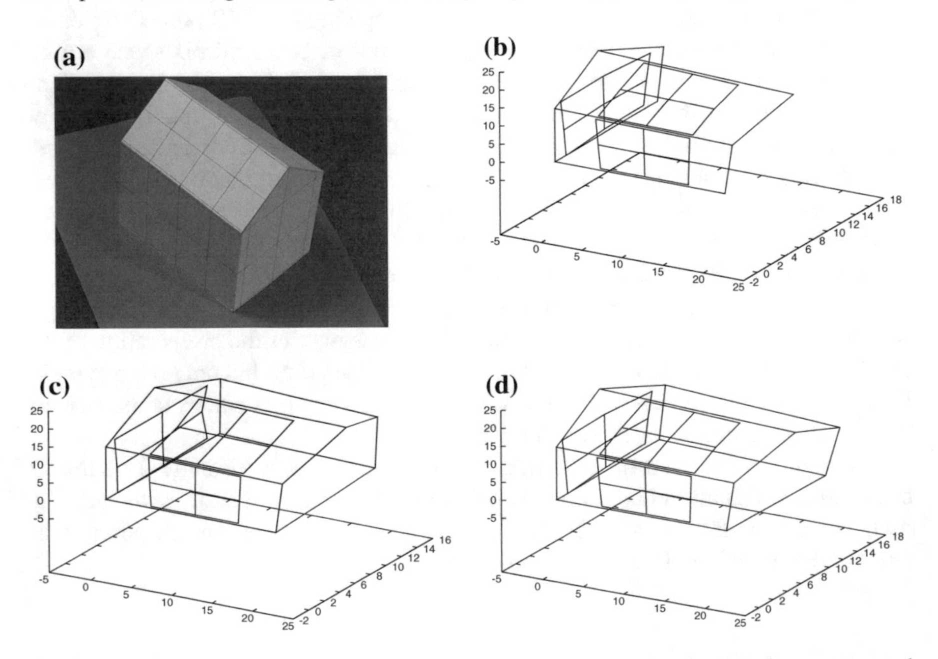

Fig. 15.16 3D reconstruction using nine images: **a** one of the nine images; **b** reconstructed incomplete house using nine images; **c** extending the join-image; **d** completing in the 3D space

15.10 Omnidirectional Vision Landmark Identification Using Projective Invariants

In Chap. 12, we explain the projective invariants using omnidirectional vision. In this section, we present a robust technique for landmark identification using the projective and permutation p^2-invariants.

A visual landmark ($\check{V}$) is a set of visual sublandmarks (v) in an image frame, where a sublandmark is a quintuple formed with five coplanar points in $\mathbb{P}^2$ [60] which are in general position. Each sublandmark is represented by a point-permutation invariant vector of points in general position. The landmark identification consists of two distinct phases: the learning and recognition phases.

15.10.1 Learning Phase

In the learning phase, potential landmarks are extracted and analyzed, and if they fulfill some constraints, they are stored in the memory. The extraction of features is performed using the Harris corner detector. Once the features have been extracted, we apply the algorithm of the previous section to project the features onto the sphere. With the features on the sphere, we are able to calculate their projective and p^2-*invariants*; using the features, we also start the formulation of legal quintuples, that is, quintuples of points in $\mathbb{P}^2$ that are in general position. These legal quintuples represent sublandmarks candidates; a set of validated sublandmarks is used to build a landmark. The necessary validations to consider a sublandmark as legal are explained next.

Collinearity test. Although collinearity is preserved under perspective transformations, quasi-collinearity is not. Thus, three quasi-collinear points in one image could be collinear in another one; therefore, this fact has to be taken into account.

Now, since the bracket (12.27) is not a reliable indicator for near singularity (i.e., quasi-collinearity) [107], we define instead, as in [36], the matrix having as columns the scalars of three vectors ($v_0, v_1, v_2 \in \mathcal{G}_{4,1}$):

$$M = \begin{pmatrix} \frac{v_0 \cdot e_1}{v_0 \cdot e_3} & \frac{v_1 \cdot e_1}{v_1 \cdot e_3} & \frac{v_2 \cdot e_1}{v_2 \cdot e_3} \\ \frac{v_0 \cdot e_2}{v_0 \cdot e_3} & \frac{v_1 \cdot e_2}{v_1 \cdot e_3} & \frac{v_2 \cdot e_2}{v_2 \cdot e_3} \\ 1 & 1 & 1 \end{pmatrix}. \tag{15.85}$$

The closeness of its smallest singular value to zero measures how collinear the three points are. The algorithm uses a threshold t_c obtained experimentally. If the smallest singular value is less than the threshold, the triplet is rejected.

Coplanarity test. The quintuples that passed the collinearity test are examined to verify the coplanarity of their constituent points. Let v and v' be the same quintuple observed in different image frames. If they are coplanar, then they must have the same projective and permutation p^2-*invariants*, thus, $\gamma = \gamma'$, where γ and γ' are the calculated quintuples in two frames. The last condition cannot be used since numerical inaccuracies may be present due to noise or small errors in corner detection. Instead we use

$$|v - v'| \leq t_o \,, \tag{15.86}$$

where t_o is a threshold value.

Convex hull test. A projective transformation that exists between two images preserves the convex hull of a coplanar point set [123]. A pair of matched quintuples puts five points in correspondence. A not sufficient but necessary condition for the match to be correct is that the two convex hulls must be in correspondence. This invariance has the following conditions [211]:

1. Corresponding points must lie either on or inside the convex hull.
2. The number of points on the convex hull must be the same.
3. For points lying on the convex hull, neighboring relations are preserved.

The convex hull can be used to correct problems with the point correspondences. If the convex hull detects two false *point pair* correspondences, then they are mutually exchanged. If more than two errors are detected, then the quintuple is rejected.

Visual landmark construction. To minimize the effect of numerical instabilities during the recognition time, we select the outliers [60]. To identify those outliers, we first calculate the mean vector $\bar{v}$ [11] as follows:

$$\bar{v} = \frac{1}{n}\sum_{i=0}^{n-1} v_i \,, \tag{15.87}$$

where the vector candidate v_i represent the ith sublandmark candidate. Then the vectors are selected based on the distance of the candidate v_i with respect to the mean vector $\bar{v}$:

$$d_i = |v_i - \bar{v}| \,. \tag{15.88}$$

The vectors with d_i greater than twice the mean of all d_i's are selected as a sublandmark of a landmark $\check{V}_j$. If such vector does not exist, then no landmark is added.

15.10.2 Recognition Phase

In the recognition phase, during navigation, a process similar to the learning phase is applied to find legal quintuples. Next we compare the detected quintuple to the stored sublandmarks. The comparison between projective and point-permutation invariant vectors is done using the Euclidean distance of the vectors.

The matched sublandmark is used as a guide for the detection of other sublandmarks that will strengthen our landmark hypothesis. Once the hypothesis is verified, the landmark is considered as recognized.

15.10.3 Omnidirectional Vision and Invariants for Robot Navigation

Experiments have been carried out in an indoor corridor environment using a mobile robot equipped with an omnidirectional vision system (Fig. 15.17). In this section, we show the results obtained with these experiments. The threshold values of the experiments have been set experimentally, in accordance with the landmark identification results. The values of the threshold in the experiments were $t_c = 0.0316$ and $t_o = 0.8$.

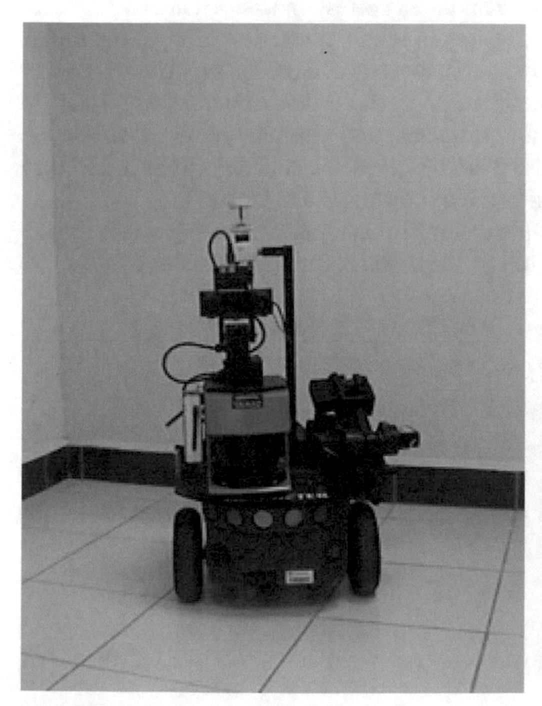

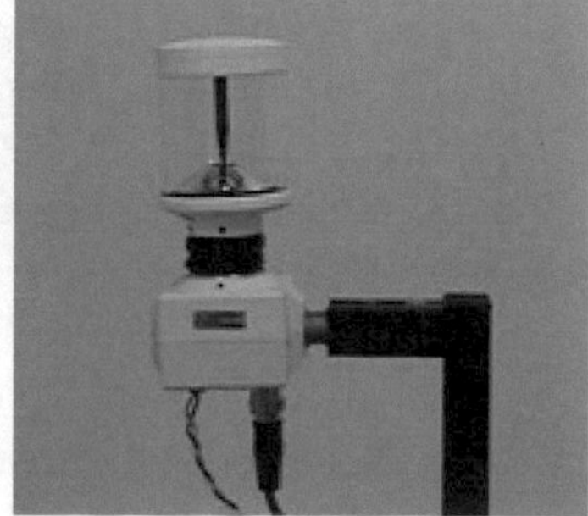

Fig. 15.17 Mobile robot and omnidirectional vision system

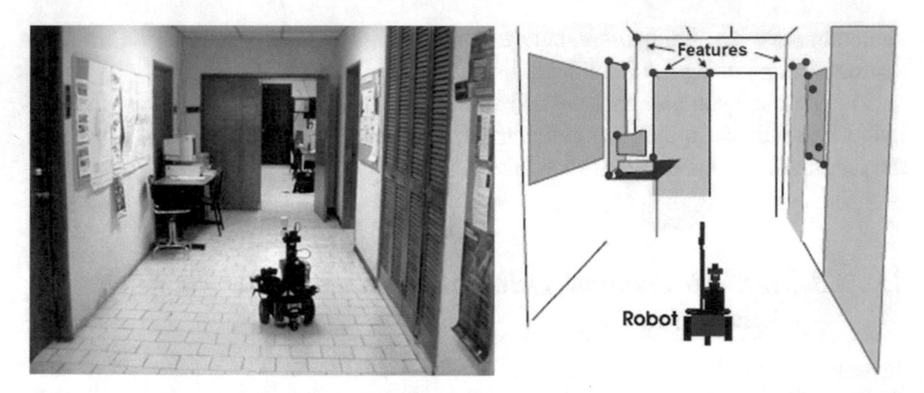

Fig. 15.18 Robot scenario and simulation of features extraction

15.10.4 Learning Phase

In this phase, we first extract the features using the Harris corner detector from two different frames F and F', and also we find their correspondences. Once we have the features and their correspondences, we begin with the creation of legal quintuples (sublandmark candidates) (Fig. 15.18). First, we project those features onto the sphere (Sect. 12.9.1). Then, for each quintuple we apply the collinearity test (Sect. 15.10.1). If it passes the test we calculate the projective and permutation p^2-*invariants* of the quintuple (Sect. 12.9.1), and with them we check the coplanarity (Sect. 15.10.1). Next, we apply the convex hull test to the quintuples (Sect. 15.10.1).

In Fig. 15.19, we show an example of four sublandmarks forming a visual landmark. The projective and permutation p^2-*invariants* of these sublandmarks are

$$
\begin{aligned}
v_0 &= 2.0035e_1 + 2.0038e_2 + 2.0000e_3 + 2.0039e_4 + 2.0041e_5,\\
v_1 &= 2.7340e_1 + 2.3999e_2 + 2.5529e_3 + 2.7517e_4 + 2.5122e_5,\\
v_2 &= 2.5096e_1 + 2.5040e_2 + 2.7769e_3 + 2.7752e_4 + 2.3379e_5,\\
v_3 &= 2.7767e_1 + 2.7461e_2 + 2.0196e_3 + 2.0059e_4 + 2.0040e_5.
\end{aligned}
$$

Finally, with the quintuples that pass such a test, we construct a visual landmark, as explained in Sect. 15.10.1.

15.10.5 Recognition Phase

Once that the robot has passed the learning phase, it navigates through the corridor and applies the same procedure for the extraction of legal quintuples to find sublandmarks. The extracted sublandmarks are compared with the stored ones.

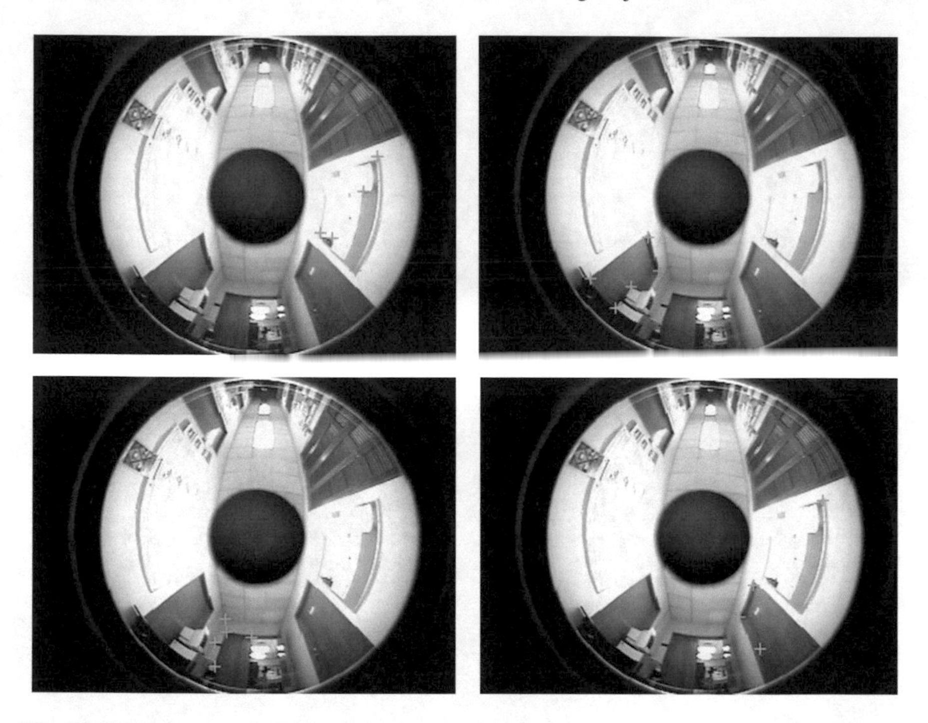

Fig. 15.19 Sequence of images acquired in the learning phase. The crosses show the extracted features to build their respective p^2-*invariants*. The images (from top to bottom) represent the sublandmark $v_0, \ldots, v_3$, respectively

In Fig. 15.20, we show an example of the four sublandmarks recognized. The values of those sublandmarks are:

$$v'_0 = 2.0804e_1 + 2.0892e_2 + 2.0069e_3 + 2.0096e_4 + 2.0002e_5,$$
$$v'_1 = 2.5744e_1 + 2.0611e_2 + 2.3071e_3 + 2.5270e_4 + 2.2623e_5,$$
$$v'_2 = 2.5771e_1 + 2.4412e_2 + 2.3406e_3 + 2.7491e_4 + 2.7916e_5,$$
$$v'_3 = 2.7266e_1 + 2.66063_2 + 2.0407e_3 + 2.0138e_4 + 2.0071e_5.$$

15.10.6 Quantitative Results

To evaluate the proposed framework quantitatively, we have set up an experiment that consists of 30 navigation trials to measure the sublandmark recognition accuracy. The results are shown in Table 15.1, where the abbreviations mean (SL) sublandmark, (CR) correct recognition, (MR) missed recognition (when one sublandmark was

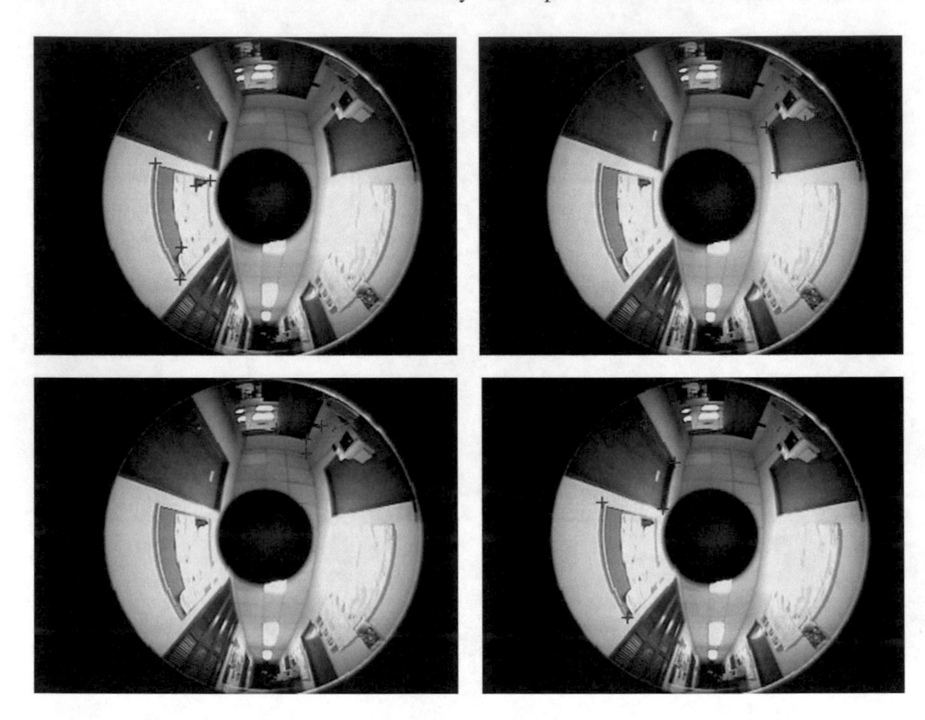

Fig. 15.20 Sequence of images acquired in the recognition phase. Observe that the images were taken with the robot navigating in the opposite direction to the learning phase. The red crosses show the extracted features to build their respective p^2-*invariants*. The images (from top to bottom) represent the sublandmark $v'_0, \ldots, v'_3$, respectively

Table 15.1 Landmark recognition results

	CR		MR		FP		FN	
SL	#	%	#	%	#	%	#	%
v_0	25	83.3	2	6.67	2	6.67	3	10
v_1	28	93.3	1	3.33	0	0	1	3.33
v_2	27	90	1	3.33	1	3.33	2	6.67
v_3	29	96.6	0	0	1	3.33	1	3.33
Mean	27.25	90.833	1	3.33	1	3.33	1.75	5.83

reported as another one), (FP) false positives (when a sublandmark was reported erroneously), (FN) false negatives (when a sublandmark was missed). As we can observe, the algorithm achieves a 90.833% of correct landmarks recognition.

15.11 Conclusions

In this chapter, we presented the applications of projective invariants using n-uncalibrated cameras. We first showed how projective invariants can be used to compute the view-center of a moving camera. We also developed geometric rules for a task of visually guided grasping. Our purpose here was to motivate the reader to apply invariants from a geometric point of view and takes advantage of the power of the algebra of incidence.

Next, using a trifocal tensor-based projective invariant, we developed a method for the computation of projective depths, which in turn were used to initiate an SVD procedure for the projective reconstruction of shape and motion. Further, we applied the rules of incidence algebra to complete the reconstruction, for example, in the critical case of occluded points.

The main contribution of this chapter is that we were able to demonstrate a simple way to unify current approaches for the computation and application of projective invariants using n-uncalibrated cameras. We formulated Pascal's theorem as a projective invariant of conics. This invariant was then used to solve the camera calibration problem. Simple illustrations, such as camera self-localization and visually guided grasping, show the potential of the use of projective invariants, points at infinity, and geometric rules developed using the algebra of incidence. The most important application given is the projective reconstruction of shape and motion. Remarkably, the use of trifocal, tensor-based, projective invariants allows for the computation of the projective depths required for the initialization of the SVD procedure to compute shape and motion. In this way, we were able to link the use of n-views-based projective invariants with SVD projective reconstruction methods.

The bracket algebra involved in the computation of the presented projective invariants is noise sensitive. We believe that to compute as we did is a promising approach for high-level geometric reasoning, especially if better ways to cope with the noise can be found. One promising approach, which should be explored further, might be to express the bracket equations as polynomials and then to look for their Gröbner basis. This helps to calculate the exact number of real solutions that satisfy certain geometric constraints arising from physical conditions.

This chapter also presents advances in theoretical issues concerning omnidirectional vision. We define our computational model without referencing any coordinate system, using only the geometric relationships between its geometric objects (i.e., the point, plane, and sphere). We have shown that we can define our model in a more general and simpler way as in the case using matrices or tensors. This allows an easier implementation in more complex applications. As an interesting application, we presented how to recover the projective invariants from a catadioptric image using the inverse projection of the UM and how to use them to construct the p^2-*invariants*. As a real application of this technique, we present a visual landmark identification application using projective and permutation p^2-*invariants* for autonomous robot navigation.

Chapter 16
Geometric Algebra Tensor Voting, Hough Transform, Voting and Perception Using Conformal Geometric Algebra

The first section presents a non-iterative algorithm that combines the power of expression of geometric algebra with the robustness of Tensor Voting to find the correspondences between two sets of 3D points with an underlying rigid transformation. In addition, we present experiments of the conformal geometric algebra voting scheme using synthetic and real images.

16.1 Problem Formulation

Using the geometric algebra of 3D space $G_{3,0,0}$, the rigid motion of a 3D point $\boldsymbol{x} = xe_1 + ye_2 + ze_3$ can be formulated as

$$\boldsymbol{x}' = \widetilde{\boldsymbol{R}}\boldsymbol{x}\boldsymbol{R} + \boldsymbol{t}, \tag{16.1}$$

where $\boldsymbol{R}$ is a rotor as described in the previous section, and $\boldsymbol{t} = t_x e_1 + t_y e_2 + t_z e_3$. For simplicity, we will represent $\boldsymbol{R}$ as rotor of the form

$$\boldsymbol{R} = q_0 + q_x e_{23} + q_y e_{31} + q_z e_{12}. \tag{16.2}$$

By left-multiplication with $\boldsymbol{R}$, the equation of rigid motion becomes

$$\boldsymbol{R}\boldsymbol{x}' = \boldsymbol{x}R + \boldsymbol{R}\boldsymbol{t}, \tag{16.3}$$

$$(q_0 + q_x e_{23} + q_y e_{31} + q_z e_{12})(x'e_1 + y'e_2 + z'e_3) = (xe_1 + ye_2 + ze_3)(q_0 + q_x e_{23} + q_y e_{31} + q_z e_{12}) + \\ (q_0 + q_x e_{23} + q_y e_{31} + q_z e_{12})(t_x e_1 + t_y e_2 + t_z e_3).$$

E. Bayro-Corrochano, *Geometric Algebra Applications Vol. I*,
https://doi.org/10.1007/978-3-319-74830-6_16

Developing products, we get

$$\begin{aligned}
& q_0x'e_1 + q_xx'e_{231} + q_yx'e_3 - q_zx'e_2 + q_0y'e_2 - q_xy'e_3 + \\
& + q_yy'e_{312} + q_zy'e_1 + q_0z'e_3 + q_xz'e_2 - q_yz'e_1 + q_zz'e_{123} = \\
& xq_0e_1 + yq_0e_2 + zq_0e_3 + +xq_xe_{123} + yq_xe_3 - zq_xe_2 - \\
& - xq_ye_3 + yq_ye_{231} + zq_ye_1 + xq_ze_2 - yq_ze_1 + zq_ze_{312} + \\
& + q_0t_xe_1 + q_xt_xe_{231} + q_yt_xe_3 - q_zt_xe_2 + +q_0t_ye_2 - \\
& - q_xt_ye_3 + q_yt_ye_{312} + q_zt_ye_1 + q_0t_ze_3 + q_xt_ze_2 - q_yt_ze_1 + q_zt_ze_{123}.
\end{aligned} \tag{16.4}$$

Rearranging terms according to their multivector basis, we obtain the following four equations:

$$\begin{aligned}
e_1 &: q_0x' + q_zy' - q_yz' = q_0x + q_yz - q_zy + q_0t_x + q_zt_y - q_yt_z, \\
e_2 &: q_0y' + q_xz' - q_zx' = q_0y + q_zx - q_xz + q_0t_y + q_xt_z - q_zt_x, \\
e_3 &: q_0z' + q_yx' - q_xy' = q_0z + q_xy - q_yx + q_0t_z + q_yt_x - q_xt_y, \\
e_{123} &: q_xx' + q_yy' + q_zz' = q_xx + q_yy + q_zz + q_xt_x + q_yt_y + q_zt_z.
\end{aligned}$$

These equations can be rearranged to express linear relationships in the joint difference and sum spaces:

$$e_1 : q_0(x' - x) - q_y(z + z') + q_z(y + y') + (q_yt_z - q_0t_x - q_zt_y) = 0, \tag{16.5}$$

$$e_2 : q_0(y' - y) + q_x(z + z') - q_z(x + x') + (q_zt_x - q_0t_y - q_xt_z) = 0, \tag{16.6}$$

$$e_3 : q_0(z' - z) - q_x(y + y') + q_y(x + x') + (q_xt_y - q_0t_z - q_yt_x) = 0, \tag{16.7}$$

$$e_{123} : q_x(x' - x) + q_y(y' - y) + q_z(z' - z) - (q_xt_x + q_yt_y + q_zt_z) = 0. \tag{16.8}$$

These equations clearly represent four 3D planes in the entries of the rotor and the translator (the unknowns). Thus, in order to estimate the correspondences due to a rigid transformation, we can use a set of tentative correspondences to populate the joint spaces $\{(x' - x), (z + z'), (y + y')\}$, $\{(y' - y), (z + z'), (x + x')\}$, $\{(z' - z), (y + y'), (x + x')\}$, and $\{(x' - x), (y' - y), (z' - z)\}$. If four planes appear in these spaces, then the points lying on them are related by a rigid transformation. However, we will show in the following section that the first three planes of Eqs. (16.5)–(16.7) are related by a powerful geometric constraint. So it is enough to find the plane described by Eq. 16.8 and verify that it satisfies this constraint. We will now show how this is done and how this geometric constraint helps in eliminating multiple matches too.

16.1.1 The Geometric Constraint

Let $(\mathbf{x}_i, \mathbf{x}'_i)$ and $(\mathbf{x}_j, \mathbf{x}'_j)$ be two points in correspondence through a rigid transformation. Then, these points satisfy Eq. (16.5):

$$q_0 d_x - q_y s_z + q_z s_y + q_k = 0, \tag{16.9}$$
$$q_0 d'_x - q_y s'_z + q_z s'_y + q_k = 0, \tag{16.10}$$

where $d_x = x'_i - x_i$, $s_z = z'_i + z_i$, $s_y = y'_i + y_i$; $d'_x = x'_j - x_j$, $s'_z = z'_j + z_j$, $s'_y = y'_j + y_j$; and $q_k = q_y t_z - q_0 t_x - q_z t_y$. If we subtract these equations, we get

$$q_0 v_x - q_y v_{sz} + q_z v_{sy} = 0, \tag{16.11}$$

where $v_x = d_x - d'_x$, $v_{sz} = s_z - s'_z$, and $v_{sy} = s_y - s'_y$. Using the definition of R, this equation can be rewritten as

$$k v_x - a_y v_{sz} + a_z v_{sy} = 0, \tag{16.12}$$

where $k = \cos(\frac{\theta}{2})/\sin(\frac{\theta}{2})$. Using a similar procedure, for the Eqs. 16.6 and 16.7, we end up with the following system of equations

$$k v_x + a_z v_{sy} - a_y v_{sz} = 0, \tag{16.13}$$
$$k v_y + a_x v_{sz} - a_z v_{sx} = 0, \tag{16.14}$$
$$k v_z + a_y v_{sx} - a_x v_{sy} = 0, \tag{16.15}$$

where v_y, v_z, and v_{sx} are defined accordingly. Note that we now have a system of equations depending on the unitary axis of rotation $[a_x, a_y, a_z]$. Since we can obtain the axis of rotation as the normal of the plane described by Eq. 16.8, we have only one unknown: k. These equations can be mixed to yield the following three constraints:

$$v_y(a_y v_{sz} - a_z v_{sy}) - v_x(a_z v_{sx} - a_x v_{sz}) = 0, \tag{16.16}$$
$$v_z(a_y v_{sz} - a_z v_{sy}) - v_x(a_x v_{sy} - a_y v_{sx}) = 0, \tag{16.17}$$
$$v_z(a_z v_{sx} - a_x v_{sz}) - v_y(a_x v_{sy} - a_y v_{sx}) = 0. \tag{16.18}$$

These equations depend only on the points themselves and on the plane spanned by them. Thus, if we populate the joint space described by Eq. (16.8) with a set of tentative correspondences and detect a plane in this space, we can verify if this plane corresponds to an actual rigid transformation by verifying that the points lie on this plane satisfy Eqs. (16.16)–(16.18). Note that these constraints never become undefined, because the factor k was removed. So this test can always be applied to confirm or reject a plane that seems to represent a rigid transformation. Furthermore, since these constraints have been derived from the original plane Eqs. (16.5)–(16.7), they are, in a sense, expressing the requirement that these points lie simultaneously on all three planes.

On the other hand, Eqs. (16.16)–(16.18) have an interesting geometric interpretation. They are in fact expressing a double cross product that is only satisfied for true correspondences. To see this, note that if $A = [a_x, a_y, a_z]^\mathsf{T}$, $V = [v_x, v_y, v_z]^\mathsf{T}$, and $V_s = [v_{sx}, v_{sy}, v_{sz}]^\mathsf{T}$, then Eqs. (16.16)–(16.18) can be rewritten as a vector equation

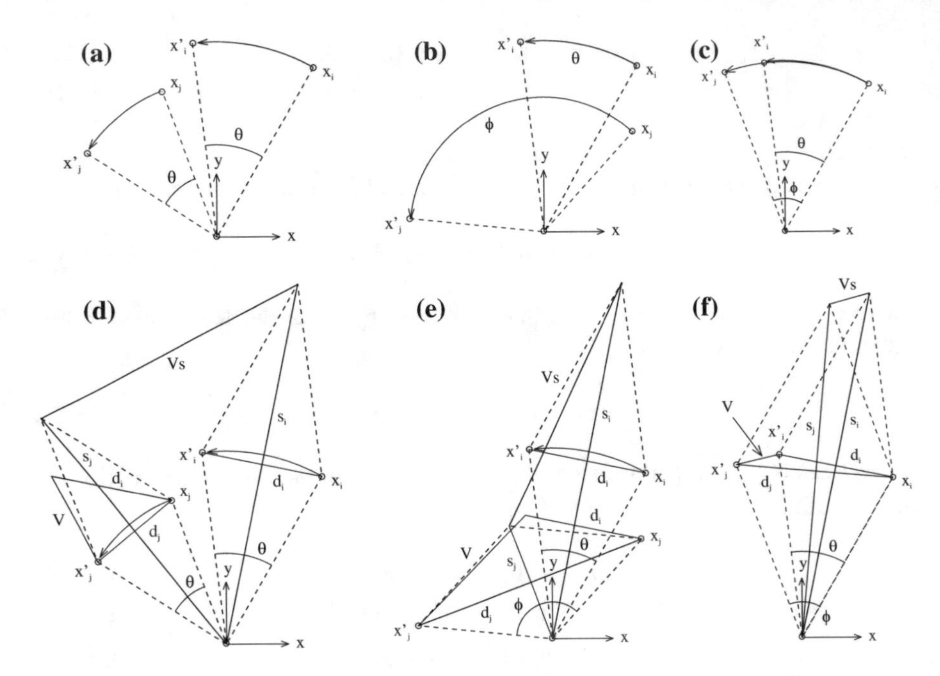

Fig. 16.1 **a** and **d** Two correspondences belonging to the same transformation and the geometry of the plane of rotation. **b** and **e** Two correspondences belonging to different transformations (different angle) and their corresponding geometry. **c** and **f** Multiple correspondence case and its geometry

of the form

$$V \times (A \times V_s) = 0. \tag{16.19}$$

This equation holds only due to an inherent symmetry that is only present for true correspondences; in other words, these equations can be used to reject false matches too. To prove this, first remember the well-known fact that any rigid motion in 3D is equivalent to a screw motion (rotation about the screw axis followed by a translation along it). Hence, without loss of generality, we can consider the case where the screw axis is aligned with the z-axis. In this case, the screw motion consists of a rotation about z followed by a translation t_z along it. Therefore,

$$v_z = d_z - d'_z = z'_i - z_i - z'_j + z_j = (z_i + t_z) - z_i - (z_j + t_z) + z_j = 0. \tag{16.20}$$

Also, note that since $A = [001]^{\mathsf{T}}$, then the first cross product of Eq. (16.19), $A \times V_s = [-v_{sy}, v_{sx}, 0]^{\mathsf{T}}$, hence the v_{sz}-component of V_s, is irrelevant in this case and can be safely disregarded. Thus, we can analyze this problem in 2D by looking only at the x- and y- components of V and V_s. Accordingly, the difference and sum vectors will only have two components, $d_i = [d_x, d_y]^{\mathsf{T}}$, $d_j = [d'_x, d'_y]^{\mathsf{T}}$, $s_i = [s_x, s_y]^{\mathsf{T}}$, and $s_j = [s'_x, s'_y]^{\mathsf{T}}$. The situation is illustrated in Fig. 16.1a and d.

Since the angle between x_i and x'_i is the same as the angle between x_j and x'_j, the parallelograms spanned by the sum and difference of these points (the dashed lines in Fig. 16.1d) are equal up to a scale factor. That is, there is a scale factor k such that $k||s_i|| = ||s_j||$ and $k||d_i|| = ||d_j||$. From whence $||s_i||/||d_i|| = ||s_j||/||d_j||$. In turn, this means that the triangle formed by the vectors d_i, d_j, and V is proportional to the triangle s_i, s_j, V_s. And since, by construction, $s_i \perp d_i$ and $s_j \perp d_j$, then $V \perp V_s$.

Now, let us return to Eq. (16.19). The cross product $A \times V_s$ has the effect of rotating V_s by 90° since $A = [0, 0, 1]^{\mathsf{T}}$ in this case. But since $V_s \perp V$, the vector $A \times V_s$ will be parallel to V, and hence, their cross product will always be 0, which is consistent with the analytic derivation.

This symmetry is broken if we have points that belong to different transformations, as shown in Fig. 16.1b and e (we assume the worst case, in which points x_i and x_j were applied the same translation but different rotation angle ϕ. If a different translation is present, the planes of motion will be different for both points, breaking the symmetry). Note how the angle between V and V_s is not orthogonal (Fig. 16.1e).

In a similar way, when we have multiple correspondences, i.e., x_i matches both x'_i and x'_j, $i \neq j$ (Fig. 16.1c), the symmetry is also broken and V is not orthogonal to V_s (see Fig. 16.1f). Hence, the constraint expressed by Eq. 16.19 can be used to reject multiple matches too.

Following this procedure, we were able to cast the problem of finding the correspondences between two sets of 3D points due to rigid transformation, into a problem of finding a 3D plane in a joint space that satisfies three geometric constraints. In order to find a 3D plane from a set of points that may contain a large proportion of outliers, several methods can be used. We have decided to use Tensor Voting because it has proven to be quite robust and because it can be used to detect general surfaces too, which in turn enables us to easily extend our method to non-rigid motion estimation. We will not explain Tensor Voting here. The reader should resort to [210] for a gentle introduction to this subject.

16.2 Tensor Voting

Tensor Voting is a methodology for the extraction of dense or sparse features from n-dimensional data. Some of the features that can be detected with this methodology include lines, curves, points of junction, and surfaces.

The Tensor Voting methodology is grounded in two elements: *tensor calculus* for data representation and *tensor voting* for data communication. Each input site propagates its information in a neighborhood (the information itself is encoded as a tensor and is defined by a predefined voting field). Each site collects the information cast there by its neighbors and analyzes it, building a saliency map for each feature type. Salient features are located at local extrema of these saliency maps, which can be extracted by non-maximal suppression.

For the present work, we found that sparse tensor voting was enough to solve the problem. Since we are only concerned with finding 3D planes, we will limit our discussion to the detection of this type of feature. We refer the interested reader to [210] for a complete description of the methodology.

16.2.1 Tensor Representation in 3D

In Tensor Voting, all points are represented as second-order symmetric tensors. To express a tensor S, we choose to take the associated quadratic form, and diagonalize it, leading to a representation based on the eigenvalues λ_1, λ_2, and λ_3 and the eigenvectors $\mathbf{e}_1$, $\mathbf{e}_2$, and $\mathbf{e}_3$. Therefore, we can write the tensor S as

$$S = \begin{bmatrix} \mathbf{e}_1 & \mathbf{e}_2 & \mathbf{e}_3 \end{bmatrix} \begin{bmatrix} \lambda_1 & 0 & 0 \\ 0 & \lambda_2 & 0 \\ 0 & 0 & \lambda_3 \end{bmatrix} \begin{bmatrix} \mathbf{e}_1^T \\ \mathbf{e}_2^T \\ \mathbf{e}_3^T \end{bmatrix}. \tag{16.21}$$

Thus, a symmetric tensor can be visualized as an ellipsoid where the eigenvectors correspond to the principal orthonormal directions of the ellipsoid and the eigenvalues encode the magnitude of each of the eigenvectors (see Fig. 16.2).

For the rest of this chapter, we will use the convention that the eigenvectors have been arranged so that $\lambda_1 > \lambda_2 > \lambda_3$. In this scheme, points are encoded as *ball tensors* (i.e., tensors with eigenvalues $\lambda_1 = \lambda_2 = \lambda_3 \geq 1$); *curvels* as *plate tensors* (i.e., tensors with $\lambda_1 = \lambda_2 = 1$, and $\lambda_3 = 0$, tangent direction given by $\mathbf{e}_3$); and *surfels* as *stick tensors* (i.e., $\lambda_1 = 1$, $\lambda_2 = \lambda_3 = 0$, normal direction given by $\mathbf{e}_1$).

A ball tensor encodes complete uncertainty of direction, a plate tensor encodes uncertainty of direction in two axes, but complete certainty in the other one, and a stick tensor encodes absolute certainty of direction. Tensors that lie between these three extremes encode differing degrees of direction certainty. The pointness of any given tensor is represented by λ_3, the curveness is represented by $\lambda_2 - \lambda_3$, and the surfaceness by $\lambda_1 - \lambda_2$. Also, note that a second-order tensor only encodes direction, but not *orientation*, i.e., two vectors $\mathbf{v}$ and $-\mathbf{v}$ will be encoded as the same second-order tensor.

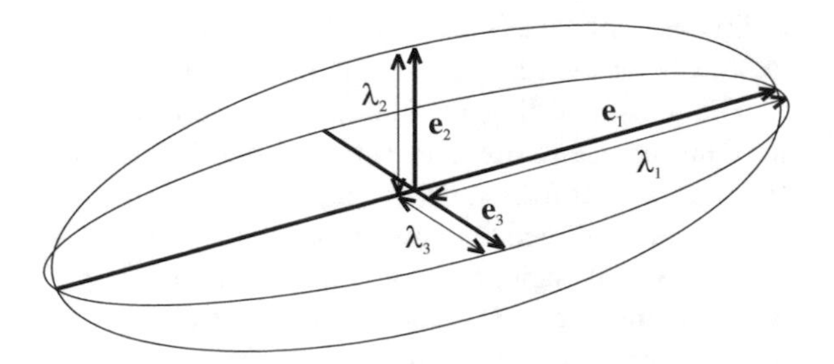

Fig. 16.2 Graphic representation of a second-order 3D symmetric tensor

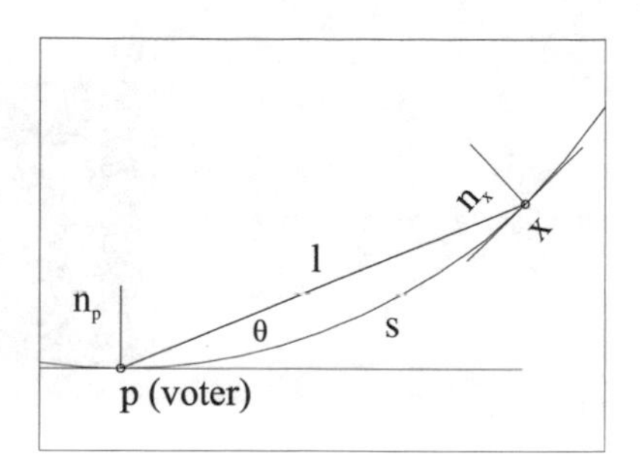

Fig. 16.3 Osculating circle and the corresponding normals of the voter and votee

16.2.2 Voting Fields in 3D

We have just seen how the various types of input data are encoded in Tensor Voting. Now, we will describe how these tensors communicate between them. The *input* usually consists of a set of sparse points. These points are encoded as ball tensors if no information is available about their direction (i.e., identity matrices with $\lambda_1 = \lambda_2 = \lambda_3 = 1$). If only a tangent is available, the points are encoded as plate tensors (i.e., tensors with $\lambda_1 = \lambda_2 = 1$, and $\lambda_3 = 0$). Finally, if information about the normal of the point is given, it is encoded as a stick tensor (i.e., tensors with only one nonzero eigenvalue: $\lambda_1 = 1, \lambda_2 = \lambda_3 = 0$). Then, each *encoded input point*, or *token*, communicates with its neighbors using either a ball voting field (if no orientation is present), a plate voting field (if local tangents are available), or a stick voting field (when the normal is available). The voting fields themselves consist of various types of tensors ranging from stick to ball tensors.

These *voting fields* have been derived from a 2D fundamental voting field that encodes the constraints of surface continuity and smoothness, among others. To see how the fundamental voting field was derived, suppose that we have a voter p with an associated normal $\mathbf{n}_p$. At each votee site x surrounding the voter p, the direction of the fundamental field $\mathbf{n}_x$ is determined by the normal of the osculating circle at x that passes through p and x and has normal $\mathbf{n}_p$ at p (see Fig. 16.3).

The saliency decay function of the fundamental field $DF(s, \kappa, \sigma)$ at each point depends on the arc length $s = \frac{l\theta}{\sin\theta}$ and curvature $\kappa = \frac{2\sin\theta}{l}$ between p and x (see Fig. 16.3) and is given by the following Gaussian function:

$$DF(s, \kappa, \sigma) = e^{-\left(\frac{s^2 + c\kappa^2}{\sigma^2}\right)}, \tag{16.22}$$

where σ is a scale factor that determines the overall rate of attenuation and c is a constant that controls the decay with high curvature. Note that the strength of the field becomes negligible beyond a certain distance; in this way, each voting field has an effective neighborhood associated with it given by σ. The shape of the fundamental field can be seen in Fig. 16.4. In this figure, the direction and strength fields are displayed separately. The direction field shows the eigenvectors with the largest associated eigenvalues for each tensor surrounding the voter (center). The strength

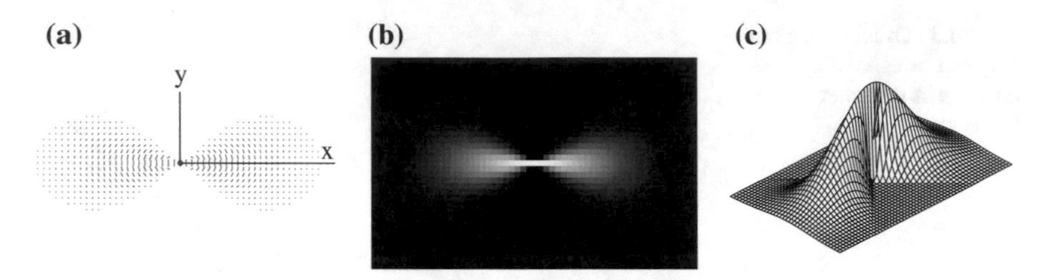

Fig. 16.4 Fundamental voting field. **a** Shape of the direction field when the voter p is located at the origin and its normal is parallel to the y-axis (i.e., $n_p = y$). **b** Strength at each location of the previous case. White colors denote high strength, and black denotes no strength. **c** 3D display of the strength field, where the strength has been mapped to the z-coordinate

field shows the value of the largest eigenvalue around the voter: White denotes a strong vote, and black denotes no intensity at all (zero eigenvalue).

Finally, both orientation and strength are encoded as a *stick tensor*. In other words, each site around this voting field, or votee, is represented as a *stick tensor* with varying strength and direction.

Communication is performed by the addition of the stick tensor present at the votee and the tensor produced by the field at that site. To exemplify the voting process, imagine we are given an *input point* x and a voter located at the point p with associated normal $\mathbf{n}_p$. The input point (votee) is first encoded as a ball tensor ($\lambda_1 = \lambda_2 = \lambda_3 = 1$). Then, the vote generated by p on x is computed. This vote is, in turn, a stick tensor. To compute this vote, Eq. 16.22 is used to compute the strength of the vote (λ_1). The direction of the vote ($\mathbf{e}_1$) is computed through the osculating circle between the voter and the votee (using the voter's associated normal $\mathbf{n}_p$). The other eigenvalues (λ_2 and λ_3) are set to zero, and then, the stick tensor is computed using Eq. (16.21). Finally, the resulting stick tensor vote is added with ordinary matrix addition to the encoded ball tensor at x. Since, in general, the stick vote has only one nonzero eigenvalue, the resulting addition produces a non-identity matrix with one eigenvalue much larger than the others ($\lambda_1 > \lambda_2$ and $\lambda_1 > \lambda_3$). In other words, the ball tensor at x *becomes an ellipsoid* in the direction given by the stick tensor. The larger the first eigenvalue of the voter is, the more pronounced this ellipsoid becomes.

To speed things up, however, these calculations are not done in practice. Instead, the set of votes surrounding the stick voting field is precomputed and stored using a discrete sampling of the space. When the voting is performed, these precomputed votes are just aligned with the voter's associated normal $\mathbf{n}_p$ and the actual vote is computed by linear interpolation.

Note that the stick voting field (or fundamental voting field) can be used to detect surfaces. In order to detect joints, curves, and other features, different voting fields must be employed. These other 3D voting fields can be generated by rotating the fundamental voting field about the x-, y- and z-axes, depending on the type of field we wish to generate. For example, if the voter is located at the origin and its normal

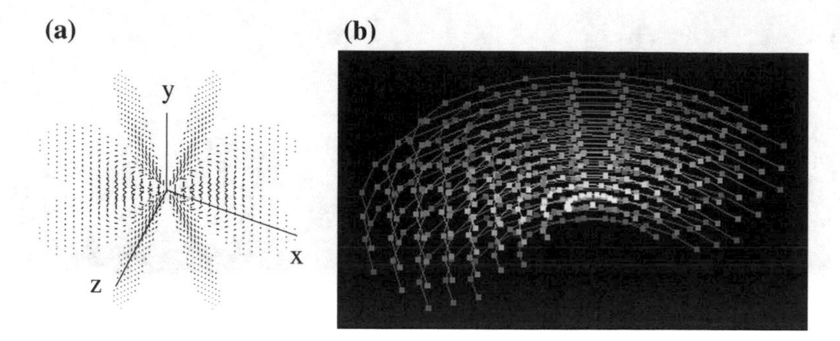

Fig. 16.5 Stick voting field in 3D. **a** The direction of the 3D stick voting field when the voter is located at the origin and its normal is parallel to the y-axis. Only the $\mathbf{e}_1$ eigenvectors are shown at several positions. **b** The strength of this field. White denotes high strength, black denotes no strength

is parallel to the y-axis, as in Fig. 16.4, then we can rotate this field about the y-axis to generate the 3D stick voting field, as shown in Fig. 16.5.

In a more formal way, let us define the general rotation matrix $\mathbf{R}_{\psi\phi\theta}$, where ψ, ϕ, and θ stand for the angles of rotation about the x-, y-, and z-axes, respectively, and let $V_S(x)$ stand for the tensor vote cast by a stick voting field in 3D at site x. Then, $V_S(x)$ can be defined as

$$V_S(x) = \int_0^{\pi} \mathbf{R}_{\psi\phi\theta} V_f(\mathbf{R}_{\psi\phi\theta}^{-1}\mathbf{p})\mathbf{R}_{\psi\phi\theta}^{\mathsf{T}} d\phi \;\; (\psi = \theta = 0), \tag{16.23}$$

where $V_f(x)$ stands for the vote cast by the Fundamental Voting Field in 2D at site x.

The stick voting field can be used when the normals of the points are available. However, when no orientation is provided, we must use the *ball voting field.* This field is produced by rotating the stick field about all the axes and integrating the contributions at each site surrounding the voter. For example, for the field depicted in Fig. 16.4, the 2D ball voting field is generated by rotating this field about the z-axis (as shown in Fig. 16.6). This 2D ball voting field is further rotated about the y-axis to generate the 3D ball voting field. In other words, the 2D ball voting field $V_b(x)$ at site x can be defined as

$$V_b(x) = \int_0^{\pi} \mathbf{R}_{\psi\phi\theta} V_f(\mathbf{R}_{\psi\phi\theta}^{-1}\mathbf{p})\mathbf{R}_{\psi\phi\theta}^{\mathsf{T}} d\theta \;\; (\psi = \phi = 0), \tag{16.24}$$

and the 3D ball voting $V_B(x)$ field can thus be further defined as

$$V_B(x) = \int_0^{\pi} \mathbf{R}_{\psi\phi\theta} V_b(\mathbf{R}_{\psi\phi\theta}^{-1}\mathbf{p})\mathbf{R}_{\psi\phi\theta}^{\mathsf{T}} d\phi \;\; (\psi = \theta = 0), \tag{16.25}$$

or, alternatively, as

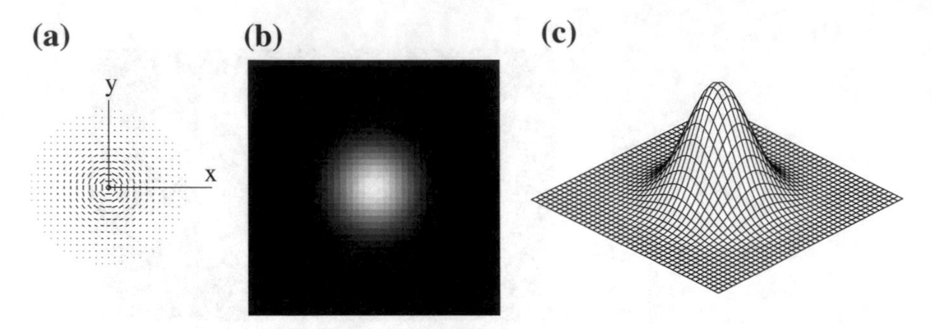

Fig. 16.6 Ball voting field. **a** Direction field when the voter is located at the origin. **b** Strength of this field. White denotes high strength, and black denotes no strength. **c** 3D display of the strength field with the strength mapped to the z-axis

$$V_B(x) = \int_0^{\pi}\int_0^{\pi} \mathbf{R}_{\psi\phi\theta} V_f(\mathbf{R}^{-1}_{\psi\phi\theta}\mathbf{p})\mathbf{R}^{\mathsf{T}}_{\psi\phi\theta} d\theta d\phi \quad (\psi = 0). \tag{16.26}$$

Note that by rotating the stick voting field about all axes and adding up all vote contributions, the shape of the votes in the ball voting field varies smoothly from nearly stick tensors at the edge ($\lambda_1 = 1$, $\lambda_2 = \lambda_3 = 0$), to ball tensors near the center of the voter ($\lambda_1 = \lambda_2 = \lambda_3 = 1$). Thus, this field consists of ellipsoid-type tensors of varying shape. This is the reason why this field is not simply a "radial stick tensor field" with stick tensors pointing radially away from the center. However, the added complexity of rotating the stick tensor voting field to generate this field does not impact the implementation. As discussed previously, in practice, this field is precomputed in discrete intervals and linearly interpolated when necessary.

Finally, the ball voting field can be used to infer the preferred orientation (normals) at each input point if no such information is present to begin with. After voting with the ball voting field, the eigensystem is computed at each input point and the eigenvector with the greatest eigenvalue is taken as the preferred normal direction at that point. With the normals at each input point thus computed, a further stick voting step can be used to reinforce the points which seem to lie in a surface. Surface detection is precisely what we need in order to solve the original 3D point registration problem. We will now describe the process of surface detection used in our approach.

16.2.3 Detection of 3D Surfaces

Now that we have defined the tensors used to encode the information and the voting fields, we can describe the process used to detect the presence of a 3D surface in a set of points. We will limit our discussion of this process to the stages of Tensor Voting that are relevant to our work. We refer the interested reader to [210] for a full account on feature inference through Tensor Voting.

As described previously, the input to our algorithm is a 3D space populated by a set of putative correspondences between two point sets. To avoid confusion, we will refer to the points in the joint space simply as "tokens." Each of these tokens is encoded as unitary ball tensor (i.e., with the identity matrix $I_{3\times3}$). Then, we place a ball voting field at each input token and cast votes to all its neighbors. Since the strength of the ball voting field becomes negligible after a certain distance (given by the free parameter σ in Eq. (16.22)), we only need to cast votes to the tokens that lie within a small neighborhood about each input token. To cast a vote, we simply add the tensor present at the votee with the tensor produced by the ball voting field at that position. This process constitutes a sparse ball voting stage. Once this stage is finished, we can examine the eigensystem left at each token and thus extract the preferred normals at each site. The preferred normal direction is given by the eigenvector $\mathbf{e}_1$, and the saliency of this orientation is given by $\lambda_1 - \lambda_2$.

After this step, each token has an associated normal. The next step consists of using the 3D stick voting field to cast votes to the neighbors so that the normals are reinforced. In order to cast a stick vote, the 3D stick voting field is first placed on the voter and oriented to match its normal. Once again, the extent of this field is limited by the parameter σ, so we only need to cast votes to the tokens that lie within a small neighborhood of the voter. After the votes have been cast, the eigensystem at each token is computed to obtain the new normal orientation and strength at each site. This process constitutes a sparse stick voting stage.

In ordinary Tensor Voting, the eigensystem at each token is used to compute different saliency maps: pointness λ_3, curveness $\lambda_2 - \lambda_3$, and surfaceness $\lambda_1 - \lambda_2$. Then, the derivative of these saliency maps is computed and non-maximal suppression is used to locate the most salient features. After this step, the surfaces are polygonized using a marching cubes algorithm (or similar). However, our objective in this case was not the extraction of polygonized surfaces, but simply the location of the most salient surface. Hence, a simple thresholding technique was used instead. The token with the greatest saliency is located, and the threshold is set to a small percentage of this saliency. Thus, for example, tokens with a small λ_1 relative to this token are discarded. In a similar fashion, tokens with a small surfaceness ($\lambda_1 - \lambda_2$) with respect to this token are also deleted.

After the sparse stick voting is performed, only the tokens that seem to belong to surfaces (i.e., λ_1 is not small and $\lambda_1 - \lambda_2$ is high) cast votes to its neighbors to further reinforce the surfaceness of the tokens. Input tokens that do not belong to surfaces are discarded (set to $0_{3\times3}$). This process is repeated a fixed number of times with increasing values of σ in order to make the surface(s) grow. In this way, there is a high confidence that the tokens that have not been discarded after the repeated application of the sparse stick voting stage belong to a surface.

16.2.4 Estimation of 3D Correspondences

Given two sets of 3D points $\mathtt{X}_1$ and $\mathtt{X}_2$, we are expected to find the correspondences between these two sets assuming a rigid transformation has taken place, and we have an unspecified number of outliers in each set. No other information is given.

In the absence of better information, we populate the joint space $(x' - x)$, $(y' - y)$, $(z' - z)$ by matching all points from the first set with all the points from the second set. Note that this allows us to detect any motion regardless of its magnitude, but the amount of outliers present in the joint space is multiplied 100-fold by this matching scheme.

The tokens in the joint space thus populated are then processed with Tensor Voting in order to detect the most salient plane, as described in the previous section. The plane thus detected is further tested against the constraint given by Eq. (16.19). This constraint requires the specification of two different tokens. In practice, we use the token on the plane with the highest saliency and test it against the rest of the points on the plane. If any pair of tokens does not satisfy the constraint, we remove it from the plane. If not enough points remain after this pruning is completed, we reject the plane. Remember that the enforcement of this constraint also avoids the presence of false or multiple matches. So the output is a one-to-one set of correspondences.

As can be easily noted, Eq. (16.8) collapses for the case of pure translation. However, in this case, all points that belong to a rigid transformation will tend to cluster together in a *single token* in the joint space. This cluster is easy to detect after the sparse ball voting stage because these tokens will have a large absolute saliency value at this stage. If such a cluster is found, we stop the algorithm and produce the correspondences based on the tokens that were found clustered together.

Following this simple procedure, we can detect any rigid transformation. Note, however, that using the simple matching scheme mentioned earlier, the number of outliers in the joint space is multiplied by 100-fold. When the number of outliers in the real space is large (for example, on the order of 90%), this can complicate the detection of surfaces in the joint space. When a situation like this arises, we have adopted a scheme where several rotation angles and axes are tested in a systematic fashion in order to make the detection process simpler. In this variation of the algorithm, the set $\mathtt{X}_1$ is first rotated according to the current axis and angle to be tested, and the joint space specified by Eq. (16.8) is populated again. We then run the detection process using Tensor Voting. If a plane with enough support is found, we stop the algorithm and output the result. Otherwise, the next angle and rotation axis are tested until a solution is found, or all the possibilities have been tested. The whole algorithm for the detection of correspondences between two 3D point sets under a single rigid transformation is sketched below. Finally, note that this variation of the algorithm is only needed when large numbers of outliers are present in the input, as stated previously.

Algorithm

- 1. Initialize the rotation angle $\alpha = 0°$, and axis $A = [0, 0, 1]^{\mathsf{T}}$.
- 2. Rotate the set X_1 according to α and A. Populate the voting space with tokens generated from the candidate correspondences.
- 3. Initialize all tokens to ball tensors.
- 4. Perform sparse ball voting and extract the preferred normals.
- 5. Check for the presence of a set of tokens clustered about a single point in space. If this cluster is found, finish and output the corresponding translation detected.
- 6. Perform sparse stick voting using the preferred normals. Optionally, repeat this step a fixed number of times to eliminate outliers. After each iteration, increase the reach of the votes slightly, so as to make the plane grow.
- 7. Obtain the equation of the plane described by the tokens with the highest saliency. Enforce the constraint of Eq. (16.19) and delete the tokens that do not satisfy it.
- 8. If a satisfactory plane is found, output the correspondences. Otherwise, increment α and A, and repeat steps 2 to 7 until all angles and axes of rotation have been tested.

Finally, we recognize that the scheme proposed here is far from perfect. The exhaustive search of all angles and rotation axes can be quite time-consuming and appears to be a little simplistic. Unfortunately, the density of the plane we are seeking varies with the angle of the rotation applied to the set of points. That is, the density of this plane is minimum (the plane spans the full voting space) when the rotation is 180°, and it becomes infinite when we have pure translation (all the points of the plane cluster in a single location in space). Hence, there does not seem to be some type of heuristic or constraint we can apply to prune the search. An alternative to this is to use the other three search spaces as described in Eqs. (16.5)–(16.7) and perform Tensor Voting to detect these planes to help improve the search method. This is a matter for future research.

However, this disadvantage is only apparent if the magnitude of the transformation is unbounded. Algorithms like the ICP require that the transformation be relatively small. If we use the same limitation in our method, we do not need this exhaustive search, and our method works without the iterative scheme. Finally, we have shown in [247] that our algorithm has a complexity of $O(n^2)$ in the worst case, where n is the number of tokens in the voting space (which never occurs in practice because this implies that each token casts a vote to every other token).

16.3 Experimental Analysis of Tensor Votings Using Geometric Algebra

When dealing with real data, the input points usually have some amount of noise. This noise in turn affects the shape and thickness of the plane that has to be detected in the joint space. Thus, instead of producing an ideal plane in the joint space, points

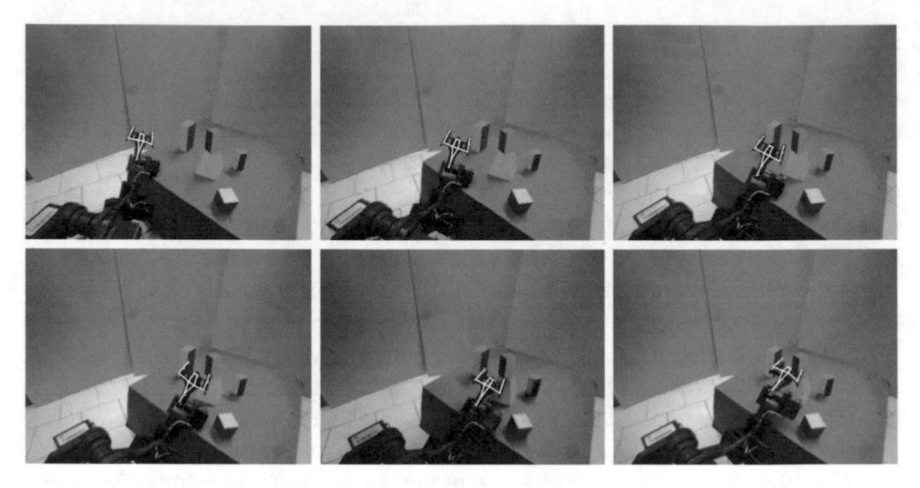

Fig. 16.7 Sequence of (left) images from a stereo camera showing the position of the reprojected arm (in white lines). This is not a video

with noise yield a "fuzzy" plane that has small variations over its surface. However, even in spite of this, Tensor Voting can be used successfully to detect these "fuzzy" surfaces. Also, the constraint given by Eq. (16.19) has to be relaxed in order to avoid rejecting points that do not seem to be in correspondence due to the noise. This relaxation is accomplished by checking that the equation yields a small absolute value, instead of zero.

16.3.1 Correspondences Between 3D Points by Rigid Motion

We performed the following experiments. First, we followed the position of a robotic arm in 3D in a sequence of stereo pairs (Fig. 16.7 shows only the left images of the sequence). In this case, the problem of 3D reconstruction is considered already solved and the input to our algorithm is the 3D points of this reconstruction. In practice, ordinary camera calibration and stereo matching (through cross-correlation) were performed to achieve the 3D reconstruction.

The model of the object was picked by hand from the first reconstruction, and then, we computed the position of the arm in the subsequent reconstructions using an optimized version of our algorithm (namely, it only consisted of two stages: sparse ball voting and sparse stick voting, and no iterations were used). Note that the sequence of images does not form a video, hence, the features cannot be tracked between successive frames due to the relatively large differences between the snapshots. After the motion was computed, the position of the arm in 3D was reprojected on the images (drawn in white) as shown in Fig. 16.7.

In a second experiment, we made a reconstruction of a Styrofoam model of a head using a stereo camera. The two reconstructions are shown in Fig. 16.8a–c. The aligned sets can be seen in Fig. 16.8d–f. In this case, however, another optimization was used. Since the sets are close to each other, and the points provide enough structure, we used Tensor Voting to compute the preferred normals at each site.

The computation of the normals procccds as in standard sparse tensor voting. First, we initialized each point to a ball tensor. Then, we placed a normal ball voting field on each point and cast votes to all the neighbors. Then, the preferred normal at each site is obtained by computing the eigensystem at each point and selecting the eigenvector with the greatest eigenvalue. A close-up of the surface of the model and some of the normals found by this method is shown in Fig. 16.8g. We used this information to prune the candidate matches to those that shared a relatively similar orientation only.

Also, note that in this case, there are non-rigid differences between both sets. This can be noted in places like the chin, where the alignment could not be made simply because the size of this section of the reconstruction differs slightly between both sets (the overall height of the head is slightly larger in the second set). Hence, it is impossible to match all the points at the same time. However, even in spite of this, our algorithm does yield a reasonable solution. In practice, two different surfaces are formed in the joint space, one corresponds to a rigid transformation that matches the forehead and the nose, and the other corresponds to the transformation that aligns the chins of both models. We have chosen to display the first solution, where the upper part of the head is correctly aligned – this solution also corresponds to the largest surface in the joint space. The main point of this experiment is to show that our algorithm still works even when the input cannot be perfectly aligned with a single rigid transformation.

Finally, in our last experiment, we aligned a model of a Toyota car taken with a laser range scanner and aligned it with a noisy reconstruction performed with a stereo camera. The noisy target is shown in Fig. 16.9a, the model and the target are shown in Fig. 16.9b, and the final alignment in Fig. 16.9c. The procedure is the same as in the previous case. Again, since the data sets provided structure, we used it to our advantage by computing the preferred normals using Tensor Voting and pruning the candidate matches as described previously (Fig. 16.9d).

16.3.2 Multiple Overlapping Motions and Non-rigid Motion

Another advantage our method has over ICP and similar methods is the ability to simultaneously detect multiple overlapping motions. This is also true for the 3D case. In this case, each different motion simply produces another plane in the voting space. There are limitations to the motions that can be differentiated, though. A quick analysis of Eq. (16.8) reveals that if two different motions share the same axis of rotation and same overall translation, then they will span the same 3D plane in the

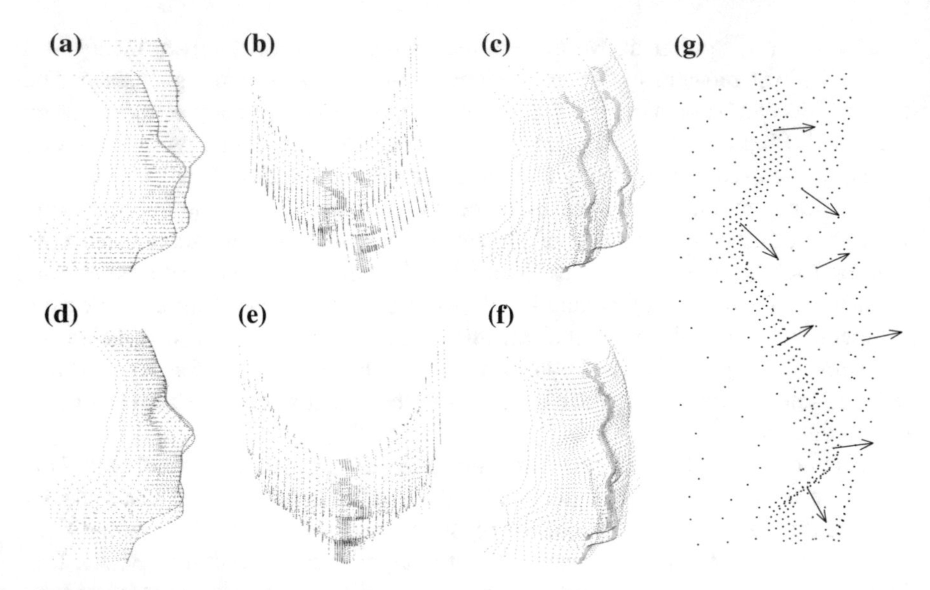

Fig. 16.8 **a–c** Sets to be aligned. **d–f** Sets after alignment

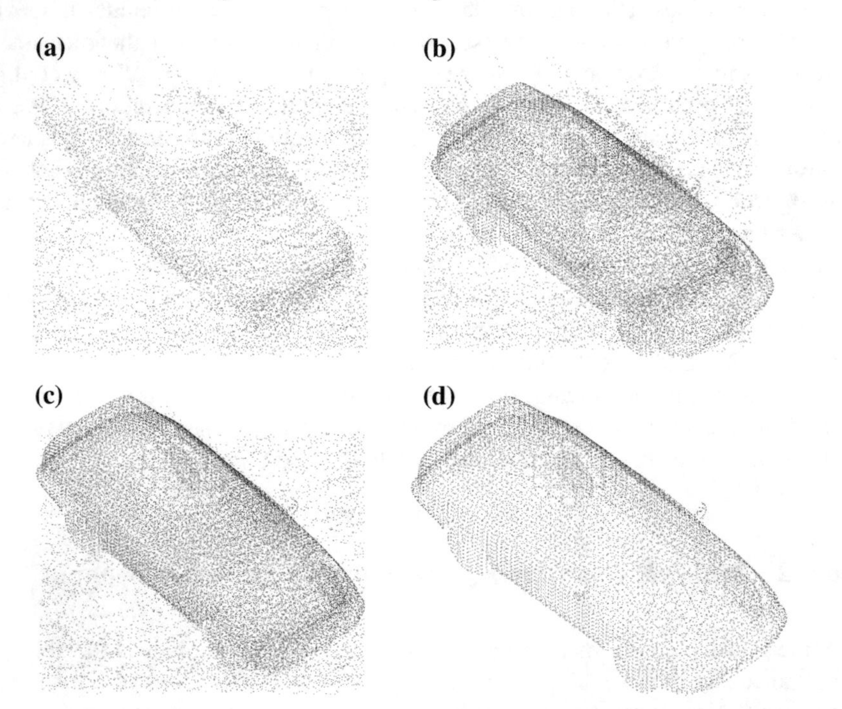

Fig. 16.9 **a** Target for alignment, note the noisy surface. **b** Model displayed over the target. **c** Model and data after alignment. **d** Close-up of the surface of the model showing some of the normals computed with Tensor Voting

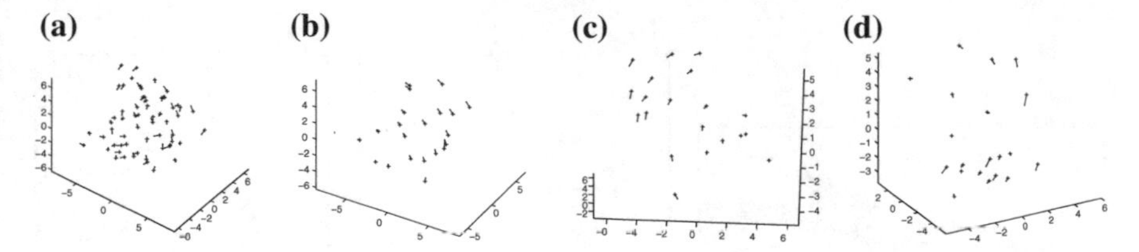

Fig. 16.10 **a** Three overlapping rigid motions in 3D. **b–d** The different motions as detected by our algorithm

voting space. However, in these circumstances, it suffices to analyze the other three voting spaces (Eqs. (16.5)–(16.7)) to disambiguate this case.

To illustrate this, we present a synthetic example where three overlapping motions with different axes of rotation, angles, and translations were generated in a $10 \times 10 \times 10$ cube centered at the origin (see Fig. 16.10). Our algorithm is applied as described in the algorithm of Sect. 16.2.4. However, after the first plane was detected, we removed its tokens from the voting space and the process was repeated until no more planes were found. This is, of course, the naive implementation of the solution. However, the algorithm can be modified to account for the presence of multiple planes. In that case, only the final stage, where the constraint from Eq. (16.19) is enforced, would be executed separately for each set of points.

16.3.3 Extension to Non-rigid Motion

While it can still be argued that, with some work, the Hough Transform might also be used to detect the same plane we obtain through Tensor Voting, there is another advantage to using the latter over the former: Tensor Voting enables us to find general surfaces. This means that we can also detect certain non-rigid motions that produce non-planar surfaces in the voting spaces.

To illustrate this, we generated a synthetic plane and then applied a twist transformation to it (see Fig. 16.11a). This transformation produces a curved surface in the voting space (clearly visible in the center of Fig. 16.11b–c, a close-up of the surface is also presented in Fig. 16.12). The surface is easily detected using Tensor Voting, and the resulting correspondences, from two different viewpoints, can be seen in Fig. 16.11d–e.

In order to detect this surface, we had to modify our algorithm as follows. The first two stages (sparse ball voting and sparse stick voting) are performed as usual. However, in the last stage, Eq. (16.19) was not enforced globally, but only locally around each active token. In other words, we enforced the presence of rigid transformations only on a local level. It must be remembered that Eq. (16.19) depends on two points. Therefore, for each token that was verified, we used the closest active

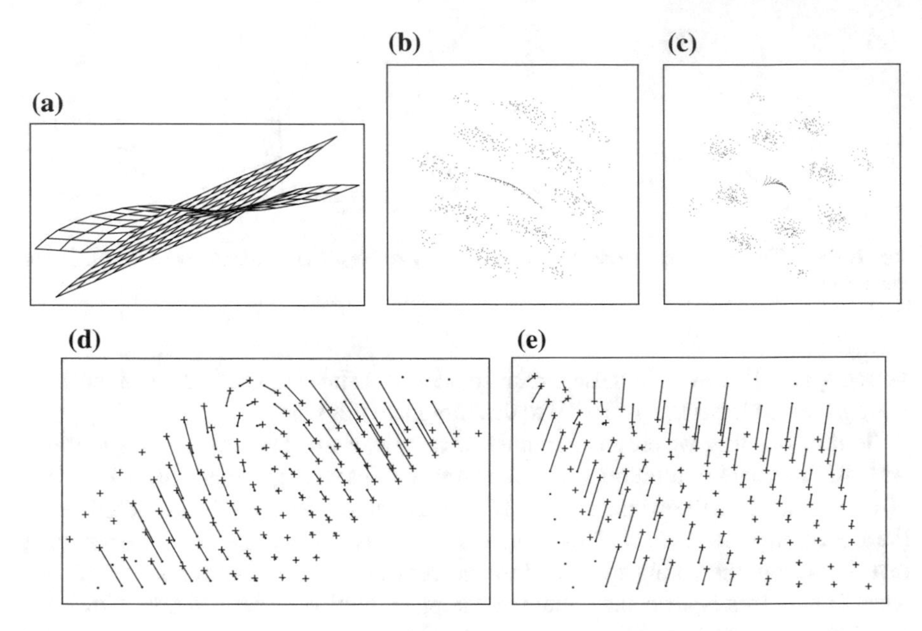

Fig. 16.11 **a** Non-rigid motion applied to a 3D plane. **b** and **c** The curved surface that was generated in the voting space from two different view points. **d** and **e** The resulting correspondences found with our algorithm seen from two different viewpoints

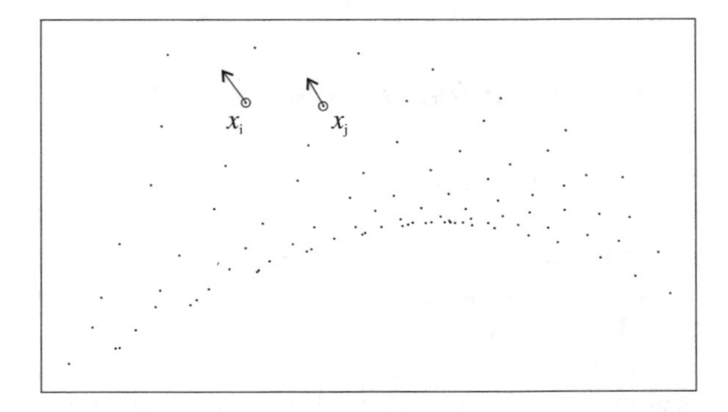

Fig. 16.12 A close-up of the surface corresponding to an elastic motion. The constraints of Eq. (16.19) are only verified locally between the closest point pairs. In this figure, token x_i is verified with its closest neighbor, x_j. Other pairs to be verified are also highlighted in the figure

neighbor. We illustrate this in Fig. 16.12. In that figure, the token x_i is being verified using its closest neighbor, x_j. The normals of the tokens are also shown.

A simpler version of this algorithm was also used to solve the correspondence problem in catadioptric images. In this case, the 2D images are mapped to a 3D sphere. In this space, the corresponding corners in the 2D images form a curved 3D surface that is easily detected using Tensor Voting. The resulting correspondences can be seen in Fig. 16.13.

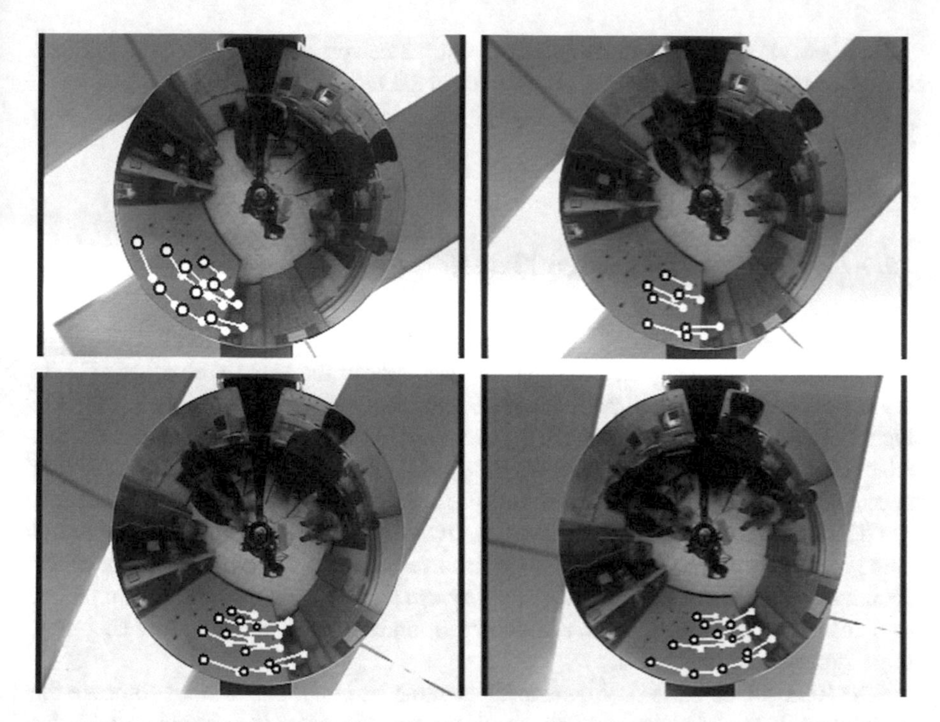

Fig. 16.13 Solving the correspondence problem in catadioptric images via a spherical mapping. White circles were successfully registered into the black-and-white circles

16.4 Hough Transform in Conformal Geometric Algebra

This section presents a method to apply the Hough transform to 2D and 3D cloud points using the conformal geometric algebra framework. The objective is to detect geometric entities, with the use of simple parametric equations and the properties of the geometric algebra. We show with real images and RGB-D data that this new method is very useful to detect lines and circles in 2D and planes and spheres in 3D. This section is based on the work done by G. López-González, N. Arana-Daniel, and E. Bayro-Corrochano [194].

The Hough transform is an algorithm for feature extraction used in image analysis, computer vision, and digital image processing [268]. This technique collects imperfect instances of objects within a certain class of shapes by a voting procedure in accumulators or cells. This voting scheme is carried out in a parameter space, where object candidates are obtained as local maximal in the accumulator space. The selection of the maxima of possible clusters is identified by a type of K-means algorithm. The Hough transform was developed for the identification of lines in the image, but later works extended the Hough transform to identifying positions of different shapes for example circles or ellipses [8, 80].

In this work, using the conformal geometric framework, we extend the randomized Hough transform to detect lines and circles in 2D cloud points of images and lines, planes, circles, and spheres in 3D cloud points obtained by 3D scanners and RGB-D sensors.

16.4.1 Randomized Hough Transform

This work is an extension of the randomized Hough transform, RHT, [167]. The RHT is an algorithm proposed to solve the problems of the Hough transform, HT. In the HT algorithm for each image pixel not only the cells of the possible entities are increased, but also of many other. This creates a problem to find the local maxima. Also the accumulator array is predefined by windowing and sampling the parameter space. For the correct detection of the entities, we need a good parameter resolution. For this, we need a big array that takes too much storage and computing time. Without some previous knowledge of the image, it is very hard to determine the size of the accumulator array. A bad accumulator array can lead to the next problems: a failure to detect some specific entities, difficulties in finding local maxima, low accuracy, large storage, and low speed.

The RHT solves these problems using parametric equations to only compute the possible entities and a dynamic accumulator array to solve the problems with the storage. By doing this, the storage space is greatly reduced. Other concept added is a scalar δ used as tolerance for similitude. When the difference between two computed entities is smaller than δ, then we consider that the two are the same. This scalar is used to set the resolution of the accumulator array. If we increase δ, the resolution and storage space will be lower.

The steps of the algorithm are:

(1) Randomly take n points from the set, being n the number of points needed to define the entity.

(2) Solve the parametric equations to get a candidate.

(3) Search for the candidate in the dynamic accumulator array. If the candidate is found, increase the accumulator by one. Otherwise add a new cell for the candidate and set its accumulator to one.

(4) If an accumulator surpasses a threshold, we check if the entity exists in the image. If it exists, we add it to a list of detected entities and delete from the set all the points that belong to it.

Finally, we must note that the RHT is a stochastic method in which its performance depends on the selection of δ and the randomized selection of the points.

16.4.2 CGA Representation of Lines, Circles, Planes, and Spheres

In conformal geometric algebra, the Euclidean space is embedded in a higher dimension. Because $\mathbb{G}_{2,1}$ is possible to visualize, we use it to exemplify this process. First the basis e_+ expands the space by one dimension. In this dimension, we can draw a unitary sphere centered in the origin. The basis e_- allows us to lift the bottom of the sphere to the origin of the space. Now, we make a line from the top of the sphere to the Euclidean point x_E. The intersection of this line with the sphere is the conformal point x_c. The conformal point x_c can be obtained with:

$$x_c = x_E + \frac{x_E \cdot x_E}{2} + e_0 = Xe_1 + Ye_2 + Ze_3 + \frac{X^2 + Y^2 + Z^2}{2} e_\infty + e_0. \quad (16.27)$$

Now a sphere in $\mathbb{R}^0$, a point, is a blade of grade 1. As defined before, the wedge product is a grade-increasing operation. If we take 2 spheres of grade 1, points, the result is a sphere of grade 2, a point pair. Therefore, the wedge product of 3 points is the circle, and with 4 points, we get a sphere.

There exist 2 special cases to consider: The first one occurs when the points are not in general position (GP); this means that there are 3 collinear points or 4 coplanar points. In this case, the result will be a line with 3 points and a plane with 4 points. The line and plane are a circle and a sphere, respectively, with infinite radius. The other is known as the dual property of the conformal space: The result of the wedge product between an entity and a point will be 0 if the point lies on the entity.

By using the wedge product results and the dual property yields, one can describe the Outer Product Null Space ($\mathbb{OPNS}$), representation. A visualization of this is presented in the next table.

Entity	Notation	$\mathbb{OPNS}$ representation	Grade
Point	x_c	$x_E + \frac{x_E^2}{2} e_\infty + e_0$	1
Point Pair	$\mathbb{ON}(PP)$	$x_{c1} \wedge x_{c2}$	2
Circle	$\mathbb{ON}(Z)$	$x_{c1} \wedge x_{c2} \wedge x_{c3}$; in GP	3
Line	$\mathbb{ON}(L)$	$x_{c1} \wedge x_{c2} \wedge x_{c3}$	3
Sphere	$\mathbb{ON}(S)$	$x_{c1} \wedge x_{c2} \wedge x_{c3} \wedge x_{c4}$; in GP	4
Plane	$\mathbb{ON}(P)$	$x_{c1} \wedge x_{c2} \wedge x_{c3} \wedge x_{c4}$	4

In $\mathbb{G}_{4,1}$ there exists an alternate representation. This representation is called the Inner Product Null Space, $\mathbb{IPNS}$. To change from $\mathbb{OPNS}$ to $\mathbb{IPNS}$ representations one multiply by the unit pseudoscalar I_c. This multiplication is called the dual operation. Because $I_c^2 = -1$, its inverse is $-I_c$, so to return to $\mathbb{OPNS}$, we multiply by $-I_c$. This is done to avoid a change of sign between both representations.

$$\mathbb{ON}(X)(I_c) = \mathbb{IN}(X) \quad \mathbb{IN}(X)(-I_c) = \mathbb{ON}(X), \tag{16.28}$$

A special case is the point representation x_c. The $\mathbb{OPNS}$ representation of the conformal point is also valid in $\mathbb{IPNS}$.

The combinations of these two representations allow us to obtain the information of the blades that define the geometric entities, as it is shown next along with the equations to obtain the parameters.

- Sphere

$$\mathbb{IN}(S) = \pm\alpha\left(C_c - \frac{r^2}{2}e_\infty\right), \tag{16.29}$$

where C_c is the conformal point of the center C_E, r is its radius, and α is some scale. Because of the fact that the blade e_0 can only be equal to 1 in the sphere case, the scale can be obtained as:

$$\pm\,\alpha = -\mathbb{IN}(S)\cdot e_\infty. \tag{16.30}$$

The equations to compute the parameters of the normalized sphere are

$$C_E = (\mathbb{IN}(S)\cdot(-I_E))I_E, \tag{16.31}$$

$$r^2 = \mathbb{IN}(S)^2. \tag{16.32}$$

- Plane

$$\mathbb{IN}(P) = \pm\alpha(n_E + de_\infty), \tag{16.33}$$

where n_E is the normal of the plane, d is the distance to the origin, and α is a scale factor equal to $|n_E|$. The equations to compute the parameters of the normalized plane are:

$$n_E = (\mathbb{IN}(P)\cdot(-I_E))I_E, \tag{16.34}$$

$$d = -\mathbb{IN}(P)\cdot e_0. \tag{16.35}$$

- Circle

To obtain the plane in which the circle lies, we use:

$$\mathbb{ON}(P) = e_\infty \wedge \mathbb{ON}(Z). \tag{16.36}$$

The norm of $\mathbb{ON}(Z)$ is the same of $\mathbb{ON}(P)$. The sphere with the same center and radius as the circle is obtained by:

$$\mathbb{IN}(S) = \mathbb{ON}(Z)\cdot\mathbb{ON}(P)^{-1}. \tag{16.37}$$

- Line

$$\alpha = \sqrt{\mathbb{ON}(L) \cdot \mathbb{ON}(L)}, \tag{16.38}$$

where α is thc scale of factor used to normalize the line. Once we have normalized the line, we can get its direction d_E, momentum m and closes point to the origin O_E.

$$d_E = \mathbb{ON}(L) \cdot E, \tag{16.39}$$

$$m = (0.5e_\infty - e_0) \cdot \mathbb{IN}(L), \tag{16.40}$$

$$O_E = -d_E \cdot mI_E. \tag{16.41}$$

16.4.3 Conformal Geometric Hough Transform

The steps for the algorithm are the same as the RHT. These steps can be described as follows:

(1) Transform the Euclidean points in the cloud to conformal points. Then randomly take sets of 4 points, x_{ci}.

(2) Do the wedge product between the first 3 points to get a circle $\mathbb{ON}(Z)$. Do $\mathbb{ON}(S) = \mathbb{ON}(Z) \wedge x_{c4}$.

(2.1) If $\mathbb{ON}(S)$ is 0, then x_{c4} lies on $\mathbb{ON}(Z)$. Do the wedge product between $\mathbb{ON}(Z)$ and e_∞. If the result is 0, $\mathbb{ON}(Z)$ is a line, otherwise a circle.

(2.2) If $\mathbb{ON}(S)$ is not 0, then do the wedge product between $\mathbb{ON}(S)$ and e_∞. If the result is 0, $\mathbb{ON}(S)$ is a plane, otherwise a sphere.

(3) After we detect the entity that the points x_{ci} form, we must eliminate two ambiguities. The first one is the scale factor, and the second is a variant sign caused by the anticommutative behavior of the wedge product between 2 vectors. To eliminate these ambiguities, we work in $\mathbb{IPNS}$ for the sphere and plane. A division by the $\pm\alpha$ obtained in Eq. 16.30 solves both ambiguities for the sphere. For the plane, we can get $\alpha = |n_E|$ to solve the scale factor. For the variant sign, we use the function atan2 with n_E, because it can distinguish between diametrically opposite directions. The function atan2 has the interval [-π, π], so the sign of the angle obtained is used to eliminate the variant sign. In the only exception to this method, where $X = Y = 0$, we use the sign of Z. To eliminate these ambiguities, we work in $\mathbb{OPNS}$ for the line and circle. For the line, we get α with Eq. 16.38. We also solve the variant sign with the function atan2 and its direction d_E. The circle can be converted to a plane and use the same steps.

Once we have discarded the ambiguities, we search for the candidate in its corresponding dynamic accumulator array. If the candidate is found, increase the accumulator by one. Otherwise, add a new cell for the candidate and set its accumulator to one.

(4) If an accumulator surpasses a threshold k, we check if it exists in the cloud. If it exists, we added it to a list of detected entities and deleted all the points that

belong to it from the set. To eliminate those points, we compute their distance to the entity to see if they are close enough with the next formulas:

$$\begin{aligned} D &= ||C_E - x_E| - r|, \\ D &= |x_C \cdot \mathbb{IN}(P)|. \end{aligned} \tag{16.42}$$

Equation 16.42 is also valid for the line. With the circle we use both.

There are some minimal changes to apply this algorithm in the planar case. The first one is that we only detect lines and circles, so we take sets of 3 points instead of 4. We only have 2 coordinates, X and Y, so Z is set to 0 by default. The elimination of the ambiguities of the circle will be different, because all the circles lie on the image plane $\mathbb{IN}(Pimg) = 0e_1 + 0e_2 + e_3 + 0e_\infty$. If we obtain the plane of the circle, this will be like $\mathbb{IN}(P) = \pm\alpha e_3$; then, both ambiguities can be discarded with a division by $\pm\alpha$.

16.4.4 Detection of Lines, Circles, Planes, and Spheres Using the Conformal Geometric Hough Transform

One good reason to use CGA to attack this problem is the parametric equations. They are more simple than the regular equations. For example, the equations used for circles in 2D with center (a, b) and radius r with 3 given points are:

$$\begin{aligned} m_a &= \frac{y_2 - y_1}{x_2 - x_1}, \qquad m_b = \frac{y_3 - y_2}{x_3 - x_2}, \\ a &= (m_a(m_b(y_1 - y_3)) + m_b(x_1 + x_2) - \frac{m_a(x_2 + x_3)}{2(m_b - m_a)}, \\ b &= \frac{-a}{m_a} + \frac{x_1 + x_2}{2m_a} + \frac{y_1 + y_2}{2}, \\ r &= \sqrt{(x - a)^2 + (y - b)^2}. \end{aligned} \tag{16.43}$$

These equations are more complicated than the equation used to represent a circle in $\mathbb{OPNS}$. Other advantage is that the $\mathbb{OPNS}$ representations are related between them. For instance, the equations of the circle and line are the same, the result only depends on the position of the points, and this also occurs with the sphere and plane. Also the result of the equation with tree points partially solves the one with four. As has been stated in other works [191], to maintain this advantage we need an optimal way to implement the equations in CGA. For this purpose, we use GAALOP [143]. GAALOP is a precompiler for C/C++ and OpenCL. This tool allows to make an efficient implementation. It reduces the storage space by using only the space needed to store the nonzero coefficients of the multivectors and also ignores all the unnecessary operations.

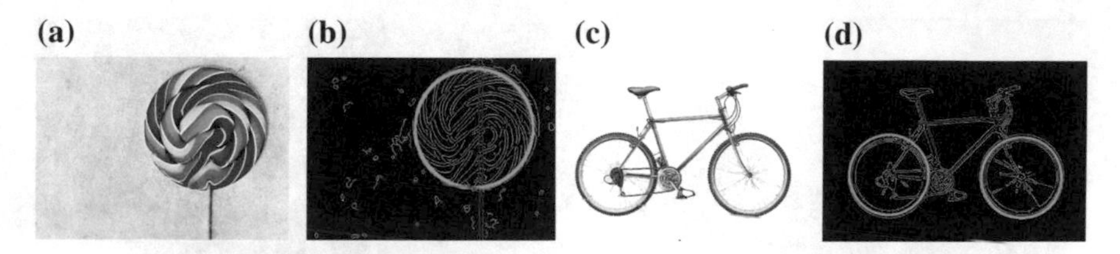

Fig. 16.14 Experiments with the candy and icicle images. Detected lines are in red and circles in green

Fig. 16.15 (left) RGB image obtained with the sensor and (right) the depth image

In the worst-case scenario, this algorithm has a computational complexity of $\mathcal{O}(\binom{n}{s})$, where n is the numbers of points in the cloud, and s is the number of points in each set, 4 for 3D cloud points and 3 in the planar case. For comparison, this is similar to $\mathcal{O}(n^s)$ but discarding permutations and repetitions. This is reduced by only taking a sample of the points from the cloud and eliminating those that already has been detected. In the 3D case, other viable option will be to use a deep segmentation and then apply the algorithm to each zone.

For the first experiments, we used 2D images. In the first, we take the picture of a candy; see Fig. 16.14a, to detect the circle-shaped candy and its stick. We first apply a Gaussian filter to reduce noise and then the Canny [47] algorithm to detect borders and then our algorithm. In Fig. 16.14b, we observe the detected circle in color green and the stick lines in color red.

For the next image, we use the picture of a bicycle; see Fig. 16.14c. The original image shows a icicle in a white background. The algorithm was able to detect the 2 wheels and 3 lines that form the frame.

In the last experiment, we used a RGB-D sensor, see Fig. 16.15.

As proposed in the analysis section, we use depth segmentation to reduce the computational cost; see Fig. 16.16. We also use the Canny algorithm in the RGB image to reduce the number of points; see Fig. 16.16. By doing this, we have been able to detect a sphere, the ball, and two planes, the table and the wall. To delimit the table, the quality of the points was not enough to detect a circle, but by using the parameters of the detected plane and the points that are close to it as constraints, we

Fig. 16.16 Depth segmentation and edge image used for the detection process

Fig. 16.17 (Left) Detection of the geometric entities. (Middle) Angle used to show the error in the approximation of the table. (Right) Rectification of the table

can get an approximation. The final result is visualized using the software Webots [213]; see Fig. 16.16.

Although the approximation of the table is good, we observe that it overlaps with the wall and ball. We can correct this with the use of conformal transformations. In order to do this, we construct a translator

$$T = 1 + \frac{\lambda}{2} n_E e_\infty, \tag{16.44}$$

where n_E is the direction of the translation, in this case the norm of the planes, and λ is the overlapping distance. We can see the result in Fig. 16.17.

16.5 Conformal Geometric Algebra Voting Scheme for Extraction of Circles and Lines in Images

This section is based on the works of G. Altamirano-Gómez and E. Bayro-Corrochano [2, 3]. The goal of this application is to find a representation of the contour of objects in images; hence, the input to our algorithm is an edge image. In addition, our method considers an image as a vector space $\mathbb{R}^2$, and each edge pixel is represented as a conformal point of CGA $G_{3,1}$. The output is a set of perceptual structures, i.e., circles and lines, associated with a saliency value. These structures are represented as elements of CGA $G_{3,1}$, according to Table 16.1.

Table 16.1 Representation of geometric entities in CGA $G_{n+1,1}$

Entity	IPNS	OPNS
Point (p_c)	$p_e + 0.5p_e^2 e_\infty + e_0$	$\bigwedge_{i=1}^{n+1} S_i$
Pair of points (PP)	$\bigwedge_{i=1}^{n} S_i$	$p_{c1} \wedge p_{c2}$
Hypersphere (S)	$c_c - 0.5\rho^2 e_\infty$	$\bigwedge_{i=1}^{n+1} p_{ci}$
Hyperplane (π)	$n_e + d_H e_\infty$	$e_\infty \bigwedge_{i=1}^{n} p_{ci}$

16.5.1 Euclidean Distance

A convenient property of CGA is that we can recover Euclidean metric using the inner product. Let x and y be vectors of $\mathbb{R}^{n+1,1}$, then its inner product is defined as follows:

$$x \cdot y = \frac{xy + yx}{2}, \tag{16.45}$$

where xy is the geometric product between x and y.

So that, the Euclidean distance between two Euclidean points $p_{ei}, p_{ej} \in \mathbb{R}^n$ is computed as follows:

$$|d| = \sqrt{-2\left(p_{ci} \cdot p_{cj}\right)}, \tag{16.46}$$

where p_{ci} and p_{cj} are the conformal representation of the Euclidean points.

In addition, the Euclidean distance between a point and a hyperplane is computed as follows:

$$|d| = \pi_c \cdot p_c, \tag{16.47}$$

where p_c is a point and π_c is an hyperplane of $\mathbb{R}^n$ in conformal representation.

Finally, the Euclidean distance between a point and an hypersphere is computed by:

$$|d| = \sqrt{\rho^2 - 2\left(S_c \cdot p_c\right)} - \rho, \tag{16.48}$$

where p_c is a point, and S_c is an hypersphere with radius ρ, in conformal representation.

16.5.2 Local Voting

In this stage, we select an edge pixel, denoted by p_0 and defines a neighborhood, denoted by P_0. Without loss of generality, we set p_0 as origin of the coordinate system; then, a pixel p_i has image coordinates (u_i, v_i).

Next, each pixel in the neighborhood p_0 casts a vote in the form of a perceptually salient line:

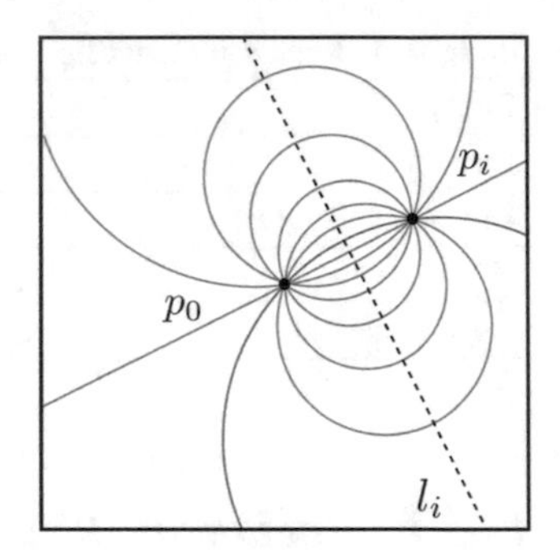

Fig. 16.18 For a pencil of circle, in which all circles meet in two real base points, p_0 and p_i, the centers of the circles lie on a line, l_i

$$\bar{l}_i = \{p_c : p_c \cdot l_i = 0\}, \ W : \mathcal{R}^m \to \mathcal{R}, \tag{16.49}$$

where

$$l_i = -\frac{u_i}{|p_{ie}|}e_1 - \frac{v_i}{|p_{ie}|}e_2 + \frac{|p_{ie}|}{2}e_\infty, \tag{16.50}$$

Since p_0 and p_i are the base points of a pencil of circles, the center of all possible circles passing through p_0 and p_i lies on l_i, as Fig. 16.18 shows. In addition, the perceptual saliency of l_i is assigned using the decay function of the Tensor Voting algorithm [209, 293]:

$$W(s, \rho, c, \sigma) = \exp\left(-\frac{s^2 + c\rho^2}{\sigma^2}\right), \tag{16.51}$$

where c controls the degree of decay with curvature, σ determine the neighborhood size of the voting, s represents the arc length, and ρ the curvature. The values of c and σ are taken as input parameters, while s and ρ are computed as follows:

$$s = \frac{\theta d}{\sin\theta}, \ \rho = \frac{2\sin\theta}{d}, \tag{16.52}$$

where d represents the Euclidean distance between p_0 and p_i computed using (16.46), and θ represents the angle between the tangent of the circle at p_0 and the line defined by p_0 and p_i, given by the equation:

$$\theta = \arctan\left(\frac{v_i}{u_i}\right) - \arctan\left(-\frac{c_x}{c_y}\right). \tag{16.53}$$

Once we have mapped each point p_i in P_0 to a perceptually salient line, l_i, we obtain a set of lines denoted by L_0. The intersection of each pair of lines in L_0 is a pair of points, composed by the base e_∞ and the center of the circle passing through points p_0, p_i, and p_j. Hence, we compute their intersection using the following equation:

$$c_{0ij} \wedge e_\infty = (l_i \wedge l_j)^* \tag{16.54}$$

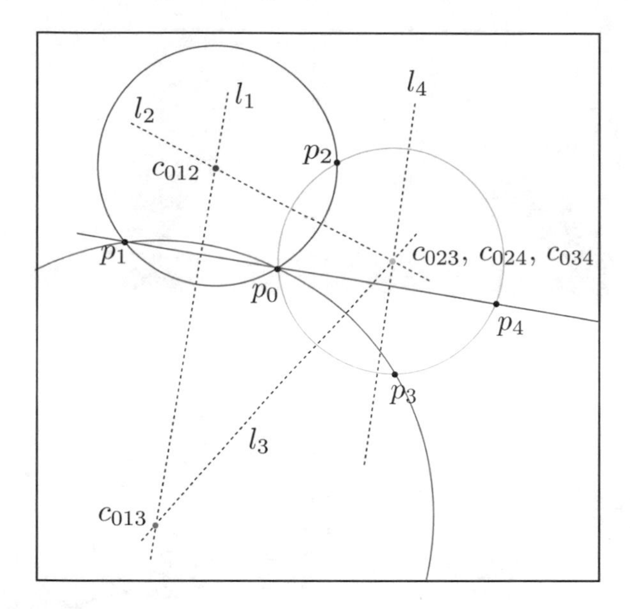

Fig. 16.19 Local voting for a set of 4 points. Each point p_i, $i = 1, \ldots, 4$, is mapped to a line l_i. Each pair of lines intersects in the center of a circle, c_{0ij}

and from this result we extract c_{0ij}. Figures 16.19 and 16.20 show an example of the local voting process using 4 points.

If the result of (16.54) is zero, then lines meet at a point at infinity; hence, p_0, p_i, and p_j are collinear. In this case, the vote is the line passing through p_0, p_i, and p_j:

$$n_{0ij} = (p_0 \wedge p_i \wedge e_\infty)^*. \tag{16.55}$$

In the voting space, each c_{0ij} or n_{0ij} is a point, and its saliency value is given by:

$$W_{0ij} = W(s_i, \rho_i, c, \sigma) + W(s_j, \rho_j, c, \sigma). \tag{16.56}$$

Once we have computed all possible intersection of lines in set L_0, we cluster resulting points using DBSCAN algorithm. The next step is to compute the perceptual saliency of each cluster using (12.170). Finally, we select clusters that surpass a threshold value, or the cluster with maximum perceptual saliency. In each case, we compute the weighted mean of points in the selected cluster, for circles:

$$\bar{c}_{0ij} = \frac{1}{\bar{W}} \sum_i \sum_j \left(W_{0ij} c_{0ij} \right). \tag{16.57}$$

and the radius is given by the magnitude of $\bar{c}_{0ij}$. On the other hand, if the winner of the voting process is a line, we compute:

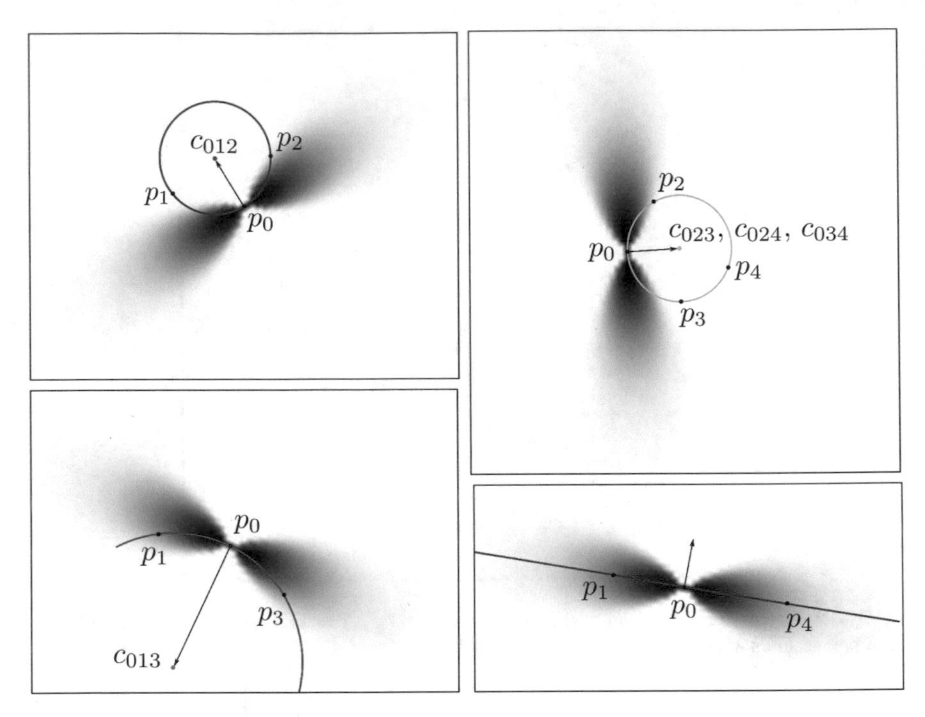

Fig. 16.20 Saliency assignment in the local voting stage. For each geometric structure, a saliency field is constructed using (16.51). Each point in the geometric structure casts a vote, with a magnitude given by the saliency field

$$\bar{n}_{0ij} = \frac{1}{\bar{W}} \sum_i \sum_j \left(W_{0ij} n_{0ij} \right) \tag{16.58}$$

and the Hesse distance or foot is zero.

16.5.3 Global Voting

Let $\hat{S}$ and $\hat{L}$, be the output of the local voting process, i.e., a set of circles and lines respectively; then, we apply DBSCAN algorithm to separate in clusters: circles with the same center and radius, and lines with the same normal and Hesse distance. Thus, $\hat{S}$ is partitioned in subsets $\hat{S}_1, \ldots, \hat{S}_q$, where each $\hat{S}_k$ is a cluster of circles, obtained by DBSCAN. In the same way, $\hat{L}$ is partitioned in subsets $\hat{L}_1, \ldots, \hat{L}_q$, where each $\hat{L}_k$ is a cluster of lines, obtained by DBSCAN.

Global Voting for Circles. Let $\hat{S}_k = \{S_1, \ldots, S_r\}$, be a cluster of circles obtained by DBSCAN, then we compute its perceptual saliency, $\bar{W}_k$, using (12.170). Afterward, we compute the weighted mean of the cluster:

$$\bar{S}_k = \bar{c}_{ek} + \frac{\bar{c}_{ek}^2 - \bar{\rho}_k^2}{2} e_\infty + e_0, \tag{16.59}$$

where

$$\bar{c}_{ek} = \frac{1}{\bar{W}_k} \sum_i^r (W_i c_{ei}), \; \bar{\rho}_k = \frac{1}{\bar{W}_k} \sum_i^r (W_i \rho_i). \tag{16.60}$$

In addition, we recompute the perceptual saliency of circle $\bar{S}_k$, and the idea is to count how many pixels in the image lies on it; in this way, we obtain a saliency value that is supported with global information. Thus, the perceptual saliency of $\bar{S}_k$ is recomputed as follows:

$$W(\bar{S}_k, D) = \sum_{p_{ic} \in D} \exp\left(-\frac{|d_i|^2}{\sigma^2}\right), \tag{16.61}$$

where $D = \{p_{c1}, \ldots, p_{cm}\}$ is the set of all edge pixels in the image represented as conformal points, $|d_i|$ is the Euclidean distance from circle $\bar{S}_k$ to each point in D, which is computed using (16.48), and σ is a parameter that sets the maximum allowed distance from a pixel to the circle. Finally, we select circles that surpass a threshold value.

Global Voting for Lines. A similar procedure is applied to lines, where for each cluster of lines, $\hat{L}_k = \{L_1, \ldots, L_r\}$, obtained by DBSCAN, we compute its perceptual saliency, $\bar{W}_k$, using (12.170). Afterward, we compute the weighted mean of the elements of cluster $\hat{L}_k$:

$$\bar{L}_k = \frac{1}{\bar{W}_k} \sum_i^r (W_i L_i). \tag{16.62}$$

Next, we recompute the perceptual saliency of line $\bar{L}_k$ using (16.61), but in this case $|d_i|$ is the Euclidean distance from line $\bar{L}_k$ to each point in D, which is computed using (16.47). Finally, we select lines that surpass a threshold value.

16.5.4 Complexity Analysis

Our voting scheme takes as input a preprocessed image with a set of k edge pixels and gives as output a set of geometric entities in conformal representation.

The local voting step selects an edge pixel, and define a neighborhood. Let m be the size of the neighborhood, then the computing of intersections takes $O(m^2)$ and produces m^2 geometric entities. Their clustering is done with DBSCAN algorithm, implemented with R-trees [114], and takes $O(m^2\, lg(m^2))$. In addition, extracting the mean of the cluster with highest density takes $O(m^2)$. Since local voting is executed for each edge pixel in the image, then the local voting step takes $O(km^2\, lg(m^2))$.

In addition, the global voting step takes as input a set of k' geometric entities, at least one for each pixel. Clustering with DBSCAN algorithm takes $O(k'\, lg(k'))$, and computing the mean of each cluster takes $O(k')$. Thus, our algorithm has a complexity of $O(km^2\, lg(m^2) + k'\, lg(k'))$.

16.5.5 Conformal Geometric Algebra Voting Scheme for Detection of Bilateral Symmetry

This section is based on the works of G. Altamirano-Gómez and E. Bayro-Corrochano [2, 3]. Using the CGA voting scheme as building blocks, we have designed a three-level architecture for detection of bilateral symmetry in images. Figure 16.21 shows an overview of the architecture. The goal of the first two levels at the right side of the architecture is to extract symmetry axis with local support, while the left side extracts salient circles and lines, and map them to symmetry axis. The third level uses these geometric entities to compute symmetry axis with local and global support. For an implementation of these algorithms using FPGA, see the works [286, 287].

Local Voting. In this stage, we select an edge pixel, denoted by p_0, and define a neighborhood. Without loss of generality, we set the coordinates of pixel p_0 as the origin. Then, each pixel in the neighborhood p_0 casts a vote in the form of a perceptually salient line:

$$\bar{l}_i = \{p_c : p_c \cdot l_i = 0\},\ W : \mathcal{R}^m \to \mathcal{R}. \tag{16.63}$$

Let p_i be a pixel in the neighborhood of p_0; then, we can reflect p_i through a line to obtain point p_0. So that, the symmetry axis defined by this pair of points is represented by the line:

$$l_i = -\frac{u_i}{|p_{ie}|}e_1 - \frac{v_i}{|p_{ie}|}e_2 + \frac{|p_{ie}|}{2}e_\infty, \tag{16.64}$$

where $p_{ie} = (u_i, v_i)$ are the coordinates of pixel p_i in the image. So that, for extracting symmetry axis, the vote of a pixel p_i on p_0 is the line l_i, with a perceptual saliency given by:

$$W(l, D) = \sum_{p_{ic} \in D} \exp\left[\frac{(l\, p_{ic}\, \tilde{l}) \cdot p_{jc}}{\sigma^2}\right], \tag{16.65}$$

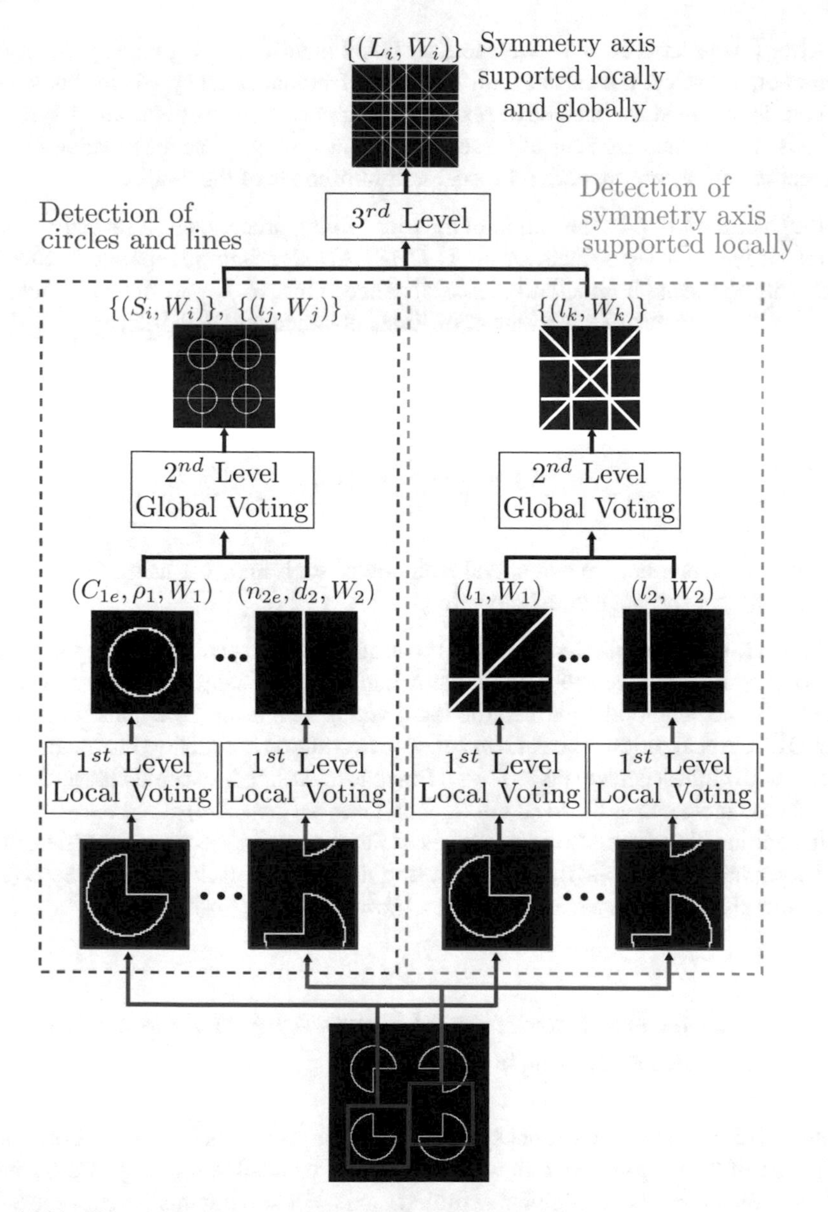

Fig. 16.21 Three-level architecture for extraction of symmetry axis

where $D = \{p_{1c}, \ldots, p_{kc}\}$, is the set of all edge pixels in the image represented as conformal points, l is the symmetry axis and is applied as a versor transformation on point p_{ic}, p_{jc} is the nearest neighbor to point $l\ p_{ic}\ \tilde{l}$, and σ is a parameter that sets the size of the window in which we search for point p_{jc}.

After this process is done, we use DBSCAN algorithm for grouping the votes; thereafter, we select the cluster with highest perceptual saliency. Since this voting process is executed for each edge pixel, the output of this level is a set of symmetry axis, where each element of these sets has an associated saliency value. These geometric entities are supported in local neighborhoods of the image.

Global Voting. Let $\hat{L}$ be the output of the local voting process, i.e., a set of lines that represent symmetry axis; then, we apply DBSCAN algorithm to separate in clusters lines with the same normal and Hesse distance. Thus, $\hat{L}$ is partitioned in subsets $\hat{L}_1, \ldots \hat{L}_q$, where each $\hat{L}_k$ is a cluster of lines, obtained by DBSCAN.

Next, for each cluster of lines, $\hat{L}_k = \{L_1, \ldots, L_r\}$, obtained by DBSCAN; we compute its perceptual saliency using (12.170). The weighted mean of the elements of cluster $\hat{L}_k$ is given by:

$$\bar{L}_k = \frac{1}{\bar{W}_k} \sum_{i}^{r} (W_i L_i) . \tag{16.66}$$

Finally, we recompute the perceptual saliency of each line, $\bar{L}_k$, using (16.65), and select lines that surpass a threshold value.

Computational Complexity. Let k be the number of edge pixels in the input image, the local voting step selects one of them and defines a neighborhood. Let m be the size of the neighborhood, then the local voting step takes $O(m)$, and clustering with DBSCAN algorithm takes $O(m\ lg(m))$. In addition, extracting the mean of the cluster with highest density takes $O(m)$. Since local voting is executed for each edge pixel in the image, then the local voting step takes $O(km\ lg(m))$.

In addition, the global voting step takes as input a set of k' geometric entities, their clustering takes $O(k'\ lg(k'))$, and computing the mean of each cluster takes $O(k')$. Thus, our algorithm has a complexity of $O(km\ lg(m) + k'\ lg(k'))$.

16.5.6 Third-Level Extraction of Symmetry Axis from Salient Structures of Shape

Section 16.5.5 allows us to extract symmetry axis, supported in local neighborhoods; if the size of the neighborhood in which we apply the local voting step is too small, this algorithm cannot detect global symmetry axis. For solving this issue, we use the circles and lines computed at the beginning of Sect. 16.5.

In this stage, each pair of circles and lines casts a vote in the form of a perceptually salient line:

$$\bar{L}_i = \{p_c : p_c \cdot L_i = 0\},\ W : \mathcal{R}^m \rightarrow \mathcal{R}, \tag{16.67}$$

which represents a symmetry axis.

Then, the vote cast by each pair of circles is computed as:

$$L = [(S_1 \wedge S_2)^* \wedge e_\infty]^*, \tag{16.68}$$

where S_i and S_j are two different circles with equal radius.

On the other hand, for each pair of parallel lines l_i and l_j, the geometric structure of the vote is given by:

$$L_d = T_d \, l_i \, \tilde{T_d}, \tag{16.69}$$

where $d = |(l_i \wedge l_j)^*| n_i$, n_i is the normal of the lines, and T is a translator of CGA $G_{3,1}$.

Finally, for lines that intersect with each other, we have two symmetry axis; thus, the geometric structure of the vote is given by:

$$L_\theta = R_\theta \, l_i \, \tilde{R}_\theta, \tag{16.70}$$

where R is a rotor of CGA $G_{3,1}$, and θ is the angle between the lines, computed as follows:

$$\theta_1 = \arctan\left(\frac{|l_i \wedge l_j|}{|l_i \cdot l_j|}\right), \ \theta_2 = \theta_1 + \pi. \tag{16.71}$$

After that, we assign a perceptual saliency to each symmetry axis using (16.65). At the end of this process, we use DBSCAN algorithm for clustering symmetry axis with similar parameters, recompute the perceptual saliency of symmetry axis,, and using a threshold value select clusters with maximum perceptual saliency.

Computational Complexity. Let m_1, m_2, and m_3 be the cardinality of the sets of circles, lines, and symmetry axis, respectively; then, after mapping circles and lines to symmetry axis, we obtain a set of size $m = C(m_1, 2) + C(m_2, 2) + m_3$. Since we apply DBSCAN algorithm to this set, the total complexity of this level is $O(m \ \lg m)$.

16.5.7 Relationship with Hough Transform, Tensor Voting, and Fitting Techniques

Our method is a flexible technique that extends previous voting schemes. Then, changing some parameters in the local voting process, we obtain equivalent results to Hough transform, Tensor Voting, and fitting techniques. For example, if we set the saliency field to a constant value, the output of our algorithm is equivalent to that obtained by Hough transform, applied in a local neighborhood.

Moreover, from lines and circles obtained in the local voting stage, we can compute the normal of the curve passing through each pixel. This result is equivalent to that obtained with Tensor Voting, and the global voting step should be replace by a marching cubes algorithm [195] as is done by the tensor voting framework; alternatively, we can apply directly a marching hyperspheres algorithm [29] to the output of the local voting stage. Even though Tensor Voting algorithm has lower computational complexity than ours, the use of a clustering technique allows our method to extract multiple features in a local neighborhood, since we are grouping features by its geometric similarity. In addition, we obtain a representation of complex objects in the image in terms of geometric entities.

Furthermore, for a set of tokens, if we use a fitting method to estimate the must likely geometric structure, by minimizing (12.161), we will face nonlinearity problems, and a bias due to outliers. Using our geometric approach, we take the outlier pixels out using a saliency field and compute the geometric structure using the rest of the points.

16.5.8 Experimental Analysis of Conformal Geometric Algebra Voting

In this section, we present a set of experiments with images to illustrate the viability of our method.

Application for Detection of Circles and Lines. Figure 16.22 shows four images for which we have applied our voting geometric scheme. Input images contain incomplete data (Fig. 16.22), noisy data[1] (Fig. 16.22), an image with illusory contours (Fig. 16.22), and an image with noise and an incomplete geometric entity (Fig. 16.22). Output of our algorithm is shown in Fig. 16.22, where in each case, as opposite to any current algorithms, our method recovers simultaneously the equations of the circles and lines that are perceptually salient, with subpixel error.

Figure 16.23 shows a set of images that contain real objects, and the input images were taken from image databases [152, 311]. For each input image, we apply a preprocessing step, which consists in a Canny edge detector [47], and a mean filter; after that, we use our algorithm to extract salient geometric structures. The second row of Fig. 16.23 shows the output of our algorithm for each image; salient structures have been superimposed to the edge images obtained after the preprocessing step. This set of images contains objects which contours describe nonlinear curves. Experimental results show that our algorithm makes a non-uniform sampling with circles and lines, in order to describe this objects; note that each geometric entity obtained by our algorithm is a *local descriptor of shape*, and the total output describes the object by a sort of an expansion of spherical wavelets [16].

[1] Each pixel has a 0.09 probability of being an erroneous site.

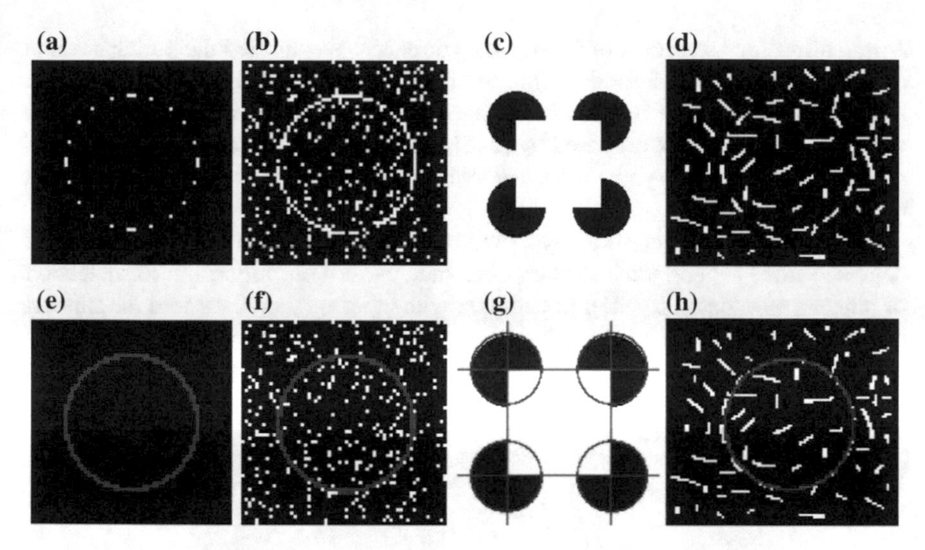

Fig. 16.22 Synthetic images (a, b, c, d) and the results obtained by our method (e, f, g, h)

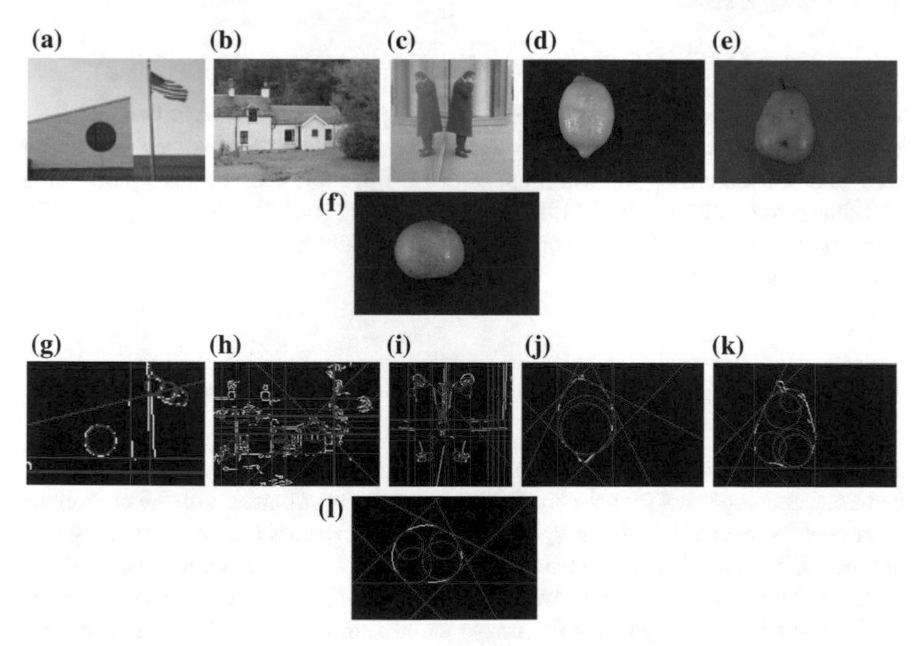

Fig. 16.23 Images that contain real objects (a, b, c, d, e, f) and the results obtained by our method (g, h, i, j, k, l)

Application for Detection of Bilateral Symmetry. We applied the voting scheme described in Sect. 16.5.5 for detection of bilateral symmetry in images that contain real objects, as well as in synthetic images. For comparison with other methods, we used the benchmark proposed by [246], which contains a set of 30 test images, divided in 4 categories: synthetic vs. real images, and images with a single vs. multiple reflection axis.

For each input image, we apply a preprocessing step, which consists in a Canny edge detector [47] and a mean filter; after that, we use our algorithm for detection of bilateral symmetry. In order to compare with other methods, we find the support region of each symmetry axis using the algorithm presented below. The output is shown in Fig. 16.24, where the detected symmetry axis is superimposed to the input image.

To evaluate our algorithm, we calculate the precision rate:

$$P = \frac{TP}{TP + FP}, \tag{16.72}$$

and the recall rate:

$$R = \frac{TP}{TP + FN}, \tag{16.73}$$

where TP, FP, and FN are the number of true positives, false positives, and false negatives, respectively. Figure 16.25 shows a comparison between precision and recall rates obtained by our voting scheme and the bilateral symmetry detection algorithms of Loy and Eklund [198], Mo and Draper [208], and Kondra et al. [162]. Our algorithm has a similar performance to state-of-the-art methods and is slightly better than three of them.

16.6 Egomotion, 3D Reconstruction, and Relocalization

Robots equipped with a stereo rig for 3D vision require a hand–eye calibration procedure so that the robot global-coordinate system is coordinated with the coordinate system of the stereo rig. However, due to unmodeled parameters of the pan–tilt unit and noise, it is still difficult to have precise coordination between those coordinate systems. After various attempts, we decided that, on the one hand, we should resort to Kalman filter techniques for the image stabilization and, on the other hand, we should carry out the tracking by applying sliding-mode-based nonlinear control techniques. Consequently, we can get a more precise perception and tracking system. In this section, we will not give the details of our algorithms for the nonlinear control of the pan–tilt unit; they can be found in [12].

Fig. 16.24 Detection of bilateral symmetry using our voting method

16.6.1 Relocalization and Mapping

In order to perceive the environment autonomously using stereo vision, our robot uses simultaneous localization and mapping (SLAM). To implement this, we used the approach based on the extended Kalman filter (EKF) proposed by Davison [63]. We extended the monocular EKF approach for stereo vision to improve the perception accuracy. We will briefly summarize the monocular SLAM and then give additional details for stereo SLAM. For the sake of simplicity, the EKF is formulated to work

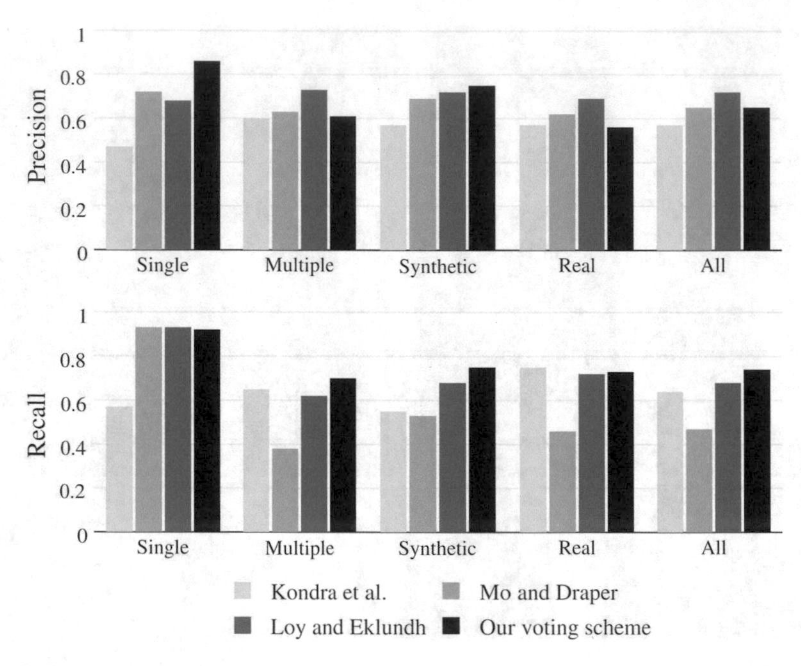

Fig. 16.25 Recall and precision rates for symmetry axis detection

in the 3D Euclidean geometric algebra $G_{3,0,0}$, which is a subalgebra of conformal geometric algebra $G_{4,1}$ for 3D space. The estimated state and covariance of the digital camera are given by

$$\hat{\mathbf{x}} = \begin{bmatrix} \hat{\mathbf{x}}_v \\ \hat{\mathbf{y}}_1 \\ \hat{\mathbf{y}}_2 \\ \vdots \end{bmatrix}, \qquad \mathrm{P} = \begin{bmatrix} \mathrm{P}_{xx} & \mathrm{P}_{xy_1} & \mathrm{P}_{xy_2} & \cdots \\ \mathrm{P}_{y_1x} & \mathrm{P}_{y_1y_1} & \mathrm{P}_{y_1y_2} & \cdots \\ \mathrm{P}_{y_2x} & \mathrm{P}_{y_2y_1} & \mathrm{P}_{y_2y_2} & \cdots \\ \vdots & \vdots & \vdots & \end{bmatrix}, \tag{16.74}$$

where the camera state vector $\hat{\mathbf{x}}_v$ comprises a 3D translation vector $\boldsymbol{t}^W$, an orientation rotor $\boldsymbol{R}^{WR}$ (isomorphic to a quaternion), a velocity vector $\mathbf{v}^W$, and an angular velocity vector w^W. This amounts to a total of 13 parameters. Feature states $\hat{\mathbf{y}}_i$ correspond to 3D points of objects or 3D points in the environment. In order to correct the rotor $\boldsymbol{R}$ and the translation vector $\boldsymbol{t}$, we rectify them using motors (rotor and dual rotor). Since a motor is given by

$$\boldsymbol{M} = \boldsymbol{T}\boldsymbol{R} = \left(1 + I\frac{\boldsymbol{t}}{2}\right)\boldsymbol{R} = \boldsymbol{R} + I\frac{\boldsymbol{t}}{2}\boldsymbol{R} = \boldsymbol{R} + I\boldsymbol{R}', \tag{16.75}$$

the rotor $\boldsymbol{R}$ and dual rotor $\boldsymbol{R}'$, which involves the translation, must fulfill the following constraints:

Algorithm 1 Computation of the support region of a symmetry axis.

Require: An image with a set of k edge-pixels:
$I = \{p_{1c}, \ldots, p_{kc}\}$,
and a symmetry axis:
$l = n_x e_1 + n_y e_2 + d_H e_\infty$.
Ensure: The initial and end points of the symmetry axis, (p_{1c}, p_{2c}).
1: Define a set $P = \{\}$.
2: **for** each edge-pixel $p_{ic} \in I$ **do**
3: $p'_{ic} = l\, p_{ic} \tilde{l}$.
4: p_{jc} =nearest edge-pixel to point p'_{ic}.
5: **if** $\sqrt{-2(p_{jc} \cdot p'_{ic})} < threshold$ **then**
6: Add (p_{ic}, p_{jc}) to set P.
7: **end if**
8: **end for**
9: Define a set $P' = \{\}$.
10: **for** each edge-pixel $p_{ic} \in P$ **do**
11: $\theta = -\arctan(n_x/n_y)$
12: Define a rotor: $R_\theta = \cos\left(\frac{\theta}{2}\right) - \sin\left(\frac{\theta}{2}\right) e_1 \wedge e_2$.
13: $p'_{ic} = R_\theta (p_{ic}) \tilde{R}_\theta$
14: Add p'_{ic} to set P'
15: **end for**
16: Select $(p_{ic}, p_{jc}) \in P$ such as $p'_{ic} \cdot e_2$ is minimum.
17: $p_{1c} \wedge e_\infty = [(p_{ic} \wedge p_{jc} \wedge e_\infty)^* \wedge l]^*$
18: Select $(p_{ic}, p_{jc}) \in P$ such as $p'_{ic} \cdot e_2$ is maximum.
19: $p_{2c} \wedge e_\infty = [(p_{ic} \wedge p_{jc} \wedge e_\infty)^* \wedge l]^*$
20: **return** (p_{1c}, p_{2c})

$$M\widetilde{M} = 1, \tag{16.76}$$
$$R\widetilde{R} = 1, \tag{16.77}$$
$$R\widetilde{R}' + \widetilde{R}R' = 0, \tag{16.78}$$

where the last equation indicates that $\boldsymbol{R}$ has to be orthogonal to the dual rotor $\boldsymbol{R}'$ and is valid up to a scalar. Unfortunately, in practice, the rotor $\boldsymbol{R}$ estimated by the EKF is usually not truly orthogonal to the estimated $\boldsymbol{R}' = \frac{t}{2}\boldsymbol{R}$. We adjust both rotors for orthogonality using the simple technique suggested in [28], and the translation vector is recomputed as follows: $\boldsymbol{t} = 2\boldsymbol{R}'\widetilde{\boldsymbol{R}}$.

In the EKF prediction step, a model for smooth motion is used involving the Gaussian-distributed perturbations $\mathbf{V}^W$ and Ω^R, which affect the camera's linear and angular velocities, respectively. The explicit process for motion in a time-step Δt is given by

$$\mathbf{x}_v = \begin{bmatrix} \mathbf{t}^W_{new} \\ \boldsymbol{R}^{WR}_{new} \\ \mathbf{v}^W_{new} \\ \mathbf{w}^R_{new} \end{bmatrix} = \begin{bmatrix} \mathbf{t}^W + (\mathbf{v}^W + \mathbf{V}^W)\Delta t \\ \boldsymbol{R}^{WR}\boldsymbol{R}((w^R + \Omega^R)\Delta t) \\ \mathbf{v}^W + \mathbf{V}^W \\ w^W + \Omega^R) \end{bmatrix}. \tag{16.79}$$

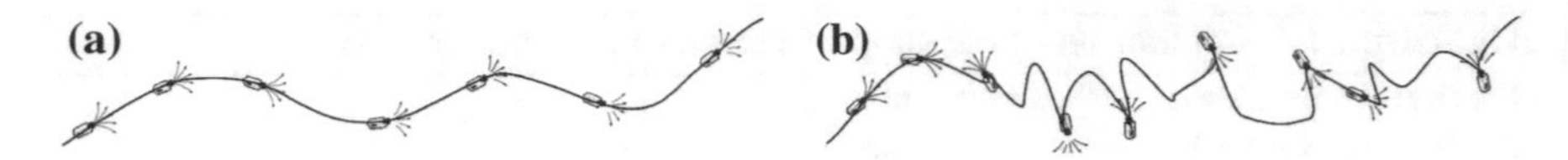

Fig. 16.26 Visualization of the camera's motion: **a** constant velocity model for smooth motion. **b** Nonzero acceleration by shaking motion

Figure 16.26a depicts the potential deviations from a constant-velocity trajectory, and Fig. 16.26b depicts a shaking motion during robot maneuvers.

The EKF implementation demands computations of the Jacobians of this motion function with respect to both $\mathbf{x}_v$ and the perturbation vector.

16.6.2 Machine Learning to Support the Robot's Spatial Orientation and Navigation

A landmark literally is a geographic feature used by explorers and others to find their way back or to move through a certain area. In the 3D map-building process, a robot can use these landmarks to remember where it was before, while it explores its environment. Also, the robot can use landmarks to find its position in a map previously built, therefore facilitating its relocalization. Since we are using a camera stereo system, the 3D position and pose of any object can be obtained and represented in the 3D virtual environment of the map. Thus, a natural or artificial landmark located in the environment can greatly help the mobile robot to know its actual position and relative pose with respect to both the map and the environment.

Viola and Jones introduced a faster machine learning approach to face detection based on the so-called *AdaBoost algorithm* [304]. This approach can be used to detect our static landmarks. Once the landmarks have been selected and trained, the robot can utilize them to navigate in the environment. This navigation is supported by the 3D virtual map, and only one camera (left camera) is used to perform the Viola and Jones algorithm. If a landmark is found, we get a subimage I_L from the left camera image. This I_L is the region of the image where the landmark was found; see Fig. 16.27a.

When a landmark is identified in one image (left camera), we must be sure that the landmark is present in the image in the right camera in order to get the 3D position. The landmark in the right image is also detected by the Viola and Jones algorithm, and its region is identified by a subimage I_R.

16.6.3 Landmark Position and Pose Estimation

When we talk about landmark position estimation, we are looking strictly for the 3D location of these landmarks in the environment, not for the pose (position and orientation) of the object found. To do this, we precalculated the depth using the

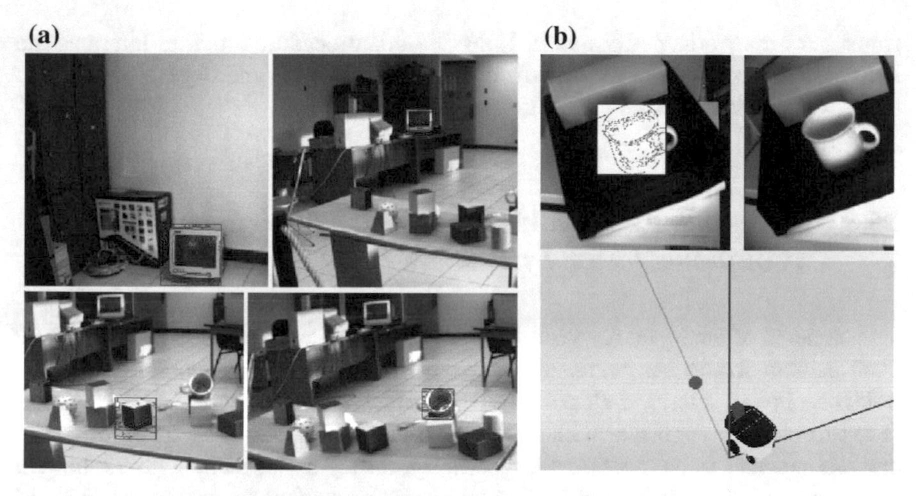

Fig. 16.27 **a** Four landmarks found in the left camera image using the Viola and Jones algorithm. The images show the point of interest. **b** Identification of an object and determining its 3D position. It is represented in the 3D virtual environment using a sphere

disparity of one object's fixed point. After getting the landmark identified in both images, we proceed to calculate the points of interest. To do this, we use Canny edge detection operator on I_L. To find the correspondence of these points, we use the epipolar geometry [126] and the *zero-mean normalized cross-correlation* (ZNCC).

Correspondences of an image patch are searched for along the epipolar line by calculating the ZNCC only in a given interval $(d_{min}, \ldots, d_{max})$ of so-called disparities [86]. A small disparity represents a large distance to the camera, whereas a large value indicates a small distance (parallax).

When all the points are matched in both images, we proceed to calculate its 3D position using the stereo triangulation. Then, we integrate this set of points to get its center of gravity and place the center of a virtual sphere on it. The radius of the covering sphere is calculated taking the highest number of points of the landmark within the sphere. The sphere is stored in the 3D virtual map using CGA and labeled as a landmark; see Fig. 16.27b.

Since the robot is using the EKF filter to estimate the position and pose of its binocular head, we use the 3D landmarks to also compute the pose relative to the binocular head with respect to the 3D landmark location. Once the object has been recognized using the Viola and Jones method, we use stereo to get the relative 3D pose. For that, a very simple method is utilized; namely, a plane of the object is used as reference and the robot computes the gravity center and its normal. This information is fed into the state vector of the EKF in order to reduce the pose and position error. Note that through time the EKF covariance increases due to the robot drifting and noise. However, by using the pose gained from the 3D landmarks, we can reduce the covariance greatly. In addition, if the robot has carried out a shaking motion, the covariance increases as well. Thus, using the pose of the landmarks, the robot can improve its performance. Note that this procedure provides intermediate help for the robot both for the initialization of the EKF filter and to get a much more robust relocalization in long robot tours.

16.6.4 Integration of SLAM and Machine Learning in the Conformal Vision System

The cameras of the binocular vision system work in a master–slave fashion, while each camera fixates on points of interest to estimate its own pose and the 3D points. Points of interest are tracked through time. Note that with the EKF the 3D reconstruction is monocular without the need to establish frame by frame the point correspondences satisfying epipolar geometry. It may be the same with the human visual system, because it appears that point correspondence between the left and right eyes is not necessarily established. If some points get out of the field of view, the EKF keeps in its state vector for a while those 3D points that are not visible. The cameras use the stereo depth for the EKF initialization only at the start. When the robot uses a landmark found with machine learning and stereo to get the disparity, the robot actualizes the 3D points in its state vector; consequently, the covariance will be diminished rapidly in fewer than 10 frames.

In Fig. 16.28a, we see different points of interest on objects and on the wall. Figure 16.28b depicts the cyclopean eye, where the 3D points estimated by the EKF lie on the Vieth–Müller circles or locus of zero of disparity (ZOD); namely, all points lying in the same circle produce the same disparity with respect to the egocenter. You can appreciate that these circles lie on spheres, which varies their radius toward infinity. Figure 16.28c shows the EKF estimated rotation angles of the camera.

In Fig. 16.29a, we see points lying on the edge of a staircase of interest on objects and on the wall. Figure 16.28b depicts the cyclopean eye, where the staircase 3D points were estimated by the EKF. The cyclopean view can be used by a humanoid to climb the staircase.

In Fig. 16.30a, we see a pair of stereo sequences showing points of interest on objects and the wall. The head of the robot was tilted slowly from the floor to the ceiling, tracking these points of interest, and simultaneously the cameras had a vergence [KCN7] motion from the inside to the outside, i.e., from fixating near to fixating far away. Figure 16.30b shows the estimated angles of the left and right camera views compared with the ground truth of the angles supplied by the motor encoders (curves at the right). You can see acceptable angle estimation by the EKF.

In Fig. 16.31a, we see detected points of interest on a cup of the set of objects lying on the table shown in Fig. 16.28a. Using features from a small window, the robot uses the Viola and Jones machine learning procedure to recognize the cup using just one camera; then, the related 3D object information is retrieved, which is useful for grasping and manipulating, as depicted in Fig. 16.31b. Here, the stereo vision can help further to get the object's pose.

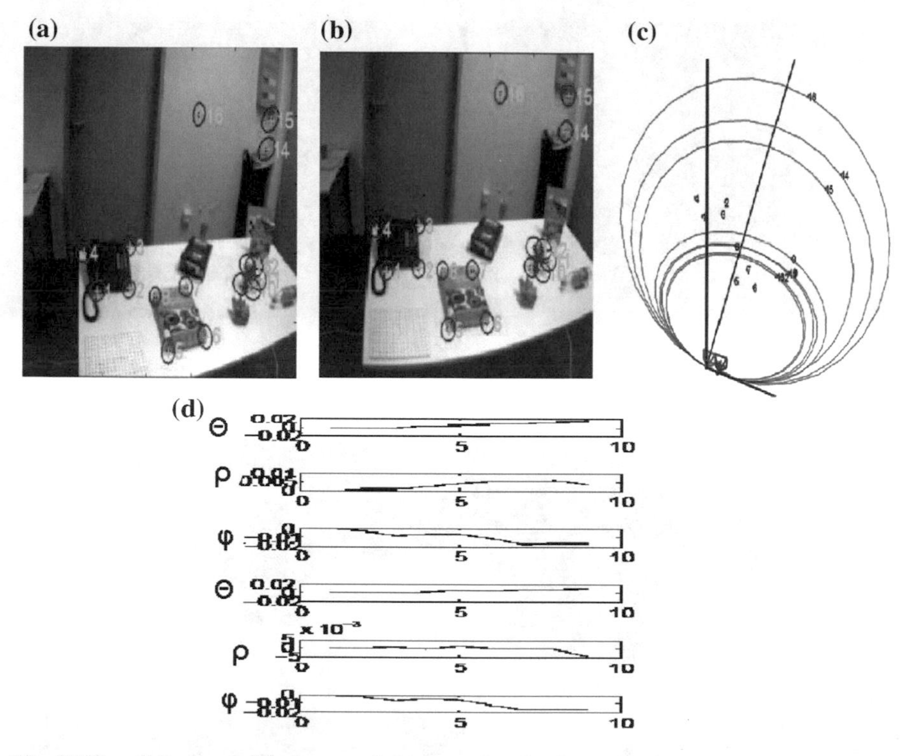

Fig. 16.28 **a**, **b** Left and right images of objects and wall points. **c** Cyclopean view of 3D space. **d** Estimation of the rotation angles of left and right cameras

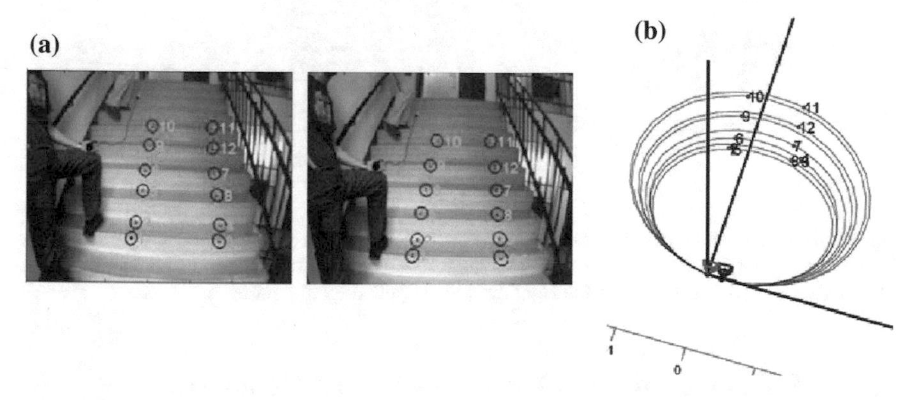

Fig. 16.29 **a** Left and right images of points of interest on a staircase. **b** Cyclopean view of the staircase

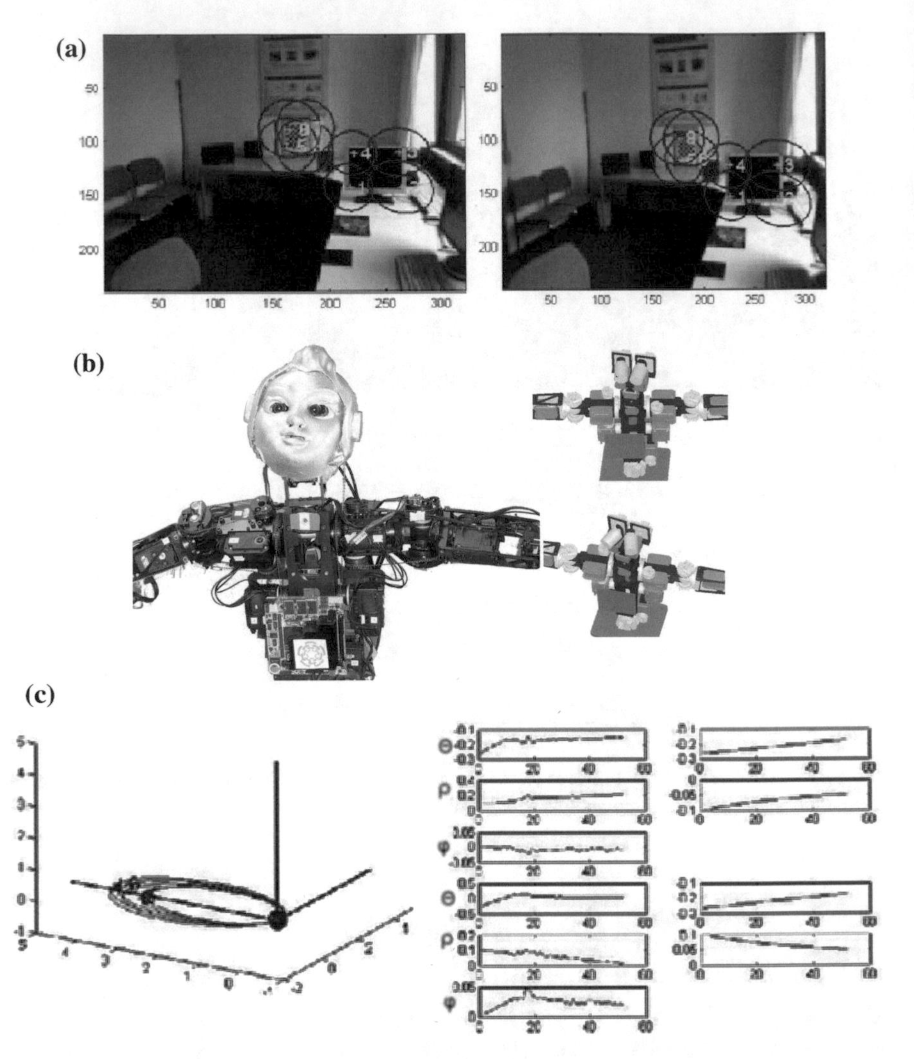

Fig. 16.30 **a** Left and right images of objects and wall points. **b** Vergence in the humanoid. **c** Cyclopean view of 3D space and estimation of the rotation angles of the left and right cameras: The left curves are EKF estimates, and the right ones are supplied by the motor encoders

16.7 3D Pose Estimation Using Conformal Geometric Algebra

Next, we describe a procedure to obtain the 3D pose estimation for tracking and manipulation of the object. The perception system consists of a stereo vision system (SVS) mounted in a pan–tilt unit (PTU) as shown in Fig. 16.32a. First, a calibration process for the SVS is realized. This process consists of retrieving the position and

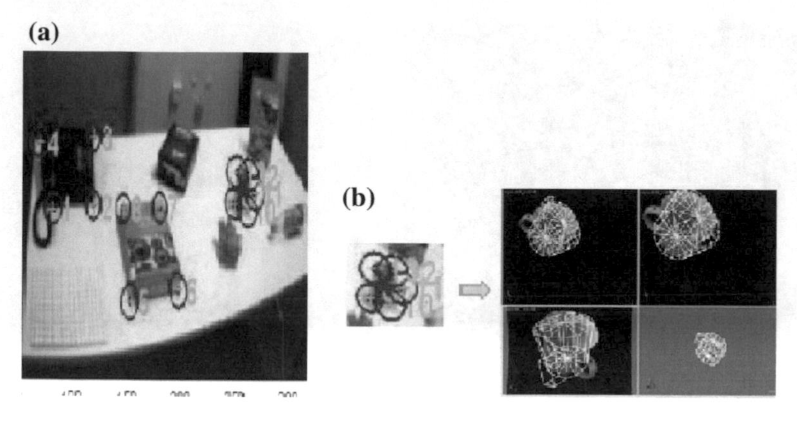

Fig. 16.31 **a** Selected object. **b** Monocular recognition of a cup using the Viola and Jones method. This retrieves 3D information, which can be used for grasping and manipulation

orientation of the principal axis of the right camera with respect to the principal axis of the left camera.

16.7.1 Line Symmetry Estimation via Reflectional Symmetry

The procedure for pose estimation is based on a *Fast Reflectional Symmetry Detection* algorithm proposed in [189]. In [190], the symmetry detection is combined with a block motion detector to find the symmetry axis of an object in motion, which is limited because it is impossible to detect static objects. Instead of this, a color-based segmentation is proposed to solve this issue. The color segmentation is performed in HSV color space (hue-saturation-value); this color space is chosen due to its relative robustness to changes in the environment's illumination. Once the image segmentation is obtained in each camera, it is converted to an edge image using an appropriate edge filter, and after that, the fast reflectional symmetry detection is applied, obtaining a parameterization of the line given by its angle and the radius or distance perpendicular to the line symmetry (θ, r) (see Fig. 16.33).

Using the data obtained from camera calibration (from both cameras), the line symmetry is transformed from the image coordinates to camera coordinates, i.e., from $\mathbb{R}^2 \rightarrow \mathbb{R}^3$, and then creates the line in conformal space as follows

$$L = \cos(\theta)e_{23} + \sin(\theta)e_{31} + re_3 \wedge e_\infty - fe_\infty(\cos(\theta)e_2 - \sin(\theta)e_1), \qquad (16.80)$$

where f is the focal distance from the origin of the coordinate frame of the camera and the camera plane.

Notice that from Eq. 16.80, line L lies in a parallel plane to xy plane. Since a rigid transformation $[R, t]$ relates both cameras, we need to define the line in both

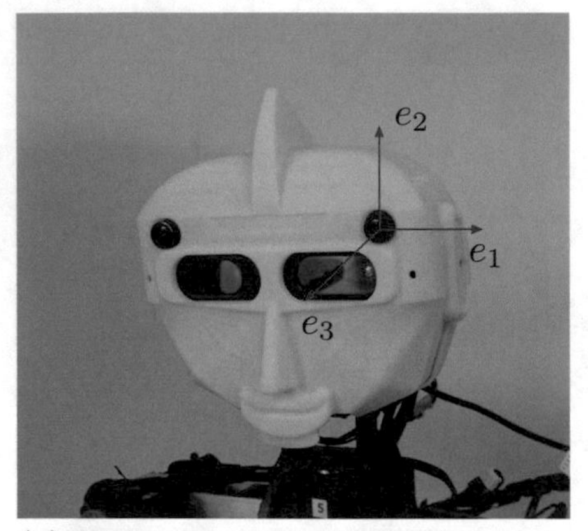

(a) SVS mounted in a PTU used to apply the proposed algorithm. The frame reference is fixed in left camera.

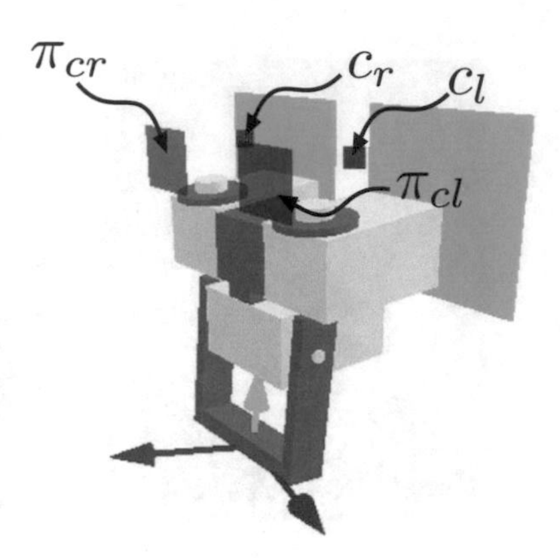

(b) Geometric entities in PTU

Fig. 16.32 Pan–tilt unit

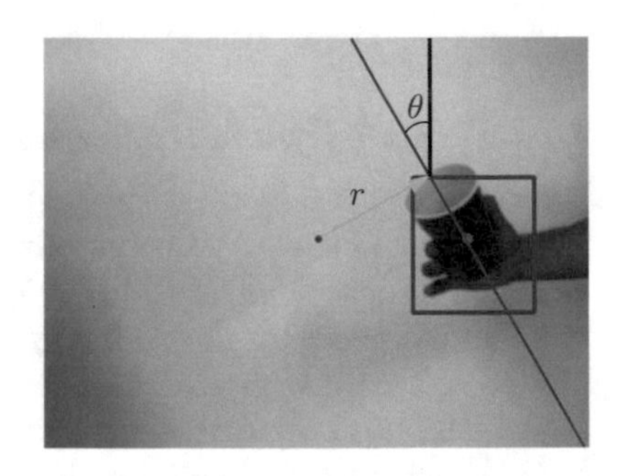

Fig. 16.33 Parameters obtained from fast reflectional symmetry detection

camera planes π_{cl} and π_{cr} (Fig. 16.32b). For that, a motor M is constructed with this transformation that relates the cameras. The lines are then defined as follow:

$$L_l = L, \tag{16.81}$$

$$L_r = ML\widetilde{M}, \tag{16.82}$$

where L_l is the line of the left camera plane π_{cl} and L_r is from the right camera plane π_{cr}.

To obtain the 3D line symmetry, it is necessary to get the points at camera center which can be calculated from calibration matrix. Without loss of generality, it is possible to define the camera center of the left camera c_l as the origin and the right camera center c_r as a rigid transformation defined before applied to the point at the origin, i.e.,

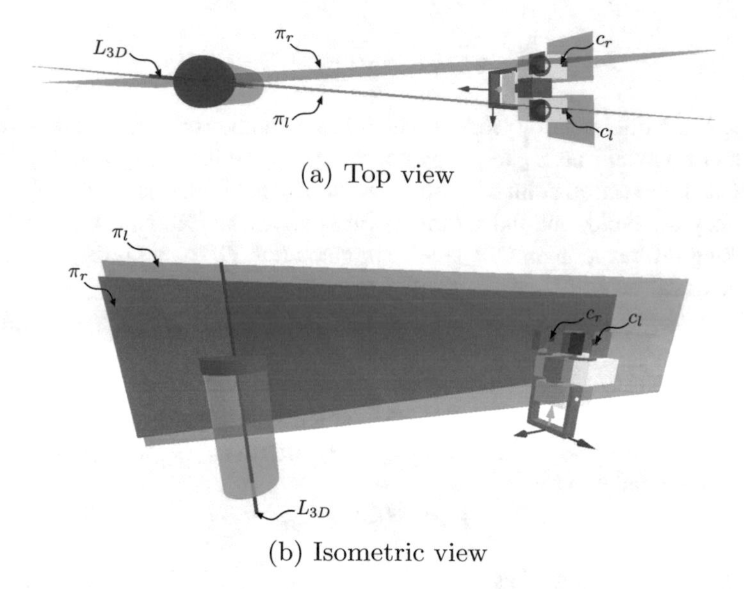

Fig. 16.34 Schematic representation of 3D line symmetry estimation

$$c_l = e_0, \tag{16.83}$$

$$c_r = Me_0\widetilde{M}. \tag{16.84}$$

With the lines and the image center points obtained, we create two planes as follow

$$\pi_l^* = c_l \wedge L_l^*, \tag{16.85}$$

$$\pi_r^* = c_r \wedge L_r^*. \tag{16.86}$$

Finally, the symmetry line in 3D L_{3D} is created by intersecting the planes

$$L_{3D} = \pi_l \wedge \pi_r. \tag{16.87}$$

Figure 16.34 shows a general scheme of the idea presented.

To create the reference for the object manipulation, a point pair will be used. To obtain this point pair, it is necessary to calculate a point that lies in the line symmetry of the object, which will be found calculating its mass center. In each segmented image, the mass center is computed and the 3D point is calculated using the parameters obtained in calibration. Once this point is obtained, it is transformed in its conformal point representation x_{ref}. This point will give us the position of the object, and for the orientation, we defined a second point using the symmetry line. This point will have to lie in a line orthogonal to L_{3D} and a line L_{ref} constructed with the mass center x_{ref} and the left image center c_l, which will be defined later,

$$L_\perp = \frac{1}{2}(L_{3D}L_{ref} - L_{ref}L_{3D}) = L_{3D}\underline{\times}L_{ref}, \tag{16.88}$$

where $\underline{\times}$ is the commutator product which has a geometric interpretation that is to build an orthogonal line $L_\perp$ to the plane created by the lines L_{3D} and L_{ref} passing through the intersection point of these lines. Notice that both lines must intersect due to how they are build, and indeed this point is x_{ref}. The line L_{ref} will be used for the tracking reference later. Once the orthogonal line $L_\perp$ is obtained, a motor M_{p_2} is created as follows

$$M_{p_2} = T_r T_o \widetilde{T}_r, \tag{16.89}$$

where $T_r = 1 - \frac{1}{2}x_{e,ref}e_\infty$ is a translator constructed with the Euclidean part $x_{e,ref}$ of x_{ref} and $T_o = 1 - \frac{1}{2}d_p n_\perp^e \wedge e_\infty$ is a translator that defines the distance d_p between the point pair and $n_\perp^e$ is defined as $n_\perp^e = n_\perp I_e$, where $n_\perp$ is the line orientation of $L_\perp$. So, the second point is

$$p_2 = M_{p_2} x_{ref} \widetilde{M}_{p_2} \tag{16.90}$$

and the point pair is formed as

$$P^*_{p\,ref} = x_{ref} \wedge p_2. \tag{16.91}$$

In order to create the tracking reference of the object, a line is defined using the mass center x_{ref} and left camera center c_l, as mentioned previously

$$L^*_{ref} = x_{ref} \wedge c_l \wedge e_\infty. \tag{16.92}$$

16.7.2 Real-Time Results

The results obtained by the real-time implementation are presented in this section. The algorithm was developed using C++ and a computer vision library called IVT (Integrating Vision Toolkit) [299]. In order to obtain the line symmetry in the plane images, the line is projected in this plane as follow

$$L_{i,l} = (L_{3D} \cdot \pi_{cl})\pi_{cl}^{-1}, \tag{16.93}$$
$$L_{i,r} = (L_{3D} \cdot \pi_{cr})\pi_{cr}^{-1}, \tag{16.94}$$

where $L_{i,l}$ and $L_{i,r}$ are the line symmetry in the image plane of the left and right camera, respectively. Figure 16.35 shows the 3D line symmetry projected in the image plane obtained using different objects.

Using this reference and the differential kinematics model for point pairs, it is possible to define a control law using references defined in CGA. A previous work of this kind of control algorithm is presented in [109]. The implementation of a complete

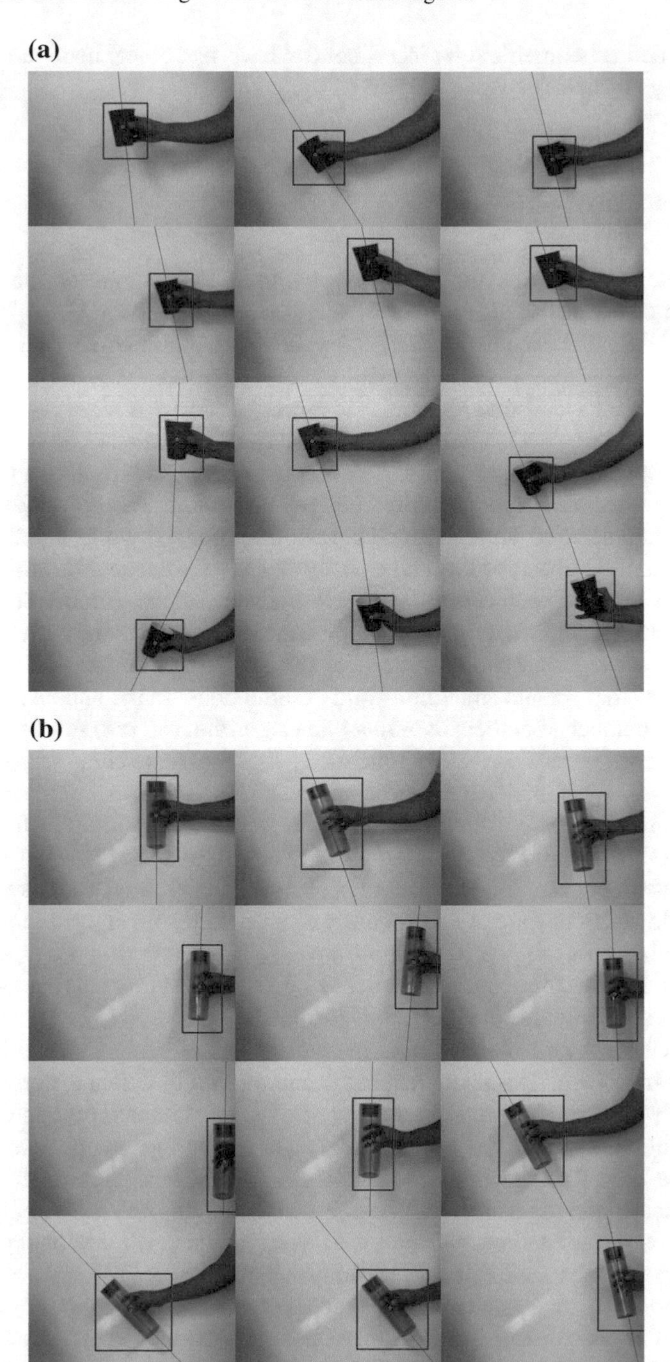

Fig. 16.35 Sequence of left camera images from real-time implementation

scheme to define geometric references, control laws, and object manipulation in the conformal geometric framework is part of the future work.

16.8 Conclusions

This chapter presented a non-iterative algorithm that combines the power of expression of geometric algebra with the robustness of Tensor Voting to find the correspondences between two sets of 3D points with an underlying rigid transformation. This algorithm was also shown to work with excessive numbers of outliers in both sets. We have also used geometric algebra to derive a set of constraints that serves a double purpose: On the one hand, it let us decide whether or not the current plane corresponds to a rigid motion; and on the other hand, it allows us to reject multiple matches and enforce the uniqueness constraint. The algorithm does not require an initialization (though it can benefit from one). It works equally well for large and small motions. And it can be easily extended to account for multiple overlapping motions and even certain non-rigid transformations. It must be noted that our algorithm can detect multiple overlapping motions, whereas the current solutions only work for one global motion. We have also shown that our algorithm can work with data sets that present small non-rigid deformations. In the unconstrained case, with a large number of outliers (83–90%), the algorithm can take several minutes to finish. However, in most real-life applications, these extreme circumstances are not found, and a good initialization can be computed. When these conditions are met, our algorithm can be rather fast. Another aspect of our 3D algorithm is that we must rotate one of the sets of points in order to make more dense the plane we are looking for in the voting space. We are currently exploring other ways to make this more efficient. However, in spite of this apparent disadvantage, our algorithm works even without initialization, unlike other algorithms like ICP. Moreover, our algorithm can be used to initialize subsequent refinement stages with ICP, thus solving the problem of having a good initialization for that algorithm.

In this chapter, we have presented the advantages of using CGA to implement the Hough transform. This mathematical framework let us detect different kinds of shapes with very simple equations even in 3D. We must remark the representations of the entities are vectors and also have an algebraic and geometric interpretations that can be used for detection algorithms at higher levels of complexity. Future development of the algorithm will be focused in solving the high computational cost. Other extension to be developed is to work in higher-dimensional algebras to detect more complex entities, e.g., $\mathbb{G}_{6,3}$ to detect ellipses.

Furthermore, in this chapter we introduced a generalization of voting schemes in the conformal geometric algebra framework using the concept of inner products with respect to a flag. Considering the Eq. (12.161), we see that we can design suitable flags for detection of complex structures involving transformation as well; in addition, we use functions to apply constraints on the flag, according to perceptual properties. Moreover, our voting scheme can be used to design a multilevel architecture, as is shown in the applications subsections. Note that the principles of Gestalt cannot be easily put into mathematics, in the above experimental analysis we show a successful attempt to do so, and we have introduced the use of geometric algebra to model perceptual properties, like symmetry, as well as geometric structures. Future works should focus on the design of new perceptual saliency functions and flags, the implementation of new stages of the architecture for high-level tasks, and the integration of other information channels like color, movement, disparity.

We proposed also a conformal model for 3D visual perception. In our model, the two views are fused in an extended 3D horopter model. For visual simultaneous localization and mapping (SLAM), an extended Kalman filter (EKF) technique is used for 3D reconstruction and determination of the robot head pose. In addition, the Viola and Jones machine learning technique is applied to improve the robot relocalization. The 3D horopter, the EKF-based SLAM, and the Viola and Jones machine learning technique are key elements for building a strong real-time perception system for robot humanoids.

Finally, a methodology for object pose estimation via reflectional symmetry of objects, modeling of robotic manipulators was developed using the conformal geometric algebra framework. The pose for the manipulator was defined using a single geometric entity: the point pair. This is impossible using vector calculus, which shows the potential of the CGA. As future work, it is proposed to implement a control law using the obtained pose reference and the algorithm developed in this work of the differential kinematics.

Chapter 17
Modeling and Registration of Medical Data

17.1 Background

In medical image analysis, the availability of 3D models is of great interest to physicians because it allows them to have a better understanding of the situation, and such models are relatively easy to build. However, sometimes and in special situations (such as surgical procedures), some structures (such as the brain or tumors) suffer a (non-rigid) transformation and the initial model must be corrected to reflect the actual shape of the object. In the literature, we can find the Union of Spheres algorithm [244], which uses the spheres to build 3D models of objects and to align or transform it over time. In our approach, we also use the spheres, but we use the marching cubes algorithm's ideas to develop an alternative method, which has the advantage of reducing the number of primitives needed; we call our method *marching spheres*.

17.1.1 Union of Spheres

This algorithm was proposed in [244] and can be summarized as follows:

- Given a set of boundary points (borders of the 3D volumetric data), calculate the Delaunay tetrahedrization (DT).
- Compute the circumscribing sphere to each tetrahedron.
- Verify from the original data which spheres are inside the defined object.
- Simplify the dense sphere representation by clustering and eliminating redundant or nonsignificant spheres.

This algorithm has the worst-case complexity of $O(n^2)$ in both time and number of primitives, where n is the number of boundary points. However, in [244] it is noted that the highest number of primitives observed in experiments was $4n$, approximately. This number, although good, could be computationally heavy for big n.

E. Bayro-Corrochano, *Geometric Algebra Applications Vol. I*,
https://doi.org/10.1007/978-3-319-74830-6_17

To register models based on spheres, the authors first match a sufficient number of spheres from the Union of Spheres representation of one object to the other, and then from the matches find the most likely transformation using some method like least squares.

17.1.2 The Marching Cubes Algorithm

The basis of the marching cubes algorithm is to subdivide the space into a series of small cubes. The algorithm then moves (or "marches") through each of the cube testing the corner points to determine whether or not they are on/inside the surface and replacing the cube with an appropriate set of polygons. The resulting set of polygons will be a surface that approximates the original one. To explain the algorithm, let us look at a two-dimensional equivalent. Suppose we want to approximate a 2D shape like the one in Fig. 17.1a, then we create a grid of squares (equivalent to the cubes for the 3D version of the algorithm). The first task is to determine which of the corners of these squares are inside (or on) the shape (see Fig. 17.1b). Then we insert new vertices that are positioned halfway between each inside and outside corner connected to each other by an edge of the grid (see Fig. 17.1c). Finally, we join

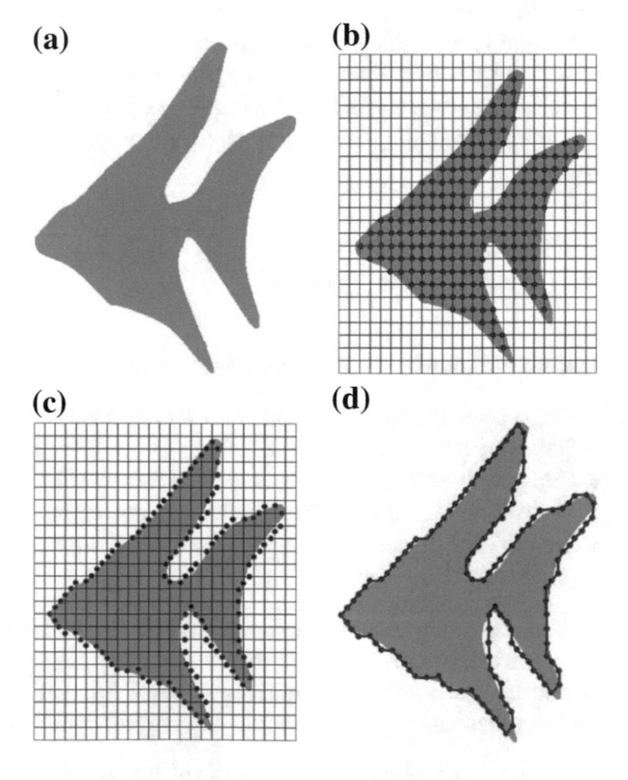

Fig. 17.1 2D equivalent of marching cubes. **a** The shape we want to represent; **b** determining which corners of the grid are inside the shape; **c** new vertices inserted; **d** approximation to the surface by joining the vertices

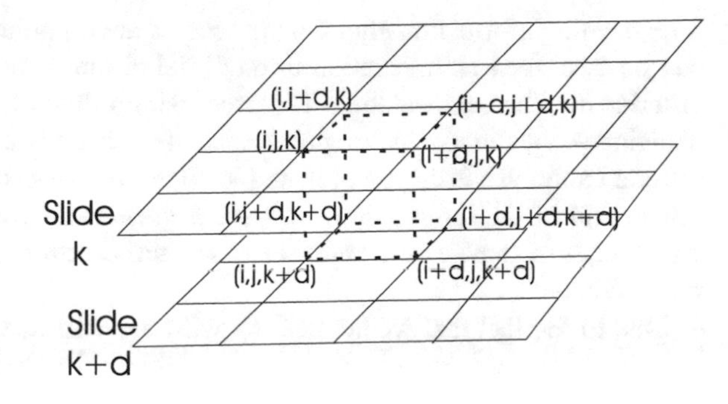

Fig. 17.2 Basic marching cubes algorithm and the order of vertices

the inserted vertices using lines and obtain an approximation to the original surface (Fig. 17.1d). Obviously, the finer the grid, the better the approximation to the shape.

For the 3D case, suppose we have a set of m images (slides) containing the shape information of the object we want to represent. Then we create a set of *logical cubes*, with each cube created from eight pixels: four of slide k and four of slide $k + 1$. The algorithm determines how the cube intersects the surface and then moves (or marches) to the next cube. At each cube, we assign a one to a cube's vertex if this vertex is inside (or on) the surface, and a zero if this vertex is outside the surface (Fig. 17.2).

Since there are eight vertices in each cube and two states for each one, there are $2^8 = 256$ ways the surface can intersect the cube. In [195], it is shown how these 256 cases can be reduced to only 15 for clarity in explanation, although a permutation of these 15 cases using complementary and rotational symmetry produces the 256 cases. With this number of basic cases, it is then easy to create predefined polygon sets for making the appropriate surface approximation.

Before we proceed to explain our proposed method for volume representation and registration based on spheres, let us introduce the basic concepts of geometric algebra and conformal geometric algebra, because we will use the representation of spheres in such algebras for such tasks.

17.2 Segmentation

Before we can proceed to model the objects, we need to segment them from the images. Segmentation techniques can be categorized in three classes [98, 221]: (a) thresholding, (b) region-based, and (c) boundary-based. Due to the advantages and disadvantages of each technique, many segmentation methods are based on the integration of information obtained by two techniques: the region and boundary information. Some of them embed the integration in the region detection, while others integrate the information after both processes are completed. Embedded integration can be described as integration through definition of new parameters or decision criterion for the segmentation. Postprocessing integration is performed after both techniques (boundary- and region-based) have been used to process the image. A different approach is the use of dynamic contours (snakes). Within each category for integration of the information, we have a great variety of methods; some of them work better in some cases, some need user initialization, some are more sensitive to noise, etc.

Due to the fact that we are dealing with medical images, we also need another important strategy: texture segmentation. Textural properties of the image can be extracted using statistical features, spatial frequency models, etc. A texture operator describes the texture in an area of the image. So, if we use a texture operator over the whole image, we obtain a new "texture feature image." In such an image, the texture of a neighborhood around each pixel is described. In most cases, a single operator does not provide enough information about the texture, and a set of operators needs to be used. This results in a set of "texture feature images" that jointly describe the texture around each pixel. The main methods for texture segmentation are Laws' texture energy filters, co-occurrence matrices, random fields, frequency domain methods, and perceptive texture features [52, 295].

Simple segmentation techniques such as region growing, split and merge, or boundary segmentation cannot be used alone to segment tomographic images due to the complexity of the computer tomographic images of the brain. For this reason, we decide not only to combine boundary and region information (as typically is done), but to integrate information obtained from texture segmentation methods with boundary information and embed that in a region-growing strategy. A block diagram of our proposed approach is shown in Fig. 17.3.

In order to obtain the texture information, we use the Laws' texture energy masks. Laws' texture energy measures are a set of filters designed to identify specific primitive features such as spots, edges, and ripples in a local region. Laws' masks are obtained from three basic vectors, which correspond to a Gaussian, and its first and second derivatives:

$$L_3 = [1 \ 2 \ 1], \qquad E_3 = [-1 \ 0 \ 1], \qquad S_3 = [1 \ -2 \ 1]. \tag{17.1}$$

Convolution of these three vectors with themselves and with one another generates five 5×1 vectors:

$$\begin{aligned} L5 = [1 \ 4 \ 6 \ 4 \ 1];\ E5 = [-1 \ -2 \ 0 \ 2 \ 1],\\ S5 = [-1 \ 0 \ 2 \ 0 \ -1];\ W5 = [-1 \ 2 \ 0 \ -2 \ 1],\\ R5 = [1 \ -4 \ 6 \ -4 \ 1], \end{aligned} \tag{17.2}$$

which identify certain types of feature: level, edge, spot, wave, and ripple. Multiplying these five vectors by themselves and by one another produces a set of 25 unique 5×5 masks (see some examples in Fig. 17.4). Convolving these masks with an image, we

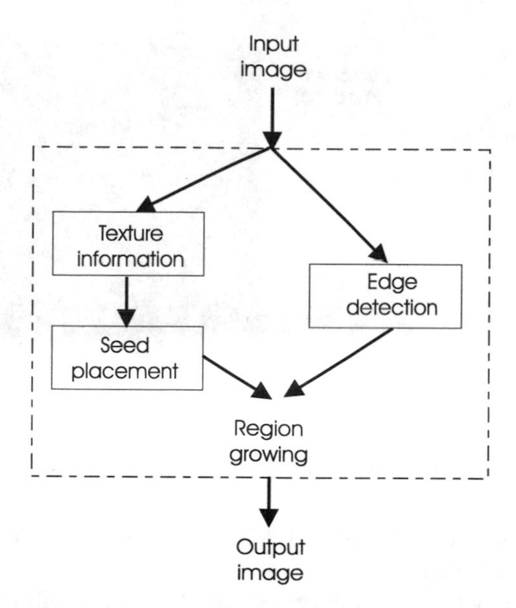

Fig. 17.3 Block diagram of our proposed approach to segment tumors in computer tomographic images

obtain the so-called texture images, which contain the response of each pixel in the image to the convoluted mask.

Our objective is to build a characteristic vector for each pixel in the image, so we take the absolute value for each pixel for each mask's result and fix the value in a position of such vector with 1 or 0, depending on whether the value is greater than zero or zero, respectively. The characteristic vector will have k values; one of them is only to identify if the pixels correspond to the background (value set to zero) or to the patient's head (value set to one), the patient's head could be skin, bone, brain, etc., while the other values are determined according to the set of masks used. If we name the characteristic vector as V_{xy} and identify each of its k coordinates as $V_{xy}[i]$, $i = 1, \ldots, k$, the procedure could be summarized as follows:

(a)

-1	-4	-6	-4	-1
-2	-8	-12	-8	-2
0	0	0	0	0
2	8	12	8	2
1	4	6	4	1

(b)

1	-4	6	-4	1
-4	16	-24	16	-4
6	-24	36	-24	6
-4	16	-24	16	-4
1	-4	6	-4	1

(c)

-1	-4	-6	-4	-1
0	0	0	0	0
2	8	12	8	2
0	0	0	0	0
-1	-4	-6	-4	-1

(d)

-1	-2	0	2	1
0	0	0	0	0
2	4	0	-4	-2
0	0	0	0	0
-1	-2	0	2	1

Fig. 17.4 Energy texture masks used to characterize each pixel over the whole image (these masks are convolved with the image

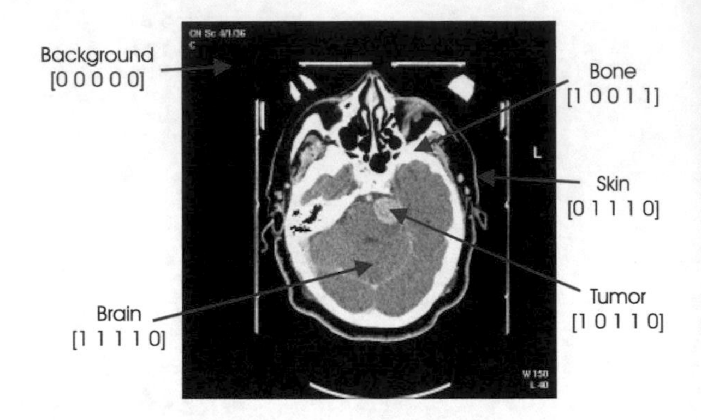

Fig. 17.5 Characteristic vectors for the main structures present in a computer tomographic image

1. For each pixel p in the CT image: if $p_{ij} \in background$, then $V[1] = 0$; else $V[1] = 1$,
2. Convolve the image with each of the texture masks to obtain a total of k "characteristic texture images."
3. For each position (x, y) in the image and for each result of the convolution of the masks, take the absolute value $absval_{xy} = \|val_{ij}\|$. If $absval_{xy} > 0$, then $V_{xy}[i] = 1$; else $V_{xy}[i] = 0$, where $i = 2, \ldots, k+1$ corresponds to the k different masks.

As a result, each structure (tissue, bone, skin, background) on the medical images has a typical vector; an example using only four masks is shown in Fig. 17.5. Obviously, the more masks used, the better the object will be characterized. It is important to note that not all the pixels belonging to the same object (i.e., the tumor) have the desired vector because of variations in values of neighboring pixels, but a good enough quantity of them do. So we can use the pixels having the characteristic vector of the object we want to extract and to establish them as seed points in a region-growing scheme.

The growing criterion (how to add neighboring pixels to the region) and the stopping criterion are as follows: compute the mean μ_{seeds} and standard deviation σ_{seeds} of the pixels fixed as seeds; then, for each neighboring pixel, to determine whether or not it is being added to the region

$$\text{If } \sum_{1}^{k+1} (V_{xy} \oplus V_{seed}) = 0 \quad \text{or} \quad \sum_{1}^{k+1} (V_{xy} \oplus V_{seed}) = 1 \qquad (17.3)$$
$$\text{and } I(x, y) = \pm 2 * \sigma_{seeds}, \text{ then } (x, y) \in R_t,$$

where $V_{xy} \oplus V_{seed}$ is like a XOR operator ($a \oplus b = 1$ if and only if $a \neq b$), $\sum_{1}^{k+1} (V_{xy} \oplus V_{seed})$ acts as a counter of different values in V_{xy} vs V_{seed} (to belong to a specific region, it must differ at most in one element), and R_t is the tumor's region. Example: suppose we use only four masks; let $V_{xy} = [0\ 1\ 0\ 1\ 1]$ and $V_{seed} =$

[0 1 1 1 1], then $V_{xy} \oplus V_{seed} = [0\ 0\ 1\ 0\ 0]$; therefore, $\sum_{1}^{k+1} (V_{xy} \oplus V_{seed}) = 1$, and the point (x, y) is included in that region.

The region grows in all directions, but when a boundary pixel is found, the grow in that direction is stopped. Note that the boundary information helps to avoid the inclusion of pixels outside the object boundary; if we use only the texture information, wrong pixels can be added to the region, but boundary information reduces the risk of doing that.

Figure 17.6 shows an example of the process explained before. Figure 17.6a shows the original image. Fig. 17.6b shows the seed points fixed. Fig. 17.6c shows the final result, highlighting the segmented tumor in the original image.

The overall process takes only a few seconds per image and could be used to segment any of the objects; but in our case, we focus our attention on the extraction of the tumor. After that, the next step is to model the volumetric data by some method. An example is shown in Fig. 17.7; however, in Sect. 17.1 we present a method that uses the spheres as basic entities for modeling, and in Sect. 17.3 we present our proposed approach.

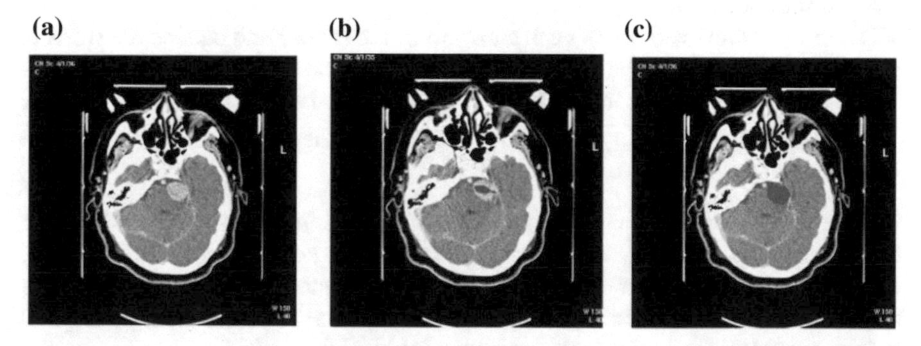

Fig. 17.6 Results for the segmentation of tumor in CT images. **a** One of the original images; **b** seed points; **c** result of segmentation.

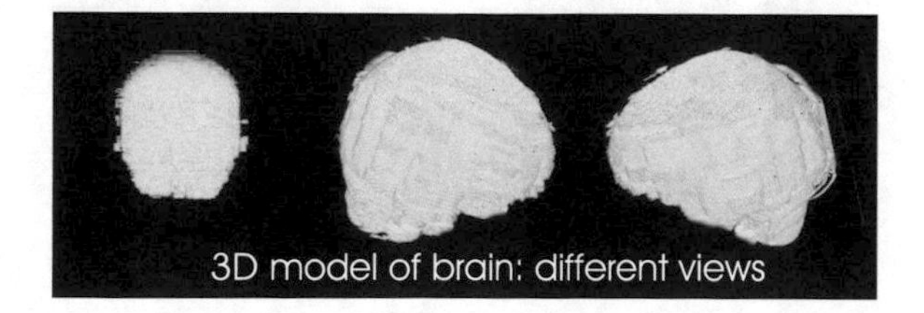

Fig. 17.7 After segmentation, 3D models are constructed. Here we show the 3D model of the brain segmented from a set of 200 MR images

17.3 Marching Spheres

Once we segment the object we are interested in, next step is to model it in 3D space. In our approach for object modeling, we will use the spheres as the basic entities, and we will follow the ideas of the marching cubes algorithm. First we will see the 2D case. We want to approximate the shape shown in Fig. 17.1a, and then the process is

- Make a grid over the image and determine the points inside (or on) and outside the surface (same as in Fig. 17.1a, b).
- Draw circles centered at each inside corner connected with others out of the surface.
- Finally, draw circles centered at vertices that are connected with two vertices inside the surface.
- As a result, we obtain an approximation to the original surface (Fig. 17.8). Obviously, the finer the grid, the better the approximation.

For the 3D case, given a set of m slides, do:

- Divide the space into logical cubes (each cube contains eight vertices, four of slide k and four of slide $k + 1$).
- Determine which vertices of each cube are inside (or on) and outside the surface.
- Define the number of spheres of each cube according to Fig. 17.9 taking the indices of the cube's corners as in Fig. 17.2. The magnitude of the radius is: $r_{p_i} = \frac{d}{2}$ for the smallest spheres $S^j_{p_i}$, $r_{m_i} = \frac{d}{2}$ for medium size spheres $S^j_{m_i}$, and $r_{g_i} = d$ for the biggest spheres $S^j_{g_i}$.

Note that we use the same 15 basic cases of the marching cubes algorithm because a total of 256 cases can be obtained from this basis. Also note that instead of triangles, we define spheres and that our goal is not to have a good render algorithm, but to have a representation of the volumetric data based on spheres, which, as we said before, could be very useful in the process of object registration.

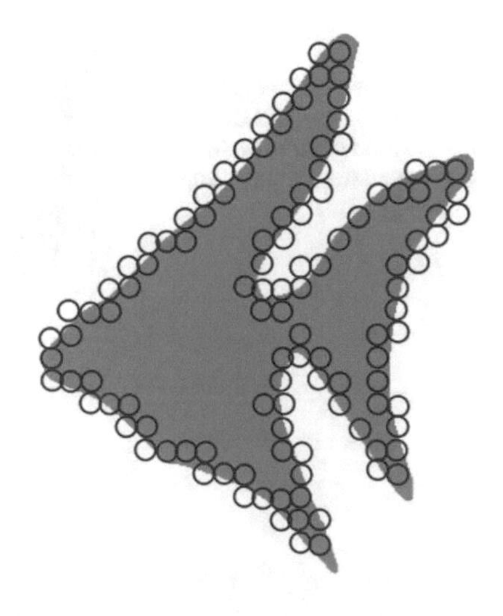

Fig. 17.8 2D equivalent of the modified marching cubes obtaining a representation based on circles

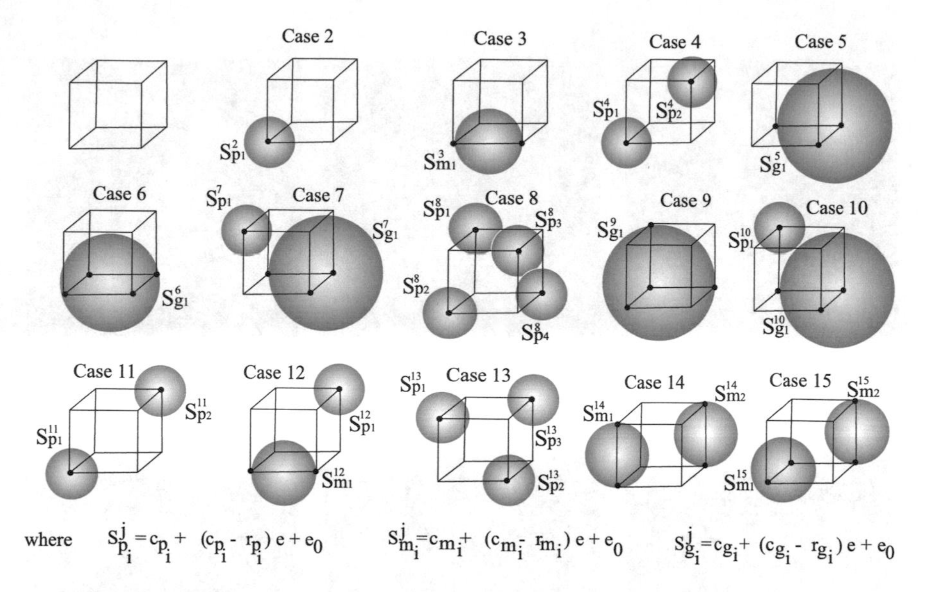

Fig. 17.9 Fifteen basic cases of surface intersecting cubes. The cases are numbered in this figure starting with the upper left corner, from 1 to 15

17.3.1 Experimental Results for Modeling

To test the algorithm, we used several images: 26 images of a human skull containing balloons to simulate the brain, 36 images of a real patient, 16 images of a real tumor in a patient, and other different CT images. The first step is to segment the brain (tumor) from the rest of structures. Once the brain (tumor) is properly segmented, the images are binarized to make the process of determining if each pixel is inside or outside the surface easier. The Canny edge detector is used to obtain the boundary points of each slide; by this way, we can compare the number of such points (which we call n) with the number of spheres obtained, also giving a comparison with the number of primitives obtained with the Union of Spheres algorithm.

Figure 17.10 shows one CT image of the skull with balloons, the segmented object, and the approximation of the surface by circles using the 2D version of our approach. Figures 17.11 and 17.12 show the CT of a real patient and tumor segmented. Figure 17.13a shows the results for the 3D case modeling the brain of our human head model (balloons) extracted from CT images, while Fig. 17.13b shows the 3D model of the tumor from the real patient (extracted from 16 CT images). Table 17.1 is a comparison between the results of the Union of Spheres and our approach for the case of brain being modeled. The first row shows the worst case with both approaches; second row shows the number of spheres with improvements in both algorithms (reduction of spheres in DT is done by grouping spheres in a single one that contains the others, while such a reduction is done using a displacement of $d = 3$ in our approach). The number of boundary points was $n = 3370$ in both cases.

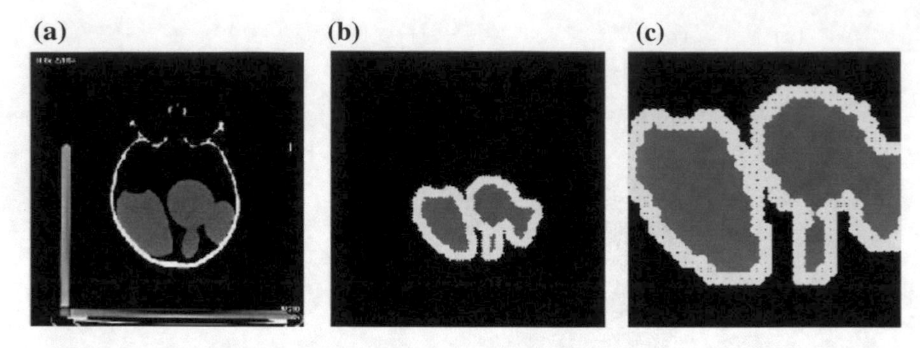

Fig. 17.10 Simulated brain: **a** original of one CT slide; **b** segmented object and its approximation by circles according the steps described in Sect. 17.3; **c** Zoom of **b**

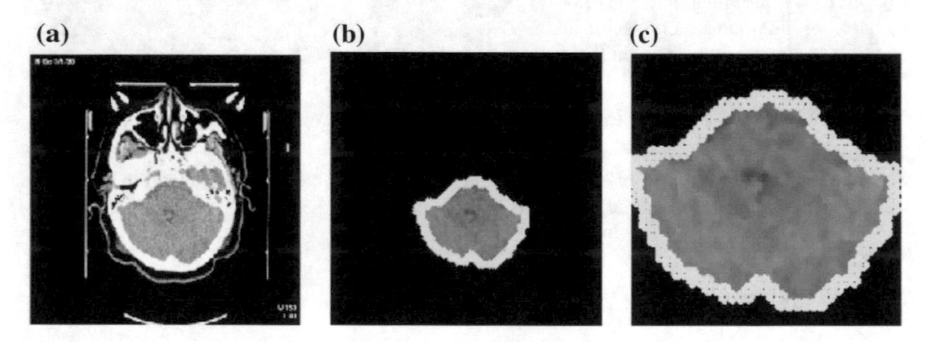

Fig. 17.11 Real patient: **a** original of one CT slide; **b** brain segmented and its approximation by circles; **d** zoom of **C**

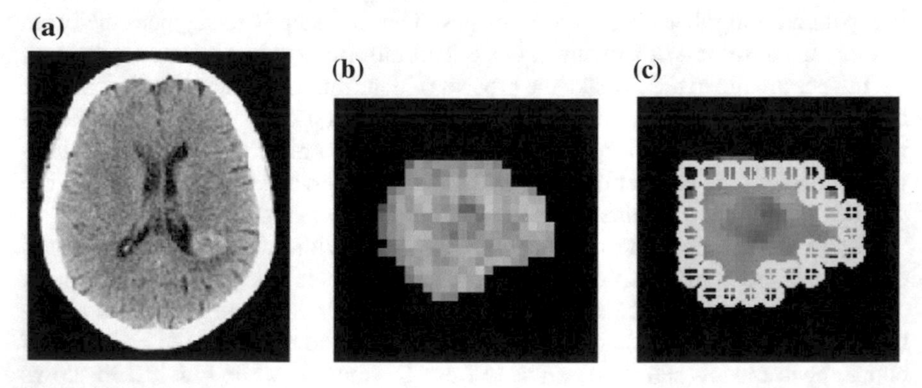

Fig. 17.12 Real patient: **a** Original of one CT slide; **b** zoom of the tumor segmented; (c) tumor approximation by circles

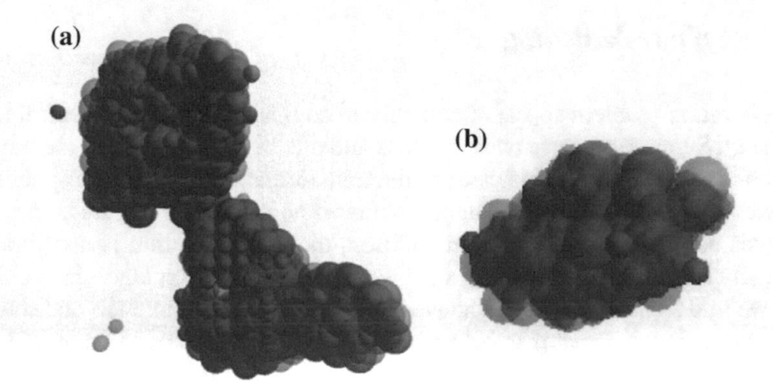

Fig. 17.13 Approximation of shape of three-dimensional objects with marching spheres; **a** approximation of the brain structure extracted from CT images (synthetic data); **b** approximation of the tumor extracted from real patient data

Table 17.1 Comparison between the number of spheres using the approach based on Delaunay tetrahedrization and our approach based on the marching cubes algorithm (called marching spheres); n is the number of boundary points; d is the distance between vertices in logical cubes of our approach)

n/d	No. of spheres in DT approach	No. of spheres in marching spheres
3370 / 1	13480	11866
3370 / 3	8642	2602
10329 / 2	25072	8362
2641 / 1	6412	5008

Note that our approach manages to reduce the number of primitives maintaining well the shape representation.

17.4 Registration of Two Models

In [56], a comparison between a version of the popular Iterative Closest Point (ICP) [320] and the Thin-Plate Spline Robust Point Matching (TPS-RPM) algorithms for non-rigid registration appears. TPS-RPM performed better because it can avoid getting trapped in local minima, and it aligns the models even in the presence of a large amount of outliers. However, the authors had used only sets of 2D and 3D points. Now we have spheres modeling the object, and these spheres have not only different centers, but also different radii. This fact prompted us to look at the representation of spheres using the conformal geometric algebra (CGA), which is a geometric algebra of five dimensions. In next section, we explain how to register two models based on spheres using their representation in CGA.

17.4.1 Sphere Matching

The registration problem appears frequently in computer vision and medical image processing. Suppose we have two point sets and one of the results from the transformation of the other, but we do not know the transformation nor the correspondences between the points. In such a situation, we need an algorithm that finds these two unknowns as well as possible. If, in addition, the transformation is non-rigid, the complexity increases enormously. In the variety of registration algorithms existing today, we find some that assume knowledge of one of these unknowns and solve for the other one; but there are two that solve for both: Iterative Closest Point (ICP) and Thin-Plate Spline Robust Point Matching (TPS-RPM). It has been shown [25] that TPS-RPM gives better results than ICP; therefore, we will adapt such an algorithm for the alignment of models based on spheres. It is important to note that previous works only deal with 2D and 3D points, but now we have spheres in the 5D space of the CGA.

The notation is as follows: $\mathbf{V} = \{S_j^I\}$, $j = 1, 2, \ldots, k$, is the set of spheres of the first model, or model at time t_1; $\mathbf{X} = \{S_i^F\}$, $i = 1, 2, \ldots, n$, is the set of spheres of the second model, or model at time t_2 (expected model); $\mathbf{U} = \{S_i^E\}$, $i = 1, 2, \ldots, n$, is the set of spheres resulting after the algorithm finishes (estimated set). The superindex denotes if the sphere S belongs to the initial set (I), the expected set (F), or the estimated set (E).

To solve the correspondence problem, one has to find the correspondence matrix Z, Eq. (17.4), which indicates which sphere of the set $\mathbf{V}$ is transformed in which sphere of the set $\mathbf{X}$ by some transformation f. Thus, if we correctly estimate the transformation f and apply it to X ($U = f(V)$), then we should obtain the set X. To deal with outliers, the matrix Z is extended as in (17.4). The inner part of Z (with size $k \times n$) defines the correspondences.

$$Z = \begin{array}{cccccccc} z_{ji} & S_1^F & S_2^F & S_3^F & S_4^F & \ldots & S_n^F & outlier_{n+1}^F \\ S_1^I & 1 & 0 & 0 & 0 & \ldots & 0 & 0 \\ S_2^I & 0 & 1 & 0 & 0 & & 0 & \ldots \\ S_3^I & 0 & 0 & 1 & 0 & \ldots & 0 & \ldots \\ \ldots & \ldots & \ldots & \ldots & \ldots & \ldots & \ldots & \ldots \\ S_k^I & 0 & 0 & 0 & 0 & \ldots & 1 & 0 \\ outlier_{k+1}^I & 0 & 0 & 0 & 1 & \ldots & 0 & \end{array} \quad (17.4)$$

However, to deal with outliers using binary correspondence matrices can be very cumbersome [56]. For this reason, TPS-RPM uses two techniques to solve the correspondence problem:

Soft assign: the basic idea is allow the matrix of correspondences M to take continuous values in the interval [0,1]; this "fuzzy" correspondence improves gradually during optimization without jump in the space of permutations of binary matrices.

Deterministic annealing: technique used to control the fuzzy correspondences by means of an entropy term in the cost function (called energy function), introducing a parameter T of temperature that is reduced in each stage of the optimization process beginning at a value T_0 and reducing by a factor r until some final temperature is reached.

Using these two techniques, the problem is to minimize the function:

$$E(M, f) = \sum_{i=1}^{n}\sum_{j=1}^{k} m_{ji} \left| S_i^F - f(S_j^I) \right|^2 + \lambda |Lf|^2 + T \sum_{i=1}^{n}\sum_{j=1}^{k} m_{ji} \log m_{ji} - \varsigma \sum_{i=1}^{n}\sum_{j=1}^{k} m_{ji}, \quad (17.5)$$

where m_{ji} satisfy the condition $\Sigma_{i=1}^{n+1} m_{ji} = 1$ for $j \in \{1, 2, \ldots, k+1\}$, $\Sigma_{j=1}^{k+1} m_{ji} = 1$ for $i \in \{1, 2, \ldots, n+1\}$, and $m_{ji} \in [0, 1]$. The parameter λ is reduced in an annealing scheme $\lambda_i = \lambda_{init} T$. The basic idea in this heuristic is that more global, and rigid transformations should be first favored with large values of λ. We will follow the next two steps:

Update the correspondences: for the spheres S_j^I, $j = 1, 2, \ldots, k$, and $S_i^F, i = 1, 2, \ldots, n$, modify m_{ji}:

$$m_{ji} = \frac{1}{T} e^{-\frac{(S_i^F - f(S_j^I))^2}{T}}, \quad (17.6)$$

for outliers $j = k+1$ e $i = 1, 2, \ldots, n$:

$$m_{k+1,i} = \frac{1}{T_0} e^{-\frac{(S_i^F - f(S_{k+1}^I))^2}{T_0}}, \quad (17.7)$$

and for outliers $j = 1, 2, \ldots, k$ e $i = n+1$:

$$m_{j,n+1} = \frac{1}{T_0} e^{-\frac{(S_{n+1}^F - f(S_j^I))^2}{T_0}}. \quad (17.8)$$

Update transformation: To update the centers, we use

$$\begin{aligned} E_{tps}(D, W) = {} & \|Y_c - V_c D - \Phi_c W\|^2 + \lambda_1 (W^T \Phi_c W), \\ & + \lambda_2 [D - I]^T [D - I], \end{aligned} \quad (17.9)$$

where V_c is the matrix given by $V_c = \{(S_j^I \wedge E) \cdot E\}$, $j = 1, 2, \ldots, k$, Y_c is interpreted as the new estimated position of the spheres, Φ_c is constructed by $\phi_b = (S_b^I \wedge E) \cdot E - (S_a^I \wedge E) \cdot E$; therefore, Φ_c contains information about the structure of the sets; D, W represent the rigid and non-rigid deformations, respectively, affecting the centers of the spheres and are obtained by solving

$$f((S_a^I \wedge E) \cdot E | D, W) = ((S_a^I \wedge E) \cdot E) D + \phi((S_a^I \wedge E) \cdot E) W. \quad (17.10)$$

The radii are updated using a dilator, which is defined as

$$D_\lambda = e^{\frac{-\log(\lambda)\wedge E}{2}} = e^{-\frac{\log(\rho_{jmax}/\rho_{imax})\wedge E}{2}}, \tag{17.11}$$

where $\rho_{j_{max}}$ and $\rho_{i_{max}}$ are the radii of the more corresponding pair of spheres, according to matrix M. That is, the index i of the initial set, and j of the expected set where the ith row of M has a maximum value $\max(m_{ij})$. Dilators are applied as

$$S'^{I}_{j} = D_\lambda S^I_j D_\lambda. \tag{17.12}$$

The annealing process helps to control the dual update process. Parameter T is initialized in T_{init} and is gradually reduced as $T_i = T_{init}r$, where r is the annealing rate until some final temperature is reached. Parameters λ_1 and λ_2 follow an annealing scheme as mentioned earlier.

17.4.2 Experimental Results for Registration

Figures 17.14 and 17.15 show two examples registering sets of spheres. Figures 17.14a and 17.15a are the initial set or representation at time t_1; Figure 17.14b and 17.15b are the deformed or expected set, or representation at time t_2. These two representations should be registered. Figure 17.14c shows the results of the registration process. Note that researchers usually use TPS-RPM with 2D or 3D vectors because they cannot go beyond such a dimension; in contrast, using conformal geometric algebra we have a homogeneous representation that preserves isometries and uses the sphere as the basic entity. Note that the algorithm adjusted the radius, as expected (this is not possible using only 3D vectors).

We carried out other experiments. Table 17.2 shows the average errors between the corresponding spheres' centers, measured with the expected and resulting sets. This error is measured as

$$\epsilon = \frac{\sum_{i,j}\sqrt{c(i,j)(S^F_i - S^E_i)^2}}{N}, \tag{17.13}$$

where N is the number of pairs of corresponding spheres; $c(i,j) = 1$ if S^F_i corresponds to S^E_i, or $c(i,j) = 0$ if they do not correspond.

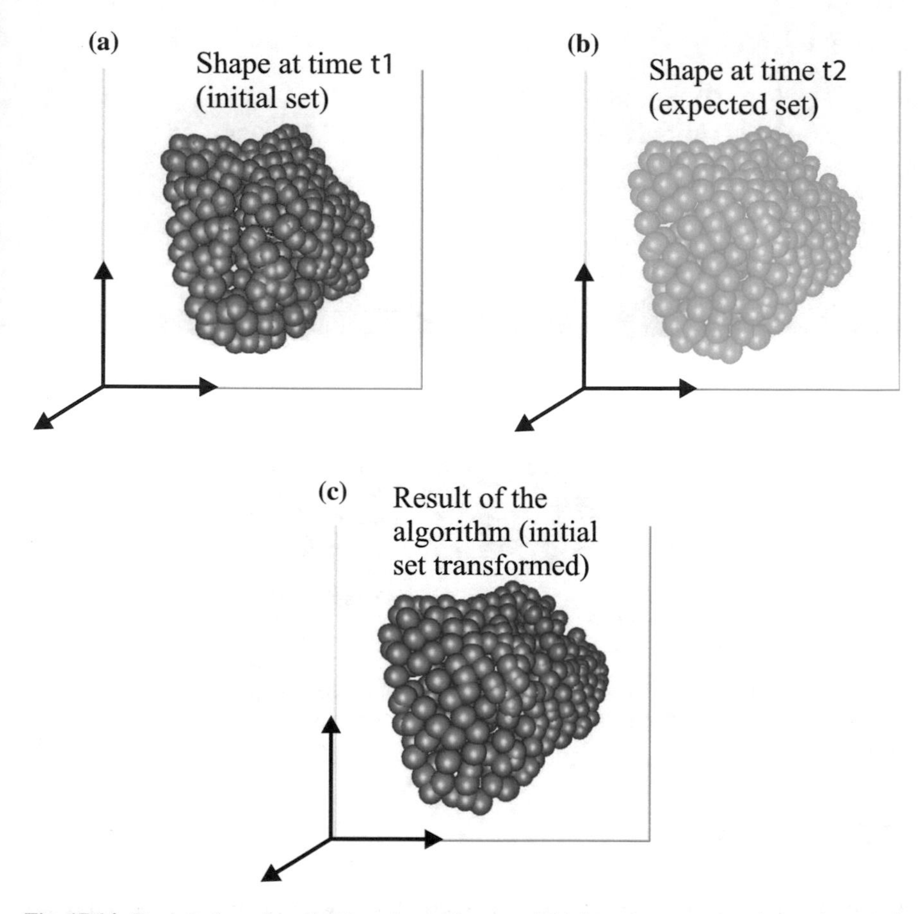

Fig. 17.14 Registration of models based on spheres: **a** initial set (or representation at time t_1); **b** expected set (or representation at time t_2); **c** result of the deformation of the initial model to match the second one

Table 17.2 Average errors measured as the distance (in voxels) between the centers of corresponding spheres, calculated with the expected and resulting sets, and according to (17.13)

Number of spheres of the initial set	Number of spheres of the expected set	Average error according to (17.13)
60	60	12.86
98	98	0.72
500	500	2.80
159	143	18.04
309	280	15.63
340	340	4.65

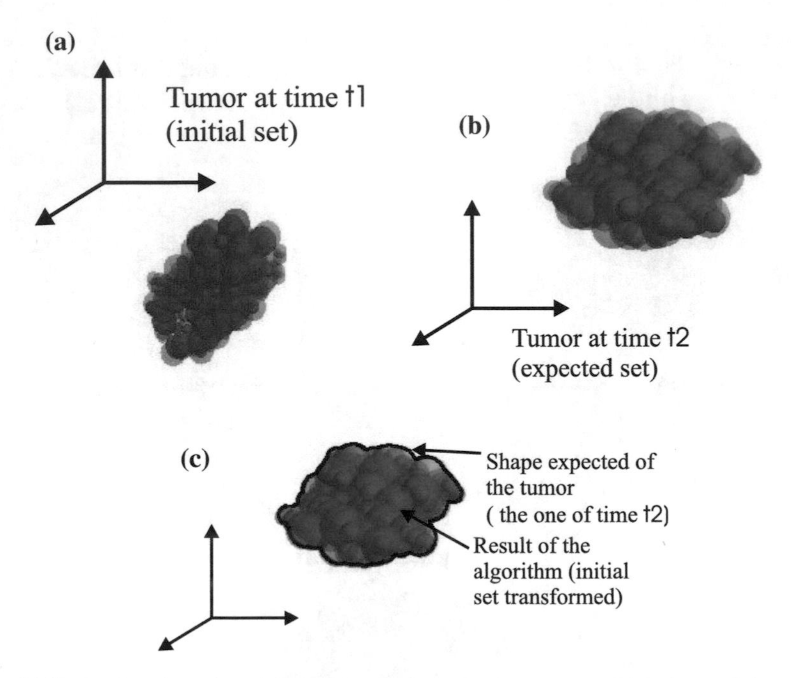

Fig. 17.15 Registration of models based on spheres for a tumor: **a** initial shape of the tumor; **b** expected shape of the tumor (representation at time t_2); **c** result after registration (transformed initial model

17.5 Conclusions

The chapter first presents an approach for medical image segmentation that combines texture and boundary information and embeds it into a region-growing scheme, having the advantage of integrating all the information in a simple process. We also show how to obtain a representation of volumetric data using spheres. Our approach, called *marching spheres*, is based on the ideas exposed in the marching cubes algorithm but it is not intended for rendering purposes or displaying in real time, but rather for reducing the number of primitives modeling the volumetric data. With this saving, it makes a better registration process and better in the sense of using fewer primitives, which reduces the registration errors. Also, the chapter shows how to represent these primitives as spheres in the conformal geometric algebra, which are five-dimensional vectors that can be naturally used with the principles of TPS-RPM. Experimental results seem to be promising.

Part VI
Applications of GA in Machine Learning

Chapter 18
Applications in Neurocomputing

In this chapter, we present a series of experiments in order to demonstrate the capabilities of geometric neural networks. We show cases of learning of a high nonlinear mapping and prediction. In the second part experiments of multiclass classification, object recognition, and robot trajectories interpolation using CSVM are included.

18.1 Experiments Using Geometric Feedforward Neural Networks

We begin with an analysis of the XOR problem by comparing a real-valued MLP with bivector-valued MLPs. In the second experiment, we used the encoder–decoder problem to analyze the performance of the geometric MLPs using different geometric algebras. In the third experiment, we used the Lorenz attractor to perform step-ahead prediction.

18.1.1 Learning a High Nonlinear Mapping

The power of using bivectors for learning is confirmed with the test using the XOR function. Figure 18.1 shows that geometric nets $\text{GMLP}_{0,2,0}$ and $\text{GMLP}_{2,0,0}$ have a faster convergence rate than either the MLP or the P-QMLP—the quaternionic multilayer perceptron of Pearson [227], which uses the activation function given by Eq. (13.5). Figure 18.1 shows the MLP with two- and four-dimensional input vectors. Since the MLP(4), working also in 4D, cannot outperform the GMLP, it can be claimed that the better performance of the geometric neural network is due not to

E. Bayro-Corrochano, *Geometric Algebra Applications Vol. I*,
https://doi.org/10.1007/978-3-319-74830-6_18

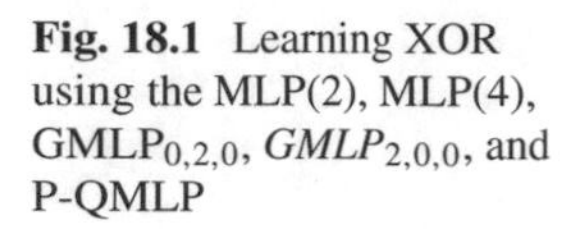
Fig. 18.1 Learning XOR using the MLP(2), MLP(4), $\mathrm{GMLP}_{0,2,0}$, $GMLP_{2,0,0}$, and P-QMLP

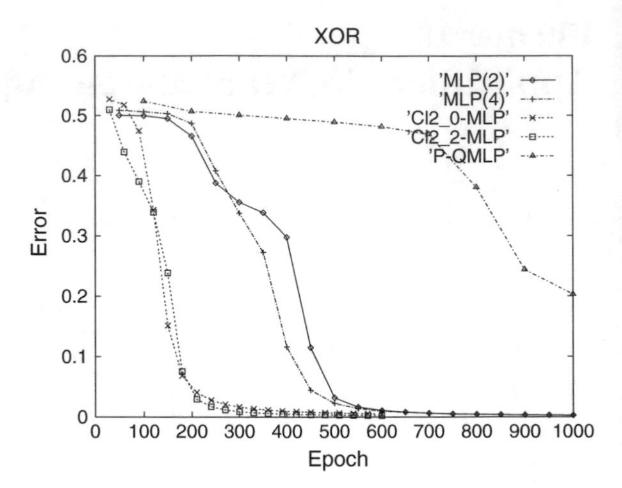

the higher-dimensional quaternionic inputs but rather to the algebraic advantages of the geometric neurons of the net.

18.1.2 Encoder–Decoder Problem

The *encoder–decoder problem* is also an interesting benchmark test to analyze the performance of three-layer neural networks. The training input patterns are equal to the output patterns. The neural network learns in its hidden neurons, a compressed binary representation of the input vectors, in such a way that the neural network can decode it at the output layer.

We tested real- and multivector-valued MLPs using sigmoid transfer functions in the hidden and output layers. Two different kinds of training sets consisting of one input neuron and of multiple input neurons were used (see Table 18.1). Since the sigmoids have asymptotic values of 0 and 1, the used output training values were numbers near to 0 or 1. Figure 18.2 shows the mean square error (MSE) during the training of the G00 (working in $G_{0,0,0}$, a real-valued 8–8–8 MLP, and the geometric MLPs—G30 working in $G_{3,0,0}$, G03 in $G_{0,3,0}$, and G301 in the degenerated algebra $G^{+}_{3,0,1}$ (algebra of the dual quaternions). For the one input case, the multivector-valued MLP network is a three-layer network with one neuron in each layer, i.e., a 1–1–1 network. Each neuron has the dimension of the used geometric algebra. For example, in the figure G03 corresponds to a neural network working in $\mathcal{G}_{3,0,0}$ with eight-dimensional neurons.

For the case of multiple input patterns, the network is a three-layer network with three input neurons, one hidden neuron, and one output neuron, i.e., a 3–1–1 network. The training method used for all neural nets was the batch momentum learning rule. We can see that in both experiments, the real-valued MLP exhibits the worst

Table 18.1 Test of real- and multivector-valued MLPs using (top) one input and one output, and (middle) three inputs and one output

Input	0.97 0.03 0.03 0.03	0.03 0.03 0.03 0.03
Output	0.97 0.03 0.03 0.03	0.03 0.03 0.03 0.03
Input	0.03 0.97 0.03 0.03	0.03 0.03 0.03 0.03
Output	0.03 0.97 0.03 0.03	0.03 0.03 0.03 0.03
Input	0.03 0.03 0.97 0.03	0.03 0.03 0.03 0.03
Output	0.03 0.03 0.97 0.03	0.03 0.03 0.03 0.03
Input	0.03 0.03 0.03 0.97	0.03 0.03 0.03 0.03
Output	0.03 0.03 0.03 0.97	0.03 0.03 0.03 0.03
Input	0.03 0.03 0.03 0.03	0.97 0.03 0.03 0.03
Output	0.03 0.03 0.03 0.03	0.97 0.03 0.03 0.03
Input	0.03 0.03 0.03 0.03	0.03 0.97 0.03 0.03
Output	0.03 0.03 0.03 0.03	0.03 0.97 0.03 0.03
Input	0.03 0.03 0.03 0.03	0.03 0.03 0.97 0.03
Output	0.03 0.03 0.03 0.03	0.03 0.03 0.97 0.03
Input	0.03 0.03 0.03 0.03	0.03 0.03 0.03 0.97
Output	0.03 0.03 0.03 0.03	0.03 0.03 0.03 0.97
Input	0.97 0.03 0.03 0.03	0.03 0.03 0.03 0.03
	0.97 0.03 0.03 0.03	0.03 0.03 0.03 0.03
	0.97 0.03 0.03 0.03	0.03 0.03 0.03 0.03
Output	0.97 0.03 0.03 0.03	0.03 0.03 0.03 0.03
Input	0.03 0.97 0.03 0.03	0.03 0.03 0.03 0.03
	0.03 0.97 0.03 0.03	0.03 0.03 0.03 0.03
	0.03 0.97 0.03 0.03	0.03 0.03 0.03 0.03
Output	0.03 0.97 0.03 0.03	0.03 0.03 0.03 0.03
Input	0.03 0.03 0.03 0.03	0.03 0.03 0.03 0.97
	0.03 0.03 0.03 0.03	0.03 0.03 0.03 0.97
	0.03 0.03 0.03 0.03	0.03 0.03 0.03 0.97
Output	0.03 0.03 0.03 0.03	0.03 0.03 0.03 0.97

performance. Since the MLP has eight inputs and the multivector-valued networks have effectively the same number of inputs, the geometric MLPs are not being favored by a higher-dimensional coding of the patterns. Thus, we can attribute the better performance of the multivector-valued MLPs solely to the benefits of the Clifford geometric products involved in the pattern processing through the layers of the neural network.

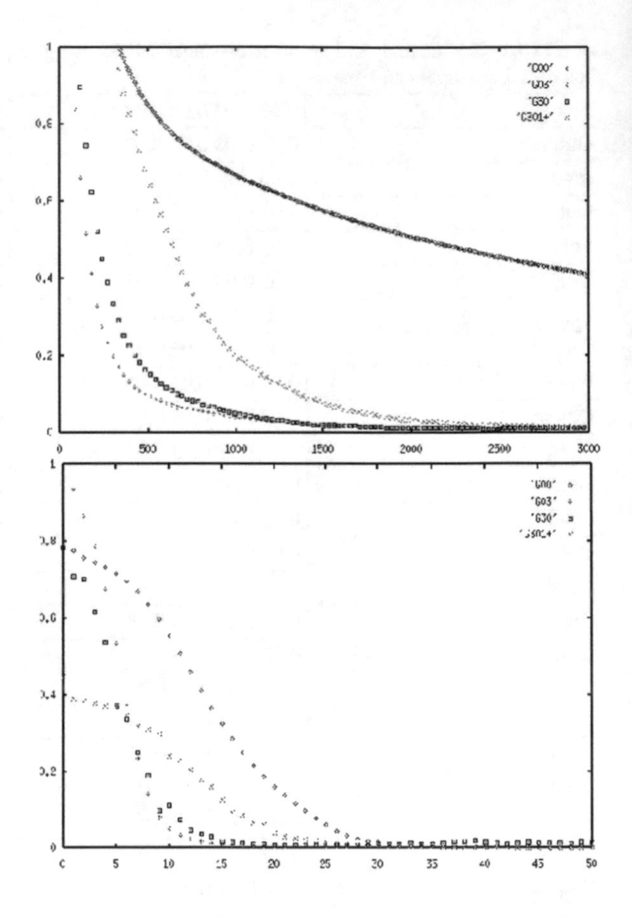

Fig. 18.2 MSE for the encoder–decoder problem with (top) one input neuron and (bottom) three input neurons

18.1.3 Prediction

Let us show another application of a geometric multilayer perceptron to distinguish the geometric information in a chaotic process. In this case, we used the well-known Lorenz attractor ($\sigma = 3$, $r = 26.5$, and $b = 1$), with initial conditions of [0,1,0] and a sample rate of 0.02 seconds. A 3–12–3 MLP and a 1–4–1 $GMLP_{0,2,0}$ were trained in the interval 12–17 s to perform an 8 τ step-ahead prediction. The next 750 samples unseen during training were used for the test. Figure 18.3a shows the error during training; note that the $GMLP_{0,2,0}$ converges faster than the MLP. It is interesting to compare the prediction capability of the nets. Figure 18.3b, c shows that the $GMLP_{0,2,0}$ predicts better than the MLP. By analyzing the covariance parameters of the MLP [0.96815, 0.67420, 0.95675] and of the $GMLP_{0,2,0}$ [0.9727, 0.93588, 0.95797], we see that the MLP requires more time to acquire the geometry involved in the second variable, and that is why the convergence is slower. The result is that the

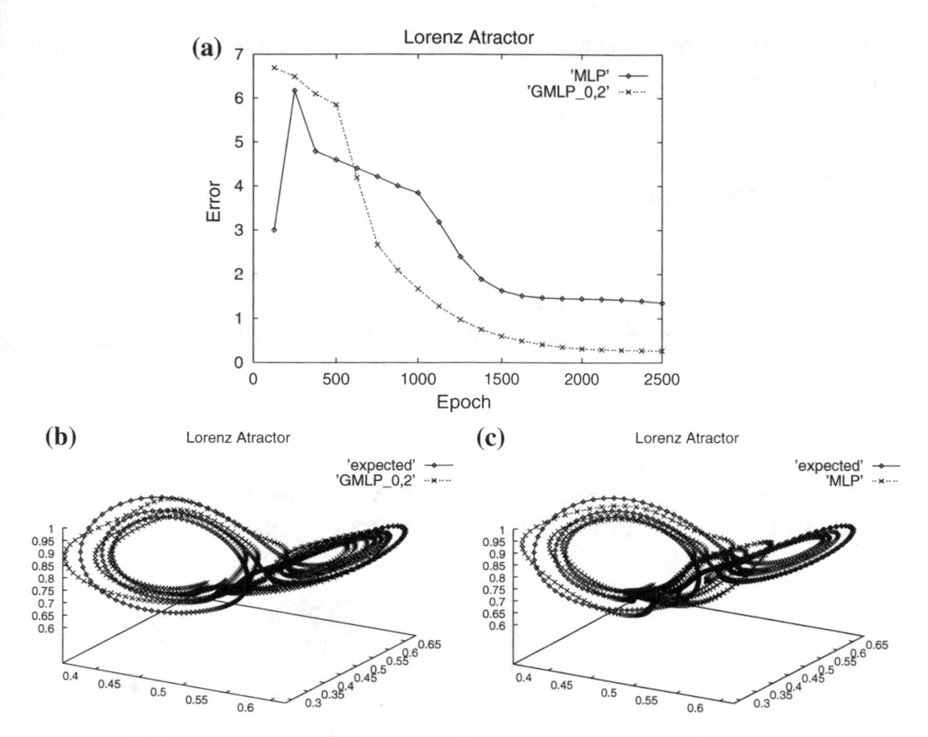

Fig. 18.3 **a** Training error; **b** prediction by $\mathrm{GMLP}_{0,2,0}$ and expected trend; **c** prediction by MLP and expected trend

MLP loses its ability to predict well in the other side of the looping (see Fig. 18.3b). In contrast, the geometric net is able to capture the geometric characteristics of the attractor from an early stage, so it cannot fail, even if it has to predict at the other side of the looping.

18.2 Recognition of 2D and 3D Low-Level Structures

The recognition of low-level structures can be seen as the recognition of geometric objects (e.g., lines, planes, and circles) and the recognition of Lie groups, which in geometric algebra are represented as versors, such as a rotor, a motor, or a dilator. These operators also have a geometric interpretation and are considered geometric objects as well.

In this section, we present the use of the geometric RBF network, first for the detection of 2D visual structures or geometric primitives in patterns. And in the second part, we show the use of the GRBFN for the detection of a rotor as the geometric operator governing the relation of pairs in sequences of 3D lines. Finally, we present the use of the SRBN for detecting in soft surfaces the underlying 3D structure together and for the interpolation at the output layer of the elasticity associated with any surface point.

18.2.1 Recognition of 2D Low-Level Structures

We tested our model with three images. We worked only with oriented lines extracted with the *Qup* filter; however, we show the capability of the model to extract lines or edges and to encode the geometric structure constructed by lines. Figure 18.4 shows the convolution of the *Qup* filter on the letters L, H, and a chessboard (Qup, which was rotated 45°) images. We see a selective detection of lines in the horizontal and vertical orientations, particularly in i and j. Figure 18.5 shows the lines' and edges' image and quaternion phases' profiles. The phase ϕ only responds to edges and white lines, while θ responds with black and white lines considering each edge. The GRBFN encodes the rotation transformation between lines and then works like

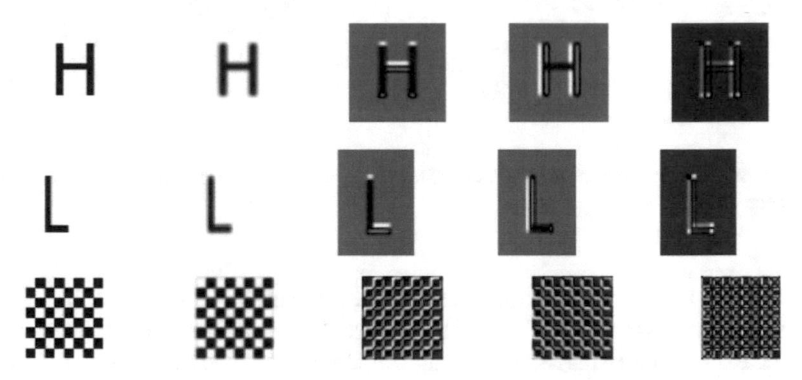

Fig. 18.4 Filtered images of H, L, and chessboard. From left to right: image, real part, i-part, j-part, k-part

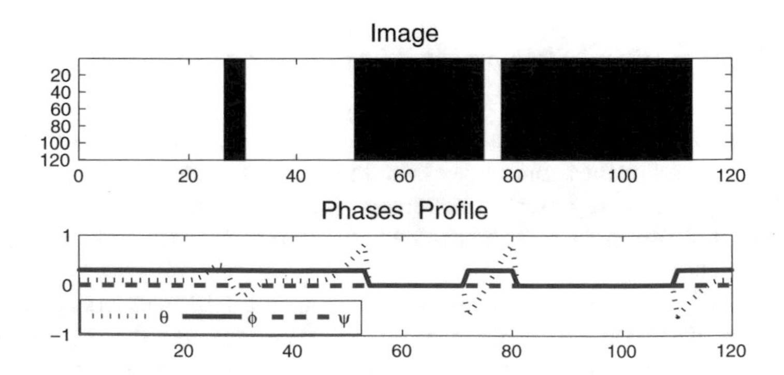

Fig. 18.5 From top to bottom: line and edge image, its phase's profile

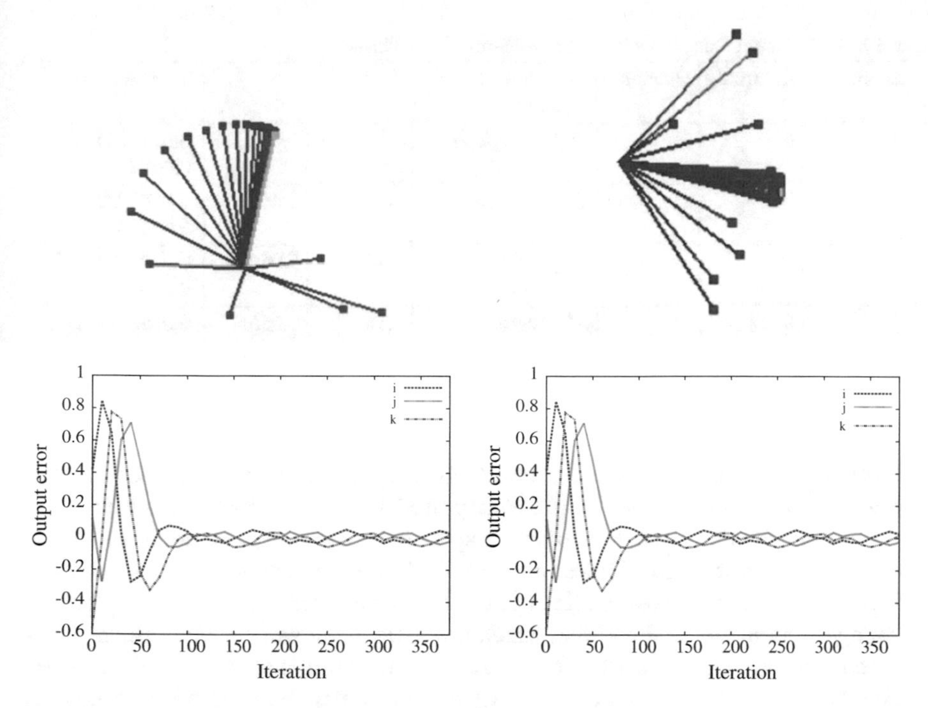

Fig. 18.6 Estimation of transformation between lines during learning process of networks 2 (left), 4 (right). Output evolution (above, in blue lines), output error (below)

a classifier of our basic geometric structures. For now, GRBFN only allows one input and one output. If the geometric structure is more complex, then it is possible to use a set of filters and networks to detect the full structure. Input pairs (for training and testing) of lines were extracted from the filtered images using the AF. Different filters give different oriented lines (see Figs. 18.4, 18.5). Figure 18.6 shows the evolution of the network output until the true transformation is finally encoded between two orthogonal lines from letter L. It can be seen that in the first iterations, i, j, k are changing in a bigger range than in the last ones. At the end, i, j, k change in a very small range. This shows that rotors converge to the expected value. Another GRBFN shows a similar result using orthogonal lines extracted from the image of the letter H. We have used four networks for experiments, whose parameters appear in Table 18.2. Networks 1, 2, and 3 were used to find the transformation using data from the H, L, and chessboard images, respectively. Network 4 used non-orthogonal lines from the chessboard image. Figure 18.6 shows that, at first, the network was completely unaware of the transformation (blue lines) between the input orientation (red line) and output[1] orientation (green line, which is considered the target orientation). However, while more pairs of orientations are fed to the network, it can approximate the real

[1]The network output is expressed as $X = xe_1e_2 + ye_2e_3 + ze_3e_1$.

Table 18.2 Parameters related to test four geometric networks

Inputs	Centroids	Outputs	Geometric structure	η	ρ	Testing error
1	6	1	Right-angle H	0.14	0.060	0.0213211 ± 0.00821
1	6	1	Right-angle L	0.74	0.060	0.0463731 ± 0.00424
1	6	1	Right-angle chessboard	0.74	0.060	0.0297221 ± 0.01327
1	6	1	45-Degree chessboard	0.15	0.055	0.0351754 ± 0.00534

rotation between the geometric entities. Considering noise in image processing, this grade of accuracy is very useful to identify these kinds of geometric patterns.

Once a network is trained to detect a specific structure, it is tested using similar structures (with orthogonal lines in the case of a network trained using orthogonal lines and with non-orthogonal lines in the case of network 4), and then we get the testing errors shown in Table 18.2, which indicates that it works well. If the network has a big error, bigger than 0.15 (defined experimentally), then it does not recognize the tested structure. Error is defined using the Euclidean distance between the target orientation and GRBFN output orientation. It is computed according to what the training algorithm step **d** indicates.

18.2.2 Recognition of 3D Low-Level Structures

Due to the noise in robot vision, we rely on observed lines and planes to track the spatial relationship between these geometric entities. In this work, we will limit ourselves to only rotation. The next experiments show that it is possible to detect rotations between lines and between planes using GRBFNs. In this approach, we use data available from a stereo vision system. Each line orientation can be represented using a bivector; this is also true for the plane orientation. We consider two points, **p** and **q**, to compute the direction **n**. It is possible to use our GRBFN to find the rotation between two line and two plane orientations. For simultaneous training, we construct an ensemble of two GRBFNs, as shown in Fig. 18.7. Input pairs (for training and testing) of line and plane orientations were generated using a disk with uniformly distributed marks, which represent one point of the two needed to construct the line, while the other point (origin) was fixed. By changing one point, we generate a set of orientations that are in different positions in 3D space (see Fig. 18.8). This gives us a set of line orientations. Similarly, we generate a set of plane orientations. These orientations constitute the training set for both networks. We want to show that our scheme is a computing block that can be used in a system to help us learn other kinds

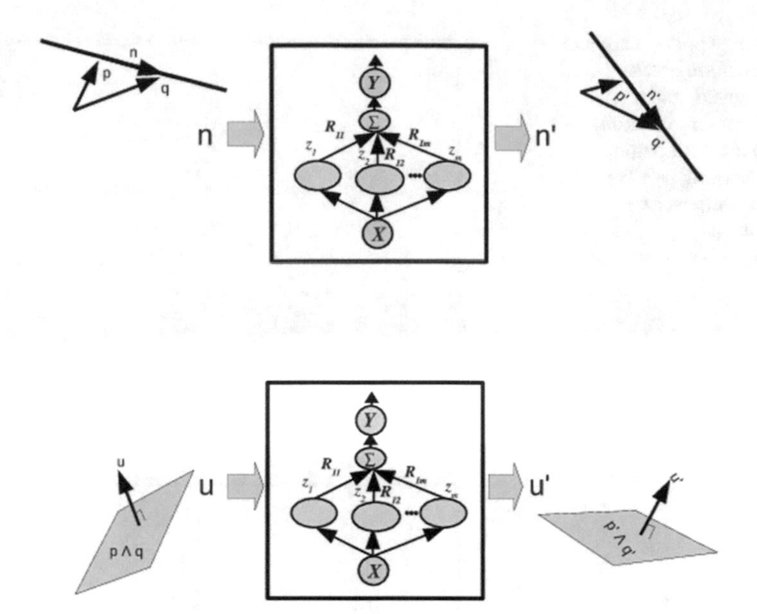

Fig. 18.7 GRBFN ensemble

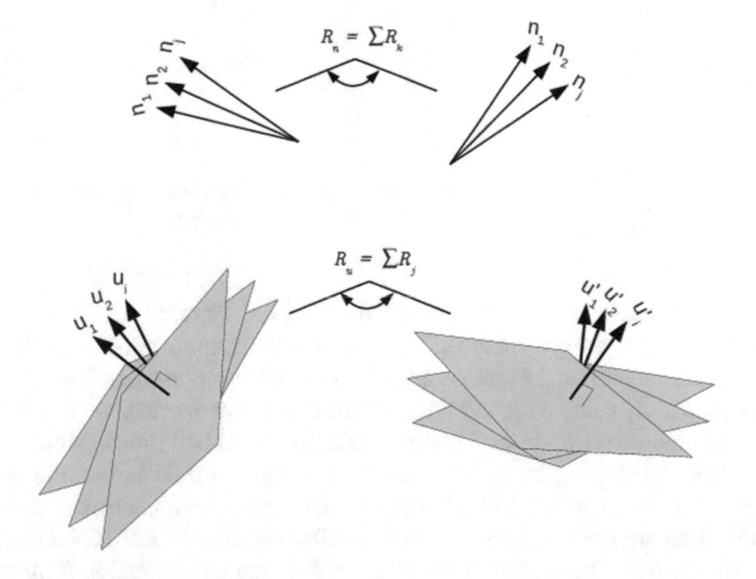

Fig. 18.8 Sets of line and plane orientations

of more complex transformations. To generate the testing set, we proceed similarly. As Fig. 18.8 shows, the first GRBFN approximates the mapping between $\mathbf{n_i}$ and $\mathbf{n'_i}$. In a similar way, the second network does the same task between $\mathbf{u_i}$ and $\mathbf{u'_i}$.

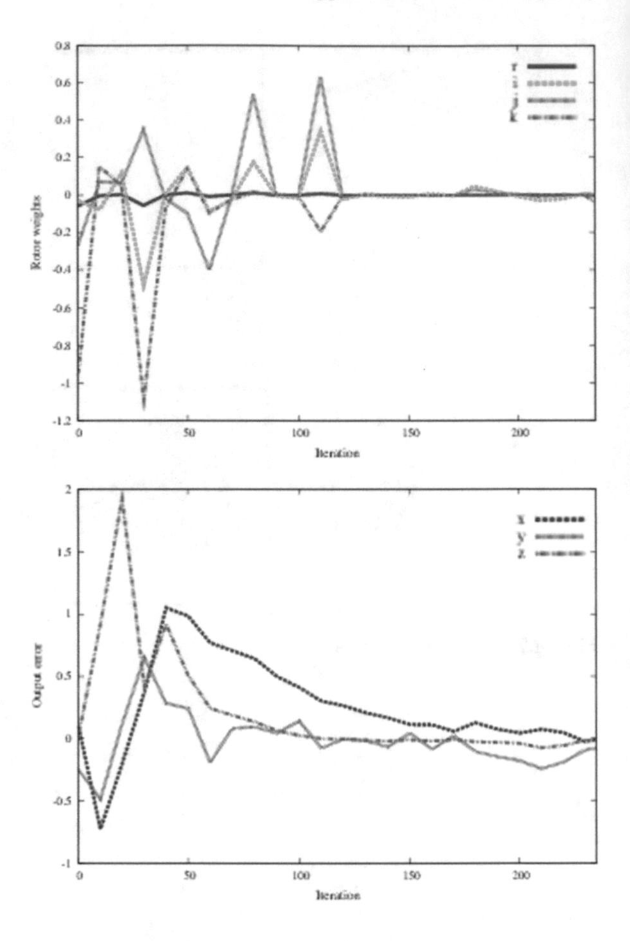

Fig. 18.9 Approximation of transformation between line orientations $\mathbf{n_i}$ and $\mathbf{n'_i}$. Rotors' vector evolution (above). Error at output during learning process (below). Outputs are expressed as $X = xe_1e_2 + ye_2e_3 + ze_3e_1$

We use these data sets to train and test the GRBFN. Figure 18.9 shows the rotor error with respect to the ground truth. This shows that rotors converge to the expected value. A similar result can be observed in Fig. 18.6 for the approximation of mapping between $\boldsymbol{u}_i$ and $\boldsymbol{u}'_i$ using a second GRBFN. In these experiments, we use bivectors to code the line direction $\boldsymbol{n}_i$ and plane orientation $\boldsymbol{u}_i$. Also, rotors (which are the weights) are coded using bivectors. We use $\eta = 0.9$ $\rho = 0.05$ in the first network and $\eta = 0.6$ $\rho = 0.1$ in the second one; in both cases, three centroids, one input, and one output are used. Figures 18.9 and 18.10 show that the GRBFN ensemble is able to approximate the needed transformation to reach an orientation from another one. Note that the first time the network is completely unaware of the transformation between the input orientation and the output orientation. However, when more pairs of orientations are fed to the network, it can approximate the real rotation between the geometric entities, which changes smoothly. In robotics, this grade of accuracy is very useful to guide a robot. We have seen how the network can help to approximate the real transformation between certain geometric entities. Our scheme converges

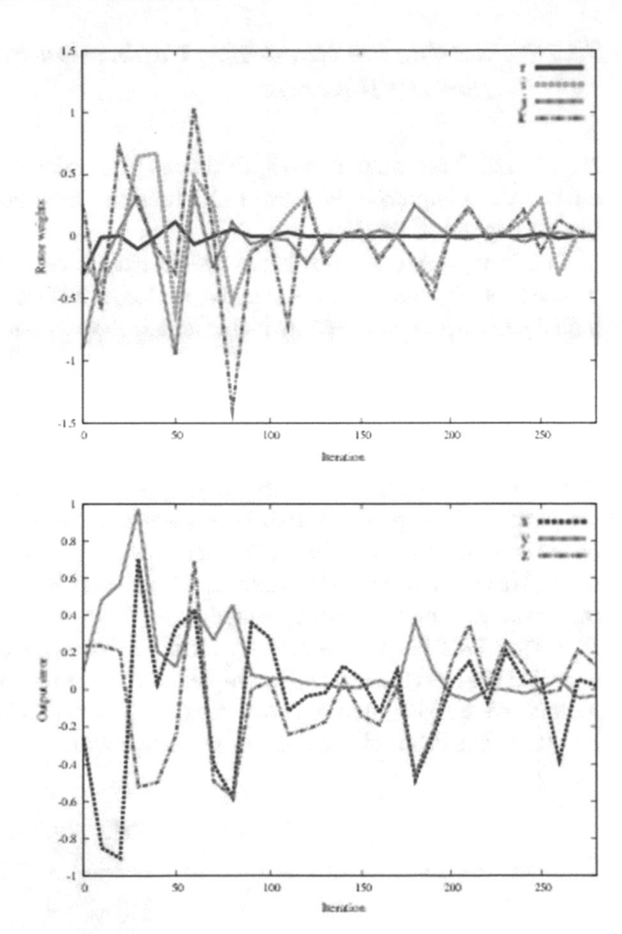

Fig. 18.10 Approximation of transformation between plane orientations $\mathbf{u_i}$ and $\mathbf{u'_i}$. Error at output during learning process (below). Outputs are expressed as $X = xe_1e_2 + ye_2e_3 + ze_3e_1$

so that by using different instances of training and testing data, we get an error of 0.0740 in the first network and 0.0627 in the second one. This error is defined using an average of the Euclidean distance between target outputs and GRBFN outputs. This indicates that our scheme works well.

Geometric algebra helps here to define geometric entities as circles, spheres, or planes so that we can have transformations between orientations of different types of geometric entities expressed using multivectors. They can be used to teach a robot the necessary transformations to move its arms or legs. Using powerful computing equipment and external cameras for the robot, we can achieve this structure and the robot will know in real time which transformation to apply to move a part of its arm or leg. Also, the robot could move an object to another orientation using its arms.

18.2.3 Solving the Hand–Eye Calibration Problem Using the GRFB Network

Now, to estimate the unknown rigid motion, we employ the GRFBN, which is actually a multivector approach. We have n-3D motions described by lines, where each one has six degrees of freedom.

In this approach, we use data available from a stereo camera. We have an object whose axis of rotation is marked by two colored, well-distributed points. With vision techniques, these two points are identified in 3D space, and then the orientation line of the object is obtained. These lines' axes help us to know the camera orientation lines. On the other hand, we construct orientation lines using movements of the end-effector. Then, we have a target set l_{Ai} and another input set l_{Bi} of line orientations, as we observe in Fig. 18.11a. Let L_{Ai} be an orientation of an object and L_{Bj} the orientation of the end-effector, which are expressed using six-dimensional multivectors. Then we construct input pairs (for training and testing) using the orientation lines obtained by the end-effector and object movements. Now we have defined L_{Ai} for X_i and L_{Bj} for Y_i. We use these data sets to train and test the GRBFN. Figure 18.12 shows that the motor values converge to the expected value; this is also observed in Fig. 18.13, which shows that GRBFN outputs are very close to a value range. In these experiments, we are using six-dimensional hypercomplex numbers to code orientations of objects and motors. We consider that the error must be less than 0.0001; also, we use $\eta = 0.4$, $\rho = 0.01$, five centroids, one input, and one output.

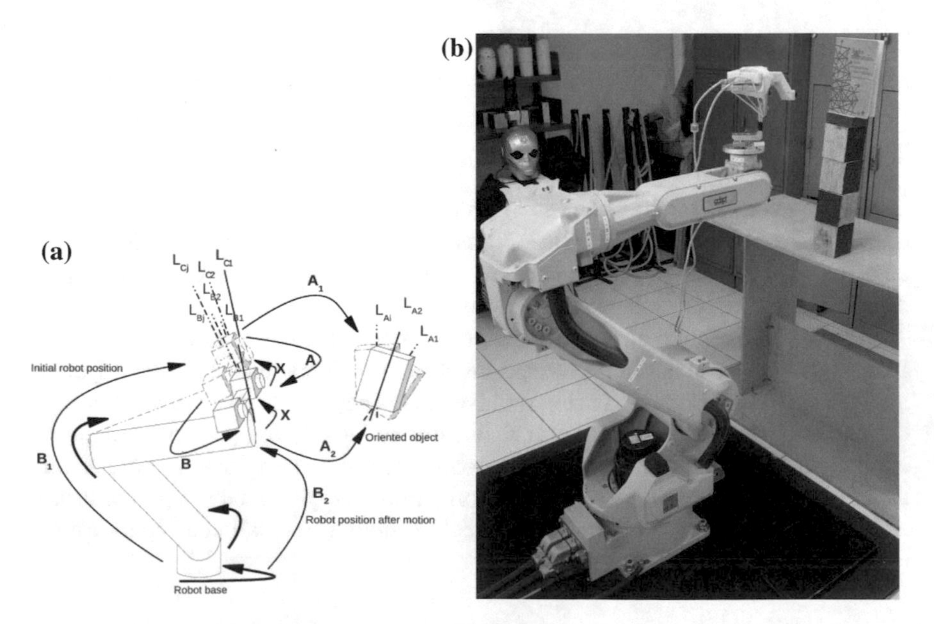

Fig. 18.11 Stereo system mounted at the end-effector of the robot arm. **a** A representative scheme; **b** real workspace

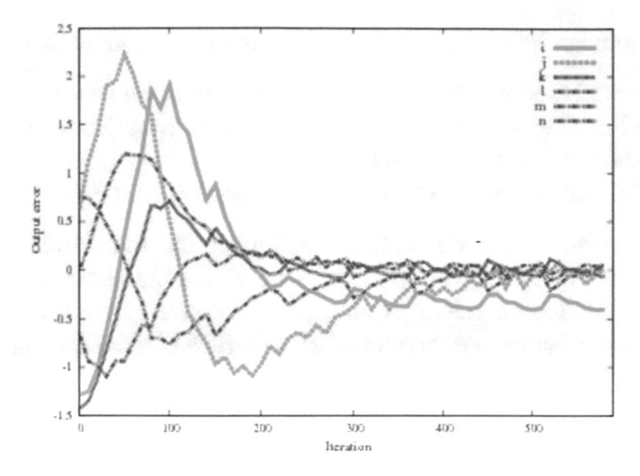

Fig. 18.12 Output error during learning process

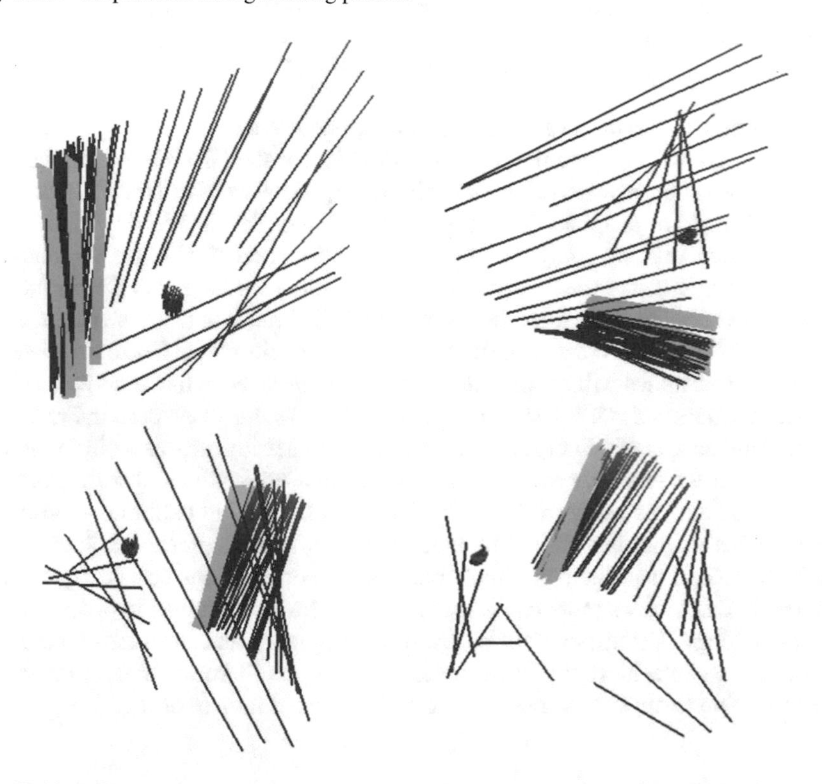

Fig. 18.13 Different views of learning process. Output network is shown (blue line). The network encodes the needed geometric transformation between the set of red lines and set of green ones

The algorithm can be summarized as consisting of the following procedure:

1. Consider n hand motions (b_i) and their corresponding camera motions (a_i). Then we extract the line directions and moments of the screw axes' lines, and we construct the input set and the target set of lines.
2. Apply GRBFN using these sets to get the motor M_X.

Figure 18.12 shows how our network estimates the transformation needed to achieve the target orientation from an initial one. As the network learns, its output is closer to the desired goal. Also, our scheme converges so that using different instances of testing data, we get an error of 0.039, which is defined using an average of the Euclidean distance between target outputs and GRBFN outputs. The M_X found by the GRBFN is one of the motors that are needed to transform an entity in the camera's coordinate system into the world coordinate system. If we know that M_w is the motor, we need to transform the end-effector in the world system, and if M_c is the motor needed to transform an entity in the camera coordinate system, then we need the total motor:

$$M_f = M_w M_X M_c \tag{18.1}$$

to transform from the camera to the world coordinate system.

Then, to know how good the transformation is that the GRBFN found, we have defined a line using our stereo system; furthermore, we have defined the exact position in the real world of this line. We apply a translation of this line to the origin of the camera system, then we apply the M_x transformation, and, finally, we translate this line to the origin of the robot coordinate system. Now we have this line in the real world; if we apply the transformation that the real line must be from the origin of the robot coordinate system (real world), then we have results that can be appreciated in Fig. 18.13. This indicates that our scheme works having an error that is defined using a Euclidean distance, and it is just 15 mm larger. We show our workspace in Fig. 18.11.

The geometric algebra helps here to define a geometric entity as a circle, line, or plane, so that we can have transformations between orientations of different types of geometric entities using multivectors. This can be used to calibrate a variety of systems where a problem similar to hand–eye calibration is presented, such as that of a humanoid, where there are many parts whose orientations can be represented by lines. In this sense, a robot can learn the needed transformations in order to move its arms or legs. Using powerful computing equipment and external cameras for the robot, we can achieve this structure and the robot will know in real time which transformation to apply to move the appropriate part of its arm or leg.

18.3 Experiments Using Clifford Support Vector Machines

In this section, we present three interesting experiments. The first one shows a multi-class classification using CSVM with a simulated example. Here, we discuss also the number of variables computed by each approach and a time comparison between

CSVM and two approaches to do multiclass classification using real SVM. The second is about object multiclass classification with two types of training data: Phase (a) artificial data and Phase (b) real data obtained from a stereo vision system. We also compared the CSVM against MLP's (for multiclass classification). The third experiment presents a multiclass interpolation.

18.3.1 3D Spiral: Nonlinear Classification Problem

We extended the well-known 2D spiral problem to the 3D space. This experiment should test whether the CSVM would be able to separate five 1D manifolds embedded in $\mathbb{R}^3$. On this application, we used a quaternion-valued CSVM whit geometric algebra $G_{0,2,0}$.[2] This allows us to have quaternion inputs and outputs, and therefore, with one output quaternion we can represent as many as 2^4 classes. The functions were generated as follows:

$$
\begin{aligned}
f_1(t) &= [x_1(t), y_1(t), z_1(t)] \\
&= [z_1 * cos(\theta) * sin(\theta), z_1 * sin(\theta) * sin(\theta), z_1 * cos(\theta)], \\
f_2(t) &= [x_2(t), y_2(t), z_2(t)] \\
&= [z_2 * cos(\theta) * sin(\theta), z_2 * sin(\theta) * sin(\theta), z_2 * cos(\theta)], \\
f_3(t) &= [x_3(t), y_3(t), z_3(t)] \\
&= [z_3 * cos(\theta) * sin(\theta), z_3 * sin(\theta) * sin(\theta), z_3 * cos(\theta)], \\
f_4(t) &= [x_4(t), y_4(t), z_4(t)] \\
&= [z_4 * cos(\theta) * sin(\theta), z_4 * sin(\theta) * sin(\theta), z_4 * cos(\theta)], \\
f_5(t) &= [x_5(t), y_5(t), z_5(t)] \\
&= [z_5 * cos(\theta) * sin(\theta), z_5 * sin(\theta) * sin(\theta), z_5 * cos(\theta)],
\end{aligned}
$$

where the program code reads $\theta = linspace(0.2 * pi, 32)$, $z_1 = 4 * linspace(0, 10, 32) + 1$, $z_2 = 4 * linspace(0, 10, 32) + 10$, $z_3 = 4 * linspace(0, 10, 32) + 20$, $z_4 = 4 * linspace(0, 10, 32) + 30$ and $z_3 = 4 * linspace(0, 10, 32) + 40$. To depict these vectors, they were normalized by 10. In Fig. 18.14, one can see that the problem is highly nonlinearly separable. The CSVM uses for training 50 input quaternions of each of the five functions, since these have three coordinates we use simply the bivector part of the quaternion, namely $x_i = x_i(t)\sigma_2\sigma_3 + y_i(t)\sigma_3\sigma_1 + z_i(t)\sigma_1\sigma_2 \equiv [0, x_i(t), y_i(t), z_i(t)]$. The CSVM used the kernel given by (13.62). Note that the CSVM indeed manage to separate the five classes.

Comparisons using the 3D spiral. According to [147], the most-used methods to do multiclass classification are: one-against-all [40], one-against-one [159], DAGSVM [236], and some methods to solve multiclass in one step, known as all together methods [307]. Table 18.3 shows a comparison of number of variables computing per approach, considering also CSVM.

---

[2]The dimension of this geometric algebra is $2^2 = 4$.

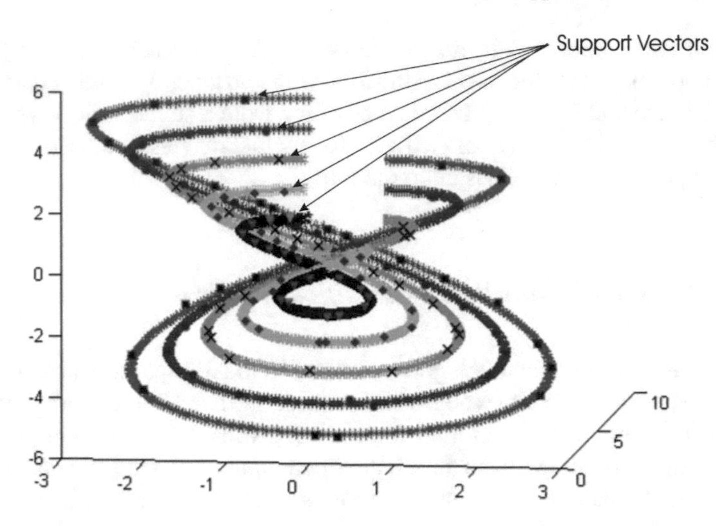

Fig. 18.14 3D spiral with five classes. The marks represent the support multivectors found by the CSVM

Table 18.3 Number of variables per approach

Approach	NQP	NVQP	TNV
CSVM	1	D*N	D*N
One-against-all	K	N	K*N
One-against-one	K(K − 1)/2	2*N/K	N(K − 1)
DAGSVM	K(K − 1)/2	2*N/K	N(K − 1)
A method by considering all data at once	1	K*N	K*N

NQP Number of quadratic problems to solve
NVQP Number of variables to compute per quadratic problem
TNV Total number of variables
D Training input data dimension
N Total number of training examples
K Number of classes

The experiments shown in [147] indicate that "one-against-one and DAG methods are more suitable for practical use than the other methods," we have chosen to implement the one-against-one and the earliest implementation for SVM multiclass classification one-against-all approach to do comparisons between them and our proposal CSVM. Table 18.4 shows three cases where CSVM was compared with three different approaches for multiclass classification; see [148, 153]. The comparisons were made using the 3D spiral toy example and the quaternion CSVM shown in the past subsection. The number of classes was increased on each experiment, and we started with K = 3 classes and 150 training inputs for each class. Since the training inputs have three coordinates, we use simply the bivector part of the quaternion for CSVM approach, namely $x_i = x_i(t)\sigma_2\sigma_3 + y_i(t)\sigma_3\sigma_1 + z_i(t)\sigma_1\sigma_2 \equiv [0, x_i(t), y_i(t), z_i(t)]$; therefore, CSVM computes $D * N = 3 * 150 = 450$ variables. The approaches one-against-all

Table 18.4 Time training per approach (seconds)

Approach	K = 3, N = 150 (Variables)	K = 5, N = 250 (Variables)(Variables)	K = 16, N = 800 (Variables)(Variables)
CSVM	0.10	2.12	22.83
C = 1000	(450)	(750)	(3200)
One-against-all	0.18	10.0	152.5
(C;σ) = (1000, 2^{-3})	(450)	(1250)	(12800)
One-against-one	0.11	3.42	42.73
(C;σ) = (1000, 2^{-2})	(300)	(1000)	(12000)
DAGSVM	0.11	4.81	48.55
(C;σ) = (1000, 2^{-3})	(300)	(1000)	(12000)

K = number of classes, N = number of training examples,
Used kernels $K(x_i, x_j) = e^{-\sigma||x_i - x_j||}$ with parameters taken from $\sigma = \{2, 2^0, 2^{-1}, 2^{-2}, 2^{-3}\}$ and costs $C = \{1, 10, 100, 1000, 10000\}$. From these $5 \times 5 = 25$ combinations, the best result was selected for each approach

and one-against-one compute 450 and 300 variables, respectively; however, the training times of CSVM and one-against-one are very similar in the first experiment. Note that when we increase the number of classes the performance of CSVM is much better than the other approaches because the number of variables to compute is greatly reduced. For K = 16 and N = 800, the CSVM uses 3200 variables in contrast with the others equal or more than 12000, i.e., four times more. In general, we can see that the multiclass SVM approaches described in [179, 301, 307] focus on real-valued data without any assumption on intrinsic geometric characteristics as ours do. Their training methods are based on either a sequential optimization of k-class functions or minimizing the misclassification rate, which often involves often a very demanding optimization. In contrast, the CSVM classifier extends in a natural manner the binary real valued to a MIMO CSVM without increasing the complexity of the optimization. Next we improved the computational efficiency of all these algorithms, namely we accelerated the computation of the $Gramm$ matrix by utilizing the "decomposition method" [148] and the "shrinking technique" [153]. We can see in Table 18.5 that the CSVM using a quarter of the variables is still faster with around a quarter of the processing time of the other approaches The classification performance of the four approaches is presented in Table 18.6; see [148, 153]. We used during training and test 50 and 20 vectors per class, respectively. We can see that the CSVM for classification has overall the best performance.

18.3.2 Object Recognition

In this subsection, we present an application of Clifford SVM for multiclass object classification. We use only one CSVM with a quaternion as input and a quaternion as output, that allow us to have up to $2^4 = 16$ classes. Basically, we packed in a feature

Table 18.5 Time training per approach (seconds) using the acceleration techniques

Approach	K = 3, N = 150 (Variables)	K = 5, N = 250 (Variables)	K = 16, N = 800 (Variables)
CSVM	0.07	0.987	10.07
C = 1000	(450)	(750)	(3200)
One-against-all	0.11	8.54	131.24
(C;σ) = (1000, 2^{-3})	(450)	(1250)	(12800)
One-against-one	0.09	2.31	30.86
(C;σ) = (1000, 2^{-2})	(300)	(1000)	(12000)
DAGSVM	0.10	3.98	38.88
(C;σ) = (1000, 2^{-3})	(300)	(1000)	(12000)

K = number of classes, N = number of training examples,
Used kernels $K(x_i, x_j) = e^{-\sigma||x_i - x_j||}$ with parameters taken from $\sigma = \{2, 2^0, 2^{-1}, 2^{-2}, 2^{-3}\}$ and costs $C = \{1, 10, 100, 1000, 10000\}$. From these $5 \times 5 = 25$ combinations, the best result was selected for each approach

Table 18.6 Performance in training and test using the acceleration techniques

Approach	Ntrain = 150 Ntest = 60 K = 3	Ntrain = 250 Ntest = 100 K = 5	Ntrain = 800 Ntest = 320 K = 16
CSVM	**98.66**	**99.2**	**99.87**
C = 1000	(**95**)	(98)	(**99.68**)
One-against-all	96.00	98.00	99.75
(C;σ) = (1000, 2^{-3})	(90)	(96)	99.06
One-against-one	98.00	98.4	**99.87**
(C;σ) = (1000, 2^{-3})	(**95**)	(**99**)	(99.375)
DAGSVM	97.33	98.4	**99.87**
(C;σ) = (1000, 2^{-3})	(**95**)	(97)	(**99.68**)

NTrain = number of training vectors, Ntest = number of test vectors,
K = classes number,
Percent of accuracy in training figure above and and below in brackets the percent by test

quaternion one 3D point (which lies in surface of the object) and the magnitude of the distance between this point and the point which lies in the main axis of the object in the same level curve. Figure 18.15 depicts the 4 features taking by the object :

$$X_i = \delta_i s + x_i e_2 e_3 + y_i e_3 e_1 + z_i e_1 e_2 \tag{18.2}$$

For each object, we trained the CSVM using a set of several feature quaternions obtained from different level curves that means that each object is represented by several feature quaternions and not only one. Due to this, the order in which the feature quaternions are given to the CSVM is important: we begin to sample data from the

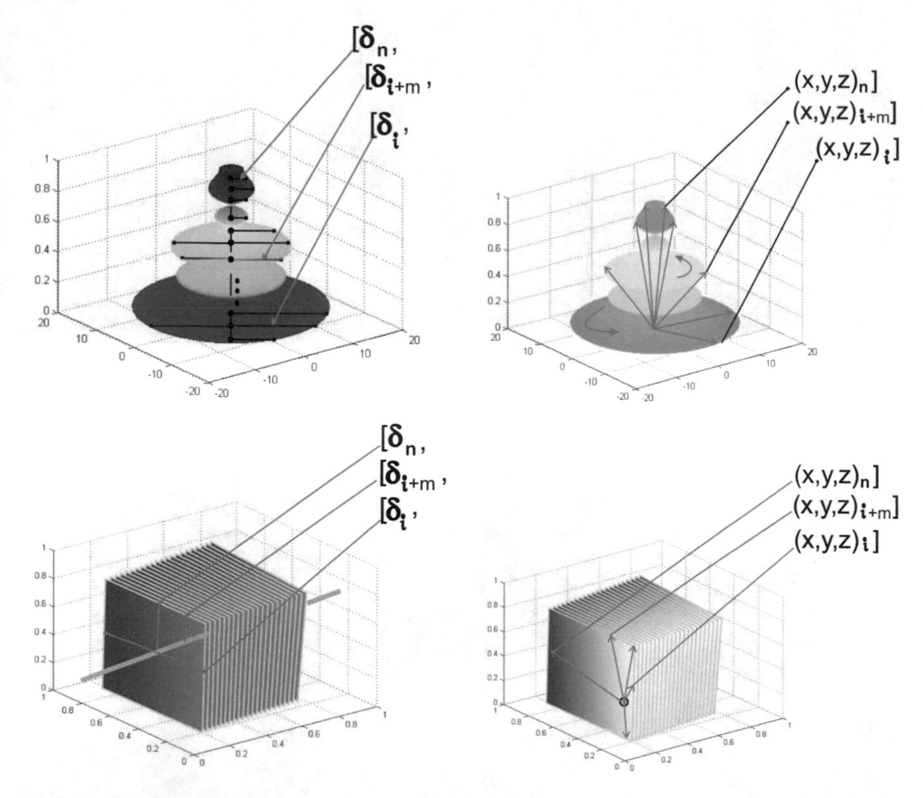

Fig. 18.15 Geometric characteristics of two training object. The magnitude is δ_i, and the 3D coordinates (x_i, y_i, z_i) to build the feature vector: $[\delta_i, (x_i, y_i, z_i)]$

bottom to the top of the objects, and we give to the CSVM the training and test data maintaining this order. We processed the sequence of input quaternions one by one and accumulated their outputs by a counter that computes after the sequence which class fires the most and thus decides which class the object belongs to; see Fig. 18.16a Note carefully, that this experiment is anyway a challenge for any algorithm for object recognition, because the feature signature is sparse. We will show later that using this kind of feature vectors the CSVM's performance is superior to that of MLP's one and the real-valued SVM-based approaches. Of course, if you spend more effort trying to improve the quality of the feature signature, the CSVM's performance will increase accordingly. It is important to note that all the objects (synthetic and real) were preprocessed in order to have a common center and the same scale. Therefore, our learning process can be seen as centered and scale invariant.

Phase (a) Synthetic data. In this experiment, we used data training obtained from synthetic objects, the training set is shown in Fig. 18.17. Note that we have six different objects, which means a six-classes classification problem, and we solve it with only one CSVM making use of its multioutput characteristic. In general,

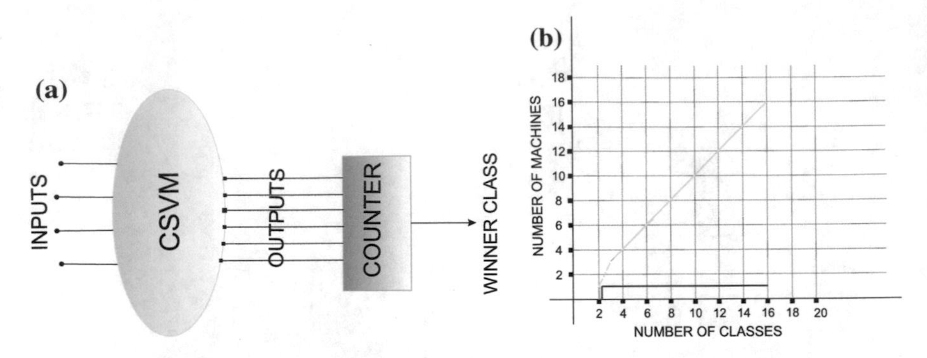

Fig. 18.16 **a** (left) After we get the outputs, these are accumulated using a counter to calculate which class the object belongs. **b** (right) Lower line represents CSVM, while the increasing line the real SVM one versus all approach

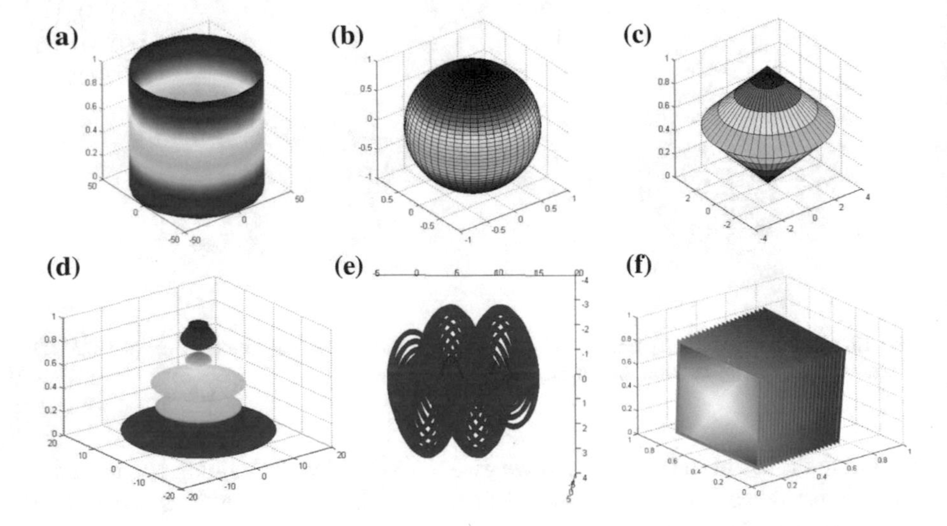

Fig. 18.17 Training synthetic object set

for the "one versus all" approach one needs n SVMs (one for each class). In contrast, the CSVM needs only one machine because its quaternion output allows us to have 16 class outputs; see Fig. 18.16b. For the input data coding, we used the non-normalized 3D point which is packed into the σ_{23}, σ_{31}, σ_{12} basis of the feature quaternion and the magnitude was packed in the scalar part of the quaternion (see Eq. (18.2)). Figure 18.18 shows the 3D points sampled from the objects. We compared the performance of the following approaches: CSVM, a 4-7-6 MLP and the real-valued SVM-based approaches one-against-one, one-against-all, and DAGSVM. The results in Tables 18.7 and 18.8 (see [148, 153]) show that CSVM has better generalization and less training errors than the MLP approach and the real-valued SVM-based approaches one-against-one, one-against-all, and DAGSVM. Note that all methods

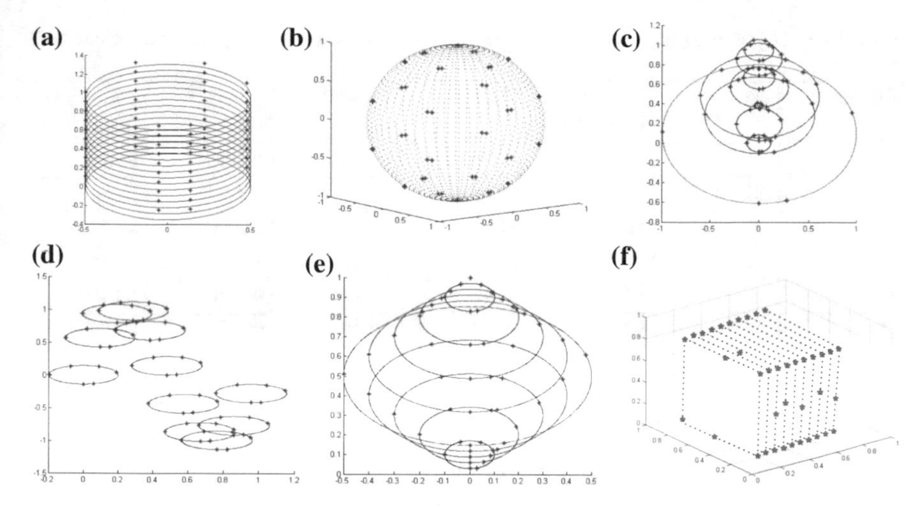

Fig. 18.18 Sampling of the training synthetic object set

Table 18.7 Object recognition performance in percent $\%(\cdot)$ during training using the acceleration techniques

Object	NTS	CSVM C = 1200	MLP	1-versus-all (a)	1-vs-1 (b)	DAGSVM (c)
C	86	**93.02**	48.83	87.2	90.69	90.69
S	84	89.28	46.42	89.28	(**90.47**)	(**90.47**)
F	84	**85.71**	40.47	83.33	84.52	83.33
W	86	91.86	46.51	90.69	91.86	(**93.02**)
D	80	**93.75**	50.00	87.5	91.25	90.00
C	84	**86.90**	48.80	82.14	83.33	84.52

C = cylinder, S = sphere, F = fountain, W = worm, D = diamond, C = cube
NTS = number of training vectors.
Used kernels $K(x_i, x_j) = e^{-\sigma||x_i - x_j||}$ with parameters taken from $\sigma = \{2^{-1}, 2^{-2}, 2^{-3}, 2^{-4}, 2^{-5}\}$ and costs $C = \{150, 1000, 1100, 1200, 1400, 1500, 10000\}$. From these $8 \times 5 = 40$ combinations, the best result was selected for each approach.
(a) $(2^{-4}$,1500), (b) $(2^{-3}$,1200) and (c) $(2^{-4}$,1400)

were speed up using the acceleration techniques [148, 153], and this extra procedure enhanced even more the accuracy of the CSVM classificators (see also Tables 18.4 and 18.5). The authors think that the MLP presents more training and generalization errors because the way we represent the objects (as feature quaternion sets) makes the MLP gets stuck in local minima very often during the learning phase, whereas the CSVM is guaranteed to find the optimal solution to the classification problem because it solves a convex quadratic problem with global minima. With respect to the real-valued SVM-based approaches, the CSVM takes advantage of the Clifford product which enhances the discriminatory power of the classificator itself unlike the other approaches which are based solely on inner products.

Table 18.8 Object recognition accuracy in percent ($\%(\cdot)$) during test using the acceleration techniques

Object	NTS	CSVM C = 1200	MLP	1-vs-all (a)	1-vs-1 (b)	DAGSVM (c)
C	52	94.23	80.76	90.38	(**96.15**)	**96.15**
S	66	**87.87**	45.45	83.33	(84.84)	86.36
F	66	**90.90**	51.51	83.33	86.36	84.84
W	66	**89.39**	57.57	86.36	83.33	86.36
D	58	**93.10**	55.17	**93.10**	**93.10**	**93.10**
C	66	**92.42**	46.96	89.39	90.90	89.39

C = cylinder, S = sphere, F = fountain, W = worm, D = diamond, C = cube
NTS = number of test vectors.
$K(x_i, x_j) = e^{-\sigma||x_i - x_j||}$, $\sigma = \{2^{-1}, 2^{-2}, 2^{-3}, 2^{-4}, 2^{-5}\}$, $C = \{150, 1000, 1100, 1200, 1400, 1500, 10000\}$.
(a) (2^{-4},1500), (b) (2^{-3},1200) and (c) (2^{-4},1400)

Fig. 18.19 Stereo vision system and experiment environment

Phase (b) Real data. In this phase of the experiment, we obtained the training data using our robot "Geometer," it is shown in Fig. 18.19. We take two stereoscopic views of each object: one frontal view and one 180 rotated view (w.r.t. the frontal view), after that, we applied the Harry's filter on each view in order to get the objects corners and then, with the stereo system, the 3D points (x_i, y_i, z_i) which laid on the object surface and to calculate the magnitude δ_i for the quaternion equation (18.2). This process is illustrated in Fig. 18.20, and the whole training object set is shown in Fig. 18.21. We follow the method explained before and proceed like in phase a.1) (for synthetic objects), because we obtained better results than in phase a.2), that is, we take the non-normalized 3D point for the bivector basis $\sigma_{23}, \sigma_{31}, \sigma_{12}$ of the feature quaternion in (18.2)

After the training, we tested with a set of feature quaternions that the machine didnot see during its training and we used the approach of "winner take all" to decide

Fig. 18.20 **a** Frontal left and right views and their sampling views. **b** 180 rotated and sampling views. We use big white cross for the depiction

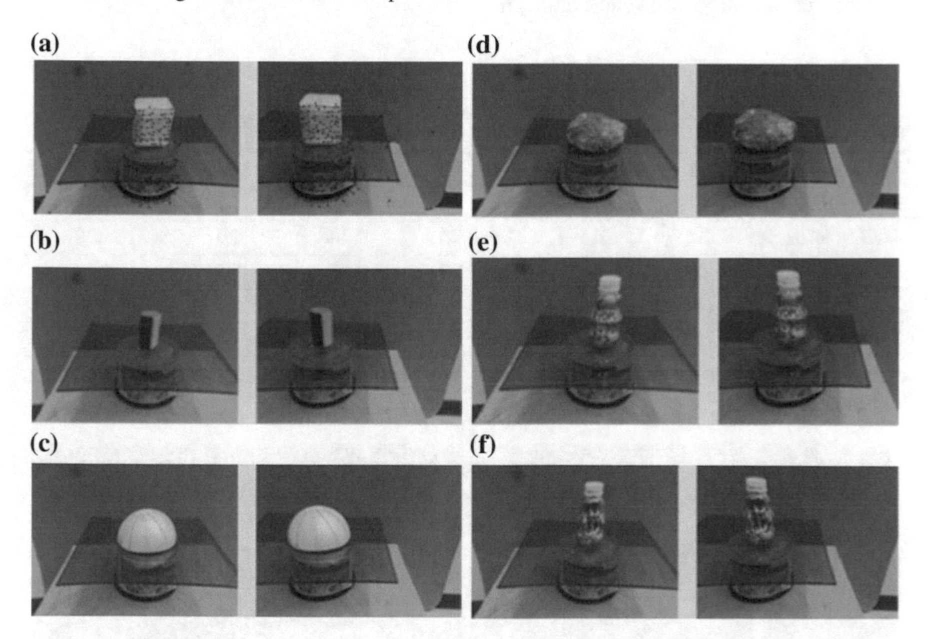

Fig. 18.21 Training real object set, stereo pair images. We include only the frontal views

which class the object belongs to. The results of the training and test are shown in Table 18.9 and in Fig. 18.22. We trained the CSVM with equal number of training data for each object, that is, 90 feature quaternions for each object, but we tested with different number of data for object. Note that we have two pairs of objects that are very similar to each other; the first pair is composed by the half sphere shown in Fig. 18.21c and the rock in Fig. 18.21d, in spite of their similarities, we got very good accuracy in the test phase for both objects: 65.9% for the half sphere and 84% for the rock. We think we got better results for the rock because this object has a lot of texture that produces many corners which in turn capture better the irregularities; therefore, we have more test feature quaternions for the rock than for half sphere (75 against 44 respectively). The second pair composed by similar objects is shown in Fig. 18.21e and f, and these are two equal plastic bottles of juice, but one of them (Fig. 18.21f) is burned, which makes the difference between them and gives the CSVM enough distinguishing features to make two object classes, shown in Table 18.9. We got 60% of correct classified test samples for the bottle in Fig. 18.21e against 61% for the burned bottle in Fig. 18.21f. The lower learn rates in the last objects (Fig. 18.21c, e and f) is because the CSVM is mixing a bit the classes due to the fact that the feature vectors are not large and reach enough.

Table 18.9 Experimental results using real data

Object label	NTS	NES	CTS	%
Cube (Fig. 18.21a) [1, 1, 1, 1]	90	50	38	76.00
Prism (Fig. 18.21b) [−1, −1, 1, 1]	90	43	32	74.42
Half sphere (Fig. 18.21c) [−1, −1, −1, −1]	90	44	29	65.90
Rock (Fig. 18.21d) [−1, −1, −1, −1]	90	75	63	84.00
Plastic bottle 1 (Fig. 18.21e) [−1, −1, −1, −1]	90	65	39	60.00
Plastic bottle 2 (Fig. 18.21f) [−1, −1, −1, −1]	90	67	41	61.20

NTS: number of training samples, NES: number of test samples, CTS: number of correct classified test samples

Fig. 18.22 Robot takes the recognized object

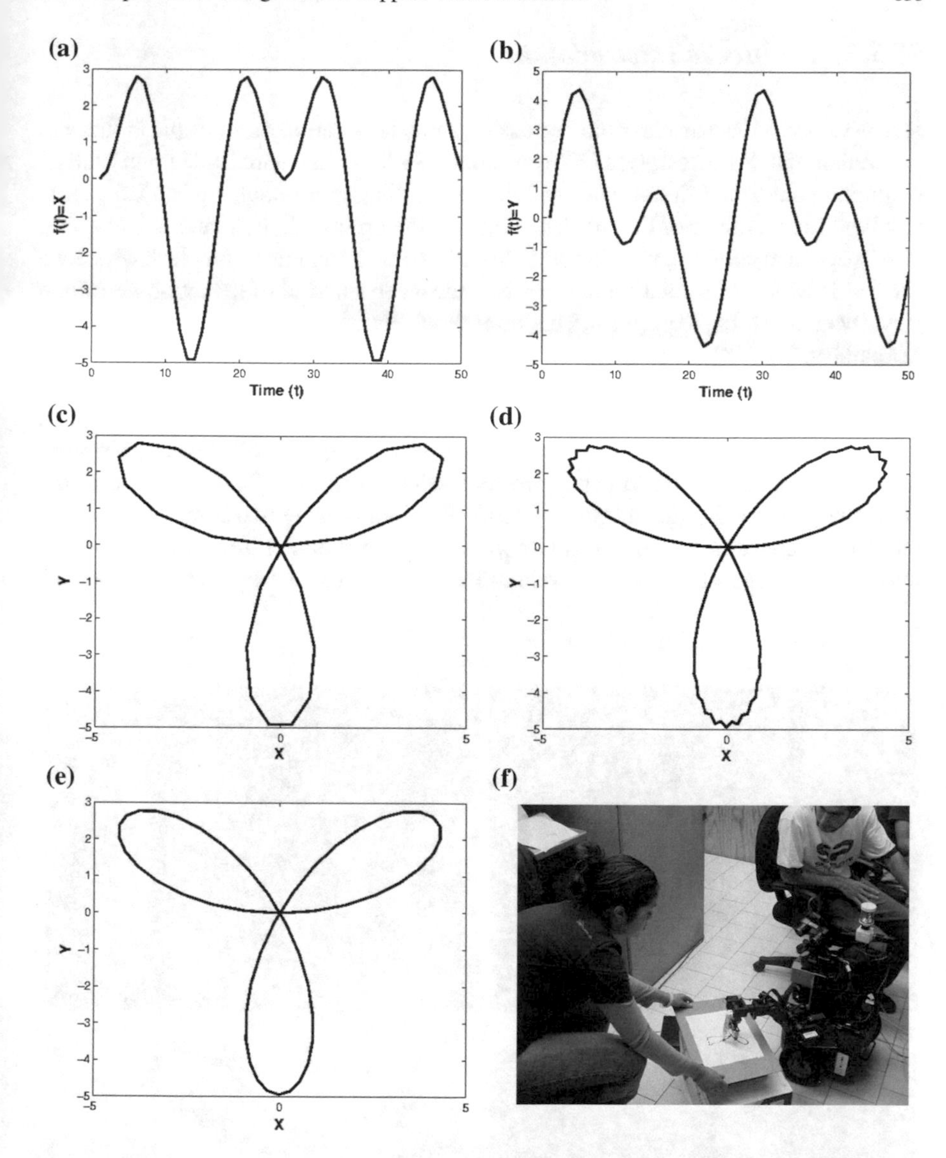

Fig. 18.23 **a** and **b** Continuous curves of training output data for axes x and y (50 points). **c** 2D result of combing axes x and y (50 points). **d** 2D result by testing with 100 input data. **e** 2D result by testing with 400 input data. **f** Experiment environment

18.3.3 Multicase Interpolation

A real-valued SVM can carry out regression and interpolation for multiple inputs and one real output. Surprisingly, a Clifford-valued SVM can have multiple inputs and 2^n outputs for a n-dimensional space or $\mathbb{R}^n$. For handling regression, we use $1.0 > \varepsilon > 0$, where the diameter of the tube surrounding the optimal hyperplane is $2 \times \epsilon$. For the case of interpolation, we use $\varepsilon = 0$. We have chosen an interesting task where we use a CSVM for interpolation in order to code a certain kind of behavior we want a visually guided robot to perform. The robot should autonomously draw a complicated 2D pattern. This capacity should be coded internally in long-term memory (LTM), so that the robot reacts immediately without the need of reasoning. Similar to a capable person who reacts in milliseconds with incredible precision to accomplish a very difficult task, for example a tennis player or tango dancer. For our purpose, we trained off line a CSVM using two real-valued functions. The CSVM used the geometric algebra G_3^+ (quaternion algebra). Two inputs using two components of the quaternion input and two outputs (two components of the quaternion output). The first input u and first output x coded the relation $x = a * sin(3 * u) * cos(u)$ for one axis.

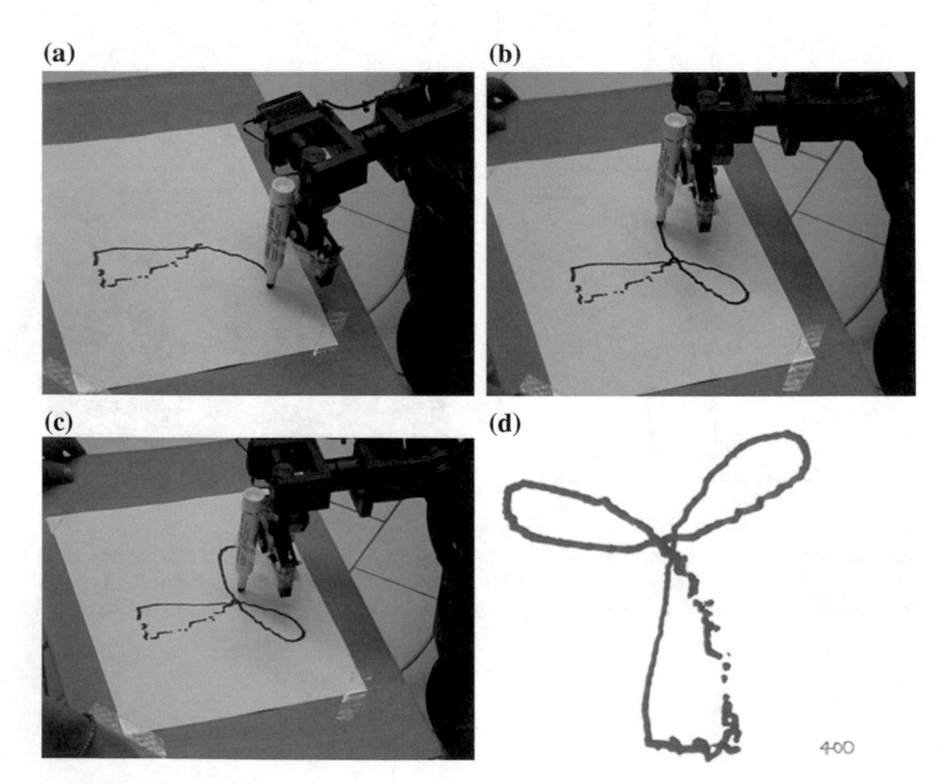

Fig. 18.24 **a, b, c** Image sequence while robot is drawing **d** Robot's draw. Result by testing with 400 input data

The second input v and second output y coded the relation $y = a * sin(3 * v) * sin(v)$ for another axis; see Fig. 18.23a, b. The 2D pattern can be drawn using these 50 points generated by functions for x and y; see Fig. 18.23c. We tested if the CSVM could interpolate well enough using 100 and 400 unseen input tuples $\{u, v\}$; see respectively Fig. 18.23d–e. Once the CSVM was trained, we incorporated it as part of the LTM of the visually guided robot shown in Fig. 18.23f. In order to carry out its task, the robot called the CSVM for a sequence of input patterns. The robot was able to draw the desired 2D pattern as we see in Fig. 18.24a–d. The reader should bear in mind that this experiment was designed using the equation of a standard function, in order to have a ground truth. Anyhow, our algorithm should be able also to learn 3D curves which do not have explicit equations.

18.4 Conclusion

This chapter generalizes the real-valued MLPs and SVMs to Clifford-valued MLPs and SVMs, and they are used for classification, regression, and interpolation. The CSVM accepts multiple multivectors inputs and multivector outputs like a MIMO architecture that allows us to have multiclass applications. We can use CSVM over complex, quaternion, or hypercomplex numbers according to our needs. The application section shows experiments in pattern recognition and visually guided robotics which illustrates the power of the algorithms and help the reader understand the Clifford SVM and use it in various tasks of complex and quaternion signal and image processing, pattern recognition, and computer vision using high-dimensional geometric primitives. The extension of the real-valued SVM to the Clifford SVM appears promising particularly in geometric computing and their applications like graphics, augmented reality, robot vision, and humanoids.

Chapter 19
Neurocomputing for 2D Contour and 3D Surface Reconstruction

In geometric algebra, there exist specific operators named *versors* to model rotations, translations, and dilations, and are called rotors, translators, and dilators, respectively. In general, a versor $\boldsymbol{G}$ is a multivector which can be expressed as the geometric product of non-singular vectors

$$G = \pm \boldsymbol{v}_1 \boldsymbol{v}_2 ... \boldsymbol{v}_k. \tag{19.1}$$

In conformal geometric algebra, such operators are defined by (19.2), (19.3), and (19.4), being R the rotor, T the translator, and D_λ the dilator.

$$\boldsymbol{R} = e^{\frac{1}{2}\theta \boldsymbol{b}}, \tag{19.2}$$

$$\boldsymbol{T} = e^{-\frac{t e_\infty}{2}}, \tag{19.3}$$

$$\boldsymbol{D}_\lambda = e^{\frac{-\log(\lambda)\wedge E}{2}}, \tag{19.4}$$

where $\boldsymbol{b}$ is the bivector dual to the rotation axis, θ is the rotation angle, $t \in \mathcal{E}^3$ is the translation vector, λ is the factor of dilation, and $\boldsymbol{E} = e \wedge e_0$.

Such operators are applied to any entity of any dimension by multiplying the entity by the operator from the left and by the reverse of the operator from the right. Let be $\boldsymbol{X}_i$ any entity in CGA; then to rotate it, we compute $\boldsymbol{X}_1' = \boldsymbol{R}\boldsymbol{X}_1\tilde{\boldsymbol{R}}$, to translate $\boldsymbol{X}_2' = \boldsymbol{T}\boldsymbol{X}_2\tilde{\boldsymbol{T}}$ and to dilate $\boldsymbol{X}_3' = \boldsymbol{D}_\lambda \boldsymbol{X}_3 \tilde{\boldsymbol{D}}_\lambda$.

19.1 Determining the Shape of an Object

To determine the shape of an object, we can use a topographic mapping that uses selected points of interest along the contour of the object to fit a low-dimensional map to the high-dimensional manifold of this contour. This mapping is commonly achieved by using self-organized neural networks such as Kohonen's self-organizing

E. Bayro-Corrochano, *Geometric Algebra Applications Vol. I*,
https://doi.org/10.1007/978-3-319-74830-6_19

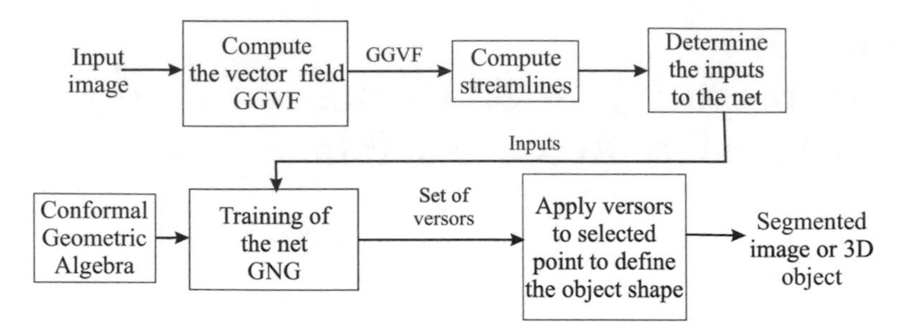

Fig. 19.1 A block diagram of our approach

maps (SOM) or neural gas (NG) [212]; however, if we desire a better topology preservation, we should not specify the number of neurons of the network a priori (as specified for neurons in SOM or NG, together with its neighborhood relations), but allow the network to grow using an incremental training algorithm, as in the case of the Growing Neural Gas (GNG) [97]. In this work, we follow the idea of growing neural networks and present an approach based on the GNG algorithm to determine the shape of objects by means of applying versors of the CGA, resulting in a model easy to handle in postprocessing stages; a scheme of our approach is shown in Fig. 19.1. The neural network has versors associated with its neurons, and its learning algorithm determines the parameters that best fit the input patterns, allowing us to get every point on the contour by interpolation of such versors.

Additionally, we modify the acquisition of input patterns by adding a preprocessing stage that determines the inputs to the net; this is done by computing the generalized gradient vector flow (GGVF) and analyzing the *streamlines* followed by particles (points) placed on the vertices of small squares defined by dividing the 2D/3D space in such squares/cubes. The streamline or the path followed by a particle that is placed on $\mathbf{x} = (x, y, z)$ coordinates will be denoted as $S(\mathbf{x})$. The information obtained with GGVF is also used in the learning stage, as explained further.

19.1.1 Automatic Sample Selection Using GGVF

In order to select the input patterns automatically, we use the GGVF [315], which is a *dense vector field* derived from the volumetric data by minimizing a certain energy functional in a variational framework. The minimization is achieved by solving linear partial differential equations that diffuses the gradient vectors computed from the volumetric data. To define the GGVF, the edge map is defined at first as

$$f(\mathbf{x}) : \Omega \rightarrow \mathcal{R}. \tag{19.5}$$

For the 2D image, it is defined as $f(x, y) = -|\nabla G(x, y) * I(x, y)|^2$, where $I(x, y)$ is the gray level of the image on pixel (x, y), $G(x, y)$ is a 2D Gaussian function (for robustness in the presence of noise), and ∇ is the gradient operator.

With this edge map, the GGVF is defined to be the vector field $\mathbf{v}(x, y, z) = [u(x, y, z), v(x, y, z), w(x, y, z)]$ which minimizes the energy functional

$$\mathcal{E} = \int\int g(|\nabla f|)\nabla^2\mathbf{v} - h(|\nabla f|)(\mathbf{v} - \nabla f), \tag{19.6}$$

where

$$g(|\nabla f|) = e^{-\frac{|\nabla f|}{\mu}} \quad \text{and} \quad h(|\nabla f|) = 1 - g(|\nabla f|) \tag{19.7}$$

and μ is a coefficient. An example of such a dense vector field obtained in a 2D image is shown in Fig. 19.2a, while an example of the vector field for a volumetric data is shown in Fig. 19.2b. In Fig. 19.2d, the points were selected according to Eq. (19.8). Observe the large range of capture of the forces in the image. Due to this large capture range, if we put particles (points) on any place over the image, they can be guided to the contour of the object. The automatic selection of input patterns is done by analyzing the *streamlines* of points on a 3D grid topology defined over the volumetric data. It means that the algorithm follows the streamlines of each point of the grid, which will guide the point to the more evident contour of the object; then, the algorithm selects the point where the streamline finds a peak in the edge map and gets its conformal representation $X = \mathbf{x} + \frac{1}{2}\mathbf{x}^2 e_\infty + e_0$ to make the input pattern set. In addition to X (conformal position of the point), the inputs have the vector $\mathbf{v_I} = [u,\ v,\ w]$, which is the value of the GGVF in such pixel and will be used in the training stage as a parameter determining the amount of energy the input has to attract neurons. This information will be used in the training stage together with the position $\mathbf{x}$ for learning the topology of the data. Summarizing, the input set $\mathbf{I}$ will be

$$\mathbf{I} = \{\zeta_\mathbf{k} = X_{\zeta_k}, \mathbf{v}_{\zeta_k} | \mathbf{x}_\zeta \in S(\mathbf{x}') \text{ and } f(\mathbf{x}_\zeta) = 1\}, \tag{19.8}$$

where X_ζ is the conformal representation of $\mathbf{x}_\zeta$; $\mathbf{x}_\zeta \in S(\mathbf{x}')$ means that $\mathbf{x}_\zeta$ is on the path followed by a particle placed in $\mathbf{x}'$, and $f(\mathbf{x}_\zeta)$ is the value of the edge map in position $\mathbf{x}_\zeta$ (assuming it is binarized). As some streamlines can carry to the same point or very close points, we can add constraints to avoid very close samples; one very simple restriction is that the candidate to be included in the input set must be at least at a fixed distance d_{thresh} of any other input.

Figure 19.2c shows the streamlines according to the vector field shown in Fig. 19.2a, and the input patterns selected as described before are shown in Fig. 19.2.

19.1.2 Learning the Shape Using Versors

It is important to note that although we will be explaining the algorithm using points, the versors can be applied to any entity in GA that we had selected to model the object. The network starts with a minimum number of versors (neural units), and new units are inserted successively. The network is specified by

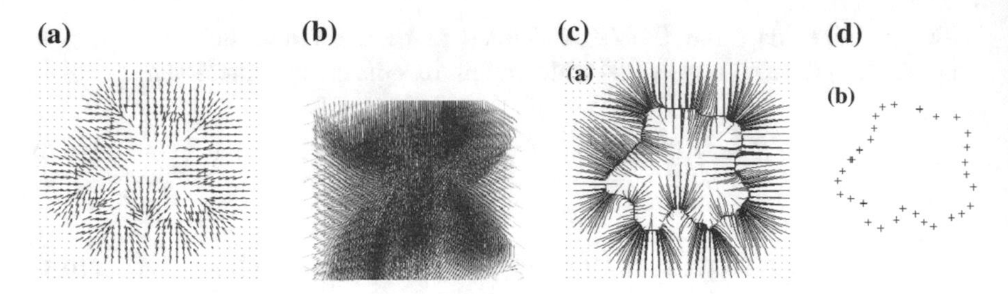

Fig. 19.2 Example of the dense vector field called GGVF (only representative samples of a grid are shown rather than the entire vector field). **a** Samples of the vector field for a 2D image; **b** samples of the vector field for volumetric data; **c** example of streamlines for particles arranged in a 32×32 grid according to the vector field shown in **a**; **d** points selected as input patterns

- A set of units (neurons) named N, where each $\mathbf{n}_l \in N$ has its associated versor $M_{\mathbf{n}_l}$; each versor is the transformation that must be applied to a point to place it in the contour of the object. The set of transformations will ultimately describe the shape of the object.
- A set of connections between neurons defining the topological structure.

Also, take into account that

- There are two learning parameters: e_w and e_n, for the winner neuron and for the direct neighbors of it; such parameters remain constant during all the process.
- Each neuron $\mathbf{n_l}$ will be composed of its versor $M_{\mathbf{n_l}}$, the *signal counter* sc_l, and the *relative signal frequency* rsf_l. The signal counter sc_l is incremented for the neuron $\mathbf{n_l}$ every time if it is a winner neuron. The relative signal frequency rsf_l is defined as

$$rsf_l = \frac{sc_l}{\sum_{\forall \mathbf{n_j}}} sc_j. \tag{19.9}$$

This parameter will act as an indicator to insert new neural units.

With these elements, we define the learning algorithm of the GNG to find the versors that will define the contour as follows:

1. Let P_0 be a fixed initial point over which the transformations will be applied. This point corresponds to the conformal representation of p_0, which can be a random point or the centroid defined by the inputs. The initial transformations will be expressed as $M = e^{-\frac{t}{2}e_\infty}$ in the conformal geometric algebra. The vector t initially is a random displacement.
2. Start with the minimal number of neurons, which have associated random motors M as well as a vector $\mathbf{v_l} = [u_l,\ v_l,\ w_l]$ whose magnitude is interpreted as the capacity of learning for such neuron.
3. Select one input ζ from the inputs set $\mathbf{I}$, and find the winner neuron; find the neuron n_l having the versor M_l that moves the point P_0 closer to such input:

$$M_{win} = \min_{\forall M} \sqrt{(X_\zeta - M P_0 \tilde{M})^2}. \tag{19.10}$$

4. Modify M_{win} and all other versors of neighboring neurons M_l in such a way that the modified M will represent a transformation moving the point P_0 nearer the input. Note that each motor is composed of a rotation ΔR and a translation ΔT. The rotation is computed as in (19.2), where θ is the angle between the actual position a and $a' = a + \mathbf{v}$ ($\mathbf{v}$ is the GGVF vector value in such position); the bivector dual to the rotation axis is computed as $\boldsymbol{b} = I_E \mathbf{b}$; the rotors and translators are defined as

$$\Delta R_{win} = e^{\frac{e_w\ \eta(\mathbf{v}_\zeta, \mathbf{v_{win}})}{2}\theta \boldsymbol{b}}, \tag{19.11}$$
$$\Delta T = e^{-\frac{\Delta t_{win}}{2} e_\infty}, \tag{19.12}$$
$$\Delta t_{win} = e_w\ \eta(\mathbf{v}_\zeta, \mathbf{v_{win}})\ (\mathbf{x}_\zeta - p_0), \tag{19.13}$$

for the winner neuron, and

$$\Delta R_n = e^{\frac{e_n\ \eta(\mathbf{v}_\zeta, \mathbf{v_n})}{2}\theta \boldsymbol{b}}, \tag{19.14}$$
$$\Delta T = e^{-\frac{\Delta t_n}{2} e_\infty}, \tag{19.15}$$
$$\Delta t_n = e_n\ \eta(\mathbf{v}_\zeta, \mathbf{v_n})\ (\mathbf{x}_\zeta - p_0), \tag{19.16}$$

for its direct neighbors, to obtain $\Delta M = \Delta T \Delta R$. Finally, the new motor is

$$M_{l_{new}} = \Delta M M_{l_{old}}. \tag{19.17}$$

ϕ is a function defining the amount a neuron can learn according to its distance from the winner one (defined as in Eq. (19.18)), and $\eta(\mathbf{v}_\zeta, \mathbf{v_l})$ is defined as in (19.19):

$$\phi = e^{-\frac{(M_{win} P_0 \tilde{M}_{win} - M_l P_0 \tilde{M}_l)^2}{2\sigma}}, \tag{19.18}$$
$$\eta(\mathbf{v}_\zeta, \mathbf{v_l}) = \|\mathbf{v}_\zeta - \mathbf{v_l}\|^2, \tag{19.19}$$

which is a function defining a quantity of learning depending on the strength of the input ζ to teach and the capacity of the neuron to learn, given in $\mathbf{v}_\zeta$ and $\mathbf{v_l}$, respectively. Also, update

$$\mathbf{v_{win}^{new}} = [u_{win}^{new}\ \ v_{win}^{new}\ \ w_{win}^{new}]^T, \tag{19.20}$$
$$\mathbf{v_n^{new}} = [u_n^{new}\ \ v_n^{new}\ \ w_n^{new}]^T, \tag{19.21}$$

where $u_{win}^{new} = (u_{win} + e_w\ u_{win})$, $v_{win}^{new} = (v_{win} + e_w\ v_{win})$, $w_{win}^{new} = (w_{win} + e_w\ w_{win})$, $u_n^{new} = (u_n + e_n\ u_n)$, $v_n^{new} = (v_n + e_n\ v_n)$, $w_n^{new} = (w_n + e_n\ w_n)$.

5. Every certain number λ of iterations determine the neuron with the rsf_l with highest value. Then, if any of the direct neighbors of that neuron is at a distance larger than c_{max}, do the following:

   – Determine neighboring neurons $\mathbf{n_i}$ and $\mathbf{n_j}$.
   – Create a new neuron n_{new} between $\mathbf{n_i}$ and $\mathbf{n_j}$ whose associated M and $\mathbf{v_l}$ will be

$$M_{n_{new}} = \frac{M_i + M_j}{2}, \quad \mathbf{v_{l\_new}} = \frac{\mathbf{v_i} + \mathbf{v_j}}{2}. \tag{19.22}$$

   The new units will have the values $sc_{new} = 0$ and $rsf_{new} = 0$.
   – Delete old edge connecting $\mathbf{n_i}$ and $\mathbf{n_j}$, and create two new edges connecting n_{new} with $\mathbf{n_i}$ and $\mathbf{n_j}$.

6. Repeat steps 3 to 5 if the stopping criterion is not achieved. The stop criterion is when a maximum number of neurons is reached or when the learning capacity of neurons approaches zero (is less than a threshold c_{min}), whichever happens first will stop the learning process.

Training the network, we find the set of M defining positions on a trajectory; such positions minimize the error measured as the average distance between X_ς and the result of $M_\varsigma P_0 \tilde{M}_\varsigma$:

$$\chi = \frac{\sum_{\forall\varsigma}(\sqrt{(M_\varsigma P_0 \tilde{M}_\varsigma - X_\varsigma)^2}}{N}, \tag{19.23}$$

where M_ς moves P_0 closer to input X_ς and N is the number of inputs.

19.2 Experiments

Figure 19.3 shows the result when the algorithm is applied to a magnetic resonance image (MRI); the goal is to obtain the shape of the ventricle. Figure 19.3a shows the original brain image and the region of interest (ROI); Fig. 19.3b shows the computed vector field for the ROI; Fig. 19.3c shows the streamlines in the ROI defined for particles placed on the vertices of a 32×32 grid; Fig. 19.3d shows the initial shape as defined for the two initial random motors M_a *and* M_b; Fig. 19.3e shows the final shape obtained; and, finally, Fig. 19.3f shows the original image with the segmented object.

Figure 19.4 shows an image showing that our approach can also be used for automated visual inspection tasks; the reader can observe that such an image contains a very blurred object. That image is for the inspection of hard disk head sliders. Figure 19.4a shows the original image and the region of interest (ROI); Fig. 19.4b shows the computed vector field of the ROI; Fig. 19.4c shows the streamlines defined for particles placed on the vertices of a 32x32 grid; Fig. 19.4d shows the inputs selected according to the streamlines and the initial shape as defined for the two

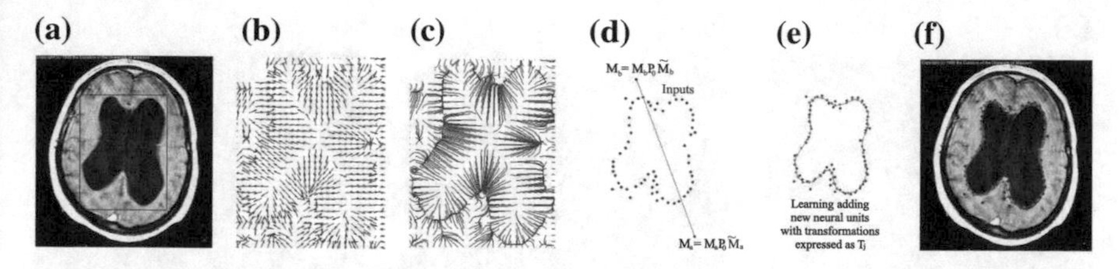

Fig. 19.3 **a** Original image and region of interest (ROI); **b** zoom of the dense vector field of the ROI; **c** zoom of the streamlines in ROI; **d** inputs and initial shape; **e** final shape defined according to the 54 estimated motors; **f** image segmented according to the results

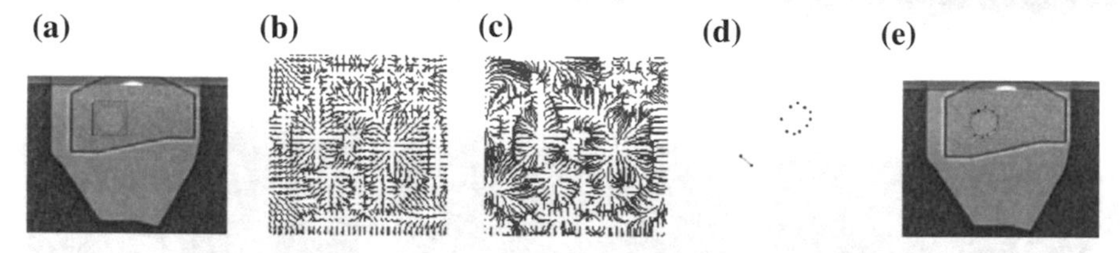

Fig. 19.4 Application in visual inspection tasks: **a** original image and the region of interest (ROI); **b** zoom of the dense vector field of the ROI; **c** zoom of the streamlines in ROI; **d** inputs and initial shape according to the two initial random transformations M_a and M_b; **e** final shape defined according to the 15 estimated motors (original image with the segmented object)

initial random motors M_a *and* M_b; Fig. 19.4e shows the final shape obtained overlapped with the original image, showing that the algorithm gives good results if it is used for segmentation.

Figure 19.5 shows the application of the ggvf-snakes algorithm in the same problem. It is important to note that although the approaches of Figs. 19.4 and 19.5 use GGVF information to find the shape of an object, the estimated final shape is better using the neural approach than the one using active contours; the second approach (see Fig. 19.5) fails to segment the object whether the initialization of the snake is given inside, outside, or over the contour we are interested in. Additionally, the fact of expressing such a shape as a set of motors allows us to have a model best suited for use in further applications that can require the deformation of the model, specially if such a model is not based on points but on other GA entities, because we do not need to change the motors (remember that they are applied in the same way to any other entity).

The proposed algorithm was applied to different sets of medical images. Figure 19.6 shows some images of such sets. The first row of each figure shows the original image and the region of interest, while the second row shows the result of the proposed approach. Table 19.1 shows the errors obtained with our approach using and not using the GGVF information. We can observe that the inclusion of GGVF information improves the approximation of the surface.

To compare our algorithm, we use the GNG with and without GGVF information, as well as a growing version of SOM, also using and not using the GGVF

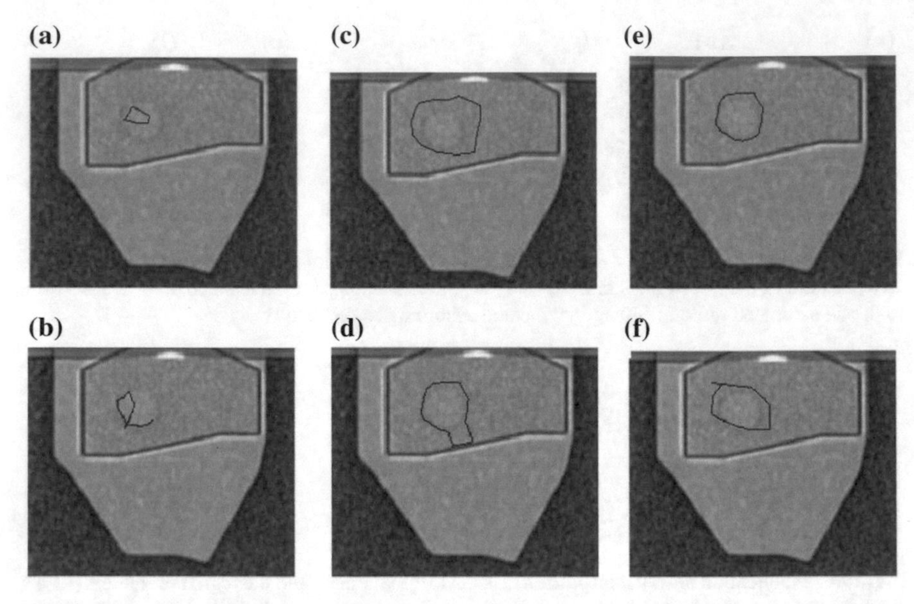

Fig. 19.5 Result obtained when using the active contour approach to segment the object in the same image as in Fig. 19.4. **a** Initialization of snake inside the object; **b** final result obtained with initialization showed in **a**; **c** initialization of snake outside the object; **d** final result obtained with initialization showed in **c**; **e** initialization of snake over the contour; **f** final result obtained with initialization shown in **e**

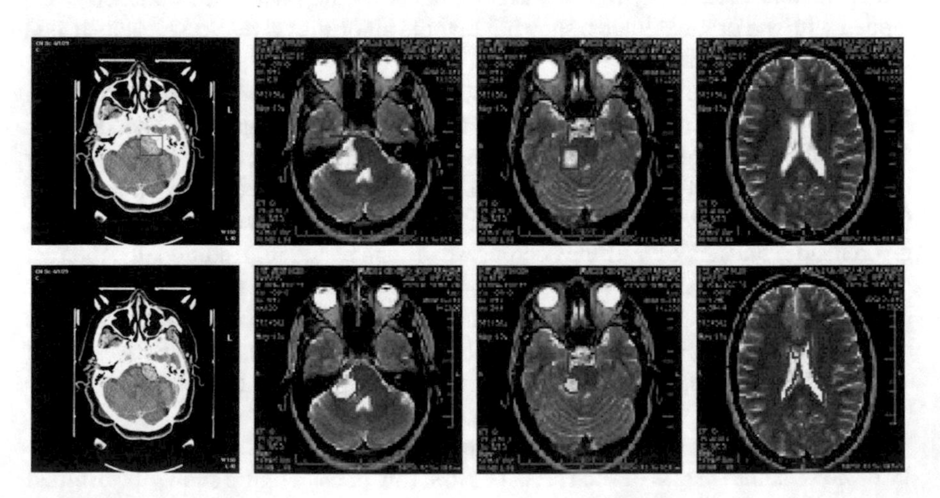

Fig. 19.6 First row (upper row): original image and the region of interest. Second row: result of segmentation

Table 19.1 Errors obtained by the algorithm with and without the GGVF information. ϵ_1: error without GGVF; ϵ_2: error with GGVF

Example	ϵ_1	ϵ_2	Example	ϵ_1	ϵ_2
Ventricle 1	3.29	2.51	Eye 1	7.63	6.8
Eye 2	3.43	2.98	Column disk 1	4.65	4.1
Tumor 1	3.41	2.85	Tumor 2	2.95	2.41
Free-form curve	2.84	1.97	Column disk 2	2.9	2.5

information. These algorithms were applied to a set of 2D medical images (some obtained with computer tomography (CT) and some with magnetic resonance (MR)). Figure 19.7a shows the average errors when GSOM stops for different examples: segmenting a ventricle, a blurred object, a free-form curve, and a column disk. Note that with the GGVF information, the error is reduced. That means that using the GGVF information, as we propose, allows a better approximation of the object shape to be obtained. Figure 19.7b shows the average errors obtained for several examples but using the GNG with and without the GGVF information. Note that, again, the GGVF contributes to obtaining a better approximation of the object's surface. Also, note that the average errors obtained with the GNG algorithm are smaller than the errors obtained with the GSOM, as can be seen in Fig. 19.7c, and that both are improved with GGVF information, although GNG gives better results.

It is necessary to mention that the whole process is quick enough; in fact, the computational time required for all the images shown in this work took only a few seconds. The computation of the GGVF is the most time-consuming task in the algorithm, but it only takes about 3 s for 64×64 images, 20 s for 256×256 images, and 110 s for 512×512 images. This is the reason why we decided not to compute it for the whole image, but for selected region of interest. The same criterion was applied to 3D examples.

Figure 19.8a shows the patient head with the tumor whose surface we need to approximate; Fig. 19.8b shows the vectors of the dense GGVF on a 3D grid arrangement of size $32\times32\times16$; Fig. 19.8c shows the inputs determined by GGVF and edge map, and also shows the initialization of the net GNG; Fig. 19.8d–f shows some stages of the adaptation process, while the net is determining the set of transformations M (Fig. 19.8f is the final shape after training has finished with a total of 170 versors M (associated with 170 neural units)).

Figure 19.9 shows other 3D examples, corresponding to a pear, and the surface is well approximated. Figure 19.9b shows the inputs and the initialization of the net with nine neural units (the topology of the net is defined as a sort of pyramid around the centroid of input points); Fig. 19.9c shows the result after the net has reached the maximum number of neurons, which was fixed to 300; and finally, Fig. 19.9d shows the minimization of the error according to (19.23).

Another useful application of the algorithm using the gradient information of the GGVF during the training of the GNG neural net in the geometric algebra framework

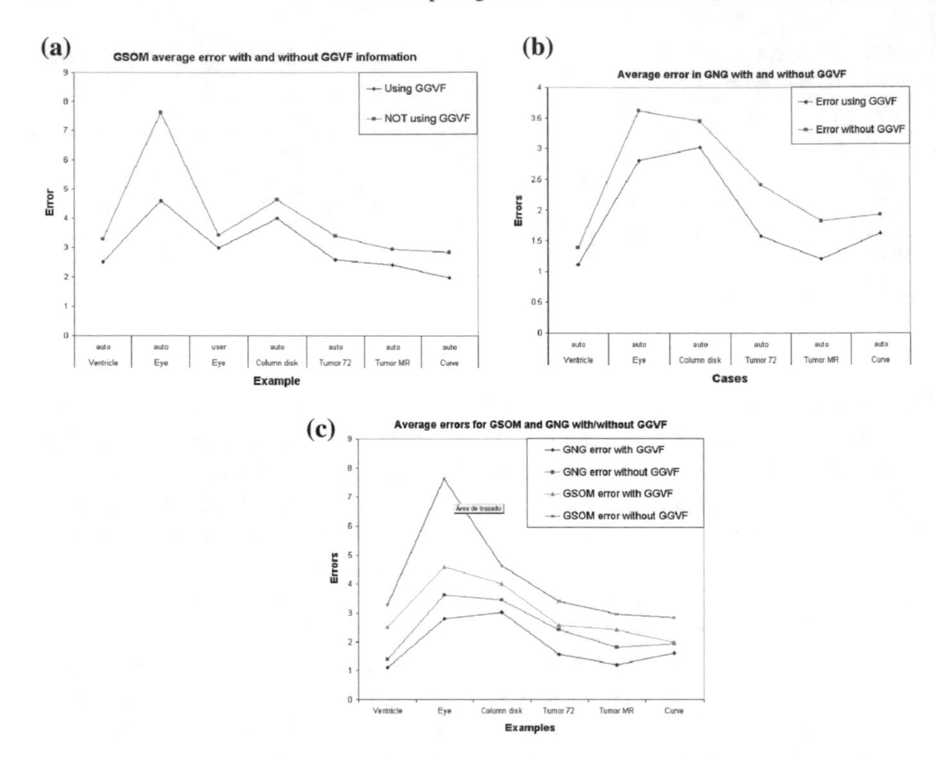

Fig. 19.7 **a** Average errors for different examples using the GSOM algorithm with and without GGVF information; **b** average errors for different examples using the GNG algorithm with and without GGVF information; **c** comparison between the errors obtained with GSOM and GNG using and without using GGVF information. Note that both are improved with GGVF information, although GNG gives better results

is the transformation of one model obtained at time t_1 into another obtained at time t_2 (a kind of morphing of 3D surfaces).

Figure 19.10a shows the initial shape, which will be transformed into the one showed in Fig. 19.10b; Fig. 19.10c–f shows some stages during the process. Note that Fig. 19.10f looks like Fig. 19.10b, as expected.

In the case shown in Fig. 19.11, we have one 3D model with an irregular shape that will be transformed into a shape similar to a pear; Fig. 19.11a shows the initial shape that will be transformed into the one shown in Fig. 19.11b; Fig. 19.11c–f shows some stages during the process. Again, the resulting volume looks like the one expected (Fig. 19.11b).

To illustrate the application of the presented algorithm in cases having models based on entities different than the points, in Fig. 19.12 we show models based on spheres [244]. The goal is the same: morphing the model shown in Figs. 19.12a and 19.12d into the ones shown in Figs. 19.12b and 19.12e, respectively. The results are shown in Figs. 19.12c and 19.12f .

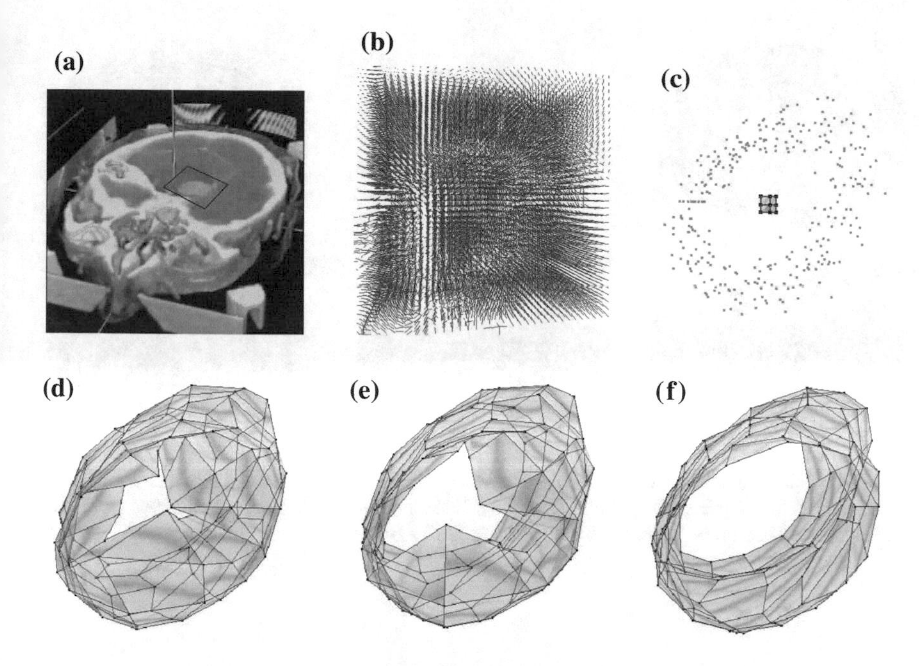

Fig. 19.8 Algorithm for 3D object's shape determination. **a** 3D model of the patient's head containing a tumor in the marked region; **b** vectors of the dense GGVF on a 3D grid arrangement of 32×32×16; **c** inputs determined by GGVF and edge map and the initialization of the net GNG; **d–e** two stages during the learning; **g** final shape after training has finished with a total of 170 versors M (associated with 170 neural units)

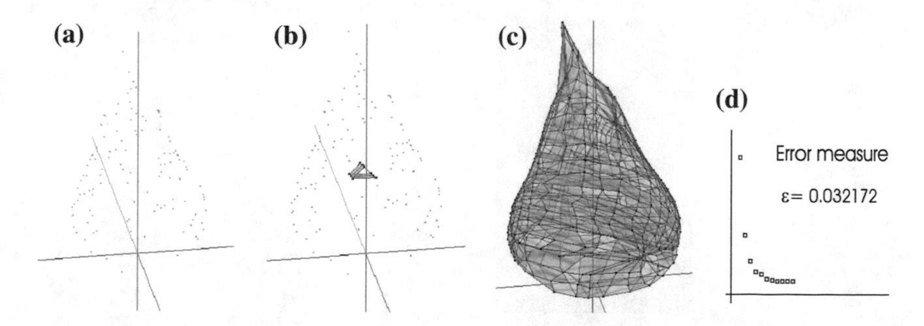

Fig. 19.9 3D object shape definition for the case of a pear. **a** Inputs to the net selected using GGVF and streamlines; **b** inputs and the initialization of the net with nine neural units; **c** result after the net has been reached the maximum number of neurons (300 neurons); **d** error measurement using Eq. (19.23)

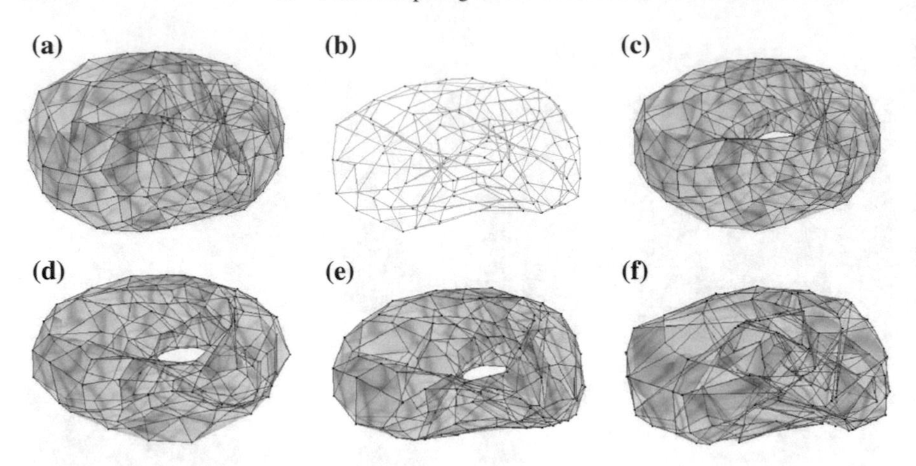

Fig. 19.10 The 3D surface shown in **a** is transformed into the one shown in **b**. Different stages during the evolution are shown in **c** to **f**, where **f** is the final shape (i.e., the final shape of a) after finishing the evolution of the net, which should look like **b**

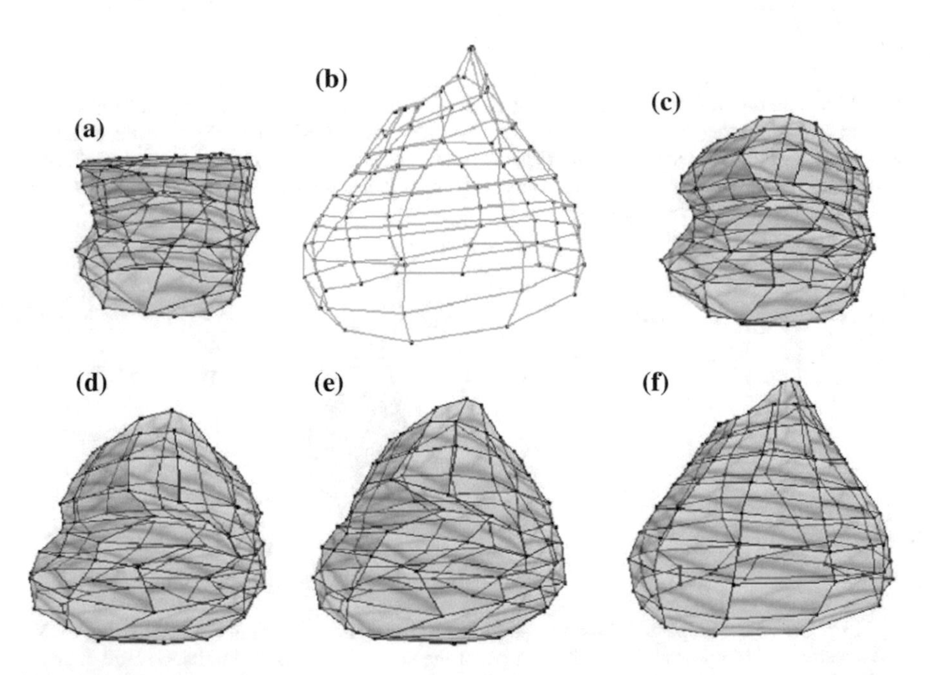

Fig. 19.11 The 3D surface shown in **a** is transformed into the one shown in **b**. Different stages during the evolution are shown in **c–f**, where **f** is the final shape (i.e., the final shape of a) after finishing the evolution of the net, which should look like **b**

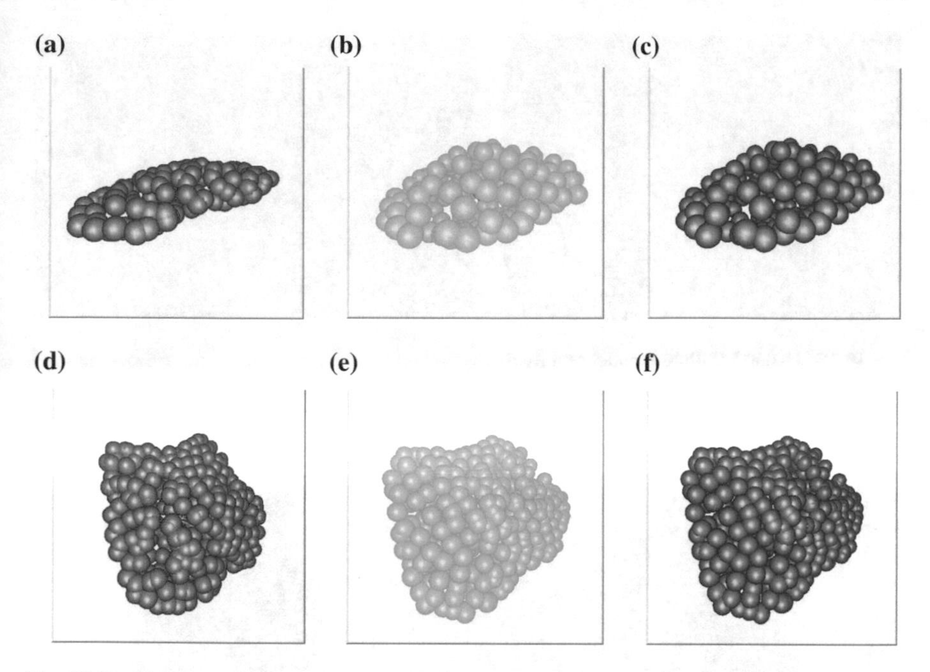

Fig. 19.12 The 3D models based on spheres shown in **a** and **d** are transformed into the ones shown in **b** and **e**, respectively, resulting in the models shown in **c** and **f**, respectively

19.2.1 Spherical Radial Basis Function Network for 3D data

We first tested our spherical networks using a 3D model as the ground truth. The National Library of Medicine has a project called the Visible Human. It has already produced computed tomography, magnetic resonance imaging, and physical cross sections of a human cadaver. It uses surface connectivity and isosurface extraction techniques to create polygonal models of the skin, bone, muscle, and bowels. The goal of such a project is the creation of complete, anatomically detailed, three-dimensional representations of typical male and female human bodies. The acquisition of transverse CT, MR, and cryosection images of representative male and female cadavers has been completed. The male was sectioned at 1 mm intervals, the female at 1/3 mm intervals. Figure 19.13 shows an example of the cloud points and the surface corresponding to a human liver. Since these 3D data are not accompanied by the elasticity coefficients at each point, we use our spherical radial basis function network (SRBFN), described in Sect. 13.6, with a coefficient of elasticity constant everywhere, and thus we did not need to apply the coupled layer for the interpolation. In the next experiment, we show simulated data from this interpolation.

This image was taken from the digital National Library of Medicine [306].

The model of the liver is in 3D. In order to better visualize the results, we start by cutting the liver into two surfaces, the top and bottom. This is achieved

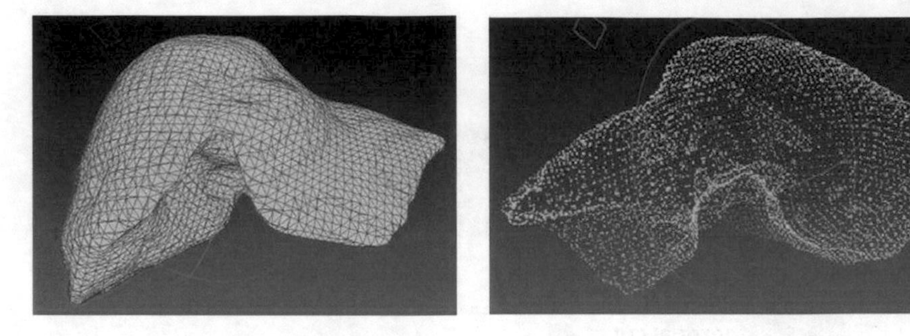

Fig. 19.13 Ground truth 3D model of a liver

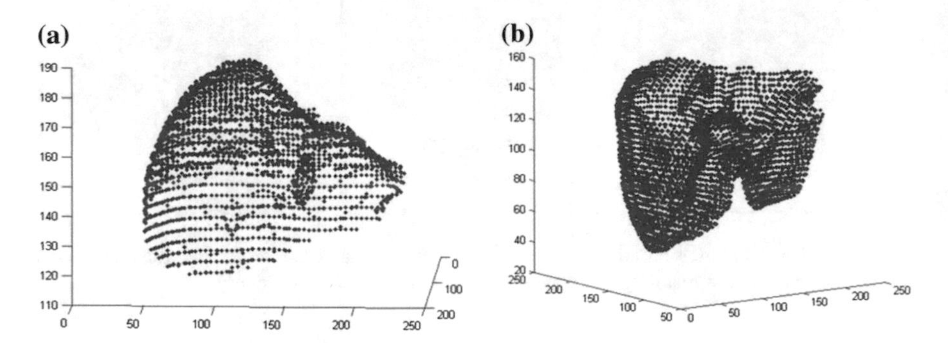

Fig. 19.14 **a** Point cloud of the top surface; **b** point cloud of the bottom surface

by "cutting" the liver with planes. To get the equation of the plane, we take three points (these points were selected visually): $p1(43, 51, 88)$, $p2(220, 56, 117)$, and $p3(103, 200, 158)$. The equation of this plane in conformal geometry in the OPNS representation is

$$p1 \wedge p2 \wedge p3 \wedge e_\infty. \tag{19.24}$$

Then, we take all points up the plane to form the top surface and the others to form the bottom surface. These surfaces are shown in Fig. 19.14.

Now, we reconstruct a visual cut on the top surface of the liver using the SRBFN. We will name a sagittal cut to one perpendicular to the x-axis and a coronal cut to one perpendicular to the y-axis, as shown in Fig. 19.15a. The goal of this is to see how many points would be sufficient for the SRBFN to fit the surface along these cuts. The experiment was performed by taking random points on each slice and calculating the normalized mean square error (MSE) between the reconstruction and the cloud of points. In each cut, we took different numbers of random points. The number of points and their resulting MSE are shown in Fig. 19.15b.

(a)

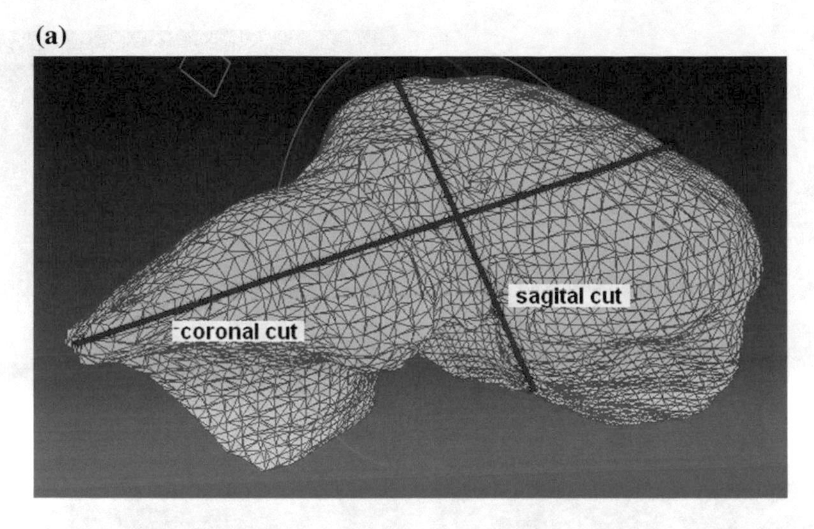

(b)

Saggital cut x=100		Coronal cut y=100	
# puntos	Normalized MSE	# puntos	Normalized MSE
5	11.2901	5	9.2
8	9.0583	8	6.9275
12	2.7492	12	1.5358
15	2.6981	15	1.4268
20	1.8539	20	1.4084

Fig. 19.15 **a** Sagittal and coronal cuts; **b** the fit of the surface cuts with the SRBFN

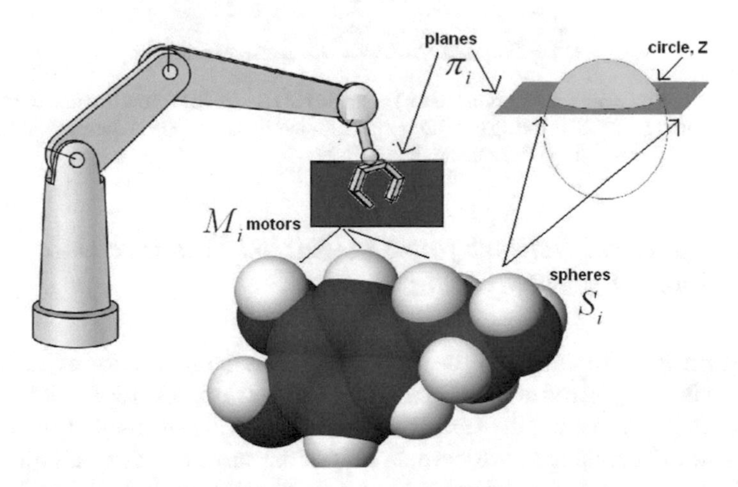

Fig. 19.16 Using the same geometric mathematical framework, we can relate the end-effector of a robot to the individual spheres with an elasticity that belongs to the surface represented via the SRBFN and GRBFN models

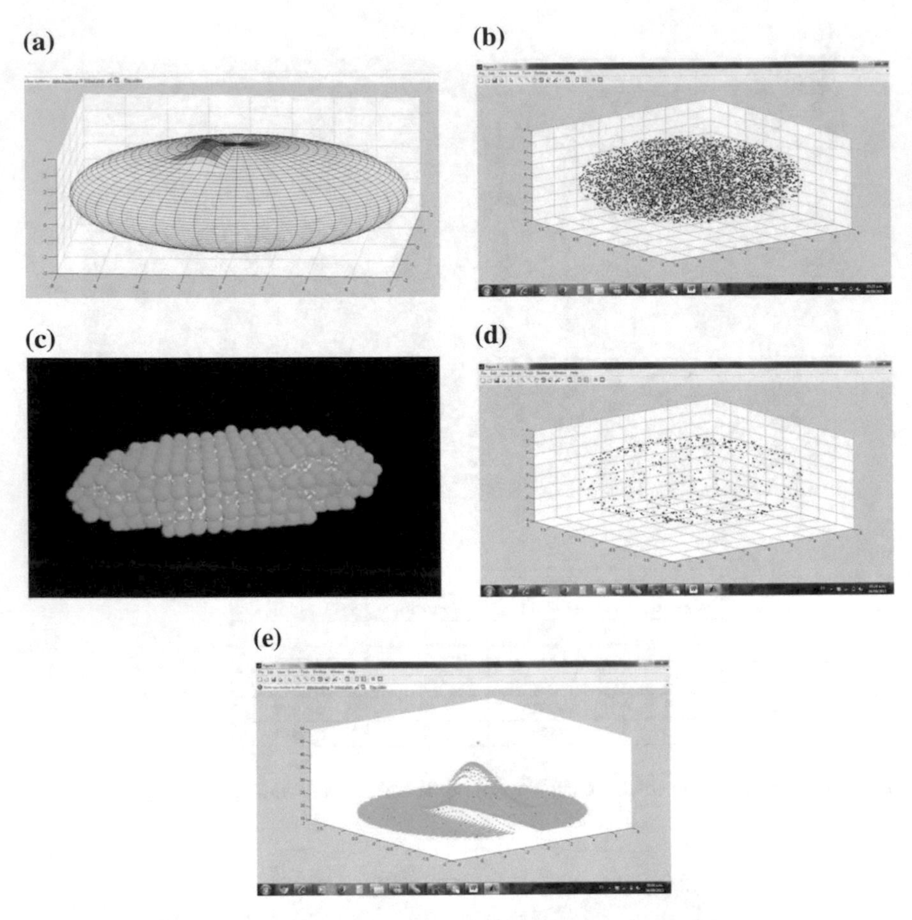

Fig. 19.17 **a** Surface with calcification bump; **b** sampled 3D data; **c** 3D reconstruction with spheres at the surface using the SRBFN; **d** sparse 3D samples for interpolation; **e** local interpolated stiffness of the bump at a patch of the surface using the GRBFN

19.2.2 Spherical Network for 4D Data: 3D Structure Data and 1D Elasticity Data

This experiment, with simulated 4D data, aims to show the application of the SRBFN together with the geometric radial basis function network (GRBFN), described in Sect. 13.5.2, to reconstruct the surface and interpolate the stiffness of the surface: $k = |\mathbf{F}|/|\mathbf{d}|$. For simplicity, we have an ellipsoidal surface with a bump function, which in a human organ could resemble a calcification; see Fig. 19.17a. First, the SRBF manages to reconstruct the surface with spheres allocated only on the surface, and then the GRBFN interpolates the data. In Fig. 19.17, we show a surface with constant elasticity everywhere except on the bump. In Fig. 19.17, this bump is not the surface; it just represents the elasticity in this small region. The function has a maximum elasticity of 50 N/m and a spread of $\sigma = 0.3$. In a real scenario, the surgeon touches the surface via a haptic device. In our simulation, only on this small area does the elasticity change, while in the other parts of the ellipsoid, the elasticity is constant. Thus, the GRBFN has to interpolate the elasticity function of the bump just in this small area. As a result, after the interpolation, the model will assign to each underlying sphere an orthogonal vector weighted with the interpolated elasticity coefficient k. So, a surgical instrument requires the position and elasticity vector for 4D data; the sphere, with its normal vector pointing out of the sphere, gives this information. In this way, the model based on the SBRFN and GRBFN is a sort of interface between the 3D visual space with elasticity and a surgical instrument held by the robot manipulator, which is commanded via a haptic device; see Fig. 19.16.

This sphere gives the 3D position and curvature at a certain point of the surface. In addition, the output layer interpolates the elasticity at this sphere, so in conformal geometric algebra, one can use the information of this sphere and its associated elasticity to relate this data with a robotic device, for instance a surgical instrument being applied on the surface of a human organ.

The results presented in Fig. 19.17b, c confirm the effectiveness of the combined SRBFN and GRBFN model.

19.3 Conclusion

In this chapter, the authors showed how to incorporate geometric algebra techniques in an artificial neural network approach to approximate 2D contours or 3D surfaces. In addition, it demonstrated the use of the dense vector field named generalized gradient vector flow (GGVF) not only to select the inputs to the neural network GNG, but also as a parameter guiding its learning process. This network was used to find a set of transformations expressed in the conformal geometric algebra framework, which move a point by means of a versor along the contour of an object, in this way defining by the shape of the object. This has the advantage that versors of the conformal geometric algebra can be used to transform any entity exactly in the same way: multiplying the entity from the left by M and from the right by $\tilde{M}$.

Some experiments show the application of the proposed method in medical image processing and also for automated visual inspection tasks. The results obtained show that by incorporating the GGVF information, we can automatically get the set of inputs to the net, and we also improve its performance.

When dealing with the 3D case, we presented two different applications: surface approximation and the transformation of a model at time t_1 onto another at time t_2, obtaining good results even using models based on spheres of the conformal geometric algebra.

Chapter 20
Clifford Algebras and Related Algebras

20.1 Clifford Algebras

Clifford algebras were created and classified by William K. Clifford (1878–1882) [57–59], when he presented a new multiplication rule for vectors in Grassmann's exterior algebra $\bigwedge \mathbb{R}$ [112]. In the special case of the Clifford algebra CL_3 for $\mathbb{R}^3$, the mathematical system embodied Hamilton's quaternions [119].

In Chap. 2, Sect. 2.2, we explain a further geometric interpretation of the Clifford algebras developed by David Hestenes and called geometric algebra. Throughout this book, we will utilize geometric algebra.

20.1.1 *Basic Properties*

A Clifford algebra is a unital associative algebra which contains and it is generated by a vector space V equipped with a quadratic form Q. The Clifford algebra $Cl(V, Q)$ is generated by V subject to the condition

$$v^2 = Q(v) \qquad \texttt{for all } v \in V. \tag{20.1}$$

This is a fundamental Clifford identity. If the characteristic of the ground field F is not 2, this condition can be rewritten in the following form

$$uv + vu = 2 < u, v > \qquad \texttt{for all } u, v \in V, \tag{20.2}$$

where $< u, v >= \frac{1}{2}(Q(u + v) - Q(u) - Q(v))$ is the symmetric bilinear form associated to Q.

Regarding the quadratic form Q, one can notice that Clifford algebras are closely related to exterior algebras. As a matter of fact, if $Q = 0$ then the Clifford algebra

E. Bayro-Corrochano, *Geometric Algebra Applications Vol. I*,
https://doi.org/10.1007/978-3-319-74830-6_20

$Cl(V, Q)$ is just the exterior algebra $\Lambda(V)$. Since the Clifford product includes the extra information of the quadratic form Q, the Clifford product is fundamentally richer than the exterior product.

20.1.2 Definitions and Existence

A pair (A, α) consisting of an R-algebra A and a homomorphism of R-modules $\alpha : V \to A$ is compatible with V if

$$\alpha(v)^2 = Q(v)1_A \qquad \texttt{for all } v \in V. \tag{20.3}$$

This equation applied to $v + u$ implies that

$$\alpha(v)\alpha(u) + \alpha(u)\alpha(v) = 2 < u, v > \qquad \texttt{for all } u, v \in V. \tag{20.4}$$

In this regard, it an be can say that a Clifford algebra $Cl(V, Q)$ is a universal with respect to the pair (A, α). Thus a Clifford algebra of V is a pair $Cl(V, Q) = (Cl(V, Q), \gamma)$ such that

(i) $Cl(V, Q)$ is an R-algebra,
(ii) $\gamma : V \to Cl(V, Q)$ is and R-module map that satisfies $\gamma(v)^2 = Q(v)1_A$ and $\gamma(u)\gamma(v) + \gamma(v)\gamma(u) = 2 < u, v >$ for all $u, v \in V$, and
(iii) if (A, α) is any pair compatible with V, then there is a unique R-algebra homomorphism $\phi : Cl(V, Q) \to A$ such that the following diagram commutes, see Fig. 20.1.

A Clifford algebra described as above always exist and it can be constructed as follows: begin with the most general algebra that contains V, namely the tensor algebra $T(V)$ and then enforce the fundamental identity by taking a suitable quotient. In this regard, one takes the two-sided ideal I_Q in $T(V)$ which is generated by all elements of the form.

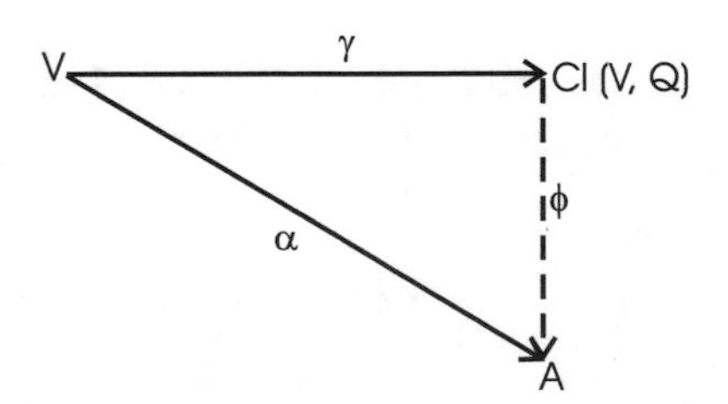

Fig. 20.1 Diagram for the definition of existence

$$v \otimes v - Q(v)1 \quad \texttt{for all } v \in V, \tag{20.5}$$

an define $Cl(V, Q)$ as the quotient

$$Cl(V, Q) = T(V)/I_Q. \tag{20.6}$$

20.1.3 Real and Complex Clifford Algebras

The most important Clifford algebras equipped with non-degenerated quadratic forms are those over the real and complex vectors spaces.

Every non-degenerated quadratic form on a finite real vector space of dimension n is equivalent to the standard diagonal form given by

$$Q(v) = v_1^2 + \cdots + v_p^2 - v_{p+1}^2 - \cdots v_{p+q}^2, \tag{20.7}$$

where $p + q = n$ corresponds to the dimension of the space V. The pair of integers (p, q) is known as the signature of the quadratic form $Q(v)$. The real vector space associated with this quadratic form $Q(v)$ is often denoted by $\mathbb{R}^{p,q}$ and the Clifford algebra on $\mathbb{R}^{p,q}$ is denoted by $Cl_{p,q}(\mathbb{R})$. The notation $Cl_n(\mathbb{R})$ means either $Cl_{n,o}(\mathbb{R})$ or $Cl_{0,n}(\mathbb{R})$. depending on the user preference for a positive definite or negative definite spaces.

The standard orthonormal basis e_i for $\mathbb{R}^{p,q}$ consists of $n = p + q$ mutually orthogonal unit vectors, p of which have the norm to 1 and q to -1. Thus, the Clifford algebra $Cl_{p,q}(\mathbb{R})$ has p base vectors which square to $+1$ and q to -1. The set of the orthonormal bases fulfills

$$< e_i, e_j >= 0 \qquad i \neq j. \tag{20.8}$$

The fundamental Clifford identity of Eq. (20.1) implies that for an orthonormal basis

$$e_i e_j = -e_j e_i \qquad i \neq j. \tag{20.9}$$

In terms of these unit vectors, the so-called unit pseudoscalar in $Cl_{p,q}(\mathbb{R})$ is defined as follows

$$I_n = e_1 e_2 \dots e_n. \tag{20.10}$$

It's square is given by

$$I_n^2 = (-1)^{\frac{n(n-1)}{2}}(-1)^q = (-1)^{\frac{(p-q)(p-q-1)}{2}} = \begin{cases} +1 \; p - q \equiv 0, 1 \; mod \; 4 \\ -1 \; p - q \equiv 2, 3 \; mod \; 4 \,. \end{cases} \tag{20.11}$$

Next, we will derive the basis of a Clifford algebra in terms of the basis $e_1, e_2, \ldots, e_n$ of the related vector space $\mathbb{R}^{p,q}$. Using the canonical order of the power set $\mathcal{P}(\{1, \ldots, n\})$, we first derive the index set

$$\mathcal{B} := \{\{b_1, \ldots, b_k\} \in \mathcal{P}(\{1, \ldots, n\}) | 1 < b_1 < \cdots < b_k < n\}. \quad (20.12)$$

Then by defining for all $B \in \mathcal{B}$

$$e_B = e_{b_1} \ldots e_{b_k}, \quad (20.13)$$

one obtains the expected basis $\{e_B | B \in \mathcal{B}\}$ of the entire Clifford algebra $Cl_{p,q}(\mathbb{R})$. Here the empty product (for k=0) is defined as the multiplicative identity element. Note that each element $x \in Cl_{p,q}(\mathbb{R})$ can then be written as

$$x = \sum_{B \in \mathcal{B}} x_B e_B. \quad (20.14)$$

For every $k \in \{0, \ldots, 2^{p+q} - 1\}$ the set $\{e_B | B \in \mathcal{B}, |B| = k\}$ spans a linear subspace of $Cl_{p,q}(\mathbb{R})$. Each e_B is called a blade of k grade. An element of this linear subspace is called a k-vector which in turn is a linear combination of e_B blades of k grade. For running k from 0 to 3, a k-vector is usually called a scalar, vector, bivector, or trivector, respectively. All even k-vectors form the even part $Cl(V, Q)^+$ of the Clifford algebra $Cl(V, Q)$. Note that $Cl(V, Q)^+$ is a subalgebra, whereas the odd part $Cl(V, Q)^-$ formed by all odd k-vectors is not. Depending of k there are n choose k basis elements, so that the total dimension of $Cl_{p,q}(\mathbb{R})$ is

$$\mathtt{dim}\ Cl(V, Q) = \sum_{k=0}^{n} \binom{n}{k} = 2^n. \quad (20.15)$$

The *center of an algebra* consists of those elements which commute with all the elements of this algebra. As examples, the center $Cen(Cl_3) = \mathbb{R} \oplus \bigwedge^3 \mathbb{R}^3$ (scalar and pseudoscalar) of Cl_3 is isomorphic to $\mathbb{C}$.

The algebra $Cl_{0,0}(\mathbb{R})$ is isomorphic to R, it has only scalars and no base vectors. $Cl_{0,1}(\mathbb{R})$ corresponds to a two-dimensional algebra generated by a single base vector which squares to -1 and therefore is isomorphic to $\mathbb{C}$ (the field of the complex numbers). The four-dimensional algebra $Cl_{0,2}(\mathbb{R})$ is spanned by $\{1, e_1, e_2, I_2 = e_1 e_2\}$. The latter three base elements square to -1 and all anticommute. This algebra is isomorphic to the quaternion algebra of Hamilton $\mathbb{H}$. The eight-dimensional algebra $Cl_{0,3}(\mathbb{R})$ spanned by $\{1, e_1, e_2, e_3, e_2e_3, e_3e_1, e_1e_2, I_3 = e_1e_2e_3\}$ is isomorphic to the Clifford biquaternions (direct sum of $\mathbb{H}$).

The Clifford algebras on complex spaces denoted as $Cl_n(\mathbb{C})$ are also of great interest. Every non-degenerate quadratic form on a n-dimensional complex space is equal to the standard diagonal form

$$Q(z) = z_1^2 + z_2^2 + \cdots + z_n^2. \tag{20.16}$$

An algebra $Cl_n(\mathbb{C})$ may be obtained by complexification of the algebra $Cl_{p,q}(\mathbb{R})$ as follows

$$Cl_n(\mathbb{C}) \cong Cl_{p,q}(\mathbb{R}) \otimes \mathbb{C} \cong Cl(\mathbb{C}^{p+q}, Q \otimes \mathbb{C}), \tag{20.17}$$

where Q is the real quadratic form of signature (p, q). Note in this equation, that the complexification does not depend on the signature. One computes easily $Cl_0(\mathbb{C}) = \mathbb{C}$, $Cl_1(\mathbb{C}) = \mathbb{C} \oplus \mathbb{C}$ and $Cl_2(\mathbb{C}) = M_2(\mathbb{C})$.

Each algebra $Cl_{p,q}(\mathbb{R})$ or $Cl_n(\mathbb{C})$ is isomorphic to a matrix algebra over $\mathbb{R}$, $\mathbb{C}$ or $\mathbb{H}$ or a direct sum of two such algebras.

20.1.4 Involutions

Given the linear map $v \to -v$ for all $v \in V$ which preserves the quadratic form $Q(V)$ and taking into account the universal propriety of the Clifford algebras, this mapping can be extended to an algebra automorphism

$$\alpha : CL(V, Q) \to CL(V, Q). \tag{20.18}$$

When α squares to the identity it is called an *involution*. There are also other two important antiautomorphism in the Clifford algebras. To elucidate these involutions consider the antiautomorphism of the tensor algebras that reverses the order in all products as follows

$$v_1 \otimes v_2 \otimes \cdots \otimes v_k \to v_k \cdots \otimes v_2 \otimes v_1. \tag{20.19}$$

Since the ideal I_Q of Eq. (20.6) is invariant under this operation of reversion, such operation descends to an automorphism of $CL(V, Q)$ and its called the *transpose* or *reversal* operation and denoted by x^t. The transpose of a product is $(xy)^t = y^t x^t$. The second antiautomorphism is composed by α and the transpose as follows

$$\hat{x} = \alpha(x^t) = \alpha(x)^t. \tag{20.20}$$

These three operations depend only on the degree module 4, i.e., if x is a k-blade then

$$\alpha(x) = \pm x \qquad x^t = \pm x \qquad \hat{x} = \pm x, \tag{20.21}$$

where the signs are given in Table 20.1.

Table 20.1 Signs in involutions depending on k mod 4

k mod 4	0	1	2	3	
$\alpha(x)$	+	−	+	−	$(-1)^k$
x^t	+	+	−	−	$-1^{k(k-1)/2}$
$\hat{x}$	+	−	−	+	$-1^{k(k+1)/2}$

20.1.5 Structure and Classification of Clifford Algebras

The structure of Clifford algebras can be worked out using a simple methodology. For this purpose, let us consider the space V with a quadratic form Q and another vector space U of an even dimension and a non-singular bilinear form with discriminant d. The Clifford algebra of $V + U$ is isomorphic to the tensor product of the Clifford algebras of $(-1)^{\frac{dim(U)}{2}} d\ V$ and U. This means that the space V with its quadratic form is multiplied by the factor $(-1)^{\frac{dim(U)}{2}} d$. Over the reals, one can write the following formulas

$$\begin{aligned} Cl_{p+2,q}(\mathbb{R}) &= M_2(\mathbb{R}) \otimes Cl_{q,p}(\mathbb{R}), \\ Cl_{p+1,q+1}(\mathbb{R}) &= M_2(\mathbb{R}) \otimes Cl_{q,p}(\mathbb{R}), \\ Cl_{p,q+2}(\mathbb{R}) &= \mathbb{H} \otimes Cl_{q,p}(\mathbb{R}), \end{aligned} \tag{20.22}$$

which in turn can be used to find the structure of all real Clifford algebras. Here $M_2(\mathbb{R})$ stands for the algebra of 2×2 matrices over $\mathbb{R}$. In the theory of non-degenerated quadratic forms on real and complex vector spaces, the finite-dimensional Clifford algebras have been completely classified. In each case, the Clifford algebra is isomorphic to a matrix algebra over $\mathbb{R}$, $\mathbb{C}$ or $\mathbb{H}$ or to a direct sum of two such algebras, but not in canonical way. Next, we will denote by $\mathcal{K}(n)$ the algebra of $n \times n$ matrices with entries in the division algebra $\mathcal{K}$. The direct sum of algebras will be denoted by $\mathcal{K}^2(n) = \mathcal{K}(n) \otimes \mathcal{K}(n)$.

The Clifford algebra on $\mathbb{C}^n$ with the quadratic form given by Eq. (20.16) will be denoted by $Cl_n(\mathbb{C})$. When n is even the algebra $Cl_n(\mathbb{C})$ is central simple and according to the Artin–Wedderbum theorem is isomorphic to a matrix algebra over $\mathbb{C}$. For the case n is odd the center included not only the scalars but the pseudoscalar I_n as well. Depending on p and q by $Cl_{q,p}(\mathbb{R})$, one can find always a pseudoscalar which squares to one ($I_n^2 = 1$). Using the pseudoscalar I_n, let us define the following operators

$$P_\pm = \frac{1}{2}(1 \pm I_n). \tag{20.23}$$

These two operators form a complete set of orthogonal idempotents. Since they are central, they allow a decomposition of $Cl_n(\mathbb{C})$ into a direct sum of two algebras

Table 20.2 Classification of complex Clifford algebras

n	$Cl_n(\mathbb{C})$
2m	$\mathbb{C}(2^m)$
$2m+1$	$\mathbb{C}(2^m) \oplus \mathbb{C}(2^m)$

$$Cl_n(\mathbb{C}) = Cl_n^+(\mathbb{C}) \oplus Cl_n^-(\mathbb{C}), \tag{20.24}$$

where $Cl_n^{\pm}(\mathbb{C}) = P_{\pm}Cl_n(\mathbb{C})$. This decomposition is called a *grading* of $Cl_n(\mathbb{C})$. The algebras $Cl_n^{\pm}(\mathbb{C})$ are just the positive and negative eigenspaces of I_n and the operators $P_{\pm}$ are just the projection operators. Since I_n is odd these algebras are isomorphic related with each other via the automorphism α as follows

$$\alpha(Cl_n^{\pm}(\mathbb{C}) = Cl_n^{\mp}(\mathbb{C}). \tag{20.25}$$

These two isomorphic algebras are each central simple and thus isomorphic to a matrix algebra over $\mathbb{C}$. The sizes of the matrices can be determined by the dimension 2^n of the $Cl_n(\mathbb{C})$. As a consequence of these considerations one can classify straightforwardly the complex Clifford algebras as it is shown in Table 20.2.

For the real case, the classification is a bit more difficult as the periodicity is 8 rather than 2. According the Artin–Wedderburn theorem, if n (or $p-q$) is even the Clifford algebra $Cl_{p,q}(\mathbb{R})$ is central simple and isomorphic to a matrix algebra over $\mathbb{R}$ or $\mathbb{H}$, but if n (or $p-q$) is odd is no longer central simple but rather it has a center which includes the pseudoscalar as well as the scalars. If n is odd and the pseudoscalar $I_n^2 = +1$, then as in the complex case the Clifford algebra can be decomposed into a direct sum of isomorphic algebras as follows

$$Cl_{p,q}(\mathbb{R}) = Cl_{p,q}^+(\mathbb{R}) \oplus Cl_{p,q}^-(\mathbb{R}). \tag{20.26}$$

Each of which is central simple and thus isomorphic to the matrix algebra over $\mathbb{R}$ or $\mathbb{H}$. Now, if n is odd and the pseudoscalar $I_n^2 = -1$, then the center of $Cl_{p,q}(\mathbb{R})$ is isomorphic to $\mathbb{C}$, thus it can be considered as a complex algebra which is central simple and so isomorphic to a matrix algebra over $\mathbb{C}$ Summarizing, there are in fact three properties which are necessary to determine the class of the algebra $Cl_{p,q}(\mathbb{H})$, namely

- n is even/odd
- $I^2 = \pm 1$
- the Brauer class of the algebra (n even) or even subalgebra (n odd) is $\mathbb{R}$ or $\mathbb{H}$.

Each of these properties depends only on the signature $p-q$ modulo 8. The size of the matrices is determined by the consequence that $Cl_{p,q}(\mathbb{R})$ has dimension 2^{p+q}.

The complete classification is given in Table 20.3, where the size of the matrices is determined by he dimension of 2^{p+q} of the involved Clifford algebra $Cl_{p,q}(\mathbb{R})$

Table 20.3 Complete Classification of real Clifford algebras

$p - q$ mod8	I_n^2	$Cl_{p,q}(\mathbb{R})$ $(p+q=2m)$	$p - q$ mod 8	I_n^2	$Cl_{p,q}(\mathbb{R})$ $(p+q=2m+1)$
0	+	$\mathbb{R}(2^m)$	1	+	$\mathbb{R}(2^m) \oplus \mathbb{R}(2^m)$
2	−	$\mathbb{R}(2^m)$	3	−	$\mathbb{C}(2^m)$
4	+	$\mathbb{H}(2^{m-1})$	5	+	$\mathbb{H}(2^{m-1}) \oplus \mathbb{H}(2^{m-1})$
6	−	$\mathbb{H}(2^{m-1})$	7	−	$\mathbb{C}(2^m)$

Table 20.4 Classification of real Clifford algebras for $p + q \leq 5$

	5	4	3	2	1	0	−1	−2	−3	−4	−5
0						$\mathbb{R}$					
1					$\mathbb{R}^2$		$\mathbb{C}$				
2				$\mathbb{R}(2)$		$\mathbb{R}(2)$		$\mathbb{H}$			
3			$\mathbb{C}(2)$		$\mathbb{C}^2(2)$		$\mathbb{C}(2)$		$\mathbb{H}^2$		
4		$\mathbb{H}(2)$		$\mathbb{R}(4)$		$\mathbb{R}(4)$		$\mathbb{H}(2)$		$\mathbb{H}(2)$	
5	$\mathbb{H}^2(2)$		$\mathbb{C}(4)$		$\mathbb{R}^2(4)$		$\mathbb{C}(4)$		$\mathbb{H}^2(2)$		$\mathbb{C}(4)$
I^2	+	+	−	−	+	+	−	−	+	+	−

Table 20.4 shows the results of this classification for $p + q \leq 5$. In this table, $p + q$ runs vertically and $p - q$ runs horizontally. For example, the Clifford algebra $Cl_{3,1}(\mathbb{R} \cong \mathbb{R}(4)$ is found in row $p + q = 4$ and column $p - q = 2$. Note the symmetry about the columns 1,5,−3.

20.1.6 *Clifford Groups, Pin and Spin Groups and Spinors*

In this subsection, we will assume that the space V is of finite dimension and the bilinear form Q is non-singular. The Clifford group $\Gamma(V)$ (or Lipschitz group) is defined to be the set of invertible elements g of the Clifford algebra $Cl_{p,q}(V)$ such that the map

$$v \mapsto g v \widehat{g}^{-1} \in V \quad \texttt{for all} \quad v \in V, \tag{20.27}$$

is an orthogonal automorphism of V. In fact, this equation represents the Clifford group action on the vector space V that preserves the norm of Q and also gives a homomorphism from the Clifford group to the orthogonal group. Since the universal algebra $Cl_{p,q}(V)$ is uniquely defined up to isomorphism, thus $\Gamma(V)$ is also defined up to isomorphism. The Clifford group $\Gamma(V)$ contains all elements of nonzero norm r of

V which act on V by reflections in the hyperplane $(R\{r\})^{\pm}$ mapping v to $v - \frac{<v,r>r}{Q(r)}$. By the case of a characteristic 2, these mappings are called orthogonal transversions rather than reflections. Every orthogonal automorphism of V like Eq. (20.27) is the composite of a finite number of hyperplane reflections.

An element g of $\Gamma(V)$ represents a *rotation* of V if and only if g is representable as the product of an even number of elements of V. The set of such elements is denoted by $\Gamma^0 = \Gamma^0(V)$. If g of $\Gamma(V)$ is representable as the product of an odd number of elements of V, then g represents an *antirotation* of V. The set of such elements will denote $\Gamma^1 = \Gamma^1(V)$. One can then see the Clifford group as the disjoint union of the subset Γ^0 (the subgroup of index 2) and the subset Γ^i (elements of degree i in Γ). If the space V is finite-dimensional with a non-degenerated bilinear form then the Clifford group maps onto the orthogonal group of V and its kernel consists of the nonzero elements of the field F. This leas to the following exact sequences

$$
\begin{aligned}
&1 \to F^* \to \Gamma \to O_V(F) \to 1, \\
&1 \to F^* \to \Gamma^0 \to O_V(F) \to 1.
\end{aligned} \tag{20.28}
$$

By an arbitrary characteristic, the spinor norm Q on the Clifford group is defined as follows

$$
Q(g) = g^t g, \tag{20.29}
$$

This is an homomorphism from the Clifford group to the group F^* of nonzero element of the field F. The nonzero elements of F have a spinor norm in the group F^{*2} (squares of nonzero elements of the field F). When V is of finite-dimensional and non-singular, one gets an induced map from the orthogonal group of V to the group F^*/F^{*2} also known as the *spinor norm*. The spinor norm of the reflection of a vector p has image $Q(p)$ in F^*/F^{*2}. This property is uniquely defined on the orthogonal group. This provides the following exact sequences

$$
\begin{aligned}
&1 \to \pm 1 \to Pin_V(F) \to O_V(F) \to F^*/F^{*2}, \\
&1 \to \pm 1 \to Spin_V(F) \to SO_V(F) \to F^*/F^{*2}.
\end{aligned} \tag{20.30}
$$

The reader should note that in characteristic 2 the group $\{\pm 1\}$ has just one element.

Next, we explain in the context of Clifford groups, the Pin, and Spin groups. The Pin group denoted by $Pin_V(F)$ is the subgroup of the Clifford group $\Gamma(V)$ of elements of spinor norm 1. Similarly, the spin group denoted by $Spin_V(F)$ is the subgroup of elements of Dickson invariant 0 in $Pin_V(F)$. Usually, the spin group has index 2 in the pin group. As it is said before, there is an homomorphism from the Clifford group $\Gamma(V)$ onto the orthogonal group, thus the special orthogonal group is defined as the image of Γ^0. There is also an homomorphism from the Pin group to the orthogonal group. The image consists of these elements of spinor norm $1 \in F^*/F^{*2}$. The kernel comprises of the elements +1 and −1 and it has order 2 unless F has a characteristic 2. Similarly, it exists an homomorphism from the Spin group to the special orthogonal group of V. When $V = \mathbb{R}^{p,q}$ the notations for Γ, Γ^0, Pin_V and $Spin_V$ will be $\Gamma(p,q)$, $\Gamma^0(p,q)$, $Pin_V(p,q)$ and $Spin_V(p,q)$. Since $\mathbb{R}^0_{q,p} \cong \mathbb{R}^0 p, q$, $\Gamma^0(q,p) \cong \Gamma^0(p,q)$ and $Spin_V(q,p) \cong Spin_V(p,q)$, respectively. Often $\Gamma^0(0,n)$ and $Spin(0,n)$ are abbreviated as $\Gamma^0(n)$ and $Spin(n)$, respectively. The groups $Pin_V(p,q)$, $Spin_V(p,q)$, $Spin_{V+}(p,q)$ are twofold covering groups of $O(p,q)$, $SO(p,q)$, $SO_+(p,q)$. As an example, let us consider the spin group $Spin(3)$. The traceless Hermitian matrices $x\sigma_1 + y\sigma_2 + z\sigma_3$ with $x, y, z \in \mathbb{R}$ represent vectors $\mathbf{v} = xe_1 + ye_2 + ze_3 \in \mathbb{R}^3$. The group of unitary and unimodular matrices given by

$$SU(2) = \{U \in \mathtt{Mat}(2, \mathbb{C}) | U^\dagger U = I,\ det\, U = I\} \tag{20.31}$$

represents the spin group $Spin(3) = \{u \in Cl_3 | u\tilde{u} = 1,\ u\hat{u} = 1\}$ or $Spin(3) = \{u \in Cl_3^+ | u\tilde{u} = 1\}$. The groups $Spin(3)$ and $SU(2)$ both are isomorphic with the group of unit quaternions $S^3 = \{\boldsymbol{q} \in \mathbb{H} | \boldsymbol{q}\hat{\boldsymbol{q}} = 1\}$. Taking an element $u \in Spin(3)$, the mapping $\mathbf{v} \rightarrow u\mathbf{v}\tilde{u}$ corresponds to a rotation of $\mathbb{R}^3$. Thus, every element of $SO(3)$ can be represented by an element in $Spin(3)$. Next, we explain what are spinors.

The Clifford algebras $Cl_{p,q}(\mathbb{C})$ with $p + q = 2n$ even are isomorphic to matrix algebras with complex representation of dimension 2^n. Restricting our analysis to the group $Pin_{p,q}(\mathbb{R})$, one obtains a complex representation of the pin group of the same dimension which is known as *spinorrepresentation*. By restricting this to the spin group $Spin_{p,q}(\mathbb{R})$, then it splits as the sum of two half spin representations (or Weyl representations) of dimension 2^{n-1}. Now, if $p + q = 2n + 1$ is odd the Clifford algebra $Cl_{p,q}(\mathbb{C})$ is a sum of two matrix algebras, each of which has a representation of 2^n dimension, and these are also both representations of the pin group $Pin_{p,q}(\mathbb{R})$. By restricting us to the spin group $Spin_{p,q}(\mathbb{R})$ these become isomorphic, thus the spin group has a complex spinor representation of dimension 2^n. Generally speaking, spinor groups and pin groups over any field F have similar representations where their exact structure depends on the structure of the corresponding Clifford algebras $Cl_{p,q}(\mathbb{C})$. If a Clifford algebra has a factor that is a matrix algebra over some division algebra, the corresponding representation of the pin and spin groups are consequently over that division algebra.

In order to describe the real spin representations, one needs to know how the spin group resides in its Clifford algebra. Let us consider the pin group, $Pin_{p,q}$, it is the set of invertible elements in $Cl_{p,q}(\mathbb{C})$ which can be expressed simply as a product of unit vectors as follows

$$Pin_{p,q} = \{v_1 v_2 \ldots v_r |\ \mathtt{for\ all}\ ||v_i|| = \pm 1\}. \tag{20.32}$$

In fact, the pin group corresponds to the product of arbitrary number of reflections and it is a cover of the full orthogonal group $O(p,q)$. Now the spin group consists of such elements of $Pin_{p,q}$ which are built by multiplying an even number of unit vectors. So, according the Cartan–Dieudonné theorem the *Spin* is a cover of the group of the proper rotations $SO(p, q)$. As a consequence, the classification of the pin representations can be done straightforwardly using the already existing classification of the Clifford algebras $Cl_{p,q}(\mathbb{C})$. The spin representations are representations of the even subalgebras. Now in order to realize the spin representations in signature (p, q) as pin representations of signature $(p, q-1)$ or $(q, p-1)$, one can make use of either of the following isomorphisms

$$\begin{aligned} Cl^{+}_{p,q} &\cong Cl_{p,q-1}, \ \texttt{for}\ q > 0, \\ Cl^{+}_{p,q} &\cong Cl_{q,p-1}, \ \texttt{for}\ p > 0, \end{aligned} \tag{20.33}$$

where $Cl^{+}_{p,q}$ is an even subalgebra.

20.2 Related Algebras

20.2.1 Gibbs Vector Algebra

Josiah Willard Gibbs (1839–1903) and independently Oliver Heavisied (1850–1925) laid the foundations of a mathematical system called vector calculus to deal at that time with challenging engineering and physic problems. In 1879, Gibbs delivered a course in vector analysis with applications to electricity and magnetism and in 1881 he let print a private version of the first half of his “Elements of Vector Analysis”; the second half appeared in 1884 [106] The first paper in which Heavisied introduced vector methods was in his 1882–1883 paper titled “The relation between magnetic force and electric current” [93, 127]. In the preface of to the third edition of his “Treatise on Quaternions” (1890) [294], Peter Guthrie Tait showed his disappointment at “how little progress has recently been made with the development of Quaternions.” He further remarked “Even Prof. Willard Gibbs must be ranked as one of the retarders of Quaternion progress, in virtue of his pamphlet on vector analysis; a sort of hermaphrodite monster, compounded of the notations of Hamilton and Grassmann.” We can accept certainly the Tait’s remark about Gibbs as correct; Gibbs indeed retarded quaternion progress, however his “pamphlet” *Elements of Vector Analysis* marked undoubtedly the beginning of modern vector analysis.

Basically, vector calculus consists of a three dimensional linear vector space $\mathbb{R}^3$ with two associated operations: first the *scalarproduct*

$$\alpha = \mathbf{a} \cdot \mathbf{b} \qquad \alpha \in \mathbb{R} \tag{20.34}$$

which computes a scalar or the projection of vector **a** toward the vector **b**. The second operation is the *cross product*

$$\mathbf{c} = \mathbf{a} \times \mathbf{b} \qquad \mathbf{a}, \mathbf{b}, \mathbf{c} \in \mathbb{R}^3. \tag{20.35}$$

Since the two vectors **a** and **b** lie on a plane their cross product generates a vector **c** orthogonal to this plane. Note that this scalar product is the *inner product* within the Clifford algebra. The crossproduct can be reformulated in geometric algebra via the concept of duality as follows

$$\mathbf{a} \times \mathbf{b} = (\mathbf{a} \wedge \mathbf{b})^* = (\mathbf{a} \wedge \mathbf{b})I_3^{-1}, \tag{20.36}$$

where $\mathbf{a}, \mathbf{b} \in G_3$, $\wedge$is the wedge product and $I_3 = e_1e_2e_3$ is the pseudoscalar of the 3D Euclidean geometric algebra G_3.

There are some identities of the Gibb's vector calculus which can be straightforwardly rewritten in geometric algebra, the triple scalar product of three vectors $\mathbf{a}, \mathbf{b}, \mathbf{c} \in \mathbb{R}^3$

$$\begin{aligned} \mathbf{a} \cdot (\mathbf{b} \times \mathbf{c}) &= \mathbf{a} \cdot (\mathbf{b} \wedge \mathbf{c})^* = \mathbf{a} \cdot ((\mathbf{b} \wedge \mathbf{c}) \cdot I_3^{-1}) = (\mathbf{a} \wedge \mathbf{b} \wedge \mathbf{c}) \cdot I_3^{-1}, \\ &= det([\mathbf{a}, \mathbf{b}, \mathbf{c}]), \end{aligned} \tag{20.37}$$

the resulting determinant is a scalar representing the parallelepiped expanded by three noncoplanar vectors. Another useful identity is the triple vector product which can be computed in geometric algebra applying successively duality and the generalized inner product concept of Eq. (2.54) as follows

$$\begin{aligned} \mathbf{a} \times (\mathbf{b} \times \mathbf{c}) &= (\mathbf{a} \wedge ((\mathbf{b} \wedge \mathbf{c})^*))^*, \\ &= (\mathbf{a} \wedge ((\mathbf{b} \wedge \mathbf{c})I_3^{-1}))I_3^{-1} = \mathbf{a} \cdot (((\mathbf{b} \wedge \mathbf{c})I_3^{-1})I_3^{-1}), \\ &= -\mathbf{a} \cdot (\mathbf{b} \wedge \mathbf{c}) = \mathbf{b}(\mathbf{a} \cdot \mathbf{c}) - \mathbf{c}(\mathbf{a} \cdot \mathbf{b}). \end{aligned} \tag{20.38}$$

Note that the wedge product in Clifford algebra is valid in any dimension, whereas the crossproduct is only defined in a 3D vector space.

After 1898s many operations of differential geometry where defined using Gibb's vector calculus, again all of them can be reformulated in Clifford algebra and even generalized beyond the 3D vector space. Gibbs vector calculus was useful for the development of certain areas like electrical engineering. But unfortunately, it slowed down the development in Physics. From a much more broad perspective, the progress would have been faster if instead researchers and engineers would had adopted for their work not only the quaternions but even better the powerful Clifford algebra framework. This claim is still far more valid at present times due to the increasing complexity of the problems and to the fortunate fact that researchers have now very convenient computational resources at lower cost.

20.2.2 Exterior Algebras

The exterioralgebra is the algebra of the exterior product and it is also called an alternating algebra or Grassmann Algebra, after Hermann Grassmann [111, 112].

In 1844, the Grassmann's brilliant contribution appeared under the full title " Die lineale Ausdehnungslehre, ein neuer Zweig der Mathematik dargestellt und durch Anwendungen auf die übrigen Zweige der Mathematik, wie auch auf die Statik, Mechanik, die Lehre vom Magnetismus und die Krystallonomie erläutert". Grassmann was unrecognized at that time and he worked as a teacher at the Friedrich-Wilhelms-Schule (high school) at Stettin. An idea of the reception of his contribution can be obtained by the following quotation extracted from a letter written to Grassmann in 1876 by the publisher: "Your book Die Ausdehnungslehre has been out of print for some time. Since your work hardly sold at all, roughly 600 copies were in 1864 used as waste paper and the remainder, a few odd copies, have now been sold with the exception of one copy which remains in our library".

The exterior algebra $\bigwedge(V)$ over a vector field V contains V as a subspace and its multiplication product is called the exterior product or wedge product $\wedge$ which is associative and a bilinear operator. The wedge product of two elements of V is defined by

$$x \wedge y = x \otimes y \ \ (mod I). \tag{20.39}$$

This product is anticommutative on elements of V. The exterior algebra $\bigwedge(V)$ for a vector V is constructed by forming monomials via the wedge product: x, $x_1 \wedge x_2$, $x_1 \wedge x_2 \wedge x_3$, etc. A monomial is called a decomposable k-vector, because it is built by the wedge product of k linearly independent vectors $x_1, x_2, \ldots, x_k \in V$. The sums formed from linear combinations of the monomials are the elements of an exterior algebra.

The exterior algebra for a vector space V can be also described as the quotient vectors space

$$\bigwedge^k V := \bigotimes^k V / W_k, \tag{20.40}$$

where W_k is the subspace of k-tensors generated by transpositions such as $W_2 = (x \otimes y + y \otimes x)$ and $\otimes$ denotes the vector space tensor product. Thus, the equivalence class $[x_1 \otimes \cdots \otimes x_k]$ or k-vector is denoted as said above $x_1 \wedge x_2 \wedge \cdots \wedge x_k$. For instance

$$x \wedge y + y \wedge x = 0, \tag{20.41}$$

since the representatives add to an element of W_2, thus

$$x \wedge y = -y \wedge x. \tag{20.42}$$

More generally, if $x_1, x_2, x_3, \ldots, x_k \in V$ and σ is a permutation of integers $[1, \ldots, k]$ then

$$x_{\sigma(1)} \wedge x_{\sigma(2)} \wedge \cdots \wedge x_{\sigma(k)} = sgn(\sigma) x_1 \wedge x_2 \wedge \cdots \wedge x_k, \tag{20.43}$$

where $sgn(\sigma)$ is the signature of the permutation σ. If any of $x_1, x_2, \ldots, x_k \in V$ is linear dependent, then

$$x_1 \wedge x_2 \cdots \wedge x_k = 0. \tag{20.44}$$

The subspace spanned by all possible decomposable k-vectors is also called the kth exterior power of V and it is denoted by $\bigwedge^k(V)$. Exterior powers are commonly used in differential geometry to define the differential forms and to compute their wedge products. The exterior product of a k-vector and p-vector yields a $(k+p)$-vector and symbolically the wedge of the two correspondent subspaces reads

$$\left(\bigwedge^k(V)\right) \wedge \left(\bigwedge^p(V)\right) \subset \bigwedge^{k+p}(V). \tag{20.45}$$

Thus, the exterior algebra is a *graded algebra* built by the direct sum of kth exterior powers of V

$$\bigwedge(V) = \bigwedge^0(V) \oplus \bigwedge^1(V) \oplus \bigwedge^2(V) \oplus \cdots \oplus \bigwedge^n(V) = \bigoplus_{k=0}^{n}(V), \tag{20.46}$$

where $\bigwedge^0(V) = F$ and $\bigwedge^1(V) = V$. Each of this spaces is spanned by $\binom{n}{k}$ k-vectors, where $\binom{n}{k} := \frac{n!}{(n-k)!k!}$. Thus $\bigwedge(V)$ is spanned by $\sum_{k=0}^{n} \binom{n}{k} = 2^n$ elements.

The k-vectors have a clear geometric interpretation for example the 2-vector or bivector $x_1 \wedge x_2$ represents a planar space spanned by the vectors x_1 and x_2 and weighted by scalar representing the area of the oriented parallelogram with sides x_1 and x_2. In an analog way, the 3-vector or trivector represents the spanned 3D space and weighted by the volume of the oriented parallelepiped with edges x_1, x_2 and x_3.

If V^* denotes the dual space to the vector space V, then for each $\vartheta \in V^*$, on can define an *antiderivation* on the algebra $\bigwedge(V)$

$$i_\vartheta : \bigwedge^k V \rightarrow \bigwedge^{k-1} V. \tag{20.47}$$

Consider $x \in \bigwedge^k V$. Then x is a multilinear mapping of V^* to $\mathbb{R}$, thus it is defined by its values on the k-fold Cartesian product $V^* \times V^* \times \cdots \times V^*$. If $y_1, y_2, \ldots, y_{k-1}$ are $k-1$ elements of V^*, then define

$$(i_\vartheta X)(y_1, y_2, \ldots, y_{k-1}) = X(\vartheta, y_1, y_2, \ldots, y_{k-1}). \tag{20.48}$$

In case of a pure scalar $g \in \bigwedge^0 V$, it is clear that $i_\vartheta g = 0$.

The interior product fulfills the following properties:

i. For each k and each $\vartheta \in V^*$, $i_\vartheta : \bigwedge^k V \to \bigwedge^{k-1} V$. By convention $\bigwedge^{-1} = 0$,
ii. If $x \in \bigwedge^1$ (=V) then $i_\vartheta x = \vartheta(x)$ is the dual paring between the elements of V and V^*.
iii. For each $\vartheta \in V^*$, i_ϑ is a *graded derivation* of degree -1: $i_\vartheta(x \wedge y) = (i_\vartheta x) \wedge y + (-1)^{deg\vartheta} \vartheta \wedge (i_\vartheta y)$.

These three properties suffice to characterize the interior product as well as define it in the general infinite-dimensional case. Other properties of the interior product follow: $i_\vartheta \circ i_\vartheta = 0$ and $i_\vartheta \circ i_\varsigma = -i_\varsigma \circ i_\vartheta$.

Suppose that V has finite dimension n, then the interior product induces a canonical isomorphism of vector spaces, namely

$$\bigwedge^{k}(V^*) \bigotimes \bigwedge^{n}(V) = \bigwedge^{n-k}(V). \tag{20.49}$$

A nonzero element of the top exterior power $\bigwedge^n(V)$ (which is a one-dimensional space) is sometimes called a volume or oriented form. Given a volume form θ, the isomorphism is given explicitly by

$$\vartheta \in \bigwedge^{k}(V^*) \to i_\vartheta \theta \in \bigwedge^{n-k}(V). \tag{20.50}$$

Now, if in addition to a volume form, the vector space V is equipped with an inner product which identifies V and V^*, then the pertinent isomorphism

$$* : \bigwedge^{k}(V) \to \bigwedge^{n-k}(V) \tag{20.51}$$

is called the *Hodge dual* or more commonly the *Hodge star operator*.

Suppose that U and V are a pair of vector spaces and $f : U \to V$ is a linear transformation, or in other words

$$\bigwedge(U)|_{\bigwedge^1(U)} = f : U = \bigwedge^{1}(U) \to V = \bigwedge^{1}(V), \tag{20.52}$$

then by the universal construction, there exists a unique homomorphism of the graded algebras, namely

$$\bigwedge(f) : \bigwedge(U) \to \bigwedge(V). \tag{20.53}$$

Note that $\bigwedge(f)$ preserves homogeneous degree. A k-graded element of $\bigwedge(f)$ is given by decomposable elements transformed individually

$$\bigwedge(f)(x_1 \wedge \cdots x_k) = f(x_1) \wedge \cdots \wedge f(x_k). \tag{20.54}$$

Consider

$$\bigwedge^k(f) = \bigwedge(f)_{\bigwedge^k(U)} : \bigwedge^k(U) \to \bigwedge^k(V). \tag{20.55}$$

The transformation $\bigwedge(f)$ relative to a basis U and V is equal to the matrix of $k \times k$ minors of f. In the case that U is of finite dimension n and $U = V$, then $\bigwedge^n(f)$ is a mapping of one-dimensional vector space $\bigwedge^n$ to itself, thus it is given by a scalar: the determinant of f.

If F is a field of characteristic 0, then the exterior algebra of a vector space V can be canonically identified with the vector subspace of $T(V)$ consisting of antisymmetric tensors. According to Eq. (20.40), the exterior algebra is the quotient of $T(V)$ by the ideal I generated by $x \otimes x$. Let be $T^r(V)$ the space of homogeneous tensors of rank r. This space is spanned by decomposable tensors

$$x_1 \otimes \cdots \otimes x_r, \quad x_i \in V. \tag{20.56}$$

The antisymmetrization also called the skew-symmetrization of a decomposable tensor is defined by

$$Alt(x_1 \otimes \cdots \otimes x_r) = \frac{1}{r!} \sum_{\sigma \in \Theta_r} sgn(\sigma)_{\sigma(1)} \otimes \cdots \otimes x_{\sigma(r)}, \tag{20.57}$$

where the sum is taken over the symmetric group of permutations on the symbols 1,...,r. By linearity and homogeneity, this extends to an operation, also denoted by Alt, on the full tensor algebra $T(V)$. The image of $Alt(T(V))$ is call the *alternating tensor algebra* and it is denoted by $A(V)$. Note that $A(V)$ is a vector subspace of $T(V)$ and it inherits the structure of a graded vector space from that on $T(V)$. $A(V)$ has an associative graded product $\hat{\otimes}$ defined as follows

$$x \hat{\otimes} y = Alt(x \otimes y). \tag{20.58}$$

Even though that this product is different to the tensor product, under the assumption that the field F has characteristic 0, the kernel of Alt is precisely the ideal I and the following isomorphism exists $A(V) \cong \bigwedge(V)$.

In physics, the exterior algebra is an archetypal example of the so called super-algebras which are essential for instance in physical theories concerning fermions and supersymmetry. The exterior algebra has remarkable applications in differential geometry for defining differential forms. One can intuitively interpret differential forms as a function on weighted subspaces of the tangent space of a differentiable manifold. Consequently, there is a natural wedge product for differential forms. The differential forms play a crucial role in diverse areas of differential geometry.

20.2.3 Grassmann–Cayley Algebras

The Grassmann–Cayley algebra is based on the work by Hermann Grassmann on exterior algebra and the work by the British mathematician Arthur Cayley (1821–1895) on matrices and linear algebra. It is also known as *double algebra*. The Grassmann–Cayley algebra is a sort of a modeling algebra for projective geometry. This mathematical system utilizes subspaces (brackets) as basic computational elements. This framework facilitates the translation of synthetic projective statements into invariant algebraic statements in the bracket ring, which is the ring of the projective invariants. Furthermore, this mathematical system is useful for the treatment with geometric insight of tensor mathematics and for the modeling of conics, quadrics among other forms.

The bracket ring. Let S be a finite set of points $\{e_1, e_2, \ldots, e_n\}$ in $(d-1)$-dimensional projective space over a field F. By using homogeneous coordinates, each point is represented by a d-tuple, which corresponds to a column in this matrix

$$X = \begin{pmatrix} x_{1,1} & x_{1,2} & \ldots & x_{1,n} \\ x_{2,1} & x_{2,2} & \ldots & x_{2,n} \\ \cdot & \cdot & & \cdot \\ \cdot & \cdot & & \cdot \\ \cdot & \cdot & \ldots & \cdot \\ x_{d,1} & x_{d,2} & \ldots & x_{d,n} \end{pmatrix}. \tag{20.59}$$

Assume now that the entries of matrix X are algebraically independent indeterminate over F, so we can define a *bracket* as follows

$$[e_{i_1}, e_{i_2}, \ldots, e_{i_d}] = det \begin{vmatrix} x_{1,i_1} & x_{1,i_2} & \ldots & x_{1,i_d} \\ \ldots & \ldots & \ldots & \ldots \\ x_{d,i_1} & x_{d,i_2} & \ldots & x_{d,i_d} \end{vmatrix}. \tag{20.60}$$

The *bracket ring* B of S (over F in rank d) is the subring of the polynomial ring $F[x_{1,1}, x_{1,2}, \ldots, x_{d,n}]$ generated by all possible brackets. For the projective group, the first theorem of invariant theory states that the projective invariants of the set of points S are strictly the elements of B or bracket polynomials which are homogeneous with respect to different values may take the elements of S. The equation of a typical projective invariant reads

$$[x_1, x_2, x_3][x_4, x_5, x_6][x_1, x_4, x_7] - 3[x_1, x_2, x_4][x_3, x_5, x_7][x_1, x_4, x_6]. \tag{20.61}$$

Note that this is a purely symbolic expression in terms of points in certain geometric configuration and coefficients belonging to the field F. Surprisingly, the coordinates are not explicit and the invariants have a geometric meaning which is completely coordinate-free. The use of coordinate-free formulas is of major relevance for representing and computing complex physical relations. The advantages of the use of coordinate-free symbolic algebraic expressions for representing geometric conditions and constraints are that it resembles to the way we humans regard geometry and that this algebra is conceptually much closer to the geometry essence than the straightforward algebra of coordinates. In fact, we can translate synthetic geometric statements into the bracket algebra by using Grassmann–Cayley algebra, and we can try to translate back invariant algebraic statements, or homogeneous bracket equations. The drawback of this procedure is that the bracket algebra is more complicated than the straightforward polynomial algebra in the coordinates themselves; this is because the brackets are not algebraically independent. They satisfy following relations:

i. $[x_1, x_2, \ldots, x_k] = 0$ *if* any $x_i = x_j$ $i \neq j$.
ii. $[x_1, x_2, \ldots, x_k] = sign(\sigma)[x_{\sigma(1)}, x_{\sigma(2)}, \ldots, x_{\sigma(k)}]$.
iii. $[x_1, x_2, \ldots, x_k][y_1, y_2, \ldots, y_k] = \sum_{j=1}^{k}[x_1, x_2, \ldots, x_{k-1}, y_j][y_1, y_2, \ldots, y_{j-1}, a_k, y_{j+1}, \ldots, y_k]$.

The relations of the type (iii) are called Grassmann–Plücker relations or *syzygies* which correspond to the generalized Laplace expansions in the ring B.

The second fundamental theorem of invariant theory for projective invariants states that all relations among brackets are results involving relations of the type i–iii.

Plücker coordinates. Consider$k$ independent columns $c_{j1}, c_{j2}, \ldots, c_{jk}$ of the matrix X of Eq. (20.59), pick k out of the d rows and index them in ascendant manner by $i_1, i_2, \ldots, i_k$, then the Plücker coordinate is defined as the determinant of a minor as follows

$$P_{i_1,i_2,\ldots,i_k} = \begin{vmatrix} x_{i_1,j_1} & x_{i_1,j_2} & \cdots & x_{i_1,j_k} \\ x_{i_2,j_1} & x_{i_2,j_2} & \cdots & x_{i_2,j_k} \\ \cdots & \cdots & \cdots & \cdots \\ x_{i_k,j_1} & x_{i_k,j_2} & \cdots & x_{i_k,j_k} \end{vmatrix}. \tag{20.62}$$

The Plücker coordinate vector

$$P = c_{j1} = c_{j1} \vee c_{j2} \vee c_{jk} = P_{i_1,i_2,\ldots,i_k} \tag{20.63}$$

is a vector over F of length $\binom{d}{k}$ and depends, up to a nonzero scalar, only on the subspace $U = span(c_{j1}, c_{j2}, \ldots, c_{jk})$ of $V = F^d$. As an illustration consider a line U in $\mathbb{R}^3$ spanned by two points X_1 and X_2, then using $k = 2, d = 4$ its Plücker vector is

$$\begin{aligned} P = X_1 \vee X_1 &= \begin{pmatrix} X_{11} \\ X_{12} \\ X_{13} \\ 1 \end{pmatrix} \vee \begin{pmatrix} X_{21} \\ X_{22} \\ X_{23} \\ 1 \end{pmatrix} = (P_{01}, P_{02}, P_{03}, P_{12}, P_{31}, P_{23})^t, \\ &= (N, R \times N)^t, \end{aligned} \tag{20.64}$$

where 1 corresponds to the homogeneous coordinate in each 3D-point, N stands for the orientation of the line and R is any point from the origin touching the line U.

Extensors, join and meet operations. The previous section gives an introduction to exterior algebra $\bigwedge(V)$, now in this subsection, we will extend $\bigwedge(V)$ to the Grassmann–Cayley algebra. Among the essentially equivalent definitions of exterior algebra let us mention three: the definition as a universal object for alternating multilinear maps on V, the definition as the quotient of the tensor algebra on V by the ideal generated by all tensors of the form $x \otimes x$ for $x \in V$ and an interesting one which assumes that V is a *Peano space* (earlier called Cayley space), meaning that V is vector space endowed with a non-degenerated alternating d-linear form, or *bracket*. Then one can define the exterior algebra $\bigwedge(V)$ as the quotient of the free associative algebra over V by its ideal generated by all expressions resulting of linear combinations, for all $k \leq d$, of k-products

$$\sum_i = \alpha_i x_{1,i} x_{2,i} \ldots x_{k,i}, \tag{20.65}$$

with $\alpha_i \in F$, $x_{j,i} \in V$ for all i, j such that for all $y_1, y_2, \ldots, y_{d-k}$,

$$\sum_i = \alpha_i [x_{1,i}, x_{2,i}, \ldots, x_{k,i}, y_1, y_2, \ldots, y_{d-k}] = 0. \tag{20.66}$$

Instead of the usual symbol for the exterior product $\wedge$ in $\bigwedge(V)$, one uses $\vee$ and refers to it as the *join operation*. This product is associative, distributive over addition, and antisymmetric. The exterior algebra is a *graded algebra* built by the direct sum of kth exterior powers of V

$$\bigwedge(V) = \bigoplus_{k=0}^{n} \bigwedge^{k}(V). \tag{20.67}$$

If one chooses a basis $\{e_1, \ldots, e_d\}$ of V over F, then a basis for $\bigwedge^k(V)$ over F is

$$\{e_{i_1} \vee e_{i_2} \vee \cdots \vee e_{i_k} | 1 \leq i_1 \leq i_2 ... \leq i_k \leq d\}. \tag{20.68}$$

Note that in the exterior algebra $\bigwedge(V)$, there is no need to choose an explicit basis for V, one has a coordinate-free symbolic algebra which in fact helps to mimic coordinate-free geometric operations in the $(d-1)$-dimensional projective space corresponding to V, in other words in an affine space embedded in the projective space.

Let $x_1, x_2, \ldots, x_k \in V$ and compute the join of these k vectors: $X = x_1 \vee x_2 \vee, \ldots \vee x_k$ which can be written simply as $X = x_1 x_2, \ldots x_k$. If $X \neq 0$, then the involved k vectors are linearly independent. For this case X will called an *extensor of step* k or a *decomposable* $k - vector$. Let $Y = y_1 y_2 \dots y_l$ an another extensor of step l, then $X \vee Y = x_1 \vee x_2 \vee, .. \vee x_k \vee y_1 \vee y_2 \vee, .. \vee y_l = x_1 x_2 \dots, x_k y_1 y_2 \dots y_l$ is an extensor of step $k + l$. In fact if $X \bigvee Y \leq 0$ if only if $x_1, x_2, \dots, x_k, y_1, y_2, \dots, y_l$ are distinct and linearly independent.

Grassmann–Cayley algebra Next, we endow the exterior algebra $\bigwedge(V)$ with a second operation $\wedge$ called the *meet* operation. If $X = x_1 x_2, ..x_k$ and $Y = y_1 y_2 \dots y_l$ with $k + l \geq d$ then

$$X \wedge Y = \sum_{\sigma} sgn(\sigma)[x_{\sigma(1)}, \dots, x_{\sigma(d-l)}, y_{\sigma(1)}, \dots, y_{\sigma(l)}]x_{\sigma(d-l+1)}, \dots, x_{\sigma(k)}, \tag{20.69}$$

where the sum is taken over all permutations σ of $\{1, 2, \dots, k\}$, such that $\sigma(1) < \sigma(2) < \cdots < \sigma(d - l)$ and $\sigma(d - l + 1) < \sigma(d - l + 2) < \cdots < \sigma(k)$. These permutations are called *shuffles* of the $(d - l, k - (d - l))$ split of X.

A meet of the extensor X of step k and extensor Y of step l, if is an extensor of step $|l - k|$. The meet is associative and anticommutative in the following sense:

$$X \wedge Y = (-1)^{(d-k)(d-l)} Y \wedge X. \tag{20.70}$$

The meet is dual to the join, where duality exchanges vectors with covectors or extensors of step $n - 1$. The definitions of join and meet are extended to arbitrary elements of $\bigwedge(V)$ by distributivity. The extended operations remain well-defined and associative. The meet operation corresponds to lattice meet of subspaces: $X \bar{\vee} Y = \bar{X} \cap \bar{Y}$ only if $\bar{X} \cup \bar{Y}$ spans V. The Grassmann–Cayley algebra is the vector space $\bigwedge(V)$ equipped with the operations join $\vee$ and meet $\wedge$. In the Grassmann–Cayley algebra framework, one can translate geometric incidence theorems or incidence relations of the projective geometry into a conjunction of Grassmann–Cayley statements. Provided that those statements involve only join and meet operations and not additions, they can be relatively easy translated back to projective geometry. Furthermore, Grassmann–Cayley statements may be in turn expanded into bracket statements by the definitions and properties of join and meet. Every simple Grassmann–Cayley statement is equivalent to a finite conjunction of bracket statements. Conversely, writing a bracket statement as a simple Grassmann–Cayley statement, it is not always

Fig. 20.2 Cayley factorization: bracket algebra → geometry

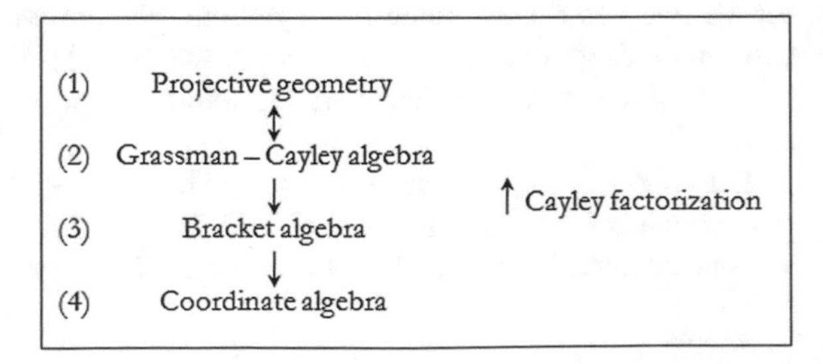

a simple task. This problem is called *Cayley factorization*. In this manner, one works with an invariant language with respect to the projective general linear group. However, one can introduce a further step away from the coordinate-free geometric computations by introducing vector coordinates, as a result the statements in a larger algebra may include non-invariant expressions. The Cayley factorization as depicted in Fig. 20.2 plays a key role in automated geometric-theorem proving like the proof of theorems in projective and Euclidean geometry [10, 55, 172, 309].

Appendix A
Glossary

This section includes the list of the notation with a brief description.

$\mathbb{R}$	The real numbers
$\mathbb{S}^n$	The unit sphere in $\mathbb{R}^{n+1}$
$\mathbb{C}$	The complex numbers
$\mathbb{H}$	The quaternion algebra
$\mathbb{R}^n$	A vector space of dimension n over the field $\mathbb{R}$ with Euclidean signature
$\mathbb{R}^{p,q}$	A vector space of dimension $n = p + q$ over the field $\mathbb{R}$ with signature (p, q)
$\mathbb{R}^{m \times n}$	The direct product $\mathbb{R}^m \otimes \mathbb{R}^n$
G_n	The geometric algebra over $\mathbb{R}^n$
$G_{p,q}$	The geometric algebra over $\mathbb{R}^{p,q}$
$G_{p,q,r}$	The special or degenerated geometric algebra over $\mathbb{R}^{p,q}$
$G^k_{p,q}$	The k-vector space of $G_{p,q}$
x	Scalar element
$\mathbf{x}$	Vector of $\mathbb{R}^n$
$\boldsymbol{x}$, $\boldsymbol{X}$	Vector and multivector of geometric algebra
$\boldsymbol{X}_k$	A blade of grade k
$\mathcal{F}$	Flag, a set of geometric entities of an object frame
$a \cdot \partial F = F_a$	a-derivative of $F = F(x)$
$A * \partial F = F_A$	A-derivative de $F = F(x)$
$\underline{F} = \underline{F}(a) = F_a$	First differential of $F = F(x)$
$\underline{\underline{F}} = \underline{F}(A) = F_A$	differential of $F = F(x)$
$\overline{f} = \overline{f}(a')$	Adjoint of $f = f(x)$
$\overline{f}(A)$	Adjoint outermorphism of $f = f(x)$
$\underline{f}(A)$	Differential outermorphism of $f = f(x)$

E. Bayro-Corrochano, *Geometric Algebra Applications Vol. I*,
https://doi.org/10.1007/978-3-319-74830-6

Operator symbols

$\boldsymbol{XY}$	Geometric product of the multivectors $\boldsymbol{X}$ and $\boldsymbol{Y}$
$\boldsymbol{X} * \boldsymbol{Y}$	Scalar product of $\boldsymbol{X}$ and $\boldsymbol{Y}$
$\boldsymbol{X} \cdot \boldsymbol{Y}$	Inner product of $\boldsymbol{X}$ and $\boldsymbol{Y}$
$\boldsymbol{X} \wedge \boldsymbol{Y}$	Wedge product of $\boldsymbol{X}$ and $\boldsymbol{Y}$
$\boldsymbol{x} \overline{\times} \boldsymbol{y}$	Anticommutator product
$\boldsymbol{x} \underline{\times} \boldsymbol{y}$	Commutator product
$\boldsymbol{XY} = \boldsymbol{X} \overline{\times} \boldsymbol{Y} + \boldsymbol{X} \underline{\times} \boldsymbol{Y}$	
$\widehat{\boldsymbol{X}}_r$	Grade involution
$\widetilde{\boldsymbol{X}}_r$	Reversion
$\langle \mathbf{X}_r \rangle_k^{\dagger}$	Conjugate
$\overline{\boldsymbol{X}}_r = \widetilde{\widehat{\boldsymbol{X}}}_r$	Clifford conjugation
$\boldsymbol{X} \bigvee \boldsymbol{Y}$	Meet operation of $\boldsymbol{X}$ and $\boldsymbol{Y}$
$\boldsymbol{X} \bigwedge \boldsymbol{Y}$	Join product of $\boldsymbol{X}$ and $\boldsymbol{Y}$
$\boldsymbol{X}^{-1}$	Inverse of $\boldsymbol{X}$
$< \boldsymbol{X} >_k$	Projection of $\boldsymbol{X}$ onto grade k
$\boldsymbol{X}^*$	Dual of $\boldsymbol{X}$
$\widetilde{\boldsymbol{X}}$	Reverse of $\boldsymbol{X}$
$\boldsymbol{X}^{\dagger}$	Conjugate of $\boldsymbol{X}$
$\lvert\lvert \boldsymbol{X} \rvert\rvert$	Norm of $\boldsymbol{X}$
$P_{Y_l}^{\parallel}(< \boldsymbol{X} >_k)$	Projection of $< \boldsymbol{X} >_k$ onto $< \boldsymbol{Y} >_l$
$P_{Y_l}^{\perp}(< \boldsymbol{X} >_k)$	Projection of $< \boldsymbol{X} >_k$ onto $< \boldsymbol{Y} >_l$
$(\boldsymbol{H} *_l \boldsymbol{F})(\boldsymbol{x})$	Left convolution
$(\boldsymbol{F} *_r \boldsymbol{H})(\boldsymbol{x})$	Right convolution
$(\boldsymbol{H} \star_l \boldsymbol{F})(\boldsymbol{x})$	Left correlation
$(\boldsymbol{F} \star_r \boldsymbol{H})(\boldsymbol{x})$	Right correlation

Appendix B
Useful Formulas for Geometric Algebra

Geometric Product

$$
\begin{aligned}
&\boldsymbol{xy} = \boldsymbol{x} \cdot \boldsymbol{y} + \boldsymbol{x} \wedge \boldsymbol{y}\\
&\boldsymbol{yx} = \boldsymbol{x} \cdot \boldsymbol{y} - \boldsymbol{x} \wedge \boldsymbol{y}\\
&\boldsymbol{x} \cdot \boldsymbol{X}_r = \tfrac{1}{2}(\boldsymbol{x}\boldsymbol{X}_r - (-1)^r \boldsymbol{X}_r \boldsymbol{x})\\
&\boldsymbol{x} \wedge \boldsymbol{X}_r = \tfrac{1}{2}(\boldsymbol{x}\boldsymbol{X}_r + (-1)^r \boldsymbol{X}_r \boldsymbol{x})
\end{aligned}
$$

Multivectors

$$
\begin{aligned}
&\boldsymbol{X} = < \boldsymbol{X} >_0 + < \boldsymbol{X} >_1 + \cdots = \textstyle\sum_r < \boldsymbol{X} >_r\\
&\boldsymbol{X}_r \boldsymbol{Y}_s = < \boldsymbol{X}_r \boldsymbol{Y}_s >_{|r-s|} + < \boldsymbol{X}_r \boldsymbol{Y}_s >_{|r-s|+2} + \cdots + < \boldsymbol{X}_r \boldsymbol{Y}_s >_{r+s}\\
&\boldsymbol{X}_r \cdot \boldsymbol{Y}_s = < \boldsymbol{X}_r \boldsymbol{Y}_s >_{|r-s|}\\
&\boldsymbol{X}_r \cdot \lambda = 0 \quad \text{for scalar } \lambda\\
&\boldsymbol{X}_r \wedge \boldsymbol{Y}_s = < \boldsymbol{X}_r \boldsymbol{Y}_s >_{r+s}\\
&< \boldsymbol{XY} > = < \boldsymbol{YX} >\\
&\boldsymbol{X}_r \cdot (\boldsymbol{Y}_s \cdot \boldsymbol{Z}_t) = (\boldsymbol{X}_r \wedge \boldsymbol{Y}_s) \cdot \boldsymbol{Z}_t \quad r + s \leq t \;\text{ and }\; r, s > 0\\
&\boldsymbol{X}_r \cdot (\boldsymbol{Y}_s \cdot \boldsymbol{Z}_t) = (\boldsymbol{X}_r \cdot \boldsymbol{Y}_s) \cdot \boldsymbol{Z}_t \quad r + t \leq s\\
&\boldsymbol{X} \times (\boldsymbol{Y} \times \boldsymbol{Z}) + \boldsymbol{Z} \times (\boldsymbol{X} \times \boldsymbol{Y}) + \boldsymbol{Y} \times (\boldsymbol{Z} \times \boldsymbol{X}) = 0
\end{aligned}
$$

Products of frequent use of vectors $\boldsymbol{x}$, $\boldsymbol{y}$, $\boldsymbol{z}$, $\boldsymbol{w}$ and bivectors $\boldsymbol{X}$, $\boldsymbol{Y}$

$$
\begin{aligned}
&\boldsymbol{x}\overline{\times}\boldsymbol{y} = \tfrac{1}{2}(\boldsymbol{xy} + \boldsymbol{yx}) \qquad \text{anticommutator product}\\
&\boldsymbol{x}\underline{\times}\boldsymbol{y} = \tfrac{1}{2}(\boldsymbol{xy} - \boldsymbol{yx}) \qquad \text{commutator product}\\
&\boldsymbol{XY} = \tfrac{1}{2}(\boldsymbol{XY} + \boldsymbol{YX}) + \tfrac{1}{2}(\boldsymbol{XY} - \boldsymbol{YX}) = \boldsymbol{X}\overline{\times}\boldsymbol{Y} + \boldsymbol{X}\underline{\times}\boldsymbol{Y}\\
&\boldsymbol{x} \cdot (\boldsymbol{y} \wedge \boldsymbol{x}) = \boldsymbol{x} \cdot \boldsymbol{yz} - \boldsymbol{x} \cdot \boldsymbol{zy}\\
&(\boldsymbol{x} \wedge \boldsymbol{y}) \cdot (\boldsymbol{z} \wedge \boldsymbol{w}) = \boldsymbol{x} \cdot \boldsymbol{wy} \cdot \boldsymbol{z} - \boldsymbol{x} \cdot \boldsymbol{zy} \cdot \boldsymbol{w}\\
&\boldsymbol{x} \cdot (\boldsymbol{Y} \cdot \boldsymbol{X}) = (\boldsymbol{x} \wedge \boldsymbol{y}) \cdot \boldsymbol{X}\\
&(\boldsymbol{x} \wedge \boldsymbol{y}) \times (\boldsymbol{z} \wedge \boldsymbol{w}) = \boldsymbol{y} \cdot \boldsymbol{zx} \wedge \boldsymbol{w} - \boldsymbol{x} \cdot \boldsymbol{zy} \wedge \boldsymbol{w} + \boldsymbol{x} \cdot \boldsymbol{wy} \wedge \boldsymbol{z} - \boldsymbol{y} \cdot \boldsymbol{wx} \wedge \boldsymbol{z}\\
&(\boldsymbol{x} \wedge \boldsymbol{y}) \times \boldsymbol{X} = (\boldsymbol{x} \cdot \boldsymbol{X}) \wedge \boldsymbol{y} + \boldsymbol{x} \wedge (\boldsymbol{y} \cdot \boldsymbol{X})
\end{aligned}
$$

Bivectors

$$
\begin{aligned}
&\boldsymbol{XY} = \boldsymbol{X} \cdot \boldsymbol{Y} + \boldsymbol{X} \times \boldsymbol{Y} + \boldsymbol{X} \wedge \boldsymbol{Y}\\
&\boldsymbol{YX} = \boldsymbol{X} \cdot \boldsymbol{Y} - \boldsymbol{X} \times \boldsymbol{Y} + \boldsymbol{X} \wedge \boldsymbol{Y}
\end{aligned}
$$

E. Bayro-Corrochano, *Geometric Algebra Applications Vol. I*,
https://doi.org/10.1007/978-3-319-74830-6

$$\boldsymbol{Y} \times \boldsymbol{X}_r = \langle \boldsymbol{Y} \times \boldsymbol{X}_r \rangle_r$$
$$\boldsymbol{Z} \times (\boldsymbol{X}\boldsymbol{Y}) = \boldsymbol{Z} \times \boldsymbol{X}\boldsymbol{Y} + \boldsymbol{X}\boldsymbol{Z} \times \boldsymbol{Y}, \quad \forall \boldsymbol{X}, \boldsymbol{Y}$$

Generalized inner product

For two blades $\boldsymbol{X}_r = \boldsymbol{x}_1 \wedge \boldsymbol{x}_2 \wedge \cdots \wedge \boldsymbol{x}_r$, and $\boldsymbol{Y}_s = \boldsymbol{y}_1 \wedge \boldsymbol{y}_2 \wedge \cdots \wedge \boldsymbol{y}_s$:

$$\boldsymbol{X}_r \cdot \boldsymbol{Y}_s = \begin{cases} ((\boldsymbol{x}_1 \wedge \boldsymbol{x}_2 \wedge \cdots \wedge \boldsymbol{x}_r) \cdot \boldsymbol{y}_1) \cdot (\boldsymbol{y}_2 \wedge \boldsymbol{y}_3 \wedge \cdots \wedge \boldsymbol{y}_s) & \text{if } r \geq s \\ (\boldsymbol{x}_1 \wedge \boldsymbol{x}_2 \wedge \cdots \wedge \boldsymbol{x}_{r-1}) \cdot (\mathbf{x}_r \cdot (\boldsymbol{y}_1 \wedge \boldsymbol{y}_2 \wedge \cdots \wedge \boldsymbol{y}_s)) & \text{if } r < s. \end{cases}$$

right contraction

$$(\boldsymbol{x}_1 \wedge \boldsymbol{x}_2 \wedge \cdots \wedge \boldsymbol{x}_r) \cdot \boldsymbol{y}_1 = \sum_{i=1}^{r} (-1)^{r-i} \mathbf{x}_1 \wedge \cdots \wedge \mathbf{x}_{i-1} \wedge (\boldsymbol{x}_i \cdot \boldsymbol{y}_1) \wedge \boldsymbol{x}_{i+1} \wedge \cdots \wedge \boldsymbol{x}_r$$

left contraction

$$\mathbf{x}_r \cdot (\boldsymbol{y}_1 \wedge \boldsymbol{y}_2 \wedge \cdots \wedge \boldsymbol{y}_s) = \sum_{i=1}^{s} (-1)^{i-1} \boldsymbol{y}_1 \wedge \cdots \wedge \boldsymbol{y}_{i-1} \wedge (\boldsymbol{x}_r \cdot \boldsymbol{y}_i) \wedge \boldsymbol{y}_{i+1} \wedge \cdots \wedge \boldsymbol{y}_s.$$

Involution, reversion, and conjugation operations

For an r-grade multivector $\boldsymbol{A}_r = \sum_{i=0}^{r} \langle \boldsymbol{A}_r \rangle_i$

$$\text{Grade involution: } \widehat{\boldsymbol{A}}_r = \sum_{i=0}^{r} (-1)^i \langle \boldsymbol{A}_r \rangle_i,$$
$$\text{Reversion: } \widetilde{\boldsymbol{A}}_r = \sum_{i=0}^{r} (-1)^{\frac{i(i-1)}{2}} \langle \boldsymbol{A}_r \rangle_i,$$
$$\text{Conjugate: } \langle \boldsymbol{A}_r \rangle_k^{\dagger} = (\boldsymbol{a}_1 \wedge \boldsymbol{a}_2 \wedge \ldots \wedge \boldsymbol{a}_k)^{\dagger} = \boldsymbol{a}_k^{\dagger} \wedge \boldsymbol{a}_{k-1}^{\dagger} \wedge \ldots \wedge \boldsymbol{a}_1^{\dagger}$$
$$e_{\boldsymbol{A}}^{\dagger} = (-1)^r \tilde{e}_{\boldsymbol{A}}, \quad r := \texttt{grade of } e_{\boldsymbol{A}}$$
$$\text{Clifford conjugation: } \overline{\boldsymbol{A}}_r = \widetilde{\widehat{\boldsymbol{A}}}_r = \sum_{i=0}^{r} (-1)^{\frac{i(i+1)}{2}} \langle \boldsymbol{A}_r \rangle_i.$$

Magnitude of a multivector

$$\begin{aligned} ||\boldsymbol{M}|| &= <\widetilde{\boldsymbol{M}}\boldsymbol{M}>_0^{\frac{1}{2}} \\ &= (|| <\boldsymbol{M}>_0 ||^2 + || <\boldsymbol{M}>_1 ||^2 + || <\boldsymbol{M}>_2 ||^2 + \ldots + || <\boldsymbol{M}>_n ||^2)^{\frac{1}{2}} \\ &= \sqrt{\sum_{r=0}^{n} || <\boldsymbol{M}>_r ||^2}. \end{aligned}$$

Pseudoscalar I

$$X_r \cdot (Y_s I) = X_r \bigwedge Y_s I \quad r+s \leq n$$
$$X_r \bigwedge (Y_s I) = X_r \cdot Y_s I \quad r \leq s$$

Linear Algebra

$$\mathtt{f}(x \wedge y \wedge \cdots \wedge w) = \mathtt{f}(x) \wedge \mathtt{f}(y) \wedge \cdots \wedge \mathtt{f}(w)$$
$$\mathtt{f}(I) = \mathtt{det}(\mathtt{f}) I$$
$$\bar{\mathtt{f}}(y) = e_k \mathtt{f}(e^k) \cdot y = \partial_x \mathtt{f}(x) \cdot y$$
$$\mathtt{f}(X_r) \cdot Y_s = \mathtt{f}[X_r \cdot \bar{\mathtt{f}}(Y_s)] \quad r \geq s$$
$$X_r \cdot \bar{\mathtt{f}}(Y_s) = \bar{\mathtt{f}}[\mathtt{f}(X_r) \cdot Y_s] \quad r \leq s$$
$$\mathtt{f}^{-1}(X) = \mathtt{det}(\mathtt{f})^{-1} I \bar{\mathtt{f}}(I^{-1} X)$$
$$\bar{\mathtt{f}}^{-1}(X) = \mathtt{det}(\mathtt{f})^{-1} I \mathtt{f}(I^{-1} X)$$

Simplex

Tangent

$$X_r \equiv x_0 \wedge x_1 \wedge x_2 \wedge \ldots \wedge x_r = x_0 \wedge \bar{X}_r \; \bar{X}_r \equiv (x_1 - x_0) \wedge (x_2 - x_0) \wedge \cdots \wedge (x_r - x_0)$$

Face

$$\mathcal{F}_i^r X_r \equiv X_r^i \equiv x^i \cdot X_r = (-1)^{i+1} x_0 \wedge x_1 \wedge \cdots \wedge \check{x}_i \wedge \cdots \wedge x_r$$

Boundary

$$\Omega_b X_r = \sum_{i=0}^r X_r^i = \sum_{i=0}^r \mathcal{F}_i X_r$$

Frames and Contractions

$$e^i \cdot e_j = \delta_j^i$$
$$e^k = (-1)^{k+1} e_1 \wedge \cdots \wedge \check{e}_k \wedge \cdots \wedge e_n E_n^{-1}$$
$$E_n = e_1 \wedge e_2 \wedge \cdots \wedge e_n$$
$$e_k e^k \cdot X_r = \partial_x x \cdot X_r = r X_r$$
$$e_k e^k \wedge X_r = \partial_x x \wedge X_r = (n-r) X_r$$
$$e_k X_r e^k = \dot{\partial}_x X_r \dot{x} = (-1)^r (n-2r) X_r$$

The multivector integral

$$\int_{E^n} F(x)|dx| = \lim_{\substack{|\Delta x_j| \to 0 \\ n \to \infty}} \sum_{j=1}^{n} F(x_j e_j) \Delta x_j,$$

Convolution and correlation of scalar fields

$$(h * f)(x) = \int_{E^n} h(x') f(x - x') dx' \qquad \text{convolution}$$
$$(h \star f)(x) = \int_{E^n} h(x') f(x + x') dx' \qquad \text{correlation}$$

Clifford convolution and correlation

For multivector field F and a multivector-valued filter H, convolution

$$(H *_l F)(x) = \int_{E^n} H(x') F(x - x') |d\, x'| \qquad \text{left}$$
$$(F *_r H)(x) = \int_{E^n} F(x - x') H(x') |d\, x'| \qquad \text{right}$$

discrete
$(\boldsymbol{H} *_l \boldsymbol{F})_{i,j,k} = \sum_{i=-d}^{d} \sum_{j=-d}^{d} \sum_{k=-d}^{d} \boldsymbol{H}_{r,s,t} \boldsymbol{F}_{i-r,j-s,k-t}$
correlation
$(\boldsymbol{H} \star_l \boldsymbol{F})(\boldsymbol{x}) = \int_{E^n} \boldsymbol{H}(\boldsymbol{x}')\boldsymbol{F}(\boldsymbol{x}+\boldsymbol{x}')|d\ x'|$ left
$(\boldsymbol{F} \star_r \boldsymbol{H})(\boldsymbol{x}) = \int_{E^n} \boldsymbol{F}(\boldsymbol{x}+\boldsymbol{x}')\boldsymbol{H}(\boldsymbol{x}')|d\ x'|.$ right

Differentiation, Linear, and Multilinear Functions (Chap. 3)

$\partial_\epsilon F(\epsilon) = \frac{\partial F(\epsilon)}{\epsilon} = \lim_{\delta\epsilon \to 0} \frac{F(\epsilon+\delta\epsilon)-F(\epsilon)}{\delta\epsilon}$
$\boldsymbol{v} \cdot \partial F = \frac{1}{2}(\boldsymbol{v}\partial F + \dot{\partial}\boldsymbol{v}\dot{F})$

c.f. [[138], 2-(1.32–1.37)]
$\partial|\boldsymbol{v}|^2 = \partial \boldsymbol{v}^2 = 2P(\boldsymbol{v}) = 2\boldsymbol{v}$
$\partial \wedge \boldsymbol{v} = 0$
$\partial \boldsymbol{v} = \partial \cdot \boldsymbol{v} = n$
$\partial|\boldsymbol{v}|^k = k|\boldsymbol{v}|^{k-2}\boldsymbol{v}$
$\partial\left(\frac{\boldsymbol{v}}{|\boldsymbol{v}|^k}\right) = \frac{n-k}{|\boldsymbol{v}|^k}$
$\partial log|\boldsymbol{v}| = \boldsymbol{v}^{-1}$
If $\boldsymbol{V} = P(\boldsymbol{V}) =< \boldsymbol{V} >_r$, c.f. [[138], 2-(1.38–1.40)]
$\dot{\partial}(\dot{\boldsymbol{v}} \cdot \boldsymbol{V}) = \boldsymbol{V} \cdot \partial \boldsymbol{v} = r\boldsymbol{V}$
$\dot{\partial}(\dot{\boldsymbol{v}} \wedge \boldsymbol{V}) = \boldsymbol{V} \wedge \partial \boldsymbol{v} = (n-r)\boldsymbol{V}$
$\dot{\partial}\boldsymbol{V}\dot{\boldsymbol{v}} = \sum_k \boldsymbol{a}^k \boldsymbol{V} \boldsymbol{a}_k = (-1)^r(n-2r)\boldsymbol{V}$
$\boldsymbol{v} \cdot (\partial \wedge \boldsymbol{u}) = \boldsymbol{v} \cdot \partial \boldsymbol{u} - \dot{\partial}\dot{\boldsymbol{u}} \cdot \boldsymbol{v}$
Lie bracket $[\boldsymbol{u}, \boldsymbol{v}] = \partial \cdot (\boldsymbol{u} \wedge \boldsymbol{v}) - \boldsymbol{v}\partial \cdot \boldsymbol{u} + \boldsymbol{u}\, \partial \cdot \boldsymbol{v}$
Jacobi identity $\boldsymbol{u} \cdot (\boldsymbol{v} \wedge \boldsymbol{w}) + \boldsymbol{v} \cdot (\boldsymbol{w} \wedge \boldsymbol{u}) + \boldsymbol{w} \cdot (\boldsymbol{u} \wedge \boldsymbol{v}) = 0$
$\dot{\boldsymbol{F}} \wedge \dot{\partial} \wedge \dot{\boldsymbol{G}} = \boldsymbol{F} \wedge \partial \wedge \boldsymbol{G} + (-1)^{r+s(r+1)} \boldsymbol{G} \wedge \partial \wedge \boldsymbol{F}$
$\boldsymbol{U} * \partial = \sum_p \boldsymbol{U} * \partial_{\hat{p}} = \sum_p \boldsymbol{U}_{\hat{p}} * \partial_{\hat{p}}.$
$\partial_X = \partial_U \boldsymbol{U} * \partial_X$
$\partial F \equiv \partial_X F(\boldsymbol{X}) = \partial_U \underline{F}(\boldsymbol{X}, \boldsymbol{U}) \equiv \underline{\partial}\, \underline{F}$
$\underline{\boldsymbol{X}}(\lambda) = \lambda * \partial \boldsymbol{X} = \lambda \partial_p \boldsymbol{X} = \lambda \frac{d\boldsymbol{X}}{d\tau}$
c.f. [[138], 2-(2.28.a–2.35)]
$\boldsymbol{U} * \partial_X \boldsymbol{X} = \dot{\partial}_X \dot{\boldsymbol{X}} * \boldsymbol{U} = \boldsymbol{U}^{||}$
$\boldsymbol{U} * \partial_X \boldsymbol{X}^\dagger = \dot{\partial}_X \dot{\boldsymbol{X}}^\dagger * \boldsymbol{U} = \boldsymbol{U}^{||\dagger}$
$\partial_X \boldsymbol{X} = d$
$\partial_X |\boldsymbol{X}|^2 = 2\boldsymbol{X}^\dagger$
$\boldsymbol{U} * \partial_X \boldsymbol{X}^k = \boldsymbol{U}^{||}\boldsymbol{X}^{k-1} + \boldsymbol{X}\boldsymbol{U}^{||}\boldsymbol{X}^{k-2} + \cdots + \boldsymbol{X}^{k-1}\boldsymbol{U}^{||}$
$\partial_X |\boldsymbol{X}|^k = k|\boldsymbol{X}|^{k-2}\boldsymbol{X}^\dagger$
$\partial_X log|\boldsymbol{X}| = \frac{\boldsymbol{X}^\dagger}{|\boldsymbol{X}|^2}$
$\boldsymbol{U} * \partial_X \{|\boldsymbol{X}|^k \boldsymbol{X}\} = |\boldsymbol{X}|^k \left\{ \boldsymbol{U}^{||} + k \frac{\boldsymbol{U}\boldsymbol{X}^\dagger \boldsymbol{X}}{|\boldsymbol{X}|^2} \right\}$
$\partial_X \{|\boldsymbol{X}|^k \boldsymbol{X}\} = |\boldsymbol{X}|^k \left\{ d + k \frac{\boldsymbol{X}^\dagger \boldsymbol{X}}{|\boldsymbol{X}|^2} \right\}$

Adjoints and Inverses

$\boldsymbol{a} \cdot \underline{f}(\boldsymbol{b}) = \overline{f}(\boldsymbol{a}) \cdot \boldsymbol{b}$
$\overline{f}(\boldsymbol{a}) = \overline{f}(\boldsymbol{a}) \cdot e_k e^k = \boldsymbol{a} \cdot \underline{f}(e_k) e^k$

$\overline{f}(\boldsymbol{a}\wedge\boldsymbol{b}) = \frac{1}{2}e_i\wedge e_j(\boldsymbol{a}\wedge\boldsymbol{b})\cdot\underline{f}(e^j\wedge e^i)$
$\boldsymbol{A}_p\cdot\overline{f}(\boldsymbol{B}_p) = \underline{f}(\boldsymbol{A}_p)\cdot\boldsymbol{B}_p$
$\underline{f}(\boldsymbol{A}_p)\cdot\boldsymbol{B}_q = \underline{f}[\boldsymbol{A}_p\cdot\overline{f}(\boldsymbol{B}_q)] \quad p\geq q$
$\boldsymbol{A}_p\cdot\overline{f}(\boldsymbol{B}_q) = \overline{\overline{f}}[\underline{f}(\boldsymbol{A}_p)\cdot\boldsymbol{B}_q] \quad p\leq q$

Eigenvectors and Eigenblades

$f(\boldsymbol{v}) = \alpha\boldsymbol{v}$
$\underline{f}(\boldsymbol{V}) = \alpha\boldsymbol{V}\in\bigwedge^k\mathbb{R}^n$
$\overline{\overline{f}}(\boldsymbol{U}) = \beta\boldsymbol{U}$
$\underline{F}(\boldsymbol{V}_k) = C_{f_k}(\lambda)\boldsymbol{V}_k$
$C_f(\lambda) = C_{f_1}(\lambda)C_{f_2}(\lambda)\cdots C_{f_m}(\lambda)$
$\beta\underline{f}(\boldsymbol{V}\cdot\boldsymbol{U}) = \alpha\boldsymbol{V}\cdot\boldsymbol{U}$, if grade $\boldsymbol{V}\geq$ grade $\boldsymbol{U}$
$\alpha\overline{\overline{f}}(\boldsymbol{V}\cdot\boldsymbol{U}) = \beta\boldsymbol{V}\cdot\boldsymbol{U}$, if grade $\boldsymbol{U}\geq$ grade $\boldsymbol{V}$

Traction Operators: traction, protraction, contraction

$F(\boldsymbol{V}) =< F(\boldsymbol{V}) >_q$ traction
$\partial_x\cdot F(\boldsymbol{v}\wedge\boldsymbol{U})$ contraction
$\partial_v\wedge F(\boldsymbol{v}\wedge\boldsymbol{U})$ protraction
$\partial_v F(\boldsymbol{v}\wedge\boldsymbol{U}) = \partial_v\cdot F(\boldsymbol{v}\wedge\boldsymbol{U}) + \partial_v\wedge F(\boldsymbol{v}\wedge\boldsymbol{U})$
$F(\boldsymbol{v}\wedge\boldsymbol{u}) = \frac{1}{2}(\boldsymbol{v}\wedge\boldsymbol{u})\cdot(\boldsymbol{V}\wedge\boldsymbol{V}) + (\boldsymbol{v}\wedge\boldsymbol{u})\cdot(\boldsymbol{V}\boldsymbol{V})$
$\partial F = \partial_V F(\boldsymbol{V}) = \partial\cdot F + \partial\times F + \partial\wedge F$ bivector derivative
$\partial F = \frac{1}{2}\partial_v\wedge\partial_u F(\boldsymbol{u}\wedge\boldsymbol{v}) = \frac{1}{2}\partial_v\partial_u F(\boldsymbol{u}\wedge\boldsymbol{v})$
c.f. [[138], 3-9.18.a-c]
$\partial_V\boldsymbol{V}\cdot\boldsymbol{Q} = \frac{k(k-1)}{2}\boldsymbol{Q} = \boldsymbol{Q}\cdot\partial_V\boldsymbol{V}$ for $k\geq 2$
$\partial_V\boldsymbol{V}\times\boldsymbol{Q} = k(n-k)\boldsymbol{Q} = \boldsymbol{Q}\times\partial_V\boldsymbol{V}$ for $k\geq 1$
$\partial_V\boldsymbol{V}\wedge\boldsymbol{Q} = \frac{(n-k)(n-k-1)}{2}\boldsymbol{Q} = \boldsymbol{Q}\wedge\partial_V\boldsymbol{V}$ for $k\geq 0$

Tensors in Geometric Algebra

$\tau(\boldsymbol{v}_1,\boldsymbol{v}_2,\ldots,\boldsymbol{v}_{r+s}) = (\boldsymbol{v}_{r+1},\boldsymbol{v}_{r+2},\ldots,\boldsymbol{v}_{r+s})\cdot T(\boldsymbol{v}_1,\boldsymbol{v}_2,\ldots,\boldsymbol{v}_r)$ T a s,r form
$(\boldsymbol{v}_1,\boldsymbol{v}_2,\ldots,\boldsymbol{v}_r) = (s!)^{-1}\partial_{u_s}\wedge\cdots\partial_{u_1}\tau(\boldsymbol{v}_1,\boldsymbol{v}_2,\ldots,\boldsymbol{v}_r,\boldsymbol{u}_1,\boldsymbol{u}_2,\ldots,\boldsymbol{u}_s)$
$= \partial_U U * T(\boldsymbol{v}_1,\boldsymbol{v}_2,\ldots,\boldsymbol{v}_r)$ using simplicial derivative
$\partial_{\boldsymbol{v}_k}T(\boldsymbol{v}_1,\boldsymbol{v}_2,\ldots,\boldsymbol{v}_p)$ traction of T
$T(\boldsymbol{v}_1,\ldots,\boldsymbol{v}_{j-1},\partial_{\boldsymbol{v}_k},\boldsymbol{v}_{j+1},\ldots,\boldsymbol{v}_p) = \partial_{\boldsymbol{v}_j}\cdot\partial_{\boldsymbol{v}_k}T(\boldsymbol{v}_1,\boldsymbol{v}_2,\ldots,\boldsymbol{v}_p)$ T degree $p-2$
$\gamma(\boldsymbol{v},\boldsymbol{u},\boldsymbol{w}) = \nu^i\mu^j\omega^k\gamma_{ijk} = \nu_i\mu_j\omega_k\gamma^{ijk} = \nu_i\mu_j\omega^k\gamma^{ij}_k$
$\gamma^{ij}_k f^k_r = \gamma(\overline{f}(\boldsymbol{v}'^i),\overline{f}(\boldsymbol{v}'^j),\underline{f}(\boldsymbol{v}_r)) = f^i_p f^j_q\gamma(\boldsymbol{v}'^p,\boldsymbol{v}'^q,\boldsymbol{v}'_r) = f^i_p f^j_q\gamma^{p'q'}_{r'}$
$\gamma'(\boldsymbol{v}',\boldsymbol{u}') = \gamma(\underline{f}^{-1}(\boldsymbol{v}'),\overline{f}(\boldsymbol{u}'))$
$(AB)_{ij} = A_{ik}B_{lj}g^{kl}$ metric tensor
$A_{\mu\nu} = f_{\mu i}f_{\nu j}A^{ij}$ non-orthonrmal frame $\{f_\mu\}$
$A^\nu_\mu = f^i_\mu f^\nu_j A^j_i$

Geometric Calculus (Chap. 4)

$\nabla\wedge\nabla = 0$

$\nabla(\boldsymbol{x}\cdot\boldsymbol{y}) = \boldsymbol{y}$
$\nabla\boldsymbol{x}^2 = 2\boldsymbol{x}$
$\nabla\cdot F = \frac{\partial}{\partial x^k}e^k\cdot F = \frac{\partial F^k}{\partial x^k} = \partial_k F^k$ divergence of F
$\nabla\cdot\boldsymbol{x} = \frac{\partial x^k}{\partial x^k} = n$
$\nabla\wedge F = e^i\wedge(\partial_i F) = e^i\wedge e^j\partial_i F_j$ exterior derivative
$\nabla F = \nabla\cdot F + \nabla\wedge F$ full vector derivative: codivergence and cocurl
$\nabla\cdot F = \frac{1}{2}(\nabla F + \dot{F}\dot{\nabla})$
$\nabla\wedge F = \frac{1}{2}(\nabla F - \dot{F}\dot{\nabla})$
$\nabla\boldsymbol{F} = e^k\partial_k\boldsymbol{F}$
$\nabla\cdot\boldsymbol{F}_m = <\nabla\boldsymbol{F}_m>_{m-1},$ $\nabla\wedge\boldsymbol{F}_m = <\nabla\boldsymbol{F}_m>_{m+1}$
$\dot{\nabla}\boldsymbol{F}\dot{\boldsymbol{G}} = e^k\boldsymbol{F}\partial_k\boldsymbol{G}$
$\nabla(\boldsymbol{F}\boldsymbol{G}) = \nabla\boldsymbol{F}\boldsymbol{G} + \dot{\nabla}\boldsymbol{F}\dot{\boldsymbol{G}}$
$\dot{\nabla}\dot{f}(\boldsymbol{x}) = \nabla f(\boldsymbol{x}) - e^k f(\partial_k\boldsymbol{x})$
$\nabla\boldsymbol{x}\cdot\boldsymbol{X}_m = e^k e_k\cdot X_m$
$\nabla\boldsymbol{x}\cdot\boldsymbol{X}_m = m\boldsymbol{X}_m$
$\nabla\boldsymbol{x}\wedge\boldsymbol{X}_m = (n-m)\boldsymbol{X}_m$
$\dot{\nabla}\boldsymbol{X}_m\dot{\boldsymbol{x}} = (-1)^m(n-2m)\boldsymbol{X}_m$
$\partial^2 F = (\partial\cdot\partial)F + (\partial\wedge\partial)\cdot F + (\partial\wedge\partial)\times F + (\partial\wedge\partial)\wedge F$
$\partial\cdot\partial = \sum_k(e^k\frac{\partial}{\partial x^k})(e^k\frac{\partial}{\partial x^k}) = \nabla^2$ Laplacian linear operator
$\triangle_M = \partial\cdot\partial = \sum_k(\tau^k\frac{\partial}{\partial x^k})(\tau^k\frac{\partial}{\partial x^k})$ Laplace–Beltrami operator
$P_u(v) = u\cdot\dot{\partial}\dot{\mathcal{P}}(v) = u\cdot\partial P(v) - P(u\cdot\partial v)$ differential of the projection
$\mathcal{S}(F) = \dot{\partial}\dot{\mathcal{P}}(F) = \partial_u P_u(F)$ shape operator
$S_v \equiv \dot{\partial}\wedge\dot{P}(v) = S(P(v))$ curl tensor
$N \equiv \dot{P}(\dot{\partial}) = \partial_v\cdot S_v = \partial_v S_v$ spur of the tensor $P_u(v)$
$P(\partial\wedge\partial) = I^{-1}I\cdot(\partial\wedge\partial) = 0$ integrability condition
$\partial F = \nabla F + S(F)$ derivative and coderivative
$\nabla\wedge F = P(\partial\wedge F)$ cocurl
$\nabla\cdot F = [\nabla\wedge(FI)]I^{-1}$ codivergence and cocurl relation
$\partial\wedge\partial = \partial\wedge P(\partial) = P(\partial\wedge\partial) + S(\partial)$
$\partial\wedge\partial F = S(\partial)F$
$\triangle_L F = \partial\cdot(\partial\wedge F) + \partial\wedge(\partial\cdot F) = \partial^2 F + (\partial\wedge\partial)\times F$ blade Laplace operator
$\triangle_L F = \frac{1}{2}(\partial^2 F + F\partial^2)$
$\triangle_L F = \triangle_{\mathcal{M}} F + (\partial\wedge\partial)\times F$
$\triangle_L F = \triangle_{\mathcal{M}} F + S(\dot{\partial})\times\dot{F}$
$\partial^2 F = \triangle_L F + \partial\wedge\partial F = \triangle_L F + S(\partial F)$ second derivative
$ih\frac{\partial\Psi}{\partial t} = \frac{1}{2m}\pi^2\Psi - \frac{he}{2m}\mathbf{B}\Psi e_3 - eV\Psi$ spinor Schrödinger equation

Lie Algebras, Lie Groups, and Algebra of Incidence (Chap. 5)

$\boldsymbol{R}(\alpha)\boldsymbol{v}\tilde{\boldsymbol{R}}(\alpha) = e^{-\frac{\alpha\boldsymbol{B}}{2}}\boldsymbol{v}e^{\frac{\alpha\boldsymbol{B}}{2}} = \boldsymbol{v} + \boldsymbol{v}\cdot\boldsymbol{B} + \frac{1}{2!}(\boldsymbol{v}\cdot\boldsymbol{B})\cdot\boldsymbol{B} + \cdots$
$\boldsymbol{R}\boldsymbol{v}\tilde{\boldsymbol{R}} = e^{-\boldsymbol{J}\frac{\theta}{2}}\boldsymbol{v}e^{\boldsymbol{J}\frac{\theta}{2}} = cos\theta\boldsymbol{v} + sin\theta\boldsymbol{v}\cdot\boldsymbol{J}$
$\boldsymbol{J}_i = e_i f_i$ $(i = 1, ..., n)$
$\boldsymbol{E}_{ij} = e_i e_j + f_i f_j$ $(i < j = 1, ..., n)$
$\boldsymbol{F}_{ij} = e_i f_j - f_i e_j$ $(i < j = 1, ..., n)$
$\{w_i, \bar{w}_i\}$ Witt bases

$N = span\{w_1, ..., w_n\} \quad \overline{N} = span\{\overline{w}_1, ..., \overline{w}_n\}$, reciprocal null cones.
$G_{n,n} = G_N \otimes G_{\overline{N}} = gen\{w_1, ..., w_n, \overline{w}_1, ..., \overline{w}_n\}$
$\underline{E}_{p,q}(\boldsymbol{x}) = (\boldsymbol{x} \cdot E_{p,q})E_{p,q}^{-1} = \frac{1}{2}[\boldsymbol{x} - (-1)^{p+q}E_{p,q}E_{p,q}\boldsymbol{x}E_{p,q}^{-1}]$
$[\boldsymbol{x}\wedge\boldsymbol{y} - (\boldsymbol{x} \cdot \boldsymbol{K})(\boldsymbol{y} \cdot \boldsymbol{K})] \times \boldsymbol{K} = 0$
$\boldsymbol{K}_i = \frac{1}{2}\boldsymbol{F}_{ij} = e_i\bar{e}_i$
$\boldsymbol{E}_{ij} = e_ie_j - \bar{e}_i\bar{e}_j \qquad (i < j = 1...n)$
$\boldsymbol{F}_{ij} = e_i\bar{e}_j - \bar{e}_ie_j \qquad (i < j = 1...n)$
$\boldsymbol{B} = \sum_{i<j} \alpha^{ij}e_{ij} + \sum_{i<j} \beta^{ij}\bar{e}_{ij}$
$\boldsymbol{R} = e^B = e^{\frac{1}{2}\sum_{i<j}\alpha^{ij}E_{ij}}e^{\frac{1}{2}\sum_{i<j}\beta^{ij}F_{ij}}e^{\frac{1}{2}\sum_{i<j}\gamma^{ij}K_{ij}}$
$\boldsymbol{E}_{ij} = e_ie_j - \bar{e}_i\bar{e}_j$
$\boldsymbol{F}_{ij} = e_i\bar{e}_j + \bar{e}_ie_j$
$G_{3,3} = G_N \otimes G_{\overline{N}} = gen\{w_1, w_2, w_3, \overline{w}_1, \overline{w}_2, \overline{w}_3\}$

Translator

$$
\begin{aligned}
\boldsymbol{T} = e^{\boldsymbol{B}_t} &= e^{\frac{t_3\bar{e}_1e_2 - t_2\bar{e}_1e_3 + t_1\bar{e}_2e_3}{2}} \\
&= e^{\frac{t_3\bar{w}_1\wedge\bar{w}_2 - t_2\bar{w}_1\wedge\bar{w}_3 + t_1\bar{w}_2\wedge\bar{w}_3}{2}} \\
&= 1 + \frac{1}{2}(t_3\bar{e}_1e_2 - t_2\bar{e}_1e_3 + t_1\bar{e}_2e_3) \\
&= 1 + \frac{1}{2}(t_3\bar{w}_1\wedge w_2 - t_2\bar{w}_1\wedge\bar{w}_3 + t_1\bar{w}_2\wedge\bar{w}_3)
\end{aligned}
$$

Perspector

$$
\begin{aligned}
\boldsymbol{P} = e^{\boldsymbol{B}_f} &= e^{\frac{-f_3e_1\bar{e}_2 + f_2e_1\bar{e}_3 - f_1e_2\bar{e}_3}{2}} \\
&= e^{\frac{-f_3w_1\wedge w_2 + f_2w_1\wedge w_3 - f_1w_2\wedge w_3}{2}} \\
&= 1 + \frac{1}{2}(-f_3e_1\bar{e}_2 + f_2e_1\bar{e}_3 - f_1e_2\bar{e}_3) \\
&= 1 + \frac{1}{2}(-f_3w_1\wedge w_2 + f_2w_1\wedge w_3 - f_1w_2\wedge w_3)
\end{aligned}
$$

Rotator

$$
\begin{aligned}
\boldsymbol{R} &= e^{\boldsymbol{B}_R} \\
&= e^{\frac{(-e_2e_3+\bar{e}_2\bar{e}_3)+(-e_1e_3+\bar{e}_1\bar{e}_3)+(-e_1e_2+\bar{e}_1\bar{e}_2)}{2}} \\
&= e^{\frac{(w_2\wedge\bar{w}_3-\bar{w}_2\wedge w_3)+(w_1\wedge\bar{w}_3-\bar{w}_1\wedge w_3)+(w_1\wedge\bar{w}_2-\bar{w}_1\wedge w_2)}{2}}
\end{aligned}
$$

Lorentor

$$
\begin{aligned}
\boldsymbol{R}_L &= e^{\boldsymbol{B}_L} \\
&= e^{\frac{(e_2\wedge e_3-\bar{e}_2\wedge\bar{e}_3)+(e_1\wedge e_3-\bar{e}_1\wedge\bar{e}_3)+(e_1\wedge e_2-\bar{e}_1\wedge\bar{e}_2)}{2}} \\
&= e^{\frac{(\bar{w}_2\wedge w_3-w_2\wedge\bar{w}_3)+(\bar{w}_1\wedge w_3-w_1\wedge\bar{w}_3)+(\bar{w}_1\wedge w_2-w_1\wedge\bar{w}_2)}{2}}
\end{aligned}
$$

Sheartor

$$
\begin{aligned}
\boldsymbol{S} &= e^{\boldsymbol{B}_S} \\
&= e^{\frac{a_{23}(e_2\wedge e_3-\bar{e}_3\wedge\bar{e}_2)+a_{13}(e_1\wedge e_3-\bar{e}_3\wedge\bar{e}_1)+a_{12}(e_1\wedge e_2-\bar{e}_2\wedge\bar{e}_1)}{2}} \\
&\qquad e^{\frac{a_{32}(e_3\wedge e_2-\bar{e}_2\wedge\bar{e}_3)+a_{31}(e_3\wedge e_1-\bar{e}_1\wedge\bar{e}_3)+a_{21}(e_2\wedge e_1-\bar{e}_1\wedge\bar{e}_2)}{2}} \\
&= e^{\frac{a_{23}(\bar{w}_2\wedge w_3-w_3\wedge\bar{w}_2)+a_{13}(\bar{w}_1\wedge w_3-w_3\bar{w}_1)+a_{12}(\bar{w}_1\wedge w_2-w_2\wedge\bar{w}_1)}{2}} \\
&\qquad e^{\frac{a_{32}(\bar{w}_3\wedge w_2-w_2\wedge\bar{w}_3)+a_{31}(\bar{w}_3\wedge w_1-w_1\bar{w}_3)+a_{21}(\bar{w}_2\wedge w_1-w_1\wedge\bar{w}_2)}{2}}
\end{aligned}
$$

Dilator

$$
\begin{aligned}
\boldsymbol{D} = &\ e^{\boldsymbol{B}_D} \\
= &\ e^{\frac{1}{2}\{(1/0,\pm)\bar{e}_1\wedge e_1+(0/1,\mp)\bar{e}_2\wedge e_2+(0/1,\mp)\bar{e}_3\wedge e_3\}} \\
&\ e^{\frac{1}{2}\{(-1/0,\pm)e_1\wedge\bar{e}_1+(0/-1,\mp)e_2\wedge\bar{e}_2+(0/-1,\mp)e_3\wedge\bar{e}_3\}} \\
= &\ e^{\frac{1}{2}\{(1/0,\pm)\bar{w}_1\wedge w_1+(0/1,\mp)\bar{w}_2\wedge w_2+(0/1,\mp)\bar{w}_3\wedge w_3\}} \\
&\ e^{\frac{1}{2}\{(-1/0,\pm)w_1\wedge\bar{w}_1+(0/-1,\mp)w_2\wedge\bar{w}_2+(0/-1,\mp)w_3\wedge\bar{w}_3\}}
\end{aligned}
$$

$\mathcal{H}_w^{p,q}(R^{p+1,q+1}) = \{\frac{1}{2}\boldsymbol{x}_h\overline{w}\boldsymbol{x}_h|\boldsymbol{x}_h \in \mathcal{A}_w(\boldsymbol{R}^{p,q})\} \in R^{p+1,q+1}$ horosphere
$\mathcal{A}_w^2 = \{\boldsymbol{x}_h|\boldsymbol{x}_h = \boldsymbol{x} + w, \boldsymbol{x} \in \boldsymbol{R}^2\}$ affine plane of $\mathbb{R}^2$
$\mathcal{H}_w^2 = \{\boldsymbol{x}_c = \frac{1}{2}\boldsymbol{x}_h\overline{w}\boldsymbol{x}_h|\boldsymbol{x}_h \in \mathcal{A}_w^2\}$ the horosphere $\mathbb{R}^2$

Generators of the 2D affine Lie group

$$
\begin{aligned}
\boldsymbol{\mathcal{L}}_x &= bivector(\mathcal{L}_x) = w_1\wedge\bar{w}_3, \\
\boldsymbol{\mathcal{L}}_y &= bivector(\mathcal{L}_y) = w_2\wedge\bar{w}_3, \\
\boldsymbol{\mathcal{L}}_s &= bivector(\mathcal{L}_s) = w_1\wedge\bar{w}_1 + w_2\wedge\bar{w}_2, \\
\boldsymbol{\mathcal{L}}_r &= bivector(\mathcal{L}_r) = w_2\wedge\bar{w}_1 - w_1\wedge\bar{w}_2, \\
\boldsymbol{\mathcal{L}}_b &= bivector(\mathcal{L}_b) = w_1\wedge\bar{w}_1 - w_2\wedge\bar{w}_2, \\
\boldsymbol{\mathcal{L}}_B &= bivector(\mathcal{L}_\phi) = w_1\wedge\bar{w}_2 + w_2\wedge\bar{w}_1
\end{aligned}
$$

Reflections, Rotations

$\boldsymbol{X}_r \rightarrow (-1)^r n\boldsymbol{X}_r n$
$\boldsymbol{X}_r \rightarrow \boldsymbol{R}\boldsymbol{X}_r\widetilde{\boldsymbol{R}}$
$\boldsymbol{R} = e^{-\boldsymbol{B}/2} = kl\cdots mn$
$\boldsymbol{R}\widetilde{\boldsymbol{R}} = \widetilde{\boldsymbol{R}}\boldsymbol{R} = 1$

Motor Algebra: Rotors, Translators, and Motors (Chap. 6)

$\boldsymbol{T} = 1 + I\frac{1}{2}\boldsymbol{t} = e^{-\frac{t}{2}I}$
$\boldsymbol{R} = cos(\frac{\theta}{2}) + sin(\frac{\theta}{2})\boldsymbol{n}$
$\boldsymbol{R}_s = cos(\frac{\theta}{2}) + sin(\frac{\theta}{2})\boldsymbol{l}$ where $\boldsymbol{l} = \boldsymbol{n} + I\boldsymbol{m}$
$\boldsymbol{M} = \boldsymbol{T}_s\boldsymbol{R}_s = (1 + I\frac{\boldsymbol{t}_s}{2})e^{\boldsymbol{l}\frac{\theta}{2}} = e^{\boldsymbol{l}\frac{\theta}{2}+I\frac{\boldsymbol{t}_s}{2}}$
$\boldsymbol{M} = \boldsymbol{T}_s\boldsymbol{R}_s = (1 + I\frac{\boldsymbol{t}_s}{2})\mathbf{R}_s = \boldsymbol{R}_s + I\frac{\boldsymbol{t}_s}{2}\boldsymbol{R}_s = \boldsymbol{R}_s + I\boldsymbol{R}_s'$

$$
\begin{aligned}
&\boldsymbol{M} = cos(\tfrac{\theta}{2} + I\tfrac{d}{2}) + sin(\tfrac{\theta}{2} + I\tfrac{d}{2})\boldsymbol{l} \\
&|\boldsymbol{M}| = \boldsymbol{M}\widetilde{\boldsymbol{M}} = 1 \\
&\boldsymbol{R}_s\widetilde{\boldsymbol{R}}_s = 1,\ \widetilde{\boldsymbol{R}}_s\boldsymbol{R}'_s + \widetilde{\boldsymbol{R}}'_s\boldsymbol{R}_s = 0,\ \boldsymbol{t}_s = 2\boldsymbol{R}'_s\widetilde{\boldsymbol{R}}_s \\
&\boldsymbol{M} = (a_0 + \boldsymbol{a}) + I(b_0 + \boldsymbol{b}) = \boldsymbol{T}_s\boldsymbol{R}_s \\
&\widetilde{\boldsymbol{M}} = (a_0 - \boldsymbol{a}) + I(b_0 - \boldsymbol{b}) = \widetilde{\boldsymbol{R}}_s\widetilde{\boldsymbol{T}}_s \\
&\bar{\boldsymbol{M}} = (a_0 + \boldsymbol{a}) - I(b_0 + \boldsymbol{b}) = \boldsymbol{R}_s\widetilde{\boldsymbol{T}}_s \\
&\widetilde{\bar{\boldsymbol{M}}} = (a_0 - \boldsymbol{a}) - I(b_0 - \boldsymbol{b}) = \widetilde{\boldsymbol{R}}_s\boldsymbol{T}_s \\
&a_0 = \tfrac{1}{4}(\boldsymbol{M} + \widetilde{\boldsymbol{M}} + \bar{\boldsymbol{M}} + \widetilde{\bar{\boldsymbol{M}}}) \\
&Ib_0 = \tfrac{1}{4}(\boldsymbol{M} + \widetilde{\boldsymbol{M}} - \bar{\boldsymbol{M}} - \widetilde{\bar{\boldsymbol{M}}}) \\
&a = \tfrac{1}{4}(\boldsymbol{M} - \widetilde{\boldsymbol{M}} + \bar{\boldsymbol{M}} - \widetilde{\bar{\boldsymbol{M}}}) \\
&I\boldsymbol{b} = \tfrac{1}{4}(\boldsymbol{M} - \widetilde{\boldsymbol{M}} - \bar{\boldsymbol{M}} + \widetilde{\bar{\boldsymbol{M}}})
\end{aligned}
$$

Exponentials of Bivectors

Exponentials of bivectors can be expanded in power series [197] and factorized in a compact manner in terms of trigonometric functions (see [78] Appendix A). For a bivector $\boldsymbol{B}^2 = 0$

$$e^{\tau \boldsymbol{B}} = 1 + \tau \boldsymbol{B},$$

for $\boldsymbol{B}^2 = -1$

$$e^{\theta \boldsymbol{B}} = \cos(\theta) + \sin\theta \boldsymbol{B},$$

for a bivector $\boldsymbol{B}^3 = -\boldsymbol{B}$

$$
\begin{aligned}
e^{\theta \boldsymbol{B}} &= 1 + \frac{1}{1!}\theta\boldsymbol{B} + \frac{1}{2!}\theta^2\boldsymbol{B}^2 + \frac{1}{3!}\theta^3\boldsymbol{B}^3 + \frac{1}{4!}\theta^4\boldsymbol{B}^4 + \cdots \\
&= 1 + \boldsymbol{B}^2 + (\frac{1}{1!}\theta\boldsymbol{B} - \frac{1}{3!}\theta^3 + \cdots)\boldsymbol{B} - (1 - \frac{1}{2!}\theta^2 + \frac{1}{4!}\theta^4 + \cdots)\boldsymbol{B}^2 \\
&= 1 + sin(\theta)\boldsymbol{B} + (1 - cos(\theta))\boldsymbol{B}^2,
\end{aligned}
$$

for a bivector $\boldsymbol{B}^2 = 1$

$$e^{\alpha \boldsymbol{B}} = \cosh(\alpha) + \sinh(\alpha)\boldsymbol{B}$$

for a bivector $\boldsymbol{B}^3 = \boldsymbol{B}$

$$
\begin{aligned}
e^{\alpha \boldsymbol{B}} &= 1 + \frac{1}{1!}\alpha\boldsymbol{B} + \frac{1}{2!}\alpha^2\boldsymbol{B}^2 + \frac{1}{3!}\alpha^3\boldsymbol{B}^3 + \frac{1}{4!}\alpha^4\boldsymbol{B}^4 + \cdots \\
&= 1 - \boldsymbol{B}^2 + (\frac{1}{1!}\alpha + \frac{1}{3!}\alpha^3 + \cdots)\boldsymbol{B} + (1 + \frac{1}{2!}\alpha^2 + \frac{1}{4!}\alpha^4 + \cdots)\boldsymbol{B}^2 \\
&= 1 + \sinh(\alpha)\boldsymbol{B} + (cos(\alpha) - 1)\boldsymbol{B}^2,
\end{aligned}
$$

Incidence Algebra (Chap. 12)

$$PI^{-1} = \alpha II^{-1} = \alpha \equiv [P]$$
$$[\boldsymbol{x}_1\boldsymbol{x}_2\boldsymbol{x}_3...\boldsymbol{x}_n] = [\boldsymbol{x}_1\wedge\boldsymbol{x}_2\wedge\boldsymbol{x}_3\wedge...\wedge\boldsymbol{x}_n] = (\boldsymbol{x}_1\wedge\boldsymbol{x}_2\wedge\boldsymbol{x}_3\wedge...\wedge\boldsymbol{x}_n)I^{-1}$$
$$P = \mathbf{X}_1\wedge\mathbf{X}_2\wedge\mathbf{X}_3\wedge\mathbf{X}_4 = W_1W_2W_3W_4\langle(1+\boldsymbol{x}_1)(1-\boldsymbol{x}_2)(1+\boldsymbol{x}_3)(1-\boldsymbol{x}_4)\rangle_4$$
$$A^* = AI^{-1}$$
$$A\cdot B^* = (A\wedge B)^* = [A\wedge B]$$
$$J = A\cup B = A\wedge B$$
$$(A\cap B)^* = A^*\cup B^* \text{ for If } A = A'C \text{ and } B = B'C$$
$$A\cap B = (A^*\cdot B)$$

Direct distance

$$\overline{e}\cdot A^h = \overline{e}\cdot(a_1^h\wedge a_2^h\wedge\ldots\wedge a_k^h) = (a_2-a_1)\wedge(a_3-a_2)\wedge\ldots\wedge(a_k-a_{k-1})$$
$$d[a_1^h\wedge\ldots a_k^h, b^h] \equiv [\{\overline{e}(\cdot a_1^h\wedge\ldots\wedge a_k^h)\}\,(\overline{e}\cdot b^h)]^{-1}[\overline{e}\cdot(a_1^h\wedge\ldots\wedge a_k^h\wedge b^h)] = [(a_2-a_1)\wedge\ldots\wedge(a_k-a_{k-1})]^{-1}[(a_2-a_1)\wedge\ldots\wedge(a_k-a_{k-1})\wedge(b-a_k)]$$
$$d[a_1^h\wedge\ldots\wedge a_r^h, b_1^h\wedge\ldots\wedge b_s^h] \equiv \{\overline{e}\cdot(a_1^h\wedge\ldots\wedge a_r^h)\}\wedge\{\overline{e}\cdot(b_1^h\wedge b_2^h\wedge\ldots\ldots\wedge b_s^h)\}]^{-1} \quad [\overline{e}\cdot(a_1^h\wedge\ldots\wedge a_r^h\wedge b_1^h\wedge\ldots\wedge b_s^h)] = [(a_2-a_1)\wedge\ldots\wedge(a_r-a_{r-1})\wedge(b_2-b_1)\wedge\ldots\wedge(b_s-b_{s-1})]^{-1}[(a_2-a_1)\wedge\ldots\wedge(a_r-a_{r-1})\wedge(b_1-a_r)\wedge(b_2-b_1)\wedge\ldots\wedge(b_s-b_{s-1})]$$

Projective Geometry (Chap. 12)

$$\mathbf{X}e_{n+1} = \mathbf{X}\cdot e_{n+1} + \mathbf{X}\wedge e_{n+1} = \mathbf{X}\cdot e_{n+1}(1+\tfrac{\mathbf{X}\wedge e_{n+1}}{\mathbf{X}e_{n+1}}) \qquad \text{projective split}$$
$$X\gamma_4 = \mathbf{X}\cdot\gamma_4 + \mathbf{X}\wedge\gamma_4 = X_4\left(1+\tfrac{\mathbf{X}\wedge\gamma_4}{X_4}\right) \equiv X_4(1+\boldsymbol{x})$$
$$L\cap\Phi = (\mathbf{X}_1\wedge\mathbf{X}_2)\cap(\mathbf{Y}_1\wedge\mathbf{Y}_2\wedge\mathbf{Y}_3) = L^*\cdot\Phi$$
$$L\cap\Phi = [\mathbf{X}_1\mathbf{X}_2\mathbf{Y}_2\mathbf{Y}_3]\mathbf{Y}_1 + [\mathbf{X}_1\mathbf{X}_2\mathbf{Y}_3\mathbf{Y}_1]\mathbf{Y}_2 + [\mathbf{X}_1\mathbf{X}_2\mathbf{Y}_1\mathbf{Y}_2]\mathbf{Y}_3$$
$$L_1\wedge L_2 = 0 \quad \text{coplanar lines}$$
$$L_1\cap L_2 = [\mathbf{X}_1\mathbf{X}_2\mathbf{Y}_1]\mathbf{Y}_2 - [\mathbf{X}_1\mathbf{X}_2\mathbf{Y}_2]\mathbf{Y}_1 \quad \text{intersecting lines}$$
$$L = \Phi_1\cap\Phi_2 = (\mathbf{X}_1\wedge\mathbf{X}_2\wedge\mathbf{X}_3)\cap(\mathbf{Y}_1\wedge\mathbf{Y}_2\wedge\mathbf{Y}_3) = [\mathbf{X}_1\mathbf{X}_2\mathbf{X}_3\mathbf{Y}_1](\mathbf{Y}_2\wedge\mathbf{Y}_3) + [\mathbf{X}_1\mathbf{X}_2\mathbf{X}_3\mathbf{Y}_2](\mathbf{Y}_3\wedge\mathbf{Y}_1) + [\mathbf{X}_1\mathbf{X}_2\mathbf{X}_3\mathbf{Y}_3](\mathbf{Y}_1\wedge\mathbf{Y}_2)$$

Projective Invariants

$$Inv_1 = \frac{(\mathbf{X}_3\wedge\mathbf{X}_1)I_2^{-1}(\mathbf{X}_4\wedge\mathbf{X}_2)I_2^{-1}}{(\mathbf{X}_4\wedge\mathbf{X}_1)I_2^{-1}(\mathbf{X}_3\wedge\mathbf{X}_2)I_2^{-1}} = \frac{(t_3-t_1)(t_4-t_2)}{(t_4-t_1)(t_3-t_2)}$$
$$Inv_2 = \frac{(\mathbf{X}_5\wedge\mathbf{X}_4\wedge\mathbf{X}_3)I_3^{-1}(\mathbf{X}_5\wedge\mathbf{X}_2\wedge\mathbf{X}_1)I_3^{-1}}{(\mathbf{X}_5\wedge\mathbf{X}_1\wedge\mathbf{X}_3)I_3^{-1}(\mathbf{X}_5\wedge\mathbf{X}_2\wedge\mathbf{X}_4)I_3^{-1}} = \frac{A_{543}A_{521}}{A_{513}A_{524}}$$
$$Inv_3 = \frac{(\mathbf{X}_1\wedge\mathbf{X}_2\wedge\mathbf{X}_3\wedge\mathbf{X}_4)I_4^{-1}(\mathbf{X}_4\wedge\mathbf{X}_5\wedge\mathbf{X}_2\wedge\mathbf{X}_6)I_4^{-1}}{(\mathbf{X}_1\wedge\mathbf{X}_2\wedge\mathbf{X}_4\wedge\mathbf{X}_5)I_4^{-1}(\mathbf{X}_3\wedge\mathbf{X}_4\wedge\mathbf{X}_2\wedge\mathbf{X}_6)I_4^{-1}} = \frac{V_{1234}V_{4526}}{V_{1245}V_{3426}}$$

Conformal Geometric Algebra (Chap. 8)

$$e_i^2 = 1,\ i = 1..,n;\ e_0 = \frac{(e_- - e_+)}{2},\ e_\infty = e_- + e_+$$
$$E = e_\infty\wedge e_0 = e_+\wedge e_- = e_+e_- \quad \text{Minkowsky plane}$$
$$\boldsymbol{x}_c = \mathbf{x}_e + \alpha e_0 + \beta e_\infty \quad \text{conformal split}$$
$$P_E(\boldsymbol{x}_c) = (\boldsymbol{x}_c\cdot\boldsymbol{E})\boldsymbol{E} = \alpha e_0 + \beta e_\infty \in \mathbb{R}^{1,1}$$
$$P_E^\perp(\boldsymbol{x}_c) = (\boldsymbol{x}_c\cdot\boldsymbol{E}^*)\widetilde{\boldsymbol{E}^*} = (\boldsymbol{x}_c\wedge\boldsymbol{E})\boldsymbol{E} = \mathbf{x}_e \in \mathbb{R}^n$$
$$\boldsymbol{x}_c = P_E(\boldsymbol{x}_c) + P_E^\perp(\boldsymbol{x}_c).$$
$$\boldsymbol{x}_c = \boldsymbol{x}_c\boldsymbol{E}^2 = (\boldsymbol{x}_c\wedge\boldsymbol{E} + \boldsymbol{x}_c\cdot\boldsymbol{E})\boldsymbol{E} = (\boldsymbol{x}_c\wedge\boldsymbol{E})\boldsymbol{E} + (\boldsymbol{x}_c\cdot\boldsymbol{E})\boldsymbol{E}$$

$\boldsymbol{x}_c = (\boldsymbol{x}_c \wedge \boldsymbol{E})\boldsymbol{E} + (\boldsymbol{x}_c \cdot \boldsymbol{E})\boldsymbol{E} = \mathbf{x}_e + e_0 + \frac{1}{2}(k_1 + k_2)e_\infty = \mathbf{x}_e + \frac{1}{2}\mathbf{x}_e^2 e_\infty + e_0$
$\boldsymbol{x}_c = \mathbf{x} + \frac{1}{2}\mathbf{x}^2 e_\infty + e_0$ point
$\boldsymbol{x}_c^* = \boldsymbol{s}_1 \wedge \boldsymbol{s}_2 \wedge \boldsymbol{s}_3 \wedge \boldsymbol{s}_4$ dual point
$\boldsymbol{s} = \mathbf{p} + \frac{1}{2}(\mathbf{p}^2 - \rho^2)e_\infty + e_0$ sphere
$\boldsymbol{s}^* = \boldsymbol{a} \wedge \boldsymbol{b} \wedge \boldsymbol{c} \wedge \boldsymbol{d}$ dual sphere
$\boldsymbol{x}_c \wedge \boldsymbol{s}^* = 0$
$\pi = \boldsymbol{n} I_E - d e_\infty$ $\boldsymbol{n} = (\boldsymbol{a} - \boldsymbol{b}) \wedge (\boldsymbol{a} - \boldsymbol{c})$ $d = (\boldsymbol{a} \wedge \boldsymbol{b} \wedge \boldsymbol{c}) I_E$ plane
$\pi^* = e_\infty \wedge \boldsymbol{a} \wedge \boldsymbol{b} \wedge \boldsymbol{c}$ dual plane
$\boldsymbol{L} = \pi_1 \wedge \pi_2$, $\boldsymbol{L} = \boldsymbol{n} I_E - e_\infty \boldsymbol{m} I_E$, $\boldsymbol{n} = (\boldsymbol{a} - \boldsymbol{b})$, $\boldsymbol{m} = (\boldsymbol{a} \wedge \boldsymbol{b})$ line
$\boldsymbol{L}^* = e_\infty \wedge \boldsymbol{a} \wedge \boldsymbol{b}$ dual line
$z = \boldsymbol{s}_1 \wedge \boldsymbol{s}_2 = \boldsymbol{s}_1 \wedge \pi_2$, circle
$z^* = \boldsymbol{a} \wedge \boldsymbol{b} \wedge \boldsymbol{c}$, dual circle
$\boldsymbol{PP} = \boldsymbol{s}_1 \wedge \boldsymbol{s}_2 \wedge \boldsymbol{s}_3$, $\boldsymbol{PP} = \boldsymbol{s} \wedge \boldsymbol{L}$ point pair
$\boldsymbol{PP}^* = \boldsymbol{a} \wedge \boldsymbol{b}$ dual point pair
$\boldsymbol{s} = \boldsymbol{A}_r \boldsymbol{A}_{r+1}^{-1}$
$\boldsymbol{s} = z\pi^{-1}$, $\boldsymbol{s} = \boldsymbol{PPL}^{-1}$, $\boldsymbol{PP} = \boldsymbol{sL} = \boldsymbol{s} \wedge \boldsymbol{L}$

Lie Algebra of the Conformal Group

$$[\boldsymbol{x} \wedge \boldsymbol{y} - (\boldsymbol{x} \cdot \boldsymbol{E})(\boldsymbol{y} \cdot \boldsymbol{E})] \times \boldsymbol{E} = 0$$
$$\boldsymbol{E}_i = e_+ e_-,$$
$$\boldsymbol{B}_{ij} = e_i e_j \qquad (i < j = 1...n),$$
$$\bar{\boldsymbol{E}}_{ij} = e_i e_\pm$$

Conformal Transformations

$s(\boldsymbol{x}_c) = -\boldsymbol{s}\boldsymbol{x}_c\boldsymbol{s}^{-1}$ inversion
$\boldsymbol{x}' = -\pi \boldsymbol{x} \pi^{-1}$ reflection
$\boldsymbol{K}_{\mathbf{b}} = e_+ \boldsymbol{T}_{\mathbf{b}} e_+ = (e_\infty - e_0)(1 + \mathbf{b} e_\infty)(e_\infty - e_0) = 1 + \mathbf{b} e_0$ transversions
$\boldsymbol{T}_a = 1 + \frac{1}{2} a e_\infty = e^{-\frac{a}{2} e_\infty}$ translator
$\boldsymbol{R}_\theta = n_2 n_1 = cos(\frac{\theta}{2}) - sin(\frac{\theta}{2}) l = e^{-\frac{\theta}{2} l}$ rotor
$\boldsymbol{M}_\theta = \boldsymbol{T}\boldsymbol{R}\widetilde{\boldsymbol{T}} = cos(\frac{\theta}{2}) - sin(\frac{\theta}{2}) L = e^{-\frac{\theta}{2} L}$ motor
$Q' = \prod_{i=1}^{n} M_i Q \prod_{i=1}^{n} \widetilde{M}_{n-i+1}$
$\boldsymbol{D}_\rho = (1 + \boldsymbol{E})\rho + (1 - \boldsymbol{E})\rho^{-1} = e^{\boldsymbol{E}\phi}$, dilator
$\boldsymbol{E}(\mathbf{x}_e + \mathbf{x}_e^2 e_\infty + e_0)\boldsymbol{E} = -(-\mathbf{x}_e + \frac{1}{2}\mathbf{x}_e^2 e_\infty + e_0)$, involution
$\boldsymbol{G} = \boldsymbol{K}_{\mathbf{b}} \boldsymbol{T}_{\mathbf{a}} \boldsymbol{R}_\alpha$ conformal transformation
$g(\boldsymbol{x}_c) = \boldsymbol{G}\boldsymbol{x}_c(\boldsymbol{G}^*)^{-1} = \sigma \boldsymbol{x}_c'$

Differential Kinematics

$$x_p' = \prod_{i=1}^{n} M_i x_p \prod_{i=1}^{n} \widetilde{M}_{n-i+1}$$
$$dx_p' = \sum_{j=1}^{n} \left[x_p' \cdot L_j' \right] dq_j$$
$$dL_p' = \sum_{j=1}^{n} \left[L_p' \underline{\times} L_j' \right] dq_j$$

Dynamics

$$\mathcal{M}\ddot{q} + C\dot{q} + G = \tau$$
$$\mathcal{M} = V^T mV + \delta I$$
$$C = V^T m\dot{V}$$
$$G = V^T m\dot{V} = V^T ma$$

$$V = \begin{pmatrix} x_1' & 0 & \cdots & 0 \\ 0 & x_2' & \cdots & 0 \\ \vdots & \vdots & \ddots & \vdots \\ 0 & 0 & \cdots & x_n' \end{pmatrix} \begin{pmatrix} L_1' & 0 & \cdots & 0 \\ L_1' & L_2' & \cdots & 0 \\ \vdots & \vdots & \ddots & \vdots \\ L_1' & L_2' & \cdots & L_n' \end{pmatrix} = XL$$

$$(V^T mV + \delta I)\ddot{q} + V^T m\dot{V}\dot{q} + V^T F = \tau$$
$$\delta I\ddot{q} + V^T(mV\ddot{q} + m\dot{V}\dot{q} + F) = \tau$$
$$\delta I\ddot{q} + V^T m(V\ddot{q} + \dot{V}\dot{q} + a) = \tau$$

The Geometric Algebras $G^+_{6,0,2}$, $G_{6,3}$, $G^+_{9,3}$, $G^+_{6,0,6}$ (Chap. 9)

The double motor algebra $G^+_{6,0,2} = \texttt{gen}\{1, e_ie_j, e_ie, f_if_j, f_if, I\}$

screw

$$\boldsymbol{s} = w_x e_2e_3 + w_y e_3e_1 + w_z e_1e_2 + v_x e_1e + v_y e_2e + v_z e_3e = \boldsymbol{w} + I_e\boldsymbol{v}$$

co-screw

$$\boldsymbol{s}_f = w_x f_2f_3 + w_y f_3f_1 + w_z f_1f_2 + v_x f_1f + v_y f_2f + v_z f_3f = \boldsymbol{w} + I_f\boldsymbol{v}$$

The Lie bracket

$$[\boldsymbol{s}_1, \boldsymbol{s}_2] = \tfrac{1}{2}(\boldsymbol{s}_1\boldsymbol{s}_2 - \boldsymbol{s}_2\boldsymbol{s}_1)$$

wrenche

$$\boldsymbol{w} = \tau_x e_2e_3 + \tau_y e_3e_1 + \tau_z e_1e_2 + F_x f_3f_1 + F_y f_3f_1 + F_z f_1f_2 = \boldsymbol{\tau} + I_e\boldsymbol{F}$$

Kinetic energy

$$E_K = \frac{1}{2}\mathcal{P}\cdot\boldsymbol{s}$$
$$= \frac{1}{2}(\boldsymbol{w}\cdot\boldsymbol{q} + \boldsymbol{v}\cdot\boldsymbol{p}) = \frac{1}{2}(w_xq_x + w_ldyq_y + w_zq_z + v_xp_x + v_yp_y + v_zp_z)$$

Inertia Matrix

$$\mathtt{N} = \begin{bmatrix} \mathtt{I} & m[\mathtt{c}]_\times \\ m[\mathtt{c}]^{\mathtt{T}}_\times & m\mathtt{I}_3 \end{bmatrix}$$

$$\mathtt{N}: \boldsymbol{s} \to \mathcal{P} = \boldsymbol{q} + I_f\boldsymbol{p} = (\mathtt{I}\boldsymbol{w} + \mathtt{m}(\boldsymbol{c}\times\boldsymbol{v})) + \mathtt{I_e m}(\boldsymbol{c}\times\boldsymbol{w} + \boldsymbol{v})$$

Shuffle product

$$\boldsymbol{A}\vee\boldsymbol{B} = \sum_\sigma sign(\sigma)\det(a_{\sigma(1)},\cdots,a_{\sigma(n-k)},b_1,\cdots,b_k)a_{\sigma(n-k+1)}\wedge\cdots\wedge a_{\sigma(j)}$$

co-screw

$$\{\boldsymbol{s}, \mathcal{P}\} = \left\{ \begin{pmatrix} \boldsymbol{w} \\ \boldsymbol{v} \end{pmatrix}, \begin{pmatrix} \boldsymbol{q} \\ \boldsymbol{p} \end{pmatrix} \right\} = \begin{pmatrix} \boldsymbol{w} \times \boldsymbol{q} + \boldsymbol{v} \times \boldsymbol{p} \\ \boldsymbol{w} \times \boldsymbol{p} \end{pmatrix}$$

Dynamic equation
$\mathrm{N}\frac{d}{dt}\boldsymbol{s} + \{\boldsymbol{s}, N\boldsymbol{s}\} = \mathrm{N}\frac{d}{dt}\boldsymbol{s} + \{\boldsymbol{s}, \mathcal{P}\} = \boldsymbol{\tau}$

Inertia in terms of bivector

$$N = d_x f_1 f e_1 e + d_y f_2 f e_2 e + d_z f_3 f_2 e_3 e + m f_2 f_3 e_2 e_3 + m f_3 f_1 e_3 e_1 + m f_1 f_2 e_1 e_2,$$

$\boldsymbol{N} \rightarrow (\boldsymbol{M}\boldsymbol{M}_f)\boldsymbol{N}(\widetilde{\boldsymbol{M}}_f\widetilde{\boldsymbol{M}})$
$\boldsymbol{s} \rightarrow (\boldsymbol{M}\boldsymbol{M}_f)\boldsymbol{s}(\widetilde{\boldsymbol{M}}_f\widetilde{\boldsymbol{M}})$
$\mathcal{P} \rightarrow (\boldsymbol{M}\boldsymbol{M}_f)\mathcal{P}(\widetilde{\boldsymbol{M}}_f\widetilde{\boldsymbol{M}})$

$$\begin{aligned} \{\boldsymbol{s}_f, \mathcal{P}\} &= \frac{1}{2}(\boldsymbol{s}_f\mathcal{P} - \mathcal{P}\boldsymbol{s}_f) \\ \{\boldsymbol{s}_f, \boldsymbol{I}_f \vee (\boldsymbol{N}\wedge\boldsymbol{s})\} &= \frac{1}{2}(\boldsymbol{s}_f(\boldsymbol{I}_f \vee (\boldsymbol{N}\wedge\boldsymbol{s})) - (\boldsymbol{I}_f \vee (\boldsymbol{N}\wedge\boldsymbol{s}))\boldsymbol{s}_f) \end{aligned}$$

Dynamic equation
$\boldsymbol{N}\wedge\dot{\boldsymbol{s}} + \frac{1}{2}((\boldsymbol{s}_f\boldsymbol{N} - \boldsymbol{N}\boldsymbol{s}_f)\wedge\boldsymbol{s}) = \boldsymbol{\tau}\boldsymbol{I}_e$
$G_{6,3} = \text{gen}\{1, e_i, e_ie_j, e_ie_je_k, ..., I = e_1e_2...e_9\}$

$$\begin{aligned} e_{01} &= \frac{(e_7 - e_4)}{2}, \quad e_{\infty 1} = e_7 + e_4, \\ e_{02} &= \frac{(e_8 - e_5)}{2}, \quad e_{\infty 2} = e_8 + e_5, \\ e_{03} &= \frac{(e_9 - e_6)}{2}, \quad e_{\infty 3} = e_9 + e_6, \end{aligned}$$

Point
$\boldsymbol{x}_w = xe_1 + ye_2 + ze_3 + \frac{1}{2}(x^2e_{\infty 1} + y^2e_{\infty 2} + z^2e_{\infty 3}) + e_0$

Ellipsoid

$$Q = \frac{p}{a^2}e_1 + \frac{q}{b^2}e_2 + \frac{s}{c^2}e_3 + \frac{1}{2}(\frac{p^2}{a^2} + \frac{q^2}{b^2} + \frac{s^2}{c^2} - 1)e_\infty + (\frac{1}{a^2}e_{01} + \frac{1}{b^2}e_{02} + \frac{1}{c^2}e_{03})$$

$Q^* = QI = \boldsymbol{x}_{w1}\wedge\boldsymbol{x}_{w2}\wedge\boldsymbol{x}_{w3}\wedge\boldsymbol{x}_{w4}\wedge\boldsymbol{x}_{w5}\wedge\boldsymbol{x}_{w6}$

Sphere

$$\begin{aligned} S &= \frac{p}{r^2}e_1 + \frac{q}{r^2}e_2 + \frac{s}{r^2}e_3 + \frac{1}{2}(\frac{p^2}{r^2} + \frac{q^2}{r^2} + \frac{s^2}{r^2} - 1)e_\infty + (\frac{1}{r^2}e_{01} + \frac{1}{r^2}e_{02} + \frac{1}{r^2}e_{03}), \\ &= pe_1 + qe_2 + se_3 + \frac{1}{2}(p^2 + q^2 + s^2 - r^2)e_\infty + (e_{01} + e_{02} + e_{03}), \\ &= \boldsymbol{x}_e + \frac{1}{2}(\boldsymbol{x}_e^2 - r^2)e_\infty + e_0 \end{aligned}$$

Cylinder

$C = \frac{p}{a^2}e_1 + \frac{q}{b^2}e_2 + \frac{1}{2}(\frac{p^2}{a^2} + \frac{q^2}{b^2} - 1)e_\infty + (\frac{1}{a^2}e_{01} + \frac{1}{b^2}e_{02})$

$C^* = CI = \boldsymbol{x}_{w1} \wedge \boldsymbol{x}_{w2} \wedge \boldsymbol{x}_{w3} \wedge \boldsymbol{x}_{w4} \wedge \boldsymbol{x}_{w5} \wedge \boldsymbol{e}_{\infty 3}$

Pair of points

$$PP_{yz} = \frac{p}{a^2}e_1 + \frac{1}{2}(\frac{p^2}{a^2} - 1)e_\infty + \frac{1}{a^2}e_{01},$$
$$= pe_1 + \frac{1}{2}(p^2 - a^2)e_\infty + e_{01}$$

$PP^*_{yz} = \boldsymbol{x}_{w1} \wedge \boldsymbol{x}_{w2} \wedge \boldsymbol{x}_{w3} \wedge \boldsymbol{x}_{w4} \wedge \boldsymbol{e}_{\infty 2} \wedge \boldsymbol{e}_{\infty 3}$

Plane

$\pi = n + de_\infty$

$\pi^* = \boldsymbol{x}_{w1} \wedge \boldsymbol{x}_{w2} \wedge \boldsymbol{x}_{w3} \wedge e_{\infty 1} \wedge \boldsymbol{e}_{\infty 2} \wedge \boldsymbol{e}_{\infty 3} = \boldsymbol{x}_{w1} \wedge \boldsymbol{x}_{w2} \wedge \boldsymbol{x}_{w3} \wedge \pi_\infty$

Line

$L = nI_e + e_\infty m$

$L^* = \pi^* = \boldsymbol{x}_{w1} \wedge \boldsymbol{x}_{w2} \wedge e_{\infty 1} \wedge \boldsymbol{e}_{\infty 2} \wedge \boldsymbol{e}_{\infty 3} = \boldsymbol{x}_{w1} \wedge \boldsymbol{x}_{w2} \wedge \pi_\infty$

Intersection of surfaces

$C^* = S_1^* \wedge S_2^*$

$\boldsymbol{Q} = \boldsymbol{A}_r \boldsymbol{A}_{r+1}^{-1}$

$\boldsymbol{Q} = \boldsymbol{q}\pi^{-1}$ elliposoid intersection of an ellipse and a plane

$\boldsymbol{PP} = \boldsymbol{sL} = \boldsymbol{s} \wedge \boldsymbol{L}$ ellipsoid intersection of a sphere and a line

$\boldsymbol{PP} = \boldsymbol{QL} = \boldsymbol{Q} \wedge \boldsymbol{L}$ elliposoid intersection of an ellipsoid and a line

$\boldsymbol{PP}^* = \boldsymbol{x}_{w_1} \wedge \boldsymbol{x}_{w_2}$

Transformations of $G_{6,3}$

$$\boldsymbol{K}_{\mathbf{b}} = 1 + \mathbf{b}e_o,$$
$$\boldsymbol{T} = e^{-\frac{t}{2}e_\infty} = 1 + \frac{1}{2}\boldsymbol{t}e_\infty,$$
$$\boldsymbol{R} = e^{-\frac{\theta}{2}\boldsymbol{n}} = cos(\frac{\theta}{2}) - sin(\frac{\theta}{2})\boldsymbol{n}$$
$$\boldsymbol{M} = \boldsymbol{TR} = e^{-\frac{\theta}{2}\boldsymbol{L}}$$

Perspector

$$\boldsymbol{P} = e^{\boldsymbol{B}_f} = e^{\frac{f_3e_4e_8 - f_2e_4e_9 + f_1e_5e_9}{2}}$$
$$= e^{\frac{f_3e_{01} \wedge e_{02} - f_2e_{01} \wedge e_{03} + f_1e_{02} \wedge e_{03}}{2}}$$
$$= 1 + \frac{1}{2}(f_3e_4e_8 - f_2e_4e_9 + f_1e_5e_9)$$
$$= 1 + \frac{1}{2}(f_3e_{01} \wedge e_{02} - f_2e_{01} \wedge e_{03} + f_1e_{02} \wedge e_{03})$$

Lorentor

$$
\begin{aligned}
\boldsymbol{R}_L &= e^{\boldsymbol{B}_L} \\
&= e^{\frac{(e_5e_6-e_8e_9)+(e_4e_6-e_7e_9)+(e_4e_5+e_7e_8)}{2}} \\
&= e^{\frac{(e_{02}\wedge e_{\infty 3}-e_{02}\wedge e_{\infty 3})+(e_{\infty 1}\wedge e_{03}-e_{01}e_{\infty 3})+(e_{\infty 1}\wedge e_{02}-e_{01}\wedge e_{\infty 1})}{2}}
\end{aligned}
$$

Sheartor

$$
\begin{aligned}
\boldsymbol{S} &= e^{\boldsymbol{B}_S} \\
&= e^{\frac{a_{23}(e_5e_6-e_9e_8)+a_{13}(e_4e_6-e_9e_7)+a_{12}(e_4e_5-e_8e_7)}{2}} \\
&\qquad e^{-\frac{a_{32}(e_6e_5-e_8e_9)+a_{31}(e_6e_4-e_7e_9)+a_{12}(e_4e_5-e_8e_7)}{2}} \\
&= e^{\frac{a_{23}(e_{\infty 2}\wedge e_{03}-e_{02}\wedge e_{\infty 3}+a_{13}(e_{\infty 1}\wedge e_{03}-e_{01}e_{\infty 3})+a_{12}(e_{\infty 1}\wedge e_{02}-e_{01}\wedge e_{\infty 2})}{2}} \\
&\qquad e^{\frac{a_{32}(e_{\infty 3}\wedge e_{02}-e_{03}\wedge e_{\infty 2})+a_{31}(e_{\infty 3}\wedge e_{01}-e_{03}e_{\infty 1})+a_{21}(e_{\infty 2}\wedge e_{01}-e_{02}\wedge e_{\infty 1}}{2}}
\end{aligned}
$$

$\mathcal{R} = \boldsymbol{S}\mathcal{C}\tilde{\boldsymbol{S}}$ strech a cube into a romboid

Dilator

$$
\begin{aligned}
\boldsymbol{D} &= e^{\boldsymbol{B}_D} \\
&= e^{\frac{(1/0,\pm)e_7e_4+(0/1,\mp)e_8e_5+(0/1,\mp)e_9e_6+(-1/0,\pm)e_4e_7(0/-1,\mp)e_5e_8+(0/-1,\mp)e_6e_9}{2}} \\
&= e^{\frac{(1/0,\pm)e_{\infty 1}e_{01}+(0/1,\mp)e_{\infty 2}e_{02}+(0/1,\mp)e_{infty3}e_{03}}{2}} \\
&\qquad e^{\frac{(-1/0,\pm)e_{01}\wedge e_{\infty 1}+(0/-1,\mp)e_{02}\wedge e_{\infty 2}+(0/-1,\mp)e_{03}\wedge e_{\infty 3}}{2}}
\end{aligned}
$$

$\mathcal{Q} = \boldsymbol{D}\mathcal{S}\tilde{\boldsymbol{D}}$ transform a sphere into an elliposoid

The geometric subalgebras $G^+_{9,3}$ and $G^+_{6,0,6}$

$G^+_{9,3} = \mathtt{gen}\{1, e_ie_j, e_ie, f_if_j, f_if, \cdots, \boldsymbol{I}\}$

Versors for Inertia and m $\mathtt{I}_3$

$$
\begin{aligned}
\boldsymbol{I}_I &= e^{\boldsymbol{B}_I} \\
&= e^{-\frac{i_{11}e\wedge\bar{e}+i_{12}e\wedge\bar{f}+i_{13}e\wedge\bar{g}+i_{21}f\wedge\bar{e}+i_{22}f\wedge\bar{f}+i_{23}f\wedge\bar{g}+i_{31}g\wedge\bar{e}+i_{32}g\wedge\bar{f}+i_{33}g\wedge\bar{g}+m\bar{e}\wedge e+m\bar{f}\wedge f+m\bar{g}\wedge g}{2}}.
\end{aligned}
$$

$$
\boldsymbol{I}_{mc} = e^{\boldsymbol{B}_{mc}} = e^{m\frac{c_3\bar{e}\wedge\bar{f}-c_2\bar{e}\wedge\bar{g}+c_3\bar{f}\wedge\bar{g}-c_1f\wedge g+c_2e\wedge g+c_3e\wedge f}{2}}
$$

Versor for Inertia

$$
\boldsymbol{N} = \boldsymbol{I}_I\boldsymbol{I}_{mc} = e^{\boldsymbol{B}_I}e^{\boldsymbol{B}_{mc}} = e^{\boldsymbol{B}_I+\boldsymbol{B}_{mc}}
$$

Dynamic

$$
\boldsymbol{N}\dot{\boldsymbol{s}}\tilde{\boldsymbol{N}} + \{\boldsymbol{s}, \boldsymbol{N}\boldsymbol{s}\tilde{\boldsymbol{N}}\} = \boldsymbol{N}\dot{\boldsymbol{s}}\tilde{\boldsymbol{N}} + \{\boldsymbol{s}, \mathcal{P}\} = \boldsymbol{\tau}
$$

References

1. Ablamowicks R. eCLIFFORD Software packet using Maple for Clifford algebra. computations. http://math.tntech.edu/rafal
2. Altamirano-Gómez, G.E., and E. Bayro-Corrochano. 2014. Conformal geometric method for voting. In *Proceedings of the Iberoamerican Conference on Pattern Recognition, CIARP 2014*, LNCS, vol. 8827, 802–809.
3. Altamirano-Gómez, G.E., and E. Bayro-Corrochano. 2016. Conformal geometric algebra method for detection of geometric primitives. In *IEEE Xplore Proceedings of the IAPR International Conference on Pattern Recognition, ICPR'2016*, 802–809, Dec 2–8, Cancun.
4. Altaisky, M. 2001. arXiv preprint quant-ph/0107012.
5. Arena, P., R. Caponetto, L. Fortuna, G. Muscato, and M.G. Xibilia. 1996. Quaternionic multilayer perceptrons for chaotic time series prediction. *IEICE Transactions Fundamentals* E79-A(10): 1–6.
6. Ashdown, M.A.J. 1998. Maple code for geometric algebra. http://www.mrao.cam.ac.uk/~maja
7. Balasubramanian, M., J. Polimeni, and E.L. Schwartz. 2002. The V1-V2-V3 complex quasiconformal dipole maps in primate striate and extra-striate cortex. *Neural Networks* 15(10): 1157–1163.
8. Ballard, D.H. 1981. Generalizing the Hough transform to detect arbitrary shapes. *Pattern Recognition, Elsevier* 13(2): 111–122.
9. Baker, S., and S. Nayar. 1998. A theory of catadioptric image formation. In *Proceedings of the International Conference on Computer Vision*, 35–42. Bombay, India.
10. Barnabei, M., A. Brini, and G.-C. Rota. 1985. On the exterior calculus of invariant theory. *Journal of Algebra* 96: 120–160.
11. Barnett, V. 1976. The ordering of multivariate data. *Journal of Royal Statistical Society A* 3: 318–343.
12. Bayro-Corrochano, E., and D. Gonzalez-Aguirre. 2008. Like vision using conformal geometric algebra. In *Proceedings of the International Conference on Robotics and Automation*, ICRA'2008, 1299–1304, May 19–23, Pasadena, CA.
13. Bayro-Corrochano, E. 2001. *Geometric computing for perception action systems*. New York: Springer Verlag.
14. Bayro-Corrochano, E. 2006. Theory and use of the quaternion wavelet transform. *Journal of Mathematical Imaging and Vision* 24: 19–35.
15. Bayro-Corrochano, E. 2003. Modeling the 3D kinematics of the eye in the geometric algebra framework. *Pattern Recognition* 36(12): 2993–3012.
16. Bayro-Corrochano, E. 2010. *Geometric computing for wavelet transforms, robot vision, learning control and action*. London: Springer.

E. Bayro-Corrochano, *Geometric Algebra Applications Vol. I*,
https://doi.org/10.1007/978-3-319-74830-6

17. Bayro-Corrochano, E. 1996. Clifford self-organizing neural network, Clifford wavelet network. In *Proceedings of the 14th IASTED International Conference on Applied Informatics*, 271–274, Feb 20–22, Innsbruck, Austria.
18. Bayro-Corrochano, E. 2005. Robot perception and action using conformal geometry. In *Handbook of Geometric Computing. Applications in Pattern Recognition, Computer Vision, Neurocomputing and Robotics*, ed. E. Bayro-Corrochano, Chap. 13, 405–458. Heidelberg: Springer.
19. Bayro-Corrochano, E., S. Buchholz, and G. Sommer. 1996. Self-organizing Clifford neural network. In *IEEE ICNN'96*, 120–125. Washington: DC, June.
20. Bayro-Corrochano, E., K. Daniilidis, and G. Sommer. 1997. Hand-eye calibration in terms of motions of lines using geometric algebra. *10th Scandinavian Conference on Image Analysis*, vol. I, 397–404. Finland: Lappeenranta.
21. Bayro-Corrochano, E., K. Daniilidis, and G. Sommer. 2000. Motor algebra for 3D kinematics. The case of the hand-eye calibration. *International Journal of Mathematical Imaging and Vision* 13(2): 79–99.
22. Bayro-Corrochano, E., and J. Lasenby. 1998. Geometric techniques for the computation of projective invariants using n uncalibrated cameras. In *Proceedings of the Indian Conference on Computer Vision and Image Processing*, 95–100, Dec 21–23, New Delhi, India.
23. Bayro-Corrochano, E., J. Lasenby, and G. Sommer. 1996. Geometric algebra: A framework for computing point and line correspondences and projective structure using n uncalibrated cameras. In *IEEE Proceedings of ICPR'96*, vol. I, 334–338, Vienna, Austria.
24. Bayro-Corrochano, E., and C. López-Franco. 2004. Omnidirectional vision: Unified model using conformal geometry. In *Proceedings of the European Conference on Computer Vision*, Prague, Czech Republic, 536–548.
25. Bayro-Corrochano, E., and J. Rivera-Rovelo. 2004. Non-rigid registration and geometric approach for tracking in neurosurgery. *International Conference on Pattern Recognition*, 717–720, Cambridge, UK.
26. Bayro-Corrochano, E., and B. Rosenhahn. 2000. Analysis and computation of the intrinsic camera parameters using Pascal's theorem. In *Geometric Computing with Clifford Algebra*, ed. G. Sommer, Chap. 16, 593–414. Heidelberg, Germany: Springer.
27. Bayro-Corrochano, E., and G. Sobczyk. 2000. Applications of Lie algebras and the Algebra of Incidence. In *Geometric algebra applications with applications in science and engineering*, ed. E. Bayro-Corrochano, and G. Sobczyk, 252–277. Boston: Birkhäuser.
28. Bayro-Corrochano, E., and Y. Zhang. 2000. The motor extended Kalman filter: A geometric approach for 3D rigid motion estimation. *Journal of Mathematical Imaging and Vision* 13(3): 205–227.
29. Bayro-Corrochano, E., and J. Rivera-Rovelo. 2009. The use of geometric algebra for 3D modeling and registration of medical data. *Journal of Mathematical Imaging and Vision* 34(1): 48–60.
30. Bauer, M. 1995. *General regression neural network for technical use*. Master's thesis, University of Wisconsin–Madison. http://digital.library.wisc.edu/1793/7779
31. Belinfante, J., and B. Kolman. 1972. *Lie groups and lie algebras: With applications and computationa methods*. Philadelphia: SIAM.
32. Bell, I. $C++$ MV 1.3.0 to 1.6 sources supporting $N \leq 63$. http://www.iancgbell.clara.net/maths/index.htm
33. Benosman, R., and S. Kang. 2000. *Panoramic vision*. New York: Springer.
34. Bernard, C. 1997. Discrete wavelet analysis for fast optic flow computation. *Applied and Computational Harmonic Analysis* 11(1): 32–63.
35. Bernd, J. 1993. *Digital image processing*. Springer.
36. Biglieri, E., and K. Yao. 2000. Some properties of singular value decomposition and their application to digital signal processing. *Signal Processing* 18: 277–289.
37. Bigun, J. 2006. *Vision with direction*. Springer.
38. Blaschke, W. 1960. *Kinematik und Quaternionen*. Berlin: VEB Deutscher Verlag der Wissenschaften.
39. Bloch, F. 1946. Nuclear induction. *Physical Review* 70: 460–474.

40. Bottou, L., C. Ortes, J. Denker, H. Drucker, I. Guyon, L. Jackel, Y. LeCun, U. Müller, E. Sackinger, P. Simard, and V. Vapnik. 1994. Comparison of classifier methods: A case study in handwriting digit recognition. In *International Conference on Pattern Recognition*, 77–87. IEEE Computer Society Press.
41. Bottou, L. 2012. Stochastic gradient descent tricks. In *Neural networks: Tricks of the trade*, 2nd ed., ed. G. Montavon, G.B. Orr, and K.-R. Mller, 421–436. Berlin, Heidelberg: Springer.
42. Brannan, D., M. Esplen, and J. Gray. 2002. *Geometry*. New York: Cambridge University Press.
43. Bruna J., S. Chintala, Y. LeCun, S. Piantino, A. Szlam, and M. Tygert. 2015. A theoretical argument for complex-valued convolutional networks, CoRR, vol. abs/1503.03438. http://arxiv.org/abs/1503.03438.
44. Bülow, T. 1999. Hypercomplex Fourier transforms. Ph.D. thesis, Computer Science Institute, Christian Albrechts Universität, Kiel.
45. Burges, C.J.C. 1998. A tutorial on support vector machines for pattern recognition. *Knowledge Discovery and Data Mining* 2(2): 1–43. Kluwer Academic Publishers.
46. Cafaro, C., and S.. Mancini. 2010. A geometric algebra perspective on quantum computational gates and universality in quantum computing. arXiv:1006.2071v1 [math-ph] 10 June 2010.
47. Canny, J. 1986. A computational approach to edge detection. *IEEE Transactions on Pattern Analysis and Machine Intelligence* 8(6): 679–698.
48. Carlsson, S. 1994. The double algebra: An effective tool for computing invariants in computer vision. In *Applications of Invariance in Computer Vision*, Lecture Notes in Computer Science, vol. 825; Proceedings of the 2nd Joint Europe-U.S. Workshop, AICV1993, Azores, 145–164, Oct 1993. Springer.
49. Carlsson, S. 1998. Symmetry in perspective. In *Proceedings of the European Conference on Computer Vision*, 249–263, Freiburg, Germany.
50. Cao, M., and P. Li. 2014. Quantum-inspired neural networks with applications. *International Journal of Computer and Information Technology* 3(01): 83–92.
51. Cerejeiras, P., M. Ferreira, U. Kähler, and F. Sommen. 2007. Continuous wavelet transform and wavelet frames on the sphere using Clifford analysis. *Communication on Pure and Applied Analysis* 6(3): 619–641.
52. Chantler, M.J. 1994. *The effect of variation in illuminant direction on texture classification*. Ph.D. thesis, Dept. of Computing and Electrical Engineering, Heriot-Watt University.
53. Chappell, J.M. 2011. *Quantum Computing, Quantum Games and Geometric Algebra*. Ph.D. thesis, The School of Chemestry amd Physics, University of Adelaide, Australia.
54. Chernov, V.M. 1995. Discrete orthogonal transforms with data representation in composition algebras. *Scandinavian Conference on Image Analysis*, 357–364. Sweden: Uppsala.
55. Chou, S., W. Schelter, and J. Yang. 1987. Characteristic sets and Gröbner bases in geometry theorem proving. In *Computer-aided geometric reasoning*, ed. H. Crapo, 29–56. Rocquencourt, France: INRIA.
56. Chui, H., and A. Rangarajan. 2000. A new point matching algorithm for non-rigid registration. *IEEE Conference on Computer Vision and Pattern Recognition (CVPR)*, vol. 2, 44–51.
57. Clifford, W.K. 1873. Preliminary sketch of bi-quaternions. *Proceedings of the London Mathematical Society* 4: 381–395.
58. Clifford, W.K. 1878. Applications of Grassmann's extensive algebra. *American Journal of Mathematics* 1: 350–358.
59. Clifford, W.K. 1882. On the classification of geometric algebras. In *Mathematical Papers by William Kingdon Clifford*, ed. R. Tucker. London: Macmillan. [Reprinted by Chelsea, New York, 1968; Title of talk announced already in Proceedings of the London Mathematical Society 7(1876): 135.
60. Colios, C., and P.E. Trahanias. 2001. A framework for visual landmark identification based on projective and point-permutation invariant vectors. *Robotics and Autonomous Systems Journal* 35: 37–51.
61. Csurka, G., and O. Faugeras. 1998. Computing three dimensional project invariants from a pair of images using the Grassmann-Cayley algebra. *Journal of Image and Vision Computing* 16: 3–12.

62. Cybenko, G. 1989. Approximation by superposition of a sigmoidal function. *Mathematics of Control, Signals and Systems* 2: 303–314.
63. Davison, A.J., I. Reid, N. Molton, and O. Stasse. 2007. Monoslam: Real time single camera slam. *IEEE Transactions on Pattern Analysis and Machine Intelligence* 6: 1052–1076.
64. Daubechies I. 1992. *Ten lectures on wavelets*. SIAM.
65. Deavours C.A. 1973. The quaternionic calculus. *The American Mathematical Monthly* 80: 995–1008. New York.
66. Deng, J., W. Dong, R. Socher, L.-J. Li, K. Li, and L. Fei-fei. 2009. Imagenet: A large-scale hierarchical image database. *IEEE International Conference on Computer Vision and Pattern Recognition, CVPR 2009*, 248–255, June 20–25, Miami, FL, USA.
67. Dierkes U., S. Hildebrandt, and A.J. Tromba. 2010. *Regularity of minimal surgaces* Rev. and enlarged 2nd ed., Grundlehren der matematische Wissenschafen, 340, Springer.
68. Dierkes, U., S. Hildebrandt, and A.J. Tromba. *Global analysis of minimal surfaces.* Rev. and enlarged 2nd ed., Grundlehren der matematische Wissenschafen, 341. Springer.
69. Dodwell, P.C. 1983. The Lie transformation group model of visual perception. *Perception and Psychophysics* 34(1): 1–16.
70. Doran, C., A. Lasenby, and S. Gull. 1993. States and operators in the space time algebra. *Foundation Physics* 23(9): 1239–1264.
71. Doran, C., A. Lasenby, S. Gull, S. Somaroo, and A. Challinor. 1996. Spacetime algebra and electron physics. *Advances in Imagin Electron Physics* 95: 271–386.
72. Doran, C., D. Hestenes, F. Sommen, and N. Van Acker. 1993. Lie Groups and spin groups. *Journal of Mathematical Physics* 34(8): 3642–3669.
73. Doran, C. 1994. *Geometric algebra and its applications to mathematical physics*. Ph.D. thesis, University of Cambridge.
74. Doran, C., and A. Lasenby. 2005. *Geometric algebra for physicists*. Cambridge University Press.
75. Dorst, L., D. Fontjine, and T. Mann. 2008. GAIGEN 2: Generates fast $C++$ or JAVA sources for low dimensional geometric algebra. http://www.science.uva.nl/ga/gaigen/
76. Dorst, L., D. Fontjine, and S. Mann. 2007. *Geometric algebra for computer science. An object-oriented approach to geometry*. Cambridge, MA: Morgan Kaufmann Series in Computer Science.
77. Dorst, L., S. Mann, and T. Bouma. 1999. GABLE: A Matlab tutorial for geometric algebra. http://www.carol.wins.uva.nl/~gable
78. Dorst, L. 2015. 3D Projective geoemtry through versors of $\mathbb{R}^{3,3}$. *Advances in Applied Clifford Algebras* 26: 1137–1172.
79. Dress, A., and T. Havel. 1993. Distance geometry and geometric algebra. *Foundations of Physics* 23(10): 1357–1374.
80. Duda, R.O., and P.E. Hart. 1972. Use of the Hough transformation to detect lines and curves in pictures. *Communications of ACM* 15: 11–15.
81. Eaester, R.B., and E. Hitzer. 2017. Double conformal geometric algebra. *Advances in Applied Clifford Algebras* 27: 2175–2199.
82. Everingham, M., L. Van Gool, C.K.I. Williams, J. Winn, and A. Zisserman. 2010. The pascal visual object classes (voc) challenge. *International Journal of Computer Vision* 88(2): 303–338.
83. Ebling, J., and G. Sheuermann. 2005. Clifford Fourier transform on vector fields. *IEEE Transactions on Visualization and Computer Graphics* 11(4): 469–479.
84. Einsten, A., B. Podolsky, and N. Rosen. 1935. Can quantum-mechanical description of physical realilty be considered complete? *Physical Review* 47(10): 777.
85. Ell, T.A. 1992. *Hypercomplex spectral transformations*. Ph.D. thesis, University of Minnesota.
86. Faugeras, O., B. Hotz, H. Mathieu, T. Viville, Z. Zhang, P. Fua, E. Thron, L. Moll, G. Berry, and J. Vuillemin. 1993. Real-time correlation-based stereo: Algorithm, implementation and applications, *INRIA Technical Report no. 2013*.
87. Faugeras, O. 1995. Stratification of three-dimensional vision: projective, affine and metric representations. *Journal of the Optical Society of America A*: 465–484.

88. Federer, H. 1969. *Geometric measure theory*. Die Grundlehren der mathematischen Wissenschaften, 153. New York: Springer.
89. Felsberg, M. 1998. *Signal processing using frequency domain methods in Clifford algebra*. M.S. thesis, Computer Science Institute, Christian Albrechts Universität, Kiel.
90. Felsberg, M. 2002. *Low-level image processing with the structure multivector*. Ph.D. thesis, Christian-Albrecht, Kiel University, Kiel, Germany.
91. Fleet, D.J., and A.D. Jepson. 1990. Computation of component image velocity from local phase information. *International Journal of Computer Vision* 5: 77–104.
92. Fletcher, R. 1987. *Practical methods of optimization*, 2nd ed. New York: Wiley.
93. FitzGerald, G.F. 1902. Review of Heaviside's electrical papers. In *Scientific writings of the late George Francis FitzGerald*, ed. J. Lamour, Dublin.
94. Fontijne, D. 2007. *Efficient implementation of geometric algebra*. Ph.D. thesis, University of Amsterdam. http://www.science.uva.nl/~fontjine/phd.html
95. Franchini, S., A. Gentile, M. Grimaudo, C. Hung, S. Impastato, F. Sorbello, G. Vassallo, and S. Vitabile. 2007. A sliced coprocessor for native Clifford algebra operations. In *Proceedings of the 10th IEEE Euromicro Conference on Digital System DesignArchitectures, Methods and Tools*, DSD07, Lbeck, 436–439.
96. Frederick, C., and E.L. Schwartz. 1990. Conformal image warping. *IEEE Computer Graphics and Applications* 10: 54–61.
97. Fritzke, B. 1995. A growing neural gas network learns topologies. In *Advances in Neural Information Processing Systems*, vol. 7, 625–632. Cambridge, MA: MIT Press.
98. Fu, K.S., and J.K. Mui. 1980. A survey on image segmentation. *Pattern Recognition* 12: 395–403.
99. Fulton, W., and J. Harris. 1991. *Representation theory: A first course*. New York: Springer.
100. Fukushima, K. 1980. Neocognitron: A self-organizing neural net- work model for a mechanism of pattern recognition unaffected by shift in position. *Biological Cybernetics* 36(4): 193–202.
101. Gabor, D. 1946. Theory of communication. *Journal of the IEE* 93: 429–457.
102. Gelf'and, M.I., M.I. Graev, and N.Y. Vilenkin. 1996. *Generalized functions*, vol. 5. Integral Geometry and Representation Theory: Academic Press.
103. Gentile, A., S. Segreto, F. Sorbello, G. Vassallo, S. Vitabile, and V. Vullo. 2005. CliffoSor, an innovative FPGA-based architecture for geometric algebra. In *Proceedings of 45th Congress of the European Regional Science Association (ERSA)*, 211–217, Aug 23–27, Vrije, Amsterdam.
104. Georgiou, G.M., and C. Koutsougeras. 1992. Complex domain backpropagation. *IEEE Transactions on Circuits and Systems*: 330–334.
105. Geyer, C., and K. Daniilidis. 2000. A unifying theory for central panoramic systems and practical implications. In *Proceedings of the European Conference on Computer Vision*, 445–461. Dublin.
106. Gibbs, J.W. 1884. *Elements of vector analysis*. Privately printed in two parts, 1881 and 1884. New Haven, CT: Yale University Press. Reprinted in 1961. *The Scientific Papers of J. Willard Gibbs*, vol. 2, 84–90, Dover, New York.
107. Golub, G.H., and C.F. van Loan. 1989. *Matrix computations*. Baltimore, MD: Johns Hopkins University Press.
108. Gorshkov, A., V.F. Kravchenko, and V.A. Rvachev. 1992. Estimation of the discrete derivative of a signal on the basis of atomic functions. Izmer. *Tekhnika* 1(8): 10.
109. González-Jiménez, L., O. Carbajal-Espinosa, A. Loukianov, and E. Bayro-Corrochano. 2014. Robust pose control of robot manipulators using conformal geometric algebra. *Advances in Applied Clifford Algebras* 24(2): 533–552.
110. Granlund, G.H., and H. Knutsson. 1995. *Signal processing for computer vision*, Kluwer Academic Publishers.
111. Grassmann, H.G. 1844. *Die Lineale Ausdehnungslehre*. Leipzig: Wiegand.
112. Grassmann, H. 1877. Der Ort der Hamilton'schen Quaternionen in der Ausdehnungslehre. *Mathematische Annalen* 12: 375.

113. Guberman, N. 2016. *On complex valued convolutional neural networks*. Israel: The Hebrew University of Jerusalem.
114. Guttman, A. 1984. R-trees: A dynamic index structure for spatial searching. In *Proceedings of the ACM International Conference on Management of Data*, SIGMOD '84, Boston, Massachusetts, 47–57.
115. Guyaev, Y.V., and V.F. Kravchenko. 2007. A new class of WA-Systems of Kravchenko-Rvachev functions. *MOSCOW Doklady Mathematics* 75(2): 325–332.
116. Hahn, S.L. 1992. Multidimensional complex signals with single-orthant spectra. *Proceedings of the IEEE* 80(8): 1287–1300.
117. Hahn, S.L. 1996. *Hilbert Transforms in Signal Processing*. Boston: Artech House.
118. Hänsch Rand Hellwich, O. 2010. Complex-valued convolutional neural networks for object detection in polsar data. In *8th European Conference on Synthetic Aperture Radar (EUSAR)*, July, 877–880.
119. Hamilton, W.R. 1853. *Lectures on Quaternions*. Dublin: Hodges and Smith.
120. Hamilton, W.R. 1866. *Elements of Quaternions, Longmans Green, 1969*. New York: London; Chelsea.
121. Havel, T.F., and C. Doran. 2002. Geometric algebra in quantum information processing. In *Quantum computation and quantum information science*, ed. S. Lomonaco, 81–100. AMS Contemporary Math series.
122. Hartley, R.I. 1997. Lines and points in three views and the trifocal tensor. *International Journal of Computer Vision* 22(2): 125–140.
123. Hartley, R. 1993. Chirality invariants. In *DARPA Image Understanding Workshop*, 745–753.
124. Hartley, R.I. 1994. Projective reconstruction and invariants from multiple images. *IEEE Transactions on PAMI* 16(10): 1036–1041.
125. Hartley, R. 1998. The quadrifocal tensor. *ECCV98*, 20–35. LNCS: Springer.
126. Hartley, R.I., and A. Zissermann. 2003. *Multiple view geometry in computer vision*, 2nd ed. Cambridge: Cambridge University Press.
127. Heaviside, O. 1892. *Electrical papers*, 2 vols. London.
128. Hestenes, D. 1966. *Space-time algebra*. London: Gordon and Breach.
129. Hestenes, D. 1975. Observables operators, and complex numbers in the Dirac theory. *Journal of Mathematical Physics* 16: 556.
130. Hestenes, D. 1986. *New foundations for classical mechanics*. Dordrecht: D. Reidel.
131. Hestenes, D. 1991. The design of linear algebra and geometry. *Acta Applicandae Mathematicae* 23: 65–93.
132. Hestenes, D. 1993. Invariant body kinematics I: Saccadic and compensatory eye movements. *Neural Networks* 7: 65–77.
133. Hestenes, D. 1993. Invariant body kinematics II: Reaching and neurogeometry. *Neural Networks* 7: 79–88.
134. Hestenes, D. 2001. Old wine in new bottles: A new algebraic framework for computational geometry. In *Geometric algebra applications with applications in science and engineering*, ed. E. Bayro-Corrochano, and G. Sobczyk, 3–17. Boston: Birkhäuser.
135. Hestenes, D. 2011. The shape of differential geometry in geometric calculus. *Guide to geometric algebra in practice*, 393–410. London: Springer.
136. Hestenes, D. 2003. Orsted Medal Lecture 2002: Reforming the mathematical language of physics. *American Journal of Physics* 71(2): 104–121.
137. Hestenes, D. 2009. New tools for computational geometry and rejuvenation of screw theory. In *Geometric algebra computing for engineering and computer science*, ed. E. Bayro-Corrochano, and G. Sheuermann, 3–33. London: Springer.
138. Hestenes, D., and G. Sobczyk. 1984. *Clifford algebra to geometric calculus: A unified language for mathematics and physics*. Dordrecht: D. Reidel.
139. Hestenes, D., and R. Ziegler. 1991. Projective geometry with Clifford algebra. *Acta Applicandae Mathematicae* 23: 25–63.
140. Hestenes, D. 1993. Differential forms in geometric calculus. In *Clifford algebra and their applications in mathematical physics*, ed. F. Brackx, et al., 269–285. Dordrecht/Boston: Kluwer.

141. Hestenes, M. 1963. *Calculus of variations and optimal control theory*. New York: Wiley.
142. Hildenbrand, D., J. Pitt, and A. Koch. 2009. High-performance geometric algebra computing using Gaalop. In *Geometric algebra computing for engineering and computer science*, ed. E. Bayro-Corrochano, and G. Sheuermann, 477–494. London: Springer.
143. Hildenbrand, D., J. Pitt, and A. Koch. 2010. Gaalop high performance parallel computing based on conformal geometric algebra. In *Geometric algebra computing for engineering and computer science*, ed. E. Bayro-Corrochano, and G. Sheuermann, Chap. 22, 477–494. Springer.
144. Hitzer, E., and B. Mawardi. 2007. Uncertainty principle for the Clifford geometric algebra $Cl_{n,0}$ n = 3(mod 4) based on Clifford Fourier transform. In *Wavelet analysis and applications*. Applied and Numerical Harmonic Analysis, ed. T. Qian, M.I. Vai, and X. Yuesheng, 45–54. New York: Springer.
145. Hoffman, W.C. 1966. The Lie algebra of visual perception. *Journal of Mathematical Psychology* 3: 65–98.
146. Hornik, K. 1989. Multilayer feedforward networks are universal approximators. *Neural Networks* 2: 359–366.
147. Hsu, C.W., and C.J. Lin. 2001. A comparison of methods for multi-class support vector machines. *Technical report*, National Taiwan University, Taiwan.
148. Hsu, C.W., and C.J. Lin. 2002. A simple decomposition method for support vector machines. *Machine Learning* 46: 291–314.
149. Hubel, D.H., and T.N. Wiesel. 1968. Receptive fields and functional architecture of monkey striate cortex. *Journal of Physiology, London* 195: 215–243.
150. Izhikevich, E.M. 2004. Which model to use for cortical spiking neurons? *IEEE Transactions on Neural Networks*: 1063–1070.
151. Jancewicz, B. 1990. Trivector Fourier transformation and electromagnetic field. *Journal of Mathematical Physics* 31(8): 1847–1852.
152. Jianxiong, X., J. Hays, K.A. Ehinger, A. Oliva, and A. Torralba. 2010. SUN database: Large-scale scene recognition from abbey to zoo. In *IEEE Conference on Computer Vision and Pattern Recognition, CVPR'2010* , 3485–3492.
153. Joachims, T. 1998. Making large-scale SVM learning practical. In *Advances in Kernel methods-support vector learning*, ed. B. Schölkopf, C.J.C. Burges, and A.J. Smola. Cambridge, MA: MIT Press. *Journal of Machine Learning Research* 5: 819–844.
154. Johnson, Barnabas D. The Cybernetics Of Society The Governance of Self and Civilization. www.jurlandia.org/writings.htm
155. Kaiser, G. 1994. *A friendly guide to wavelets*. Boston: Birkhäuser.
156. Kantor, I.L., and A.S. Solodovnikov. 1989. *Hypercomplex numbers: An elementary introduction to algebras*. New York: Springer.
157. Kingsbury, N. 1999. Image processing with complex wavelets. *Philosophical Transactions of the Royal Society of London A* 357: 2543–2560.
158. Klawitter, D. 2014. A Clifford algebraic approach to line geometry. *Advances in Applied Clifford Algebra* 24: 713–736.
159. Knerr, S., L. Personnaz, and G. Dreyfus. 1990. Single-layer learning revisited: A stepwise procedure for building and training a neural network. In *Neurocomputing: Algorithms, architectures and applications*, ed. J. Fogelman, 51–60. New York: Springer.
160. Koenderink, J.J. 1990. The brain: A geometry engine. *Psychological Research* 52: 122–127.
161. Kolodyazhnya, V.M., and V.A. Rvachev. 2007. Atomic functions: Generalization to the multivariable case and promising applications. *Cybernetics and Systems Analysis* 43(6): 893–911.
162. Kondra, S., A. Petrosino, and S. Iodice. 2013. Multi-scale Kernel operators for reflection and rotation symmetry: Further achievements. *The IEEE Conference on Computer Vision and Pattern Recognition (CVPR) Workshops*, June, 217–222.
163. Kovesi, P. 1997. Symmetry and asymmetry from local phase. In *Proceedings of the AI'97, Tenth Australian Joint Conference on Artificial Intelligence*, 185–190, Dec 2–4.
164. Kovesi, P. 1999. Image features from phase congruency. *Videre: A Journal of Computer Vision Research* 1(3 Summer). MIT Press.

165. Kovesi P. 1996. *Invariant measures of image features from phase information*. Ph.D. thesis, University of Western Australia.
166. Kouda, N., N. Matsui, and H. Nishimura. 2014. A multialyered feed-forward network based on qubit neuron model. *Systems and Coputers in Japan* 3(13): 641–648.
167. Kultanen, P., L. Xu, and E. Oja. 1990. Randomized Hough transform (RHT). In *Proceedings of the 10th International Conference on Pattern Recognition*, ICPR'1990, Atlantic City, USA, vol. 1, 631–635, June 16–21.
168. Kravchenko, V.F., A.S. Gorshkov, and V.A. Rvachev. 1992. Estimation of the discrete derivative of a signal on the basis of atomic functions, Izmer. *Tekhnika* 1: 8–10.
169. Kolodyazhny, V.M., and V.A. Rvachov. 2008. Atomic radial basis functions in numerical algorithms for solving boundary-value problems for the Laplace equation. *Cybernetics and Systems Analysis* 44: 4.
170. Kravchenko, O. Approximate Solutions of a Functional Differential Equation From the Wolfram Demonstrations Project http://demonstrations.wolfram.com/ApproximateSolutionsOfAFunctionalDifferentialEquation/
171. Krizhevsky, A., I. Sutskever, and G.E. Hinton. 2012. Imagenet classification with deep convolutional neural networks. In *Advances in neural information processing systems*, 1106–1114.
172. Kutzler, B., and S. Sifter. 1986. On the application of Buchberger's algorithm to automated geometry theorem proving. *Journal of Symbolic Computation* 2: 389–398.
173. La Poutré, H., S.H. Bohte, and J.N. Kok. 2002. Error-backpropagation in temporally encoded networks of spiking neurons. *Neurocomputing* 48: 17–37.
174. Lasenby, A.N. 1994. A 4D Maple package for geometric algebra manipulations in spacetime. http://www.mrao.cam.ac.uk/~clifford
175. Lasenby, A., C. Doran, and S. Gull. 1993. 2-Spinors. In *twistors and supersymmetry in the space time algebra*, ed. I.Z. Oziewicz, et al., 233–245. Dordrecht: Spinos, Twistors, Clifford Algebras and Quantum Deoformations, Kluwer Academic.
176. Lasenby, J., E.J. Bayro-Corrochano, A. Lasenby, and G. Sommer. 1996. A new methodology for computing invariants in computer vision. In *IEEE Proceedings of the International Conference on Pattern Recognition (ICPR'96)*, Vienna, Austria, vol. I, 393–397.
177. Lasenby, J., and E. Bayro-Corrochano. 1997. Computing 3D projective invariants from points and lines. In *7th International Conference on Computer Analysis of Images and Patterns*, CAIP'97, ed. G. Sommer, K. Daniilidis, and J. Pauli, 82–89. Kiel: Springer.
178. Lasenby, J., and E. Bayro-Corrochano. 1999. Analysis and computation of projective invariants from multiple views in the geometric algebra framework. In *Special Issue on Invariants for pattern recognition and classification*, ed. M.A. Rodrigues. *International Journal of Pattern Recognition and Artificial Intelligence* 13(8): 1105–1121.
179. Lee, Y., Y. Lin, and G. Wahba. 2001. Multicategory support vector machines. Technical report 1043, University of Wisconsin, Department of Statistics, 10–35.
180. LeCun, Y. 1989. Generalization and network design strategies. University of Toronto. Connectionis Research Group, Toronto, Ontario, Canada, Technical Report CRG-TR-89-4, 1–19, June.
181. LeCun, Y., L. Bottou, G.B. Orr, and K.R. Mller. 1998. Efficient backpropagation. In *Neural networks: Tricks of the trade*, ed. G.B. Orr, K.-R. Mller, 9–48. Berlin, Heidelberg: Springer Berlin Heidelberg.
182. Levenberg, K. 1944. A method for the solution of certain non-linear problems in least squares. *Quarterly of Applied Mathematics* 2: 164–168.
183. Lennart, W. 2008. *Local feature detection by higher order Riesz transforms on images*. Master's thesis, Christian-Albrecht, Kiel University, Kiel Germany.
184. Lewintan, P. 2016. Geometric Calculus of the Gaussian map. *Journal of Advances on Applications of Clifford Algebras* 27: 503–521.
185. Li, H., D. Hestenes, and A. Rockwood. 2001. Generalized homogeneous coordinates for computational geometry. In *Geometric computing with Clifford algebra*, ed. G. Sommer, 27–59. New York: Springer.

186. Li, H., D. Hestenes, and A. Rockwood. 2001. Generalized homogeneous coordinates for computational geometry. In *Geometric computing with Clifford algebra*, ed. G. Sommer, 27–52. New York: Springer.
187. Li, H., and L. Zhang. 2011. Line geometry in terms of the null geometric algebra over $\mathbb{R}^{3,3}$, and applications to the inverse singularity of generalized Stewart plataforms. In *Guide to geometric algebra in practice*, ed. L. Dorst, and J. Lasenby, 253–272. New York: Springer.
188. Li, H., L. Huang, Ch. Shao, and L. Dong. 2015. Three-dimensional projective geometry with geometric algebra. arXiv:1507.06634v1.
189. Li, W.H., A. Zhang, and L. Kleeman. 2005. *Fast global reflectional symmetry detection for robotic grasping and visual tracking*. In *Proceedings of Australasian Conference on Robotics and Automation* (ACRA '05), 147–151, Sydney, Australia.
190. Li, W.H., and L. Kleeman. 2006. Real time object tracking using reflectional symmetry and motion. In *Proceedings of the IEEE International Conference on Intelligent Robots and Systems* (IROS '06), 2798–2803, Beijing, China.
191. Li, Z., X. Hong, and Y. Liu. 2011. Detection geometric object in the conformal geometric algebra framework. In *CADGRAPHICS 11 Proceedings of the 2011 12th International Conference on Computer-Aided Design and Computer Graphics*, 198–201, USA.
192. Light, W.A. 1992. Some, and aspects of radial basis function approximation. In *Approximation theory*, ed. S.P. Sing, 163–190. Dordrecht: Spline Function and Applications, Kluwer Academic Publishers.
193. Lina, J.-M. 1997. Complex Daubechies wavelets: Filters design and applications, 95–112. In *ISAAC Conference*. University of Delaware.
194. López-González, G., G. Altamirano-Gómez, and E. Bayro-Corrochano. 2013. Geometric entities voting schemes in the conformal geometric algebra framework. *Advances in Applications of Clifford Algebra* 26(3): 1045–1059.
195. Lorensen, W., and H. Cline. 1987. Marching cubes: A high resolution 3D surface construction algorithm. *Computer Graphics* 21(4): 163–169.
196. Lounesto, P. 1987. *CLICAL software packet and user manual*, A248. Research report. Helsinki University of Technology of Mathematics.
197. Lounesto, P. 1997. *Clifford algebras and spinors*. Cambridge: Cambridge University Press.
198. Loy, G. and J.O. Eklundh. 2006. Detecting symmetry and symmetric constellations of features. *Proceedings of the 9th European Conference on Computer Vision - Volume Part II*, ECCV'06, 508–521, Graz, Austria, LNCS. Springer.
199. Luong, Q.T., and O.D. Faugeras. 1996. The fundamental matrix. *International Journal of Computer Vision* 17(1): 43–75.
200. Magarey, J.F.A., and N.G. Kingsbury. 1998. Motion estimation using a complex-valued wavelet transform. *IEEE Transactions on Image Processing* 6: 549–565.
201. Mackintosh, A.R. 1983. The Stern-Garlach experiment, electron spin and intermediate quantum mechanics. *European Journal of Physics* 4: 97.
202. Mallat, S. 1989. A theory for multiresolution signal decomposition: The wavelet representation. *IEEE Transactions on Pattern Analysis and Machine Intelligence* 11(7): 674–693.
203. Mallat, S. 2001. *A wavelet tour of signal processing*, 2nd ed. San Diego: Academic Press.
204. Maybank, S.J., and O.D. Faugeras. 1992. A theory of self-calibration of a moving camera. *International Journal of Computer Vision* 8(2): 123–151.
205. Mawardi, B., and E. Hitzer. 2006. Clifford Fourier transformation and uncertainty principle for the Clifford geometric algebra $Cl_{3,0}$. *Advances in Applications of Clifford Algebras* 16(1): 41–61.
206. Mawardi, B., and E. Hitzer. 2007. Clifford algebra $Cl_{3,0}$-valued wavelet transformation, Clifford wavelet uncertainty inequality and Clifford Gabor wavelets. *International Journal of Wavelets, Multiresolution and Information Processing* 5(6): 997–1019.
207. McCulloch W. and W. Pitts. A logical calculus of the ideas immanent in nervous activity. *Bulletin of Mathematical Biophysics* 5: 115–133.
208. Mo, Q., and B. Draper 2011. Detecting bilateral symmetry with feature mirroring. *The IEEE Conference on Computer Vision and Pattern Recognition (CVPR) Workshops*, 1715–1724, June, Colorado, USA.

209. Mordohai, P., and G. Medioni. 2006. *Tensor Voting. A perceptual organization approach to computer vision and machine learning*. Morgan and Claypool Publishers, Synthesis Lectures on Image, Video, and Multimedia Processing, Morgan and Claypool.
210. Medioni, G., M. Lee, and C. Tang. 2000. *A computational framework for segmentation and grouping*. Burlington, MA: Elsevier Science.
211. Meer, P., R. Lenz, and S. Ramakrishna. 1998. Efficient invariant representation. *International Journal of Computer Vision* 26: 137–152.
212. Mehrotra, K., C.K. Mohan, and S. Ranka. 1997. *Elements of Artificial Neural Networks*. Cambridge, MA: MIT Press.
213. Michel, O. 2004. Webots: Professional mobile robot simulation. *International Journal of Advanced Robotic Systems* 1: 39–42.
214. Miller, W. 1968. *Lie theory and special functions*. New York: Academic Press.
215. Mishra, B., and P. Wilson. 2005. Hardware implementation of a geometric algebra processor core. In *Proceedings of IMACS International Conference on Applications of Computer Algebra*, Nara, Japan. http://eprints.ecs.soton.ac.uk/10957/.
216. Mitrea, M. 1994. *Clifford wavelets, singular integrals and hardy spaces*. Lecture Notes in Mathematics, vol. 1575, New York: Springer.
217. Moya-Sánchez, E.U., and E. Bayro-Corrochano. 2013. Hilbert and Riesz transform using atomic function for quaternionic phase computation. *Advances on Applications of Clifford Algebra* 23: 929–949.
218. Moya-Sánchez, E.U., and E. Bayro-Corrochano. 2013. Quaternion local phase for low-level image processing using atomic functions. In *Quaternion and Clifford-Fourier transforms and wavelets, trends in matehmatics*, ed. E. Hitzer, and S.J. Sangwine, 57–83, Birkhauser.
219. Moya-Sánchez, E.U., and E. Bayro-Corrochano. 2014. Symmetry feature extraction Advances on Applications of Clifford. *Algebra* 24: 333–354.
220. Mundy, J., and A. Zisserman (eds.). 1992. *Geometric invariance in computer vision*. Cambridge, MA: MIT Press.
221. Muñoz, X. 2002. *Image segmentation integrating color, texture and boundary information.* Ph.D. thesis, University of Girona, Girona, Spain.
222. Nair, V., and G.E. Hinton. 2010. Rectified linear units improve restricted boltzmann machines. In *Proceedings of the 27th International Conference on Machine Learning (ICML-10)*, ed. J. Frnkranz, and T. Joachims, 807–814, Omnipress.
223. Nielsen, A., and I.L. Chuang. 2000. *Quantum computation and information*. Cambridge Press.
224. Nguyen, V., S. Gächter, A. Martinelli, N. Tomatis, and R. Siegwart. 2007. A comparison of line extraction algorithms using 2D range data for indoor mobile robotics. *Autonomous Robots* 23(2): 97–111.
225. Pan, H.-P. 1996. Uniform full information image matching complex conjugate wavelet pyramids. In *Proceedings of the 18th International Society for Photogrametry and Remote Sensing ISPRS Congress*, Vienna, Austria, vol. XXXI, 619–625, July 9–19.
226. Passino, K.M. 1998. *Fuzzy control*. Reading, MA: Addison-Wesley.
227. Pearson, J.K., and D.L. Bisset. 1992. Back propagation in a Clifford algebra. *Artificial neural networks*, ed. I. Aleksander, and J. Taylor, vol. 2, 413–416.
228. Pellionisz, A., and R. Llinàs. 1980. Tensorial approach to the geometry of brain function: Cerebellar coordination via a metric tensor. *Neuroscience* 5: 1125–1136.
229. Pellionisz, A., and R. Llinàs. 1985. Tensor network theory of the metaorganization of functional geometries in the central nervous system. *Neuroscience* 16(2): 245–273.
230. Perantonis, S.J., and P.J.G. Lisboa. 1992. Translation, rotation, and scale invariant pattern recognition by high-order neural networks and moment classifiers. *IEEE Transactions on Neural Networks* 3(2): 241–251.
231. Pérez-Meana H., V.F. Kravchenko, and V.I. Ponomaryov. 2010. *Adaptive digital processing of multidimensional signals with applications.* Moscow Fizmatlit.
232. Perwass, C.B.U. 2000. *Applications of geometric algebra in computer vision*. Ph.D. thesis, University of Cambridge.

233. Perwass, C.B.U. 2009. *Geometric algebra with applications in engineering*. Berline, Heidelberg: Springer.
234. Perwass, C.B.U. 2006. CLUCal. http://www.clucal.info/.
235. Perwass, C., C. Gebken, and G. Sommer. 2003. Implementation of a Clifford algebra co-processor design on a field programmable gate array. In *Clifford Algebras: Application to mathematics, physics, and engineering*, ed. R. Ablamowicz. *6th International Conference on Clifford Algebras and Applications*, Cookeville, TN, Progress in Mathematical Physics, 561–575, Birkhäuser, Boston.
236. Platt, J.C., N. Cristianini, and J. Shaw-Taylor. 2000. Large margin DAGs for multiclass classification. In *Advances in neural information processing systems*, vol. 12, 547–553. Cambridge, MA: MIT Press.
237. Poelman, C.J., and T. Kanade. 1994. A paraperspective factorization method for shape and motion recovery. In *European Conference on Computer Vision*, ed. J.-O. Eklundh, 97–108, Stockholm.
238. Poggio, T., and F. Girosi. 1990. Networks for approximation and learning. *IEEE Proceedings* 78(9): 1481–1497.
239. Porteous, R.I. 1995. *Clifford algebras and the classical groups*. Cambridge: Cambridge University Press.
240. Pozo, J.M., and G. Sobczyk. 2001. Realizations of the conformal group. In *Geometric algebra with applications in science and engineerings*, ed. E. Bayro-Corrochano, and G. Sobczyk, 42–60. Boston: Birkhäuser.
241. Press, W.H., S.A. Teukolsky, W.T. Vetterling, and B.P. Flannery. 1994. *Numerical Recipes in C*. New York: Cambridge University Press.
242. Quan, L. 1994. Invariants of 6 points from 3 uncalibrated images. *Proceedings of the European Conference on Computer Vision* II: 459–470.
243. Radon, J. 1917. Über die Bestimmung von Funktionen durch ihre Integralwerte längs gewisser Mannigfaltigkeiten, Berichte Sächsische Akademie der Wissenschaften. *Leipzig, Mathematisch-Physikalische Klasse* 69: 262–277.
244. Ranjan, V., and A. Fournier. 1995. Union of spheres (UoS) model for volumetric data. In *Proceedings of the 11th Annual Symposium on Computational Geometry*, 402–403, Vancouver, Canada, C2-C3.
245. Rvachev, V.A. 1990. Compactly supported solution of functional-differential equations and their applications. *Russian Mathematical Survey* 45(1): 87–120.
246. Rauschert, I., K. Brocklehurst, S. Kashyap, J. Liu, and Y. Liu. 2011. *First symmetry detection competition: Summary and results*, 1–17. The Pennsylvania State University, PA, CSE11-012, Oct.
247. Reyes-Lozano, L., G. Medioni, and E. Bayro-Corrochano. 2007. Registration of 3D points using geometric algebra and tensor voting. *Journal of Computer Vision* 75(3): 351–369.
248. Riesz, M. 1958. *Clifford Numbers and Spinors*, ed. E.F. Bolinder, and P. Lounesto. Lecture Series no. 38, University of Maryland. Reprinted as facsimile. Dordrecht: Kluwer. 1993.
249. Robinson, A.J., and F. Fallside. 1987. The utility driven dynamic error propagation network. Technical report CUED/F-INFENG/TR. 1, Cambridge University Engineering Department.
250. Rivera-Rovelo, J., and E. Bayro-Corrochano. 2009. The use of geometric algebra for 3D modelling and registration of medical data. *Journal of Mathematical Imaging and Vision* 34(1): 48–60.
251. Rooney, J. 1978. On the three types of complex number and planar transformations. *Environment and Planning B* 5: 89–99.
252. Rosenhan, B., C. Perwass, and G. Sommer. 2005. Pose estimation of 3D free-form curves. *Journal of Computer Vision* 62(3): 267–289.
253. Rosenblueth, Arturo, Norbert Wiener, and Julian Bigelow. 1943. Behavior. Purpose and teleology. Philosophy of Science 10(1): 21.
254. Ruder, S. 2016. An overview of gradient descent optimization algorithms. CoRR, vol. abs/1609.04747, 2016. http://arxiv.org/abs/1609.04747.

255. Rumelhart, D.E., and J.L. McClelland. 1986. *Parallel distributed processing: Explorations in the microstructure of cognition*. Cambridge, MA: MIT Press.
256. Russakovsky, O., J. Deng, H. Su, et al. 2014. Imagenet large scale visual recognition challenge. *CoRR*, vol. abs/1409.0575. http://arxiv.org/abs/1409.0575.
257. Sabata, R., and J.K. Aggarwal. 1991. Estimation of motion from a pair of range images: A review. *CVGIP: Image Understanding* 54: 309–324.
258. Salomon, J., S. King, and M. Osborne. 2002. Framewise phone classification using support vector machines. Spoke Language Processing, Denver: In *Proceedings of the International Conference*.
259. Sander, J., M. Ester, H.P. Kriegel, and X. Xu. 1998. Density-based clustering in spatial databases: The algorithm GDBSCAN and its applications. *Data Mining and Knowledge Discovery* 2(2): 169–194.
260. Sangwine, S.J. 1996. Fourier transforms of colour images using quaternion or hypercomplex numbers. *Electronic Letters* 32(21): 1979–1980. evoke
261. Schmidhuber, J., M. Gagliolo, D. Wierstra, and F. Gomex. 2005. Recurrent support vector machines. Technical report IDSIA 19-05.
262. Schmidhuber, J., D. Wierstra, and F.J. Gómez. 2005. Evolino: Hybrid neuroevolution optimal linear search for sequence prediction. In *Proceedings of the 19th International Joint Conference on Artificial Intelligence (IJCAI)*, Morgan Kaufmann, 853–858.
263. Selig, J.M. 1999. Clifford algebra of points, lines and planes. Technical report SBU-CISM-99-06, South Bank University, School of Computing, Information Technology and Maths.
264. Selig, J. 2000. Robotics kinematics and flags. In *Advances in geometric algebra with applications in science and engineering*, ed. E. Bayro-Corrochano, and G. Sobczyk, 211–234. Boston: Birkhäuser.
265. Selig, J. 2005. *Geometric fundamentals of robotics*. New York: Springer.
266. Selig, J., and E. Bayro-Corrochano. 2010. Rigid body dynamics uisng Clifford algebra. *Advances in Applications of Clifford algebras* 20: 141–154.
267. Semple, J.G., and G.T. Kneebone. 1985. *Algebraic projective geometry*. Oxford: Oxford University Press. Reprinted by Oxford Science Publications.
268. Shapiro, L., and G. Stockman. 2001. *Computer vision*. Prentice-Hall, Inc.
269. Shashua, A. 1994. Projective structure from uncalibrated images: Structure from motion and recognition. *IEEE Transations on Pattern Analysis and Machine Intelligence* 16(8): 778–790.
270. Shashua, A., and M. Werman. 1995. Trilinearity of three perspective views and its associated tensor. In *Proceedings ICCV'95*, Los Alamitos, 920–925, June 12–14. MIT Press.
271. Shimodaira, H., K.-I. Noma, M. Nakai, and S. Sagayama. 2002. Dynamic time-alignment kernel in support vector machine. In *Advances in neural information processing systems*, vol. 14, ed. T.G. Dietterich, S. Becker, and Z. Ghahamani. Cambridge, MA: MIT Press.
272. Shiu, Y.C., and S. Ahmad. 1989. Calibration of wrist-mounted robotic sensors by solving homogeneous transform equations of the form $AX = XB$. *IEEE Transactions on Robotics Automation* 5: 16–27.
273. Siadat, A., A. Kaske, S. Klausmann, M. Dufaut, and R. Husson. 1997. An optimized segmentation method for a 2D laser-scanner applied to mobile robot navigation. In *Proceedings of the 3rd IFAC Symposium on Intelligent Components and Instruments for Control Applications*, 153–158.
274. Siciliano, B. 1990. *Robot kinematics*. Università degli di napoli, Italy, vol. 1, 1–50. CRC Press Handbook.
275. Simons, J. 1968. Minimal varieties in riemannian manifolds. *Annals of Mathematics* 2(88): 62–105.
276. Simonyan, K., and A. Zisserman. 2014. Very deep convolutional networks for large-scale image recognition. *CoRR*, vol. abs/1409.1556.
277. Sobczyk, G. 1997. The generalized spectral decomposition of a linear operator. *The College Mathematics Journal* 28(1): 27–38.
278. Sobczyk, G. 1997. Spectral integral domains in the classroom. *Aportaciones Matemáticas, Serie Comunicaciones* 20: 169–188.

279. Sobczyk, G. 2001. Universal geometric algebra. In *Advances in geometric algebra with applications in science and engineering*, ed. E. Bayro-Corrochano, and G. Sobczyk, 18–41. Boston: Birkhäuser.
280. Somaroo, S.S., A. Lasenby, and C. Doran. 1999. Geometric algebra and the causal approach to multiparticle quantum mechanics. *Journal of Mathematical Physics* 40: 3327.
281. Somaroo, S.S., D.G. Cory, and T.F. Havel. 1998. Expressing the operations of quantum computing in multiparticle geometric algebra. *Physics Letters* 240: 1–7.
282. Sommen, F., and N. Van Acker. 1993. SO(m)-invariant differential operators on Clifford algebra-valued functions. *Foundations of Physics* 23(11): 1491–1519.
283. Sommer, G., and C. Perwass. 2004. Implementation of a Clifford algebra co-processor design on a field-programmable gate array. In *Clifford Algebras: Applications to mathematics, physics, and engineering*. Progress in Mathematical Physics.
284. Sparr, G. 1994. Kinetic depth. *In Proceedings of the European Conference on Computer Vision*, Vol. II, pp. 471–482.
285. Specht, D.F. 1991. A general regression neural network. *Transactions on Neural Networks* 2(6): 568–576.
286. Soria-García, G., G. Altamirano-Gómez, S. Ortega-Cisneros, and E. Bayro-Corrochano. 2017. Conformal geometric algebra voting scheme implemented in reconfigurable devices for geometric entities extraction. *IEEE Transactions on Industrial Electronics*.
287. Soria-García, G., G. Altamirano-Gómez, S. Ortega-Cisneros, and E. Bayro-Corrochano. 2017. FPGA Implementation of a geometric voting scheme for the extraction of geometric entities from images. *Advances in Applications of Clifford Algebras* 27: 685–705.
288. Stark, H. 1971. An extension of the Hilbert transform product theorem. *Proceedings of the IEEE* 59: 1359–1360.
289. Steinman, S. 1994. *Binocular vision module the empirical Horopter*. Addison-Wesley Reading.
290. Study, E. 1903. *Geometrie der Dynamen*. Leipzig.
291. Svensson, B. 2008. *A multidimensional filtering framework with applications to local structure analysis and image enhancement*. Ph.D. thesis, Linköping University, Linköping, Sweden.
292. Szegedy, C., W. Liu, Y. Jia, P. Sermanet, S. Reed, D. Anguelov, D. Erhan, V. Vanhoucke, and A. Rabinovich. 2015. Going deeper with convolutions. In *IEEE Conference on Computer Vision and Pattern Recognition (CVPR 2015)*, 1–9.
293. Tai-Pang, Wu, Sai-Kit Yeung, Jiaya Jia, Chi-Keung Tang, and G. Medioni. 2012. A closed-form solution to tensor voting: theory and applications. *IEEE Transactions on Pattern Analysis and Machine Intelligence* 34(8): 1482–1495.
294. Tait, P.G. 1890. *Elementary treatise on quaternions*, 3rd ed., vi. Cambridge.
295. Tamura, H., S. Mori, and T. Yamawaki. 1980. Textural features corresponding to visual perception. *IEEE Transactions on Systems, Man and Cybernetics* 8(6): 460–473.
296. Tomasi, C., and T. Kanade. 1992. Shape and motion from image streams under orthography: A factorization method. *International Journal on Computer Vision* 9(2): 137–154.
297. Introduction to Tensor Flow. https://www.tensorflow.org/
298. A course on Torch. http://torch.ch/
299. The Integrating Vision Toolkit. http://ivt.sourceforge.net/
300. Triggs, W. 1995. Matching constraints and the joint image. In *IEEE Proceedings of the International Conference on Computer Vision (ICCV'95)*, 338–343, Boston.
301. Tsochantaridis, I., T. Hofmann, T. Joachims, and Y. Altun. 2004. Support vector machine learning for interdependent and structured output spaces. In *21st Proceedings of the International Conference on Machine Learning (ICML-04)*, 104–112, July 04–08, Alberta, Canada.
302. Van Rullen, R., A. Delorme, J. Gautrais, and S. Thorpe. 1999. SpikeNET: A simulator for modeling large networks of integrate and fire neurons. *Neurocomputing*: 989–996.
303. Vapnik, V. 1998. *Statistical learning theory*. New York: Wiley.
304. Viola, P., and M. Jones. 2001. R+++++++apid object detection using a boosted cascade of simple features. In *IEEE Computer Society Conference on Computer Vision and Pattern Recognition*, 511–518.

305. Visual, Needham T. 2003. *Complex analysis*. New York: Oxford University Press. Reprinted.
306. http://www.nlm.nih.gov/research/visible/visible_human.html
307. Weston, J., and C. Watkins. 1998. Multi-class support vector machines. In *Proceedings of ESANN99*, ed. M. Verleysen. Brussels, D. Facto Press. Technical Report CSD-TR-98-04, Royal Holloway, University of London.
308. White, N. 1994. Grassmann-Cayley algebras in robotics. *Journal of Intelligent Robot Systems* 11: 97–187.
309. White, N. 1991. Multilinear Cayley factorization. *Journal of Symbolic Computation* 11: 421–438.
310. Wiaux, Y., L. Jacques, and P. Vandergheynest. 2005. Correspondence principle between spherical and Euclidean wavelets. *Astrophysical Journal* 632(1): 15–28.
311. Williams, L.R., and K.K. Thornber. 1998. A comparison of measures for detecting natural shapes in cluttered backgrounds. *International Journal of Computer Vision* 34(2–3): 81–96.
312. Wilmankski, M., C. Kreuchner, and A. Hero. 2016. Complex input convolutional neural networks for wide angle sar atr. In *2016 IEEE Global Conference on Signal and Information Processing*, 1037–1041.
313. Wiener, Norbert. 1948. *Cybernetics, or control and communication in the animal and the machine*. Cambridge: MIT Press.
314. Wulfram, G., and M.K. Werner. 2002. *Spiking neuron models, single neurons, population. Plasticity*. Cambridge: MIT Press.
315. Xu, C. 1999. *Deformable models with applications to human cerebral cortex reconstruction from magnetic resonance images*, 14–63. Ph.D. thesis, Johns Hopkins University.
316. Yaglom, M. 1968. *Complex numbers in geometry*. Leicester, UK: Academic Press.
317. Yeshurun, Y., and E.L. Schwartz. 1989. Cepstral filtering on a columnar image architecture: A fast algorithm for binocular stereo segmentation. *IEEE Transactions on Pattern Analysis and Machine Intelligence* II(5)July: 759–767.
318. Zamora-Esquivel, J. 2014. $G_{6,3}$ Geometric algebra: Decription and implementation. *Advances in Applied Clifford Algebras* 24: 493–514.
319. Zeiler, M.D., and R. Fergus. 2014. Visualizing and understanding convolutional networks. In *Proceedings, Part I, Computer Vision ECCV 2014: 13th European Conference, Zurich, Switzerland*, ed. D. Fleet, T. Pajdla, B. Schiele, and T. Tuytelaars, 818–833, Sept 6–12. Cham: Springer International Publishing.
320. Zhang, Z. 1992. Iterative point matching for registration of free-form curves. Technical report 1658, INRIA.
321. Zhang, Z., and O. Faugeras. 1992. *3-D Dynamic scene analysis*. New York: Springer.
322. Zwicker, E. 1998. *Psychoacoustics: Facts and models*, 2nd ed. New York: Springer.

Index

E. Bayro-Corrochano, *Geometric Algebra Applications Vol. I*,
https://doi.org/10.1007/978-3-319-74830-6

C

Printed in the United States
By Bookmasters